Binding, Transport and Storage of Metal Ions in Biological Cells

RSC Metallobiology Series

Editor-in-Chief:
Professor C. David Garner, *University of Nottingham, UK*

Series Editors:
Professor Hongzhe Sun, *University of Hong Kong, China*
Professor Anthony Wedd, *University of Melbourne, Australia*
Professor Barry P. Rosen, *Florida International University, USA*

Titles in the Series:
1: Mechanisms and Metal Involvement in Neurodegenerative Diseases
2: Binding, Transport and Storage of Metal Ions in Biological Cells

How to obtain future titles on publication:
A standing order plan is available for this series. A standing order will bring delivery of each new volume immediately on publication.

For further information please contact:
Book Sales Department, Royal Society of Chemistry, Thomas Graham House, Science Park, Milton Road, Cambridge, CB4 0WF, UK
Telephone: +44 (0)1223 420066, Fax: +44 (0)1223 420247,
Email: booksales@rsc.org
Visit our website at www.rsc.org/books

Binding, Transport and Storage of Metal Ions in Biological Cells

Edited by

Wolfgang Maret
School of Medicine, King's College London, London, UK

and

Anthony Wedd
School of Chemistry, University of Melbourne, Melbourne, Australia

THE QUEEN'S AWARDS
FOR ENTERPRISE:
INTERNATIONAL TRADE
2013

RSC Metallobiology Series No. 2

ISBN: 978-1-84973-599-5
eISBN: 978-1-84973-997-9
ISSN: 2045-547X

A catalogue record for this book is available from the British Library

Published by The Royal Society of Chemistry,
Thomas Graham House, Science Park, Milton Road,
Cambridge CB4 0WF, UK

Registered Charity Number 207890

For further information see our website at www.rsc.org

Printed and bound in Great Britain by CPI Group (UK) Ltd, Croydon, CR0 4YY

Preface

Biochemistry is more than merely the chemistry of hydrogen, carbon, nitrogen, oxygen, sulfur and phosphorus. A significant number of additional elements of the periodic table are essential for life. In fact, life would not have evolved without them. Yet, after so many years of research, considerable uncertainty persists as to the complete list of elements that are essential to humans.

Non-essential elements are not innocent by-standers. Some interact with living systems and with essential elements, some have beneficial or pharmacological effects and yet others are toxic. Though we rarely monitor their concentrations, many of them are present in living systems, some of them at higher concentrations than essential elements, and our knowledge of how they affect biological function remains limited. Due to new industrial applications and manufacturing processes, we continue to introduce new chemical species and rare elements into the environment. This poses a considerable risk to unprotected biological systems, the stability of eco-systems and, of course, our health. Nanomaterials are one but not the only example. Thus, the study of the non-essential elements, traditionally an area of toxicology, gains importance similar to that of the essential elements, traditionally an area of nutrition.

Homeostatic control maintains the essential elements at their required concentrations. This control involves a large number of processes. It avoids negative effects associated with either deficiency or overload. We need to understand the chemistry underlying metal ion overloads and deficiencies in order to define pathophysiological mechanisms. This area is receiving increased attention because perturbation of homeostatic control of some elements is associated with causation, symptoms or progression of genetic, infectious and degenerative diseases, including neurodegeneration, cancer and diabetes. Detoxification of some non-essential elements also involves

RSC Metallobiology Series No. 2
Binding, Transport and Storage of Metal Ions in Biological Cells
Edited by Wolfgang Maret and Anthony Wedd
© The Royal Society of Chemistry 2014
Published by the Royal Society of Chemistry, www.rsc.org

biological control and specific chemical and biological mechanisms. Such processes can be mistaken for homeostatic control. For yet other non-essential elements with chemical characteristics similar to essential elements, detoxification mechanisms may not be in place and these elements can have pleiotropic actions.

This book focuses on the binding, storage and transport of metal ions in biological systems. All these functions require specific chemistries, coordination dynamics and selectivity in transient binding of metal ions.

Metals are involved in the biochemistry of gases including those involving nitrogen (N_2, NH_3, NO), carbon (CH_4, CO, CO_2), hydrogen (H_2), oxygen (O_2) and sulfur (SO, SO_2). This includes the fixation of these elements in biological matter and geo-biochemical cycles. Metal ions support fundamental biological functions in catalysis, structure and regulation, including the establishment of electrochemical potentials across membranes for energy generation and communication. This inorganic aspect of biochemistry is as important as its organic aspects. About 50% of all proteins contain a metal cofactor. The protein machinery that controls metal ions in cells is extensive, and how those ions are bound, stored and transported is critical to all their diverse functions. Cells are loaded with metal ions and even some considered as trace elements occur at remarkably high concentrations. But there are also pools of non-protein bound metal ions that are reactive and can be catalytically or biologically active.

This book does not address all the functions of metal ions in biology but rather how they are handled and controlled. We took the periodic table as a basis for the book and the line between metals and non-metals, which led to the exclusion of certain metalloids. Some metals are not discussed (Li, Be, Sn, lanthanides) due to a lack of molecular information. We plan to include them in future editions. We also largely excluded specific processes that biology employs to scavenge metal ions outside the cell, *e.g.* siderophores for iron and similar compounds for other metal ions, plus cellular processes that involve complex cofactors for metal ions, *e.g.* haem. Each chapter addresses the biochemistry of one particular element, with the exception of the chapter on metallothioneins, which discusses their involvement in the metabolism of several metals. Regarding biology, we focus in a textbook-like way on general aspects of the inorganic biochemistry of either prokarya or eukarya. Some authors have chosen to discuss plants. The usage of essential elements differs in organisms and while some are essential in some organisms they are not in others. Such aspects of biological variety are rarely considered and neither are protection mechanisms that also vary between organisms.

The knowledge presented here is necessary for defining nutritional requirements, thresholds for toxicity and disease-causing mechanisms. It is also the key to the mechanisms of action of metallodrugs and to the interactions of metal ions with biomolecules when used in the imaging of cells and tissues. The contributing authors are working in different scientific disciplines. We believe that, in addition to the inherently inter- and multidisciplinary nature of the subject matter, the different views of the authors provide the reader with broader perspectives. We also hope that the book will raise the interest of scientists from many disciplines. We noticed that investigations have become so narrowly focused that "independent" fields develop unique terminologies for each metal ion. This book attempts to surmount this issue by taking a metallomics approach. We asked the authors to discuss the same broad questions but they came up with different and sometimes conflicting answers. We attempted an approach that transcends investigations of the interaction between metals and biomolecules *in vitro* as we were asking the authors to address how metal ions are controlled in the cell and whether or not the phenomena observed are physiologically relevant and meaningful.

We acknowledge advice from David Giedroc, Robert Haussinger, Barry Rosen and Dennis Winge.

Contents

RSC Metallobiology Series No. 2
Binding, Transport and Storage of Metal Ions in Biological Cells
Edited by Wolfgang Maret and Anthony Wedd
© The Royal Society of Chemistry 2014
Published by the Royal Society of Chemistry, www.rsc.org

Chapter 23 Cadmium   **695**
Jean-Marc Moulis, Jacques Bourguignon and Patrice Catty

Chapter 26 Actinides in Biological Systems   **800**
Gerhard Geipel and Katrin Viehweger

Chapter 27 Aluminium   **833**
Christopher Exley

CHAPTER 1

Overview

ROBERT J. P. WILLIAMS

Inorganic Chemistry Laboratory, University of Oxford, South Parks Road,
Oxford, OX1 3QR, UK
Email: bob.williams@chem.ox.ac.uk

This book has the objective of describing the binding, transport and storage, mainly of metal ions, and many of their functions in organisms. It does not cover the link between these functions and enzyme action. The widest possible range of metal ions and metalloids are described. The functions include intracellular and extracellular activities but the book does not look at the transport into prokaryote cells through siderophores and similar molecules. Great stress is placed on healthy rather than diseased conditions.

Looking first at binding by all kinds of biological ligands we must remember that in an organism selection between the metal ions and their ligands is essential. The selection is constrained differently inside different compartments by the limitations imposed by transport of metal ions through membranes. The specific transport by inward or outward pumps may give either high concentrations of a given element in a vesicle (for example, in the message vesicles of zinc and calcium) or low concentrations. We must also note that binding in any compartment can be either weak or strong depending on the ligand. For example adenosine triphosphate, ATP, binds weakly to all the metal ions in cells. Hence ATP is only bound by an ion of the element magnesium, which is in high free ion concentration, much higher than that of any other ions. ATP and Mg^{2+} are overwhelmingly retained in one compartment, the cytoplasm. There are further factors governing stronger metal ligand selective binding at equilibrium. In addition to controlling metal ion concentration the ligand can also be controlled in its

RSC Metallobiology Series No. 2
Binding, Transport and Storage of Metal Ions in Biological Cells
Edited by Wolfgang Maret and Anthony Wedd
© The Royal Society of Chemistry 2014
Published by the Royal Society of Chemistry, www.rsc.org

concentration by transport through membranes into vesicles. Clearly the two can then give selectivity in bound states. Examples are the small molecule organic anions in zinc vesicles and the specific proteins in calcium vesicles. An additional control on ligand concentration is synthesis, to which we turn later. As oxidative/reductive potentials are fixed differently in certain compartments so the resultant binding at equilibrium is different and ligands are selected for this purpose. We note that ferrous and cuprous states are usually found inside the cytoplasm of cells whereas in the extracellular fluids and the environment, ferric and cupric states prevail. With those metal ions that are responsible for activity in membranes there must be transport to them and selective binding in the membranes. Membranes then form separate compartments and in them there is little interaction between fixed sites and free ions except at pumps.

Worthy of note are the orders of selection of metal ions with ligands at equilibrium. In the cases of Mg^{2+} and Ca^{2+} the binding selectivity between the two is dependent upon the sizes of the metal ions and the cavity provided by the ligand. For a small cavity magnesium ions bind most strongly but for a cavity that cannot close around the small magnesium its hydration makes binding weak. However, calcium may fit such a constrained space very well when the order of binding strength becomes $Ca^{2+} > Mg^{2+}$. The possibility of selection is very different in the cytoplasm where Mg and K ions are accumulated while the other two ions of Na and Ca are rejected from this space. This makes gradients across membranes; see later. While here the Mg/Ca selectivity is related to different ion size fittings, similarly so is that of Na^+ relative to K^+. This discrimination based on size is not very powerful in the equilibrium binding of the transition metal ions in the observed Irving–Williams series of their complexes:

$$Mn^{2+} < Fe^{2+} < Co^{2+} < Ni^{2+} < Cu^{2+} > Zn^{2+}$$

The ions here are of similar though declining size but the order arises from the electron affinity of the ions. This order is found *in vitro* with almost every ligand and it is very probably true in biological compartments to which there is access to all of them. All these ions are present to some extent in the cytoplasm. Fe^{2+} ions are not found in vesicles but Mn^{2+} is.

It is important to observe that the series is broken by change of spin state, to some degree by steric demands of ligands (crystal field effects) and of course by redox state changes of some ions. The spin-state changes are observed for example in iron *ortho*-phenanthroline complexes *in vitro* and in porphyrins *in vivo*. The binding of low-spin ions is then such that $Fe^{2+} = Cu^{2+}$. Steric factors influence the relative binding of some pairs of ions. For example Zn^{2+} prefers tetrahedral but Ni^{2+} prefers octahedral coordination. Thus the order of binding can be $Ni^{2+} > Zn^{2+}$ or $Zn^{2+} > Ni^{2+}$ according to the geometric disposition of liganding groups. A discussion of all the above bindings is given in ref. 1.

We turn from thermodynamic binding in a compartment to the kinetic factors generating selection. Most obviously the production of protein ligands from translated RNA is controlled, started or stopped, often by transcription factors and promoters bound to DNA. These factors and promoters are often activated themselves by the binding of metal ions or the action of small organic molecules. This control leads to limiting concentrations of specific protein ligands. Proteins can also be transferred to selected binding sites and to compartments. Since most proteins are not very stable they are constantly slowly degraded and must be replaced. A flow system in steady state is then a limiting factor on such concentrations. Some ions in proteins are selective transcription factors such as zinc fingers bound to DNA or are hydrolases such as metallophosphatases giving rise to phosphate, both of which control gene expression.

Apart from these conditions of the kinetics of ligand concentrations there are also kinetic limitations on exchange of metal ions from binding sites. This is particularly true for very strongly bound ions such as those of copper and zinc. These ions are found with different protein partners inside and outside cells. This condition is brought about by transport extending from across membranes to carrier proteins, often called metallochaperones. The first observed example of the second kind of transport is that across the cytoplasm of epithelial cells of calcium by the intestinal calcium binding protein. Subsequently many other metallochaperones have been found. Different chaperones convey a given metal ion to a target site where it may be virtually irreversibly bound. A very good example that has been known for some time is the range of chelatases, protein ligands, which carry different metal ions to different porphyrins in proteins. Six metal ions are involved, ferrous, cobalt, nickel, magnesium and, much more rarely, copper and zinc. Recognition of insertion sites is by selective protein/protein binding. Porphyrin binding of metal ions is not exchangeable and is then irreversible. Irreversible binding cannot be included in the Irving–Williams series, which only applies to reversible equilibrium binding. This does not mean, however, that selection at early steps by chaperones and by pumps is also excluded. In most cases of chelatases binding it is reversible. This may be true of other chaperones and many of the ion pumps where binding, though different on the two different sides of a membrane, is reversible with the contents of the two compartments which are separated by the membrane. It may be that free ion concentration (and binding) is selective in the cytoplasm following the Irving–Williams series, common to all cells.

Our analysis so far has largely concerned the cytoplasm of all cells. A different kind of selection is apparent in quite different compartments including the internal extracellular media of multicellular organisms. This book is not concerned with the binding of elements in the environment even where it is known that molecules, even proteins, are exported to scavenge for a particular ion, for example ferric ion by siderophores. On the other hand scavenging by ligands between cells in multicellular organisms is included in the chapters.

We must now turn to the functions of selective handling of metal ions in organisms. The most obvious is that strong selection by ligand sites allows subsequent strict selection of enzyme action. In no way must acid/base catalysis, mainly a function of zinc and magnesium ions, be confused with redox activity dominated by iron and copper ions. It is very noticeable that many enzyme sites are associated with one metal only, allowing experimental extraction and fractionation of specific metalloenzymes and proteins, such as copper oxidases and zinc fingers. This is not universally true and certain sites may be bound by one cation but be replaceable by another. A recent example are forms of carbonic anhydrase in which zinc can be replaced by cadmium in one organism and perhaps by cobalt in several. This is very different from the finding of different metal ions for a given function in very different organisms, for example there are iron, manganese, nickel and zinc/copper superoxide dismutases even in different cell compartments. The most obvious use of metal ion binding to proteins, RNA or DNA, is to give them structure. The structures can remain mobile between large domains or very locally. The mobility allows the proteins in particular to act as triggers of actions. Valence states or simple spin-state changes can adjust structure to induce uptake cooperativity as of oxygen in haemoglobin.

Returning to vesicles, or similar containers together with the extracellular fluids inside multicellular organisms, their redox and/or acidity is usually different from those of the fluids in the cytoplasm. Certain vesicles for the uptake of iron are acidic and the extracellular fluids and these and other vesicles are controlled. In these compartments iron is ferric not ferrous and copper is cupric not cuprous. There are exceptions as the circulating cupric oxygen carrier in some organisms, haemocyanin, is a cuprous complex but another oxygen carrier is a ferric protein. Of course all metal ions have to be circulated and apart from the free ions of Na, K, Mg and Ca, more strongly bound ions, such as Cu, are usually transferred in complexes or by proteins such as albumins.

Many cells have vesicles or other structures for storage even within the cytoplasm of cells. The well-known examples are the storage of iron in ferritin, of haem iron in haemoglobin and myoglobin in muscle tissue. The haemoglobin iron in cells is reduced iron, needed for oxygen binding. Many organisms store somewhat poisonous metal ions in vesicles rejecting the vesicles later. In this way stores can act in purification. Plants that can survive in nickel-rich soils carry the nickel to vesicles in the leaves and then it is thrown away as the leaf falls. Cadmium is stored in the animal multi-metal ion protein, metallothionein, in the kidney cells, while plants and more primitive organisms contain it in a peptide, glutathione. Other kinds of vesicle storage are the precursors of external biominerals of which the most obvious are calcium carbonates or phosphates, which on export are placed in the outer surface of lower organisms. The most studied example is that of the coccolith. By way of contrast animals store calcium phosphates in bone and plants store silica in small granules, concentrated in large numbers internally. While the silica is very free from other ions bone is not. There are many other biomineral storages such as magnetite (Fe), Sr and Ba in sulfates, and even Zn in the teeth of some organisms. Of course transport is common to movement of vesicles or of ions by chaperones.

Finally binding of ions can give structure to proteins, DNA and RNA. Particularly protein folding often occurs only on binding calcium hence calcium acts as a trigger messenger of action. The structure of ribosomes is dependent on the presence of magnesium and some part of DNA depends on potassium binding.

A way of understanding the vast range of metalloproteins, even excluding enzymes, is to analyse their evolution through their changing DNA sequences with the changing environment through the ages.[2] This analysis reveals relatively sudden changes of the numbers of such proteins at the two major stages of organization of cells. The first is the evolution of single-cell eukaryotes from prokaryotes while the second is that of multicellular organisms. Different functions of the metalloproteins can be seen in the three different groups of organisms.

In conclusion the book then covers a very wide range of the various metal ion functions quite apart from in enzyme catalysis. These functions are of equal importance with those of catalysis and like it they have evolved over billions of years.

References

1. J. J. R. Fraústo da Silva and R. J. P. Williams, *The Biological Chemistry of the Elements*, Oxford University Press, 2nd edn, 2001.
2. R. J. P. Williams and R. E. M. Rickaby, *Evolution's Destiny: Co-Evolving Chemistry of the Environment and Life*, Royal Society of Chemistry, Cambridge, 2012.

CHAPTER 2

Sodium. Its Role in Bacterial Metabolism

MASAHIRO ITO*[a] AND BLANCA BARQUERA*[b]

[a] Faculty of Life Sciences, Toyo University, 1-1-1 Izumino, Itakura-machi, Oura-gun, Gunma 374-0193, Japan; [b] Department of Biology and Center for Biotechnology and Interdisciplinary Studies, Rensselaer Polytechnic Institute, Rm 2239 Biotech., 110 8th Street, Troy, NY 12180, USA
*Email: masahiro.ito@toyo.jp; barqub@rpi.edu

2.1 Sodium Ions in Nature

Sodium and potassium are the two elements in Group 1 of the periodic table that display evolved biological activity. As monovalent cations, they are central to the chemistry of life: they serve as counter ions to cellular buffers, electrolytes and DNA, and as solutes for exchange and transport processes.[1–3],[†] The three other stable elements in the group (lithium, rubidium and caesium) are involved in very few if any biological processes, in spite of their natural abundance.

Na$^+$ is found in a wide range of concentrations in the natural environment. It ranges from micromolar to molar and living organisms occur throughout this range.[4] In a function that is absolutely essential for survival, both prokaryotic and eukaryotic cells exhibit homeostatic mechanisms that maintain constant but relatively low internal concentrations of Na$^+$. In many bacteria and archaea, these homeostatic mechanisms are sufficiently robust

---

[†] For further comparison of the chemistries of Li, Na, K, Mg and Ca, see Chapter 3, Section 3.1.1, and Chapter 5, Section 5.2.

RSC Metallobiology Series No. 2
Binding, Transport and Storage of Metal Ions in Biological Cells
Edited by Wolfgang Maret and Anthony Wedd
© The Royal Society of Chemistry 2014
Published by the Royal Society of Chemistry, www.rsc.org

that the organisms can tolerate external Na^+ concentrations that are significantly higher than those tolerated by eukaryotic cells.[5–8]

In organisms such as marine- or blood-borne bacteria that have adapted to high-Na^+ environments, ionic balance is maintained by the energy-dependent extrusion of the ion.[8,9] Interestingly, in some of these organisms, the resulting electrochemical gradient of Na^+ across the cell membrane can be used for energy production, in place of the proton motive force;[10] this topic will be discussed in a later section of this review. In addition to pumping Na^+ out of the cell, some halophilic organisms have membranes with anionic phospholipids to prevent dehydration. They are able to synthesize small molecules, known as osmolytes, that help to maintain osmotic balance with the highly saline extracellular medium.[11]

2.1.1 The Coordination Chemistry of the Sodium Ion

Group IA elements carry positive charges but exhibit low electron affinities and favourable hydration energies. The study of low-molar-mass ionophores for monovalent cations (M^+) has revealed several key principles relevant to the interactions of these ions with proteins. Typically, the cation is bound in a nest of oxygen atoms, provided by the side-chains of Asp and Glu and the polypeptide backbone. For the latter, there is a slight preference for the backbone oxygens of Ala, Gly, Leu, Ile, Val, Ser and Thr. In the case of Na^+, the most common stereochemistry is six-coordinate (octahedral).[12–14]

2.1.2 Small Molecule Ligands

There are a number of small molecules that bind monovalent cations with high selectivity and affinity. Some of these compounds function as ionophores, enabling the bound ion to pass through a membrane, thus dissipating the ion gradient. Since these ion gradients are essential for cell function, many ionophores have antibiotic activity. For example, monensin, which comes from the soil bacterium *Streptomyces cinnamonensis*, has a flexible structure that allows it to bind Na^+ tightly. Facilitated diffusion of the complex through the membrane collapses the ion gradient causing the death of the cell. It is also worth noting that there are also two classes of synthetic organic compounds known as crown ethers and cryptands that bind alkali cations with high affinity and high specificity (Figure 2.1).[15–17]

2.2 Sodium *versus* Hydrogen Ions in Bacteria

In general, living organisms use H^+ gradients as the media of primary energy production. Essentially all cells utilize H^+ pumps to convert energy from light or metabolic processes into a proton motive force that drives production of ATP.[18] In eukaryotic cells, these primary H^+ pumps are confined to the membranes of subcellular organelles (chloroplasts and mitochondria). However, at the cell membrane, the interface with the external

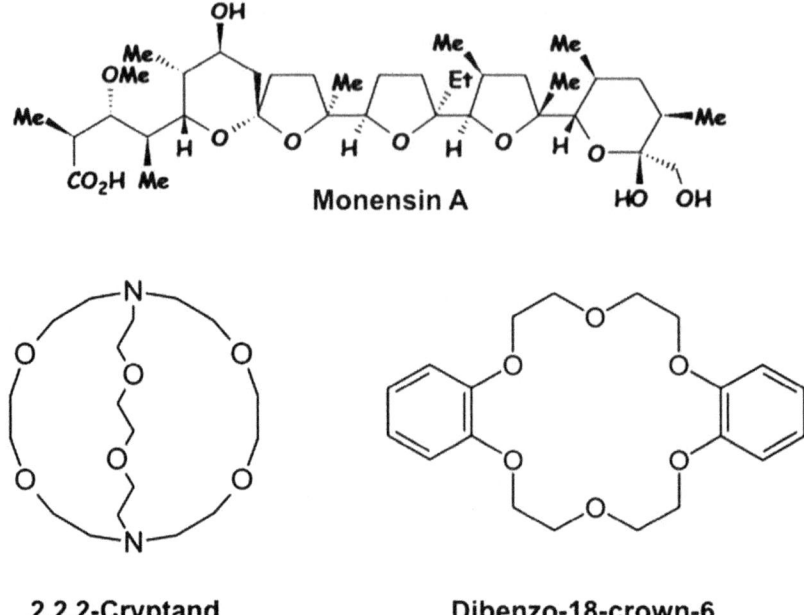

Figure 2.1   Small molecules that bind sodium: ionophores (Monensin A), cryptands (2.2.2-cryptand) and crown ethers (dibenzo-18-crown-6).

environment, generation of ion gradients depends on ATP-driven pumps, which typically transport Na^+ rather than protons.[19] H^+ gradients across the cell membrane are typically generated by secondary transporters that convert the Na^+ gradient into a H^+ gradient *via* Na^+/H^+ antiporters.[20] In contrast, in bacteria, the ion pumps responsible for primary energy production are themselves located in the cell membrane. In many cases, these pumps generate proton gradients, and it is the Na^+ gradient that is secondary and produced by Na^+/H^+ antiporters. However, some bacteria are able to generate the Na^+ gradient directly using the several different mechanisms described below.

2.2.1   The Sodium Ion Gradient as a Distributed Power Source

Since controlling internal Na^+ concentration is a central aspect of cellular homeostasis, maintaining the Na^+ gradient is itself important for the cell. However, this sodium-motive-force also serves as a distributed power source to drive energy-dependent processes throughout the cell membrane.

There are several groups of bacteria that rely on the Na^+ gradient for metabolic work, including the transport of nutrients. These include the halophiles, marine bacteria, rumen bacteria, alkaliphiles and alkali-tolerant bacteria. The formation and use of the Na^+ gradient in these bacteria seems to be a mechanism of adaptation to the high-salt environments in which these organisms live.

2.2.1.1 Alkaliphilic Bacteria

Although they live in alkaline rather than high-salt environments, these bacteria use the Na^+ gradient for some of their energetic processes. A likely reason is that, in an alkaline external environment, the cells must keep their cytoplasm at a lower pH than is present in the external medium. This is the reverse of the usual situation in which the internal pH is higher than the external pH. Even though proton pumps can create an electrical membrane potential, the external H^+ concentration is too low for protonic chemiosmosis to be feasible kinetically. In these circumstances, the Na^+ gradient is of increased importance since even a minimal external Na^+ concentration will be far higher than the H^+ concentration.[21]

2.2.1.2 Halophilic Bacteria

These organisms live in high salt concentrations in which $[Na^+] = 2$ to 5 M. They are able to form very large Na^+ gradients that are used to transport nutrients into the cell. The relevant transporters exhibit K_m values $\sim$1–3 M for the substrate Na^+.[21]

2.2.1.3 Marine Bacteria

All marine bacteria are dependent on Na^+ for growth. This dependence is due in part to the use of Na^+ transporters for the acquisition of nutrients, including amino acids and sugars. Most transporters have K_m values $\sim$50–500 mM. Many pathogenic bacteria exhibit similar Na^+ transport properties. They include *Vibrio cholerae*, *Pseudomonas aeruginosa*, *Yersinia pestis* and *Neisseria gonorrhoeae*.[22–24]

2.2.1.4 Rumen Bacteria

These organisms also require Na^+ for growth but at lower concentrations relative to those required by marine bacteria. Some Na^+-dependent transporters from these bacteria have been identified but little is known about their properties.[25]

2.3 Co-transport of Nutrients Using Sodium Ions

Many bacteria that live in normal Na^+ environments (1 to 200 mM Na^+) use the Na^+ gradient to supply energy for transport of nutrients into the cell. In *Escherichia coli*, uptake of amino acids is driven by a Na^+ gradient generated by the Na^+/H^+ antiporter that is, in turn, driven by respiratory H^+ pumping.

The proline/Na$^+$ symporter is an extensively studied example of these transporters. This enzyme transports one proline per Na$^+$ and is able to use Li$^+$ instead of Na$^+$ but not H$^+$. Since the enzyme is active at the low concentrations of Na$^+$ where *E. coli* live, its affinity for Na$^+$ must be in the micromolar range.[26,27]

2.4 Role of Sodium/Proton Antiporters

Monovalent/proton antiporters have important roles in homeostasis of pH, monovalent cation and volume in bacterial and eukaryotic cells and organelles.[28–33] These antiporters are categorized into at least eight transporter protein families based on sequence similarity.[34,35] The two main clusters are the CPA (monovalent-cation/proton antiporter) families CPA1, CPA2 and CPA3 and the Nha (Na$^+$/H$^+$ antiporter) families NhaA, NhaB, NhaC and NhaD. Some calcium/cation antiporter (CaCA) family members are also capable of Na$^+$/H$^+$ antiporter activity (that is, they are 2$^+$(Na$^+$)(K$^+$)/H$^+$ antiporters.[36] Most monovalent-cation/proton antiporters are single hydrophobic gene products, a few of which are functionally active as homo-oligomers. The most structurally complex antiporters are multiple resistance and pH adaptation (Mrp) transporters (CPA3).

Members of the Nha family catalyse active Na$^+$ efflux in electrogenic exchange for H$^+$. The number of H$^+$ ions entering is greater than the number of Na$^+$ exiting in each turnover; *e.g.* the *E. coli* NhaA (Ec-NhaA) and NhaB (Ec-NhaB) isoforms have stoichiometries of Na$^+$/2H$^+$ and 2Na$^+$/3H$^+$, respectively.[37,38] This property enables alkaliphilic bacteria to accumulate cytoplasmic H$^+$ against a concentration gradient, offsetting proton ejection due to respiration, while preserving a large fraction of the electrical component of the proton motive force (PMF)[39] in the resulting electrochemical sodium gradient.[40] Sodium/proton antiporters are ubiquitous membrane proteins and the prokaryotic NhaA and eukaryotic Na$^+$/H$^+$ exchanger (NHE) families have been studied comprehensively.[41–44] Their major physiological roles are maintenance of intracellular pH homeostasis and Na$^+$ efflux. Many bacteria use sodium motive force (SMF) and PMF simultaneously for energetic purposes. Therefore, Na$^+$ efflux through sodium/proton antiporters is critical in Na$^+$ circulation inside and outside the cell.[41,44–46] Mammalian NHE family members are 12 transmembrane-spanning segment (TMS) proteins with a number of different isoforms.[42–44] The transporter proteins NHE1–NHE5 and NHE6–NHE9 are located on the plasma membrane and organelle membranes, respectively.[45] NHE family members have a C-terminal hydrophilic domain exposed to the cytoplasmic side of the membrane and a Ca^{2+} binding protein is indispensable for monovalent cation transport activity of the NHE proteins.[46] This mammalian family has high sequence homology with the prokaryotic NhaP family which belongs to CPA1, but poor homology with the NhaA family.[45,47] The NhaP family also has a long C-terminal hydrophilic domain.

Bacteria generally have several sodium/proton antiporters that can be deployed differently in response to growth and to environmental stresses.[48–50] In *E. coli*, expression levels of Ec-NhaA and Ec-NhaB depend on stresses such as alkaline pH and elevated sodium levels.[28,36,41,51] Ec-NhaA functions best at alkaline pH while Ec-NhaB is most active in neutral conditions. Consequently, Ec-NhaA is critical for adaptation to an alkaline environment. The *E. coli* Ca^{2+}/H^+ antiporter ChaA and multidrug transporter MdfA also display sodium/proton antiporter activity.[52,53]

Ec-NhaA is a prototypical pH-regulated antiporter that has been studied extensively.[54,55] Orthologues are found in a wide range of Gram-negative bacteria including *Helicobacter pylori*, *V. cholerae*, *Salmonella typhimurium* and *E. coli*.[38,56,57] Analyses of X-ray crystal structures and site-specific amino acid substitution experiments have identified both the coupling-ion binding sites as well as a "pH sensor" domain (Figure 2.2).[58–63] Ec-NhaA is inactive below pH 6.5 but as the pH is increased to 8.5, V_{max} increases over three orders of magnitude and is then constant above pH 9. Thus, an alkaline pH shift converts Ec-NhaA from an inactive to an active molecular structure.[54,63,67] In the crystal structure of NhaA, residues Glu-78, Arg-81, Glu-82, Glu-252, His-253 and His-256, on one side of the cytoplasmic domain, are located in close proximity to the funnel entrance of the antiporter. They are suggested to play a role in the pH sensor activity.[63–67] The ion translocation stoichiometry of Ec-NhaA is $1Na^+/2H^+$, which indicates that the enzyme consumes the membrane potential while exchanging cations. In this way, it can function in an alkaline environment despite the intracellular pH being close to neutral and the membrane proton concentration gradient being reversed.[37]

2.4.1 Role of Sodium/Proton Antiporters at Alkaline pH

In alkaliphilic bacteria, sodium/proton antiporters have an essential role in supporting cytoplasmic pH homeostasis.[28] Antiporter-dependent H^+ accumulation is crucial for maintenance of a cytoplasmic pH that is approximately 2.3 units below the pH of the external medium (pH 10–11).[68,69] Their active Na^+ efflux activity in alkaliphilic bacteria concomitantly helps to prevent accumulation of toxic levels of Na^+. In particular, the Mrp antiporter in alkaliphilic *Bacillus* spp. plays a central role in adaption to highly alkaline environments.

2.4.2 The Sodium/Proton Antiporter Activity of Mrp

The *mrp* operon, which occurs in a wide range of bacteria and archaea (Figure 2.3) encodes a unique multi-subunit monovalent cation/proton antiporter whose role is to exchange extracellular H^+ for intracellular Na^+.[68] Mrp antiporters are hetero-oligomeric complexes, with multiple membrane-spanning subunits that appear to be essential for activity.[72] This property contrasts with that of other prokaryotic monovalent cation/proton

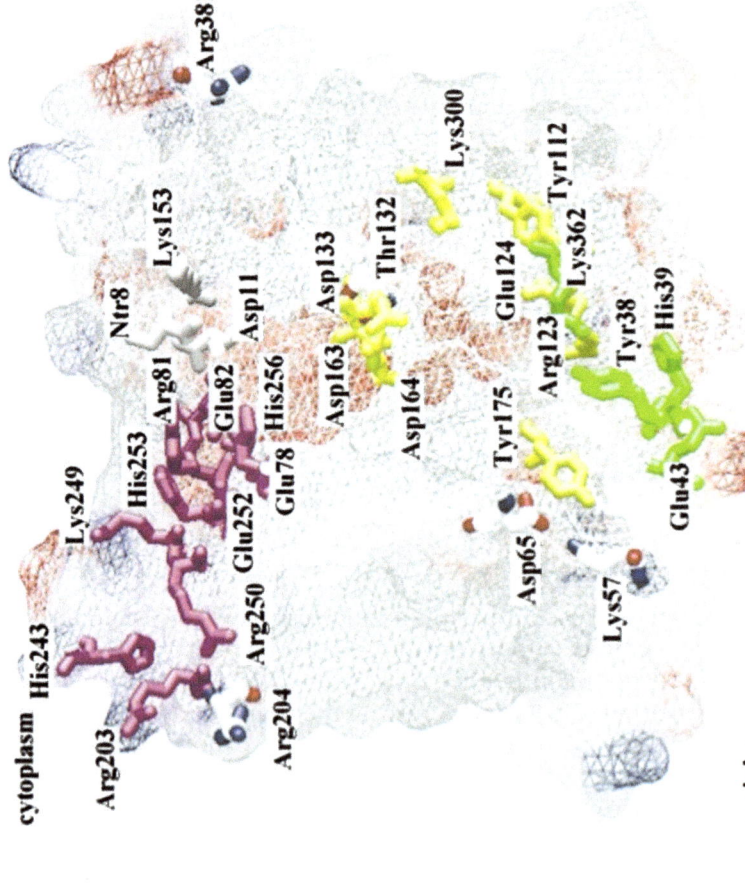

Figure 2.2 Clusters of ionizable groups that show strong electrostatic interactions in the NhaA antiporter. Zones of negative and positive potential are coloured red and blue, respectively. Four clusters of interacting residues are shown in grey, green, magenta and yellow.

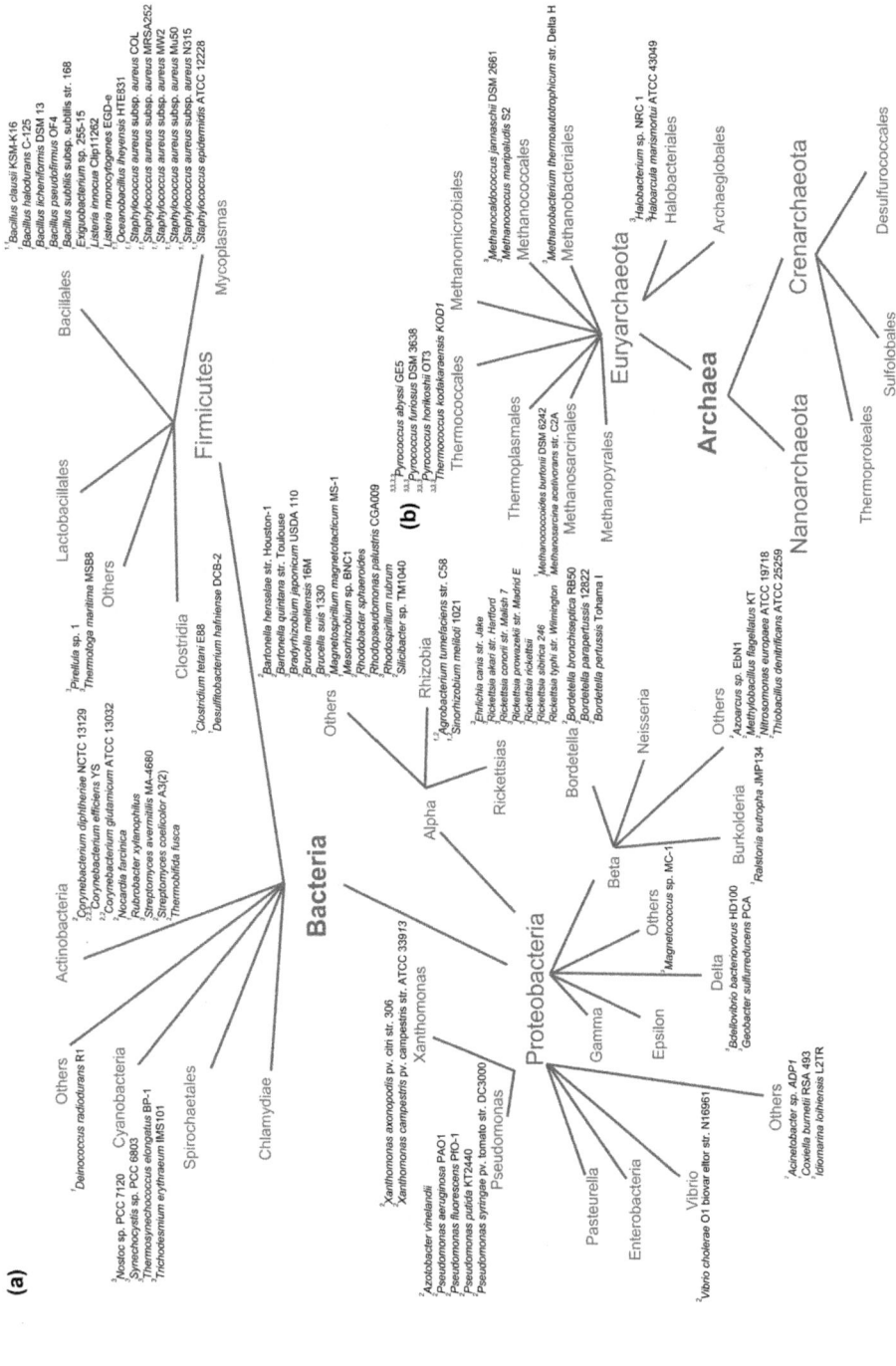

Figure 2.3 Groups of *mrp* operons in diverse prokaryotes.

antiporters that are single gene products.[67,77,78] Because of their structural features, the Transporter Database (http://www.membranetransport.org/) classifies the Mrp antiporters as a discrete family (CPA3).[79] Mrp antiporters (Figure 2.3)[68] often play indispensable roles in adaptation to alkaline or saline environments and in pathogenicity.[74] For almost all known aerobes, the Mrp antiporter comprises seven subunits that are the products of the *mrpABCDEFG* operon.[68,73] A few other aerobic bacteria have a six-subunit version in which MrpA and MrpB are fused.[70] The Mrp complexes from anaerobes and archaea exhibit larger differences. In many cases, MrpA is missing while, in others, there is more than one copy in the operon of other Mrp subunits, particularly of MrpB and MrpD. There are also differences in the gene order within the operons.[68] Some aerobes and anaerobes have more than one Mrp antiporter locus and, in others, the gene products have different cation specificities.[68] To date, there are Mrp antiporters known to catalyse Na^+, Li^+ and K^+ efflux in different combinations. In alkaliphilic *B. pseudofirmus* OF4, *B. subtilis* and *Staphylococcus aureus*, a Mrp antiporter functions as a $Na^+(Li^+)/H^+$ antiporter.[68,72,76] Conversely, the *Sinorhizobium meliloti* Mrp (Pha1)[80] and the *V. cholerae* Mrp (Vc-Mrp) transport K^+ as well as Na^+.[71,75]

The transporter present in *Thermomicrobium roseum*, a thermophilic Gram-negative bacterium, can transport Ca^{2+} instead of Na^+ or K^+.[81] A novel Nha sodium/proton antiporter, Nt-Na, from the anaerobe *Natranaerobius thermophilus* has been reported to consist of homologues of the largest two Mrp subunits, MrpA and MrpD, encoded at a single gene locus (Figure 2.4).[82]

2.5   Generation of Sodium Ion Gradients

2.5.1   Redox-driven Sodium Ion Pumps

The Na^+-NQR and RNF family. Many marine and pathogenic bacteria have a primary respiratory pump, the Na^+-dependent NADH-quinone oxidoreductase (Na^+-NQR) that translocates Na^+ across the cell membrane using the redox energy of metabolism. A related enzyme RNF ("*Rhodobacter* nitrogen fixation" protein) also appears to be a redox-linked Na^+ pump.[84,85] Na^+-NQR and RNF constitute a superfamily of membrane proteins that are found in many marine and pathogenic bacteria.[85–87]

2.5.1.1   *Na^+-NQR: Distribution and Physiological Role*

This enzyme is found exclusively in prokaryotes and serves as the entry point for electrons into the respiratory chain of many marine and pathogenic bacteria, including *V. cholerae*, *Y. pestis*, *N. gonorrhoeae*, *V. alginolyticus* and *V. harveyi*.[88–90] It oxidizes NADH while reducing ubiquinone and uses the free energy of this redox reaction to selectively translocate sodium across the cell membrane, creating a SMF that is used by the cell for metabolic

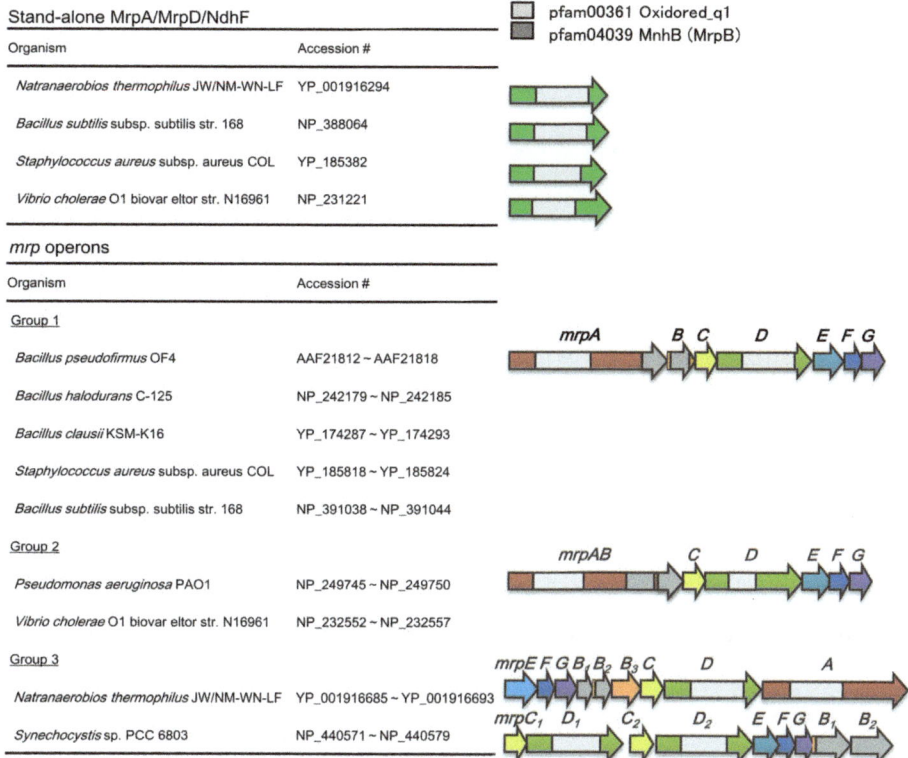

Figure 2.4 Diagrammatic representations of the genes encoding MrpA/MrpD/ NdhF-like proteins encoded outside CPA3 operons and operons of different CPA3 groups, showing the oxidoreductase and MnhB (MrpB) domains.

work.[91-93] Na$^+$-NQR carries out the same redox reaction as the H$^+$-pumping NADH:quinone oxidoreductase (complex I) found in mitochondria and many bacteria.[94] However, these two enzymes share no homology.

2.5.1.1.1 Subunit Composition and Redox Cofactors. Na$^+$-NQR is an integral membrane complex made up of six subunits: NqrA-F with a total molar mass of ∼200 kDa. Electrons enter the enzyme from NADH at a binding site located in NqrF and are carried through the enzyme by means of five redox cofactors: an FAD, a 2Fe-2S centre,[‡] two FMNs and a riboflavin. The two FMN centres are attached by phosphoester bonds to conserved threonines in NqrB (T236) and in NqrC (T225), and are named FMNB and FMNC, respectively. Na$^+$-NQR is unique in that it contains a molecule of riboflavin as a true redox cofactor (Figure 2.5A).[83,88,95-99]

[‡]The properties of iron-sulfur centres are discussed in Chapter 13.

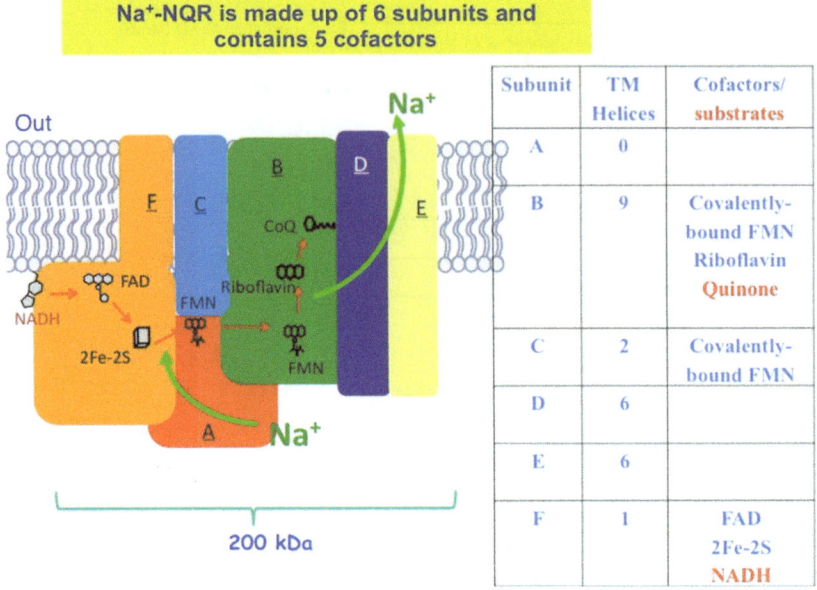

Figure 2.5A Diagrammatic representation of Na$^+$-NQR. The diagram includes the six subunits of the complex and the redox cofactors.

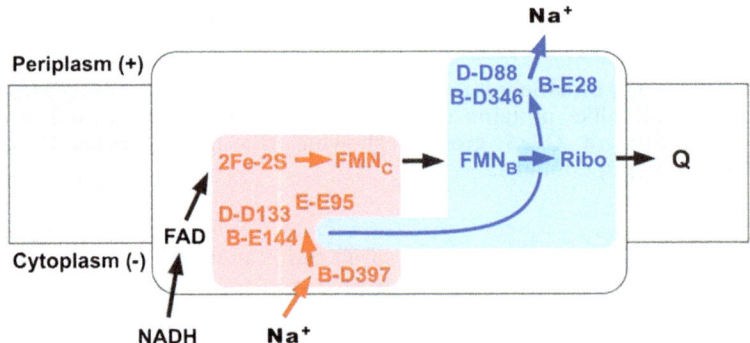

Figure 2.5B Diagrammatic representation of Na$^+$-NQR coupling sites. The red box includes the redox reaction and cytoplasmic-side acidic residues that participate in Na$^+$ uptake. The blue box includes the redox reaction and the periplasmic-side acidic residues that participate in Na$^+$ release.

The current model of the enzyme reaction is summarized in Figure 2.5B. The electron transfer chain can be represented as <u>NADH</u> → FAD → 2Fe-2S → FMN$_C$ → FMN$_B$ → riboflavin → quinone.[102] Two different steps in this redox reaction are linked to different parts of the Na$^+$ transport process (see red and blue boxes in Figure 2.5B). During one of these steps (2Fe-2S → FMN$_C$; red box), Na$^+$ is taken up from solution and binds to a group of acidic residues near the cytoplasmic side of the membrane (see below). This Na$^+$ uptake step does not contribute to the membrane potential. In the other redox step (FMN_B → riboflavin; blue box), the Na^+ ion is transported across the membrane, thereby generating a potential, and is released. A second group of acidic residues located near the periplasmic side of the membrane seems to be involved in this process.

The two redox steps involved in the Na^+ transport do not have a cofactor in common. This suggests that Na^+-NQR does not operate by a simple electroneutrality or "redox-Bohr" mechanism, in which reduction and oxidation of a single site drives uptake and release of the substrate.[99,101]

2.5.1.1.2 Mechanism of Sodium Ion Translocation.

The two groups of acid residues that form the Na^+ uptake and putative exit sites are located on opposite sides of the membrane. The cytoplasmic group (in red in Figure 2.5B; NqrB-E144, NqrB-D397, NqrD-D133 and NqrE-E95) participates in the ion uptake/binding. Replacement of any of these residues by an aliphatic one results in a significant increase in $K_{m(app)}$ for Na^+, consistent with lower Na^+ affinity. In these mutants, electron flow is inhibited at the 2Fe-2S → FMN_C step. The residues of the "periplasmic group" (NqrB-E28, NqrB-D346 and NqrD-D88) are located on the opposite side of the membrane and are involved in the Na^+ translocation process. Replacement of any of these residues by an aliphatic one results in slower turnover but no significant change in $K_{m(app)}$ for Na^+. In these mutants, electron flow is inhibited at the FMN_B → riboflavin step. On the basis of these results, it has been hypothesized that the "cytoplasmic group" residues constitute the Na^+ uptake/binding site of the pump while the "periplasmic group" forms the release site.[100,101,103]

It has also been suggested that Na^+ translocation operates by a delocalized mechanism in which the redox reactions are coupled to conformational changes that in turn control binding and release of Na^+. As described above, at least four acid residues in Na^+-NQR play important roles in Na^+ binding, while three additional Glu or Asp residues participate in later steps of transport. The proposed mechanism is based on the fact that Na^+ uptake and transport across the membrane are each linked to a separate redox step. In addition, a locally coupled mechanism would not be consistent with the generally understood requirements of Na^+ binding to proteins and cation selectivity in pumps. In all cases where structures are known, Na^+ binding sites consist of six negatively charged or polar amino acids arranged octahedrally. Localized coupling does not appear to be consistent with the highly hydrophilic environments in which Na^+ typically binds. Furthermore, the cofactor redox potentials are not dependent on pH, suggesting that the cofactors are sequestered away from the aqueous environment.

Although there is now evidence that the redox potential of FMN_C is dependent on the Na^+ concentration, the magnitude of this dependence is not sufficient to account for Na^+ translocation. It is thus likely that coupling between redox and Na^+ reactions has significant kinetic character.

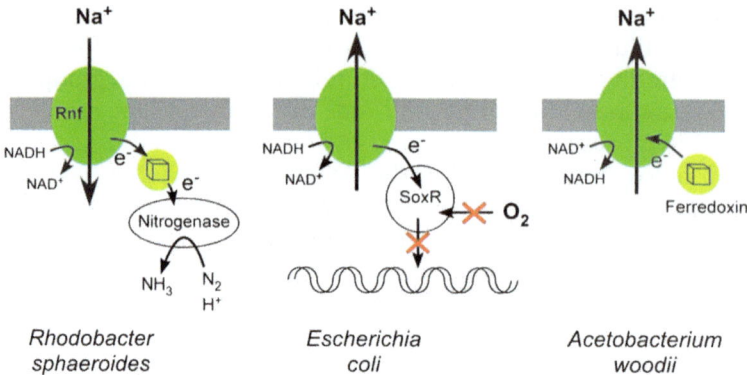

Figure 2.6 Diagrammatic representation of RNF complexes. The diagram illustrates the three possible functions of the complex in different bacteria.

2.5.1.2 Rnf

The Rnf family of membrane proteins is little understood but shares many properties with Na^+-NQR, including structural similarities and the fact that some or all Rnf proteins function as Na^+ pumps (Figure 2.6). However, the Rnf family exhibits more structural and functional diversity than Na^+-NQR. The enzyme was originally found in *Rhodobacter capsulatus sphaeroides* where it delivers low potential reducing equivalents to nitrogenase to supply energy for the fixation of nitrogen. It was therefore named the "Rhodobacter nitrogen fixation" protein.[104–109] Rnf has been identified in the genomes of a number of bacteria and archaea including *V. cholerae, Y. pestis, Porphyromonas gingivalis and E. coli*. In the latter, Rnf is responsible for maintaining the oxygen radical sensor SoxR in a ready state.[109] In the anaerobic bacteria *Clostridium tetanomorphum* and *Acetobacterium woodii*, Rnf pumps Na^+ out of the cell. The driving force is provided by electron transfer from a 4Fe-4S ferredoxin to NAD^+.[87] Rnf has also been identified in *Methanosarcina acetivorans*, but its function in that organism has not been determined clearly.

The Rnf enzymes from *A. woodii* and *M. acetivorans* are able to translocate Na^+ in membrane vesicles.[86,111] However, there is a recent report that the Rnf complex from *Clostridium* pumps H^+ instead of Na^+.[110] It is worth noting that the conserved residues involved in Na^+ pumping in Na^+-NQR are also found in RNF, which suggests a common function.

2.5.2 Decarboxylases that Pump Sodium Ions

These enzymes remove the carboxyl moiety from carboxylic acids producing CO_2, and use the energy liberated to transport Na^+ across the cell membrane. These enzymes are of special importance for pathogens (such as *Vibrio* and *Klebsiella*) that use citrate as a carbon source.[111–113] The Na^+-dependent decarboxylases export one or two Na^+ ions and include oxaloacetate decarboxylase, methylmalonyl CoA decarboxylase, glutaconyl CoA decarboxylase and malonate decarboxylase.[114-116]

Oxaloacetate decarboxylase was the first to be studied and was demonstrated to be a sodium pump. It plays a central role in anaerobic metabolism of citrate. The latter anion is transported into the cell by a sodium-dependent transporter and then cleaved into oxaloacetate and acetate by citrate lyase. Oxaloacetate is then decarboxylated to pyruvate by oxaloacetate decarboxylase and the free energy of this reaction used to pump Na^+ out of the cell. The electrical component of the Na^+ gradient created by this enzyme drives the synthesis of ATP by the H^+ F_1F_0 ATP-synthase.[117,118]

The enzyme consists of three subunits, alpha, beta and gamma. Briefly, the alpha subunit is peripherally bound to the membrane on the cytoplasmic site and contains the carboxyl transferase domain near the N-terminus and a biotin-binding domain near the C-terminus. The two domains are connected by a Pro-Ala peptide. The beta subunit is a transmembrane protein involved in Na^+ translocation. The gamma subunit also resides in the membrane and interacts with the other two to maintain the stability of the complex. The gamma subunit contains a binding site for Zn^{2+}.

The mechanism of the enzyme has been defined. Oxaloacetate binds at the alpha subunit and one of its carboxyl groups (Asp-203) binds to biotin near the beta subunit. The decarboxylation site in the beta subunit acts upon this carboxybiotin fragment in a reaction coupled to the transport of an outside proton to biotin, the release of CO_2 into the cytosol and the pumping of two Na^+ ions to the periplasm. These reactions are accompanied by conformational changes that allow flipping of the biotin between the alpha and beta subunits and the opening and closing of sites for sodium uptake and sodium release. Mutagenesis studies on the beta subunit have identified the amino acids involved in Na^+ transport. Asp-203 in helix IIIa is absolutely necessary for both the decarboxylase activity and the Na^+ translocation. It plays a role in Na^+ binding and also in the transport of the H^+ ion required for the decarboxylase reaction. Mutations of Ser-382 (in helix VIII) also affect both reactions. These properties indicate that these groups deliver a H^+ to the catalytic site of the decarboxylation reaction. Mutants S382E and S383C are inactive, indicating that the length of the side-chain at this position is important for H^+ release. Ser-382 also plays a role in sodium binding since the mutant S382E inhibits Na^+ translocation, possibly because a larger side-chain at this position disrupts Na^+ binding. Ser-383 plays also a double role in the mechanism. There is also evidence for the presence of two Na^+ binding sites in the enzyme. The rate of oxaloacetate decarboxylation depends on the concentration of Na^+ in a cooperative fashion: both Na^+ binding sites have to be occupied for the decarboxylation reaction to start.[120]

A model has been proposed for the mechanism of oxaloacetate decarboxylase on the basis of this mutagenesis work.[119-123] The enzyme contains two Na^+ binding sites in the beta subunit, located on opposite sides of the membrane, one involving Asp-203 and the other involving Ser-383. The enzyme adopts at least two conformations; a Na^+ site is accessible from the cytosol in one conformation and another is accessible from the periplasm in the second conformation.

The carboxybiotin group moves from the alpha catalytic site to the beta subunit, allowing two Na^+ ions to access their binding sites. The first binds at Asp-203 and the second at Ser-383. A conformational change occurs that allows the two Na^+ to be released to the periplasm. The displacement of Na^+ from Ser-383 allows a proton to enter from the periplasm restoring the hydroxyl group at S383. This mechanism uses residues in the hydrophobic part of the membrane as binding sites for opposite movement of Na^+ from the cytosol to the periplasm and the opposite movement of H^+ from the periplasm to the cytosol.

2.5.3 The Sodium Ion-translocating Methyl Transferase from Methanogenic Archaea

Methanogenic archaea obtain energy from an anaerobic disproportionation reaction in which acetic acid or other C1 compounds are converted to CO_2 and methane (CH_4). This reaction is dependent on Na^+ and the methyl transfer from N5-methyl-tetrahydromethanopterin to co-enzyme M is the sodium-dependent step, catalysed by a sodium-pumping membrane-bound methyltransferase.

The N5-methyl-tetrahydromethanopterin:co-enzyme M methyltransferase (Mtra-H) has been purified from membrane fractions of *M. mazei* and *M. marburgensis*. The enzyme is a subunit complex of 670 kDa that features 5-hydroxybenzimidazolyl cobalamide as a cofactor. The reconstituted enzyme pumps Na^+ with an experimental stoichiometry of 1.7 mol Na^+ translocated per mol of methyl group transferred. The mechanism of Na^+ translocation is not understood: it has been speculated that subunit MtrE participates in Na^+ translocation but no evidence was provided.[124,125]

2.6 The Sodium Ion-translocating ATPase/ATP Synthase

2.6.1 ATPase as a Rotary Motor

Ion-translocating rotary ATPases are classified into three groups (F-, V- and A-ATPases) based on their function and phylogenetic origin.[126,127] They are distinguished from P-type ATPases, such as Ca^{2+}-ATPases and Na^+/K^+-ATPases, by their rotational mechanism, and by the fact that there are no phosphorylated intermediates in their catalytic cycles. F-ATPases are found in the membranes of mitochondria, chloroplasts and respiring bacteria where they synthesize ATP from ADP and phosphate using energy derived from a H^+ or Na^+ gradient.[128] V-ATPases have a similar mechanism but physiologically they operate in the opposite sense, functioning as proton pumps in acidic organelles and eukaryotic plasma membranes. A-ATPases, so named because they are found in Archaea, play the same physiological role as F-ATPases, synthesizing ATP, but are more similar in structure and subunit composition to V-ATPases.[129]

The rotary ATPases consist of a hydrophilic catalytic domain (F_1, V_1 or A_1) and a membrane-embedded ion-translocating domain (F_0, V_0 or A_0), which interacts with ADP, inorganic phosphate and ATP and mediates the movement of either protons or sodium ions across the membrane, respectively.

Homologues of eukaryotic V-ATPase have been identified in some eubacteria, including *Thermus thermophilus* (Tt-V-ATPase) and *Enterococcus hirae* (Eh-V-ATPase). Tt-V-ATPase synthesizes ATP under physiological conditions and is thus sometimes classified as an A-ATPase. Eh-V-ATPase acts as a primary ion pump but translocates Na^+ and Li^+ instead of protons.

2.6.2 Structure of the Sodium Ion-translocating V-ATPase

A schematic diagram of the Na^+-translocating Eh-V-ATPase (Figure 2.7) shows the nine subunits (Eh-A, -B, -d, -D, -E, -F, -G, -a and -c) arranged in ATPase-driven (V_1) and ion-translocating motor (V_0) domains. Each subunit has homology to a corresponding subunit of the eukaryotic V-ATPase.[130] However, yeast V-ATPase is composed of 14 subunits making the composition of the eukaryotic V-ATPase more complex than that of prokaryotic V-ATPase.[131] The core of the V_1 domain consists of a hexameric arrangement of alternating A and B subunits, which have homology with the α and β

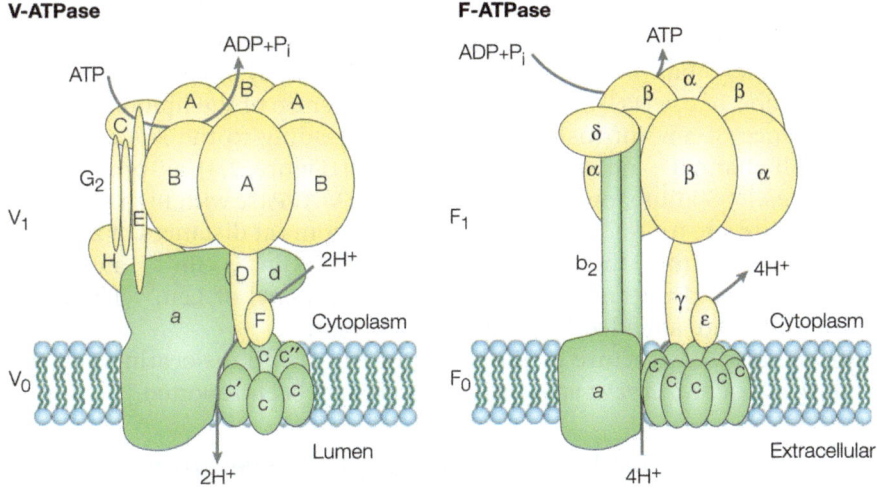

Figure 2.7 Structural comparison of the V- and F-ATPases. For the V-ATPases, ATP hydrolysis by the peripheral V_1 domain (shown in yellow) drives proton transport through the integral V_0 domain (shown in green) from the cytoplasm to the lumen. In contrast, the F-ATPases (or ATP synthases) normally function in the opposite direction (that is, in ATP synthesis), coupling the downhill movement of protons through F_0 to the synthesis of ATP in F_1. Both enzymes are thought to operate through a rotary mechanism.

subunits of the F_1-ATPase.[132] The V_0 domain, in which rotational energy drives Na$^+$ translocation, is composed of oligomers of the 16-kDa c and a subunits.[133,134] The c subunits, each of which has an ion-binding site, form the rotor ring. The V_1 and V_0 domains are connected by a central stalk and two peripheral stalks, which are composed of E and G subunits and D, F and d subunits, respectively (Figure 2.7). The V_1-G subunit is homologous to the F_1-b subunit. The V_0-a, V_0-d, V_1-D, V_1-E and V_1-F subunits have no homologous protein in the F-ATPase. In addition, the eukaryotic V-ATPase includes a C subunit and an H subunit that may regulate enzymatic activity. ATP hydrolysis induces rotation of the central stalk (DFd complex) and an attached membrane ring c that induces ion pumping at the interface between the c ring and the a subunit.[133]

2.6.3 Sodium Ion Binding Site in the K Ring

ATP synthases/ATPases display ion selectivity at their membrane-embedded domain, particularly at their rotor ring (K ring). The ion selectivity and rotary action of the ion-translocating F_0-ATPase domain have been extensively studied. In particular, a detailed reaction mechanism has been proposed for F_1-ATPase, in which ATP synthesis and hydrolysis reactions are catalysed.[135] A high-resolution crystallographic structure of the rotor ring (K ring) of the Na$^+$-translocating V-ATPase has been determined, and a detailed mechanism of transport has been proposed.[133,136] A structure of the rotor ring of V-type Na$^+$-ATPase from *E. hirae* at 2.1-Å resolution has also been reported. This latter structure unequivocally shows the ring to have 10-fold symmetry, and it is likely that this symmetry persists in the fully assembled V-ATPase. The position of the Na$^+$-binding site near the centre of the lipid bilayer, and the absence of any intrinsic K ring channel leading to the site supports a "two-half-channel" ion translocation model (Figure 2.8) as proposed for F-ATPases.[137] Sodium ions bind to the *E. hirae* V-ATPase with high affinity.[130] The binding site for Na$^+$ includes five oxygen atoms at distances of 2.2–2.3 Å, four of them from the side-chains of Thr^{63}, Gln^{64}, Gln^{108} and Glu^{137} (Figure 2.9), and the fifth from a backbone carbonyl of Leu.^{60} The bound sodium ions are shielded by a Glu^{137} side-chain.

In addition, a structure of the c ring of the Na$^+$-translocating F-ATPase from *Ilyobacter tartaricus* at 2.4 Å resolution has been reported.[138]

2.6.4 The Sodium Ion-translocating F-ATPase from Polyextremophilic *Natranaerobius thermophilus*

Microorganisms that can tolerate two or more extreme conditions are defined as "polyextremophiles". Anaerobic *N. thermophilus* is a halophilic and alkalithermophilic eubacterium that exhibits optimum growth at Na$^+$ concentrations in the range 3.3–3.9 M, pH 9.5 and 53 °C.[139] The F-ATPase of *N. thermophilus* (Nt-F-ATPase) has been studied to understand how its ATP synthesis mechanism is adapted to these harsh environmental conditions. It

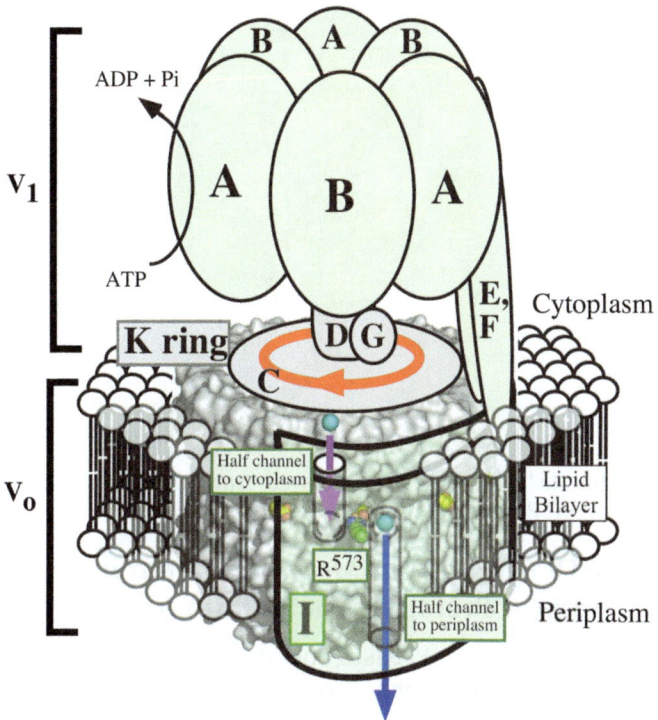

Figure 2.8   A schematic structure model for ion translocation by the V-ATPase of *E. hirae*. The V-ATPase is composed of a hydrophilic catalytic moiety (V_1) and an integral membrane moiety (V_0) for Na^+ translocation. ATP hydrolysis in A3B3 generates rotation of the central stalk (D, G and C subunits) and an attached membrane ring (K ring) as indicated by the red arrow. Na^+ ions are pumped across the membrane at the interface between the K ring and I-subunit. This ion translocation model is based on the Na^+-bound K ring structure that supports the two-channel model of F-ATPase.[137] The K ring is shown in space-filling representation with essential Glu-139 (yellow and red for oxygen atoms). Sodium ions are shown as blue spheres. The I-subunit (structure unknown) possesses two half-channels and an essential Arg-573 that are required for Na^+ translocation across the membrane. Clockwise rotation of the ring (red arrow) driven by ATP hydrolysis and an electrostatic Arg-573–Glu-139 interaction enables the release of bound Na^+ into a half-channel connected to the periplasm as indicated by the blue arrow and the binding of Na^+ from a second half-channel connected to the cytoplasm as shown by the purple arrow.

acts as a Na^+ pump and plays a critical role.[139] In addition, the ATP synthesis activity of Nt-F-ATPase was found to be extremely low, though it may be regulated by other factors. The complete ε subunit of Nt-F-ATPase is hypothesized to control excessive ATP consumption and to preserve the intracellular levels of Na^+ required by electrogenic Na^+ (K^+)/proton antiporters that are crucial for cytoplasmic acidification in the obligatory alkaliphilic *N. thermophilus*.

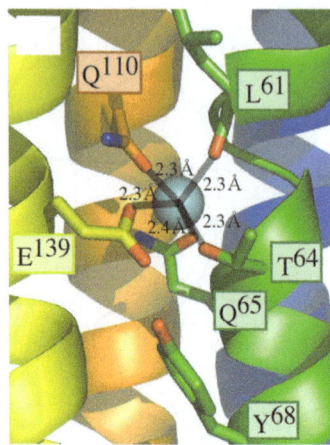

Figure 2.9   The Na$^+$-binding pocket viewed from outside the K ring of V-type Na$^+$-ATPase from *Enterococcus hirae*.

2.6.5   The ATP Synthase in *Methanosarcina acetivorans* is Driven by a Proton and a Sodium Ion Gradient

Until 2012, all known rotary ATPases operated with a single coupling ion (H$^+$ or Na$^+$). However, it was then reported that the A-ATPase of Archaea *Methanosarcina acetivorans* can drive ATP synthesis under physiological conditions using transmembrane gradients of *either* H$^+$ or Na$^+$.[140] Such dual ion specificity has been observed in other biological systems, *i.e.* some cation/solute symporters,[141,142] and the flagellar motors of alkaliphilic microorganisms.[143,144] This finding may facilitate the identification of other cation coupling capacities among the rapidly growing number of bacteria whose genomic sequences are available.

2.6.6   Future Prospects

V-ATPase is expressed in the cell membranes of osteoclasts and cancer cells. The acid environmental abnormality associated with this enzyme is one of the causes of the metastasis in osteoporosis in combination with cancer. A specific V-ATPase repressor would be expected to be a novel drug for those diseases and a target for antifungal drugs.[128,145]

2.7   A Voltage-gated Sodium Ion Channel, NaChBac

Voltage-gated Na$^+$ (Na$_V$) channels play an important role in the rapid de-polarization of mammalian neurons and muscle cells, and have become the target of innovative drug development.[146] A Na$_V$ channel was discovered in the alkaliphilic *Bacillus halodurans* (NaChBac) in 2001.[147] This was selective for sodium, even though its sequence is similar to that of voltage-gated Ca^{2+} channels. Pore-forming subunits of mammalian Na$_V$ and voltage-gated Ca^{2+}

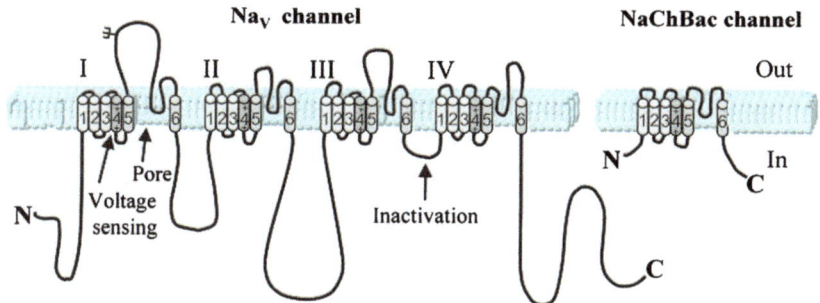

Figure 2.10 Schematic diagram of eukaryotic Na_V channel and prokaryotic NaCh-Bac channel.

channels are encoded by four repeated domains (called the α domain) of six transmembrane (6TM) segments. In contrast, NaChBac is encoded by one 6TM segment and four monomers forming a tetrameric channel (Figure 2.10). NaChBac is blocked by the dihydropyridine class (nifedipine and nimodipine) of L-type voltage-gated Ca^{2+} channel blockers.[148] NaChBac superfamily members have been identified in diverse bacteria whose common habitats include marine and/or alkaline niches.[149]

Since its discovery, several research groups worldwide have tried to crystallize NaChBac because its structure is simpler. The crystal structure was reported in 2011 for the NaChBac (Na_VAb) derived from the eubacterium *Arcobacter butzleri* (Figure 2.11).[150] The crystal structure of an orthologue (Na_VRh) derived from the eubacterium *Rickettsiales* sp. HIMB114 was reported at about the same time.[151] As shown in Figure 2.11, the segments 5 and 6 of each repeat of the α subunit gather and form a pore domain, and a voltage sensor composed of segments 1, 2, 3 and 4 (S1, S2, S3 and S4) is located in the fourth corner. Ion selectivity is determined in the portion called the ion selectivity filter, for which the pore diameter is narrowest. The Na_V channel has some permeability for ions of large ionic radius, including guanidinium ions. The width of the ion selective filter of the Na_V channel is larger than the diameter of Na^+. Therefore, it was predicted that the pore selects hydrated sodium ions.[151,152] The pore diameter at the narrowest part of the ion-selective filter of the Na_VAb crystal structure is identical to that of hydrated Na^+.[150]

2.7.1 The Physiological Roles of NaChBac in Bacteria

2.7.1.1 Re-entry Routes for Sodium Ions are Required for Optimal pH Homeostasis in Extremely Alkaline Conditions

It has been hypothesized that the Na_V channel provides an important reentry route for Na^+ to support the critical function of alkaliphile pH homeostasis such as under conditions where solutes whose uptake is

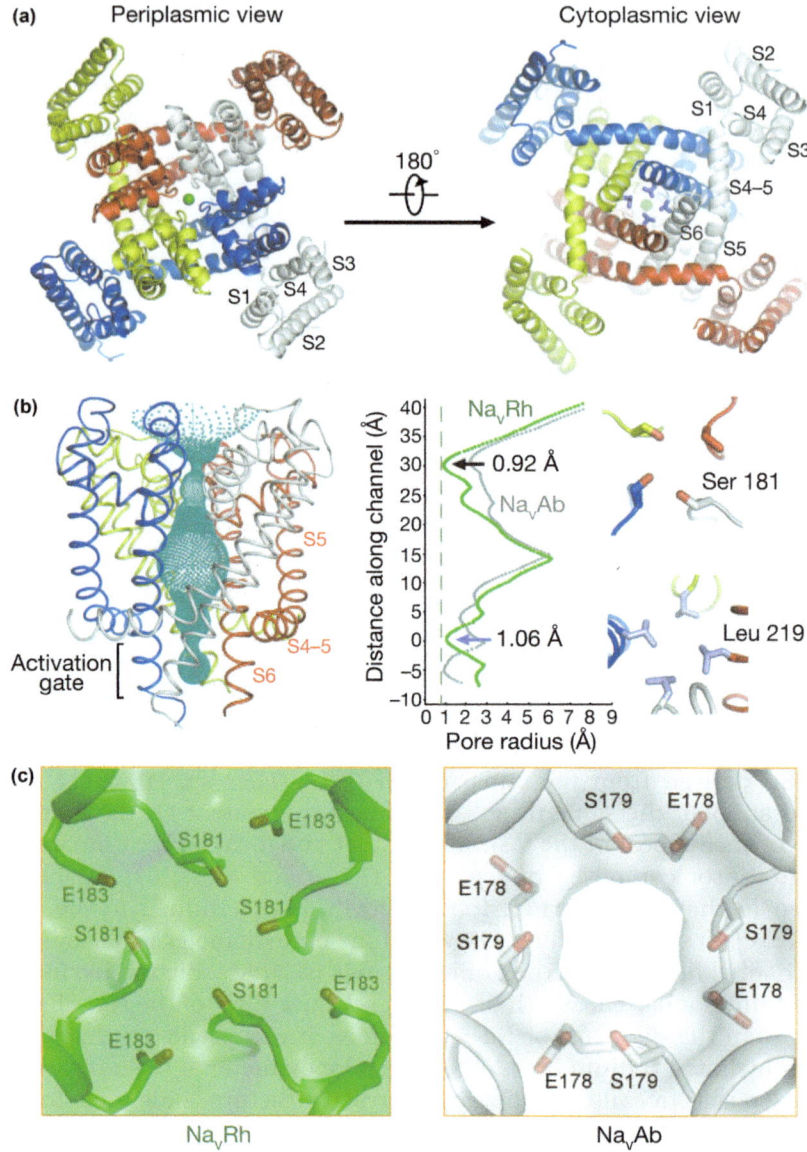

Figure 2.11 The structure of Na$_V$Rh exhibits a closed conformation. **a)** Na$_V$Rh crystallized as an asymmetric tetramer. The green sphere indicates the bound ion within the selectivity filter. Leu-219, which occludes the activation gate, is shown in light purple sticks in the cytoplasmic view. **b)** Na$_V$Rh is closed at the activation gate and the entrance to the selectivity filter. The channel passage (left panel) is indicated by cyan dots. The pore radii (right panel) of Na$_V$Rh (green) are compared with those of Na$_V$Ab (grey). The residues that constitute the constriction sites, Ser-181 at the entrance to the selectivity filter and Leu-219 at the activation gate, are shown in sticks in periplasmic and cytoplasmic views, respectively. **c)** A semi-transparent surface illustration of the periplasmic entrance to the selectivity filter in Na$_V$Rh and Na$_V$Ab.

Table 2.1 Mutational loss of Na$_V$BP and MotPS affect pH homeostasis in alkaline-shifted cells. *ncbA* encodes Na$_V$BP belonging to the Na$_V$Bac superfamily of bacterial voltage-gated Na$^+$-selective ion channels. The values are mean values from at least two independent experiments in which cells equilibrated at pH 8.5 were subjected to shift of pH$_{out}$ to 10.5. Mutants lacking functional Na$_V$BP were also defective in pH homeostasis in response to a sudden alkaline shift in external pH under conditions in which cytoplasmic [Na$^+$] is limiting for this crucial process. The role of MotPS in pH homeostasis at high pH is apparently masked in the single *motPS* mutant by compensatory Na$_V$BP activity. (This table is a modified version of Table 3 from ref. 148.)

Strain (phenotype)	Cytoplasmic pH, 10 min after pH shift from 8.5 to 10.5	
	100 mM added Na$^+$, pH$_{in}$	*2.5 mM added Na$^+$, pH$_{in}$*
B. pseudofirmus OF4-811M (wild-type)		
SC34 (811M Δ*ncbA*::Spc^R)	8.46	9.03
SC34-R (SC34 *ncbA* restored)	**8.66**	**9.43**
Mot6 (811M Δ*motPS*::Cm^R)	8.48	9.02
SC34/Mot6 (811M Δ*ncbA*::Spc^R Δ*motPS*::Cm^R)	8.52	9.03
	8.72	**9.51**

coupled to Na$^+$ are scarce.[153,154] A Na$^+$ channel that opens at high pH was predicted to be an important alternative Na$^+$ re-entry route to support cytoplasmic pH homeostasis.[153,154] The Na$^+$-translocating Mot channel associated with the flagellar rotor was a prime candidate for such a channel because alkaliphile motility is restricted to alkaline pH values.[154,155] However, no pH homeostasis defect was revealed upon disruption of the genes encoding MotPS, the Na$^+$-translocating stator-channel proteins required for motility of the alkaliphilic *B. pseudofirmus* OF4[148] (Table 2.1).

2.7.1.2 The Physiological Functions of NaChBac (Na$_V$BP) in Alkaliphilic B. pseudofirmus OF4

Studies are reported only for the alkaliphilic *B. pseudofirmus* OF4.[148,156,158] The gene named *ncbA* encodes Na$_V$BP, which belongs to the NaChBac superfamily of bacterial Na$_V$ channels. To investigate the physiological functions of this Na$_V$ channel, a NaChBac (Na$_V$BP)-defective mutant SC34 (Δ*ncbA*) was constructed from *B. pseudofirmus* OF4. Wild-type *B. pseudofirmus* OF4, the channel mutant SC34 and the SC34-R complemented strain (in which *ncbA* was restored) all exhibited growth doubling times ranging from 95–110 min. at pH 7.5 and 70–75 min at pH 10.5, respectively. The growth yield of SC34 was reduced ≦20% relative to wild-type on complex medium at pH 10.5. SC34 was also defective in pH homeostasis in response to a sudden alkaline shift in external pH, under conditions in which cytoplasmic Na$^+$ is limiting for this crucial process (Table 2.1). The role of MotPS, a second Na$^+$ channel in *B. pseudofirmus* OF4, in pH homeostasis at high pH is apparently

Table 2.2 Inverse chemotaxis behaviour of up-motile *ncbA* mutant. Chemotaxis behaviour was assessed by a capillary assay of pH 10.5-grown strains of up-motile wild-type, SC34 and SC34-R that swim well enough for use of this assay. The low nutrient condition compromises pH homeostasis, so a pH of 8.5 was used instead of pH 10.5. Chemical effectors were added at 1 mM and the buffer pH inside the capillary was 8.5 for aspartate, proline and glucose, and 10.5 for malate; pH 10.5 was also tried as a variable without malate. The buffer pH outside the capillary was always pH 8.5.[148] (This table is a modified version of Table 2 from ref. 148.)

Strain	Attractants	Repellants
Wild-type SC34-R	Aspartate (pH 8.5) Proline (pH 8.5) Glucose (pH 8.5) Malate (pH 10.5)	pH 10.5
SC34	pH 10.5	Aspartate (pH 8.5) Proline (pH 8.5) Glucose (pH 8.5) Malate (pH 10.5)
Wild-type (with 50 mM nifedipine treatment)		Aspartate (pH 8.5)

masked in the single *motPS* mutant Mot6 by compensatory $Na_V BP$ activity. Therefore, $Na_V BP$ is a dominant Na^+ channel for pH homeostasis at high pH. In addition, the *ncbA* mutant exhibited greatly reduced motility at pH 10.5 that was reversed by restoration of the native channel.[148] $Na_V BP$ has an additional role in chemotaxis: mutants in $Na_V BP$ exhibited inverse chemotaxis, as did the wild-type alkaliphile when treated with the $Na_V BP$ channel inhibitor nifedipine (Table 2.2).

2.7.2   Localization of NaChBac in the Cell

Immunofluorescence microscopy (IFM) and fluorescent protein fusion studies have revealed that an alkaliphile protein (designated McpX) that cross-reacts with antibodies raised against *Bacillus subtilis* chemoreceptor McpB co-localizes with $Na_V BP$ at the cell poles of *B. pseudofirmus* OF4 (Figure 2.12). In a mutant in which the $Na_V BP$-encoding *ncbA* is deleted, the content of McpX was similar to the wild-type level but McpX was significantly delocalized. A Δ*cheAW* mutant of *B. pseudofirmus* OF4 was constructed to assess whether this mutation impaired McpX polar localization, as expected from studies in *E. coli* and, if so, whether $Na_V BP$ would be similarly affected. Polar localization of both McpX and $Na_V BP$ decreased in the *cheAW* mutant. The results suggest interactions between McpX and $Na_V BP$ that affect their co-localization. The inverse chemotaxis phenotype of *ncbA* mutants may result, in part, from MCP delocalization. The cytoplasmic C-terminus of $Na_V BP$ contains four aspartate residues that could serve as potential phosphor-acceptors for the CheA kinase; changes in C-terminal phosphorylation of eukaryotic channels can profoundly affect channel gating.[157] Further work

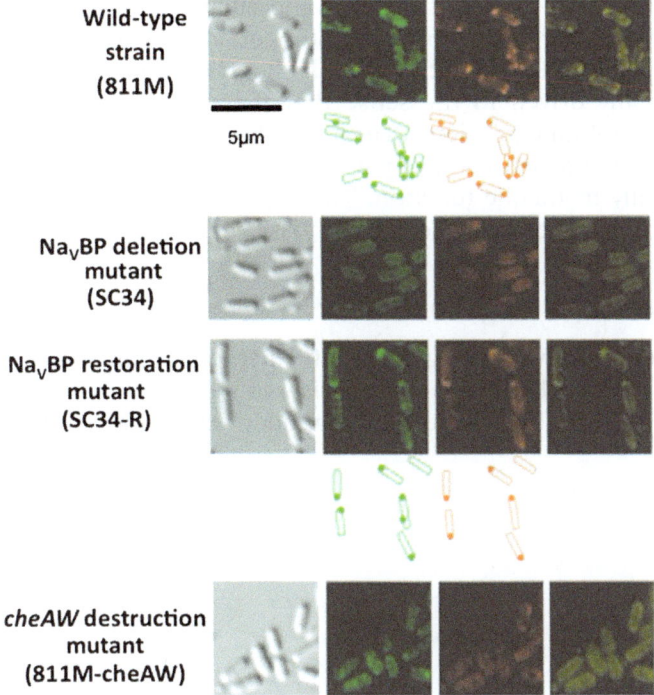

Figure 2.12 Immunofluorescence microscopy analyses of the cellular localization of Na$_V$BP and McpX in *B. pseudofirmus* OF4. The strains are indicated on the left side of the figure and the image analyses are indicated at the top: (A) Wild-type (811M), SC34, SC34-R and 811M-CheAW; (B) Nifedipine-treated wild-type 811M. First (left) column of images: DIC, differential-interference contrast microscopy. Second column of images: Na$_V$BP, Immunofluorescence microscopy for Na$_V$BP (Alexa Fluor 488 rabbit anti-mouse IgG is green). Third column of images: McpX, Immunofluorescence microscopy for McpX (Alexa Fluor 546 goat anti-rabbit IgG is red). Fourth column of images: Overlay images of the images in columns 2 (Na$_V$BP) and 3 (McpX). Bar, 5 μm. Underneath the photographs, cartoons are provided to indicate the interpretation of each image that was used for purposes of quantification of localization.

will be required to define the nature of MCP and channel interactions in the alkaliphile, and the effect of such interactions on channel function and chemotaxis.

2.8 Sodium Ion-coupled Flagellar Motility

Living organisms have molecular-scale motors that are similar in many ways to the motors made by human technology. They include both linear and rotary motors, and play important roles in locomotion, transport and metabolism. The sliding of myosin and actin filaments in muscle tissues, and the tubulin-dynein system of flagellar and ciliary movement are among the linear motors. Four types of rotary motors exist in nature: the F-ATPase, the V-ATPase, the A-ATPase and the bacterial flagellar motor. All of these motors are electrically driven in the sense that they obtain energy from electro-chemical membrane gradients. However, the ATPases all use rotary motion to transmit energy between two parts of an enzyme complex that carries out an essentially metabolic function. The flagellar motor is the only naturally occurring rotary motor whose function is locomotion.

2.8.1 Bacterial Flagellar Motor

Many bacteria can swim using flagella, filamentous organelles extending from the cell surface. A flagellar motor consists of the filament (helical propeller), the hook (universal joint) and the basal body (rotary motor) (Figure 2.13).[160]

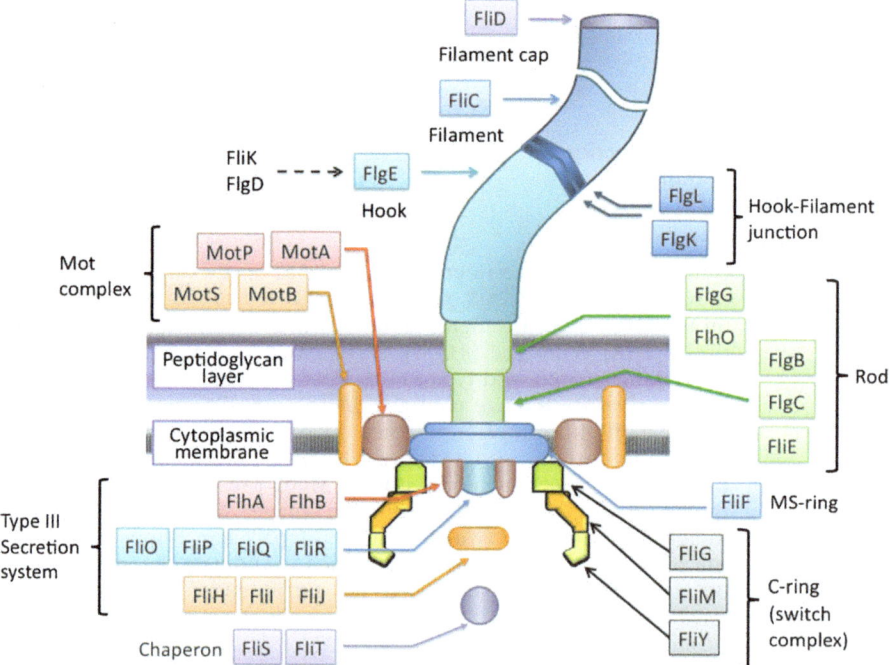

Figure 2.13   The flagellum in Gram-positive bacteria that are conserved in many *Bacillus* species including alkaliphilic *Bacillus*. Gram-negative species like *Escherichia coli* have the LP-ring assembly. In *Bacillus* species, there is an FlhO protein that has been reported to substitute for FlgF (Proximal rod), FlgI (P ring) and FlgH (L ring). An additional flagellar assembly factor FliW was reported in *B. subtilis* and *Campylobacter jejuni*. *B. subtilis* has the orthologues FliM and FliG but instead of FliN this species has FliY, which is homologous to FliN only in its C terminus. *B. subtilis* has two kinds of flagellar stator complexes, MotAB and MotPS. Some alkaliphilic *Bacillus* species have only a sodium-coupled MotPS type stator instead of proton-coupled MotAB type.

The motor is embedded in the cell envelope and is powered by an electro-chemical gradient of H^+ or Na^+ across the cytoplasmic membrane.

2.8.2 Proton-coupled Flagellar Motors

E. coli and *Salmonella enterica* both have H^+-driven flagella. In these motors, five proteins (MotA, MotB, FliG, FliM and FliN) appear to be responsible for torque generation. Of these, only MotA, MotB and FliG are thought to be directly involved in the energy conversion that generates rotation from the flow of ions through the membrane.[159-163] MotA and MotB make up a H^+ channel complex (Figure 2.14). MotA comprises four transmembrane segments with a large cytoplasmic loop between the second and third segments. MotB has a single transmembrane segment and a C-terminal domain that is believed to bind to the peptidoglycan layer and anchor the MotA/MotB complex as a stator around the basal body[164-166] (Figure 2.14). According to electron microscopy, FliG, FliM and FliN proteins are components of the "switch complex" ring structure similar to that formed by the c subunits of the ATPase.[167,168] These three proteins are necessary for flagellar assembly, torque generation and rotation control (counter-clockwise or clockwise).[169,170] The MotA/MotB complex converts the H^+ flux into mechanical power for flagellar rotation. The ion flow induces a conformational change in the stator complex that modulates interactions between the cytoplasmic loop regions of MotA and FliG, to drive the motor.

2.8.3 Sodium Ion-coupled Flagellar Motor

The halophilic *V. cholerae* and the extremely alkaliphilic *Bacillus pseudofirmus* OF4 are the only bacteria that have been shown to rely solely on Na^+-coupled motors for motility, but some other alkaline-resistant bacteria

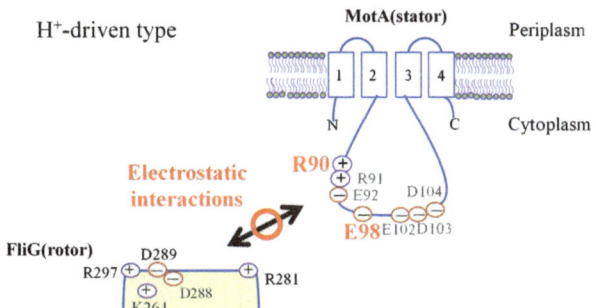

Figure 2.14 Interaction between rotor and stator in torque generation in *E. coli*. In *E. coli*, there are conserved charged residues in the cytoplasmic loop of MotA that interact with the conserved charged residues of the C-terminal domain of FliG.

possess both Na$^+$- and H$^+$-driven flagellar motors.[171–177] *V. parahaemolyticus* and *V. alginolyticus* constitutively express a single Na$^+$-driven polar flagellum and conditionally express multiple H$^+$-driven lateral flagella. Distinct sets of genes encode these polar and lateral flagella. Genomic evidence suggests that many other alkaline-resistant bacteria have hybrid motility systems. In each bacterial setting, the two systems appear to optimize motility under different conditions.[172,173] However, unlike the two examples above, few of these are encoded by entirely distinct sets of genes coding for motors that use different coupling ions.

2.8.4   Dual Motility Systems of Bacterial Flagellar Motors

B. subtilis possesses two stators, the H$^+$-coupled MotA/MotB and the Na$^+$-coupled MotP/MotS, but possess only one set of flagellar rotor genes. Thus, the two different stators apparently interact with a single rotor form FliG.[177–179] *Shewanella oneidensis* MR-1 also has two different stator systems, a H$^+$-coupled MotA/MotB and a Na$^+$-coupled PomA/PomB that drive a single polar flagellum.[180] In *B. subtilis*, mutations in the large *fla/che* operon (that encodes FliG and the other proteins of the rotor switch complex) abolish all motility.[181] This system will help to clarify whether the two stators with different ion coupling properties interact differently with a single FliG.

2.8.5   Proposed Mechanisms of Bacterial Flagellar Rotation

Figure 2.14 illustrates a schematic model for MotA–FliG interactions during rotation of H$^+$-driven flagellar motors.[182] Residues in the loop between the second and third transmembrane helices of MotA and the C-terminal FliG region have been analysed by mutagenesis. Residues Arg-90 and Glu-98 in MotA and Lys-264, Arg-281, Glu-288, Glu-289 and Arg-297 in FliG were shown to be important for motility.[169,182] Crystal structure analysis of *Thermotoga maritima* FliG indicated that these residues are localized at the outer edges, and on this basis it was proposed that electrostatic interactions between residues on the stator cytoplasmic loop and the C terminus of FliG drive motor rotation.[183,184] *V. alginolyticus* PomA shares two conserved charged residues with MotA (Arg-88 and Glu-96). When these residues in MotA are substituted with neutral residues, motor rotation is prevented. However, substitutions of charged residues in PomA do not affect the rotation of the Na$^+$-driven flagellar motor.[185] This difference suggests that electrostatic interactions in the Na$^+$-driven PomA/PomB-type flagellar motor are different or that other charged residues are involved.[185,186] However, it is worth noting that charged residues in the *V. alginolyticus* rotor and stator proteins are essential for motor rotation, when the proteins function in the setting of the *E. coli* motor.[187]

2.8.6 The Energy Source for the Flagellar Stator Protein from Alkaliphilic Bacteria

The motility of alkaliphilic *Bacillus* species depends on SMF. The genomic sequence of alkaliphilic *B. halodurans* C-125[171] revealed one set of genes encoding a MotA/MotB-like pair of proteins downstream of a putative *ccpA*, which encodes a central regulator of carbon metabolism in *Bacillus* species and other Gram-positive bacteria.[176,188] When *B. subtilis* homologues of this pair (MotPS: for pH and salt) were first described, they did not appear to have a role in motility. These results led to the role of MotPS in this alkaliphile to be re-assessed.

To date, based on the properties of each flagellar stator, there are at least four groups of flagellar motors in alkaliphilic *Bacillus* species (Figure 2.15):

Group 1: organisms including *B. pseudofirmus* and *B. halodurans*, which have a Na^+-coupled stator (MotPS), and use SMF for flagellar rotation;

Group 2: organisms including *Oceanobacillus iheyensis* and the neutralophilic *B. subtilis*, which have two stators in the same motor, the H^+-coupled MotAB and the Na^+-coupled MotPS. They simultaneously use both SMF and PMF for rotational energy;

Group 3: organisms including *B. clausii* KSM-K16 that have a MotAB variant that uses protons at near-neutral pH, sodium at high pH and both ions at intermediate pH 9;

Group 4: organisms including *B. alcalophilus* Vedder1934, which can use both K^+ and Na^+ to drive flagellar rotation. These bacteria may have both Na^+- and K^+-dependent pH homeostasis at pH 10.[189,190] This was the first report describing a flagellar motor that can use both K^+ and Rb^+ as coupling ions. These findings will increase our understanding of the operating principles of flagellar motors and the molecular mechanisms of ion selectivity; they will also enhance our understanding of the evolution of the ability of bacteria to respond to changes in environment and stress and will also have ramifications for nanotechnology.

2.9 Final Remarks

This chapter reviews some of the remarkable functions of proteins that transport Na^+ in prokaryotes. In bacteria and archaea, Na^+ ions are transported by highly selective enzymes that allow the interior of the cells to maintain a constant low-Na^+ concentration. Most of these enzymes convert energy from one form to another. Some, such as antiporters, interconvert H^+ and Na^+ gradients, while others draw energy from the sodium gradient for important metabolic functions including transport of nutrients, ATP synthesis and motility. Thus, the mechanisms of these enzymes go well beyond simple catalysis. A great deal remains to be learned about the structure and function of these enzymes, as well as their role in the physiology of prokaryotes. High-resolution structures are available only in a few cases.

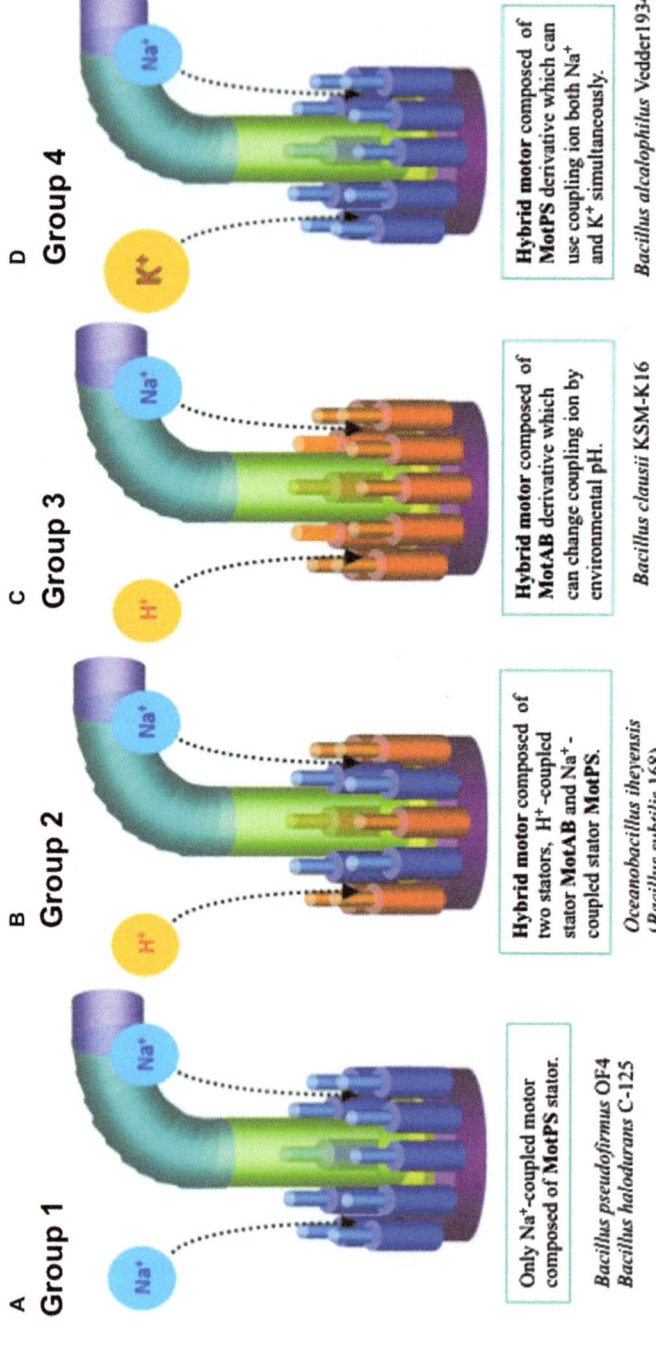

Figure 2.15   A schematic diagram of a classification of alkaliphilic *Bacillus* flagella based on differences in the property of the stator. A: A conserved pair of stator proteins, MotPS, that are homologues of MotAB mediate Na[+]-coupled motility of extreme alkaliphilic *B. halodurans* C-125 and *B. pseudofirmus* OF4. B: Moderately alkaliphilic *O. iheyensis* and neutralophilic *B. subtilis* exhibited dual ion-coupling capacity using MotPS and MotAB as stators for a single rotor. C: Extreme alkaliphilic *B. clausii* has a novel form of dual ion coupling to motility in which a single MotAB stator uses both sodium and protons at different ranges of pH values. D: *B. alcalophilus* Vedder1934 can use both K[+] and Na[+] for flagellar rotation.

Mechanistic and structural studies are complicated by the fact that, because of their function, all of these enzymes must be membrane proteins. However, the study of sodium transporters offers to reveal important insights into the physiology of prokaryotes and into the fundamentals of the chemistry of enzyme catalysis.

References

1. M. J. Page and E. Di Cera, *Physiol. Rev.*, 2006, **86**, 1049–1092.
2. B. Halle and V. P. Denisov, *Biopolymers*, 1998, **48**, 210–233.
3. E. Gouaux and R. Mackinnon, *Science*, 2005, **310**, 1461–1465.
4. M. O. Andreae, J. R. Asmode, P. Foster and L. Van't Dack, *Anal. Chem.*, 1981, **53**, 1766–1771.
5. T. Unemoto, H. Tokuda and M. Hayashi, in *Bacterial Energetics*, ed. T. A. Krulwich, Academic Press, San Diego, California, 1990, vol. XII, pp. 33–54.
6. S. Saum and V. J. Muller, *Bacteriol.*, 2007, **189**, 6968–6975.
7. T. A. Krulwich, G. Sachs and E. Padan, *Nat. Rev. Microbiol.*, 2011, **9**, 330–343.
8. J. P. Dimroth and M. Schmid, *Biochim. Biophys. Acta*, 2001, **1505**, 1–14.
9. L. Costernaro, G. Zaccai and C. Ebel, *Biochemisty*, 2002, **41**, 13245–13252.
10. H. Tokuda and T. Unemoto, *J. Biol. Chem.*, 1984, **259**, 7785–7790.
11. J. M. Wood, *Microbiol. Mol. Biol. Rev.*, 1999, **63**, 230–262.
12. M. M. Harding, *Acta Crystallograph. D, Biol. Crystallogr.*, 2001, **60**, 849–859.
13. M. M. Harding, *Acta Crystallograph. D, Biol. Crystallogr.*, 2001, 57, 401–411.
14. M. M. Harding, *Acta Crystallograph. D, Biol. Crystallogr.*, 2002, 58, 872–874.
15. F. Bauer and G. Dost, *Antimcrob. Agents Chemother.*, 1965, **5**, 749–752.
16. J. M. Lehn, *Science*, 1985, **227**, 849–857.
17. V. V. Martin, A. Rothe and K. R. Gee, *Bioorg. Med. Chem. Lett.*, 2005, **15**, 1851–1855.
18. B. L. Trumpower and R. B. Gennis, *Annu. Rev. Biochem.*, 1994, **63**, 675–716.
19. M. Futai and T. Tsuchiya, in *Ion Transport in Prokaryotes*, ed. B. Rosen and S. Silver, Academic Press, San Diego, 1987, pp. 4–83.
20. E. Padan, K. Herz and A. Rimon, *J. Exp. Biol.*, 2009, **12**, 1593–1603.
21. T. A. Krulwich and D. Mack Ivey, in *Bacterial Energetics*, ed. T. A. Krulwich, Academic Press, San Diego, 1990, vol. XII , pp. 417–447.
22. C. C. Hase, N. D. Fedorova, M. Y. Galperin and P. A. Dibrov, *Microbiol. Mol. Biol. Rev.*, 2001, **65**, 353–370.
23. S. Kato and I. Yumoto, *Can. J. Microbiol.*, 2000, **46**, 325–332.
24. H. Tokuda and K. Kogure, *J. Gen. Microbiol.*, 1989, **135**, 703–709.
25. J. Russell and H. Strobel, *Appl. Environ. Microbiol.*, 1989, **55**, 1–6.

26. J. Russell, H. Strobel, A. Driessen and W. Konings, *J. Bacteriol.*, 1988, **170**, 3531–3536.

27. S. R. Malloy, in *Bacterial Energetics*, ed. T. A. Krulwich, Academic Press, San Diego, 1990, vol. XII , pp. 203–224.

28. H. Jung, D. Hilger and M. Raba, *Front. Biosci.*, 2012, **17**, 45–59.

29. E. Padan, E. Bibi, M. Ito and T. A. Krulwich, *Biochim. Biophys. Acta*, 2005, **1717**, 67–88.

30. J. L. Slonczewski, M. Fujisawa, M. Dopson and T. A. Krulwich, *Adv. Microb. Physiol.*, 2009, **55**, 1–79.

31. H. Hahne, U. Mader, A. Otto, F. Bonn, L. Steil, E. Bremer, M. Hecker and D. Becher, *J. Bacteriol.*, 2010, **192**, 870–882.

32. L. Counillon and J. Pouyssegur, *J. Biol. Chem.*, 2000, **275**, 1–4.

33. C. L. Brett, M. Donowitz and R. Rao, *Am. J. Physiol. Cell. Physiol.*, 2005, **288**, C223–239.

34. G. S. Casey and J. Orlowski, *Nat. Rev. Mol. Cell Biol.*, 2010, **11**, 50–61.

35. Q. Ren, K. Chen and I. T. Paulsen, *Nucleic Acids Res.*, 2007, **35**, D274–279.

36. M. H. Saier, Jr., C. V. Tran and R. D. Barabote, *Nucleic Acids Res.*, 2006, **34**, D181–186.

37. E. B. Goldberg, T. Arbel, H. J. Chen, R. Karpel, G. A. Mackie, S. Schuldiner and E. Padan, *Proc. Natl Acad. Sci. USA*, 1987, **84**, 2615–2619.

38. D. Taglicht, E. Padan and S. Schuldiner, *J. Biol. Chem.*, 1993, **268**, 5382–5387.

39. E. Pinner, Y. Kotler, E. Padan and S. Schuldiner, *J. Biol. Chem.*, 1993, **268**, 1729–1734.

40. I. R. Booth, *Microbiol. Rev.*, 1985, **49**, 359–378.

41. R. M. Macnab and A. M. Castle, *Biophys. J.*, 1987, **52**, 637–647.

42. D. Zilberstein, V. Agmon, S. Schuldiner and E. Padan, *J. Biol. Chem.*, 1982, **257**, 3687–3691.

43. S. Wakabayashi, M. Shigekawa and J. Pouyssegur, *Physiol. Rev.*, 1997, 77, 51–74.

44. J. Orlowski and S. Grinstein, *Pflugers Arch.*, 2004, **447**, 549–565.

45. J. Orlowski and S. Grinstein, *J. Biol. Chem.*, 1997, **272**, 22373–22376.

46. R. Waditee, T. Hibino, Y. Tanaka, T. Nakamura, A. Incharoensakdi and T. Takabe, *J. Biol. Chem.*, 2001, **276**, 36931–36938.

47. X. S. T. Pang, S. Wakabayashi and M. Shigekawa, *J. Biol. Chem.*, 2001, **276**, 17367–17372.

48. C. T. Resch, J. L. Winogrodzki, C. C. Hase and P. Dibrov, *Biochem. Cell Biol.*, 2011, **89**, 130–137.

49. K. Herz, S. Vimont, E. Padan and P. Berche, *J. Bacteriol.*, 2003, **185**, 1236–1244.

50. T. A. Krulwich, A. A. Guffanti and M. Ito, *Novartis Found Symp.*, 1999, **221**, 167–182.

51. E. Padan, Y. Gerchman and N. Dover, *Biochim. Biophys. Acta*, 2001, **1505**, 144–157.

52. E. Padan, M. Ito and T. A. Krulwich, *Biochim. Biophys. Acta*, 2005, **1717**, 67–88.
53. J. A. O. Lewinson, G. J. Poelarends, P. Mazurkiewicz, A. J. Driessen and E. Bibi, *Proc. Natl Acad. Sci. USA*, 2003, **100**, 1667–1672.
54. T. Y. T. Shijuku, H. Ohashi, H. Saito, T. Kakegawa, M. Ohta and H. Kobayashi, *Biochim. Biophys. Acta*, 2002, **1556**, 142–148.
55. E. Padan, K. Herz, L. Kozachkov, A. Rimon and L. Galili, *Biochim. Biophys. Acta*, 2004, **1658**, 142–148.
56. E. Padan, K. Herz, L. Kozachkov, A. Rimon and L. Galili, *Biochim. Biophys. Acta*, 2004, **1658**, 2–13.
57. H. Inoue, T. Sakurai, S. Ujike, T. Tsuchiya, H. Murakami and H. Kanazawa, *FEBS Lett.*, 1999, **443**, 11–16.
58. S. Vimont and P. Berche, *J. Bacteriol.*, 2000, **182**, 2937–2944.
59. E. Olkhova, C. Hunte, E. Screpanti, E. Padan and H. Michel, *Proc. Natl. Acad. Sci. USA*, 2006, **103**, 2629–2634.
60. E. Padan, *Trends Biochem. Sci.*, 2008, **33**, 435–443.
61. Y. Gerchman, A. Rimon and E. Padan, *J. Biol. Chem.*, 1999, **274**, 24617–24624.
62. Y. Gerchman, Y. Olami, A. Rimon, D. Taglicht, S. Schuldiner and E. Padan, *Proc. Natl Acad. Sci. USA*, 1993, **90**, 1212–1216.
63. A. Rimon, C. Hunte, H. Michel and E. Padan, *J. Mol. Biol.*, 2008, **379**, 471–481.
64. A. Rothman, Y. Gerchman, E. Padan and S. Schuldiner, *Biochemistry*, 1997, **36**, 14572–14576.
65. Y. Olami, A. Rimon, Y. Gerchman, A. Rothman and E. Padan, *J. Biol. Chem.*, 1997, **272**, 1761–1768.
66. C. Hunte, E. Screpanti, M. Venturi, A. Rimon, E. Padan and H. Michel, *Nature*, 2005, **435**, 1197–1202.
67. T. H. Swartz, S. Ikewada, O. Ishikawa, M. Ito and T. A. Krulwich, *Extremophiles*, 2005, **9**, 345–354.
68. T. Hamamoto, M. Hashimoto, M. Hino, M. Kitada, Y. Seto, T. Kudo and K. Horikoshi, *Mol. Microbiol.*, 1994, **14**, 939–946.
69. J. L. Slonczewski, M. Fujisawa, M. Dopson and T. A. Krulwich, *Adv. Microb. Physiol.*, 2009, **55**, 1–79.
70. P. Putnoky, A. Kereszt, T. Nakamura, G. Endre, E. Grosskopf, P. Kiss and A. Kondorosi, *Mol. Microbiol.*, 1998, **28**, 1091–1101.
71. T. Hiramatsu, K. Kodama, T. Kuroda, T. Mizushima and T. Tsuchiya, *J. Bacteriol.*, 1998, **180**, 6642–6648.
72. M. Ito, A. A. Guffanti, B. Oudega and T. A. Krulwich, *J. Bacteriol.*, 1999, **181**, 2394–2402.
73. S. Kosono, S. Morotomi, M. Kitada and T. Kudo, *Biochim. Biophys. Acta*, 1999, **1409**, 171–175.
74. S. Kosono, K. Haga, R. Tomizawa, Y. Kajiyama, K. Hatano, S. Takeda, Y. Wakai, M. Hino and T. Kudo, *J. Bacteriol.*, 2005, **187**, 5242–5248.
75. J. Dzioba-Winogrodzki, O. Winogrodzki, T. A. Krulwich, M. A. Boin, C. C. Hase and P. Dibrov, *J. Mol. Microbiol. Biotechnol.*, 2009, **16**, 176–186.

76. Y. Kajiyama, M. Otagiri, J. Sekiguchi, S. Kosono and T. Kudo, *J. Bacteriol.*, 2007, **189**, 7511–7514.
77. M. Morino, S. Natsui, T. H. Swartz, T. A. Krulwich and M. Ito, *J. Bacteriol.*, 2008, **190**, 4162–4172.
78. M. H. Saier, Jr., B. H. Eng, S. Fard, J. Garg, D. A. Haggerty, W. J. Hutchinson, D. L. Jack, E. C. Lai, H. J. Liu, D. P. Nusinew, A. M. Omar, S. S. Pao, I. T. Paulsen, J. A. Quan, M. Sliwinski, T. T. Tseng, S. Wachi and G. B. Young, *Biochim. Biophys. Acta*, 1999, **1422**, 1–56.
79. T. H. Swartz, M. Ito, T. Ohira, S. Natsui, D. B. Hicks and T. A. Krulwich, *J. Bacteriol.*, 2007, **189**, 3081–3090.
80. T. Yamaguchi, F. Tsutsumi, P. Putnoky, M. Fukuhara and T. Nakamura, *Microbiology*, 2009, **155**, 2750–2756.
81. M. Morino and M. Ito, *FEMS Microbiol. Lett.*, 2012, **335**, 26–30.
82. N. M. Mesbah, G. M. Cook and J. Wiegel, *Mol. Microbiol.*, 2009, **74**, 270–281.
83. O. Juárez and B. Barquera, *Biochim. Biophys. Acta*, 2012, **1817**, 1823–1832.
84. V. Muller, F. Imkamp, E. Biegel, S. Schmidt and S. Dilling, *Ann. NY Acad. Sci.*, 2008, **1125**, 137–146.
85. E. Biegel and V. Muller, *Proc. Natl Acad. Sci. USA*, 2010, **107**, 18138–18142.
86. S. Schmidt, E. Biegel and V. Muller, *Biochim. Biophys. Acta*, 2009, **1787**, 691–696.
87. B. Barquera, P. Hellwig, W. Zhou, J. E. Morgan, C. C. Hase, K. K. Gosink, M. Nilges, P. J. Bruesehoff, A. Roth, C. R. Lancaster and R. B. Gennis, *Biochemistry*, 2002, **41**, 3781–3789.
88. M. Hayashi, Y. Nakayama and T. Unemoto, *Biochim. Biophys. Acta*, 2001, **1505**, 37–44.
89. A. V. Bogachev and M. I. Verkhovsky, *Biochemistry (Mosc.)*, 2005, **70**, 143–149.
90. P. A. Dibrov, V. A. Kostryko, R. L. Lazarova, V. P. Skulachev and I. A. Smirnova, *Biochim. Biophys. Acta*, 1986, **850**, 449–457.
91. P. A. Dibrov, R. L. Lazarova, V. P. Skulachev and M. L. Verkhovskaya, *Biochim. Biophys. Acta*, 1986, **850**, 458–465.
92. P. Dimroth, *Biochim. Biophys. Acta*, 1997, **1318**, 11–51.
93. R. Baradaran, J. M. Berrisford, S. G. Minhas and L. A. Sazanov, *Nature*, 2013, **494**, 443–448.
94. B. Barquera, W. Zhou, J. E. Morgan and R. B. Gennis, *Proc. Natl Acad. Sci. USA*, 2002, **99**, 10322–10324.
95. B. Barquera, J. E. Morgan, D. Lukoyanov, C. P. Scholes, R. B. Gennis and M. J. Nilges, *J. Am. Chem. Soc.*, 2003, **125**, 265–275.
96. B. Barquera, M. J. Nilges, J. E. Morgan, L. Ramirez-Silva, W. Zhou and R. B. Gennis, *Biochemistry*, 2004, **43**, 12322–12330.
97. O. Juarez, M. J. Nilges, P. Gillespie, J. Cotton and B. Barquera, *J. Biol. Chem.*, 2008, **283**, 33162–33167.

98. M. Hayashi, Y. Nakayama, M. Yasui, M. Maeda, K. Furuishi and T. Unemoto, *FEBS Lett.*, 2001, **488**, 5–8.

99. Y. Nakayama, M. Yasui, K. Sugahara, M. Hayashi and T. Unemoto, *FEBS Lett.*, 2000, **474**, 165–168.

100. O. Juarez, J. E. Morgan, M. J. Nilges and B. Barquera, *Proc. Natl Acad. Sci. USA*, 2010, **107**, 12505–12510.

101. O. Juárez, J. E. Morgan and B. Barquera, *J. Biol. Chem.*, 2009, **284**, 8963–8972.

102. O. Juárez, M. E. Shea, G. I. Makhatadze and B. Barquera, *J. Biol. Chem.*, 2011, **286**, 26383–26390.

103. O. Juarez, K. Athearn, P. Gillespie and B. Barquera, *Biochemistry*, 2009, **48**, 9516–9524.

104. H. S. Jeong and Y. Jouanneau, *J. Bacteriol.*, 2000, **182**, 1208–1214.

105. Y. Jouanneau, H. S. Jeong, N. Hugo, C. Meyer and J. C. Willison, *Eur. J. Biochem.*, 1998, **251**, 54–64.

106. H. Kumagai, T. Fujiwara, H. Matsubara and K. Saeki, *Biochemistry*, 1997, **36**, 5509–5521.

107. K. Saeki and H. Kumagai, *Arch. Microbiol.*, 1998, **169**, 464–467.

108. A. Saaf, M. Johansson, E. Wallin and G. von Heijne, *Proc. Natl Acad. Sci. USA*, 1999, **96**, 8540–8544.

109. M. S. Koo, J. H. Lee, S. Y. Rah, W. S. Yeo, J. W. Lee, K. L. Lee, Y. S. Koh, S. O. Kang and J. H. Roe, *EMBO J.*, 2003, **22**, 2614–2622.

110. K. Schlegel, C. Welte, U. Deppenmeier and V. Müller, *FEBS J.*, 2012, **279**, 4444–4452.

111. P. Dahinden, Y. Auchli, T. Granjon, M. Taralczak, M. Wild and P. Dimroth, *Arch. Microbiol.*, 2005, **183**, 121–129.

112. R. Studer, P. Dahinden, W. W. Wang, Y. Auchli, X. D. Li and P. Dimroth, *J. Mol. Biol.*, 2007, **367**, 547–557.

113. M. Bott, *Arch. Microbiol.*, 1997, **167**, 78–88.

114. A. V. Demirev, A. Khanal, N. P. Hanh, K. T. Nam and D. H. Nam, *J. Microbiol.*, 2011, **49**, 407–412.

115. D. Kress, D. Brügel, I. Schall, D. Linder, W. Buckel and L. O. Essen, *J. Biol. Chem.*, 2009, **284**, 28401–28409.

116. C. D. Boiangiu, E. Jayamani, D. Brügel, G. Herrmann, J. Kim, L. Forzi, R. Hedderich, I. Vgenopoulou, A. J. Pierik, J. Steuber and W. Buckel, *J. Mol. Microbiol. Biotechnol.*, 2005, **10**, 105–119.

117. G. D. Repizo, V. S. Blancato, P. Mortera, J. S. Lolkema and C. Magni, *Appl. Environ. Microbiol.*, 2013, **79**, 2882–2890.

118. W. Buckel, *Biochim. Biophys. Acta*, 2001, **1505**, 15–27.

119. P. K. Dahinden and P. Dimroth, *FEBS J.*, 2005, **272**, 846–855.

120. M. R. Wild, K. M. Pos and P. Dimroth, *Biochemistry*, 2003, **42**, 11615–11624.

121. M. Schmid, T. Vorburger, K. M. Pos and P. Dimroth, *Eur. J. Biochem.*, 2002, **269**, 2997–3004.

122. M. Schmid, M. R. Wild, P. Dahinden and P. Dimroth, *Biochemistry*, 2002, **41**, 1285–1292.

123. P. Dimroth and M. Schmid, *Biochim. Biophys. Acta*, 2001, **1505**, 1–14.
124. G. Gottschalk and R. Thauer, *Biochim. Biophys. Acta*, 2001, **1505**, 28–36.
125. U. Deppenmeier, *Prog. Nucleic Acid Res. Mol. Biol.*, 2002, **71**, 223–283.
126. V. Muller and G. Gruber, *Cell. Mol. Life Sci.*, 2003, **60**, 474–494.
127. G. Gruber, H. Wieczorek, W. R. Harvey and V. Muller, *J. Exp. Biol.*, 2001, **204**, 2597–2605.
128. M. Forgac, *Nat. Rev. Mol. Cell Biol.*, 2007, **8**, 917–929.
129. G. Schafer, M. Engelhard and V. Muller, *Microbiol. Mol. Biol. Rev.*, 1999, **63**, 570–620.
130. T. Murata, K. Igarashi, Y. Kakinuma and I. Yamato, *J. Biol. Chem.*, 2000, **275**, 13415–13419.
131. P. M. Kane, *Microbiol. Mol. Biol. Rev.*, 2006, **70**, 177–191.
132. T. Murata, K. Takase, I. Yamato, K. Igarashi and Y. Kakinuma, *J. Biochem.*, 1999, **125**, 414–421.
133. T. Murata, I. Yamato, Y. Kakinuma, A. G. Leslie and J. E. Walker, *Science*, 2005, **308**, 654–659.
134. T. Murata, I. Yamato, Y. Kakinuma, M. Shirouzu, J. E. Walker, S. Yokoyama and S. Iwata, *Proc. Natl Acad. Sci. USA*, 2008, **105**, 8607–8612.
135. D. Okuno, R. Iino and H. Noji, *J. Biochem.*, 2011, **149**, 655–664.
136. S. Arai, S. Saijo, K. Suzuki, K. Mizutani, Y. Kakinuma, Y. Ishizuka-Katsura, N. Ohsawa, T. Terada, M. Shirouzu, S. Yokoyama, S. Iwata, I. Yamato and T. Murata, *Nature*, 2013, **493**, 703–707.
137. W. Junge, H. Lill and S. Engelbrecht, *Trends Biochem. Sci.*, 1997, **22**, 420–423.
138. T. Meier, P. Polzer, K. Diederichs, W. Welte and P. Dimroth, *Science*, 2005, **308**, 659–662.
139. N. M. Mesbah and J. Wiegel, *Appl. Environ. Microbiol.*, 2012, **78**, 4074–4082.
140. K. Schlegel, V. Leone, J. D. Faraldo-Gómez and V. Müller, *Proc. Natl Acad. Sci. USA*, 2012, **109**, 947–952.
141. T. Pourcher, S. Leclercq, G. Brandolin and G. Leblanc, *Biochemistry*, 1995, **34**, 4412–4420.
142. B. Tolner, T. Ubbink-Kok, B. Poolman and W. N. Konings, *Mol. Microbiol.*, 1995, **18**, 123–133.
143. H. Terashima, S. Kojima and M. Homma, *Int. Rev. Cell Mol. Biol.*, 2008, **270**, 39–85.
144. N. Terahara, M. Sano and M. Ito, *PLoS ONE*, 2012, 7, e46248.
145. X. Zhang, M. Xia, Y. Li, H. Liu, X. Jiang, W. Ren, J. Wu, P. DeCaen, F. Yu, S. Huang, J. He, D. E. Clapham, N. Yan and H. Gong, *Cell Res.*, 2013, **23**, 409–422.
146. M. Mantegazza, G. Curia, G. Biagini, D. S. Ragsdale and M. Avoli, *Lancet Neurol.*, 2010, **9**, 413–424.
147. D. Ren, B. Navarro, H. Xu, L. Yue, Q. Shi and D. E. Clapham, *Science*, 2001, **294**, 2372–2375.
148. M. Ito, H. Xu, A. A. Guffanti, Y. Wei, L. Zvi, D. E. Clapham and T. A. Krulwich, *Proc. Natl Acad. Sci. USA*, 2004, **101**, 10566–10571.

149. R. Koishi, H. Xu, D. Ren, B. Navarro, B. W. Spiller, Q. Shi and D. E. Clapham, *J. Biol. Chem.*, 2004, **279**, 9532–9538.
150. J. Payandeh, T. Scheuer, N. Zheng and W. A. Catterall, *Nature*, 2011, **475**, 353–358.
151. X. Zhang, W. Ren, P. DeCaen, C. Yan, X. Tao, L. Tang, J. Wang, K. Hasegawa, T. Kumasaka, J. He, D. E. Clapham and N. Yan, *Nature*, 2012, **486**, 130–134.
152. B. Hille, *Proc. Natl Acad. Sci. USA*, 1971, **68**, 280–282.
153. T. A. Krulwich, *Mol. Microbiol.*, 1995, **15**, 403–410.
154. S. Sugiyama, *Mol. Microbiol.*, 1995, **15**, 592.
155. M. G. Sturr, A. A. Guffanti and T. A. Krulwich, *J. Bacteriol.*, 1994, **176**, 3111–3116.
156. S. Fujinami, T. Sato, J. S. Trimmer, B. W. Spiller, D. E. Clapham, T. A. Krulwich, I. Kawagishi and M. Ito, *Microbiology*, 2007, **153**, 4027–4038.
157. S. Y. Park, B. Lowder, A. M. Bilwes, D. F. Blair and B. R. Crane, *Proc. Natl Acad. Sci. USA*, 2006, **103**, 11886–11891.
158. S. Fujinami, N. Terahara, T. A. Krulwich and M. Ito, *Future Microbiol.*, 2009, **4**, 1137–1149.
159. L. E. Bakeeva, K. M. Chumakov, A. L. Drachev, A. L. Metlina and V. P. Skulachev, *Biochim. Biophys. Acta*, 1986, **850**, 466–472.
160. H. C. Berg, *Annu. Rev. Biochem.*, 2003, **72**, 19–54.
161. D. F. Blair, *FEBS Lett.*, 2003, **545**, 86–95.
162. R. M. Macnab, *J. Bacteriol.*, 1999, **181**, 7149–7153.
163. M. D. Manson, P. Tedesco, H. C. Berg, F. M. Harold and C. Van der Drift, *Proc. Natl Acad. Sci. USA*, 1977, **74**, 3060–3064.
164. S. Khan, *J. Bacteriol.*, 1993, **175**, 2169–2174.
165. D. F. Blair and H. C. Berg, *Cell*, 1990, **60**, 439–449.
166. B. Stolz and H. C. Berg, *J. Bacteriol.*, 1991, **173**, 7033–7037.
167. D. L. Marykwas, S. A. Schmidt and H. C. Berg, *J. Mol. Biol.*, 1996, **256**, 564–576.
168. H. Tang, T. F. Braun and D. F. Blair, *J. Mol. Biol.*, 1996, **261**, 209–221.
169. S. A. Lloyd and D. F. Blair, *J. Mol. Biol.*, 1997, **266**, 733–744.
170. S. Yamaguchi, H. Fujita, A. Ishihara, S. Aizawa and R. M. Macnab, *J. Bacteriol.*, 1986, **166**, 187–193.
171. H. Takami, K. Nakasone, Y. Takaki, G. Maeno, R. Sasaki, N. Masui, F. Fuji, C. Hirama, Y. Nakamura, N. Ogasawara, S. Kuhara and K. Horikoshi, *Nucleic Acids Res.*, 2000, **28**, 4317–4331.
172. L. McCarter, *J. Mol. Microbiol. Biotechnol.*, 2004, 7, 18–29.
173. C. C. Hase, *Microbiology*, 2001, **147**, 831–837.
174. T. Atsumi, L. McCarter and Y. Imae, *Nature*, 1992, **355**, 182–184.
175. L. L. McCarter, *J. Bacteriol.*, 2005, **187**, 1207–1209.
176. N. Hirota and Y. Imae, *J. Biol. Chem.*, 1983, **258**, 10577–10581.
177. S. Matsuura, J. Shioi and Y. Imae, *FEBS Lett.*, 1977, **82**, 187–190.
178. J. I. Shioi, Y. Imae and F. Oosawa, *J. Bacteriol.*, 1978, **133**, 1083–1088.
179. S. Matsuura, J. I. Shioi, Y. Imae and S. Iida, *J. Bacteriol.*, 1979, **140**, 28–36.

180. A. Paulick, A. Koerdt, J. Lassak, S. Huntley, I. Wilms, F. Narberhaus and K. M. Thormann, *Mol. Microbiol.*, 2009, **71**, 836–850.
181. J. Zhou and D. F. Blair, *J. Mol. Biol.*, 1997, **273**, 428–439.
182. J. Zhou, S. A. Lloyd and D. F. Blair, *Proc. Natl Acad. Sci. USA*, 1998, **95**, 6436–6441.
183. S. A. Lloyd, F. G. Whitby, D. F. Blair and C. P. Hill, *Nature*, 1999, **400**, 472–475.
184. P. N. Brown, M. A. Mathews, L. A. Joss, C. P. Hill and D. F. Blair, *J. Bacteriol.*, 2005, **187**, 2890–2902.
185. T. Yorimitsu, A. Mimaki, T. Yakushi and M. Homma, *J. Mol. Biol.*, 2003, **334**, 567–583.
186. T. Yorimitsu, Y. Sowa, A. Ishijima, T. Yakushi and M. Homma, *J. Mol. Biol.*, 2002, **320**, 403–413.
187. T. Yakushi, J. Yang, H. Fukuoka, M. Homma and D. F. Blair, *J. Bacteriol.*, 2006, **188**, 1466–1472.
188. N. Hirota, M. Kitada and Y. Imae, *FEBS Lett.*, 1981, **132**, 278–280.
189. N. Terahara, T. A. Krulwich and M. Ito, *Proc. Natl Acad. Sci. USA*, 2008, **105**, 14359–14364.
190. N. Terahara, M. Sano and M. Ito, *PloS One*, 2012, **7**, e46248.

CHAPTER 3
Potassium

DAVID M. MILLER[a,b] AND JACQUELINE M. GULBIS*[a,b]

[a] The Walter and Eliza Hall Institute of Medical Research; [b] Department of
Medical Biology, The University of Melbourne, Parkville, Victoria 3052,
Australia
*Email: jgulbis@wehi.edu.au

3.1 Eukaryotes

3.1.1 Introduction

Potassium cuts a diminutive figure in the chemical literature yet plays an
incomparable role in living systems. Elemental potassium does not occur in
nature, although the soft, silvery metal can be produced by electrochemical
methods and was, historically, the first known metal to be isolated in this
manner. Sir Humphry Davy demonstrated the electrolysis of potassium salt
solutions experimentally at the Royal Institution, London, in 1806. Potassium is a member of the Group 1 alkali metals (Li, Na, K, Rb and Cs) that
immediately follow the noble gases in the periodic table. Consequently, loss
of the single outer shell electron yields a stable closed shell configuration.
The resulting monovalent cations occur naturally as salts, the three most
geologically abundant being Na^+, K^+ and Li^+, in that order. The propensity
to ionize is central to the physicochemical properties of these elements, and,
in the case of potassium, even synthetic organo-potassium compounds exhibit essentially ionic behaviour.[1] When the dioxygen and water present in
air contact alkali metals, oxidation occurs by vigorous exothermic reactions,
turning any residual water alkaline. Students of chemistry will be familiar
with the explosive reaction of metallic potassium in water and the

RSC Metallobiology Series No. 2
Binding, Transport and Storage of Metal Ions in Biological Cells
Edited by Wolfgang Maret and Anthony Wedd

characteristic purple flames accompanying its spontaneous ignition. The coordination chemistry of *ionic* potassium thus underpins its physiological role(s) and has provided the evolutionary blueprint for transport molecules that are able to discriminate between dissolved K^+, Na^+ and other physiologically relevant cations.

Mineral ions have an array of biological functions, including protein cofactors and cellular messengers. Potassium is the predominant intracellular cationic species, yet with few exceptions its importance in living cells is as an essential macronutrient and lies in its bulk properties. Prokaryotes, for instance, often exist in harsh conditions incurring extremes of pH, heat or salts. Key to their survival is maintenance of a controlled intracellular environment with respect to pH, hydrostatic pressure, osmotic balance and cell volume. All of these fundamental qualities are contingent on controlled K^+ transport across the bounding membranes. Multicellular organisms share these needs, as well as additional requirements for intercellular communication. In animals, K^+ is involved in maintaining tissue electrolyte balance in addition to participating in a complex network of regulatory and cell signalling processes. Sodium, the most abundant cation in blood, is another alkali metal ion important in regulation of blood volume and osmotic balance. In a physiological sense, neither Na^+ nor K^+ can be considered in isolation. Lithium appears to be essential as a trace element, but its physiological role is as yet unclear.[2] Importantly, the capacity of transport proteins to render cell membranes selectively permeable to specific ions is used to effect charge separation, creating the transmembrane electrochemical gradients that are utilized in signalling. In higher eukaryotes, stringent maintenance of differential concentrations of potassium ions on either side of an excitable membrane fulfils a key requirement for intercellular transmission.

Potassium ions are actively accumulated in living cells and tissues and, as a consequence, K^+ is recycled continuously through biological systems and food chains. While the natural abundance of K^+ in seawater ($0.39\,\mathrm{g}\,l^{-1}$ or 10 mM) is very low relative to that of Na^+ ($10.7\,\mathrm{g}\,l^{-1}$ or 465 mM), the relationship is reversed in living cells. Inside cells, K^+ is present at significantly higher levels. *Extra*cellular K^+ concentrations in nerve and muscle, for example, are maintained within the range 3.8–5.0 mM whereas resting *intra*cellular levels are held within 120–140 mM.[3–5] Sodium, by contrast, is maintained at low concentrations within cells with respect to the external environment. Lithium, the lightest of the Group 1 elements, is also present in all biological systems. Its concentration is not under stringent control, however, and intra- and extracellular Li^+ levels are comparable.

The cellular enrichment of K^+ in plant and animal material ensures unrefined foods are premium sources of dietary K^+ for humans. Plants require sufficient K^+ for growth, flowering and setting seed, and soil supplementation of K^+ *via* organic and inorganic fertilizers is commonplace in both intensive agricultural and domestic situations. A major utilizable source of K^+ for use in agricultural fertilizers is potash ("pot ash"), a complex mixture of salts from which the name potassium is derived. The ash of burnt vegetation is a rich mineral resource from which K^+ (typically as the carbonate K_2CO_3) can be leached, re-crystallized and harvested. On a commercial scale, however, potash is most often mined from mineralized marine deposits that are rich in potassium salts such as KCl and saltpetre KNO_3.

In a similar vein, grains and vegetables are also a source of Li^+, as is drinking water, although Li^+ is not actively taken up and its presence in plant matter appears to reflect regional levels in soil or growth media. Lithium is present at low concentrations in all human tissue.[2] There is evidence that Li^+ is retained in some tissues (*e.g.* the endocrine glands: adrenal, thyroid and pituitary) more strongly than others, irrespective of dietary levels.[6,7] Unlike Na^+ and K^+, however, its role has remained obscure. Lithium is best known for its indispensability as an approved medicine for psychiatric conditions including mania, bipolar disorder and depression, initially utilized for this purpose in 1949.[8] It has proven to be particularly effective in curbing suicidal tendencies in patients[9] (reviewed in 2012[10]). There are some side effects, however, which were comprehensively reviewed in 2012.[11] Recent attention has focussed on the utility of lithium as a neuroprotective agent in cases of acute brain injury or neurodegenerative conditions (reviewed in 2013[12]).

Selective accumulation of K^+ within living cells is of great consequence for cellular chemistry. Why K^+ and not Na^+ is the primary ion species sequestered by cells remains a perplexing question but is the case in organisms from bacteria through to man, with the exception of some of the halophiles. The choice of K^+ thus came early on an evolutionary time-scale. The ability to select one ion over another is undoubtedly the important factor in generating potential differences across membranes. The precise identity of the selectable ion may be a secondary consideration. A simple explanation consistent with the ionic properties of the Group 1 elements is that subtly different coordination chemistry allows biomolecules to select K^+ with far higher fidelity than Na^+ or Li^+ ions. Such questions notwithstanding, control of the abundance of K^+ ions equips cell membranes to mediate electrical and charge driven processes and requires tight regulation of K^+ homeostasis. To even begin to understand how K^+ features so prominently in biological systems requires some knowledge of its solution chemistry. This includes the comparative size and charge density of ionic and hydrated K^+ species, the preferential coordination geometry of K^+ and its bulk hydration properties.

3.1.1.1 Coordination Chemistry of Potassium

The coordination chemistry of K^+ in solution is central to its biology. Although K^+ ions can bind both nitrogen- and oxygen-based ligands, K^+ in biological molecules appears to be exclusively coordinated *via* oxygen atoms. Ligand moieties include, but are not exclusive to, carbonyls, side-chain hydroxyls and water molecules. Collectively, crystal structures of small inorganic molecules containing K^+ show that when nitrogen ligands are bridging two K^+ sites, interatomic distances between K^+ ions can be as short as 3.5 Å. When oxygen is the bridging ligand, however, the inter-potassium distances are invariably >4 Å, comparable to the distances observed in protein structures.

Examples of K^+ bound in protein structures are sparse. In the pores of K^+ channels, the ions are octa-coordinate, bound by eight ligands regularly arranged at the vertices of a distorted square antiprism (Figure 3.1a). The stereochemistry is similar in DNA quadruplexes (Figure 3.1b). Interestingly, when telomeric DNA sequences are crystallized in the presence of Na^+ instead of K^+, the multimers alter substantially and in a way that is consistent with their respective coordination numbers and geometric preferences (Figure 3.2). The Na/K^+-ATPase is a less specific K^+ binding protein than K^+ channels and binds K^+ at two sites 4.1 Å apart. Neither site exhibits a

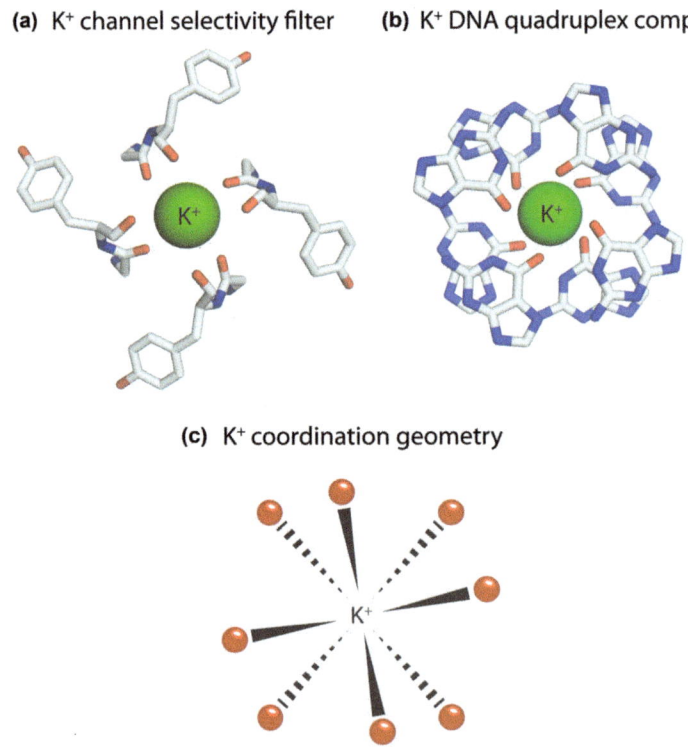

(a) K⁺ channel selectivity filter   **(b)** K⁺ DNA quadruplex complex

(c) K⁺ coordination geometry

Figure 3.1   Coordination of K^+ ions by biological molecules. **a.** A K^+ ion in the selectivity filter of a potassium channel is bound by eight oxygen ligands (peptide carbonyls and threonine hydroxyls). **b.** A DNA quadruplex also binds K^+ in octa-coordinate geometry. In this case, the ligands are carbonyls and anionic C-O moieties. **c.** In both **a** and **b** the coordination geometry is intermediate between cubic and square antiprismatic.

(a) K⁺ liganded   **(b)** Na⁺ liganded

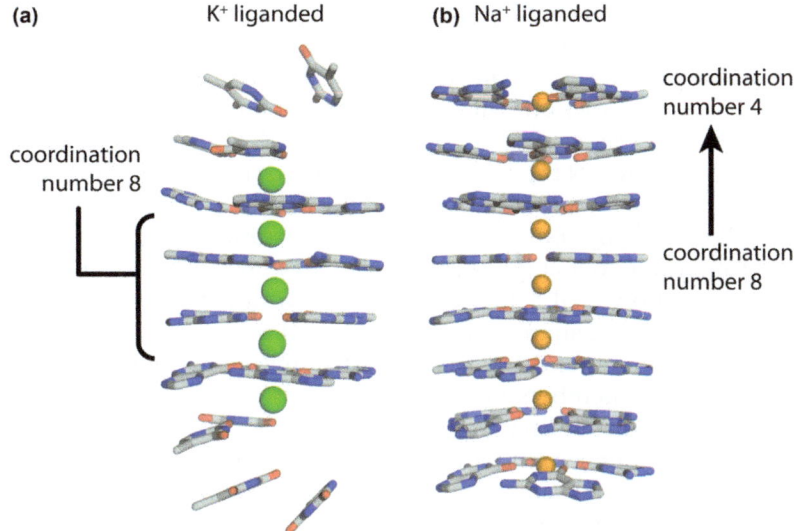

coordination number 4

coordination number 8

coordination number 8

Figure 3.2 Contrasting coordination of K^+ and Na^+ by DNA quadruplexes. **a.** Octa-coordinate K^+ ions present in the crystal structure of a DNA quadruplex. The K^+ ions intercalated between the tiers of bases exhibit distorted square antiprismatic geometry. This is invariably the case with K^+-containing DNA complexes. **b.** Variation in the coordination number of Na^+ in the crystal structure of a Na^+-containing DNA quadruplex. Unlike in **a**, the stereochemistry varies from octa-coordinate at the central Na^+ ion to tetra-coordinate at terminal Na^+ ions. Protein data bank entries are 2GWQ and 352D, respectively.

classical ligand geometry: one K^+ is five-coordinate and the other six-coordinate (distorted octahedral). An unexpected example of a protein showing apparently specific K^+ binding is yeast Sec12, a guanine nucleotide exchange factor that initiates assembly of Coat Protein II onto the endoplasmic reticulum membrane prior to vesicle budding and fission. Potassium apparently enhances exchange factor activity. The β-propeller molecular architecture of Sec12 contains a glycine-rich linker near the N-terminus that chelates K^+ in a distorted octahedral geometry, employing as ligands four backbone carbonyl oxygens, one asparagine side-chain and a bound water molecule.[13]

3.1.1.2 Potassium Hydration

Charge densities of alkali and alkali earth ions differ considerably, with significant consequences for their hydration properties. The monovalent ions K^+ and Na^+ have larger ionic and smaller hydration radii than their divalent counterparts Ca^{2+} and Mg^{2+}.[†] The resultant lower charge densities

---

[†]For further comparison, see Chapter 5, Section 5.2 and Tables 5.2 and 5.3.

result in comparatively weaker hydration forces. In addition, potassium salts are usually highly soluble and the energetic cost of displacing hydrating waters in the primary solvation shell of K^+ relatively low. Both are important factors in its interaction with transport proteins. Comparative single-ion data for the free energy of hydration are -338 kJ mol^{-1} for K^+ *versus* -411 kJ mol^{-1} for Na^+, based on measured ionization potentials and electron affinities.[14] Opinions differ as to whether K^+ has one or more hydration shells.[15,16] By contrast, Li^+, the smallest member of Group 1, binds water strongly due to its higher charge density and there is clear evidence of a second hydration shell.[16] Its free energy of hydration is -516.7 kJ mol^{-1}.[14] A detailed study of alkali metal hydration found a robust correlation between coordination number and $M-OH_2$ bond distance and verified or revised the ionic radii for all members of the group except the short-lived Francium.[17] However, favoured coordination numbers and the nature of hydration shells remain somewhat uncertain for the alkali metals, chiefly because of inherent technical issues associated with making comparable measurements over the entire set of Group 1 ions.

3.1.1.3   Selection of Potassium by Biological Molecules

In terms of cell biology, the hydration state of an ion is particularly relevant to its transport. Gel-sieving studies on the Group 1 monovalent cations and NH_4^+ revealed a transition from strong to weak hydration occurring at an ionic radius of 1.06 Å.[18] Smaller "pores" appear to select for K^+, whilst larger ones preferentially allow passage of Na^+ and Li^+,[19] indicating that the lighter ions are more likely to maintain their hydration mantle. Experiments using nano-filtration membranes indicate a correlation between membrane permeability and the bulk hydration properties of cationic species.[20] Moreover, molecular dynamics simulations exploring energetic barriers to permeation of anions through nano-pores have suggested a non-linear correlation of pore size with energetic barrier, asserting that the dominant contributing factor is the difference between the radius of the pore and that of the ion in question.[21,22] The major implication is that the energetic cost of permeation is contingent upon whether or not an ion must dehydrate (fully or partially) to enter a pore or other binding cavity. It follows that ion selectivity depends upon a critical trade-off: while dehydration raises the energetic barriers, it has the advantage that protein oxygen atoms can directly bind to the ion, allowing the geometry and character of the binding site to determine selectivity. On the other hand, if ions remain fully hydrated, thermodynamic costs would be lessened but close approaches compromised such that permeating ions are subject only to rudimentary molecular sieving. Transportation of hydrated K^+ ions across membranes would require pore diameters too large to discriminate between K^+ and a variety of other ionic species. Sodium channels, which transport hydrated ions, are less stringently selective than K^+ channels. Stripping K^+ of its primary hydration shell substantially decreases its size and allows it to be captured by ligand sets that have evolved to bind it in preference to other monovalent cations.

Aside from these factors, electrostatic considerations include ligand-ligand electrostatic repulsion[23] and charge transfer between ligand and K^+ that attenuates the charge on the cation and polarizes the ligand.[24] Molecular simulation data suggest that higher ligand numbers can be used to selectively enhance K^+ binding; for example, a site with eight ligands would preferentially bind K^+ over Na^+.[25] This study argues that the chemical properties of ligands (in this case, carbonyl *versus* water) may be less important than topological constraints, with the qualification that a peptide backbone can naturally impose a specific geometry whereas free water molecules cannot. However, this remains an open question. A related issue is to what extent intrinsic physical and electrostatic differences between ligand types influence ion selectivity. For instance, does the presence of a single high-field carboxylate ligand shift ion selectivity to favour Na^+ over K^+?[26]

3.1.1.4 Introduction to Potassium Transport

The two main classes of molecule facilitating potassium homeostasis operate on vastly different principles. The "uphill" separation of ionic species across bounding membranes, facilitated by molecular exchangers and pumps, sets up a potential difference across the insulating lipid bilayer. The pumps (ATPases) couple the movement of K^+ ions with the energy released by ATP hydrolysis. On the other hand, potassium-specific pores (K^+ channels) that allow controlled "downhill" diffusion are responsible for fine-tuning the membrane potential set up by the ion exchangers. The cell potential at equilibrium relies on attaining a balance between opposing diffusive and electrical forces. Ion channels simply serve as gates to diffusion and, while K^+ channels play the leading role in setting the cell potential, *all* ion channels utilize the stored membrane potential *via* the electrochemical gradient.

Despite increasingly complex regulatory requirements as the evolutionary tree is scaled, the specialized transporters and channels governing K^+ accumulation and homeostasis in the higher eukaryotes share common origins with those found in prokaryotes. The conserved protein cores are responsible for governing ion specificity, which is exquisitely tailored to the distinctive properties of the transported ions. Cationic species are differentiated with respect to their ionic radii, charge and ligand and stereochemical preferences. Although penetrating insights have arisen from structural and functional studies, the precise hierarchy of molecular factors determining selectivity by K^+ transport proteins has not been determined unequivocally. In addition, adaptations to the core protein folds tailoring the functional attributes (*e.g.* responsiveness and kinetics) to cellular requirements are only partially understood at this time.

3.1.2 Uptake and Excretion

In mammals, K^+ and other electrolytes are ingested as food and absorbed through intestinal epithelia. The necessity to maintain extracellular K^+ levels within stringent tolerance limits is of overriding concern during the digestive process. In addition, uptake and excretion must equate precisely, necessitating temporal correspondence of the relevant biological pathways. While in adult humans the daily dietary intake of K^+ exactly matches the amount excreted, between infancy and adolescence small positive increments in the amount of K^+ retained maintain homeostatic levels as cell number and body mass increase. In a physiological context, a good western diet provides approximately 90 millimoles (3.5 g) of K^+ per day.[27] The K^+ content of a nutritious meal can be as much as 50–100% of the total K^+ present in extracellular body fluids at any given time. For instance, at a K^+ concentration of 4 mM, the 15 litres or so of extracellular aqueous fluid present in a 70 kg adult contain approximately 2.3 g of K^+. The comparison vividly illustrates the acuteness of the rise in extracellular K^+ concentration at mealtimes and the reason why, if homeostatic mechanisms were not in place, a failure to swiftly dispel peaking K^+ levels could lead rapidly to a doubling of extracellular concentrations.

Approximately 90% of dietary K^+ is *absorbed* passively in the small intestine through an as yet unidentified mechanism[27] (Figure 3.3), noting that a range of specialized K^+ channels exist in gastrointestinal epithelia. The remaining 10% is absorbed actively in the colon.[28] On the other hand, the bulk of excess K^+ is *secreted via* the distal kidney tubules and expelled from the body in the urine.[29] A small remaining fraction, which amounts to 10% or so, is excreted from the colon in faecal matter.[30] Under normal conditions, a degree of spatial compartmentalization of the colonic K^+ absorption and secretion processes occurs. However, this is not absolute and, while the predominant process in the distal colon is absorption and in the proximal is secretion,[31–33] the situation can alter under high or low K^+ stress as ancillary pathways come into play. Potassium homeostasis and regulation are discussed later in this chapter. Lithium also is absorbed in the small intestine, where it enters through epithelial Na^+ channels.

3.1.2.1 Colonic Potassium Handling

In the epithelial membranes of the colon, uptake and excretion of electrolytes (including Na^+ and Cl^- as well as K^+) are mediated by a selection of co-transporters and channels. The walls of the colonic lumen display distinctive morphological features, including invaginations opening to the mucosal surface that are referred to formally as crypts (Figure 3.4a). A *crypt* is a central duct surrounded by goblet and columnar cells, distinct from the *surface epithelium*. All epithelial cells lining the colon are polarized, *i.e.* the portion of the cell membrane facing the lumen of the gastrointestinal tract, known as the apical membrane, differs in constitution from the

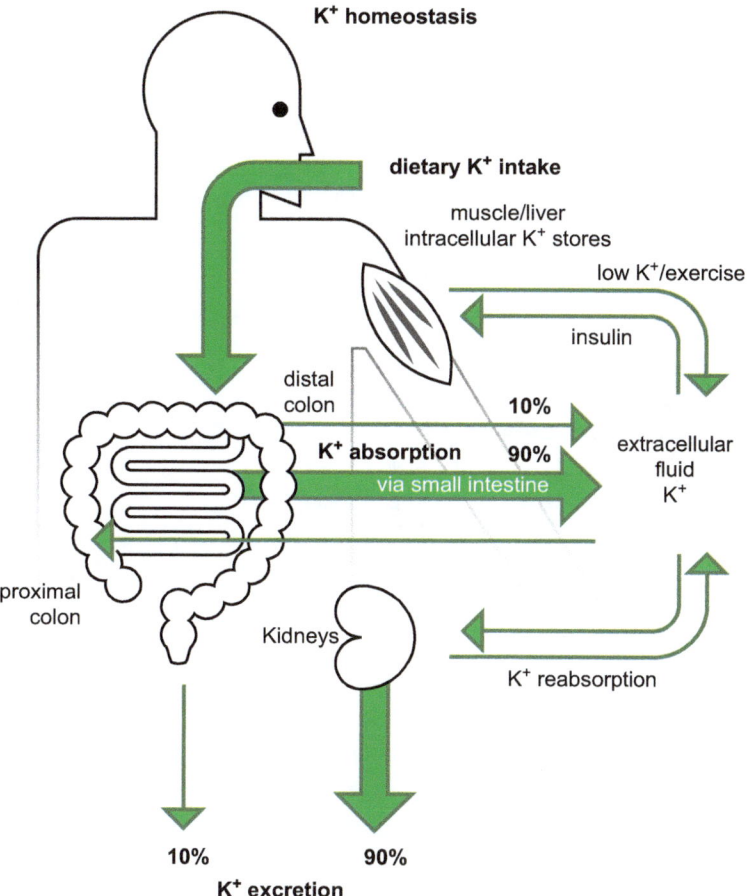

Figure 3.3 Spatial compartmentalization of K^+ absorption, secretion, excretion and an overview of homeostatic relationships.

remaining portions that are collectively known as the basolateral membrane (Figure 3.4b). Both membrane "types" contain a specific cohort of ion channels and pumps and, while transport proteins can move fluidly *within* either membrane, they are thought not to migrate *between* the apical and basolateral zones. Thus the transporter populations in each membrane can be considered more-or-less separately. Basolateral ion transport pathways serve multiple functions including tuning the membrane potential, maintaining pH and regulating cell volume.[34] In so doing, they underpin the primary processes of electrolyte uptake through the apical membrane, as reviewed in 2011.[35]

Colonic crypt cells also express small-, intermediate- and large-conductance Ca^{2+}-sensitive channels implicated in K^+ secretion; these have not been found in surface cells.[36–38] The large conductance Ca^{2+}-sensitive K^+

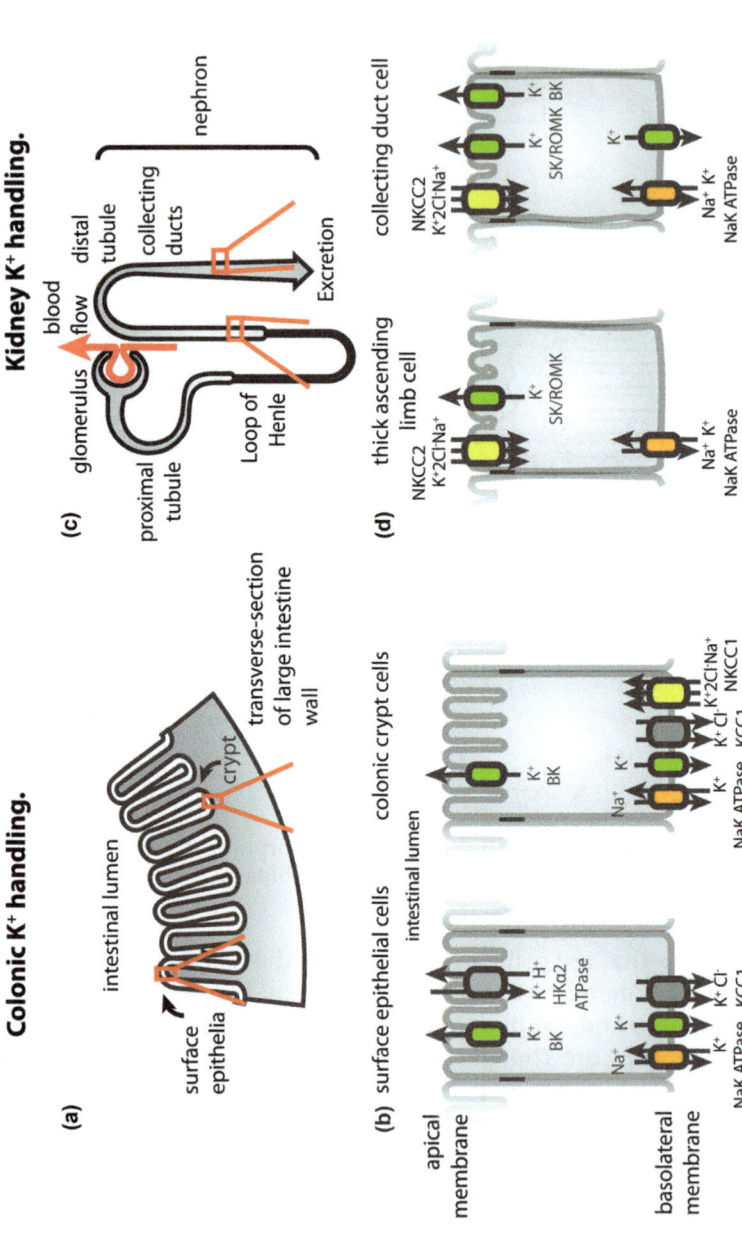

Figure 3.4   The apical and basolateral membranes of epithelial cells carry ion channels and pumps involved in K⁺ uptake or secretion/excretion. Refer to the text for details. **a.** Potassium handling in the colon: transverse section of part of the intestinal wall. **b.** An epithelial cell on the convoluted intestinal surface of the colon (left) has a different complement of ion transporters to colonic crypt cells (right). **c.** Microscopic structure of the kidney nephron. **d.** Ion channels and transporters in the thick ascending limb of the loop of Henle (left) and in the collecting ducts (right).

channel BK (*syn* Maxi-K, *Slo1*) in the apical membrane of crypt cells appears to be the principal secretory K^+ channel.[39,40]

The interdependency of Na^+, K^+ and Cl^- uptake and secretion is evident in the way in which local environmental affronts to epithelia are corrected. The Na^+/K^+ ATPase pump in the basolateral membrane, for instance, cycles rapidly between Na^+ expulsion and K^+ import in a respective ratio of 3 : 2, which could result in an imbalance in Cl^- in the absence of contemporaneous Cl^- regulation. Fluctuations in cell volume are countered with direct modification of the solute balance by basolateral $K:Cl$ and $Na:(K):2Cl$ cotransporters (*e.g.* NKCC1), themselves regulated by specific serine-threonine kinases.[41,42] The former couple K^+ and Cl^- transport, mediating net efflux of Cl^-, while the latter couple the movement of Na^+ and Cl^- (with K^+ optional) resulting in net Cl^- uptake. Small-conductance Ca^{2+}-sensitive K^+ channels in the basolateral membrane promote K^+ recycling, hyper-polarizing the cell and providing the driving force for Cl^- exodus through apical Cl^- channels.[37]

A non-gastric epithelial H^+/K^+ ATPase (HKα2) located in the distal colon provides an additional route for K^+ import from the lumen.[43] It is situated on the apical membrane of surface epithelia and has not been found in colonic crypts (Figure 3.4b). HKα2 is closely related to the acid-producing gastric ATPase HKα1 (H^+ out/K^+ in) in the apical membrane of gastric parietal cells, which acts in collaboration with a Clc-2 Cl^- channel and two K^+ channels (the inward rectifier KCNJ10, or $K_{ir}4.1$, and the voltage-gated K^+ channel KCNQ1, or $K_v7.1$) to achieve secretion of HCl into gastric mucosa.[44,45] HKα2 is an example of a pump that responds to local conditions. In instances where dietary K^+ is low, active potassium absorption is partially chloride dependent, inferring a functional coupling between HKα2 and a cellular Cl^- transporter, analogous to the gastric ATPase HKα1 system. In the case of HKα2, however, the identity of the partner for basolateral Cl^- exit remains uncertain. Candidates include an electrically neutral $K:Cl$ cotransporter KCC1 in the basolateral membrane that is up-regulated under K^+ deprivation,[46] a Clc-2 chloride channel[47] and the HCO_3^-/Cl^- anion antiporter.[48] It should be noted that the overview presented here represents a simplified picture of colonic ion transport.

3.1.2.2 Renal Potassium Handling

The remaining 90% of dietary K^+ (plus other electrolytes) is eliminated through the *kidneys* (Figure 3.4c). Water and ions are continuously filtered at the glomerulus and re-absorbed in the proximal tubule and the loop of Henle, particularly its thick ascending limb where the $Na:(K):2Cl$ transporter NKCC2 is active.[49] The filtrate volume can be of the order of 180 litres per day, of which approximately 1% is excreted as urine. The fine balancing of K^+ in response to hormonal levels occurs by K^+ secretion (or re-absorption, under conditions of K^+ deprivation) in the distal tubule and collection ducts. Similarly to the intestine, the epithelial cells of the nephron are polarized. Here K^+ enters the cell *via* basolateral Na^+/K^+ ATPases and undergoes passive efflux into the lumen through K^+ channels, specifically *via* the ATP-sensitive inward rectifier ROMK (IRK1) (Figure 3.4d).[50] Lithium takes a similar route, but is in large part reabsorbed by apical epithelial Na^+ channels in the principal cells of the collecting duct (Section 3.1.5.2).[51] Additional K^+ channels implicated in K^+ secretion (under conditions of elevated K^+ intake) are the large conductance BK channels,[52] which contribute to urinary K^+ excretion, and the ROMK-like small conductance channel SK[53] responsible for baseline secretion and recycling.

3.1.3 Coordination Chemistry of the Ion in the Blood

The plasma concentration of K^+ is maintained between 3.5 and 5 mM and largely consists of fully hydrated K^+ ions circulating in the bloodstream and extracellular fluid as a constituent of the bulk electrolyte. The latter may additionally include K^+ and other monovalent cations electrostatically associated with amino-acid side-chains on the surfaces of proteins. Continuous bidirectional exchange diffusion between blood and extracellular fluids has been demonstrated by radioactive tracer ion experiments.[54]

Weak associations with oxygen ligands, coupled with a preference for high coordination numbers (seven to eight), mean that K^+ in biological systems is almost exclusively coordinated by specialized channels and transporter molecules (a notable exception being pyruvate kinase). The bloodstream is no exception and transporters are found in erythrocyte membranes, most notably the electrically neutral K:Cl co-transporters, which instigate K^+ efflux following cellular swelling. Electroneutral K:Cl co-transport is a major factor regulating the cellular volume of reticulocytes during maturation.[55] Of the four known co-transporters of this type in human tissue, three are expressed in red blood cells.[56] Co-transporter activity results in net transport of Cl^- out of the red blood cells; the activity is highest in reticulocytes, diminishing with the age of the cell. Elevated K:Cl co-transporter levels also feature in sickle-cell anaemia where they have been implicated in erythrocyte dehydration.[57] Potassium channels present in blood T-lymphocytes include the voltage-gated K^+ channel $K_v1.3$ and the intermediate conductance Ca^{2+}-activated K^+ channel SK4. Both feature prominently in immune cell development, maintaining the membrane potential required to drive activity of store-operated Ca^{2+} channels.

An example of a non-transporter K^+-binding protein is pyruvate kinase. Its enzymatic function is to catalyse the breakdown of phosphoenolpyruvate during ATP catabolism. The enzyme requires both K^+ and Mg^{2+} (or Mn^{2+}) as cofactors. The crystal structure of pyruvate kinase from human erythrocytes (pdb code 2VGB) reveals K^+ and Mn^{2+} ions bound 5.5 Å apart. In the structure, the ions are bridged by a 2-phosphoglycolic acid molecule (occupying the phosphoenolpyruvate binding site) and coordinated by backbone carbonyls and oxygen-containing side-chains.[58] The geometry of K^+ in pyruvate kinase is irregular four-coordinate, or possibly five-coordinate (a low residual peak in the electron density of 2VGB near the K^+ ion appears to correspond to a water molecule).

A non-natural example of K^+ binding in the blood is the aptamer "TBA". Designer targeting agents such as nucleic acid or peptide aptamers hold considerable potential for biotechnological and pharmaceutical applications. An important role has been attributed to potassium in aptamer recognition of human thrombin, a serine protease in human blood that promotes blood coagulation. Specifically, α-thrombin cleaves fibrinogen to release fibrin, sparking the clotting process. TBA is an antiparallel G-quadruplex deoxyribonucleic acid that interacts directly with the fibrinogen-binding site of the protease. Potassium ions bind to the quadruplex with distorted square antiprismatic geometry, not unlike that observed in K^+ channels. The K^+ ions connect the two constituent tetrads and enhance both the affinity and inhibitory ability of TBA for α-thrombin (relative to its affinity when Na^+ is bound).[59] The changes induced in the quadruplex suggest that the nature of the ion binding site, and hence the quaternary structure, is governed by physicochemical properties of the ion, rather than the other way around.

3.1.4 Transport through Membranes

The principal, or certainly the most thoroughly documented, purpose of sequestering K^+ in human cells is to facilitate the electrical transmission responsible for activities of the central nervous system and vital organs, without which all breathing and sensation would cease. The role of cytoplasmic K^+ in electrical signalling is contingent upon intertwined processes that accumulate and rapidly discharge it in response to cellular or electrical signals. The specialized membrane transporter proteins mediating these tasks fall into two broad classes: ATPases (energy-driven pumps) and K^+ channels (gated pores). Painstaking work over several decades has built up a picture of ion transport across cell membranes, although knowledge of the molecular mechanisms remains sketchy. This section elaborates more deeply upon the modes of K^+ transport and provides a molecular context. While acknowledging that a minority of K^+ channels exists in organellar membranes (*e.g.* mitochondrial), the emphasis here is placed on transport across the plasma membrane.

3.1.4.1 Active K^+ Transport

Active ion transport corresponds to "appropriating and amassing", and occurs where movement of a charged species across the membrane is driven by energy release, such as the hydrolysis of a bound cofactor, in the absence of, or opposing, an electrochemical driving force. The cofactor for ion exchangers and transporters (collectively known as pumps) is invariably adenosine triphosphate (ATP), which serves as a form of cellular energy currency. Movement of an ion of interest may also be coupled in parallel or antiparallel to that of prescribed transport counter-ions (*e.g.* Na^+, Cl^-) or cellular molecules (*e.g.* H^+ or glutamate). ATP-driven K^+ pumps feature cycles in which ATP is hydrolysed on one stroke and regenerated on the next. Ion transport is thus the consequence of a sequence of linked energetically favourable and unfavourable events, in this case, ATP hydrolysis to a process of ion translocation against a concentration gradient. Both biophysical[60] and structural analyses of ATPases indicate that they adopt discrete alternative conformations during their functional cycle. This lends credence to the thesis that the ability to assume different conformational states underpins mechanism, where binding, hydrolysis and/or release of ATP is linked to conformational change.[61]

The first description of the Na^+/K^+ ATPase (Figure 3.5a) was published in 1957 by Jens Skou,[62] for which he received (jointly) the 1997 Nobel Prize in Chemistry. This ATPase utilizes ATP hydrolysis to facilitate the exchange of Na^+ and K^+ against their respective concentration gradients (Figure 3.5b). Unlike Na^+ channels, the Na^+/K^+ ATPase has a low affinity for Li^+. While the precise numbers vary a little according to the source, the intracellular concentrations of Na^+ and K^+ are set at around 12 and 140 mM, respectively, whereas extracellular concentrations of Na^+ and K^+ approximate 140 and 5 mM, respectively. Continual regeneration of ATP in the cytoplasm of living cells provides the impetus to maintain this energetically un-favourable ion balance across the cellular membrane. The physiological worth of active transportation can be estimated by the cellular quotient of energy equivalents expended on transporting ions against a concentration gradient. This -has been estimated at 20% of total metabolic energy[63] and possibly more than 50% in excitable cells such as neurons that utilize the potential difference as stored energy for the propagation of nerve action potentials.[64]

In setting up an electrochemical gradient to drive electrical signalling and cellular chemistry, the principal Na^+/K^+ ATPase transports two K^+ ions into and three Na^+ ions out of the cell for each ATP that is hydrolysed. This is not only expensive energetically but, in comparison to efflux during electrical signalling, it is also a very slow process. Potassium ions are pumped in through the ATPase at a steady influx rate of around 60 to 170 ions sec^{-1} or 30–85 cycles sec^{-1}.[65] By comparison, passive efflux through K^+ channels occurs approximately 10 000 times faster at $\sim 10^7$ ions sec^{-1}. As a con-sequence, maintaining sufficiently high intracellular levels of K^+ to support signalling requires individual animal cells to express anywhere from 80 000 to 30 million Na^+/K^+ pumps, depending on cell type.[65,66] In electric eels, for example, a packed array of Na^+/K^+ ATPases is required to set the resting membrane potential of the cells (electrocytes) of the electric organ. Upon depolarization, the net electric discharge from batteries comprising thou-sands of electrocytes produces high voltage shockwaves: 500 V has been recorded for a medium size (0.5 m long) eel – a significant voltage that can cause human death, albeit usually as a result of the shocked individual drowning.

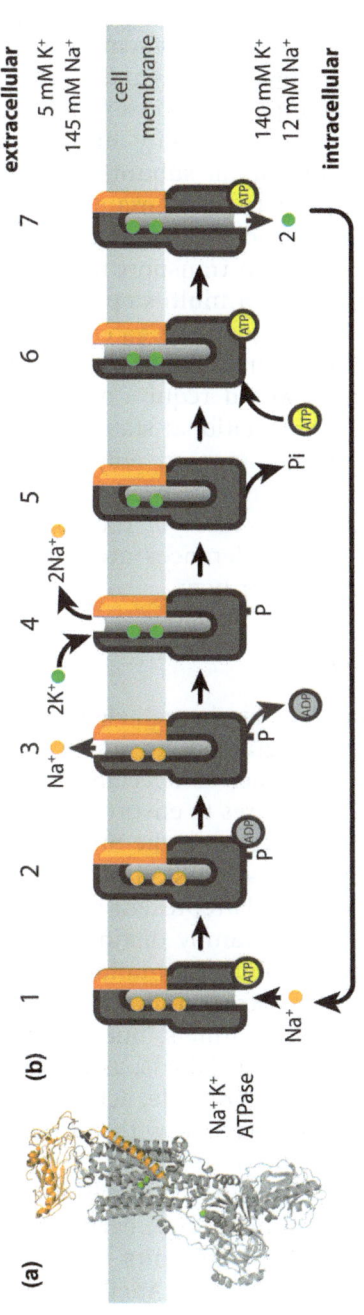

Figure 3.5 Structure and simplified model of the exchange mechanism of the Na^+/K^+-ATPase pump. **a.** The crystal structure of the Na^+/K^+-ATPase pump in the K^+ bound state. **b.** Mechanism of action of the pump. The pump cycles between multiple states according to the Post–Albers scheme **1:** 3 Na^+ and ATP bound; **2:** The pump is phosphorylated and ADP is bound; **3:** One Na^+ is released to the extracellular environment in conjunction with release of ADP; **4:** Two K^+ ions from the extracellular solution exchange with the two remaining Na^+ on the phosphorylated pump; **5:** The phosphate is hydrolysed; **6:** ATP binds in the ATP binding site; **7:** K^+ is released.

Molecular pumps are classified into different categories on mechanistic criteria, and the Na^+/K^+ pump is a member of the P-type ATPase family that is distinguished by autophosphorylation and dephosphorylation of a catalytic aspartate during cycling. Other family members include the sarcoplasmic reticulum Ca^{2+} ATPase pump (SERCA) and the H^+/K^+ ATPase. The Na^+/K^+ ATPase consists of a functional catalytic α-subunit, as well as β- and γ-subunits that provide conformational stability and fine-tune ion affinity. Tissue-specific isoforms of the individual subunits combine into Na^+/K^+ ATPase isozymes with differing properties; for example, α-subunit isoforms differ with respect to ion affinity[67] and turnover rates,[68] as reviewed.[69] According to the Post–Albers scheme, ion transport involves cycling between distinct K^+ and Na^+ bound states in a multi-step reaction cycle in which sequential ion uptake and release occurs by gates on either face of the membrane (Figure 3.5b).[70,71] Characterization of the structural changes occurring during the reaction cycle would require a complement of crystal structures describing discrete conformational states, each representing an individual kinetic step. At present, however, only four structures representing two states of a Na^+/K^+ ATPase have been determined.[72–75] In contrast, there are over 20 separate structures of the closely related SERCA pump.[76] So, despite some notable differences around the binding sites, SERCA remains the model of choice for now.

3.1.4.2   *Passive K^+ Transport*

Ion channels correspond to cellular "discharge" elements, utilizing the energy stored in the electrochemical gradient to transport ions across cell membranes. These molecular pores act as specialized gates to diffusion, possessing well-honed molecular features to ensure that only a specific type of ion (Na^+, K^+, Ca^{2+} or Cl^-) is able to cross the membrane and that other types cannot bind productively. Non-specific classes of cation and anion channels also exist. Potassium channels represent a large and naturally diverse subdivision of the ion channel family distinguished by extreme selectivity for K^+, a consequence of evolutionary tailoring to its coordination chemistry. Potassium channels are structurally related to both the Na^+ channels and some classes of Ca^{2+} channels, but are more rigorously selective than either. The secret to K^+ selectivity is, as discussed earlier, *direct* interaction of the ion with the protein molecule, achieved by stripping away a shell of hydrating waters and replacing it with an octet of polypeptide backbone carbonyls buried within the protein.

Humans have 87 genes known to encode potassium channels. They are classified on the basis of sequence homology, each grouping readily distinguishable by intrinsic domains and subunits that respond to specific environmental or chemical cues (Figure 3.6). Ancillary regulatory domains typically span the membrane or protrude into the cytoplasm. The voltage-sensing domain of one class, known as voltage-gated K^+ (K_v) channels (Figure 3.6a), is an example of the former, while the N-terminal T1 domain

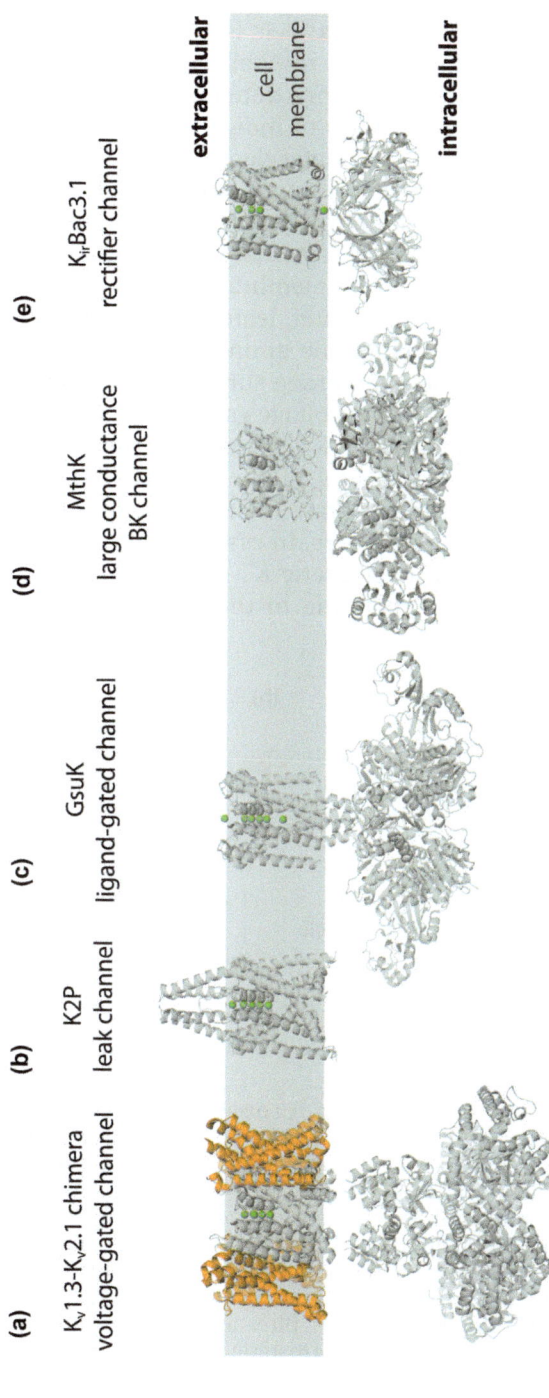

Figure 3.6 Potassium channels are classified into families and subfamilies on the basis of regulatory domains attached to the central channel. **a.** A voltage-gated potassium channel. The structure is that of a chimeric molecule constructed for the purpose of crystallization $K_V1.3$-$K_V2.1$ (PDB 2R9R). The voltage sensor domain (orange) has four membrane-spanning segments. **b.** A twin-pore K2P potassium leak channel with large extracellular protrusions (PDB 3UKM). **c.** A nucleotide-gated Ca^{2+}-inhibited K^+ channel from *Geobacter sulfurreducens* GsuK (PDB 4GX5). **d.** MthK from *Methanococcus jannaschii* is a prokaryotic homologue of the large-conductance BK channel, except that it lacks a voltage-sensing domain in the membrane (PDB 1LNQ). **e.** An inwardly rectifying potassium channel K_{ir}Bac3.1 from *Magnetospirillum magnetotacticum* (PDB 2WLM) is in the same family as RomK ($K_{ir}1.1$). Potassium ions are depicted in green.

is an example of the latter. In some groups, another protein (such as the β-subunit of K_v channels) may be permanently associated with a K^+ channel over its lifetime. Each major class of K^+ channels is identifiable by its electrophysiological signature, but the factors responsible are less clear and contributions from both pore and regulatory domains appear likely. The differences underpin the physiological roles of individual types of K^+ channels, some of which will be discussed in greater depth within this chapter.

Architecturally, the conserved pore-forming domain of K^+ channels comprises two anti-parallel transmembrane helices (the "inner" and "outer" helices), along with a shorter "pore helix" spanning only one leaflet of the lipid bilayer (Figure 3.7a).[77] In the plasma membrane, K^+ channels are invariably oriented such that this is the outer leaflet. A lesser-known but seemingly general feature is an amphipathic amino-terminal extension of the outer helix located at the cytosol-membrane surface. This is termed the S4-S5 linker in K_v channels (and other voltage-gated ion channels) and the "slide helix"[78] in other K^+ channels. A K^+ channel consensus sequence T(V/I)G(Y/F)G connects the C-terminal end of the pore helix (positioned midway through the membrane) to the extracellular surface of the channel.

Potassium channels are obligate tetramers, in essence at least. The 4-fold symmetry is not always exact, and heteromeric K^+ channels are known, as are pseudo-tetrameric dimers corresponding to the so-called "twin pore"

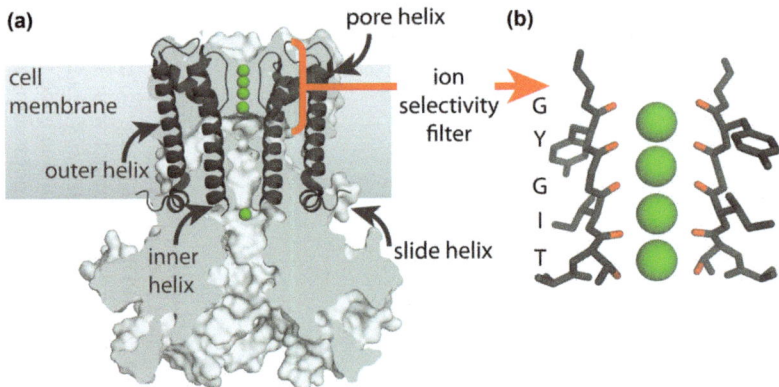

Figure 3.7 Potassium channel architecture. **a.** Central transection through the molecular surface of a K_{ir} channel with the protein backbone of the pore shown as a dark ribbon. The key elements of the canonical fold depicted include the inner, outer, pore and slide helices. The molecular surface reveals the depth and dimensions of the ion conduction pathway. **b.** The ion selectivity filter corresponds to the K^+ channel signature sequence T(V/I)G(Y/F)G. Potassium ions are liganded by peptide carbonyl oxygen atoms and a threonine hydroxyl oxygen. The four ion sites depicted within the selectivity filter are not all occupied simultaneously due to unfavourable electrostatic repulsion. During conduction K^+ rapidly transfers between sites to achieve net flow down an electrochemical gradient.

or K2P channels. Importantly, individual monomeric subunits are non-functional as the ion permeation pathway forms, not within them, but at a common interface coincident with the tetramer axis. The relative orientation and placement of the four subunits thus decides the critical dimensions of the conduction pathway through the membrane, and hence its capacity to act as a high-fidelity conduit for ion translocation.

Surprisingly, despite the nomenclature and unlike Na^+ channels, K^+ channels do not have a *bona fide* pore spanning the entire lipid bilayer. In the tetramer, extended polypeptide linkers encoded by the consensus sequence create an "ion-selectivity filter" perforating only the outer leaflet of the bilayer, wherein rings of carbonyl oxygen atoms coordinate K^+ at discrete sites by providing each ion with eight oxygen ligands (Figure 3.7b). This singular feature of K^+ channels ensures fidelity by optimizing coordination of K^+ in preference to other monovalent cations. It is possible only because the signature glycine residues confer sufficient conformational flexibility on the polypeptide backbone to allow the carbonyl oxygen atoms and threonine hydroxyl in the motif to align appropriately when K^+ is present. Only in crystal structures where the selectivity filter is distorted, for example by crystallization under conditions where K^+ is depleted, or as a consequence of mutagenesis, are the carbonyl oxygen atoms visibly out of alignment. Thus the selectivity filter represents four internal coordination sites linked in series along the molecular axis. The approaches of the four consensus linkers of the selectivity filter are so close that molecular surface calculations show no cavitation, and hence no "pore". The pore as such is a deep invagination traversing only the inner leaflet of the bilayer (Figure 3.7a).

The precision placement of oxygen ligands in the selectivity filter is the crux of a geometrically tailored environment that favours K^+ over Na^+ binding. The design is successful because of the high free energy penalty that dictates that K^+ must always be ligand-bound. Consequently, substitution of peptide carbonyls for the water ligands stripped from each K^+ ion as it enters the bilayer is absolutely critical to ion permeation. By the same token, the capacity to release K^+ once it has passed through the selectivity filter is contingent upon the availability of free water to complete its coordination sphere so that at all times – before, during and after transit – K^+ ions exist in a stable coordination complex. Exiting K^+ ions must therefore undergo a reversal of the entry process, exchanging transient interactions with eight carbonyl oxygens of the selectivity filter for a similar number of ligand water molecules.

3.1.4.3 Electrical Excitability

While the purpose of this chapter is to define routes for the uptake and homeostasis of K^+ rather than to elaborate on its role in excitation, much of the chemistry described herein is directed toward facilitation of electrical signalling through the body when and where it is needed. Ion channels are the vehicles by which action potentials in the central nervous system (CNS) are generated, the heartbeat perpetuated and non-electrical signals by second messenger pathways initiated. Figure 3.8 briefly depicts the principles underlying electrophysiological research, but the interested reader is referred to an excellent text to gain a greater understanding of electrical phenomena in higher eukaryotes.[79]

3.1.5   Sensing, Buffering and other Control Mechanisms (Cellular Homeostasis)

3.1.5.1   Overview

The membrane potential essential for excitation of nerve and muscle is utterly dependent upon maintaining the steep concentration gradient of K^+ across cellular membranes within tight strictures. An additional requirement is that the daily whole-body intake of K^+ must exactly balance the amount excreted. Potassium homeostasis is thus active at both tissue-specific and systemic levels, and entails detecting K^+ intake, quantitatively adjusting intra- and extracellular fluids and eliminating excess, all of which is managed through a suite of acute and medium-term measures (reviewed in 2009[80]). Failure to balance K^+ uptake and excretion can lead to serious derangements in the form of hypo- or hyper-kalaemia, as well as to a swag of associated conditions. As the respective names imply, hypo-kalaemia occurs under conditions of K^+ deprivation (such as during fasting or diarrhoeal episodes) and can lead to draw down of muscle K^+, while hyper-kalaemia occurs when K^+ is in excessive supply (*e.g.* if insulin secretion is compromised). Both are cardiovascular risk factors and can trigger deleterious physiological outcomes. Primary control of K^+ homeostasis is exerted in the distal nephrons of the kidney and the colon while fine adjustments to extracellular and intracellular K^+ concentrations are effected in organs and tissues (Figure 3.3). In the central nervous system, for example, where neurotransmission causes rapid fluctuations in K^+ levels, coexisting pathways for K^+ uptake and spatial buffering act in synchrony to manipulate local K^+ levels and maintain electrical neutrality.

Knowledge of the interconnections that govern K^+ homeostasis is slowly building. For example, like those of glucose, K^+ levels are regulated by insulin, although this territory is poorly charted as yet. Moreover, in addition to the familiar feedback loops that operate in biological systems, there is significant evidence for anticipatory (feed-forward) control mechanisms. As an example, excretion of K^+ in the urine after a meal occurs *before* any measurable rise in extracellular K^+ levels, suggesting rapid, pre-emptive dissipation. The first instance of this phenomenon in the literature appears to be a report of "early K^+ sensing" in the visceral organs.[81] Cogent support derived from evidence that partaking of K^+ *with* a meal enhances the renal efficiency of K^+ clearance *prior* to any rise in plasma concentration (*i.e.* preceding any feedback response).[82] The as yet unidentified sensor has been branded a regulatory "gut-factor". Glucagon has also been implicated

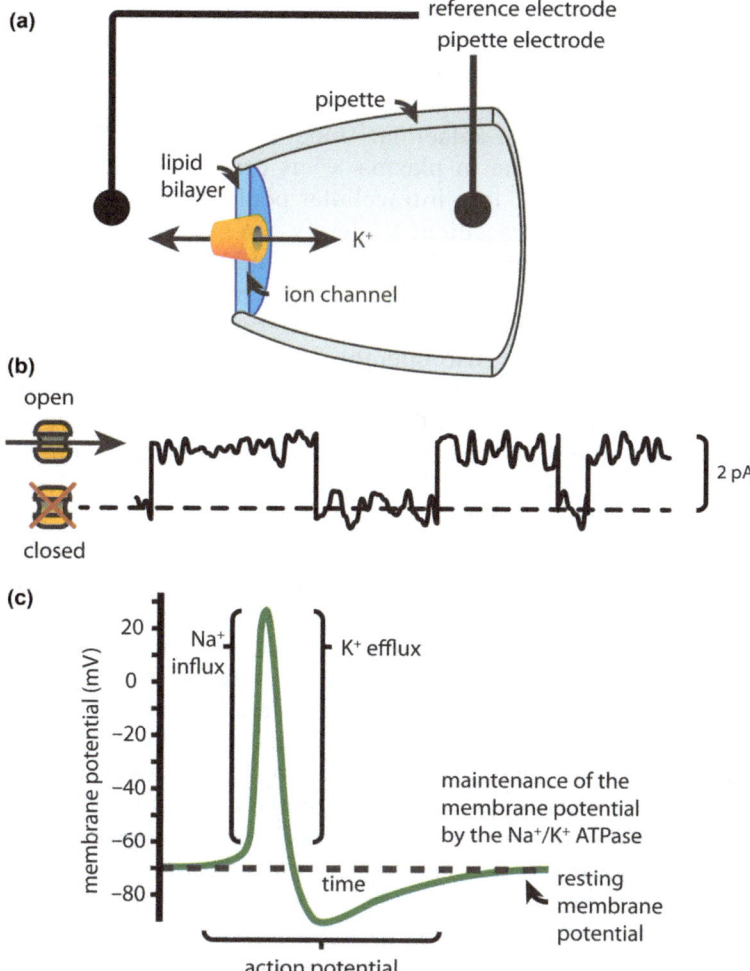

Figure 3.8 Basis of electrophysiological experiments. **a.** An "inside-out" single channel patch clamp setup. A borosilicate pipette (grey) that was initially pressed against a clean cell membrane until it formed a tight seal has been pulled away, excising a membrane patch (pale blue) containing a single K^+ channel. The seal between membrane and pipette must be extremely tight (resistance > 1 GΩ) or ion leakage at the edges will drown out any K^+ signal through the K^+ channel. The composition of the inner and outer solutions can be manipulated. The voltage within the pipette is clamped relative to a reference electrode. **b.** A current of picoampere (pA) amplitude can be measured from a single K^+ channel. **c.** Electrical potentials in nerve cells are due to conduction through Na^+ and K^+ channels. Impulses in nerve axons occur as a result of a combination of the activities of a number of ion channels that result in the reversal of the resting membrane potential. An action potential results from a wave of spatiotemporal changes in the permeability of Na^+ (NaV voltage gated channels). Potassium channels (K_v voltage gated channels) act in the recovery phase of an action potential. During action potentials an influx of Na^+ is closely followed by efflux of K^+ ions. Adapted from Hille.[79]

in feed-forward control, increasing renal excretion of K$^+$ after a protein-containing meal in response to elevated levels of the second messenger cAMP.[83,84] Overall, feed-forward regulation appears to play a principal role in acute non-renal modulation of K$^+$ at mealtimes, when fast K$^+$ depletion is required to forestall hyper-kalaemia. Under normal dietary conditions, a rapid post-prandial decline in plasma K$^+$ is caused by insulin-stimulated uptake of extracellular K$^+$ into intracellular pools in muscle and liver cells, temporarily buffering extracellular K$^+$ levels.[85] A slow, protracted release of K$^+$ back into the extracellular fluid is promoted by muscle contractions, allowing a quantity of K$^+$ almost precisely matching the amount consumed to pass concomitantly to the kidneys for excretion.

Classical feedback loops also operate. A rise in extracellular K$^+$ concentrations brings about an increase in K$^+$ secretion into the collecting ducts of the kidney nephrons, whereas a decrease causes the converse. Inter-dependency of K$^+$ and Na$^+$ levels reflects overlap between their nominally distinct transport pathways, and this is nowhere more obvious than in these feedback loops, where up or down regulation of renal Na$^+$/K$^+$ ATPase and epithelial Na$^+$ channels are primary mechanisms driving K$^+$ modulation.

As well as general dietary fluctuations, a multitude of stress conditions, in the form of hibernation, starvation, muscular exertion, cellular acid-base disturbances, excessively K$^+$-rich diets or even compromised blood supply to organs (*e.g.* brain ischaemia) can affect K$^+$ balance in mammals. Under unusually high or low K$^+$ loads, the homeostatic system must compensate and physiological studies have revealed a network of adaptive strategies. During short-term K$^+$ deprivation, for example, sensitivity to insulin decreases such that insulin-dependent K$^+$ uptake is minimal or absent.[86]

Regulation of K$^+$ homeostasis in mammals is thus stratified with feed-forward and feedback loops operating on different time scales, some with immediate effect and others markedly slower. The slower pathways include mechanisms for modulating the membrane concentrations of ATPases, epithelial Na$^+$ channels and K$^+$ channels specifically involved in homeostatic K$^+$ secretion. These operate by transcriptional control (*e.g.* of epithelial Na$^+$ channels by the mineralocorticoid hormone aldosterone), by post-translational modification (*e.g.* RomK phosphorylation), by redistribution (*e.g.* from vesicles to the apical membrane) and/or by elimination.

3.1.5.2   Whole Body Homeostasis: Secretion and Excretion of K$^+$ via the Kidneys

Renal homeostasis is controlled in the distal nephron, a segmented organ incorporating the cortical and medullary collecting ducts and comprising a range of specialized cell types. These include the *intercalated* cells (K$^+$ secreting) that differentiate into specialized α (acid secreting) and β (base secreting) subtypes[87,88] and also the *principal* cells (K$^+$ absorbing) which carry the epithelial Na$^+$ channel. Like the colon, the kidney is a prime site of action of aldosterone, a mineralocorticoid hormone produced in response to elevated K^+ levels and associated with Na^+ reabsorption and enhanced K^+ secretion. As it turns out, these are two of many sites and recent studies attesting to the widespread off-target activities of aldosterone are prompting re-examination and review of its role (reviewed in 2012[89]). In the nephron, adjustment of water and electrolyte levels plus pH occurs under the influence of paracrine signalling. Intercalated cells represent the major source of paracrine factors, cooperate with the principal cells in mediating K^+ homeostasis (reviewed in 2012[90]) and play a central role in maintaining electrolyte balance, plasma K^+ levels and blood pressure.

Regulated secretion of excess K^+ from the collecting tubules is followed by excretion from the kidneys. The process of elimination is under purinergic (*i.e.* ATP) control. It has been proposed that hexameric hemichannels of connexin-30 in the apical membrane[91,92] facilitate ATP release into the lumen[93] where the nucleotide acts on specific K^+ channels to prevent secretion. However, another hemichannel-forming protein in the plasma membrane, known as a pannexin, was implicated recently in ATP release,[94] suggesting that the jury is still out on this issue. Independent of these transport issues, luminal ATP directly inhibits the ROMK-like small-conductance inward rectifier K^+ channel SK. Preliminary evidence of coupled ATP and K^+ efflux from intercalated cells[95] suggests ATP may also modulate the activity of the large-conductance Ca^{2+}-sensitive BK K^+ channels.

3.1.5.3 Local Homeostasis: Spatial K^+ Buffering and Uptake in the Central Nervous System

The CNS of animals presents specific challenges for K^+ homeostasis: firstly, at 15 to 30% (typically 20%) of the volume of a healthy adult brain[96] the extracellular space is very limited and secondly, neuronal activity causes significant temporal fluctuations in extra- and intracellular K^+ levels. These factors affect membrane polarity, local chemistry, neurotransmitter release and, importantly, neuronal excitability post-impulse. Measurements on feline cortex and thalamus have revealed extracellular K^+ concentrations plateauing significantly above the homeostasis level of approximately 3 mM. Under normal conditions, K^+ plateaus at 10–12 mM[97,98] but under pathophysiological duress this increases dramatically to 50–80 mM.[99,100] Left uncorrected, the consequences of *ad hoc* compensation would be detrimental, at a minimum, and potentially neurotoxic in extreme circumstances where the actions of other proteins are invoked (*e.g.* glutamate excitotoxicity is caused by glutamate release resulting from the reverse action of the Na^+/glutamate co-transporter during brain ischaemia[101]). Any mechanism for correction of such large fluctuations must preserve electroneutrality and pH while stabilizing a vacillating cell volume. Fine-tuning a rapidly changing K^+ balance thus appears to be a complex issue. Complementary mechanisms underpinning local K^+ homeostasis in the CNS have been identified as *K$^+$*

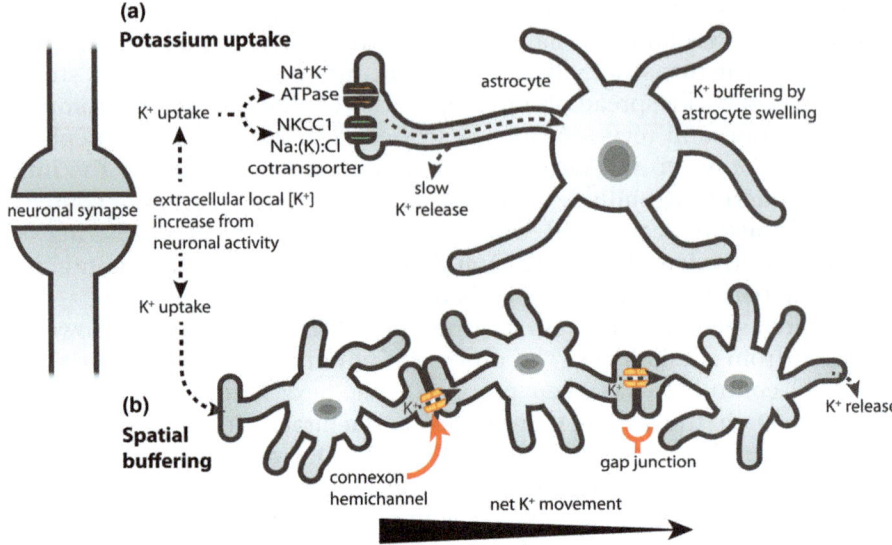

Figure 3.9   Potassium uptake and spatial buffering by glial cells results in a net movement of K$^+$ away from areas of high concentration such as the neuronal synapse. **a.** Buffering as a result of active K$^+$ uptake and sequestration in the astrocyte. Potassium is imported *via* the Na$^+$/K$^+$ ATPase and NKCC1. The movement of ions into the astrocyte also results in uptake of water resulting in astrocyte swelling. **b.** Spatial buffering resulting in movement of K$^+$ away from the site of neuronal activity to distant regions to maintain a low *extracellular* K$^+$ concentration. In this case no significant swelling occurs. Potassium entering the astrocyte passively moves away from the regions of high K$^+$ concentration through gap junctions consisting of adjoining connexon hemichannels, into neighbouring astrocytes. This is followed by release at distant sites where the extracellular potassium concentration is low. The mechanisms for K$^+$ uptake and spatial buffering can operate in unison.

uptake and *spatial buffering*,[102–104] both drawing on an intricate web of interrelationships involving channels, active transporters and other receptors (Figure 3.9). As yet, it remains unclear whether and how much functional overlap occurs and what factors determine the mechanism utilized (*e.g.* developmental phase, cell type).

3.1.5.4   Potassium Uptake

In the context of homeostasis, K$^+$ uptake refers to the temporary accumulation of excess K$^+$ into glial cells, of which astrocytes, or star-shaped glia, are the predominant cell type in the CNS. The roles of astrocytes in K$^+$ clearance in the CNS and in the ischaemic brain have been reviewed.[101,105,106] Both active and passive modes of uptake occur and the main mediators have been identified along with the auxiliary co-transport systems that maintain electrical balance and regulate astrocyte volume. The central participants in K^+ clearance, which is accompanied by astrocyte swelling,[107] are a glial isoform of the Na^+/K^+ ATPase[108] and the NKCC1 isoform of the $Na:(K):Cl$ co-transporter.[109] As extracellular K^+ levels return to normal, K^+ is released back into the extracellular space from whence it came. In astrocytes, the intracellular Na^+ concentrations may be insufficient to furnish uptake of K^+ *via* the Na^+/K^+ ATPase alone, indicating a further reason for employing co-transporters. Indeed fluorescence ratio imaging experiments on cultured rat hippocampal astrocytes have measured baseline intracellular Na^+ concentrations in the order of only 15 ± 5 mM.[110] Co-transporters, including the $Na:(K):Cl$ and $Na-HCO_3$ carriers, serve to bring in additional Na^+ to facilitate Na^+/K^+ ATPase action[110] (Figure 3.9a).

3.1.5.5 *Spatial Buffering*

Spatial buffering, sometimes termed gap-junction coupling, similarly involves clearance of excess K^+ into glial cells, but in this case without resulting in a net accumulation of K^+ or water. As each K^+ ion is conveyed into a glial cell another simultaneously departs from a cell some distance away. Potassium ions diffuse away from sites of high extracellular K^+ through an extended syncytium connected *via* connexon hemichannels on adjoining cells in the glial membranes (Figure 3.9b). Evidence for gap junction coupling between astrocytes[111] or between astrocytes and oligodendrocytes[112] has been experimentally obtained either by injection of dyes into astrocytes, which diffuse to nearby cells, or by using optical or electrophysiological means. In mammals, the syncytium can cover tens or hundreds of cells, and diffusion can be modulated by K^+ or neurotransmitter application.[113] Potassium ions traverse variable distances through these continuous cell-to-cell networks, from sites of elevated concentration to remote, K^+-depleted areas where they are released. A return current of Na^+ and Cl^- through the extracellular space completes the loop (reviewed in 2000[105]).

3.1.5.6 *Potassium Siphoning*

In retinal astrocytes (Müller cells), which differ from other types of glia in size and shape, the buffering process has specialized into "K^+ siphoning", first hypothesized in 1984 (Figure 3.10).[114] The term refers to the rapid release and dispersal of K^+ into a reservoir of gelatinous fluid (the vitreous humour). It has been proposed that this is achieved by passive K^+ currents involving K^+ channels,[115] and that siphoning is facilitated by a high concentration of K^+ channels located in the specialized cellular processes (endfeet) of Müller cells that border the humour.[116,117]

Potassium channels from the inward-rectifier K^+ (K_{ir}) family are strongly implicated in siphoning.[118] Glial cells throughout the CNS are highly permeable to K^+, less so to Na^+, and characterized by resting potentials

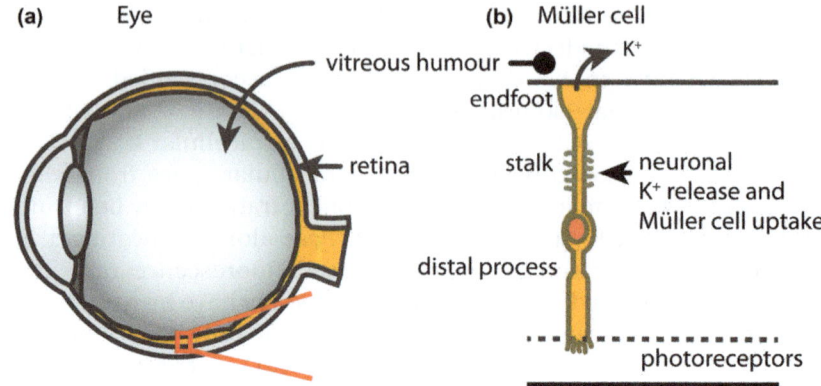

Figure 3.10   Potassium siphoning by Müller cells is a specialized form of spatial buffering. **a.** Relevant morphology of the eye. **b.** A rapid increase in extracellular K^+ concentrations as a result of neuronal activity is dispersed by the action of K_{ir} K^+ channels localized in the endfeet of specialized glia called Müller cells. Excess extracellular K^+ is taken up by Müller cells and siphoned rapidly into the vitreous humour, resulting in net movement of K^+ ions away from firing neurons.

approximately 20 mV more negative than neighbouring cell-types.[119] The hyper-polarized membranes favour activity of K_{ir} channels, which are most active at hyper-polarized potentials whereas most other K^+ channels are not. The major physiological role of K_{ir} channels is to maintain the resting potential of cells near the reversal potential of K^+ (E_K) by permitting small outward fluxes of K^+ at slightly depolarizing potentials (just above E_K). They were named for their ability to rectify (or stem) outward K^+ currents, in accordance with a phenomenon first noted in 1949.[120] Rectification is achieved by "open channel block", meaning that under strongly depolarizing conditions endogenous cellular polyamines (polycations) diffuse into and occlude the conduction pathway, preventing K^+ efflux.

Retinal Müller cells retain a high concentration of $K_{ir}4.1$ channels in the endfeet. The water channel aquaporin-4 (AQP4) exhibits a similarly polarized distribution in CNS glia, where it is enriched in the endfeet bordering microvessels or subarachnoidal spaces. Their co-localization in the glia[121] formed the basis of early suggestions that, by coupling ion and water movement across membranes, the channels might collaborate in reducing cell swelling associated with elevated external K^+.[122] Since then it has been shown that $K_{ir}4.1$ and AQP4 are associated within a macromolecular complex in rat retina,[123] but there is no direct evidence for a regulatory relationship. In fact, the available evidence favours the converse[124] and so the rationale for association remains to be established. More recently, AQP4 was shown to associate with the Na^+/K^+ ATPase in glia[125] while studies on AQP4 knock-out mice have implicated impaired K^+ uptake through the glial Na^+/K^+ ATPase in AQP4-deficient hippocampus. These observations suggest that AQP4 may directly influence Na^+/K^+ ATPase function.[126]

Although the twin phenomena of K^+ uptake and spatial buffering described have gained general acceptance, the identities of the proteins, channels and transporters involved in homeostasis, and specific details of their associations and roles, remain under investigation.

3.1.6 Conclusions

Potassium is present in high abundance within living tissues. In humans and higher eukaryotes, both intracellular and extracellular K^+ levels are subject to continuous monitoring and fine-tuning. The complexity within the system is that the activities of K^+, Na^+ and Cl^- in tissues and organs are not independent and the electrolyte balance as a whole must be concurrently adjusted and maintained. Knowledge of the physiological processes governing uptake and homeostasis is continually being refined as new interconnections between pathways come to light. While their spectroscopic silence has somewhat disadvantaged studies of the Group 1 ions in the past, a revolution in visualization technologies and structural biology is paving the way to elucidating the molecular interactions governing the intricate series of events in K^+ (and electrolyte) homeostasis. These emerging scientific frontiers hold enormous potential to enhance understanding of K^+ physiology and its impact on cell biology. Key issues include, for example, how electrical signalling feeds into cellular Ca^{2+} signalling cascades and the role of the membrane potential in a diversity of cellular processes that require it (*e.g.* protein translocation).

3.2 Prokaryotes

Eukaryotes, either multicellular or unicellular, are characterized by compartmentalizing membranes that give definition to the nuclei and organelles. Prokaryotes, which include the *eubacteria* and *archaea*, are much simpler, at their most advanced exhibiting only rudimentary intracellular membrane systems. An example is the thylakoid membrane system internal to the plasma membrane of cyanobacteria. The nature of the evolutionary relationship between eubacteria and archaea is as complex as that between either and the eukaryotes (see, for example,[127,128]), involving a combination of vertical and horizontal genetic transfer. The cellular structure and organization of the two prokaryotic groups differs markedly, even to the lipids present in their membranes. Lateral (horizontal) gene transfer between archaea and eubacteria is well documented. In general, indications are that genetic material can be accepted or discarded in response to environmental pressures and there is even a suggestion that viruses may act as intermediaries in some circumstances. Moreover, endosymbiotic associations resulting in lateral gene transfer from bacteria to eukaryotic hosts are increasingly coming to light.[129,130] So far, transfer in the opposite direction is unknown. The complexity of processes contributing to heredity explains in part why the origins of some K^+ transport proteins discussed here are far from obvious.

As a bulk electrolyte, K^+ is essential to the life cycle and survival of prokaryotic cells. Gram-negative bacteria (*e.g. Escherichia coli*) are the best-studied examples and a brief overview of their features is provided for context. Beneath an outer capsid, twin phospholipid membranes are physically separated by a periplasmic space that incorporates a peptidoglycan layer of high tensile strength that contributes to the morphology of the bacterium (Figure 3.11). The outer membrane is perforated by proteins known as porins, which adopt a β-barrel topology, rendering it permeable to particles up to 2 nm. This imbues it with the qualities of a semi-permeable

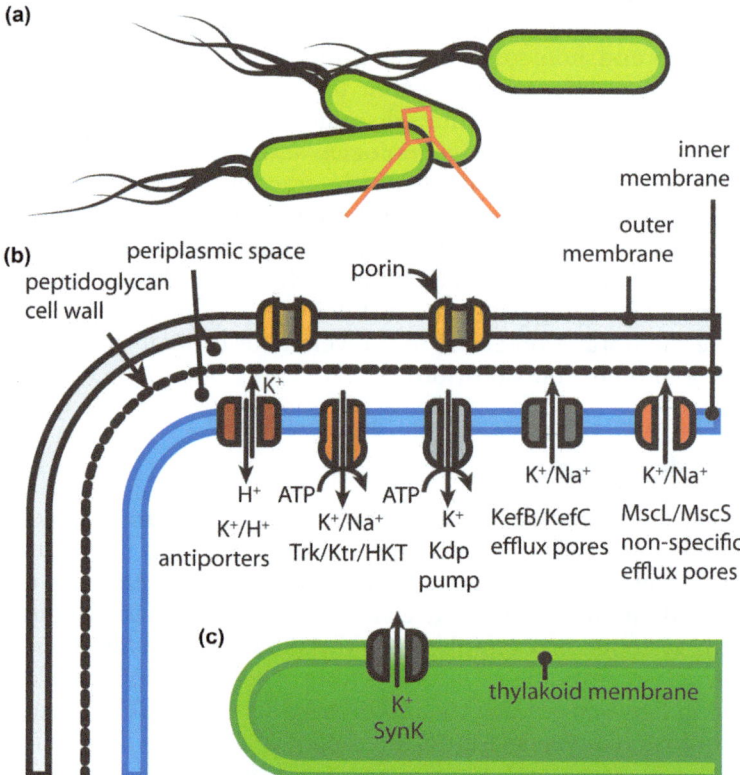

Figure 3.11   Bacteria (*e.g. Escherichia coli*) harvest K^+ to maintain their osmotic potential. **a.** Bacteria are simple unicellular organisms with species-specific morphologies. **b.** Gram-negative bacteria are bounded by dual phospholipid membranes separated by a periplasmic space. A peptidoglycan layer between the membranes is present to maintain cell morphology. Beta-barrel proteins known as porins perforate the outer membrane. The inner plasma membrane harbours K^+ (and other ion) channels and transporters specific to the organism. **c.** The thylakoid membrane (green) is present only in cyanobacteria, which are distantly related to Gram-positive bacteria.

membrane, in that solutes and ions can permeate but periplasmic macro-molecules cannot. The consequence is an unequal distribution of ionic species across the outer membrane (Donnan equilibrium), some being as-sociated with macromolecules, generating an electrical Donnan potential. With this crude solute filter, bacteria depend upon efflux systems with components in both outer and inner membranes for rapid elimination of permeating drugs and toxins. The inner (plasma) membrane surrounding the cytoplasm is significantly less penetrable and, by contrast, actively takes up K^+ to generate a highly negative membrane potential.

The shape of cells and their ability to divide and grow are dependent upon osmoregulation, in which K^+ plays a key role.[131,132] The interior of a cell has a different osmotic constitution from its surroundings, and balance is maintained by hydrostatic pressure (turgor) across the cell envelope. While this gives the cells form and firmness, any significant perturbations to the osmotic equilibrium place the cell at risk. The gross morphological features of bacteria are also of relevance. Isolated cells have a high surface area to volume ratio and the variety of distinctive cell species morphologies elevate the ratio even further such that maintaining cell volume in the face of os-motic shock presents a significant challenge. Bacteria require means by which cellular integrity can be preserved under extreme conditions, and the ace up their sleeves is the ability to rapidly manipulate the levels of ions and osmotic solutes. The role of K^+ can be illustrated by the ultrafast responses of cells to an increase in the external ionic strength, or hyper-osmotic shift, entailing a switch to K^+ uptake accompanied by corresponding synthesis of the physiological counterion/osmolyte (*e.g.* glutamate in *E. coli*) to restore turgor. This is somewhat simplifying the real situation, however. Nominally unicellular bacteria typically exist in self-adherent colonies, which can take the form of small asexually produced mounds of cells (such as on agar plates) or, encased in a polymeric matrix, as biofilms that may contain one or a multiplicity of species (reviewed in 2002[133]). Evidence for cell-to-cell communication within biofilms is growing: horizontal gene transfer leading to greater phenotypic diversity is a notable instance[134] with quorum sensing being another. Some recent indications are that K^+ ions may act as signal-ling molecules in these processes.

3.2.1 Transport through the Membrane

3.2.1.1 K^+ Influx

In the same manner as eukaryotes, their simple cousins actively accumulate K^+ to high intracellular concentrations, resulting in a negative resting po-tential. Acidophilic bacteria represent an exception in that a positive po-tential across the cell membrane is maintained. A limited number of K^+ transporters have been identified in bacteria, including influx systems used to accumulate K^+ and efflux systems that alleviate the toxicity com-mensurate with excessive K^+ levels. Constitutively expressed members of the

Trk/Ktr/HKT "transporter" family are involved in K$^+$ and Na$^+$ uptake and its members are found widely in prokaryotes and plants. This was reviewed in 2010[135]. Curiously, this class of putative ATP-fuelled transporters is more closely related to ion channels than to ATPases. The crystal structure of KtrAB[136,137] reveals an ion selectivity filter similar to that of the ion channel NaK[138] that conducts both Na$^+$ and K$^+$ but not divalent ions. In addition, an octamer of domains on the cytoplasmic side of the membrane is related structurally to that of the Ca^{2+}-sensitive K$^+$ channel BK, albeit somewhat differently constructed and assembled. Ion channel activity has been confirmed by measurement of single channel cation (but not anion) currents in TrkH from *Vibrio parahaemolyticus*. Moreover a significant increase in the *open probability* (the relative time TrkH is open, or conducting) in the presence of ATP is reproduced when non-hydrolysable ATP analogues are substituted, indicating that cation transport is not an energy-dependent process. This indicates that Trk mediates passive, rather than active, K$^+$ transport, consistent with it being a channel.[137] Another group of K$^+$ transporters, the Kup family, also appears to be constitutive.

A class of transport protein with higher (micromolar) affinity for K$^+$ was identified from K$^+$ transport mutants of *E. coli*.[139] These P-type ATPase pumps are now known as Kdp transporters. Encoded by the KdpFABC operon, the multisubunit ATPases differ from their eukaryotic counterparts in that ATP hydrolysis and substrate binding/release are coupled but performed by different subunits.[140] Additionally, K$^+$ influx does not appear to be linked to the transport of another ion.[141] Kdp pumps are unusual prokaryotic transport systems, being regulated at the transcriptional level in response to environmental stressors and local K$^+$ levels and possibly to hydrostatic pressure (although this remains controversial;[142,143] reviewed in 2010[144]). Expression of KdpFABC is regulated by the KdpD (histidine kinase) and KdpE (response regulator) subunits associated with Gram-negative bacterial Kdp complexes (but not present in some archaeal extremophile counterparts[145]). Recent studies indicate that the transcriptional regulation is due to specific effects of K$^+$ on the histidine kinase KdpD.[146,147]

3.2.1.2   K$^+$ Efflux

Significantly less is known about prokaryotic K$^+$ efflux systems. The major bacterial efflux pores KefB and KefC act independently of K$^+$ uptake systems to enhance cellular recovery from toxic insults.[148] Inactive in the presence of cellular glutathione, they are activated by the glutathione adducts of toxic electrophiles such as *N*-ethylmaleimide, switching on K$^+$ efflux and Na$^+$ influx, decreasing the cytoplasmic pH and, in so doing, promoting cell survival.

A consequence of reduced osmotic concentration in the surrounding environment is that the hydrostatic pressure across the bounding envelope of the bacterium exceeds normal levels. The immediate risk to the cells is osmotic shock. When this occurs, the mechanosensitive cation channels

MscS and MscL come into play. They are particularly important in extreme conditions such as rapid bacterial rehydration and instantaneously activate in response to high hydrostatic pressure, correcting the hypo-osmotic shock. While they are relatively non-specific ion channels, K^+, as the predominant cation present within cells, is the major ionic species extruded through the pore. The rapid elimination of cytoplasmic K^+ under hypo-osmotic conditions is essential to reduce the cell turgor and to prevent rupture.

The K^+/H^+ antiporters are another class long predicted to play a role in K^+ efflux because, with K^+ as the main intracellular cation, they would provide convenient mediators for control of cellular pH. While early demonstrations of K^+/H^+ exchange in everted *E. coli* membrane vesicles[149] appeared to support this idea, little corroborative data has since been forthcoming. A K^+/H^+ antiporter active at high pH was identified in *Enterococcus hirae*,[150,151] but ensuing literature is sparse. More recently, the *Vibrio cholerae* Na^+/H^+ antiporter was shown to exhibit K^+/H^+ exchange activity under some conditions[152] and there is preliminary evidence of a K^+/H^+ antiporter in *Streptomyces* species.[153]

3.2.1.3 K^+ Channels

Given their relatively simple lifestyle requirements, just why prokaryotic cells possess highly specialized K^+ channels remains an open question. Not a great number have been identified, which seems incongruous with the idea of prokaryotes as the ancestral gene pool, but those that have been are representative of the major K^+ channel classes found in eukaryotes. The answers may lie in specialized functions. SynK, for instance, is a recently discovered K^+ channel with six membrane-spanning segments per subunit that is localized to the thylakoid membranes of *Synechocystis* cyanobacteria.[154] A SynK knock-out does not affect cell viability but the cells are unable to build up a strong proton motive force across the thylakoid, indicating a (non-essential) role in the photosynthetic process.

3.2.2 Coordination Chemistry of the Potassium Ion

Valinomycin is a prokaryotic K^+ chelating agent well noted for its experimental utility (*e.g.* dissipation of membrane potential). A highly specific K^+ binding ionophore $(K^+ : Na^+ = 10\,000 : 1)$, it was first isolated from the actinobacterium *Streptomyces fulvissimus* in 1955. Like other natural products from *Streptomyces* strains valinomycin has antibiotic properties. Resembling crown ethers in form, valinomycin is a cyclic dodecadepsipeptide with alternating amide and ester linkages (Figure 3.12). The functional groups enable it to exist in both polar and non-polar environments and as such to act as a transmembrane carrier. The crystal structure of a K^+-valinomycin complex reveals a central potassium ion with regular octahedral geometry coordinated by six ester carbonyls.[155]

(a)

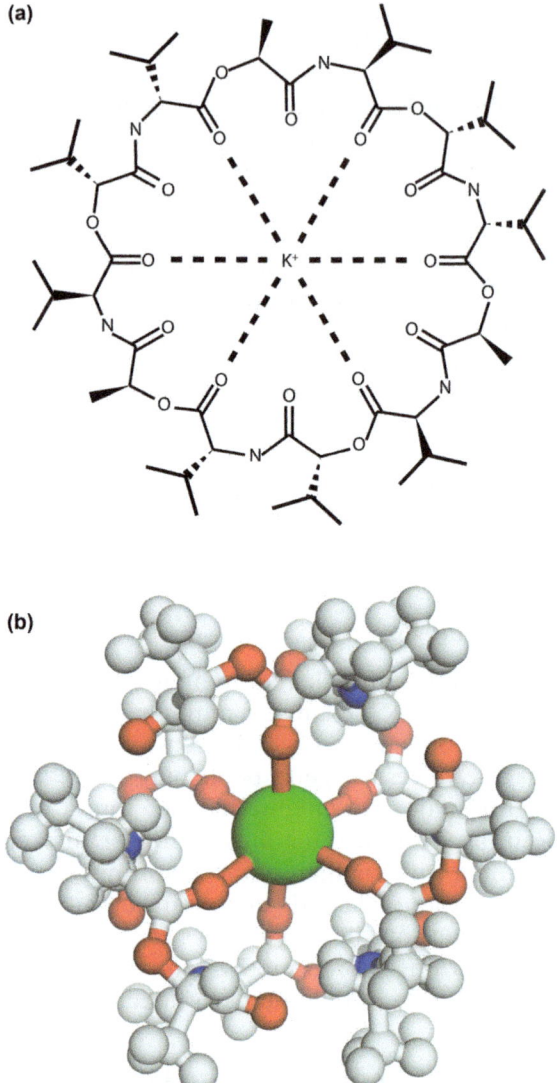

(b)

Figure 3.12 Valinomycin is a K^+ ionophore isolated from actinobacteria that is widely utilized in biology laboratories. **a.** Chemical structure showing the arrangement of the 12 units of D- and L-valine and L-lactic acid. **b.** The molecular structure obtained by single crystal X-ray crystallographic analysis of valinomycin contains a central K^+ ion (green) with six coordinating ester carbonyl ligands arranged in octahedral geometry.[155]

3.2.3 Sensing, Buffering and other Control Mechanisms (Cellular Homeostasis)

As described above, the osmotic status of prokaryotic cells is closely tied to potassium regulation. Recent studies suggest that K^+ may have a more complex role in biofilms and other manifestations of "multicellularity". *Bacillus subtilis* has proved to be the organism of choice for such studies.[156] Quorum sensing is a phenomenon where life decisions are linked to population density and can be mediated by receptor binding of molecules secreted from within a cohort. A recent report found that a variety of molecules inducing K^+ leakage across the plasma membrane trigger a cascade of events that promote biofilm formation *via* synthesis of the extracellular matrix.[157] The authors concluded that this was a new form of sensing in that bacteria are receptive to a chemical (K^+) released in response to the actions of chemically diverse molecules. It could also, however, be thought of as paralleling second messenger pathways in higher eukaryotes.

3.2.4 Conclusions

A pivotal difference between the prokaryotic and eukaryotic K^+ systems, unicellular algae and yeasts aside, is that in prokaryotes only the internal environment can be regulated with any degree of exactitude. This is the realm of ion pumps and channels located in the plasma membrane that, whilst homologous to eukaryotic transport proteins, have evolved distinctive character in keeping with the specific environmental challenges faced by prokaryotes. Mechanisms controlling electrolyte balance thus take precedence over other cellular processes, and an internal stockpile of K^+ provides a ready source of ammunition for use in situations of osmotic or toxic affront.

3.3 Epilogue

The natural propensity of the Group 1 elements to ionize and form coordination complexes is fundamental to their role in biological systems. The separation of charge across cell membranes responsible for instating and maintaining the membrane potential is at the crux of cellular preservation and communication. It is achievable only because of the physiochemical attributes that distinguish potassium from the other naturally abundant Group 1 ions, sodium and lithium. The lower charge density of K^+ confers upon it the capacity to readily interchange ligands, while its preference for a high coordination number offers greater possibilities for stringent selectivity and hence for control over the electrolyte balance.

Acknowledgements

The authors acknowledge the Australian NHMRC for fellowship and project funding. They thank Brian J. Smith, Oliver B. Clarke, Jamie Vandenberg, Katrina Black and Pauline Crewther for informative discussions that contributed to the manuscript.

References

1. F. A. Cotton and G. Wilkinson, *Advanced Inorganic Chemistry*. Wiley Interscience Publishers, New York, 3rd edn, 1972.
2. G. N. Schrauzer, *J. Am. Coll. Nutr.*, 2002, **21**, 14–21.
3. O. M. Sejersted and G. Sjøgaard, *Physiol. Rev.*, 2000, **80**, 1411–1481.
4. J. I. Medbø and O. M. Sejersted, *J. Physiol. (Lond.)*, 1990, **421**, 105–122.
5. C. B. Thompson, C. Choi, J. H. Youn and A. A. McDonough, *Am. J. Physiol.*, 1999, **276**, C1411–1149.
6. E. L. Patt, E. E. Pickett and B. L. O'Dell, *Bioinorg. Chem.*, 1978, **9**, 299–310.
7. E. E. Pickett and B. L. O'Dell, *Biol. Trace Elem. Res.*, 1992, **34**, 299–319.
8. J. F. J. Cade, *Med. J. Aust.*, 1949, **2**, 349–352.
9. L. Tondo and R. J. Baldessarini, *Epidemiol. Psichiatr. Soc.*, 2009, **18**, 179–183.
10. A. M. A. Nivoli, A. Murru, J. M. Goikolea, J. M. Crespo, J. M. Montes, A. González-Pinto, P. García-Portilla, J. Bobes, J. Sáiz-Ruiz and E. Vieta, *J. Affect. Disord.*, 2012, **140**, 125–141.
11. R. F. McKnight, M. Adida, K. Budge and S. Stockton, *The Lancet*, 2012, **379**, 721–728.
12. B. K. Kishore and C. M. Ecelbarger, *Am. J. Physiol. Renal Physiol.*, 2013, **304**, F1139–49.
13. C. McMahon, S. M. Studer, C. Clendinen, G. P. Dann, P. D. Jeffrey and F. M. Hughson, *J. Biol. Chem.*, 2012, **287**, 43599–43606.
14. H. L. Friedman, C. V. Krishnan, Thermodynamics of ion hydration, in *Aqueous Solutions of Simple Electrolytes*, ed. F. Franks, Water: A comprehensive treatise, Springer, USA, 1973, vol. 3, ch. 1, pp. 1–118.
15. S. Varma and S. B. Rempe, *Biophys. Chem.*, 2006, **124**, 192–199.
16. S. Ansell, A. C. Barnes, P. E. Mason, G. W. Neilson and S. Ramos, *Biophys. Chem.*, 2006, **124**, 171–179.
17. J. Mähler and I. Persson, *Inorg. Chem.*, 2012, **51**, 425–438.
18. K. D. Collins, *Proc. Natl Acad. Sci. USA*, 1995, **92**, 5553–5557.
19. M. Carrillo-Tripp, M. L. San-Román, J. Hernańdez-Cobos, H. Saint-Martin and I. Ortega-Blake, *Biophys. Chem.*, 2006, **124**, 243–250.
20. B. Tansel, J. Sager, T. Rector, J. Garland, R. F. Strayer, L. Levine, M. Roberts, M. Hummerick and J. Bauer, *Sep. Purif. Technol.*, 2006, **51**, 40–47.
21. L. A. Richards, A. I. Schäfer, B. S. Richards and B. Corry, *Phys. Chem. Chem. Phys.*, 2012, **14**, 11633–11638.
22. L. A. Richards, A. I. Schäfer, B. S. Richards and B. Corry, *Small*, 2012, **8**, 1701–1709.
23. S. Y. Noskov and B. Roux, *Biophys. Chem.*, 2006, **124**, 279–291.
24. D. Bucher, S. Raugei, L. Guidoni, M. Dal Peraro, U. Rothlisberger, P. Carloni and M. L. Klein, *Biophys. Chem.*, 2006, **124**, 292–301.
25. D. L. Bostick and C. L. Brooks, *Proc. Natl Acad. Sci. USA*, 2007, **104**, 9260–9265.

26. S. Y. Noskov and B. Roux, *J. Gen. Physiol.*, 2007, **129**, 135–143.
27. R. Agarwal, R. Afzalpurkar and J. S. Fordtran, *Gastroenterology*, 1994, **107**, 548–571.
28. J. P. Hayslett, J. Halevy, P. E. Pace and H. J. Binder, *Am. J. Physiol.*, 1982, **242**, G209–14.
29. G. Giebisch and W. Wang, *Kidney Int.*, 1996, **49**, 1624–1631.
30. J. P. Hayslett and H. J. Binder, *Am. J. Physiol.*, 1982, **243**, F103–12.
31. G. Rechkemmer, R. A. Frizzell and D. R. Halm, *J. Physiol. (Lond.)*, 1996, **493**(Pt 2), 485–502.
32. J. H. Sweiry and H. J. Binder, *J. Clin. Invest.*, 1989, **83**, 844–851.
33. J. H. Sweiry and H. J. Binder, *J. Physiol. (Lond.)*, 1990, **423**, 155–170.
34. O. Bachmann, B. Riederer, H. Rossmann, S. Groos, P. J. Schultheis, G. E. Shull, M. Gregor, M. P. Manns and U. Seidler, *Am. J. Physiol. Gastrointest. Liver Physiol.*, 2004, **287**, G125–33.
35. O. Bachmann, M. Juric, U. Seidler, M. P. Manns and H. Yu, *Acta Physiol. (Oxf.)*, 2011, **201**, 33–46.
36. M. Bleich, N. Riedemann, R. Warth, D. Kerstan, J. Leipziger, M. Hör, W. V. Driessche and R. Greger, *Pflugers Arch.*, 1996, **432**, 1011–1022.
37. G. I. Sandle, C. M. McNicholas and R. B. Lomax, *Lancet*, 1994, **343**, 23–25.
38. M. V. Sorensen, J. E. Matos, M. Sausbier, U. Sausbier, P. Ruth, H. A. Praetorius and J. Leipziger, *J. Physiol. (Lond.)*, 2008, **586**, 4251–4264.
39. M. V. Sorensen, M. Sausbier, P. Ruth, U. Seidler, B. Riederer, H. A. Praetorius and J. Leipziger, *J. Physiol. (Lond.)*, 2010, **588**, 1763–1777.
40. M. V. Sorensen, J. E. Matos, H. A. Praetorius and J. Leipziger, *Pflugers Arch.*, 2010, **459**, 645–656.
41. C. Lytle and T. McManus, *Am. J. Physiol. Cell Physiol.*, 2002, **283**, C1422–31.
42. K. T. Kahle, J. Rinehart and R. P. Lifton, *Biochim. Biophys. Acta*, 2010, **1802**, 1150–1158.
43. A. Dörge, F. X. Beck and G. Rechkemmer, *Pflugers Arch.*, 1998, **436**, 280–288.
44. A. M. Sherry, D. H. Malinowska, R. E. Morris, G. M. Ciraolo and J. Cuppoletti, *Am. J. Physiol. Cell Physiol.*, 2001, **280**, C1599–606.
45. R. Warth and J. Barhanin, *J. Membr. Biol.*, 2003, **193**, 67–78.
46. P. Sangan, S. R. Brill, S. Sangan, B. Forbush and H. J. Binder, *J. Biol. Chem.*, 2000, **275**, 30813–30816.
47. G. Peña-Münzenmayer, M. Catalán, I. Cornejo, C. D. Figueroa, J. E. Melvin, M. I. Niemeyer, L. P. Cid and F. V. Sepúlveda, *J. Cell Sci.*, 2005, **118**, 4243–4252.
48. M. Ikuma, J. Geibel, H. J. Binder and V. M. Rajendran, *Am. J. Physiol. Cell Physiol.*, 2003, **285**, C912–21.
49. S. Gurkan, G. K. Estilo, Y. Wei and L. M. Satlin, *Pediatr. Nephrol.*, 2007, **22**, 915–925.

50. J. Z. Xu, A. E. Hall, L. N. Peterson, M. J. Bienkowski, T. E. Eessalu and S. C. Hebert, *Am. J. Physiol.*, 1997, **273**, F739–48.

51. B. M. Christensen, A. M. Zuber, J. Loffing, J.-C. Stehle, P. M. T. Deen, B. C. Rossier and E. Hummler, *J. Am. Soc. Nephrol.*, 2011, **22**, 253–261.

52. M. A. Bailey, A. Cantone, Q. Yan, G. G. MacGregor, Q. Leng, J. B. O. Amorim, T. Wang, S. C. Hebert, G. Giebisch and G. Malnic, *Kidney Int.*, 2006, **70**, 51–59.

53. M. Lu, G. G. MacGregor, W. Wang and G. Giebisch, *J. Gen. Physiol.*, 2000, **116**, 299–310.

54. R. Katzman and H. M. Pappius, Hypernatremia and hyperosmolarity, in *Brain Electrolytes and Fluid Metabolism*, Williams & Wilkins, Baltimore, 1973, ch. 13, pp. 278–290.

55. M. Canessa, M. E. Fabry, N. Blumenfeld and R. L. Nagel, *J. Membr. Biol.*, 1987, **97**, 97–105.

56. S. C. Crable, S. M. Hammond, R. Papes, R. K. Rettig, G.-P. Zhou, P. G. Gallagher, C. H. Joiner and K. P. Anderson, *Exp. Hematol.*, 2005, **33**, 624–631.

57. M.-O. Quarmyne, M. Risinger, A. Linkugel, A. Frazier and C. Joiner, *Blood Cells Mol. Dis.*, 2011, **47**, 95–99.

58. G. Valentini, L. R. Chiarelli, R. Fortin, M. Dolzan, A. Galizzi, D. J. Abraham, C. Wang, P. Bianchi, A. Zanella and A. Mattevi, *J. Biol. Chem.*, 2002, **277**, 23807–23814.

59. I. Russo Krauss, A. Merlino, A. Randazzo, E. Novellino, L. Mazzarella and F. Sica, *Nucleic Acids Res.*, 2012, **40**, 8119–8128.

60. S. Geibel, J. H. Kaplan, E. Bamberg and T. Friedrich, *Proc. Natl Acad. Sci. USA*, 2003, **100**, 964–969.

61. M. Hilge, G. Siegal, G. W. Vuister, P. Güntert, S. M. Gloor and J. P. Abrahams, *Nat. Struct. Biol.*, 2003, **10**, 468–474.

62. J. C. Skou, *Biochim. Biophys. Acta*, 1957, **23**, 394–401.

63. W. G. Siems, H. Schmidt, S. Gruner and M. Jakstadt, *Cell Biochem. Funct.*, 1992, **10**, 61–66.

64. D. Attwell and S. B. Laughlin, *J. Cereb. Blood Flow Metab.*, 2001, **21**, 1133–1145.

65. L. J. Boardman, J. F. Lamb and D. McCall, *J. Physiol. (Lond.)*, 1972, **225**, 619–635.

66. M. Liang, J. Tian, L. Liu, S. Pierre, J. Liu, J. Shapiro and Z.-J. Xie, *J. Biol. Chem.*, 2007, **282**, 10585–10593.

67. R. Zahler, M. Lufburrow, M. Manor, R. Shenoy, D. Fornasari, M. Romana and W. Sun, *Ann. NY Acad. Sci.*, 1997, **834**, 687–689.

68. G. Crambert, U. Hasler, A. T. Beggah, C. Yu, N. N. Modyanov, J. D. Horisberger, L. Lelièvre and K. Geering, *J. Biol. Chem.*, 2000, **275**, 1976–1986.

69. K. Geering, *Curr. Opin. Nephrol. Hypertens.*, 2008, **17**, 526–532.

70. R. W. Albers, *Annu. Rev. Biochem.*, 1967, **36**, 727–756.

71. R. L. Post, C. Hegyvary and S. Kume, *J. Biol. Chem.*, 1972, **247**, 6530–6540.

72. J. P. Morth, B. P. Pedersen, M. S. Toustrup-Jensen, T. L. M. Sørensen, J. Petersen, J. P. Andersen, B. Vilsen and P. Nissen, *Nature*, 2007, **450**, 1043–1049.

73. T. Shinoda, H. Ogawa, F. Cornelius and C. Toyoshima, *Nature*, 2009, **459**, 446–450.

74. H. Ogawa, T. Shinoda, F. Cornelius and C. Toyoshima, *Proc. Natl Acad. Sci. USA*, 2009, **106**, 13742–13747.

75. L. Yatime, M. Laursen, J. P. Morth, M. Esmann, P. Nissen and N. U. Fedosova, *J. Struct. Biol.*, 2011, **174**, 296–306.

76. A.-M. L. Winther, M. Bublitz, J. L. Karlsen, J. V. Møller, J. B. Hansen, P. Nissen and M. J. Buch-Pedersen, *Nature*, 2013, **495**, 265–269.

77. D. A. Doyle, J. Morais Cabral, R. A. Pfuetzner, A. Kuo, J. M. Gulbis, S. L. Cohen, B. T. Chait and R. Mackinnon, *Science*, 1998, **280**, 69–77.

78. A. Kuo, J. M. Gulbis, J. F. Antcliff, T. Rahman, E. D. Lowe, J. Zimmer, J. Cuthbertson, F. M. Ashcroft, T. Ezaki and D. A. Doyle, *Science*, 2003, **300**, 1922–1926.

79. B. Hille, *Ion Channels of Excitable Membranes*, Sinauer Associates Inc, Sunderland, MA, 3rd edn, 2001.

80. J. H. Youn and A. A. McDonough, *Annu. Rev. Physiol.*, 2009, **71**, 381–401.

81. L. Rabinowitz, *Am. J. Physiol.*, 1988, **254**, R381–8.

82. F. N. Lee, G. Oh, A. A. McDonough and J. H. Youn, *Am. J. Physiol. Renal Physiol.*, 2007, **293**, F541–7.

83. M. Ahloulay, M. Déchaux, C. Hassler, N. Bouby and L. Bankir, *J. Clin. Invest.*, 1996, **98**, 2251–2258.

84. L. Bankir, M. Ahloulay, P. N. Devreotes and C. A. Parent, *Am. J. Physiol. Renal Physiol.*, 2002, **282**, F376–92.

85. R. A. DeFronzo, P. Felig, E. Ferrannini and J. Wahren, *Am. J. Physiol.*, 1980, **238**, E421–7.

86. C. S. Choi, C. B. Thompson, P. K. Leong, A. A. McDonough and J. H. Youn, *Am. J. Physiol. Renal Physiol.*, 2001, **280**, F95–F102.

87. G. J. Schwartz, S. Tsuruoka, S. Vijayakumar, S. Petrovic, A. Mian and Q. Al-Awqati, *J. Clin. Invest.*, 2002, **109**, 89–99.

88. X. Gao, D. Eladari, F. Leviel, B. Y. Tew, C. Miró-Julià, F. H. Cheema, F. Cheema, L. Miller, R. Nelson, T. G. Paunescu, M. McKee, D. Brown and Q. Al-Awqati, *Proc. Natl Acad. Sci. USA*, 2010, **107**, 21872–21877.

89. L. Shavit, M. D. Lifschitz and M. Epstein, *Kidney Int.*, 2012, **81**, 955–968.

90. D. Eladari, R. Chambrey and J. Peti-Peterdi, *Annu. Rev. Physiol.*, 2012, **74**, 325–349.

91. F. Hanner, M. Schnichels, Q. Zheng-Fischhöfer, L. E. Yang, I. Toma, K. Willecke, A. A. McDonough and J. Peti-Peterdi, *Cell Commun. Adhes.*, 2008, **15**, 219–230.

92. F. McCulloch, R. Chambrey, D. Eladari and J. Peti-Peterdi, *Am. J. Physiol. Renal Physiol.*, 2005, **289**, F1304–12.

93. A. Sipos, S. L. Vargas, I. Toma, F. Hanner, K. Willecke and J. Peti-Peterdi, *J. Am. Soc. Nephrol.*, 2009, **20**, 1724–1732.

94. F. Hanner, L. Lam, M. T. X. Nguyen, A. Yu and J. Peti-Peterdi, *Am. J. Physiol. Renal Physiol.*, 2012, **303**, F1454–9.

95. J. D. Holtzclaw, R. J. Cornelius, L. I. Hatcher and S. C. Sansom, *Am. J. Physiol. Renal Physiol.*, 2011, **300**, F1319–26.

96. E. Syková and C. Nicholson, *Physiol. Rev.*, 2008, **88**, 1277–1340.

97. U. Heinemann and H. D. Lux, *Brain Res.*, 1977, **120**, 231–249.

98. M. J. Gutnick, U. Heinemann and H. D. Lux, *Electroencephalogr. Clin. Neurophysiol.*, 1979, **47**, 329–344.

99. G. G. Somjen, *Annu. Rev. Physiol.*, 1979, **41**, 159–177.

100. F. Vyskocil, N. Kritz and J. Bures, *Brain Res.*, 1972, **39**, 255–259.

101. J. A. Leis, L. K. Bekar and W. Walz, *Glia*, 2005, **50**, 407–416.

102. R. K. Orkand, J. G. Nicholls and S. W. Kuffler, *J. Neurophysiol.*, 1966, **29**, 788–806.

103. S. W. Kuffler, J. G. Nicholls and R. K. Orkand, *J. Neurophysiol.*, 1966, **29**, 768–787.

104. H. Marrero and R. K. Orkand, *Glia*, 1996, **16**, 285–289.

105. W. Walz, *Neurochemistry International*, 2000, **36**, 291–300.

106. P. Kofuji and E. A. Newman, *Neuroscience*, 2004, **129**, 1045–1056.

107. B. A. MacVicar, D. Feighan, A. Brown and B. Ransom, *Glia*, 2002, **37**, 114–123.

108. A. Reichenbach, A. Henke, W. Eberhardt, W. Reichelt and D. Dettmer, *Can. J. Physiol. Pharmacol.*, 1992, **70**(Suppl.), S239–47.

109. W. Walz and E. C. Hinks, *Neuroscience*, 1987, **20**, 341–346.

110. C. R. Rose and B. R. Ransom, *J. Physiol. (Lond.)*, 1996, **491**(Pt 2), 291–305.

111. B. W. Connors, L. S. Benardo and D. A. Prince, *J. Neurosci.*, 1984, **4**, 1324–1330.

112. H. Kettenmann and B. R. Ransom, *Glia*, 1988, **1**, 64–73.

113. M. O. Kristian Enkvist and K. D. McCarthy, *J. Neurochem.*, 1994, **62**, 489–495.

114. E. A. Newman, D. A. Frambach and L. L. Odette, *Science*, 1984, **225**, 1174–1175.

115. C. J. Karwoski, H. K. Lu and E. A. Newman, *Science*, 1989, **244**, 578–580.

116. E. A. Newman, *Ann. NY Acad. Sci.*, 1986, **481**, 273–286.

117. H. Brew and D. Attwell, *Biophys. J.*, 1985, **48**, 843–847.

118. E. A. Newman, *J. Neurosci.*, 1993, **13**, 3333–3345.

119. G. G. Somjen, *Neuroglia*, Oxford University Press, New York, 1995, pp. 319–331.

120. B. Katz, *Arch. Sci. Physiol.*, 1949, **3**, 285–299.

121. E. A. Nagelhus, Y. Horio, A. Inanobe, A. Fujita, F. M. Haug, S. Nielsen, Y. Kurachi and O. P. Ottersen, *Glia*, 1999, **26**, 47–54.

122. E. A. Nagelhus, T. M. Mathiisen and O. P. Ottersen, *Neuroscience*, 2004, **129**, 905–913.

123. N. C. N. Connors and P. P. Kofuji, *Glia*, 2006, **53**, 124–131.

124. J. Ruiz-Ederra, H. Zhang and A. S. Verkman, *J. Biol. Chem.*, 2007, **282**, 21866–21872.

125. N. B. Illarionova, E. Gunnarson, Y. Li, H. Brismar, A. Bondar, S. Zelenin and A. Aperia, *Neuroscience*, 2010, **168**, 915–925.
126. S. Strohschein, K. Hüttmann, S. Gabriel, D. K. Binder, U. Heinemann and C. Steinhäuser, *Glia*, 2011, **59**, 973–980.
127. C. R. Woese and N. Goldenfeld, *Microbiol. Mol. Biol. Rev.*, 2009, **73**, 14–21.
128. N. Goldenfeld and C. Woese, *Nature*, 2007, **445**, 369.
129. J. C. Dunning Hotopp, M. E. Clark, D. C. S. G. Oliveira, J. M. Foster, P. Fischer, M. C. M. Torres, J. D. Giebel, N. Kumar, N. Ishmael, S. Wang, J. Ingram, R. V. Nene, J. Shepard, J. Tomkins, S. Richards, D. J. Spiro, E. Ghedin, B. E. Slatko, H. Tettelin and J. H. Werren, *Science*, 2007, **317**, 1753–1756.
130. J. C. Dunning Hotopp, *Trends Genet.*, 2011, **27**, 157–163.
131. A. L. Koch, *Adv. Microb. Physiol.*, 1983, **24**, 301–366.
132. S. Morbach and R. Krämer, *ChemBioChem*, 2002, **3**, 384–397.
133. P. Stoodley, K. Sauer, D. G. Davies and J. W. Costerton, *Annu. Rev. Microbiol.*, 2002, **56**, 187–209.
134. N. Chia, C. R. Woese and N. Goldenfeld, *Proc. Natl Acad. Sci. USA*, 2008, **105**, 14597–14602.
135. C. Corratgé-Faillie, M. Jabnoune, S. Zimmermann, A.-A. Véry, C. Fizames and H. Sentenac, *Cell. Mol. Life Sci. (CMLS)*, 2010, **67**, 2511–2532.
136. R. S. Vieira-Pires, A. Szollosi and J. H. Morais-Cabral, *Nature*, 2013, **496**, 323–328.
137. Y. Cao, Y. Pan, H. Huang, X. Jin, E. J. Levin, B. Kloss and M. Zhou, *Nature*, 2013, **496**, 317–322.
138. N. Shi, S. Ye, A. Alam, L. Chen and Y. Jiang, *Nature*, 2006, **440**, 570–574.
139. D. B. Rhoads, F. B. Waters and W. Epstein, *J. Gen. Physiol.*, 1976, **67**, 325–341.
140. E. T. Buurman, K. T. Kim and W. Epstein, *J. Biol. Chem.*, 1995, **270**, 6678–6685.
141. K. Fendler, S. Dröse, K. Altendorf and E. Bamberg, *Biochemistry*, 1996, **35**, 8009–8017.
142. J. Gowrishankar, *J. Bacteriol.*, 1985, **164**, 434–445.
143. L. Sutherland, J. Cairney, M. J. Elmore, I. R. Booth and C. F. Higgins, *J. Bacteriol.*, 1986, **168**, 805–814.
144. R. Heermann and K. Jung, *FEMS Microbiol. Lett.*, 2010, **304**, 97–106.
145. D. Kixmüller, H. Strahl, A. Wende and J.-C. Greie, *Extremophiles*, 2011, **15**, 643–652.
146. R. Heermann, A. Weber, B. Mayer, M. Ott, E. Hauser, G. Gabriel, T. Pirch and K. Jung, *J. Mol. Biol.*, 2009, **386**, 134–148.
147. D. Lüttmann, R. Heermann, B. Zimmer, A. Hillmann, I. S. Rampp, K. Jung and B. Görke, *Mol. Microbiol.*, 2009, **72**, 978–994.
148. G. P. Ferguson, Y. Nikolaev, D. McLaggan, M. Maclean and I. R. Booth, *J. Bacteriol.*, 1997, **179**, 1007–1012.

149. R. N. Brey, B. P. Rosen and E. N. Sorensen, *J. Biol. Chem.*, 1980, **255**, 39–44.
150. Y. Kakinuma and K. Igarashi, *J. Biol. Chem.*, 1988, **263**, 14166–14170.
151. Y. Kakinuma and K. Igarashi, *J. Bacteriol.*, 1999, **181**, 4103–4105.
152. C. T. Resch, J. L. Winogrodzki, C. T. Patterson, E. J. Lind, M. J. Quinn, P. Dibrov and C. C. Häse, *Biochemistry*, 2010, **49**, 2520–2528.
153. J. Y. Song, Y. B. Seo, S.-K. Hong and Y. K. Chang, *J. Microbiol. Biotechnol.*, 2013, **23**, 149–155.
154. V. Checchetto, A. Segalla, G. Allorent, N. La Rocca, L. Leanza, G. M. Giacometti, N. Uozumi, G. Finazzi, E. Bergantino and I. Szabò, *Proc. Natl Acad. Sci. USA*, 2012, **109**, 11043–11048.
155. K. Neupert-Laves and M. Dobler, *Helv. Chim. Acta*, 1975, **58**, 432–442.
156. C. Aguilar, H. Vlamakis, R. Losick and R. Kolter, *Curr. Opin. Microbiol.*, 2007, **10**, 638–643.
157. D. López, M. A. Fischbach, F. Chu, R. Losick and R. Kolter, *Proc. Natl Acad. Sci. USA*, 2009, **106**, 280–285.

Magnesium

ANDREA M. P. ROMANI

Department of Physiology and Biophysics, School of Medicine,
Case Western Reserve University, 10900 Euclid Avenue, Cleveland, OH,
44106-4970, USA
Email: amr5@po.cwru.edu

4.1 The Chemistry of Magnesium

Magnesium is the eighth most abundant element on Earth, the fourth most abundant element in vertebrates and, after potassium, the second most abundant cation within cells. A look at the basic chemical properties of the magnesium ion Mg^{2+} documents its special position amongst biological cations. Its ionic radius is 0.65 Å (65 pm), about one-third smaller than those of Na^+ and Ca^{2+} (0.95 and 0.99 Å, respectively), and about half that of K^+ (1.38 Å).[1] Following hydration to Mg^{2+}_{aq}, however, the effective radius increases approximately 400-fold, becoming twice as large as that of K^+_{aq}, and 1.7 and 1.6 times larger than those of Na^+_{aq} and Ca^{2+}_{aq}, respectively.[1]

Moreover, Mg^{2+} compounds are consistently hexa-coordinate, forming a rigid octahedron with minimal angular deviation from the characteristic 90° angle. Consequently, the Mg-O bond distances in proteins or small molecules vary minimally compared to other cations. This rigidity reflects the co-enzyme nature of Mg^{2+} that normally binds ATP, trimeric G-proteins or other enzymes in the catalytic pocket. In the case of ATP, binding to the trisphosphate side-chain is essential for hydrolysis and energy transfer. When bound into a catalytic site, Mg^{2+} normally retains one or two aqua ligands. These ligands are integral components as they either contribute to the stereochemistry required for reaction, or at least one of them is utilized

RSC Metallobiology Series No. 2
Binding, Transport and Storage of Metal Ions in Biological Cells
Edited by Wolfgang Maret and Anthony Wedd
© The Royal Society of Chemistry 2014
Published by the Royal Society of Chemistry, www.rsc.org

directly in the reaction itself. The high charge density of partially hydrated Mg^{2+} makes the cation particularly suitable to serve as a Lewis acid in processes that involve hydrolysis of phosphate esters, phosphoanhydride bonds and phosphoryl transfer.[2]

Lastly, because the number of solvent water molecules associated with Mg^{2+} is significantly higher than for other cations,[2,3] Mg^{2+}-based macromolecular complexes are larger. This has major implications for the ability of Mg^{2+} to bind to proteins and/or to move within a solvent or a cell. This solvation state becomes important for several glycolytic enzymes in which Mg^{2+} acts as a co-enzyme.[3] These enzymes usually require two metal ions to act cooperatively: one ion binds at an allosteric regulatory site whereas the second plays a catalytic role. The same is true for several of the substrates that bind to the activated enzyme. The substrates *per se* do not have a high affinity for Mg^{2+} but the partially hydrated metal molecule stabilizes the structure and favours the progression of the catalytic process.[2,3] In this respect, Mg^{2+} is more adaptable than Ca^{2+} or Mn^{2+} despite slow ligand exchange rates and the relative low affinities of substrates for the cation.[†]

4.2   Magnesium in Plants

4.2.1   Distribution and Transport

Plant viability is tightly associated with the maintenance of Mg^{2+} homeostasis. Mg^{2+} is abundant but highly compartmentalized within the cytoplasm. It plays an essential role in numerous plant enzymes and factors including chlorophyll, ATP, ATPases, protein kinases and phosphatases. Chloroplast enzymes, as well as mammalian mitochondrial enzymes, are markedly impacted by changes in Mg^{2+} content within the chloroplast itself or in the cytoplasm.

Magnesium ion uptake occurs *via* the root. The radial movement of solutes from the soil to the stele, the innermost component of the endodermis, occurs through two parallel routes. The first is through the cell wall and the extracellular space (apoplasmic route). The second is through plasmodesmata (symplasmic route). These are cytoplasmic strands that connect two adjacent cells by penetrating and crossing the two cell membranes.[4] In the case of Mg^{2+}, the apoplasmic route is approximately two orders of magnitude faster than the symplasmic pathway.[4] Once in the stele, Mg^{2+} has to be released into the stelar apoplasm. This release (aka xylem loading) is mediated by xylem parenchyma cells that line the xylem vessels. Hence, two transport processes need to be in place: Mg^{2+} uptake into the root symplasm followed by its release to the xylem. Currently, no information is available about the proteins and transport mechanisms involved in these steps.

The presence of proton pumps in the cell membrane promotes H^+ movement from the cytoplasm into the extracellular space (apoplasm).[4]

[†]For further comparison of the chemistries of Li, Na, K, Mg and Ca, see Chapter 2, Section 2.1, Chapter 3, Section 3.1.1 and Chapter 5, Section 5.2.2.

Hence, the increase in alkalinity and negative charge within the cytoplasm favours Mg^{2+} entry down its electrochemical potential gradient. The specific mechanism responsible for Mg^{2+} entry into the cell, however, has not been identified clearly. It has been proposed that the ion may enter through a homologue of the *rca* channel.[4] The latter is not specific for Mg^{2+} and transports several mono- and divalent cations.[4] How it might select for Mg^{2+} is not understood. As is the case in mammalian cells (Section 4.4), it is conceivable that Mg^{2+} movement across the plasma membrane of plant cells depends on the membrane potential $\Delta\psi$, which in turn depends on the transmembrane movements and gradients of other ions.[4] This potential and the outward-oriented proton motive force regulate the Mg^{2+} concentration within the plant cell cytoplasm in the range between 2 and 10 mM.[5] Most of Mg^{2+} ($\geq$90%) is complexed to ATP, leaving the free cytoplasmic concentration as $\sim$0.4 mM.[5]

A significant percentage of cellular Mg^{2+} is bound to the cell wall or sequestered within the vacuoles, discrete compartments within the plant cell. They occupy 80% to 90% of plant cell volume and are essential to maintain turgor pressure against the cell wall allowing support of the structure of leaves and flowers. The vacuoles are filled with water containing inorganic ions and organic molecules including enzymes. They are surrounded by a membrane called the tonoplast or vacuolar membrane that separates the vacuolar contents from the cytoplasm. Proton pumps are also present in this membrane, and the movement of H^+ towards the lumen of the vacuole helps to stabilize the cytoplasmic pH.[4] It also creates a proton motive force that favours the transport of nutrients and ions including Mg^{2+} in and out of the vacuole. The vacuole is, in fact, the main plant cell organelle involved in maintaining Mg^{2+} content within the cytoplasm and chloroplasts.[4,5] Concentrations of Mg^{2+} ranging between 3 and 60 mM have been estimated to be present within vacuoles in different plants, or in different parts of the same plant.[4,6] A significant portion of the vacuolar Mg^{2+} is assumed to be bound to the vacuolar membrane and to vacuolar anionic compounds or enzymes, but no precise measure of free vacuolar Mg^{2+} concentration is available.

The main mechanism responsible for the movement of Mg^{2+} from the cytoplasm into the vacuole lumen involves an Mg^{2+}/H^+ exchanger. Amiloride and imipramine both inhibit this exchanger, with IC_{50} values of 0.3 and 0.12 mM, respectively. The relevant gene *AtMHX* in *Arabidopsis thaliana* has been cloned.[7] AtHMX was the first Mg^{2+} transporter to be cloned from a multicellular organism, and shows no homology to transporters cloned from bacteria or archaea (Section 4.3). It does share a limited homology with the mammalian Na^+/Ca^{2+} exchanger, isoform 1 (NCX1). Because of its localization, AtHMX is reputed to be essential in determining the availability of Mg^{2+} in the plant cytoplasm while sequestering the excess into a pool that can be drained under specific pathophysiological conditions (*e.g.* deficiency).

Interestingly, overexpression of AtHMX in tobacco plants generates necrotic lesions in the plant leaves that occur in the absence of significant changes in total cellular Mg^{2+} content.[8] These lesions occur in the absence of significant changes in total cellular magnesium content. Due to its specific localization in the vacuolar membrane, it is possible that this transporter can affect the intracellular distribution of Mg^{2+} between the cytoplasm, chloroplasts and vacuoles. An increase in Mg^{2+} concentration within the vacuole can impact cytoplasmic pH (more protons moved into the cytoplasm from the vacuole) while a decrease in concentration within the chloroplast can limit ATP production at a time when more ATP is required to restore pH balance through H^+-ATPases.

While AtMHX is considered responsible for Mg^{2+} accumulation within the vacuole, the mechanism responsible for its extrusion from the vacuole into the cytoplasm remains undefined. One likely candidate is the Slow Activation Vacuolar (SV) ion channel.[4] Mg^{2+} is not only transported by the channel but also regulates it.

AtMRS2 is another Mg^{2+} transporter in plants.[9] About ten homologues of this gene have been identified in the *Arabidopsis* genome. Localized in the inner mitochondrial membrane of yeast or plant cells, this protein is essential to: 1) maintain a physiological level of Mg^{2+} within the organelle; 2) maintain a functional respiratory system in yeast mitochondria and 3) control group II intron RNA splicing.[10] *Arabidopsis* AtMRS2 and yeast MRS2 are structurally and functionally related to CorA, the bacterial Mg^{2+} transporter (Section 4.3).[11]

4.2.2   Role of Mg^{2+} in Plant Physiology

Maintenance of proper Mg^{2+} homeostasis is essential to regulate several K^+ and Na^+ transport pathways in the cell membrane. In most cases, the inhibitory effect of Mg^{2+} can be attributed to its ability to screen negative charges on the membrane surface (limiting the electrostatic attraction for other cations) or to its inhibition of the passage of mono- and divalent cations through channels. In addition, Mg^{2+} can regulate the activity of the H^+-ATPase (V-ATPase) and the inorganic pyrophosphatase (V-PP$_i$ase) enzymes in plant vacuoles by forming a complex with the specific substrates for these two pumps (*i.e.* Mg*ATP and Mg*PPi, respectively).[4] In addition, the ion can act as an allosteric activator of the V-PP$_i$ase[4] *via* the binding of two Mg^{2+} ions to anionic residues in its cytosolic loop. As a result, the Mg^{2+} concentrations eliciting half-maximal and maximal pump activity have been estimated to be 42 μM and 150 μM, respectively.[4] The physiological concentration of Mg^{2+} measured within plant cells far exceeds these concentrations, raising the question as to whether Mg^{2+} does have any real effect in regulating V-PP$_i$ase activity under *in vivo* conditions. On the other hand, two channels in plant cell membranes do appear to be regulated by changes in Mg^{2+} concentration. One of these is the previously mentioned SV channel. The second one is the Fast-activating Vacuolar (FV) ion channel that is also present ubiquitously in plant membranes. This channel sustains a continuous leakage of K^+ out of the vacuole. Free Mg^{2+} concentrations of 1–2 mM appear to close this channel from the lumen side of both the cytoplasm and the vacuole.[4] In addition, Mg^{2+} also modulates the transport of cationic amino acids[4] and the operation of the NH_4^+ permease.[4]

Photosynthesis is one of the most important plant processes requiring Mg^{2+}. It is the essential metal ion in chlorophyll II, the key molecule in the inner membrane of the chloroplast responsible for capturing light. Replacement of Mg^{2+} with other divalent cations impairs the photosynthetic process. Its presence is essential for proton pumping from the stroma into the chloroplast, promoting the conversion of ADP and $NADP^+$ to ATP and NADPH (light step).[4] This proton gradient is compensated by the movements of anions, and by the movement of Mg^{2+} ions from the lumen of the chloroplast to the cytoplasm.[4] Changes in Mg^{2+} concentration and the proton gradient (pH) affect the activity of the ATPase complex and promote ATP production. Following ADP to ATP conversion, Mg^{2+} binds to the same sites undergoing protonation during the light step, effectively inactivating the enzymatic complex. In addition, recent evidence indicates that adenylyl cyclase is essential to maintain plant mitochondrial respiration and to regulate the flow of energetic substrates into the organelle.[4] This process is promoted by Mg^{2+} ions. By forming an energetically favourable complex with ATP, the Mg^{2+} content within mitochondria appears to play a key role in optimizing the interaction between adenylyl cyclase and Mg·ATP and in maintaining an optimal Mg·ATP/Mg·ADP ratio within the organelle.[4] The high-energy compounds produced during the light cycle (ATP and NADPH) are then used by enzymes to convert CO_2 to carbohydrates (dark reaction). Mg^{2+} is essential to activate several of the enzymes involved in this conversion including fructose 1,6-bisphosphatase and sedo-heptulose 1,7-bisphosphatase.[2,4] Maintenance of a physiological level of Mg^{2+} within the plant cells is key to avoiding inhibition of photosynthesis (Mg^{2+} excess) or inhibition of carbohydrate production by blocking the H^+ATPase (Mg^{2+} deficiency). In addition, Mg^{2+} deficiency results in the dissociation of ribosome subunits, *de facto* inhibiting proton synthesis.[4]

The complex interaction between Mg*ATP, Mg*PP_i and adenylyl cyclase mentioned above is pivotal to regulation of plant bioenergetics under anoxic or hypoxic conditions. A decrease in ATP content as registered under either condition results in an increase in free Mg^{2+} content, and the activation of several enzymes normally inhibited by low Mg^{2+} concentration. An increase in Mg^{2+} concentration displaces adenylyl cyclase from its binding site, and favours a 2–3-fold increase in the citrate/isocitrate ratio, with consequent efflux of citrate from the mitochondria into the cytoplasm. Furthermore, elevation of the Mg^{2+} concentration activates PP_i-dependent glycolysis (a specific process occurring in plants) and V-PP_iase activity.[12] The equilibrium between free PP_i and $MgPP_i$ complexes is under pH control, and assessment of hydrolysis energy for PP_i and ATP indicates that the energy of hydrolysis for these two moieties is comparable at acidic pH.[12] In mammalian tissues, the adenylyl cyclase equilibrium is linked to the equilibrium catalysed by the creatine kinase, but in plants it appears to depend on the equilibrium between pyruvate kinase and pyruvate phosphate dikinase. Expression of the latter increases several fold under anoxia and the enzyme increases the rate of conversion of ADP + P_i to AMP + PP_i, with PP_i becoming the energetic substrate of choice under anoxia. The complex interplay between these substrates revolves around the concentration of Mg^{2+}. Its increase, as observed under anoxia, favours the shift of the Mg-ADP/Mg-P_i ratio towards Mg-P_i.[12] The take-home message here is that Mg^{2+} availability dictates the direction of the phosphorylation/dephosphorylation reaction catalysed by pyruvate phosphate dikinase. Dephosphorylation of the enzyme is promoted by millimolar Mg^{2+} (anoxic condition). Ultimately, the PP_i formed from ADP and P_i enters glycolysis at the level of UDP-glucose pyrophosphorylase, generating UTP and glucose 1-phosphate while saving ATP.[12]

4.3   Magnesium and Prokaryotic Cells

Our knowledge of magnesium transport and homeostasis in prokaryotes is significantly more advanced than that in eukaryotes. Three classes of Mg^{2+} transporters have been identified and cloned from these monocellular organisms: CorA, MgtA/B and MgtE.[11,13] They were originally cloned from *Salmonella enterica typhimurium* but their presence has been reported in almost all bacteria and archaea assessed to date.

4.3.1   The CorA Pump

The Mg^{2+} channel from *S. typhimurium* was the first one to be cloned.[14] It is present in nearly half of all sequenced microbial genomes. Its absence is usually observed in species with small genomes and is compensated for by the presence of MgtE (see below). In some species, multiple CorA-like sequences have been discovered that can be assigned to two subclasses. One is termed MPEL for a highly conserved sequence between the second and third transmembrane domain. The second is termed CorA II, and the members of this subfamily exhibit less homology.[14] This second subclass appears to mediate the efflux of unidentified ions but not of Mg^{2+}. Because there is no correlation between the molecular phylogeny of CorA and the bacterial phylogeny as derived by the 16S rRNA sequence, it has been suggested that CorA is a very ancient protein, antecedent to the divergence of bacteria and archaea.

Recently, three independent crystal structures of CorA from *Thermotoga maritima* have been published.[15–17] CorA is a homopentamer with two transmembrane regions per monomer (Figure 4.1).[15–18] Both the N- and C-termini are on the cytoplasmic side. The large N-terminus comprises seven parallel/antiparallel β-sheets between two sets of three α-helices each. A seventh α-helix (α7) forms the stalk that connects this domain to the first transmembrane domain and extends for 100 Å into the cytoplasm.[18] It participates in forming the inner wall of a funnel-like structure that enlarges from a diameter of 5 Å at the membrane-cytosol interface to 20 Å diameter at its cytoplasmic mouth. Several anionic residues or hydroxyl groups line the

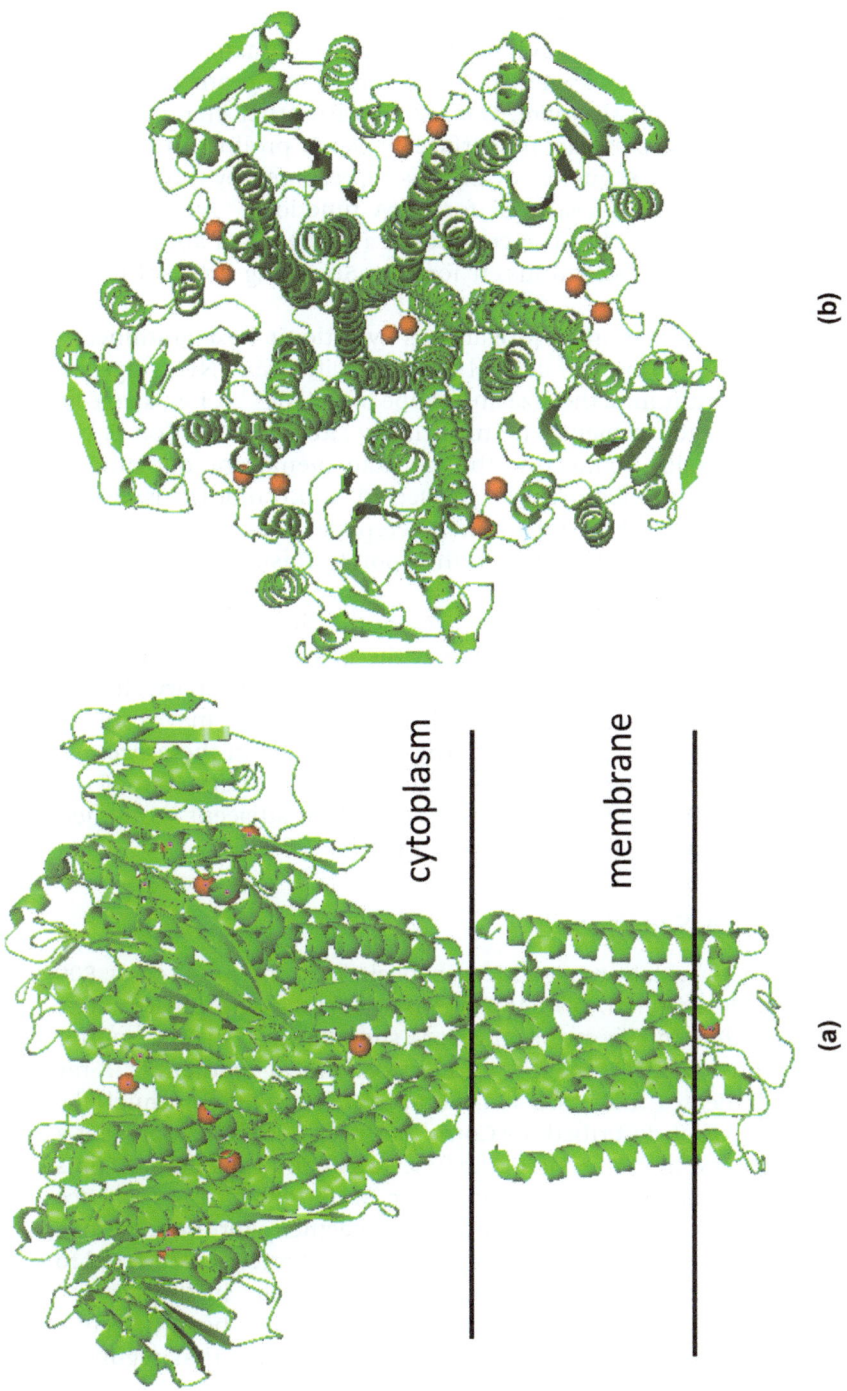

(b)

(a)

Figure 4.1 CorA Structure. Ribbon diagram of the CorA pentameric complex. (a) The narrow part of the "funnel" corresponds to the membrane-bound structure of the protein whereas the larger part of the "funnel" is in the cytoplasm. (b) View from the intracellular region. The putative magnesium ions between monomers or inside the pore region are rendered in red. Interested readers are referred to refs. 15 and 18 for further information. This figure was generated by the PyMol program from PDB file 4I0U.

inner face of the funnel, likely interacting with the transported Mg^{2+} ions.[18] Each CorA monomer possesses two transmembrane domains, for a total of ten segments per channel. The five transmembrane domains-1 form the actual pore of the channel pore whereas the five transmembrane domains-2 form an outer ring around the pore. Consistent with other channel structures, the pore near the periplasmic surface presents proline and glycine residues. The "signature" sequence of CorA is represented by a five amino acid string (YGMNE) that is essential for CorA function.[18] Another well-conserved amino acid string in CorA consists of a sequence of six arginines and/or lysines within the C-terminus. It forms a 50 Å ring external to the pore.[15–18]

Inhibition of Mg^{2+} transport in prokaryotes by cationic hexa-ammines,[18] *i.e.* molecules formed by a central metal with six NH_3 ligands (*e.g.* $[Co(NH_3)_6]^{3+}$) that mimic the size and shape of the $Mg^{2+}{}_{aq}$ ion, suggests that CorA binds that ion moiety. The most likely site of interaction is the loop of nine amino acids connecting the two transmembrane domains in each monomer. This loop features the conserved MPEL motif, suggesting that it may represent a selectivity filter. Discrepancies in modelling this loop,[17] however, do not fully support this point, raising the need for additional study.

Interestingly, no charged residue is present in the first and second transmembrane domains of CorA, hinting at the possibility that the backbone carbonyls present within the pore structure are the primary sites of interaction during passage of Mg^{2+} through the pore.[18] This may result in the outward and/or upward movement of the "outer ring" in the process.[18] Interaction of Mg^{2+} with the stalk helix would promote interfacing of the stalk helix of one monomer with the N-terminus of an adjacent monomer, thus stabilizing the protein in a closed conformation.[15–18]

4.3.1.1   The CorA Superfamily

The CorA protein is widely represented in bacteria, with more than 800 sequences currently detected. A BLAST homology search and phylogenetic analysis indicates that ZntB is part of the CorA superfamily. This transporter mediates the efflux of Zn^{2+} against its electrochemical gradient.[19] X-ray crystallography indicates that ZntB is a homopentamer with a funnel-like structure similar to that described for CorA.[‡] Both CorA and ZntB present a Gly-Met-Asn (GMN) motif at the end of the first transmembrane domain plus two transmembrane domains in each monomer. Mutations in this sequence in CorA abolish Mg^{2+} transport while providing specificity for other cations such as Zn^{2+} (GVN) or Cd^{2+} (GIN). Hence, this class of proteins is referred to as 2TM-GxN.[20] In addition to ZntB, the CorA superfamily includes some yeast proteins for which a limited homology with CorA has been detected. These include MRS2p, a mitochondrial RNA splicing factor that regulates

---

‡See Chapter 24.

Mg^{2+} transport across the inner mitochondrial membrane of yeast, eukaryotes and plant cells,[10] and ALR1 and ALR2, two proteins that confer resistance to Al^{3+} in yeast.[21] The last two proteins can form either homo- or hetero-oligomers,[22] but because of the replacement of Glu by Arg in the MPEL sequence, they are unable to transport Mg^{2+}.

4.3.2 MgtE

MgtE is the primary Mg^{2+} transport protein in the 50% of bacteria that do not possess CorA. Both CorA and MgtE are expressed in these organisms. In addition to Mg^{2+}, MgtE can transport Co^{2+} but not Ni^{2+}, which actually inhibits the transporter.[18] The crystal structure of this protein has been determined in both the absence and the presence of Mg^{2+}.[23] It is a homodimer, with each monomer possessing five transmembrane domains,[18,23] *i.e.* a total of ten transmembrane domains as in CorA. The cytoplasmic N-terminus of each monomer presents two subdomains. The first consists of 10 α-helices, and is structurally homologous to the soluble domains of the bacterial flagellum motor component FliG.[23] It is followed by two cystathionine-β-synthase domains (CBS), a feature common to other transporters.[24] In addition, each monomer possesses five transmembrane helices connected by five loops. Loop-1 and loop-4 each contain a short helix termed H1b and H4b, respectively.[18,23] As for other channels, Pro and Gly residues are present in several of these transmembrane helices, contributing to the natural "kinking" of the helices. Transmembrane helices-2 and -5 of each monomer form the MgtE ion pore. The periplasmic entrance to the pore is formed by the H1b helix and its natural continuation into loop-1, which forms a highly conserved hydrophobic, non-helical portion of transmembrane domain-2. The latter features a VVILA specific sequence. While the periplasmic entrance to the pore is 15 Å in diameter, the cytoplasmic pore entrance is ~ 6 Å. While the cytoplasmic exit of CorA does not present charged amino acids, that for the MgtE pore contains several hydrophilic residues including a conserved Asp.

The crystal structure of MgtE presents five regions of high electron density (Figure 4.2), which have been interpreted as Mg^{2+} binding regions.[18,23] One of these (Mg1) is located in the conducting pore, suggesting that Mg^{2+} binds to the Asp carboxyl group present in transmembrane helix-5, and with the carbonyl group of an Ala residue present one turn of the helix away. The second area of density (Mg2) has been assigned to an Mg^{2+} ion coordinating with one CBS domain and with two residues of the connecting helix. A third area of density (Mg3) appears to coordinate with the CBS domain and with a residue of the connecting helix plus a residue between transmembrane helices-4 and -5. Hence, it appears that Mg^{2+} itself binds at key points to maintain the channel in the closed conformation by inducing the connecting helix to interact with H4 and with transmembrane helices-2 and -5. The other two regions of high electron density (Mg4 and Mg5) appear to connect the CBS domain with the N-terminal domain.[18,23] Such a

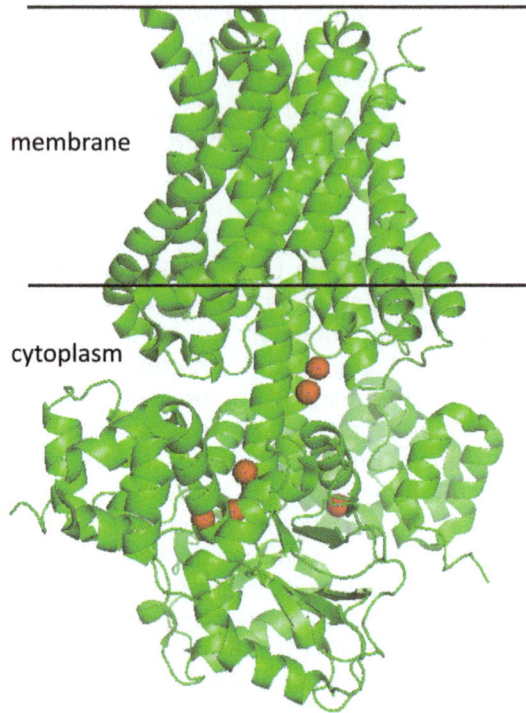

Figure 4.2   MgtE Structure: A ribbon representation of the MgtE dimer in the plane of the membrane. The membrane surface is indicated. The putative magnesium ions bound within the structure are rendered in red. Interested readers are referred to refs. 18 and 23 for further information. This figure was generated by the PyMol program from PDB file 2ZY9.

distribution would guarantee that the flexibility of the cytoplasmic domain increases in the absence of Mg^{2+}. The N domain could then rotate away from the adjacent CBS domain and increase the distance between the two CBS domains. Consequently, the helices forming the pore will change in apposition, resulting in opening of the pore. A cursory comparison of MgtE and CorA suggests that, in both cases, the cytoplasmic component of the channel acts as a Mg^{2+} sensor and as a gate for the ion-conducting channel.

4.3.2.1   The MgtE Family

Phylogenetic analysis of MgtE proteins in bacteria points to a single family.[25] Eukaryotic homologues of MgtE have been identified in the solute carrier family SLC41, which comprises three isoforms (SLC41A1, A2 and A3) and is discussed in more detail below. The SLC41 transporters differ structurally from MgtE in that their N-terminus is shorter and lacks any similarity to the N-terminal and CBS domains of MgtE. A second level of difference is represented by MgtE being involved in Mg^{2+} entry while the members of the SLC41 family appear to predominantly mediate Mg^{2+} extrusion.

4.3.3   MgtA and MgtB Mg^{2+} Transporters

This class of Mg^{2+} transporters is not well defined. While *mgtA* encodes for an endogenous P-type ATPase transporter, *mgtB* has all the characteristics of an acquired gene. Their different nature leads to a different phylogenetic distribution. Although not ubiquitous, MgtA is present mostly in Gram-negative *Enterobacteriaciae*. In contrast, MgtB is present in only three *Enterobacteriaciae* species,[26] and is ≥50% homologous to MgtA. It is noteworthy that neither of the MgtA- nor MgtB-related P-type ATPases is present in the majority of bacterial and archaeal genomes.

Structurally, MgtA and MgtB are similar to yeast H^+-ATPases and eukaryotic Ca^{2+}-ATPases.[§] MgtB was the first P-type ATPase for which some structural information became available.[27] It features ten transmembrane domains and both the N- and C-termini are on the cytoplasmic side. The presence of these transmembrane domains is quite unusual as all other P-type ATPases are shorter, lacking the four transmembrane domains proximal to the C-terminus. A second difference is the direction of Mg^{2+} transport. Both MgtA and MgtB mediate Mg^{2+} influx across the periplasmic membrane[27] whereas all other mammalian or bacterial P-Type ATPases mediate ion efflux from the cytoplasm. Further, mammalian P-Type ATPases possess six highly conserved and negatively charged residues in the transmembrane domains involved in cation binding. Only two of these residues are conserved in MgtA and MgtB, and one of them is uncharged.[14]

While CorA and MgtE are constitutively expressed in bacteria, MgtA and MgtB are induced by Mg^{2+} deprivation, and repressed as the Mg^{2+} concentration in the extracellular milieu increases.[18] The induction of these proteins is mediated by the PhoPQ system whose genes are involved in mediating bacterial virulence and are derepressed following the invasion of host cells. MgtA and MgtB, however, do not appear to play a major role in bacterial pathogenesis.[26] In terms of transport, K_m (Mg^{2+}) for both these transporters is in the 5–20 μM range, which is not dissimilar to that reported for CorA.[26] Small differences exist between MgtA and MgtB in terms of pH sensitivity and cation transport but their main difference is in thermal sensitivity: MgtA becomes inactive at 4 °C but MgtB is already inactive upon cooling to 20 °C. Because of the apparent K_m, it is difficult to attribute to MgtA and MgtB the role of subsidiary transporters upon substitution of the primary CorA. It is possible that Mg^{2+} influx is not their primary role, and that Mg^{2+} is accumulated in exchange for another cation or a yet-to-be determined substrate that is extruded from the bacteria. This hypothesis is supported by the phylogenetic distribution of MgtA and MgtB, and by

[§]See Chapter 5, Section 5.2.2.

the observation that under no circumstances do MgtA or MgtB represent the only transporters responsible for maintaining Mg^{2+} homeostasis in a bacterial strain.

4.3.4   The MgtCB Operon

As indicated above, *mgtB* appears not to be an endogenous bacterial gene but rather an acquired one, part of the *mgtCB* operon responsible for controlling bacterial virulence. This operon comprises three genes, of which *mgtB* is the second. The first is *mgtC* while the third is *orf X*, for which no specific function or homology gene has been identified. Because it is part of this operon, *mgtC* is also regulated by extracellular Mg^{2+} through the PhoPQ system.[28] Homologues of MgtC have been observed in several bacterial strains and species, but their distribution does not allow for any phylogenetic conclusion.[26] The protein expressed by this gene has a molecular mass of ~23 kDa, is very hydrophobic and presents five or six transmembrane domains. While its function is not clearly identified, there is significant, albeit not unanimous, evidence that *mgtC* is essential for both bacterial virulence and bacterial growth under conditions in which extracellular magnesium is limited.[26] It has also been proposed that, at least in strains like *S. thyphimurium* and *Mycobacterium tuberculosis*, MgtC is a fourth Mg^{2+} transporter.[26] However, other data indicate that, in the absence of other bacterial transport mechanisms, MgtC alone is not sufficient to fully restore Mg^{2+} homeostasis and bacterial growth.[29] Overall, it remains to be elucidated whether these differences are related to differences in gene expression between diverse bacterial strains.

4.4   Magnesium and Eukaryotic Cells

In eukaryotes, magnesium is the second most abundant cation after potassium. Various analytical techniques consistently indicate that total Mg^{2+} concentrations range between 17 and 20 mM in most eukaryotic cells[30,31] and between 15 and 18 mM within the nucleus, mitochondria and endo-(sarco)-plasmic reticulum of a given cell.[30] The persistence of such high Mg^{2+} concentrations within these organelles has been rationalized by the binding of Mg^{2+} to nucleic acids, chromatin, nucleotides, phospholipids and proteins. Although the specific nature of this binding has not been fully investigated, experimental evidence confirms that the concentration of *free* Mg^{2+} is ≤ 1 mM within the cytoplasm or in the lumen of the above organelles.[30,32] This is supported by estimates of 0.8 and 1.2 mM *free* $[Mg^{2+}]$, respectively, in the cardiac and liver mitochondrial matrices.[30,32] In contrast, no data on luminal free $[Mg^{2+}]$ are available for the nucleus and the endo-(sarco)-plasmic reticulum. Due to the porous structure of the nuclear envelope, it is reasonable to assume that intranuclear levels are similar to those in the cytoplasm. On the other hand, the presence of millimolar concentrations of Ca^{2+} inside the endo-(sarco)-plasmic reticulum[30–32] combined with the high affinity of Mag-Fura or Mag-Indo fluorescent dyes for Ca^{2+} ($K_d = 50$ μM) as compared to Mg^{2+} ($K_d = 1.5$ mM)[30–32] prevents the accurate estimation of the *free* $[Mg^{2+}]$ within the organelle lumen. Additional cellular pools of Mg^{2+} are in the form of complexes with ATP, phosphonucleotides and phosphometabolites.[30–33]

Because of its abundance (~ 5 mM) and affinity for Mg^{2+} ($K_d = 78$ μM), ATP constitutes the largest metabolic pool able to bind Mg^{2+} within the cytoplasm, and the mitochondrial matrix as well.[30–33] Because of this distribution, cytosolic *free* $[Mg^{2+}]$ represents less than 5% of the total cellular Mg^{2+} content in eukaryotic cells,[30] resulting in the presence of a very limited chemical gradient of Mg^{2+} across the cell membrane and across the membranes of cellular organelles.

The detected Mg^{2+} concentrations are essential for regulation of numerous cellular functions and enzymes, including ion channels, metabolic cycles and signalling pathways. However, our understanding of the regulation of Mg^{2+} homeostasis within eukaryotic cells remains incomplete for a variety of conceptual and methodological reasons. The absolute abundance of cellular Mg^{2+} (both total and *free*), its slow turnover across biological membranes and the presence of small changes, if any, in cytoplasmic free $[Mg^{2+}]$ following metabolic or physiological stimuli have all generated the assumption that Mg^{2+} content within eukaryotic cells remains largely constant at a level commensurate with that of a cofactor for cellular enzymes and proteins. Having been accepted until the last decade of the twentieth century, this assumption led to a limited interest in developing techniques and methodologies to properly quantify cellular Mg^{2+} content and its changes. Since the 1990s, steadily increasing numbers of experimental and clinical observations support a regulatory role for Mg^{2+} in a large variety of cellular functions and the presence of major fluxes of Mg^{2+} through the eukaryote cell membrane. Genetic, electrophysiological and molecular biology approaches in both prokaryotes and eukaryotes have identified several Mg^{2+} transport mechanisms operating in the cell membrane or in the membrane of cellular organelles (*e.g.* mitochondria and the *trans*-Golgi network). New fluorescent indicators have been developed to better measure cytosolic free $[Mg^{2+}]$. All together, these approaches and techniques have led to renewed interest in Mg^{2+} homeostasis and its effect upon specific physiological and pathological conditions in human patients. The picture that emerges from the available information is that Mg^{2+} is highly compartmentalized within the cellular organelles and cytoplasm and that a limited gradient exists across the cell membrane and those of the cellular organelles. Moreover, the concentrations of free Mg^{2+} in the cytoplasm and the mitochondrial matrix under resting conditions and their variations following hormonal and metabolic stimuli are consistent with a key regulatory role in several cellular functions.

This section highlights our current understanding of Mg^{2+} regulation and homeostasis in mammalian cells and aims to provide the reader with a useful framework to relate changes in cellular Mg^{2+} content with variations in specific cellular functions under physiological conditions.

4.4.1 Uptake of Mg^{2+} in Intestinal Cells and Release into the Blood

Diet is the main source of Mg^{2+} intake. The daily recommended dose is approximately 300 mg for men and 250 mg for women, and corresponds to the amount eliminated daily through the urinary and digestive systems.[34] Magnesium absorption occurs at the apical side of intestinal cells of the small intestine. The cation is transported through these cells, and released into the bloodstream at the basolateral side. This series of events requires the presence and cooperation of specific entry and exit mechanisms for Mg^{2+} on the two sides of the cell membrane.

Apical Side: Limited information is available about the mechanism of Mg^{2+} uptake from the intestinal lumen. Specific accumulation mechanisms have been observed in brush border cells of rabbit ileum and rat duodenum and jejunum.[35] The saturable Mg^{2+} uptake occurring in rabbit ileum is Na^+-dependent ($K_m = 16$ mM) but becomes inoperative when $[Na^+]_i < [Na^+]_o$ as the process is not reversible.[35] Further, the transport is electroneutral and is not inhibited by amiloride analogues but is modulated by intravesicular anions (*e.g.* Cl^- and SCN^-) and antagonists of anion transport such as 4,4'-diisothiocyanatostilbene-2,2'-disulfonic acid (DIDS).[35]

As for Mg^{2+} entry into rat duodenum and jejunum, the main difference is that a single Mg^{2+} transporter operates in the duodenum ($K_m = 0.8$ mM) while two transporters operate in the jejunum ($K_m = 0.15$ and 2.4 mM, respectively).[35] In both models, Mg^{2+} accumulation is reduced by alkaline phosphatase inhibitors and blocked by amiloride but not by Ca^{2+} channel antagonists.[35] Furthermore, Mg^{2+} accumulation is stimulated by an intravesicular electronegative potential or alkaline pH_o,[35] similar to what is observed in rabbit ileum.[35] The effect of external pH, however, is lost when $[Mg^{2+}]_o > 1$ mM.[35] Under the latter condition, Mg^{2+} accumulation is enhanced by the presence of Na^+ or K^+, and inhibited by the presence of divalent cations ($Co^{2+} > Mn^{2+} > Ca^{2+} > Ni^{2+} > Ba^{2+} > Sr^{2+}$) in the extravesicular space.[35]

More recently, attention has been paid to the distribution and operation of apical Mg^{2+} channels. Two channels elicit Mg^{2+} accumulation in eukaryotes: TRPM6[36] and TRPM7.[37] They are members of the melastatin subfamily of transient receptor potential (TRP) channels.[38] While TRPM6 is specifically located in the colon and also in the distal convolute tubule of the nephron,[36] TRPM7 is ubiquitously expressed in the majority of mammalian cells,[37] including the various segments of the small intestine.[39]

Both TRPM6 and TRPM7 are tetramers, and present an α-kinase domain at their C-terminus. This kinase is functionally homologous to eEF2-kinase, and phosphorylates serine and threonine residues within a α-helical structure.[40] Removal of the kinase domain does not abolish entirely the channel activity but affects the extent to which the channel is regulated by intracellular free Mg^{2+} or the Mg·ATP complex.[40] The best known target phosphorylated by the TRPM6 kinase domain is TRPM7.[40]

TRPM6 expression and activity are modulated *in vivo* by diet and estrogens. The 17β-estradiol markedly up-regulates TRPM6 mRNA in both colon and kidney while having no effect on TRPM7 mRNA.[40] In the absence of estrogens, the repressor of estrogen receptor activity (REA) binds to the sixth, seventh and eighth β-sheets of the TRPM6 kinase domain in a phosphorylation-dependent manner, and inhibits TRPM6 activity.[40] Short-term estrogen administration leads to dissociation of REA from TRPM6 and increases channel activity.[40] Expression of TRPM6 mRNA increases in colon and kidney following restriction of dietary Mg^{2+}.[40] In contrast, an Mg^{2+}-enriched diet up-regulates TRPM6 mRNA only in the colon, increasing intestinal absorption.[40] In contrast, neither restriction nor enrichment of dietary Mg^{2+} affects the expression of TRPM7 mRNA.[40] Thus, it appears that genetic factors and dietary levels control TRPM6 expression and activity in the large intestine to favour Mg^{2+} absorption. On the other hand, renal Mg^{2+} resorption occurs only following restriction of dietary Mg^{2+}.[40]

As indicated previously, an increase in intracellular ATP levels specifically decreased TRPM6 and TRPM7 currents,[40] in a manner independent of α-kinase autophosphorylation activity.[40] The site of inhibition resides in the conserved ATP-binding motif GXG(A)XXG within the α-kinase domain.[40] Full deletion of the kinase domain or point mutations within the ATP-binding motif ($G^{1955}D$) completely prevents the inhibitory effect of intracellular ATP.

TRPM6 and possibly TRPM7 activity is also modulated by cellular signalling molecules. RACK1 (receptor for activated protein kinase C) directly binds to the α-kinase domain of TRPM6, and possibly TRPM7 due to the high homology (>84%) between the two kinase domains.[41] The RACK1 binding site is restricted to the region between amino acids 1857 and 1885 (sixth, seventh and eighth β-sheets) of TRPM6, the same β-sheets involved in REA regulation.[40] Accessibility analysis suggests that 18 of the indicated 28 amino acids are localized at the surface of the TRPM6 α-kinase domain.[41] As a result of this interaction, the channel activities of TRPM6 and TRPM7 are inhibited. RACK1's inhibitory effect specifically depends on autophosphorylation of threonine 1851 (T^{1851}) within the kinase domain. This residue is located at the end of the fourth α-helix adjacent to the RACK1 binding site. Mutation of T^{1851} to Ala (A) or to Asp (D) significantly decreases TRPM6 autophosphorylation while leaving RACK1 binding unaffected. The inhibitory effect of RACK1 on the channel activity is completely abolished only in the case of the $T^{1851}A$ mutation while it persists in the case of $T^{1851}D$ mutation.[41] Interestingly, $T^{1851}D$ autophosphorylation strongly depends on the Mg^{2+} concentration, steadily increasing for concentrations between 0.1 and 1 mM. In contrast, the $T^{1851}A$ mutant is less sensitive to intracellular Mg^{2+} concentrations as compared to the wild-type (IC_{50} ~0.7 *vs.* 0.5 mM, respectively). Activation of protein kinase C, the natural ligand for RACK1, completely prevents the inhibitory effect of RACK1 on TRPM6 channel activity,[41] suggesting a competition between TRPM6 and RACK1 for PKC.

Cellular Transport: In contrast to a total cellular Mg^{2+} concentration of 15–20 mM,[30–32] the cytoplasmic Mg^{2+} concentration is approximately 0.5–1 mM.[30–32] This suggests that entry of Mg^{2+} into the cell is rapidly buffered by ATP, phosphonucleotides and proteins.

It has been suggested that cytosolic proteins contribute to buffering cytosolic Mg^{2+} levels in intestinal cells while also transporting Mg^{2+} through the cytoplasm to the basolateral side where it enters the circulation. Limited information is available about the proteins that bind Mg^{2+} within the cell. Aside from calmodulin, troponin C, parvalbumin, S100 protein and the proteins that use Mg^{2+} as a cofactor, no indication is available about other cytoplasmic or intraorganellar proteins that specifically bind and buffer cellular Mg^{2+}.[30–32,40] More than 40 years ago, it was reported that two proteins bind Mg^{2+} with high affinity/low capacity and high capacity/low affinity, respectively, in the intermembrane space of the mitochondrion.[42] However, these proteins have not been identified as yet. Several other cellular and organellar proteins appear to have potential binding sites for Mg^{2+}, but it is not clear whether any of them acts under physiological conditions. In the case of the kidney, it has been proposed that parvalbumin and cal-bindin-D_{28k}, two proteins abundantly present within cells of the distal convolute tubule of the nephron, favour the transcellular transport of Mg^{2+} that has accumulated at the apical domain of the cell, thereby accelerating the rate of delivery to the basolateral domain. Both proteins are present in intestinal cells, in which they contribute to Ca^{2+} binding and transport upon vitamin D stimulation. Whether these proteins operate in a similar manner for Mg^{2+} in the intestinal epithelium is presently undefined. The physiological relevance of cytoplasmic proteins to bind and transport Mg^{2+} from the apical to the basolateral side of intestinal cells is further questioned because parvalbumin *null* mice do not exhibit hypomagnesaemia or significant changes in tissue Mg^{2+} handling and homeostasis.[43]

Basolateral Side: Intestinal cells extrude Mg^{2+} into the bloodstream through a Na^+-dependent pathway, currently identified as a Na^+/Mg^{2+} exchanger.[44]

Na^+-dependent Mg^{2+} Extrusion Mechanism: The first evidence for the operation of a Na^+-dependent Mg^{2+} transport in mammalian cells was provided in 1984.[44] In sequential studies,[44,45] the *modus operandi* of the Mg^{2+} extrusion was detailed as well as its sensitivity to inhibition by amiloride. The initial observation has been confirmed in a variety of mammalian cell types.[40] Additional observation has provided compelling evidence that this putative Na^+-dependent, amiloride-inhibited Mg^{2+} extrusion transporter is phosphorylated by cAMP. Stimulation by β-adrenergic agonists, glucagon or prostaglandin E2, or administration of forskolin or cell-permeant cyclic-AMP analogues, all activate Na^+-dependent Mg^{2+} extrusion.[30–32,40] In contrast, pre-treatment of cells with inhibitors of adenylyl cyclase (*e.g.* Rp-cAMP) prevents Mg^{2+} mobilization.[30–32]

The Mg^{2+} extrusion depends strictly on the presence of extracellular Na^+ suggesting a Na^+/Mg^{2+} exchanger.[30–32] Controversy persists as to whether this exchanger is electrogenic ($1Na^+ : 1Mg^{2+}$)[30–32] or electroneutral ($2Na^+$:

$1Mg^{2+}$).[30–32,40] Present results consistently indicate a K_m for Na^+ between 15 and 20 mM,[30,31,40] similar to that reported for the Mg^{2+} transporter in rabbit ileum.[30–32] No specific inhibitor has been identified. Amiloride, imipramine and quinidine represent the three most widely utilized inhibitors of Na^+-dependent Mg^{2+} extrusion.[40,46] Their limited specificity, however, does not indicate whether they inhibit the Na^+/Mg^{2+}-exchanger directly or indirectly by affecting other Na^+ transport processes (including Na^+ and possibly K^+ channels), by altering the cellular membrane potential or by decreasing the driving force for Mg^{2+} transport across the plasma membrane.

Despite extensive research, uncertainty remains about the actual nature and structure of the Na^+/Mg^{2+} exchanger. Using monoclonal antibodies, evidence for a protein of ~ 70 kDa was obtained.[47] The Mg^{2+} transporter known as SLC41 is thought not to be an Mg^{2+} importer,[48] but an exporter similar to the putative Na^+/Mg^{2+} exchanger, as it displays cAMP-dependent activation.[49]

Na^+-independent Mg^{2+} Extrusion Mechanism: Under conditions in which no or limited extracellular Na^+ is available to exchange for cellular Mg^{2+}, a Na^+-independent Mg^{2+} extrusion is observed. The specifics of this transporter, however, remain uncertain. Different cations (*e.g.* Ca^{2+}, Mn^{2+} and choline) but also anions (*e.g.* HCO_3^-, Cl^-) appear to be utilized by this transporter.[50] It is unclear whether it is an antiporter for cations or a symporter for cations and anions dependent on experimental conditions. It has been proposed that the Na^+-independent Mg^{2+} export may occur *via* the choline transporter, a suggestion based upon its quite specific inhibition by cinchona alkaloids.[51]

Lastly, it remains controversial whether the Na^+-dependent or the Na^+-independent transporters require ATP, with experimental evidence that is both in favour and against it.[40]

4.4.2   Coordination Chemistry of Mg^{2+} in the Blood

In humans, the physiological plasma Mg^{2+} level is 1.7–2.1 mg dL^{-1} (0.7–1.2 mM or 1.4–1.7 mEq L^{-1}).[52] Eighty percent of plasma Mg^{2+} is *free* or complexed to filterable ions (*e.g.* oxalate, phosphate, citrate) and thus available for glomerular filtration in the kidney. The remaining 20% is protein-bound (60%–70% associated with albumin and the rest bound to globulins).[53] The concentration of free Mg^{2+} in the serum ranges between 0.54 and 0.67 mM, narrower than that for calcium.[53] When these values are compared to cytosolic *free* $[Mg^{2+}]$ (estimated to be between 0.5–0.6 and 0.8–1.2 mM), most eukaryotic cells appear not to have an inward gradient favouring Mg^{2+} entry across the cell membrane or the biomembranes of cellular organelles (*e.g.* mitochondria). Moreover, the levels of cytosolic *free* $[Mg^{2+}]$ are almost two orders of magnitude lower than the value calculated using the Nernst equation.[54] This supports the notion that eukaryotic cells must possess specific and powerful transporters and buffering systems to maintain total Mg^{2+} concentrations and the cytosolic *free* levels at the values estimated experimentally.

Serum and extracellular Mg^{2+} content represents $\sim 1\%$ of total body Mg^{2+} content. Presently, it is unknown whether serum Mg^{2+} undergoes circadian fluctuations under physiological conditions (*e.g.* fasting, exercise, lactation or changes in hormone levels). The infusion of catecholamine or isoproterenol in conscious humans or in ewes has generated contrasting results.[55] In rats, however, the infusion of these agents resulted in a dose-dependent increase in circulating Mg^{2+} concentration that persisted up to two hours following infusion.[55] The effect is independent of the haemo-dynamics changes (increase in heart rate and decrease in mean arterial pressure) elicited by β-adrenergic stimulation.[55] The persistent increase in circulating Mg^{2+} level suggests that the systemic stimulation of β-adrenoceptors mobilizes Mg^{2+} from different organs and tissues into the circulation promoting at the same time Mg^{2+} resorption in the Henle's loop to prevents net Mg^{2+} loss *via* renal excretion.[55]

Consistent with the predominance of $β_2$ *versus* $β_1$ adrenoceptors in the whole body, the increase in serum Mg^{2+} level appears to occur *via* specific $β_2$-adrenoceptor activation.[55] The amplitude of the increase also suggests that adrenergic agonists mobilize Mg^{2+} from various tissues.[55] The bone Mg^{2+} reservoir may be involved in the process based on the inhibitory effect of carbonic anhydrase blockers under similar experimental conditions.[55]

Several factors may explain the inconsistency in the reported serum Mg^{2+} changes. They include the relative expression of β-adrenoceptor subtypes in different experimental models, the ability of adrenergic agonists to stimulate different adrenoceptors subtypes with various potencies and the modality, rate and duration of drug infusion. These inconsistencies have also limited our understanding of the physiological significance to the increase in serum Mg^{2+} level following catecholamine infusion.

While an increase in calcaemia results in detectable symptoms such as muscle weakness and arrhythmia in humans, an increase in serum Mg^{2+} level appears to be tolerated well. In rats, the infusion of Mg^{2+} boluses that increase serum Mg^{2+} level by 50% results in an increase in coronary artery blood flow in the absence of significant changes in systemic hemo-dynamics.[56] At the same time, infusion of pharmacological doses of Mg^{2+} sufficient to prevent epinephrine-induced cardiac arrhythmias in baboons markedly attenuates the epinephrine-induced increase in mean arterial pressure and systemic vascular resistance.[55] Thus, it appears that an increase in circulating Mg^{2+} level can regulate catecholamine release from peripheral and adrenal sources,[55] and modulate cardiac contractility.[57] Altogether, these observations suggest that the increase in serum Mg^{2+} level elicited by adre-nergic agonists can: 1) act as a feedback mechanism to limit catecholamine release and activity, and 2) contribute to improve coronary blood flow and O_2 delivery to the heart at a time when energy production is expected to increase.

The serum Mg^{2+} level decreases significantly under several chronic dis-eases.[58] Hypertension, diabetes (type-I and type-II), hyperaldosteronism and alcoholism are among the prominent pathological conditions accompanied by a decrease in circulating Mg^{2+} level.[58] Clinically or experimentally, the term hypomagnesaemia is used to indicate a reduction in circulating Mg^{2+} concentration below 1 mEq L^{-1}. More properly, two conditions can be recognized based upon the absence or the presence of clear clinical symptoms: one in which the circulating Mg^{2+} level is below 0.5 mM, and a second in which the circulating Mg^{2+} level falls below 0.4 mM.[58] Although symptoms tend to appear once the serum Mg^{2+} falls below 1 mEq L^{-1}, most patients with mild Mg^{2+} deficiency (~ 0.5 mM hypomagnesaemia) are asymptomatic. Further, there is no direct correlation between serum Mg^{2+} and the severity of the symptoms. When present, the symptoms are non-specific and include tiredness, neuromuscular irritability, muscle cramps (usually in legs), tetany, rapid heartbeats, elevated blood pressure, tachycardia and/or ventricular arrhythmias, vertigo, ataxia, depression and seizure activity.[58]

Hypomagnesaemia is often but not always associated with tissue Mg^{2+} deficiency. Patients with abnormal Mg^{2+} homeostasis typically fall into one of the following three categories:

1) Low total body Mg^{2+} content (*i.e.* Mg^{2+} deficiency) *and* a resultant hypomagnesaemia;
2) Hypomagnesaemia in the *absence* of Mg^{2+} deficiency (*i.e.* normal total body Mg^{2+} content);
3) Low total body Mg^{2+} content (Mg^{2+} deficiency) *but* no hypomagnesaemia.

Clinicians routinely assess serum Mg^{2+} level as an indicator of total body Mg^{2+} content. However, because of the inconsistent relationship between circulating Mg^{2+} levels and tissue Mg^{2+} content and the relative percentages involved (1% in serum *vs.* 99% within tissues), this assessment is limited and incomplete. In addition, suitable methodologies are lacking for timely and effective screening of the population. As a result, Mg^{2+} deficiency goes largely undetected. It is currently estimated that approximately 40% of the US population has less than optimal serum and/or tissue Mg^{2+} level, or is actually Mg^{2+} deficient.[58]

From a dietary viewpoint, the recommended daily Mg^{2+} allowance (RDA) in adults is 4.5 mg kg^{-1} day^{-1}, although it is recognized that the RDA should be increased in pregnancy, lactation and following a debilitating illness.[58] Magnesium is abundant in green leafy vegetables, cereals, grains, nuts and legumes, and low in dairy products, while chocolate, vegetables, fruits, meat and fish have intermediate values.[58] Drinking water is another important source of magnesium, especially "hard water" that can contain up to 30 mg L^{-1} Mg. Food refining or processing may deplete magnesium content by nearly 85%. Cooking, especially boiling of magnesium-rich foods, results in significant loss. Altogether, these dietary limitations may explain the low magnesium intake and serum/tissue level observed in many populations by a recent dietary survey that concluded that the average daily magnesium intake in many western countries is approximately 35% to 45% lower than the RDA.[58]

4.4.3 Mg^{2+} Transport through Cellular Membranes

Magnesium is transported across the cell membrane or the membrane of cellular organelles through channels and exchangers. While channels are predominantly involved in Mg^{2+} accumulation, exchangers essentially mediate Mg^{2+} extrusion.

4.4.3.1 Channels

Channels transporting Mg^{2+} into cells were originally reported for pro-karyotes,[14,59] including protozoans.[60] Recently, Mg^{2+} entry through channels or channel-like pathways has been reported for eukaryotic cells. Several exhibit a relatively high specificity for Mg^{2+} while others transport a variety of divalent cations. The majority are located in the cell membrane although one was found in the mitochondrial membrane and another in the Golgi cisternae. As the identification and characterization of these Mg^{2+} channels is an on-going process, information relative to their regulation is fragmentary. In addition, apparent redundancy raises the question of whether different mechanisms cooperate to modulate Mg^{2+} entry or whether one of them predominates in a specific cell under well-defined conditions.

TRPM6 and TRPM7: These were the first Mg^{2+} channels identified con-temporaneously in mammalian cells.[36,37] TRMP6 is specifically localized in the colon and the distal convolute tubule of the nephron, a distribution that emphasizes its role in controlling whole body (systemic) Mg^{2+} homeostasis *via* intestinal absorption and renal resorption. On the other hand, TRPM7 appears to be present in all mammalian cells, suggesting a role in the control of cellular Mg^{2+} homeostasis.

TRPM6: The gene *TRPM6* was identified originally as the site of various mutations responsible for Hypomagnesaemia with Secondary Hypocalcaemia (HSH). This is a rare autosomal recessive disease charac-terized by Mg^{2+} and Ca^{2+} wasting, the symptoms of which are ameliorated by massive intravenous administrations of Mg^{2+} followed by oral Mg^{2+} supplementation.[36] However, while this treatment alleviates the hypo-calcaemia completely, serum Mg^{2+} levels remain around 0.5–0.6 mM in these patients, roughly half the physiological level.[36] Because the primary defect in these patients is at the level of the TRPM6 expressed in the intes-tine,[36] the excess Mg^{2+} supplementation is rapidly filtered at the glomerular level and increases passive renal absorption *via* paracellin-1 (see below). Transcellular absorption *via* renal TRPM6 remains depressed and unable to restore physiological serum levels.[36]

Available evidence suggests that this channel forms a tetramer within the plasma membrane. Several TRPM6 mutations have been identified that re-sult in the expression of a truncated and non-functional channel.[40] TRPM6 *null* mice have been developed. The heterozygous $Trpm6^{+/-}$ has a normal electrolyte profile if one excludes a modest decrease in plasma Mg^{2+} (~ 0.67 mM *vs.* 0.75 mM in the wild-type).[40] The majority of the homozygous *Trpm6⁻/⁻* animals die by embryonic day 12.5. Among the few animals that survive to term, many present neural tube defects such as exencephaly and spina bifida occulta. Administration of a high Mg^{2+} diet to dams improves offspring survival to weaning.[40]

Most of the available information about TRPM6 was provided in Section 4.1. An important addition is the role of epidermal growth factor (EGF) as an autocrine/paracrine magnesiotropic hormone.[7] By engaging its receptor in the basolateral domain of the distal convolute tubule, EGF activates TRPM6 at the apical domain of the cell and induces cellular Mg^{2+} accumulation. A point mutation in the pro-EGF ($P^{1070}L$) molecule retains EGF secretion to the apical membrane and disrupts this cascade of events, ultimately resulting in the Mg^{2+} wasting typical of isolated recessive renal hypo-magnesaemia (IRH). An alteration of the axis EGF/TRPM6/Mg^{2+} leads to reabsorption/renal Mg^{2+} wasting in cancer patients undergoing treatment with anti-EGFR antibodies.[40] The antibodies antagonize the stimulation of TRPM6 activity *via* EGF. EGF modulates TRPM6 activity and expression through ERK1/2 signalling *via* activator protein-1 (AP-1). The process is prevented by antagonists for integrin $\alpha_v\beta3$ or MEK1/MEK2 activity, or by siRNA for TRPM6.[40] How exactly EGF, integrin and ERK1/2 interact to enhance TRPM6 expression needs further elucidation. It is unclear whether the activation of this signalling axis removes RACK1-mediated inhibition of TRPM6 activity.[41]

TRPM7: Although this channel carries Mg^{2+} and Ca^{2+} preferentially,[37] it also transports other divalent cations such as Ni^{2+} and Zn^{2+}.[40] Under physiological conditions, the transport of divalent cations blocks the inward movement of monovalent cations (*e.g.* Na^+).

Located at the locus 15q21 of human chromosome 15, the gene encodes for a protein containing 1865 amino acids with ten transmembrane domains and both the C- and N-termini at the cytoplasmic side. The protein is expressed ubiquitously, albeit to varying extents, in all mammalian cells tested so far. The functional structure is thought to be a tetramer but disagreement exists as to whether it is formed by four identical monomers or by a combination of TRPM7 and TRPM6 arranged with a varying stoichiometry in different portions of the cell membrane or in different cell types.[40] A detailed functional characterization demonstrated that TRPM6 and TRPM7 form a chimeric heterotetramer, and that TRPM6, TRPM7 and TRPM6/7 constitute three distinct ion channels.[40] These display different divalent cation permeabilities, pH sensitivities and characteristic single channel conductances. In addition, the calcium channel inhibitor 2-aminoethoxydiphenyl-borate (2-APB) can differentially regulate the activities of the three channels, markedly increasing Mg^{2+} and Ca^{2+} entry through TRPM6.[40] While this observation strongly suggests the possibility that the three channels are expressed differently in distinct tissues and play diverse roles under a variety of physiological or pathological conditions, a detailed map of homomeric TRPM7 *versus* heteromeric TRPM6/7 channel distribution within tissues is still lacking. Consequently, their relative roles remain largely undefined.

In terms of regulation, the TRPM7-mediated inward current is enhanced by protons that compete with Ca^{2+} and Mg^{2+} for binding sites at the level of the channel pore, and release the blockade by divalent cations on inward monovalent currents.[40] This observation suggests that, at physiological pH values, Mg^{2+} binds to TRPM7 and inhibits monovalent cation currents while, at higher pH, the affinity of TRPM7 for Mg^{2+} decreases, allowing monovalent cations to permeate the channel.[61] A second level of regulation is provided by the phospholipid phosphatidyl inositol bisphosphate (PIP_2). The depletion in PIP_2 levels that follows phospholipase C (PLC)-activation by different hormones[40,62] counteracts TRPM7 activation, and accelerates the progressive decrease in current (channel "run-down"). In contrast, inhibition of PLC or addition of exogenous PIP_2 decreases the channel "run-down".[40] An accelerated channel run-down is also observed following addition of ATP or non-hydrolysable GTP analogues.[40] It is noteworthy that TRPM7 activation occurs only in the presence of physiological $[Mg^{2+}]_i$. Reducing this concentration below its physiological level with ethylenediaminetetraacetic acid-acetoxymethylester (EDTA-AM) results in PLC-mediated TRPM7 inactivation, most likely *via* PIP_2 depletion.[40] A regulatory role by cAMP has also been reported,[40] but no information is available about the mechanism(s) behind this regulation. The interplay between phosphatidylinositol metabolites and TRPM7 is further supported by the observation that a functional TRPM7 is required for sustained phosphoinositide-3-kinase (PI3K) mediated signalling and growth,[63] to the extent that TRPM7-deficient cells exit the cell cycle and enter quiescence.

TRPM7, like TRPM6, presents an α-kinase at the C-terminus that specifically phosphorylates serines and threonines located in an alpha-helix.[40] While not strictly necessary for channel activation and gating, the absence of the kinase domain results in the inability of the channel to properly phosphorylate and regulate downstream cellular components upon activation. Autophosphorylation of the kinase domain is essential for target recognition and to increase the rate of substrate phosphorylation by the kinase domain.[40] Phosphomapping has identified 37 out of 46 autophosphorylation sites in a Ser/Thr rich region immediately preceding the kinase catalytic domain. Deletion of this region does not affect the intrinsic catalytic activity of the kinase but prevents substrate phosphorylation, confirming the role of this region in substrate recognition. Although this Ser/Thr region is poorly conserved in TRPM6, the kinase domain of the TRPM6 channel appears to require a similar massive autophosphorylation of Ser/Thr residues for effective substrate recognition and target phosphorylation.[40]

Presently, only annexin-I, myosin IIA heavy chain and calpain have been identified unambiguously as phosphorylation substrates for the TRPM7 kinase domain.[40] Hence, it appears that TRPM7 plays two key roles within the cell: it regulates Mg^{2+} homeostasis, and cellular functions centred on cell adhesion, contractility and (anti)-inflammatory processes. This double role in smooth muscle cells is emphasized by the observation that aortic segments of mice presenting low intracellular Mg^{2+} levels have increased medial cross-section and increased TRPM7 expression, but decreased levels of annexin-I expression.[64] Because annexin-I has a major anti-inflammatory role, these results point to TRPM7 as a regulator of both inflammation and vascular structure and integrity.

Removal of the kinase domain results in embryonic lethality in homozygous mice, and hypomagnesaemia in heterozygous mice due to a defect in intestinal Mg^{2+} absorption.[40] Moreover, embryonic stem cells lacking the TRPM7 kinase domain are growth-arrested and can be rescued by Mg^{2+} supplementation.[63]

Claudins: *PCLN-1* and its gene product paracellin 1 (PCLN-1 or claudin 16) was the first Mg^{2+} transport-related gene and protein to be discovered.[65] It was identified through genetic analysis of patients affected by Familial Hypomagnesaemia with Hypercalciuria and Nephrocalcinosis (FHHNC). This disease is characterized by massive renal Mg^{2+} and Ca^{2+} wasting that leads rapidly and irreversibly to renal failure and does not respond to Mg^{2+} supplementation. Proteins of the claudin family are involved in tight junction formation. They present four transmembrane spans coordinated by two extracellular loops, with both C- and N-termini on the cytoplasm side. More than 20 mutations affecting claudin 16 trafficking or permeability have been identified.[66]

Claudin-16 mediates paracellular Ca^{2+} and Mg^{2+} fluxes throughout the nephron seemingly by increasing paracellular Na^+-permeation. This, in turn, generates a positive potential within the lumen that acts as the driving force for Mg^{2+} and Ca^{2+} resorption.[40] Results in cultured kidney cells support a decrease in Na^+-permeability and an increase in Mg^{2+}-permeability. It is unclear whether this effect depends on the experimental conditions or the differing origins of the cell lines utilized in the two studies. Nevertheless, claudin-16 expression is modulated positively by the Mg^{2+} concentration present in the extracellular medium.[67]

Functionally, claudin-16 delivered to the tight junction interacts with the scaffolding protein ZO-1,[40,67] with their association and dissociation being regulated *via* PKA-mediated phosphorylation of Ser^{217} in claudin-16. Activation of the Calcium Sensing Receptor (CaSR) dephosphorylates this residue and results in dissociation of claudin-16 from ZO-1 and its accumulation in the lysosomal compartment.[40,67] Claudin-19 appears to be involved in mediating Mg^{2+} and Ca^{2+} resorption by forming a head-to-head cation-selective complex with claudin-16 at the tight junction level.[68] In this regard, claudin-19 is indispensable for recruitment of claudin-16 and to switch the channel selectivity from anions to cations.[68] The association between claudin-16 and claudin-19 is affected by point mutations in claudin-16 ($L^{145}P$, $L^{151}F$, $G^{191}R$, $A^{209}T$ and $F^{232}C$) and in claudin-19 ($L^{90}P$ and $G^{123}R$). All these mutations abolish the synergism between the two proteins and result in the development of FHHNC.

MagT1: This protein was identified based upon its up-regulation in human epithelial cells following exposure to low $[Mg^{2+}]_o$.[69] It has an estimated molecular mass of 38 kDa and presents five transmembrane domains in its premature form and four transmembrane domains in its mature form due to cleavage of the first transmembrane segment located near the C-terminus. Unlike other transporters such as SLC41 and Mrs2 (Section 4.3), MagT1 is not homologous to prokaryotic Mg^{2+} transporters but shows some similarity to the oligosaccharide transferase complex OST3/OST6 that regulates protein glycosylation in the endoplasmic reticulum in yeast.[40,69] The murine orthologue of MagT1 is highly expressed in liver, heart, kidney and colon, with detectable levels in lung, brain and spleen.[69] For the most part, MagT1 protein levels in these tissues are consistent with the mRNA levels. The only exception is the liver, in which a low protein level is detected.[69] MagT1 appears to possess high specificity for Mg^{2+} ($K_m = 0.23$ mM), and the ion current can be inhibited specifically by the Ca^{2+}-channel blocker nitrendipine at ~ 10 μM.[69] N33 is a second member of the MagT1 family and can also transport Mg^{2+} but with a much lower specificity than MagT1.[69]

Because of its high selectivity for Mg^{2+}, MagT1 is considered essential in regulating Mg^{2+} homeostasis in mammalian cells. Knock-out of MagT1, and its human homologue TUSC3, in HEK-293 cells markedly reduces cellular Mg^{2+} content.[70] On the other hand, decreasing $[Mg^{2+}]_o$ increases the mRNA levels of MagT1 but not TUSC3 while incubation with high $[Mg^{2+}]_o$ has no effect on the cellular expression of either protein.[70]

4.4.3.2 Transporters

Magnesium transport mechanisms other than channels have also been identified in eukaryotic cells. Two of these transporters were described in Section 4.2 as Na^+-dependent and -independent Mg^{2+} exchangers based upon the electrochemical requirements favouring Mg^{2+} extrusion.

What follows is a list of novel Mg^{2+} transport mechanisms of murine or human origin identified as a result of diet restriction (*i.e.* Mg^{2+}-deficient diet) or medium restriction (*i.e.* low $[Mg^{2+}]_o$).

SLC41: This family of Mg^{2+} transporters includes three members (A1, A2 and A3) that are distantly related to the prokaryotic MgtE channel (Section 4.3.2). The discussion will focus on A1 and A2 isoforms of SLC41 since the A3 isoform has not been investigated functionally.

SLC41A1 (~ 56 kDa) was the first member to be identified.[25] The hydrophobic profile of the protein predicts the presence of ten transmembrane domains, two of which have a discrete level of homology with MgtE. The *SLC41A1* gene is widely distributed among tissues, with an abundance that varies markedly (high in heart and testis, low in haematopoietic cells).[25] Under basal conditions, this gene is expressed modestly in the renal cortex but is up-regulated markedly following several days of exposure to a low Mg^{2+} diet.[71] Functional expression of murine SLC41A1 in *Xenopus laevis* oocyte indicates that this protein transports Mg^{2+} and also Fe^{2+}, Cu^{2+}, Zn^{2+} and Cd^{2+} but not Ca^{2+}.[71] Originally, it was thought to operate as a channel.[71] However, recent experimental evidence strongly suggests that it is an Mg^{2+} exporter,[72] forming high molar mass complexes within the cell membrane with masses ranging from 360 to 1236 kDa.[72] Whether this observation indicates that the SLC41A1 monomer forms large multimeric complexes and/or interacts with auxiliary proteins needs further clarification. It also remains to be determined whether the discrepancy in *modus operandi* between the highly homologous (>90%) murine[71] and human orthologue[72] depends on point mutations that dramatically alter SLC41A1 ion specificity and function. The SLC41A2 isoform also transports Mg^{2+} in addition to other divalent cations,[73] but at variance with SLC41A1, it is inhibited by Ca^{2+}.[73] Widely expressed in mammalian tissues, SLC41A2 expression is unaltered by a low Mg^{2+} diet.[73] At the structural level, SLC41A2 exhibits >70% homology with SLC41A1, and is organized in ten transmembrane domains.[73] This structural arrangement is not universally accepted.[74]

ACDP2: Mapped to the region q23–q24 of human chromosome 10, the *ACDP* gene family comprises four isoforms differentially distributed in human tissues.[75] The term ancient conserved domain protein derives from the fact that all isoforms have in common one domain that appears to be phylogenetically conserved from bacteria to man. ACDP1 is restricted to the brain. ACDP2 is expressed most highly in the brain while being absent in skeletal muscles. ACDP3 and ACDP4 are expressed ubiquitously, with the highest expression in the heart.[75] The murine distribution of ACDP isoforms is similar to that observed in humans.[75] Overexpression of ACDP2 in *Xenopus* oocytes indicates that this protein can transport a broad range of divalent cations including Mg^{2+}, Co^{2+}, Mn^{2+}, Sr^{2+}, Ba^{2+}, Cu^{2+} and Fe^{2+} while Zn^{2+} inhibits its activity.[76] Mg^{2+} transport *via* ACDP2 is voltage-dependent, displays a $K_m \sim 0.5$ mM and does not require Na^+ or Cl^- ion transport. As for SLC41A1, overexpression of the *ACDP2* gene is induced by a diet low in Mg^{2+}.[76]

NIPA: Located in the SPG6 locus of human chromosome 15q11-q13, the *NIPA1* gene is labelled "non-imprinted in Prader–Willi/Angelman syndrome", a complex developmental and multisystem disorder. Located among about 30 genes linked to this disease, *NIPA1* has been implicated in autosomal dominant hereditary spastic paraplegia (HSP). The human and mouse genome contain four members of the NIPA family, NIPA1-4, with 40% similarity overall. Homology between the individual human and mouse proteins is ∼98%. Both NIPA1 and NIPA2 operate as Mg^{2+} transporters in a saturable fashion, with different K_m values and Mg^{2+} specificity.[77,78] NIPA1 can transport other divalent cations but NIPA2 is specific for Mg^{2+}.[77] The other two isoforms transport a variety of divalent cations but not Mg^{2+}.[77] Certain point mutations in NIPA1 ($G^{100}R$ or $T^{45}R$) promote the development of autosomal dominant HSP.[40] While conserved among NIPA1 orthologues in different species, these glycine and threonine residues are absent in NIPA2-4, implying that these proteins fold differently. NIPA2 is apparently normal in HSP patients, and yet it cannot replace functionally NIPA1 to ameliorate HSP symptoms nor can NIPA3 or NIPA4 substitute for defective NIPA1. It is not known if the Prader–Willi syndrome presents with an alteration in Mg^{2+} homeostasis.

Huntingtin: Huntingtin-interacting protein 14 (HIP14) and its related protein HIP14-like (HIP14L) are both three-fold up-regulated under low $[Mg^{2+}]_o$ conditions.[79] Consisting of ~ 532 amino acids arranged in six transmembrane domains, HIP14 exhibits 69% homology to HIP14L plus a strong sequence similarity to the ankyrin repeat protein Akr1p. Also, HIP14 possesses a cytoplasmic DHHC (Asp-His-His-Cys) domain that confers palmitoyl-acyltransferase activity for palmitoylating membrane components. Mg^{2+} accumulation *via* HIP14 and HIP14L appears to be electrogenic, voltage-dependent and saturable with very similar K_m values. Inhibition of palmitoylation activity by 2-Br-palmitate, or deletion of the Asp-His-His-Cys domain decreases HIP14-mediated Mg^{2+} accumulation by $\sim 50\%$, suggesting that palmitoylation is not required for basal Mg^{2+} transport. Because of its widespread tissue distribution and intracellular localization (nuclear and perinuclear regions, Golgi complex, mitochondria, microtubules, endosomes, clathrin-coated and non-coated vesicles and plasma membrane), HIP14 is implicated in numerous cellular processes from transcriptional regulation to mitochondrial bioenergetics, scaffolding, vesicle trafficking, endocytosis and dendrite formation.[79] The primary location of the protein is in the Golgi and post-Golgi vesicles. The accumulation of Mg^{2+} elicited by this protein may be linked to its physiological role. At the molecular level, Huntington disease and its progressive neurodegeneration, cognitive deficits and choreic movements are linked to the abnormal expansion of glutamine residues from <34 to >37 at the amino acid in position 18.[79] It is presently unknown which mechanism is responsible for the expression of these defects, how poly-glutamine expansions affect Mg^{2+} transport and whether perturbation of Mg^{2+} homeostasis plays any role in the onset and uprising of the neuronal defects typical of Huntington disease.

4.4.4 Mg^{2+} Transport through Subcellular Membranes

Recent evidence indicates the presence of Mg^{2+} transporters in the biological membrane of mitochondria and Golgi cisternae.

Mrs2: The protein specifically operates in the inner mitochondrial membrane. The specifics of this protein were described in Section 4.2.

MMgTs: This gene family contains two proteins termed MMgT1 and MMgT2 (for membrane Mg^{2+} transporter 1 and 2),[80] located on chromosome XA5 (MMgT1) and 11B2 (MMgT2) in the mouse, and on Xq36 (MMgT1) and 10q23 (MMgT2) in the rat. In humans, MMgT1 is located on Xq26.3. Both MMgT1 and MMgT2 are predominantly localized to the Golgi complex and post-Golgi vesicles, where they may contribute to the regulation of Mg^{2+}-dependent enzymes involved in protein assembly and glycosylation.[80] Widely distributed in tissues, MMgT1 and MMgT2 respectively contain 131 and 123 amino acids assembled into two transmembrane domains. This structure suggests that they can form homo-oligomeric and possibly hetero-oligomeric channels to favour Mg^{2+} permeation. Mg^{2+} uptake is saturable with a $K_m \sim 1.5$ mM for MMgT1 and ~ 0.6 mM for MMgT2. These values do not vary significantly with membrane potential. Neither isoform is specific for Mg^{2+}. Expression of either transporter in the kidney is increased about three-fold following exposure to a low-Mg^{2+} diet.[80]

4.4.5 Mg^{2+} Sensing, Buffering and other Control Mechanisms

Sensing: The observation that serum Mg^{2+} levels increase under certain conditions implies that specific organs or tissues are able to *sense* these changes. However, no specific Mg^{2+}-sensing mechanism has been identified. One potential candidate is the Ca^{2+}-sensing receptor (CaSR).[81] This receptor usually detects changes in circulating Mg^{2+} levels in a range of concentrations far higher than those of Ca^{2+},[81] but consistent with the increases in serum Mg^{2+} levels reported.[40,81] Luminal Ca^{2+} and Mg^{2+} activate the CaSR with similar sensitivity in distal convoluted tubule cells of the mouse.[82] This observation has implications for whole body physiology. Activation of CaSR would inhibit Mg^{2+} reabsorption in the nephron,[82] favouring its urinary elimination. This activation would rationalize the clinical and experimental evidence that hypermagnesaemia and hypercalcaemia inhibit hormone-stimulated cAMP-mediated resorption of Mg^{2+} and Ca^{2+} along the different segments of the nephron.[82] Also, it would represent a distal regulatory mechanism for restoration of physiological Mg^{2+} levels following the increase in circulating Mg^{2+} in animals infused with adrenergic agonists.[40,81]

ATP and Mg^{2+} Homeostasis: In describing cellular Mg^{2+} distribution, ATP was indicated as one of the main complexing agents within the cytoplasm and mitochondria.[30–32] Treatment of cells with agents that decrease cellular ATP levels results in Mg^{2+} losses. Cyanide, mitochondrial uncouplers, fructose, ethanol or hypoxia are some of the agents that impact ATP levels and consequently cellular Mg^{2+} homeostasis.[40] As ATP represents the major buffering component for Mg^{2+} within the cell,[30] a decrease in ATP results in an increase in cytosolic free $[Mg^{2+}]_i$, and ultimately in an extrusion of Mg^{2+} from the cell.[40] This occurs despite the additional buffering provided by proteins or cellular organelles. In addition, cellular ATP appears to play a key role in regulating Mg^{2+} extrusion from erythrocytes and hepatocytes through the Na^+-independent Mg^{2+} extrusion process.[40] The exact role of ATP in regulating the process, however, is unclear as the extrusion does not appear to be mediated by an ATPase.

Regulation of Mg^{2+} Transport and Homeostasis: While mammalian cells retain constant basal Mg^{2+} content under resting conditions, compelling evidence supports the ability of different hormones to induce significant movement of Mg^{2+} in either direction across the cell membrane. As a result, changes in serum and total cell levels have been observed, as well as in *free* Mg^{2+} (to a lesser extent). The picture that emerges relates changes in total cellular Mg^{2+} content to the utilization of energetic metabolites (*e.g.* glucose, see below). In addition, changes within discrete portions of cellular organelles translate into modulation of activity of specific enzymes located within these compartments (see below).

cAMP-Dependent Extrusion: Catecholamine, glucagon and other hormones or agents that increase cellular cAMP level all induce Mg^{2+} extrusion from various cell types that affects the Mg^{2+} pools in the cytoplasm and cellular compartments to variable extents. The mechanism ultimately responsible for the extrusion of Mg^{2+} across the cell membrane is primarily the Na^+/Mg^{2+} exchanger, with some contribution from the Na^+-independent pathway.

This cAMP-dependent Mg^{2+} extrusion is quantitatively significant (≥ 2–8% of total cellular Mg^{2+} content), ubiquitous (see ref. 35 for a comprehensive list of cells and tissues) and fast, reaching the maximum within eight minutes after application of the stimulus irrespective of the hormone or agent utilized.[35,40] Conversely, hormones or agents that decrease cAMP production or prevent PKA activation inhibit extrusion. The time frame of Mg^{2+} extrusion suggests a mobilization from a cellular Mg^{2+} pool that can be rapidly depleted. This notion is supported by the fact that submaximal doses of agonist infused sequentially within a few minutes elicit Mg^{2+} extrusions of progressively decreasing amplitudes.[40] Under all these conditions, limited changes in cytosolic free $[Mg^{2+}]_i$ are observed.[35,40] The Mg^{2+} extrusion induced by cAMP occurs *via* phosphorylation of the Na^+/Mg^{2+} exchanger. Removal of extracellular Na^+ or addition of Na^+ transport inhibitors blocks Mg^{2+} extrusion through this transporter and results in a more sustained rise in cytosolic free $[Mg^{2+}]_i$.[40,46] Overall, these data indicate that export of cellular Mg^{2+} depends on the transmembrane movement of Na^+ while Mg^{2+} mobilization from intracellular pools does not.

cAMP-Independent Extrusion: Addition of the α_1-adrenergic agonist phenylephrine also elicits Mg^{2+} extrusion from liver cells.[40] The author has confirmed this observation and provided evidence that the stimulation of α_1- and β-adrenergic receptors is not alternative but additive, and that complementary processes elicit Mg^{2+} extrusion from liver cells and cardiac ventricular myocytes, especially when catecholamine release stimulates both classes of adrenergic receptors.[40] On the other hand, pre-treatment with insulin abolishes the Mg^{2+} extrusion mediated by β-adrenergic receptor agonists or cAMP analogues, but leaves the Mg^{2+} mobilization mediated *via* α_1-adrenergic receptors unaffected. The effect of insulin has been ascribed to inhibition of β-adrenergic receptors and stimulation of the cytosolic phosphodiesterase that degrades cAMP.[40] These effects are distinct from the direct modulatory effect of insulin on the Na^+/Mg^{2+} exchanger.

The phenylephrine-induced Mg^{2+} extrusion appears to depend on the activation of capacitative Ca^{2+} entry.[40] Inhibition of IP_3-induced Ca^{2+} release from the ER, chelation of cytosolic Ca^{2+} or inhibition of Ca^{2+} entry at the plasma membrane all prevent phenylephrine from inducing Mg^{2+} extrusion from the hepatocyte. It is unclear, however, whether the extruded Mg^{2+} is mobilized from the ER or displaced from cytosolic binding sites following Ca^{2+} entry across the hepatocytes' cell membrane. It is noteworthy that extracellular Na^+ and Ca^{2+} are both required for the phenylephrine-induced Mg^{2+} extrusion to occur.[40] The amplitude of Mg^{2+} extrusion decreases by 15–20% in the absence of extracellular Ca^{2+}: the remaining 80–85% depends on extracellular Na^+. It is possible that extracellular Na^+ is required to maintain the membrane potential and to facilitate Ca^{2+} entry across the hepatocellular membrane. In the absence of receptor activation, administration of thapsigargin mimics phenylephrine stimulation and elicits Mg^{2+} extrusion from the hepatocyte even in the absence of extracellular Ca^{2+}.[40] Hence, it appears that an optimal level of cytosolic Ca^{2+} has to be attained for Mg^{2+} extrusion to occur, possibly *via* displacement from cellular binding sites or *via* a Ca^{2+}-calmodulin-activated mechanism.

Mg^{2+} Homeostasis and Glucose: The redundancy in Mg^{2+} extrusion mechanisms and modalities of activation raises the following question: what is the physiological significance of Mg^{2+} mobilization in mammalian cells? In the case of cardiac myocytes, an increase in extracellular Mg^{2+} level appears to modulate the probability of opening L-type Ca^{2+}-channels and elicit a temporary decrease in the action potential of the sinus-atrial (SA) pacemaker cells.[83] In the case of liver cells, Mg^{2+} transport appears to be associated with glucose transport and utilization.[84] Catecholamine, glucagon and phenylephrine all elicit Mg^{2+} extrusion together with glucose output (through glycogenolysis) from liver cells.[40,84] Inhibition of Mg^{2+} extrusion by amiloride or imipramine blocks hepatic glucose output. The converse is also true: inhibition of glucose transport by phloretin inhibits Mg^{2+} extrusion from liver cells.[84] The presence of a functional "link" between glucose and Mg^{2+} homeostasis is supported by the observation that overnight starvation depletes hepatic glycogen stores, and markedly decreases total Mg^{2+} content (by 15%) as a consequence of the endogenous activation of the pro-glycemic hormones (catecholamine and glucagon).[85] This decrease in hepatic Mg^{2+} content is quantitatively equivalent to the Mg^{2+} loss elicited in fed livers perfused with the same hormones,[85] or that observed in livers of type-I diabetic rats.[40] This functional link between glucose and Mg^{2+} homeostasis is also observed following insulin-induced glucose accumulation in cardiac ventricular myocytes or pancreatic beta cells.[40] In both experimental models, the amount of Mg^{2+} accumulated within the cells is directly proportional to the amplitude of glucose accumulation. Conversely, decreasing extracellular Mg^{2+} concentration directly reduces the amount of glucose accumulated within the cells.[84]

The decrease in cellular Mg^{2+} content observed in diabetic animals provides an additional albeit indirect proof of the relationship between glucose and Mg^{2+}. Observed originally in erythrocytes, the decrease also occurs in various other tissues including muscles, liver and cardiac myocytes.[40,86] Interestingly, Mg^{2+} extrusion *via* β-adrenergic receptor stimulation remains operative and is actually up-regulated in hepatocytes from diabetic animals while it is markedly inhibited in cardiac myocytes from the same animals.[40] This might reflect a differential operation and modulation of β_2-adrenergic receptors in liver cells *vs.* β_1-adrenergic receptors in cardiac cells. Both cell models exhibit a defective Mg^{2+} entry that can be observed in liver plasma membrane vesicles.[40,87] The decrease in total cellular Mg^{2+} content is associated with a decrease in protein and ATP content. Administration of exogenous insulin for at least two weeks restores these two parameters as well as Mg^{2+} homeostasis and extrusion.[40] Insulin appears to modulate Mg^{2+} homeostasis by controlling: 1) glucose homeostasis and accumulation, 2) the release of pro-glycemic hormones like glucagon and 3) the Na^+/Mg^{2+} exchanger directly.[40] Conversely, Mg^{2+} deficiency prevents the insulin receptor from properly phosphorylating the insulin receptor substrate (IRS) and from propagating insulin signalling within the cell but does not affect autophosphorylation of the insulin receptor.[88] This result has been primarily observed in skeletal muscle cells, and it could contribute to explaining the decrease in glucose accumulation observed in skeletal muscles under diabetic conditions. In turn, a decrease in insulin-mediated signalling would further affect Mg^{2+} homeostasis within tissues and explain why Mg^{2+} deficiency is commonly associated with both type I and type II diabetes in human patients and in experimental models.[86]

Mg^{2+} Accumulation: The identification of several channels for Mg^{2+} influx strongly supports the hypothesis that cellular Mg^{2+} is dynamically maintained through both entry and exit pathways with differential regulation by hormones and metabolic conditions. In the case of Mg^{2+} extrusion, a good understanding of the activating signalling has been achieved but there is a lack of specific structural information. In contrast, structural information for proteins involved in Mg^{2+} entry is available but for the most part detailed information about the activating signalling is lacking.

Role of Protein Kinase C: Eukaryotic cells accumulate 1–2 mM Mg^{2+} following stimulation by hormones like carbachol, vasopressin, angiotensin-II or insulin (see ref. 35 for a list of responsive cells). In addition to inhibiting cAMP production, several of these hormones activate protein kinase C (PKC) as part of their cellular signalling. A role for PKC in mediating Mg^{2+} accumulation has been observed in S49 lymphoma cells, thymocytes, cardiac myocytes and hepatocytes.[35,40] Consistent with this observation, down-regulation of PKC completely abolishes the ability of cardiac and liver cells to accumulate Mg^{2+} while leaving the response of these cells to adrenergic agonists unaffected.[89] Mg^{2+} accumulation is also blocked by administration of the PKC inhibitors calphostin or staurosporine.[35,40] Blunted responses in accumulating Mg^{2+} due to alteration in PKC distribution and activity have been observed in arterial smooth muscle cells and hepatocytes of animals exposed to alcohol, or in liver cells of diabetic animals.[35,40]

Activation of PKC is associated with Ca^{2+} transients, and it is part of the signalling of hormones like angiotensin-II or vasopressin. Yet, the effect of Ca^{2+} on Mg^{2+} accumulation is poorly defined. Liver cells loaded with ligands that chelate cytosolic Ca^{2+} are unable to extrude and accumulate Mg^{2+} following stimulation by α_1-adrenergic agonists.[90] When thapsigargin is used to increase cytosolic Ca^{2+}, Mg^{2+} accumulation is also prevented and, in fact, Mg^{2+} extrusion from the hepatocyte is observed when thapsigargin is applied for more than 3–5 minutes.[90] The different time-scale and amplitude of changes in cellular Ca^{2+} and Mg^{2+} content make it difficult to properly correlate the changes in Ca^{2+} and Mg^{2+}.[90] Cytosolic free Ca^{2+} transiently increases by one order of magnitude over its basal level while basal cytosolic free Mg^{2+} is already three orders of magnitude higher than Ca^{2+} (0.5–0.7 mM Mg^{2+} *versus* ~ 0.1 μM Ca^{2+}) but increases by ~ 10–15% at the most.[40] While these changes are small in percent terms when compared to those of Ca^{2+}, nevertheless they result in detectable variations in Mg^{2+} levels within organelles, especially mitochondria, with major repercussions for cellular bioenergetics. The physiological significance of these changes for the cell and the organism in coordinating a response to the hormones is discussed in Section 4.7.

Activation of α_1-adrenoceptors by phenylephrine also activates PKC and IP_3/Ca^{2+} signalling. Yet, this agent induces Mg^{2+} extrusion rather than Mg^{2+} accumulation in liver cells.[40] The reason for this inconsistency is unclear. One possibility is that different stimuli activate different PKC isoforms. For example, hepatocytes possess three classical and at least two novel PKC isoforms. Thus, it is possible that one isoform (or class of isoforms) is involved in mediating Mg^{2+} accumulation whereas another is involved in modulating Mg^{2+} extrusion. Consistent with this hypothesis, PKCε appears to be essential for Mg^{2+} accumulation.[40] Preventing the expression or the translocation of this isoform to the cell membrane (*e.g.* by ethanol administration) abolishes Mg^{2+} accumulation in liver cells. Interestingly, this PKC isoform has the highest affinity for Mg^{2+} among all PKC isoenzymes ($K_m \sim 1$ mM),[91] close to the physiological *free* $[Mg^{2+}]_i$ measured under basal conditions in the cytoplasm of eukaryotic cells (including the hepatocyte).[30–32,40] The amount of Mg^{2+} accumulated in the hepatocyte following hormonal stimuli is 1–2 mM total Mg^{2+} with a ~ 15% increase in cytosolic free $[Mg^{2+}]$.[30–32,40] Such an increase would be meaningful only by invoking Mg^{2+} compartmentalization (undetected by the fluorescent indicators for Mg^{2+} currently available) or by postulating weak Mg^{2+} buffering within the cytoplasm.

Role of MAPKs: Data in the literature indicate that MAPK signalling is involved in mediating Mg^{2+} accumulation.[35,40] Similar results are observed in hepatocytes in which inhibition of ERK1/2 and p38 MAPKs abolishes PKC-mediated Mg^{2+} accumulation.[40] Inhibition of MAPKs also hampers cyclin activity and cell cycle progression. This effect may occur *via* changes in nuclear functions directly regulated by Mg^{2+} or by ERK2, as the latter depends on Mg^{2+} to properly dimerize, translocate and activate specific nuclear targets.[40] Increased phosphorylation of ERK1/2 together with increased TRPM6 expression has been observed upon EGF administration to renal epithelial cells.[40] The role of MAPKs, however, needs further elucidation as these kinases are also involved in Mg^{2+} extrusion.

Role of EGF: Direct and indirect evidence indicates that EGF controls proper TRPM6 operation in the apical domain of renal epithelial cells to promote Mg^{2+} accumulation.[40] Point mutations in the EGF sequence affect

TRPM6 function and Mg^{2+} accumulation within the cells. EGF modulates TRPM6 *via* ERK1/2 signalling and activator protein-1 (AP-1).

4.4.6 The Physiological Role of Intracellular Mg^{2+}

Magnesium has an indispensable role in regulating various enzymes, phosphometabolites and channel activities.[30,92] Several glycolytic enzymes, including hexokinase, phosphofructokinase, phosphoglycerate mutase, phosphoglycerate kinase, enolase and pyruvate kinase, are activated at low and inhibited at high Mg^{2+} concentrations.[93,94] Adenylyl cyclase is another enzyme directly regulated by Mg^{2+}.[95] The regulation of all these enzymes occurs at Mg^{2+} concentrations between 0.5 and 1 mM, which are well within the fluctuations in free $[Mg^{2+}]_i$ measured in the cytoplasm of various cells including hepatocytes.[40] With the exception of glycolytic enzymes, studies attempting to demonstrate a regulatory role of Mg^{2+} for cytosolic enzymes have been inconclusive because Mg^{2+} was used in high millimolar concentrations that are physiologically not significant. Under these conditions, an increase or decrease in cytosolic Mg^{2+} levels such as observed for $[Ca^{2+}]_i$ will go largely undetected by commonly employed fluorescence or ^{31}P-NMR techniques. Under conditions in which hormones elicit major fluxes of Mg^{2+} across the cell plasma membrane in either direction to change total cellular Mg^{2+} content by 1 to 2 mM (or 5–10% of total cellular concentration), changes in cytosolic free $[Mg^{2+}]_i$ of 100–150 μM have been observed.[35,40] Although small, these changes persist well beyond the time frame of Ca^{2+} oscillation.[35,40] Hence, Mg^{2+} should not be considered a transient regulator of enzymes like Ca^{2+} but rather like a long-term regulator of enzymes within the cytoplasm and organelles. Due to the specific compartmentalization of Mg^{2+} within the cell, it is reasonable to assume that the source or destination of transported Mg^{2+} is ultimately a cellular compartment. Consequently, Mg^{2+} will regulate cellular functions mainly within organelles. This situation would also apply to the plasma where Mg^{2+} concentration can increase or decrease more than 20%.[35,40]

The following discussion will highlight the known regulatory effects of extra- or intracellular Mg^{2+} on cation channels in the plasma membrane, on volume and respiration rate in mitochondria, and in specific functions of the endoplasmic reticulum.[40]

Ca^{2+}- and K^+-Channels: Increasing intracellular free $[Mg^{2+}]_i$ from 0.3 to 3.0 mM by internal perfusion decreased the amplitude of I_{Ca} elevated by cAMP-dependent phosphorylation by more than 50% while having a small effect on the basal L-type Ca^{2+}-channel current. Mg^{2+} appears to affect directly the phosphorylated channel or the channel dephosphorylation rate. The block induced by Mg^{2+} on the Ca^{2+} current depends on a direct effect on the inactivation state of the channel as the block persisted in the absence and in the presence of cAMP, and was not reversed by elevating extracellular Ca^{2+} concentration or adding catecholamine. An effect of Mg^{2+} on Ca^{2+}-channels has been reported in vascular smooth muscle cells and

endothelial cells from human placenta in which Mg^{2+} regulates Ca^{2+} influx through voltage-gated Ca^{2+} channels by acting on the channels at an extracellular site. Extracellular Mg^{2+} exerts a similar block on T-type Ca^{2+}-channels. The modulatory effect of Mg^{2+} appears to take place at the EF-hand motif of the COOH-terminus of $Ca_v1.2$.

Extracellular Mg^{2+} also modulates the activity of store-operated Ca^{2+} channels, mimicking nifedipine in preventing or reversing the vaso-constriction elicited store-operated Ca^{2+} entry. These results may explain some of the modifications in vascular myotone observed under hypertensive conditions, in which a decrease in plasma Mg^{2+} has often been reported.

Intracellular Mg^{2+} also affects the operation of store-operated calcium re-lease-activated Ca^{2+} (CRAC) channels.[40] These channels are highly Ca^{2+}-se-lective under physiological conditions, but removal of extracellular divalent cations makes them freely permeable to monovalent cations, in particular Na^+. Intracellular Mg^{2+} also modulates the activity and selectivity of these channels, thus affecting monovalent cation permeability. However, the chan-nels modulated by intracellular Mg^{2+} may not be CRAC channels, but a dif-ferent class of channels (Mg^{2+}-inhibited cation or MIC) that open when Mg^{2+} is washed out of the cytosol. These channels possess distinctive functional parameters in terms of inhibition, regulation, ion permeation and selectivity.

Potassium channels are also targets of Mg^{2+}. The presence of Mg^{2+} on the cytoplasmic side of the inwardly rectifying K^+ channel blocks the outward current without affecting the inward current.[40] Mg^{2+} blocks with a binding constant of 1.7 µM. When the Mg^{2+} concentration increases to 2–10 µM, the outward current fluctuates between two intermediate values distinct from those of the open and closed states. Because these inhibitory concentrations are far from the physiological range of Mg^{2+} concentrations, it is difficult to envision such a regulatory effect under normal conditions without invoking Mg^{2+} compartmentation. K_v channels in vascular smooth muscle cells are also regulated by intracellular Mg^{2+}. In this case, Mg^{2+} slows down the kinetics of activation of the channel, causing inward rectification at positive membrane potentials, and shifting the voltage-dependent inactivation. These effects occur in a range of intracellular Mg^{2+} concentrations that are consistent with the physiological variations detected in many cells. Overall, this represents a novel mechanism for the regulation of K_v channels in the vasculature.

Intracellular Mg^{2+} also modulates large-conductance (BK-type) Ca^{2+}-dependent K^+ channels either by blocking the pore of BK channels in a voltage-dependent manner, or by activating the channels independently of changes in Ca^{2+} and voltage, binding to the channel's open conformation at a site different from Ca^{2+}-binding sites. Interestingly, Mg^{2+} may also bind to Ca^{2+} sites and competitively inhibit Ca^{2+}-dependent activation.

The inhibitory effect of Mg^{2+} is not restricted to channels in the plasma membrane, as Mg^{2+} in the mitochondrial matrix modulates gating and conductance of mitochondrial K_{ATP} channels that play a key role in ischemia/reperfusion.[40]

Mitochondrial Dehydrogenases: Mitochondria represent one of the major cellular pools for Mg^{2+} with a total concentration of 14–16 mM.[1-3,96] Circumstantial evidence suggests that Mg^{2+} can be mobilized from mitochondria under various conditions including hormone-mediated increase in the cytosolic cAMP level. The underlying mechanism has not been fully elucidated but it appears to involve the adenine nucleotide translocase.[40] Several reviews have analysed in detail how Mg^{2+} homeostasis is regulated in the organelle. This section focuses on the evidence for a role for intra- and extramitochondrial Mg^{2+} on the activity of specific dehydrogenases and proteins.

Catalysis of mitochondrial dehydrogenases and the respiration rate are regulated by changes in matrix Ca^{2+}.[97] A similar role has been suggested for Mg^{2+} since the activity of several mitochondrial dehydrogenases increases within minutes from the application of hormonal or metabolic stimuli in the absence of a detectable increase in mitochondrial Ca^{2+}.[40] The role of matrix Mg^{2+} in regulating mitochondrial dehydrogenases and respiration has been investigated by measuring the activity of several dehydrogenases in mitochondria under conditions in which matrix Ca^{2+} or Mg^{2+} concentration were varied. From such data, it appears that Mg^{2+} removal within mitochondria increases the activities of succinate and glutamate dehydrogenases by several-fold while having no effect on the α-ketoglutarate and pyruvate dehydrogenases.[40] This evidence indicates that changes in matrix Mg^{2+} content modulate mitochondrial respiration. In this respect, mitochondrial Mg^{2+} content appears to change significantly during transition from respiration state 3 to 4, with dioxygen consumption being directly proportional to Mg^{2+} content.

The regulatory effect of Mg^{2+} on mitochondrial function also affects the anion channel present in the mitochondrial membrane and the opening of the permeability transition pore (PTP).[40]

The mitochondrial inner membrane anion channel (IMAC) transports various anions, and is involved in regulating the organelle volume in conjunction with the K^+/H^+ antiporter. Although its fine regulation is not completely elucidated, experimental evidence suggests that matrix Mg^{2+} and protons inhibit IMAC, maintaining the channel in its closed state.[40] Kinetic studies indicate that the main role of Mg^{2+} is to maintain the channel in a condition that would allow its fine modulation by small changes in pH and proton distribution under physiological conditions.[40]

The PTP is a proteinaceous pore that opens in the inner mitochondrial membrane following a decrease in mitochondrial $\Delta\psi$, resulting in the rapid redistribution and equilibration of matrix and extramitochondrial solutes down their concentration gradient. While an increase in mitochondrial Ca^{2+} content facilitates PTP opening, an increase in mitochondrial Mg^{2+} antagonizes it. According to one proposed model, creatine kinase can regulate PTP opening by tightly associating to the mitochondrial membrane and remaining in an active state.[40] Both processes are Mg^{2+}-dependent, and Mg^{2+} removal from the extramitochondrial environment decreases creatine kinase activity and favours the PTP opening.

Considering the effect of Mg^{2+} on mitochondrial functions and channels, it appears that Mg^{2+} plays more than one role within this organelle by regulating: 1) mitochondrial volume, 2) ion composition, 3) ATP production and 4) metabolic interaction with the hosting cell.

Endoplasmic Reticulum: The ER is one of the major cellular Mg^{2+} pools with a total concentration range of 14–18 mM.[30-32,40] Limited information, however, is available about the role luminal Mg^{2+} plays on reticular functions in addition to regulating protein synthesis. Whether luminal Mg^{2+} is buffered or chelated by ER proteins in a manner similar to luminal Ca^{2+} is also undefined, as are the transport mechanisms involved in maintaining such a large Mg^{2+} concentration within the organelle.

Work from different laboratories suggests a major role for cytosolic and luminal Mg^{2+} concentrations in limiting Ca^{2+} uptake into the ER and its release from the organelle via IP_3 and the ryanodine receptor.[40] While a direct effect of Mg^{2+} on the ryanodine receptor has been reported, whether Mg^{2+} has a similar effect on the IP_3 receptor is less clear. Recently, it has been shown that cytosolic Mg^{2+} can regulate the activity of reticular glucose 6-phosphatase (G6Pase) in liver cells.[98] Cytoplasmic glucose 6-phosphate (G6P) is specifically transported into the lumen of the ER. The G6P transport component of the G6Pase is the limiting step of the process, and it is at this level that cytosolic Mg^{2+} plays its regulatory effect in a dynamic and biphasic manner. The optimal stimulatory effect requires ~ 0.5 mM $[Mg^{2+}]_i$ with an inhibitory setting at higher Mg^{2+} concentrations.[98]

Cell pH and Volume: Cellular acidification, as it occurs following exposure to cyanide, fructose, hypoxia or ethanol, decreases cellular ATP content and favours a major Mg^{2+} extrusion from the cell as a result of the decrease in Mg^{2+} buffering capacity.[40] At the same time, changes in intracellular Mg^{2+} modulate the electrogenic Na^+-HCO_3^- co-transporter NBCe1-B. This effect is exerted by Mg^{2+} directly, and requires a functional N-terminus on the NBCe1-B transporter. It is unclear whether Mg^{2+} binds to the N-terminus of the transporter, or affects it indirectly *via* Mg^{2+}-modulated regulatory protein(s). A cytosolic Mg^{2+} concentration of ~ 1 mM (a physiological level measured within various cells[92]) inhibits the NBCe1-B current by $\sim 50\%$, while no detectable current can be measured when the free Mg^{2+} concentration is raised to ≥ 3 mM.

Increasing cellular Mg^{2+} content also stimulates the expression of aquaporin 3 in CaCo-3 cells.[40] This isoform of aquaporin is highly expressed in the gastro-intestinal tract, where it absorbs water, glycerol and urea. The effect of Mg^{2+} on aquaporin mRNA expression involves cAMP/PKA/CREB signalling, as well as MEK1/2 and MSK1,[40] suggesting short- and long-term regulation of the protein's activity and expression. Aquaporin 3 is also expressed in brain, erythrocytes, kidney and skin. Hence, a modulatory role of Mg^{2+} on protein expression in these tissues may be relevant for various physiological and/or pathological conditions. No information is available as to whether Mg^{2+} also regulates the expression and activity of other aquaporin isoforms. Taken together, these two sets of information suggest a major regulatory role for Mg^{2+} on cellular pH, volume and cation concentration, especially Na^+.

Cell Cycle: Cell cycle, cell proliferation and cell differentiation are all associated with the maintenance of an optimal cellular Mg^{2+} level.[92–99] Under conditions of restricted cellular Mg^{2+} accessibility, cell proliferation and cell cycle progression become impaired. Accordingly, a decrease in extracellular Mg^{2+} content also affects cell differentiation.[99] A decrease in cellular Mg^{2+} affects these cellular processes through defective MAPKs and p27 signalling, increased oxidative stress and decreased MgATP levels.[40] Because the cellular MgATP level is optimal for protein synthesis, any alteration will impact the proper functioning of the cell. In addition, extracellular Mg^{2+} levels regulate integrin signalling, modulating the interaction among cells, and between cells and the extracellular matrix.[40] In agreement with the proposed long-term regulatory function of Mg^{2+},[92] all these observations underscore the important role of Mg^{2+} on cell cycle progression and proper morphology while avoiding the undesired progression towards cell death or neoplasm.[100]

4.5 Conclusions

Although still incomplete and limited as compared to the knowledge on hand for other ions such as Ca^{2+}, H^+, K^+ or Na^+, our understanding of cellular and whole body Mg^{2+} homeostasis has advanced at a momentous pace in the last two decades. The picture that emerges is that of a cation that has a small inward-oriented gradient across all biological membranes. It is tightly regulated in its transport across these membranes through a combination of channels for entry and exchangers for extrusion, all regulated by a variety of signalling molecules. In addition, while most of the Mg^{2+} transported in and out of the cell is rapidly buffered in the cytoplasm, the extracellular space and the serum experience changes in the order of 15–20% in Mg^{2+} levels. Although these changes are smaller compared to those reported for Ca^{2+}, they are still effective in modulating channels or enzymes within the cytoplasm and the cellular organelles. This distribution and regulation support the notion put forward more than 25 years ago that Mg^{2+} should be considered a long-term regulator of cellular function as opposed to the short-term regulation exerted by Ca^{2+}.[92]

As the knowledge of Mg^{2+} channels and transport mechanisms in biological membranes and the signalling pathways behind them increases, we anticipate that new tools will become available to address important questions about the physiological role Mg^{2+} plays inside the cell and in the whole body under both physiological and pathological conditions.

References

1. M. E. Maguire and J. A. Cowan, *Biometals*, 2002, **15**, 203–210.
2. J. A. Cowan, *Biometals*, 2002, **15**, 225–235.

3. J. A. Cowan, *The Biological Chemistry of Magnesium*, ed. J. A. Cowan, VCH, New York, 1995.

4. O. Shaul, *Biometals*, 2002, **15**, 309–323.

5. R. A. Leigh and R. G. Wyn Jones, in *Advances in Plant Nutrition* 2 ed. B. Thinker and A. Lauchli, Praeger Scientific, New York, 1986. pp. 249–279.

6. H. Marschner, *Mineral Nutrition of Higher Plants*, Academic Press, London, San Diego, 1995.

7. Z. Amalou, R. Gibrat, P. Trouslot and J. d'Auzac, *Plant Physiol.*, 1992, **100**, 255–260.

8. O. Shaul, D. W. Hilgemann, J. Almeida-Engler, M. Van Montagu, D. Inzé and G. Galili, *EMBO J.*, 1999, **18**, 3973–3980.

9. I. Schock, J. Gregan, S. Steinhauser, R. Schweyen, A. Brennicke and V. Knoop, *Plant J.*, 2000, **24**, 489–501.

10. D. M. Bui, J. Gregan, E. Jarosch, A Ragnini and R. J. Schweyen, *J. Biol. Chem.*, 1999, **274**, 20438–20443.

11. R. L. Smith and M. E. Maguire, *Mol. Microbiol.*, 1998, **28**, 217–226.

12. A. U. Igamberdiev and L. A. Kleczkowski, *Biochem J.*, 2011, **437**, 373–379.

13. R. L. Smith and M. E. Maguire, in *The Biological Chemistry of Magnesium*, ed. J. A. Cowan, VCH Publishing, London, 1995, pp. 211–234.

14. D. G. Kehres, C. H. Lawyer and M. E. Maguire, *Microb. Compar. Genomics*, 1998, **43**, 151–169.

15. S. Eshaghi, D. Niegowski, A. Kohl, M. D. Martinez, S. A. Lesley and P. Nordlund, *Science*, 2006, **313**, 354–357.

16. W. Lunin, E. Dobrovetsky, G. Khutoreskaya, R. Zhang, A. Joachimiak, D. A. Doyle, A. Bochkarev, M. E. Maguire, A. M. Edwards and C. M. Koth, *Nature*, 2006, **440**, 833–837.

17. J. Payandeh and E. F. Pai, *EMBO J.*, 2006, **25**, 3762–3773.

18. A. S. Moomaw and M. E. Maguire, *Physiology*, 2008, **23**, 275–285.

19. A. M. Caldwell and R. L. Smith, *J. Bacteriol.*, 2003, **185**, 374–376.

20. V. Knoop, M. Gorth-Malonek, M. Gebert, K. Eifler and K. Weyand, *Mol. Genet. Genomics*, 2005, **274**, 205–226.

21. C. W. MacDiarmid and R. C. Gardner, *J. Biol. Chem.*, 1998, **273**, 1727–1732.

22. M. Wachek, M. C. Aichinger, J. A. Stadler, R. J. Schweyen and A. Graschopf, *FEBS J.*, 2006, **273**, 4236–4249.

23. M. Hattori, Y. Tanaka, S. Fukai, R. Ishitani and O. Nureki, *Nature*, 2007, **448**, 1072–1075.

24. S. Ignoul and J. Eggermont, *Am. J. Phyisol.*, 2005, **289**, C1369–1378.

25. T. Wabakken, E. Rian, M. Kveine and H. C. Aashein, *Biochem. Biophys. Res. Commun.*, 2003, **306**, 718724.

26. A. B. Blanc-Potard and E. A. Groisman, *EMBO J.*, 1997, **16**, 5376–5385.

27. D. L. Smith, T. Tao and M. E. Maguire, *J. Biol. Chem.*, 1993, **268**, 22469–22479.

28. E. A. Groisman, E. Chiao, C. J. Lipps and F. Heffron, *Proc. Natl Acad. Sci. USA*, 1989, **86**, 7077–7081.

29. M. B. C. Moncrief and M. E. Maguire, *Infect. Immun.*, 1998, **66**, 3802–3809.
30. A. Romani and A. Scarpa, *Arch. Biochem. Biophys.*, 1992, **298**, 1–12.
31. F. I. Wolf, A. Torsello, S. Fasanella and A. Cittadini, *Mol. Asp. Med.*, 2003, **24**, 11–26.
32. F. I. Wolf and A Cittadini, *Mol. Asp. Med.*, 2003, **24**, 3–9.
33. A. Scarpa and F. J. Brinley, *Fed. Proc.*, 1981, **40**, 2646–2652.
34. M. E. Shils, in *Modern Nutrition in Health and Disease*, ed. M. E. Shils, J. A. Olson, M. Shike and A. C. Ross, Lippincott, Williams, and Wilkins, 9th edn, 1999, pp. 169–192.
35. A. Romani and A. Scarpa, *Front. Biosci.*, 2000, **5**, D720–D734.
36. K. P. Schlingmann, S. Weber, M. Peters, N. L. Niemann, H. Vitzthum, K. Klingel, M. Kratz, E. Haddad, E. Ristoff, D. Dinour, M. Syrrou, S. Nielsen, M. Sassen, S. Waldegger, H. W. Seyberth and M. Konrad, *Nat. Genet.*, 2002, **31**, 166–170.
37. M. J. Nadler, M. C. Hermosura, K. Inabe, A. L. Perraud, Q. Zhu, A. J. Stokes, T. Kurosaki, J. P. Kinet, R. Penner, A. M. Scharenberg and A. Fleig, *Nature*, 2001, **411**, 590–595.
38. B. Nilius and T. Voets, *Pflugers Arch. Eur. J. Physiol.*, 2005, **451**, 1–10.
39. B. J. Kim, K. J. Park, H. W. Kim, S. Choio, J. Y. Jun, I. Y. Chang, J.-H. Jeon, I. So and S. J. Kim, *World J. Gastroenterol.*, 2009, **14**, 5799–5804.
40. A. M. Romani, *Met. Ions Life Sci.*, 2013, **12**, 69–118.
41. G. Cao, S. Thébault, J. van der Wijst, A. van der Kemp, E. Lasonder, R. J. Bindels and J. G. Hoenderop, *Curr. Biol.*, 2008, **18**, 168–176.
42. K. Bogucka and L. Wojtczak, *Biochem. Biophys. Res. Commun.*, 1971, **44**, 1330–1337.
43. H. Belge, P. Gailly, B. Schwaller, J. Loffing, H. Debaix, E. Riveira-Munoz, R. Beauwens, J. P. Devogelaer, J. G. Hoenderop, R. J. Bindels and O. Devuyst, *Proc. Natl Acad. Sci. USA*, 2007, **104**, 14849–14854.
44. T. Gunther, J. Vormann and R. Forster, *Biochem. Biophys. Res. Commun.*, 1984, **119**, 124–131.
45. T. Gunther and J. Vormann, *Biochem. Biophys. Res. Commun.*, 1985, **130**, 540–545.
46. T. Gunther, *Magnes. Bull.*, 1996, **18**, 2–6.
47. M. Schweigel, J. Vormann and H. Martens, *Am. J. Physiol.*, 2000, **278**, G400–G408.
48. A. Goytain and G. A. Quamme, *Physiol. Genom.*, 2005, **21**, 337–342.
49. M. Koliske, A. Nestler, J. Vorman and M. Schweigel-Rontgen, *Am. J. Physiol.*, 2012, **302**, 318–326.
50. T. Günther, *Miner. Electrolyte Metab.*, 1993, **19**, 259–265.
51. H. Ebel, M. Hollstein and T. Gunther, *Biochim. Biophys. Acta*, 2002, **1559**, 135–144.
52. Geigy Scientific Tables, ed. C. Lentner, Ciba-Geigy, Basel, Switzerland, 1984.
53. T. B. Drueke and B. Lacour, in *Comprehensive Clinical Nephrology*, ed. J. Feehally, J. Fleoge and R. J. Johnson, Mosby, Philadelphia, PA, 3rd edn, 2007, pp. 136–138.

54. P. W. Flatman, *J. Membr. Biol.*, 1984, **80**, 1–14.
55. A. Romani and M. E. Maguire, *Biometals*, 2002, **15**, 271–283.
56. D. J. Dipette, K. Simpson and J. Guntupalli, *Magnesium*, 1987, **6**, 136–149.
57. F. C. Howarth, J. Waring, B. I. Hustler and J. Singh, *Magnes. Res.*, 1994, **7**, 187–197.
58. A. Romani, *Brit. Med. J., Point of Care*, www.pointofcare.bmj.com, 2011.
59. M. B. Moncrief and M. E. Maguire, *J. Biol. Inorg. Chem.*, 1999, **4**, 523–527.
60. R. R. Preston, *Science*, 1990, **250**, 285–288.
61. J. Jiang, M. Li and L. Yue, *J. Gen. Physiol.*, 2005, **126**, 137–150.
62. R. M. Touyz, Y. He, A. C. I. Montezano, G. Yao, V. Chubanov, T. Gudermann and G. E. Callera, *Am. J. Physiol.*, 2006, **290**, R73–R78.
63. J. Sahni and A. M. Scharenberg, *Cell Metab.*, 2008, **8**, 84–93.
64. T. M. Paravicini, A. Yogi, A. Mazur and R. M. Touyz, *Hypertension*, 2009, **53**, 423–429.
65. D. B. Simon, Y. Lu, K. A. Choate, H. Velazquez, E. Al-Sabban, M. Praga, G. Casari, A. Bettinelli, G. Colussi, J. Rodrigues-Soriano, D. McCredie, D. Milford, S. Sanjad and R. P. Lifton, *Science*, 1999, **285**, 103–106.
66. P. J. Kausalya, S. Amasheh, D. Gunzel, H. Wurps, D. Muller, M. Fromm and W. Hunziker, *J. Clin. Invest.*, 2006, **116**, 878–891.
67. E. Efrati, J. Arsentiev-Rozenfeld and I. Zelikovic, *Am. J. Physiol.*, 2005, **288**, F272–F283.
68. J. Hou, A. Renigunta, A. S. Gomes, M. Hou, D. L. Paul, S. Waldegger and D. A. Goodenough, *Proc. Natl Acad. Sci. USA*, 2009, **106**, 15350–15355.
69. A. Goytain and G. A. Quamme, *BMC Genomics*, 2005, **6**, 48.
70. H. Zhou and D. E. Clapham, *Proc. Natl Acad. Sci. USA*, 2009, **106**, 15750–15755.
71. A. Goytain and G. A. Quamme, *Physiol. Genomics*, 2005, **21**, 337–342.
72. M. Kolisek, P. Launay, A. Beck, G. Sponder, N. Serafini, M. Brenkus, E. M. Froschauer, H. Martens, A. Fleig and M. Schweigel, *J. Biol. Chem.*, 2008, **283**, 16235–16247.
73. A. Goytain and G. A. Quamme, *Biochem. Biophys. Res. Commun.*, 2005, **330**, 701–705.
74. J. Sahni, B. Nelson and A. M. Scharenberg, *Biochem. J.*, 2007, **401**, 505–513.
75. C. Y. Wang, J. D. Shi, P. Yang, P. G. Kumar, Q. Z. Li, Q. G. Run, Y. C. Su, H. S. Scott, K. J. Kao and J. X. She, *Gene*, 2003, **306**, 37–44.
76. A. Goytain and G. A. Quamme, *Physiol. Genomics*, 2005, **22**, 382–389.
77. A. Goytain, R. M. Hines, A. El-Husseini and G. A. Quamme, *J. Biol. Chem.*, 2007, **282**, 8060–8068.
78. A. Goytain, R. M. Hines and G. A. Quamme, *Am. J. Physiol.*, 2008, **295**, C944–C953.
79. A. Goytain, R. M. Hines and G. A. Quamme, *J. Biol. Chem.*, 2008, **283**, 33365–33374.

80. A. Goytain and G. A. Quamme, *Am. J. Physiol.*, 2008, **294**, C495–502.
81. E. M. Brown, G. Gamba, D. Riccardi, M. Lombardi, R. Butters, O. Kifor, A. Sun, M. A. Hediger, J. Lytton and S. C. Herbert, *Nature*, 1993, **366**, 575–580.
82. B. W. Bapty, L. J. Dai, G. Ritchie, F. Jirik, L. Canaff, G. N. Hendy and G. A. Quamme, *Kidney Intern.*, 1998, **53**, 583–592.
83. F. C. Howarth, J. Waring, B. I. Hustler and J. Singh, *Magnes. Res.*, 1994, **7**, 187–197.
84. T. E. Fagan and A. Romani, *Am. J. Physiol.*, 2000, **279**, G943–G950.
85. L. M. Torres, J. Youngner and A. Romani, *Am. J. Physiol.*, 2005, **288**, G195–G206.
86. M. Barbagallo and L. J. Dominguez, *Arch. Biochem. Biophys.*, 2007, **458**, 40–47.
87. C. Cefaratti, A. McKinnis and A. Romani, *Mol. Cell. Biochem.*, 2004, **262**, 145–154.
88. A. Suarez, N. Pulido, A. Casla, B. Casanova, F. J. Arrieta and A Rovira, *Diabetologia*, 1995, **38**, 1262–1270.
89. A. Romani, C. Marfella and A. Scarpa, *FEBS Lett.*, 1992, **296**, 135–140.
90. A. Romani, C. Marfella and A. Scarpa, *J. Biol. Chem.*, 1993, **268**, 15489–15495.
91. Y. Konno, S. Ohno, Y. Akita, H. Kawasaki and K. Suzuki, *J. Biochem.*, 1989, **106**, 673–678.
92. R. D. Grubbs and M. E. Maguire, *Magnesium*, 1987, **6**, 113–127.
93. D. Garfinkel and L. Garfinkel, *Magnesium*, 1988, 7, 249–261.
94. M. Otto, R. Heinrich, B. Kuhn and G. Jacobasch, *Eur. J. Biochem.*, 1974, **49**, 169–178.
95. M. E. Maguire, *Trends Pharmacol. Sci.*, 1984, **5**, 73–77.
96. T. Gunther, *Magnesium*, 1986, 5, 53–59.
97. R. G. Hansford, *J. Bioenerg. Biomembr.*, 1994, **26**, 495–508.
98. L. Doleh and A. Romani, *Arch. Biochem. Biophys.*, 2007, **467**, 283–290.
99. M. E. Maguire, *Ann. NY Acad. Sci.*, 1988, **551**, 215–217.
100. F. I. Wolf, A. R. Cittadini and J. A. Maier, *Cancer Treat. Rev.*, 2009, **35**, 378–382.

CHAPTER 5

Calcium

J. A. COWAN

Department of Chemistry and Biochemistry, Evans Laboratory,
The Ohio State University, 100 West 18th Avenue, Columbus,
OH 43210-1173, USA
Email: cowan@chemistry.ohio-state.edu

5.1 Summary

The focus of this chapter is on the biological chemistry of calcium, building from a basic understanding of the unique physicochemical and structural properties that both differentiate it from other competing metal cations, as well as lending it the required functional characteristics to promote biological activity. Following a brief introduction to key kinetic, thermodynamic and coordination chemistry properties of calcium, pathways of uptake will be described for intestinal cells and ultimately serum for general cellular distribution. This will include a review of various categories of membrane transporters (both active and passive) and the role of vitamin D and its derivatives and other hormones in regulating calcium transport, as well as the important role for bone as a central store of calcium and mechanisms of release into the blood. Subsequently, the speciation of calcium in blood as a hydrated cation, as albumin-bound forms, and complexed to other specialized proteins will be described, as well as the important role for calcium-activated proteins in blood clotting and cell signalling mechanisms. The latter topic will bring us back to the intracellular role of calcium as a second messenger and a review of intracellular depositories of calcium in the endoplasmic and sarcoplasmic reticula and in mitochondria. Also discussed are the mechanisms of membrane transporters that facilitate the rapid

RSC Metallobiology Series No. 2
Binding, Transport and Storage of Metal Ions in Biological Cells
Edited by Wolfgang Maret and Anthony Wedd
© The Royal Society of Chemistry 2014
Published by the Royal Society of Chemistry, www.rsc.org

increase or decrease in cytosolic calcium levels that underlie its role as a second messenger. These include voltage- and receptor-gated channels, as well as symports and antiports involving co-transport of sodium and hydrogen ions. This provides an opportunity to understand further the role of proteins in buffering intracellular calcium levels, as well as in promoting signalling pathways. Finally, the similarities in eukaryotic and prokaryotic calcium biology will be detailed, from the viewpoint of their roles in transport, buffering and signalling. A comment will cover the chemistry of ionophores secreted/excreted by prokaryotae, which can equilibrate calcium concentrations across membranes, killing target organisms and serving as a defence mechanism for select bacteria. A final section will detail some important questions and topics that remain to be addressed. The reader will find additional general details and insights on several of the topics and proteins discussed in this chapter in several texts that are worthy of review.[1-6]

5.2 Introduction to Calcium Chemistry

5.2.1 Distribution

Calcium was first identified by Sir Humphrey Davy in 1808. It is the fifth most abundant element in the Earth's crust by mass, as well as by solution concentration in the oceans, and is the fourth most abundant in the plasma and other extracellular fluids. By contrast, the intracellular concentrations of divalent calcium (Ca^{2+}) are low, on the order of nM, but can rapidly increase with an appropriate stimulus, reflecting the dominant role of Ca^{2+} as a species that promotes and is involved in cellular signalling pathways. Table 5.1 summarizes calcium concentrations in a variety of environments and cellular organelles.

The bulk of calcium ion ($\sim 99\%$) in the human body is found in bones and teeth where it forms solid support structures through biomineralization. These also provide a buffer medium to maintain homeostatic balance within the serum. An excess or deficiency of serum Ca^{2+} (hypercalcemia or hypocalcemia, respectively) underlie a variety of medical conditions. While the mineral component represents the dominant physiological form of calcium, it is the solution chemistry of the cation that more fully represents the diverse functional and physiological roles of the metal ion and so forms the

Table 5.1   Calcium concentrations in a variety of environments.[4]

Environment	Concentration (mM)
Sea water	10
Fresh water	0.02–2 (depending on source)
Human serum	2.5
Milk	70
Bone	1
Cytoplasm of eukaryotic cell	0.0001
Cytoplasm of prokaryotic cell	0.0001

focus of the chapter.[7] Much of this chemistry involves either membrane transport or protein binding and release, and so an appreciation of the magnitude of some basic solution physicochemical parameters will provide significant insights on the reactivity of the cation.

5.2.2 Physicochemical Properties

The alkali and alkaline earth ions (Na^+, K^+, Mg^{2+}, Ca^{2+}) are the most abundant metal ions in biological systems.[8,9] The distinct biological roles for the alkali and alkaline earth ions (Table 5.2) reflect fundamental differences in their coordination chemistries and include variations in structural, thermodynamic and kinetic properties (Table 5.3).[†] The functional role of each ion is also reflected to a great extent by the intra- and extracellular distribution of the alkali and alkaline earth ions (Table 5.2). It is this distribution that dictates both availability and the necessary concentration gradients that underlie many important cellular functions of these ions. While intra- and extracellular concentrations of magnesium are similar, the distribution of the other ions is not, with K^+ most abundant inside the cell, and Na^+ and Ca^{2+} found in higher concentrations outside of the cell.

The basic physicochemical solution properties of divalent calcium are reflected by the specific roles it plays in bioenergetics, catalysis, bio-structural chemistry (both solution molecules and minerals) and as a regulator or

Table 5.2 Distribution of intracellular and extracellular ions in and around a typical mammalian cell system and the biochemical roles associated with physiologically relevant alkali and alkaline earth cations.[3]

Ion	$[M^{n+}]_{intra}$, total (mM)	$[M^{n+}]_{intra}$, free (mM)	$[M^{n+}]_{extra}$, total (mM)	Role Intracellular	Transmembrane	Extracellular
Na^+	12	8	145	Regulation of intracellular calcium	Osmotic balance, nerve conduction	Electrolyte
K^+	140	120	5	Ribosomes, enzyme activation (structural)	Osmotic balance, nerve conduction	No major role
Mg^{2+}	30	0.3	1	Enzyme activation (catalytic/ structural), ribosomes, complex with NTPs	Regulation of intracellular levels through Na^+/Mg^{2+} antiporters	Ill defined, but various health attributes
Ca^{2+}	3	0.0001	4	Second messenger, muscle activation, skeletal mass	Gradients underlie signalling and regulatory roles	Enzyme activation, (catalytic/ structural)

[†]For further comparison, see Chapter 2, Section 2.1, Chapter 3, Section 3.1.1 and Chapter 4, Section 4.1.

Table 5.3 Physicochemical properties of the physiologically relevant s-block cations.

Ion	Ionic radius (Å)	Hydrated radius (Å)	Hydrated volume (Å^3)	Coordination number	$k_{ex}(H_2O)$, s^{-1}
Na$^+$	0.95	2.75	88.3	6	8×10^8
K$^+$	1.38	2.2	52.5	6 to 8	10^9
Mg^{2+}	0.65	4.76	453	6	10^5
Ca^{2+}	0.99	2.95	108	6 to 8	3×10^8

The hydrated volume reflects the number of inner and outer sphere associated solvent molecules, providing a measure of the electrostatic ordering around the cation.

promoter of enzyme or protein function.[4,5] In turn, these properties provide pathways for control of the molecular mechanisms that underpin the physiology of neurotransmission, muscle contraction, maintenance of the action potential in cardiac muscle and blood clotting. To achieve a measure of control over these roles, there is a need for both intra- and extracellular recognition of calcium, relative to other abundant competitive cations, to allow for selective transport and binding.

Extending from this, the binding affinity of biological ligands (large and small) typically correlates with the available concentration of free metal ions.[6,8] For example, binding affinities for intracellular magnesium ligands show $K_D \sim 0.5$ mM, and $K_D \sim 2$ to 4 mM for extracellular calcium ligands. In contrast, the K_D values for intracellular calcium ligands are on the order of tens of nanomolar in concentration. In turn, such variations in cellular distribution also underlie mechanisms of selectivity, since abundant cations are selectively bound relative to other ions of much lower abundance even if their binding affinities are intrinsically greater. Consequently, intracellular calcium does not inhibit magnesium enzymes even if it binds more tightly.

Calcium is exclusively bound by oxygen ligands (including water, hydroxide, backbone carbonyl, carboxylate and occasional alkoxyl (Figure 5.1) and usually adopts a distorted octahedral or seven-coordinate geometry.[10] With carboxylate ligation, both monodentate and bifurcating coordination is possible (Figure 5.1). As noted earlier, the relatively low charge density for the larger calcium ion results in relatively weak solvation, and so, accounting for associated waters in the secondary coordination sphere, *the hydrated calcium is effectively smaller than the hydrated magnesium ion* (see the hydrated volume entry in Table 5.3). This phenomenon has significant implications for selectivity by membrane cation pores inasmuch as both the "smaller" size and lower dehydration energy strongly favour Ca^{2+} transport. Also the relatively low charge density of the calcium ion is more comparable to that of the alkali metal ions. Consequently, it exhibits water exchange rate constants ($\sim 10^9$ s^{-1}) closer to the diffusion limited values ($\sim 10^{10}$ s^{-1}) for hydrated sodium and potassium ions, relative to the smaller value ($\sim 10^5$ to 10^6 s^{-1}) exhibited by the high charge density of the magnesium ion.

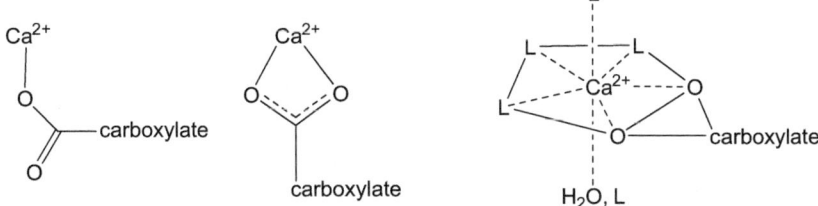

Figure 5.1 Monodentate (left) *versus* bifurcating (middle) coordination of carboxylate to Ca^{2+}. A standard pentagonal bipyramidal coordination geometry for calcium bound to an EF-hand motif is shown (right), where L ligands are typically monodentate carboxylate oxygen, backbone carbonyl or occasionally Ser/Thr hydroxyl. One of the axial ligands is often water, but can be a protein residue. For magnesium coordination, the bifurcating carboxylate switches to monodentate mode, with a consequent change to octahedral coordination.

It is clear that the structural and kinetic parameters associated with divalent calcium and the other alkali and alkaline earth ions are strictly controlled by charge and size effects. These properties are important since they underpin many of the functional roles of calcium. These include the need for rapid exchange kinetics (by virtue of its low charge density and fast exchange rates) to facilitate its role as a second messenger in cellular signalling, plus an acceptable binding affinity (by virtue of the divalent charge) to promote selective recognition of proteins, membrane sites, *etc.*

5.2.3 Cellular Roles

Relative to the other alkali and alkaline earth metals, divalent calcium is characterized by higher charge density than monovalent ions (promoting tighter binding to cognate proteins), a larger radius (promoting faster ligand exchange than magnesium ion) and an expandable coordination geometry (allowing for selective binding). Indeed the combination of rapid exchange kinetics and strong ligand binding is consistent with Ca^{2+} rather than Mg^{2+} being a more effective triggering ion for the activation of biological reactions, and therefore a better intracellular second messenger. For this reason, the intracellular concentration of Ca^{2+} is maintained at a low level (Table 5.2) so that rapid influx to the cytosol can trigger cellular processes with a rapidity promoted by the rapid exchange of calcium-bound ligands. This phenomenon will be reflected in transport of the ion and the sensory roles described in later sections. An additional consequence of the low intracellular calcium levels is the absence of any clear example of a calcium-dependent enzyme that would be engaged in regular cellular metabolism. Such an enzyme would require an extremely high affinity for Ca^{2+} under the conditions typically prevalent in the cell, and presumably such functions have been taken up by the more abundant magnesium ion to avoid the evolutionary development of a specialized calcium binding site.

As noted earlier, the ability of calcium to adopt expanded coordination numbers and more asymmetric geometries also allows for recognition of calcium over magnesium both inside the cell, where the free calcium levels are typically several orders of magnitude lower than magnesium, and in the extracellular spaces where the concentrations are more comparable. For example, hydrolytic enzymes such as phospholipase, protein kinase C and staphylococcal nuclease show a high degree of specificity for Ca^{2+}. If these enzymes were readily activated by Mg^{2+}, it would be more difficult to control their activity.

Binding constants for Ca^{2+} vary from the high-affinity ($K_D \leq 10^{-6}$ M) storage and regulatory proteins (such as parvalbumin, calsequestrin, calmodulin and troponin C, discussed in later sections) to the low-affinity enzymes or storage proteins ($K_D \sim 10^{-3}$ M) noted in Table 5.4. The latter low-affinity proteins are typically located either in serum or in specific cellular organelles, where local calcium concentrations are high. A comparison of calcium-binding EF hands *versus* magnesium-dependent enzymes shows that the Ca^{2+} binding sites are highly size-selective and tunable through a combination of size selectivity, coordination number and geometry.[11,12] In cases such as parvalbumin, higher affinity is achieved by retention of all protein ligands (excluding water coordination; Figure 5.1), thereby fixing the coordination number and controlling the radius of the substrate cavity (Figure 5.2).[13] The plasticity of proteins and their ability to fine tune coordination environments (relative to the more rigid scaffold provided by ligands such as nucleotides and ribozymes, or mineral structures *etc.*) also serves to allow preferential association with calcium rather than magnesium ion as required by the specific physiological role.

Aside from the need to maintain low calcium concentrations to facilitate cellular signalling, this again avoids the inhibitory effect of Ca^{2+} on many of the Mg^{2+}-dependent enzymes used in cellular metabolism and eliminates

Table 5.4 Range of affinities and roles for calcium-binding proteins.[3]

	Ca^{2+} sites	K_D (M)
EF-hand proteins (sensory, buffering, regulatory)		
Parvalbumin	2	10^{-9}
Calbindin-9K	2	10^{-6}
Calmodulin	4	10^{-6}
Troponin C (skeletal)	4	10^{-6}
Extracellular digestive enzymes		
Staphylococcus nuclease	1	10^{-5}
Phospholipase A_2	2	$10^{-5}, 10^{-3}$
Trypsin	1	10^{-4}
Structural/storage proteins		
Thrombin	Many	10^{-3}
Phosphodentine (material in teeth)	Many	10^{-3}
Calsequestrin (Ca^{2+} storage in the sarcoplasmic reticulum)	40	10^{-3}

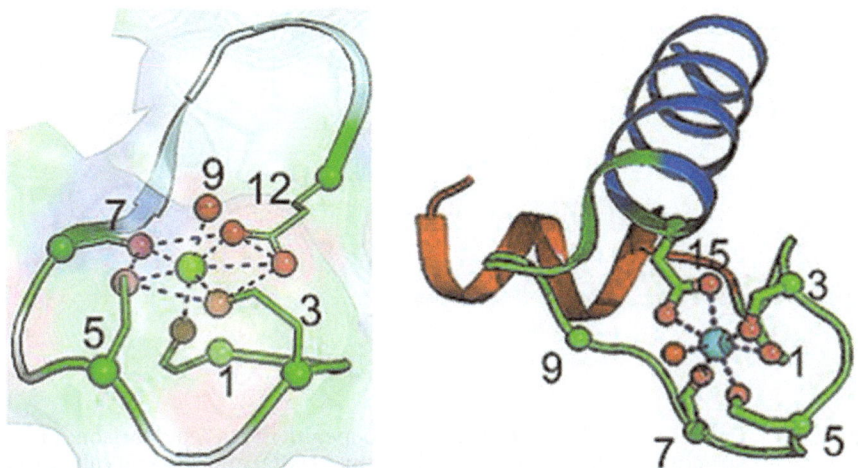

Figure 5.2 A prototypical EF-hand. Sites 1, 3 and 5 are typically ligated by aspartate, although in this case a Ser hydroxyl coordinates at position 5. A backbone carbonyl coordinates at position 7 and the glutamate carboxylate at site 9 binds at the axial position (although this is often taken by water). The bifurcating glutamate switches to monodentate when magnesium binds. The additional bound water and enhanced chelate effect provide even higher affinity than is commonly observed for calcium binding, even for a cytosolic protein.
Reproduced with permission from ref. 14.

problems from the low solubility of calcium salts. To allow for the tight regulation of intracellular calcium, the cell makes use of intracellular buffering proteins such as the parvalbumins that show high affinity binding for Ca^{2+} ($K_D \sim 10$ nM).

5.2.4   EF Hands

The role of Ca^{2+} as a second messenger, and an on/off switch for calcium biochemistry, is primarily executed through the use of the EF hand motif. For example, this allows troponin C to initiate the direct mechanical response of muscle contraction following Ca^{2+} binding and for calmodulin to serve as a general switch for cellular metabolism by activating key enzymes. The various calcium regulatory proteins all initiate a cellular response following binding of the calcium ion through the formation or breakdown of contacts with other cellular proteins.[16,17] The signalling is based on structural change induced by calcium binding to a conserved structural feature found in each of these regulatory proteins (Figure 5.3).[18] In particular, each Ca^{2+} is bound in a helix-loop-helix site named after the E and F helices that comprise part of the structure (Figure 5.3). The EF nomenclature derives from the notation used to describe the six α-helical segments of carp muscle

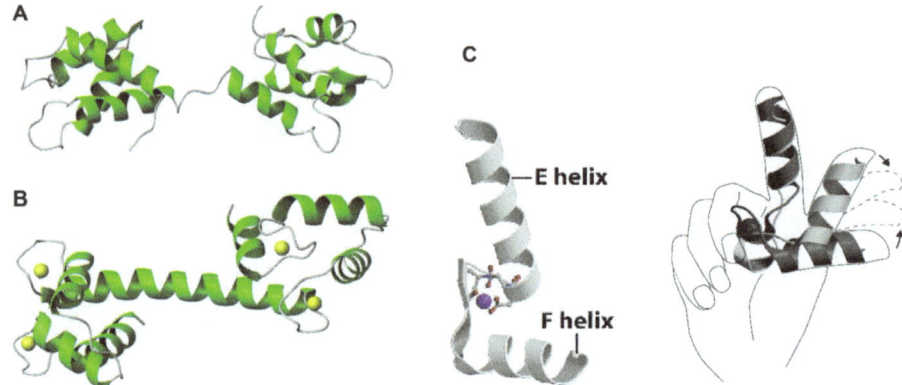

Figure 5.3 Calcium coordination controls cooperativity. A comparison of the structures of (A) *apo*-calmodulin (pdb: 1DMO) and (B) *holo*-calmodulin (calcium loaded; pdb: 1CLL), illustrating the structural changes that accompany cooperative uptake of calcium ions. Adapted from ref. 15. (C) A depiction of the classical EF-hand motif. As shown right, binding of calcium induces relative movement of the helices that underpins both the cooperativity in binding between calcium sites, and the switching on and off of binding to protein partners.
Adapted from ref. 16.

parvalbumin B and, in particular, the solvent accessible loop that lies between two helical sequences (Figure 5.3).

The term EF-hand domain has been extensively used to define Ca^{2+}-binding domains that are structurally homologous to that identified in parvalbumin, and is a central structural feature of all the calcium-binding proteins involved in second messenger pathways. Typically, two calcium-binding sites are found in close proximity and binding of the metal ions results in perturbation of the bonding interactions between the two principal elements of secondary structure (Figure 5.3). Such perturbations underlie the cooperativity often observed in proteins carrying such motifs. By contrast, no distinct structural motif is used by calcium-dependent proteins or enzymes that do not function through allosteric protein-protein contacts, such as nucleases and phosphatases. As expected, the ligands that bind Ca^{2+} are always oxygen atoms from side-chain carboxylates, backbone carbonyls, alkoxyls or water molecules.

5.3 Uptake in the Intestinal Cells and Release into the Blood

5.3.1 Absorption

A healthy adult diet requires the daily intake of approximately 1 g of calcium to maintain homeostatic balance. Calcium is absorbed across the epithelial cells that form the wall of the intestine *via* a special membrane termed the brush border membrane as a result of its bristle-like texture. Key details are summarized in Figure 5.4 and in ref. 4. Transport across intestinal epithelial cells is mediated by voltage-gated ion channels (such as TRPV6) that sense intracellular calcium levels and facilitate uptake when the concentration is low, since opening and closing of the channel is sensitive to the resulting transmembrane potential. Uptake is also driven by ATPase-driven pumps or by vesicle/vesicular transport involving endocytotic and exocytotic pathways. In the latter pathways, portions of the cellular membrane encapsulate local calcium solutions and either break off from the main cell membrane for uptake of calcium, or blend into the cell membrane to release calcium.[19] Because the cytosolic calcium levels must remain low, effective and rapid transport from intestine to serum is facilitated by carrier proteins such as calbindin-D9k and -D28k, which contain two of the EF-hands described earlier. Export from the cytosol across the basolateral membrane to serum is energetically uphill against a pronounced concentration gradient and is driven primarily by a Ca-ATPase pump or by a Ca^{2+}-$3Na^+$ antiporter that

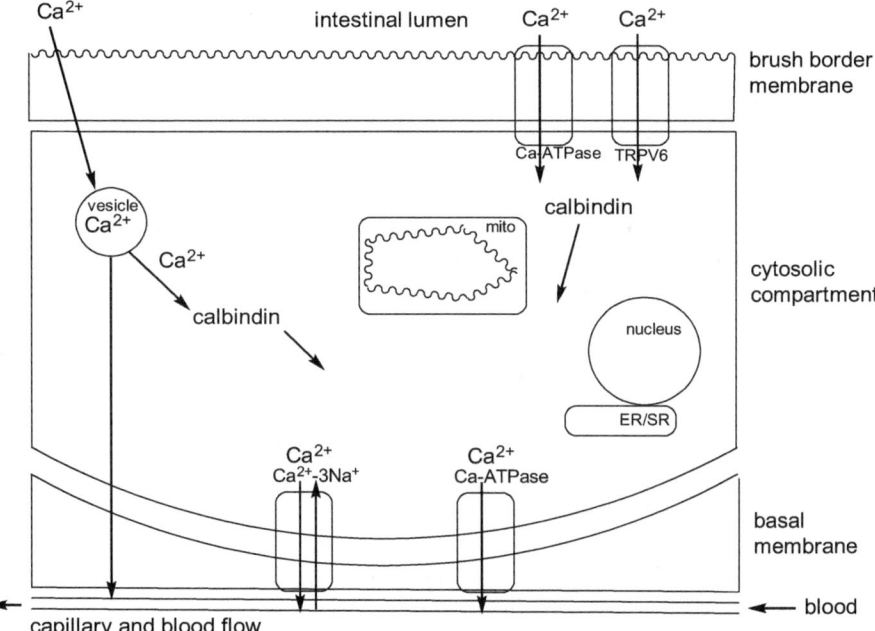

Figure 5.4   General aspects of calcium transport in intestinal cells. Uptake of calcium is mediated by both voltage-gated passive transporters and active ATP-driven pumps, as well as vesicular migration. Release to blood serum through the plasma membrane is an active transport process driven by sodium/calcium antiporter or a distinct plasma membrane-associated ATPase. The abbreviations ER/SR denote the endoplasmic/sarcoplasmic reticulum, and mito represents the mitochondrion.

couples a favourable sodium ion gradient to drive transmembrane calcium movement (3 Na^+ per Ca^{2+}).

5.3.2 Conservation

While the kidney excretes a significant amount of calcium to the urine, most of this is reabsorbed into the blood. Serum calcium levels can also be increased by hormone stimulation that causes the release of calcium from bone, and increases the uptake from the intestine (Section 5.6.3).

5.4 Coordination Chemistry of Calcium in the Blood

5.4.1 Serum Calcium

Serum calcium is one of the most regulated species in the human body with variations of only 1 or 2% allowed and concentrations typically poised at 2.45 $\pm$ 0.05 mM.[4] Significant increases or decreases in serum calcium levels are reflected by serious health conditions. Approximately half of the available calcium circulates in the free hydrated ion form, while the other half is complexed to serum albumin.

Calcium binds and activates a variety of serum, or serum-exposed, proteins and enzymes. Because there is a higher free calcium concentration in blood, the affinities of such proteins for calcium are typically on the order of 0.1 to 1 mM K_D. Moreover, since such proteins are normally enzymes or structural components, they typically do not use the sensory EF-hand binding motif. Instead they have metal binding pockets that reflect the functional role of the protein and the chemistry required at the metal cofactor (structural or catalytic).

5.4.2 Role of Calcium in Coagulation

Calcium plays a key role in multiple steps of blood coagulation during wound repair.[20] When external tissue or internal blood vessels are damaged, a sequence of events is initiated that triggers an increased level of intracellular cytosolic calcium that in turn activates protein kinase C and phospholipase A2, which subsequently modify cellular membrane proteins and enhances binding by fibrinogen. A key part of that response is the exposure of the protein tissue factor (Factor III) to recruit plasma coagulation proteins that then trigger a cascade of events that involve formation of multiple active protein factors (normally designated by Roman numerals) by cleavage of zymogens (inactive precursor proteins). This "coagulation cascade" ultimately leads to formation of active thrombin through the action of a calcium-dependent Factor Xa enzyme on prothrombin, where prothrombin itself is post-translationally modified to convert 10 Glu amino acids to calcium binding Gla (Figure 5.5) residues that promote calcium-dependent membrane association at the wound site. Thrombin promotes cleavage of the

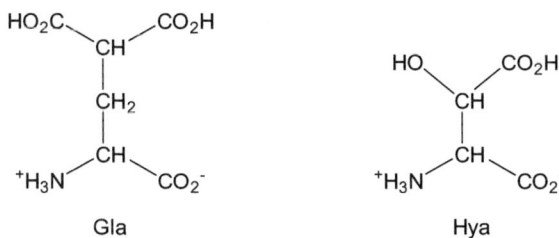

Figure 5.5 Specialized calcium-binding amino acids. Structures of two modified amino acids (γ-carboxyglutamic acid and β-hydroxyaspartic acid) commonly formed post-translationally for calcium-binding proteins.

soluble fibrinogen (2–3% of plasma protein) to insoluble fibrin fibres, as well as converting Factor XIII to Factor XIIIa that in turn cross-links the fibrin strands to form a clot. This clot, along with blood platelets, blocks the damaged site and traps other cells that assist in repair. The extracellular calcium levels remain constant and calcium is taken up by proteins such as prothrombin and Factor XIII as they are needed. Clearly calcium is a central cofactor in the design of the coagulation cascade.

Thrombin itself is activated through a series of protein cleavage events where inactive zymogens (precursor proteins) are converted to an active form by protease action with deletion of inhibitory domains.[20] Many of these proteases are calcium-activated, and in several cases post-translational modification is required to activate a protein domain for Ca^{2+} binding. This is mediated by vitamin K serving as a redox cofactor for iron-dependent oxidases that promote the conversion of glutamic acid to γ-carboxyglutamic acid (Gla, Figure 5.5). A deficiency of vitamin K or introduction of the anticoagulant drug warfarin inhibits the production of γ-carboxyglutamic acid residues and diminishes activation of the coagulation process. Other modified amino acids, β-hydroxyaspartic acid (Hya, Figure 5.5) or β-hydroxyasparagine (Hyn), are formed through the enzymatic oxidation of aspartic acid or asparagines, respectively, again by iron-dependent oxidases with an alpha-ketoglutarate redox coreagent. Both Hya and Hyn are found in epidermal growth factor domains and in some cases are implicated with calcium binding.

5.4.3 Role of Calcium in Membrane Biochemistry

Another family of calcium-activated proteins promotes the repair of membrane surfaces.[5] In particular, lysosomal synaptotagmin mediates the fusion of lysosomal vesicles with the cellular plasma membrane to repair damage.[21] The function of this protein in calcium-promoted vesicle fusion is discussed further in Section 5.6.4. Calcium-dependent phospholipases mediate lipid transformations by catalysing the hydrolysis of the 2-acyl ester bond in 1,2-diacylglycero-3-phosphoglycerate derived lipids that allow remodelling of membrane surfaces and dynamics.[22] The calcium binding site for phospholipase A2 enzymes is heptacoordinate and does not readily bind Mg^{2+}, which prefers a regular octahedral geometry. Again, Ca^{2+} binds to the phosphate group and stabilizes the increased negative charge that arises in the transition state following hydroxide attack at the phosphate ester. Calcium also activates serine protease activity, kinase C and phospholipases involved in membrane repair.[19,23] Many of these membrane-associated proteins are post-translationally modified to contain Gla and Hya that selectively bind calcium and promote interaction with phospholipid bilayers and membrane repair.

5.4.4 Role of Calcium in Cell Adhesion and Membrane Signalling

Other important extracellular processes that require Ca^{2+} include complement activation (the immune system), cell surface receptor/ligand interactions, cell adhesion, formation of bone and connective tissue and degradative enzymes (Table 5.5).[24] There is greater involvement by calcium in catalysis, stabilization of protein structure and fostering protein-protein contacts as well as interactions with cell surfaces. The latter also involves calcium coordination to oligosaccharides. In these roles, the flexibility of calcium coordination is evident, with variable coordination number and coordination asymmetry. However, the preference for carboxylate and carbonyl ligands is retained.[25] Extracellular calcium proteins also tend to be more modular in structure. Examples include the Gla-rich domains associated with proteins involved in coagulation, the C-type lectin domain where calcium is involved in binding carbohydrates,[26] the calcium-binding domains located on cadherins (calcium-dependent adhesion molecules) and epidermal growth factor (EGF).[24] The EGF domain consists of ~40–45 amino acids and contains three disulfide bonds. Calcium-binding EGF domains (cbEGF) are often subject to post-translational modifications of

Table 5.5 Extracellular and membrane-associated calcium-dependent proteins.[23]

Enzyme class	Function
Selectins	Cell adhesion glycoproteins that bind to sugars and promote binding of other cell types. E-selectins are associated with epithelial cells and recruit T lymphocytes as part of the inflammatory response.
Osteonectins	Matrix glycoproteins that promote bone mineralization and adhesion to collagen fibres.
Integrins	Mediate signals between the cell and extracellular matrix.
Cadherins	Promote adhesion of cells and tissue in a calcium-dependent manner.
Fibrillin	A calcium binding glycoprotein that is incorporated into connective tissue to assist with elasticity.

amino acids and glycosylation that promote both calcium binding and interactions with other proteins or cells.

The annexin family is another important class of proteins that function primarily inside the cell, but there exist a few examples of extracellular annexins modulating cell fusion events. Annexins associate with the inner membranes through protein domains that bind multiple calcium ions and serve to promote binding to the phospholipid membrane. Calcium binding also displaces an N-terminal protein domain that makes contact with other cellular proteins which are part of the cellular signalling and regulatory apparatus.[27] The calcium binding domains are modified to contain Gla and Hya residues that recruit the divalent ion.

5.5 Transport through Subcellular Membranes

5.5.1 Endoplasmic/Sarcoplasmic Reticulum and Mitochondrial Transport

We have already seen how calcium can be taken up or removed from the cell by transport proteins that promote uptake from the intestine, or mediate import/export from the serum (Figure 5.4). Cytosolic calcium levels can also be increased by transport from intracellular stores (which include the endoplasmic/sarcoplasmic reticulum or mitochondria; Figure 5.6) and by

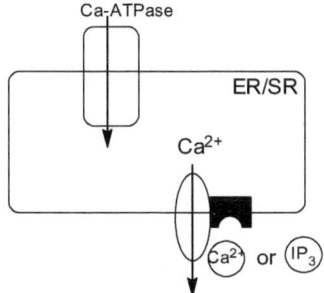
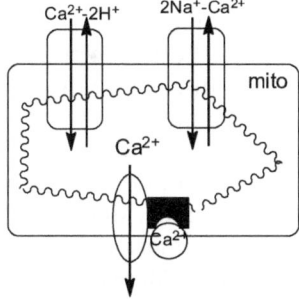

Figure 5.6 Calcium uptake and release from cellular calcium stores is highly regulated. Both the endoplasmic and sarcoplasmic reticula and the mitochondria contain high levels of calcium that are released following triggering by inositol triphosphate (IP₃), calcium ion or transmembrane potentials linked to sodium ions. The sarcoplasmic reticulum is more closely associated with muscle contraction. Following release of calcium into the cytosol, there is an almost immediate return of calcium to the cellular store, driven by calcium ATPase pumps, or by mitochondrial proton gradients. The possible existence of a nucleoplasmic reticulum, associated inside the nucleus with the nuclear envelope, has also been invoked, although details remain uncertain.[29,30] It could increase nuclear calcium levels and activate nuclear protein kinase C and IP₃ signalling pathways while also impacting transcriptional regulation of select genes.

direct uptake from the plasma. The significantly higher levels of Ca^{2+} in the ER/SR and mitochondria could approach 30 mM if all of the ion were soluble, and has the potential to result in osmotic shock. The possibility of this happening is reduced by precipitation of Ca^{2+} as phosphate or oxalate salts, or by binding to buffering proteins such as calsequestrin.[28]

Release of calcium from the endoplasmic/sarcoplasmic reticulum is most commonly promoted by the messenger molecule inositol triphosphate (IP_3) or by direct coupling to cytosolic calcium (Figure 5.6).[17,19] For mitochondrial export, the favourable release of calcium can be coupled to the electroneutral uptake of two sodium ions, or be triggered by cytosolic calcium. Because cytosolic levels of calcium are very low (Table 5.2), release of calcium into the cytosol can occur passively, from either extracellular or intracellular pools. Triggers for the stimulation of increasing cytosolic calcium include extracellular hormone action through membrane-bound receptors, or small molecule signalling within a cell (where release may be either hormone- or calcium-controlled). The rapid return of cytosolic levels is essential and must involve active transport pathways, including those driven by ATPase activity or coupling to favourable proton gradients (Figure 5.6).

5.5.2   Ca-ATPases

The sarcoplasmic/endoplasmic reticulum ATPases (SERCA) pump two Ca^{2+} per ATP hydrolysed. This is one of the best characterized examples of an active calcium pump, having been crystallographically defined to 2.6 Å resolution (Figure 5.7).[31] In common with other ion pumps, the protein belongs to the P-type transporter family. The exposed cytosolic surface includes the nucleotide binding domain that promotes phosphorylation of an adjacent domain.

Two calcium binding sites have been identified in the transmembrane spanning domain which are populated in one conformation open to the cytoplasmic side, and become exposed to the interior following an ATP-driven conformational change with a resulting lowering of affinity that favours release. There appears to be a substantial rearrangement of the various domains when comparing the calcium-bound and -free states.[31] The calcium ions lie adjacent to each other, are bridged through an aspartate side-chain that is important for the cooperative movement of the two Ca^{2+} ions and are surrounded by four helices, akin to a 4-helix bundle except that two helices are slightly unwound to allow attainment of an effective coordination geometry for the Ca^{2+} at site II.[31] The unwound domain contains a PEGL motif that has been observed in other heavy metal transporters where the E is replaced with C or H residues.[32]

5.5.3 Na^+/Ca^{2+}-Exchangers

The molecular details behind the function of these transporters remain to be elucidated. The energetics is such that at least three Na^+ ions must move

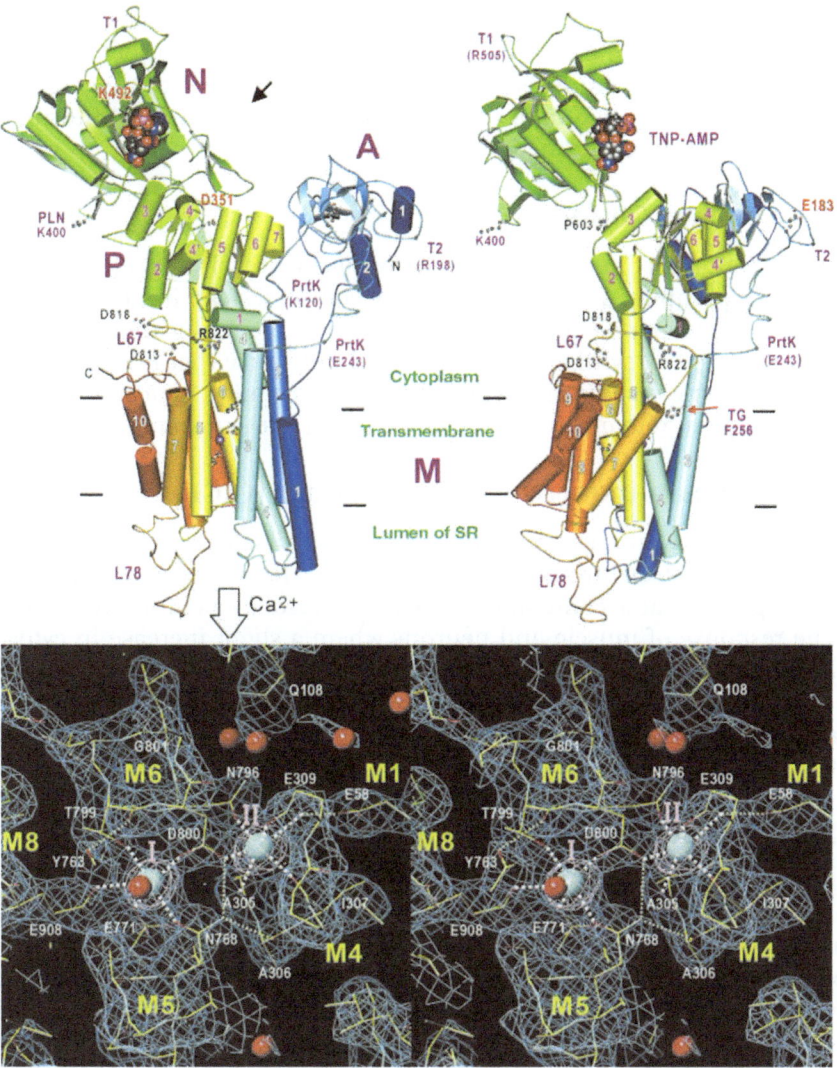

Figure 5.7 Crystallographically defined structure of the skeletal muscle Ca²⁺-ATPase. (Top) The cytoplasmic domains (labelled A, N and P) are shown, as well as the binding site for 2′,3′-O-(2,4,6-trinitrophenyl)adenosine 5′-monophosphate (TNP-AMP), a probe of the nucleotide binding site. The structure on the right is rotated 50 degrees around the M5 helix. Two purple spheres represent calcium ions bound in the transmembrane spanning domain M. (Bottom) A closer view of the calcium binding sites shows site I with a bound water molecule, but otherwise the ligands are oxygens from backbone carbonyl or side-chains, and a carboxylate bridge to site II that presumably underlies the cooperative binding and movement of the two calcium ions. The coordination around these ions is illustrated in stereo mode.
Reproduced with permission from ref. 31.

along a typical sodium gradient ($[Na^+]_{out}/[Na^+]_{in} \sim 10$) to push Ca^{2+} against a concentration gradient of $\sim 10^3$. In the case of the mitochondrial exchanger, where calcium export is driven passively, the energetically favourable flow of calcium is used to pump two sodium ions into the mitochondrion in an electroneutral process.

5.5.4   Voltage Gating

In mitochondria, H^+ gradients formed by normal respiratory chemistry and the proton pumping action of respiratory complexes can in turn be used to drive Ca^{2+} transport (Figure 5.6) in addition to their typical mitochondrial role of driving ATP synthesis. The potential difference that is established by proton pumping is sufficient to push Ca^{2+} into the mitochondrion so long as the H^+ gradient remains intact from the action of the respiratory apparatus.

5.5.5   Calcium-induced Calcium Release (CICR)

This describes a process where the calcium ion itself serves as the effector that activates release from intracellular stores. It further accelerates a cellular response that accompanies increased cytosolic calcium levels, such as in the response of muscle and neurons where a slight increase in cytosolic calcium stimulates a "wave" of calcium release. Many of these channels are described as ryanodine receptors,[33,34] so called after the name of an alkaloid that shows high affinity binding to the open form of the channel and can cause violent muscle contractions and convulsions.

5.5.6   Receptor-driven Pathways

IP$_3$ is a potent intracellular messenger that binds to receptors on the endoplasmic/sarcoplasmic reticulum and opens passive transport channels for Ca^{2+}. Other signalling molecules that are used include calcium itself (as noted previously), cyclic AMP and sphingolipids.[19] Varied mechanisms communicate extracellular signals to these receptors, including the uptake of calcium from voltage-gated channels to increase cytosolic calcium levels and plasma-membrane-associated enzymes that result in the release of IP$_3$, sphingolipids or cAMP (Figure 5.8). In certain cases, the receptor complexes can be regulated by interactions with G-coupled proteins and other components of the cell-signalling apparatus that can both up- or down-regulate activity.[34,35]

5.6   Sensing, Buffering and Transport (Cellular Homeostasis)

The cellular chemistry of calcium represents great diversity of structure and function, as reflected in Figure 5.9. This section, more than any other, best represents the various roles of calcium in cellular physiology.

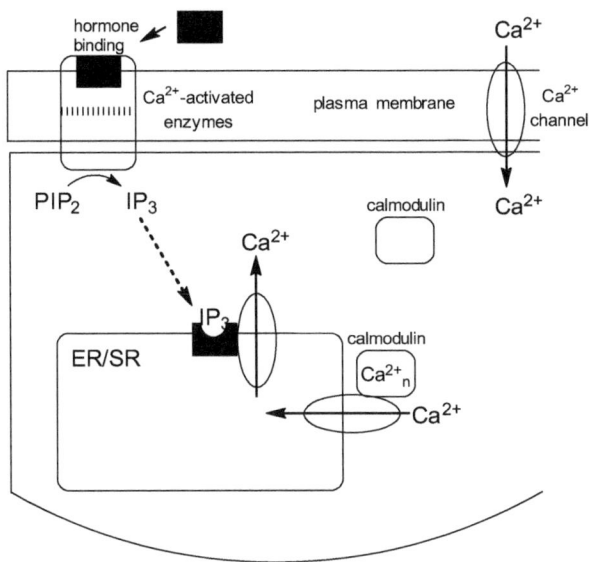

Figure 5.8 Triggering change in cytosolic calcium levels. The increase of cytosolic calcium is promoted by activation of IP_3 and Ca^{2+} second messenger pathways as a result of IP_3 production mediated by calcium-dependent enzymes, or various chemically or voltage-gated Ca^{2+} channels. In turn, IP_3 activates channels that allow passive transport and rapid release of Ca^{2+} from the endoplasmic/sarcoplasmic reticula or mitochondrial pools. This is a transient increase because the increased calcium levels promote binding to calmodulin and this calcium-activated protein stimulates the action of Ca^{2+}-ATPases that remove the divalent ion back to the ER/SR or mitochondria.

5.6.1   Calcium as a Second Messenger

Calcium plays a key role in the regulation of a number of intracellular events where it serves as a second messenger, either through binding to proteins and eliciting conformational changes, or *via* transport pathways across membranes where biochemical processes are driven by concentration gradients. To achieve signalling *via* these mechanisms, the intracellular resting concentration of Ca^{2+} is held at ~ 100 nM. During activation, cytosolic calcium levels can increase to tens of micromolar. Calcium proteins exhibit several Ca^{2+} binding sites of varying affinity (Table 5.4). Distinct sets of sites are populated at differing concentrations of calcium (basal *versus* stimulated), thereby activating the protein toward distinct functional roles. Intracellular calcium proteins, particularly those serving sensory or regulatory roles, possess EF-hand calcium binding motifs that activate the protein following metal-induced conformational changes that open up a target binding site either on the metal itself (in the case of an enzyme) or on the protein surface.[15,19,34,36,37]

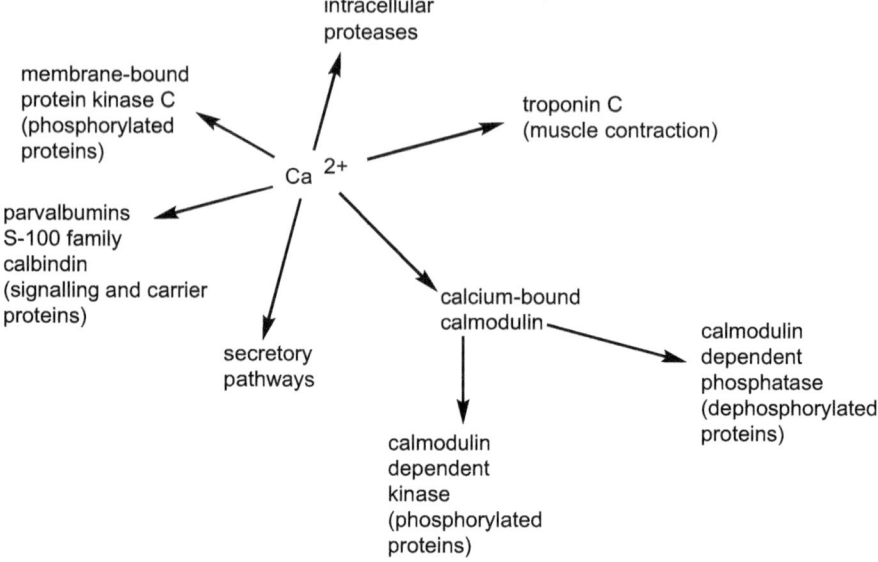

Figure 5.9 Diverse cellular roles for calcium in the activation of proteins and enzymes. Both the free calcium ion and calcium-bound proteins (such as calmodulin) regulate cellular function.

   In contrast to a primary messenger that either acts as an extracellular agent and promotes communication between cells or triggers an intracellular response, the second messenger functions within the cell and regulates change within the cell. Because the action of a second messenger relies on a rapid response to an external stimulus, calcium must satisfy certain functional criteria to serve this role. First, there must be target molecules (effectors) within the cell that remain passive until Ca^{2+} levels increase and, following rapid binding, the calcium-effector complex then stimulates subsequent cellular chemistry. Effector molecules for calcium are normally proteins that either initiate an immediate cellular response after binding Ca^{2+} (*e.g.* calcium binding to troponin C results in muscle contraction) or activate a target protein (*e.g.* calmodulin or calpain) that subsequently initiates further cellular responses. These protein targets in mammalian cells are localized in the cytosol and on membrane surfaces open to the cytosol (Figure 5.8, and later in Figure 5.10). Binding of the calcium ion causes changes in protein conformation that allow for selective interactions with other cellular proteins or molecular substrates. Second, the concentration of calcium ions must remain low ($\sim$100 nM) in an unstimulated cell to avoid interaction with effector proteins. These proteins must be highly selective for Ca^{2+} since they must recognize this ion in the presence of high intracellular levels of Mg^{2+} and typically bind Ca^{2+} with $K_D<1$ μM (and >1 mM for Mg^{2+}). Third, during stimulation the intracellular calcium levels must increase rapidly. Subsequently, the calcium levels must be reduced to normal

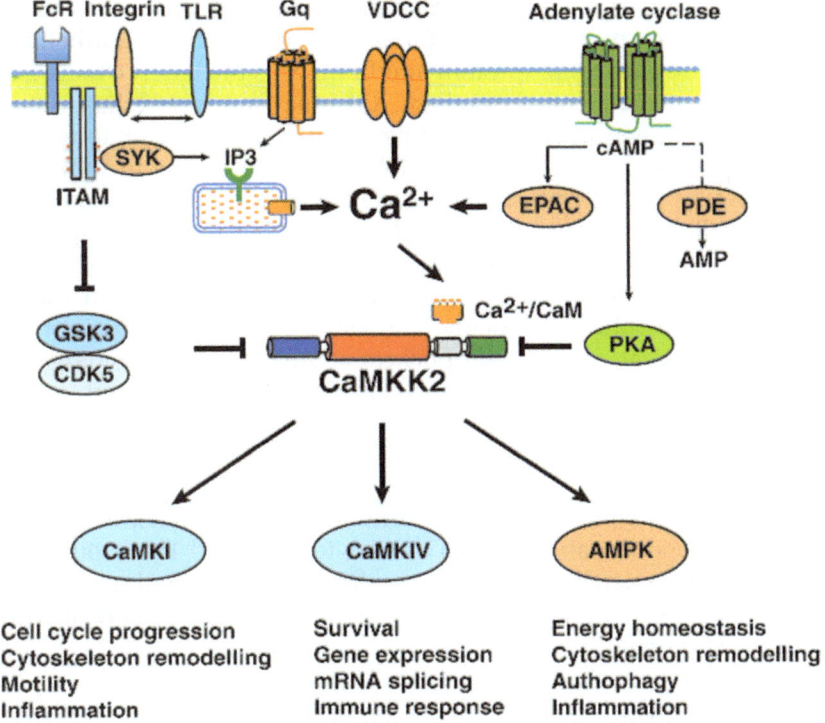

Figure 5.10 Calcium as a regulator of biocellular chemistry. A schematic illustration of the important role for a calmodulin-activated kinase (CaMKK2) in the regulation of a variety of cellular responses prompted by external signals that include activation through G-coupled receptors, IP₃-receptors or voltage-dependent calcium channels (VDCC). Other abbreviations include CDK5, cyclin-dependent kinase 5; EPAC, exchange protein directly activated by cAMP; FcR, Fc receptor; GSK3, glycogen synthase kinase 3; ITAM, immunoreceptor tyrosine-based activation motif; PDE, phosphodiesterase; PKA, protein kinase A; TLR, toll-like receptors. Reproduced with permission from ref. 35.

levels. Such rapid changes are made possible by the rapid exchange kinetics of the calcium ion (Section 5.2.2).

5.6.2 Buffering and Sensory Proteins

Such proteins typically display high affinity binding for Ca^{2+}, which binds to EF-hand motifs to prompt a transition to an active role.[38] They include parvalbumins and calbindins as ubiquitous buffering proteins, as well as more specialized calcium binding proteins such as members of the S-100 protein family (in brain) and calcimedins in smooth muscle.

Calbindin-D9k and -D28k production within the cell is regulated by vitamin D. Each protein contains EF-hands and serves as a carrier and buffering agent for calcium in a variety of cell types, being primarily involved in the transport of calcium across cells for further metabolism. Epithelial cells primarily utilize the former protein and brain cells the latter. Otherwise, the proteins have no sequence homology. A comparison can perhaps be made with the cooperativity of oxygen binding within a haemoglobin tetramer that ensures the rapid uptake or release of dioxygen, according to solution conditions. Similarly, cooperative binding of calcium can promote a prompt and regulated cellular response that reflects and further induces change in calcium levels that drives other cellular chemistry and pathways.

Calsequestrin is a calcium storage protein that is located in the endoplasmic and sarcoplasmic reticula, and can bind up to 50 calcium ions.[28] To achieve this feat, it binds *via* pockets of surface negative charge rather than EF-hand type motifs. The protein lacks a pronounced structure, although calcium binding does induce a greater degree of secondary and tertiary structure organization, as well as higher order calcium-promoted polymerization.

Parvalbumin is typically found in muscle and brain cells, is structurally related to calmodulin and serves a similar function.[2,13] The S100 family of proteins are also related structurally to calmodulin, but regulate protein and enzyme activities in a narrower range of neural cells and cells related to the immune response.[5]

5.6.3 Calmodulin Signalling Pathways

The calmodulin family of proteins occurs in a diverse array of cell types and regulates multiple cellular activities. As explained in Section 2.4, the regulation of calmodulin function is mediated by calcium binding to EF hand motifs. This provides a functional on/off switch that promotes calmodulin binding to partner proteins through induced structural changes, as illustrated earlier in Figure 5.3, Calmodulin-$(Ca^{2+})_n$ serves as an intracellular receptor that binds to and activates many proteins within the cell (Table 5.6), including proteins such as kinases, phosphatases and ion transporters.[15,39] Kinases are an important class of enzymes that specifically phosphorylate serine or threonine residues and regulate intracellular signalling pathways that control a variety of cellular functions. These include learning and memory, Ca^{2+} homeostasis, neurotransmission, chloride transport and T-cell activation. It is the structural changes that result from phosphorylation that produce a physiological response. Figure 5.10 illustrates the role for calmodulin in activating protein kinases involved in the regulation of cellular processes (as suggested by Figure 5.8). In addition to kinases, these cells also possess cyclic nucleotide phosphodiesterases that catalyse the hydrolysis of the second messengers cAMP and cGMP. These enzymes are also activated by calmodulin-$(Ca^{2+})_n$, emphasizing the close biochemical relationship among the various families of second messengers.

Calmodulin is also a key regulator of cellular mechanisms that remove Ca^{2+} from the intracellular space, through close association with both the endoplasmic/sarcoplasmic reticula calcium transport proteins (Ca^{2+}-ATP-ases, Figure 5.8), as well as the plasma membrane Ca^{2+}-ATPase (PMCA) pump (Figures 5.8 and 5.11).[37,40] The latter is also activated by Ca^{2+} binding to calmodulin following displacement of an auto-inhibitory peptide. The

Table 5.6 Calmodulin-dependent enzymes and their cellular functions.[38]

Enzyme	Function
Adenylate cyclase	Synthesis of cyclic adenosine monophosphate (cAMP)
Guanylate cyclase	Synthesis of cyclic guanosine monophosphate (cGMP)
Ca^{2+}-dependent phosphodiesterase	Hydrolysis of cAMP and cGMP
Ca^{2+}-adenosine-triphosphatase	Ca^{2+}-pump
Nicotinamide adenine dinucleotide kinase	Synthesis of nicotinamide adenine dinucleotide phosphate
Phosphorylase kinase	Glycogen degradation
Inositol triphosphate kinase	Phosphoinositol metabolism
Myosin light-chain kinase	Contractility and motility
Calmodulin-dependent protein kinase	Phosphorylation of various proteins
Calmodulin-dependent protein phosphatase	Dephosphorylation of various proteins

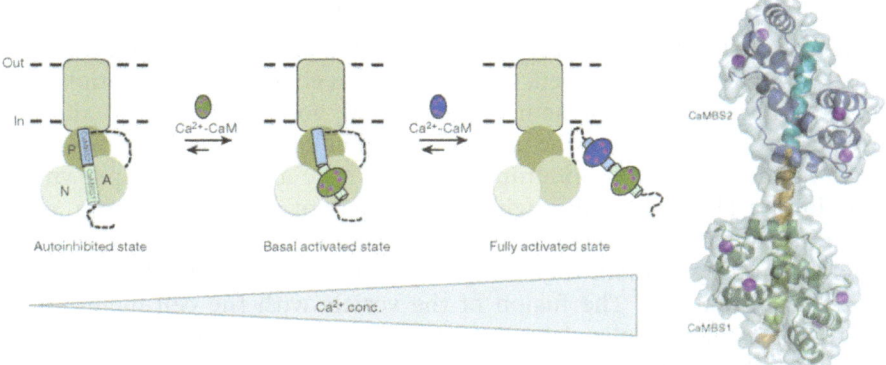

Figure 5.11 Activation of membrane pumps. The plasma membrane Ca^{2+}-ATPase (PMCA) pump is activated by calmodulin binding, which displaces an inhibitory peptide that belongs to the N- or C-terminus of the pump (for plant or mammalian system, respectively). The PMCA pump is phosphorylated by cAMP-dependent kinase and protein kinase C. At low calcium levels one calmodulin-$(Ca^{2+})_n$ complex binds to the peptide and effects a partial release that allows for basal pump activity. When intracellular calcium levels increase further, a second calmodulin-$(Ca^{2+})_n$ complex binds with rapid release of the inhibitory peptide domain that generates the fully active pump.
Adapted from ref. 37.

activity varies with the concentration of intracellular calcium. The calmodulin-$(Ca^{2+})_n$-induced basal activity at low cytosolic concentrations increases greatly when prompted by higher calcium levels that, in turn, prompt binding of a second calmodulin-$(Ca^{2+})_n$ complex that fully activates the pump and rapidly reduces calcium levels to the resting state. This allows for tuning of calcium homeostasis and signalling over a wide concentration range. The PMCA pump is also phosphorylated by cAMP-dependent kinase (PKA) and protein kinase C (PKC), each of which is also calcium activated. Other membrane-associated proteins that are activated by calmodulin include several that mediate formation of other second messengers (see Figures 5.8 and 5.9).

Activated calmodulin also influences transcriptional regulation by blocking the activity of calcineurin,[17] which then loses its capacity to dephosphorylate transcription factors. This prevents their entry to the nucleus, inhibiting their role in regulating gene expression. Clearly calcium has a central role in cellular signalling pathways.

5.6.4   Calcium Channels as Regulatory Elements

Some of the most interesting problems in biological science lie in the field of neurochemistry (Figures 5.10 and 5.12). Ca^{2+} is involved in the release of neurotransmitters from vesicles at nerve terminals and provides a useful illustration of how this connects with the chemistry of troponin C and muscle contraction discussed later. The relevant chemical pathways are illustrated in Figure 5.12,[3] but are substantially expanded upon in a recent review.[41] We have already seen (Figure 5.8) that an external stimulus results in an influx of Ca^{2+} to the cell. In response to a change in polarization at the axon terminal of a motor neuron (nerve cell), voltage-gated Ca^{2+} channels open and Ca^{2+} flows into the axon. The higher calcium levels stimulate the fusion of acetylcholine-containing vesicles in the axon with the presynaptic membrane. This occurs through the involvement of synaptotagmin, a protein that embeds itself in the vesicle membrane, but has a separate cytosolic calcium-binding domain that can promote interaction with the membrane of the neuronal cell. The fusion of the vesicle with the cell membrane is promoted by a specialized protein complex termed SNARE that works in tandem with calcium-bound synaptotagmin to promote the exocytotic event. The positively-charged calcium facilitates binding to the neuronal membrane and underlies the mechanism of calcium-induced vesicle exocytosis with and subsequent release of acetylcholine into the synaptic cleft. Basically, the action potential that travels along the nerve cell ultimately results in a change of membrane potential that switches on voltage-gated calcium channels and results in a rapid influx of calcium to the synapse. In turn this promotes rapid calcium-promoted fusion of the synaptic vesicles to the membrane and release of the neurotransmitter acetylcholine, all of which well illustrates the multifunctional roles for Ca^{2+}, including transport through voltage-gated ion channels and activation of membrane-associated

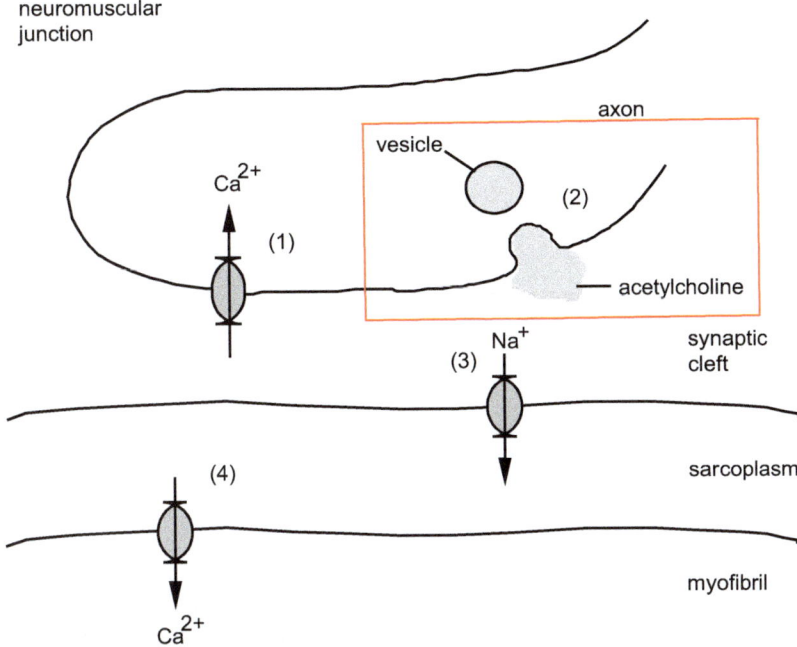

Figure 5.12 Schematic drawing of a neuromuscular junction. The axon is part of a nerve cell (neuron) that interacts with the muscle fibre. The region between the nerve and muscle cells is called the synapse. Signal transduction at the neuromuscular junction follows the following sequence of events: (1) Influx of Ca^{2+} to the axon of the motor neuron through voltage-gated calcium channels. (2) Release of acetylcholine following calcium-promoted fusion of synaptic vesicles with the pre-synaptic membrane by use of vesicle-embedded synaptotagmin protein and the SNARE protein complex. The calcium-dependent chemistry for the vesicle membrane fusion (boxed red) is illustrated in greater detail in Figure 5.13. (3) Influx of Na^+ to the muscle cell following binding of acetylcholine to the acetylcholine-gated sodium channel on the sarcoplasmic reticulum. (4) Depolarization of the sarcoplasmic reticulum triggers the release of Ca^{2+} through another voltage-gated calcium channel, which ultimately activates muscle contraction as described in Section 5.6.2. Adapted from ref. 3.

enzyme activity (kinases) and lipid association. Subsequently, these neuro-transmitters cross the synaptic cleft and bind to acetylcholine receptors. The change in receptor protein conformation allows the influx of Na^+ into the muscle cell and the resulting depolarization of the membrane potential triggers the release of Ca^{2+} from the sarcoplasmic reticulum into the cytoplasm, which triggers muscle contraction as described for troponin C (Section 5.6.5). Production of the enzyme acetylcholine esterase hydrolyses acetylcholine and allows the synaptic junction to return to pre-stimulus state. Interestingly, this enzyme is also under the control of cellular calcium,

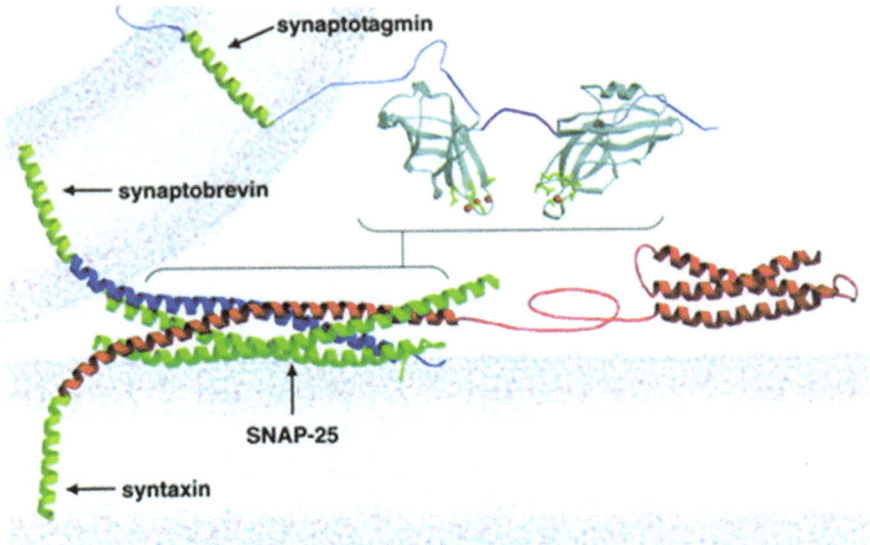

Figure 5.13   Role for calcium in vesicular membrane fusion. Calcium binds to the
EF-hand domains of synaptotagmin (red spheres) and promotes
interaction with the membrane of the neuronal cell by a specialized
protein complex termed SNARE (including syntaxin, SNAP-25 and
synaptobrevin) that works in tandem with calcium-bound synaptotagmin to facilitate the exocytotic event.
Reproduced with permission from ref. 21.

which up-regulates transcription of the gene and stabilizes the mRNA.[42] The
formation of the SNARE protein complex that facilitates "zippering" of the
synaptic vesicle to the neuronal membrane is also dependent on phosphorylation events promoted by protein kinase C and calmodulin-dependent
kinases.

5.6.5   Troponin C and Muscle Activation

The triggering of muscle action provides a useful illustration of the action of
calcium as a second messenger. The sarcoplasmic reticulum is a special
form of the endoplasmic reticulum that surrounds the fibres within the
muscle cell and, in response to a nerve stimulus, calcium ions are released
from this intracellular store. The close proximity of the sarcoplasmic reticulum to the muscle fibre minimizes the diffusional time for calcium to
reach troponin C (Tn C) on the muscle fibre (Figure 5.14). Four Ca^{2+} ions
bind to EF-hands on Tn C and induce a large conformational change (similar
to that exhibited for calmodulin in Figure 5.3) that releases an inhibitory
complex (Tn I) and allows direct interactions between the thick and thin
filaments, resulting in muscle contraction (Figure 5.14). The muscle relaxes

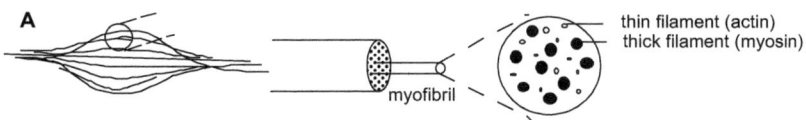

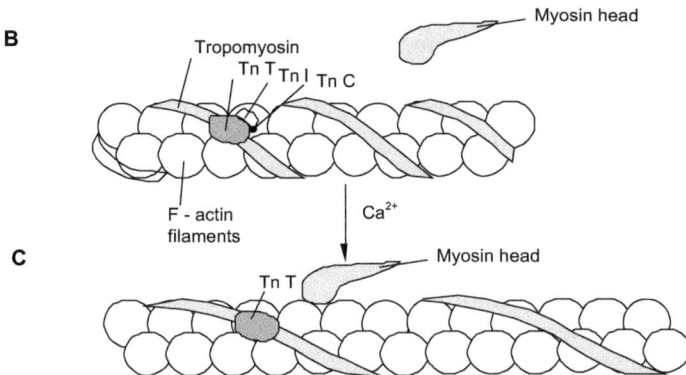

Figure 5.14 Role for calcium in the mechanics of muscle motion. (A) Schematic illustration of a muscle fibre showing the network of thick and thin filaments. (B) Actin filaments are formed from G-actin, a globular protein that polymerizes to form the filamentous F-actin complex. Tropomyosin is a helical dimer that wraps around the actin filament. Movement of the tropomyosin chain, relative to the actin filament, is responsible for muscle contraction. The interaction between the myosin heads in the thick filament, and actin, is regulated by the trimeric troponin complex consisting of Tn T, Tn I and Tn C. Calcium binds to Tn C, which results in loss of the inhibitory Tn I subunit and allows the myosin head to make intimate contact with the actin filament. The relative motion of the myosin and actin filaments results in a contraction. ATP phosphorylation of proteins provides the energy source for myosin head movement. After release of Ca^{2+} from the cell, the system reverts back to its original state. Adapted from ref. 3.

when Ca^{2+} is removed from the cell by the action of the Ca^{2+}-ATPase as described earlier (Figures 5.2 and 5.6).

5.6.6   Activation of Cellular Enzymes

We have already seen how many extracellular enzymes require calcium for full activation – particularly in the thrombin cascade. Intracellular proteases also serve important functional roles and include a number of enzymes that function on membrane proteins and promote membrane fusion.[23] Others, such as the calpains, are calcium-dependent cysteine proteases that recognize substrates on the basis of tertiary structure motifs rather than a specific amino acid substrate sequence, and have been implicated in a variety of cellular roles that include cell mobility and the cell cycle.

5.6.7   Hormonal Regulation

Calcium regulation is principally orchestrated by two hormones. Parathyroid hormone (PTH) is an 84-amino acid peptide that serves to increase calcium levels in the blood while calcitonin, a 32-amino acid peptide that is produced by the parafollicular C-cells of the thyroid, does the opposite, resulting in a lowering of blood calcium levels. PTH release from the parathyroid gland is dependent on low levels of plasma calcium and serves to enhance serum calcium by a number of discrete pathways. One involves stimulation of calcium release from bone. In another pathway, PTH up-regulates the cellular production of cytochrome P450 in the liver that hydroxylates vitamin D and converts it to a more active di-hydroxy form (25-hydroxyvitamin D, or calcidiol), which in turn is a pre-hormone that is subsequently oxidized by a kidney enzyme to produce calcitriol as the active form of the hormone. In turn, calcitriol promotes expression of the transport channel proteins that promote uptake of calcium from the intestine and transport calcium across the plasma membrane to the serum. This activity is additional to that of calbindin, a carrier protein that facilitates the uptake and movement of calcium from the intestine to the serum (Figure 8.4). In a third pathway, PTH stimulates reabsorption of calcium from urine *via* the kidney. When the levels of serum calcium return to normal, the production of PTH is switched off by use of a negative feedback loop involving G-protein-coupled calcium receptors that convert extracellular stimuli into an intracellular response. These receptors stimulate the activity of phospholipase C in the parathyroid gland that, in turn, promotes cellular conversion of PIP_2 into the second messenger IP_3. This prompts an increase in cytosolic calcium levels (Figure 5.8) that inhibits the exocytosis of PTH hormone. Calcitonin essentially serves to decrease blood calcium levels by inhibiting these three pathways for enhancing blood calcium by blocking the breakdown of bone tissue and inhibiting the uptake of calcium through the intestine and liver.

5.6.8   Transcriptional Regulation

As described earlier, there is increasing evidence for a calcium pool associated with the nuclear envelope, a discrete nucleoplasmic reticulum that is separate from the ER/SR and can provide a transient increase in nuclear calcium levels.[29,30] This pathway would be distinct from the nuclear pores that mediate passive diffusion of calcium to or from the nucleus. Nuclear calcium can activate the nuclear-localized protein kinase C and an independent nuclear IP_3 signalling pathway that regulates the production and release of nuclear protein kinases, some of which are directed to the nuclear envelope and exported to the cytosol.

There is substantial evidence for calcium indirectly regulating gene expression by phosphorylation and dephosphorylation of transcription factors through the action of calmodulin and calcineurin, among other proteins.[17,19,36] Other mechanisms for transcriptional regulation include binding of S100B (an EF-hand protein) to tumour suppressor proteins such as p53, a regulatory protein that is involved in control of cell cycle arrest, apoptosis, DNA repair and other changes in metabolism. A direct role for calcium in the regulation of gene transcription has been uncovered through the action of another EF-hand protein, the so-called DREAM protein, which binds to regulatory sequences of DNA and serves as a transcriptional repressor.[17]

5.7 Prokaryotes

5.7.1 Cytoplasmic Calcium

Because much of the cellular chemistry involving Ca^{2+} is regulatory in nature, a less important role might be expected for simpler prokaryotes and that indeed is the case.[43,44] Overall there appears to be substantial homology between prokaryotic and eukaryotic ion channels and ATP-driven pumps.[44,45] Intracellular calcium concentrations in bacteria are very similar to eukaryotic cytosolic concentrations (~ 100 nM). However, with the exception of a role in cell movement through the action of chemotactic proteins,[11] there appears to be no significant specialized role for this cation. Calcium is known to accumulate in bacteria during spore formation. A protein called protein S that is implicated in this process bears a strong homology to calmodulin. Since calmodulin is ubiquitous in eukaryotes it is likely that this, and other bacterial calmodulin homologues,[46] are the forerunners of eukaryotic calmodulin.

5.7.2 Bacterial Defence Mechanisms

Certain prokaryotes do arm themselves with ionophores (ion-carrying chelate ligands) that selectively bind divalent metals.[47] However, as there is no pronounced gradient for intra- and extracellular levels of magnesium, the imbalance of calcium concentration is critical for function and provides a target for therapeutic intervention. Accordingly, excretion of calcium-binding ionophores can serve as a defence mechanism for select bacteria by equilibrating calcium concentrations across membranes, thereby killing the target organisms. The absence of a magnesium gradient also diminishes possible side effects from magnesium coordination. Representative examples of calcium-binding ionophores demonstrate a preference for oxygen coordination and apparently a degree of asymmetry that promotes selectivity for calcium ion over magnesium (Figure 5.15).

Bacterial pathogens also impact host organisms by releasing toxins that can act at a distance from the site of infection and induce an increase in cytosolic calcium levels in epithelial cells increasing the production of pro-inflammatory cytokines or triggering muscle or cytoskeletal activity.[36] Such

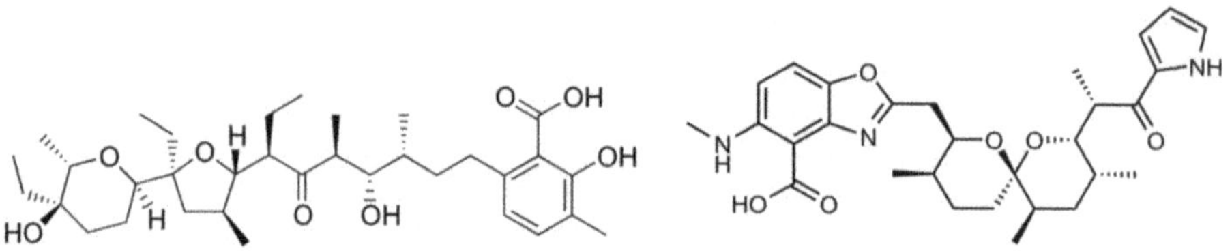

Figure 5.15   Calcium binding ionophores are effective antibiotics. (Left) Lasalocid A (X537A) from *Streptomyces lasaliensis* and (right) calcimycin (A23187) from *Streptomyces chartreusensis*.

responses may also divert the host's natural defences against the invading pathogen. Alternatively, calcium-dependent scavenging of essential nutrient metal ions by antimicrobial proteins within a host can impact the viability of pathogenic organisms.[48]

5.8   Conclusions with Open Questions

Clearly the broad scope of cellular calcium biochemistry is apparent. While new calcium-activated proteins continue to be identified, these normally fit into existing classes and functional roles. Advances in electrophysiology provide experimental tools to understand channel activity, and various microscopies and improved methods of sample preparation allow definition of membrane-associated protein complexes. With this wealth of data, many of the key questions that remain to be addressed focus on the molecular level understanding of structure and function. Much remains to be done in the field of high-resolution structural definition of the various classes of calcium channels, and a molecular understanding of how they are regulated and serve to selectively transport calcium. The molecular mechanisms of signalling within the cell are now most certainly poised for more detailed elaboration with the tools of molecular biology and solution spectroscopies now at hand. Further elaboration of the chemistry of calcium-dependent enzymes will surely provide a new collection of therapeutic targets for cardiac, muscular and neurodegenerative disease.

Areas that appear particularly ripe for future development include understanding the spatiotemporal dependence of calcium concentrations on physiological response. It is becoming increasingly clear that subcellular localization of calcium is not only focused on internal organelles, but even at the suborganellar level, with evidence for local concentration spikes and uneven concentration distributions within the cytosol, nucleus, *etc*. The latter are of specific relevance to developmental biology, cellular replication and neuronal activities such as memory, pain, addiction and neuroprotection pathways. There will undoubtedly be further advances in the molecular level understanding of the role of calcium in nuclear regulatory chemistry, as well as in understanding its chemistry as a neural effector.[41]

References

1. E. Carafoli and J. Krebs, ed., *Calcium Homeostasis*, Springer, New York, 2000.
2. M. R. Celio, T. Pauls and B. Schwaller, ed., *Guidebook to the Calcium-Binding Proteins*, Oxford University Press, Oxford, 1996.
3. J. A. Cowan, *Inorganic Biochemistry. An Introduction*, Wiley-VCH, New York, 2nd edn, 1997.
4. S. Forsen and J. Kordel, in *Bioinorganic Chemistry*, ed. I. Bertini, H. B. Gray, S. J. Lippard and J. Valentine, University Science Books, Mill Valley, CA, 1994, pp. 107–166.
5. A. Muranyi and B. E. Finn, in *Handbook on Metalloproteins*, ed. I. Bertini, H. Sigel and A. Sigel, Marcel Decker, New York, 2001, pp. 93–152.
6. J. A. Cowan, *Chem. Rev.*, 1998, **98**, 1067–1087.
7. R. J. P. Williams, *Biochim. Biophys. Acta*, 2006, **1763**, 1139–1146.
8. C. B. Black, W.-W. Huang and J. A. Cowan, *Coord. Chem. Rev.*, 1994, **135/136**, 165–202.
9. M. E. Maguire and J. A. Cowan, *BioMetals*, 2002, **15**, 203–210.
10. J. A. Cowan, in *Comprehensive Coordination Chemistry II*, Pergamon Press, New York, 2004, pp. 123–140.
11. J. J. Falke, S. K. Drake, A. L. Hazard and O. B. Peersen, *Q. Rev. Biophys.*, 1994, **27**, 219–290.
12. W. Yang, H.-W. Lee, H. Hellinga and J. J. Yang, *Proteins: Struct. Funct. Genet.*, 2002, **47**, 344–356.
13. S. Cates, M. L. Teodoro and G. N. Phillips, Jr., *Biophys. J.*, 2002, **82**, 1133–1146.
14. Y. Zhou, F. K. Frey and J. J. Yang, *Cell Calcium*, 2009, **46**, 1–17.
15. N. V. Valeyev, D. G. Bates, P. Heslop-Harrison, I. Postlethwaite and N. V. Kotov, *BMC Systems Biology*, 2008, **2**, 48–65.
16. A. Lewit-Bentley and S. Réty, *Curr. Opin. Struct. Biol.*, 2000, **10**, 637–643.
17. M. Ikura, M. Osawa and J. B. Ames, *BioEssays*, 2002, **24**, 625–636.
18. Y. Chen, S. Xue, Y. Zhou and J. J. Yang, *Sci. China: Chem.*, 2010, **53**, 52–60.
19. R. Rizzuto and T. Pozzan, *Nat. Genet.*, 2003, **34**, 135–141.
20. E. W. Davie and J. D. Kulman, *Semin. Thromb. Hemost.*, 2006, **32**(1), 3–15.
21. J. T. Littleton, J. Bai, B. Vyas, R. Desai, A. E. Baltus, M. B. Garment, S. D. Carlson, B. Ganetzky and E. R. Chapman, *J. Neurosci.*, 2001, **21**, 1421–1433.
22. J. C. Lee, A. Simonyi, A. Y. Sun and G. Y. Sun, *J. Neurochem.*, 2011, **116**, 813–819.
23. R. L. Mellgren, *FASEB J.*, 1987, **1**, 110–115.
24. A. K. Downing, P. A. Handford and I. D. Campbell, in *Topics in Biological Inorganic Chemistry Calcium Homeostasis.*, ed. E. Carafoli and J. Krebs, Springer, New York, 2000, vol. 3, pp. 83–99.
25. J. Evenas, A. Malmendal and S. Forsen, *Curr. Opin. Chem. Biol.*, 1998, **2**, 293.
26. A. N. Zelensky and J. E. Gready, *FEBS J.*, 2005, **272**, 6179–6217.

27. V. Gerke, C. E. Creutz and S. E. Moss, *Nat. Rev. Molec. Cell Biol.*, 2005, **6**, 449–461.

28. S. Wang, W. R. Trumble, H. Liao, C. R. Wesson, A. K. Dunker and C. Kang, *Nat. Struct. Biol.*, 1998, **5**, 476–483.

29. R. R. Resende, L. M. Andrade, A. G. Oliveira, E. S. Guimarães, S. Guatimosim and M. F. Leite, *Cell Commun. Signaling*, 2013, **11**, 14.

30. D. A. Gomes, M. F. Leite, A. M. Bennet and M. H. Nathanson, *Can. J. Physiol. Pharmacol.*, 2006, **84**, 325–332.

31. C. Toyoshima, M. Nakasako, H. Nomura and H. Ogawa, *Nature*, 2000, **405**, 647–655.

32. J. V. Mùller, B. Juul and M. Le Maire, *Biochim. Biophys. Acta*, 1996, **1286**, 1–51.

33. D. Lee and M. Michalak, *Anim. Cells Syst.*, 2012, **16**, 269–273.

34. J. J. Mackrill, *Biochem. J.*, 1999, **337**, 345–361.

35. T. Yamauchi, *Biol. Pharm. Bull.*, 2005, **28**, 1342–1354.

36. G. T. Nhieu, C. Clair, G. Grompone and P. Sansonetti, *Biol. Cell*, 2004, **96**, 93–101.

37. H. Tidow, L. R. Poulsen, A. Andreeva, M. Knudsen, K. L. Hein, C. Wiuf, M. G. Palmgren and P. Nissen, *Nature*, 2012, **491**, 468–473.

38. W. Y. Cheung, *Science*, 1980, **207**, 19–27.

39. L. A. Onek and J. R. Smith, *J. Gen. Microbiol.*, 1992, **138**, 1039–1049.

40. F. D. Leva, T. Domi, L. Fedrizzi, D. Lim and E. Carafoli, *Arch. Biochem. Biophys.*, 2008, **476**, 65–74.

41. H. Bading, *Nat. Rev. Neurosci.*, 2013, **14**, 593–608.

42. W. Gao, H. Zhu, J. Y. Zhang and X. J. Zhang, *Cell. Mol. Life Sci.*, 2009, **66**, 2181–2193.

43. J. Michiels, C. Xi, J. Verhaert and J. Vanderleyden, *Trends Microbiol.*, 2002, **10**, 87–93.

44. D. C. Dominguez, *Molec. Microbiol.*, 2004, **54**, 291–297.

45. I. V. Shemarova and V. P. Nesterov, *J. Evol. Biochem. Physiol.*, 2005, **41**, 12–19.

46. K. Yang, *J. Mol. Microbiol. Biotechnol.*, 2001, **3**, 457–459.

47. M. R. Truter, in *Calcium in Biological Systems,* 30th Symposium of the Society for Experimental Biology, ed. C. J. Duncan, Cambridge University Press, Cambridge, 1976, vol. 30, pp. 19–40.

48. M. B. Brophy, J. A. Hayden and E. M. Nolan, *J. Am. Chem. Soc.*, 2012, **134**, 18089–18100.

CHAPTER 6

Vanadium

EUGENIO GARRIBBA*[a] AND DANIELE SANNA[b]

[a] Dipartimento di Chimica e Farmacia, Università di Sassari, Via Vienna 2, I-07100 Sassari, Italy; [b] Istituto CNR di Chimica Biomolecolare, Trav. La Crucca 3, I-07040, Sassari, Italy
*Email: garribba@uniss.it

6.1   Introduction

If things had gone differently this chapter would be entitled *panchromium* or, more likely, *erythronium*, but it would still describe the element with atomic number 23. Vanadium was discovered in 1801 by Andrés Manuel del Rio, chair of mineralogy at the Royal College of Mines in Mexico City, in *plomo pardo* (brown lead), a mineral extracted from the mine of Purisima del Cardonal near Zimapán.[1] Since the chemistry of the element isolated from this mineral was unlike that of any element known to him, he decided that he had discovered a new element. Del Rio convinced himself later that the new element was actually the recently discovered chromium. Samples of minerals were sent to the Institut de France and analysed in 1804 by Hyppolyte Victor Collet-Descotils, who wrote in his report that "*this ore contains nothing of new metal*". In 1831 the Swedish chemist Nils Gabriel Sefström discovered a new element and named it *vanadium*, after Vanadis, another name for Freya, the Norse goddess of love and beauty. Friedrich Wöhler, after analysing samples of brown lead from Zimapán, confirmed the presence of vanadium but supported Sefström's priority in the discovery. The brown lead of Zimapán is now recognized as *vanadinite*, $Pb_5Cl(VO_4)_3$.

RSC Metallobiology Series No. 2
Binding, Transport and Storage of Metal Ions in Biological Cells
Edited by Wolfgang Maret and Anthony Wedd
© The Royal Society of Chemistry 2014
Published by the Royal Society of Chemistry, www.rsc.org

Vanadium (symbol V) is the 22nd most abundant element in the Earth's crust (0.013%), being more abundant than better known elements such as copper and zinc. Of the 60 known V minerals, only patronite (VS_4), roscoelite ($K(V,Al,Mg)_2AlSi_3O_{10}(OH)_2$), carnotite ($K_2(UO_2)_2V_2O_8 \cdot 3(H_2O)$) and vanadinite are commercial sources. Small amounts of V (ranging from 0.1 to 1%) occur as porphyrin complexes in fossil fuels from Venezuela, Angola, California, Iran, Iraq and Kuwait.

Burning of oil and coal fuels provides about 90% of the 64 000 tons of V that are emitted each year to the atmosphere from natural and anthropogenic origins. This may constitute a health problem especially in populated areas near power plants burning V-rich fossil fuels. The element enters the atmosphere as fly ash and its absorption in the lungs is much more efficient than that ingested with food. In sea water, V as vanadate(V) is the 2nd most abundant soluble transition metal. It occurs at a concentration of 35 nM, exceeded only by molybdenum (100 nM).

Vanadium can assume oxidation states ranging from $-I$ to $+V$ and $+III$, $+IV$ and $+V$ have been found in biological systems.[†] V(III) has a $3d^2$ electronic configuration and is normally hexa-coordinate with octahedral geometry. However, its relatively large ionic radius allows it to form hepta-coordinate complexes of distorted pentagonal bipyramidal structure. At biologically relevant concentrations ($<1~\mu M$), V(III) exists as the aqua ion $[V^{III}_{(aq)}]^{3+}$ only in acidic solutions. At micromolar concentrations, $[V^{III}(H_2O)_5(OH)]^{2+}$ and $[V^{III}(H_2O)_4(OH)_2]^+$ are formed with pK_a values of 2.7 and 4.0.[2]

The electronic configuration $3d^1$ for V(IV) makes it an ideal candidate for study with electron paramagnetic resonance spectroscopy (EPR). In an acidic aqueous solution, V(IV) exists as the oxidovanadium(IV) cation $[V^{IV}O_{(aq)}]^{2+}$ (vanadyl) featuring a VO multiple bond.[‡] At higher pH, hydrolysis products $[V^{IV}O(H_2O)_n(OH)_m]^{(2-m)+}$ are formed before precipitation of the hydroxide $V^{IV}O(OH)_2$. The latter is soluble in alkaline solution forming anions $[(V^{IV}O)_2(OH)_5]^-$ and $[V^{IV}O(OH)_3]^-$.[3,§] $[V^{IV}O(H_2O)_5]^{2+}$ and $[V^{IV}O(OH)_3]^-$ are EPR-active, while the other hydrolysed species and the hydroxide are EPR-silent, evidently due to spin coupling induced by polymerization.

The coordination geometry of $[V^{IV}O]^{2+}$ complexes is square pyramidal when they are penta-coordinate, or distorted octahedral when hexa-co-ordinate. Regular square pyramidal complexes are observed rarely and, when

---

[†]According to the IUPAC nomenclature, the sign "plus" for positive oxidation states must be omitted. In this chapter, however, this is used to avoid confusion with the atomic symbol of vanadium, V.

[‡]Historically, $V^{IV}O^{2+}$ ion has been called vanadyl or oxovanadium(IV) ion. Now, IUPAC nomenclature recommends oxidovanadium(IV). The bond order between V^{4+} and O^{2-} anion is 2, if the valence bond theory is used.[2] This description has been adopted for many years and will be used in this chapter. However, the VO bond should be correctly treated as a triple bond with one σ plus two π bonds.[2]

[§]The relative and absolute concentrations of $[(VO)_2(OH)_5]^-$ and $[VO(OH)_3]^-$ at pH 7.4 depend on the total vanadium concentration. $[(VO)_2(OH)_5]^-$ is the only anionic species in solution for concentrations around 1 mM, but the importance of $[VO(OH)_3]^-$ increases as the total concentration decreases to the biologically relevant values (1–100 nM).

the ligands are bidentate with two different donors, a distortion toward trigonal bipyramidal geometry is detected. The extent of this distortion can be measured by a parameter $\tau = (\beta - \alpha)/60$, where β and α are the angles between the two pseudo-axial and two pseudo-equatorial donors. The so-called "bare" V^{IV} species are very rare because the oxido ligand of $[V^{IV}O]^{2+}$ can be lost only under particular conditions, usually leading to the formation of hexa-coordinate complexes of stoichiometry $[V^{IV}L_3]^{2-}$. L^{2-} must be a ligand such as catecholate, 1,2-benzenedithiolate or maleonitriledithiolate.[2] In these cases, the coordination geometry is intermediate between the octahedron and the trigonal prism, depending on the ligand bite.

V(V) has no 3d electrons and can be studied with ^{51}V nuclear magnetic resonance (NMR) spectroscopy. It exists in alkaline solutions (pH > 13) in the form of vanadate $[V^{V}O_4]^{3-}$ that can be protonated to form $[HV^{V}O_4]^{2-}$ and $[H_2V^{V}O_4]^{-}$ (mononuclear vanadate(V) ions can be generically indicated by V1). An increase of concentration and a decrease of pH leads to the formation of oligonuclear condensation products. Di- ($[V^{V}_2O_7]^{4-}$, $[HV^{V}_2O_7]^{3-}$ and $[H_2V^{V}_2O_7]^{2-}$ or V2), tetra- ($V_4O_{12}^{4-}$ or V4), penta- ($V_5O_{15}^{5-}$ or V5) and deca-vanadates ($[V_{10}O_{28}]^{6-}$, $[HV_{10}O_{28}]^{5-}$, $[H_2V_{10}O_{28}]^{4-}$ and $[H_3V_{10}O_{28}]^{3-}$, V10) are formed depending on the pH, total concentration and ionic strength. Consequently, speciation can be very complicated. VO_4 tetrahedra share corners with neighbouring units in a number of these structures, but the coordination geometry is variable. It can be tetrahedral, as in the case of $[V^{V}O_4]^{3-}$ and its oligomers, penta-coordinate (with geometries intermediate between square pyramidal and trigonal bipyramidal) or distorted octahedral. Around pH 7.4, uncomplexed V(V) forms only anionic species. The most important are $[H_2V^{V}O_4]^{-}$ and $[HV^{V}O_4]^{2-}$ (for concentrations below 1 μM). On the other hand, the cationic forms $[V^{V}O]^{3+}$ and $[V^{V}O_2]^{+}$ are significantly stabilized by complexation.[2] In an alternative description, V(V) complexes can be represented by the formal replacement of one or two hydroxido or oxido ligands of $[H_2V^{V}O_4]^{-}$ (indicated with V^{-}) by the ligand donors.[4,¶]

Under aerobic conditions, $[V^{IV}O_{(aq)}]^{2+}$ and $[V^{III}_{(aq)}]^{3+}$ are readily oxidized by O_2 (O_2/H_2O, $E^{\circ} = 1.23$ V) to V^{V}. However, in the anaerobic environment of the cytoplasm, specific reducing agents can reduce $[V^{V}O_{2(aq)}]^{+}$ to $[V^{IV}O_{(aq)}]^{2+}$, especially if vanadium(V) is in an uncomplexed form. The reduction of $[H_2V^{V}O_4]^{-}$ is less easy, whereas that of $[V^{IV}O_{(aq)}]^{2+}$ to $[V^{III}_{(aq)}]^{3+}$ does not occur under common conditions (Table 6.1).

6.2 Nutritional Aspects and Absorption of Vanadium

Vanadium has been classified as an ultra-trace metal *i.e.* a human nutritional requirement of less than 1 mg kg^{-1} diet and present in tissues in

---

¶Using the notation of ref. 4, the formation of mono- and bis-chelated V(V) species with a bidentate ligand LH can be described with the reactions: $[H_2V^{V}O_4]^{-} + LH \leftrightarrows [HV^{V}O_3(L)]^{-} + H_2O$ (with $[HV^{V}O_3(L)]^{-} = VL^{-}$) and $[HV^{V}O_3(L)]^{-} + LH \leftrightarrows [V^{V}O_2(L)_2]^{-} + H_2O$ (with $[V^{V}O_2(L)_2]^{-} = VL_2^{-}$).

Table 6.1 Redox potentials for inorganic vanadium species.[a]

Redox reaction	$E°$	E (pH = 7)
$[V^VO_{2(aq)}]^+ + 2H^+ + e^- = [V^{IV}O_{(aq)}]^{2+} + H_2O$	+1.02	+0.19
$[H_2V^VO_4]^- + 4H^+ + e^- = [V^{IV}O_{(aq)}]^{2+} + 3H_2O$	+1.31	−0.34
$[V^{IV}O_{(aq)}]^{2+} + 2H^+ + e^- = [V^{III}_{(aq)}]^{3+} + H_2O$	+0.36	−0.46

[a]Potentials measured in V (volts) *versus* the normal hydrogen electrode (NHE).

the range of micrograms per kg. However, even if multiple lines of evidence suggest that vanadium is beneficial to human health, its mechanism of action remains obscure. In the seventh and eighth decades of the twentieth century, possible signs of V deficiency were described for some animals. Goats fed with less than 10 ng day^{-1} diet had difficulty in conceiving and exhibited a high rate of spontaneous abortion. Other experiments with goats demonstrated that a V-deficient diet causes skeletal deformations, convulsion and death within 90 days of birth. In rats, V deprivation resulted in an increase of thyroid weight and a decrease in growth; it also affected the activity of pancreatic amylase and serum lactate dehydrogenase. These findings indicate that physiological amounts of vanadium affect thyroid hormone and carbohydrate metabolism, providing evidence that it is an essential element for higher forms of life.[5]

A balance of the daily intake of V from the diet, the amount excreted in the faeces and the concentration found in urine indicates that less than 5% of the intake is absorbed in the gastrointestinal tract and that the remainder is excreted.[6] However, the absorption of orally ingested V is dependent on a variety of factors, including whether the metal is complexed to a ligand. Most studies on animals confirm the conclusion that V is poorly absorbed, although other evidence indicates that V absorption can reach values as high as 40%, depending on the chemical form and the dietary conditions.[7] Vanadium intake in the US diet is thought to be in the range 10–60 μg day^{-1}, with many diets containing <15 μg.[7] The latter diets cause no obvious signs of deficiency. Therefore, a daily dietary intake of 10 μg V probably meets the requirements of humans. An upper safe nutritional intake for V is difficult to define because humans apparently are more tolerant to high concentrations than animals such as rats.

Foods rich in vanadium (more than 40 ng g^{-1}) include shellfish, mushrooms, parsley, dill seed and black pepper and their high consumption can increase the intake considerably. Cereals, liver and fish tend to have intermediate amounts of vanadium (from 5 to 40 ng g^{-1}). Beer and wine have high concentrations (25–45 ng g^{-1}), whereas other beverages, fats, oils, fresh fruits and fresh vegetables generally contain less than 5.0 ng g^{-1}, and often less than 1.0 ng g^{-1}.[5]

Commercially available preparations containing $V^{IV}OSO_4$, such as *Vanadyl Fuel*, are nowadays popular among body builders because it is supposed that they help to increase muscle mass.[8] In Japan, a mineral water from the Mount Fuji region is on the market as *Vanadium Water* and is supposed to act as a general tonic. It contains *ca.* 50 μg of V per litre in the form of hydrogenvanadates(V), $[H_2V^VO_4]^-$ and $[HV^VO_4]^{2-}$.[2]

6.3 Uptake in Intestinal Cells and Release in the Blood

The pathways for distribution of vanadium compounds in the body were reviewed recently (Figure 6.1).[9] Absorption is oral or through the lungs after inhalation. In the lungs, a fraction of the absorbed particles containing V (mainly as V_2O_5) undergo solubilization and conversion to vanadium(V) that then enters the bloodstream. The absorption from the lungs depends on the particle size and solubility of the compounds. Humans exposed to vanadium have higher levels of vanadium in urine and blood, but the V concentration drops rapidly after the exposure has ceased.

Most of the orally ingested vanadium is reduced to $[V^{IV}O]^{2+}$ in the stomach and remains in this form after passing into the duodenum. *In vitro* studies suggest that absorption depends upon oxidation state: vanadium(V)

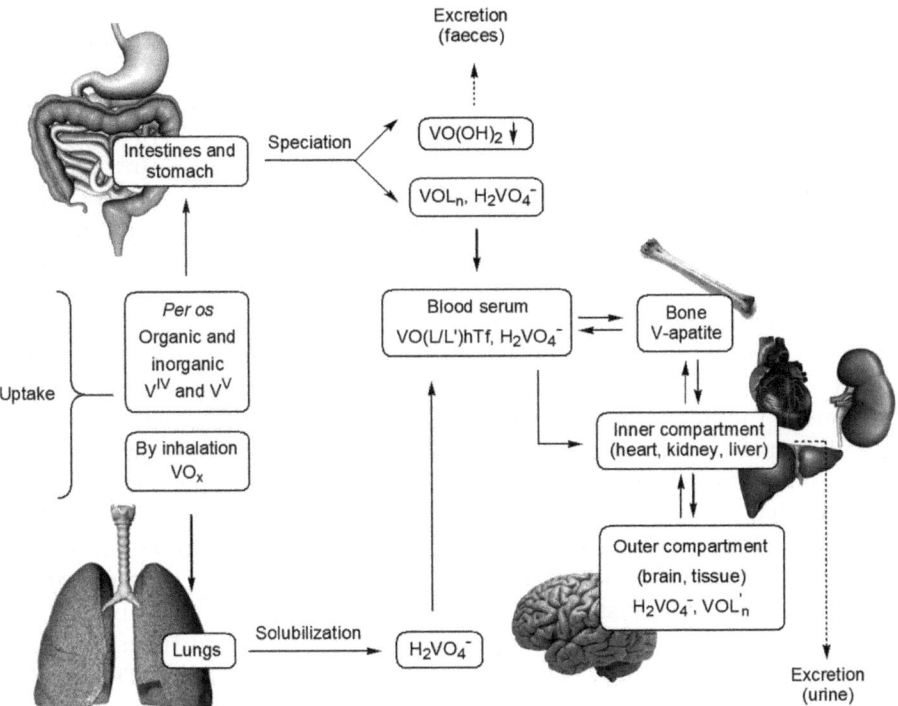

Figure 6.1 The excretion forms and the target tissues of vanadium in humans. L and L' denote possible bio-ligands (*e.g.* lactate, citrate, glutathione); hTf denotes human serum transferrin.
Adapted from ref. 9 with permission from Future Science Ltd.

appears to be absorbed three to five times more effectively than is $[V^{IV}O]^{2+}$. It is proposed that the vanadate(V) anion can cross the cellular membranes using available anion transport systems, for example, that of phosphate.[10–13] Caution is needed before assuming that ingested inorganic vanadium is poorly absorbed because the nature of the diet can greatly affect its absorption when administered during meals;[7] for example, when $V^{IV}O$ compounds such as $[V^{IV}OL_2]$ (L^- is a monoanionic bidentate ligand) are orally administered, the absorption is considerably increased in comparison to salts such as $V^{IV}OSO_4$. Therefore, a number of substances can affect the amount of vanadium absorbed,[6] including ligands that promote reduction of vanadium(V) to oxidovanadium(IV) and stabilize $[V^{IV}O]^{2+}$ through coordination.

6.4   Transport of Vanadium in the Blood

6.4.1   Speciation

The relevant oxidation states in biological systems are +III, +IV and +V.[2,14,15] However, in humans and higher animals, V^{III} is susceptible to oxidation and is less important. The only exceptions are ascidians and polychaeta worms (see Section 6.9.1) where this is the predominant oxidation state.[16,17]

Under physiological conditions, V(IV) should precipitate as $V^{IV}O(OH)_2$ (with partial formation of soluble hydrolysis products $[(V^{IV}O)_2(OH)_5]^-$ and $[V^{IV}O(OH)_3]^-$). However, bio-ligands in the blood can suppress the hydrolysis processes and maintain $[V^{IV}O]^{2+}$ in solution. On the other hand, V(V) should form $[H_2V^VO_4]^-$ (the main species at pH 7.4) and $[HV^VO_4]^{2-}$ in the absence of bio-ligands, or coordination compounds featuring $[V^VO]^{3+}$ and $[V^VO_2]^+$ centres (Section 6.1).

No consensus exists on the nature of the species nor on the oxidation state of vanadium in blood. Regardless of the form initially injected, interconversion of V^{IV} and V^V occurs until a constant ratio between the concentrations of the two states is reached.[18] *In vivo* EPR studies monitoring blood circulation in rats showed that, when administered as $V^{IV}OSO_4$, at least 90% is present in oxidation state +IV in nearly all organs.[19] This contrasts with the fact that oxidation of $[V^{IV}O]^{2+}$ ion to V^V is expected to occur rapidly in aqueous solution at physiological pH, with a half-time of 5–13 minutes at room temperature.[20] The presence of reducing agents such as ascorbate, catecholamines and cysteine plus the binding of $[V^{IV}O]^{2+}$ by bio-ligands stabilizes the +IV state and suppresses oxidation, at least in part. In this regard, complexation of $[V^{IV}O]^{2+}$ by human serum transferrin (hTf) seems to be particularly important. However, availability of molecular oxygen dissolved in the blood should be sufficient to oxidize the available V^{IV} to V^V.[18] Therefore, a definitive description must take into account interconversion of oxidation states +IV and +V and the fact that an equilibrium distribution of V^{IV} and V^V is probably reached in the bloodstream.

Divergent values have been reported for the concentration for V in human serum and blood[||] (from 0.45 to 15 700 nM). A critical re-examination of the available data concluded that 0.45–18.4 nM was more realistic and that values in the range 1–10 nM may be considered normal for blood and urine.[21] Rehder proposed that the value could be larger, around 200 nM.[2] Further analysis of results indicates that higher values for V concentration were determined in blood rather than in serum. This, in turn, suggests a distribution of vanadium between the serum and erythrocytes.

6.4.2 Distribution of Vanadium between Blood Serum and Erythrocytes

Many studies have examined the mechanism of transport across the erythrocyte membrane to determine the ratio of V between the plasma and erythrocytes. Vanadate(V) crosses the cell wall easily, *via* anion channels, and penetrates the membrane much faster than do V(IV) salts.[22] Vanadate(V) transport is inhibited by 4,4′-diisothiocyanostilbene-2,2′-disulfonic acid (DIDS), a specific inhibitor of anion channels.[23] Inside the cells, vanadate(V) is reduced to V(IV) by the intact erythrocyte at the expense of cellular glutathione (GSH) without detection of the glutathione thiyl radical GS•. The reduction is nearly quantitative after 23 h. Observed inhibition of vanadate(V) uptake in the presence of the GSH depletor diethyl maleate supports the proposed mechanism of reduction.[22] Further evidence is the observed delay of uptake by red blood cells when these are incubated with V(IV) instead of V(V). The effect is attributed to the time required for partial oxidation of V(IV) prior to possible transport.[11,18] It has been proposed that the divalent metal transporter 1 protein (DMT1, also known as Nramp2, that carries Fe(II) into the cells of the gastrointestinal system and out of endosomes in the transferrin cycle) may transport $[V^{IV}O]^{2+}$ ion through the erythrocyte membrane.[24] On the whole, the concentration of V in erythrocytes is determined by the balance of the efflux and influx rates. The latter is dependent primarily on the extracellular concentration of vanadate(V), tempered by that of oxidovanadium(IV) and their rates of conversion.[11]

Several hypotheses have been advanced about the distribution of vanadium between blood serum and erythrocytes but none appears to be definitive. The ratio depends on the erythrocyte V^V/V^{IV} redox reactions (reduction to V^{III} is usually neglected), on binding of both V^{IV} and V^V to available bio-ligands and on rates of influx and efflux across the cell membrane.[18] It may also depend on the presence or absence of human transferrin (hTf) in the experiments as hTf stabilizes V^{IV} in the serum, slowing oxidation to V^V.[5,22] On the basis of all available data, it is suggested that in

---

[||]Blood serum is distinguished from blood plasma by the absence of fibrinogen and other proteins that promote coagulation. For the present purposes, blood serum and blood plasma can be considered as synonyms, even if this is not technically correct.

mammals the distribution of vanadium between plasma and erythrocytes is in the range between 70 : 30 and 95 : 5 and does not seem to be dependent on the chemical form initially present.[18,25] In addition, neutral $[V^{IV}OL_2]$ complexes can diffuse passively through the erythrocyte membrane and strongly affect the ratio V(plasma)/V(erythrocytes).[23]

Most of the experimental studies indicate that the $[V^{IV}O]^{2+}$ ion is bound mainly to haemoglobin in erythrocytes, even though neither V^{IV} nor V^V have been shown to bind to haemoglobin *in vitro*.[26] The level of $[V^{IV}O]^{2+}$ bound to haemoglobin was diminished by addition of adenosine 5′-triphosphate (ATP) and 2,3-diphosphoglycerate, suggesting that intracellular bio-ligands compete for the V binding to haemoglobin.[27]

6.4.3 Vanadium Transport in Blood Plasma

About 90% of vanadium present in the blood is proposed to be associated with the plasma fraction (Table 6.2)[15] and exclusively in oxidation states +IV and +V. When vanadium is absorbed in the gastrointestinal tract and passes into the plasma, it encounters many bio-ligands of low molecular mass (*l.m.m.* components) such as citrate, lactate, phosphates, amino acids, *etc.* and others of high molecular mass (*h.m.m.* components), such as human serum transferrin (hTf), albumin (HSA) and immunoglobulins (Ig). Its final form depends on its affinity for the various bio-ligands and on their relative concentrations. For example, albumin is present in much larger concentration than transferrin (approximately 20 times), but transferrin has a higher affinity for vanadium. Consequently, the balance between thermodynamic stability of the species formed and the relative concentration of HSA and hTf determines the distribution between the two proteins.

6.4.4 Binding of Vanadium(IV) and Vanadium(V) to Components of Low Molecular Mass

On the basis of the high affinity of V(IV) and V(V) for oxygen donors and their concentration in the plasma, lactate (lact), citrate (citr), oxalate (ox), phosphate

Table 6.2 Concentrations of the most important bio-ligands of blood serum.[a,b]

Bio-ligand	Concentration	Bio-ligand	Concentration
Albumin (HSA)	630 μM	**Histidine (His)**	77.0 μM
Carbonate (HCO_3^-/ CO_3^{2-})	24.9 mM	**Immunoglobulin G (IgG)**	84.0 μM
Cystine (Cys-Cys)	10.9 μM	**Lactate (lact⁻)**	1.51 mM
Cysteine (Cys)	33.0 μM	**Oxalate (ox²⁻)**	9.20 μM
Citrate (citr³⁻)	99.0 μM	**Phosphate ($H_2PO_4^-$/ HPO_4^{2-})**	1.10 mM
Glycine (Gly)	2.30 mM	Sulfate (SO_4^{2-})	330 μM
Glutamate (Glu)	60.0 μM	**Transferrin (hTf)**	37 μM

[a]Values taken from refs 28 and 29.
[b]The bio-ligands with highest affinity for V(IV) and V(V) are indicated in bold.

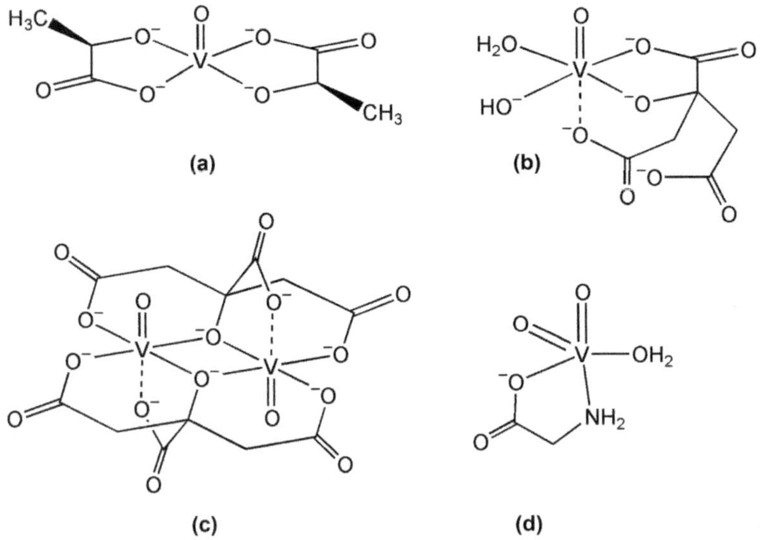

Figure 6.2 The most important complexes of vanadium(IV) and vanadium(V) formed under physiological conditions by *l.m.m.* bio-ligands of the blood: (a) $[V^{IV}O(lactH_{-1})_2]^{2-}$; (b) $[V^{IV}O(citrH_{-1})(OH)]^{3-}$; (c) $[(V^{IV}O)_2(citrH_{-1})_2]^{4-}$ and (d) $[V^{V}O_2(H_2O)(Gly)]$.

($H_2PO_4^-$ and HPO_4^{2-}), glycine (Gly) and histidine (His) have been identified as the most important *l.m.m.* bio-ligands (listed in bold in Table 6.2).[30,31] The binding strength of other serum components such as sulfate and carbonate can be considered as negligible. EPR evidence indicates that only lactate and citrate interact with $[V^{IV}O]^{2+}$ under physiological conditions,[31] the complexation being favoured for the $lactH_{-1}^{2-}$ and $citrH_{-1}^{4-}$ forms, where the subscript −1 refers to deprotonation of the alcohol groups. The most important complexes formed by lactate and citrate under such conditions are $[VO(lactH_{-1})_2]^{2-}$, $[V^{IV}O(citrH_{-1})(OH)]^{3-}$ (at V concentration of 10 nM) and $[(VO)_2(citrH_{-1})_2]^{4-}$ (100 nM) (see Figure 6.2a–c for molecular structures). For vanadate(V), models based on thermodynamic stability constants suggest that, in the absence of proteins, the main species are $[H_2V^{V}O_4]^-$ (major) and $[HV^{V}O_4]^{2-}$ (minor) for $[V^{V}_{total}] = 10–100$ nM and that, among the six most important *l.m.m.* bio-ligands, only glycine contributes appreciably to the metal speciation forming $[V^{V}O_2(H_2O)(Gly)]$ (VGly; Figure 6.2d; see ref. 4.[30]

6.4.5 Binding of Vanadium(III), Vanadium(IV) and Vanadium(V) to Human Serum Transferrin

The principal biochemical function of hTf is the transport of ferric ions in the blood *via* reversible binding. However, since only 30% of the four binding sites in the homo-dimer are occupied by this ion under normal conditions, hTf is available to bind a wide variety of metal cations, both essential and toxic.[32,33]

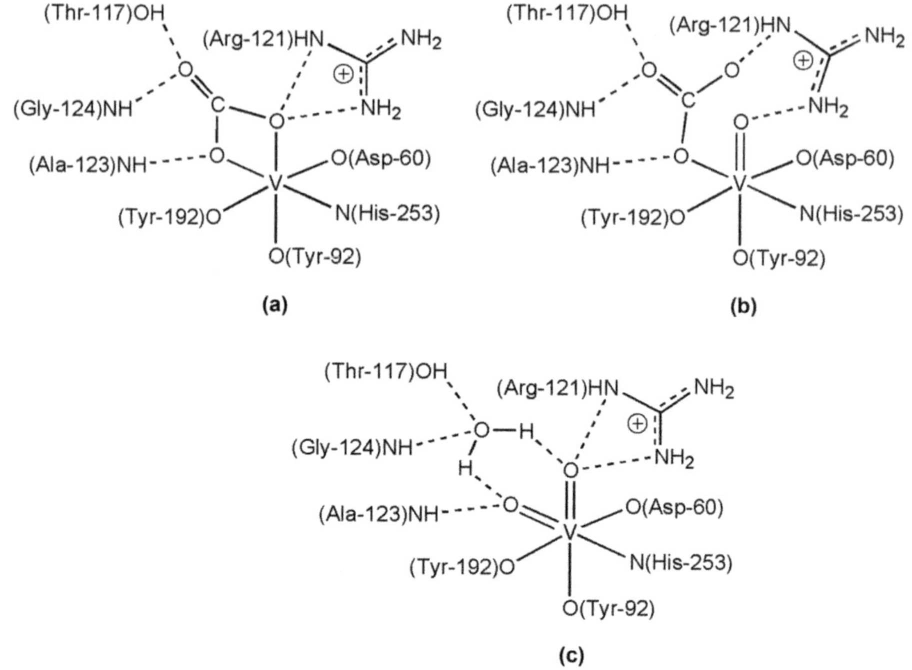

Figure 6.3   Coordination environment of V bound to hTf$_C$ site of transferrin: (a) V^{III}; (b) [V^{IV}O]$^{2+}$ and (c) [V^{V}O$_2$]$^+$. The proposed networks of hydrogen bonds involving carbonate and V=O groups are also shown.

[V^{IV}O]$^{2+}$ binds to the same sites occupied by Fe^{3+}, one in the *N*-lobe (hTf$_N$) and the second in the *C*-lobe (hTf$_C$).[33] A single conformer corresponding to binding to hTf$_C$ and two conformers corresponding to the binding to hTf$_N$ are observed. As does Fe^{III}, binding of [V^{IV}O]$^{2+}$ requires HCO$_3^-$ or CO$_3^{2-}$, anchored in place by electrostatic and hydrogen bonding interactions with positively charged arginine residues. They can be replaced by other (synergistic**) anions that have the features (carboxylate and an electron-withdrawing group) that favour metal binding.[32,33] In the absence of an adequate synergistic anion, the positively charged arginine side-chain and the *N*-terminus of helix 5 may inhibit metal binding to the specific sites.

Spectroscopic results indicate that the [V^{IV}O]$^{2+}$ ion in the hTf$_C$ and hTf$_N$ sites occupies the same coordination environment as does Fe^{III} (Figure 6.3b):†† a monodentate carbonate, a monodentate carboxylate (Asp-60 for hTf$_C$ and Asp-63 for hTf$_N$), two tyrosinate oxygens (one equatorial [Tyr-192 for hTf$_C$ and Tyr-188 for hTf$_N$, the other axial [Tyr-92 for hTf$_C$ and

---

**An anion is defined as synergistic if it promotes the binding of iron(III) to the hTf$_C$ and hTf$_N$ sites.
††See Chapter 11 for the structure of transferrin.

Tyr-95 for hTf$_N$]) and a histidine nitrogen (His-253 for hTf$_C$ and His-249 for hTf$_N$).[34] The rather high value of the ^{51}V hyperfine coupling constant (A_z) is consistent with one of the two Tyr-O ligands occupying the axial position *trans* to the oxido ligand (Figure 6.3b). Analogously to the Fe^{3+}-hTf structure, the binding of [V^{IV}O]$^{2+}$ in the pocket is presumed to be stabilized by a hydrogen bond with N–H of Arg-121 in hTf$_C$ and Arg-124 in the hTf$_N$ lobe, which in their turn would be bound to the carbonate ligand. A recent theoretical study indicates that the most stable structure for the hTf$_N$ site is characterized by bond distances V=O, 1.59 Å; V-O(Tyr-188), 1.92 Å; V-O(Asp-63), 1.92 Å and V-N(His-249), 2.13 Å. It suggests that O(Tyr-95) is protonated and interacting weakly at 3.28 Å from V and that the coordination of CO$_3^{2-}$ is favoured with respect to that of HCO$_3^-$. The plane of the carbonate ligand appears to be essentially parallel to the V=O vector (dihedral angle of –5.6°), with V-O(carbonate), 1.79 Å.[35]

The first estimates of the binding constants of [V^{IV}O]$^{2+}$ to the two lobes of human serum transferrin were based on a linear free energy relationship between log K_1(hTf) and the stability constant for the binding of hydroxide ion, log K_{OH}: 13.2 ± 1.6 and 12.2 ± 1.6 for log K_1 and log K_2, with a rather high degree of uncertainty. Subsequently, competition equilibria between hTf and anionic ligands (1,2-dimethyl-3-hydroxy-4(1*H*)-pyridinonate or acetylacetonate) for [V^{IV}O]$^{2+}$ led to more precise estimates (Table 6.3).[36,37] The values indicate that hTf exhibits a high affinity for [V^{IV}O]$^{2+}$, comparable to those reported for divalent metal ions.[32]

Among the biologically relevant anions, oxalate, phosphate, lactate (only at the N-terminal site) and citrate are synergistic. Under physiological conditions, the (hydrogen)carbonate concentration (24.9 mM) should be high enough to supply a synergistic anion. Only lactate seems to be capable of replacing carbonate at the hTf$_N$ site to form the species (V^{IV}O)hTf(lact) (Table 6.3).

Vanadium(V) also occupies the Fe^{III} binding sites of transferrin but a synergistic anion is not required (Figure 6.3c). It is probable that the two oxido groups of the [V^{V}O$_2$]$^+$ ion prevent access of hydrogencarbonate.

The proposed structure is consistent with the fact that the V^V stability constants are 3–5 orders of magnitude higher than those for inorganic

Table 6.3 Log K for the species formed by V^{III}, [V^{IV}O]$^{2+}$ and [V^{V}O$_2$]$^+$ with blood plasma proteins.

Protein	Species	log K	Ref.	Protein	Species	log K	Ref.
hTf	(V^{III})hTf	20.0 ± 1.5	15	hTf	(V^{V}O$_2$)$_2$hTf	6.6 ± 0.3	38
hTf	(V^{IV}O)hTf	13.0 ± 0.5	36	HSA	(V^{IV}O)$_x$HSA	9.1 ± 1.0	36
hTf	(V^{IV}O)hTf	13.4 ± 0.2	37	HSA	(V^{IV}O)$_x$HSA	9.1 ± 0.4	39
hTf	(V^{IV}O)$_2$hTf	12.5 ± 0.5	36	HSA	(V^{IV}O)$_2$HSA	20.9 ± 1.0	36
hTf	(V^{IV}O)$_2$hTf	11.8 ± 0.4	37	HSA	(V^{IV}O)$_2$HSA	20.6 ± 0.4	39
hTf	(V^{IV}O)hTf(lact)	14.5 ± 0.8	36	HSA	(V^{V}O$_2$)HSA	1.8 ± 0.3	40
hTf	(V^{V}O$_2$)hTf	6.0 ± 0.1	37	HSA	(V^{V}O$_2$)HSA	3.0	41
hTf	(V^{V}O$_2$)hTf	7.5 ± 0.2	38	IgG	(V^{IV}O)IgG	10.3 ± 1.0	42
hTf	(V^{V}O$_2$)$_2$hTf	5.5 ± 0.2	37				

anions such as carbonate, phosphate and sulfate.[15,43] For example, log $K_1 = 7.45$ and log $K_2 = 6.6$ for V(V) and log $K_1 = 2.66$ and log $K_2 = 1.8$ for carbonate.[43] NMR results suggest that there is no competition between HCO_3^- and $[H_2V^VO_4]^-$ for the anion binding site of hTf, in agreement with the structure proposed.[40]

A last comment concerns the behaviour of V^{III}, which has little relevance for human physiology. Vanadium(III) binds strongly to hTf, with a co-ordination mode very similar to that of Fe^{III}. The order of affinity of the three oxidation states of V to hTf is $V^{III} > V^{IV} > V^V$ in the presence of carbonate and $V^{III} \approx V^{IV} > V^V$ in its absence.[44] Although it has been suggested that vanadium can be transferred to cells as V^{III}-hTf species, it is unlikely that V is transported in this oxidation state.

6.4.6 Binding of Vanadium(IV) and Vanadium(V) to Human Serum Albumin

HSA is the most abundant protein in the blood $(\sim 0.63$ mM$)$,[29] and plays a major role in the transport of Cu^{II}, Zn^{II}, Co^{II} and Ca^{II}, and toxic Ni^{II} and Cd^{II}. For example, 5–10% of serum copper and 95% of serum nickel is bound to albumin.[45]

It contains a primary metal binding site, the N-terminal site (NTS or site I),[45] and a secondary site, the multi-metal binding site (MBS or site II).[46] The NTS is the specific site for metal ions with a high affinity for nitrogen, such as Cu^{II} and Ni^{II} ions. It supplies a square planar geometry through the donor set (NH_2, N^-, N^-, His-N) where NH_2 represents the terminal amino, N^- the deprotonated amide groups of residues 2 and 3 and His-N the imidazole nitrogen of His-3. The MBS site in HSA is located at the interface of domains I and II and is the primary binding site for Zn^{II}. It also binds Cu^{II} and Ni^{II} with similar affinity, and Cd^{II} with an affinity that is one order of magnitude lower.[46] The metal ions at the MBS are coordinated by His-67, His-247, Asn-99 and Asp-249. Two additional sites have been identified: site B (or CdB) that is specific for cadmium (one His and several carboxylate residues)[47] and Cys-34, the only thiol not involved in disulfide bridges. The accessibility of Cys-34 is limited and it displays a certain specificity for gold and platinum compounds.

The binary system $[V^{IV}O]^{2+}$-HSA has been studied with several techniques. Competition experiments between Cu_{aq}^{2+} and $[V^{IV}O]^{2+}$ ions for bovine and porcine serum albumin (the latter lacks His3 in the NTS) established that $[V^{IV}O]^{2+}$ binds to the MBS more strongly than does Cu^{II}.[48] A further indication that the $[V^{IV}O]^{2+}$ ion does not bind at the NTS was through comparison of ^{51}V A_z values for $[V^{IV}O]^{2+}$-HSA with that of the complex formed by peptide GlyGlyHis, a model of the NTS site.

The behaviour of the system $[V^{IV}O]^{2+}$-HSA is significantly different when studied at different molar ratios. In particular, EPR spectra of an equimolar solution are characterized by the presence of the signals attributable to a dinuclear species, denoted as $(V^{IV}O)_2HSA$, in which two $[V^{IV}O]^{2+}$ ions are magnetically coupled.[49] When the metal to protein ratio is increased to 2 : 1 or higher, the EPR resonances are attributable to a multinuclear complex $(V^{IV}O)_xHSA$ ($x = 5$ or 6) in which the metal ions are not coupled and have similar coordination environments.[49] Competition experiments between Zn^{II} and $[V^{IV}O]^{2+}$ suggest that $[V^{IV}O]^{2+}$ has two types of binding sites, one of them corresponding to the MBS. Zn^{II} is able to displace $[V^{IV}O]^{2+}$ from this site, but not from the other sites.[39] Glu-252 is at ~ 5 Å distance from Asp-249 (one of the donors of the MBS) and might be involved in the coordination to V in the dinuclear form.[39]

It has been possible to estimate average log K values for $(V^{IV}O)_xHSA$ and $(V^{IV}O)_2HSA$ (Table 6.3).[‡‡] The interaction between V(V) is weak and adventitious (Table 6.3), probably involving surface carboxylic functions. Ultrafiltration, ^{51}V NMR spectroscopy and fluorimetry experiments confirm the weak interaction.[40,50]

6.4.7 Binding of Vanadium(IV) to Immunoglobulins

Several earlier investigations of V^{III}, $[V^{IV}O]^{2+}$ and $[H_2V^VO_4]^-$ binding to immunoglobulin G found no evidence for interaction. The system $[V^{IV}O]^{2+}$-IgG (human) was recently re-assessed.[42] The ion was found to distribute among at least three distinct coordination sites 1, 2 and 3, forming species with stoichiometry $(V^{IV}O)_xIgG$ ($x = 1-3$). ^{51}V A_z values indicated that affinities are in the order $1 > 2 > 3$. The most likely candidates for ligands in site 1 would be Ser-O$^-$/Thr-O$^-$ residues rather than a Tyr-O$^-$ or Cys-S$^-$. The number of surface Ser in the F_{ab} region is higher than that of Tyr and Cys, and the coordination of a deprotonated serine has been proposed for the active site of other V-proteins, such as the reduced form of vanadium bromoperoxidase. Site 2 may feature His-N and Asp-O$^-$ or Glu-O$^-$ donors that are abundant on the surface of the F_{ab} and F_c regions (3–5, 4 and 3, respectively, in F_{ab}, and 8, 8 and 5 in F_c).[§§] These conclusions are supported by simulations using DFT (density functional theory) methods.[42]

A log K estimate for $[V^{IV}O]^{2+}$-IgG is 10.3 ± 1.0 (Table 6.3). Since the stabilities of the three sites in IgG are comparable, only a mean value for the association constant of the species $(V^{IV}O)IgG$ can be calculated.[42] The higher thermodynamic stability of the $[V^{IV}O]^{2+}$-hTf over $[V^{IV}O]^{2+}$-IgG species is because hTf contains specific sites for metal ion binding.

---

‡‡It has been noticed that the values of log β_x for the five to six equilibrium steps in the $[V^{IV}O]^{2+}$-HSA system should differ only slightly from each other (not more that 0.1–0.2 log units) and from the mean value (around 0.3–0.4 log units).

§§IgG is a "Y"-shaped protein. The arms of the "Y" contain the sites that can bind the antigens (F_{ab}, Fragment antigen binding region), whereas the base of the "Y" plays a role in modulating immune cell activity (F_c, Fragment crystallizable region).

6.4.8 Distribution of Vanadium(IV) and Vanadium(V) in Blood Serum

The distribution of vanadium between the components of blood serum depends on several factors, including the relative proportions of vanadium(IV) and vanadium(V), their relative affinity for the *l.m.m.* components and the plasma proteins and the capacity to form ternary complexes. As a first approximation, we take the amount of V^{III} in blood to be negligible. The most important bio-ligands of blood plasma are proteins, hTf and HSA, in particular. The data of Table 6.3 identify hTf as the strongest bio-ligand for $[V^{IV}O]^{2+}$ and $[V^{V}O_2]^+$ ions. The values of the stability constants for the binding of $[V^{IV}O]^{2+}$ and $[V^{V}O_2]^+$ to *l.m.m.* and *h.m.m.* components allow prediction of the speciation of vanadium(IV) and vanadium(V) in blood plasma. Total V concentrations in the range 1–100 nM and a ratio V^{IV}/V^V of 80 : 20 were considered to construct a species distribution (Table 6.4).

In the concentration range 1–100 nM most of the V is bound to hTf. Since the hTf concentration in the blood is much higher than that of $V^{IV}O$ and V^V, the species $(V^{IV}O)$hTf and (V^VO_2)hTf predominate. The V ions are coordinated in the hTf_C site, and the contribution of the species with both iron-binding sites occupied is negligible. The total V bound as $(V^{IV}O)$hTf and (V^VO_2)hTf is around 95%; the remainder is coordinated mainly to lactate in the ternary complex (VO)hTf(lact) where lactate replaces CO_3^{2-}. For V^V, the presence of free $[H_2V^VO_4]^-$ at physiological conditions has been proposed.[30]

Table 6.4 Speciation of vanadium in blood plasma at different physiological concentrations of vanadium.[a,b]

Species	1 nM [c]			10 nM [d]			100 nM [e]		
	$V^{IV}O$	V^V	Tot.	$V^{IV}O$	V^V	Tot.	$V^{IV}O$	V^V	Tot.
$(V^{IV}O)$hTf/(V^VO_2)hTf	0.76	0.19	0.95	7.55	1.89	9.45	75.34	18.89	94.23
$(V^{IV}O)_2$hTf/$(V^VO_2)_2$hTf	0.00	0.00	0.00	0.00	0.00	0.00	0.22	0.06	0.27
$(V^{IV}O)$hTf(lact)	0.04	–	0.04	0.36	–	0.36	3.60	–	3.60
$(V^{IV}O)_2$HSA[d]	0.00	–	0.00	0.00	–	0.00	0.00	–	0.00
$(V^{IV}O)$HSA/(V^VO_2)HSA	0.00	0.00	0.00	0.02	0.00	0.03	0.23	0.03	0.26
$(V^{IV}O)$IgG	0.00	–	0.00	0.05	–	0.05	0.49	–	0.49
$[V^{IV}O(lactH_{-1})_2]^{2-}$	0.00	–	0.00	0.01	–	0.01	0.06	–	0.06
$[V^{IV}O(citrH_{-1})]^{3-}$	0.00	–	0.00	0.00	–	0.00	0.03	–	0.03
$[V^VO_2(H_2O)(Gly)]$[f]	–	0.00	0.00	–	0.01	0.01	–	0.14	0.14
$[HV^VO_3(lactH_{-1})]^{2-}$[g]	–	0.00	0.00	–	0.00	0.00	–	0.01	0.01
$[H_2V^VO_4]^-$	–	0.01	0.01	–	0.07	0.07	–	0.74	0.74
$[HV^VO_4]^{2-}$	–	0.00	0.00	–	0.01	0.01	–	0.12	0.12

[a]Data based on the thermodynamic stability constants reported in the literature.
[b]For hTf a concentration of 25.9 μM was used to take into account the fraction of transferrin involved in iron complexation (*ca.* 30%).
[c]Values reported in nM; the sum of the species concentration is 1 nM.
[d]Values reported in nM; the sum of the species concentration is 10 nM.
[e]Values reported in nM; the sum of the species concentration is 100 nM.
[f]VGly with the notation used for potentiometry (ref. 4).
[g]Vlact[2-] with the notation used for potentiometry (ref. 4).

Binding of vanadium to HSA and IgG is almost negligible and, for the given concentrations, remains below 1%. HSA could bind $V^{IV}O$ and V^V at higher metal ion concentrations (≥ 50 μM) when the hTf sites are saturated. At lower concentrations, $(V^{IV}O)_2$HSA does not form at all and $(V^{IV}O)_x$HSA is formed only in very small amounts.

The results confirm that for total vanadium concentrations <10 μM, only hTf binds vanadium in both oxidation states +IV and +V.[37,51] Of course, besides Fe^{III}, other metal ions such as Cr^{III}, Mn^{III} and Al^{III} may be available and will reduce the sites available for binding of vanadium.

6.5 Transport of Vanadium across the Cell Membrane

Cellular uptake depends on the V(IV) and V(V) species reaching the cell membrane. V compounds employ a number of different mechanisms for entering cells.[9]

Inorganic V(V) species, mainly $[H_2VO_4]^-$ and $[HVO_4]^{2-}$, enter *via* phosphate channels.[23] Cells accumulate twice the amount of vanadium when incubated with V(V) salts such as NaV^VO_3 as opposed to V(IV) salts such as $V^{IV}OSO_4$.[22] Apparently, oxidation of $V^{IV}O$ to V^V takes place before uptake occurs. Since most of the V is present in the plasma as the binary species of transferrin $(V^{IV}O)$hTf and (V^VO_2)hTf, binding to transferrin would appear to be necessary for uptake *via* hTf receptor-mediated endocytosis.[9,52] In fact, only Fe_2-hTf is recognized and correctly internalized by the transferrin receptor (TfR). Apo-hTf or hTf with bound vanadium exhibit different conformations and are not recognized. A recent study demonstrates that $[V^{IV}O]^{2+}$ and some of its insulin-enhancing compounds can bind also to holo-hTf and so endocytosis of holo-hTf could be a mechanism for transport of $[V^{IV}O]^{2+}$ and its complexes inside the cell.[53,¶¶]

As discussed for erythrocytes (Section 6.4.2), the divalent metal transporter DMT1 may be involved in transport of the $[V^{IV}O]^{2+}$ ion through the endosomal membrane into the cytosol.[24] If a lipophilic and sufficiently stable complex was formed in the blood, it could cross the cellular membrane *via* passive diffusion. This mechanism has been observed in Caco-2 cells.[23] Diffusion of $[VO(acac)_2]$ through human erythrocyte membranes was not inhibited by DIDS while uptake of vanadate(V) was completely inhibited.[23] This mechanism seems to be secondary for inorganic vanadium compounds but is relevant for pharmacologically active complexes, for example the antidiabetic $[V^{IV}O]^{2+}$-pyridinone complexes.[9] In the experiments with human erythrocyte ghosts, the interaction with protein sulfhydryl groups has been demonstrated.[54] Other studies with inside-out vesicle (IOV) suggested that the inner membrane proteins of the erythrocytes delay the oxidation of

---

¶¶At the physiological conditions, vanadium(IV) exists as a cation $[V^{IV}O]^{2+}$, whereas vanadium(V) exists as an anion $[H_2V^VO_4]^-$ or $[HV^VO_4]^{2-}$, structurally similar to phosphate. The mechanism of cellular uptake is related to this difference.

$[V^{IV}O]^{2+}$ compounds to a greater extent than the right-side-out vesicle (ROV), confirming an interaction with the membrane proteins from the cytoplasmatic side.[23]

Decavanadate ($[V_{10}O_{28}]^{6-}$ or V10) is well known to interact with several proteins and to have many biological activities.[55] It is taken up through anion channels, but also binds to membrane proteins (Figure 6.4). Furthermore, it can be formed inside the mitochondria due to a local increase of V concentration.[55]

Recent studies have explored how vanadium compounds associate with interfaces of membrane model systems, in particular micelles and reverse micelles obtained from mixtures of a non-polar solvent, a surfactant and a polar solvent (usually H_2O). The type of interaction depends on the chemical species examined. For example, the association with the hydrophilic side of $[V^{V}O_2(maltolato)_2]^-$ – the oxidation product of the potential anti-diabetic compound $[V^{IV}O(maltolato)_2]$ – results in the formation of hydrolytic species and loss of the coordinated ligand.[56] The anion $[V^{V}O_2(dipic)]^-$ (dipic = 2,6-pyridinedicarboxylate) readily penetrates the artificial membrane despite the high solubility of the complex in aqueous environments and the negatively charged interface. The anionic species resides in the interface between the hydrophobic and the hydrophilic regions; this fact is surprising and suggests that a charged polar compound, depending on its structure, may have affinity for hydrophobic membrane environments.[57] Even if these results do not necessarily apply to the cellular membranes, these possibilities must be taken into account. The chemical transformation of V complexes in a hydrophobic environment may yield redox products that can exert different

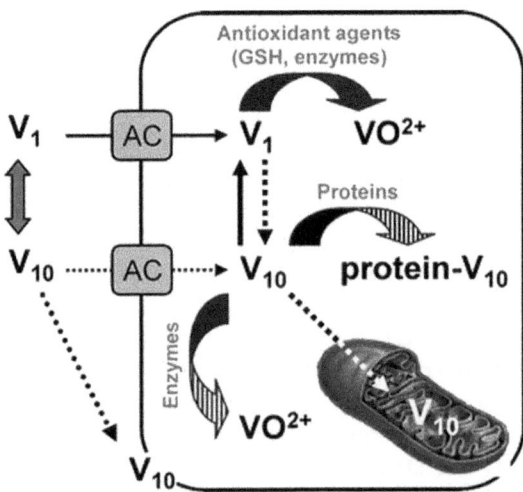

Figure 6.4   Uptake and cellular targets of decavandate. AC indicates the anion channels, V1 monomeric vanadate and V10 decavanadate.
Reproduced from ref. 55 with permission from The Royal Society of Chemistry.

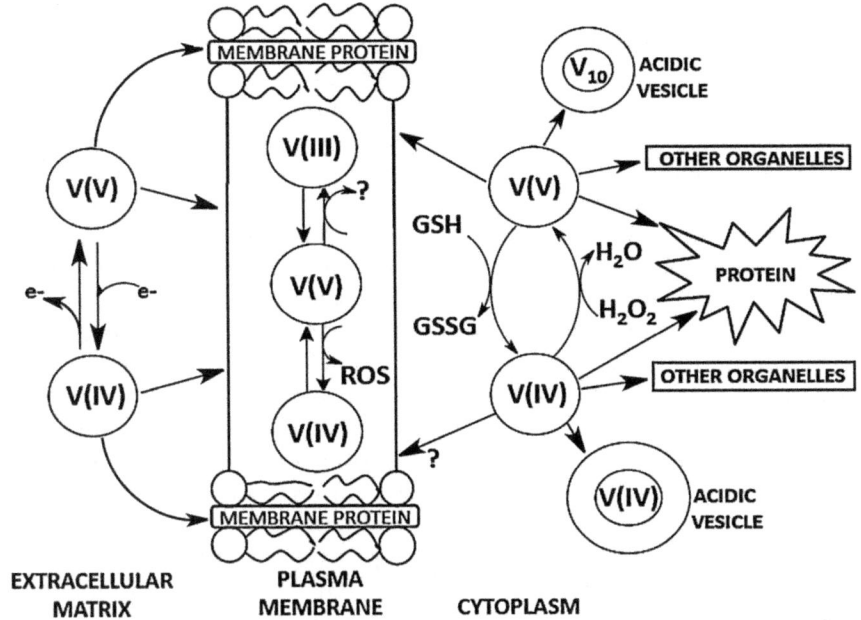

Figure 6.5 The pathway for vanadium(IV) and vanadium(V) compounds after passive diffusion into the membrane.
Reproduced from ref. 57 with permission from Elsevier.

types of biochemical action. Ligand substitution reactions can result in complicated redox processes including the formation of ROS (Figure 6.5). The two-electron reduction of V^V to V^{III} proposed is similar to that observed for a range of different alcohols.[58]

This description proposed in Figure 6.5 highlights the importance of the interaction with the membranes and can explain the toxic effects of vanadium by the formation of organic radicals but also suggest its beneficial activities scavenging the radicals inside the cells through the reduction of V(V) to V(IV). The final balance, beneficial or toxic, may depend on the specific redox cycle and control of the chemistry in these different environments. This description may also help in understanding how the same complex induces beneficial effects in some cases and is toxic in others.

6.5.1 Redox Reactions and Complexation Processes in the Intracellular Environment

It is necessary to postulate the presence of ligand exchange and subsequent redox processes of V^{IV} and V^V complexes to account for the cellular effects of vanadium. Cytoplasmic reductants GSH and ascorbate plus cofactors NADH and NADPH are available, and so are oxidants such as dioxygen and peroxide. The V^V/V^{IV} one-electron couple can promote Fenton- and

Haber–Weiss-like chemistries to produce ROS. For example, reduction of O_2 and H_2O_2 lead to the radicals superoxide $O_2^{\bullet-}$ and $^{\bullet}OH$, respectively. $O_2^{\bullet-}$ can be subsequently reduced by NADH generating H_2O_2.

The hydrolysis and subsequent oxidation of $[V^{IV}O]^{2+}$ species to V(V) must also be considered (Figure 6.6). The final form of vanadium in cells is the result of the balance of all these factors and the stabilization of V^V/V^{IV} species through complexation by cellular bio-ligands. In human erythrocytes and rat adipocytes, GSH reduces V^V and bound $[V^{IV}O]^{2+}$ ions are found.[22,27] *In vitro*, reduction of vanadate(V) by glutathione reductase and NADPH at pH 7.2 has been observed. The intracellular excess of GSH (2.3 mM[59]) over vanadium (maximum concentration ~ 100 nM) is at least four orders of magnitude and highly favours reduction of V^V to $V^{IV}O$ with concomitant formation of oxidized glutathione (GSSG) through the formation of a short-lived glutathionato-V^V species.[60] The excess of GSH can rapidly bind the $[V^{IV}O]^{2+}$ ion. The relatively high concentrations of reducing agents such as NADH (~ 0.1–1.0 mM),‖‖ NADPH (~ 0.01–0.1 mM) and ascorbate (1–10 mM) have led to the suggestion that further reduction to V^{III} may occur but no definitive data exist.[60] The viability of this process would also depend on the relative stabilization of both the V^{III} and $[V^{IV}O]^{2+}$ species *via* complexation.

The strongest intracellular *l.m.m.* binders are GSH, GSSG, ATP, NADH, NADPH, NAD^+, $NADP^+$ and ascorbate. The nature of the $[V^{IV}O]^{2+}$ species formed at pH 7.4 can be inferred from potentiometric and spectroscopic data for the binary systems.

GSH. For an excess of GSH in the pH range 5.0–6.5, the bis-chelate $[V^{IV}O(GSH)_2]$ with a $2\times(COO^-, NH_2)$ donor set is the main species. At the physiological pH, the thiolate function is involved in the metal coordination (Figure 6.6).[61]

GSSG. As observed with other first-row transition metal ions, the 1:1 complex uses a $2\times(COO^-, NH_2)$ donor set, analogous to that of an amino acid.[62]

ATP. From a comparison of the coordinating strength of GSH and ATP, it was concluded that ATP is the more efficient ligand for $[V^{IV}O]^{2+}$ over the whole pH range. ATP employs its terminal phosphate donor(s) at neutral pH.[51]

NAD^+ and $NADP^+$. $[V^{IV}O]^{2+}$ binds NAD^+ and $NADP^+$ under physiological conditions. The dominant 1:2 complex with NAD^+ features diphosphate oxygens and two ribose alcoholate groups as ligands. At pH 7.4, $NADP^+$ forms a di-μ-hydroxide dinuclear species, in which the first ligand is coordinated through the (O^-, O^-) ribose moiety and the second one through the 2′-phosphate group.[63]

‖‖ The total amount of NAD^+ and NADH is approximately 1 μmole per gram of wet weight and is about 10 times the concentration of $NADP^+$ and NADPH in the same cells. The actual concentration of NADH is not easy to measure because it depends also on the [NADH]/[NAD^+] ratio. In the light of the recent estimates on animal cells, a value in the range 0.1–1.0 mM seems to be reasonable.

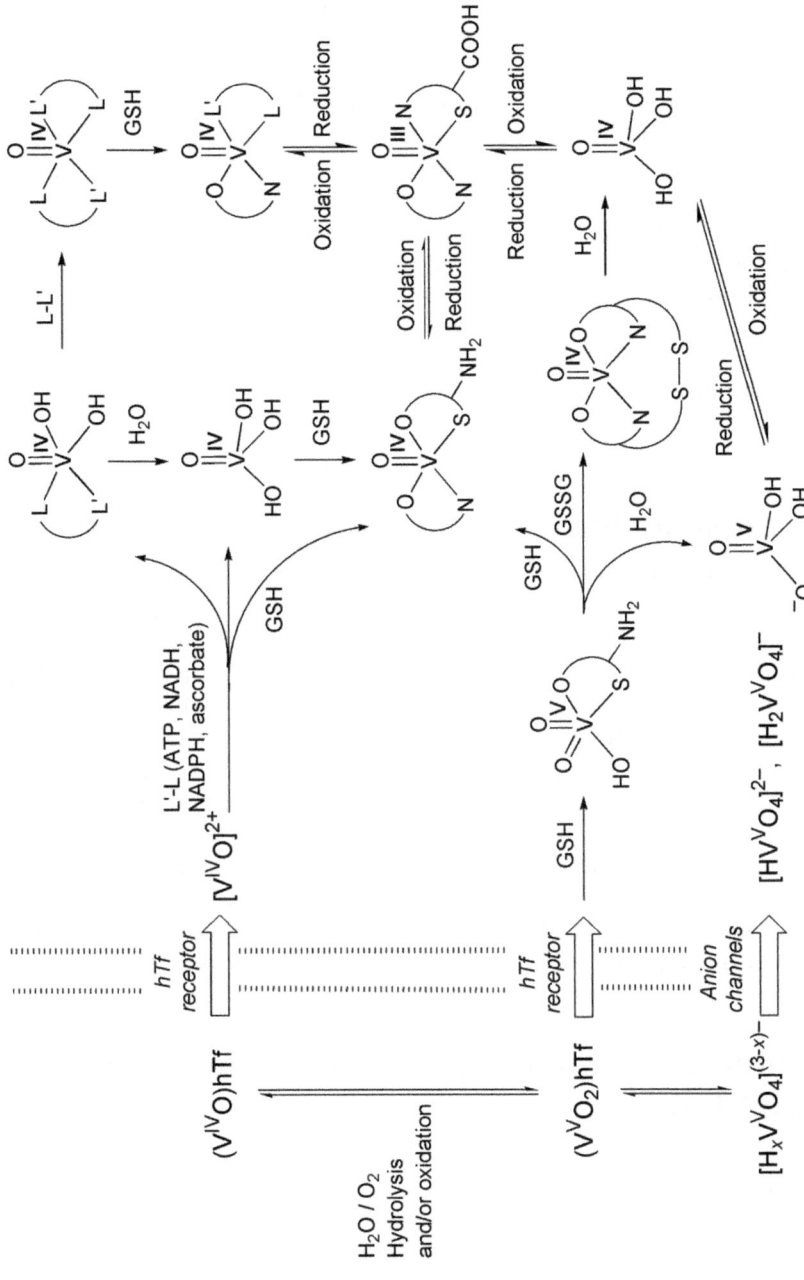

Figure 6.6 Possible biotransformations of V in the blood and in the cell. Adapted from ref. 60 with permission from Springer.

Ascorbate. Ascorbate can exert two effects on intracellular vanadium. On the one hand, it can reduce vanadate(V) at pH 7.4–8.0. On the other, both ascorbate and its oxidized form dehydroascorbate can complex the $[V^{IV}O]^{2+}$ ion. Three physiologically relevant complexes have been described: one 1 : 1 and one 1 : 2 species containing bound ascorbate and one species containing bound dehydroascorbate.[64]

X-ray absorption spectroscopic studies have shown that, in a biological system, both V^{IV} and V^V are present in the cells treated with $V^{IV}O$ complexes.[65] Other studies have shown the formation of V^{IV} after addition of V^V in several biological systems.[8,13,27] On the whole, the oxidation state of V inside the cells depends on the specific cellular compartment and on the nature and amount of complexing, reductant and oxidant agents; therefore, both vanadium(IV) and vanadium(V) can exist inside the cell. Local oxidizing conditions in some compartment, due to the presence of reactive oxygen species, could convert $[V^{IV}O]^{2+}$ to $[H_2V^VO_4]^-$ and $[HV^VO_4]^{2-}$. Even if a redox cycle has not been demonstrated and the relative amount of V^{IV} and V^V species is not known, an equilibrium between the two oxidation states in the cellular environment is established.[2,8]

6.6   Transport through Subcellular Membranes

Few data are available on this aspect. When the yeast *Saccharomyces cerevisiae* was treated with a solution containing vanadate(V) at pH 6.5 and 4.0, little V(V) was observed in the cells at pH 4.0 but V10 was found at pH 6.5.[66] However, *in vitro*, V10 is mainly formed at the lower pH. Thus, V10 may form in the subcellular organelles with vesicular structures where pH reaches values near 5.0. Therefore, equilibration with oligonuclear vanadate(V) species may be part of the cellular chemistry. Vanadium also accumulates in subcellular organelles, mainly mitochondria.[55,67] The consequence may be that a chemical species that is not stable at ambient conditions can be formed and found inside the cell and subcellular compartments. *In vitro* studies have clearly demonstrated that V10 affects mitochondrial oxygen consumption and membrane potential more strongly than does mononuclear V1. It appears to act as a potent inhibitor of the respiratory chain.[55,67]

6.7   Inhibition and Stimulation of Cellular Enzymes and Proteins

Once vanadium is inside the cell, it can interact with a number of components. In particular, it has been shown that, at the cellular level, vanadate(V) inhibits a wide variety of enzymes, including phosphatases, ribonucleases and ATPases. The discovery that vanadate(V) inhibits (Na^+, K^+)-ATPase dates back to 1977.[68]

Phosphatases can be divided on the basis of the residue in the active site: Ser in alkaline phosphatase; His in acid phosphatase; Cys in protein tyrosine

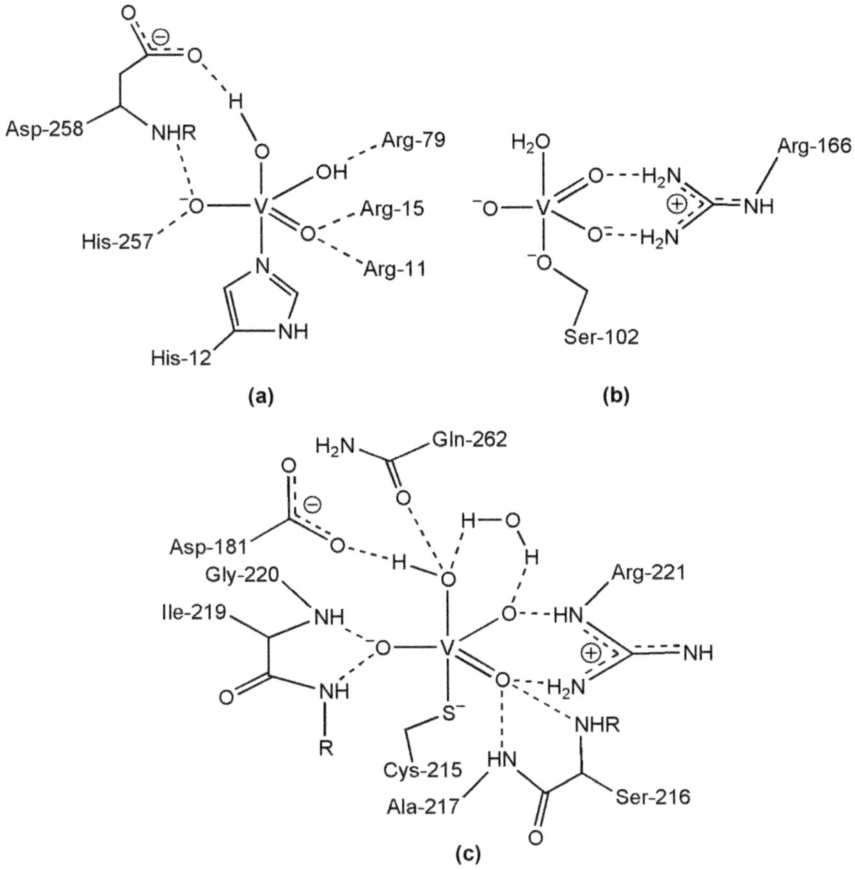

Figure 6.7 V coordination environment in (a) an acid phosphatase, (b) an alkaline phosphatase and (c) a PTP.
Adapted from ref. 2 with permission from John Wiley & Sons.

phosphatase. They catalyse the hydrolysis of phosphate ester bonds according to the reaction $RO\text{-}PO_3H^- + H_2O \rightarrow ROH + H_2PO_4^-$. The reaction mechanism involves formation of a five-coordinate trigonal bipyramidal intermediate, $RO\text{-}PO_3H(donor^{aa})$, where $donor^{aa}$ is Ser-O$^-$, His-N or Cys-S$^-$. The inhibition by vanadate(V) is due to the fact that it is recognized as phosphate and binds irreversibly at the active site. The crystal structure of vanadate-inhibited rat acid phosphatase is shown as an example in Figure 6.7a. The His-12 side-chain acts as an axial while the three equatorial oxygen atoms are hydrogen bonded to adjacent residues.[69] In *Escherichia coli* alkaline phosphatase, the axial site of V is occupied by a Ser-O$^-$ rather than a His-N and there is a different pattern of hydrogen bonds to the equatorial oxygen atoms (Figure 6.7b).[70]

Vanadate(V) is a strong inhibitor of protein tyrosine phosphatases (PTPs) that regulate tyrosine phosphorylation. They influence numerous cellular processes such as mitosis and insulin-receptor signalling.[2] The structure of vanadate-inhibited PTP-1B was solved recently by single-crystal X-ray diffraction and is shown in Figure 6.7c.[71] The overall features are similar to the vanadate phosphatase structures with a Cys-S$^-$ ligand occupying the axial position.

The observation that $V^{IV}OSO_4$ is a potent inhibitor for some phosphatases (*e.g. E. coli* alkaline phosphatase) is not fully understood. The effect may be associated with the hydrolysis of $[V^{IV}O]^{2+}$ near neutral pH values to give species such as $[(V^{IV}O)_2(OH)_5]^-$ and $[V^{IV}O(OH)_3]^-$. These may behave analogously to vanadate(V).[14] In particular, as recently suggested, $[V^{IV}O(OH)_3]^-$ can be interpreted as a vanadate(IV), $[H_3V^{IV}O_4]^-$.[72]

Ribonucleases (RNases) are also inhibited by vanadate(V) and the adducts formed with RNase-A, RNase-T$_1$ and T-DNA-PDase have been characterized.*** The geometry around V is trigonal bipyramidal in each case but comparisons are particularly interesting. In the RNase-A structure, V^V is coordinated to the oxido groups in position 2' and 3' of the substrate uridine. However, in RNase-T$_1$, there are no V-protein covalent bonds and vanadate is fixed to the catalytic site through a network of hydrogen bonds to Tyr, His, Glu and Arg residues and to water molecules. Finally, in T-DNA-PDase, V is bound to an equatorial desoxyribose-O and axial Tyr-O$^-$ and His-N donors.[2] It has been observed that this state resembles the active centre in vanadate-dependent haloperoxidases.[2] The decrease in the energy of stabilization with respect to the analogous species with phosphate has been attributed to the longer V–O bonds *versus* P–O bonds (around 15%). This results in an imperfect matching of V in the pocket of the active site and a distortion of the optimal trigonal bipyramidal structure (that corresponding to the transition state with phosphate) toward square pyramidal.[8]

Vanadium also inhibits enzymes in metabolic pathways. These include phosphoglycerate mutase, which catalyses the conversion of 3-phosphoglycerate to 2-phosphoglycerate through a phosphate transfer, with 2,3-diphosphoglycerate being an intermediate.[8] The inhibition of vanadate(V) is explained with the formation of structures analogous to the enzyme transition-state, 2-vanadio-3-phosphoglycerate and 3-vanadio-2-phosphoglycerate. Similar results were found for phosphoglucomutase, which catalyses the transformation (or mutation) of glucose-1-phosphate into glucose-6-phosphate *via* a phosphate shift.

A different role for vanadium is observed in the stimulation of enzymes that catalyse the transfer of phosphate groups. It is based on the formation of esters between vanadate(V) and substrates such as sugars. As it has been noted, however, vanadate esters are not distinguishable from phosphate

---

***RNase-A catalyses the hydrolysis of phosphate ester linkages in RNA, RNase-T1 promotes the specific cleavage of RNA at guanosine to yield oligonucleotides with terminal guanosine-3'-phosphate and tyrosyl-DNA phosphodiesterase catalyses the hydrolysis of a phosphodiester bond between a tyrosine side-chain and a DNA 3'-phosphate.

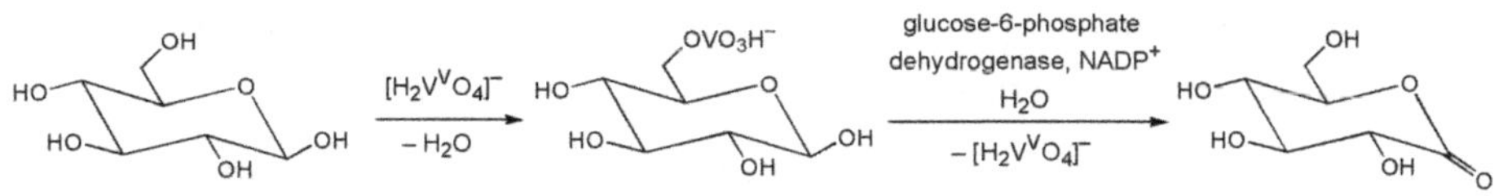

Figure 6.8 Vanadate-activated transformation of glucose into gluconolactone.

esters and can represent alternative target species for dehydrogenases and isomerases.[2] The replacement of phosphate by vanadate stabilizes the transition state of the enzyme because of the higher coordination number of the metal ion. Glucose-6-phosphate dehydrogenase, adenylate cyclase, glyceraldehyde-3-phosphate dehydrogenase, tyrosine phosphatase and glycogen synthase are examples of enzymes stimulated by vanadate(V) action. In this regard, the formation of glucose-6-vanadate has been proposed to explain the activation exerted by vanadate(V) on the oxidation of glucose by $NADP^+$ catalysed by glucose-6-phosphate dehydrogenase. Glucose-6-vanadate is recognized by the enzyme, rapidly oxidized and hydrolysed to gluconate (Figure 6.8).[2]

Vanadate(V) esters formed with sugar-OH groups are known to influence other reactions. For example, nicotine adenine dinucleotide vanadate (NADV), an analogue of NADP, is considered to be a potent cofactor in reductases otherwise dependent on NADP.[73] Adenosine monophosphate vanadate (AMPV) and adenosine diphosphate vanadate (ADPV), analogous of ADP and ATP, are formed under physiological conditions but are not cleaved by phosphatases. The explanation for this behaviour suggests that the V–P bond is weaker than the P – P bond (~ 10 *vs.* 30 kJ mol^{-1}) and that mixed vanadate-phosphate derivatives do not bind Mg^{2+} effectively, a condition for an efficient turnover.

Mononuclear and oligonuclear vanadium(V) species exert different effects on the proteins and enzymes. For example, in contrast to mononuclear V1, dinuclear V2 stimulates the 2,3-diphosphoglycerate phosphatase activity of phosphoglycerate mutase. V2 acts as a mimic of monophosphoglycerate and binds to the enzyme. It has been postulated that the two anionic moieties of V2 occupy the phosphate and carboxylate binding site.[74] V1 and V10 affect muscle contraction/relaxation differently and, in particular, the actions of the sarcoplasmic reticulum (SR) Ca^{2+}-ATPase transmembrane transport system and molecular motor (myosin and actomyosin) ATPase activities.[55] V1 is a well-known specific inhibitor of the sarcoplasmic reticulum Ca^{2+}-ATPase, forming a transition state similar to the phosphorylated intermediate. V10, instead, interacts with this calcium pump at a site distinct from that of phosphorylation and, unlike V1, is able to inhibit calcium accumulation coupled with the ATP hydrolysis in SR vesicles. The mechanisms of inhibition of myosin ATPase are also different. V1 mimics the transition state for γ-phosphate hydrolysis while V10 induces the formation of an intermediate myosin-MgATP-V10 complex, blocking the contractile cycle presumably in the pre-hydrolysis state.[67]

6.7.1 Pharmacological Effects of the Interactions with Cellular Components

The first *in vitro* studies on the insulin-mimetic action of vanadium(IV) and vanadium(V) inorganic salts were published more than thirty years ago. The observed effects were the uptake and degradation of glucose by adipocytes (fat cells), the stimulation of hepatic glycogenesis and the inhibition of hepatic gluconeogenesis and lipolysis.[14,75] The inhibition of PTPs by vanadate(V) discussed above is the main cause of the anti-diabetic action of vanadium. Therefore, the current prevailing opinion is that the V complexes are sources of an uncomplexed form of vanadium(V), which causes the insulin-enhancing effects. However, the possibility that $[V^{IV}O(OH)_3]^-$, which has a structure similar to $[H_3V^VO_4]$, may be the inhibiting species must be taken into account.[14,72] The active species is vanadate(V), independent of the specific vanadium salt administered (V^{IV} or V^V). In the mechanism of glucose uptake, a signal transduction cascade is induced by insulin, stimulating the tyrosine kinase activity of the subunit β of the insulin receptor (IR) and phosphorylation of insulin receptor substrates (IRS). Phosphorylation of IRS induces the signal transduction, the glucose transporter Glut4 is translocated to the membrane and the glucose is transported into the cell. The signal transduction is suppressed, in the case of the absence of insulin or insufficient insulin response, by PTP, which catalyses the hydrolysis of the phospho-ester bonds of IRS. Vanadate(V) inhibits PTP and hinders the dephosphorylation of IRS forming a bond with a cysteine residue of the active site of protein tyrosine phosphatase. Therefore, V re-activates the signal cascade and, as a consequence, the uptake of glucose by the cell is re-established.[2] The inhibition by vanadate(V) of glucose-6-phosphatase results in the normalization of the serum glucose concentration, promoting glycogenesis and inhibiting glycogenolysis and gluconeogenesis in the liver. Finally, vanadate(V) inhibits free fatty acid release in adipocytes (lipolysis) through interaction with a phosphodiesterase involved in the reduction of the amount of cyclic adenosine monophosphate (cAMP) present in the cells.

Vanadium compounds have been found to be effective in the treatment of several variants of leukaemia, tumour and carcinoma pathologies. While different types of complexes have been tested recently, the simplest compounds used in these studies were $[V^{IV}O]^{2+}$ and $[V^VO_2]^+$ salts and the peroxovanadate(V) anion. The anti-cancer action of vanadium appears to be related to the inhibition of cellular protein tyrosine phosphatases. This leads to an increase in phosphorylation of tyrosine residues in proteins, a condition which modulates the signal transduction pathways leading to apoptosis or activation of tumour suppressor genes. It has been proposed that regulation of other functions of phosphate-metabolizing enzymes by vanadate can also be involved.[76] An additional feature of vanadium compounds in cancer treatment is detoxification of alkylating agents, which are known to induce DNA damage due to alkylation and/or oxidative processes.

These agents can be detoxified by transfer of the electrophilic alkyl to a nucleophile such as $[H_2V^VO_4]^-$. The toxic compound is thus consumed and DNA damage prevented. This reaction has been modelled using the trivanadate $[V_3O_9]^{3-}$ as nucleophile.[2] Vanadium complexes such as vanadocene $(Cp)_2V^{IV}Cl_2$ bind to DNA.[77] Their hydrolysis products interact in aqueous solution at physiological pH with the phosphate moiety of desoxyadenosyl phosphate to form outer-sphere complexes with a V···P distance of *ca.* 5.5 Å.

6.8 Vanadium Metabolism

The human body contains *ca.* 100 µg of vanadium. An equilibrium exists between the amount of vanadium excreted and absorbed (up to some tens of µg daily). The existence of homeostatic mechanisms has been postulated but has not been demonstrated. The concentration of vanadium inside cells rarely exceeds 0.2 µM and rises to 10–20 µM in animals and 1–5 µM in humans only at the moment of administration of vanadium.[78] These concentrations depend on the absorption, excretion and storage processes that remove vanadium from the bloodstream as well as from cells. Thus, as for other ultratrace elements, the main mechanism controlling the concentration of vanadium is the amount of metal absorbed in the gastrointestinal tract.

Accumulation was reported to occur primarily in bones, kidneys and, to a smaller degree, in spleen and liver.[22,79] Use of the radioactive isotope ^{48}V to follow tissue uptake of $[^{48}V^{IV}O(maltolato)_2]$ (^{48}V-BMOV) and $^{48}V^{IV}OSO_4$ demonstrated a different localization pattern at 24 h after the gavage.[80] The ratio found in bone : kidney : liver was 8 : 3 : 2 for ^{48}V-BMOV and 6 : 4 : 1 for $^{48}V^{IV}OSO_4$. Other studies with V^{IV} or V^V salts indicate that three days after administration 10–25% of V was found in bone, 4–5% in liver and kidney, 0.2% in the testes and 0.1% in spleen.[81]

Rats treated with $V^{IV}OSO_4$ and the anti-diabetic compound $[V^{IV}O(picolinato)_2(H_2O)]$ confirmed that the target organs depend on the form of vanadium administered. For $V^{IV}OSO_4$, the order of accumulation is kidney > liver > bone > pancreas > spleen ≫ blood (serum), whereas $[V^{IV}O(picolinato)_2(H_2O)]$ accumulates mostly in bones, followed by kidney, spleen, heart, liver, lung and serum.[82] Bones seem to be the major sink for retained vanadium. It is most likely incorporated into bone hydroxyapatite $Ca_{10}(PO_4)_6(OH)_2$ as $[V^VO_4]^{3-}$, consistent with the hypothesis that the bone-accumulated V is gradually released and transported to the other organs as occurs with Ca^{2+}. The high skeletal retention of vanadate(V) can probably be related to its rapid exchange with bone phosphate, based on the strong structural analogies between vanadate and phosphate.[12] In bone, residence times of 31 days for ^{48}V-BMOV and 11 days for $V^{IV}OSO_4$ were reported.[80]

Since absorbed vanadium is excreted with urine, accumulation in kidney seems unavoidable. Elevated levels in the liver most likely reflect endogenous excretion of absorbed vanadium *via* the bile. Brain levels were found to be considerably lower than in other organs, suggesting that the tested compounds did not pass the blood-brain barrier. Vanadium compounds can pass the placenta, though their levels are usually higher in the placenta than in the foetus.[81]

Residence lifetime measurements after an extensive feeding regime have shown that the half-life for elimination of vanadium from the kidney of rats is about 12 days, but the bulk of the element is removed within two days of intake.[8] In a healthy young man, orally administered sodium metavanadate (12.5 mg day^{-1} for 12 days) was completely recovered, largely unabsorbed, in the faeces (87.6%) and the remainder (12.4%) in the urine.[81] This is another indication of the low absorption of V compounds in the gastrointestinal tract. When working with ^{48}V, only the absorbed vanadium is excreted *via* the urine, whereas the unabsorbed fraction is excreted *via* the faeces.[80] Excretion through the bile seems to be a secondary route, as suggested by the low concentration of V found in the human bile (about 1.0 ng g^{-1}).[5] There is no agreement in the literature on the identity of the compounds excreted, although it has been suggested that low molecular mass $[V^{IV}O]^{2+}$ complexes may be the most plausible form.[12] All the processes discussed are summarized in Figure 6.1.

6.9   Uptake and Transport of Vanadium in other Living Systems

6.9.1   Vanadium in Ascidians and Polychaete Worms

One hundred years ago, Henze reported that the haemocytes of the ascidian *Phallusia mamillata* contained a high concentration of vanadium. He identified a "chromogen" (because of the colour change) that turned red-brown when vanadocytes are haemolysed and blue after contact with dioxygen.[2] The highest amount of vanadium was found for *Ascidia gemmate*, where the concentration can reach 350 mM, *i.e.* more than 10^7 times the concentration found in the sea water (35 nM). It is considered the highest degree of accumulation of a metal in any living organism.

In 1979, one of the compounds responsible for the colour change was isolated from vanadocytes of *Ascidia nigra* and named "tunichrome". Subsequently, several tunichromes have been isolated: these contain dopa (3,4-dihydroxyphenylalanine) and its derivatives dehydrodopa and dehydrodopamine, topa (3,4,5-trihydroxyphenylalanine) and its derivatives dehydrotopa and dehydrotopamine. The presence of tunichromes in ascidians has transiently enforced the notion that they are the primary reducing agents for vanadate(V), and the primary ligands to stabilize $[V^{IV}O]^{2+}$ formed by reduction. As a matter of fact, the tunichromes are not associated with the vanadocytes, the signet ring cells that have a large vacuole containing vanadium (Figure 6.9). The role of the tunichromes in the ascidians has not been established and remains hypothetical. It has been proposed to be connected to antibacterial and adhesive/polymerization properties.

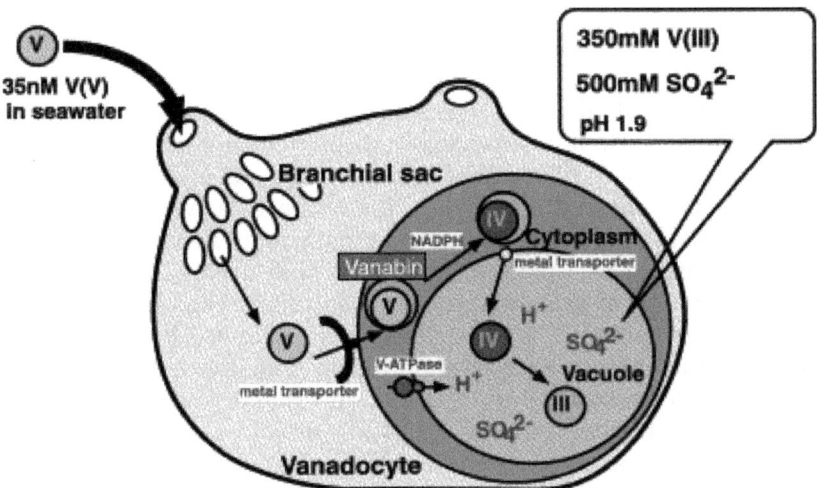

Figure 6.9 Representation of vanadium uptake, transport and accumulation in the ascidians.
Reproduced from ref. 83 with permission from Elsevier.

In ascidians, vanadate(V) likely enters the organism through anion channels, migrates into the cytoplasm of the vanadocytes through the phosphate channels and/or is translocated by a metal transporter. There it is reduced to $[V^{IV}O]^{2+}$ by NADPH[83] and bound subsequently to vanabins (vanadium-binding proteins). Another metal transporter conveys the $[V^{IV}O]^{2+}$ ions into the vacuole where the metal is concentrated and subsequently reduced to the oxidation state +III (Figure 6.9). At the low pH (1.9) inside the vacuoles, V^{III} forms aqua and (hydrogen)sulfate complexes,[†††] such as $[V^{III}(HSO_4)(H_2O)_5]^{2+}$, $[V^{III}(HSO_4)_2(H_2O)_4]^+$ and $[V^{III}(SO_4)(H_2O)_3(OH)_2]^-$.[2,16]

Five vanabins have been identified in *Ascidia sydneiensis samea* and *Ciona intestinalis* and named vanabin1–4 and vanabinP. Four are contained in the vanadocytes.[16] Vanabin1–3 are small cytosolic proteins (vanabin2 has 91 amino acids), whereas vanabin4 is located in the cytoplasmic membrane. VanabinP is one of the constituents of the coelomic fluid. Vanabin1 and 2 are metallochaperone proteins transporting $[V^{IV}O]^{2+}$ from the cytoplasm to the vacuoles and can bind up to 20 metal ions. The vanadium centres are isolated and not interacting. The binding sites of $[V^{IV}O]^{2+}$ are thought to be the lysine residues and are exclusively located on one side of the protein (14 for vanabin2).[84] VanabinP proteins are carriers of $[V^{IV}O]^{2+}$ in the coelomic fluid and can bind a maximum of 13 ions.[85]

The characterization of the vanadium content in polychaetes tissues is limited to a few species. The sabellid *Eudistylia vancouveri* was the first marine worm characterized in terms of its capability to accumulate vanadium from the environment. Concentrations ranged from about 90 to

---

[†††]An approximate concentration of 500 mM of (hydrogen)sulfate is present in the vacuoles.

almost 800 μg g^{-1} in the whole tissues.[86] High vanadium concentrations have been recently reported for other sabellid polychaetes such as *Pseudopotamilla occelata* and *Perkinsiana littoralis*. About 90% of the overall V is found in the branchial crown, concentrated in vacuoles of the outer part of the epidermal cells; it is associated with high amounts of sulfur, predominantly present in the form of sulfate. Analysis of *P. occelata* revealed an oxidation state +III and a VO_6 coordination, with a (V–O) bond length of about 2 Å, suggesting mixed aqua-(hydrogen)sulfate species.[2]

The mechanism of vanadium reduction and accumulation within the vacuoles is thought to be similar to that in ascidians: vanadium accumulated from sea water is bound by a specific protein, reduced in the cytosol from V(V) to V(IV) in the presence of glutathione, glutathione reductase and an electron transfer cascade from NADPH, and finally transported into vacuoles, where it is further reduced to V(III) by unknown reductants.

The function of vanadium in the ascidians and polychaetes is not known: it has been proposed that it may facilitate absorption of dioxygen and could be used in chemical defence against predation, fouling, overgrowth and bacteria.[86]

6.9.2   Vanadium in Amanita

Another example of accumulation of vanadium is given by *Amanita muscaria*, a discovery that dates back to 1931. Its vanadium content exceeds 400 times the value normally detected in other species of the same *genus* and is independent of the vanadium content in the soil.[‡‡‡] This mushroom, known as fly agaric, accumulates a vanadium compound named amavadin that was first isolated in 1972.[87] Subsequently, amavadin was also found, in comparably high concentrations, in two other species belonging to the same genus *Amanita*, *A. regalis* and *A. velatipes*.

Amavadin is a low-molecular-mass, anionic non-oxido V^{IV} complex derived from the bio-ligand *N*-hydroxyimino-2,2′-diisopropionic acid, (*S,S*)-H$_3$hidpa (Figure 6.10a), in its anionic form hidpa(3–).[2] Initially, spectroscopic data were consistent with a classic $[V^{IV}O]^{2+}$ complex as solutions containing amavadin are light blue.[87] But this interpretation was re-examined when a very large value for log β_2 of the species formed by V^{IV} with H$_3$hidpa and its analogue *N*-hydroxyiminodiacetic acid (H$_3$hida) was measured. Further studies assigned an eight-coordinate structure.[88,§§§] The X-ray structure of the calcium salt of amavadin [Ca(H$_2$O)$_5$][V((*S,S*)-hidpa)$_2$]·2H$_2$O showed that the V^{IV} coordination sphere involves two η^2-N,O$^-$ groups and four monodentate carboxylates (Figure 6.10b).[89] Oxidation in aqueous solutions and pH 7 of $[VL_2]^{2-}$ (L = hida(3–) or (*S,S*)-hidpa(3–)) is a single-electron reversible

---

[‡‡‡]There is no apparent relationship between amavadin and muscimol, the real psychoactive compound responsible for the mushroom's toxicity.

[§§§]As it is known, the chemistry of vanadium(IV) is dominated by penta- or hexa-coordinate species formed by the $[V^{IV}O]^{2+}$ ion. The peculiarity of amavadin is that this is a non-oxido (or "bare"), eight-coordinated vanadium(IV) complex.

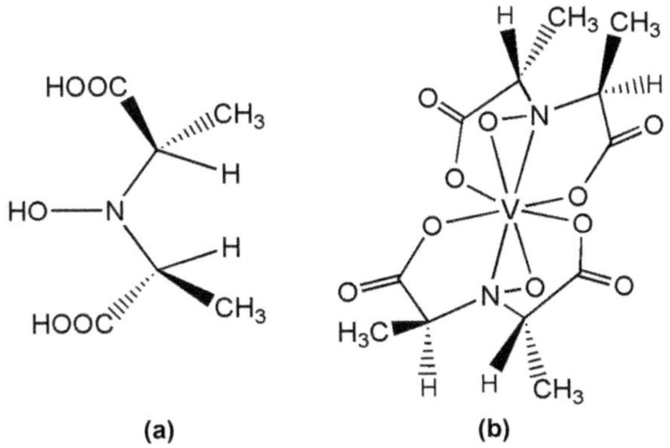

Figure 6.10   (a) The ligand present in amavadin, (*S,S*)-H₃hidpa, and (b) the structure of the V^{IV} complex of amavadin.
Adapted from ref. 2 with permission from John Wiley & Sons.

process (at *ca.* 0.8 V *vs.* NHE) that does not affect significantly the molecular structure of the complexes.[90]

The role of amavadin in *A. muscaria* remains unclear. On the one hand, (*S,S*)-hidpa(3–) forms a very stable complex with vanadium(IV) that is a necessary condition for efficient uptake of the metal from vanadium-poor environments. On the other hand, the anionic charge of the complex in both oxidation states +IV and +V does not favour crossing of cellular membranes. This is in contrast with other bio-ligands such as the siderophores, important in uptake of iron.[¶¶¶] The capacity of amavadin to occur in two different oxidation states with minor differences only in structure and the high value of the redox potential suggest a role in electron transfer. The ability of amavadin to act as an electron-transfer mediator in the oxidation of some biological thiols may support a role as a cofactor in oxidoreductases, a function usually performed by other transition metal ions.[90]

6.9.3   Vanadium in Siderophores and Bacteria

Many bacteria release strong iron-binding compounds (siderophores) for iron acquisition, and the same may happen with nitrogen-fixing bacteria acquiring V (and Mo).[‖‖‖] Recently, it was shown that *Azotobacter vinelandii* produces compounds that form strong complexes with vanadate(V) (and molybdate(VI)) and that these species are available for metal uptake.[91] These were named "vanadophores" in analogy with siderophores and have been identified as 2,3-dihydroxybenzoic acid, protochelin (a tris-catechol ligand) and, in lower concentrations, azotochelin (a bis-catechol ligand). A study with *Pseudomonas aeruginosa* showed that V^{IV}OSO₄ inhibits the growth of this Gram-negative bacterium, especially in the presence of Fe^{III} chelators.

¶¶¶See Chapter 10 for a discussion of siderophores.
‖‖‖See Chapter 8 for a discussion of uptake of molybdenum.

This interference implies that V may partially replace Fe in siderophore complexes and affect the iron uptake.[14]

The low level of vanadium present in the charged form in water around neutrality has been explained in terms of the activity of bacteria that transform soluble vanadate(V) species into insoluble oxidovanadium(IV) hydroxides, a process followed by incorporation into minerals.[2] For bacteria vanadate(V), as other high valent metal ions, may be an electron acceptor in place of O_2, used by higher animals and plants. *Saccharomyces cerevisiae* and *Shewanella oneidensis* are examples of organisms that can metabolize vanadium.[92,93] It has been proposed that some of these bacteria may be used for the detoxification of waters with elevated concentrations of vanadium through the transfer of electrons to V^V;[2] for example *Geobacter metallireducens*, a bacterium able to reduce metal ions in high oxidation states, can convert V^V in concentrations up to 5 mM to $[V^{IV}O]^{2+}$ by respiratory reduction. In this process acetate is the source of electron delivery.[94]

6.10   Future Developments

In spite of the great number of papers published over the last years and the progresses in the comprehension of V chemistry, many aspects of the biological functions of this metal in higher animals remain unclear. Only for certain organisms, such as ascidians, polychaete worms and *Amanita* mushrooms, the essential role of vanadium has been unambiguously demonstrated. On this basis, the future developments will concern several aspects of the uptake and transport in the biological fluids. First of all, it must be demonstrated if a homeostatic mechanism that controls the V concentration in human organism exists, beyond the low absorption in the gastrointestinal tract. The role in the higher animals is still not fully known and, even if many evidences suggest that vanadium is beneficial to health, its mechanism of action remains obscure. The inhibition of several enzymes and proteins explains most of the pharmacological activity, but the only examples of V-containing enzymes — vanadate-dependent haloperoxidases and vanadium nitrogenases — were isolated from fungi, lichens, marine algae and bacteria. One of the challenges of the next years will be to relate all the signs of V deficiency observed in animals (difficulty of conceiving, skeletal deformations, decrease in growth, convulsion and death) to some specific physiological function. The discovery of the presumable functions of V in higher animals must be connected with the study of the transport in the biological fluids, of the cellular uptake and the biotransformations in the cytosol. Whereas the transport in the blood, where the binding to the free ferric sites of transferrin is decisive and appears to be (almost) completely clear, little information is available about the interaction with the cellular membranes and biospeciation inside the cell. The versatility and complexity of V chemistry, such as the capability to interconvert the oxidation states between +IV and +V (and possibly also +III), to form oxido and non-oxido complexes in the +IV state and to be in the cationic or anionic forms, play a key role and, probably, were an obstacle for the full characterization of the biospeciation of the V^{III}, V^{IV} and V^V compounds in the cellular environment. Nevertheless, these factors influence the association with the membrane, the uptake in cells and the subsequent biotransformation after the interaction with the components of the cytosol, proteins and low-molecular-mass bio-ligands, and the knowledge of the biological role of V in the higher animals will be subordinated to the comprehension of these processes.

References

1. R. C. Lyman, *Bull. Hist. Chem.*, 2003, **28**, 35–41.
2. D. Rehder, *Bioinorganic Vanadium Chemistry*, John Wiley & Sons, Ltd, Chichester, 2008.
3. L. F. Vilas Boas and J. Costa Pessoa, in *Comprehensive Coordination Chemistry*, ed. G. Wilkinson, R. D. Gillard and J. A. McCleverty, Pergamon Press, Oxford, 1987, vol. 3, pp. 453–583.
4. L. Pettersson and K. Elvingson, in *Vanadium Compounds. Chemistry, Biochemistry, and Therapeutic Applications*, ACS Symposium Series, 1998, vol. 711, ch. 2, pp. 30–50.
5. F. H. Nielsen, in *Vanadium Compounds. Chemistry, Biochemistry, and Therapeutic Applications*, ACS Symposium Series, 1998, vol. 711, ch. 23, pp. 297–307.
6. F. H. Nielsen, in *Handbook of Nutritionally Essential Mineral Elements*, ed. B. L. O'Dell and R. A. Sunde, Marcel Dekker, 1997, ch. 22, pp. 619–630.
7. J. L. S. Sareen, S. Gropper and James L. Groff, *Advanced Nutrition and Human Metabolism*, Wadsworth, Cengage Learning, 2009.
8. A. S. Tracey, G. R. Willsky and E. S. Takeuchi, *Vanadium: Chemistry, Biochemistry, Pharmacology and Practical Applications*, CRC Press, Boca Raton, FL, 2007.
9. D. Rehder, *Future Med. Chem.*, 2012, **4**, 1823–1837.
10. L. C. Cantley, M. D. Resh and G. Guidotti, *Nature*, 1978, **272**, 552–554.
11. A. Heinz, K. A. Rubinson and J. J. Grantham, *J. Lab. Clin. Med.*, 1982, **100**, 593–612.
12. E. J. Baran, *Chem. Biodiversity*, 2008, **5**, 1475–1484.
13. M. Garner, J. Reglinski, W. E. Smith, J. McMurray, I. Abdullah and R. Wilson, *J. Biol. Inorg. Chem.*, 1997, **2**, 235–241.
14. D. C. Crans, J. J. Smee, E. Gaidamauskas and L. Yang, *Chem. Rev.*, 2004, **104**, 849–902.
15. J. Costa Pessoa and I. Tomaz, *Curr. Med. Chem.*, 2010, **17**, 3701–3738.
16. T. Ueki and H. Michibata, *Coord. Chem. Rev.*, 2011, **255**, 2249–2257.
17. H. Michibata and T. Ueki, in *Vanadium. Biochemical and Molecular Biological Approaches*, ed. H. Michibata, Springer, Netherlands, 2012, pp. 51–71.
18. W. R. Harris, S. B. Friedman and D. Silberman, *J. Inorg. Biochem.*, 1984, **20**, 157–169.

19. H. Yasui, K. Takechi and H. Sakurai, *J. Inorg. Biochem.*, 2000, **78**, 185–196.
20. N. D. Chasteen, J. K. Grady and C. E. Holloway, *Inorg. Chem.*, 1986, **25**, 2754–2760.
21. E. Sabbioni, J. Kuèera, R. Pietra and O. Vesterberg, *Sci. Total Environ.*, 1996, **188**, 49–58.
22. T. V. Hansen, J. Aaseth and J. Alexander, *Arch. Toxicol.*, 1982, **50**, 195–202.
23. X. Yang, K. Wang, J. Lu and D. C. Crans, *Coord. Chem. Rev.*, 2003, **237**, 103–111.
24. A. J. Ghio, C. A. Piantadosi, X. Wang, L. A. Dailey, J. D. Stonehuerner, M. C. Madden, F. Yang, K. G. Dolan, M. D. Garrick and L. M. Garrick, *Am. J. Physiol.*, 2005, **289**, L460–L467.
25. E. Sabbioni, E. Marafante, L. Amantini, L. Ubertalli and C. Birattari, *Bioinorg. Chem.*, 1978, **8**, 503–515.
26. L. C. Cantley and P. Aisen, *J. Biol. Chem.*, 1979, **254**, 1781–1784.
27. I. G. Macara, K. Kustin and L. C. Cantley, Jr., *Biochim. Biophys. Acta*, 1980, **629**, 95–106.
28. R. G. Hamilton, *Clin. Chem.*, 1987, **33**, 1707–1725.
29. W. R. Harris, *Clin. Chem.*, 1992, **38**, 1809–1818.
30. A. Gorzsás, I. Andersson and L. Pettersson, *Eur. J. Inorg. Chem.*, 2006, 3559–3565.
31. D. Sanna, G. Micera and E. Garribba, *Inorg. Chem.*, 2009, **48**, 5747–5757.
32. H. Sun, M. Cox, H. Li and P. Sadler, in *Struct. Bonding (Berlin)*, ed. H. Hill, P. Sadler and A. Thomson, Springer, Heidelberg, 1997, vol. 88, pp. 71–102.
33. D. N. Chasteen, *Coord. Chem. Rev.*, 1977, **22**, 1–36.
34. C. A. Smith, E. W. Ainscough and A. M. Brodie, *J. Chem. Soc. Dalton Trans.*, 1995, 1121–1126.
35. G. Justino, E. Garribba and J. Costa Pessoa, *J. Biol. Inorg. Chem.*, 2013, **18**, 803–813.
36. D. Sanna, P. Buglyó, G. Micera and E. Garribba, *J. Biol. Inorg. Chem.*, 2010, **15**, 825–839.
37. T. Jakusch, J. Costa Pessoa and T. Kiss, *Coord. Chem. Rev.*, 2011, **255**, 2218–2226.
38. D. C. Harris, *Biochemistry*, 1977, **16**, 560–564.
39. I. Correia, T. Jakusch, E. Cobbinna, S. Mehtab, I. Tomaz, N. V. Nagy, A. Rockenbauer, J. Costa Pessoa and T. Kiss, *Dalton Trans.*, 2012, **41**, 6477–6487.
40. T. Jakusch, A. Dean, T. Oncsik, A. C. Benyei, V. Di Marco and T. Kiss, *Dalton Trans.*, 2010, **39**, 212–220.
41. D. C. Crans, R. L. Bunch and L. A. Theisen, *J. Am. Chem. Soc.*, 1989, **111**, 7597–7607.
42. D. Sanna, G. Micera and E. Garribba, *Inorg. Chem.*, 2011, **50**, 3717–3728.
43. W. R. Harris, *Biochemistry*, 1985, **24**, 7412–7418.
44. M. H. Nagaoka, T. Yamazaki and T. Maitani, *Biochem. Biophys. Res. Commun.*, 2002, **296**, 1207–1214.

45. C. Harford and B. Sarkar, *Acc. Chem. Res.*, 1997, **30**, 123–130.
46. W. Bal, J. Christodoulou, P. J. Sadler and A. Tucker, *J. Inorg. Biochem.*, 1998, **70**, 33–39.
47. A. J. Stewart, C. A. Blindauer, S. Berezenko, D. Sleep and P. J. Sadler, *Proc. Natl Acad. Sci. USA*, 2003, **100**, 3701–3706.
48. H. Yasui, Y. Kunori and H. Sakurai, *Chem. Lett.*, 2003, **32**, 1032–1033.
49. D. Sanna, G. Micera and E. Garribba, *Inorg. Chem.*, 2009, **49**, 174–187.
50. G. Heinemann, B. Fichtl, M. Mentler and W. Vogt, *J. Inorg. Biochem.*, 2002, **90**, 38–42.
51. T. Kiss, T. Jakusch, D. Hollender, A. Dornyei, E. A. Enyedy, J. Costa Pessoa, H. Sakurai and A. Sanz-Medel, *Coord. Chem. Rev.*, 2008, **252**, 1153–1162.
52. Á. Dörnyei, S. Marcão, J. Costa Pessoa, T. Jakusch and T. Kiss, *Eur. J. Inorg. Chem.*, 2006, 3614–3621.
53. D. Sanna, G. Micera and E. Garribba, *Inorg. Chem.*, 2013, **52**, 11975–11985.
54. B. Zhang, L. Ruan, B. Chen, J. Lu and K. Wang, *BioMetals*, 1997, **10**, 291–298.
55. M. Aureliano, *Dalton Trans.*, 2009, 9093–9100.
56. D. Crans, S. Schoeberl, E. Gaidamauskas, B. Baruah and D. Roess, *J. Biol. Inorg. Chem.*, 2011, **16**, 961–972.
57. D. C. Crans, A. M. Trujillo, P. S. Pharazyn and M. D. Cohen, *Coord. Chem. Rev.*, 2011, **255**, 2178–2192.
58. S. K. Hanson, R. T. Baker, J. C. Gordon, B. L. Scott, L. A. P. Silks and D. L. Thorn, *J. Am. Chem. Soc.*, 2010, **132**, 17804–17816.
59. *Williams Hematology, Eighth Edition*, The McGraw-Hill Companies, Inc., China, 2010.
60. D. Rehder, J. Costa Pessoa, C. Geraldes, M. Castro, T. Kabanos, T. Kiss, B. Meier, G. Micera, L. Pettersson, M. Rangel, A. Salifoglou, I. Turel and D. Wang, *J. Biol. Inorg. Chem.*, 2002, 7, 384–396.
61. J. Costa Pessoa, I. Tomaz, T. Kiss, E. Kiss and P. Buglyó, *J. Biol. Inorg. Chem.*, 2002, 7, 225–240.
62. J. Costa Pessoa, I. Tomaz, T. Kiss and P. Buglyó, *J. Inorg. Biochem.*, 2001, **84**, 259–270.
63. G. Micera, D. Sanna, E. Kiss, E. Garribba and T. Kiss, *J. Inorg. Biochem.*, 1999, **75**, 303–309.
64. P. C. Wilkins, M. D. Johnson, A. A. Holder and D. C. Crans, *Inorg. Chem.*, 2006, **45**, 1471–1479.
65. J. B. Aitken, A. Levina and P. A. Lay, *Curr. Top. Med. Chem.*, 2011, **11**, 553–571.
66. G. R. Willsky, A. B. Goldfine and P. J. Kostyniak, in *Vanadium Compounds. Chemistry, Biochemistry, and Therapeutic Applications*, ACS Symposium Series, 1998, vol. 711, ch. 22, pp. 278–296.
67. M. Aureliano, *World J. Biol. Chem.*, 2011, **2**, 215–225.
68. L. C. Cantley, L. Josephson, R. Warner, M. Yanagisawa, C. Lechene and G. Guidotti, *J. Biol. Chem.*, 1977, **252**, 7421–7423.

69. Y. Lindqvist, G. Schneider and P. Vihko, *Eur. J. Biochem.*, 1994, **221**, 139–142.
70. K. M. Holtz, B. Stec and E. R. Kantrowitz, *J. Biol. Chem.*, 1999, **274**, 8351–8354.
71. T. A. S. Brandão, A. C. Hengge and S. J. Johnson, *J. Biol. Chem.*, 2010, **285**, 15874–15883.
72. D. Rehder, *Dalton Trans.*, 2013, **42**, 11749–11761.
73. K. Elvingson, D. C. Crans and L. Pettersson, *J. Am. Chem. Soc.*, 1997, **119**, 7005–7012.
74. P. J. Stankiewicz and A. S. Tracey, in *Metal Ions in Biological Systems*, ed. H. Sigel and A. Sigel, Marcel Dekker, New York, 1995, vol. 31, ch. 8, pp. 249–285.
75. K. H. Thompson and C. Orvig, *Coord. Chem. Rev.*, 2001, **219–221**, 1033–1053.
76. A. M. Evangelou, *Crit. Rev. Oncol. Hematol.*, 2002, **42**, 249–265.
77. P. Ghosh, O. J. D'Cruz, R. K. Narla and F. M. Uckun, *Clin. Cancer Res.*, 2000, **6**, 1536–1545.
78. A. Goc, *Cent. Eur. J. Biol.*, 2006, **1**, 314–332.
79. K. De Cremer, M. Van Hulle, C. Chéry, R. Cornelis, K. Strijckmans, R. Dams, N. Lameire and R. Vanholder, *J. Biol. Inorg. Chem.*, 2002, 7, 884–890.
80. I. A. Setyawati, K. H. Thompson, V. G. Yuen, Y. Sun, M. Battell, D. M. Lyster, C. Vo, T. J. Ruth, S. Zeisler, J. H. McNeill and C. Orvig, *J. Appl. Physiol.*, 1998, **84**, 569–575.
81. B. J.-S. Lagerkvist and A. Oskarsson, in *Handbook on the Toxicology of Metals (Third Edition)*, ed. G. F. Nordberg, B. A. Fowler, M. Nordberg and L. T. Friberg, Academic Press, Burlington, Vermont, 2007, pp. 905–923.
82. K. Fukui, Y. Fujisawa, H. Ohya-Nishiguchi, H. Kamada and H. Sakurai, *J. Inorg. Biochem.*, 1999, 77, 215–224.
83. H. Michibata, N. Yamaguchi, T. Uyama and T. Ueki, *Coord. Chem. Rev.*, 2003, **237**, 41–51.
84. T. Hamada, M. Asanuma, T. Ueki, F. Hayashi, N. Kobayashi, S. Yokoyama, H. Michibata and H. Hirota, *J. Am. Chem. Soc.*, 2005, **127**, 4216–4222.
85. M. Yoshihara, T. Ueki, T. Watanabe, N. Yamaguchi, K. Kamino and H. Michibata, *Biochim. Biophys. Acta*, 2005, **1730**, 206–214.
86. D. Fattorini and F. Regoli, in *Vanadium. Biochemical and Molecular Biological Approaches*, ed. H. Michibata, Springer, Netherlands, 2012, ch. 4, pp. 73–92.
87. H. Kneifel and E. Bayer, *Angew. Chem. Int. Ed.*, 1973, **12**, 508–508.
88. P. Krauss, E. Bayer and H. Kneifel, *Z. Naturforsch. B*, 1984, **39**, 829–832.
89. R. E. Berry, E. M. Armstrong, R. L. Beddoes, D. Collison, S. N. Ertok, M. Helliwell and C. D. Garner, *Angew. Chem., Int. Ed.*, 1999, **38**, 795–797.
90. J. A. L. Silva, J. J. R. F. Silva and A. J. L. Pombeiro, in *Vanadium. Biochemical and Molecular Biological Approaches*, ed. H. Michibata, Springer, Netherlands, 2012, pp. 35–49.

91. J. P. Bellenger, T. Wichard, A. B. Kustka and A. M. L. Kraepiel, *Nature Geosci.* , 2008, **1**, 243–246.
92. G. R. Willsky, D. A. White and B. C. McCabe, *J. Biol. Chem.*, 1984, **259**, 13273–13281.
93. W. Carpentier, L. De Smet, J. Van Beeumen and A. Brigé, *J. Bacteriol.*, 2005, **187**, 3293–3301.
94. I. Ortiz-Bernad, R. T. Anderson, H. A. Vrionis and D. R. Lovley, *Appl. Environ. Microbiol.*, 2004, **70**, 3091–3095.

CHAPTER 7

Chromium

PETER A. LAY* AND AVIVA LEVINA

School of Chemistry, The University of Sydney, Sydney NSW 2006,
Australia
*Email: peter.lay@sydney.edu.au

7.1 Biological and Environmental Chemistry of Chromium

Chromium (Cr, atomic number 24) is a first-row transition element of Group
6 of the periodic table and is among the most abundant of metals on Earth.
Its outer electronic configuration $3d^5 4s^1$ allows it to form compounds in
oxidation states from II− to VI +, among which Cr(0) (free metal), Cr(III) and
Cr(VI) are the most stable ones. The main sources of human exposure to Cr
compounds in the environment, in industrial settings and in everyday life,
are summarized in Scheme 7.1; a detailed description is available in several
books[1,2] and internet resources.[3] The most abundant Cr-containing mineral,
chromite ($Cr_2O_3 \cdot FeO$), is the main source of Cr(III) in soil and in ground
waters while Cr(VI) (chromate, $[CrO_4]^{2-}$) is predominant in sea water.[1,2]
Water-soluble Cr(III) complexes, formed by the reactions of chromite with
natural ligands such as humic acids, are absorbed by plants and enter the
food chain.[1,4] The low content of organic matter in many soils and the
presence of oxidizing minerals (primarily MnO_2) facilitate Cr(III) oxidation
to highly water-soluble $[CrO_4]^{2-}$.[2] Another major source of both Cr(III) and
Cr(VI) in the environment is fly ash formed during the burning of Cr(III)-
containing fossil fuels (Scheme 7.1).[5]

Chromium is a vital element for modern civilization, mainly as an essential component of stainless steel, as well as of a number of chemicals

RSC Metallobiology Series No. 2
Binding, Transport and Storage of Metal Ions in Biological Cells
Edited by Wolfgang Maret and Anthony Wedd
© The Royal Society of Chemistry 2014
Published by the Royal Society of Chemistry, www.rsc.org

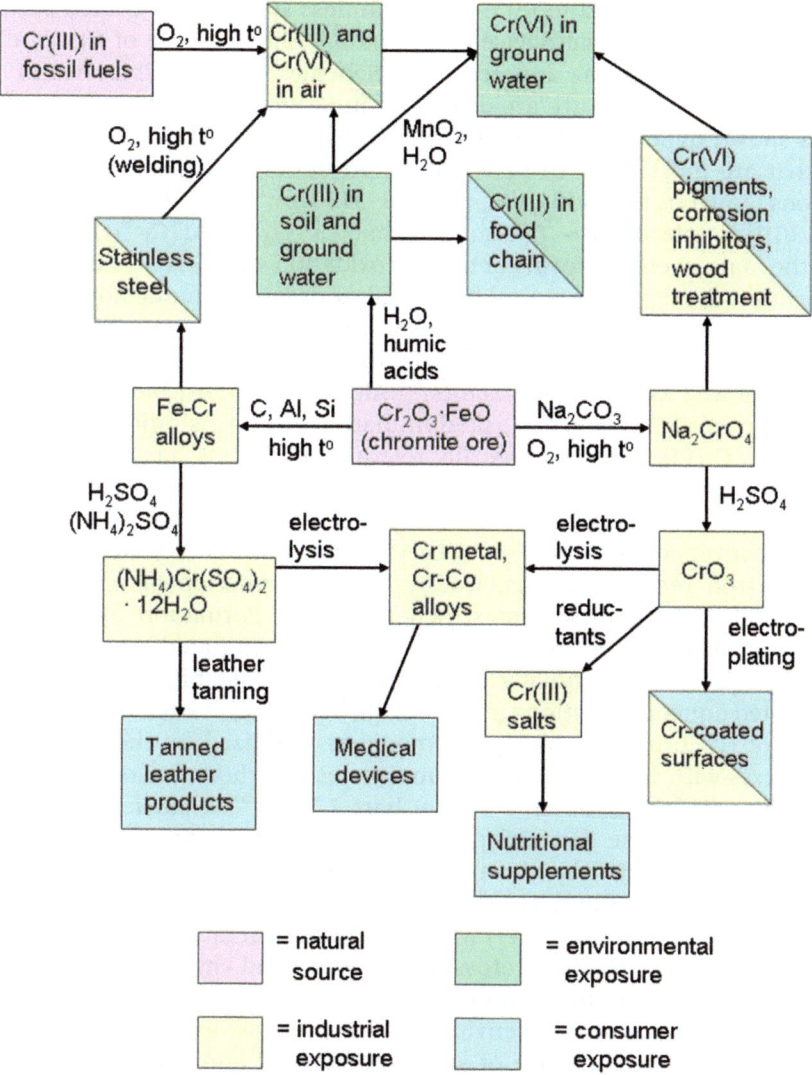

Scheme 7.1 Main chemical transformations of chromium in industry and the environment (based on published data).[1-3] The symbol t° refers to temperature.

(Scheme 7.1).[3] Primary processing of chromite is performed either by thermal reduction leading to Cr(0) (in alloys with iron), or by thermal oxidation leading to Cr(VI) (Na_2CrO_4, Scheme 7.1). Secondary products include stainless steel and other Cr alloys (Cr(0)), leather tanning solutions (Cr(III)), electroplating solutions (Cr(VI)), Cr/As/Cu wood preservatives (containing chromate) and pigments (water-insoluble Cr(VI) salts such as $PbCrO_4$). The most likely exposure of the general population to Cr compounds (apart from traces in air, water and food) include stainless steel and Cr-coated surfaces, tanned leather, medical devices (such as prostheses made of Cr-Co alloys) and Cr(III) nutritional supplements (Scheme 7.1). Possible health consequences of Cr exposure are discussed in Section 7.2.

As shown in Scheme 7.1, Cr(III) and Cr(VI) are the two oxidation states of Cr that are most likely to be encountered under biologically relevant conditions (aqueous solutions, pH $\sim$ 7, 293–313 K).[1,2] The redox potential of the Cr(VI/III) couple at pH $\sim$ 7 (assuming that $[Cr^{VI}O_4]^{2-}$ and $[Cr^{III}(OH)_2(OH_2)_4]^+$ are the main chemical forms of the two oxidation states) is $\sim +0.5$ V,[6] which lies within the range for typical biological oxidants and reductants (from -0.5 to $+1.0$ V).[7] These values can be used for guidance only, since the redox properties of the Cr(VI/III) couple are strongly affected by complexation of Cr(III), as well as by kinetic factors.[6,8] Although in most biological systems reduction of Cr(VI) to Cr(III) predominates,[8,9] the feasibility of Cr(III) oxidation to Cr(VI) under biologically relevant conditions has also been demonstrated (see Sections 7.5.1 and 7.6).[10,11] The likely biological reductants and oxidants for Cr(VI) and Cr(III), respectively, are listed in Scheme 7.2.[11,12]

Interconversions of Cr(III) and Cr(VI) occur through a series of one- and two-electron redox reactions, leading to the formation of highly reactive Cr(V) and Cr(IV) intermediates (Scheme 7.2).[12] Formation of Cr(II) intermediates under biological conditions has been postulated by some authors, but their presence is unlikely due to the extreme instability of these species in aerated aqueous solutions at pH $\sim$ 7 (Scheme 7.2).[12] Each of the oxidation states of Cr formed in biological media is stabilized by the formation of complexes with certain types of biomolecules, *e.g.* those listed in Scheme 7.2 and illustrated by structures **1–13** (Chart 7.1).[13–25] Of particular biological importance are the formation of thiolato-Cr(VI) complexes (**1**);[13,26] the stabilization of Cr(V) by the ubiquitous biological ligands carbohydrates and glycoproteins (**3** and **4**);[15,27] the formation of peroxido-Cr(V) complexes during the reduction of Cr(VI) with ascorbate in aerated aqueous solutions (**5**);[16] the ability of Cr(III) to cross-link proteins and small molecules to DNA (**13**),[28] as discussed in Section 7.5. Formation of Cr(VI) species both in biological media and in the environment is favoured under basic conditions (pH > 7)[6,10] in the absence of strong reductants and chelators for Cr(III). Cr(V) and Cr(IV) intermediates are generally more stable in weakly acidic media (pH = 4–6; *e.g.* in acidic lysosomes within mammalian cells),[8] and most Cr(III) complexes with biological ligands are stable over the whole biological pH range (in the absence of other strong ligands, Scheme 7.2).[29]

The vast majority of Cr(III) complexes is characterized by octahedral geometries (see **8–13** in Chart 7.1) and by slow rates of ligand exchange (kinetic inertness).[30] For instance, the mean time of a particular water molecule binding to a Cr(III) ion in aqueous solutions of $[Cr^{III}(OH_2)_6]^{3+}$ at 298 K is $\sim 5 \times 10^5$ s ($\sim$6 days), compared with $\sim$1 s for Al(III), $\sim 10^{-2}$ s for Fe(III), $\sim 10^{-7}$ s for Fe(II) and $\sim 10^{-9}$ s for Cu(II).[31] Biological consequences of the formation of kinetically inert Cr(III) complexes with biomolecules, including DNA and proteins, are discussed in Sections 7.5.1 and 7.6.

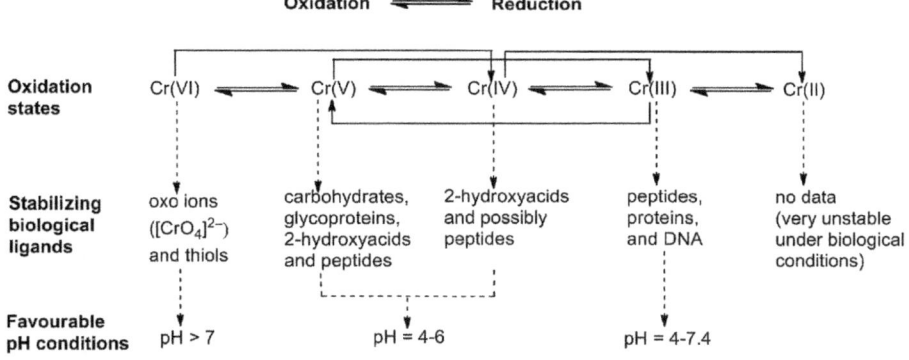

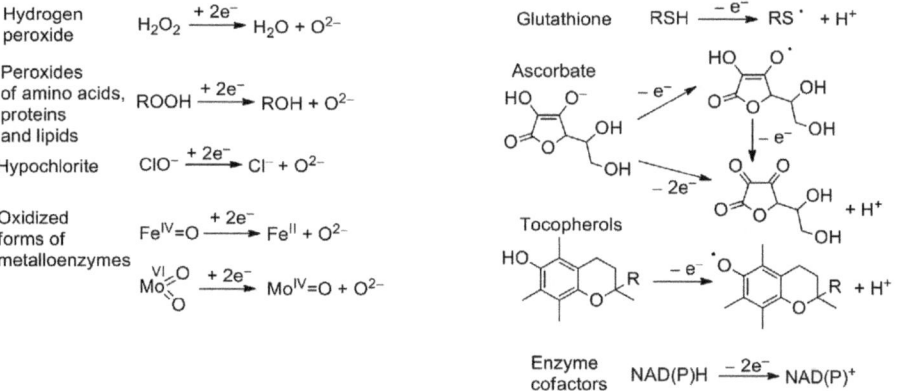

Scheme 7.2 An overview of Cr redox chemistry in biological systems (adapted from published data).[12] The O^{2-} in the Scheme refers to the oxido ligand transfered to a Cr atom in a two-electron oxidation of the Cr centre.

7.2 Biological Activities of Chromium

7.2.1 Toxicity

A detailed description of the toxic effects of Cr compounds in humans is given in recent reviews[32–34] and on a public internet site.[3] Briefly, Cr(VI) compounds were among the first chemicals for which a link between industrial exposure and an increased incidence of cancers was noticed (at the end of the nineteenth century), and are currently listed among well-established (Group 1) human carcinogens.[35] The carcinogenicity of Cr(VI) has been associated primarily with an increased risk of lung and bronchial cancers in exposed industrial workers, while the data about other cancer types (*e.g.* gastrointestinal) and non-industrial exposure are inconclusive.[35] The highest risk occupations involving Cr(VI) are the manufacture and use of Cr(VI)-containing pigments, chrome electroplating and manual welding of

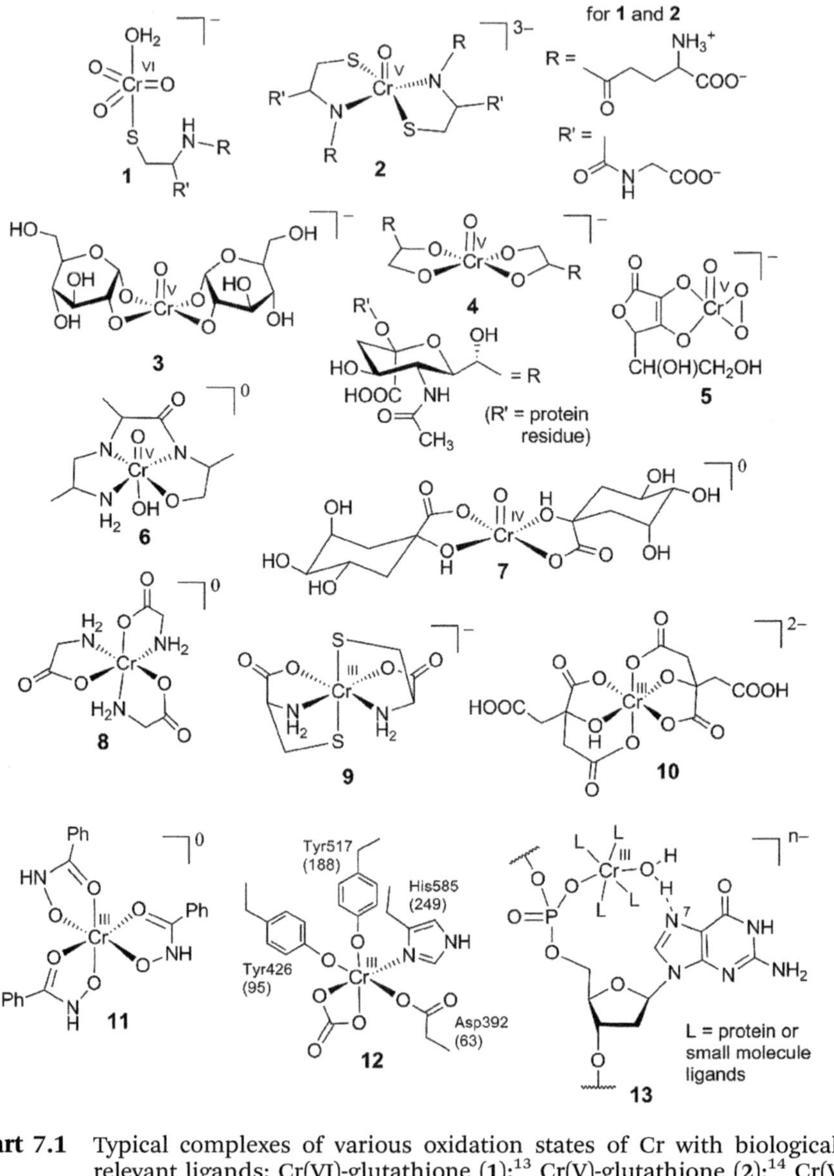

Chart 7.1 Typical complexes of various oxidation states of Cr with biologically relevant ligands: Cr(VI)-glutathione (1);[13] Cr(V)-glutathione (2);[14] Cr(V)-D-glucose (3);[15] Cr(V)-sialoglycoprotein (4);[15] Cr(V)-ascorbato-peroxido complex (5);[16] Cr(V)-Ala$_3$ (6);[17] Cr(IV)-quinic acid (7);[18] Cr(III)-glycine (8);[19] Cr(III)-L-cysteine (9);[20] Cr(III)-citric acid (10);[21] Cr(III)-benzhydroxamic acid (a model of siderophores) (11);[22] Cr(III)-transferrin (12, numbers denote amino acid residues in the C- and N-terminal lobes, respectively; the Cr(III) binding modes at both sites are identical);[23] and DNA-Cr(III)-amino acid (protein) (13).[24] The structures of 8–11 were determined by X-ray crystallography, and those of 1–7 and 12–13 were assigned from spectroscopic studies.

stainless steel that produces Cr(VI)-containing vapour (Scheme 7.1).[35] Moderately water-soluble chromates (Ca(II), Sr(II) and Zn(II) salts used as pigments) are considered more carcinogenic than highly water-soluble Na^+, K^+ and NH_4^+ salts or highly insoluble Pb(II) and Ba(II) salts.[35] The possible carcinogenicity of Cr(VI) in drinking water (originating from either natural or anthropogenic sources, Scheme 7.1) has caused considerable controversy,[2] but recent data suggest that this does present a cancer risk.[36,37] Natural Cr(VI) content in ground water can reach 50–100 ppb, and it is likely to increase further during the disinfecting treatments (chlorination or ozonation) that lead both to Cr(III) oxidation to Cr(VI) and to destruction of Cr(VI)-reducing organic matter.[2,37] Importantly, Cr(VI) can act as a co-carcinogen in conjunction with UV radiation or with aromatic polycyclic hydrocarbons.[38,39] Likely mechanisms of Cr(VI)-induced carcinogenicity are discussed in Sections 7.5 and 7.6 and in several recent reviews.[33,34,40]

Although there is currently no conclusive evidence on the carcinogenicity of Cr(III) or Cr(0) compounds to humans,[35] data from cell and animal experiments have pointed to the potential carcinogenicity of some Cr(III) nutritional supplements, as well as to an increased cancer risk in offspring due to parental exposure to Cr(III).[29,41–43] Recent epidemiological studies have suggested that exposure to high levels of Cr(III) in the environment (*e.g.* near leather tanneries) increases the risk of a range of other diseases, including respiratory illness, diabetes and gastrointestinal and dermal problems.[44] The use of Cr-Co alloys in hip prostheses and other medical implants is now recognized as a source of increased metal concentrations in the blood of patients, which can potentially lead to carcinogenicity and other health problems.[45] In addition, allergies caused by Cr(0), Cr(III) or Cr(VI) compounds remain a major cause of concern both in Cr-related industries and among the general population (1–3% of which are allergic to Cr compounds).[3,46] There is also ample evidence for Cr(VI) toxicity in animals, plants and microorganisms, while Cr(III) compounds are considered relatively non-toxic due to their poor bioavailability (Section 7.7).[35,47]

7.2.2 Anti-diabetic Activity and the Problem of Chromium Essentiality

The history and current status of Cr(III) as a potential cofactor in insulin signalling and glucose metabolism are comprehensively described in two recent books[43,48] and several reviews.[49–51] Claims of Cr(III) essentiality for humans are often found in the nutritional literature[52] and date back to the reports of Mertz and co-workers in the 1950s on the isolation of a Cr(III)-containing "glucose tolerance factor" from brewer's yeast.[53] However, the subsequent 50 years of research has led to the conclusion that *Cr(III) is probably not an essential element for human nutrition.*[43,49,51,54,55] First, developments in analytical techniques in the 1980s led to decreases by several orders of magnitude in estimated Cr concentrations in normal human tissues and in typical food sources. The early data were affected by Cr contamination from stainless steel tools and from the environment, as well as by instrumental artefacts.[56,57] These findings led to the realization that the amounts of Cr compounds used in the early research designed to demonstrate the effects on glucose metabolism were clearly pharmacological, rather than nutritional.[43] Second, the generation of chromium deficiency states in experimental animals has proven to be extremely difficult[55,56] and no genetic Cr deficiency disease has been identified as yet.[43] Third, despite some reports in the literature, there is currently no conclusive evidence of regulated Cr homeostasis in animals or humans (Section 7.3).[43] Fourth, despite decades of extensive research, no functional Cr(III)-containing biomolecule has been characterized unambiguously and several hypotheses on their possible structures have been refuted (see Section 7.5.2).[43,51,56] Finally, there is no conclusive evidence for Cr essentiality in microorganisms, plants or animals (see Section 7.7).[47,58,59]

Despite this uncertainty about the nutritional status of Cr(III), there is a steady consumer demand for Cr(III)-containing nutritional supplements, primarily Cr(III) picolinate (**14** in Chart 7.2),[60,61] which are claimed to assist weight loss and improve body condition.[62–64] These claims are not supported by several large-scale clinical trials that have shown no significant metabolism-enhancing activity of Cr(III)-nutritional supplements in non-diabetic humans.[65–68] Moreover, there is growing concern over possible adverse health effects, including cancers and birth defects, caused by uncontrolled use of **14** and other Cr(III) nutritional supplements.[11,29,41] Although recent animal experiments did not provide compelling evidence for carcinogenicity or teratogenicity of **14**,[69,70] these studies cannot be directly extrapolated to long-term human exposure, because of the long latency of Cr-induced cancers in humans[35] and likely effects of co-carcinogens, such as UV radiation and oxidative stress, in diabetic patients.[38,50]

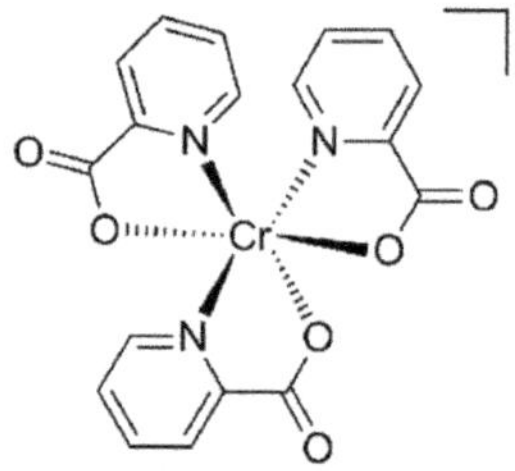

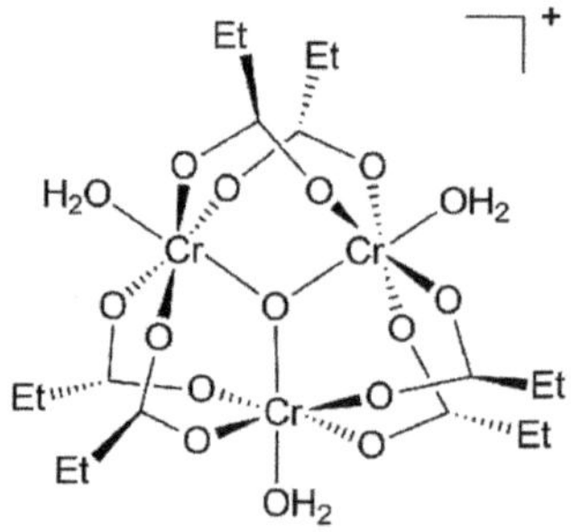

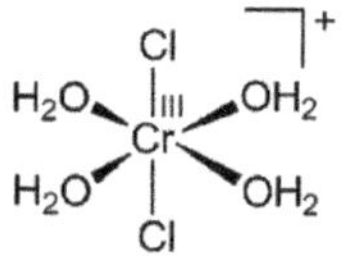

Chart 7.2   Structures of typical Cr(III) complexes used in biological studies: $[Cr(pic)_3]$ (**14**, where pic = picolinato = 2-pyridinecarboxylato), $[Cr_3O(OCOEt)_6(OH_2)_3]^+$ (**15**) and *trans*-$[CrCl_2(OH_2)_4]^+$ (aquated $CrCl_3$, **16**).

The efficacy of high doses of Cr(III) nutritional supplements (*e.g.* those exceeding normal Cr dietary intakes by at least ten times)[71] in the treatment of type 2 (non-insulin dependent)[72] diabetes and related metabolic disorders is currently under debate.[43] A recent large-scale and highly controlled clinical trial has shown that only about a half of diabetic patients responded to the treatment with **14**, and those in the responding group had significantly higher initial values of fasting blood glucose and related parameters (insulin resistance, haemoglobin glycation) compared with the non-responding group.[73] Importantly, the decreases of these parameters in responding patients were not to normal levels, but to those of the non-responding group.[73] These data suggest that **14** is only active in patients with poorly controlled diabetes that is associated with chronic oxidative stress,[74] in agreement with the oxidative hypothesis of biological activity of Cr(III) (Section 7.6).[10,29,50]

A number of Cr(III)-containing supplements, including complexes with propionate (predominantly in the trinuclear form; **15** in Chart 7.2),[75,76] amino acids or nicotinate, have been proposed as supposedly safer and more bioavailable alternatives to **14**.[43] However, all these compounds are capable of producing toxic Cr(VI) under biologically relevant conditions (Sections 7.1 and 7.6), and their anti-diabetic activity in animal experiments correlate with their oxidation rates in blood.[10,11,29,50,77] The ability of Cr(III) complexes with N-donor ligands to be oxidized to relatively stable and genotoxic Cr(V) complexes under biologically relevant conditions has also been demonstrated.[78–80] Experiments with animal models of type 2 diabetes have shown some activity of Cr(III) complexes in improving blood lipid parameters but the glucose levels were largely unaffected.[43,50]

In summary, Cr(III) complexes do not show great promise as *long-term oral* anti-diabetic agents, due to their limited activity but also due to the safety concerns over their prolonged use.[43,50] This conclusion is consistent with the current policy of the American Diabetes Association that advises against the use of Cr(III) compounds in the treatment of glucose intolerance or diabetes.[81] Still, several independent reports appeared recently on dramatic improvements in glucose tolerance of critically ill patients by *short-term intravenous* injections of Cr(III) solutions ($CrCl_3 \cdot 6H_2O$ in physiological saline; see Sections 7.3 and 7.4).[82–84] These data show that the studies on the possible mechanisms of anti-diabetic activity of Cr(III) should be continued (Section 7.6).

7.3 Gastrointestinal Absorption and Metabolism of Chromium

As described in detail elsewhere,[3] human exposure to Cr occurs *via* ingestion, inhalation or dermal contact. Since the last two routes are mostly limited to high-level exposures to anthropogenic Cr sources in industrial settings (Section 7.2.1),[3] this chapter concentrates on the gastrointestinal metabolism of naturally occurring Cr compounds in the general population.

Given the uncertain nutritional status of Cr (Section 7.2.2),[43,50] it is hard to determine the physiological requirements (if any) for this element. The currently accepted adequate Cr intake levels (25–35 µg day^{-1} for adults)[71] are based on the measurements of Cr content in typical food sources.[85] Note that the previously reported higher values (50–200 µg day^{-1})[86] were based on unreliable analytical data (see Section 7.1).[85] It is believed that Cr is ingested mainly in the form of Cr(III) carboxylato (*e.g.* oxalato or citrato) or amino acid complexes in food of plant or animal origin. However, knowledge of Cr speciation in food sources is very limited.[47] Breakfast cereals, broccoli and some brands of beer have been reported as relatively rich sources of dietary Cr although it is unclear if Cr leakage from stainless steel utensils also contributes significantly to dietary Cr intake.[85,86]

The chemical form of Cr(III) is a crucial factor that determines the extent of its gastrointestinal absorption.[3,87] Early studies using $CrCl_3 \cdot 6H_2O$ (mostly *trans*-$[CrCl_2(OH_2)_4]Cl_2 \cdot 2H_2O$; **16** in Chart 7.2)[88] showed low absorption efficiency ($\sim$0.4%) in humans.[89] This value may not be representative of natural Cr metabolism since **16** is a non-biological and highly acidic source of Cr(III), which rapidly hydrolyses under intestinal conditions with the formation of insoluble oligomeric hydroxido complexes or of Cr(III)-biomolecule adducts.[29,50] Formation of insoluble Cr(III) phosphates in the acidic environment of the stomach is also likely to contribute to the poor oral bioavailability of Cr(III) compounds, as has been shown also for Al(III).[90] Nutritional supplements, such as Cr(III) picolinate or nicotinate, which were developed with the goal of increased oral bioavailability, have provided only a modest increase (to 1–3%) in gastrointestinal absorption of Cr(III).[87] Significantly higher absorption values (up to $\sim$20%) were reported for some amino acid complexes.[91] Exceptionally high intestinal absorption in rats (40–60%) was reported for **15** (Chart 7.2),[92] but these data were not verified by direct comparison with other Cr(III) complexes. Notably, a detailed recent study with the use of ^{51}Cr(III) tracers suggested that the absorption of Cr(III) from various types of supplements (including picolinato, nicotinato, propionato and amino acid complexes) was lower than previously thought (up to 0.24% in rats and up to 1.0% in humans).[93] Thus, the extent of Cr(III) absorption from nutritional sources remains controversial. There are also controversial reports on the influence of food components, such as starch or ascorbic acid, on the gastrointestinal absorption of Cr(III).[3,87,91]

Gastrointestinal metabolism of Cr(III) complexes that are intended for use as oral anti-diabetics can be crucial for their biological activities.[50] Typical Cr(III) nutritional supplements, **14** and **15**, remained mostly (>75%) unchanged in simulated gastrointestinal media in the presence or absence of food components.[77,94] However, studies in mice and rats, using ^{14}C-labelled **14**, have shown that a large proportion of the complex decomposed on the intestinal walls, followed by significant absorption of the ligand (26–56%), but poor absorption of Cr (1–5%).[95] Compound **14** was also metabolized (with the release of the ligand) *in vitro* by liver enzymes.[96] These findings raise the question as to whether some of the reported pharmacological effects of **14** are due to the ligand and its metabolites (such as *N*-picolinoylglycine),[95] rather than due to Cr(III).[43] Such decomposition can be expected to occur to an even larger extent during the gastrointestinal absorption of other Cr(III) nutritional supplements which are generally more reactive *in vitro* compared with **14**.[29,77]

No active mechanisms of Cr(III) uptake and release by intestinal epithelial cells (such as those recently described for Fe(III))[97] are currently known.[98] Studies using monolayers of cultured human intestinal cells suggested that Cr(III) from **14** or **16** (or rather from their reaction products with cell culture media)[29,77] is transported through the intestinal epithelium *via* passive diffusion.[99] By contrast, it is well known that $[Cr^{VI}O_4]^{2-}$ is taken up actively by various types of mammalian cells through anion channels. This active transport is due to its structural similarity to essential SO_4^{2-} and HPO_4^{2-} ions (see Section 7.5.2 for details).[8,9,100] The Cr(V) complexes that are formed as oxidation products of Cr(III) are also highly permeable to cells, although the mechanisms of their uptake are not yet clear.[78]

The metabolism of ingested Cr(VI) is at the centre of the current controversy over the possible carcinogenicity of Cr(VI) in drinking water (Section 7.2.1).[2,37,101] The acidic gastric environment has a high capacity to reduce Cr(VI) to Cr(III), particularly in the presence of food components.[100,102–104] The ability of intestinal microflora to reduce Cr(VI), and to facilitate the excretion of the resultant Cr(III) products, has recently been recognized as an important factor in detoxification of ingested Cr(VI) (see also Section 7.7.3).[105] Nevertheless, numerous studies have demonstrated higher systemic uptakes of ingested Cr(VI), compared with equivalent doses of Cr(III), in animals and humans (see the recent reviews).[3,100] Therefore, contrary to earlier claims,[102,103,106] traces of Cr(VI) in drinking water (Section 7.2.1) that escape reduction in the stomach and pass into the intestines may form a significant part of total Cr intake in humans.[37]

There is also an ongoing controversy on whether the amount of Cr absorbed from food sources is regulated by the organism (homeostasis), and whether the disruptions in such regulation can contribute to the development of type 2 diabetes.[43] One study from 1985 reported that Cr absorption levels in healthy humans were inversely proportional to dietary intake, which suggests the presence of an active regulatory mechanism.[107] However, no independent verification of this finding has appeared since its publication and such studies are extremely difficult to perform due to the low concentrations of Cr in food and in human tissues.[43] No significant correlations between dietary Cr intake and absorption levels were found in rats.[108] There are numerous reports on decreased plasma and increased urinary Cr levels in type 2 diabetics,[109] but these changes can represent one of the effects of diabetes, rather than its cause.[110] Changes in Cr(III) metabolism in type 2 diabetes can also be linked to changes in Fe(III/II) metabolism, since these two metals are likely to share some metabolic pathways (Sections 7.4 and 7.5.2).[43]

7.4 Chromium Speciation in Blood and Other Biological Fluids

Although most studies are performed using isolated plasma or serum, red blood cells (RBC) can be equally or even more important in effecting the transport and speciation of metal complexes in the bloodstream.[111] Most of the Cr speciation studies in whole blood used Cr(VI) (*e.g.* Na_2CrO_4) as a source of Cr. For humans, measurable amounts in the blood can occur as a result of the following: (i) exposure to Cr(VI) compounds by inhalation (Section 7.2.1);[3,112] (ii) corrosion of Cr-containing medical implants;[113] (iii) regular ingestion of Cr(VI)-containing water (Section 7.3);[37] and (iv) oxidation of Cr(III) complexes taken as anti-diabetics under the conditions of oxidative stress that is characteristic of advanced diabetes (Section 7.6).[10,29,50] The uptake rate of Cr(VI) by RBC is three orders of magnitude higher than that of Cr(III) complexes of amino acids.[114] The RBC fraction is mostly responsible for the rapid reduction of Cr(VI) to Cr(III) by fresh whole blood, while Cr(VI) persists for hours in isolated plasma.[115] Cellular thiols (essentially glutathione) have been implicated as the most important Cr(VI) reductants in RBCs, and the resultant Cr(III) is predominantly bound to haemoglobin as the most abundant RBC protein.[116,117] Since the resultant Cr(III) is trapped inside the RBC as long as the cellular membrane is intact (due to its kinetic inertness, see Section 7.1), accumulation of Cr in RBC is used as a biomarker of systemic Cr(VI) uptake and an early sign that Cr(VI)-related toxicity is likely to have occurred.[118] In contrast to Cr(VI), oral or intravenous administration of Cr(III) compounds to animals resulted in extensive Cr(III) binding to plasma proteins, and little if any Cr uptake by RBC.[119]

On the basis of perfusion experiments in rats,[120] complementary roles in Cr(III) transport have been proposed for two of the main metal-binding blood proteins albumin and transferrin (Tf).[111,†] In these experiments, the small intestines of the animals were filled with a liquid meal containing a ^{51}Cr(III) tracer; the surrounding blood vessels were filled with a protein-depleted buffer solution. These conditions led to a drastic decrease of Cr(III) transport from the intestines to the blood vessels, which could be restored to the normal level by the addition of subphysiological concentrations of either albumin or *apo*-Tf to the buffer solution.[120] On the other hand, distribution studies of a ^{51}Cr(III) tracer in the blood of humans undergoing peritoneal dialysis indicated that Cr(III) binds preferentially to Tf compared with albumin.[121] A low-molecular-mass (LMM, usually referred as low-molecular-weight, LMW $\sim$5 kDa) Cr-containing fraction was also prominent in human plasma, particularly at long incubation times, but this was not characterized further.[121] However, the results of these studies can be affected by the use of a non-biological source of Cr(III) (**16** in Chart 7.2).[29]

---

†The biological role and molecular properties of the ferric ion metallochaperone transferrin are discussed in Chapter 11.

Administration of a synthetic radiolabelled ^{51}Cr(III)-Tf complex to rats *via* intravenous injection resulted in rapid and insulin-sensitive Cr(III) uptake into tissues (similar to that for Fe(III)-Tf).[43,122] Although Fe(III) transport is the primary biological function of Tf, its Fe-binding sites are only ~30% saturated under normal conditions.[‡] This creates opportunities for the transport of other metal ions, particularly those such as Cr(III) that are chemically similar to Fe(III).[123,124] Early studies by EPR spectroscopy,[125] corroborated by more recent data,[23] indicated that Cr(III) binds strongly (overall effective binding constant ~10^{15} M^{-2}) and selectively to the two Fe(III)-binding sites of isolated apo-Tf (**12** in Chart 7.1).[124] In summary, currently available data point to the role of Tf as a major Cr(III) transporter in blood[124] but the roles of albumin[120] and of LMM binders such as small peptides,[126] citrate[127] or phosphate[128] are still to be elucidated.

Studies of the reaction products of proposed oral anti-diabetic drugs, **14** and **15** (Chart 7.2), with bovine serum were carried out by X-ray absorption spectroscopy.[77] Each complex underwent extensive ligand-exchange reactions in serum (reaction times, 6–24 h at 310 K) that led to similar products consisting predominantly of Cr(III)-protein adducts (binding to albumin and transferrin could not be distinguished by this method) plus Cr(III) complexes with amino acid and aqua/hydroxido ligands.[77] Importantly, these reaction products (unlike **14**) were susceptible to oxidation to Cr(VI) by biological oxidants, such as H_2O_2 (Scheme 7.2).[77] These observations are consistent with rapid dissociation of radioactively labelled Cr(III) ions and the ligands in **14** (^{51}Cr, ^{3}H) and **15** (^{51}Cr, ^{14}C) when either of the complexes was administered intravenously to rats.[129,130] These results, together with the data on decomposition of **14** during intestinal absorption (Section 7.3),[95] indicate that Cr(III) complexes act as pro-drugs, *i.e.* they release potentially active components during interactions with biological media.[50]

Administration of either Cr(III) or Cr(VI) compounds to experimental animals resulted in excretion of LMM Cr-containing species in the urine (~1.5 kDa according to gel-filtration chromatography).[131,132] Species of similar molar mass containing Cr(III) and several amino acids (predominantly Asp, Glu, Gly and Cys) were isolated from the livers of Cr(VI)-treated animals[133] or from the *ex vivo* reactions of bovine liver homogenates with Cr(VI).[134] They were represented as a natural biologically active Cr-binding substance, termed low-molecular-weight chromium binding substance (LMWCr) or chromodulin in the literature (referred to as LMMCr in the following text).[43,132] However, it is not yet known whether LMMCr is a unique compound or a mixture, and the chemical nature and biological roles of Cr(III) species in urine remain unknown (see Section 7.5).[51] Binding of Cr(III) by LMM ligands in biological fluids is likely to play a protective role against Cr-induced toxicity.[54,135] Modern chromatographic and

---

[‡]The biological role and molecular properties of the ferric ion metallochaperone transferrin are discussed in Chapter 11.

mass-spectrometric techniques allow for the selective determination of both Cr(III) and Cr(VI) in human urine, and the presence of the latter can be used as an indicator of high-level occupational exposure to Cr(VI).[136]

Speciation of Cr in synovial fluid has been studied[137] in relation to the corrosion and toxicity of Cr-containing medical implants.[45] Paramagnetic broadening of ^{1}H NMR spectra due to binding of Cr(III) to biomolecules indicated that the ion binds predominantly to amino acids in human synovial fluid that was aspirated from patients not wearing metallic implants.[137] A limitation of this study was the use of high concentrations of an external, non-biological source of Cr (**16** in Chart 7.2).[29] Incubation of medical Co-Cr alloys with synovial fluid *in vitro* led to the detection of Cr ions in the fluid, but oxidation states and coordination environments could not be determined due to their low concentrations.[138] Distribution of Cr between RBC and blood plasma can be used as an indicator of the oxidation state of Cr formed during the corrosion of Cr-containing implants.[118] Recent human studies showed that Cr was localized almost exclusively in the plasma of implant-wearing patients,[139] indicating that the metal is released as Cr(III). However, earlier animal studies showed that corroded stainless steel implants led to significant Cr accumulation in the RBC fraction,[113] suggesting that Cr(VI) can be released from the implants under unfavourable conditions (including inflammation that locally leads to up to millimolar concentrations of biological oxidants, such as H_2O_2 or ClO^-).[140]

Inhalation of Cr(VI)-containing vapours or dusts (*e.g.* in electroplating or paint manufacturing; see Scheme 7.1) results in Cr(VI) interactions with a range of extracellular fluids (saliva, nasal mucus, bronchial and lung fluids), as well as with mucin-producing epithelial cells.[15,103,141] Despite the significant reductive capacity of these media, detoxification of inhaled Cr(VI) is expected to be less efficient compared with the ingested forms (Section 7.3).[103] This is particularly true for poorly soluble particles (such as $PbCrO_4$ used in paints) that adsorb on the surface of lung or bronchial epithelial cells and slowly dissolve with the formation of high local concentrations of Cr(VI).[142,143] Furthermore, interactions with sialoglycoproteins and mucins that are abundant in such fluids can lead to the formation of relatively stable and highly genotoxic Cr(V) species, such as **4** in Chart 7.1.[15,141] Consequently, inhalation of Cr(VI) (particularly of its poorly water-soluble salts) is the predominant exposure route by which Cr(VI) can cause cancer (Section 7.2.1).[3,35]

7.5 Uptake and Metabolism of Chromium by Mammalian Cells

7.5.1 Chromium(VI)

Metabolism of Cr(VI) by animal cells has been studied extensively, starting from the early 1980s with the uptake-reduction model of Wetterhahn and co-workers[9] that was subsequently developed by other researchers.[8,34,144]

The ability of Cr(VI) compounds (both soluble and insoluble) to penetrate mammalian cells is among the most efficient that is known for metal complexes.[8] For instance, incubation of non-cancerous human bronchial cells with Na_2CrO_4 (10 μM) or $PbCrO_4$ (5.0 mg per cm^2 of the cell monolayer) for 24 h resulted in intracellular Cr concentrations of ∼4 mM and ∼2 mM, respectively.[145] Recent experimental data[100] confirmed an earlier hypothesis[9] that soluble $[CrO_4]^{2-}$ enters cells *via* a general anion transport mechanism, primarily through a $Na^+/[SO_4]^{2-}$ co-transporter. Poorly soluble chromates ($MCrO_4$, where $M = Ca(II)$, $Zn(II)$ or $Pb(II)$) are likely to absorb on the surface of cells and dissolve slowly, creating high local concentrations of $[CrO_4]^{2-}$.[146] Alternatively, the cells can engulf such particles by phagocytosis.[143]

The ability of X-ray absorption spectroscopy (XAS) to clearly distinguish between various oxidation states of Cr has been used to show that Cr(III) is the predominant (>95%) form of Cr in Cr(VI)-treated cells.[147,148] Analysis of XAS data with the use of libraries of model compounds showed that the majority of this Cr(III) is bound to carboxylato, amine and imine residues of proteins.[25] These reaction products were practically identical over a wide range of human or animal cells that originated from various tissues.[25,77] The resultant Cr(III) is very slow to efflux from the cell due to the kinetic inertness of the adducts (Section 7.1), which rationalizes the ability of cells to accumulate large amounts of Cr.[34] The fast cellular reduction of Cr(VI) to Cr(III) (which is probably limited by the rate of Cr(VI) transport into the cell) also maintains the Cr(VI) concentration gradient between extra- and intracellular spaces and thus facilitates uptake.[8]

Studies of Cr distribution in single Cr(VI)-treated human lung cancer cells using microprobe X-ray fluorescence mapping techniques[147] suggested that, at the initial stages of Cr(VI) uptake (∼20 min at $[Cr(VI)] = 0.10$ mM), a protective mechanism operates in which metal is confined to small areas in the cytoplasm and strongly co-localized with S, Cl, K and Ca.[149] Longer treatments (>1 h) led to the distribution of Cr throughout the cell and to a loss of cell viability. This corresponded to significant Cr(VI) uptake by the nucleus.[149] The latter may be facilitated by electrostatic binding of $[CrO_4]^{2-}$ to the positively charged residues of nuclear proteins, such as histones that migrate from the cytoplasm to the nucleus and carry structurally similar phosphate and sulfate ions.[150] Reduction of Cr(VI) in the vicinity of DNA inside the nucleus is likely to lead to the formation of bulky DNA adducts, such as **13** in Chart 7.1. These may disrupt the transcription process and cause genotoxicity and malignant transformations of the cell.[28,34,151] Other important cellular targets for Cr(VI) and Cr(V) complexes include RNA,[27,152] zinc finger proteins[153,154] and cysteine moieties of regulatory proteins (such as tyrosine phosphatases).[10,50] These processes are likely to lead to disruptions in gene regulation, DNA repair and cell signalling, all of which are potential steps on the pathway to carcinogenicity.[34,155,156]

The likely cellular reductants of Cr(VI) include both small molecules (predominantly ascorbate and glutathione)[8] and proteins, such as nitric oxide synthase,[157] cytochrome b_5[158] and metallothioneins[159] (Scheme 7.2).

Conversely, the resultant Cr(III) complexes can be oxidized to Cr(VI) (both intra- and extracellularly) by biological oxidants such as H_2O_2, ClO^- or oxidase enzymes (Scheme 7.2).[29,50,140] The reactive Cr(VI), Cr(V) and Cr(IV) intermediates (all of which are bound to biological ligands such as thiols or carbohydrates; Scheme 7.2 and **1–7** in Chart 7.1) are likely to form in both processes.[13–18,27] These species have been generated and characterized *in vitro*, as discussed in recent reviews.[144,147] In addition, the formation of carbohydrate-Cr(V) complexes, similar to **3** and **4** in Chart 7.1, has been demonstrated in living Cr(VI)-treated cells[25,160,161] and animals.[117,162,163] Such complexes are known to damage DNA.[164] Apart from reactive Cr intermediates, the redox reactions of Cr(VI) and Cr(III) in the presence of O_2 produce reactive oxygen species (ROS) that would contribute to cellular oxidative stress.[165,166]

A recent publication has attempted to provide mathematical modelling of the intracellular reactions of species in the available range of Cr oxidation states with a variety of oxidants and reductants in order to determine changes in the *in vivo* concentrations of these species and of various radicals.[167] While this paper comes to some interesting conclusions, it has not included all of the potentially important reactions discussed above and further development of this modelling is required to more accurately reflect all of the potentially important reactions of both the chromium and radical species. A multi-compartment physiologically based pharmacokinetic (PBPK) model that was developed to describe the behaviour of Cr(III) and Cr(VI) in humans is of particular interest.[168] The PBPK model included the reactions and compartmentalization of Cr in gastrointestinal lumen, oral mucosa, stomach, small intestinal tissue, blood, liver, kidney, bone and a combined compartment for other tissues. While the authors acknowledge that there is a degree of uncertainty in the modelling, the Cr compartmentalization results are in reasonable agreement with *in vivo* observations and further development of the model is likely to inform health authorities on risk factors associated with different levels of exposure.

7.5.2 Chromium(III)

Cellular metabolism of Cr(III) under physiologically relevant conditions is much less understood than that of toxic Cr(VI), primarily due to the low concentrations of the Cr(III) species involved.[49–51] Although Tf has long been implicated as the main Cr(III) transporter in the blood (Section 7.4), only indirect evidence existed until recently that Cr(III)-Tf complexes are able to interact with Tf receptors at the cell surface and to be taken into cells by endocytosis, similarly to the well-established mechanism for Fe(III)-Tf.[124,†] Recently, using cultured mouse muscle cells, it has been shown that additions of physiologically relevant concentrations of Tf (25 μM) in the form of radiolabelled, Cr(III)-saturated Tf ($^{51}Cr_2Tf$) to cell culture media resulted in Cr(III) uptake levels in the cells that were comparable ($\sim$ 2-fold lower) with Fe(III) uptake from $^{59}Fe_2Tf$ under the same conditions.[169] These results are consistent with selective Cr(III) binding to the Fe(III)-binding sites of Tf (Section 7.4; **11** in Chart 7.1),[23,124,125] which does not prevent the resultant complex from being recognized by cellular Tf receptors. Analysis of the subcellular distribution of ^{51}Cr showed that it initially concentrated in the microsomal fraction of the cell lysate, which suggested that Cr_2Tf, similarly to Fe_2Tf, is transported from the cell membrane into endosomes within the cytoplasm.[169] The mechanisms of ^{51}Cr release from the endosomes and of its further distribution within the cell remain unknown. These processes for Fe(III) rely on reduction to Fe(II) but Cr(III) is not expected to be reduced to Cr(II) under cellular conditions.[169] However, acid-catalysed aquation of Cr(III) bound to carboxylato donors in Tf or to the proposed carbonato co-ligand[170] may be important within these acidic organelles.[171] Cellular metabolic pathways of Cr(III) are likely to be shared with those of other redox-inactive trivalent metal ions with no known biological function, such as Al(III) and Ga(III), which are only beginning to be unravelled.[124,172] It is not yet known whether anionic Cr(III) complexes featuring bound citrate (such as **10** in Chart 7.1) or other small biomolecules can be taken up by cells *via* the numerous organic anion transporter proteins,[169] as has been suggested for Al(III).[173,174] Certainly, cellular uptake by some Cr(III) complexes with small peptides is relatively high.[175]

A role for Cr(III) in regulation of insulin signalling has been originally suggested by Mertz and co-workers.[53,56] Subsequently, it was proposed that cellular uptake of Cr(III) *via* the Tf pathway is followed by selective Cr(III) binding to a specific small molecule (LMMCr, Section 7.4).[176,177] The resultant LMMCr was purported to trigger the cascade of cytoplasmic kinase enzymes in response to insulin binding to the extracellular part of the insulin receptor and so result in enhancement of insulin signalling.[176,177] The authors of this hypothesis are currently re-evaluating its likelihood in light of the growing evidence that Cr(III) is unlikely to be an essential element (Section 7.2).[43] Furthermore, the existence and biological role of LMMCr as a unique Cr(III)-binding compound remain controversial.[51] It could be either an artefact of the isolation procedure[25,178] or representative of a class of metal-binding peptides that serve for detoxification of heavy metal ions.[54,124] Interestingly, isolation of a LMM (~ 1.0–1.5 kDa) Fe(III)-binding compound (that contained $\sim 10\%$ Cr(III) relative to $\sim 90\%$ Fe(III)) from mouse liver was reported simultaneously with, but independently from, LMMCr.[179] This compound was tentatively assigned as a multinuclear assembly of Fe(III) ions bound to the carboxylato residues of glutamine[179] and corresponds to the reported characteristics of LMMCr.[126,180] However, no further characterization of the LMMFe compound was reported, and it can now be considered as an artefact of the multitude of small Fe(II)-binding ligands that are present in the cell (labile iron pool).[181,182] This analogy emphasizes the difficulty in isolation and characterization of complexes with LMM biological ligands for any metal, including Cr(III).[25]

There is controversy in the literature on whether the shared metabolic pathways of Cr(III) and Fe(III) in mammalian cells can contribute to the toxicity or beneficial effects of Cr.[54] Some of the earlier studies suggested that increased dietary Cr intake alleviated the oxidative stress that is linked to excessive Fe levels in tissues, which in the long term was expected to decrease the chances of developing diabetes, cardiovascular diseases and cancer.[54] More recent data showed that even pharmacological doses of Cr (0.20–1.0 mg Cr as **14** daily) in non-diabetic humans did not significantly affect blood Fe parameters, including haemoglobin levels and Fe(III)-Tf binding.[183] Furthermore, synthetic Cr(III) complexes are known to inhibit some of the Fe-related metabolic pathways that protect cells from oxidative stress, such as induction of haem oxygenase 1 (HO-1, a general stress response enzyme)[184] and sequestration of Fe(III) by ferritin (the main Fe storage protein).[185] Blocking the production of HO-1 with a Cr(III)-porphyrin complex was shown directly to increase oxidative stress and decrease glucose tolerance in rat models of diabetes.[185] Therefore, current data do not support the hypothesis that interference with Fe metabolism is the primary reason for the anti-diabetic activity of Cr(III).[54] Cultured human cells (particularly those in the mitotic phase) respond to Cr(VI) treatments with the formation of a Fe-rich ring around the nucleus.[186] This may point to the involvement of Fe-dependent enzymes in the protective reaction against Cr(VI) toxicity.[149]

Cellular metabolism of typical Cr(III) nutritional supplements (**14** and **15** and also **16** [$CrCl_3 \cdot 6H_2O$, a common source of Cr(III) in *in-vitro* biochemical assays]) was studied by X-ray absorption spectroscopy.[77] An insulin-sensitive rat skeletal muscle cell line (L6) was chosen in connection with the proposed role of Cr(III) in insulin signalling and glucose metabolism (skeletal muscle tissue is responsible for most of the insulin-dependent glucose metabolism in mammals).[77] Addition of freshly prepared solutions of either **15** or **16** to the cell cultures resulted in extensive adsorption of Cr(III) hydrolysis products onto the surface of the cells.[77] This was confused in earlier work as "efficient uptake" of Cr(III) by mammalian cells.[187] Pre-equilibration of the Cr(III) complexes with cell culture media led to extensive changes in co-ordination environments of the complexes, and to drastic reduction in Cr(III) interactions with the cell surface.[77] Contrary to previous suggestions,[43,130] none of the complexes studied entered cells intact.[77] The metabolic products both in the cells and in the media consisted predominantly of Cr(III)-amino acid and Cr(III)-protein complexes. These were somewhat different from the Cr(III) products formed as a result of Cr(VI) uptake by the same cells (Section 7.5.1).[77]

7.6 Mechanisms of Toxicity and Anti-diabetic Activity of Chromium

Scheme 7.3 summarizes the likely metabolic pathways of Cr(VI) and Cr(III) compounds in mammals (Sections 7.3–7.5). Cr(III) complexes with chelating ligands that are likely to be present in food sources[47] as well as a popular nutritional supplement **14**[61] appear to feature amino or hydroxido

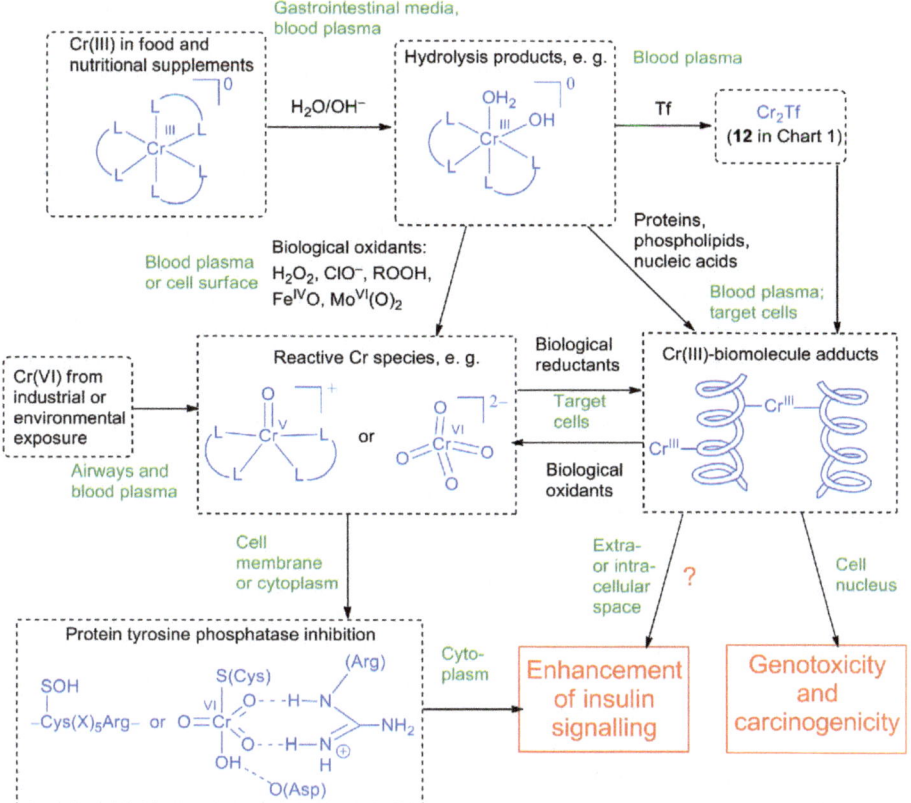

Scheme 7.3 Proposed main metabolic pathways and biological effects of Cr(III) and Cr(VI) compounds in mammals, adapted from published data.[29,51] The proposed structures of Cr metabolites are shown in blue, and the corresponding biological compartments are indicated in green. Designations: L are generic biological ligands for Cr(III) or Cr(V) (see Chart 7.1 for typical examples) and Tf is apo-transferrin.

carboxylato ligands. These compounds decompose at least partially in a gastrointestinal environment (Section 7.3)[95] and even more extensively in blood plasma (Section 7.4).[77] This leads to the formation of Cr(III) complexes with biomolecules, including transferrin (Scheme 7.3 and **12** in Chart 7.1).[23,124] Possible forms of Cr(III) transport into the cell include the complexes with transferrin as well as those with other blood proteins and small-molecule ligands (Scheme 7.3; see Section 7.5.2 for details).

Soluble (Na^+, K^+) and insoluble (Ca(II), Zn(II), Pb(II)) chromate salts are taken up by inhalation and ingestion from polluted environments (Section 7.2.1).[2,3] Extra- and intracellular reduction of these Cr(VI) compounds in the presence of ubiquitous biological 1,2-diols (carbohydrates and glyco-proteins) lead to the formation of relatively stable and potentially genotoxic Cr(V) intermediates (Schemes 7.2 and 7.3; 3 and 4 in Chart 7.1)[15,27,188] en route to reduction to kinetically inert Cr(III)-biomolecule complexes (Scheme 7.3).[25,28]

Dynamic redox equilibria between reduced (Cr(III)) and oxidized (Cr(IV), Cr(V) and Cr(VI)) forms of Cr are likely to exist in biological media and will be dependent on the balance between biological oxidants and reductants in particular compartments (Schemes 7.2 and 7.3).[7,74] Of particular importance is the oxidation of Cr(III) to Cr(V) and Cr(VI) species in blood plasma under the conditions of oxidative stress caused by the presence of poorly controlled diabetes or chronic inflammation.[10,29,50,77] These lead to localized formation of high micromolar to low millimolar concentrations of strong oxidants (including H_2O_2, ClO^- and lipid peroxide; Scheme 7.2),[7,140] which is consistent with the conditions used for the *in vitro* oxidation of biologically relevant Cr(III) complexes to Cr(VI).[10,11] The resultant oxidized Cr species can easily enter cells (Section 7.5.1) and react with Cys residues of protein tyrosine phosphatases (PTPs) in the cell membrane or cytoplasm. These interactions can lead either to reversible inhibition of these enzymes (*via* $[CrO_4]^{2-}$ binding mimicking $[PO_4]^{3-}$ binding) or to irreversible inhibition due to oxidation of Cys residues (Scheme 7.3).[10,29,50] The proposed structure of a Cr(VI)-PTP complex (Scheme 7.3) is based on the X-ray crystal structure of a V(V)-PTP complex[189] (V(V) is isoelectronic with Cr(VI))[§] and the structures of Cr(VI) thiolato complexes in aqueous solutions (determined by X-ray absorption spectroscopy).[13] Since several PTPs (such as PTP1B) provide negative regulation of the insulin signalling cascade,[190] their inhibition will lead to insulin-enhancing activity.[50] In agreement with this hypothesis, subcytotoxic concentrations of Cr(VI) caused increases in glucose uptake rates in cultured cells[191] and subcutaneous injections of Cr(VI) caused severe, but short-term, decreases in blood glucose levels in non-diabetic rats.[192] On the other hand, Cr(V) and Cr(VI) interactions with PTPs that are involved in other cell signalling pathways could contribute to the toxicity of Cr compounds.[29,193,194] Penetration of Cr(VI) into the cell nucleus and formation of Cr(III) adducts with DNA or nuclear proteins (Scheme 7.3) lead to genotoxicity that can develop into cancer (Section 7.5.1).[28,29,34,151]

It is possible that the extra- or intracellular formation of certain types of Cr(III)-biomolecule complexes can also lead to insulin-enhancing activity[43,54] although no unambiguous evidence for such action is available yet (indicated with a question mark in Scheme 7.3). Mertz and co-workers hypothesized that some biologically relevant Cr(III) complexes can bind to insulin and stabilize it against enzymatic hydrolysis[53,56] but it was only much later that this was demonstrated in *in vitro* experiments.[195–197] Similar reactions may account for the rapid increase in insulin sensitivity in critically ill patients following intravenous injections of **16** ($CrCl_3 \cdot 6H_2O$ in physiological saline; see Section 7.2).[82–84] This reactive Cr(III) complex is expected to bind rapidly to various biomolecules in the blood plasma,

---

[§]See Chapter 6 for further discussion of this V-PTP complex.

including insulin.[29] Another proposed mechanism of anti-diabetic activity involves reactions of **14** with the cell surface, which would lead to reorganization of lipid bilayers of cell membranes.[198] However, this scenario is less likely since **14** and similar complexes are not expected to reach the target cells intact.[77,95] In summary, the same chemical intermediates are likely to be responsible for both the anti-diabetic and carcinogenic actions of Cr compounds (Scheme 7.3).[29,51]

7.7 Chromium Metabolism in Plants and Microorganisms

7.7.1 Plants

Despite some early reports of beneficial effects of Cr(III) compounds in plants (summarized in a review),[199] both Cr(III) and Cr(VI) compounds are generally considered to be toxic for plants.[200] The stimulating effects of low doses of Cr(VI) on the growth of some plants[201] can be explained in terms of hormesis (beneficial effects of low doses of toxic chemicals).[54,202] For plants growing in a natural environment, it is difficult to distinguish between the effects of Cr(VI) and Cr(III)[199,203] since these oxidation states are readily interconvertible in soils and ground waters (Section 7.1).[2,204] It is likely that Cr is taken up by plant roots predominantly in the form of highly soluble $[CrO_4]^{2-}$ (through sulfate channels, in a similar manner to that observed for mammalian cells; see Section 7.5.1). Subsequently, the chromate is rapidly reduced to Cr(III) compounds that bind to root cell walls. Consequently, relatively little Cr would be transported up the xylem.[203,205,206] In support of this hypothesis, Cr is predominantly accumulated by sulfur-loving plant species such as cauliflower, kale and cabbage.[205]

Predictably, interactions of Cr(VI) with carbohydrate-rich plant cells lead to the formation of relatively stable, Cr(V) carbohydrato intermediates (Section 7.1; **3** in Chart 7.1).[207,208] These species represent a health hazard for animals grazing on Cr(VI)-contaminated soil, and a potential danger in the use of plants for bioremediation.[8,15] Potential applications of plants and microorganisms in bioremediation of Cr(VI)-contaminated soils have been reviewed.[209,210] Speciation and distribution of Cr metabolism products in plants have been studied by X-ray absorption spectroscopy.[205,211,212] Apart from confirming that Cr(VI) is reduced efficiently to Cr(III) by plants (predominantly in the roots),[205,212] these studies indicated that different Cr(III) species are likely to form in different plant compartments (attributed to aetate and oxalate complexes in roots and leaves, respectively, of *Brassica juncea*).[211]

7.7.2 Yeasts, Fungi and Algae

In the last two decades, the yeast *Saccharomyces cerevisiae* has become the most widely used model in studies of metal ion uptake and metabolism in eukaryotes, due to its well-characterized genome and the availability of a wide range of gene knock-outs.[213] Although significant progress has been achieved in understanding of metal (particularly Cu and Fe) metabolism in the yeast, no specific metabolic pathways for Cr have been found so far.[213] The available data on the metabolism and biological effects of Cr(III) and Cr(VI) in various species of yeasts and fungi have been reviewed recently.[58]

Metabolism of Cr(VI) in yeasts shows many similarities to that in mammalian cells (Section 7.5.1) and includes: (i) active uptake of $[CrO_4]^{2-}$ through sulfate channels; (ii) rapid intracellular reduction of Cr(VI) to Cr(III); (iii) high accumulation of Cr(III) within the cells due to the concentration gradient with extracellular Cr(VI) and kinetic inertness of the Cr(III) products; (iv) Cr(VI)-induced genotoxicity and mutagenicity.[58] Yeast cells can be used as complementary models for the studies of Cr(VI) toxicity in humans, in addition to cultured human cancer cells that are usually used for this purpose. The antioxidant defence system of normal human cells seems to be closer to that of yeasts than to that of human cancer cells in culture.[58] The most significant difference from mammalian cells is the important role of cell walls in the reduction of Cr(VI) and adsorption of the resultant Cr(III) products.[214] This observation has led to numerous applications of dead biomass from yeasts, fungi, algae and plants for the removal of Cr(VI) from waste waters, as reviewed recently.[215,216] As mentioned in Section 7.7.1, a potential danger of this approach is the formation of genotoxic Cr(V) intermediates.[15,217] Prolonged treatments of the unicellular green algae *Micrasterias dentriculata* with non-cytotoxic doses of Cr(VI) caused the formation of solid particles, rich and Cr, Fe in O, along the inner side of the cell walls.[218] This observation supports earlier suggestions[203] that interference with Fe(III) metabolism is one of the sources of Cr toxicity in microorganisms.

The role of cell walls in the metabolism of Cr(III) compounds by yeasts is thought to be even more important than in the case of Cr(VI). Such compounds are likely to be adsorbed rapidly onto the cell surface, a process followed by slow internalization.[219] Such adsorption is likely to be responsible for the high uptake values reported for Cr(III) citrato or histidinato complexes in *S. cerevisiae* (comparable to or higher than Cr(VI) uptake under similar conditions).[220] These observations are in marked contrast to the relative uptake of Cr(III) and Cr(VI) by mammalian cells (Section 7.4).[114] High Cr(III) uptake by *S. cerevisiae* and the associated increase in glucose uptake values were interpreted previously as a sign of Cr(III) essentiality for the yeast, related to the "glucose tolerance factor".[221] No firm evidence has yet been found for the existence of such a factor (see Section 7.2.2 for details),[49] while the increase in glucose uptake is likely to be a part of the general stress response of the cells.[54] Genotoxicity in *S. cerevisiae* was observed for Cr(III) as well as for Cr(VI) compounds.[222]

7.7.3 Prokaryotes

There are currently no convincing data that Cr is essential for any prokaryotes. The ability of strains of some anaerobic bacteria to use $[CrO_4]^{2-}$ as their sole electron acceptor in respiration is considered to be an adaptation to Cr(VI)-polluted environments.[200] Studies of bacterial Cr metabolism have mostly been limited to Cr(VI), with the view of using bacteria for bioremediation of Cr(VI) pollution.[209,210,223] Stainless steel production and other industries using Cr compounds generate millions of tonnes of Cr(VI)-containing residues per year, as such, Cr(VI) leaching into ground waters has become a serious public health concern in many areas (Section 7.2.1).[210]

As for eukaryotes (Sections 7.5.1, 7.7.1 and 7.7.2), prokaryotes are most likely to take up $[CrO_4]^{2-}$ through sulfate channels.[203] Unlike eukaryotes, a distinct pathway of Cr(VI) efflux has been characterized in several species of bacteria and is considered to be one of the main mechanisms of bacterial resistance to Cr(VI).[200] Such efflux involves specific transmembrane proteins (called ChrA) that are also likely to be sulfate exporters, although this function has yet to be proven.[200]

Most bacteria are capable of reducing Cr(VI) *via* one-electron pathways with the formation of Cr(V) intermediates (predominantly 1,2-diolato complexes; see Scheme 7.2 and structures 3 and 4 in Chart 7.1).[15,217] Both enzymatic reduction and non-enzymatic reductants appear to be employed.[200,224,225] However, formation of Cr(V) leads to redox reactions that regenerate Cr(VI) (Scheme 7.2) and produce cell-damaging ROS.[12,224,225] Therefore, bioremediation studies are focused on bacteria capable of direct three-electron reduction of Cr(VI) to Cr(III) in either an anaerobic or an aerobic environment.[223]

Shewanella oneidensis strain MR-1 is the most studied anaerobic Cr(VI)-reducing organism. It is also capable of reducing other biologically non-essential metal ions such as U(VI), Tc(VII) or Au(III).[226–228] The reduction is thought to occur predominantly on the cell surface, as it leads to encrustation of the bacterial cells with insoluble Cr(III) compounds.[227,229] The enzymes responsible for this process are multi-haem membrane proteins related to cytochrome *c*,[227] which normally serve for electron transfer to Fe(III) and Mn(IV) oxides that are used as terminal acceptors in anaerobic respiration.[230] However, recent single-cell imaging studies have shown that Cr(VI) reduction can also occur in the cytoplasm of *S. oneidensis*, including the formation of Cr(VI)-containing pockets within the cell.[228] The reduction products were previously thought to consist mainly of $Cr(OH)_3$ but the formation of polynuclear Cr(III) complexes featuring small biomolecules as ligands has been demonstrated.[231] In addition, transient formation of Cr(II) and Cr(IV) (but not Cr(V)) species has been detected.[229,232]

Aerobic bacteria capable of direct Cr(VI) reduction to Cr(III) include such ubiquitous species as *Escherichia coli*.[224] The chromate reductase (ChrR) from the latter organism has recently been characterized by X-ray crystallography and belongs to the flavodoxin superfamily that features a flavin mononucleotide cofactor.[225] Its main physiological role is the two-electron reduction of quinones to hydroquinones by NADH (nicotinamide adenine dinucleotide, reduced form), but an oligomeric form of the enzyme is capable of concerted transfer of four electrons (three to reduce Cr(VI) to Cr(III) and one to an organic substrate).[225] A related Cr(VI) reductase from *Gluco-nacetobacter hansenii* has also been characterized crystallographically.[233] Reduction of Cr(VI) by ChrR occurs in the cytoplasm. The reaction products are likely to include a soluble Cr(III)-NAD$^+$ complex.[234] Apart from Cr(VI), these enzymes are capable of reducing a range of abiological substrates including $[Fe(CN)_6]^{3-}$, U(VI) and quinone anti-cancer drugs. Their capability of reducing Cr(VI) is clearly an adaptation to a Cr(VI)-polluted environment, rather than a sign of Cr essentiality.[200] Other types of aerobic Cr(VI)-reducing enzymes are likely to exist in bacteria and include a putative Cu-dependent membrane-associated reductase from the *Amphibacillus* genus.[235] The bioremediating capacity of Cr(VI)-reducing bacteria can be improved by genetic engineering but the application of these bacteria to environmental remediation has been hampered by the reluctance to release genetically modified organisms into the environment.[210,224]

In addition to the enzymatic mechanisms described above, some anaerobic bacteria reduce Cr(VI) to Cr(III) extracellularly *via* the release of small-molecule reductants (such as H_2S or Fe(II)) into the environment.[210] A variation of this mechanism is the release of siderophores, ligands of low molar mass used by microorganisms to harvest essential metal ions from the environment.[236,237] Some of the siderophores have reducing properties, such as pyridine-2,6-bis(thiocarboxylic acid) produced by *Pseudomonas stutzeri*, and are capable of both reducing Cr(VI) and binding the resultant Cr(III).[238] Typical Fe(III)-binding siderophores feature hydroxamato or catechol donor groups[237] and are also strong Cr(III) binders (as illustrated by the model compound **11** in Chart 7.1).[22] They are likely to promote adventitious uptake of Cr(III) by microorganisms.[239] Hydroxamato and catecholato groups of siderophores are also expected to stabilize potentially dangerous Cr(IV) and Cr(V) intermediates.[240–242]

7.8   Conclusions

Chromium compounds, originating from both natural sources and anthropogenic activity, are widespread in the environment (Scheme 7.1).[1,3] Despite its high abundance, *no proven natural biological functions of Cr have emerged so far*, despite long-standing claims on the essentiality of Cr(III).[43,51] However, pharmacological doses of Cr(III) compounds can act as anti-diabetics, although their efficacy and safety are currently under dispute (Section 7.2).[43,50] A crucial chemical property that is likely to prevent Cr(III) from being a biometal is its kinetic inertness, which is even more pronounced than that for another highly abundant non-biological metal ion, Al(III) (Section 7.1).[31,243,¶] Together with relative redox inertness (Section 7.5.2), this property means that Cr(III) is unlikely to function as either a catalytic or a structural metal ion in biological systems, in contrast to other first-row transition elements.[51] On the other hand, the kinetic inertness of Cr(III) (as a

---

¶See Chapter 25 for a discussion of the biological effects of aluminium.

product of intracellular reduction of Cr(VI)) is a key factor in determining the toxicity of Cr(VI), due to its accumulation in cells and the ability to cross-link biological macromolecules (Section 7.5.1).[28,29] Another relevant property of Cr(III) is the similarity of its chemistry to that of Fe(III). This includes specific binding to the main Fe(III) transport protein, transferrin (Section 7.5.2).[124] Sharing Fe(III) uptake pathways is likely to be the main mechanism of Cr(III) uptake from food sources, although the biological significance of such uptake has not yet been proven.[124]

A crucial biologically relevant property of $[CrO_4]^{2-}$ (chromate; the main form of soluble Cr(VI) at pH $\geq$ 7) is its structural similarity to the essential anions $[SO_4]^{2-}$ and $[HPO_4]^{2-}$. This aspect is responsible for the high uptake of Cr(VI) by most cell types (Sections 7.5.1 and 7.7).[100] The resemblance of $[CrO_4]^{2-}$ to $[HPO_4]^{2-}$ also leads to its binding to the active sites of phosphate-metabolizing enzymes, which, in turn, leads to alterations in cell signalling that are likely to contribute to both the toxic and anti-diabetic effects of Cr compounds (Section 7.6).[29,50] Similarly, the isostructural nature of these ions is likely to be important in the transport of $[CrO_4]^{2-}$ into the nucleus by electrostatic binding to histone proteins.[150] Extra- and intracellular reduction of Cr(VI) by both enzymatic and non-enzymatic reductants (Scheme 7.2)[12] leads, in most instances, to the formation of Cr(V) intermediates that are stabilized by highly abundant biological 1,2-diols that include carbohydrates and glycoproteins.[15,27] These reactive intermediates are likely to contribute both to Cr-induced damage of the genetic apparatus of the cell (including DNA, RNA and zinc finger proteins) and to the inhibition of phosphatase enzymes (and hence to the anti-diabetic effects of Cr; see Scheme 7.3 and Section 7.6).[29,50]

Studies of speciation of Cr compounds in biological media, including the oxidation states and the ligand environment (as illustrated in Scheme 7.3) are required in order to understand its biological activity. For instance, the widespread use of highly acidic Cr(III) salts (*e.g.* $CrCl_3 \cdot 6H_2O$, **16** in Chart 7.2) in biological studies of Cr(III) is likely to lead to results that are not biologically relevant. Such salts rapidly hydrolyse in biological media with the formation of either insoluble oligomeric hydroxido complexes or Cr(III)-biomolecule adducts.[29] The characterization of Cr(III) products formed in Cr(VI)- or Cr(III)-treated cells is complicated by the chemical changes occurring during cell lysis and chromatographic separation,[25] which so far have prevented a definitive description of any natural Cr(III)-containing factors.[43,51] These studies have so far been limited to the use of supra-physiological concentrations of Cr but further developments in detection techniques may allow the use of near-physiological concentrations of metal ions.[147]

In summary, attempts to prove the essentiality of Cr(III) and to characterize functional Cr(III)-containing biomolecules have been problematical so far, but they have taught scientists many useful lessons on the methodologies of studies of metal ion metabolism.[49,244] On the other hand, studies of Cr(VI) metabolism in mammalian cells has led to significant progress in the understanding of general mechanisms of metal-induced cancers (Sections 7.5.1 and 7.6).[34] Studies of Cr(VI) metabolism by plants and microorganisms has led to development of potential Cr(VI) bioremediation systems (Section 7.7).[209,210]

Acknowledgements

The authors are grateful for support from Australian Research Council (ARC) grants to PAL for an ARC Professorial Fellowship (DP0984722) and for an ARC Senior Research Associate position for AL (DP1095310).

References

1. J. O. Nriagu and E. Nieboer, ed., *Chromium in the Natural and Human Environments*, John Wiley & Sons, New York, 1988.
2. J. Guertin, J. A. Jacobs and C. P. Avakian, ed., *Chromium(VI) Handbook*, CRC Press, Boca Raton, FL, 2005.
3. Agency for Toxic Substances and Disease Registry (USA), *Toxicological profile for chromium*, ATSDR, Washington, DC, 2008, http://www.atsdr.cdc.gov/ToxProfiles/TP.asp?id=62&tid=17.
4. J. G. Farmer, R. P. Thomas, M. C. Graham, J. S. Geelhoed, D. G. Lumsdon and E. Paterson, *J. Environ. Monit.*, 2002, **4**, 235.
5. P. Shah, V. Strezov and P. F. Nelson, *Fuel*, 2012, **102**, 1.
6. F. Y. Saleh, T. F. Parkerton, R. V. Lewis, J. H. Huang and K. L. Dickson, *Sci. Total Environ.*, 1989, **86**, 25.
7. B. Halliwell and J. M. C. Gutteridge, *Free Radicals in Biology and Medicine*, Oxford University Press, Oxford, UK, 4th edn, 2007.
8. A. Levina, R. Codd, C. T. Dillon and P. A. Lay, *Prog. Inorg. Chem.*, 2003, **51**, 145.
9. P. H. Connett and K. E. Wetterhahn, *Struct. Bonding (Berlin)*, 1983, **54**, 93.
10. I. Mulyani, A. Levina and P. A. Lay, *Angew. Chem. Int. Ed.*, 2004, **43**, 4504.
11. A. Levina, I. Mulyani and P. A. Lay, *Redox Chemistry and Biological Activities of Chromium(III) Complexes*, in *The Nutritional Biochemistry of Chromium(III)*, ed. J. B. Vincent, Elsevier, Amsterdam, 2007, Ch. 11, pp. 225–256.
12. A. Levina and P. A. Lay, *Coord. Chem. Rev.*, 2005, **249**, 281.
13. A. Levina and P. A. Lay, *Inorg. Chem.*, 2004, **43**, 324.
14. A. Levina, L. Zhang and P. A. Lay, *Inorg. Chem.*, 2003, **42**, 767.
15. R. Codd, J. A. Irwin and P. A. Lay, *Curr. Opin. Chem. Biol.*, 2003, 7, 213.
16. L. Zhang and P. A. Lay, *J. Am. Chem. Soc.*, 1996, **118**, 12624.
17. H. A. Headlam and P. A. Lay, *Inorg. Chem.*, 2001, **40**, 78.
18. R. Codd, P. A. Lay and A. Levina, *Inorg. Chem.*, 1997, **36**, 5440.
19. R. F. Bryan, P. T. Greene, P. F. Stokely and E. W. Wilson, *Inorg. Chem.*, 1971, **10**, 1468.
20. P. De Meester, D. J. Hodgson, H. C. Freeman and C. J. Moore, *Inorg. Chem.*, 1977, **16**, 1494.

21. M. Quiros, D. M. L. Goodgame and D. J. Williams, *Polyhedron*, 1992, **11**, 1343.
22. K. Abu-Dari, J. D. Ekstrand, D. P. Freyberg and K. N. Raymond, *Inorg. Chem.*, 1979, **18**, 108.
23. C. D. Quarles, R. K. Marcus and J. L. Brumaghim, *J. Biol. Inorg. Chem.*, 2011, **16**, 913.
24. A. Zhitkovich, Y. Song, G. Quievryn and V. Voitkun, *Biochemistry*, 2001, **40**, 549.
25. A. Levina, H. H. Harris and P. A. Lay, *J. Am. Chem. Soc.*, 2007, **129**, 1065.
26. P. H. Connett and K. E. Wetterhahn, *J. Am. Chem. Soc.*, 1985, **107**, 4282.
27. L. F. Sala, J. C. Gonzalez, S. I. Garcia, M. I. Frascaroli and D. S. Van, *Adv. Carbohydr. Chem. Biochem.*, 2011, **66**, 69.
28. A. Zhitkovich, *Chem. Res. Toxicol.*, 2005, **18**, 3.
29. A. Levina and P. A. Lay, *Chem. Res. Toxicol.*, 2008, **21**, 563.
30. P. A. Lay and A. Levina, *Chromium*, in *Comprehensive Coordination Chemistry II: From Biology to Nanotechnology*, ed. J. A. McCleverty and T. J. Meyer, Elsevier Science, Amsterdam, The Netherlands, 2004, vol. 4, pp. 313–413.
31. L. Helm and A. E. Merbach, *Chem. Rev.*, 2005, **105**, 1923.
32. A. Chiu, X. L. Shi, W. K. P. Lee, R. Hill, T. P. Wakeman, A. Katz, B. Xu, N. S. Dalal, J. D. Robertson, C. Chen, N. Chiu and L. Donehower, *J. Environ. Sci. Health C*, 2010, **28**, 188.
33. K. P. Nickens, S. R. Patierno and S. Ceryak, *Chem.-Biol. Interact.*, 2010, **188**, 276.
34. P. A. Lay and A. Levina, Metal Carcinogens in *Comprehensive Inorganic Chemistry II*, ed. J. Reedijk and K. Poeppelmeier, Elsevier, Oxford, 2013, vol. 3, pp. 835–856.
35. International Agency for Research on Cancer, *IARC Monographs on the Evaluation of Carcinogenic Risks to Humans. Vol. 100. A Review of Human Carcinogens. Part C: Arsenic, Metals, Fibres, and Dusts*, IARC, Lyon, France, 2012.
36. R. M. Sedman, J. Beaumont, T. A. McDonald, S. Reynolds, G. Krowech and R. Howd, *J. Environ. Sci. Health, Part C*, 2006, **24**, 155.
37. A. Zhitkovich, *Chem. Res. Toxicol.*, 2011, **24**, 1617.
38. T. Davidson, T. Kluz, F. Burns, T. Rossman, Q. Zhang, A. Uddin, A. Nadas and M. Costa, *Toxicol. Appl. Pharmacol.*, 2004, **196**, 431.
39. Y. Fan, J. L. Ovesen and A. Puga, *J. Trace Elem. Med. Biol.*, 2012, **26**, 188.
40. N. McCarroll, N. Keshava, J. Chen, G. Akerman, A. Kligerman and E. Rinde, *Environ. Mol. Mutagen.*, 2010, **51**, 89.
41. D. M. Stearns, *Evaluation of Chromium (III) Genotoxicity with Cell Culture and In Vitro Assays*, in *The Nutritional Biochemistry of Chromium(III)*, ed. J. B. Vincent, Elsevier, Amsterdam, 2007, Ch. 10, pp. 209–224.
42. Y.-H. Shiao, R. M. Leighty, C. Wang, X. Ge, E. B. Crawford, J. M. Spurrier, S. D. McCann, J. R. Fields, L. Fornwald, L. Riffle, C. Driver, K. S. Kasprzak, O. A. Quinones, R. E. Wilson, G. S. Travlos, W. G. Alvord and L. M. Anderson, *Environ. Mol. Mutagen.*, 2012, **53**, 392.

43. J. B. Vincent, *The Bioinorganic Chemistry of Chromium*, John Wiley & Sons, Chichester, UK, 2013.
44. F. H. Khan, K. Ambreen, G. Fatima and S. Kumar, *Sci. Total Environ.*, 2012, **430**, 68.
45. D. Cohen, *Br. Med. J.*, 2012, **344**, e1410.
46. J. P. Thyssen and T. Menne, *Chem. Res. Toxicol.*, 2010, **23**, 309.
47. D. E. Kimbrough, Y. Cohen, A. M. Winer, L. Creelman and C. Mabuni, *Crit. Rev. Environ. Sci. Technol.*, 1999, **29**, 1.
48. J. B. Vincent, ed., *The Nutritional Biochemistry of Chromium(III)*, Elsevier, Amsterdam, 2007.
49. J. B. Vincent, *Dalton Trans.*, 2010, **39**, 3787.
50. A. Levina and P. A. Lay, *Dalton Trans.*, 2011, **40**, 11675.
51. P. A. Lay and A. Levina, *Chromium: Biological Relevance*, in *Encyclopedia of Inorganic and Bioinorganic Chemistry*, ed. R. A. Scott, John Wiley & Sons, Chicester, UK, 2012. doi: 10.1002/9781119951438.eibc0040.pub2. http://onlinelibrary.wiley.com/doi/10.1002/9781119951438.eibc0040.pub2/otherversions (published online 15.06.12).
52. P. Mason, *Dietary Supplements*, Pharmaceutical Press, London, 2007.
53. K. Schwarz and W. Mertz, *Arch. Biochem. Biophys.*, 1959, **85**, 292.
54. D. M. Stearns, *BioFactors*, 2000, **11**, 149.
55. K. R. Di Bona, S. Love, N. R. Rhodes, D. McAdory, S. Halder Sinha, N. Kern, J. Kent, J. Strickland, A. Wilson, J. Beaird, J. Ramage, J. F. Rasco and J. B. Vincent, *J. Biol. Inorg. Chem.*, 2011, **16**, 381.
56. W. Mertz, *J. Am. Coll. Nutr.*, 1998, **17**, 544.
57. C. Veillon and K. Y. Patterson, *J. Trace Elem. Exp. Med.*, 1999, **12**, 99.
58. B. Poljsak, I. Pocsi, P. Raspor and M. Pesti, *J. Basic Microbiol.*, 2010, **50**, 21.
59. S. Hayat, G. Khalique, M. Irfan, A. S. Wani, B. N. Tripathi and A. Ahmad, *Protoplasma*, 2012, **249**, 599.
60. D. M. Stearns and W. H. Armstrong, *Inorg. Chem.*, 1992, **31**, 5178.
61. G. W. Evans and D. J. Pouchnik, *J. Inorg. Biochem.*, 1993, **49**, 177.
62. Anon., *Chromium Picolinate and Weight Loss*, 2011, http://chromiumpicolinateandweightloss.org/.
63. F. Mirasol, *Chromium picolinate market sees robust growth and high demand.* 2000, http://www.icis.com/Articles/2000/02/14/105617/chromium-picolinate-market-sees-robust-growth-and-high.html.
64. J. Dalli, *The European Commision decision of 27 May 2011, Authorising the placing on the market of Chromium Picolinate as a novel food ingredient*, 2011, http://eur-lex.europa.eu/LexUriServ/LexUriServ.do?uri = OJ:L:2011:143:0036:0037:EN:PDF.
65. M. H. Pittler, C. Stevinson and E. Ernst, *Int. J. Obes.*, 2003, **27**, 522.
66. N. Iqbal, S. Cardillo, S. Volger, L. T. Bloedon, R. A. Anderson, R. Boston and P. O. Szapary, *Metab. Syndr. Relat. Disord.*, 2009, 7, 143.
67. Y. Yazaki, Z. Faridi, Y. Ma, A. Ali, V. Northrup, V. Y. Njike, L. Liberti and D. L. Katz, *J. Altern. Complement. Med.*, 2010, **16**, 291.
68. U. Masharani, C. Gjerde, S. McCoy, B. A. Maddux, D. Hessler, I. D. Goldfine and J. F. Youngren, *BMC Endocr. Disord.*, 2012, **12**, 31.

69. M. D. Stout, A. Nyska, B. J. Collins, K. L. Witt, G. E. Kissling, D. E. Malarkey and M. J. Hooth, *Food Chem. Toxicol.*, 2009, **47**, 729.

70. D. McAdory, N. R. Rhodes, F. Briggins, M. M. Bailey, K. R. Bona, C. Goodwin, J. B. Vincent and J. F. Rasco, *Biol. Trace Elem. Res.*, 2011, **143**, 1666.

71. National Research Council (USA), *Dietary Reference Intakes for Vitamin A, Arsenic, Boron, Chromium, Copper, Iodine, Iron, Manganese, Molybdenum, Nickel, Silicon, Vanadium, and Zinc. A Report of the Panel on Micronutrients, Subcommittee on Upper Reference Levels of Nutrients and of Interpretations and Uses of Dietary Reference Intakes, and the Standing Committee on the Scientific Evaluation of Dietary Reference Intakes*, National Academy of Sciences, Washington, DC, 2002.

72. National Institutes of Health (USA), *Type 2 diabetes*, 2010, http://www.nlm.nih.gov/medlineplus/ency/article/000313.htm.

73. W. T. Cefalu, J. Rood, P. Pinsonat, J. Qin, O. Sereda, L. Levitan, R. A. Anderson, X. H. Zhang, J. M. Martin, C. K. Martin, Z. Q. Wang and B. Newcomer, *Metab. Clin. Exp.*, 2010, **59**, 755.

74. M. A. Abdul-Ghani and R. A. DeFronzo, in *Oxidative Stress in Aging: from Model Systems to Human Diseases*, ed. S. Miwa, K. B. Beckman and F. L. Muller, Springer, New Jersey, 2008, pp. 191–211.

75. A. S. Antsyshkina, M.-A. Porai-Koshits, I. V. Arkhangel'skii and I. N. Diallo, *Russ. J. Inorg. Chem.*, 1987, **32**, 2928; *Engl. Trans.*, 1700.

76. C. M. Davis, A. C. Royer and J. B. Vincent, *Inorg. Chem.*, 1997, **36**, 5316.

77. A. Nguyen, I. Mulyani, A. Levina and P. A. Lay, *Inorg. Chem.*, 2008, **47**, 4299.

78. C. T. Dillon, P. A. Lay, A. M. Bonin, M. Cholewa and G. J. F. Legge, *Chem. Res. Toxicol.*, 2000, **13**, 742.

79. C. L. Weeks, A. Levina, C. T. Dillon, P. Turner, R. R. Fenton and P. A. Lay, *Inorg. Chem.*, 2004, **43**, 7844.

80. J. M. Carraher and A. Bakac, *Chem. Res. Toxicol.*, 2010, **23**, 1735.

81. American Diabetes Association, *Diabetes Care*, 2009, **32**, S13.

82. M. Via, C. Scurlock, J. Raikhelkar, G. Di Luozzo and J. I. Mechanick, *Nutr. Clin. Pract.*, 2008, **23**, 325.

83. T. C. Drake, K. D. Rudser, E. R. Seaquist and A. Saeed, *Endocr. Pract.*, 2012, **18**, 394.

84. S. R. Surani, I. Ratnani, B. Guntupalli and S. Bopparaju, *World J. Diab.*, 2012, **3**, 170.

85. B. J. Stoecker, *Basis for Dietary Recommendations for Chromium*, in *The Nutritional Biochemistry of Chromium(III)*, ed. J. B. Vincent, Elsevier, Amsterdam, 2007, Ch. 2, pp. 43–55.

86. R. A. Anderson, N. A. Bryden and M. M. Polansky, *Biol. Trace Elem. Res.*, 1992, **32**, 117.

87. D. W. Lamson and S. M. Plaza, *Altern. Med. Rev.*, 2002, 7, 218.

88. I. G. Dance and H. C. Freeman, *Inorg. Chem.*, 1965, **4**, 1555.

89. R. A. Anderson, M. M. Polansky, N. A. Bryden, K. Y. Patterson, C. Veillon and W. H. Glinsmann, *J. Nutr.*, 1983, **113**, 276.

90. J. Poirier, H. Semple, J. Davies, R. Lapointe, M. Dziwenka, M. Hiltz and D. Mujibi, *Neuroscience*, 2011, **193**, 338.

91. R. A. Anderson, M. M. Polansky and N. A. Bryden, *Biol. Trace Elem. Res.*, 2004, **101**, 211.

92. B. J. Clodfelder, C. Chang and J. B. Vincent, *Biol. Trace Elem. Res.*, 2004, **98**, 159.

93. N. Laschinsky, K. Kottwitz, B. Freund, B. Dresow, R. Fischer and P. Nielsen, *Biometals*, 2012, **25**, 1051.

94. B. Gammelgaard, K. Jensen and B. Steffansen, *J. Trace Elem. Med. Biol.*, 1999, **13**, 82.

95. A. R. Jeffcoat, *[¹⁴C] Chromium Picolinate Monohydrate: Disposition and Metabolism in Rats and Mice*, National Institute of Environmental Health Sciences, USA, Research Triangle Park, NC, 2002.

96. S. A. Kareus, C. Kelley, H. S. Walton and P. R. Sinclair, *J. Haz. Mat.*, 2001, **84**, 163.

97. B. K. Fuqua, C. D. Vulpe and G. J. Anderson, *J. Trace Elem. Med. Biol.*, 2012, **26**, 115.

98. E. J. Martinez-Finley, S. Chakraborty, S. J. B. Fretham and M. Asshner, *Metallomics*, 2012, **4**, 593.

99. L.-Y. Zha, Z.-R. Xu, M.-Q. Wang and L.-Y. Gu, *J. Anim. Physiol. Anim. Nutr.*, 2008, **92**, 131.

100. B. J. Collins, M. D. Stout, K. E. Levine, G. E. Kissling, R. L. Melnick, T. R. Fennell, R. Walden, K. Abdo, J. B. Pritchard, R. A. Fernando, L. T. Burka and M. J. Hooth, *Toxicol. Sci.*, 2010, **118**, 368.

101. J. E. McLean, L. S. McNeill, M. Edwards and J. L. Parks, *J. Am. Water Works Assoc.*, 2012, **104**, 35.

102. B. D. Kerger, B. L. Finley, G. E. Corbett, D. G. Dodge and D. J. Paustenbach, *J. Toxicol. Environ. Health*, 1997, **50**, 67.

103. S. De Flora, A. Camoirano, M. Bagnasco, C. Bennicelli, G. Corbett and B. D. Kerger, *Carcinogenesis*, 1997, **18**, 531.

104. D. M. Proctor, M. Suh, L. L. Aylward, C. R. Kirman, M. A. Harris, C. M. Thompson, H. Gurleyuk, R. Gerads, L. C. Haws and S. M. Hays, *Chemosphere*, 2012, **89**, 487.

105. M. Monachese, J. P. Burton and G. Reid, *Appl. Environ. Microbiol.*, 2012, **78**, 6397.

106. S. De Flora, *Carcinogenesis*, 2000, **21**, 533.

107. R. A. Anderson and A. S. Kozlovsky, *Am. J. Clin. Nutr.*, 1985, **41**, 1177.

108. R. A. Anderson and M. M. Polansky, *Biol. Trace Elem. Res.*, 1995, **50**, 97.

109. B. W. Morris, S. MacNeil, C. A. Hardisty, S. Heller, C. Burgin and T. A. Gray, *J. Trace Elem. Med. Biol.*, 1999, **13**, 57.

110. N. R. Rhodes, D. McAdory, S. Love, K. R. Di Bona, Y. Chen, K. Ansorge, J. Hira, N. Kern, J. Kent, P. Lara, J. F. Rasco and J. B. Vincent, *J. Inorg. Biochem.*, 2010, **104**, 790.

111. J. C. Pessoa and I. Tomaz, *Curr. Med. Chem.*, 2010, **17**, 3701.

112. S. A. Katz, B. Ballantyne and H. Salem, *The Inhalation Toxicology of Chromium Compounds*, in *Inhalation Toxicology*, ed. H. Salem and

S. A. Katz, CRC Press, Cambridge, UK, 2nd edn, 2006, Ch. 23, pp. 543–564.

113. K. Merritt and S. A. Brown, *J. Biomed. Mater. Res.*, 1995, **29**, 627.
114. A. Kortenkamp and D. Beyersmann, *Toxicol. Environ. Chem.*, 1987, **14**, 23.
115. G. E. Corbett, D. G. Dodge, E. O'Flaherty, J. Liang, L. Throop, B. L. Finley and B. D. Kerger, *Environ. Res. A*, 1998, **78**, 7.
116. J. Aaseth, J. Alexander and T. Norseth, *Acta Pharmacol. Toxicol.*, 1982, **50**, 310.
117. H. Sakurai, K. Takechi, H. Tsuboi and H. Yasui, *J. Inorg. Biochem.*, 1999, **76**, 71.
118. B. L. Finley, B. D. Kerger, M. W. Katona, M. L. Gargas, G. C. Corbett and D. J. Paustenbach, *Toxicol. Appl. Pharmacol.*, 1997, **142**, 151.
119. Y. Sayato, K. Nakamuro, S. Matsui and M. Ando, *J. Pharmacobio-Dyn.*, 1980, **3**, 17.
120. H. J. Dowling, E. G. Offenbacher and F. X. Pi-Sunyer, *Nutr. Res.*, 1990, **10**, 1251.
121. F. Borguet, R. Cornelis, J. Delanghe, M.-C. Lambert and N. Lameire, *Clin. Chim. Acta*, 1995, **238**, 71.
122. B. J. Clodfelder and J. B. Vincent, *J. Biol. Inorg. Chem.*, 2005, **10**, 383.
123. H. Sun, H. Li and P. J. Sadler, *Chem. Rev.*, 1999, **99**, 2817.
124. J. B. Vincent and S. Love, *Biochim. Biophys. Acta*, 2012, **1820**, 362.
125. P. Aisen, R. Aasa and A. G. Redfield, *J. Biol. Chem.*, 1969, **244**, 4628.
126. Y. Chen, H. M. Watson, J. Gao, S. H. Sinha, C. J. Cassady and J. B. Vincent, *J. Nutr.*, 2011, **141**, 1225.
127. M. Hémadi, G. Miquel, P. H. Kahn and J.-M. El Hage Chahine, *Biochemistry*, 2003, **42**, 3120.
128. R. J. Hilton, M. C. Seare, N. D. Andros, Z. Kenealey, C. M. Orozco, M. Webb and R. K. Watt, *J. Inorg. Biochem.*, 2012, **110**, 1.
129. D. D. D. Hepburn and J. B. Vincent, *Chem. Res. Toxicol.*, 2002, **15**, 93.
130. A. A. Shute, N. E. Chakov and J. B. Vincent, *Polyhedron*, 2001, **20**, 2241.
131. A. Yamamoto, O. Wada and T. Ono, *Toxicol. Appl. Pharmacol.*, 1981, **59**, 515.
132. B. J. Clodfelder, J. Emamaullee, D. D. D. Hepburn, N. E. Chakov, H. S. Nettles and J. B. Vincent, *J. Biol. Inorg. Chem.*, 2001, **6**, 608.
133. A. Yamamoto, O. Wada and T. Ono, *Eur. J. Biochem.*, 1987, **165**, 627.
134. C. M. Davis and J. B. Vincent, *Arch. Biochem. Biophys.*, 1997, **339**, 335.
135. D. M. Stearns, *Multiple hypotheses for chromium (III) biochemistry: why the essentiality of chromium (III) is still questioned*, in *The Nutritional Biochemistry of Chromium(III)*, ed. J. B. Vincent, Elsevier, Amsterdam, 2007, Ch. 3, pp. 57–70.
136. H.-J. Wang, X.-M. Du, M. Wang, T.-C. Wang, O.-Y. Hong, B. Wang, M.-T. Zhu, Y. Wang, G. Jia and W.-Y. Feng, *Talanta*, 2010, **81**, 1856.
137. C. J. L. Silwood and M. Grootveld, *J. Inorg. Biochem.*, 2005, **99**, 1390.
138. A. C. Lewis, M. R. Kilburn, I. Papageorgiou, G. C. Allen and C. P. Case, *J. Biomed. Mater. Res. Part A*, 2005, **73A**, 456.

139. A. W. Newton, L. Ranganath, C. Armstrong, V. Peter and N. B. Roberts, *J. Orthop. Res.*, 2012, **30**, 1640.

140. C. C. Winterbourn, M. B. Hampton, J. H. Livesey and A. J. Kettle, *J. Biol. Chem.*, 2006, **281**, 39860.

141. R. Codd and P. A. Lay, *J. Am. Chem. Soc.*, 2001, **123**, 11799.

142. J. Singh, D. E. Pritchard, D. L. Carlisle, J. A. Mclean, A. Montaser, J. M. Orenstein and S. R. Patierno, *Toxicol. Appl. Pharmacol.*, 1999, **161**, 240.

143. H. Xie, A. L. Holmes, S. S. Wise, N. Gordon and J. P. Wise, *Chem. Res. Toxicol.*, 2004, **17**, 1362.

144. P. A. Lay and A. Levina, *Chromium toxicity - high valent chromium*, in *Encyclopedia of Metalloproteins*, ed. V. Uversky, R. H. Kretsinger and E. A. Permyakov, Springer, New York, 2013, doi: 10.1007/978-1-4614-1533-6.

145. S. S. Wise, A. L. Holmes, Q. Qin, H. Xie, S. P. Katsifis, W. D. Thompson and J. P. Wise, *Chem. Res. Toxicol.*, 2010, **23**, 365.

146. H. Xie, A. L. Holmes, S. S. Wise, S. Huang, C. Peng and J. P. Wise, *Am. J. Respir. Cell Mol. Biol.*, 2007, **37**, 544.

147. J. B. Aitken, A. Levina and P. A. Lay, *Curr. Top. Med. Chem.*, 2011, **11**, 553.

148. C. T. Dillon, P. A. Lay, M. Cholewa, G. J. F. Legge, A. M. Bonin, T. J. Collins, K. L. Kostka and G. Shea-McCarthy, *Chem. Res. Toxicol.*, 1997, **10**, 533.

149. H. H. Harris, A. Levina, C. T. Dillon, I. Mulyani, B. Lai, Z. Cai and P. A. Lay, *J. Biol. Inorg. Chem.*, 2004, **10**, 105.

150. A. Levina, H. H. Harris and P. A. Lay, *J. Biol. Inorg. Chem.*, 2006, **11**, 225.

151. H. Arakawa, M.-W. Weng, W.-C. Chen and M.-S. Tang, *Carcinogenesis*, 2012, **33**, 1993.

152. R. Krupa, M. Stanczak and Z. Walter, *Z. Naturforsch. C*, 2002, **57**, 951.

153. A. Levina, A. M. Bailey, G. Champion and P. A. Lay, *J. Am. Chem. Soc.*, 2000, **122**, 6208.

154. A. Hartwig, T. Schwerdtle and W. Bal, *Methods Mol. Biol.*, 2010, **649**, 399.

155. D. Beyersmann and A. Hartwig, *Arch. Toxicol.*, 2008, **82**, 493.

156. A. Arita and M. Costa, *Metallomics*, 2009, **1**, 222.

157. R. Porter, M. Jáchymová, P. Martásek, B. Kalyanaraman and J. Vásquez-Vivar, *Chem. Res. Toxicol.*, 2005, **18**, 834.

158. G. R. Borthiry, W. E. Antholine, J. M. Myers and C. R. Myers, *Chem. Biodivers.*, 2008, **5**, 1545.

159. D. Krepkiy, W. E. Antholine and D. H. Petering, *Chem. Res. Toxicol.*, 2003, **16**, 750.

160. M. Sugiyama, K. Tsuzuki and R. Ogura, *J. Biol. Chem.*, 1991, **266**, 3383.

161. C. Witmer, E. Faria, H.-S. Park, N. Sadrieh, E. Yurkow, S. O'Connell, A. Sirak and H. Schleyer, *Environ. Health Perspect.*, 1994, **102**, 169.

162. K. J. Liu, X. Shi, J. Jiang, F. Goda, N. Dalal and H. M. Swartz, *Ann. Clin. Lab. Sci.*, 1996, **26**, 176.

163. K. J. Liu, K. Mäder, X. Shi and H. M. Swartz, *Magn. Reson. Med.*, 1997, **38**, 524.

164. R. Bartholomäus, J. A. Irwin, L. Shi, S. Meejoo, A. Levina and P. A. Lay, *Inorg. Chem.*, 2013, **52**, 4282.

165. P. A. Lay and A. Levina, *J. Am. Chem. Soc.*, 1998, **120**, 6704.

166. A. Hartwig, *Free Radic. Biol. Med.*, 2013, **55**, 63.

167. A. Lamb, G. Evans and J. R. King, *Bull. Math. Biol.*, 2013, **75**, 1472.

168. C. R. Kirman, L. L. Aylward, M. Suh, M. A. Harris, C. M. Thompson, L. C. Haws, D. M. Proctor, S. S. Lin, W. Parker and S. M. Hays, *Chem.-Biol. Interact.*, 2013, **204**, 13.

169. N. R. Rhodes, P. A. LeBlanc, J. F. Rasco and J. B. Vincent, *Biol. Trace Elem. Res.*, 2012, **148**, 409.

170. M. A. W. Lawrence, P. T. Maragh and T. P. Dasgupta, *J. Coord. Chem.*, 2010, **63**, 2517.

171. J. A. Mindell, *Annu. Rev. Physiol.*, 2012, **74**, 69.

172. J.-M. El Hage Chahine, M. Hemadi and N.-T. Ha-Duong, *Biochim. Biophys. Acta*, 2012, **1820**, 334.

173. R. A. Yokel, M. Wilson, W. R. Harris and A. P. Halestrap, *Brain Res.*, 2002, **930**, 101.

174. K. Nagasawa, S. Ito, T. Kakuda, K. Nagai, I. Tamai, A. Tsuji and S. Fujimoto, *Toxicol. Lett.*, 2005, **155**, 289.

175. H. A. Headlam, *The role of Cr(III) and Cr(V) peptide and amino acid complexes in Cr-induced carcinogenesis*, Ph.D. Thesis, The University of Sydney, 1998.

176. J. B. Vincent, *J. Am. Coll. Nutr.*, 1999, **18**, 6.

177. J. B. Vincent, *Acc. Chem. Res.*, 2000, **33**, 503.

178. D. Dinakarpandian, V. Morrissette, S. Chaudhary, K. Amini, B. Bennett and J. D. Van Horn, *BMC Chem. Biol.*, 2004, **4**, doi: 10.1186/1472-6769-1184-1182.

179. N. Deighton and R. C. Hider, *Biochem. Soc. Trans.*, 1989, **17**, 490.

180. L. Jacquamet, Y. Sun, J. Hatfield, W. Gu, S. P. Cramer, M. W. Crowder, G. A. Lorigan, J. B. Vincent and J.-M. Latour, *J. Am. Chem. Soc.*, 2003, **125**, 774.

181. O. Kakhlon and Z. I. Cabantchik, *Free Radic. Biol. Med.*, 2002, **33**, 1037.

182. R. C. Hider and X. L. Kong, *BioMetals*, 2011, **24**, 1179.

183. H. C. Lukaski, W. A. Siders and J. G. Penland, *Nutrition*, 2007, **23**, 187.

184. J. Fomusi Ndisang, N. Lane, N. Syed and A. Jadhav, *Endocrinology*, 2010, **151**, 549.

185. C. M. Barnes, E. C. Theil and K. N. Raymond, *Proc. Natl Acad. Sci. USA*, 2002, **99**, 5195.

186. J. B. Aitken, E. A. Carter, H. Eastgate, M. J. Hackett, H. H. Harris, A. Levina, Y.-C. Lee, C.-I. Chen, B. Lai, S. Vogt and P. A. Lay, *Rad. Phys. Chem.*, 2010, **79**, 176.

187. S. A. Blankert, V. H. Coryell, B. T. Picard, K. K. Wolf, R. E. Lomas and D. M. Stearns, *Chem. Res. Toxicol.*, 2003, **16**, 847.

188. A. Levina, L. Zhang and P. A. Lay, *J. Am. Chem. Soc.*, 2010, **132**, 8720.

189. M. Zhang, M. Zhou, R. L. Van Etten and C. V. Stauffacher, *Biochemistry*, 1997, **36**, 15.

190. A. R. Saltiel and R. C. Kahn, *Nature*, 2001, **414**, 799.
191. M. J. Gonçalves, A. C. C. Santos, C. F. D. Rodrigues, P. Coelho, A. N. Costa, A. J. Guiomar, M. S. Santos, M. C. Alpoim and A. M. Urbano, *Toxicol. Environ. Chem.*, 2011, **93**, 1202.
192. E. Kim and K. J. Na, *Toxicol. Appl. Pharmacol.*, 1991, **110**, 251.
193. E. J. Yurkow and G. Kim, *Molec. Pharmacol.*, 1995, **47**, 686.
194. Y. Qian, B. Jiang, D. C. Flynn, S. S. Leonard, S. W. Wang, Z. Zhang, J. P. Ye, F. Chen, L. Y. Wang and X. L. Shi, *Molec. Cell. Biochem.*, 2001, **222**, 199.
195. K. Govindaraju, T. Ramasami and D. Ramaswamy, *J. Inorg. Biochem.*, 1989, **35**, 127.
196. R. Sreekanth, V. Pattabhi and S. S. Rajan, *Biochem. Biophys. Res. Commun.*, 2008, **369**, 725.
197. P. Mackowiak, Z. Krejpcio, M. Sassek, P. Kaczmarek, I. Hertig, J. Chmielewska, T. Wojciechowicz, D. Szczepankiewicz, D. Wieczorek, H. Szymusiak and K. W. Nowak, *Mol. Med. Rep.*, 2010, **3**, 347.
198. A. Al-Qatati, P. W. Winter, A. L. Wolf-Ringwall, P. B. Chatterjee, A. K. Van Orden, D. C. Crans, D. A. Roess and B. G. Barisas, *Cell. Biochem. Biophys.*, 2012, **62**, 441.
199. S. Samantaray, G. R. Rout and P. Das, *Acta Physiol. Plant.*, 1998, **20**, 201.
200. M. I. Ramírez-Díaz, C. Díaz-Pérez, E. Vargas, H. Riveros-Rosas, J. Campos-García and C. Cervantes, *Biometals*, 2008, **21**, 321.
201. R. O. Castro, M. M. Trujillo, J. L. Bucio, C. Cervantes and J. Dubrovsky, *Plant Science*, 2007, **172**, 684.
202. E. J. Calabrese, *Toxicol. Appl. Pharmacol.*, 2005, **204**, 1.
203. C. Cervantes, J. Campos-García, S. Devars, F. Gutiérrez-Corona, H. Loza-Tavera, J. C. Torres-Guzmán and R. Moreno-Sánchez, *FEMS Microbiol. Rev.*, 2001, **25**, 335.
204. D. A. Brose and B. R. James, *Environ. Sci. Technol.*, 2010, **44**, 9438.
205. A. Zayed, C. M. Lytle, J.-H. Qian and N. Terry, *Planta*, 1998, **206**, 293.
206. M. Schiavon, E. A. H. Pilon-Smits, M. Wirtz, R. Hell and M. Malagoli, *J. Environ. Qual.*, 2008, **37**, 1536.
207. G. Micera and A. Dessí, *J. Inorg. Biochem.*, 1988, **34**, 157.
208. K. J. Appenroth, M. Bischoff, H. Gabrys, J. Stoeckel, H. M. Swartz, T. Walczak and K. Winnefeld, *J. Inorg. Biochem.*, 2000, **78**, 235.
209. N. Das and L. Mathew, *Environ. Pollut.*, 2011, **20**, 297.
210. Y. Cheng, H.-Y. Holman and L. Zhang, *Elements*, 2012, **8**, 107.
211. S. Bluskov, J. M. Arocena, O. O. Omotoso and J. P. Young, *Int. J. Phytoremediation*, 2005, **7**, 153.
212. Y. Zhao, J. G. Parsons, J. R. Peralta-Videa, M. L. Lopez-Moreno and J. L. Gardea-Torresdey, *Metallomics*, 2009, **1**, 330.
213. M. R. Bleackley and R. T. A. MacGillivray, *Biometals*, 2011, **24**, 785.
214. D. Park, Y.-S. Yun, J. H. Jo and J. M. Park, *Water Res.*, 2005, **39**, 533.
215. S. S. Ahluwalia and D. Goyal, *Bioresour. Technol.*, 2007, **98**, 2243.
216. M. N. Sahmoune, K. Louhab and A. Boukhiar, *Environ. Prog. Sustainable Energy*, 2011, **30**, 284.

217. R. Codd, P. A. Lay, N. Y. Tsibakhashvili, T. L. Kalabegishvili, I. G. Murusidze and H.-Y. N. Holman, *J. Inorg. Biochem.*, 2006, **100**, 1827.
218. S. Volland, C. Luetz, B. Michalke and U. Luetz-Meindl, *Aquat. Toxicol.*, 2012, **109**, 59.
219. P. Kaszycki, D. Fedorovych, H. Ksheminska, L. Babyak, D. Wojcik and H. Koloczek, *Microbiol. Res.*, 2004, **159**, 11.
220. N. Chatterjee, Z. Luo, S. Malghani, J. J. Lian and W. L. Zheng, *Chem. Speciation Bioavailability*, 2009, **21**, 245.
221. N. Mirsky, A. Weiss and Z. Dori, *J. Inorg. Biochem.*, 1981, **15**, 275.
222. Z. Kirpnick-Sobol, R. Reliene and R. H. Schiestl, *Cancer Res.*, 2006, **66**, 3480.
223. K. H. Cheung and J.-D. Gu, *Int. Biodeterior. Biodegrad.*, 2007, **59**, 8.
224. Y. Barak, D. F. Ackerley, C. J. Dodge, L. Banwari, C. Alex, A. J. Francis and A. Matin, *Appl. Environ. Microbiol.*, 2006, **72**, 7074.
225. S. Eswaramoorthy, S. Poulain, R. Hienerwadel, N. Bremond, M. D. Sylvester, Y.-B. Zhang, C. Berthomieu, D. Van Der Lelie and A. Matin, *PLoS One*, 2012, **7**, e36017.
226. C. Liu, Y. A. Gorby, J. M. Zachara, J. K. Fredrickson and C. F. Brown, *Biotechnol. Bioeng.*, 2002, **80**, 637.
227. S. M. Belchik, D. W. Kennedy, A. C. Dohnalkova, Y. Wang, P. C. Sevinc, H. Wu, Y. Lin, H. P. Lu, J. K. Fredrickson and L. Shi, *Appl. Environ. Microbiol.*, 2011, 77, 4035.
228. S. P. Ravindranath, K. L. Henne, D. K. Thompson and J. Irudayaraj, *ACS Nano*, 2011, **5**, 4729.
229. T. L. Daulton, B. J. Little, J. Jones-Meehan, D. A. Blom and L. F. Allard, *Geochim. Cosmochim. Acta*, 2007, **71**, 556.
230. L. Shi, T. C. Squier, J. M. Zachara and J. K. Fredrickson, *Molec. Microbiol.*, 2007, **65**, 12.
231. G. J. Puzon, L. Xun, R. Tokala, Z. Zhang, S. Clark, B. Peyton and D. Yonge, *IAHS Publ.*, 2008, **324**, 420.
232. A. L. Neal, K. Lowe, T. L. Daulton, J. Jones-Meehan and B. J. Little, *Appl. Surf. Sci.*, 2002, **202**, 150.
233. H. Jin, Y. Zhang, G. W. Buchko, S. M. Varnum, H. Robinson, T. C. Squier and P. E. Long, *PLoS One*, 2012, 7, e42432.
234. G. J. Puzon, J. N. Petersen, A. G. Roberts, D. M. Kramer and L. Xun, *Biochem. Biophys. Res. Comm.*, 2002, **294**, 76.
235. A. S. S. Ibrahim, M. A. El-Tayeb, Y. B. Elbadawi, A. A. Al-Salamah and G. Antranikian, *Extremophiles*, 2012, **16**, 659.
236. A. Garenaux and C. M. Dozois, *Nat. Chem. Biol.*, 2012, **8**, 680.
237. I. J. Schalk, M. Hannauer and A. Braud, *Environ. Microbiol.*, 2011, **13**, 2844.
238. A. M. Zawadzka, R. L. Crawford and A. J. Paszczynski, *BioMetals*, 2007, **20**, 145.
239. A. Braud, F. Hoegy, K. Jezequel, T. Lebeau and I. J. Schalk, *Environ. Microbiol.*, 2009, **11**, 1079.

240. D. I. Pattison, A. Levina, M. J. Davies and P. A. Lay, *Inorg. Chem.*, 2001, **40**, 214.
241. A. Levina, G. J. Foran, D. I. Pattison and P. A. Lay, *Angew. Chem. Int. Ed.*, 2004, **43**, 4504.
242. S. Gez, R. Luxenhofer, A. Levina, R. Codd and P. A. Lay, *Inorg. Chem.*, 2005, **44**, 2934.
243. R. J. P. Williams, *J. Inorg. Biochem.*, 1999, **76**, 81.
244. D. Stallings and J. B. Vincent, *Curr. Top. Nutraceutical Res.*, 2006, **4**, 89.

Molybdenum and Tungsten

MANUEL TEJADA-JIMÉNEZ[a] AND GUENTER SCHWARZ[*a,b,c]

[a] Institute of Biochemistry, Department of Chemistry, University of Cologne, Zuelpicher Str. 47, Cologne, 50674, Germany; [b] Center for Molecular Medicine Cologne, University of Cologne, Robert-Koch Str. 21, Cologne, 50931, Germany; [c] Cluster of Excellence in Ageing Research, CECAD Research Center, Joseph-Stelzmann-Str. 26, Cologne, 50931, Germany
*Email: gschwarz@uni-koeln.de

8.1 Molybdenum Distribution, Availability and Biological Relevance

Molybdenum (Mo) is the only transition metal of the fifth row of the periodic table that has been confirmed as an essential element. The element is found in nature in different oxidation states ranging from zero to six and forms complexes with coordination numbers from four to eight.

Due to the solubility of the molybdate anion $[MoO_4]^{2-}$ and its polynuclear hydrolysis products,[1] Mo is the 25th most abundant element in sea water at an average concentration of 100 nM. Its concentration in continental water is much lower (5 nM). In the geosphere, Mo is the 54th most abundant element: its overall concentration is about 3 mg kg^{-1} but this can increase up to 300 mg kg^{-1} when the content of organic matter is high.[2,3] There are several mineral forms including wulfenite (MoS_2), molybdenite ($PbMoO_4$) and ferrimolybdenite ($Fe_2(MoO_4)_3$).[3] In soil, Mo forms complexes with catechol groups in tannin-like compounds present in natural organic matter.[4]

RSC Metallobiology Series No. 2
Binding, Transport and Storage of Metal Ions in Biological Cells
Edited by Wolfgang Maret and Anthony Wedd
© The Royal Society of Chemistry 2014
Published by the Royal Society of Chemistry, www.rsc.org

The metal is present at low concentrations in living organisms. Its content is about 10^5 atoms per cell in micro-algae[5] and about 0.2 mg kg^{-1} dry weight in higher plants.[6] In humans, the highest levels are found in kidney, liver, small intestine and adrenal glands.[7] In serum, the concentration is about 0.6 ng ml^{-1} but depends on dietary intake.[8,9]

Mo and its Group 6 congener tungsten (W) share many chemical properties due to i) equivalent valence electron configurations and ii) essentially identical atomic radii.[1] However there are important variations regarding chemical reactivity and detailed electronic structure. These arise from different ionization and redox potentials that relate, for example, to differences in stoichiometry for polynuclear polyoxometalate anions in aqueous solution.[1] In aqueous solution at pH higher than 6, the predominant species $[M^{VI}O_4]^{2-}$ (M = Mo or W) have virtually the same protonation constant.[1] Below pH 6, these molybdate and tungstate anions undergo condensation reactions to form iso-polyoxometalate anions. Equilibration of condensation is faster for molybdate than for tungstate, a consequence of a lower substitutional lability and stronger M^{VI}–O bonds for W.[1] The stereochemistry of $M^{IV,V,VI}$ complexes with the same ligand donor set is the same with minor differences in bond angles.[1] Essentially equal bond lengths pertain for a given oxidation state.

The simple oxyanion $[Mo^{VI}O_4]^{2-}$ is the only known form of Mo that cells can take up from the environment. In solution, it can be chemi-adsorbed onto positively charged surfaces of iron, aluminium or manganese oxides.[10] Its availability increases as the pH increases, mainly due to decreased association with these metal oxides.

The importance of Mo for the living world was recognized in 1939 when Arnon and Stout demonstrated its essentiality for plants using a fully defined nutrient solution.[11] Tomato plants grown in its absence developed a number of characteristic symptoms including mottling and necrotic leaves, which also showed involution of the laminae. Fruiting was also impaired. All of these deficiency symptoms were rescued upon the addition of trace amounts of Mo.

The discovery of molybdenum in enzymes such as nitrogenase, nitrate reductase (NR) and xanthine oxidase (XO) highlighted its catalytic role.[12] Finally, the identification of de-activated enzyme mutants confirmed its essentiality in metabolism.[13] Since then, more than 50 different Mo-dependent enzymes have been found in all kingdoms of life.[14] Most of them are of bacterial origin and, except for nitrogenase, they all bind the ion *via* a pterin-based prosthetic group forming the so-called molybdenum cofactor (Moco; Figure 8.1A). This is composed of a fully reduced pterin backbone with a C6-substituted four-carbon side-chain forming a pyran ring that hosts a dithiolene group (for chelation of Mo) and a terminal phosphate. In bacterial enzymes, this phosphate can be linked to certain nucleotide monophosphates and the Mo atom can be bound by the dithiolenes of two separate pterins.[12] In nitrogenase, Mo is chelated by one of the most complex metal centres produced by nature, the iron-molybdenum cofactor

(FeMo-co) that catalyses the multistep reduction of dinitrogen to ammonia.[15]

Tungsten is able to replace molybdenum in Moco due to its similar atomic radius and coordination chemistry.[16,17] The tungsten cofactor has been identified as the prosthetic group in enzymes from (hyper)thermophilic anaerobic archaea and may be the evolutionary ancestor of Mo cofactors and enzymes.[17]

8.2 Molybdenum- and Tungsten-containing Enzymes

More than 50 Mo-containing enzymes have been found in bacteria and were classified for the first time by R. Hille resulting into four families based on the nature of the cofactor and other ligands bound in their active centres in the oxidized Mo(VI) forms (Figure 8.1A).[18] The first group is termed the xanthine oxidase (XO) family and features a single pterin cofactor and terminal sulfido and oxido-ligands.[19] The second group comprises the sulfite oxidase (SO) family and carries a single cofactor, a cysteine ligand and two oxido-ligands.[20] The third group is the dimethyl sulfoxide (DMSO) reductase family and harbours a bis-pyranopterin guanine dinucleotide cofactor and two oxido-ligands plus, and depending on the particular enzyme, an extra cysteine, seleno-cysteine, serine or aspartate ligand.[19] The fourth group comprises the archaeal aldehyde oxidoreductase family that harbours a bis-pyranopterin cofactor preferentially bound to a W atom.[21]

It is still not clear why some organisms use W and others do not. The relative abundance of W over Mo may answer this question partially. The ability to use W could be an advantage for organisms exposed to environments with a high W/Mo ratio.[17] The ability to use W also confers an advantage for catalysing reactions with extremely low electrochemical potentials. Indeed almost all known tungsten-containing enzymes catalyse reactions at potentials near or below that of the standard hydrogen electrode.[22]

The biological relevance of W was revealed in 1973 by the positive influence of tungstate on the growth rate of different *Clostridia* strains.[23] The first identified W-containing enzyme was formate dehydrogenase from *Clostridia*,[24] and subsequently a number of enzymes harbouring W have been purified and characterized mainly derived from thermophilic bacteria.[25] Later W-containing enzymes were also found in mesophilic bacteria and aerophilic organisms while W-containing enzymes have not been found in eukaryotes.[17]

Five molybdo-enzymes are found in eukaryotes and belong to only two of the families (Figure 8.1B): nitrate reductase (NO), SO and the amidoxime-reducing component (mARC) are members of the SO family, while XO/xanthine dehydrogenase (XDH) and aldehyde oxidase (AO) form part of the XO family.[26] Molybdo-enzymes (mainly bacterial) are involved in key processes in the global carbon, nitrogen and sulfur cycles, such as nitrate reduction, sulfite detoxification and purine catabolism.

A

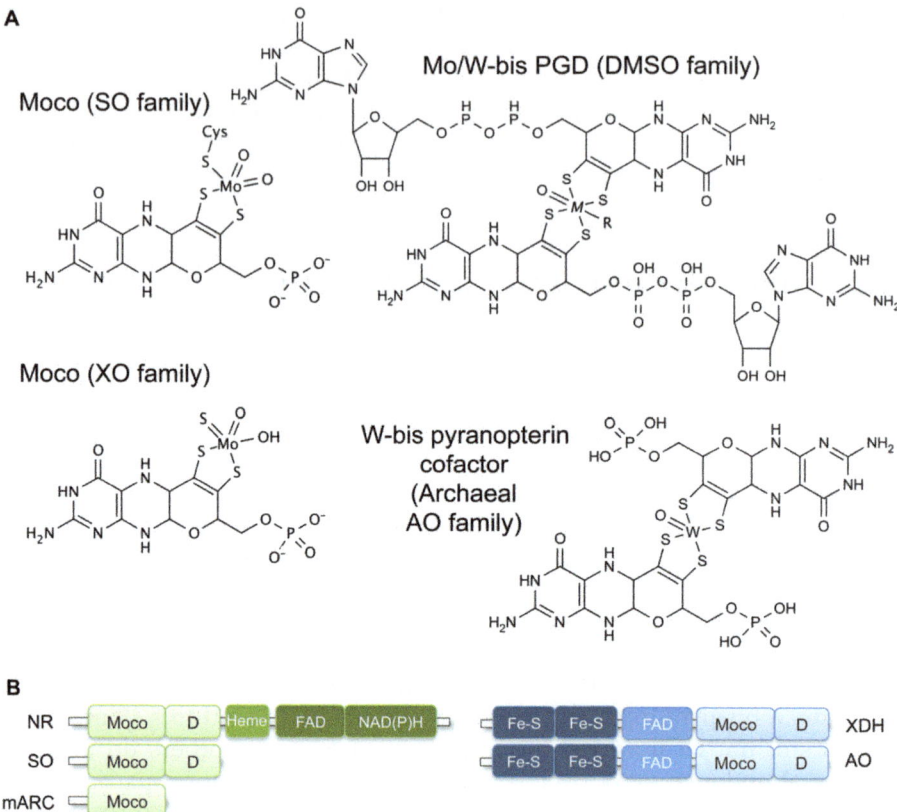

Figure 8.1 Molybdenum- and tungsten-containing cofactors and enzymes. (A) Chemical structures of molybdenum cofactor (Moco), Mo/W-bis pyranopterin guanosine dinucleotide (PGD) and W-bis pyranopterin cofactor. (B) Domain structures of eukaryotic Mo enzymes of the xanthine oxidase (XO) and sulfite oxidase (SO) families. D, dimerization domain.

8.2.1 Nitrate Reductase

Eukaryotic NR (EC 1.6.6.1) is a cytoplasmic, water-soluble enzyme involved in the reduction of nitrate to nitrite as the first step of assimilation of nitrogen in plants, algae and fungi. It is a homo-dimeric protein of about 200–220 kDa. Each monomer folds into five functional domains (Figure 8.1B) of which three are involved in the binding of single prosthetic groups: the N-terminal Moco domain, the central cytochrome b_5-type haem domain and the C-terminal FAD domain. Each monomer is completed by a dimerization domain and a NAD(P)H-binding domain.[27] The dimerization, haem and FAD domains are connected by protease-sensitive and solvent-exposed loop sequences called hinge-I and -II. NADH (in most higher plants and algae) or NADPH (in fungi) provides the reducing equivalents for nitrate reduction.

An intramolecular electron transfer chain transfers reducing equivalents and starts with the reduction of FAD by NAD(P)H followed by two one-electron transfer steps *via* the cytochrome b_5 domain to the Mo centre where nitrate undergoes two-electron reduction to nitrite.[28] In plants and algae, the product nitrite is translocated to the chloroplast where it is reduced to ammonia by the separate enzyme nitrite reductase.[29] However, NR has also been shown to reduce nitrite to nitric oxide, a reaction crucial to plant development.[30]

NR activity is tightly regulated by both transcriptional and post-transcriptional mechanisms: nitrate, light, CO_2, phyto-hormones and other metabolites of nitrogen and carbon are the main effectors.[31] A conserved phospho-serine residue in hinge-I is involved in post-translational regulation of plant NR. Upon the phosphorylation-dependent binding of the so-called 14-3-3 proteins to NR,[32] the enzyme activity is strongly decreased due to a reduced intramolecular electron transfer rate between the haem and Moco domains.[33,34] The crystal structure of the Mo-domain has been determined for the enzyme from *Pichia angusta* and shows that the active site contains a conserved binding pocket for nitrate,[35] an extended, mostly unstructured N-terminus and an overall structure remarkably similar to that of SO.[20,36]

8.2.2 Sulfite Oxidase

SO (EC 1.8.3.1) is essential in sulfur catabolism. In vertebrates, it catalyses the two-electron oxidation of sulfite to sulfate coupled to the reduction of two molecules of cytochrome c.[37] Sulfite oxidation is the terminal step in the oxidative degradation of cysteine. Vertebrate SO is a homo-dimeric protein and each monomer harbours an individual N-terminal cytochrome b_5-type haem, Moco and C-terminal dimerization domains.[20]

SO is localized in the mitochondrial intermembrane space where electrons derived from sulfite oxidation are transferred to the physiological electron acceptor cytochrome c.[38] Recently, the mechanism of maturation of mammalian SO has been clarified. A conventional leader sequence-based translocation mechanism is combined with the folding trap mechanism for which the presence of Moco is strictly required.[39] The catalytic mechanism of SO involves transfer of two electrons from sulfite to Moco ($Mo^{VI} \rightarrow Mo^{IV}$) followed by two one-electron transfer steps *via* the haem domain to cytochrome c.

While SO from the green algae *Chlamydomonas reinhardtii* shows the same primary structure as in animal SO, the enzyme in higher plants lacks the haem (Figure 8.1B).[40] It is localized in peroxisomes and uses molecular oxygen as physiological electron acceptor with the release of hydrogen peroxide as secondary product.[41–43] Recent studies have proposed that plant SO controls the sulfate-sulfite cycle and so contributes to the overall sulfur homeostasis in plants. The enzyme protects the plant from sulfite overload, particularly when exposed to atmospheric SO_2.[44]

8.2.3   Xanthine Dehydrogenase and Oxidase

Eukaryotic XDH/XO (EC 1.17.1.4) systems participate in the degradation of purines by oxidation of hypoxanthine to xanthine and xanthine to uric acid. The enzyme can function either as a dehydrogenase using NAD^+ as electron acceptor or, upon reversible cysteine oxidation, as an oxidase using dioxygen as terminal electron acceptor. In contrast, proteolytic cleavage of XDH converts the enzyme irreversibly into the XO form.[45,46]

XDH/XO are homo-dimeric enzymes harbouring three cofactor-binding domains (Fe-S clusters, FAD and Moco) and an additional domain important for dimerization (Figure 8.1B). Hydroxylation of purine substrates causes two-electron reduction of Moco and the electrons are transferred singly *via* two [2Fe-2S] clusters to the FAD cofactor where either NAD^+ or dioxygen is reduced. The latter co-substrate produces superoxide anions.[47] XDH from *Arabidopsis* exhibits an alternative activity, being able to oxidize both NADH (to NAD^+) and superoxide.[48] This finding points to a role for plant XDH in the metabolism of reactive oxygen species and may impact plant-pathogen interactions, hypersensitivity response, starvation stress and natural senescence.[26,47]

8.2.4   Aldehyde Oxidase

AO enzymes (EC 1.2.3.1) originate from a duplication of the *xdh* gene in eukaryotes before the origin of multicellularity.[49] Consequently, both enzymes contain the same cofactor-binding domains (Fe-S clusters, FAD and Moco) as well as a dimerization domain (Figure 8.1B). The human genome contains a single *AOX* gene together with two pseudo-genes clustered in chromosome 2q. However, mouse and rat genomes express four *aox* genes giving rise to four iso-enzymes.[50] The crystal structure of AOX3 from mouse exhibits high structural similarity to XDH/XO.[51] As expected, the overall mechanism of action of AO is similar to that of XO.[51]

AO exhibits a broad substrate specificity that includes heterocycles, aldehydes, purines and pteridines.[52] As for XDH/XO, animal AO produces superoxide and hydrogen peroxide as secondary products but plant AO generates hydrogen peroxide only. The physiological role of plant AO is crucial for the metabolism of the phytohormone abscisic acid (ABA), which is formed by oxidation of abscisic aldehyde.[53] ABA is crucial for plant development and stress adaptation.[54] In contrast, the physiological function of animal AO remains enigmatic. Since AOs have broad substrate specificity, they are believed to function in the metabolism of drugs and xenobiotics. In addition, they may participate in the oxidation of endogenous products involved in various metabolic pathways such as: neurotransmitters (*i.e.* serotonin); the conversion of the hydroquinone precursor gentisate aldehyde into gentisate; the catabolism of valine, leucine or isoleucine and also vitamins (nicotinamide and pyridoxal).[52]

8.2.5 The Amidoxime-reducing Component

The mitochondrial amidoxime-reducing component (mARC) has been identified recently as a novel Mo-containing enzyme in mammals.[55] In addition, a homologous protein also has been identified and characterized in the green algae *C. reinhardtii*.[56] Furthermore, similar proteins are encoded in plant and prokaryotic genomes. In some organisms, two mARC isoforms have been identified. Mouse mARC1 is localized in the outer mitochondrial membrane.[57] The enzyme is targeted by a bipartite N-terminal signal peptide leading to tail-anchored integration into the membrane with cytosolic exposure of the majority of the protein. In contrast, the homologous protein (CrARC) in *C. reinhardtii* lacks any targeting sequence, which suggests a different subcellular localization.[56]

Besides a role for mARC2 in the regulation of nitric oxide synthesis,[58] no primary physiological substrates are known for the mARC proteins. The human versions may be involved in pro-drug metabolism, given their ability to reduce several N-hydroxylated substances commonly used as pro-drugs.

All mARC and ARC proteins contain a C-terminal Moco-binding domain that forms a novel class of molybdo-enzymes in plants.[59] In contrast, studies of the ARC from *C. reinhardtii* have demonstrated that it contains a highly conserved cysteine residue (also present in mARC proteins), which is essential for catalytic activity. Given that enzymes of the SO family are characterized by a covalent cysteine link between the Mo and the enzyme one can argue that also mARC and ARC proteins belong to the SO family of Mo-enzymes (Figure 8.1B).[56] So far, for catalytic activity, all characterized mARC/ARC proteins require formation of complexes with cytochrome b_5 and with a NADH/cytochrome b_5 reductase.[60]

8.3 Molybdate and Tungstate Transport in Bacteria

Both pro- and eukaryotes acquire Mo as the oxyanion molybdate. In the former, molybdate enters the cell through the action of proteins belonging to the ATP-binding cassette (ABC) transporter family.[61] Molybdate and tungstate can be transported into cells by the same transport systems due to their similarities in charge, shape and size.

In *E. coli*, Mo is taken up mainly by the molybdate transport system ModABC that exhibits both high affinity and specificity.[62] However, other membrane transporters operate in various organisms as non-specific and low-affinity molybdate transporters (Table 8.1). These include the high-affinity sulfate permease CysUWA[63] (a non-specific anion transport system with a very low affinity for molybdate[64]) and PerO, a permease identified in *Rhodobacter capsulatus*.[65] PerO belongs to the broadly distributed ArsB/NhaD family whose members transport ions as varied as arsenite, citrate and sodium. PerO also exhibits sulfate, tungstate and vanadate transport activities, suggestive of a general role in oxyanion transport.[65]

Table 8.1 Transport systems involved in molybdate uptake.

Transport system	Specific anion	K_m	Other substrates	Organism
ModABC	MoO_4^{2-}	50 nM	WO_4^{2-} SO_4^{2-}	Bacteria
CysUWA	SO_4^{2-}	2 μM	SeO_4^{2-} SeO_3^{2-} MoO_4^{2-}	Bacteria
Non-specific system	$-^a$	$-^a$	SeO_3^{2-} MoO_4^{2-}	Bacteria
PerO	$-^a$	$-^a$	MoO_4^{2-} WO_4^{2-} SO_4^{2-} VO_4^{3-}	Bacteria
MOT1	MoO_4^{2-}	7–20 nM	WO_4^{2-}	Algae, plants, fungi, bacteria
MOT2	MoO_4^{2-}	~500 nM	WO_4^{2-}	Algae, plants, animals
SHST1	SO_4^{2-}	10 μM	MoO_4^{2-}	Plants

^{a}Information not available.

In addition, W can be taken up *via* specific tungstate transporters belonging to the ABC family. The <u>T</u>ungsten <u>u</u>ptake <u>p</u>rotein (TupABC) and the W-transport protein (WtpABC) are the two members that have been identified so far in *Eubacterium acidaminophilum* and *Pyrococcus furiosus* respectively. Proteins A of these transporters show differences in sequence and in metal-binding affinity.[17] Estimates of dissociation constants (K_D) suggest that WtpA ($K_D = 17$ pM)[66] has a higher affinity for tungstate than does TupA ($K_D = 0.2$ nM).[67] Furthermore, WtpA exhibits a binding preference for tungstate over molybdate as it supplies two extra carboxylate ligands (Asp + Glu) to expand the metal coordination sphere from four (tetrahedral) to six (octahedral).[67]

8.3.1 The ModABC Molybdate Transporter

This transporter is the main carrier of Mo in bacteria. It is encoded in *E. coli* by the *mod*ABCD operon (Figure 8.2) that comprises the structural genes needed to encode a fully functional ModABC system (*mod*A, *mod*B and *mod*C), together with a fourth gene *mod*D whose function is not known. Mutants in *mod*A, *mod*B or *mod*C, but not in *mod*D, show a loss of all molybdo-enzyme activities. These can be rescued by the addition to the medium of non-physiological high concentrations of molybdate. Consequently, these mutants display reduced molybdate uptake and accumulation.[61] Direct transport studies revealed that the molybdate transport process mediated by ModABC is energy dependent and bi-phasic.[68] Initial fast uptake is followed by a lower rate of transport (K_m is 50 nM at pH 7.0).[61]

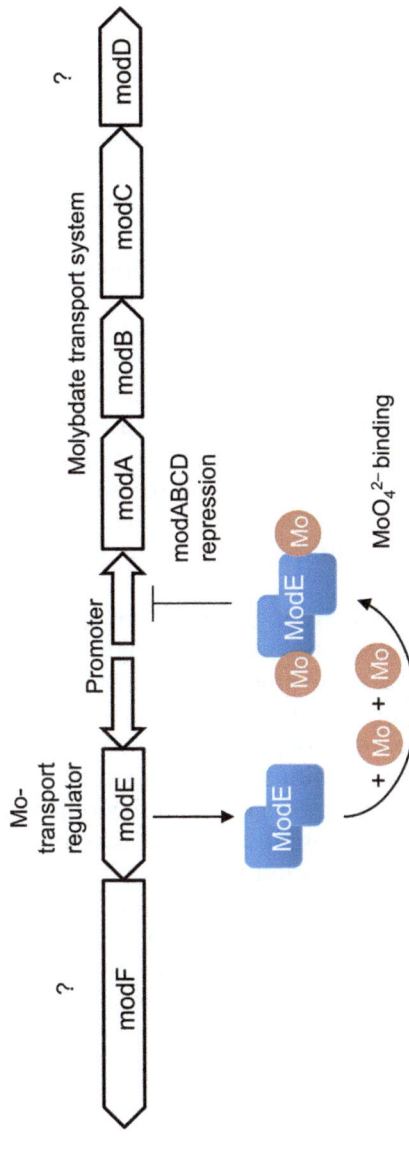

Figure 8.2 Organization of genes involved in molybdate transport and regulation in *E. coli*. Four different open reading frames organized in two divergent operons are required for transport. The *modABCD* operon encodes for the three transport components ModA, ModB and ModC while the function of ModD is unknown. The *modEF* operon encodes for the regulatory protein ModE and for ModF of unknown function. Two molybdate anions bind to one ModE dimer inducing conformational changes. The ModE-molybdate complex binds to the *modABCD* operon and prevents its transcription. The ModE protein structure was generated with PyMol 1.5 using the deposited coordinates (RCSB protein data bank 1B9M).

The *modA* gene encodes ModA, a periplasmic protein (257 residues) involved in molybdate binding.[69] It carries a leader sequence of 24 residues that is cleaved upon secretion into the periplasm. It binds molybdate and tungstate with K_D values of 0.13 and 0.17 nM, respectively, suggesting that ModA cannot discriminate between these oxyanions.[67] A mobility shift assay in non-denaturing electrophoresis gels and a change in isoelectric point suggest that conformational changes occur after substrate binding.[69] ModB acts as a homo-dimeric membrane channel.[62] This organization contrasts with the membrane channels of other ABC anion transporters which are hetero-dimeric.[70] ModC is a typical ATPase, providing the energy for molybdate transport.

Crystal structures of the intact ModABC systems from *Archaeoglobus fulgidus* and *Methanosarcina acetivorans* have been determined at high resolution.[71,72] Both are molybdate/tungstate transporters although, phylogenetically, they are more closely related to tungstate transporters.[17] The *A. fulgidus* structure shows a single ModA protein with bound tungstate attached to the $ModB_2C_2$ complex (Figure 8.3). In addition, crystal structures of ModA proteins from different bacteria are available, including structures of ModA containing either bound molybdate or tungstate.[71,73,74] They all consist of two folded domains connected *via* the anion-binding site. Selectivity for molybdate and tungstate compared to other tetrahedral anions such as sulfate or phosphate is achieved *via* seven hydrogen bonds involving both domains.[73] Once bound, neither MoO_4^{2-} nor WO_4^{2-} exchange with anions in solution.[67]

ModB features six membrane-spanning helices that lead to a dodecameric anion-conduction pore in the functional dimer (Figure 8.3).[71] This organization differs from known ABC transporters. Each ModB subunit builds a gate formed by four converging protein sections located in transmembrane helices 3 and 5. Two extracellular loops make contacts with two lobes of ModA that surround the substrate-binding site (Figure 8.3).

The ModB and ModC dimers interact *via* ModB cytoplasmic helices and grooves on the surface of ModC (Figure 8.3). The two ModC subunits are oriented in a "head-to-tail" organization with the phosphate-binding loop next to the conserved ABC motif (LSGGQ) of the other ModC subunit.[71] ATP binding and hydrolysis carried out by ModC provides critical conformational changes that are transduced to the ModB protein, powering effective transport of molybdate or tungstate from the external medium to the cytoplasm against the concentration gradient.[71]

8.3.2 Regulation of Molybdate Transport

The *mod* operon is bi-directionally transcribed to produce translation products, ModE and ModF (*modEF* transcript) as well as ModABC (Figure 8.2). ModE is a transcription factor that binds intracellular molybdate and controls transcription of the *modABCD* operon as well as that of another operon involved in the synthesis of Moco.[75] The function of ModF is

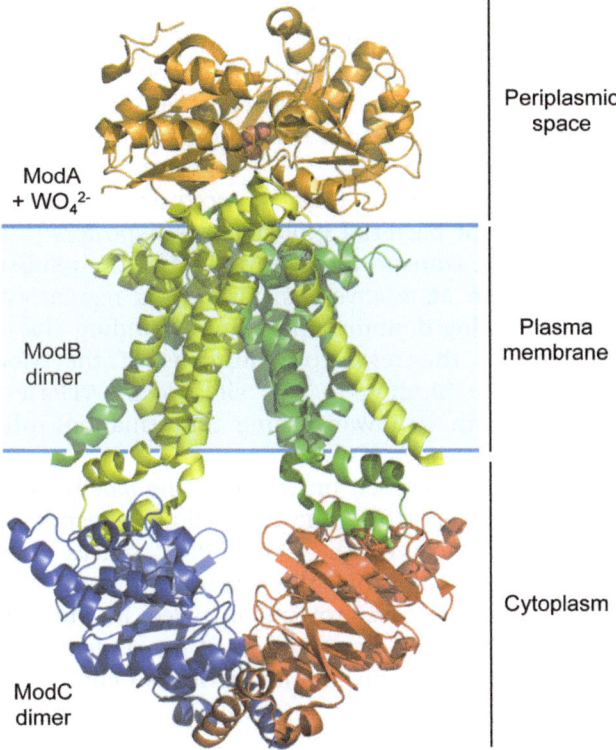

ModA + WO$_4^{2-}$

ModB dimer

ModC dimer

Periplasmic space

Plasma membrane

Cytoplasm

Figure 8.3 Structure of the ModABC transport system from *Archaeoglobus fulgidus*.[81] The ribbon representation of the ModAB$_2$C$_2$ complex shows the molybdate/tungstate binding protein ModA in orange (bound tungstate in red), the channel components ModB$_2$ in yellow and green and the ModC subunits in red and blue. The structure was generated with PyMol 1.5 using the deposited coordinates (Protein data bank 2ONK).

unknown and *modF* mutations have no effect on Mo metabolism.[62] Under standard conditions, transcription levels of *modABCD* are very low; however, a significant increase in mRNA levels is detectable in *modA*, *modB* or *modC* mutants. This activation can be reversed by adding molybdate to the growth medium.[62] Nevertheless, in a *modE* mutant, *modABC* transcription is activated even in the presence of high levels of molybdate.[61]

The crystal structure of ModE shows that each monomer consists of two major domains.[76] The N-terminal domain displays a helix-turn-helix motif that interacts with DNA and drives ModE dimerization. The C-terminal domain contains the molybdate-binding site.

Intact ModE is a homo-dimer that binds two equivalents of molybdate with high affinity ($K_D = 0.8$ μM).[61] This binding involves hydrogen-bonding interactions of the molybdate oxygen with the residues Thr163 and Ser166 from one chain and Ser126, Arg128, Lys183 and Ala184 from the other chain.[77] This complex binds to the operator of *modABCD* ($K_D = 0.3$ nM) preventing its transcription (Figure 8.2). Molybdate-free ModE binds with lower, but still substantial, affinity ($K_D = 8$ nM).[62] These differences appear to arise from the conformational changes induced by molybdate binding. They include loss of the asymmetry between the two chains of the ModE dimer and modification of the relative orientation of the two helix-turn-helix motifs that interact with DNA.[77]

A crystal structure of ModBC from *M. acetivorans* sheds light on the regulation of turnover of bacterial molybdate transporters.[72] The turnover rate of some membrane transporters is inhibited by their substrates (trans-inhibition). ModC from *M. acetivorans* harbours a regulatory domain attached to the ATP-binding domain. An oxyanion binding site is formed *via* close contact between the regulatory domains of the ModC dimer.[72] Molybdate or tungstate binding to this site blocks ATPase activity and freezes the transporter in an inward-facing conformation, inhibiting oxyanion uptake.

A regulatory RNA is transcribed upstream of the *mod*ABCD operon. This is highly conserved and controls gene expression in response to the level of Moco (riboswitch, Moco-RNA).[78] Moco-RNA senses specifically the presence of Moco and controls the transcription of a key operon (*moaABCDE*) encoding for four proteins involved in Moco biosynthesis. Activation of Moco-RNA has been observed in a *modC* mutant background, which suggests that this motif might also control the transcription of the *mod* operon.[78]

8.4   Molybdate Transport in Eukaryotes

8.4.1   Physiological Aspects

In contrast to bacteria, molybdate transport in eukaryotes is still poorly understood as the first eukaryotic molybdate transporters were identified only recently.[79,80] In different organisms, molybdate transporters that are either homologous or unrelated have been described (Table 8.1).

Genome-wide sequence analysis has concluded that homologues of bacterial ABC-type molybdate transporters are absent in eukaryotes. Consequently, it was assumed that Mo uptake involves non-specific molybdate transport systems that might be shared with other oxyanions such as phosphate or sulfate. This view is supported by known interferences between molybdate, sulfate and phosphate transport in plant cells.[81] Thus, in tomato plants, molybdate uptake is promoted by the presence of phosphate and inhibited by sulfate.[82] However, Mo transport increases in tomato plants when phosphate is withdrawn from the medium.[83] Nevertheless, molybdate uptake is not affected when phosphate is present at physiological concentrations in the growth medium,[83] suggesting that plant phosphate transport systems are able to significantly transport molybdate only under conditions of low phosphate availability.

Sulfate transporters have also been related to molybdate transport in plants and fungi. In *Glycine max*, molybdate transport is inhibited by sulfate in the medium.[84] In addition, sulfate and molybdate co-transport has been observed in filamentous fungi.[85] However, molybdate transport has also been linked to other ions. In rice, molybdate uptake is increased in the presence of Fe^{2+}.[86] Not only in plants and fungi but also in animal cells, molybdate and the transport of other elements such as selenium are linked. In a keratinocyte model, molybdate inhibits selenide uptake in a concentration-dependent manner, suggesting that both ions are taken up by the same transport system.[87]

The first physiological data regarding the presence of eukaryotic high-affinity molybdate transporters were reported in the unicellular green algae *C. reinhardtii* and suggested at least two molybdate transport activities.[88] One of these showed high affinity but low capacity for molybdate, inhibition by sulfate and clear discrimination between molybdate and tungstate. In contrast, the other activity exhibited low affinity but high capacity and was inhibited by tungstate but not by sulfate. The high-capacity transport activity seems to be linked to increased Moco levels under conditions of high molybdate concentrations in the external medium.[88] Both transport activities were related to genetic *loci Nit5* and *Nit6* in *C. reinhardtii*.[88] However, the precise roles of these *loci* in molybdate transport remain unclear.

8.4.2 Transporters of the MOT1 Family

The transporter MOT1 (MOlybdate Transporter, type 1) was identified simultaneously in *C. reinhardtii* and *Arabidopsis*.[79,80] It exhibits both a high specificity and a high affinity for molybdate. The *C. reinhardtii* version (*Cr*MOT1) was identified on the basis of the previously described functional relationship between molybdate and sulfate transport in plants and bacteria.[63,84] MOT1 proteins belong to the Major Facilitator Superfamily (MFS) and are distantly related to the plant sulfate transporter family (SULTR). For that reason they were previously classified as sulfate transporters.[89] However, *Cr*MOT1 shows only 13% sequence identity with other *C. reinhardtii* SULTR proteins. Consequently, MOT1 proteins are significantly different from SULTR members, including the lack of the STAS domain (Sulfate Transporter and Anti-Sigma antagonist) that is crucial for sulfate transport as well as for the stability and maturation of the constituent proteins.[90] Given that no function in sulfate transport had been demonstrated for MOT1 proteins so far, their sequence-based inclusion in the SULTR family needs to be revised.

*Cr*MOT1 shows a high sequence similarity with other proteins from algae, plants, fungi and bacteria, but no homologues have been identified in animals so far (Figure 8.4). MOT1-homologous proteins contain two sequence motifs that define a new family of transport proteins (MOT1 family), likely to be related to molybdate transport (Figure 8.5).[79]

The functional importance of *Cr*MOT1 in molybdate transport in *C. reinhardtii* was demonstrated using an antisense RNA approach. *Cr*MoT1 expression was knocked down in a wild-type strain and in a *Nit5* mutant strain (21gr). The latter was previously characterized as partially deficient in

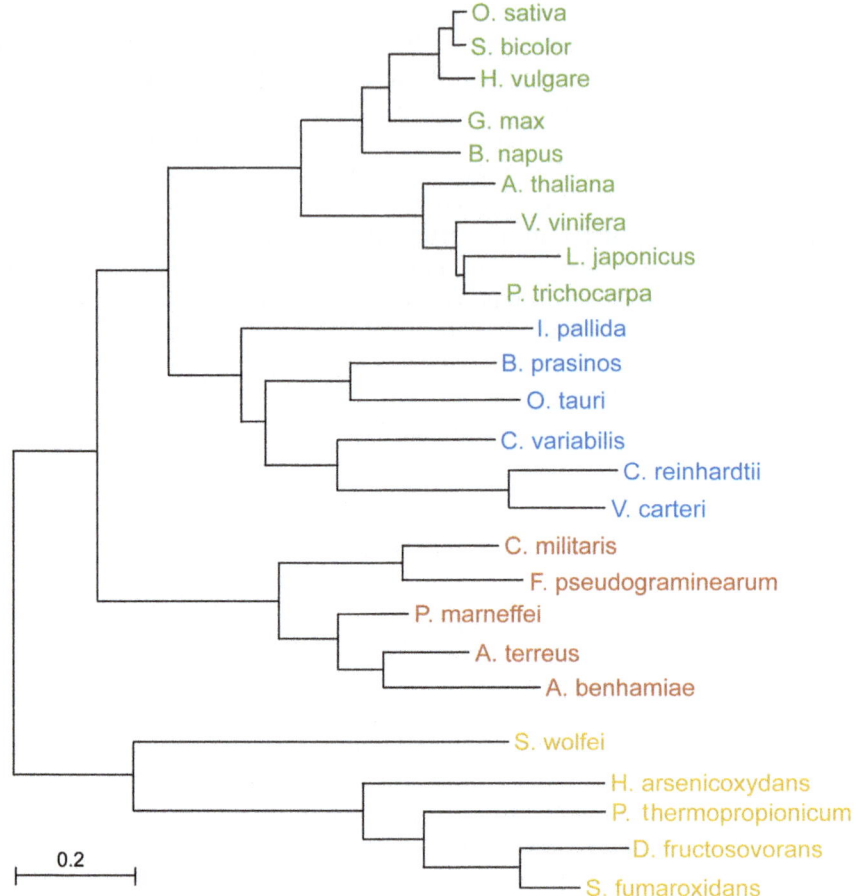

Figure 8.4   Phylogenetic tree of proteins homologous to MOT1. The tree was generated with the Mega 5 software package (http://www.megasoftware.net/). Colours represent different phyla: plants in green, algae in blue, yeasts in red and bacteria in yellow. 0.2, substitutions per site. Accession numbers: *O. sativa*, NP_001043703.1; *S. bicolor*, XP_002458231.1; *H. vulgare*, BAJ94535.1; *G. max*, XP_003531693.1; *B. napus*, CAC39421.1; *A. thaliana*, NP_180139.1; *V. vinifera*, XP_002281989.1; *L. japonicus*, AFK43331.1; *P. trichocarpa*, XP_002308631.1; *I. pallida*, YP_004180 804.1; *B. prasinos*, CCO14514.1; *O. tauri*, XP_003081530.1; *C. variabilis*, EFN52559.1; *C. reinhardtii*, A6YCJ2.1; *V. carteri*, XP_002951222.1; *C. militaris*, EGX88390.1; *F. pseudograminearum*, EKJ75971.1; *P. marneffei*, XP_002151861.1; *A. terreus*, XP_001214924.1; *A. benhamiae*, XP_0030 11208.1; *S. wolfei*, YP_753064.1; *H. arsenicoxydans*, YP_001098522.1; *P. thermopropionicum*, YP_001210746.1; *D. fructosovorans*, ZP_07333518.1 and *S. fumaroxidans*, YP_844407.1.

molybdate uptake.[88] Only in the 21gr strain a proportional decrease in molybdate transport activity was observed as a function of *CrMoT1* knockdown, suggesting that it expresses *Cr*MOT1 as the sole molybdate

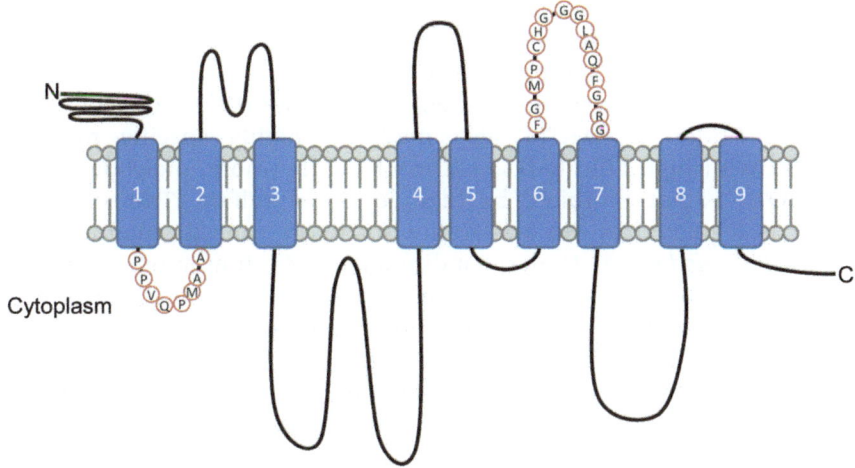

Figure 8.5 Schematic representation of the predicted transmembrane topology of *Cr*MOT1. Highlighted residues indicate the position of the conserved motifs found in the MOT1 family. Transmembrane topology has been predicted using the *Cr*MOT1 sequence with the HMMTOP software package (http://www.enzim.hu/hmmtop/).

transporter. Given the known partial deficiency in molybdate transport in 21gr cells, wild-type cells must express a second transporter that compensates for the loss of MOT1 function following the antisense RNA treatment. The effect of *CrMoT1* knockdown on intracellular Mo levels was mirrored by the activity of the NR Mo-enzyme in 21gr cells.[79]

Molybdate transport mediated by *Cr*MOT1 has high affinity and specificity. Uptake studies revealed a K_m of 7 nM,[79] a higher affinity than seen for *E. coli* ModABC (50 nM).[61] However, in contrast, *Cr*MOT1 is not regulated by intracellular molybdate but by the nitrogen source in the growth medium. Both *CrMoT1* transcription and *Cr*MOT1 transport activity are strongly activated by nitrate when that anion provides the sole nitrogen source. This suggests that transcriptional regulation of *Cr*MOT1 is predominantly controlled by nitrate under these conditions and points to metabolic links between molybdate uptake, Moco biosynthesis and nitrate assimilation.[79]

Arabidopsis MOT1 (*At*MOT1) was identified by a three-fold difference in Mo content between the two *Arabidopsis* lines Columbia (Col-0) and Landsberg *erecta* (L*er*). Genetic analysis showed that the responsible trait is linked to a single locus containing a gene previously annotated as *Sultr5;2* and finally named *AtMoT1*.[80]

Analysis of the genomic sequence of the *AtMoT1* gene revealed two differences between accessions Col-0 and L*er*. *Arabidopsis* T-DNA mutants are affected in *AtMoT1* and confirmed its involvement in molybdate transport. Transfer DNA (T-DNA) mutant lines and their F_1 progeny showed a reduced Mo concentration in shoots and roots (about 20% of wild-type uptake). In addition, when these mutants were cultivated in Mo-free media, the growth of roots and shoots was reduced by 35 to 80% as compared to plants grown in media with normal Mo levels. However, under those conditions, wild-type plants did not change their root or shoot development.[80] Therefore, AtMOT1 is believed to be necessary for efficient molybdate uptake from soil.

As for *Cr*MOT1, *At*MOT1 also mediates highly specific molybdate transport with high affinity ($K_m = 20$ nM). Furthermore, molybdate uptake by *At*MOT1 is not inhibited by high levels of sulfate. Transcription of *AtMoT1* is regulated by Mo availability but in an opposite manner to that of bacterial *modABCD*. Under conditions of Mo limitation, *AtMoT1* expression in shoots was found to be 50% of that in plants grown with adequate molybdate supply. Furthermore, roots of plants cultured under Mo starvation also had reduced *AtMoT1* expression.[80] *AtMOT1* expression is not tissue-specific as it occurs throughout the entire plant body, suggesting that the protein is of general importance for many cells and cell types in the whole plant.

A number of fusion proteins carrying the green fluorescent protein (GFP) have been expressed in plant cells to identify subcellular localization of *At*MOT1. However, an N-terminal GFP fusion to *At*MOT1 resulted in plasma membrane localization while a C-terminal GFP fusion showed mitochondrial localization.[80,91] The presence of a predicted mitochondrial target sequence within the N-terminus of *At*MOT1 supports the latter finding.[91] Since molybdate is needed in the cytosol, the plasma membrane would be an adequate compartment for localization of MOT1. Furthermore, molybdate needs to be taken up from the soil and therefore, at some point, it has to pass the plasma membrane barrier. However, given that both experimental and *in silico* data support localization of *At*MOT1 to mitochondria, additional studies are required to elucidate its function there.

A second member of the MOT1 family, originally named MOT2, occurs in *Arabidopsis*. However, since this transporter belongs to the MOT1 family and in order to avoid confusion with members of a recently described distinct molybdate transporter family, we recommend that this transporter be renamed *At*MOT1;2 (*Arabidopsis thaliana* molybdate transporter type 1; member 2).

*At*MOT1;2 was identified on the basis of its sequence similarity to *At*MOT1 (51% identity).[92] Its role in molybdate transport/accumulation was investigated by analysing a T-DNA tagged *At*MOT1;2 mutant (*mot1;2*) that shows altered Mo homeostasis in different plant tissues. While the Mo content in seeds was higher in wild-type than in the *mot1;2* mutant, wild-type leaves accumulated less Mo than *mot1;2* mutant leaves. Furthermore, withered leaves from the *mot1;2* mutant contained 15-fold more molybdate than wild-type leaves. These findings suggest that *At*MOT1;2 is also involved in molybdate transport with a specific role in its interorgan allocation.[92] NR, SO and AO activities, as well as molybdenum cofactor content, were substantially higher in *mot1;2* mutants than in wild-type plants under molybdate-limiting conditions.[92]

*At*MOT1;2-GFP fusions localized in the vacuolar membrane of epidermal onion cells and *Arabidopsis* protoplasts. Vacuolar targeting of *At*MOT1;2 is believed to be promoted by an N-terminal di-leucine motif.[92] This observation points to a role for plant vacuoles in Mo storage and homeostasis. Despite the evidence for involvement of *At*MOT1;2 in Mo uptake/accumulation, its molybdate transport activity has not yet been investigated.

8.4.3 Transporters of the MOT2 Family

In contrast to *Arabidopsis* or *Oriza sativa*, which express two and three MOT1 proteins, respectively, the *C. reinhardtii* genome contains only one copy of the *CrMOT1* gene.[79] However, this algae presents at least two different molybdate transport activities, with fundamentally different properties, suggesting a second family of molybdate transporters not related to the MOT1 family.[88,79] Consequently, *Cr*MOT2 was identified as a second transporter in *C. reinhardtii*.[93] *Cr*MOT2 showed molybdate transport comparable to *Cr*MOT1 when expressed in *Saccharomyces*. Furthermore, as seen for the *CrMoT1* knockdown, silencing *CrMoT2* transcription did not affect molybdate uptake in a wild-type strain. However, under conditions of low *Cr*MOT1 activity, knockdown of *Cr*MOT2 severely impaired molybdate transport in wild-type cells too.[93] Importantly, the *CrMoT2* gene is not affected in the *Nit5* mutants (Tejada-Jimenez *et al.* unpublished results), in which *Cr*MOT2-dependent molybdate transport should be affected due to similar uptake characteristics. It seems either that a third gene is involved in molybdate transport or that a yet unknown regulatory factor is impaired in those mutants.

While molybdate transport mediated by *Cr*MOT2 is specific, the estimated K_m value of 550 nM suggests a significantly lower affinity than for the MOT1 transporters for *C. reinhardtii* and *Arabidopsis*.[93] In addition, the presence of a tungstate concentration of 20 µM competitively inhibits *Cr*MOT2 transport activity, while a sulfate concentration of 1 mM had no effect.[93] This result is comparable to the inhibition pattern reported for MOT1 proteins.[79,80]

*Cr*MOT2 regulation depends on Mo availability since *CrMoT2* transcription is up-regulated by Mo starvation.[93] This response contrasts with *Cr*MOT1 regulation and points to different roles for these transporters in Mo homeostasis in *C. reinhardtii*. According to the regulation profiles, *Cr*MOT1 seems to be connected to NR activity whilst *Cr*MOT2 may be essential for internal Mo homeostasis.

*Cr*MOT2 belongs to the ubiquitous MFS superfamily but shows a very low sequence identity with MOT1 family members (11–14%) and contains an additional domain (DUF791) of unknown function. Proteins with a significant identity with *Cr*MOT2 (26–51%) are found in algae, plants and animals (Figure 8.6). They each share four conserved motifs that are located mainly in predicted transmembrane domains (Figure 8.7).

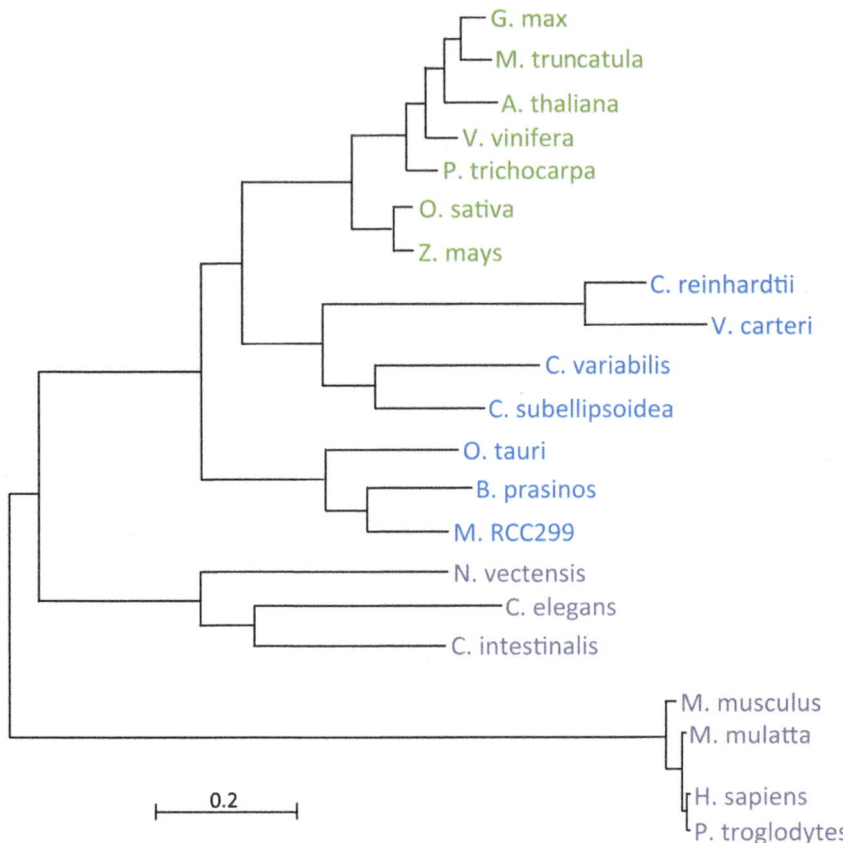

Figure 8.6 Phylogenetic tree of proteins homologous to MOT2. The tree was generated with the Mega 5 software package (http://www.megasoftware.net/). Colours represent different organisms: plants in green, algae in blue and animals in purple. 0.2, substitutions per site. Accession numbers: *G. max*, XP_003526731.1; *M. truncatula*, XP_003602305.1; *A. thaliana*, NP_567786.1; *V. vinifera*, XP_002280860.1; *P. trichocarpa*, XP_002318449.1; *O. sativa*, NP_001048746.1; *Z. mays*, DAA42909.1; *C. reinhardtii*, AEY68285.1; *V. carteri*, XP_002948878.1; *C. variabilis*, EFN51339.1; *C. subellipsoidea*, EIE22553.1; *O. tauri*, XP_003078107.1; *B. prasinos*, CCO66472.1; *M. RCC299*; XP_002506106.1; *N. vectensis*, XP_001629395.1; *C. elegans*, NP_500274.2; *C. intestinalis*, XP_00212 9829.1; *M. musculus*, NP_598861.1; *M. mulatta*, NP_001247568.1; *H. sapiens*, NP_116278.3 and *P. troglodytes*, XP_522401.3.

Saccharomyces yeasts expressing the human member of the MOT2 family (HsMOT2) show molybdate uptake levels comparable to cells expressing *Cr*MOT1 or *Cr*MOT2. Kinetics and specificity of the transport process mediated by HsMOT2 are similar to those of *Cr*MOT2 ($K_m = 546$ nM and competitive inhibition by tungstate).[93] Therefore, HsMOT2 represents the first animal protein able to facilitate molybdate transport.

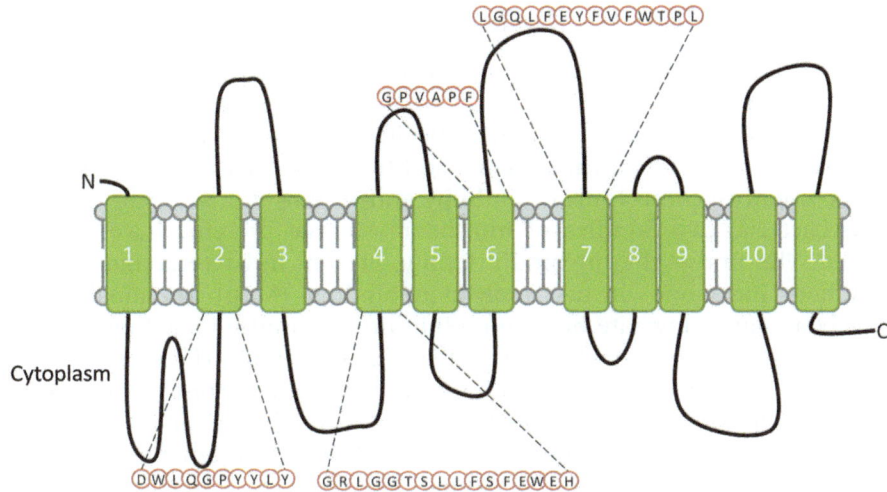

Figure 8.7 Schematic representation of the predicted transmembrane topology of *Cr*MOT2. Highlighted residues indicate the position of the conserved motifs found in the MOT2 family. Transmembrane topology has been predicted by using the CrMOT2 sequence with the HMMTOP software package (http://www.enzim.hu/hmmtop/).

8.4.4 Other Eukaryotic Molybdate Transporters

In the plant *Stylosanthes hamata*, the high-affinity sulfate transporter SHST1 promotes low-affinity molybdate uptake.[94] Following expression in a *Saccharomyces* strain, SHST1 enhances molybdate uptake at nanomolar concentrations. Molybdate transport *via* SHST1 is not inhibited by sulfate whereas sulfate transport mediated by SHST1 is affected by external molybdate. These results suggest that SHST1 transports both sulfate and molybdate at physiological concentrations, thus establishing a direct relationship between sulfate and molybdate transport in this organism.

A new molybdate transport-deficient mutant (DB6) has been identified in *C. reinhardtii*.[95] It shows impaired growth on nitrate or hypoxanthine as sole nitrogen sources due to a deficiency in the activity of the Mo-enzymes NR and XDH, respectively. This decreased growth was restored partially by adding molybdate (10 mM) to the medium suggesting that the mutant is deficient in molybdate uptake. This conclusion was supported by the fact that DB6 exhibited a reduced intracellular Mo concentration and a resistance to tungstate at a concentration of 20 mM, a phenotype that is typical of *C. reinhardtii* strains affected by reduced levels of molybdate transport (such as *21gr* mutants).[88] Given that DB6 mutants cannot be complemented by a wild-type copy of *CrMoT1*, it is possible that the gene affected in DB6 could be either *CrMoT2* or a yet-to-be-identified molybdate transporter gene or an unknown molybdate uptake regulatory gene.

8.5 Molybdenum Storage

Cellular storage systems maintain intracellular levels of essential elements such as transition metals. Proteins are often used in metal storage since they can bind and release metals under specific conditions *via* protein-protein interactions or cellular changes. In addition, proteins provide a suitable environment to control metal solubility and reactivity.

In bacteria, two families of molybdate storage proteins have been reported. The first is the molbindin family, present in bacteria and archaea. Members have one or two molybdate-binding (Mop) domains.[96] Both molybdate and tungstate bind to molbindins *via* multiple hydrogen bonds derived from protein main-chain and side-chain atoms. The second group, denoted Mo/WSto, binds both anions *via* polyoxometalate clusters.[97] Crystallization of Mo/Wsto to form *Azotobacter vinelandii* revealed a hexameric structure formed by a trimer of hetero-dimers. Such a hexamer appears to be able to store up to 100 Mo or W atoms.

In eukaryotes, Mo storage systems seem to be very different from those in bacteria since genes homologous to molbindin or Mo/WSto are yet to be found. On the other hand, proteins able to store Mo in the form of its pterin-cofactor have been described in algae and higher plants.[98,99] The Moco carrier protein (MCP) from *C. reinhardtii* (16 kDa) forms a tetramer that binds up to four Moco molecules. MCP discriminates between Moco and the Mo-free molybdopterin (MPT), suggesting specific binding and recognition by the protein of the pterin-bound molybdenum centre.[100]

Binding of Moco to MCP goes hand in hand with protection of the cofactor against oxidation and its facilitated transfer to *apo*-NR.[98] While free Moco shows a half-life of a few minutes *in vitro*, this is extended to about two days for Moco-MCP.[98,100] Transfer of Moco to apo-NR involves direct protein-protein interactions as separation of the proteins *via* a dialysis membrane strongly inhibits activation of *apo*-NR by Moco-MCP.[101] Photosynthetic organisms such *C. reinhardtii* produce a large amount of dioxygen and therefore a Moco-protective environment may be favourable for maintenance of an active pool of Moco within the cytosol.

The crystal structure of MCP revealed a typical Rossman fold with a central parallel β-sheet surrounded by five α-helices.[100] Analysis of its conserved surface residues and charge distribution together with *in silico* docking studies predicted a putative Moco-binding site. Mutations in two conserved residues located in this predicted binding pocket result in a lower half-life for Moco due to impairment of Moco binding and/or protection.[100]

MCP shows similarities with proteins that have sequence and structural homology to lysine decarboxylase-like proteins.[98,100] A family of these proteins from *Arabidopsis thaliana* has been identified on the basis of its structural similarity to *C. reinhardtii* MCP. It consists of nine members that have been described as Moco-binding proteins and named MoBP1-9.[99] In contrast to *C. reinhardtii* MCP, MoBPs do not co-purify with Moco. Nevertheless, four of them (MoBP1, MoBP3, MoBP4 and MoBP5) show moderate affinities for Moco with K_D values in the micromolar range. The lack of co-purification suggests a role apart from Moco storage. Future studies are needed to clarify the physiological role and importance of MBP proteins in plants.

In addition to the Mo storage systems described above, other processes not involving proteins have been reported in plants. Molybdate accumulation in *Brassica* ssp. seedlings seems to be related to anthocyanin levels. Plants with normal or high levels of anthocyanins accumulated molybdate while anthocyanin-deficient plants did not.[102] Furthermore, malic acid is involved in Mo storage in alfalfa (*Medicago sativa* L.). Alfalfa plants grown in a Mo-rich soil contain high levels of intracellular Mo chelated by two malate ligands.[103]

8.6 Molybdenum Cofactor Biosynthesis

Genetic studies in fungi, bacteria and plants suggested the existence of a common Mo-dependent cofactor that is formed by a multistep biosynthetic pathway.[104] Later, following the identification of the chemical structure of Moco as well as its biosynthetic intermediates, it became clear that this pathway is highly conserved in all kingdoms of life.[12,26] Characterization of Moco-deficient mutants contributed significantly to the understanding of the genetics and biochemistry of Moco biosynthesis.[105,106]

At least six gene products are involved in the biosynthesis of Moco in bacteria, fungi, plants and humans.[13,121] In *E. coli*, 16 gene products are encoded by five operons (*moa*, *mob*, *mod*, *moe* and *mog*), of which 11 are strictly required to produce active cofactor. In contrast, only four and six genes, respectively, are required for the biogenesis of Moco in humans and plants.[107] The nomenclature used for Moco biosynthetic genes and proteins differs among the organisms. Human genes and gene products follow the MOCS nomenclature (<u>Mo</u> <u>C</u>ofactor <u>S</u>ynthesis) while in plants and fungi the "old" genetically introduced CNX (<u>C</u>ofactor for <u>N</u>R and <u>X</u>DH) nomenclature has been maintained.[108]

Moco biosynthesis can be divided into three and four major steps in eukaryotes and bacteria, respectively, according to the intermediates cyclic pyranopterin monophosphate (cPMP), metal-binding pterin or MPT and Moco. Each of the steps involves the action of one or more proteins producing additional reaction intermediates (pyranopterin triphosphate,[109] thio-pyranopterin phosphate,[110] MPT-AMP) (Figure 8.8).[111]

8.6.1 Pterin Biosynthesis

Similar to the biosynthesis of other pterins and flavins, Moco synthesis starts with GTP.[112] Labelling studies in *E. coli* proposed a complex and unique rearrangement mechanism in which the C8 atom of the purine base is removed as a formyl group and subsequently inserted between the 2′- and 3′-ribose carbon atoms resulting in the formation of the four-carbon pyran

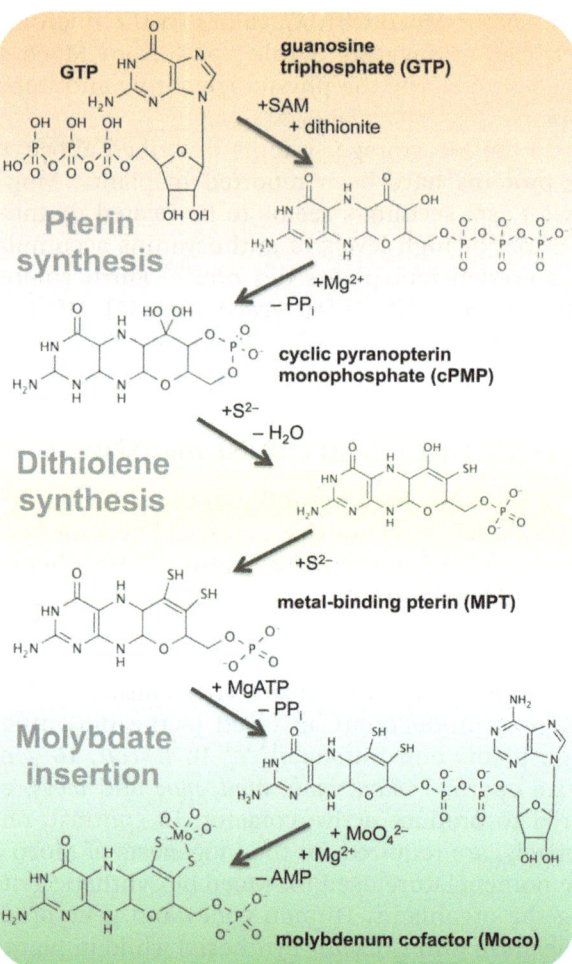

Figure 8.8 Chemical structures of molecules involved in molybdenum cofactor biosynthesis. Major intermediates are shown on the left side of the pathway and define the three stages of the biosynthesis. Reaction intermediates for each of the three steps are shown on the right side. Structures were generated with the MarvinSketch software (ChemAxon).

ring[110,113] present in both cPMP and Moco. This reaction leads to the formation of the pterin backbone of the cofactor. cPMP is the most stable biosynthetic intermediate with a half-life of several hours, depending on the environment.[114] Its structure was resolved by high-resolution mass spectrometry and ^{1}H NMR spectroscopy first and then by ^{13}C NMR, confirming the presence of a pyranopterin featuring an unusual geminal diol.[114,115]

In bacteria and plants, the respective protein pairs MoaA/MoaC and Cnx2/Cnx3 are involved in the synthesis of cPMP.[110,116] In humans, the *MOCS1* gene encodes for different splice variants producing a MOCS1A protein and a multi-domain MOCS1AB protein. These translational products cooperate in cPMP synthesis.[108] MoaA and homologous proteins (MOCS1A/Cnx2) are members of the family of *S*-adenosylmethionine (SAM)-dependent radical enzymes containing one or two [4Fe-4S] clusters.[112,117,†] The N-terminal [4Fe-4S] cluster in MoaA promotes SAM cleavage to generate a 5′-deoxyadenosyl radical. This initiates the transformation of 5′-GTP bound to the C-terminal [4Fe-4S] cluster by abstracting the 3′ proton from the ribose moiety resulting in the formation of pyranopterin triphosphate (Figure 8.8).[109] This mechanism is believed to be conserved given that plant Cnx2 and human MOCS1A are able to complement *E. coli moa*A mutants.[13] The other protein (MoaC) or domain (MOCS1B) is involved in pyrophosphate release and formation of the geminal diol.[13]

The plant proteins Cnx2 and Cnx3 localize to the mitochondrial matrix (Figure 8.9),[118] which is consistent with the respective predicted targeting signals in both pro-proteins.[116] Note that the biogenesis of Fe-S clusters within mitochondria and the high abundance of GTP might have been the driving force, thus resulting in the subcellular localization of these proteins in eukaryotes.[119,†] Consequently, the human proteins (MOCS1A and MOCS1AB) should also localize to mitochondria, given that a putative targeting signal can be found in exon 1. Because the following steps in Moco biosynthesis are localized in the cytosol (Figure 8.9), cPMP needs to cross the mitochondrial inner membrane. In *Arabidopsis*, ATM3 (<u>A</u>BC <u>T</u>ransporter of the <u>M</u>itochondria) influences the export of cPMP from the mitochondrion to the cytosol.[118] Members of the ATM family are also present in humans (ABCB6 and ABCB7) but their role in Moco biosynthesis is still unknown.[120] Given that this transporter is also important for cytosolic Fe-S cluster biogenesis,[121] future work is needed to clarify the connection between these pathways.

8.6.2 Dithiolene Synthesis

In the second step of Moco biosynthesis, two sulfur atoms are incorporated into cPMP to form the dithiolene function of MPT. The reaction is catalysed by MPT synthase, a hetero-tetrameric complex containing two small (MoaD/Cnx7/MOCS2A) and two large (MoaE/Cnx6/MOCS2B) subunits that stoichiometrically convert cPMP into MPT (Figure 8.9). The reaction mechanism involves stepwise transfer of sulfur *via* a mono-thiol intermediate (Figure 8.8) and hydrolysis of the cyclic phosphate (thio-pyranopterin phosphate).[13] Human MPT synthase is encoded by a bicistronic gene (*MOCS2*) that produces both protein subunits (MOCS2A and MOCS2B) in an approximately equimolar ratio.[122]

---

†See Chapter 12 for further information on SAM-dependent enzymes and on biogenesis of Fe-S centres.

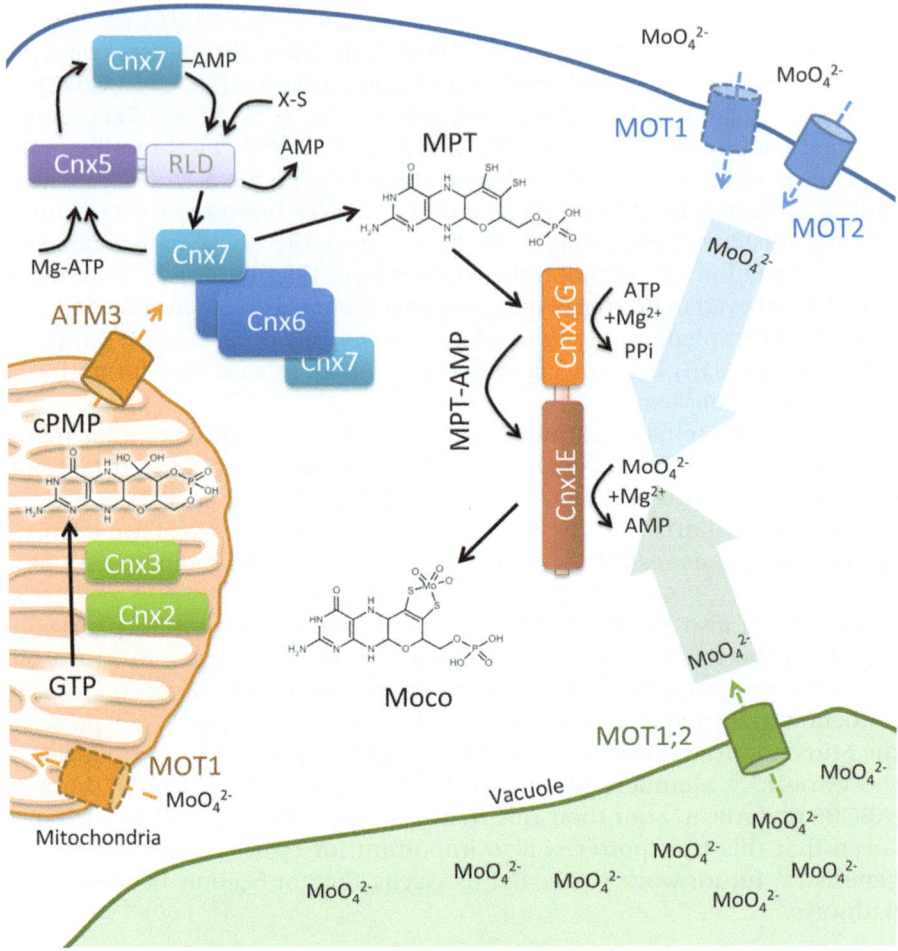

Figure 8.9   Subcellular organization of eukaryotic molybdenum cofactor biosynthesis. The basic steps are depicted for a plant cell. Firstly, GTP is
converted to cPMP in mitochondria by Cnx2 and Cnx3. The mitochondrial transporter ATM3 is involved in the export of cPMP to the cytosol
where it is converted to MPT by MPT-synthase (composed of two large
(Cnx6) and two small subunits (Cnx7)). Cnx5 is responsible for the re-
sulfuration of the MPT-synthase subunit Cnx7, a reaction that is ATP-
dependent and involves a cysteine desulfuration reaction. Given that the
final substrate for Cnx7 sulfuration is unknown, X-S has been used to
depict a putative substrate. MPT is adenylated by the G-domain of Cnx1
(Cnx1G) and MPT-AMP is subsequently transferred to the E-domain
(Cnx1E). Finally, active Moco is formed by hydrolysis of MPT-AMP and
insertion of molybdate by Cnx1E. Molybdenum is supplied by transport
proteins belonging to the MOT1 and MOT2 families. Subcellular localization identified MOT1 either in the plasma membrane or in mitochondrial membranes while MOT1.2 is localized in the vacuolar membrane
and functions as an exporter. Vacuoles have been proposed as the main
cellular compartment for molybdate accumulation in plants.

Each small subunit of MPT synthase carries a sulfur atom as thiocarbox-ylate.[123] The atom is transferred by an ATP-dependent reaction involving an adenyltransferase (MoeB/Cnx5/MOCS3). In *E. coli*, separate cysteine desul-furases provide sulfide for the sulfuration reaction following the hydrolysis of the protein-adenylate. In eukaryotes, by contrast, a rhodanese-like C-terminal domain (Cnx5/MOCS3) binds sulfur as persulfide at a conserved cysteine, which is believed to mediate the sulfuration of the small sub-unit.[124] Rhodanese enzymes are present in many organisms and catalyse the *in vitro* transfer of sulfur from thiosulfate to cyanide. However, physiological substrates of rhodaneses are still unknown. In humans, L-cysteine desul-furases have been proposed to directly donate sulfur to cytosolic MOSC3.[125]

8.6.3 Molybdate Insertion

Once MPT is formed, it is capable of binding molydenum to the dithiolene group. However, prior to this step, the MPT is activated by adenylylation (Figure 8.8). In bacteria, the MogA protein catalyses MPT adenylylation,[126] while another protein (MoeA) promotes Mo insertion.[127] In contrast, eu-karyotic organisms mostly use two-domain proteins (Cnx1 in plants/ gephyrin in mammals; Figure 8.9) in which both activities have been fused yielding multi-domain and, in some cases, multi-functional proteins.[128] The MogA-homologous G-domain binds MPT with high affinity and forms ade-nylated MPT.[111,129] Furthermore, in the structure of the Cnx1 G-domain complexed with either MPT or MPT-AMP, a copper atom was found bound at the dithiolene.[111] This copper may protect the dithiolene unit and/or might be involved in facilitating Mo insertion by providing a suitable leaving group.

Following its synthesis by the G-domain, MPT-AMP is transferred to the E-domain where it is subsequently hydrolysed in a reaction that is dependent upon molybdate plus Zn^{2+} or Mg^{2+} (Figure 8.9).[130] Tungstate and sulfate compete with molybdate in the E-domain, confirming the existence of a molybdate-binding site. Similar competitive behaviour has been ob-served in different molybdate transporters.[61,79,93] Molybdate and MPT-AMP bind to the E-domain in an equimolar and cooperative manner and initiate AMP hydrolysis in the presence of Mg^{2+} or Zn^{2+}, resulting in stoichiometric release of Moco. Adenylated molybdate has been proposed as a reaction intermediate in analogy to the adenylated sulfate in sulfate assimilation.[130] The effects of mutations in the *moeA* gene (corresponding to the E-domain) can be reversed under conditions of low sulfur supply.[131] Consequently, an "off-target" reaction can be postulated involving ATP sulfurase, which is induced under such conditions and is known to accept molybdate as substrate.[132]

Based on the domain fusion in Cnx1 and gephyrin, a product/substrate channelling has been assumed. The generation of a fully defined *in vitro* system for Moco biosynthesis has now made possible a direct comparison of the reaction rates of the individual gephyrin domains with holo-gephyrin.[133] As expected, a 300-fold increase in ATP-dependent Moco synthetic activity was found only for the fusion protein. An estimate of K_m for molybdate was close to the intracellular concentration of molybdate, a property observed for the molybdate transporters discussed above.[133]

As a side effect of the orientation of the fused domains in gephyrin, novel functions have evolved, such as gephyrin's essential role in synaptogenesis. In the central nervous system, in addition to Moco synthesis, gephyrin functions as a scaffolding protein at inhibitory synapses where it binds to glycine and γ-amino butyric acid type A receptors.[134,135]

8.6.4 Maturation of Moco

Following the release of Moco from the biosynthetic machinery, the cofactor either binds directly to enzymes of the SO family or it undergoes additional modification before binding to enzymes of the XO family (Figure 8.1A). In the latter enzymes, one of the three oxido ligands of Moco (Figure 8.8) is replaced by a terminal sulfido ligand. In contrast, in enzymes of the SO family, the third sulfur ligand (thiolate) is provided by a cysteine side-chain.

The sulfuration of Moco is catalysed by a Moco sulfurase, different versions of which have been reported in fruit fly (Ma-1), fungi (HxB), cattle (MCSU), plants (ABA3) and humans (MCSU).[13] *In vitro* studies demonstrated that ABA3 is able to sulfurate Moco in holo-enzymes of the XO family.[136] However, under *in vivo* conditions, it remains unclear whether ABA3 transfers sulfur to Moco before or after cofactor insertion into apo-enzymes. ABA3 and homologous proteins are two-domain proteins with an N-terminal cysteine desulfurase domain and a <u>Mo</u>co <u>s</u>ulfurase <u>C</u>-terminal domain (MOSC) that binds Moco and is believed to interact with the apo-enzymes.[136] Following the desulfuration of L-cysteine, a protein-bound persulfide is formed on a conserved cysteine and subsequently transferred to bound Moco.[136,137]

Activation of enzymes of the XO family by the Moco sulfurase is believed to be a regulatory mechanism. This is supported by the finding that *aba3* transcription is controlled by salt stress and ABA levels in plants.[136,138]

Moco also undergoes additional modification prior to insertion into bacterial enzymes. Enzymes of the DMSO reductase family contain a cofactor with an attached guanine resulting in a bis-MPT guanine dinucleotide cofactor.[139] Furthermore, carbon monoxide dehydrogenases harbour a cofactor modified by cytosine forming a mono-MPT cytosine dinucleotide cofactor with an additional copper ion forming a dinuclear metal centre within the enzyme.[140]

8.7 Biosynthesis of FeMo-co

FeMo-co is the prosthetic group in the enzyme nitrogenase. This enzyme is involved in the reduction of dinitrogen to ammonia representing the major contribution of reduced nitrogen into the nitrogen cycle.[14] FeMo-co comprises an inorganic Mo-7Fe-9S-C cluster and the organic acid homocitrate coordinated to the Mo atom *via* its C-2 carboxyl and hydroxyl groups.[‡] FeMo-co biosynthesis is a very complex process due to the many gene products involved in this pathway.[14] In the diazotrophic bacteria *Klebsiella pneumonia*, proteins participating in FeMo-co biosynthesis can be grouped into three classes according to their functionality: i) NifU, NifB and NifEN that act as molecular scaffolds involved in FeMo-co assembly; ii) NifX and NafY that participate as cofactor carrier between the different assembly sites; and iii) NifS, NifQ and NifV that function as sulfur, molybdenum and homocitrate supplier.[141] First the 6Fe-9S core of FeMo-co (so called NifB-co) is synthesized *via* the coordinated activity of NifS, NifU and NifB. NifB-co is subsequently transferred to NifEN *via* NifX. FeMo-co is finally formed by the incorporation of homocitrate and Mo, catalyzed by NifV and probably NifQ, respectively. Assembled FeMo-co is transferred to the apo-dinitrogenase *via* NafY.[141]

8.8   Molybdenum Deficiency and Toxicity

Mo-related deficiencies arise from three causes: i) low availability of molybdate in the external medium, ii) mutations in molybdate-specific transporters resulting in an impairment in the molybdate uptake process or iii) a defective Moco biosynthesis. In plants, Mo deficiency has been intensively studied and documented in many species presenting a wide range of phenotypes.[142] Human Moco deficiency (MoCD) is a rare hereditary metabolic disorder characterized by severe neurodegeneration resulting in death in early childhood (see discussion below).[13]

8.8.1   Defects in Molybdate Uptake

The most common phenotypic symptoms of Mo deficiency in plants are reduction of overall plant growth, poor development of seeds, altered morphology of leaves, distorted flower development, reduced pollen grain production and abnormal maturation of fruits.[142] Recently, an *Arabidopsis* AtMOT1 mutant has been reported, showing yellow and curled leaves as well as retarded growth under Mo-limiting conditions.[143] In humans, only one case of Mo deficiency has been reported, caused by prolonged omission of Mo supplementation during parenteral nutrition.[144]

Transcriptome analysis of AtMOT1 mutants in comparison to wild-type plants grown in the presence or absence of Mo identified genes involved in metabolism, transport, stress response and signal transduction of nitrogen and sulfur assimilation as well as in Moco biosynthesis. Under Mo-limiting conditions, a total of 187 genes were transcriptionally up-regulated while 203 genes were repressed in the AtMOT1 mutant.[143] Metabolites such as amino acids, sugars, organic acids and purines showed an altered profile in molybdate transport-deficient plants. It is remarkable that one of the most

---

[‡]See Chapter 12 for further discussion of the structure of FeMo-co and its biosynthesis.

responsive genes was NR1 showing up to 40-fold induction in the *AtMOT1* mutant in response to Mo limitation. Given that the intracellular nitrate concentration increased 1.4-fold only, Mo might be involved directly in the expression of NR.[143] The latter might be associated with the up-regulation of a MYB transcription factor under Mo starvation. MYB proteins belong to a superfamily of transcription factors that control developmental processes and defence responses in plants.[145]

8.8.2 Moco Deficiency

Mo deficiency ultimately leads to Molybdenum Cofactor Deficiency (MoCD) that severely affects all organisms that depend on Mo enzymes. MoCD can also be caused by defects in one of the steps in Moco biosynthesis. Plants with a defective Moco biosynthesis are not viable. The loss of NR is associated with increased intracellular nitrate concentration and decrease in plant growth.[146] Mutations in AO result in low levels of the phyto-hormone ABA leading to abnormal leave development and retarded growth.[81] Under physiological conditions, no phenotypic changes have been observed upon loss of SO, XDH and mARC activities;[26] however SO-deficient plants show an impaired growth in an atmosphere with high concentrations of sulfur dioxide.[147]

Human MoCD is a rare autosomal recessive disorder, which mostly affects neonates and is characterized by progressive brain injury leading to early childhood death.[13] Moco-deficient patients show feeding difficulties, treatment-resistant seizures, exaggerated startle reactions and unusual EEG traces. Abnormal brain morphology develops as a hallmark of rapidly progressing encephalopathy. Clinical symptoms are mainly triggered by the deficiency of SO activity, which results in the accumulation of toxic sulfite. When uncontrolled, this anion breaks disulfide bridges in proteins and in cystine and forms secondary metabolites that are believed to be neurotoxic. Mutations in three disease-causing genes (*MOCS1*, *MOCS2* and *GPHN*) have been identified in human patients suffering from MoCD.[106] No mutations were found in *MOCS3*, probably due to its additional function in tRNA thiolation.[148]

Two thirds of all MoCD patients carry mutations in the *MOCS1* gene and are therefore not able to synthesize cPMP (Figure 8.8).[108] A mouse model with a *MOCS1* gene knock-out was generated and presented symptoms resembling human MoCD patients. The average survival time of pups was 7.5 days.[149] Given the fact that cPMP is the most stable intermediate in Moco biosynthesis while free Moco has a half-life of minutes,[115,150] a substitution therapy using purified cPMP was established for the MOCS1-deficient mice.[151] Repeated intrahepatic injections of cPMP into *MOCS1* knock-out mice normalized the phenotype and reconstituted Moco biosynthesis as well as Mo-enzyme activities. Later, in a translational approach, a first human patient was exposed to cPMP treatment.[152] Intravenous injection of *E. coli*-derived cPMP markedly normalized all MoCD-specific biomarkers including sulfite, *S*-sulfocysteine, thiosulfate, xanthine and uric acid levels and was followed by a robust clinical improvement. Seizure frequencies were reduced and abnormal electro-encephalography patterns normalized. Today, several patients are treated with cPMP and in some of them an almost complete neurological development has been achieved (B. Schwahn, unpublished results). The recently reported chemical synthesis of cPMP will open new avenues for the development of cPMP as the first drug for treatment of MoCD.[153]

8.8.3 Molybdenum Toxicity

Severe Mo toxicity has been reported only in ruminants. Under conditions of high molybdate uptake, the reducing sulfide-rich gastrointestinal tract promotes the formation of the tetrathiomolybdate anion $[MoS_4]^{2-}$, which in turn readily reacts with Cu^I or Cu^{II} to form insoluble copper complexes.[154] Therefore, high Mo intake causes a secondary copper deficiency that is called molybdenosis or hypocuprosis.[155] Mo toxicity in ruminants is characterized by severe diarrhoea, anorexia, greying of hair, anaemia and leg-stiffness accompanied with infertility or sterility. These symptoms are readily reversed by copper supplementation. The ability of thiomolybdates ($[MoO_{4-x}S_x]^{2-}$, $x = 0\text{--}2$) to interact with copper has been used in the treatment of copper-dependent disorders such as Wilson's disease that result in hepatic and neurological dysfunctions due to intracellular copper accumulation.[156] Thiomolybdates are also used to inhibit metastatic cancer progression through anti-angiogenic copper chelation and inhibition of the nuclear factor κB.[154]

Monogastric animals are less sensitive to Mo toxicity. In humans, cases of Mo toxicity are extremely rare and confined to geographic areas with high amounts of Mo in drinking water or soils.[157] In some regions of Armenia, the population is exposed to high dietary Mo intake (10–15 mg d^{-1} as compared to 1–2 mg d^{-1} under normal conditions). In those individuals aching joints and symptoms resembling gout have been reported, probably due to increased XDH/XO activity. The optimum intake for Mo has been estimated to be 9 µg Mo kg^{-1} d^{-1}.[158]

8.9 Future Questions

Our knowledge in many aspects of Mo metabolism, Moco biosynthesis and Mo homeostasis has developed due to significant progress in recent years. Nevertheless, the findings raise new questions that emphasize the complex picture of Mo biology. While the basic principles of the biosynthetic pathway of Moco are understood, key remaining questions are the reaction mechanism of cPMP synthesis (Figure 8.8), the mitochondrial maturation of the human MOCS1 proteins and the translocation of cPMP from mitochondria to the cytosol (Figure 8.9). Another open question is the association of MPT with copper, a finding that can be interpreted as either non-physiological co-purification or as a functionally important association.

The recent identification of the first molybdate transporters in eukaryotes represents a major breakthrough. However, some of the reported subcellular localizations of MOT1 and MOT2 are contradictory or unclear and their physiological role needs to be studied in detail. While an additional molybdate transport activity has been reported for MOT2 from *C. reinhardtii*, members of this protein family remain to be investigated in humans. Moco carrier/binding proteins have been identified in algae and plants but not in animals. Given the large amount of MoBPs present in *Arabidopsis* and their clearly distinctive properties relative to *C. reinhardtii* MCP, future studies should address the physiological impact of these proteins on Mo and Moco homeostasis as well as on molybdo-enzyme activity. Furthermore, taking into account the severe effects of Moco-deficiency in animals together with the short half-life of the free cofactor, the existence of Moco-binding/protecting proteins could ensure a "ready to use" pool of Moco in animal cells.

Mechanisms of regulation of Moco biosynthesis are unknown in eukaryotes. Identification of signalling molecules, effectors, transcription factors and other regulatory proteins would complete our picture of Mo metabolism and its connection with other metabolic pathways. Finally, the degradation of Moco to its urinary excretion product urothione $(C_{11}H_{11}N_5O_3S_2)$ is entirely enigmatic.[159]

References

1. R. H. Holm, E. I. Solomon, A. Majumdar and A. Tenderholt, *Coord. Chem. Rev.*, 2011, **255**, 993–1015.
2. J. A. C. Fortescue, *Appl. Geochem.*, 1992, 7, 1–53.
3. K. J. Reddy, L. C. Munn and L. Wang, in *Molybdenum in Agriculture*, ed. U. C. Gupta, Cambridge University Press, Cambridge, 1997.
4. T. Wichard, B. Mishra, S. C. B. Myneni, J. P. Bellenger and A. M. L. Kraepiel, *Nat. Geosci.*, 2009, **2**, 625–629.
5. S. S. Merchant, M. D. Allen, J. Kropat, J. L. Moseley, J. C. Long, S. Tottey and A. M. Terauchi, *Biochim. Biophys. Acta*, 2006, **1763**, 578–594.
6. U. C. Gupta, *J. Plant Nutr.*, 1991, **14**, 613–621.
7. J. L. Burguera and M. Burguera, *J. Trace Elem. Med. Biol.*, 2007, **21**, 178–183.
8. J. Versieck, J. Hoste, F. Barbier, L. Vanballenberghe, J. Derudder and R. Cornelis, *Clin. Chim. Acta*, 1978, **87**, 135–140.
9. J. R. Turnlund and W. R. Keyes, *J. Nutr. Biochem.*, 2004, **15**, 90–95.
10. W. L. Lindsay, *Chemical Equilibria in Soils*, John Wiley & Sons, New York, 1979.
11. D. I. Arnon and P. R. Stout, *Plant Physiol.*, 1939, **14**, 599–602.
12. S. Leimkuhler, M. M. Wuebbens and K. V. Rajagopalan, *Coord. Chem. Rev.*, 2011, **255**, 1129–1144.
13. G. Schwarz, *Cell. Mol. Life Sci.*, 2005, **62**, 2792–2810.
14. G. Schwarz, R. R. Mendel and M. W. Ribbe, *Nature*, 2009, **460**, 839–847.
15. Y. Hu and M. W. Ribbe, *Methods Mol. Biol.*, 2011, **766**, 3–7.

16. F. H. Allen, J. E. Davies, J. J. Galloy, O. Johnson, O. Kennard, C. F. Macrae, E. M. Mitchell, G. F. Mitchell, J. M. Smith and D. G. Watson, *J. Chem. Inf. Comp. Sci.*, 1991, **31**, 187–204.
17. L. E. Bevers, P. L. Hagedoorn and W. R. Hagen, *Coord. Chem. Rev.*, 2009, **253**, 269–290.
18. R. Hille, *Chem. Rev.*, 1996, **96**, 2757–2816.
19. A. Magalon, J. G. Fedor, A. Walburger and J. H. Weiner, *Coord. Chem. Rev.*, 2011, **255**, 1159–1178.
20. C. Kisker, H. Schindelin, A. Pacheco, W. A. Wehbi, R. M. Garrett, K. V. Rajagopalan, J. H. Enemark and D. C. Rees, *Cell*, 1997, **91**, 973–983.
21. M. K. Chan, S. Mukund, A. Kletzin, M. W. Adams and D. C. Rees, *Science*, 1995, **267**, 1463–1469.
22. A. Kletzin and M. W. Adams, *FEMS Microbiol. Rev.*, 1996, **18**, 5–63.
23. J. R. Andreesen and L. G. Ljungdahl, *J. Bacteriol.*, 1973, **116**, 867–873.
24. I. Yamamoto, T. Saiki, S. M. Liu and L. G. Ljungdahl, *J. Biol. Chem.*, 1983, **258**, 1826–1832.
25. P. L. Hagedoorn, T. Chen, I. Schroder, S. R. Piersma, S. de Vries and W. R. Hagen, *J. Biol. Inorg. Chem.*, 2005, **10**, 259–269.
26. R. R. Mendel and T. Kruse, *Biochim. Biophys. Acta*, 2012, **1823**, 1568–1579.
27. W. H. Campbell, *Cell. Mol. Life Sci.*, 2001, **58**, 194–204.
28. L. Skipper, W. H. Campbell, J. A. Mertens and D. J. Lowe, *J. Biol. Chem.*, 2001, **276**, 26995–27002.
29. C. Masclaux-Daubresse, F. Daniel-Vedele, J. Dechorgnat, F. Chardon, L. Gaufichon and A. Suzuki, *Ann. Bot.*, 2010, **105**, 1141–1157.
30. M. V. Beligni and L. Lamattina, *Planta*, 2000, **210**, 215–221.
31. W. H. Campbell, *Annu. Rev. Plant Physiol. Plant. Mol. Biol.*, 1999, **50**, 277–303.
32. W. M. Kaiser and S. C. Huber, *J. Exp. Bot.*, 2001, **52**, 1981–1989.
33. I. Lambeck, J. C. Chi, S. Krizowski, S. Mueller, N. Mehlmer, M. Teige, K. Fischer and G. Schwarz, *Biochemistry*, 2010, **49**, 8177–8186.
34. I. C. Lambeck, K. Fischer-Schrader, D. Niks, J. Roeper, J. C. Chi, R. Hille and G. Schwarz, *J. Biol. Chem.*, 2012, **287**, 4562–4571.
35. K. Fischer, G. G. Barbier, H. J. Hecht, R. R. Mendel, W. H. Campbell and G. Schwarz, *Plant Cell*, 2005, **17**, 1167–1179.
36. N. Schrader, K. Fischer, K. Theis, R. R. Mendel, G. Schwarz and C. Kisker, *Structure*, 2003, **11**, 1251–1263.
37. C. Feng, G. Tollin and J. H. Enemark, *Biochim. Biophys. Acta*, 2007, **1774**, 527–539.
38. H. J. Cohen, S. Betcher-Lange, D. L. Kessler and K. V. Rajagopalan, *J. Biol. Chem.*, 1972, **247**, 7759–7766.
39. J. M. Klein and G. Schwarz, *J. Cell Sci.*, 2012, **125**, 4876–4885.
40. S. Gerin, G. Mathy, A. Blomme, F. Franck and F. E. Sluse, *Biochim. Biophys. Acta*, 2010, **1797**, 994–1003.
41. K. Nowak, N. Luniak, C. Witt, Y. Wustefeld, A. Wachter, R. R. Mendel and R. Hansch, *Plant Cell Physiol.*, 2004, **45**, 1889–1894.

42. R. Hansch, C. Lang, E. Riebeseel, R. Lindigkeit, A. Gessler, H. Rennenberg and R. R. Mendel, *J. Biol. Chem.*, 2006, **281**, 6884–6888.
43. R. S. Byrne, R. Hansch, R. R. Mendel and R. Hille, *J. Biol. Chem.*, 2009, **284**, 35479–35484.
44. D. Randewig, D. Hamisch, C. Herschbach, M. Eiblmeier, C. Gehl, J. Jurgeleit, J. Skerra, R. R. Mendel, H. Rennenberg and R. Hansch, *Plant Cell Environ.*, 2012, **35**, 100–115.
45. Y. Amaya, K. Yamazaki, M. Sato, K. Noda and T. Nishino, *J. Biol. Chem.*, 1990, **265**, 14170–14175.
46. T. Nishino, *J. Biol. Chem.*, 1997, **272**, 29859–29864.
47. Z. Yesbergenova, G. Yang, E. Oron, D. Soffer, R. Fluhr and M. Sagi, *Plant J.*, 2005, **42**, 862–876.
48. M. Zarepour, K. Kaspari, S. Stagge, R. Rethmeier, R. R. Mendel and F. Bittner, *Plant Mol. Biol.*, 2010, **72**, 301–310.
49. F. Rodriguez-Trelles, R. Tarrio and F. J. Ayala, *Proc. Natl Acad. Sci. USA*, 2003, **100**, 13413–13417.
50. E. Garattini, M. Fratelli and M. Terao, *Hum. Genomics*, 2009, **4**, 119–130.
51. C. Coelho, M. Mahro, J. Trincao, A. T. Carvalho, M. J. Ramos, M. Terao, E. Garattini, S. Leimkuhler and M. J. Romao, *J. Biol. Chem.*, 2012, **287**, 40690–40702.
52. E. Garattini, R. Mendel, M. J. Romao, R. Wright and M. Terao, *Biochem. J.*, 2003, **372**, 15–32.
53. M. Seo, A. J. M. Peeters, H. Koiwai, T. Oritani, A. Marion-Poll, J. A. D. Zeevaart, M. Koornneef, Y. Kamiya and T. Koshiba, *Proc. Natl Acad. Sci. USA*, 2000, **97**, 12908–12913.
54. P. E. Verslues and J. K. Zhu, *Biochem. Soc. Trans.*, 2005, **33**, 375–379.
55. A. Havemeyer, F. Bittner, S. Wollers, R. Mendel, T. Kunze and B. Clement, *J. Biol. Chem.*, 2006, **281**, 34796–34802.
56. A. Chamizo-Ampudia, A. Galvan, E. Fernandez and A. Llamas, *Eukaryotic Cell*, 2011, **10**, 1270–1282.
57. J. M. Klein, J. D. Busch, C. Potting, M. J. Baker, T. Langer and G. Schwarz, *J. Biol. Chem.*, 2012, **287**, 42795–42803.
58. J. Kotthaus, B. Wahl, A. Havemeyer, D. Schade, D. Garbe-Schonberg, R. Mendel, F. Bittner and B. Clement, *Biochem. J.*, 2011, **433**, 383–391.
59. B. Wahl, D. Reichmann, D. Niks, N. Krompholz, A. Havemeyer, B. Clement, T. Messerschmidt, M. Rothkegel, H. Biester, R. Hille, R. R. Mendel and F. Bittner, *J. Biol. Chem.*, 2010, **285**, 37847–37859.
60. A. Havemeyer, J. A. Lang and B. Clement, *Drug Metab. Rev.*, 2011, **43**, 524–539.
61. A. M. Grunden and K. T. Shanmugam, *Arch. Microbiol.*, 1997, **168**, 345–354.
62. W. T. Self, A. M. Grunden, A. Hasona and K. T. Shanmugam, *Res. Microbiol.*, 2001, **152**, 311–321.
63. J. K. Rosentel, F. Healy, J. A. Maupin-Furlow, J. H. Lee and K. T. Shanmugam, *J. Bacteriol.*, 1995, **177**, 4857–4864.

64. J. H. Lee, J. C. Wendt and K. T. Shanmugam, *J. Bacteriol.*, 1990, **172**, 2079–2087.

65. J. Gisin, A. Müller, Y. Pfänder, S. Leimkühler, F. Narberhaus and B. Masepohl, *J. Bacteriol.*, 2011, **192**, 5943–5952.

66. L. E. Bevers, P. L. Hagedoorn, G. C. Krijger and W. R. Hagen, *J. Bacteriol.*, 2006, **188**, 6498–6505.

67. L. E. Bevers, G. Schwarz and W. R. Hagen, *J. Bacteriol.*, 2011, **193**, 4999–5001.

68. T. Thiel, B. Pratte and M. Zahalak, *Arch. Microbiol.*, 2002, **179**, 50–56.

69. S. Rech, C. Wolin and R. P. Gunsalus, *J. Biol. Chem.*, 1996, **271**, 2557–2562.

70. A. Sirko, M. Hryniewicz, D. Hulanicka and A. Bock, *J. Bacteriol.*, 1990, **172**, 3351–3357.

71. K. Hollenstein, D. C. Frei and K. P. Locher, *Nature*, 2007, **446**, 213–216.

72. S. Gerber, M. Comellas-Bigler, B. A. Goetz and K. P. Locher, *Science*, 2008, **321**, 246–250.

73. Y. L. Hu, S. Rech, R. P. Gunsalus and D. C. Rees, *Nat. Struct. Biol.*, 1997, **4**, 703–707.

74. D. M. Lawson, C. E. M. Williams, L. A. Mitchenall and R. N. Pau, *Struct. Fold. Des.*, 1998, **6**, 1529–1539.

75. E. Aguilar-Barajas, C. Diaz-Perez, M. I. Ramirez-Diaz, H. Riveros-Rosas and C. Cervantes, *Biometals*, 2011, **24**, 687–707.

76. D. R. Hall, D. G. Gourley, G. A. Leonard, E. M. H. Duke, L. A. Anderson, D. H. Boxer and W. N. Hunter, *EMBO J.*, 1999, **18**, 1435–1446.

77. A. W. Schuttelkopf, D. H. Boxer and W. N. Hunter, *J. Mol. Biol.*, 2003, **326**, 761–767.

78. E. E. Regulski, R. H. Moy, Z. Weinberg, J. E. Barrick, Z. Yao, W. L. Ruzzo and R. R. Breaker, *Mol. Microbiol.*, 2008, **68**, 918–932.

79. M. Tejada-Jimenez, A. Llamas, E. Sanz-Luque, A. Galvan and E. Fernandez, *Proc. Natl Acad. Sci. USA*, 2007, **104**, 20126–20130.

80. H. Tomatsu, J. Takano, H. Takahashi, A. Watanabe-Takahashi, N. Shibagaki and T. Fujiwara, *Proc. Natl Acad. Sci. USA*, 2007, **104**, 18807–18812.

81. R. R. Mendel and R. Hansch, *J. Exp. Bot.*, 2002, **53**, 1689–1698.

82. P. R. Stout and W. R. Meagher, *Science*, 1948, **108**, 471–473.

83. H. Heuwinkel, E. A. Kirkby, J. Lebot and H. Marschner, *J. Plant Nutr.*, 1992, **15**, 549–568.

84. M. Singh and V. Kumar, *Soil Sci.*, 1979, **127**, 307–312.

85. J. W. Tweedie and I. H. Segel, *Biochim. Biophys. Acta*, 1970, **196**, 95–106.

86. U. Patel, M. D. Baxi and V. V. Modi, *Curr. Microbiol.*, 1988, **17**, 179–182.

87. D. Ganyc and W. T. Self, *FEBS Lett.*, 2008, **582**, 299–304.

88. A. Llamas, K. L. Kalakoutskii and E. Fernandez, *Plant Cell Environ.*, 2000, **23**, 1247–1255.

89. M. J. Hawkesford, *Physiol. Plantarum*, 2003, **117**, 155–163.

90. A. K. Sharma, A. C. Rigby and S. L. Alper, *Cell. Physiol. Biochem.*, 2011, **28**, 407–422.

91. I. Baxter, B. Muthukumar, H. C. Park, P. Buchner, B. Lahner, J. Danku, K. Zhao, J. Lee, M. J. Hawkesford, M. L. Guerinot and D. E. Salt, *PLoS Genet.*, 2008, **4**, e1000004.

92. A. Gasber, S. Klaumann, O. Trentmann, A. Trampczynska, S. Clemens, S. Schneider, N. Sauer, I. Feifer, F. Bittner, R. R. Mendel and H. E. Neuhaus, *Plant Biol.*, 2011, **13**, 710–718.

93. M. Tejada-Jimenez, A. Galvan and E. Fernandez, *Proc. Natl Acad. Sci. USA*, 2011, **108**, 6420–6425.

94. K. L. Fitzpatrick, S. D. Tyerman and B. N. Kaiser, *FEBS Lett.*, 2008, **582**, 1508–1513.

95. W. Z. Li, D. R. Fingrut and D. P. Maxwell, *Physiol. Plantarum*, 2009, **136**, 336–350.

96. D. M. Lawson, C. E. Williams, D. J. White, A. P. Choay, L. A. Mitchenall and R. N. Pau, *J. Chem. Soc. Dalton*, 1997, 3981–3984.

97. D. Fenske, M. Gnida, K. Schneider, W. Meyer-Klaucke, J. Schemberg, V. Henschel, A. K. Meyer, A. Knochel and A. Muller, *ChemBioChem*, 2005, **6**, 405–413.

98. F. S. Ataya, C. P. Witte, A. Galvan, M. I. Igeno and E. Fernandez, *J. Biol. Chem.*, 2003, **278**, 10885–10890.

99. T. Kruse, C. Gehl, M. Geisler, M. Lehrke, P. Ringel, S. Hallier, R. Hansch and R. R. Mendel, *J. Biol. Chem.*, 2010, **285**, 6623–6635.

100. K. Fischer, A. Llamas, M. Tejada-Jimenez, N. Schrader, J. Kuper, F. S. Ataya, A. Galvan, R. R. Mendel, E. Fernandez and G. Schwarz, *J. Biol. Chem.*, 2006, **281**, 30186–30194.

101. M. Aguilar, J. Cardenas and E. Fernandez, *Biochim. Biophys. Acta*, 1992, **1160**, 269–274.

102. K. L. Hale, S. P. McGrath, E. Lombi, S. M. Stack, N. Terry, I. J. Pickering, G. N. George and E. A. Pilon-Smits, *Plant Physiol.*, 2001, **126**, 1391–1402.

103. D. R. Steinke, W. Majak, T. S. Sorensen and M. Parvez, *J. Agr. Food Chem.*, 2008, **56**, 5437–5442.

104. R. R. Mendel and T. Kruse, *Biochim. Biophys.Acta*, 2012, **1823**, 1568–1579.

105. R. R. Mendel and G. Schwarz, *Met. Ions Biol. Syst.*, 2002, **39**, 317–368.

106. J. Reiss and J. L. Johnson, *Hum. Mutat.*, 2003, **21**, 569–576.

107. K. T. Shanmugam, V. Stewart, R. P. Gunsalus, D. H. Boxer, J. A. Cole, M. Chippaux, J. A. Demoss, G. Giordano, E. C. C. Lin and K. V. Rajagopalan, *Mol. Microbiol.*, 1992, **6**, 3452–3454.

108. J. Reiss, N. Cohen, C. Dorche, H. Mandel, R. R. Mendel, B. Stallmeyer, M. T. Zabot and T. Dierks, *Nat. Genet.*, 1998, **20**, 51–53.

109. A. P. Mehta, J. W. Hanes, S. H. Abdelwahed, D. G. Hilmey, P. Hanzelmann and T. P. Begley, *Biochemistry*, 2013, **52**, 1134–1136.

110. M. M. Wuebbens and K. V. Rajagopalan, *J. Biol. Chem.*, 1995, **270**, 1082–1087.

111. J. Kuper, A. Llamas, H. J. Hecht, R. R. Mendel and G. Schwarz, *Nature*, 2004, **430**, 803–806.

112. P. Hanzelmann, H. L. Hernandez, C. Menzel, R. Garcia-Serres, B. H. Huynh, M. K. Johnson, R. R. Mendel and H. Schindelin, *J. Biol. Chem.*, 2004, **279**, 34721–34732.
113. C. Rieder, W. Eisenreich, J. O'Brien, G. Richter, E. Gotze, P. Boyle, S. Blanchard, A. Bacher and H. Simon, *Eur. J. Biochem.*, 1998, **255**, 24–36.
114. J. A. Santamaria-Araujo, V. Wray and G. Schwarz, *J. Biol. Inorg. Chem.*, 2012, **17**, 113–122.
115. J. A. Santamaria-Araujo, B. Fischer, T. Otte, M. Nimtz, R. R. Mendel, V. Wray and G. N. Schwarz, *J. Biol. Chem.*, 2004, **279**, 15994–15999.
116. T. Hoff, K. M. Schnorr, C. Meyer and M. Caboche, *J. Biol. Chem.*, 1995, **270**, 6100–6107.
117. H. J. Sofia, G. Chen, B. G. Hetzler, J. F. Reyes-Spindola and N. E. Miller, *Nucleic Acids Res.*, 2001, **29**, 1097–1106.
118. J. Teschner, N. Lachmann, J. Schulze, M. Geisler, K. Selbach, J. Santamaria-Araujo, J. Balk, R. R. Mendel and F. Bittner, *Plant Cell*, 2010, **22**, 468–480.
119. R. Lill, *Nature*, 2009, **460**, 831–838.
120. D. R. Richardson, D. J. R. Lane, E. M. Becker, M. L. H. Huang, M. Whitnall, Y. S. Rahmanto, A. D. Sheftel and P. Ponka, *Proc. Natl Acad. Sci. USA*, 2010, **107**, 10775–10782.
121. D. Y. Kim, L. Bovet, S. Kushnir, E. W. Noh, E. Martinoia and Y. Lee, *Plant Physiol.*, 2006, **140**, 922–932.
122. B. Stallmeyer, G. Drugeon, J. Reiss, A. L. Haenni and R. R. Mendel, *Am. J. Hum. Genet.*, 1999, **64**, 698–705.
123. Z. Marelja, W. Stocklein, M. Nimtz and S. Leimkuhler, *J. Biol. Chem.*, 2008, **283**, 25178–25185.
124. A. Matthies, M. Nimtz and S. Leimkuhler, *Biochemistry*, 2005, **44**, 7912–7920.
125. A. Matthies, K. V. Rajagopalan, R. R. Mendel and S. Leimkuhler, *Proc. Natl Acad. Sci. USA*, 2004, **101**, 5946–5951.
126. L. E. Bevers, P. L. Hagedoorn, J. A. Santamaria-Araujo, A. Magalon, W. R. Hagen and G. Schwarz, *Biochemistry*, 2008, **47**, 949–956.
127. J. Nichols and K. V. Rajagopalan, *J. Biol. Chem.*, 2002, **277**, 24995–25000.
128. B. Stallmeyer, A. Nerlich, J. Schiemann, H. Brinkmann and R. R. Mendel, *Plant J.*, 1995, **8**, 751–762.
129. A. Llamas, R. R. Mendel and N. Schwarz, *J. Biol. Chem.*, 2004, **279**, 55241–55246.
130. A. Llamas, T. Otte, G. Multhaup, R. R. Mendel and G. Schwarz, *J. Biol. Chem.*, 2006, **281**, 18343–18350.
131. A. Hasona, R. M. Ray and K. T. Shanmugam, *J. Bacteriol.*, 1998, **180**, 1466–1472.
132. P. A. Seubert, P. A. Grant, E. A. Christie, J. R. Farley and I. H. Segel, *Ciba Found. Symp.*, 1979, **72**, 19–47.
133. A. A. Belaidi and G. Schwarz, *Biochem. J.*, 2013, **450**, 149–157.
134. B. Stallmeyer, G. Schwarz, J. Schulze, A. Nerlich, J. Reiss, J. Kirsch and R. R. Mendel, *Proc. Natl Acad. Sci. USA*, 1999, **96**, 1333–1338.

135. J. M. Fritschy, R. J. Harvey and G. Schwarz, *Trends Neurosci.*, 2008, **31**, 257–264.

136. F. Bittner, M. Oreb and R. R. Mendel, *J. Biol. Chem.*, 2001, **276**, 40381–40384.

137. L. Amrani, J. Primus, A. Glatigny, L. Arcangeli, C. Scazzocchio and V. Finnerty, *Mol. Microbiol.*, 2000, **38**, 114–125.

138. L. M. Xiong, M. Ishitani, H. Lee and J. K. Zhu, *Plant Cell*, 2001, **13**, 2063–2083.

139. J. J. Moura, C. D. Brondino, J. Trincao and M. J. Romao, *J. Biol. Inorg. Chem.*, 2004, **9**, 791–799.

140. H. Dobbek, L. Gremer, O. Meyer and R. Huber, *Proc. Natl Acad. Sci. USA*, 1999, **96**, 8884–8889.

141. L. M. Rubio and P. W. Ludden, *Annu. Rev. Microbiol.*, 2008, **62**, 93–111.

142. B. N. Kaiser, K. L. Gridley, J. N. Brady, T. Phillips and S. D. Tyerman, *Ann. Bot-London*, 2005, **96**, 745–754.

143. Y. Ide, M. Kusano, A. Oikawa, A. Fukushima, H. Tomatsu, K. Saito, M. Y. Hirai and T. Fujiwara, *J. Exp. Bot.*, 2011, **62**, 1483–1497.

144. N. N. Abumrad, *Bull. NY Acad. Med.*, 1984, **60**, 163–171.

145. C. Yanhui, Y. Xiaoyuan, H. Kun, L. Meihua, L. Jigang, G. Zhaofeng, L. Zhiqiang, Z. Yunfei, W. Xiaoxiao, Q. Xiaoming, S. Yunping, Z. Li, D. Xiaohui, L. Jingchu, D. Xing-Wang, C. Zhangliang, G. Hongya and Q. Li-Jia, *Plant Mol. Biol.*, 2006, **60**, 107–124.

146. C. Chatterjee, N. Nautiyal and S. C. Agarwala, *New Phytol.*, 1985, **100**, 511–518.

147. C. Lang, J. Popko, M. Wirtz, R. Hell, C. Herschbach, J. Kreuzwieser, H. Rennenberg, R. R. Mendel and R. Hansch, *Plant Cell Environ.*, 2007, **30**, 447–455.

148. M. M. Chowdhury, C. Dosche, H. G. Lohmannsroben and S. Leimkuhler, *J. Biol. Chem.*, 2012, **287**, 17297–17307.

149. H. J. Lee, I. M. Adham, G. Schwarz, M. Kneussel, J. O. Sass, W. Engel and J. Reiss, *Hum. Mol. Genet.*, 2002, **11**, 3309–3317.

150. S. Kramer, R. V. Hageman and K. V. Rajagopalan, *Arch. Biochem. Biophys.*, 1984, **233**, 821–829.

151. G. Schwarz, J. A. Santamaria-Araujo, S. Wolf, H. J. Lee, I. M. Adham, H. J. Grone, H. Schwegler, J. O. Sass, T. Otte, P. Hanzelmann, R. R. Mendel, W. Engel and J. Reiss, *Hum. Mol. Genet.*, 2004, **13**, 1249–1255.

152. A. Veldman, J. A. Santamaria-Araujo, S. Sollazzo, J. Pitt, R. Gianello, J. Yaplito-Lee, F. Wong, C. A. Ramsden, J. Reiss, I. Cook, J. Fairweather and G. Schwarz, *Pediatrics*, 2010, **125**, E1249–E1254.

153. K. Clinch, D. K. Watt, R. A. Dixon, S. M. Baars, G. J. Gainsford, A. Tiwari, G. Schwarz, Y. Saotome, M. Storek, A. A. Belaidi and J. A. Santamaria-Araujo, *J. Med. Chem.*, 2013, **56**, 1730–1738.

154. H. M. Alvarez, Y. Xue, C. D. Robinson, M. A. Canalizo-Hernandez, R. G. Marvin, R. A. Kelly, A. Mondragon, J. E. Penner-Hahn and T. V. O'Halloran, *Science*, 2010, **327**, 331–334.

155. G. M. Ward, *J. Anim. Sci.*, 1978, **46**, 1078–1085.
156. G. J. Brewer, F. Askari, R. B. Dick, J. Sitterly, J. K. Fink, M. Carlson, K. J. Kluin and M. T. Lorincz, *Transl. Res.*, 2009, **154**, 70–77.
157. A. Vyskocil and C. Viau, *J. Appl. Toxicol.*, 1999, **19**, 185–192.
158. T. V. Fungwe, F. Buddingh, D. S. Demick, C. D. Lox, M. T. Yang and S. P. Yang, *Nutr. Res.*, 1990, **10**, 515–524.
159. J. L. Johnson and K. V. Rajagopalan, *Proc. Natl Acad. USA*, 1982, **79**, 6856–6860.

CHAPTER 9

Manganese

SUDIPTA CHAKRABORTY,[a,c] EBANY MARTINEZ-FINLEY,[b,c] SAM CAITO,[b,c] PAN CHEN[b] AND MICHAEL ASCHNER*[b,c,d]

[a] Vanderbilt Brain Institute, Vanderbilt University Medical Center, Nashville, TN, USA; [b] Division of Clinical Pharmacology and Pediatric Toxicology, Vanderbilt University Medical Center, Nashville, TN, USA; [c] Center in Molecular Toxicology, Vanderbilt University Medical Center, Nashville, TN, USA; [d] The Kennedy Center for Research on Human Development, Vanderbilt University Medical Center, Nashville, TN, USA
*Email: michael.aschner@vanderbilt.edu

9.1   Introduction

Manganese (Mn) is an essential heavy metal that comprises about 0.1% of the Earth's crust. It is the fifth most abundant metal and twelfth most abundant element overall, usually existing in its natural form as oxides, carbonates and silicates. As erosion produces a naturally ubiquitous presence of Mn in air, soil and waterways, the human population is exposed to Mn through a variety of environmental sources. However, the primary route of Mn exposure is through dietary intake, as several Mn-containing foods are found in human diets. Legumes, rice, nuts and whole grains contain the highest Mn levels, while Mn is also found in leafy green vegetables, tea, chocolate and fruits like blueberries and acai.[1] Mn-containing nutritional supplements and vitamins are commonly taken on a daily basis, in addition to infant and neonatal formulas that contain a trace element-enriched solution. The abundant Mn-containing dietary sources allow humans to obtain the proper Mn levels (2.3 mg day^{-1} for men and 1.8 mg day^{-1} for women) necessary for several important physiological processes, including

RSC Metallobiology Series No. 2
Binding, Transport and Storage of Metal Ions in Biological Cells
Edited by Wolfgang Maret and Anthony Wedd
© The Royal Society of Chemistry 2014
Published by the Royal Society of Chemistry, www.rsc.org

development, digestion, reproduction, immune function, energy metabolism and antioxidant defences. Upon ingestion, in adults only 3–5% of Mn is absorbed into the body from the digestive tract.[2]

Outside of dietary sources, exposure to inorganic forms of Mn can occur in several industrial settings, as Mn is used in the manufacture of steel, batteries and fireworks, as well as ceramics, cosmetics, leather, glass and textiles. On the other hand, organic Mn is also highly prevalent in the environment. Methylcyclopentadienyl manganese tricarbonyl [MMT or MCMT; $(CH_3C_5H_4)Mn(CO)_3$] is a component of a widely used antiknock gasoline additive. In addition, Mn is a component of certain fungicides and pesticides (*e.g.* maneb and mancozeb), smoke inhibitors and medical magnetic resonance imaging (MRI) contrast agents.[1] Unfortunately, the impact on human health from environmental overexposure through these sources can be significantly debilitating and ultimately result in a neurotoxic condition known as "manganism". This condition leads to irreversible damage to the basal ganglia region of the brain that is also affected in Parkinson's disease (PD), resulting in similar cognitive, emotional and motor deficits.[3,4] It mostly arises from occupational exposure to Mn, affecting miners, welders, smelters and other industrial workers who handle Mn-containing steel and other manufacturing and who consequently inhale fumes containing high levels of Mn.[5,6]

In addition to industrial workers, there are other populations at risk of Mn-induced toxicity. Patients suffering from hepatic encephalopathy or any stage of liver failure are at high risk of Mn toxicity, as a properly functioning biliary system is required for proper Mn excretion.[7] Similarly, unhealthy neonates partaking in Mn-supplemented total parenteral nutrition (TPN) are also vulnerable to Mn toxicity.[8] Another significant population at risk for Mn poisoning are those suffering from iron deficiency, one of the most common nutritional deficiencies. Fe and Mn compete for the same transporters, resulting in higher Mn accumulation when Fe levels are low.[9,10] Thus, in people suffering from chronic iron deficiency (*e.g.* anaemia), low levels of Fe result in high accumulation of Mn over time.

It is apparent that several control mechanisms must exist in different life forms to closely regulate Mn homeostasis in the cell in order to thwart Mn-induced toxicity. This chapter will closely examine how Mn is taken up into intestinal cells and distributed through the blood into various cells through several different uptake and transport mechanisms. Furthermore, we will also discuss how the cell can sense and buffer Mn to maintain optimal levels, including subcellular sequestration and proper extrusion from cells.

9.2 Uptake in Intestinal Cells and Release into Blood

Ingestion is the most common route of Mn exposure. The typical adult ingests <5 mg Mn kg^{-1} of food, mostly from grains, rice, nuts, tea and chocolate. Adult humans absorb around 3–5% of ingested Mn. Radiolabeled

[54]Mn uptake studies showed that, for a meal containing 1 mg Mn, adult males absorb $1.35 \pm 0.51\%$ while adult females absorb $3.55 \pm 2.11\%$.[11,12] Mammalian tissues typically contain 0.3–2.9 µg Mn g^{-1} wet tissue weight.[2] Turnover of ingested Mn is rapid, with a mean retention of 10 days after ingestion estimated at $5.0 \pm 3.1\%$ in women.[13] Levels are tightly controlled by absorption in the gastrointestinal tract and excretion by the liver. The majority of excreted Mn is contained in bile in the liver and secreted into the intestines for elimination in the faeces.[11,14] Low levels of bile-Mn conjugates are reabsorbed in the intestines forming an enterohepatic circulation.[15] Additionally, small amounts of Mn are excreted by the pancreas or *via* urine.[11]

Molecular mechanisms of Mn uptake by intestinal cells are not well characterized. Early studies using rat brush border membrane vesicles found evidence of lactoferrin receptor-mediated uptake of Mn.[16] Studies using the Caco-2 intestinal cell line derived from human colonic carcinoma revealed a biphasic uptake process, indicating that transport falls into its steady-state condition following a brief period of equilibration between intracellular and extracellular components.[17] Moreover, *in vivo* studies using rat intestinal perfusions have found that intestinal Mn uptake involves a high-affinity, low-capacity active transport process that is rapidly saturable.[18] It is thought that Mn can enter cells through either passive diffusion or active transport *via* the divalent metal transporter 1 (DMT1).[19,20] This transporter uses the proton gradient of the cell membrane to move several divalent metals across the cell membrane, including Mn, Fe and Cu.[20] Due to this shared transport mechanism, the levels of other trace metals influence the amount of Mn absorbed. Other dietary components that alter Mn absorption include phytates and ascorbic acid.[21]

Absorption of Mn by the gastrointestinal (GI) tract is modulated by a variety of factors. The concentration in the diet influences both absorption and elimination in bile. When instances of high Mn intake occur, through either dietary or environmental exposure, the GI tract absorbs less Mn, the liver increases metabolism and there is increased biliary and pancreatic excretion.[11,22,23] Gender influences Mn uptake, with males absorbing significantly lower amounts of Mn than females.[11,12] This may reflect men's higher iron status and higher serum ferritin concentrations, which may lead to competition for transport by DMT1. Age is another determinant of Mn absorption. Younger individuals, particularly infants, absorb and retain higher levels of Mn than adults,[24,25] likely because their requirement for Mn is much higher. Data concerning the intestinal absorption of Mn in infants compared to adults are limited, but studies have examined the detrimental effects of TPN in severely ill or premature infants. These solutions are usually supplemented with a trace element solution that contains small amounts of Mn. However, unlike the minimal absorption of Mn from milk, the intravenous exposure to Mn-supplemented TPN solutions results in bypass of any intestinal control of absorption. Consequently, nearly 100% of Mn can be absorbed, resulting in conditions of toxicity in these vulnerable neonates. These infants also pass little stool, leading to even higher retention of Mn.[2]

9.3 Coordination Chemistry of Manganese Ions in the Blood

Upon absorption in the intestines, Mn enters the blood through uncharacterized mechanisms and quickly leaves the plasma for distribution to tissues. Leggett has estimated the half-life for Mn in plasma is 1min, corresponding to an outflow rate of 1000 d^{-1}.[26] The most recent biokinetic model of systemic Mn distribution shows outflow from the plasma divided up amongst the body tissues as follows: 30% to liver, 5% to kidney, 5% to pancreas, 1% to colon, 0.2% to urinary system, 0.5% to bone, 0.1% to brain, 0.02% to erythrocytes and the remaining 58% to the rest of the soft tissues.[26] Incidentally, the liver, pancreas, bone, kidney and brain retain Mn and have the highest Mn concentrations in the body.[2] This is due to the essential nature of Mn in energy production and the high-energy demands of these tissues.

Mn speciation is important for binding and transport in the blood and incorporation into metalloproteins, and as well as toxicity.[27,28] The metal exists as Mn^{2+} and Mn^{3+} in aqueous environments at physiological pH, although the latter is highly reactive and will undergo disproportionation to Mn^{2+} and Mn^{4+} unless stabilized by complex formation[28] (redox potentials for Mn are given in Table 9.1). Furthermore, Mn^{2+} can be oxidized in the blood to Mn^{3+} by ceruloplasmin,[29] although it is unknown what proportion of the total Mn pool is Mn^{3+}.[30] Studies employing electron paramagnetic resonance (EPR) and difference UV spectroscopies, supported by computer modelling, indicate that Mn^{2+} exists in different forms in the serum.[31] The major form is Mn^{2+}-albumin (84%),[32,33] but there are significant levels of Mn^{2+}-transferrin (Tf) (1%) and simple complexes (Mn^{2+}_{aq} (6%) and bicarbonate- and citrate-substitution products (6 and 2%, respectively)). A similar distribution was found using size exclusion chromatography.[34] These circulating Mn-protein complexes are essential for the distribution of Mn to tissues.

Mn^{2+} binds to albumin relatively weakly (K_A, 0.63×10^4 M^{-1}; Table 9.2) *via* imidazole nitrogen atoms of His residues and oxygen atoms of Asp residues. This site is not specific for Mn^{2+} and is able to coordinate other metal ions.[35] His and Asp/Glu dominate the bonding in several Mn-containing metalloproteins (Figure 9.1). The ferric ion metallochaperone transferrin also binds Mn^{3+} in plasma.[36] This follows from the fact that Mn and Fe

Table 9.1   Redox potentials of Mn.

Reaction	E^0 (V)
$Mn^{2+} + 2e^- \rightarrow Mn$	-1.18
$MnO_2 + 4H^+ + e^- \rightarrow Mn^{3+} + 2H_2O$	$+0.95$
$MnO_2 + 4H^+ + 2e^- \rightarrow Mn^{2+} + 2H_2O$	$+1.23$
$Mn^{3+} + e^- \rightarrow Mn^{2+}$	$+1.52$

Table 9.2   Binding constants for manganese complexes.

	Oxidation state	K_A (M^{-1})	Reference
Transferrin	2+	K_{A1}: 1.1×10^4	31
		K_{A2}: 0.91×10^3	
Albumin	2+	0.63×10^4	144
MnSOD	2+	3×10^9	145
	3+	5×10^{17}	
Arginase	2+	3×10^7	146

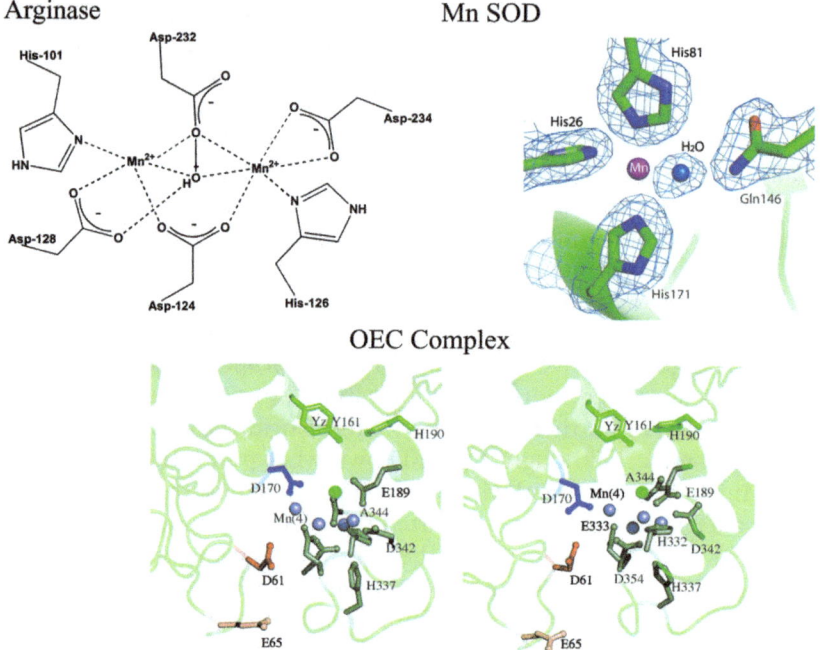

Figure 9.1   Structures of Mn coordination sites in mammalian arginase, *E. coli* Mn-superoxide dismutase (SOD) and the OEC (oxygen-evolving complex) in plants.
Adapted from Refs. 141–143.

share similar chemical properties: both are first-row transition metals with similar atomic masses, radii and electronic structures. Therefore, it is not surprising that Mn and Fe might share transport mechanisms. Tf displays two metal ion-binding sites per molecule (M$_r$, 77 000), of which Fe^{3+} occupies only 30%. Consequently, at its usual concentration in plasma (3 mg mL^{-1}), Tf has 50 μmol L^{-1} of unoccupied binding sites available for coordination by Mn^{3+}.[37] Interestingly, purified Mn^{3+}-Tf has been labelled with the fluorescent dye Alexa green. This new tool demonstrated that a significant amount of Mn^{3+} is taken up into neuronal cells by the Tf mechanism in a similar manner to Fe^{3+}-Tf transport. The Mn^{3+}-Tf transport is noticeably slower and less important than other Mn transport processes.[38]

9.4   Transport through the Cytoplasmic Membrane

Mn is an opportunistic metal and the available evidence suggests that, in mammals, all Mn-specific uptake and transport proteins have yet to be identified. Mn binds to proteins, such as Tf, that bind other metal ions. Mn enters into cells *via* DMT1 and/or Tf receptor-mediated endocytosis (Figure 9.2).[39–41] Mn^{3+} complexes with Tf and binds to the Tf receptor (TfR) allowing internalization of the Mn complex.[42] This internalization is dependent on iron levels and pH, especially in the basal ganglia.[43] Following TfR-mediated endocytosis in endothelial cells, Mn is released from the complex by endosomal acidification and the apo-Tf-TfR complex is returned to the luminal surface.[42] Endothelial Mn is taken up by neurons, oligodendrocytes and astrocytes.

A primary transport mechanism of Mn^{2+} from the extracellular space to the cytoplasm is DMT1. The role of DMT1 in Fe regulation has been examined by several groups.[44,45] The affinity of DMT1 for Mn has also been established[46] and Fe is capable of inhibiting $\sim$75% of Mn uptake in pheochromocytoma PC12 cells.[46] DMT1 expression has been shown to co-localize with TfR,[47] indicating overlap of the two systems. DMT1 is expressed

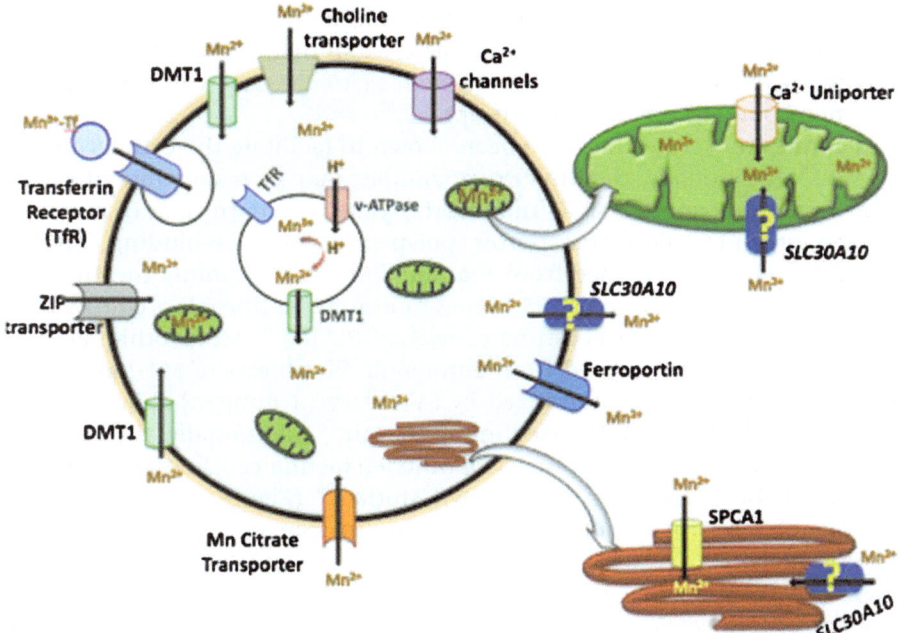

Figure 9.2   The mammalian Mn transport system, including transport across the plasma membrane, as well as into the Golgi lumen and mitochondrial matrix. Though the localization of the putative Mn transporter SLC30A10 currently remains unknown, it is proposed that the transporter either allows Mn efflux from the cell at the plasma membrane or sequesters Mn into organelles from the cytosol.

on the surface of enterocytes and in the olfactory epithelium.[48] The Mn, which is transported into the cellular endosomes by the Mn^{3+}-Tf-TfR complex, is reduced subsequently to Mn^{2+}. This conversion facilitates the transport of Mn^{2+} into the cytosol by DMT1 localized to the endosomal membrane[49] (Figure 9.2).

Mn^{2+} uptake *via* the choline transporter has also been described using Fischer-344 rats and *in situ* brain perfusion. In these studies, increasing doses of 0.4–16 mM Mn^{2+} and/or Cd^{2+} resulted in inhibition of brain choline uptake, suggesting Mn^{2+} can enter cells through the choline transporter. Cu^{2+} or Al^{3+} ions did not have the same effect, suggesting some specificity. Mn transport can also occur *via* voltage-gated and store-operated calcium channels, including ionotropic glutamate receptor Ca^{2+} channels. Uptake of Mn by calcium channels was detected using EPR spectroscopy, competition studies and calcium blockers in human erythrocytes.[50] Ca^{2+} inhibited Mn entry into cells using a hepato-carcinoma cell line.[51] Additional evidence for Mn entry through calcium channels was shown in a study where use of Ca^{2+} inhibitors resulted in inhibited entry of Mn.[52] Mn entry through store-operated Ca^{2+} channels (SOCCs) has been demonstrated in multiple systems. Permeability of Ca^{2+} channels by Mn^{2+} ions was demonstrated using Fura2 quenching in rat peritoneal mast cells.[53] Similarly, the inhibitor 2-aminoethoxydiphenyl borate (2APB) blocked store-operated calcium channels and inhibited attainment of basal Mn^{2+} levels (Figure 9.2).[54] Astrocytes are also affected, as 10 µM Ca^{2+} can competitively and non-competitively inhibit Mn uptake.[55]

A citrate transporter has also been shown to facilitate the cellular uptake of Mn.[56] Additionally, the Mn-citrate complex may be a substrate of the organic anion transporter MCT (monocarboxylate transporter), or of members of the organic anion transporter polypeptide or ATP-binding cassette families.[4] Further evidence from mammalian systems points to Mn transport *via* the Zn transporter ZIP8. Using mouse foetal fibroblast cultures, the affinity of ZIP8 for Mn was estimated as $K_m = 2.2$ µM.[57] Metallothionein-null Cd^{2+}-resistant cells (derived from embryonic fibroblasts of MT-I and MT-II knock-out mice and immortalized by SV40 large T antigen) showed downregulation of ZIP8 and decreased uptake of Mn.[58] Additionally, the transport of Mn through ZIP8 in the rat basophilic leukaemia cell line RBL-2H3 was confirmed by uptake and knockdown studies[59] (Figure 9.2). The physiological relevance of transport through the Zn transporters remains to be determined.

In lower organisms, such as bacteria and yeast, Mn functions as an important protective anti-oxidant [Mn-superoxide dismutase (SOD)], as suppression of oxidative damage is correlated with increased Mn accumulation.[60] Compared to eukaryotes, higher transporter selectivity towards Mn has been evident through several studies. There are three different classes of Mn transporters in prokaryotes: the Nramp H^+-Mn^{2+} transporters, the ABC Mn permeases and the Mn transporting P-type ATPase.[61] Mn transport is relatively simple in prokaryotic cells as there are no intracellular organelles. The Nramp Mn transporters, homologues of human DMT1 described above, are the most conserved Mn transporters from bacteria to human.[62] The first Nramp Mn transporter was found in mice susceptible to infections caused by *Salmonella typhimurium*, *Leishmania donovani* and *Mycobacterium bovis*.[63–65] The mutant mouse gene was mapped out, encoding a transporter protein expressed mainly in macrophages. When expressing the mutant gene, the macrophages lost their capability to combat intracellular pathogens.[66] Thus, this gene was named Nramp (natural resistance associated macrophage protein). The prokaryotic Nramp transporters contain about 450 amino acids, with 10 to 12 transmembrane domains, although the detailed structure remains unknown.[61] Mn transport was first demonstrated in *Bacillus subtilis*[67] and *Escherichia coli*[68] and found to be highly selective and energy dependent. About three decades later, an Nramp homologue known as MntH, as well as MntABCD (an ABC class transporter), were discovered in *S. aureus*. Both promote accumulation of Mn^{2+}, but not of Mg^{2+}, Ca^{2+}, Cu^{2+}, Fe^{2+}, Ni^{2+} or Zn^{2+} ions.[69] Later, Nramp Mn transporters were also identified in other bacteria, such as *S. typhimurium*, *Pseudomonas aeruginosa* and *Burkholderia cepacia*.[70]

The ABC Mn permeases were first reported as adhesion proteins[71] but subsequent studies showed that they were actually high-affinity Mn transporters.[72,73] A permease protein usually contains three major domains: an extracellular lipoprotein acting as a cation binder, a cytoplasmic ATP-binding protein for energy supply and an integral membrane protein regulating cation flux. These transporters seem to have higher affinities for Mn than for other cations. For example, the *S. typhimurium* ABC Mn transporter SitABCD regulates Mn^{2+} transport at $K_{0.5}$ (the cation concentration at which 50% maximal transport occurs) of 100 nM, compared with Zn^{2+} at 3 μM, Cd^{2+} at 3 μM and Fe^{2+} at 30 μM.[61] The P-type ATPase Mn transporter is rare in prokaryotic cells and will not be discussed further.

Interestingly, in yeast, a majority of cytosolic Mn is further transported into different organelles (mainly the Golgi apparatus and the vacuoles), where it is either bioactive or destined to be secreted. This process is quite different from those in prokaryotic cells, as described above. The Nramp Mn transporters have been well characterized in the budding yeast *Saccharomyces cerevisiae*. There are two proteins, Smf1p and Smf2p (suppressor of mitochondria import function) that are homologous to mammalian DMT1. When first identified in 1992, they were considered to function in mitochondrial protein processing, with no indication of a role in metal transport.[74] Three years later, Smf1p was recognized as an Nramp Mn transporter localized to the plasma membrane, with high affinity for Mn.[75] Later, Smf3p was identified as a third Nramp transporter but it transports iron, not Mn.[76] Although Smf1p has been shown to localize on the yeast plasma membrane[75] and on intracellular vacuoles,[77] how Smf1p functions as a Mn transporter remains unknown. This uncertainty arises from evidence that Mn levels remain unchanged in Smf1p deletion mutants under unstressed, Mn sufficient conditions, as did the activity of the Mn-dependent enzyme

SOD2.[78] A recent study showed that Smf1p might play a role in fighting oxidative stress rather than in Mn transport and distribution. Oxidative stress could be reversed by increasing intracellular Mn levels in yeast cells devoid of SOD1 function, but not in the *sod1*Δ & *smf1*Δ double mutant cells.[79] Regardless, it is apparent that yeast *smf2*Δ deletion mutants are less capable of accumulating Mn and that the activity of Mn-dependent enzymes are decreased, due to low intracellular Mn levels.[78]

In contrast to Smf1p, there is no clear evidence that the Smf2p protein localizes to the cell surface. One possible reason is that, since Smf2p has a high affinity for Mn, its localization to the plasma membrane is transient, rendering its levels too low for detection. Similarly, the Smf Mn transporters are also conserved in *Caenorhabditis elegans*. Three DMT1 homologues, known as SMF1, SMF2 and SMF3, have been identified in this nematode, with each transporter having a separate role in Mn homeostasis. SMF3 is the primary uptake transporter, while SMF1 has a minor role. Interestingly, SMF2 may be involved with metal content regulation, through either efflux or sequestration, or as a transcriptional mediator of the other *smf* genes in the worm. Similar to DMT1 isoforms, the differential localization of the transporters is also conserved, as SMF1 and SMF3 are primarily expressed in the gut epithelium, whereas SMF2 is expressed in the pharyngeal epithelium.[80]

The Nramp Mn transporters in yeast are up-regulated when the environmental Mn level is intermediate or low. Under conditions of excess Mn, the high-affinity cell surface phosphate transporter Pho84p becomes the primary Mn transporter.[81] *pho84*Δ yeasts are resistant to high Mn concentrations, as well as to iron, cobalt and copper ions,[81] indicating that this protein is not specifically responsible for Mn transport. However, under conditions of excess Mn, most of the Mn accumulated in yeast cells as $MnHPO_4$ is taken up through Pho84p, demonstrating that it is a major source of excessive Mn.[81]

9.5 Transport within the Cell

Mn^{2+} does not readily bind to sulfhydryl (-SH) groups and so is not transported by metallothioneins. However, Mn is an important component of several intracellular proteins. For example, binding of Mn^{2+} to glutamine synthetase (GS) under physiological conditions has been demonstrated in cultured chick glial cells, in which the absence of measureable free Mn^{2+} in the mitochondria was also reported.[82] Such metallo-enzymes depend on specific coordination of Mn ions for proper catalytic functioning. Mn superoxide dismutase (Mn-SOD) is another Mn-binding metalloenzyme and catalyses the disproportionation of superoxide into hydrogen peroxide and dioxygen. It is a tetramer, with each monomer binding a single Mn ion[83] (Figure 9.1). The superoxide radical HO_2 binds to a free site on the Mn^{3+} form of the enzyme resulting in reduction to Mn^{2+} and production of a dioxygen molecule upon protonation of a neighbouring hydroxo ligand.

A second HO_2 radical then oxidizes the Mn^{2+} form, and transfer of a proton from the Mn^{2+}-aqua ligand results in reduced product H_2O_2.[84] The active sites of Mn-SOD are indistinguishable from Fe-SOD, and both metals are coordinated to three histidines, one aspartate and one solvent molecule in distorted trigonal bipyramidal geometry.

There are also examples of interesting Mn coordination chemistry in plant cells, as Mn is a necessary component of a metal ion cluster in the protein complex Photosystem II in plants, a vital complex that produces electrons in photosynthesis from the oxidation of water to ultimately generate energy as ATP. The manganese centre of this complex is also known as the oxygen-evolving complex, or OEC.[84] The defining property of the OEC is that it produces O_2 and makes an O–O bond, the latter being the only example in biology where an O–O bond is formed from two H_2O molecules, in the absence of release of long-lived intermediates (such as H_2O_2). The OEC is composed of four Mn ions with differing oxidation states, as well as one Ca^{2+} ion, and oxidizes water to produce dioxygen and four protons *via* the transfer of four electrons from the water molecule. Though the structure of the OEC remains controversial, X-ray crystallography of the cluster has revealed a cubane-like structure[84,85] (Figure 9.1). Mn(IV) complexes are predominately found with bridging or terminal oxo-ligands, and polynuclear Mn–O clusters of mixed valence have been identified.[147] The OEC consists of an inorganic tetra-manganese penta-oxygen calcium (Mn_4O_5Ca) cluster and the nearby redox-active tyrosine residue YZ (D1-Tyr161) that couples electron transfer from the Mn_4O_5Ca cluster to P680, the photo-oxidant of pigment-protein complex photosystem II (PSII).[148] It has been recently shown that ammonia perturbs the exchangeable μ-oxo bridge of the OEC, with ammonia binding elongating the MnA4-O5 bond in the OEC, thus perturbing the μ-oxo bridge resonance.[148]

9.6 Transport through Subcellular Membranes

In mammalian cells, the secretory pathway Ca^{2+}/Mn^{2+} ATPase isoform 1 (SPCA1) is localized to the membrane of the Golgi apparatus.[86] SPCA1 is the only known P-type ATPase with high affinity for transporting Mn^{2+} from the cytosol to the Golgi lumen (Figure 9.2). Moreover, silencing of SPCA1 results in deficiency of Ca^{2+} or Mn^{2+} uptake inside the Golgi stores.[87] However, intracellular Mn is found primarily within mitochondria, the nucleus and synaptosomes.[88] Mitochondria have an existing transport system for Ca^{2+}, and due to the affinity of Mn^{2+} for Ca^{2+} binding and transport sites, the mitochondrial Ca^{2+} transport mechanisms are important. Mn^{2+} influx to the mitochondrion has been reported to occur through the Ca^{2+} uniporter[89–91] (Figure 9.2), and efflux occurs *via* Na^+-independent mechanisms,[92] which is an active (energy-requiring) mechanism. The mitochondrial Ca^{2+} uniporter is a cooperative transport mechanism with an external activation site and a transport site.[88] Ca^{2+} binding to this external activation site increases the uptake velocity of both Ca^{2+} and Mn^{2+}. Mn in the mitochondrion interferes with oxidative phosphorylation and consequently can inhibit efflux of Ca^{2+} (K_i values of 7.9 and 5.1 nmol mg^{-1} protein for Na^+-independent and -dependent efflux, respectively). This can result in disruption of membrane integrity under conditions of Mn overload,[88] characterized primarily by inhibition of State 3 mitochondrial respiration using either succinate or malate plus glutamate as substrate.[88]

In yeast, Pmr1p (plasma membrane ATPase-related) encodes a P-type Ca- and Mn-transporting ATPase localized on the Golgi membrane and responsible for transport of Mn and Ca from the cytosol into the Golgi.[93,94] This protein is important for processing and trafficking of secretory proteins as well as excretion of excessive Mn. $pmr1\Delta$ yeasts hyperaccumulated Mn and the cells were hypersensitive to Mn toxicity[93,95] due to their inability to transport excessive cytosolic Mn into the Golgi for excretion. Mn transported in the Golgi lumen is required for proper function of various Mn-dependent sugar transferases that modify glycoproteins before secretion. Mutations in its human homologue hSPCA1 cause the disorder Hailey–Hailey disease (severe skin blistering) due to improper protein glycosylation.[95] Function-structure studies have revealed that the N-terminal "EF hand" motif and the transmembrane helices 4–6 are responsible for the relative selectivity of Ca and Mn, respectively.[96,97]

While excessive Mn can be secreted through the Pmr1p-mediated secretory pathway, vacuoles also play an important role in excessive metal ion storage and detoxification in yeast. Their importance is indicated by the fact that more Mn is found in vacuoles than in the cytosol.[98,99] To date, it is clear that Ccc1p (cross complementer of csg1) functions as a vacuolar Mn transporter to import Mn from the cytosol into vacuoles. Overexpression of Ccc1p in a $pmr1\Delta$ mutant can rescue the toxicity caused by Mn overexposure.[100] The protective role of Ccc1p against Mn toxicity was later ascribed to vacuolar sequestration of Mn *via* Ccc1p, identifying it as a significant Mn transporter in vacuoles.[101]

Manganese is also an essential trace metal for plants as a cofactor of MnSOD in the mitochondrion[102] and as part of the catalytic centre of the water-splitting complex in PSII in the chloroplast.[103] Although Mn homeostasis in plants remains largely undefined, two Mn transporters have been well studied in *Arabidopsis thaliana* (*A. thaliana*). One is AtECA3, a member of the P-type ATPase family. Its expression in yeast resulted in localization in the Golgi or endosomes.[104,105] When carrying mutant *AtECA3*, *A. thaliana* suffered from severe developmental defects at low Mn conditions. The problem could be rescued by Mn supplementation.[105] Such evidence indicates that AtECA3 probably functions to transport Mn into the cytosol or endosomes.

A member of the cation diffusion facilitator (CDF) family AtMTP11 also plays a role in plant Mn transport. When carrying mutant *AtMTP11*, *A. thaliana* accumulated more Mn than wild-type,[106,107] with localization in prevacuolar[106] and *trans*-Golgi compartments.[107] This indicates a function in transport of excess Mn from the cytosol to vacuoles and Golgi for secretion.

9.7 Sensing, Buffering and other Control Mechanisms

Endogenous total Mn concentrations are relatively low compared to other metals, with an apparent subtoxic tissue threshold around 80–100 μM.[108–110] A transporter that solely transports Mn ions is yet to be identified. Consequently, a complex interplay of transport systems used also by other metals[111] may be the key to an understanding of the mechanisms behind Mn homeostasis and toxicity. In particular, Mn accumulation in the brain upon overexposure can result in conditions of neurotoxicity, with areas such as the globus pallidus, caudate and putamen, and substantia nigra showing the highest deposition.[6] In addition to the aforementioned mitochondrial and Golgi sequestration, novel findings reveal the role of calprotectin (CP), the innate immune factor in neutrophils that protects against bacterial pathogens, in sequestering Mn and Zn.[112] This protein works against pathogens and the spread of bacterial infections by promoting bacterial metal starvation through Mn and Zn sequestration. The high affinity for Mn^{2+}, even over Zn^{2+}, arises from its asymmetrical heterodimer structure that allows one Mn binding site made of six histidine residues, and this binding is necessary for CP's antimicrobial features.[113] Other Mn-associated processes include binding to Mn-dependent metallo-enzymes or -proteins, transcriptional and translational regulation of Mn transporters and efficient efflux of Mn from the cell.

As Mn is an enzymatic cofactor, proper metalloregulation can promote optimal intracellular Mn levels, with some of this regulation occurring in a cell type-specific manner. Within the brain, astroglia are thought to be a significant sink for Mn accumulation. This is primarily due to the localization of glutamine synthetase (GS) within astroglia, an enzyme necessary for the glutamate-glutamine shuttle. Nearly 80% of brain Mn is found associated with GS, as Mn is a necessary cofactor for GS activity.[114] Consequently, the abundance and selectivity of Mn accumulation in these cells can result in astroglia being the initial targets of toxicity. Similar to GS, there are several other enzymes that require Mn for proper enzymatic function. These include Mn-SOD or SOD2, the aforementioned antioxidant mitochondrial protein that converts superoxide into hydrogen peroxide and $O_{2.}$ Another Mn-containing mitochondrial protein is pyruvate carboxylase (PC), which catalyses the carboxylation of pyruvate to oxaloacetate and serves as a bridge between carbohydrate and lipid metabolism.[115] Either Mg^{2+} or Mn^{2+} can activate PC.[116] Yet another Mn-containing metalloenzyme is arginase (ARG1 and ARG2), which plays a vital role in the urea cycle, converting arginine into urea and ornithine.[117] It requires two Mn ions for activation. Finally, certain serine/threonine protein phosphatases (PPMs) are also metal-dependent (PP2C), appearing to require two Mg^{2+} or Mn^{2+} ions for their function in dephosphorylation of proteins at serine and threonine sites.[118]

Mn^{2+} does not compete well with other transition metals ions for typical protein binding sites (Irving–Williams series). One mechanism for its selection has been identified in the cyanobacterium *Synechocystis* PCC 6803. Discrimination in metal-binding and subsequent regulation of metal content in a manganese-binding cupin MncA and a copper-binding cupin CucfA is achieved by protein folding and metal uptake in different cellular compartments (cytoplasm *vs.* periplasm).[119]

Genetic regulation of Mn transporter expression also plays a role in Mn homeostasis. Mammalian systems corroborate a role of transcriptional regulation of a transporter that can regulate Mn (as well as other metal) levels, such as DMT1.[120] DMT1 has four mRNA transcripts encoding four different proteins that differ in their N- and C-termini processing. Two of the alternatively spliced isoforms differ in their transcription start sites: Exon 1A contains an AUG codon that extends its N-terminal, while Exon 1B lacks this initiator codon and, consequently, translation begins in exon 2. Two of the isoforms contain an iron responsive element (IRE) in the 3' UTR of the DMT1 mRNA. IREs serve an important sensing role. If iron levels are low (which can indirectly affect intracellular Mn concentrations), iron regulatory proteins (IRPs) bind to the IRE in the DMT1 isoforms to stabilize the IRE message.[121] The presence of such isoforms of DMT1 and the influence of iron on their expression most likely has an effect on intracellular Mn concentrations as well, though further studies are needed.

The Nramp Mn transporters Smf1p and Smf2p in yeast are usually regulated post-translationally. Excessive production of transporters is quickly eliminated based on demand. This is supported by the evidence that 90% of newly synthesized Smf1p and Smf2p undergo vacuole-mediated degradation when Mn is neither deficient nor excessively toxic.[122] During vacuolar degradation, Smf1p and Smf2p are first ubiquitinated by the E3 ubiquitin ligase Rsp5p in the Golgi, then enter multi-vesicular bodies and eventually are degraded in the vacuoles.[123,124] The ubiquitination by Rsp5p requires at least three adaptor proteins: Bsd2p (bypass SOD1 defect),[125] Tre1p and Tre2p (Tf receptor-like).[123] Rsp5p only recognizes substrates with proline-rich PY (proline-tyrosine-containing) sequences, which are absent in Smf1p and Smf2p.[123,124] Bsd2p has 2 PY motifs, while Tre1p and Tre2p each have one. Thus, binding with Bsd2p, Tre1p and Tre2p activates recognition of Smf1p and Smf2p by Rsp5p to facilitate the degradation process. This is indicated by evidence that Smf1p and Smf2p levels, as well as the Mn concentration, are increased dramatically in either one of *bsd2Δ*, *Tre1Δ* or *Tre2Δ* yeast cells.[76,77,79,123] In summary, vacuolar degradation of Smf1p and Smf2p decreases under Mn starvation conditions, increases under Mn excess conditions and remains at a basal level under physiological Mn conditions.

In contrast to the Nramp Mn transporters, the phosphate transporter Pho84p is primarily regulated transcriptionally. Pho84p levels are positively regulated by the transcription factor Pho4p, which is negatively controlled by Pho80p and Pho85p.[126,127] In *pho80Δ* and *pho85Δ* mutant yeast strains, Pho84p levels are high and Mn hyperaccumulates.[122] The regulation of the Golgi Mn transporter Pmr1p and the vacuolar Mn transporter Ccc1p remains largely unknown.

In prokaryotes, sensing and subsequent buffering can occur through transcriptional regulation of Mn transporters. MntR is the primary transcriptional suppressor of Mn transporters. When intracellular Mn levels are adequate, MntR binds to Mn^{2+}, and then translocates to the promoter regions of *mntH* and *mnt/sitABCD* genes and suppresses transcription of these transporter genes. On the other hand, when intracellular Mn levels are low, the affinity of MntR to the promoter sequence decreases, promoting detachment of MntR from the *cis*-regulatory element to allow transcription of the transporter genes.[128] In addition to MntR, iron is also able to regulate transcription of Mn transporters. For example, the transcription of *mntH* and *sitABCD* is greatly decreased by adding Fe^{2+} to the growth media of *E. coli*[129] and *S. typhimurium*.[130] There are binding sites in the promoter regions of *mntH* and *sitABCD* genes that interact with iron-binding proteins, separate from the MntR binding sites.[129] Moreover, as Mn is involved in combating oxidative stress in prokaryotic cells, OxyR and PerR also play a role in regulating Mn transporters as reactive oxygen sensors.[61]

Another example of Mn buffering is SOD2 regulation *via* chaperones. In yeast, the protein Mtm1, while not a general mitochondrial Mn transporter, is necessary for the insertion of Mn into SOD2.[131] In *E. coli*, the chaperone GroESL can increase SOD activity through an increase in SOD metallation that is post-translationally regulated.[132] Evidence has also suggested that the heat shock protein Hsp60 is necessary for proper Mn-SOD folding and function.[133] Another major mammalian regulator of Mn-SOD activity is the mitochondrial deacetylase sirtuin-3 (SIRT3), a protein that is dependent on NAD^+ and acts as a cellular metabolic regulator.[134] The ability of SIRT3 to affect SOD2 activity is thought to occur *via* the deacetylation of specific lysine residues (lysines 68 and 122).[135]

Currently, the only efflux mechanism known for Mn is through the iron exporter ferroportin (Fpn). Fpn expression decreased Mn accumulation within cells, thereby reducing Mn-induced cytotoxicity.[136] These data were supported by export of Mn in *Xenopus laevis* oocytes expressing human Fpn (SLC40A1).[137] Mn accumulation was decreased in oocytes expressing Fpn with no difference in uptake. Co-expression of Fpn and DMT1 in oocytes decreased overall accumulation to a greater extent than in oocytes expressing DMT1 alone. Export of Mn was also concentration-dependent and inhibited by lower pH, substitution of K^+ for Na^+ in the media and partially by addition of Fe, Co and Ni salts. Though additional studies must be conducted to confirm this putative Mn exporter (Figure 9.2) and to define the exact mechanism behind the extrusion, it is logical that Mn would share the same exporter as Fe in the same way the two metals share uptake mechanisms. Interestingly, recent findings have described the zinc transporter, ZnT10 (*SLC30A10*), as potentially exporting Mn as well. Using whole-genome homozygosity mapping and sequencing, alterations in the sequence of the *SLC30A10* gene in families with hyper-manganesaemia were found. To probe the effect of these alterations, human wild-type *SLC30A10*, and/or *SLC30A10* missense or nonsense mutations were introduced into the Mn-sensitive yeast strain Δ*pmr1*. It was found that Δ*pmr1*-expressing wild-type *SLC30A10* rescued growth while Δ*pmr1* expressing *SLC30A10* missense or nonsense mutations did not regain resistance to Mn, suggesting that normal *SLC30A10* is important for Mn export.[138] No localization studies have been conducted to date. Protection of ZnT10 against Mn toxicity indicates either expression at the plasma membrane to expel Mn from the cell or organellar membrane localization that would buffer Mn levels in the cytosol by bringing Mn into mitochondria or the Golgi lumen (Figure 9.2).

9.8   Conclusions with Open Questions

The intricacies surrounding intracellular Mn homeostasis reveal a complex network of regulation necessary both to maintain the optimal levels for proper functioning, while also protecting against toxicity. Though still under current examination, several uptake mechanisms have been reported. They include the DMT1 transporter, Tf-TfR-mediated endocytosis, a number of Ca^{2+} channels, a choline transporter, an organic anion transporter (*via* binding of Mn^{2+} to citrate) and the ZIP family of Zn transporters. In lower organisms, there are conserved homologues of the mammalian DMT1 transporters, in addition to ABC Mn permeases and P-type ATPase Mn transporters. Despite the abundance of studies looking at the role of these transporters in Mn uptake, further work must focus on how these transporters work together in order to control Mn levels, as compensation may mediate uptake in the case of a non-functioning transporter or group of transporters. In terms of efflux pumps, currently only ferroportin has been definitively identified. It seems disadvantageous for the cell to have many modes of uptake but few efflux mechanisms. The recent identification of the SLC30A10 as a putative Mn exporter (in addition to Zn) is a step in the right direction.

Compared to transport through the plasma membrane, subcellular Mn transport, buffering and sequestration remains more of a mystery. Currently known organellar transporters include the Golgi-specific SPCA1 transporter (Pmr1p in yeast) and the mitochondrial Ca^{2+} uniporter, with yeast also containing vacuole-specific Mn transporters (Ccc1p) to sequester Mn from the cytosol. Interplay between DMT1 (Mn^{2+}) and TfR-mediated (Mn^{3+}) uptake can be considered an endosome/vacuole-specific process in mammalians that is dependent on Mn oxidation state. The PD-related gene *ATP13A2* (*PARK9*) encoding a P-type ATPase has also been shown to suppress Mn toxicity,[139] indicating a potential role in mediating Mn homeostasis *via* lysosomal sequestration.[140]

The cell has also evolved several mechanisms to sense and buffer optimal Mn levels. These appear to include metal response elements such as the IREs in DMT1 that may indirectly affect intracellular Mn levels in response to Fe levels. However, further work is needed to examine the role of the various DMT1 isoforms in buffering Mn levels, in addition to the cell specificity in isoform-specific metal content regulation. There can also be transcriptional regulation of Mn transporters, as seen with the Pho84p in yeast, as well as post-translational regulation with Smf1p and Smf2p. In prokaryotes, transcriptional repression *via* MntR is the main form of sensing and of buffering optimal intracellular levels. Though there are several Mn-containing metalloenzymes (GS, SOD2, ARG1/2, PC, PP2C, *etc.*), Mn does not react with biological sulfhydryl groups, thereby eliminating the roles of metallothionein and glutathione in storage and detoxification that is typical of other metals. It is apparent that further work is needed to define how Mn is transported within the cell, as relatively few Mn-selective binding partners and organelle-specific transporters are known at present.

Acknowledgements

This chapter was supported in part by grants from the National Institute of Environmental Health Sciences ESR01-10563, R01-07331, ES T32-007028 and the Molecular Toxicology Center ES P30-000267.

References

1. ATSDR, *US Department of Health and Human Services, Public Service*, 2008.
2. J. L. Aschner and M. Aschner, *Mol. Aspects Med.*, 2005, **26**, 353–362.
3. J. A. Roth, *Biol. Res.*, 2006, **39**, 45–57.
4. M. Aschner, T. R. Guilarte, J. S. Schneider and W. Zheng, *Toxicol. Appl. Pharmacol.*, 2007, **221**, 131–147.
5. B. A. Racette, S. R. Criswell, J. I. Lundin, A. Hobson, N. Seixas, P. T. Kotzbauer, B. A. Evanoff, J. S. Perlmutter, J. Zhang, L. Sheppard and H. Checkoway, *NeuroToxicology*, 2012, **33**, 1356–1361.
6. S. R. Criswell, J. S. Perlmutter, J. L. Huang, N. Golchin, H. P. Flores, A. Hobson, M. Aschner, K. M. Erikson, H. Checkoway and B. A. Racette, *Occup. Environ. Med.*, 2012, **69**, 437–443.
7. H. M. Zeron, M. R. Rodriguez, S. Montes and C. R. Castaneda, *J. Trace Elem. Med. Biol.*, 2011, **25**, 225–229.
8. K. Fitzgerald, V. Mikalunas, H. Rubin, R. McCarthey, A. Vanagunas and R. M. Craig, *J. Parenter. Enteral Nutr.*, 1999, **23**, 333–336.
9. E. A. Smith, P. Newland, K. G. Bestwick and N. Ahmed, *J. Trace Elem. Med. Biol.*, 2013, **27**(1), 65–69.
10. V. A. Fitsanakis, N. Zhang, M. J. Avison, K. M. Erikson, J. C. Gore and M. Aschner, *Toxicol. Sci.*, 2011, **120**, 146–153.
11. C. D. Davis, L. Zech and J. L. Greger, *Proc. Soc. Exp. Biol. Med.*, 1993, **202**, 103–108.
12. J. W. Finley, P. E. Johnson and L. K. Johnson, *Am. J. Clin. Nutr.*, 1994, **60**, 949–955.
13. L. Davidsson, A. Cederblad, E. Hagebo, B. Lonnerdal and B. Sandstrom, *J. Nutr.*, 1988, **118**, 1517–1521.

14. E. A. Malecki, G. M. Radzanowski, T. J. Radzanowski, D. D. Gallaher and J. L. Greger, *J. Nutr.*, 1996, **126**, 489–498.
15. H. A. Schroeder, J. J. Balassa and I. H. Tipton, *J. Chronic Dis.*, 1996, **19**, 545–571.
16. L. A. Davidson and B. Lonnerdal, *Am. J. Physiol.*, 1989, **257**, G930–934.
17. G. Leblondel and P. Allain, *Biol. Trace Elem. Res.*, 1999, **67**, 13–28.
18. J. A. Garcia-Aranda, R. A. Wapnir and F. Lifshitz, *J. Nutr.*, 1983, **113**, 2601–2607.
19. J. G. Bell, C. L. Keen and B. Lonnerdal, *J. Toxicol. Environ. Health*, 1989, **26**, 387–398.
20. M. D. Garrick, K. G. Dolan, C. Horbinski, A. J. Ghio, D. Higgins, M. Porubcin, E. G. Moore, L. N. Hainsworth, J. N. Umbreit, M. E. Conrad, L. Feng, A. Lis, J. A. Roth, S. Singleton and L. M. Garrick, *Biometals*, 2003, **16**, 41–54.
21. L. Davidsson, A. Cederblad, B. Lonnerdal and B. Sandstrom, *Am. J. Clin. Nutr.*, 1991, **54**, 1065–1070.
22. A. A. Britton and G. C. Cotzias, *Am. J. Physiol.*, 1966, **211**, 203–206.
23. D. C. Dorman, M. F. Struve, R. A. James, B. E. McManus, M. W. Marshall and B. A. Wong, *Toxicol. Sci.*, 2001, **60**, 242–251.
24. C. L. Keen, J. G. Bell and B. Lonnerdal, *J. Nutr.*, 1986, **116**, 395–402.
25. S. H. Zlotkin, S. Atkinson and G. Lockitch, *Clin. Perinatol.*, 1995, **22**, 223–240.
26. R. W. Leggett, *Sci. Total Environ.*, 2011, **409**, 4179–4186.
27. S. H. Reaney, C. L. Kwik-Uribe and D. R. Smith, *Chem. Res. Toxicol.*, 2002, **15**, 1119–1126.
28. R. A. Yokel, *NeuroMol. Med.*, 2009, **11**, 297–310.
29. T. Jursa and D. R. Smith, *Toxicol. Sci.*, 2009, **107**, 182–193.
30. A. M. Scheuhammer and M. G. Cherian, *Biochim. Biophys. Acta*, 1985, **840**, 163–169.
31. W. R. Harris and Y. Chen, *J. Inorg. Biochem.*, 1994, **54**, 1–19.
32. O. Rabin, L. Hegedus, J. M. Bourre and Q. R. Smith, *J. Neurochem.*, 1993, **61**, 509–517.
33. A. C. Foradori, A. Bertinchamps, J. M. Gulibon and G. C. Cotzias, *J. Gen. Physiol.*, 1967, **50**, 2255–2266.
34. V. Nischwitz, A. Berthele and B. Michalke, *Anal. Chim. Acta*, 2008, **627**, 258–269.
35. G. Fanali, Y. Cao, P. Ascenzi and M. Fasano, *J. Inorg. Biochem.*, 2012, **117**, 198–203.
36. L. Davidsson, B. Lonnerdal, B. Sandstrom, C. Kunz and C. L. Keen, *J. Nutr.*, 1989, **119**, 1461–1464.
37. M. Aschner, G. Shanker, K. Erikson, J. Yang and L. A. Mutkus, *Neuro-Toxicology*, 2002, **23**, 165–168.
38. T. E. Gunter, B. Gerstner, K. K. Gunter, J. Malecki, R. Gelein, W. M. Valentine, M. Aschner and D. I. Yule, *NeuroToxicology*, 2013, **34**, 118–127.

39. M. Aschner and M. Gannon, *Brain Res. Bull.*, 1994, **33**, 345–349.
40. C. Au, A. Benedetto and M. Aschner, *NeuroToxicology*, 2008, **29**, 569–576.
41. K. Erikson and M. Aschner, *NeuroToxicology*, 2002, **23**, 595–602.
42. C. M. Morris, A. B. Keith, J. A. Edwardson and R. G. Pullen, *J. Neurochem.*, 1992, **59**, 300–306.
43. E. Huang, W. Y. Ong and J. R. Connor, *Neuroscience*, 2004, **128**, 487–496.
44. M. E. Conrad and J. N. Umbreit, *Blood Cells Mol. Dis.*, 2002, **29**, 336–355.
45. B. Mackenzie and M. A. Hediger, *Pflugers Arch.*, 2004, **447**, 571–579.
46. J. A. Roth and M. D. Garrick, *Biochem. Pharmacol.*, 2003, **66**, 1–13.
47. D. L. Stredrick, A. H. Stokes, T. J. Worst, W. M. Freeman, E. A. Johnson, L. H. Lash, M. Aschner and K. E. Vrana, *NeuroToxicology*, 2004, **25**, 543–553.
48. J. R. Burdo, S. L. Menzies, I. A. Simpson, L. M. Garrick, M. D. Garrick, K. G. Dolan, D. J. Haile, J. L. Beard and J. R. Connor, *J. Neurosci. Res.*, 2001, **66**, 1198–1207.
49. N. Touret, W. Furuya, J. Forbes, P. Gros and S. Grinstein, *J. Biol. Chem.*, 2003, **278**, 25548–25557.
50. C. M. Lucaciu, C. Dragu, L. Copaescu and V. V. Morariu, *Biochim. Biophys. Acta*, 1997, **1328**, 90–98.
51. J. W. Finley, *Biol. Trace Elem. Res.*, 1998, **64**, 101–118.
52. R. P. Mason, *Biochem. Pharmacol.*, 1993, **45**, 2173–2183.
53. C. Fasolato, M. Hoth and R. Penner, *J. Biol. Chem.*, 1993, **268**, 20737–20740.
54. Y. Dobrydneva and P. Blackmore, *Mol. Pharmacol.*, 2001, **60**, 541–552.
55. M. Aschner, M. Gannon and H. K. Kimelberg, *J. Neurochem.*, 1992, **58**, 730–735.
56. J. S. Crossgrove, D. D. Allen, B. L. Bukaveckas, S. S. Rhineheimer and R. A. Yokel, *NeuroToxicology*, 2003, **24**, 3–13.
57. L. He, K. Girijashanker, T. P. Dalton, J. Reed, H. Li, M. Soleimani and D. W. Nebert, *Mol. Pharmacol.*, 2006, **70**, 171–180.
58. S. Himeno, T. Yanagiya and H. Fujishiro, *Biochimie*, 2009, **91**, 1218–1222.
59. H. Fujishiro, M. Doi, S. Enomoto and S. Himeno, *Metallomics*, 2011, **3**, 710–718.
60. A. R. Reddi, L. T. Jensen, A. Naranuntarat, L. Rosenfeld, E. Leung, R. Shah and V. C. Culotta, *Free Radic. Biol. Med.*, 2009, **46**, 154–162.
61. K. M. Papp-Wallace and M. E. Maguire, *Annu. Rev. Microbiol.*, 2006, **60**, 187–209.
62. M. Cellier, G. Privé, A. Belouchi, T. Kwan, V. Rodrigues, W. Chia and P. Gros, *Proc. Natl Acad. Sci.*, 1995, **92**, 10089–10093.
63. J. Plant and A. A. Glynn, *Nature*, 1974, **248**, 345–347.
64. J. Plant and A. A. Glynn, *J. Infect. Dis.*, 1976, **133**, 72–78.
65. J. E. Plant, J. M. Blackwell, A. D. O'Brien, D. J. Bradley and A. A. Glynn, *Nature*, 1982, **297**, 510–511.

66. A. A. Glynn, D. J. Bradley, J. M. Blackwell and J. E. Plant, *The Lancet*, 1982, **320**, 151.
67. E. Eisenstadt, S. Fisher, C.-L. Der and S. Silver, *J. Bacteriol.*, 1973, **113**, 1363–1372.
68. S. Silver and M. L. Kralovic, *Biochem. Biophys. Res. Commun.*, 1969, **34**, 640–645.
69. M. J. Horsburgh, S. J. Wharton, A. G. Cox, E. Ingham, S. Peacock and S. J. Foster, *Mol. Microbiol.*, 2002, **44**, 1269–1286.
70. D. G. Kehres, M. L. Zaharik, B. B. Finlay and M. E. Maguire, *Mol. Microbiol.*, 2000, **36**, 1085–1100.
71. H. F. Jenkinson, *FEMS Microbiol. Lett.*, 1994, **121**, 133–140.
72. T. Kitten, C. L. Munro, S. M. Michalek and F. L. Macrina, *Infec. Immun.*, 2000, **68**, 4441–4451.
73. P. E. Kolenbrander, R. N. Andersen, R. A. Baker and H. F. Jenkinson, *J. Bacteriol.*, 1998, **180**, 290–295.
74. A. H. West, D. J. Clark, J. Martin, W. Neupert, F. U. Hartl and A. L. Horwich, *J. Biol. Chem.*, 1992, **267**, 24625–24633.
75. F. Supek, L. Supekova, H. Nelson and N. Nelson, *Proc. Natl Acad. Sci.*, 1996, **93**, 5105–5110.
76. M. E. Portnoy, X. F. Liu and V. C. Culotta, *Mol. Cell. Biol.*, 2000, **20**, 7893–7902.
77. X. F. Liu and V. C. Culotta, *J. Biol. Chem.*, 1999, **274**, 4863–4868.
78. E. E.-C. Luk and V. C. Culotta, *J. Biol. Chem.*, 2001, **276**, 47556–47562.
79. A. R. Reddi, L. T. Jensen, A. Naranuntarat, L. Rosenfeld, E. Leung, R. Shah and V. C. Culotta, *Free Radic. Biol. Med.*, 2009, **46**, 154–162.
80. C. Au, A. Benedetto, J. Anderson, A. Labrousse, K. Erikson, J. J. Ewbank and M. Aschner, *PloS One*, 2009, **4**, e7792.
81. L. T. Jensen, M. Ajua-Alemanji and V. C. Culotta, *J. Biol. Chem.*, 2003, **278**, 42036–42040.
82. F. C. Wedler, M. C. Vichnin, B. W. Ley, G. Tholey, M. Ledig and J. C. Copin, *Neurochem. Res.*, 1994, **19**, 145–151.
83. G. E. Borgstahl, H. E. Parge, M. J. Hickey, W. F. Beyer, Jr, R. A. Hallewell and J. A. Tainer, *Cell*, 1992, **71**, 107–118.
84. K. N. Ferreira, T. M. Iverson, K. Maghlaoui, J. Barber and S. Iwata, *Science*, 2004, **303**, 1831–1838.
85. Y. Umena, K. Kawakami, J. R. Shen and N. Kamiya, *Nature*, 2011, **473**, 55–60.
86. R. Murin, S. Verleysdonk, L. Raeymaekers, P. Kaplan and J. Lehotsky, *Cell. Mol. Neurobiol.*, 2006, **26**, 1355–1365.
87. M. R. Sepulveda, J. Vanoevelen, L. Raeymaekers, A. M. Mata and F. Wuytack, *J. Neurosci.*, 2009, **29**, 12174–12182.
88. C. E. Gavin, K. K. Gunter and T. E. Gunter, *NeuroToxicology*, 1999, **20**, 445–453.
89. Z. Drahota, P. Gazzotti, E. Carafoli and C. S. Rossi, *Arch. Biochem. Biophys.*, 1969, **130**, 267–273.

90. R. E. Gunter, J. S. Puskin and P. R. Russell, *Biophys. J.*, 1975, **15**, 319–333.
91. T. E. Gunter and J. S. Puskin, *Biophys. J.*, 1972, **12**, 625–635.
92. C. E. Gavin, K. K. Gunter and T. E. Gunter, *Biochem. J.*, 1990, **266**, 329–334.
93. P. J. Lapinskas, K. W. Cunningham, X. F. Liu, G. R. Fink and V. C. Culotta, *Mol. Cell. Biol.*, 1995, **15**, 1382–1388.
94. A. Antebi and G. R. Fink, *Mol. Biol. Cell*, 1992, **3**, 633–654.
95. V.-K. Ton, D. Mandal, C. Vahadji and R. Rao, *J. Biol. Chem.*, 2002, **277**, 6422–6427.
96. D. Mandal, T. B. Woolf and R. Rao, *J. Biol. Chem.*, 2000, **275**, 23933–23938.
97. M. Xiang, D. Mohamalawari and R. Rao, *J. Biol. Chem.*, 2005, **280**, 11608–11614.
98. M. Yang, L. T. Jensen, A. J. Gardner and V. C. Culotta, *Biochem. J.*, 2005, **386**, 479–487.
99. L. M. Ramsay and G. M. Gadd, *FEMS Microbiol. Lett.*, 1997, **152**, 293–298.
100. P. J. Lapinskas, S.-J. Lin and V. C. Culotta, *Mol. Microbiol.*, 1996, **21**, 519–528.
101. L. Li, O. S. Chen, D. M. Ward and J. Kaplan, *J. Biol. Chem.*, 2001, **276**, 29515–29519.
102. Z. Su, M. F. Chai, P. L. Lu, R. An, J. Chen and X. C. Wang, *Planta*, 2007, **226**, 1031–1039.
103. M. D. Allen, J. Kropat, S. Tottey, J. A. Del Campo and S. S. Merchant, *Plant Physiol.*, 2007, **143**, 263–277.
104. X. Li, S. Chanroj, Z. Wu, S. M. Romanowsky, J. F. Harper and H. Sze, *Plant Physiol.*, 2008, **147**, 1675–1689.
105. R. F. Mills, M. L. Doherty, R. L. Lopez-Marques, T. Weimar, P. Dupree, M. G. Palmgren, J. K. Pittman and L. E. Williams, *Plant Physiol.*, 2008, **146**, 116–128.
106. E. Delhaize, B. D. Gruber, J. K. Pittman, R. G. White, H. Leung, Y. Miao, L. Jiang, P. R. Ryan and A. E. Richardson, *Plant J.*, 2007, **51**, 198–210.
107. E. Peiter, B. Montanini, A. Gobert, P. Pedas, S. R. Husted, F. J. M. Maathuis, D. Blaudez, M. Chalot and D. Sanders, *Proc. Natl Acad. Sci.*, 2007, **104**, 8532–8537.
108. Y. Suzuki, T. Mouri, Y. Suzuki, K. Nishiyama and N. Fujii, *Tokushima J. Exp. Med.*, 1975, **22**, 5–10.
109. M. Yamada, S. Ohno, I. Okayasu, R. Okeda, S. Hatakeyama, H. Watanabe, K. Ushio and H. Tsukagoshi, *Acta Neuropathol.*, 1986, **70**, 273–278.
110. I. L. Crawford and J. D. Connor, *J. Neurochem.*, 1972, **19**, 1451–1458.
111. R. A. Yokel, *J. Alzheimer's Dis.*, 2006, **10**, 223–253.
112. B. D. Corbin, E. H. Seeley, A. Raab, J. Feldmann, M. R. Miller, V. J. Torres, K. L. Anderson, B. M. Dattilo, P. M. Dunman, R. Gerads,

R. M. Caprioli, W. Nacken, W. J. Chazin and E. P. Skaar, *Science*, 2008, **319**, 962–965.

113. S. M. Damo, T. E. Kehl-Fie, N. Sugitani, M. E. Holt, S. Rathi, W. J. Murphy, Y. Zhang, C. Betz, L. Hench, G. Fritz, E. P. Skaar and W. J. Chazin, *Proc. Natl Acad. Sci. USA*, 2013, **110**, 3841–3846.

114. F. C. Wedler, R. B. Denman and W. G. Roby, *Biochemistry*, 1982, **21**, 6389–6396.

115. S. Jitrapakdee, M. St Maurice, I. Rayment, W. W. Cleland, J. C. Wallace and P. V. Attwood, *Biochem. J.*, 2008, **413**, 369–387.

116. M. C. Scrutton, P. Griminger and J. C. Wallace, *J. Biol. Chem.*, 1972, **247**, 3305–3313.

117. Z. F. Kanyo, L. R. Scolnick, D. E. Ash and D. W. Christianson, *Nature*, 1996, **383**, 554–557.

118. A. K. Das, N. R. Helps, P. T. Cohen and D. Barford, *EMBO J.*, 1996, **15**, 6798–6809.

119. S. Tottey, K. J. Waldron, S. J. Firbank, B. Reale, C. Bessant, K. Sato, T. R. Cheek, J. Gray, M. J. Banfield, C. Dennison and N. J. Robinson, *Nature*, 2008, **455**, 1138–1142.

120. M. D. Garrick, L. Zhao, J. A. Roth, H. Jiang, J. Feng, N. J. Foot, H. Dalton, S. Kumar and L. M. Garrick, *Biometals*, 2012, **25**, 787–793.

121. H. Gunshin, B. Mackenzie, U. V. Berger, Y. Gunshin, M. F. Romero, W. F. Boron, S. Nussberger, J. L. Gollan and M. A. Hediger, *Nature*, 1997, **388**, 482–488.

122. A. R. Reddi, L. T. Jensen and V. C. Culotta, *Chem. Rev.*, 2009, **109**, 4722–4732.

123. H. E. M. Stimpson, M. J. Lewis and H. R. B. Pelham, *EMBO J.*, 2006, **25**, 662–672.

124. J. A. Sullivan, M. J. Lewis, E. Nikko and H. R. B. Pelham, *Mol. Biol. Cell*, 2007, **18**, 2429–2440.

125. E. H. Hettema, J. Valdez-Taubas and H. R. B. Pelham, *EMBO J.*, 2004, **23**, 1279–1288.

126. H. C. Hürlimann, M. Stadler-Waibel, T. P. Werner and F. M. Freimoser, *Mol. Biol. Cell*, 2007, **18**, 4438–4445.

127. U. Fristedt, R. Weinander, H.-S. Martinsson and B. L. Persson, *FEBS Lett.*, 1999, **458**, 1–5.

128. S. A. Lieser, T. C. Davis, J. D. Helmann and S. M. Cohen, *Biochemistry*, 2003, **42**, 12634–12642.

129. S. I. Patzer and K. Hantke, *J. Bacteriol.*, 2001, **183**, 4806–4813.

130. D. G. Kehres, A. Janakiraman, J. M. Slauch and M. E. Maguire, *J. Bacteriol.*, 2002, **184**, 3151–3158.

131. E. Luk, M. Carroll, M. Baker and V. C. Culotta, *Proc. Natl Acad. Sci. USA*, 2003, **100**, 10353–10357.

132. G. Hunter and T. Hunter, *Health*, 2013, **5**, 1719–1729.

133. R. Magnoni, J. Palmfeldt, J. Hansen, J. H. Christensen, T. J. Corydon and P. Bross, *Free Radic. Res.*, 2014, **48**(2), 168–179.

134. L. Jin, W. Wei, Y. Jiang, H. Peng, J. Cai, C. Mao, H. Dai, W. Choy, J. E. Bemis, M. R. Jirousek, J. C. Milne, C. H. Westphal and R. B. Perni, *J. Biol. Chem.*, 2009, **284**, 24394–24405.

135. R. Tao, A. Vassilopoulos, L. Parisiadou, Y. Yan and D. Gius, *Antioxid. Redox Signal.*, 2014, **20**(10), 1646–1654.

136. Z. Yin, H. Jiang, E. S. Lee, M. Ni, K. M. Erikson, D. Milatovic, A. B. Bowman and M. Aschner, *J. Neurochem.*, 2010, **112**, 1190–1198.

137. M. S. Madejczyk and N. Ballatori, *Biochim. Biophys. Acta*, 2012, **1818**, 651–657.

138. K. Tuschl, P. T. Clayton, S. M. Gospe, Jr., S. Gulab, S. Ibrahim, P. Singhi, R. Aulakh, R. T. Ribeiro, O. G. Barsottini, M. S. Zaki, M. L. Del Rosario, S. Dyack, V. Price, A. Rideout, K. Gordon, R. A. Wevers, W. K. Chong and P. B. Mills, *Am. J. Hum. Genet.*, 2012, **90**, 457–466.

139. A. D. Gitler, A. Chesi, M. L. Geddie, K. E. Strathearn, S. Hamamichi, K. J. Hill, K. A. Caldwell, G. A. Caldwell, A. A. Cooper, J. C. Rochet and S. Lindquist, *Nat. Genet.*, 2009, **41**, 308–315.

140. J. Tan, T. Zhang, L. Jiang, J. Chi, D. Hu, Q. Pan, D. Wang and Z. Zhang, *J. Biol. Chem.*, 2011, **286**, 29654–29662.

141. D. E. Ash, *J. Nutr.*, 2004, **134**, 2760S–2764S; discussion 2765S–2767S.

142. S. Singh, R. J. Debus, T. Wydrzynski and W. Hillier, *Philos. Trans. R. Soc. B*, 2008, **363**, 1229–1234; discussion 1234–1235.

143. M. M. Whittaker, T. F. Lerch, O. Kirillova, M. S. Chapman and J. W. Whittaker, *Arch. Biochem. Biophys.*, 2011, **505**, 213–225.

144. E. N. Chikvaidze, *Gen. Physiol. Biophys.*, 1990, **9**, 411–414.

145. K. Mizuno, M. M. Whittaker, H. P. Bachinger and J. W. Whittaker, *J. Biol. Chem.*, 2004, **279**, 27339–27344.

146. H. Hirsch-Kolb, H. J. Kolb and D. M. Greenberg, *J. Biol. Chem.*, 1971, **246**, 395–401.

147. F. A. Armstrong, *Philos. Trans. R. Soc. B*, 2008, **363**, 1263–1270.

148. M. Pérez Navarro, W. M. Ames, H. Nilsson, T. Lohmiller, D. A. Pantazis, L. Rapatskiy, M. M. Nowaczyk, F. Neese, A. Boussac, J. Messinger, W. Lubitz and N. Cox, *Proc. Natl Acad. Sci.*, 2013, **110**, 15561–15566.

CHAPTER 10

Iron in Eukarya

PAUL SHARP

Diabetes & Nutritional Sciences Division, King's College London, School of
Medicine, Franklin Wilkins Building, 150 Stamford Street, London SE1
9NH, UK
Email: paul.a.sharp@kcl.ac.uk

10.1   Introduction

Total body iron content is 3–4 g in the adult human, with approximately
60–70% contained within haemoglobin in circulating erythrocytes. A further
20–30% is stored in hepatocytes where it serves as a physiological reserve
of iron which can be mobilized rapidly to maintain homeostasis. Some
20–30 mg iron is recycled each day from effete erythrocytes *via* the reticuloendothelial macrophage system. While this system is highly efficient, approximately 0.5–2 mg iron is lost each day due to minor blood loss and
shedding of iron-containing cells. Thus, to maintain homeostasis, an
equivalent amount of iron must be absorbed from dietary sources to replace
these endogenous losses.

10.1.1   Iron in the Diet

Iron is present in the diet in two major forms; haem and non-haem (reviewed
in refs 1,2). Non-haem iron is the major form (comprising approximately
90–95% of total dietary iron) and is present in foods either as simple ferric
(Fe^{3+}) oxides and salts or more complex organic chelates, *e.g.* ferritin. In
contrast, haem forms only a small fraction of total dietary iron intake (5–10%)
and is found (almost) exclusively as haemoglobin and myoglobin in meat and

RSC Metallobiology Series No. 2
Binding, Transport and Storage of Metal Ions in Biological Cells
Edited by Wolfgang Maret and Anthony Wedd
© The Royal Society of Chemistry 2014
Published by the Royal Society of Chemistry, www.rsc.org

meat products. Haem iron is the most bioavailable form; it is estimated that the average absorption from a meat-containing meal is 20–30%. In contrast, the absorption of non-haem iron is more variable (1–10% of dietary load) and is greatly influenced by other dietary components which either increase or decrease non-haem iron bioavailability. Principal amongst these dietary regulatory factors are ascorbic acid (vitamin C), a powerful reducing agent that converts Fe^{3+} to Fe^{2+} increasing bioavailability, and the phytates (present in cereals and cereal products), which are potent iron chelators that form insoluble ferric iron chelates that cannot be absorbed. A further important influence is gastric acid, which maintains the solubility of iron in the luminal fluid and provides a source of protons to stimulate Fe^{2+} uptake *via* the duodenal apical membrane resident divalent metal transporter (DMT1; see below). Achlorhydria, due to either gastric atrophy or chemical inhibition of acid secretion, is associated with iron deficiency in some patients.[3]

10.2 Uptake of Iron by Intestinal Cells and Subsequent Release into the Blood

(a) Reduction of iron at the apical membrane: Ferric iron (the major ingested form of non-haem iron) has low bioavailability and must first be reduced to ferrous iron prior to transport across the apical membrane of duodenal enterocytes (Figure 10.1). This is achieved to a large extent by dietary reducing agents in the bulk-phase of the intestinal fluid; however, further reduction takes place at the surface of enterocytes *via* the apical membrane ferric reductase Dcytb (duodenal cytochrome *b*).[4] This is a member of the cytochrome *b*561 superfamily of electron-transporting enzymes and is a haem-containing protein with putative binding sites for ascorbate and semi-dehydroascorbate. *In vitro* studies have provided a strong link between Dcytb expression and iron uptake by cultured cells;[5,6] however, deletion of the *Cybrd1* gene (encoding Dcytb) does not lead to an iron-deficient phenotype in mice.[7] Thus the role of Dcytb in iron absorption may be species specific. Humans have an obligate requirement for ascorbic acid in the diet, whereas mice (and other rodents) are able to synthesize ascorbic acid from glucose and can secrete it into the intestinal lumen.[8] Humans may therefore have an enhanced need for a duodenal surface ferric reductase. A report linking a single nucleotide polymorphism in the *Cybrd1* gene to impaired iron metabolism in humans provides some support for a requirement for Dcytb as a functional intestinal ferric reductase.[9]

(b) Transport of ferrous iron across the apical membrane of enterocytes: Ferrous iron is a substrate for DMT1 (SLC11A2).[10–12,†] Transport of iron across the apical membrane of enterocytes is driven by a proton electrochemical

---

[†]See Chapter 6, Section 6.2 for a discussion of possible transport of the $[V^{IV}O]^{2+}$ ion by DMT1 (NRAMP2).

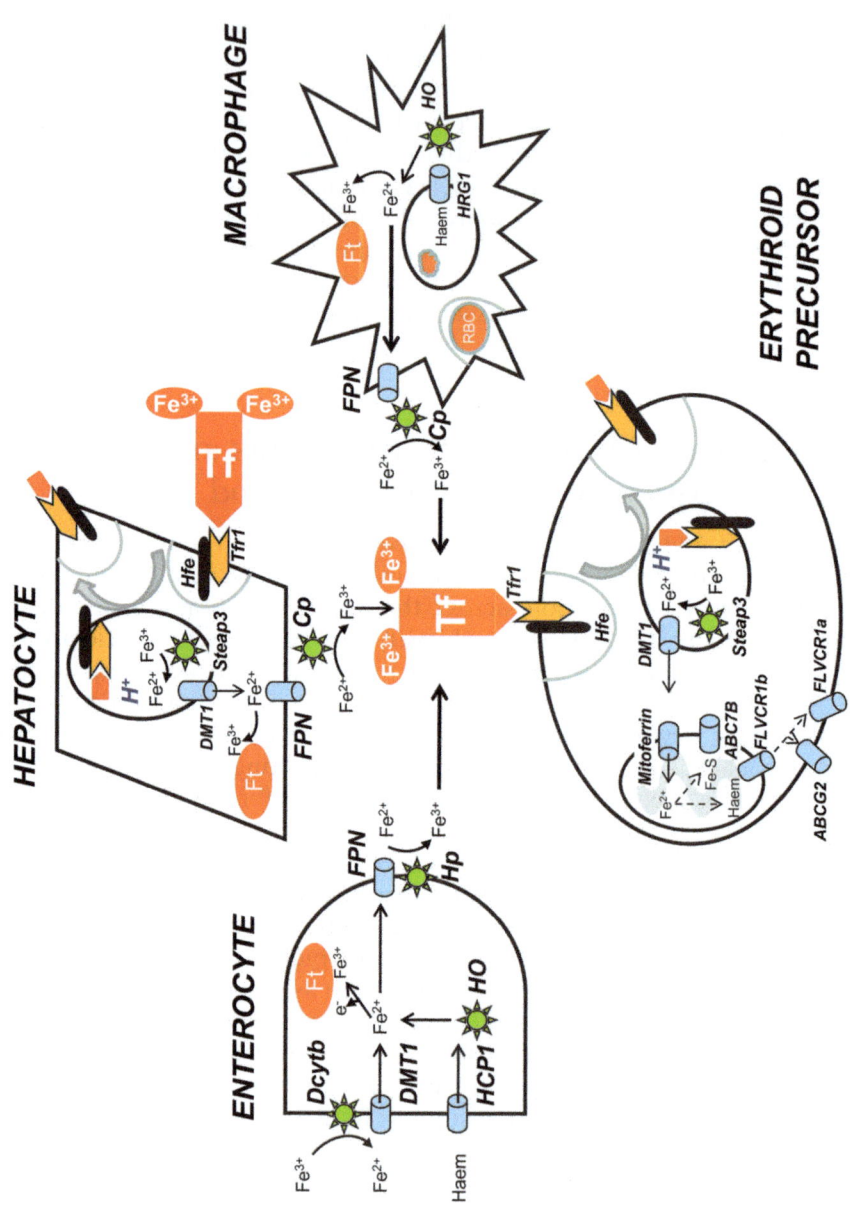

gradient; *i.e.* low extracellular pH and a cell-negative potential difference across the apical membrane of enterocytes. Luminal pH in the proximal duodenum is approximately 6.5 and the resting membrane potential across the apical membrane of the duodenal enterocyte is approximately −50 mV.[13,14]

The essential role of DMT1 in intestinal iron transport has been confirmed in mouse models with targeted disruption of the *SLC11A2* gene.[15] Cellular processing of this gene is complex and gives rise to at least four functional DMT1 isoforms. An alternate splice site in exon 16 gives rise to two variants that differ in their terminal 19–25 amino acids and their 3′ untranslated region (UTR).[16] One variant contains an iron responsive element (IRE) in its 3′ UTR whereas the other lacks this sequence. In addition, two transcription start sites have been identified at the 5′ end of the gene – in exon 1A and 1B, respectively.[17] The non-IRE expressing DMT1 isoforms are closely associated with the transferrin receptor-containing early endosomes, while the exon 1A/IRE-containing variant has been suggested to be the predominant isoform expressed in intestinal enterocytes.

(c) Haem transport: Haem released from haemoglobin and myoglobin in the intestinal lumen is absorbed intact; however, the transporter involved in its uptake remains to be elucidated. A number of candidate haem transporters have been identified including haem carrier protein 1 (HCP-1),[18] breast cancer resistance protein (ABCG2)[19] and the feline leukaemia virus C

Figure 10.1   Schematic representation of iron transport across cell membranes. Iron is acquired from the diet in either haem or non-haem forms. Non-haem iron is reduced by the ferric reductase Dcytb and transported across the apical membrane of enterocytes as Fe^{2+} *via* the proton-coupled transporter DMT1. Haem is taken up intact, possibly through the haem carrier protein HCP1 and is broken down inside the cell by haem oxygenase (HO) to liberate Fe^{2+}. Iron can either be stored in ferritin (as Fe^{3+}) or transported out of the cell *via* ferroportin (FPN). Fe^{2+} leaving enterocytes is re-oxidized to Fe^{3+} by hephaestin (Hp) and loaded onto transferrin (Tf) for onward transport. Tf-Fe binds to the transferrin receptor (TfR1) and is taken up through receptor-mediated endocytosis. Intracellular endosomes are acidified; Fe^{3+} is liberated from Tf and reduced by the endosomal reductase Steap3. Fe^{2+} exits the endosome *via* DMT1. In erythroid precursors, Fe^{2+} enters the mitochondria through mitoferrin and is used in the formation of Fe-S clusters and haem. These products exit the mitochondria *via* ABCB7 and FLVCR1b respectively. Haem can exit across cell membranes *via* ABCG2 or FLVCR1a. Iron contained within the red blood cell is recovered at the end of their circulating lifetime following phagocytosis by reticulo-endothelial macrophages; liberation of haem in the phago-lysosome, which is subsequently transported into the cytosol through HRG-1. Haem is broken down by HO and the released iron can either be stored in ferritin or can exit the macrophage *via* FPN. In macrophages and hepatocytes, Fe^{2+} exiting through FPN is re-oxidized by the serum oxidase ceruloplasmin (Cp) and loaded onto Tf for transport back to the bone marrow for synthesis of haemoglobin, or to the liver for storage as ferritin in hepatocytes.

receptor (FLVCR).[20] While each of these proteins can transport haem, ABCG2 and FLVCR are both efflux transporters, while HCP-1, which acts as a haem importer, is primarily a proton-coupled folate transporter.[21]

The release of iron from haem is catalysed by haem oxygenase 1 (HO-1), which is localized to the endoplasmic reticulum; how intact haem is delivered to this cellular locus is also unknown. HO-1 catabolism of haem yields Fe^{2+}, bile pigments (bilirubin and biliverdin) and carbon monoxide. Ferrous iron from the haem and non-haem iron pools are thought to combine at this stage, and in subsequent metabolism the body does not distinguish between iron from these two dietary sources.

(d) Iron efflux across the basolateral membrane of intestinal enterocytes: The release of iron from enterocytes into the villus microvasculature is mediated by ferroportin (SLC40A1).[22–24] To date this is the only iron efflux protein to have been identified. Deletion of the *SLC40A1* gene is embryonically lethal, highlighting the importance of ferroportin for iron efflux from a number of tissues.[25] Intestinal-specific deletion of the gene produces severe iron deficiency anaemia demonstrating the essential role of dietary iron in maintaining body iron homeostasis.[25] Recent work has established that two ferroportin transcripts are expressed in the duodenum (FPN1A and FPN1B) and that they differ in their regulation by iron.[26] However, the protein products of these transcripts are identical and it is thought that the mode of iron transport does not differ.

(e) Re-oxidation of iron at the basolateral membrane of enterocytes: In vitro expression studies have shown that ferroportin mediates selectively the efflux of Fe^{2+}.[22] For efficient transport in the circulation and delivery to its sites of utilization, iron is bound to transferrin (Tf); however, Tf preferentially binds iron in the oxidized Fe^{3+} form. Thus iron must be re-oxidized upon release from enterocytes in order to be loaded onto Tf.

Oxidation of iron at the basolateral pole of enterocytes is mediated by hephaestin, a copper-dependent ferroxidase. Hephaestin is anchored to the basolateral membrane through a glycophosphatidylinositol link, and immunocytochemical and co-immunoprecipitation studies have shown that hephaestin and ferroportin are co-expressed and interact at the basolateral membrane of enterocytes.[27] Hephaestin was first identified as the gene mutated in the sex linked anaemia (*sla*) mouse.[28] The *sla* mutation gives rise to a truncated form of hephaestin with reduced ferroxidase activity,[29] which is largely retained intracellularly. In *sla* mice, iron absorption across the apical membrane of enterocytes is normal but the efflux phase is diminished, suggesting that ferroportin and hephaestin are required to operate together as a functional unit to permit iron efflux and its loading onto Tf for onward transport in the blood.

10.3   Coordination Chemistry of Iron in the Blood

(a) Transferrin-bound iron. Iron is normally present in plasma exclusively bound to transferrin (Tf). Tf can accommodate either one (monoferric-Tf) or

two (diferric-Tf) ferric ions; and this binding is reversible and pH-dependent. At physiological pH, iron is completely associated with Tf and has a binding constant of approximately 10^{22} M^{-1}.[30] The transferrin molecule is characterized by two globular domains at the N- and C-terminal, respectively. Each of these domains contains an iron-binding site formed of two tyrosine, one histidine and one aspartic acid residues. Iron binding is facilitated by the presence of a synergistic anion, usually carbonate or hydrogen carbonate.[‡]

In healthy humans, the circulating Tf concentration is approximately 30 μM and it is 30% saturated with iron; the majority is in the monoferric form. Following the cellular uptake of iron-Tf, rapid acidification of intracellular endosomes allows iron to be released from Tf, reduced by STEAP3 (see below) and released from the endosome *via* DMT1 so that it can be utilized for cellular functions (see below).

(b) Non-transferrin bound iron. Under normal physiological conditions, virtually all serum iron is bound to Tf. However, in a number of disease states (*e.g.* haemochromatosis, thalassaemias, *etc.*), there is a significant increase in non-transferrin-bound-iron in the circulation with levels detected in the low μM range. Its precise nature is unknown but a high percentage exists as ferric citrate.

(c) Serum ferritin.[§] Small amounts of the tissue iron storage protein ferritin are also present in the circulation. Serum ferritin levels range from 30–300 μg L^{-1} in men and 15–150 μg L^{-1} in women. Serum ferritin is derived from tissue ferritin and there is a linear relationship between serum ferritin and the body iron stores with 1 μg L^{-1} serum ferritin being equivalent to approximately 8–10 mg tissue iron.[31,32] Serum ferritin is therefore a good circulating biomarker of iron status in healthy human populations. It consists largely of a glycosylated form of L-ferritin secreted from the liver[33] and other tissues, and has a low iron content.[34,35]

10.4 Transport of Iron across Cytoplasmic Membranes

(a) Transferrin receptors. Most mammalian cells acquire iron from transferrin, which enters cells following binding to transferrin receptors. Two transferrin receptors have been identified: TfR1, which is expressed ubiquitously, and TfR2, which is expressed primarily in hepatocytes.[36] While the transport mechanism for TfR1 has been well characterized, the role of TfR2 in iron uptake is less certain and this protein may act as an iron "sensor" rather than as a major transporter.

Fe-Tf binds to TfR1 at the cell surface and the receptor-transferrin complex is endocytosed *via* a clathrin-mediated mechanism (Figure 10.1). The recruitment of a V-type proton ATPase to the endosomal membrane acidifies the vesicular contents and permits the dissociation of iron (as Fe^{3+}) from

---

[‡]See Chapter 6, Section 6.5 for a discussion of vanadium binding to transferrin.
[§]See Chapter 13 for structural aspects and further functional details of ferritins.

transferrin. The apo-protein remains bound to its receptor at the acidic pH of the endosome and is recycled back to the cell surface. Following reduction by STEAP3, release of iron (Fe^{2+}) from the endosome occurs *via* DMT1 (see below) (reviewed in ref. 37).

(b) *Phagocytosis of senescent erythrocytes.* Approximately 70% of total body iron is contained in circulating erythrocytes as haemoglobin and this iron is recycled and re-utilized for new red cell production. Some 25–30 mg of iron is recycled each day through this pathway. Typically, erythrocytes are present in the circulation for approximately 120 days. The mechanisms involved in removing senescent red cells from the circulation are uncertain, but several erythrophagocytic pathways have been identified.

Following phagocytosis, red blood cells are broken down by proteolytic enzymes within the phago-lysosome. The mechanism of iron release from the phago-lysosome into the cytosol remains the subject of debate. Some studies suggest haem catabolism takes place within the phago-lysosome and iron exits in its inorganic Fe^{2+} form *via* DMT1 and Nramp1. Other data indicate that haem is released from the phago-lysosome *via* the haem transporter HRG-1[38] and interacts with haem oxygenases in the endoplasmic reticulum to release iron into the cytosol.

(c) *Zip14.* In addition to DMT1 and TfR1, non-haem iron can cross cell membranes *via* other transport pathways. Recent work has focused on a member of the zinc importer superfamily, Zip14, which is highly expressed in duodenum and liver[39] and can function as both an iron and a zinc transporter.[40] In contrast to DMT1, Zip14 is up-regulated in the liver by iron loading[41] and appears to play an important role in hepatic iron accumulation from circulating non-transferrin bound iron.[42]

(d) *Renal iron transporters.* Rates of iron excretion by the kidney are extremely low. Iron (bound to transferrin) can pass through the glomerulus and has been detected at significant levels in tubular fluid.[43] The renal tubular epithelium expresses both DMT1 and ferroportin.[44–46] While ferroportin is expressed on the basolateral membrane, DMT1 is primarily localized to endosomal vesicles rather than the apical surface of epithelial cells.[44] This has led to the development of a model for renal iron handling in which it is proposed that iron-transferrin complexes are taken up from the tubular fluid through receptor-mediated endocytosis *via* the cubulin-megalin scavenger receptor complex and is liberated from endosomes *via* DMT1. It subsequently exits the cell across the basolateral membrane through ferroportin.[47]

(e) *Ferroportin and ceruloplasmin.* Iron that exits enterocytes *via* ferroportin is re-oxidized to Fe(III) by the copper-containing ferroxidase hephaestin.[28] In other cells, oxidation of iron utilizes the serum oxidase ceruloplasmin.[48] The importance of ceruloplasmin in iron metabolism is demonstrated by human patients and mice with impaired ceruloplasmin production. Patients with aceruloplasminaemia and ceruloplasmin-null mice accumulate iron in a number of organs, including the liver and various regions of the brain, and exhibit mild anaemia.[49]

Ceruloplasmin can also be found as a membrane-bound form (glycosyl-phosphatidylinositol (GPI)-ceruloplasmin), produced by alternative splicing of the ceruloplasmin gene.[50] This form is localized primarily to the surface of astrocytes in the central nervous system (CNS)[51] where it plays a role in iron efflux from cells in the CNS.

Ceruloplasmin-like molecules are also required for the regulation of iron metabolism in lower eukaryotic species. In the yeast *Saccharomyces cerevisiae*, a copper-binding protein Fet3, which has sequence homology to ceruloplasmin, is required for high-affinity iron transport across the plasma membrane.[52,53] Like ceruloplasmin, Fet3 is a ferrioxidase, suggesting that oxidation and reduction of iron are crucial to its movement across biological membranes in yeast as well as in mammals.

(f) Iron transport proteins in the CNS. Iron handling by the CNS has been the subject of much recent work. There is evidence that imbalance in iron accumulation in the CNS is associated with a number of pathologies. It is intriguing therefore that uptake and release of iron across CNS cell membranes involves oxidation and reduction of iron by proteins associated with disease. The amyloid precursor protein (APP) has been suggested to act as a neuronal ferroxidase, controlling the rate of iron efflux.[54] A deficiency in Tau protein has been proposed to decrease APP-mediated iron efflux and contribute to toxic neuronal iron accumulation seen in pathologies such as Alzheimer's disease and Parkinson's disease.[55] Alpha-synuclein, which forms aggregates in a number of neuropathologies, including Parkinson's disease, also has cellular ferric reductase activity.[56] More recently, there is evidence that prion protein may also regulate neuronal iron transport by acting as a ferric reductase.[57] Taken together, these studies provide further evidence that impaired iron homeostasis in the CNS may contribute to the aetiology of a number of neuropathological conditions.

(g) Haem/haemoglobin scavenger receptors. In pathological situations leading to haemolysis of red blood cells, free haemoglobin and ultimately haem can be liberated in the circulation. These compounds are highly redox active and can cause significant damage at the cellular and tissue level. This potential pathological challenge is averted by the presence of scavenger mechanisms that remove free haemoglobin and haem from the circulation. Free circulating haemoglobin is bound by haptoglobin, which prevents oxidation and release of haem.[58] The haemoglobin:haptoglobin complex binds at the cell surface receptor CD163,[59] which is highly expressed in hepatocytes and macrophages, and is endocytosed and degraded intracellularly.

Liberation of free haem following oxidation of haemoglobin molecules can result in significant toxicity. Free haem can be scavenged by a number of serum proteins, but predominantly forms a high-affinity complex with hemopexin.[60,61] Haem bound to hemopexin has a limited capacity to generate free radicals.[62] The complex is readily removed from the circulation following binding to the macrophage scavenger receptor CD91.[63] Following internalization of the complex, haem is subject to intracellular degradation[64] and hemopexin is recycled back into the circulation.[63]

10.5 Transport of Iron through Subcellular Membranes

(a) Endosomal DMT1 and Steap3. The majority of cells acquire their iron *via* transferrin-transferrin receptor-mediated endocytosis. The release of iron from this pathway and its subsequent cellular utilization is dependent on the presence of DMT1 at the endosomal membrane. Targeted disruption of DMT1 in mice impairs the development of erythroid precursors into mature erythrocytes.[15] Furthermore, spontaneous mutations in the *mk/mk* mouse[65] and the Belgrade rat[66] and a number of human DMT1 mutations[67-70] are associated with the development of microcytic anaemia.

In order to exit the endosomal vesicle *via* DMT1, iron released from transferrin must first be reduced. In erythroid precursors, this is achieved by STEAP3,[71] a member of the Six Transmembrane Epithelial Antigen of the Prostate family of proteins (Figure 10.1). The role of STEAP3 in this process was confirmed in *nm1054* mice, which have a mutation in STEAP3, and in STEAP3 knock-out mice. In both models, animals exhibit hypochromic, microcytic anaemia due to the inability of erythroid precursors to utilize transferrin-bound iron. With the exception of STEAP1, all of the STEAP family of proteins act as metalloreductases *in vitro*,[72] suggesting that other STEAPs may act as intracellular reductases in different tissues.

(b) Natural Resistance Associated Macrophage Protein 1. Two mammalian members of the NRAMP metal transporter family have been characterized: NRAMP1 (SLC11A1), which confers resistance to infection by mycobacteria,[73] and NRAMP2 (DMT1 or SLC11A2), discussed above. Interestingly, NRAMP orthologues also exist in yeast (SMF1 and SMF2)[74] and in *Drosophila melanogaster* (malvolio).[75]

NRAMP1 is almost exclusively expressed at the phagosomal membrane of macrophages and neutrophils,[76] where it pumps essential metals such as manganese and iron out of the phagosome to prevent their use by invading pathogens.[77,78] A number of inbred mouse models have increased susceptibility to infections by intracellular pathogens, which is associated with a glycine to aspartic acid substitution at position 169 in NRAMP1.[79] The reason for expressing this phenotype remains unclear; however, it has been proposed that the mutation decreases the metal transport function of NRAMP1.

(c) Mitoferrin. Many of the key steps in iron metabolism including haem synthesis, the production of haemoglobin and the assembly of iron-sulfur clusters take place in the mitochondria. The mechanisms by which mitochondria accumulate and store iron have only recently been identified and the mitochondrial solute carrier family member SLC25A37 (mitoferrin) appears crucial to this role (Figure 10.1). Mitoferrin was first identified by positional cloning. It is highly expressed in haematopoietic tissue[80] and plays an essential role in haem synthesis. A mutation in the mitoferrin gene in zebrafish is responsible for the *frascati* phenotype that shows profound

hypochromic anaemia and the arrest of erythroid maturation owing to defects in mitochondrial iron uptake. Deletion of the mouse mitoferrin-1 gene is embryonically lethal and targeted disruption of mitoferrin-1 in adult mouse haematopoietic tissue leads to severe anaemia.[81] The stability of mitoferrin is governed by interaction with the mitochondrial ABC transporter Abcb10.[82] There is evidence that ferrochelatase (the enzyme responsible for the insertion of iron into protoporphyrin IX, the final step of haem biosynthesis) forms an oligomeric complex with mitoferrin-1 and Abcb10 in erythroid tissue that coordinates mitochondrial iron intake with haem biosynthesis.[83]

Mitoferrin orthologues (MRS3 and MRS4) have been identified in yeast and disruption of these genes causes defects in haemoprotein production and the mitochondrial synthesis of iron-sulfur clusters.[84–86] Expression of murine mitoferrin can rescue defective iron metabolism in the *frascati* zebrafish. Furthermore, introduction of zebrafish mitoferrin can complement the yeast MRS3/4 mutant, demonstrating that the function of the gene is highly conserved.[80]

(d) FLVCR1b. To complete the process of haem protein synthesis, it is essential that an efficient mitochondrial haem efflux pathway be in place. Recent evidence demonstrates that the feline leukaemia virus subgroup C receptor 1 (FLVCR1), implicated in haem export in the intestine,[20] is also essential for haemoprotein synthesis (Figure 10.1).[87] Deletion of FLVCR1 in mice is embryonically lethal and embryos lack the erythropoietic machinery. Post-natal deletion of FLVCR1 results in macrocytic anaemia with a blockade of pro-erythroblast maturation. Recently, a second isoform of FLVCR1 (FLVCR1b) has been identified.[88] FLVCR1b is highly expressed in mitochondria of tissues exhibiting high rates of haem synthesis. Overexpression of FLVCR1b *in vitro* results in increased cytosolic iron content, whereas knockdown of the transporter gives rise to mitochondrial haem accumulation. These data suggest that this transporter plays a key role in mitochondrial haem export and therefore erythropoiesis.

10.6 Transport of Iron within the Cell

(a) Low-molecular-mass iron binding proteins. In contrast to some metals, little is known regarding the mechanisms that transport iron across the cytosol either for intracellular storage or for efflux out of the cell. Studies in intestinal enterocytes showed that following absorption iron was rapidly associated with a low-molecular weight binding protein; however, the protein was not identified.[89] Recent evidence implicates glutathione-bound iron as a component of the cytosolic iron pool. Furthermore, it is suggested that iron-glutathione and the glutathione-binding glutaredoxins may play a critical role in the synthesis of iron sulfur clusters.[90]

(b) Mitochondrial chaperones. Frataxin is a 210-amino acid protein located predominantly within the mitochondria and is also associated with

formation of iron sulfur (Fe-S) clusters. Human metabolism is dependent on the efficient synthesis of Fe-S clusters for the activity of a number of enzymes. Frataxin has been suggested to act as a chaperone for iron in mitochondrial Fe-S cluster formation by binding directly to the Isu scaffold proteins and transferring iron to Isu during cluster assembly.[91,92]

The initial link between frataxin and iron metabolism was established in studies using the yeast frataxin homologue (Yfh1p). Yeasts lacking Yfh1p accumulate iron but exhibit decreased capability for synthesis of Fe-S clusters.[93] Frataxin can substitute for Yfh1p in yeast, suggesting that they are functional homologues.[94,95] The structure of human frataxin has been solved and a glutamate/aspartate-rich region in the N-terminal region has been identified, which is thought to be involved in metal binding (reviewed in ref. 96). However, how this contributes to frataxin function is still unclear.

While Fe-S clusters can be used directly for the formation of mitochondrial Fe-S proteins, the maturation of cytosolic Fe-S proteins requires export of the Fe-S clusters from the mitochondria into the cytosol. In mammals, Fe-S export involves the ATP-binding cassette transporter ABCB7 (Figure 10.1).[97,98] In yeast the same function is achieved using the ABCB7 homologue Atm1,[99] highlighting further that the mechanisms involved in Fe-S cluster formation are well conserved.

(c) Cytosolic chaperones. A putative cytosolic "iron chaperone" has been identified – the human poly (rC)-binding protein 1 (PCBP1), which delivers iron into ferritin.[100] PCBP1 knockdown in human cells inhibited iron loading into ferritin and increased cytosolic iron content, suggesting this protein plays a key role in coordinating cellular iron storage. Other PCBP proteins (PCBP2–4) have been identified and may also play a role in cellular iron processing. Recent data support this possibility and indicate that PCBP1 and PCBP2 both play a role in delivering iron to the prolyl hydroxylases that regulate the stability of the hypoxia-inducible factor transcription factors.[101]

10.7   Sensing, Buffering and Cellular Homeostasis

(a) Hepcidin and iron sensing. The major advance in iron metabolism research over the past ten years has arisen from the identification of a pathway that senses changes in body iron levels and, accordingly, regulates the rates of dietary iron absorption and macrophage iron recycling in order to maintain iron supply for erythropoiesis.

The hepcidin gene is highly expressed in hepatocytes and cellular processing of its mRNA gives rise to the release of a 25 amino acid peptide into the circulation.[102,103] While hepcidin was first identified as an antimicrobial peptide, it was also apparent that it played an important role in iron metabolism.[103,104] Hepcidin knock-out mice develop severe liver iron overloading that is similar to that found in human haemochromatosis patients.[105,106] Subsequent studies have generated transgenic mice overexpressing hepcidin, which exhibit severe body iron deficiency and microcytic hypochromic anaemia.[107] Taken together, these studies indicate a reciprocal relationship between hepcidin expression and iron accumulation.

The nature of the iron-sensing pathway that leads to regulation of hepcidin release has been elucidated and transferrin saturation is central to this regulatory axis (reviewed in refs 108,109). As body iron levels increase, circulating levels of holo-Tf (diferric) increase. The latter competes with HFE (the protein product of the gene mutated in hereditary haemochromatosis) for binding to TfR1. At high holo-Tf levels, HFE is displaced from TfR1 and binds instead to TfR2. The TfR2-HFE complex interacts with the bone morphogenetic protein (BMP) signalling pathway. The BMP receptor complex includes a co-receptor activator protein hemojuvelin (HJV, mutations in this gene result in juvenile haemochromatosis). The BMP signalling cascade is activated by BMP6, which is itself increased by iron loading. Activation of the BMP receptor by BMP6 leads to phosphorylation of the receptor-activated SMAD signalling proteins SMAD 1,5,8. These form a heterodimer with SMAD4, which translocates to the nucleus and binds to BMP/SMAD responsive elements located in the hepcidin promoter, thus leading to activation of gene expression of hepcidin.

The "off switch" for this iron-sensing pathway, activated during iron deficiency, is encoded by the *TMPRSS6* gene (also known as matriptase-2), which exhibits serine protease activity. TMPRSS6 is active under low iron conditions and disrupts the iron-sensing complex by cleaving HJV from the hepatocyte membrane.[110] As a result HJV, the activator protein for the BMP signalling cascade, is non-functional and hepcidin expression is suppressed. Interestingly, a number of mutations in human *TMPRSS6* have been identified. These give rise to iron-refractory iron deficiency anaemia, a situation where iron deficient individuals do not respond to iron supplementation therapy.[111–113]

(b) Hepcidin – mode of action. Hepcidin can interact directly with the iron efflux transporter ferroportin limiting cellular iron release.[114] Iron recycling macrophages are very sensitive to increased hepcidin levels and respond rapidly by decreasing ferroportin protein levels and the subsequent release of iron.[115,116] Furthermore, elevated expression of human hepcidin is associated with the anaemia of chronic disease,[117] indicating that pathological changes in hepcidin expression have severe consequences for body iron metabolism. Interestingly, the evidence that hepcidin regulates intestinal iron transport through modulation of ferroportin protein expression is less robust and suggests that the effects of hepcidin may be tissue-specific.[116,118,119]

(c) Iron sensing by intestinal enterocytes. In the absence of definitive evidence to support hepcidin as the major regulator of intestinal iron absorption, recent work has focused on local sensing pathways including iron regulatory proteins, ferritin-H and the hypoxia-inducible transcription factor HIF2α.

More than half a century ago, it was observed that administering a large oral bolus of iron to dogs could significantly reduce the absorption of subsequent smaller doses of iron given several hours later.[120] The short time interval between doses suggested that the initial dose must be having a direct effect on the mature enterocytes rather than the crypt cells, resulting in a "mucosal block" for iron absorption. The mucosal block phenomenon was associated with an increase in intestinal ferritin levels[121] and more recent studies have shown that mice with an intestinal ferritin-H gene deletion have greatly enhanced rates of intestinal iron absorption.[122] Taken together, this suggests that ferritin may be an important cellular iron sensor in addition to its role as an iron storage protein.

Many of the genes controlling iron homeostasis are tightly regulated in line with the prevailing intracellular iron levels through post-transcriptional mechanisms that involve interactions between cytosolic iron regulatory proteins (IRP-1 and -2) and iron responsive elements (IRE). The latter are stem-loop structures present in either the 5' or 3' untranslated region (UTR) of several target mRNA species. IRP-1 and IRP-2 can both bind to IRE structures when cellular iron levels are depressed.[123–127]

Under iron-replete conditions, RNA binding is quickly inactivated by either post-translational modification of IRP-1 or degradation of IRP-2. For IRP1, RNA-binding activity is controlled by the presence or absence of an iron-sulfur cluster. When cells are iron replete, the IRE-binding domain is occupied by a 4Fe-4S cluster bound by three conserved cysteine residues.¶ As cellular iron levels decrease, iron is removed from the cluster, leading to its disintegration/disassembly, which permits the *apo*-protein to bind to IRE motifs. Iron-mediated regulation of IRP2 activity is controlled *via* F-box and leucine-rich repeat protein 5 (FBXL5).[128–130] When cellular iron levels are high, iron binds to the hemerythrin-like domains of FBXL5, stabilizing the protein and permitting association with IRP2.[128] FBXL5 is a component of a larger protein complex possessing ubiquitin ligase E3 activity and IRP2-FBXL5 binding targets IRP2 for proteasomal degradation.[131,132]

IRP knock-out mice have revealed that IRP1 and IRP2 can largely replace each other functionally but that expression of both proteins is essential for normal tissue homeostasis.[133–135] Mice with intestinal IRP deficiency exhibit decreased iron absorption together with enhanced mucosal iron retention. This occurs despite elevated levels of DMT1 and ferroportin. However, IRP-deficient enterocytes also highly overexpress ferritin, which in turn can sequester absorbed iron. Taken together these data indicate that IRPs are required for efficient iron transfer across the intestinal mucosa.[136]

Hypoxia inducible factor 2α (HIF2α) transcription factor has emerged as an important iron sensor and regulator of iron transporter expression in the intestine.[137–139] Iron deficiency and hypoxia lead to stabilization of HIF2α, which dimerizes with the HIFβ subunit (ARNT) and binds to HIF-responsive

---

¶See Chapter 12 for a discussion of 4Fe-4S centres.

elements in the promoter region of target genes, including DMT1 and Dcytb. The HIF2α regulatory pathway provides an explanation for the strong up-regulation of Dcytb (which does not contain an IRE) in iron deficiency. Interestingly, while the HIF2α protein is stabilized and active in an iron-deficient environment, its mRNA expression is repressed due to the presence of an IRE in the 5′ UTR.[140] These findings suggest a complex regulatory iron-sensing network exists in enterocytes that is encompassed by IRP activity, HIF2α stability and ferritin expression. In turn, these factors control the expression of iron transporters and the rate of transepithelial flux of iron.

10.8 Conclusions and Future Perspectives

Despite the recent dramatic expansion in our understanding of the mechanisms controlling iron metabolism, there are still a number of unanswered questions. At a level of dietary iron absorption, there is still no clear consensus on the role of the duodenal ferric reductase Dcytb in non-haem iron uptake. The nature of the haem absorption pathway also remains elusive. Unlocking these mechanisms would have important implications for iron nutrition. The pathways involved in processing haem recovered from effete erythrocytes in reticuloendothelial macrophages are still poorly understood. Identifying the proteins coordinating haem catabolism and the subsequent release of iron back into the circulation will enhance our knowledge of how impaired macrophage iron turnover leads to disorders of iron metabolism. The recent identification of a range of iron-sensing proteins in different tissues and with different molecular mechanisms offers an opportunity to develop novel therapies aimed at regulating iron metabolism at the cellular level. Perhaps the greatest challenge is to understand more fully the role of iron in the brain. In particular, the reasons why brain iron content increases with age and how changes in the pattern of iron deposition in different brain structures might influence age-related neurodegeneration. Given the increasing longevity of the world's population and the cost of treating neurodegenerative disorders, this is likely to become a topic of great socio-economic importance.

References

1. P. Sharp and S. K. Srai, *World J. Gastroenterol.*, 2007, **13**, 4716–4724.
2. P. A. Sharp, *Int. J. Vitam. Nutr. Res.*, 2010, **80**, 231–242.
3. C. Hutchinson, C. A. Geissler, J. J. Powell and A. Bomford, *Gut*, 2007, **56**, 1291–1295.
4. A. T. McKie, D. Barrow, G. O. Latunde-Dada, A. Rolfs, G. Sager, E. Mudaly, M. Mudaly, C. Richardson, D. Barlow, A. Bomford, T. J. Peters, K. B. Raja, S. Shirali, M. A. Hediger, F. Farzaneh and R. J. Simpson, *Science*, 2001, **291**, 1755–1759.
5. G. O. Latunde-Dada, R. J. Simpson and A. T. McKie, *J. Nutr.*, 2008, **138**, 991–995.

6. S. Wyman, R. J. Simpson, A. T. McKie and P. A. Sharp, *FEBS Lett.*, 2008, **582**, 1901–1906.

7. H. Gunshin, C. N. Starr, C. Direnzo, M. D. Fleming, J. Jin, E. L. Greer, V. M. Sellers, S. M. Galica and N. C. Andrews, *Blood*, 2005, **106**, 2879–2883.

8. B. Atanasova, I. S. Mudway, A. H. Laftah, G. O. Latunde-Dada, A. T. McKie, T. J. Peters, K. N. Tzatchev and R. J. Simpson, *J. Nutr.*, 2004, **134**, 501–505.

9. C. C. Constantine, G. J. Anderson, C. D. Vulpe, C. E. McLaren, M. Bahlo, H. L. Yeap, D. M. Gertig, N. J. Osborne, N. A. Bertalli, K. B. Beckman, V. Chen, P. Matak, A. T. McKie, M. B. Delatycki, J. K. Olynyk, D. R. English, M. C. Southey, G. G. Giles, J. L. Hopper, K. J. Allen and L. C. Gurrin, *Br. J. Haematol.*, 2009, **147**, 140–149.

10. H. Gunshin, B. Mackenzie, U. V. Berger, Y. Gunshin, M. F. Romero, W. F. Boron, S. Nussberger, J. L. Gollan and M. A. Hediger, *Nature*, 1997, **388**, 482–488.

11. S. Tandy, M. Williams, A. Leggett, M. Lopez-Jimenez, M. Dedes, B. Ramesh, S. K. Srai and P. Sharp, *J. Biol. Chem.*, 2000, **275**, 1023–1029.

12. A. C. Illing, A. Shawki, C. L. Cunningham and B. Mackenzie, *J. Biol. Chem.*, 2012, **287**, 30485–30496.

13. D. K. O'Riordan, E. S. Debnam, P. A. Sharp, R. J. Simpson, E. M. Taylor and S. K. Srai, *J. Physiol.*, 1997, **500**, 379–384.

14. D. K. O'Riordan, P. Sharp, R. M. Sykes, S. K. Srai, O. Epstein and E. S. Debnam, *Eur. J. Clin. Invest*, 1995, **25**, 722–727.

15. H. Gunshin, Y. Fujiwara, A. O. Custodio, C. Direnzo, S. Robine and N. C. Andrews, *J. Clin. Invest*, 2005, **115**, 1258–1266.

16. P. L. Lee, T. Gelbart, C. West, C. Halloran and E. Beutler, *Blood Cells Mol. Dis.*, 1998, **24**, 199–215.

17. N. Hubert and M. W. Hentze, *Proc. Natl Acad. Sci. USA*, 2002, **99**, 12345–12350.

18. M. Shayeghi, G. O. Latunde-Dada, J. S. Oakhill, A. H. Laftah, K. Takeuchi, N. Halliday, Y. Khan, A. Warley, F. E. McCann, R. C. Hider, D. M. Frazer, G. J. Anderson, C. D. Vulpe, R. J. Simpson and A. T. McKie, *Cell*, 2005, **122**, 789–801.

19. P. Krishnamurthy, D. D. Ross, T. Nakanishi, K. Bailey-Dell, S. Zhou, K. E. Mercer, B. Sarkadi, B. P. Sorrentino and J. D. Schuetz, *J. Biol. Chem.*, 2004, **279**, 24218–24225.

20. J. G. Quigley, Z. Yang, M. T. Worthington, J. D. Phillips, K. M. Sabo, D. E. Sabath, C. L. Berg, S. Sassa, B. L. Wood and J. L. Abkowitz, *Cell*, 2004, **118**, 757–766.

21. A. Qiu, M. Jansen, A. Sakaris, S. H. Min, S. Chattopadhyay, E. Tsai, C. Sandoval, R. Zhao, M. H. Akabas and I. D. Goldman, *Cell*, 2006, **127**, 917–928.

22. A. T. McKie, P. Marciani, A. Rolfs, K. Brennan, K. Wehr, D. Barrow, S. Miret, A. Bomford, T. J. Peters, F. Farzaneh, M. A. Hediger, M. W. Hentze and R. J. Simpson, *Mol. Cell*, 2000, **5**, 299–309.

23. S. Abboud and D. J. Haile, *J. Biol. Chem.*, 2000, **275**, 19906–19912.
24. A. Donovan, A. Brownlie, Y. Zhou, J. Shepard, S. J. Pratt, J. Moynihan, B. H. Paw, A. Drejer, B. Barut, A. Zapata, T. C. Law, C. Brugnara, S. E. Lux, G. S. Pinkus, J. L. Pinkus, P. D. Kingsley, J. Palis, M. D. Fleming, N. C. Andrews and L. I. Zon, *Nature*, 2000, **403**, 776–781.
25. A. Donovan, C. A. Lima, J. L. Pinkus, G. S. Pinkus, L. I. Zon, S. Robine and N. C. Andrews, *Cell Metab.*, 2005, **1**, 191–200.
26. D. L. Zhang, R. M. Hughes, H. Ollivierre-Wilson, M. C. Ghosh and T. A. Rouault, *Cell Metab.*, 2009, **9**, 461–473.
27. O. Han and E. Y. Kim, *J. Cell. Biochem.*, 2007, **101**, 1000–1010.
28. C. D. Vulpe, Y. M. Kuo, T. L. Murphy, L. Cowley, C. Askwith, N. Libina, J. Gitschier and G. J. Anderson, *Nat. Genet.*, 1999, **21**, 195–199.
29. H. Chen, Z. K. Attieh, T. Su, B. A. Syed, H. Gao, R. M. Alaeddine, T. C. Fox, J. Usta, C. E. Naylor, R. W. Evans, A. T. McKie, G. J. Anderson and C. D. Vulpe, *Blood*, 2004, **103**, 3933–3939.
30. P. Aisen and I. Listowsky, *Annu. Rev. Biochem.*, 1980, **49**, 357–393.
31. G. O. Walters, F. M. Miller and M. Worwood, *J. Clin. Pathol.*, 1973, **26**, 770–772.
32. J. D. Cook, *Proc. Nutr. Soc.*, 1999, **58**, 489–495.
33. S. Ghosh, S. Hevi and S. L. Chuck, *Blood*, 2004, **103**, 2369–2376.
34. P. Arosio, M. Yokota and J. W. Drysdale, *Br. J. Haematol.*, 1977, **36**, 199–207.
35. M. Worwood, S. Dawkins, M. Wagstaff and A. Jacobs, *Biochem. J.*, 1976, **157**, 97–103.
36. A. S. Zhang, S. Xiong, H. Tsukamoto and C. A. Enns, *Blood*, 2004, **103**, 1509–1514.
37. M. W. Hentze, M. U. Muckenthaler and N. C. Andrews, *Cell*, 2004, **117**, 285–297.
38. C. Delaby, C. Rondeau, C. Pouzet, A. Willemetz, N. Pilard, M. Desjardins and F. Canonne-Hergaux, *PloS One*, 2012, 7, e42199.
39. J. P. Liuzzi, F. Aydemir, H. Nam, M. D. Knutson and R. J. Cousins, *Proc. Natl Acad. Sci. USA*, 2006, **103**, 13612–13617.
40. J. J. Pinilla-Tenas, B. K. Sparkman, A. Shawki, A. C. Illing, C. J. Mitchell, N. Zhao, J. P. Liuzzi, R. J. Cousins, M. D. Knutson and B. Mackenzie, *Am. J. Physiol. Cell Physiol.*, 2011, **301**, C862–871.
41. H. Nam, C. Y. Wang, L. Zhang, W. Zhang, S. Hojyo, T. Fukada and M. D. Knutson, *Haematologica*, 2013, **98**, 1049–1057.
42. C. Y. Wang and M. D. Knutson, *Hepatology (Baltimore, MD)*, 2013, **58**, 788–798.
43. M. Wareing, C. J. Ferguson, R. Green, D. Riccardi and C. P. Smith, *J. Physiol.*, 2000, **524**, 581–586.
44. M. Abouhamed, J. Gburek, W. Liu, B. Torchalski, A. Wilhelm, N. A. Wolff, E. I. Christensen, F. Thevenod and C. P. Smith, *Am. J. Physiol. Renal Physiol.*, 2006, **290**, F1525–F1533.
45. C. J. Ferguson, M. Wareing, D. T. Ward, R. Green, C. P. Smith and D. Riccardi, *Am. J. Physiol. Renal Physiol.*, 2001, **280**, F803–F814.

46. N. A. Wolff, W. Liu, R. A. Fenton, W. K. Lee, F. Thevenod and C. P. Smith, *J. Cell. Mol. Med.*, 2011, **15**, 209–219.

47. C. P. Smith and F. Thevenod, *Biochim. Biophys. Acta*, 2009, **1790**, 724–730.

48. S. Osaki and D. A. Johnson, *J. Biol. Chem.*, 1969, **244**, 5757–5758.

49. Z. L. Harris, *J. Neurol. Sci.*, 2003, **207**, 108–109.

50. B. N. Patel, R. J. Dunn and S. David, *J. Biol. Chem.*, 2000, **275**, 4305–4310.

51. B. N. Patel and S. David, *J. Biol. Chem.*, 1997, **272**, 20185–20190.

52. C. Askwith, D. Eide, H. A. Van, P. S. Bernard, L. Li, S. Davis-Kaplan, D. M. Sipe and J. Kaplan, *Cell*, 1994, **76**, 403–410.

53. A. Dancis, D. S. Yuan, D. Haile, C. Askwith, D. Eide, C. Moehle, J. Kaplan and R. D. Klausner, *Cell*, 1994, **76**, 393–402.

54. J. A. Duce, A. Tsatsanis, M. A. Cater, S. A. James, E. Robb, K. Wikhe, S. L. Leong, K. Perez, T. Johanssen, M. A. Greenough, H. H. Cho, D. Galatis, R. D. Moir, C. L. Masters, C. McLean, R. E. Tanzi, R. Cappai, K. J. Barnham, G. D. Ciccotosto, J. T. Rogers and A. I. Bush, *Cell*, 2010, **142**, 857–867.

55. P. Lei, S. Ayton, D. I. Finkelstein, L. Spoerri, G. D. Ciccotosto, D. K. Wright, B. X. Wong, P. A. Adlard, R. A. Cherny, L. Q. Lam, B. R. Roberts, I. Volitakis, G. F. Egan, C. A. McLean, R. Cappai, J. A. Duce and A. I. Bush, *Nat. Med.*, 2012, **18**, 291–295.

56. P. Davies, D. Moualla and D. R. Brown, *PloS One*, 2011, **6**, e15814.

57. A. Singh, S. Haldar, K. Horback, C. Tom, L. Zhou, H. Meyerson and N. Singh, *J. Alzheimer's Dis.*, 2013, **35**, 541–552.

58. M. Melamed-Frank, O. Lache, B. I. Enav, T. Szafranek, N. S. Levy, R. M. Ricklis and A. P. Levy, *Blood*, 2001, **98**, 3693–3698.

59. M. Kristiansen, J. H. Graversen, C. Jacobsen, O. Sonne, H. J. Hoffman, S. K. Law and S. K. Moestrup, *Nature*, 2001, **409**, 198–201.

60. M. Paoli, B. F. Anderson, H. M. Baker, W. T. Morgan, A. Smith and E. N. Baker, *Nat. Struct. Biol.*, 1999, **6**, 926–931.

61. E. Tolosano, S. Fagoonee, N. Morello, F. Vinchi and V. Fiorito, *Antioxid. Redox Signal.*, 2010, **12**, 305–320.

62. J. D. Eskew, R. M. Vanacore, L. Sung, P. J. Morales and A. Smith, *J. Biol. Chem.*, 1999, **274**, 638–648.

63. V. Hvidberg, M. B. Maniecki, C. Jacobsen, P. Hojrup, H. J. Moller and S. K. Moestrup, *Blood*, 2005, **106**, 2572–2579.

64. J. Alam and A. Smith, *J. Biol. Chem.*, 1989, **264**, 17637–17640.

65. M. D. Fleming, C. C. Trenor, III, M. A. Su, D. Foernzler, D. R. Beier, W. F. Dietrich and N. C. Andrews, *Nat. Genet.*, 1997, **16**, 383–386.

66. M. D. Fleming, M. A. Romano, M. A. Su, L. M. Garrick, M. D. Garrick and N. C. Andrews, *Proc. Natl Acad. Sci. USA*, 1998, **95**, 1148–1153.

67. M. P. Mims, Y. Guan, D. Pospisilova, M. Priwitzerova, K. Indrak, P. Ponka, V. Divoky and J. T. Prchal, *Blood*, 2005, **105**, 1337–1342.

68. C. Beaumont, J. Delaunay, G. Hetet, B. Grandchamp, M. de Montalembert and G. Tchernia, *Blood*, 2006, **107**, 4168–4170.

69. A. Iolascon, M. d'Apolito, V. Servedio, F. Cimmino, A. Piga and C. Camaschella, *Blood*, 2006, **107**, 349–354.

70. S. Lam-Yuk-Tseung, C. Camaschella, A. Iolascon and P. Gros, *Blood Cells Mol. Dis.*, 2006, **36**, 347–354.

71. R. S. Ohgami, D. R. Campagna, E. L. Greer, B. Antiochos, A. McDonald, J. Chen, J. J. Sharp, Y. Fujiwara, J. E. Barker and M. D. Fleming, *Nat. Genet.*, 2005, **37**, 1264–1269.

72. R. S. Ohgami, D. R. Campagna, A. McDonald and M. D. Fleming, *Blood*, 2006, **108**, 1388–1394.

73. S. M. Vidal, D. Malo, K. Vogan, E. Skamene and P. Gros, *Cell*, 1993, **73**, 469–485.

74. A. H. West, D. J. Clark, J. Martin, W. Neupert, F. U. Hartl and A. L. Horwich, *J. Biol. Chem.*, 1992, **267**, 24625–24633.

75. V. Rodrigues, P. Y. Cheah, K. Ray and W. Chia, *EMBO J.*, 1995, **14**, 3007–3020.

76. A. Fortier, G. Min-Oo, J. Forbes, S. Lam-Yuk-Tseung and P. Gros, *J. Leukoc. Biol.*, 2005, **77**, 868–877.

77. J. R. Forbes and P. Gros, *Trends Microbiol.*, 2001, **9**, 397–403.

78. N. Jabado, A. Jankowski, S. Dougaparsad, V. Picard, S. Grinstein and P. Gros, *J. Exp. Med.*, 2000, **192**, 1237–1248.

79. S. Vidal, P. Gros and E. Skamene, *J. Leukoc. Biol.*, 1995, **58**, 382–390.

80. G. C. Shaw, J. J. Cope, L. Li, K. Corson, C. Hersey, G. E. Ackermann, B. Gwynn, A. J. Lambert, R. A. Wingert, D. Traver, N. S. Trede, B. A. Barut, Y. Zhou, E. Minet, A. Donovan, A. Brownlie, R. Balzan, M. J. Weiss, L. L. Peters, J. Kaplan, L. I. Zon and B. H. Paw, *Nature*, 2006, **440**, 96–100.

81. M. B. Troadec, D. Warner, J. Wallace, K. Thomas, G. J. Spangrude, J. Phillips, O. Khalimonchuk, B. H. Paw, D. M. Ward and J. Kaplan, *Blood*, 2011, **117**, 5494–5502.

82. W. Chen, P. N. Paradkar, L. Li, E. L. Pierce, N. B. Langer, N. Takahashi-Makise, B. B. Hyde, O. S. Shirihai, D. M. Ward, J. Kaplan and B. H. Paw, *Proc. Natl Acad. Sci. USA*, 2009, **106**, 16263–16268.

83. W. Chen, H. A. Dailey and B. H. Paw, *Blood*, 2010, **116**, 628–630.

84. F. Foury and T. Roganti, *J. Biol. Chem.*, 2002, **277**, 24475–24483.

85. U. Muhlenhoff, J. A. Stadler, N. Richhardt, A. Seubert, T. Eickhorst, R. J. Schweyen, R. Lill and G. Wiesenberger, *J. Biol. Chem.*, 2003, **278**, 40612–40620.

86. Y. Zhang, E. R. Lyver, S. A. Knight, E. Lesuisse and A. Dancis, *J. Biol. Chem.*, 2005, **280**, 19794–19807.

87. S. B. Keel, R. T. Doty, Z. Yang, J. G. Quigley, J. Chen, S. Knoblaugh, P. D. Kingsley, I. De Domenico, M. B. Vaughn, J. Kaplan, J. Palis and J. L. Abkowitz, *Science*, 2008, **319**, 825–828.

88. D. Chiabrando, S. Marro, S. Mercurio, C. Giorgi, S. Petrillo, F. Vinchi, V. Fiorito, S. Fagoonee, A. Camporeale, F. Turco, G. R. Merlo, L. Silengo, F. Altruda, P. Pinton and E. Tolosano, *J. Clin. Invest.*, 2012, **122**, 4569–4579.

89. M. M. Kozma, G. Chowrimootoo, E. S. Debnam, O. Epstein and S. K. Srai, *Biochim. Biophys. Acta*, 1994, **1201**, 229–234.

90. R. C. Hider and X. Kong, *Dalton Trans.*, 2013, **42**, 3220–3229.
91. J. Gerber, U. Muhlenhoff and R. Lill, *EMBO Rep.*, 2003, **4**, 906–911.
92. T. Yoon and J. A. Cowan, *J. Am. Chem. Soc.*, 2003, **125**, 6078–6084.
93. O. S. Chen, S. Hemenway and J. Kaplan, *Proc. Natl Acad. Sci. USA*, 2002, **99**, 12321–12326.
94. M. Babcock, D. de Silva, R. Oaks, S. Davis-Kaplan, S. Jiralerspong, L. Montermini, M. Pandolfo and J. Kaplan, *Science*, 1997, **276**, 1709–1712.
95. P. Cavadini, C. Gellera, P. I. Patel and G. Isaya, *Hum. Mol. Genet.*, 2000, **9**, 2523–2530.
96. M. Pandolfo and A. Pastore, *J. Neurol.*, 2009, **256**, 9–17.
97. C. Pondarre, B. B. Antiochos, D. R. Campagna, S. L. Clarke, E. L. Greer, K. M. Deck, A. McDonald, A. P. Han, A. Medlock, J. L. Kutok, S. A. Anderson, R. S. Eisenstein and M. D. Fleming, *Hum. Mol. Genet.*, 2006, **15**, 953–964.
98. S. Bekri, G. Kispal, H. Lange, E. Fitzsimons, J. Tolmie, R. Lill and D. F. Bishop, *Blood*, 2000, **96**, 3256–3264.
99. G. Kispal, P. Csere, C. Prohl and R. Lill, *EMBO J.*, 1999, **18**, 3981–3989.
100. H. Shi, K. Z. Bencze, T. L. Stemmler and C. C. Philpott, *Science*, 2008, **320**, 1207–1210.
101. A. Nandal, J. C. Ruiz, P. Subramanian, S. Ghimire-Rijal, R. A. Sinnamon, T. L. Stemmler, R. K. Bruick and C. C. Philpott, *Cell Metab.*, 2011, **14**, 647–657.
102. A. Krause, S. Neitz, H. J. Magert, A. Schulz, W. G. Forssmann, P. Schulz-Knappe and K. Adermann, *FEBS Lett.*, 2000, **480**, 147–150.
103. C. H. Park, E. V. Valore, A. J. Waring and T. Ganz, *J. Biol. Chem.*, 2001, **276**, 7806–7810.
104. C. Pigeon, G. Ilyin, B. Courselaud, P. Leroyer, B. Turlin, P. Brissot and O. Loreal, *J. Biol. Chem.*, 2001, **276**, 7811–7819.
105. G. Nicolas, M. Bennoun, I. Devaux, C. Beaumont, B. Grandchamp, A. Kahn and S. Vaulont, *Proc. Natl Acad. Sci. USA*, 2001, **98**, 8780–8785.
106. J. C. Lesbordes-Brion, L. Viatte, M. Bennoun, D. Q. Lou, G. Ramey, C. Houbron, G. Hamard, A. Kahn and S. Vaulont, *Blood*, 2006, **108**, 1402–1405.
107. G. Nicolas, M. Bennoun, A. Porteu, S. Mativet, C. Beaumont, B. Grandchamp, M. Sirito, M. Sawadogo, A. Kahn and S. Vaulont, *Proc. Natl Acad. Sci. USA*, 2002, **99**, 4596–4601.
108. T. Ganz, *Cell Metab.*, 2008, 7, 288–290.
109. K. Pantopoulos, S. K. Porwal, A. Tartakoff and L. Devireddy, *Biochemistry*, 2012, **51**, 5705–5724.
110. L. Silvestri, A. Pagani, A. Nai, I. De Domenico, J. Kaplan and C. Camaschella, *Cell Metab.*, 2008, **8**, 502–511.
111. K. E. Finberg, M. M. Heeney, D. R. Campagna, Y. Aydinok, H. A. Pearson, K. R. Hartman, M. M. Mayo, S. M. Samuel, J. J. Strouse, K. Markianos, N. C. Andrews and M. D. Fleming, *Nat. Genet.*, 2008, **40**, 569–571.
112. F. Guillem, S. Lawson, C. Kannengiesser, M. Westerman, C. Beaumont and B. Grandchamp, *Blood*, 2008, **112**, 2089–2091.

113. M. A. Melis, M. Cau, R. Congiu, G. Sole, S. Barella, A. Cao, M. Westerman, M. Cazzola and R. Galanello, *Haematologica*, 2008, **93**, 1473–1479.

114. E. Nemeth, M. S. Tuttle, J. Powelson, M. B. Vaughn, A. Donovan, D. M. Ward, T. Ganz and J. Kaplan, *Science*, 2004, **306**, 2090–2093.

115. M. D. Knutson, M. Oukka, L. M. Koss, F. Aydemir and M. Wessling-Resnick, *Proc. Natl Acad. Sci. USA*, 2005, **102**, 1324–1328.

116. T. Chaston, B. Chung, M. Mascarenhas, J. Marks, B. Patel, S. K. Srai and P. Sharp, *Gut*, 2008, **57**, 374–382.

117. D. A. Weinstein, C. N. Roy, M. D. Fleming, M. F. Loda, J. I. Wolfsdorf and N. C. Andrews, *Blood*, 2002, **100**, 3776–3781.

118. B. Chung, T. Chaston, J. Marks, S. K. Srai and P. A. Sharp, *J. Nutr.*, 2009, **139**, 1457–1462.

119. C. Brasse-Lagnel, Z. Karim, P. Letteron, S. Bekri, A. Bado and C. Beaumont, *Gastroenterology*, 2011, **140**, 1261–1271, e1261.

120. P. F. Hahn, W. F. Bale, J. F. Ross, W. M. Balfour and G. H. Whipple, *J. Exp. Med.*, 1943, **78**, 169–188.

121. S. Granick, *Science*, 1946, **103**, 107.

122. L. Vanoaica, D. Darshan, L. Richman, K. Schumann and L. C. Kuhn, *Cell Metab.*, 2010, **12**, 273–282.

123. K. Iwai, S. K. Drake, N. B. Wehr, A. M. Weissman, T. LaVaute, N. Minato, R. D. Klausner, R. L. Levine and T. A. Rouault, *Proc. Natl Acad. Sci. USA*, 1998, **95**, 4924–4928.

124. T. A. Rouault, C. D. Stout, S. Kaptain, J. B. Harford and R. D. Klausner, *Cell*, 1991, **64**, 881–883.

125. B. R. Henderson, C. Seiser and L. C. Kuhn, *J. Biol. Chem.*, 1993, **268**, 27327–27334.

126. M. W. Hentze and L. C. Kuhn, *Proc. Natl Acad. Sci. USA*, 1996, **93**, 8175–8182.

127. F. Samaniego, J. Chin, K. Iwai, T. A. Rouault and R. D. Klausner, *J. Biol. Chem.*, 1994, **269**, 30904–30910.

128. S. Chollangi, J. W. Thompson, J. C. Ruiz, K. H. Gardner and R. K. Bruick, *J. Biol. Chem.*, 2012, **287**, 23710–23717.

129. T. Moroishi, M. Nishiyama, Y. Takeda, K. Iwai and K. I. Nakayama, *Cell Metab.*, 2011, **14**, 339–351.

130. J. C. Ruiz, S. D. Walker, S. A. Anderson, R. S. Eisenstein and R. K. Bruick, *J. Biol. Chem.*, 2013, **288**, 552–560.

131. A. A. Salahudeen, J. W. Thompson, J. C. Ruiz, H. W. Ma, L. N. Kinch, Q. Li, N. V. Grishin and R. K. Bruick, *Science*, 2009, **326**, 722–726.

132. A. A. Vashisht, K. B. Zumbrennen, X. Huang, D. N. Powers, A. Durazo, D. Sun, N. Bhaskaran, A. Persson, M. Uhlen, O. Sangfelt, C. Spruck, E. A. Leibold and J. A. Wohlschlegel, *Science*, 2009, **326**, 718–721.

133. B. Galy, D. Ferring-Appel, S. Kaden, H. J. Grone and M. W. Hentze, *Cell Metab.*, 2008, **7**, 79–85.

134. E. G. Meyron-Holtz, M. C. Ghosh, K. Iwai, T. LaVaute, X. Brazzolotto, U. V. Berger, W. Land, H. Ollivierre-Wilson, A. Grinberg, P. Love and T. A. Rouault, *EMBO J.*, 2004, **23**, 386–395.

135. S. R. Smith, M. C. Ghosh, H. Ollivierre-Wilson, W. Hang Tong and T. A. Rouault, *Blood Cells Mol. Dis.*, 2006, **36**, 283–287.
136. B. Galy, D. Ferring-Appel, C. Becker, N. Gretz, H. J. Grone, K. Schumann and M. W. Hentze, *Cell Rep.*, 2013, **3**, 844–857.
137. M. Mastrogiannaki, P. Matak, B. Keith, M. C. Simon, S. Vaulont and C. Peyssonnaux, *J. Clin. Invest*, 2009, **119**, 1159–1166.
138. Y. M. Shah, T. Matsubara, S. Ito, S. H. Yim and F. J. Gonzalez, *Cell Metab.*, 2009, **9**, 152–164.
139. M. Taylor, A. Qu, E. R. Anderson, T. Matsubara, A. Martin, F. J. Gonzalez and Y. M. Shah, *Gastroenterology*, 2011, **140**, 2044–2055.
140. M. Sanchez, B. Galy, M. W. Hentze and M. U. Muckenthaler, *Nat. Protoc.*, 2007, **2**, 2033–2042.

CHAPTER 11

Iron Uptake and Homeostasis in Prokaryotic Microorganisms

PIERRE CORNELIS

Department of Bioengineering Sciences, Research Group Microbiology, VIB Structural Biology, Vrije Universiteit Brussel, 1050, Brussels, Belgium
Email: pcornel@vub.ac.be

11.1 Iron: a Biologically Important, but Dangerous, Metal

For almost all microorganisms iron is an essential element involved in many important reactions involving, among others, [Fe-S] proteins and haem in cytochromes.[1] One notable exception is the Lyme disease bacterium *Borrelia burgdorferi*, which does not need iron, and uses manganese instead.[2] Under normal environmental conditions, iron presents two oxidation states, Fe^{2+} and Fe^{3+}, that are particularly suitable in oxido-reduction reactions.[1] While Fe^{2+} is the dominant form under anaerobic conditions, Fe^{3+} is the major form in oxygenated environments. This presents a problem for microorganisms with an aerobic lifestyle because of the extremely low solubility of the ferric iron.[1] For pathogens, the iron-restriction problem is also acute since the host actively limits iron availability by sequestering it within intracellular proteins such as haemoglobin, cytochromes or ferritins, or by chelating extracellular Fe^{3+} with the glycoproteins transferrin and lactoferrin.[3] Iron is also dangerous because of the Fe^{2+}-triggered Fenton/Haber–Weiss reaction that produces harmful reactive oxygen species (ROS) such as

RSC Metallobiology Series No. 2
Binding, Transport and Storage of Metal Ions in Biological Cells
Edited by Wolfgang Maret and Anthony Wedd
© The Royal Society of Chemistry 2014
Published by the Royal Society of Chemistry, www.rsc.org

superoxide (O_2^-), hydrogen peroxide (H_2O_2) and the highly destructive hydroxyl radical ($^\bullet$OH).

These ROS can cause considerable oxidative damage, including destruction of [Fe-S] clusters, protein carbonylation, Cys/Met-residue oxidation, membrane lipid peroxidation and DNA lesions.[4] A delicate balance must therefore be found between the need for iron for essential biological reactions and limiting its accumulation.

Although the title of this chapter suggests that all prokaryotes will be covered, the scarcity of information concerning Archaea prevents a detailed account. The different mechanisms of iron accumulation in bacteria will be reviewed, starting with the uptake of Fe^{2+}, moving to siderophore-mediated Fe^{3+} uptake and then haem uptake. Since regulation of iron homeostasis is important, the different iron uptake regulators and their mode of action (direct or indirect) will also be reviewed. Finally, we will look at the overlap between the regulation of oxidative stress response and the regulation of iron uptake.

11.2   Uptake of Ferrous Iron

11.2.1   The FeoABC System

The ferrous iron Fe^{2+} is highly soluble compared to the oxidized Fe^{3+} and is stable under anaerobic conditions and at low pH.[5] Soluble Fe^{2+} is transported into cells *via* a transport system termed Feo, composed of a permease FeoB and the proteins FeoA and FeoC.[5] This system is probably ancient since the primitive atmosphere of Earth was anoxic, resulting in the high abundance of this form of iron.[6] The Feo uptake system was first described in *E. coli*[7] and was later discovered in many other bacteria and, recently, even in the Archaeon *Pyrococcus furiosus*.[8] *Feo* genes are present in more than 1000 bacterial genomes.[9] The gene organization in *E. coli* is shown in Figure 11.1A. The *feoC* gene encoding a small cytoplasmic protein (78 residues) seems to be conserved only in γ-Proteobacteria.[5] Upstream of the *feoABC* operon, there are binding sites for the regulators Fur and Fnr. Their presence explains up-regulation of expression of *feo* under iron limitation (Fur) and anaerobiosis (Fnr).[5] The major component is the FeoB permease, which is a polytopic cytoplasmic membrane protein with eight transmembrane domains in the majority of cases (Figure 11.1B). The protein has an N-terminal domain in the cytoplasm with a GTPase function, which is typical of eukaryotic G-proteins such as Ras.[5,10] It possesses low affinity for GTP coupled with a low hydrolysis rate but the release rate of GDP is fast. The N-terminal FeoB domain forms a trimeric structure where the G-domain covers amino acids 1–170 while the second part of the FeoB N-domain, termed the S-domain, could act as a gate for Fe^{2+}.[11] FeoA has an SH3 domain typical of GTPase activating proteins that interact with eukaryotic G-proteins.

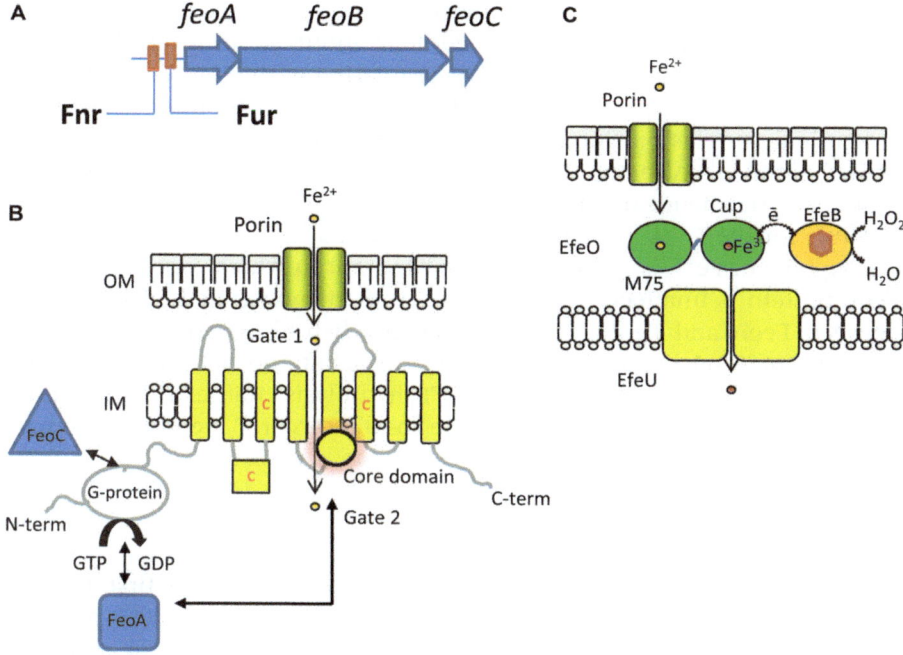

Figure 11.1 A: The *feoABC* genes involved in the Fe^{2+} uptake system. The operon is under the control of the Fur protein (induced by iron limitation) and by the Fnr regulator (induced under anaerobic conditions). B: The FeoABC system in Gram-negative bacteria, based on the *E. coli* model.[5] Fe^{2+} probably diffuses to the periplasmic space *via* a porin in the outer membrane. The large FeoB protein is an integral membrane protein with eight transmembrane domains forming two "iron gates". Conserved Cys residues are indicated in red. The N-terminal domain of FeoB resides in the cytoplasm and has a GTPase activity similar to eukaryotic G proteins. The cytoplasmic proteins FeoA and FeoC both interact with the N-terminal domain of FeoB. It has been proposed recently that FeoA also interacts with a cytoplasmic loop of FeoB called the "core domain".[14] C: The EfeUOB system in Gram-negative bacteria: EfeU is an Fe^{3+} permease localized in the cytoplasmic membrane. EfeO is a periplasmic protein with two domains, an N-terminal M75 peptidase domain and a C-terminal cupredoxin domain responsible for the oxidation of Fe^{2+} to Fe^{3+}. The donated electrons are transferred to the periplasmic EfeB haemoprotein using H_2O_2 as an electron acceptor.[16]

The C-terminal membrane domain of FeoB has two "gates" in opposite orientations, similar to those in the iron permease of yeast, Ftr1p.[12] There are also several conserved cysteine residues in transmembrane domains 3 and 6 and in the first cytoplasmic loop (Figure 11.1B). It has been proposed that the conserved Cys residues in the gates 1 and 2 transmembrane domains serve as ligands for Fe^{2+}.[5]

The small FeoC protein is predicted to contain an Fe-S cluster and it presents a LysR-regulator-like winged-helix motif at its N-terminus, suggesting that it could be involved in the regulation of the transcription of the *feo* operon.[13] Recently, it was demonstrated that FeoC also interacts with the N-terminal cytoplasmic domain of FeoB, suggesting a role in the modulation of Fe^{2+} uptake.[14] Although the *feoC* gene is found only in γ-Proteobacteria, other bacteria encode other small Fe-S cluster proteins in their *feo* operons.[5]

A role for the FeoA protein has been proposed recently based on the fact that uptake of Fe^{2+} is reduced in a *feoA* mutant in which expression of the FeoB protein is unaffected.[15] The same study demonstrated an interaction between FeoA and FeoB *via* a bacterial two-hybrid system. Recently, the structure of FeoA has been solved and the results do not support the hypothesis that FeoA assists FeoB in GTP hydrolysis. Rather they suggest a direct interaction between FeoA and the so-called "core" membrane domain of FeoB.[9]

11.2.2 The EfeUOB System

The EfeUOB Fe^{2+} transport system was identified for the first time in *Escherichia coli*.[16,17] Although the genes are present in *E. coli* K12, a frameshift in *efeU* renders the system inactive in this strain while the system is active in *E. coli* O157 : H7.[16] In contrast to the Feo system that is induced by anaerobic conditions, the EfeUOB system operates under aerobic acidic conditions where Fe^{2+} is more stable.[16] The operon is regulated by Fur and by the two-component system CpxAR, which is involved in the response to acid stress.[16–18] It has been suggested that this reflects an adaptation to acidic conditions, which can prevail outside the host.[16] A similar system, designated as YwbLMN, has been described in *Bacillus subtilis*[19] while *Staphylococcus aureus* has the FepABC system where FepC is the homologue of EfeU, FepB of EfeB and FepA of EfeO.[20]

EfeU is an inner membrane protein that shows similarity to the Frt1p iron permease from *Saccharomyces cerevisiae*,[21] suggesting that it is the iron transporter. EfeB is a periplasmic DyP peroxidase which is translocated to the periplasm *via* the TAT (twin arginine translocation) pathway, making it the only example of a haemoprotein translocated by this system.[22] EfeO is also localized in the periplasmic space and has one M75 peptidase domain and a cupredoxin domain, and it is hypothesized to bind and oxidize Fe^{2+} to Fe^{3+}, generating electrons that are likely to be accepted by EfeB.[21] In *S. aureus*, the FepB protein has the same haem-peroxidase activity as its *E. coli* counterpart EfeB and is also transported to the cell wall *via* the TAT pathway.[20] *S. aureus* FepA is a lipoprotein anchored to the membrane and is therefore present in the cell wall.[20] It has been proposed that both EfeUOB and FepABC are involved in iron extraction from haem, without destruction of the porphyrin ring, by a mechanism of deferrochelation (Section 11.4.2) that allows the protoporphyrin IX molecule to be recycled.[20,23] The EfuOB system is depicted in Figure 11.1C.

11.2.3 Other Systems

The gene encoding the p19 protein of *Campylobacter jejuni* is adjacent to the iron transporter gene *cftr1*.[24] P19 is a periplasmic protein, which binds both iron and copper and is necessary for the transport of iron.[24] The p19 gene is up-regulated in conditions of iron limitation and is regulated by Fur.[25]

Salmonella enterica has the SitABCD system, which is also present in *Shigella* and some *E. coli* strains. It is involved in the uptake of both Fe^{2+} and Mn^{2+} and the operon is regulated by both Fur and the manganese uptake regulator MntR.[26–29]

11.3 Siderophore-mediated Iron Uptake

11.3.1 Types of Siderophores and Their Biosynthesis

Siderophores are low-molar-mass molecules that are able to chelate Fe^{3+} with high affinity.[30] As ligands, they are classified into phenolate-, catecholate-, hydroxamate-, carboxylate- or mixed types. The structure of one composite peptidic siderophore, type I pyoverdine from *Pseudomonas aeruginosa*, is shown in Figure 11.2A. It has both catecholate and hydroxamate groups participating in iron chelation. A majority of siderophores are synthesized by **n**on-**r**ibosomal **p**eptide **s**ynthetases (NRPS).[31] These are large multi-modular enzymes in which each module is responsible for the incorporation of one amino acid into the peptide chain. The number and order of the modules usually dictate the number and order of amino acids in the peptide product.[31,32] Each module contains **A** (activation or adenylation), **T** (thiolation) and **C** (condensation) domains. The **A** domains activate a given amino acid to form an aminoacyl-AMP molecule that is further tethered to a **T** domain *via* a thioester bond. The **C** domains catalyze the transfer of the amino-acyl intermediate to the adjacent downstream module to form a peptide bond. In some cases, an **E** (epimerization) domain converts the ʟ-amino acid to the ᴅ-configuration. A terminal **TE** (thioesterase) domain is often present to release the peptide from the enzyme by cyclization or hydrolysis. It is now possible to predict which amino acid is to be activated by a given A domain, allowing the peptide sequence to be identified.[33] Figure 11.2B shows the organization of the domains of the different NRPS units (PvdL, PvdI, PvdJ, PvdD) involved in the biosynthesis of *P. aeruginosa* type I pyoverdine.[32] PvdL is the synthetase involved in the biosynthesis of the chromophore part of pyoverdine (Figure 11.2A).[34] It is different from the other NRPS units because its first activation domain is an acyl-CoA ligase domain, in agreement with the recent discovery that the pyoverdine precursor ferribactin is myristoylated.[35] Pyoverdines are among the most complex and diverse siderophores produced by bacteria and each species or subspecies of fluorescent pseudomonad produces its own pyoverdine with a specific peptide chain containing between 6 and 12 amino acids.[36,37]

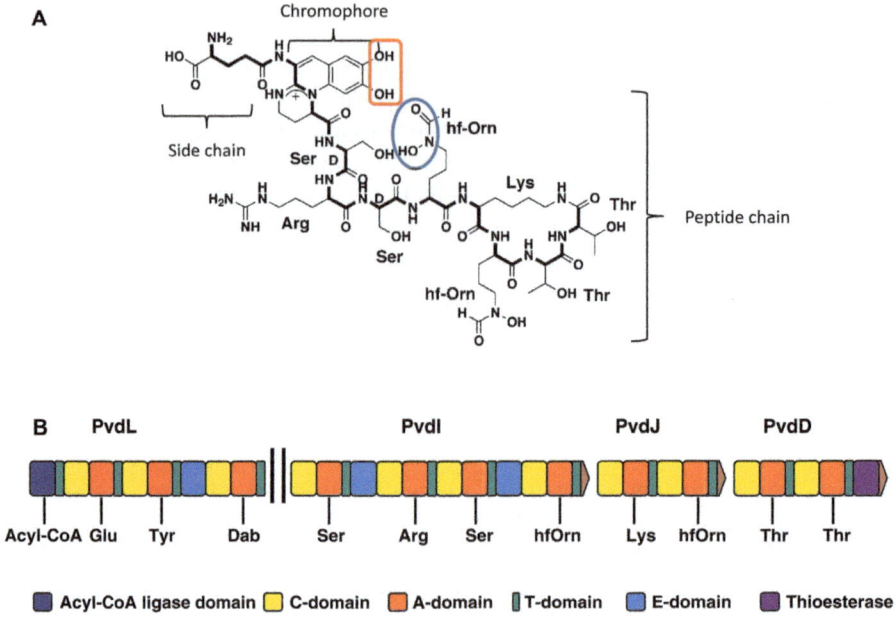

Figure 11.2 A: Structure of the *P. aeruginosa* peptidic siderophore pyoverdine. The chromophore is conserved in all pyoverdine structures and provides a catecholate group for Fe^{3+} chelation (highlighted in red). The peptidic part is variable and contains 6 to 12 amino acids. In this structure, the ligand hydroxamate is highlighted in blue. The side-chain is either a carboxylic acid or an amide. B: Non-ribosomal peptide synthetases (NRPS) involved in pyoverdine biosynthesis in *P. aeruginosa*. PvdL is involved in the synthesis of the part corresponding to the chromophore precursor. It consists in an acyl-CoA activating domain, followed by adenylation (A), thiolation (T), condensation (C) and some epimerization domains (E). The thioesterase domain releases the completed peptide from the protein. The PvdI, J and D NRPS are involved in the peptide chain formation. The *pvdL* gene is not clustered with the other NRPS genes as indicated by two bars. Note that the prediction of the activated amino acid corresponds perfectly to the sequence of the residues present in this particular pyoverdine. Dab: diamino butyric acid, hfOrn: *N*-formyl hydroxyornithine.

However, not all siderophores are synthesized *via* NRPS enzymes. Many are produced *via* what is termed the NIS biosynthesis pathway (**n**on-ribo-somal-**i**ndependent **s**iderophore **s**ynthesis).[38] A typical example is the biosynthesis of thioquinolobactin from *Pseudomonas fluorescens* ATCC17400. Steps involving the catabolism of tryptophan lead to the production of kynurenine and xanthurenic acid. The latter is activated by an AMP ligase to form AMP-xanthurenic acid that is further processed to give both the thioquinolobactin siderophore and its spontaneous hydrolysis product quinolobactin. Both molecules are able to chelate Fe^{3+}.[39–41]

11.3.2 Siderophore-mediated Iron Uptake in Gram-positive Bacteria

Gram-positive bacteria produce different types of siderophores. One example is the catecholate bacillibactin produced by *Bacillus subtilis* while the pathogen *B. anthracis* produces the two siderophores bacillibactin and petrobactin.[42-44] In the case of these siderophores, the Fe^{3+}-siderophore complex is hydrolysed by an esterase once inside the cytoplasm to liberate Fe^{3+}.[44] *S. aureus* also produces siderophores, called staphyloferrins A and B, which are both carboxylates.[45-49] Figure 11.3A shows a schematic representation of Fe^{3+}-siderophore uptake in Gram-positive bacteria. Because Gram-positive bacteria have only one cytoplasmic membrane, the Fe^{3+}-siderophores

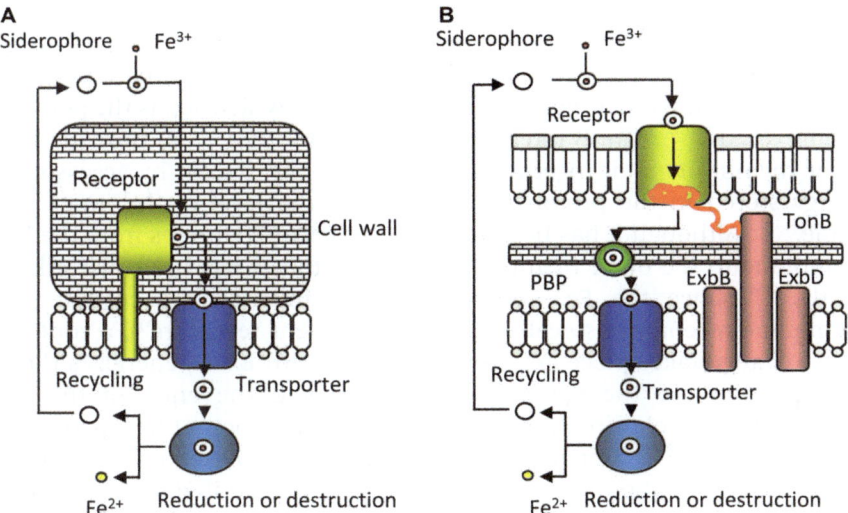

Figure 11.3 Siderophore-mediated high-affinity iron uptake. In both cases, the apo-siderophore is released to the exterior of the cell where it binds Fe^{3+}. Once in the cytoplasm, the ferric iron in the Fe^{3+}-siderophore can be removed in a non-destructive manner by reduction or by destruction of the siderophore skeleton. A: Gram-positive bacteria. The Fe^{3+}-siderophore is recognized by a receptor protein that is anchored to the cytoplasmic membrane. This protein subsequently delivers the Fe^{3+}-siderophore to a permease protein for transport to the cytoplasm. B: Gram-negative bacteria. The Fe^{3+}-siderophore is recognized by a receptor protein in the outer membrane, which forms a gated β-barrel porin. The N-terminal part of the receptor binds the periplasmic domain of the cytoplasmic membrane TonB protein, which, in association with the two membrane proteins ExbB and ExbD, relays the energy stored in the cytoplasmic membrane as the proton motive force, resulting in the opening of the gate by displacement of the plug domain of the receptor. Once in the periplasm, the Fe^{3+}-siderophore is bound to a periplasmic binding protein, which further transfers the Fe^{3+}-siderophore to a permease in the inner membrane for transport to the cytoplasm.

are first recognized by a siderophore-interacting protein anchored to the membrane, which then transfers it to an ABC transporter.[50] Once inside the cell, the iron is released from the siderophore either by degradation of the siderophore or by reductive release.[1,50]

11.3.3 Siderophore-mediated Iron Uptake in Gram-negative Bacteria

Gram-negative bacteria have two membranes, a cytoplasmic (inner) membrane and an asymmetric outer membrane. The latter is composed of an inner leaflet of phospholipids and an outer leaflet composed of lipopolysaccharides (LPS). The Fe^{3+}-siderophore complex is recognized at the outer membrane by a specific receptor, which typically is a β-barrel protein made of 22 β-strands.[30,51] These porins are gated by their own N-terminal domain that forms a plug. This is extended further into a periplasmic domain that interacts with the inner membrane protein TonB, which spans the periplasm (Figure 11.3B).[52,53] TonB interacts with the ExbD and ExbB proteins in the inner membrane and, together, they energize the system by accessing the energy stored in the proton motive force at the level of the inner membrane.[54,55] Although it has been proposed that TonB shuttles from the inner membrane to the outer membrane, this hypothesis has been discarded recently.[53,56] Once the Fe^{3+}-siderophore has touched the cognate outer membrane receptor (also called "TonB-dependent receptors"), a conformational change takes place and TonB, together with ExbD, energizes the system, resulting in the opening of the gate and the entry of the Fe^{3+}-siderophore complex into the periplasm where it is bound by a periplasmic binding protein.[30,57] Generally, the complex is transported to the cytoplasm by a permease where the iron is removed by an ill-defined reduction mechanism or by destruction of the siderophore ligand (as in the case of ferri-enterobactin).[1]

The specificity of the Fe^{3+}-siderophore uptake is conferred by the interaction with the outer membrane TonB-dependent receptor. An interesting illustration of this specific interaction is the uptake of Fe^{3+}-pyochelin in P. aeruginosa but of an enantiomeric form Fe^{3+}-enantiopyochelin in P. fluorescens Pf5.[58,59] The complexes are of stoichiometry ligand : Fe = 2 : 1. They are recognized at the level of the outer membrane by TonB-dependent receptors.[60] The structure of the FptA receptor from P. aeruginosa bound to Fe^{3+}-pyochelin has been determined.[61] The two ligands from the different species are enantiomers (Figure 11.4A). This subtle difference is enough to forbid recognition of Fe^{3+}-enantiopyochelin by the FptA receptor. Likewise, the FetA receptor of Fe^{3+}-enantiopyochelin in P. fluorescens Pf5 cannot recognize Fe^{3+}-pyochelin[58,59] (Figure 11.4B). Differences are also observed in the transport of these two ferri-siderophores through the periplasm and the inner membrane.[62] In P. aeruginosa, there is a single transporter FptX in the inner membrane that shows no specificity for either Fe^{3+}-pyochelin or

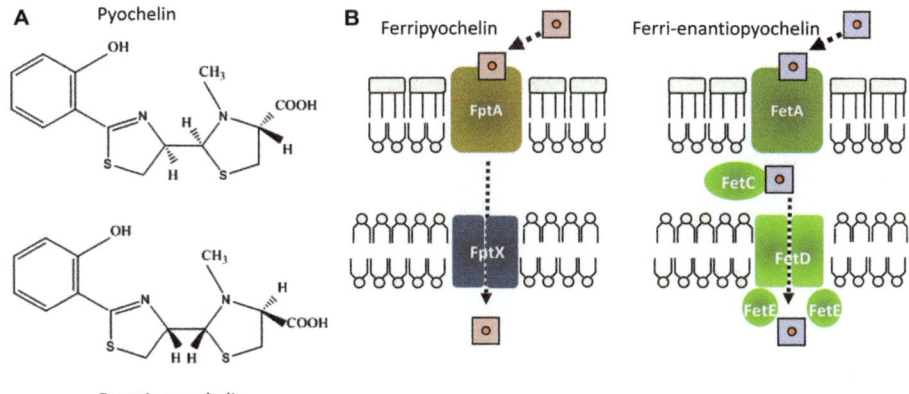

Enantio-pyochelin

Figure 11.4 A: Structures of the pyochelin siderophore from *P. aeruginosa* and the enantio-pyochelin siderophore from *P. protegens* (*P. fluorescens* Pf5). B: Mechanisms of uptake of Fe^{3+}-pyochelin and Fe^{3+}-enantio-pyochelin. Fe^{3+}-pyochelin is recognized by the FptA receptor that does not recognize Fe^{3+}-enantio-pyochelin. In turn, the Fe^{3+}-enantio-pyochelin receptor FetA does not recognize Fe^{3+}-pyochelin. A periplasmic binding protein FetC was identified for Fe^{3+}-enantio-pyochelin but not for Fe^{3+}-pyochelin. The Fe^{3+}-pyochelin permease FptX does not discriminate between the two Fe^{3+}-siderophores while the ABC transporter FetDF shows a strict specificity for Fe^{3+}-enantio-pyochelin.

Fe^{3+}-enantiopyochelin (Figure 11.4B). A periplasmic Fe^{3+}-pyochelin-binding protein is not present. On the other hand, transport of Fe^{3+}-enantiopyochelin in *P. fluorescens* Pf5 involves both a periplasmic binding protein FetC and an ABC transporter FetDE. The latter exhibits a strict specificity towards Fe^{3+}-enantiopyochelin.[62]

In *Pseudomonas*, the uptake of Fe^{3+}-pyoverdines (Figure 11.2A) is also conferred by a specific interaction with a cognate receptor in the outer membrane. The diversity of peptide chains in different pyoverdines implies that each Fe^{3+}-siderophore is recognized by a specific receptor. In turn, this suggests co-evolution of the modular NRPS enzymes (Section 11.3.1) and their receptors, as exemplified by the three different pyoverdines produced by *P. aeruginosa* and their respective receptors: FpvAI, II and III.[63–66]

The uptake of Fe^{3+}-pyoverdines is illustrated in Figure 11.5 and differs from the general plan schema presented in Figure 11.3B for the uptake of Fe^{3+}-siderophores in Gram-negative bacteria. The biosynthesis of pyoverdine takes place in the cytoplasm, probably involving a complex of NRPS and other enzymes.[67] The myristoylated precursor of pyoverdine, acylated ferribactin,[35] is transported from the cytoplasm to the periplasm by the PvdE ABC transporter.[68] The periplasmic PvdQ acylase removes the acyl chains, generating non-fluorescent ferribactin. Maturation of the fluorescent chromophore is mediated by the periplasmic proteins PvdONMP.[69–71] Pyoverdine is exported from the periplasm by the PvdT-PvdR-OpmQ type I export system

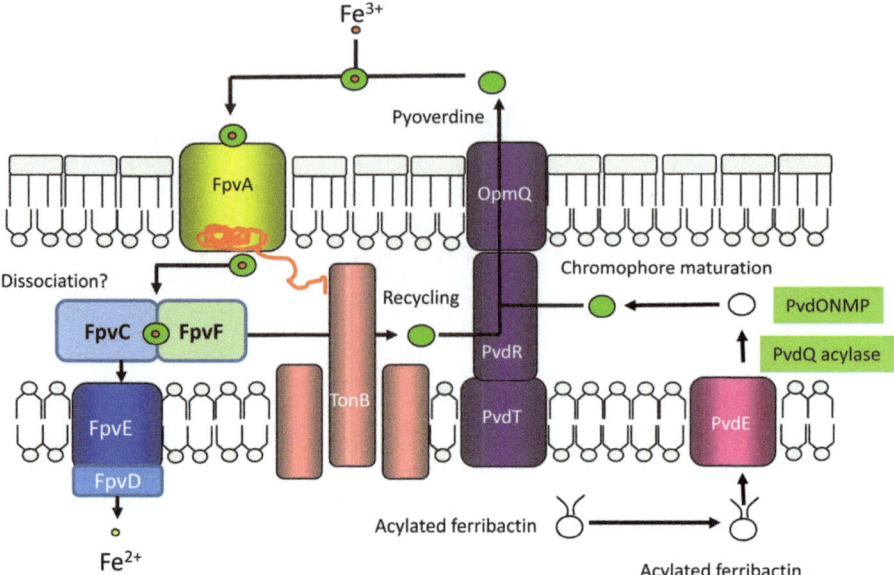

Figure 11.5 Uptake of Fe^{3+}-pyoverdine in *P. aeruginosa*: the precursor of pyoverdine, an acylated form of ferribactin (non-fluorescent), is transported to the periplasm, probably *via* the PvdE ABC transporter. In the periplasm, the PvdQ acylase removes the acyl chain and maturation of the chromophore occurs, as mediated by the periplasmic proteins PvdOMNP. Mature apo-pyoverdine is transported out of the cell by a tri-partite efflux system composed of the proteins PvdT, PvdR and OpmQ. Fe^{3+}-pyoverdine is recognized by the FpvA receptor and internalized *via* interaction with TonB. This is followed by interaction with two periplasmic proteins, FpvC and FpvF (Section 11.3.3) and the dissociated Fe^{2+} is transferred to an ABC transporter FpvED for transport to the cytoplasm. Apo-pyoverdine is rapidly recycled *via* the PvdTR-OpmQ efflux system.

but it has been suggested that another efflux system might be involved as well.[72]

The major difference with the general uptake schema of Figure 11.3 is that the reductive removal of iron from ferri-pyoverdine takes place in the periplasmic space by an unknown mechanism (Figure 11.5). The intact apo-pyoverdine is then rapidly recycled outside of the cell by the PVDRT-OpmQ pump.[72–75] Recently, it has been shown that PvdRT-OpmQ is also involved in the re-export of unwanted metal-pyoverdine chelates (metal ions other than ferric), which could be toxic for the cell.[72] Recently, the involvement of two periplasmic proteins FpvC and FpvF in the binding of ferri-pyoverdine has been described.[76] FpvC belongs to a class of periplasmic binding proteins that bind metal ions (similar to ZnuA or MntA) and could well bind Fe^{2+}. FpvF belongs to the same class of periplasmic binding proteins as FhuD, which binds hydroxamate siderophores.[76] These two periplasmic proteins could therefore be involved in the reductive dissociation of Fe^{3+}-pyoverdine.

The apo-siderophore would be recycled and Fe^{2+} transported inside the cytoplasm by an ABC transporter involving two proteins, FpvE and FpvD (Figure 11.5).[76]

11.3.4 Uptake of Xeno-siderophores

Next to the production of their own siderophores, many bacteria have evolved the capacity to use siderophores produced by other microorganisms, including bacteria and fungi. This strategy has been termed "siderophore piracy".[77] Bacteria able to colonize various niches have the capacity to utilize a wider variety of these siderophores compared to bacteria living in a defined specialized environment.[77]

As would be expected, in the case of Gram-negative bacteria, the increased capacity to utilize different siderophores is reflected by a larger repertoire of cognate TonB-dependent receptors. This is the case for pseudomonads that have between 30 and 50 TonB-dependent receptor genes in their genome[78,79] (Pierre Cornelis and Lumeng Ye, unpublished results). Interestingly, there are sometimes several ferri-pyoverdine receptor genes present in some genomes: two in *P. aeruginosa* (*fpvA* and *fpvB*[80]) and six in *P. fluorescens* Pf5 (now *P. protegens*).[81] Many Gram-negative bacteria are able to take up the Fe^{3+}-enterobactin catecholate siderophore or its degradation products.[82–84] The β-proteobacterium *Bordetella pertussis* is able to utilize human catecholamines (such as epinephrine) as sources of iron *via* three different receptors.[85,86] It should, however, be noted that not all TonB-dependent receptors are involved in the uptake of iron. Some participate in the utilization of sources of organic sulfur.[57]

11.4 Haem Uptake Systems

11.4.1 Gram-positive Bacteria

Gram-positive bacteria such as *S. aureus* are able to extract haem from different haemoproteins, such as haemoglobin or haemopexin, and use it as a source of iron.[87] Haem is not found in the unbound form because of its toxicity as a redox-active molecule.[88,89] In order to acquire haem-iron from haemoglobin, *S. aureus* secretes several haemolysins that cause lysis and release of intracellular haemoglobin.[90] The Isd system in *S. aureus* is the best characterized haem uptake system in Gram-positive bacteria (Figure 11.6A). *S. aureus* binds haemoglobin and haemoglobin-haptoglobin complexes *via* proteins termed **i**ron-regulated **s**urface **d**eterminants (Isd).[91]

The nine proteins IsdA-I bind host haemoproteins, transport haem into the bacterial cytoplasm, and degrade it to release iron. IsdB and IsdH are haemoprotein receptors that bind haemoglobin and haemoglobin-haptoglobin complexes, respectively. They feature NEAT domains (**near** iron **t**ransporter); 3 for IsdH and 2 for IsdB.[92] Domains NEAT 1 and 2 for IsdH and NEAT 1 for IsdB bind the haemoproteins and remove the haem cofactor,

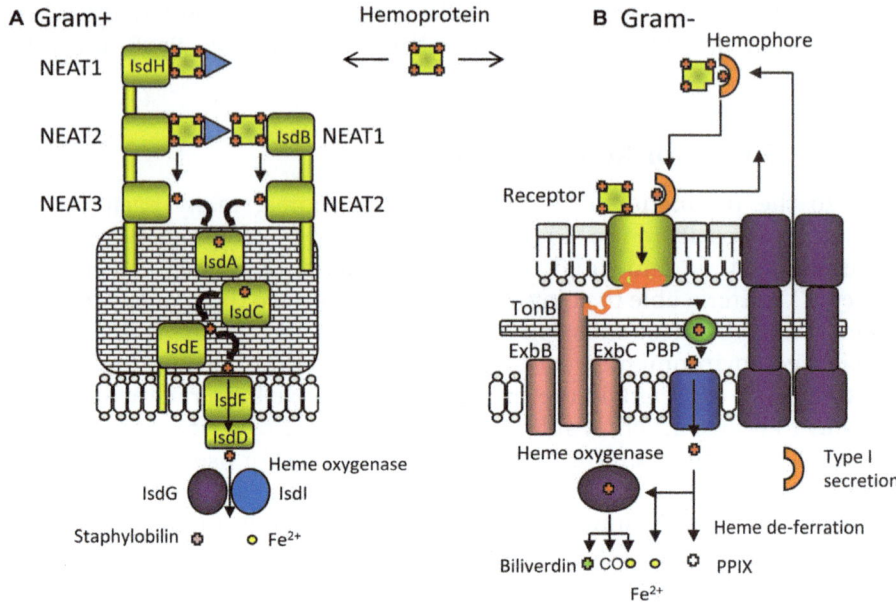

Figure 11.6   A: Haem uptake in Gram-positive bacteria taking *S. aureus* as a model:
the receptor IsdH binds haemoglobin complexed to haemopexin
while the other receptor IsdB binds haemoglobin itself (haem is
represented as a red cross). IsdH has three NEAT domains (**Near
Iron Transporter**) while IsdB has two. Haem is extracted and bound to
NEAT 3 of IsdH or NEAT 2 of IsdB before being transferred to IsdA,
which is embedded in the cell wall. Haem is further transferred to
IsdC and IsdE and, finally, to the IsdFD transporter. Within the
cytoplasm, the two subunits of the haem oxygenase, IsdG and IsdI,
remove Fe^{2+} from haem leaving the porphyrin chromogenic compound staphylobilin. B: Haem uptake in Gram-negative bacteria: two
types of haem uptake systems have been described, one involving
direct binding of the haemoprotein by the TonB-dependent receptor,
and the other involving a haemophore protein (secreted by a tripartite type I secretion system) that binds the haemoprotein outside
of the cell and then extracts haem. The haemophore-haem complex
further interacts with a receptor to which the haem is transferred. It is
transported to the periplasm *via* interaction of the receptor with TonB.
There it can bind to different periplasmic binding protein(s) before
transfer to a transporter for entry to the cytoplasm. A haem oxygenase
removes Fe^{2+} from haem leaving biliverdin and CO. Alternatively,
Fe^{2+} is removed by a de-ferration mechanism, leaving the protoporphyrin IX ligand intact.

which is then transferred to the last NEAT domain of the respective
receptors.[92,93]

Haem is then transferred to the IsdA protein (anchored in the peptidoglycan cell wall) and further on to IsdC.[94,95] All haem and haemoprotein
receptors are anchored to the peptidoglycan cell wall through a transpepti-
dation reaction catalysed by sortase A (IsdA, IsdB and IsdH), *via* an LPXTG

motif near their C-terminus of these proteins), and by sortase B, which anchors IsdC through a C-terminal NPQTN motif.[96,97] The SrtA sortase is constitutively expressed while SrtB is expressed when the concentration of iron is low.[97] IsdA and IsdC also each contain a single NEAT domain. IsdC transfers haem to IsdE, a membrane-anchored lipoprotein, which chaperones haem to IsdF (Figure 11.6A).[98–100] IsdF and IsdD together form an ABC transporter, in which IsdF is a membrane-spanning permease and IsdD is a membrane-associated ATP-binding protein.[101,102] Finally, the porphyrin ring of haem in the cytoplasm is broken down by the two haem oxygenase subunits IsdI and IsdG (64% similarity at the amino acid level), thereby releasing Fe^{2+} and the unusual chromophore staphylobilin, which differs from biliverdin.[103–105]

The Isd system is also present in other Gram-positive bacteria. In addition, similar haem uptake systems have been described for other Gram-positive bacteria (reviewed by Hood and Skaar).[50]

11.4.2 Gram-negative Bacteria

There are two major pathways for haem uptake (Figure 11.6B).[87] The first is mediated *via* an extracellular protein called a haemophore that scavenges haem from the haemoprotein and delivers it to a TonB-dependent outer membrane receptor.[87,106,107] The second pathway utilizes receptors that bind haemoproteins directly to extract the haem.[87] In the cytoplasm, iron is released by a haem oxygenase-mediated degradation of the protoporphyrin ring, resulting in the production of biliverdin and CO. Alternatively, as recently described, iron can be removed without cleavage of the tetrapyrrole ring by periplasmic or cytosolic haem-iron retrieving enzymes.[23] The haem molecule, being toxic, is bound in the cytoplasm to a chaperone protein before being delivered to the haem oxygenase.[88,89]

11.5 Regulation of Iron Uptake

11.5.1 The Fur Regulator Family

Tight control of iron acquisition is necessary since any excess of free iron in the cell can trigger Fenton chemistry. On the other hand, lack of iron affects cell metabolism, including the destruction of Fe-S clusters, among other effects.[108,109] Iron-uptake systems are up-regulated under conditions of low iron availability and stores are released from iron storage proteins. Conversely, the uptake systems are down-regulated under conditions of iron sufficiency and stores are increased[1] in three different types of proteins: ferritins, bacterioferritins and Dps (**D**NA-binding **p**roteins from **s**tarved cells).[110–112]

The control of iron homeostasis in many bacterial species (both Gram-negative and Gram-positive) is coordinated by the **f**erric **u**ptake **r**egulator (Fur) (Figure 11.7). Fur is a classical repressor employing Fe^{2+} as co-repressor

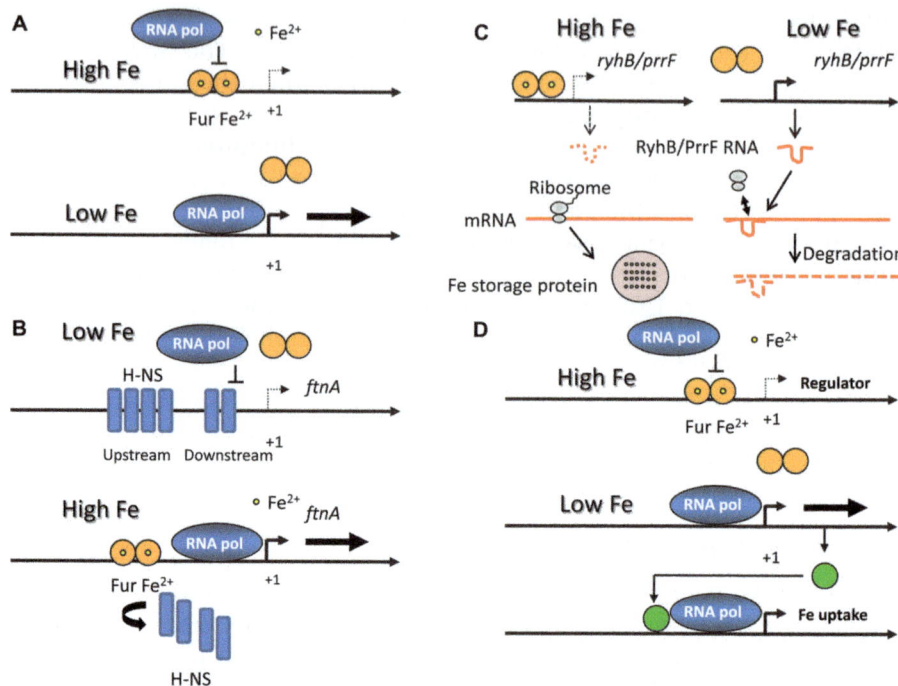

Figure 11.7   A: Mechanism of repression of iron uptake by Fe^{2+}-Fur. When iron levels are adequate, Fur will bind available Fe^{2+} and the dimer will bind to operator sequences called Fur boxes, restricting access of the RNA polymerase holo-enzyme and thereby blocking initiation of transcription of genes for the uptake of iron. When iron levels are low, apo-Fur is unable to bind to the operator sequences and iron uptake genes are expressed. B: Fe^{2+}-Fur activates the transcription of the ferritin gene *ftnA* in *E. coli*. When iron levels are low, transcription is blocked by binding of histone-like protein H-NS to several sequences upstream and downstream of the initiation point for expression of the *ftnA* gene. When iron levels are adequate, Fe^{2+}-Fur binds to the upstream sequences, displacing H-NS and allowing transcription initiation to take place. C: Indirect activation by Fe^{2+}-Fur of iron storage genes. When iron levels are adequate, Fe^{2+}-Fur represses the transcription of the small RNA *rhyB* gene while, when iron levels are low, RhyB RNA is produced. This small anti-sense RNA molecule will hybridize with the target RNA (example: bacterioferritin or SodB), promoting its degradation and preventing production of the iron-storage protein. D: Involvement of secondary regulators in iron uptake activation. Fe^{2+}-Fur represses the expression of a gene encoding an activator (ECF σ factor, AraC regulator,...), which itself activates transcription of target genes involved in specific iron uptake mechanisms (see text for details).

and switches off the transcription of iron transport and siderophore biosynthesis genes under high iron conditions.[113–115] Fe^{2+}-Fur normally represses transcription of target genes by binding to an operator sequence overlapping the promoter region, thereby blocking access by the RNA polymerase holoenzyme (Figure 11.7A). The minimal Fur-Fe^{2+} binding site (the Iron or Fur box) has 19 bp,[113] which can be subdivided into three identical and adjacent 6 bp sites.[116]

Sometimes, however, Fe^{2+}-Fur acts as an activator, either indirectly or directly, as in the case of induction of the ferritin gene (*ftnA*) of *E. coli*[117] (Figure 11.7B). When iron is low in the cell, the transcription of *ftnA* is blocked by the histone-like protein H-NS, which binds at a number of upstream and downstream sites within the *ftnA* promoter.[117] Under high-iron conditions, Fe^{2+}-Fur binds to the upstream sites, displacing H-NS from both sites and allowing transcription to take place.[117]

In response to high-iron conditions, Fe^{2+}-Fur can also induce the expression of certain iron-containing proteins (such as the iron storage protein bacterioferritin and the iron superoxide dismutase SodB) (Figure 11.7C).[1] This positive regulation is indirect and is generally mediated by Fe^{2+}-Fur-repressed small-regulatory RNA molecules, such as RyhB in *E. coli* and Prrf1 and Prrf2 in *P. aeruginosa*.[118–121] The RyhB RNA molecule, together with the RNA chaperone Hfq, hybridizes to targeted messages and acts post-transcriptionally by activating the degradation of the mRNAs by ribonuclease (RNAse E) (Figure 11.7C).

In *Helicobacter pylori*, apo-Fur is able to act as a repressor of the *sodB* and *pfr* genes that encode SodB and ferritin, respectively. However, Fe^{2+}-Fur is not able to do so, resulting in de-repression of their transcription.[122] In addition, Fe^{2+}-Fur can also function as a transcriptional activator in *H. pylori* by inducing the *oorDABC* operon, which encodes the enzyme 2-oxoglutarate oxidoreductase. It binds just upstream of the promoter.[122]

In many cases, however, the regulation of transcription of iron uptake genes is indirect. Fe^{2+}-Fur represses expression of a gene that encodes an activator of transcription (Figure 11.7D). The mechanism is explained in the following sections.

11.5.1.1 The Fur Protein and Extracytoplasmic Sigma Factors

Sigma factors are subunits of bacterial RNA polymerase that associate with the core enzyme to form the holo-enzyme, thereby allowing the recognition of promoter sequences, which is a prerequisite for initiation of the transcription process.[123] Some sigma factors are called extracytoplasmic sigma factors (ECF σ) because they are sequestered by the cytoplasmic domain of a membrane anti-sigma protein.[124,125] Generally, the genes encoding the ECF σ factors are adjacent to a gene encoding the transmembrane anti-sigma factor and a gene encoding a TonB-dependent receptor. The ECF σ factors regulate these genes according to the ferric citrate uptake FecI/FecR paradigm.[126]

In the absence of ferric citrate, the ECF σ FecI is sequestered by the FecR anti-σ, preventing its association with the core RNA polymerase enzyme.[126] In *P. aeruginosa* PAO1, there are 19 genes encoding ECF σ factors;[127] ten of them are regulated by Fur and their genes are found next to the gene of the TonB-dependent receptor they control.[128,129] In other *Pseudomonas* strains, the number of ECF σ factors controlling the expression of TonB-dependent receptors is even larger.[79] The ECF σ dependent receptors are termed TonB-dependent transducers (TBDTs), which participate in cell-surface signalling systems.[79] The unprecedented number of 25 TBDT receptors was identified recently by us in *P. fluorescens* ATCC 17400 (Lumeng Ye and Pierre Cornelis, unpublished results). A typical example is the FpvA Fe^{3+}-pyoverdine receptor of *P. aeruginosa*, which interacts with the FpvR anti-sigma factor *via* its N-terminal signalling domain (Figure 11.8).[37,129,130] The inner membrane protein FpvR interacts, *via* its cytoplasmic domain, with the two ECF σ factors PvdS and FpvI. Binding of Fe^{3+}-pyoverdine to the FpvA receptor induces proteolytic degradation of FpvR, releasing PvdS and FpvI to associate with the RNA polymerase.[130,131] PvdS is needed for transcription of the pyoverdine biosynthesis genes, but also for expression of two virulence factors genes, the exotoxin A and the exoprotease PrpL genes.[32,37,129,132] FpvI is the ECF σ involved in the transcription of the *fpvA* ferripyoverdine receptor gene.[32,37,130] This kind of regulation prevents the bacterium from expressing all the TBDT genes when iron is scarce, thereby saving cell energy since expression is only possible when the cognate Fe^{3+}-siderophore is present. It is interesting to note that a potential Fur binding site is nevertheless present upstream of some of the TBDT genes in *P. aeruginosa*. This may indicate that the ECF σ mediated regulation was superimposed later on the "primitive" Fe^{2+}-Fur regulation.[116,129]

11.5.1.2   *The Fur Protein and other Regulators*

AraC regulators are sometimes involved in the control of production and uptake of some siderophores and the expression of these activators is also subject to Fe^{2+}-Fur repression. This is the case for the pyochelin biosynthesis genes and for the *fptA* Fe^{3+}-pyochelin receptor gene, which is under the control of the PchR AraC regulator.[133] This regulator activates the *pchDCBA*, *pchEFGHI* and the *fptA* genes upon binding to the ferri-pyochelin co-activator.[133,134] The genes for the biosynthesis and uptake of enantio-pyochelin are also activated by a PchR regulator, but here again there is a strict specificity since the regulators from *P. aeruginosa* and *P. fluorescens* Pf5 are not interchangeable (Section 11.3.3).[58] In *Bordetella pertussis*, the uptake of Fe^{3+}-enterobactin *via* the BfeA receptor is regulated by the AraC regulator BfeR.[135] Likewise, the *P. fluorescens* ATCC17400 thioquinolobactin synthesis and uptake is under the control of an AraC regulator.[40]

In *P. aeruginosa*, the uptake of the xenosiderophore Fe^{3+}-enterobactin is regulated by a two-component system involving a cytoplasmic membrane sensor PfeS and a response regulator PfeR.[136] In this system, the PfeS sensor detects the Fe^{3+}-enterobactin and phosphorylates its own cytoplasmic domain. This is followed by phosphorylation of the PfeR response regulator, which activates transcription of the *pfeRS* and *pfeA* receptor genes.[129,136]

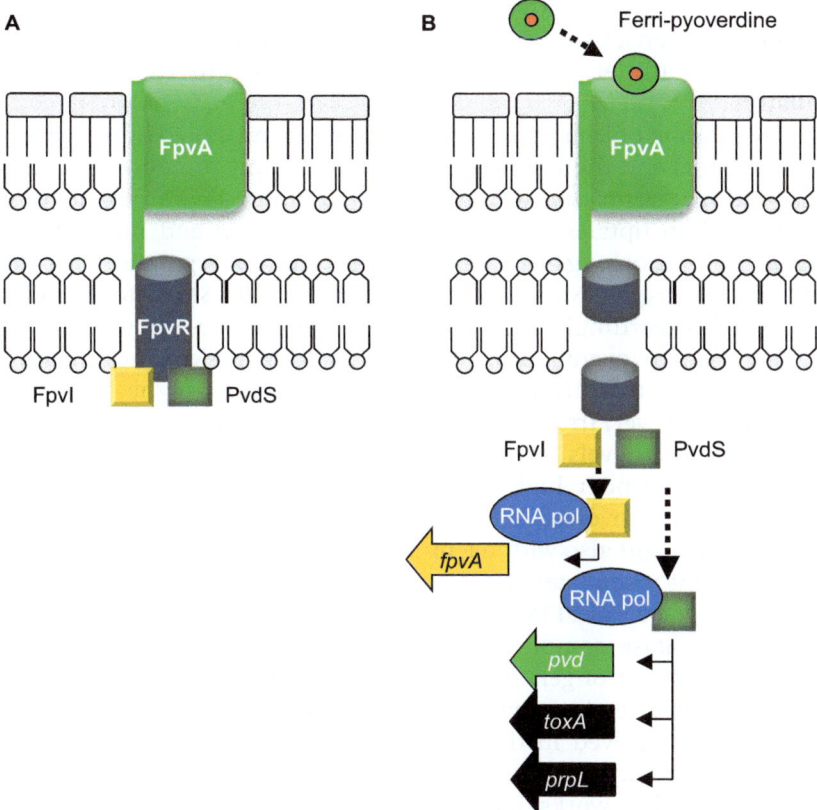

Figure 11.8 Mechanism of TonB-dependent transduction (TBDT) as exemplified by Fe^{3+}-pyoverdine uptake *via* the FpvA receptor in *P. aeruginosa*. A: When the few FpvA receptors in the outer membrane are not bound by Fe^{3+}-pyoverdine, the N-terminal domain of the receptor (the signalling domain) interacts with the FpvR membrane protein, which is an anti-sigma factor that sequesters two cytoplasmic ECF σ factors (PvdS and FpvI). The two sigma factors cannot associate with the RNA polymerase core enzyme and so the target genes (pyoverdine biosynthesis for PvdS and *fpvA* for FpvI) are not expressed. B: When Fe^{3+}-pyoverdine binds specifically to the FpvA receptor, a conformational change in the receptor protein triggers proteolytic degradation of the FpvR anti-sigma factor, releasing PvdS and FpvI, which are now able to complex with the RNA polymerase. PvdS activates the transcription not only of the pyoverdine biosynthesis genes, but also of two exoproteins, the exotoxin A and the protease PrpL (see text for details).

11.5.2 RirA, DtxR, IdeR and IscR Regulators

Although Fur is the most common general regulator of bacterial iron homeostasis, several bacteria employ other types of iron-responsive regulators.[115] On the other hand, Fur-like proteins are sometimes present, but are not involved in iron-dependent transcriptional control. This is the case for the Mur protein in Rhizobiales that responds to manganese, the Zur regulator that senses zinc, PerR, which is involved in response to peroxides, and the Nur nickel regulator. These all belong to the Fur family of regulators.[137–139]

The regulation of iron uptake in Rhizobiales is mediated by RirA (rhizobial iron regulator) and the Irr (iron response regulator) protein.[140,141] RirA is a transcriptional repressor working like Fur and it controls siderophore-dependent iron uptake in *Rhizobium leguminosarum* and in *Sinorhizobium meliloti*.[142–146] It belongs to a group of transcription regulators that include NsrR, the NO-responsive activator, and IscR, the [Fe–S] cluster biosynthesis regulator.[147] Accordingly, RirA responds to cellular [Fe–S] cluster levels and not, like Fur, to overall iron levels.[146] In Rhizobiales, the second iron-regulatory protein is Irr that, unlike RirA, belongs to the Fur family of regulators. However, Irr differs from Fur since it does not directly bind Fe^{2+} but detects the levels of haem and its precursor, protoporphyrin IX (PPIX).[140,148,149] Under high iron conditions, Irr interacts with the ferro-chelatase enzyme involved in the last step of haem biosynthesis (insertion of iron into PPIX) and also with haem, which causes Irr proteolysis. Under low iron conditions when haem concentrations are also low, there are increased levels of the precursor PPIX. Under such conditions, Irr does not bind to the PPIX-ferrochelatase complex, preventing its degradation and enabling its activity as regulator of genes involved in iron homeostasis.

In Gram-positive Actinomycetes, DtxR-like proteins (**D**iphtheria **to**xin **r**epressor) are involved in the control of iron homeostasis. These include DtxR in *Corynebacterium* and IdeR in *Mycobacterium tuberculosis*.[150–152] Interestingly, IdeR works like Fur as a repressor of expression of side-rophore biosynthesis genes, but it can also work as an activator of transcription by binding to a region of DNA upstream of the promoter of the bacterioferritin and ferritin genes.[153]

Iron-sulfur proteins include enzymes involved in the TCA cycle, proteins involved in electron transport chains and in several oxido-reduction reactions. Some regulators also have [Fe–S] clusters and include the oxidative-stress regulator SoxR, the redox sensor Fnr and IscR, the regulator of [Fe–S] cluster biogenesis.[154],† *E. coli* uses two pathways Isc and Suf[154] for the biosynthesis and insertion of [Fe–S] clusters. The Suf pathway is used only to repair oxidation-damaged [Fe–S] clusters.[154,155] IscR regulates both the *iscR-iscSUA hscBA-fdx* and the *sufABCDSE* gene clusters, as presented in Figure 11.9. Under conditions when iron is not limiting, holo-IscR (with intact [Fe–S] cluster) represses its own transcription and the transcription of the *isc hsc* genes.[154] When iron is limiting, levels of apo-IscR (without [Fe–S] cluster) increase, resulting in the production of *iscR-iscSUA hscBA-fdx* transcripts. However, these are then targeted by the small RNA RyhB mentioned above (Section 11.5.1), causing degradation of the transcript downstream of

---

†Fe-S cluster synthesis is discussed in Chapter 12.

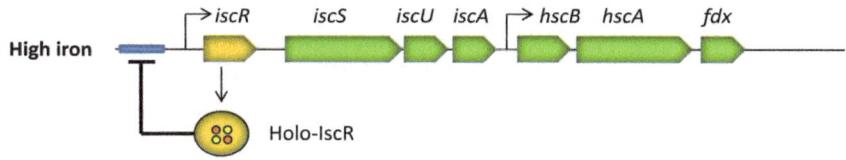

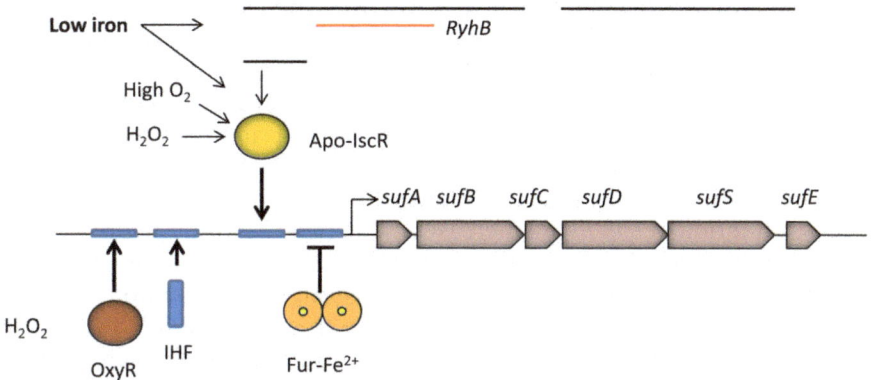

Figure 11.9 Regulation of [Fe-S] cluster assembly by the IscR regulator: when iron levels are adequate, the [Fe-S] clusters in IscR are assembled. Holo-IscR auto-regulates its own transcription by a repression mechanism and also represses transcription of the *iscSUA* and *hscBAfDX* genes. When iron levels are low, the concentration of apo-IscR rises, resulting in de-repression and increased transcription of the *iscR-iscSUA hscBAfdx* operons. The small RNA molecule RyhB binds to the transcripts, resulting in degradation of the transcript downstream of the *iscR* region. Apo-IscR positively regulates the transcription of the *sufABCDSE* genes, which are also positively regulated by OxyR and H-NS and negatively by Fe^{2+}-Fur (see text for details).

the *iscR* region (Figure 11.9). Apo-IscR positively regulates the transcription of the *sufABCDSE* genes. This regulation is quite complex since it involves the OxyR regulator, Fur and the nucleoid protein Ihf (**i**ntegration **h**ost **f**actor) (Figure 11.9). When iron levels return to normal, Fe^{2+}-Fur represses the *suf* operon.[154]

11.6 Iron Homeostasis and Oxidative Stress Regulation

11.6.1 Fur-mediated Oxidative Stress Resistance

It is not surprising that the response to oxidative stress and the regulation of iron homeostasis are intricate since excess iron in the cell can generate the production of reactive oxygen species (ROS) *via* the Fenton–Haber–Weiss reaction.[4,156] These species can damage cell processes *via* degradation of [Fe-S] clusters and DNA, protein carbonylation, Cys/Met-residue oxidation and membrane lipid peroxidation.[108,109,157] ROS are generated by aerobes *via* incomplete reduction of O_2 during respiration as well as through flavin-mediated reductive processes.[4] A mutation affecting global iron regulation can result in the accumulation of high levels of free-cytosolic iron that, in turn, catalyses production of reactive oxygen species.[158] In some bacteria, the global-iron regulator (Fur in *P. aeruginosa* and Irr in *M. tuberculosis*) cannot be fully inactivated, suggesting that they are essential for survival of the cell.[159,160] As already mentioned, Fe^{2+}-Fur directly represses the uptake of iron while indirectly (*via* RyhB RNA in *E. coli*) promoting the storage of iron. This means that *fur* inactivation simultaneously leads to uncontrolled iron uptake and diminished incorporation of internalized iron into iron proteins.[156] The consequence is increased Fenton chemistry and ROS build-up. A similar iron toxicity effect has been described for *M. smegmatis* where an *ideR* mutation resulted in increased sensitivity towards hydrogen per-oxide, decreased levels of the catalase KatG and the manganese superoxide dismutase (Mn-Sod). These observations suggested co-regulation of oxida-tive stress response and iron homeostasis in this bacterium.[161] The same phenomenon was observed in *P. aeruginosa* for mutants with decreased Fur activity as well as in other bacteria (*Xanthomonas campestris*, *Caulobacter crescentus*, *Anabaena*).[162–165] As already mentioned, the role of central iron regulator in α-Proteobacteria is taken by RirA and not by Fur. In *Agro-bacterium tumefaciens* and *Sinorhizobium meliloti*, *rirA* mutants display a growth deficiency in iron-replete media, due to enhanced accumulation of iron that causes an increased sensitivity to H_2O_2.[145,166]

11.6.2 The Fur-like Peroxide Regulator PerR

In some bacteria, the PerR protein (**per**oxide **r**egulator), a member of the Fur family, plays an essential function in orchestrating the response to oxidative stress. PerR is found in Gram-positive bacteria such as *B. subtilis*, *Clostridium acetobutylicum*, *Streptococcus* and *S. aureus*.[114,167–169] *B. subtilis* has both PerR and Fur that recognize very similar DNA-binding sites but, nevertheless, control distinct sets of target genes.[170] The PerR repressor regulates the adaptive response to peroxide stress including the catalase gene *katA*, an iron storage protein gene *mrgA*, an alkylhydroperoxide reductase gene *ahpCF*, a zinc uptake system *zosA*, haem biosynthesis genes *hemAXCDBL*, plus the *fur* gene and *perR* itself.[171] A *B. subtilis perR* knock-out mutant is resistant to hydrogen peroxide (due to the accumulation of KatA), accumu-lates a porphyrin but grows very slowly, as a result of the high expression of the two proteins KatA and Fur. This phenomenon can be traced to elevated repression of iron uptake by Fur, exacerbated by haem sequestration by the abundant KatA catalase protein.[171] Curiously, PerR is also present in two Gram-negative bacteria, *Neisseria gonorrhoeae* and in *C. jejuni*.[172–174] In the latter, the Fur and PerR regulons overlap since the *katA* gene is controlled by both regulators.[173,175] Altogether, these results suggest that Fur and PerR could derive from a common ancestor by gene duplication. One has kept its role in controlling iron homeostasis while the second evolved to specialize in the response to oxidative stress.

11.6.3 Redox-sensing Regulators

Redox-active groups (cysteine, haem, [Fe-S] or mononuclear iron) are present in redox-stress responsive transcriptional regulators. They respond *via* an oxidation-dependent conformational change that controls their interaction with target DNA sequences, influencing the expression of target genes.[176,177] The best known example is homodimeric OxyR that belongs to the large family of LysR type transcriptional regulators (LTTR).[178] Upon exposure to H_2O_2 or to change in the cytoplasmic thiol-disulfide redox status, two conserved cysteine residues in each monomer form a reversible, intramolecular disulfide bridge (S-S) that induces a conformational change and tetramer formation, allowing transcriptional activation of OxyR-dependent genes.[179–181] In *E. coli*, OxyR regulates several ROS-inactivating enzymes including the catalase KatG (hydroperoxidase I) and the alkylhydroperoxide reductase AhpCF. In addition, OxyR regulates other genes involved in oxidative stress defence, including *dps* (coding for a DNA binding mini iron-storage ferritin) and *fur* itself.[182,183] Therefore, in *E. coli*, OxyR activates *fur* transcription (which would result in decreased available iron levels during redox stress) and, together with Fur, controls *fhuF*, which encodes a ferric reductase, a [2Fe–2S]-protein required for release of iron from some hydroxamate siderophores.[183,184]

In *P. aeruginosa*, the growth of a redox-stress sensitive *oxyR* mutant is affected both by iron depletion and by the presence of high concentrations of the metal.[185,186] These properties suggest that the *oxyR* mutant is unable to cope with either iron limitation or excess. As seems to be the case for *E. coli*, levels of the Fur protein are decreased in this *P. aeruginosa oxyR* mutant, probably because of the sensitivity of Fur to oxidative damage.[187,188] A recent comprehensive analysis of the OxyR regulon in *P. aeruginosa* by chromatin immuno-precipitation revealed that the number of potential OxyR gene targets is higher than expected.[189] Interestingly, OxyR was confirmed to bind to the promotor-operator region of the *pvdS* ECF σ factor gene whose product is needed for the transcription of pyoverdine siderophore genes (Section 11.3.3).[189] As for *E. coli* again, *P. aeruginosa* OxyR binds to the promotor-operator region of the *dps* mini-ferritin gene, once more linking iron homeostasis to the response to oxidative stress.

Recently, another LysR regulator PA2206 has been described in *P. aeruginosa*. This regulator contributes significantly to the resistance of the bacterium to oxidative stress: inactivation of the gene resulted in hypersensitivity to hydrogen peroxide.[190] Like OxyR, this LysR regulator was found to activate the expression of, and to bind to, the operator-promotor sequence of pvdS.[190]

11.7 Concluding Remarks

During the evolution of life on Earth, bacteria have developed different strategies to take up the precious metal iron involved in essential metabolic reactions. Probably the most ancient and conserved system is the Feo Fe^{2+} uptake system, which was prevalent when the atmosphere was still anoxic. With the accumulation of O_2 due to cyanobacterial oxygenic photosynthesis, bacteria living in aerobic conditions and at pH around neutrality became confronted with the problem of Fe^{3+} insolubility. This led to the development of siderophore-mediated iron uptake. Competition with other microorganisms and interactions with eukaryotes also led to the evolution of xeno-siderophore uptake systems and of ways to extract and use haem as a source of iron. However, iron has two faces: it is indispensable for the majority of organisms but it is dangerous. Consequently, elaborate regulation networks had to be developed in order to maintain homeostasis of the metal and to save energy when iron is available. Because iron can generate an oxidative stress, it is not surprising that the response to oxidative stress and the control of iron levels in the cell are coordinated. Although much has been learned during the last two decades about iron uptake and homeostasis in bacteria, one would like to see more research being done about iron acquisition in Archaea. No doubt that the next decade will bring more interesting data and challenges.

References

1. S. C. Andrews, A. K. Robinson and F. Rodriguez-Quinones, *FEMS Microbiol. Rev.*, 2003, **27**, 215–237.
2. B. Troxell, H. Xu and X. F. Yang, *J. Biol. Chem.*, 2012, **287**, 19284–19293.
3. R. A. Finkelstein, C. V. Sciortino and M. A. McIntosh, *Rev. Infect. Dis.*, 1983, 5(4), S759–777.
4. J. A. Imlay, *Adv. Microb. Physiol.*, 2002, **46**, 111–153.
5. M. L. Cartron, S. Maddocks, P. Gillingham, C. J. Craven and S. C. Andrews, *Biometals*, 2006, **19**, 143–157.
6. K. Hantke, *Trends Microbiol.*, 2003, **11**, 192–195.
7. M. Kammler, C. Schon and K. Hantke, *J. Bacteriol.*, 1993, **175**, 6212–6219.
8. Y. Zhu, S. Kumar, A. L. Menon, R. A. Scott and M. W. Adams, *J. Bacteriol.*, 2013, **195**, 2400–2407.
9. C. K. Lau, H. Ishida, Z. Liu and H. J. Vogel, *J. Bacteriol.*, 2013, **195**, 46–55.
10. T. C. Marlovits, W. Haase, C. Herrmann, S. G. Aller and V. M. Unger, *Proc. Natl Acad. Sci. USA*, 2002, **99**, 16243–16248.
11. A. Guilfoyle, M. J. Maher, M. Rapp, R. Clarke, S. Harrop and M. Jormakka, *EMBO J.*, 2009, **28**, 2677–2685.
12. S. Severance, S. Chakraborty and D. J. Kosman, *Biochem. J.*, 2004, **380**, 487–496.

13. K. S. Gajiwala, H. Chen, F. Cornille, B. P. Roques, W. Reith, B. Mach and S. K. Burley, *Nature*, 2000, **403**, 916–921.
14. K. W. Hung, J. Y. Tsai, T. H. Juan, Y. L. Hsu, C. D. Hsiao and T. H. Huang, *J. Bacteriol.*, 2012, **194**, 6518–6526.
15. H. Kim, H. Lee and D. Shin, *Biochem. Biophys. Res. Commun.*, 2012, **423**, 733–738.
16. J. Cao, M. R. Woodhall, J. Alvarez, M. L. Cartron and S. C. Andrews, *Mol. Microbiol.*, 2007, **65**, 857–875.
17. C. Grosse, J. Scherer, D. Koch, M. Otto, N. Taudte and G. Grass, *Mol. Microbiol.*, 2006, **62**, 120–131.
18. A. D. Hernday, B. A. Braaten, G. Broitman-Maduro, P. Engelberts and D. A. Low, *Mol. Cell*, 2004, **16**, 537–547.
19. N. Baichoo, T. Wang, R. Ye and J. D. Helmann, *Mol. Microbiol.*, 2002, **45**, 1613–1629.
20. E. Turlin, M. Debarbouille, K. Augustyniak, A. M. Gilles and C. Wandersman, *PLoS One*, 2013, **8**, e56529.
21. M. B. Rajasekaran, S. Nilapwar, S. C. Andrews and K. A. Watson, *Biometals*, 2010, **23**, 1–17.
22. A. Sturm, A. Schierhorn, U. Lindenstrauss, H. Lilie and T. Bruser, *J. Biol. Chem.*, 2006, **281**, 13972–13978.
23. S. Letoffe, G. Heuck, P. Delepelaire, N. Lange and C. Wandersman, *Proc. Natl Acad. Sci. USA*, 2009, **106**, 11719–11724.
24. A. C. Chan, T. I. Doukov, M. Scofield, S. A. Tom-Yew, A. B. Ramin, J. K. Mackichan, E. C. Gaynor and M. E. Murphy, *J. Mol. Biol.*, 2010, **401**, 590–604.
25. K. Holmes, F. Mulholland, B. M. Pearson, C. Pin, J. McNicholl-Kennedy, J. M. Ketley and J. M. Wells, *Microbiology*, 2005, **151**, 243–257.
26. J. S. Ikeda, A. Janakiraman, D. G. Kehres, M. E. Maguire and J. M. Slauch, *J. Bacteriol.*, 2005, **187**, 912–922.
27. A. Janakiraman and J. M. Slauch, *Mol. Microbiol.*, 2000, **35**, 1146–1155.
28. D. G. Kehres, A. Janakiraman, J. M. Slauch and M. E. Maguire, *J. Bacteriol.*, 2002, **184**, 3159–3166.
29. M. Sabri, M. Caza, J. Proulx, M. H. Lymberopoulos, A. Bree, M. Moulin-Schouleur, R. Curtiss, 3rd and C. M. Dozois, *Infect. Immun.*, 2008, **76**, 601–611.
30. V. Braun and H. Killmann, *Trends Biochem. Sci.*, 1999, **24**, 104–109.
31. J. H. Crosa and C. T. Walsh, *Microbiol. Mol. Biol. Rev.*, 2002, **66**, 223–249.
32. J. Ravel and P. Cornelis, *Trends Microbiol.*, 2003, **11**, 195–200.
33. G. L. Challis, J. Ravel and C. A. Townsend, *Chem. Biol.*, 2000, 7, 211–224.
34. D. Mossialos, U. Ochsner, C. Baysse, P. Chablain, J. P. Pirnay, N. Koedam, H. Budzikiewicz, D. U. Fernandez, M. Schafer, J. Ravel and P. Cornelis, *Mol. Microbiol.*, 2002, **45**, 1673–1685.
35. M. Hannauer, M. Schafer, F. Hoegy, P. Gizzi, P. Wehrung, G. L. Mislin, H. Budzikiewicz and I. J. Schalk, *FEBS Lett.*, 2012, **586**, 96–101.
36. J. M. Meyer, *Arch. Microbiol.*, 2000, **174**, 135–142.

37. P. Visca, F. Imperi and I. L. Lamont, *Trends Microbiol.*, 2007, **15**, 22–30.
38. D. Oves-Costales, N. Kadi and G. L. Challis, *Chem. Commun. (Camb.)*, 2009, 6530–6541.
39. D. Mossialos, J. M. Meyer, H. Budzikiewicz, U. Wolff, N. Koedam, C. Baysse, V. Anjaiah and P. Cornelis, *Appl. Environ. Microbiol.*, 2000, **66**, 487–492.
40. S. Matthijs, C. Baysse, N. Koedam, K. A. Tehrani, L. Verheyden, H. Budzikiewicz, M. Schafer, B. Hoorelbeke, J. M. Meyer, H. De Greve and P. Cornelis, *Mol. Microbiol.*, 2004, **52**, 371–384.
41. S. Matthijs, K. A. Tehrani, G. Laus, R. W. Jackson, R. M. Cooper and P. Cornelis, *Environ. Microbiol.*, 2007, **9**, 425–434.
42. J. J. May, T. M. Wendrich and M. A. Marahiel, *J. Biol. Chem.*, 2001, **276**, 7209–7217.
43. E. A. Dertz, J. Xu, A. Stintzi and K. N. Raymond, *J. Am. Chem. Soc.*, 2006, **128**, 22–23.
44. M. Miethke, O. Klotz, U. Linne, J. J. May, C. L. Beckering and M. A. Marahiel, *Mol. Microbiol.*, 2006, **61**, 1413–1427.
45. J. Meiwes, H. P. Fiedler, H. Haag, H. Zahner, S. Konetschny-Rapp and G. Jung, *FEMS Microbiol. Lett.*, 1990, **55**, 201–205.
46. S. Konetschny-Rapp, G. Jung, J. Meiwes and H. Zahner, *Eur. J. Biochem.*, 1990, **191**, 65–74.
47. J. C. Grigg, J. Cheung, D. E. Heinrichs and M. E. Murphy, *J. Biol. Chem.*, 2010, **285**, 34579–34588.
48. J. Cheung, F. C. Beasley, S. Liu, G. A. Lajoie and D. E. Heinrichs, *Mol. Microbiol.*, 2009, **74**, 594–608.
49. K. P. Haley and E. P. Skaar, *Microbes Infect.*, 2012, **14**, 217–227.
50. M. I. Hood and E. P. Skaar, *Nat. Rev. Microbiol.*, 2012, **10**, 525–537.
51. V. Braun and F. Endriss, *Biometals*, 2007, **20**, 219–231.
52. J. T. Skare, B. M. Ahmer, C. L. Seachord, R. P. Darveau and K. Postle, *J. Biol. Chem.*, 1993, **268**, 16302–16308.
53. K. Postle and R. J. Kadner, *Mol. Microbiol.*, 2003, **49**, 869–882.
54. B. M. Ahmer, M. G. Thomas, R. A. Larsen and K. Postle, *J. Bacteriol.*, 1995, **177**, 4742–4747.
55. A. A. Ollis, M. Manning, K. G. Held and K. Postle, *Mol. Microbiol.*, 2009, **73**, 466–481.
56. M. G. Gresock, M. I. Savenkova, R. A. Larsen, A. A. Ollis and K. Postle, *Front. Microbiol.*, 2011, **2**, 206.
57. P. Cornelis, *Appl. Microbiol. Biotechnol.*, 2010, **86**, 1637–1645.
58. Z. A. Youard and C. Reimmann, *Microbiology*, 2010, **156**, 1772–1782.
59. Z. A. Youard, N. Wenner and C. Reimmann, *Biometals*, 2011, **24**, 513–522.
60. R. G. Ankenbauer, *J. Bacteriol.*, 1992, **174**, 4401–4409.
61. D. Cobessi, H. Celia and F. Pattus, *J. Mol. Biol.*, 2005, **352**, 893–904.
62. C. Reimmann, *Microbiology*, 2012, **158**, 1317–1324.
63. M. de Chial, B. Ghysels, S. A. Beatson, V. Geoffroy, J. M. Meyer, T. Pattery, C. Baysse, P. Chablain, Y. N. Parsons, C. Winstanley, S. J. Cordwell and P. Cornelis, *Microbiology*, 2003, **149**, 821–831.

64. E. E. Smith, E. H. Sims, D. H. Spencer, R. Kaul and M. V. Olson, *J. Bacteriol.*, 2005, **187**, 2138–2147.
65. J. Bodilis, B. Ghysels, J. Osayande, S. Matthijs, J. P. Pirnay, S. Denayer, D. De Vos and P. Cornelis, *Environ. Microbiol.*, 2009, **11**, 2123–2135.
66. B. Tummler and P. Cornelis, *J. Bacteriol.*, 2005, **187**, 3289–3292.
67. L. Guillon, M. El Mecherki, S. Altenburger, P. L. Graumann and I. J. Schalk, *Environ. Microbiol.*, 2012, **14**, 1982–1994.
68. B. J. McMorran, M. E. Merriman, I. T. Rombel and I. L. Lamont, *Gene*, 1996, **176**, 55–59.
69. E. Yeterian, L. W. Martin, L. Guillon, L. Journet, I. L. Lamont and I. J. Schalk, *Amino Acids*, 2010, **38**, 1447–1459.
70. R. Voulhoux, A. Filloux and I. J. Schalk, *J. Bacteriol.*, 2006, **188**, 3317–3323.
71. P. Nadal Jimenez, G. Koch, E. Papaioannou, M. Wahjudi, J. Krzeslak, T. Coenye, R. H. Cool and W. J. Quax, *Microbiology*, 2010, **156**, 49–59.
72. M. Hannauer, A. Braud, F. Hoegy, P. Ronot, A. Boos and I. J. Schalk, *Environ. Microbiol.*, 2012, **14**, 1696–1708.
73. I. J. Schalk and L. Guillon, *Environ. Microbiol.*, 2013, **15**, 1661–1673.
74. J. Greenwald, F. Hoegy, M. Nader, L. Journet, G. L. Mislin, P. L. Graumann and I. J. Schalk, *J. Biol. Chem.*, 2007, **282**, 2987–2995.
75. F. Imperi, F. Tiburzi and P. Visca, *Proc. Natl Acad. Sci. USA*, 2009, **106**, 20440–20445.
76. K. Brillet, F. Ruffenach, H. Adams, L. Journet, V. Gasser, F. Hoegy, L. Guillon, M. Hannauer, A. Page and I. J. Schalk, *ACS Chem. Biol.*, 2012, **7**, 2036–2045.
77. M. F. Traxler, M. R. Seyedsayamdost, J. Clardy and R. Kolter, *Mol. Microbiol.*, 2012, **86**, 628–644.
78. P. Cornelis and J. Bodilis, *Environ. Microbiol. Reports*, 2009, **1**, 256–262.
79. S. L. Hartney, S. Mazurier, T. A. Kidarsa, M. C. Quecine, P. Lemanceau and J. E. Loper, *Biometals*, 2011, **24**, 193–213.
80. B. Ghysels, B. T. Dieu, S. A. Beatson, J. P. Pirnay, U. A. Ochsner, M. L. Vasil and P. Cornelis, *Microbiology*, 2004, **150**, 1671–1680.
81. S. L. Hartney, S. Mazurier, M. K. Girard, S. Mehnaz, E. W. Davis, 2nd, H. Gross, P. Lemanceau and J. E. Loper, *J. Bacteriol.*, 2013, **195**, 765–776.
82. M. T. Anderson and S. K. Armstrong, *J. Bacteriol.*, 2006, **188**, 5731–5740.
83. C. R. Dean and K. Poole, *J. Bacteriol.*, 1993, **175**, 317–324.
84. B. Ghysels, U. Ochsner, U. Mollman, L. Heinisch, M. Vasil, P. Cornelis and S. Matthijs, *FEMS Microbiol. Lett.*, 2005, **246**, 167–174.
85. M. T. Anderson and S. K. Armstrong, *J. Bacteriol.*, 2008, **190**, 3940–3947.
86. S. K. Armstrong, T. J. Brickman and R. J. Suhadolc, *Mol. Microbiol.*, 2012, **84**, 446–462.
87. C. Wandersman and P. Delepelaire, *Annu. Rev. Microbiol.*, 2004, **58**, 611–647.
88. I. B. Lansky, G. S. Lukat-Rodgers, D. Block, K. R. Rodgers, M. Ratliff and A. Wilks, *J. Biol. Chem.*, 2006, **281**, 13652–13662.

89. E. E. Wyckoff, G. F. Lopreato, K. A. Tipton and S. M. Payne, *J. Bacteriol.*, 2005, **187**, 5658–5664.

90. G. Menestrina, M. Dalla Serra, M. Comai, M. Coraiola, G. Viero, S. Werner, D. A. Colin, H. Monteil and G. Prevost, *FEBS Lett.*, 2003, **552**, 54–60.

91. E. P. Skaar and O. Schneewind, *Microbes Infect.*, 2004, **6**, 390–397.

92. R. M. Pilpa, S. A. Robson, V. A. Villareal, M. L. Wong, M. Phillips and R. T. Clubb, *J. Biol. Chem.*, 2009, **284**, 1166–1176.

93. R. M. Pilpa, E. A. Fadeev, V. A. Villareal, M. L. Wong, M. Phillips and R. T. Clubb, *J. Mol. Biol.*, 2006, **360**, 435–447.

94. C. L. Vermeiren, M. Pluym, J. Mack, D. E. Heinrichs and M. J. Stillman, *Biochemistry*, 2006, **45**, 12867–12875.

95. V. A. Villareal, R. M. Pilpa, S. A. Robson, E. A. Fadeev and R. T. Clubb, *J. Biol. Chem.*, 2008, **283**, 31591–31600.

96. S. K. Mazmanian, G. Liu, H. Ton-That and O. Schneewind, *Science*, 1999, **285**, 760–763.

97. S. K. Mazmanian, H. Ton-That, K. Su and O. Schneewind, *Proc. Natl Acad. Sci. USA*, 2002, **99**, 2293–2298.

98. J. C. Grigg, C. L. Vermeiren, D. E. Heinrichs and M. E. Murphy, *J. Biol. Chem.*, 2007, **282**, 28815–28822.

99. M. T. Tiedemann and M. J. Stillman, *J. Biol. Inorg. Chem.*, 2012, **17**, 995–1007.

100. M. Pluym, C. L. Vermeiren, J. Mack, D. E. Heinrichs and M. J. Stillman, *Biochemistry*, 2007, **46**, 12777–12787.

101. M. L. Reniere, V. J. Torres and E. P. Skaar, *Biometals*, 2007, **20**, 333–3345.

102. S. K. Mazmanian, E. P. Skaar, A. H. Gaspar, M. Humayun, P. Gornicki, J. Jelenska, A. Joachmiak, D. M. Missiakas and O. Schneewind, *Science*, 2003, **299**, 906–909.

103. M. L. Reniere and E. P. Skaar, *Mol. Microbiol.*, 2008, **69**, 1304–1315.

104. M. L. Reniere, G. N. Ukpabi, S. R. Harry, D. F. Stec, R. Krull, D. W. Wright, B. O. Bachmann, M. E. Murphy and E. P. Skaar, *Mol. Microbiol.*, 2010, **75**, 1529–1538.

105. E. P. Skaar, A. H. Gaspar and O. Schneewind, *J. Biol. Chem.*, 2004, **279**, 436–443.

106. C. Wandersman and P. Delepelaire, *Mol. Microbiol.*, 2012, **85**, 618–631.

107. S. Letoffe, V. Redeker and C. Wandersman, *Mol. Microbiol.*, 1998, **28**, 1223–1234.

108. J. A. Imlay, S. M. Chin and S. Linn, *Science*, 1988, **240**, 640–642.

109. S. Jang and J. A. Imlay, *J. Biol. Chem.*, 2007, **282**, 929–937.

110. S. C. Andrews, *Adv. Microb. Physiol.*, 1998, **40**, 281–351.

111. E. Chiancone, P. Ceci, A. Ilari, F. Ribacchi and S. Stefanini, *Biometals*, 2004, **17**, 197–202.

112. G. Bellapadrona, M. Ardini, P. Ceci, S. Stefanini and E. Chiancone, *Free Radic. Biol. Med.*, 2010, **48**, 292–297.

113. L. Escolar, J. Perez-Martin and V. de Lorenzo, *J. Bacteriol.*, 1999, **181**, 6223–6229.
114. N. Bsat, A. Herbig, L. Casillas-Martinez, P. Setlow and J. D. Helmann, *Mol. Microbiol.*, 1998, **29**, 189–198.
115. J. W. Lee and J. D. Helmann, *Biometals*, 2007, **20**, 485–499.
116. L. van Oeffelen, P. Cornelis, W. Van Delm, F. De Ridder, B. De Moor and Y. Moreau, *Nucleic Acids Res.*, 2008, **36**, e46.
117. A. Nandal, C. C. Huggins, M. R. Woodhall, J. McHugh, F. Rodriguez-Quinones, M. A. Quail, J. R. Guest and S. C. Andrews, *Mol. Microbiol.*, 2010, **75**, 637–657.
118. E. Massé and S. Gottesman, *Proc. Natl Acad. Sci. USA*, 2002, **99**, 4620–4625.
119. E. Massé, C. K. Vanderpool and S. Gottesman, *J. Bacteriol.*, 2005, **187**, 6962–6971.
120. P. J. Wilderman, N. A. Sowa, D. J. FitzGerald, P. C. FitzGerald, S. Gottesman, U. A. Ochsner and M. L. Vasil, *Proc. Natl Acad. Sci. USA*, 2004, **101**, 9792–9797.
121. K. Prevost, H. Salvail, G. Desnoyers, J. F. Jacques, E. Phaneuf and E. Masse, *Mol. Microbiol.*, 2007, **64**, 1260–1273.
122. A. Danielli and V. Scarlato, *FEMS Microbiol. Rev.*, 2010, **34**, 738–752.
123. J. D. Helmann and M. J. Chamberlin, *Annu. Rev. Biochem.*, 1988, **57**, 839–872.
124. J. D. Helmann, *Curr. Opin. Microbiol.*, 1999, **2**, 135–141.
125. J. D. Helmann, *Adv. Microb. Physiol.*, 2002, **46**, 47–110.
126. V. Braun, S. Mahren and A. Sauter, *Biometals*, 2006, **19**, 103–113.
127. E. Potvin, F. Sanschagrin and R. C. Levesque, *FEMS Microbiol. Rev.*, 2008, **32**, 38–55.
128. M. A. Llamas, M. J. Mooij, M. Sparrius, C. M. Vandenbroucke-Grauls, C. Ratledge and W. Bitter, *Mol. Microbiol.*, 2008, **67**, 458–472.
129. P. Cornelis, S. Matthijs and L. Van Oeffelen, *Biometals*, 2009, **22**, 15–22.
130. P. A. Beare, R. J. For, L. W. Martin and I. L. Lamont, *Mol. Microbiol.*, 2003, **47**, 195–207.
131. R. C. Draper, L. W. Martin, P. A. Beare and I. L. Lamont, *Mol. Microbiol.*, 2011, **82**, 1444–1453.
132. P. J. Wilderman, A. I. Vasil, Z. Johnson, M. J. Wilson, H. E. Cunliffe, I. L. Lamont and M. L. Vasil, *Infect. Immun.*, 2001, **69**, 5385–5394.
133. L. Michel, N. Gonzalez, S. Jagdeep, T. Nguyen-Ngoc and C. Reimmann, *Mol. Microbiol.*, 2005, **58**, 495–509.
134. L. Michel, A. Bachelard and C. Reimmann, *Microbiology*, 2007, **153**, 1508–1518.
135. M. T. Anderson and S. K. Armstrong, *J. Bacteriol.*, 2004, **186**, 7302–7311.
136. C. R. Dean and K. Poole, *Mol. Microbiol.*, 1993, **8**, 1095–1103.
137. S. I. Patzer and K. Hantke, *Mol. Microbiol.*, 1998, **28**, 1199–1210.
138. T. C. Chao, A. Becker, J. Buhrmester, A. Puhler and S. Weidner, *J. Bacteriol.*, 2004, **186**, 3609–3620.

139. B. E. Ahn, J. Cha, E. J. Lee, A. R. Han, C. J. Thompson and J. H. Roe, *Mol. Microbiol.*, 2006, **59**, 1848–1858.

140. I. Hamza, S. Chauhan, R. Hassett and M. R. O'Brian, *J. Biol. Chem.*, 1998, **273**, 21669–21674.

141. A. W. Johnston, J. D. Todd, A. R. Curson, S. Lei, N. Nikolaidou-Katsaridou, M. S. Gelfand and D. A. Rodionov, *Biometals*, 2007, **20**, 501–511.

142. J. D. Todd, M. Wexler, G. Sawers, K. H. Yeoman, P. S. Poole and A. W. Johnston, *Microbiology*, 2002, **148**, 4059–4071.

143. K. H. Yeoman, A. R. Curson, J. D. Todd, G. Sawers and A. W. Johnston, *Microbiology*, 2004, **150**, 4065–4074.

144. C. Viguier, P. Ó. Cuív, P. Clarke and M. O'Connell, *FEMS Microbiol. Lett.*, 2005, **246**, 235–242.

145. P. Ngok-Ngam, N. Ruangkiattikul, A. Mahavihakanont, S. S. Virgem, R. Sukchawalit and S. Mongkolsuk, *J. Bacteriol.*, 2009, **191**, 2083–2090.

146. J. D. Todd, G. Sawers and A. W. Johnston, *Mol. Genet. Genomics*, 2005, **273**, 197–206.

147. W. S. Yeo, J. H. Lee, K. C. Lee and J. H. Roe, *Mol. Microbiol.*, 2006, **61**, 206–218.

148. M. Martinez, R. A. Ugalde and M. Almiron, *Microbiology*, 2005, **151**, 3427–3433.

149. J. D. Todd, G. Sawers, D. A. Rodionov and A. W. Johnston, *Mol. Genet. Genomics*, 2006, **275**, 564–577.

150. M. P. Schmitt, M. Predich, L. Doukhan, I. Smith and R. K. Holmes, *Infect. Immun.*, 1995, **63**, 4284–4289.

151. I. Brune, H. Werner, A. T. Huser, J. Kalinowski, A. Puhler and A. Tauch, *BMC Genomics*, 2006, **7**, 21.

152. B. Gold, G. M. Rodriguez, S. A. Marras, M. Pentecost and I. Smith, *Mol. Microbiol.*, 2001, **42**, 851–865.

153. G. M. Rodriguez, *Trends Microbiol.*, 2006, **14**, 320–327.

154. B. Py and F. Barras, *Nat. Rev. Microbiol.*, 2010, **8**, 436–446.

155. K. C. Lee, W. S. Yeo and J. H. Roe, *J. Bacteriol.*, 2008, **190**, 8244–8247.

156. P. Cornelis, Q. Wei, S. C. Andrews and T. Vinckx, *Metallomics*, 2011, **3**, 540–549.

157. J. A. Imlay and S. Linn, *Science*, 1988, **240**, 1302–1309.

158. D. Touati, *Arch. Biochem. Biophys.*, 2000, **373**, 1–6.

159. G. M. Rodriguez, M. I. Voskuil, B. Gold, G. K. Schoolnik and I. Smith, *Infect. Immun.*, 2002, **70**, 3371–3381.

160. H. A. Barton, Z. Johnson, C. D. Cox, A. I. Vasil and M. L. Vasil, *Mol. Microbiol.*, 1996, **21**, 1001–1017.

161. O. Dussurget, M. Rodriguez and I. Smith, *Mol. Microbiol.*, 1996, **22**, 535–544.

162. T. Jittawuttipoka, R. Sallabhan, P. Vattanaviboon, M. Fuangthong and S. Mongkolsuk, *Arch. Microbiol.*, 2010, **192**, 331–339.

163. D. J. Hassett, P. A. Sokol, M. L. Howell, J. F. Ma, H. T. Schweizer, U. Ochsner and M. L. Vasil, *J. Bacteriol.*, 1996, **178**, 3996–4003.

164. J. F. da Silva Neto, V. S. Braz, V. C. Italiani and M. V. Marques, *Nucleic Acids Res.*, 2009, **37**, 4812–4825.

165. S. Lopez-Gomollon, E. Sevilla, M. T. Bes, M. L. Peleato and M. F. Fillat, *Biochem. J.*, 2009, **418**, 201–207.

166. T. C. Chao, J. Buhrmester, N. Hansmeier, A. Puhler and S. Weidner, *Appl. Environ. Microbiol.*, 2005, **71**, 5969–5982.

167. M. J. Horsburgh, M. O. Clements, H. Crossley, E. Ingham and S. J. Foster, *Infect. Immun.*, 2001, **69**, 3744–3754.

168. I. Gryllos, R. Grifantini, A. Colaprico, M. E. Cary, A. Hakansson, D. W. Carey, M. Suarez-Chavez, L. A. Kalish, P. D. Mitchell, G. L. White and M. R. Wessels, *PLoS Pathog.*, 2008, **4**, e1000145.

169. F. Hillmann, C. Doring, O. Riebe, A. Ehrenreich, R. J. Fischer and H. Bahl, *J. Bacteriol.*, 2009, **191**, 6082–6093.

170. N. Baichoo and J. D. Helmann, *J. Bacteriol.*, 2002, **184**, 5826–5832.

171. M. J. Faulkner, Z. Ma, M. Fuangthong and J. D. Helmann, *J. Bacteriol.*, 2012, **194**, 1226–1235.

172. H. J. Wu, K. L. Seib, Y. N. Srikhanta, S. P. Kidd, J. L. Edwards, T. L. Maguire, S. M. Grimmond, M. A. Apicella, A. G. McEwan and M. P. Jennings, *Mol. Microbiol.*, 2006, **60**, 401–416.

173. A. H. van Vliet, M. L. Baillon, C. W. Penn and J. M. Ketley, *J. Bacteriol.*, 1999, **181**, 6371–6376.

174. K. Palyada, Y. Q. Sun, A. Flint, J. Butcher, H. Naikare and A. Stintzi, *BMC Genomics*, 2009, **10**, 481.

175. A. H. van Vliet, J. M. Ketley, S. F. Park and C. W. Penn, *FEMS Microbiol. Rev.*, 2002, **26**, 173–186.

176. J. Green and M. S. Paget, *Nat. Rev. Microbiol.*, 2004, **2**, 954–966.

177. J. M. Dubbs and S. Mongkolsuk, *J. Bacteriol.*, 2012, **194**, 5495–5503.

178. S. E. Maddocks and P. C. Oyston, *Microbiology*, 2008, **154**, 3609–3623.

179. H. Choi, S. Kim, P. Mukhopadhyay, S. Cho, J. Woo, G. Storz and S. E. Ryu, *Cell*, 2001, **105**, 103–113.

180. F. Aslund, M. Zheng, J. Beckwith and G. Storz, *Proc. Natl Acad. Sci. USA*, 1999, **96**, 6161–6165.

181. M. Zheng, F. Aslund and G. Storz, *Science*, 1998, **279**, 1718–1721.

182. M. Zheng, B. Doan, T. D. Schneider and G. Storz, *J. Bacteriol.*, 1999, **181**, 4639–4643.

183. M. Zheng, X. Wang, B. Doan, K. A. Lewis, T. D. Schneider and G. Storz, *J. Bacteriol.*, 2001, **183**, 4571–4579.

184. K. Muller, B. F. Matzanke, V. Schunemann, A. X. Trautwein and K. Hantke, *Eur. J. Biochem.*, 1998, **258**, 1001–1008.

185. T. Vinckx, S. Matthijs and P. Cornelis, *FEMS Microbiol. Lett.*, 2008, **288**, 258–265.

186. T. Vinckx, Q. Wei, S. Matthijs and P. Cornelis, *Microbiology*, 2010, **156**, 678–686.

187. S. Varghese, A. Wu, S. Park, K. R. Imlay and J. A. Imlay, *Mol. Microbiol.*, 2007, **64**, 822–830.

188. T. Vinckx, Q. Wei, S. Matthijs, J. P. Noben, R. Daniels and P. Cornelis, *Biometals*, 2011, **24**, 523–532.
189. Q. Wei, P. N. Minh, A. Dotsch, F. Hildebrand, W. Panmanee, A. Elfarash, S. Schulz, S. Plaisance, D. Charlier, D. Hassett, S. Haussler and P. Cornelis, *Nucleic Acids Res.*, 2012, **40**, 4320–4333.
190. F. J. Reen, J. M. Haynes, M. J. Mooij and F. O'Gara, *PLoS One*, 2013, **8**, e54479.

CHAPTER 12

Iron-sulfur Clusters

RICHARD CAMMACK*[a] AND JANNEKE BALK[b]

[a] King's College London, Department of Biochemistry, 150 Stamford Street, London, SE1 9NH, UK; [b] John Innes Centre and University of East Anglia, Norwich Research Park, Colney Lane, Norwich, NR4 7UH, UK
*Email: richard.cammack@kcl.ac.uk

12.1 Introduction

Iron-sulfur (Fe-S) proteins are named for their inorganic prosthetic groups, the iron-sulfur clusters, and are found in virtually all forms of life. They are divided into *ferredoxins*, which transfer electrons and contain only Fe-S clusters, and *complex iron-sulfur proteins*, which contain additional metals or prosthetic groups and often have an enzymic function.[1] Fe-S proteins are essential components in the membrane-bound electron transport chains of respiration and photosynthesis and are also responsible for the activity of hundreds of enzymes.[2–4] Genome sequencing has revealed the existence of around 150 different Fe-S proteins in the single bacterium *Escherichia coli*[5] and there are higher numbers in eukaryotes. Some examples, together with their functions and cellular locations, are listed in Table 12.1.

These Fe-S clusters comprise iron ions bridged by sulfide and they are coordinated to ligands from the protein, usually cysteines (Figure 12.1). Over the decades, a number of different types of clusters have been found. The commonest arrangements are [2Fe-2S] rhombs and [4Fe-4S] cubes, each with four cysteine thiolate ligands (Figure 12.1(a),(d)).

The clusters are generated dynamically in the cell, being formed from iron and cysteine sulfur during maturation of the proteins. There is little evidence of storage of the intact clusters. Polyferredoxins in methanogenic

RSC Metallobiology Series No. 2
Binding, Transport and Storage of Metal Ions in Biological Cells
Edited by Wolfgang Maret and Anthony Wedd
© The Royal Society of Chemistry 2014
Published by the Royal Society of Chemistry, www.rsc.org

Table 12.1 Examples of iron-sulfur enzymes and their functions.

Enzyme	Source	Function	Cofactors	Location in the cell	References
NADH:ubiquinone reductase (Complex I in mitochondria)	Most bacteria, eukaryotes (some organisms have lost Complex I, *e.g.* baker's yeast (*S. cerevisiae*))	Electron and H^+ transport	[2Fe-2S], 7[4Fe-4S], FMN	Mitochondrial inner membrane	50
Succinate:ubiquinone reductase (Complex II)	Bacteria, eukaryotes	Electron transport	[2Fe-2S], [3Fe-4S], [4Fe-4S]	Mitochondrial inner membrane	141
Ubiquinol:cyt *c* reductase (Complex III)	Bacteria, eukaryotes	Electron and H^+ transport	Rieske [2Fe-2S]	Mitochondrial inner membrane	142
Biotin synthase	Mostly universal (lost in mammals)	Sulfur insertion into dethiobiotin	[2Fe-2S], [4Fe-4S]	Cytosol in bacteria and archaea, mitochondrial matrix in lower eukaryotes	61
Lipoyl synthase	Universal	Insertion of two sulfur atoms	2[4Fe-4S]	Mitochondria, plastids	143
Nitrogenase	Nitrogen-fixing bacteria	Nitrogen assimilation	[4Fe-4S], FeMoco, P clusters = [7Fe-8S]	Cytosol	144
Nitrite reductase	Plants, cyanobacteria	Nitrogen assimilation	[4Fe-4S] coupled to sirohaem	Chloroplast stroma, cytosol of cyanobacteria	145
NAD-reducing NiFe-hydrogenase	Hydrogeno-trophic bacteria	H_2 assimilation	[2Fe-2S], [3Fe-4S], [4Fe-4S], [NiFe] centre, FMN	Membrane	146
H_2-producing hydrogenase	Anaerobic fermentative bacteria, some green algae	H_2 production	[2Fe-2S], [3Fe-4S], [4Fe-4S], H-cluster; or H-cluster only in algae	Cytosol in bacteria, chloroplast stroma in algae	147
CO-methylating acetyl-CoA synthase	Acetogenic bacteria	Synthesis of acetate	[4Fe-4S], Ni-[4Fe-4S]		148

Enzyme/protein	Function	Fe-S cluster	Distribution	Location	Ref.
CoB–CoM heterodisulfide reductase	Energy capture methanogenesis	[3Fe-4S], [4Fe-4S]	Methanogens	Cytosol	149
Isopropylmalate isomerase	Biosynthesis of leucine	[4Fe-4S]	Plants, bacteria	Cytosol	150
Anaerobic ribonucleotide reductase	Deoxyribonucleotide synthesis	[4Fe-4S]	Bacteria	Cytosol	151
Xanthine dehydrogenase	Purine degradation	2[2Fe-2S], Mo-pterin, FAD	Universal	Cytosol	152
Aconitase	Dehydratase/hydratase	[4Fe-4S]	Universal	Cytosol, mitochondrial matrix	54
Ferrochelatase	Haem biosynthesis	[2Fe-2S]	Universal, but not all bind an Fe-S cluster	Bacterial cytosol, mitochondria, plastids	153
DNA polymerase	DNA synthesis	[4Fe-4S]	Eukaryotes	Nucleus	154
Ferredoxin:thioredoxin reductase	Reduction of thioredoxin in metabolic regulation	[4Fe-4S]	Plants and cyanobacteria	Chloroplast and cytosol in cyanobacteria	155
Helicases	Unwind DNA or RNA	[4Fe-4S]	Bacteria, Archaea, eukaryotes	Bacterial cytosol, nucleus	156
RNA polymerase	RNA synthesis	[4Fe-4S]	Archaea, plants, some pathogenic eukaryotes	Cytosol, nucleus	157
RlmN and Cfr	Methylation of adenosine	[4Fe-4S]	Bacteria	Cytosol	158
Endonuclease III (*E. coli*), Nth2 (yeast)	DNA glycosylase	[4Fe-4S]	Universal	Nucleus	159
Fumarate/nitrate reductase (FNR) regulator		[4Fe-4S] or [2Fe-2S]	Bacteria	Cytosol	75

archaea[6] contain up to 12 [4Fe-4S] clusters in repeated ferredoxin-like domains and were suggested as a form of storage. However, an electron-transfer function has been described subsequently for these proteins.[7]

This chapter will focus on some of the most important iron-sulfur clusters and their properties. It will focus particularly on their biosynthesis and its regulation, plus their transport around the cell. The clusters are generally assembled in the cell compartment in which the Fe-S proteins reside. However, as will be seen, folded Fe-S proteins can be transported across membranes by the twin-arginine translocation (TAT) system. Also, Fe-S proteins are thought to move from the cytosol to the nucleus *via* the nuclear pores.

12.1.1 Chemistry of Iron-sulfur (Fe-S) Clusters

In pioneering chemical studies,[8] it was shown that "analogue" Fe-S clusters could be formed spontaneously in solution from a thiolate, sulfide and iron salts. The ones first described were [4Fe-4S] clusters with ligands such as benzenethiolate. Other types of cluster, such as [2Fe-2S], could be produced by adjusting the geometric spacing of the thiolates.[9] The compounds were nickel replacing an iron atom.[10,11] However, they did not display the catalytic activities of the molybdenum-containing nitrogenase or nickel-containing hydrogenases, which, as will be seen, require the support of highly specific protein structures.

These simple analogues showed the chemistry expected of the biological clusters and that is reflected in the *in vivo* capabilities of Fe-S proteins. They exhibit plasticity – they can undergo ligand exchange with other thiolates, and cluster rearrangements, such as conversion from [2Fe-2S] to [4Fe-4S] types.[12] Moreover, it was shown that the thiolate ligands in the compounds could be substituted by cysteine-containing peptides, creating artificial Fe-S proteins.[13] The analogue Fe-S clusters could also be transferred to the apo-forms of Fe-S proteins, such as ferredoxins.[3]

Clusters with cysteine thiolate ligands are usually stable in Fe-S proteins though, as with the analogue compounds, they tend to be sensitive to O_2 and NO. They can be assembled in *apo*-proteins by using an iron salt such as ammonium iron(II) sulfate, sulfide and a thiolate reagent such as dithiothreitol under anaerobic conditions, a process known as *reconstitution.*[14] In most cases, the *apo*-forms of ferredoxins spontaneously reconstitute the right types of [2Fe-2S] or [4Fe-4S] clusters, indicating that the proteins have an intrinsic capacity to assemble the appropriate cluster as they fold. Moreover, Fe-S clusters could be extruded from the *holo*-proteins by thiolate ligands under denaturing conditions,[15,16] and so the clusters could be transferred *in vitro* from one Fe-S protein into another. These observations are relevant to the processes of Fe-S cluster insertion and transfer *in vivo*. Stable clusters in proteins are also formed with ligands other than thiols, *e.g.* imidazole (His) (Figure 12.1(b),(c)) and carboxylate (Asp, Glu).

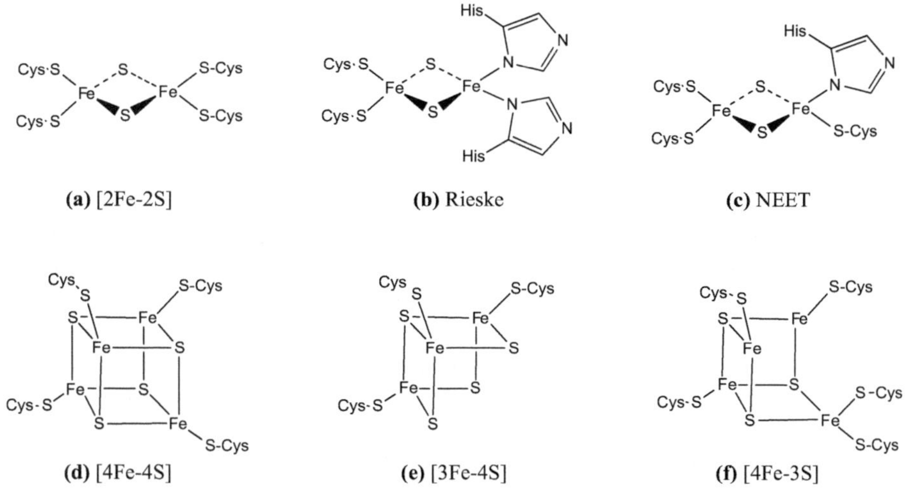

(a) [2Fe-2S]   (b) Rieske   (c) NEET

(d) [4Fe-4S]   (e) [3Fe-4S]   (f) [4Fe-3S]

Figure 12.1 Structures of Fe-S clusters. (a) Canonical [2Fe-2S]; (b) [2Fe-2S] in Rieske protein;[160] (c) [2Fe-2S] in NEET protein;[19] (d) [4Fe-4S]; (e) [3Fe-4S]; (f) [4Fe-3S] in oxygen-tolerant hydrogenase.[25,71]

12.1.2 Fe-S Clusters in Proteins

Most common are Cys-liganded [2Fe-2S] and [4Fe-4S] clusters (Figure 12.1(a),(d)). Rieske clusters, found in respiratory-chain Complex III[17] and some hydroxylases,[18] are [2Fe-2S] clusters with two histidine imidazole ligands to one iron atom, and two cysteine sulfurs to the other (Figure 12.1(b)). They tend to have less negative midpoint redox potentials than those with all-Cys coordination. The NEET proteins are dimeric proteins, each subunit of which has a [2Fe-2S] cluster coordinated by three cysteines and one histidine[19] (Figure 12.1(c)). They occur in the endoplasmic reticulum[20] and, unusually, in the outer rather than in the inner membrane of mitochondria. [3Fe-4S] clusters (Figure 12.1(e)) are naturally occurring electron carriers in enzymes such as succinate dehydrogenase,[21] NiFe-hydrogenases[22] and glutamate synthase.[23] The [4Fe-3S] cluster has recently been reported in the structures of some O_2-tolerant nickel-iron (NiFe)-hydrogenases.[24,25] The fourth iron is bound to two cysteine residues (Figure 12.1(f)).

The iron atoms in Fe-S clusters are in nearly tetrahedral coordination geometry, are high-spin but are spin-coupled. Formal oxidation states Fe^{III} and Fe^{II} can usually be assigned. For [2Fe-2S] proteins in the oxidized state, both iron atoms are Fe^{III} (net charge +2);[†] in the reduced state, they are

---

[†]In biochemical nomenclature, Fe-S clusters are described by a non-standard formalism in which the numbers of atoms of iron and sulfide sulfur are enclosed in brackets, as shown in Table 12.1. The valence states are defined by a convention which adds up the formal charges on the iron ions and sulfides, but ignoring the charges on the ligands.[1]

valence-trapped as Fe^{III} and Fe^{II} (net charge +1).[26] In most [4Fe-4S] clusters, the electronic structures are highly delocalized. In the oxidized forms, the valence states are formally $2Fe^{III} + 2Fe^{II}$ (net charge +2) and in the reduced state $Fe^{III} + 3Fe^{II}$ (net charge +1).[†]

In many systems, the clusters act as redox agents, capable of accessing two oxidation levels. At one extreme, the [4Fe-4S] cluster in the Fe-protein of nitrogenase can undergo further reduction to the all-Fe^{II} level.[27] At the other extreme, in the high-potential iron-sulfur protein (HiPIP) of photosynthetic bacteria, the [4Fe-4S] cluster can undergo oxidation to the higher $[3Fe^{III} + Fe^{II}]$ level.[28] Both have proved to be rare cases. More commonly, the action of strong oxidants such as O_2 on [4Fe-4S] clusters in proteins is to cause oxidative damage, leading to release of an iron atom and formation of a [3Fe-4S] cluster.[29] Aconitase, for example, has this cluster (Figure 12.1(e)) when isolated *in vitro*, but this form of the protein is non-physiological and inactive. Activity is regained when the fourth iron atom is bound to restore the [4Fe-4S] cluster.[30]

The clusters normally undergo oxidation and reduction between two oxidation levels. Only the [4Fe-3S] cluster in membrane-bound O_2-tolerant hydrogenases (Figure 12.1(f)) is believed to undergo facile redox transitions between three levels.[31] This would allow transfer of two electrons simultaneously to O_2 to yield peroxide, avoiding formation of toxic superoxide.

12.1.3   Fe-S Proteins

The structural simplicity of proteins such as ferredoxins and their occurrence in fundamental bioenergetic pathways, particularly in anaerobic bacteria and archaea, are indications that they evolved from ancient origins.[32] The structures of many more complex iron-sulfur proteins have been determined and they often comprise ferredoxin-like domains. Hundreds of different types have been identified and studied. Because of the redox properties discussed above, their commonest activity is one-electron oxidation/reduction. Some of the most important iron-sulfur clusters are essential components of complexes I–III of the respiratory chain located in the inner membrane of mitochondria and in the membranes of aerobic bacteria. Furthermore, Fe-S clusters have other reactivities, which are exploited by some enzymes. Representative examples are given in Table 12.1.

Methods for studying Fe-S proteins are well established (reviewed in refs 33–35). The clusters were first detected by their characteristic UV/visible absorption and electron paramagnetic resonance (EPR) spectra, which can be observed even in unpurified preparations. The EPR spectra of the [2Fe-2S] ferredoxins were the first indication that the iron atoms were spin-coupled.[36,37] In purified proteins, the different types of cluster can be identified by spectroscopy techniques such as UV/visible absorption, circular dichroism, EPR,[38] Raman, X-ray absorption and Mössbauer. More recently, dynamic electrochemistry has provided insights into the redox mechanisms of Fe-S enzymes such as hydrogenase.[39,40] Catalytic rates are measured on protein films adsorbed on electrodes. Besides measuring the midpoint potentials of clusters and other reducible centres, controlled-potential voltammetry can dissect the individual redox steps of enzymes in their reactions with substrates.[41]

In the last few decades, molecular biology approaches have been used extensively to explore the functions of Fe-S proteins. They are identified in gene sequences by conserved patterns of cysteine residues.[42] Their genes can be deleted or mutated and their expression suppressed using RNAi or increased by gene overproduction. As will be seen below, genetic methods have also thrown light on the elaborate processes of iron-sulfur cluster assembly and trafficking.

12.2 Functions of Fe-S Proteins

A wide variety of iron-sulfur proteins are employed for energy conservation, biosynthesis of cofactors, central metabolism and cell regulation, as described in recent reviews.[5,43] Many bioenergetic pathways (including methanogenesis, photosynthesis (heterotrophic and lithotrophic), fermentation and H_2 production and consumption) involve oxygen-sensitive enzymes and presumably evolved under anaerobic conditions. With the evolution of oxygenic photosynthesis, new Fe-S proteins evolved to take advantage of aerobic metabolism, including respiratory-chain ATP production and hydroxylation reactions.

12.2.1 Electron Transport

The redox properties of the clusters are tuned by the protein environment. For [2Fe-2S] and [4Fe-4S] clusters with cysteine ligands, midpoint potentials range between −500 and +100 mV *vs.* the hydrogen electrode; HiPIPs have less negative potentials, around +350 mV. Clusters with nitrogen ligands, such as the Rieske proteins, have less negative potentials (up to +350 mV). Redox changes in Fe-S clusters require low reorganization energies,[44] which means that electrons can be transferred efficiently by quantum-mechanical tunnelling over considerable distances, up to 1.4 nm.[45] This allows [4Fe-4S] clusters to participate in rapid electron capture in the primary photochemical reactions of Photosystem I.[46,47] Electrons can be exchanged with other types of prosthetic groups such as haem or flavin.

Electron transfers over longer distances are made possible by stepping-stone arrangements of the clusters to form an electron transfer chain. The most extreme example is the chain of seven clusters, extending over 9.0 nm, in the membrane-extrinsic part of complex I of the respiratory chain and its bacterial equivalent NADH:ubiquinone reductase.[48] The latter protein in *Thermus thermophilus* comprises 16 subunits of total mass 536 kDa, containing FMN and 9 Fe-S clusters.[49] The mitochondrial protein is even larger, featuring 14 subunits similar to those found in the bacterial enzyme complex plus more than 30 additional accessory subunits comprising a total molecular mass of 1 MDa with 8 Fe-S clusters.[50] The membrane-resident domain of the protein comprises transmembrane channels, predicted to be involved in pumping four protons across the cell membrane for each molecule of NADH oxidized.[51] Respiratory complex I also illustrates the observation that many large electron-transfer Fe-S proteins have a modular structure. They appear to have evolved by accretion of domains derived from other Fe-S electron transfer proteins.[52,53] Various subunits are structurally similar to other Fe-S enzymes, including NiFe- and FeFe-hydrogenases, ferredoxins and thioredoxins. This overall structure has been conserved in species from bacteria to humans, showing that, once optimized for its function, it was maintained throughout evolution.

12.2.2 Enzyme Catalysis

An important type of Fe-S protein catalyst is a [4Fe-4S] cluster bound by only three cysteine ligands, leaving the fourth ligand position on one iron atom vacant and available for coordination. Aconitase (aconitate hydratase), which catalyses the interconversion of citrate and isocitrate in the tricarboxylic acid cycle, has an iron site capable of binding citrate, isocitrate and aconitate as bidentate ligands (see, *e.g.*, citrate in Figure 12.2(a)). The cluster acts as a Lewis-acid catalyst during the enzyme reaction.[54] Other dehydratases with similar active sites include fumarate hydratase of bacteria (also part of the tricarboxylic acid cycle)[55] and isopropylmalate dehydratase of the leucine biosynthesis pathway.[56]

The "radical SAM" or radical AdoMet enzymes represent an important class of Fe-S enzymes identified in recent years. They employ a [4Fe-4S] cluster that binds *S*-adenosyl methionine (known as SAM, or AdoMet). The latter is cleaved by reduction to produce the 5'-deoxyadenosyl radical (Figure 12.2(b)), a potent oxidant which is particularly useful in anaerobic metabolism. It is used in a wide range of biochemical transformations in the biosynthesis of amino acids, nucleotides, co-enzymes and antibiotics.[57–59] Gene sequencing indicates that there are hundreds of radical-AdoMet enzymes.[60] As an example, biotin synthase from *E. coli* is a radical-AdoMet enzyme, which employs two successive molecules of AdoMet to activate two C-H groups in dethiobiotin for insertion of a bridging sulfur atom. The primary source of the sulfur is a [2Fe-2S] cluster in the enzyme itself.[61] The reaction cycle requires the Fe-S cluster assembly system to repair this cluster. In the protein structure, dethiobiotin is bound between the special [4Fe-4S] cluster and the [2Fe-2S] cluster, which has an unusual Cys$_3$Arg coordination site.[62] The protein is flexible, which allows the [2Fe-2S] cluster to be restored by the ISC system after each reaction cycle.[63] Radical-AdoMet enzymes are not restricted to anaerobic bacteria: higher eukaryotes, including man, have them in the form of lipoyl synthase and several tRNA modifying enzymes.

12.2.3 Redox Reactions of Gases Catalysed by Superclusters

Fe-S enzymes that catalyse some difficult reactions involving gases such as CO, H_2 and N_2 contain complex Fe-S clusters with nuclearities of up to eight metal ions, described here as "superclusters".[64] They include nitrogenase, which reduces N_2 to form ammonia, and hydrogenases, which produce or consume H_2. The molybdenum-containing nitrogenase comprises two proteins, the MoFe-protein and the Fe-protein. The former contains two superclusters, the P-cluster (Figure 12.2(c)) for intramolecular electron transport and the MoFe-cluster, "FeMoCo" (Figure 12.2(d)), where reduction of N_2 takes place. Each cluster can be considered to be a combination of two [4Fe-4S] clusters, one containing a bridging sulfide, the other by a unique bridging carbide.[65,66,130,131] The Fe-protein of nitrogenase is dimeric with a single [4Fe-4S] cluster bridging two identical protein subunits. In the presence of ATP, this cluster injects electrons into the MoFe-protein. The kinetics of the multistep reduction of N_2 to NH_3, with concomitant release of H_2, were determined originally by Thorneley and Lowe.[67] Recently, a model for the reaction cycle has been proposed,[68] which involves the formation of two bridging hydrides in the MoFe-cluster, their displacement by N_2, followed by a hydrazine intermediate. Other nitrogenases exist and contain similar clusters but with a vanadium or an iron atom substituting for molybdenum.[69] These alternative nitrogenases are less efficient catalytically as they release more H_2 during the reduction of N_2.

Hydrogenases are enzymes that consume or produce H_2. There are two main structural classes, which are not related phylogenetically.[22] They both contain Fe centres with cyanide and carbonyl ligands.[70] The FeFe-hydrogenases feature a supercluster (the H-cluster) comprising a [4Fe-4S] subcluster linked to a dinuclear iron site bearing CN and CO ligands plus a dithiolate ligand (Figure 12.2(e)). The NiFe-hydrogenases contain a dinuclear Ni-Fe site bridged with cysteine sulfur ligands, also with CN and CO ligands but no dithiolate ligand (Figure 12.2(f)). Both FeFe- and NiFe-hydrogenases contain Fe-S clusters that form an electron transfer chain into the active site.[22] The NiFe-hydrogenases tend to be more tolerant to O_2.[25,71]

12.2.4 Response Regulation

The protein structures of Fe-S enzymes catalysing particular reactions tend to have a high degree of evolutionary conservation. Fe-S proteins with a regulatory function are more diverse, as are their mechanisms of action. A number are employed to sense the intracellular levels of Fe-S clusters, iron and reactive molecules such as oxygen, O_2^-, H_2O_2 and NO.[72,73]

Response to cellular iron levels: The biosynthesis of Fe-S clusters themselves in *E. coli* is, in part, regulated by binding of an Fe-S cluster to the transcription factor IscR.[‡] This metalloregulator is unusual in that both the

---

[‡]The transcription factor IscR is discussed in Chapter 11, Section 11.5.2.

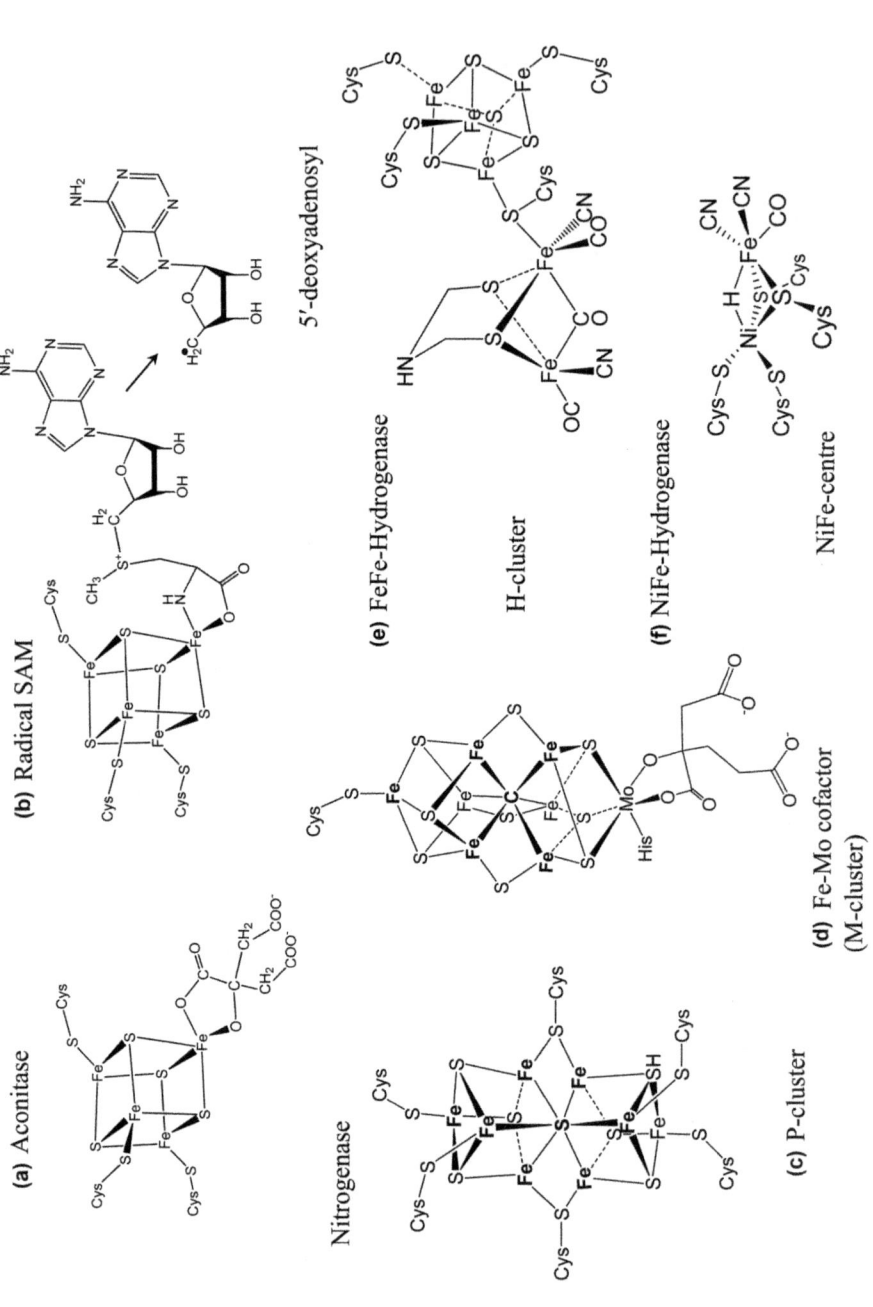

Figure 12.2 Catalytic Fe-S clusters. (a) [4Fe-4S] cluster in aconitase, with citrate bound; (b) [4Fe-4S] cluster in radical AdoMet enzymes, with 5'-deoxyadenosyl radical; (c) P-cluster in nitrogenase; (d) FeMo-cluster (FeMoco) in nitrogenase; (e) H-cluster in FeFe-hydrogenase; (f) nickel-iron centre in NiFe-hydrogenase.

apo- and Fe-S-bound forms of IscR regulate transcription and exhibit different DNA binding specificities.[74] Response to iron levels requires a reliable and selective criterion to recognize iron ions. Selective chelation properties of metal-binding proteins alone cannot discriminate sufficiently between Fe^{2+} and other divalent ions such as Mn^{2+} and Zn^{2+}. However, this distinction is easily made if the ions are selected for their ability to form Fe-S clusters. Thus a number of different transcription and translation factors bind Fe-S clusters.[75] In *S. cerevisiae*, two transcriptional activators, Aft1 and Aft2, regulate iron homeostasis. The process depends on Fe-S protein biogenesis in the mitochondria but not in the cytosol.[76] Aft1 is thought to interact with a complex consisting of a glutaredoxin, glutathione and a [2Fe-2S] cluster.[77] This complex appears to tether Aft1 in the cytosol, preventing it from initiating transcription of Fe-responsive genes in the nucleus. The Fe-S cluster is sufficiently labile that low concentrations of iron in the cell lead to its disassembly. Its loss leads to the release of Aft1, which travels to the nucleus to activate expression of its target genes. Mammals have a different system of regulation, involving the iron-regulatory proteins IRP1 and IRP2.[§] These control the translation of mRNA for proteins involved in iron uptake and storage.[78] IRP1 is identical to the *apo*-form of the iron-sulfur protein, cytosolic aconitase. When the [4Fe-4S] cluster is present, it is no longer able to bind to the iron-responsive element, and its aconitase activity is restored.[79]

Response to oxidant status: In prokaryotes, a number of transcription factors containing Fe-S clusters are now known to respond to O_2 or NO.[72,73] The FNR regulatory system employs a [4Fe-4S] cluster that regulates the activity of enzymes in response to the redox state of the cell. Under anaerobic conditions, this cluster is intact and the protein exists as a dimer. In this form it binds to specific sites in DNA and induces the genes for fumarate and nitrate reductases (hence the name FNR) and other enzymes involved in anaerobic metabolism. When it is oxidized to form a [2Fe-2S] cluster, the protein becomes monomeric and represses the enzymes.[80] The cluster features persulfide ligands in this oxidized apo-form, allowing the [4Fe-4S] cluster to be recovered by reduction without the requirement for added sulfide.[81]

12.2.5 Function Unknown – Possibly Protein Stabilization

An increasing number of enzymes have been found to contain Fe-S clusters that are essential for enzyme activity but, owing to their distance from the active site, do not appear to contribute to the catalytic function.[82] These clusters often are resistant to reduction. The rigidity of [4Fe-4S] cluster structures suggests that they could play a role in stabilizing protein structure in critical sites such as those dependent on DNA base pairing. A redox function in sensing sites of DNA damage has been suggested on the basis of

---

§See Chapter 10, Section 10.7.

electrochemical measurements of isolated proteins, but the role in the mechanism *in vivo* remains to be established.[83]

12.3   Iron-sulfur Cluster Assembly

12.3.1   Assembly in Prokaryotes

The processes of production of Fe-S clusters have been elucidated in a number of model microorganisms, largely by application of genetic techniques.[84] In bacteria, the NIF (nitrogen fixation) genes were first identified in the gene cluster dedicated to maturation of nitrogenase.[85] Subsequently two other distinct types of Fe-S cluster machinery, ISC (iron-sulfur cluster) and SUF (sulfur mobilization), were identified.[86,87] Each system is used for a wide variety of purposes but, in general, ISC appears to be involved principally in normal (housekeeping) cluster assembly while SUF is used under stress conditions (oxygen stress or low iron availability) and NIF for assembly of the complex clusters of nitrogenase.[88] All three systems have a common mechanism that involves the combined action of a pyridoxal phosphate-dependent cysteine desulfurase (encoded by NifS, IscS or SufS) for sulfide generation and scaffold proteins containing a cluster-binding site (encoded by NifU, IscU, SufU or SufB). In the desulfurase, sulfur is transferred from a free cysteine to form a persulfide (*S*-sulfanyl) group on an active-site cysteine residue leaving alanine as a product (Figure 12.3).

The mechanism of delivery of iron to the Fe-S assembly scaffolds in the NIF, ISC, SUF or CIA pathways is still speculative. Because of its toxicity, the concentration of available Fe^{2+} is maintained at a very low level that is difficult to measure. The "labile iron pool" in cells, equivalent to the intracellular pools of Ca^{2+} and Zn^{2+}, has been proposed to be available to the numerous biosynthetic systems that require iron.[89] Interest has focused on Fe^{II} complexes with small molecules such as glutathione as constituting this labile iron pool; citrate was also considered, but this is now less likely.[90]

The NifU, IscU and SufU proteins each contain three cysteine residues that bind the nascent Fe-S cluster. In IscU, the best-studied of the scaffolds, a [2Fe-2S] cluster is bound by these three cysteines and one conserved histidine. NMR spectroscopy showed that a structurally disordered form of IscU preferentially binds to IscS, and takes an ordered conformation upon cluster assembly.[91] The crystal structure of the holoform of IscU (mutant D35A) was obtained in a complex with IscS and is proposed to be a trapped intermediate.[92] It features a [2Fe-2S] cluster coordinated by the three Cys residues and, unexpectedly, by an active-site Cys residue of IscS. The latter has to undergo a 1.4 nm movement from the active site where the persulfide is initially formed, in order to add it to the cluster. It was proposed that this movement could be repeated to supply the first and second sulfide of the [2Fe-2S] cluster without the need for dissociation of the IscU-IscS complex. The persulfide from the cysteine desulfurase needs to be reduced to the oxidation level of sulfide. The reducing agent in the ISC system was identified as a ferredoxin.[93]

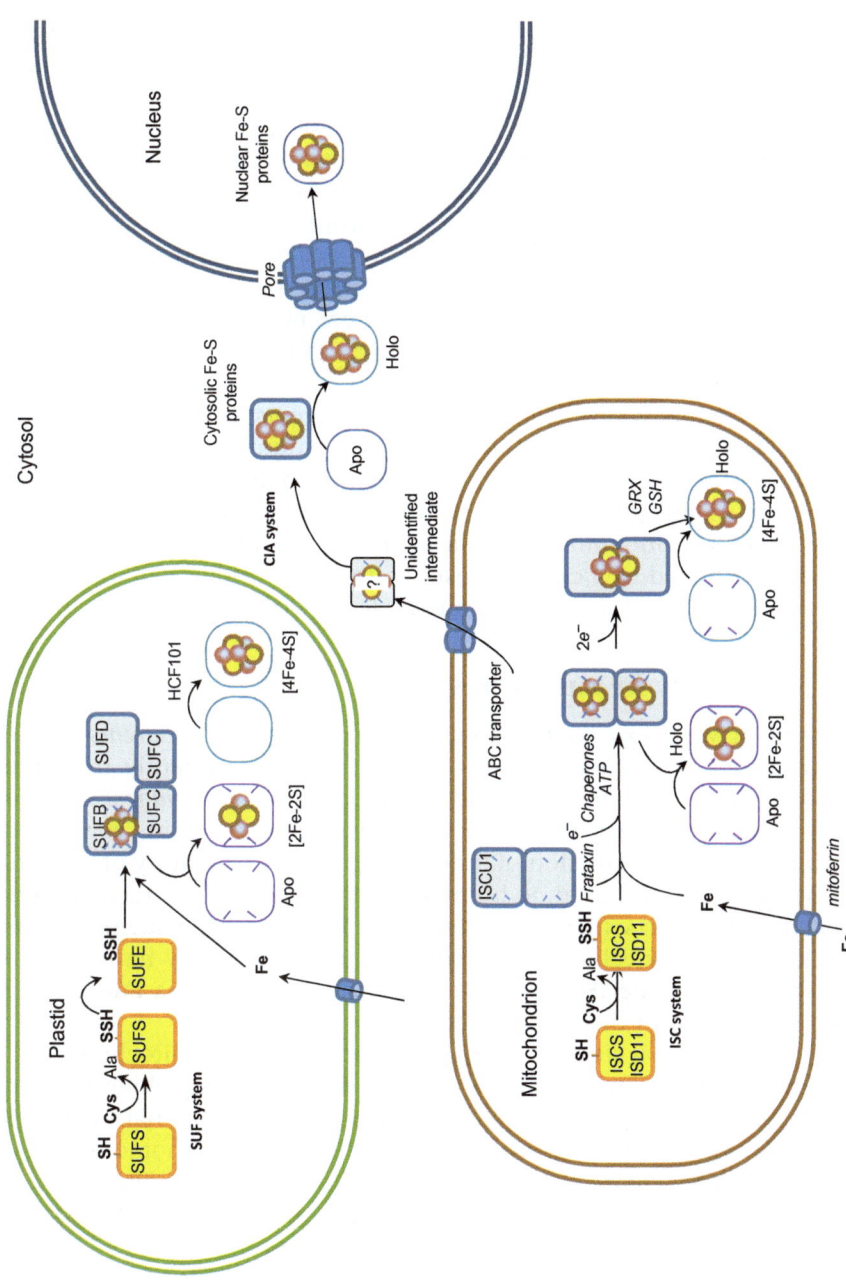

Figure 12.3 Pathways for assembly of complex Fe-S clusters in a plant cell. Pathways in yeast and animal cells are similar, except for the pathway in plastids.

Individual IscU proteins with [2Fe-2S] bound can combine to form a dimer which, on reduction, allows formation of [4Fe-4S] clusters (Figure 12.3).[94] Thus IscU is a versatile scaffold protein capable of providing clusters for the maturation of both $[2Fe-2S]^{2+}$ and $[4Fe-4S]^{2+}$ cluster-containing proteins. It can also form [3Fe-4S] clusters in enzymes such as succinate dehydrogenase where the binding site provides only three cysteine ligands. Currently little is known about additional factors that mediate these cluster conversions.

In addition to a cysteine desulfurase and a scaffold, numerous other associated proteins have been implicated in the assembly machinery (reviewed in refs 95, 96). Fe-S clusters are transferred from the *holo*-forms of U-type scaffold proteins to recipient *apo*-proteins. Cluster transfer and protein folding are assisted by chaperones.[97,98] So called A-type carrier proteins such as IscA and SufA deliver clusters to the correct position(s) in target proteins.[95] There is some evidence[99] that A-type carrier proteins are also involved in cellular iron delivery.[100]

Monothiol glutaredoxins (GRX) are small proteins containing cysteine residues that interact with glutathione. There are a number of different types that are highly abundant in cells and function in maintaining the cellular redox status.[101] Some glutaredoxins can form complexes in which a [2Fe-2S] cluster bridges two monomers.[102] These are unique in the sense that the cluster is coordinated by a cysteine residue from each monomer and by the thiolate sulfur atoms of two glutathione molecules (Figure 12.4).

12.3.2 Assembly in Eukaryotes

Much of what we know comes from studies on the yeast *Saccharomyces cerevisiae*. Other eukaryotic model systems that are frequently used are human cell lines, trypanosomes and the model plant *Arabidopsis*. When deletion of central genes for Fe-S cluster assembly is lethal, controlled transcriptional down-regulation of individual genes by regulatory promoters or by RNAi techniques is used. In humans, a number of congenital disorders associated with defects in Fe-S cluster assembly have been identified, as discussed later. Plant cells are the most complex, since they have two independent systems, the ISC/CIA systems in the mitochondria and cytosol and the SUF system in the chloroplasts.[119]

The synthesis of Fe-S clusters is a primary function of the mitochondria.[96] Even protists that lack mitochondria with a viable respiratory chain contain highly degenerate mitochondria known as mitosomes that harbour Fe-S cluster assembly machinery.[103] All these organelles contain homologues of the bacterial ISC system, supplemented by a number of additional proteins.[96] Sulfur is provided by Nfs1 when complexed to a eukaryote-specific protein called Isd11, which modulates its desulfurase activity.[104–106]

In yeast, the Cytosolic Iron-sulfur protein Assembly (CIA) system[107,108] comprises several proteins that are specifically required for the maturation of cytosolic and nuclear Fe-S proteins. Cytosolic isoforms of the cysteine desulfurase and other assembly proteins have been reported in some model organisms.[109] However, the assembly of Fe-S clusters in the eukaryotic cytosol does not appear to be completely independent of the mitochondrial ISC system, at least in yeast where the mitochondrial localization of Nfs1 appears to be critical for Fe-S cluster assembly in all cell compartments.[110] An ABC transporter in the mitochondria (ATM) has been shown to be required for assembly of cytosolic and nuclear Fe-S proteins. Mutation of the yeast Atm1p transporter also results in dramatic accumulation of iron in the mitochondria.[111] The human and plant orthologues can complement the yeast *atm1* mutant and the loss of cytosolic Fe-S clusters or enzyme activities has been shown in mutants in each organism.[112,113] Overall, the available evidence indicates that the mitochondrial ISC system produces an unidentified factor that is exported from the mitochondria. This is speculated to be a cluster intermediate or an essential sulfur compound.

The CIA proteins are highly conserved in eukaryotes, and their function has been investigated in yeast, human cell lines and plants. They include two P-loop nucleoside triphosphatases Cfd1 and Nbp35 (or Nbp35 only in plants) that are believed to perform a scaffold function. These proteins require the presence of a bridging [4Fe-4S] cluster and a nucleotide binding site (ATP or GTP).[114] The diflavin reductase Tah18 and the Fe-S protein Dre2 form a short electron transfer chain and act beforeNbp35.[115] Other proteins include: Nar1, a protein with strong sequence homology to Fe-Fe hydrogenase but no hydrogenase activity;[116] Cia1, a beta-propeller protein;[117] and Met18 and Cia2, late assembly factors that interact with the Fe-S target proteins.[118] The molecular functions of these proteins are under investigation.

Recent progress in unravelling the mechanism of iron-sensing in yeast shows that the sensor, which is located in the cytosol, depends on a [2Fe-2S] centre (see above).[77,102] Therefore, yeast cells with defects in the mitochondrial ISC pathway or Atm1p function cannot sense iron. This results in constitutive expression of iron uptake genes and uncontrolled uptake that is

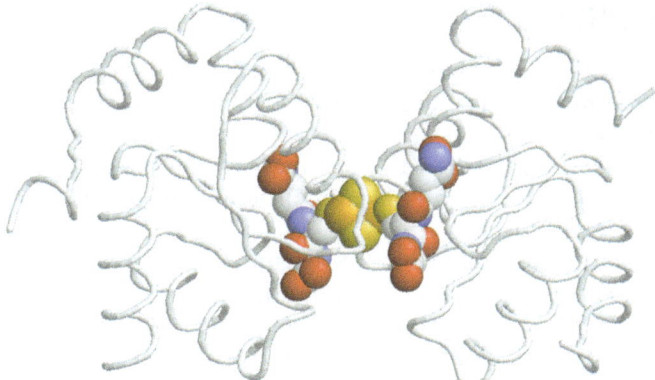

Figure 12.4   Structure of a dimeric complex of human glutaredoxin 2 (polypeptide trace) and glutathione (in CPK), binding a [2Fe-2S] cluster. Drawn from PDB file 2HT9 using RasTop 2.2.

directed to the vacuole and mitochondria to minimize damage by free ferrous ions in the cytosol. A similar mechanism involving the iron-regulatory protein IRP1 would explain mitochondrial iron accumulation in mammals. The fact that *atm3* mutants in the model plant *Arabidopsis thaliana* do not accumulate iron in the mitochondria suggests that plants have a different mechanism of iron-sensing. Also, it indicates that ATM3 does not transport iron. In *A. thaliana*, mutation of vacuolar iron transporters strongly decreases cytosolic aconitase activity (Balk, unpublished), suggesting that iron from the vacuole is directly used for cytosolic Fe-S clusters, without a detour through the mitochondria.

Plant cells have the most complex Fe-S cluster metabolism in eukaryotes, because Fe-S clusters are assembled in chloroplasts, as well as in the mitochondria and in the cytosol[119] (Figure 12.3). The chloroplasts (or plastids) contain the SUF system, which is generally found in bacteria but not in non-green eukaryotes. The plastid SUF system may be inherited from the common ancestor of plastids and cyanobacteria. All SUF genes have, in the course of evolution, been transferred to the nucleus. The only exception found thus far is a *sufB* gene on the genome of the apicoplast, a degenerate chloroplast in the malaria parasite and its relatives.[120] The eukaryotic ISC and SUF proteins, like most mitochondrial proteins, are encoded in the nuclear genome, translated in the cytosol and imported into mitochondria or plastids, respectively. For assembly of other proteins, specific proteins are involved. For example in chloroplasts, the protein HCF101 (high chlorophyll fluorescence mutant) has been shown to be essential for the production of the [4Fe-4S] clusters in Photosystem I and ferredoxin-thioredoxin reductase.[121] A simplified NIF-like system consisting of only NifS and NifU also functions in the assembly of common Fe-S clusters in certain anaerobic protozoa such as *Entamoeba histolytica*. The genes were probably acquired by lateral gene transfer.[122,123]

12.4   Trafficking of Fe-S Proteins

In eukaryotes, the majority of Fe-S proteins are encoded by nuclear genes. After translation, they are directed to the mitochondrion, plastid and nucleus by specific targeting sequences. Nuclear Fe-S proteins presumably receive their clusters from the CIA machinery in the cytosol whereas the Fe-S clusters of mitochondrial and plastidic proteins are inserted by the ISC and SUF systems, respectively.

Because of the instability of Fe-S clusters, the proteins are sometimes transferred across membranes in fully folded form, complete with metal centres. This is mediated by the TAT system,[124,125] which contrasts with the Sec system that transports proteins in their unfolded state. TAT recognizes an N-terminal signal peptide in the proteins with a twin-arginine (consensus SRRXFLK) motif. The system provides a transmembrane pathway that is large enough to allow the passage of proteins of different sizes but, remarkably, maintains the impermeability of the membrane to ions. It involves three small proteins, TatA, B and C. Firstly, in an energy-independent step, the TatBC complex binds the signal peptide of the passenger substrate protein at one side of the membrane; the twin-arginine consensus motif in the signal peptide is specifically recognized by a site in TatC. The TatA component polymerizes in the membrane bilayer to assemble an annular structure around the substrate protein. The resulting complex comprises multiple subunits of TatA, B and C, and constitutes the active translocation site. The twin-arginine signal peptide is then proteolytically cleaved by a signal peptidase at the opposite face of the membrane.

The Rieske protein in plastids is part of the b_6f complex of photosynthetic electron transport and is also thought to be transported by the TAT system across the thylakoid membrane. The motif serves to target the proteins to a specific membrane leaflet.[125,126] Interestingly, the plant mitochondrial DNA encodes a TatC homologue. It is thought that a TatB homologue is imported and forms a minimal Tat system for the plant mitochondrial Rieske protein, which has a twin-arginine signal.[126]

12.4.1 Assembly and Insertion of Superclusters

Superclusters require additional assembly proteins such as intermediate scaffolds and assembly proteins for specialized ligands. The maturation of the H-cluster of FeFe-hydrogenase requires, apart from the basic ISC or SUF pathway, three additional open reading frames: HydE, HydF and HydG. The genes were first identified by a genetic screen in the green algae *Chlamydomonas reinhardtii.*[127] This organism is unusual among eukaryotes in that it has a [FeFe]-hydrogenase which is localized to plastids. The HydE, F and G open reading frames are conserved in all prokaryotes with FeFe-hydrogenases.[128] HydF serves as the scaffold for the assembly of the dinuclear iron site and its CN, CO and dithiolate ligands. The cyanide and carbonyl ligands of the H-cluster are derived from the amino acid tyrosine by the radical-AdoMet protein HydG.[88] When fully assembled, the [FeFe] functional group is transferred to the hydrogenase protein and added to a pre-assembled [4Fe-4S] cluster. The precise role of HydE is yet to be identified.

In nitrogenase, the Fe-Mo-cofactor is assembled in the NifEN scaffold protein,[129] together with the molybdenum and homocitrate. The central carbide[130,131] (Figure 12.2(d)) is derived from the methyl carbon of S-adenosylmethionine.[132,133] The cofactor is then transferred to the NifDK protein.[64]

12.5 Human Diseases Associated with Fe-S Proteins and Fe-S Cluster Assembly

Friedreich's ataxia is a disease caused by GAA triplet expansion in an intron of the gene for the protein frataxin, which diminishes its expression. In patients and also in yeast, disruption of frataxin function results in perturbed iron homeostasis and iron accumulation in the mitochondria.

This phenotype is not specific for frataxin dysfunction, however, and mutation of any of the core ISC proteins tends to cause a similar mitochondrial iron overload. Frataxin forms a complex with cysteine desulfurase/Isd11 and the IscU/Isu scaffold protein. Recent studies showed that it affects the activity of the cysteine desulfurase and might therefore function as an allosteric regulator rather than as an iron donor.[134] In addition to Friedreich's ataxia, a considerable number of inherited diseases have been attributed to congenital defects in genes for Fe-S cluster maturation.[135,136–138] A single intronic mutation in the human *iscu* gene is associated with a myopathy resulting from deficient succinate dehydrogenase and aconitase.[139] A missense mutation in the gene for the alternative scaffold protein NFU causes a severe deficiency of respiratory chain function and oxoacid dehydrogenase complexes.[140] Other defects in cytosolic Fe-S cluster assembly are connected with fragility of DNA. Human congenital mutations have led to the discovery of other factors involved in Fe-S enzyme assembly. Mutations in the yeast MMS19 gene caused, among other effects, sensitivity to the carcinogen methyl methanesulfonate (MMS, hence the name of the gene). The MMS19 gene product is part of the CIA complex, where it facilitates iron-sulfur cluster insertion into *apo*-proteins involved in methionine biosynthesis, DNA replication, DNA repair and telomere maintenance.[118]

12.6 Conclusions

Recent genetic studies have revealed the major players in elaborate systems for Fe-S protein maturation and for their delivery to cellular locations. There are similarities in species from archaea and bacteria to plants and humans. Many proteins are involved and many of their functions are not clear at present. The systems rely on a generic cluster-assembly machinery, though many Fe-S proteins have specific systems for insertion of particular clusters. Details remain to be determined. For example, how is iron transported from its stores to the U-type scaffold proteins? What is the nature of the compound exported from the mitochondria that is essential for cytosolic Fe-S cluster assembly? Molecular details of the assembly processes are emerging as more of the components are isolated and as the structures of the protein complexes are determined by crystallography and NMR spectroscopy.

Acknowledgements

We thank Prof. Antonio Pierik (University of Kaiserslautern, Germany) for helpful comments.

References

1. G. Palmer and J. Reedijk, *Eur. J. Biochem.*, 1991, **200**, 599–611.
2. R. Cammack, *Adv. Inorg. Chem.*, 1992, **38**, 281–322.
3. H. Beinert, R. H. Holm and E. Munck, *Science*, 1997, **277**, 653–659.

4. D. C. Johnson, D. R. Dean, A. D. Smith and M. K. Johnson, *Annu. Rev. Biochem.*, 2005, **74**, 247–281.
5. B. Py and F. Barras, *Nat. Rev. Microbiol.*, 2010, **8**, 436–446.
6. R. Hedderich, S. P. J. Albracht, D. Linder, J. Koch and R. K. Thauer, *FEBS Lett.*, 1992, **298**, 65–68.
7. J. A. Vorholt, M. Vaupel and R. K. Thauer, *Eur. J. Biochem.*, 1996, **236**, 309–317.
8. R. H. Holm, *Accounts Chem. Res.*, 1977, **10**, 427–434.
9. J. J. Mayerle, R. B. Frankel, R. H. Holm, J. A. Ibers, W. D. Phillips and J. F. Weiher, *Proc. Natl Acad. Sci. USA*, 1973, **70**, 2429–2433.
10. K. D. Demadis, S. M. Malinak and D. Coucouvanis, *Inorg. Chem.*, 1996, **35**, 4038–4046.
11. S. Groysman and R. H. Holm, *Biochemistry*, 2009, **48**, 2310–2320.
12. K. S. Hagen, J. G. Reynolds and R. H. Holm, *J. Am. Chem. Soc.*, 1981, **103**, 4054–4063.
13. L. Que, J. R. Anglin, M. A. Bobrik, A. Davison and R. H. Holm, *J. Am. Chem. Soc.*, 1974, **96**, 6042–6048.
14. R. Malkin and J. C. Rabinowitz, *Biochem. Biophys. Res. Commun.*, 1966, **23**, 822–827.
15. L. Que, R. H. Holm and L. E. Mortenson, *J. Am. Chem. Soc.*, 1975, **97**, 463–464.
16. B. A. Averill, J. R. Bale and W. H. Orme-Johnson, *J. Am. Chem. Soc.*, 1978, **100**, 3034–3043.
17. J. S. Rieske, R. E. Hansen and W. S. Zaugg, *J. Biol. Chem.*, 1964, **239**, 3017–3021.
18. J. R. Mason and R. Cammack, *Annu. Rev. Microbiol.*, 1992, **46**, 277–305.
19. M. L. Paddock, S. E. Wiley, H. L. Axelrod, A. E. Cohen, M. Roy, E. C. Abresch, D. Capraro, A. N. Murphy, R. Nechushtai, J. E. Dixon and P. A. Jennings, *Proc. Natl Acad. Sci. USA*, 2007, **104**, 14342–14347.
20. A. R. Conlan, H. L. Axelrod, A. E. Cohen, E. C. Abresch, J. Zuris, D. Yee, R. Nechushtai, P. A. Jennings and M. L. Paddock, *J. Mol. Biol.*, 2009, **392**, 143–153.
21. B. A. C. Ackrell, E. B. Kearney, W. B. Mims, J. Peisach and H. Beinert, *J. Biol. Chem.*, 1984, **259**, 4019–4022.
22. J. C. Fontecilla-Camps, A. Volbeda, C. Cavazza and Y. Nicolet, *Chem. Rev.*, 2007, **107**, 4273–4303.
23. S. Ravasio, L. Dossena, E. Martin-Figueroa, F. J. Florencio, A. Mattevi, P. Morandi, B. Curti and M. A. Vanoni, *Biochemistry*, 2002, **41**, 8120–8133.
24. J. Fritsch, P. Scheerer, S. Frielingsdorf, S. Kroschinsky, B. Friedrich, O. Lenz and C. M. T. Spahn, *Nature*, 2011, **479**, 249–252.
25. Y. Shomura, K. S. Yoon, H. Nishihara and Y. Higuchi, *Nature*, 2011, **479**, 253–257.
26. W. R. Dunham, G. Palmer, R. H. Sands and A. J. Bearden, *Biochim. Biophys. Acta*, 1971, **253**, 373–384.
27. H. C. Angove, S. J. Yoo, B. K. Burgess and E. Munck, *J. Am. Chem. Soc.*, 1997, **119**, 8730–8731.

28. G. Palmer and J. Reedijk, *Biochim. Biophys. Acta*, 1991, **1060**, 599–611.

29. A. J. Thomson, A. E. Robinson, M. K. Johnson, R. Cammack, K. K. Rao and D. O. Hall, *Biochim. Biophys. Acta*, 1981, **637**, 423–432.

30. T. A. Kent, J. L. Dreyer, M. C. Kennedy, B. H. Huynh, M. H. Emptage, H. Beinert and E. Münck, *Proc. Natl Acad. Sci. USA*, 1982, **79**, 1096–1100.

31. R. M. Evans, A. Parkin, M. M. Roessler, B. J. Murphy, H. Adamson, M. J. Lukey, F. Sargent, A. Volbeda, J. C. Fontecilla-Camps and F. A. Armstrong, *J. Am. Chem. Soc.*, 2013, **135**, 2694–2707.

32. D. O. Hall, R. Cammack and K. K. Rao, *Nature*, 1971, **233**, 136–138.

33. T. G. Spiro, ed., *Iron-Sulfur Proteins*, Wiley, New York, 1982.

34. P. A. Lindahl, J. G. Morales, R. Miao and G. Holmes-Hampton, in *Methods in Enzymology, Vol 456: Mitochondrial Function, Part A: Mitochondrial Electron Transport Complexes and Reactive Oxygen Species*, ed. W. S. Allison, Academic Press, San Diego, 2009, vol. 456, pp. 267–285.

35. O. Stehling, P. M. Smith, A. Biederbick, J. Balk, R. Lill and U. Muhlenhoff, *Methods Mol. Biol.*, 2007, **372**, 325–342.

36. J. F. Gibson, D. O. Hall, J. H. M. Thornley and F. R. Whatley, *Proc. Natl Acad. Sci. USA*, 1966, **56**, 987–990.

37. J. H. M. Thornley, J. F. Gibson, F. R. Whatley and D. O. Hall, *Biochem. Biophys. Res. Commun.*, 1966, **24**, 877–879.

38. R. Cammack, *Methods Enz.*, 1988, **167**, 427–436.

39. C. Leger, S. J. Elliott, K. R. Hoke, L. J. C. Jeuken, A. K. Jones and F. A. Armstrong, *Biochemistry*, 2003, **42**, 8653–8662.

40. F. A. Armstrong, N. A. Belsey, J. A. Cracknell, G. Goldet, A. Parkin, E. Reisner, K. A. Vincent and A. F. Wait, *Chem. Soc. Rev.*, 2009, **38**, 36–51.

41. S. V. Hexter, F. Grey, T. Happe, V. Climent and F. A. Armstrong, *Proc. Natl Acad. Sci. USA*, 2012, **109**, 11516–11521.

42. J. Meyer, *J. Biol. Inorg. Chem.*, 2008, **13**, 157–170.

43. R. Lill and U. Muhlenhoff, *Annu. Rev. Biochem.*, 2008, **77**, 669–700.

44. K. P. Jensen, *J. Inorg. Biochem.*, 2006, **100**, 1436–1439.

45. C. C. Moser, J. M. Keske, K. Warncke, R. S. Farid and P. L. Dutton, *Nature*, 1992, **355**, 796–802.

46. M. C. W. Evans, C. K. Sihra, J. R. Bolton and R. Cammack, *Nature*, 1975, **256**, 668–670.

47. J. H. Golbeck, *Photosystem I: The Light-Driven Plastocyanin : Ferredoxin Oxidoreductase*, Springer, Dordrecht, 2006.

48. R. Baradaran, J. M. Berrisford, G. S. Minhas and L. A. Sazanov, *Nature*, 2013, **494**, 443–448.

49. L. A. Sazanov and P. Hinchliffe, *Science*, 2006, **311**, 1430–1436.

50. J. Hirst, *Annu. Rev. Biochem.*, 2013, **82**, 551–575.

51. R. G. Efremov and L. A. Sazanov, *Biochim. Biophys. Acta Bioenerg.*, 2012, **1817**, 1785–1795.

52. J. Carroll, I. M. Fearnley, J. M. Skehel, R. J. Shannon, J. Hirst and J. E. Walker, *J. Biol. Chem.*, 2006, **281**, 32724–32727.

53. C.-y. Yip, M. E. Harbour, K. Jayawardena, I. M. Fearnley and L. A. Sazanov, *J. Biol. Chem.*, 2011, **286**, 5023–5033.

54. H. Beinert, M. C. Kennedy and C. D. Stout, *Chem. Rev.*, 1996, **96**, 2335–2373.

55. D. H. Flint, M. H. Emptage and J. R. Guest, *Biochemistry*, 1992, **31**, 10331–10337.

56. E. Yoda, Y. Anraku, H. Kirino, T. Wakagi and T. Oshima, *FEMS Microbiol. Lett.*, 1995, **131**, 243–247.

57. M. J. Hiscox, R. C. Driesener and P. L. Roach, *Biochim. Biophys. Acta Protein Proteomics*, 2012, **1824**, 1165–1177.

58. N. D. Lanz and S. J. Booker, *Biochim. Biophys. Acta Protein Proteomics*, 2012, **1824**, 1196–1212.

59. R. U. Hutcheson and J. B. Broderick, *Metallomics*, 2012, **4**, 1149–1154.

60. K. A. Shisler and J. B. Broderick, *Curr. Opin. Struct. Biol.*, 2012, **22**, 701–710.

61. J. T. Jarrett, *Arch. Biochem. Biophys.*, 2005, **433**, 312–321.

62. R. B. Broach and J. T. Jarrett, *Biochemistry*, 2006, **45**, 14166–14174.

63. M. R. Reyda, C. J. Fugate and J. T. Jarrett, *Biochemistry*, 2009, **48**, 10782–10792.

64. J. W. Peters and J. B. Broderick, *Annu. Rev. Biochem*, 2012, **81**, 429–450.

65. M. K. Chan, J. S. Kim and D. C. Rees, *Science*, 1993, **260**, 792–794.

66. S. M. Mayer, D. M. Lawson, C. A. Gormal, S. M. Roe and B. E. Smith, *J. Mol. Biol.*, 1999, **292**, 871–891.

67. R. N. F. Thorneley and D. J. Lowe, *Biochem. J.*, 1983, **215**, 393–403.

68. B. M. Hoffman, D. Lukoyanov, D. R. Dean and L. C. Seefeldt, *Accounts Chem. Res.*, 2013, **46**, 587–595.

69. R. R. Eady, *Chem. Rev.*, 1996, **96**, 3013–3030.

70. R. P. Happe, W. Roseboom, A. J. Pierik, S. P. J. Albracht and K. A. Bagley, *Nature*, 1997, **385**, 126–126.

71. J. Fritsch, O. Lenz and B. Friedrich, *Nat. Rev. Microbiol.*, 2013, **11**, 106–114.

72. P. J. Kiley and H. Beinert, *Curr. Opin. Microbiol.*, 2003, **6**, 181–185.

73. J. C. Crack, J. Green, M. I. Hutchings, A. J. Thomson and N. E. Le Brun, *Antioxid. Redox Signal.*, 2012, **17**, 1215–1231.

74. S. Rajagopalan, S. J. Teter, P. H. Zwart, R. G. Brennan, K. J. Phillips and P. J. Kiley, *Nat. Struct. Mol. Biol.*, 2013, **20**, 740–747.

75. A. S. Fleischhacker and P. J. Kiley, *Curr. Opin. Chem. Biol.*, 2011, **15**, 335–341.

76. J. C. Rutherford, L. Ojeda, J. Balk, U. Muhlenhoff, R. Lill and D. R. Winge, *J. Biol. Chem.*, 2005, **280**, 10135–10140.

77. H. R. Li and C. E. Outten, *Biochemistry*, 2012, **51**, 4377–4389.

78. T. A. Rouault, *Nat. Chem. Biol.*, 2006, **2**, 406–414.

79. W. E. Walden, A. I. Selezneva, J. Dupuy, A. Volbeda, J. C. Fontecilla-Camps, E. C. Theil and K. Volz, *Science*, 2006, **314**, 1903–1908.

80. N. Khoroshilova, C. Popescu, E. Munck, H. Beinert and P. J. Kiley, *Proc. Natl Acad. Sci. USA*, 1997, **94**, 6087–6092.

81. B. Zhang, J. C. Crack, S. Subramanian, J. Green, A. J. Thomson, N. E. Le Brun and M. K. Johnson, *Proc. Natl Acad. Sci. USA*, 2012, **109**, 15734–15739.

82. M. F. White and M. S. Dillingham, *Curr. Opin. Struct. Biol.*, 2012, **22**, 94–100.

83. E. J. Merino, A. K. Boal and J. K. Barton, *Curr. Opin. Chem. Biol.*, 2008, **12**, 229–237.

84. R. Lill, *Nature*, 2009, **460**, 831–838.

85. M. R. Jacobson, V. L. Cash, M. C. Weiss, N. F. Laird, W. E. Newton and D. R. Dean, *Mol. Gen. Genet.*, 1989, **219**, 49–57.

86. L. M. Zheng, V. L. Cash, D. H. Flint and D. R. Dean, *J. Biol. Chem.*, 1998, **273**, 13264–13272.

87. Y. Takahashi and U. Tokumoto, *J. Biol. Chem.*, 2002, **277**, 28380–28383.

88. L. M. Rubio and P. W. Ludden, *Annu. Rev. Microbiol.*, 2008, **62**, 93–111.

89. O. Kakhlon and Z. I. Cabantchik, *Free Radic. Biol. Med.*, 2002, **33**, 1037–1046.

90. R. C. Hider and X. L. Kong, *Dalton Trans.*, 2013, **42**, 3220–3229.

91. J. H. Kim, M. Tonelli and J. L. Markley, *Proc. Natl Acad. Sci. USA*, 2012, **109**, 454–459.

92. E. N. Marinoni, J. S. de Oliveira, Y. Nicolet, E. C. Raulfs, P. Amara, D. R. Dean and J. C. Fontecilla-Camps, *Angew. Chem. Int. Ed.*, 2012, **51**, 5439–5442.

93. J. H. Kim, R. O. Frederick, N. M. Reinen, A. T. Troupis and J. L. Markley, *J. Am. Chem. Soc.*, 2013, **135**, 8117–8120.

94. K. Chandramouli, M. C. Unciuleac, S. Naik, D. R. Dean, B. H. Huynh and M. K. Johnson, *Biochemistry*, 2007, **46**, 6804–6811.

95. B. Roche, L. Aussel, B. Ezraty, P. Mandin, B. Py and F. Barras, *Biochim. Biophys. Acta Bioenerg.*, 2013, **1827**, 455–469.

96. R. Lill, B. Hoffmann, S. Molik, A. J. Pierik, N. Rietzschel, O. Stehling, M. A. Uzarska, H. Webert, C. Wilbrecht and U. Muhlenhoff, *Biochim. Biophys. Acta*, 2012, **1823**, 1491–1508.

97. R. Dutkiewicz, J. Marszalek, B. Schillke, E. A. Craig, R. Lill and U. Muhlenhoff, *J. Biol. Chem.*, 2006, **281**, 7801–7808.

98. M. A. Uzarska, R. Dutkiewicz, S. A. Freibert, R. Lill and U. Muehlenhoff, *Mol. Biol. Cell*, 2013, **24**, 1830–1841.

99. C. C. Philpott, *J. Biol. Chem.*, 2012, **287**, 13518–13523.

100. D. T. Mapolelo, B. Zhang, S. Randeniya, A. N. Albetel, H. R. Li, J. Couturier, C. E. Outten, N. Rouhier and M. K. Johnson, *Dalton Trans.*, 2013, **42**, 3107–3115.

101. A. P. Fernandes and A. Holmgren, *Antioxid. Redox Signal.*, 2004, **6**, 63–74.

102. H. R. Li, D. T. Mapolelo, S. Randeniya, M. K. Johnson and C. E. Outten, *Biochemistry*, 2012, **51**, 1687–1696.

103. K. Hjort, A. V. Goldberg, A. D. Tsaousis, R. P. Hirt and T. M. Embley, *Philos. Trans. R. Soc. B*, 2010, **365**, 713–727.

104. A. C. Adam, C. Bornhovd, H. Prokisch, W. Neupert and K. Hell, *EMBO J.*, 2006, **25**, 174–183.

105. N. Wiedemann, E. Urzica, B. Guiard, H. Muller, C. Lohaus, H. E. Meyer, M. T. Ryan, C. Meisinger, U. Muhlenhoff, R. Lill and N. Pfanner, *EMBO J.*, 2006, **25**, 184–195.

106. A. Pandey, R. Golla, H. Yoon, A. Dancis and D. Pain, *Biochem. J.*, 2012, **448**, 171–187.

107. A. K. Sharma, L. J. Pallesen, R. J. Spang and W. E. Walden, *J. Biol. Chem.*, 2010, **285**, 26745–26751.

108. D. J. A. Netz, A. J. Pierik, M. Stuempfig, U. Muhlenhoff and R. Lill, *Nat. Chem. Biol.*, 2007, **3**, 278–286.

109. W. H. Tong and T. Rouault, *EMBO J.*, 2000, **19**, 5692–5700.

110. U. Muhlenhoff, J. Balk, N. Richhardt, J. T. Kaiser, K. Sipos, G. Kispal and R. Lill, *J. Biol. Chem.*, 2004, **279**, 36906–36915.

111. G. Kispal, P. Csere, C. Prohl and R. Lill, *EMBO J.*, 1999, **18**, 3981–3989.

112. C. Pondarre, B. B. Antiochos, D. R. Campagna, E. L. Greer, K. M. Deck, A. McDonald, A. P. Han, A. Medlock, J. L. Kutok, S. A. Anderson, R. S. Eisenstein and M. D. Fleming, *Hum. Mol. Genet.*, 2006, **15**, 953–964.

113. D. G. Bernard, Y. F. Cheng, Y. D. Zhao and J. Balk, *Plant Physiol.*, 2009, **151**, 590–602.

114. D. J. Netz, A. J. Pierik, M. Stumpfig, E. Bill, A. K. Sharma, L. J. Pallesen, W. E. Walden and R. Lill, *J. Biol. Chem.*, 2012, **287**, 12365–12378.

115. D. J. A. Netz, M. Stumpfig, C. Dore, U. Muhlenhoff, A. J. Pierik and R. Lill, *Nat. Chem. Biol.*, 2010, **6**, 758–765.

116. J. Balk, A. J. Pierik, D. J. A. Netz, U. Muhlenhoff and R. Lill, *EMBO J.*, 2004, **23**, 2105–2115.

117. V. Srinivasan, D. J. A. Netz, H. Webert, J. Mascarenhas, A. J. Pierik, H. Michel and R. Lill, *Structure*, 2007, **15**, 1246–1257.

118. O. Stehling, A. A. Vashisht, J. Mascarenhas, Z. O. Jonsson, T. Sharma, D. J. A. Netz, A. J. Pierik, J. A. Wohlschlegel and R. Lill, *Science*, 2012, **337**, 195–199.

119. J. Balk and M. Pilon, *Trends Plant Sci.*, 2011, **16**, 218–226.

120. B. Kumar, S. Chaubey, P. Shah, A. Tanveer, M. Charan, M. I. Siddiqi and S. Habib, *Int. J. Parasitol.*, 2011, **41**, 991–999.

121. L. Lezhneva, K. Amann and J. Meurer, *Plant J.*, 2004, **37**, 174–185.

122. V. Ali, Y. Shigeta, U. Tokumoto, Y. Takahashi and T. Nozaki, *J. Biol. Chem.*, 2004, **279**, 16863–16874.

123. B. Maralikova, V. Ali, K. Nakada-Tsukui, T. Nozaki, M. van der Giezen, K. Henze and J. Tovar, *Cell. Microbiol.*, 2010, **12**, 331–342.

124. T. Palmer and B. C. Berks, *Nat. Rev. Microbiol.*, 2012, **10**, 483–496.

125. F. Sargent, *Microbiology*, 2007, **153**, 633–651.

126. R. Keller, J. de Keyzer, A. J. M. Driessen and T. Palmer, *J. Cell Biol.*, 2012, **199**, 303–315.

127. M. C. Posewitz, P. W. King, S. L. Smolinski, L. P. Zhang, M. Seibert and M. L. Ghirardi, *J. Biol. Chem.*, 2004, **279**, 25711–25720.

128. P. M. Vignais and B. Billoud, *Chem. Rev.*, 2007, **107**, 4206–4272.

129. J. T. Kaiser, Y. L. Hu, J. A. Wiig, D. C. Rees and M. W. Ribbe, *Science*, 2011, **331**, 91–94.

130. K. M. Lancaster, M. Roemelt, P. Ettenhuber, Y. L. Hu, M. W. Ribbe, F. Neese, U. Bergmann and S. DeBeer, *Science*, 2011, **334**, 974–977.

131. T. Spatzal, M. Aksoyoglu, L. M. Zhang, S. L. A. Andrade, E. Schleicher, S. Weber, D. C. Rees and O. Einsle, *Science*, 2011, **334**, 940–940.

132. J. A. Wiig, Y. L. Hu, C. C. Lee and M. W. Ribbe, *Science*, 2012, **337**, 1672–1675.

133. Y. Hu, M. W. Ribbe, J. A. Wiig, C. C. Lee, K. M. Lancaster, U. Bergmann, S. DeBeer, K. Rupnik and B. J. Hales, *J. Biol. Chem.*, 2013, **288**, 13173–13177.

134. A. Pastore and H. Puccio, *J. Neurochem.*, 2013, **126**, 43–52.

135. R. Lill and U. Muhlenhoff, *Annu. Rev. Biochem.*, 2008, 77, 669–700.

136. S. H. Kevelam, R. J. Rodenburg, N. I. Wolf, P. Ferreira, R. J. Lunsing, L. G. Nijtmans, A. Mitchell, H. A. Arroyo, D. Rating, A. Vanderver, C. G. M. van Berkel, T. E. M. Abbink, P. Heutink and M. S. van der Knaap, *Neurology*, 2013, **80**, 1577–1583.

137. N. A. Bolar, A. V. Vanlander, C. Wilbrecht, N. Van der Aa, J. Smet, B. De Paepe, G. Vandeweyer, F. Kooy, F. Eyskens, E. De Latter, G. Delanghe, P. Govaert, J. G. Leroy, B. Loeys, R. Lill, L. Van Laer and R. Van Coster, *Hum. Mol. Genet.*, 2013, **22**, 2590–2602.

138. A. Navarro-Sastre, F. Tort, O. Stehling, M. A. Uzarska, J. A. Arranz, M. del Toro, M. T. Labayru, J. Landa, A. Font, J. Garcia-Villoria, B. Merinero, M. Ugarte, L. G. Gutierrez-Solana, J. Campistol, A. Garcia-Cazorla, J. Vaquerizo, E. Riudor, P. Briones, O. Elpeleg, A. Ribes and R. Lill, *Am. J. Hum. Genet.*, 2011, **89**, 656–667.

139. F. Mochel, M. A. Knight, W. H. Tong, D. Hernandez, K. Ayyad, T. Taivassalo, P. M. Andersen, A. Singleton, T. A. Rouault, K. H. Fischbeck and R. G. Haller, *Am. J. Hum. Genet.*, 2008, **82**, 652–660.

140. J. M. Cameron, A. Janer, V. Levandovskiy, N. MacKay, T. A. Rouault, W. H. Tong, I. Ogilvie, E. A. Shoubridge and B. H. Robinson, *Am. J. Hum. Genet.*, 2011, **89**, 486–495.

141. G. Cecchini, I. Schroder, R. P. Gunsalus and E. Maklashina, *Biochim. Biophys. Acta Bioenerg.*, 2002, **1553**, 140–157.

142. E. Darrouzet, M. Valkova-Valchanova, C. C. Moser, P. L. Dutton and F. Daldal, *Proc. Natl Acad. Sci. USA*, 2000, **97**, 4567–4572.

143. R. M. Cicchillo, K. H. Lee, C. Baleanu-Gogonea, N. M. Nesbitt, C. Krebs and S. J. Booker, *Biochemistry*, 2004, **43**, 11770–11781.

144. L. C. Seefeldt, B. M. Hoffman and D. R. Dean, *Curr. Opin. Chem. Biol.*, 2012, **16**, 19–25.

145. U. Swamy, M. T. Wang, J. N. Tripathy, S. K. Kim, M. Hirasawa, D. B. Knaff and J. P. Allen, *Biochemistry*, 2005, **44**, 16054–16063.

146. T. Burgdorf, O. Lenz, T. Buhrke, E. van der Linden, A. K. Jones, S. P. J. Albracht and B. Friedrich, *J. Mol. Microbiol. Biotechnol.*, 2005, **10**, 181–196.

147. D. W. Mulder, E. M. Shepard, J. E. Meuser, N. Joshi, P. W. King, M. C. Posewitz, J. B. Broderick and J. W. Peters, *Structure*, 2011, **19**, 1038–1052.

148. T. I. Doukov, J. Seravalli, S. W. Ragsdale and C. L. Drennan, *Biochemistry*, 2002, **41**, 18.

149. E. C. Duin, S. Madadi-Kahkesh, R. Hedderich, M. D. Clay and M. K. Johnson, *FEBS Lett.*, 2002, **512**, 263–268.
150. D. H. Flint and R. M. Allen, *Chem. Rev.*, 1996, **96**, 2315–2334.
151. M. Fontecave and E. Mulliez, *J. Inorg. Biochem.*, 2001, **86**, 47–47.
152. C. Enroth, B. T. Eger, K. Okamoto, T. Nishino and E. F. Pai, *Proc. Natl Acad. Sci. USA*, 2000, **97**, 10723–10728.
153. K. Saha, M. E. Webb, S. E. J. Rigby, H. K. Leech, M. J. Warren and A. G. Smith, *Biochem. J.*, 2012, **444**, 227–237.
154. D. J. A. Netz, C. M. Stith, M. Stumpfig, G. Kopf, D. Vogel, H. M. Genau, J. L. Stodola, R. Lill, P. M. J. Burgers and A. J. Pierik, *Nat. Chem. Biol.*, 2012, **8**, 125–132.
155. E. M. Walters and M. K. Johnson, *Photosynth. Res.*, 2004, **79**, 249–264.
156. Y. L. Wu and R. M. Brosh, *Nucleic Acids Res.*, 2012, **40**, 4247–4260.
157. F. H. Lessner, M. E. Jennings, A. Hirata, E. C. Duin and D. J. Lessner, *J. Biol. Chem.*, 2012, **287**, 18510–18523.
158. T. L. Grove, J. S. Benner, M. I. Radle, J. H. Ahlum, B. J. Landgraf, C. Krebs and S. J. Booker, *Science*, 2011, **332**, 604–607.
159. M. M. Thayer, H. Ahern, D. Xing, R. P. Cunningham and J. A. Tainer, *EMBO J.*, 1995, **14**, 4108–4120.
160. S. Iwata, M. Saynovits, T. A. Link and H. Michel, *Structure*, 1996, **4**, 567–579.

CHAPTER 13

Ferritin and Its Role in Iron Homeostasis

ELIZABETH C. THEIL[a,b]

[a] CHORI (Children's Hospital Oakland Research Institute), 5700 Martin Luther King, Jr. Way, Oakland, CA 94609, USA; [b] Department of Molecular and Structural Biochemistry, Polk Hall, North Carolina State University, Raleigh, NC 27695-7622, USA
Email: etheil@chori.org

13.1 Introduction

Ferritins are a family of nano-cage proteins that act as iron storage sites and are crucial players in iron homeostasis. They are ferroxidases, synthesizing biominerals of stoichiometry $Fe^{III}_2O_3 \cdot H_2O$ from Fe^{2+} and O_2 using multiple sites within the ferritin protein cage. The precursor ferrous ion is the metabolic form of intracellular iron in transit. It is also the product of dissolution of the ferritin mineral that is released into the cytoplasm. In addition, Fe^{2+} is the iron species that is sensed by mRNA ribo-regulators, the non-coding regulators that control biosynthesis of the ferritin protein as well as a number of iron-trafficking proteins.

Ferritins vary dramatically among eukaryotes, bacteria and Archaea. The differences include amino acid sequence (up to 80% divergence), location and composition of the enzymatic sites, the nature of the catalytic intermediates and the order of the mineral and its phosphate content. In animal cells, ferritins are cytoplasmic but with a specific mitochondrial subclass while, in plants, they are nuclear-encoded, synthesized in the cytoplasm but

RSC Metallobiology Series No. 2
Binding, Transport and Storage of Metal Ions in Biological Cells
Edited by Wolfgang Maret and Anthony Wedd
© The Royal Society of Chemistry 2014
Published by the Royal Society of Chemistry, www.rsc.org

targeted to plastids. The microbial Dps proteins are a subclass of small or mini-ferritins that use Fe^{2+} and H_2O_2 to form the mineral precursors.

The ferritin minerals are concentrates of metabolic iron required for synthesis of iron cofactors in proteins (iron-sulfur, haem, mono- and di-iron).[†] These cofactors are important in enzymes involved in cell division, bioenergetics, dioxygen transport and the syntheses and transformation of complex organic molecules (unsaturated fatty acids, steroid hormones, vitamins, xenobiotics, toxic chemicals, *etc.*). Certain iron cofactors are specific to plants and certain bacteria, being needed for the enzymes of photosynthesis and nitrogen fixation. Ferritins exert antioxidant (catalytic) activity as the substrates utilized (Fe^{2+} and O_2/H_2O_2) are otherwise sources of Fenton chemistry and reactive oxygen species. In eukaryotes, copying of the ferritin DNA into mRNA is regulated with other genes during recovery from oxidant stress.

As a preliminary to detailed discussion of the function of the ferritins, the following outline uses the structure of the well-characterized vertebrate ferritin from frogs (PDB ID 1MFR) as the representative for the initial discussion. The *apo* or iron-free form is an approximately spherical protein shell or cage that consists of 24 subunits (maxi-ferritin). Eukaryotic ferritins are combinations of multiple types of catalytic (H) subunits except for animal ferritins found in the cytoplasm where catalytically inactive (L) subunits combine with the H subunits. The kinetics of ferritin reactions vary with the different subunit combinations, presumably reflecting different metabolic needs in the different cells of multicellular animals.

The combined molar mass of ferritin subunits in a single maxi-ferritin protein cage is about 480 000 Da; backbone ribbon structures are shown in Figures 13.1A (total molecule) and 13.1B (single 4α-helix subunit). The ferritin polypeptide subunits pack to form a hollow cage about 80 Å in diameter with walls that are about 10 Å thick (Figure 13.1C). The overall shape of ferritin protein cages is related to that of a cube. Ion channels for Fe^{2+} entry and exit are defined by the interfaces between three subunits; in maxi-ferritins, eight channels of three-fold symmetry occur at the "corners" of the *pseudo*-cube (Figure 13.1A). Sites of ferroxidase enzyme activity are centred in the middle of the α-helix bundle of each H subunit. Mineral nucleation channels exit in a cluster around the four-fold symmetry axes present at the centre of the "cube faces".

The sites of the four known activities of ferritin protein cages are depicted in Figure 13.1A–D:

A. Transport of Fe^{2+} into the protein (*via* the eight ion channels of three-fold symmetry);

B. Catalytic oxidation of Fe^{2+} by O_2 at internal di-iron sites (in regions depicted in blue in Figures 13.1A,B);

[†]The nature of some of these cofactors is discussed in Chapters 10–12.

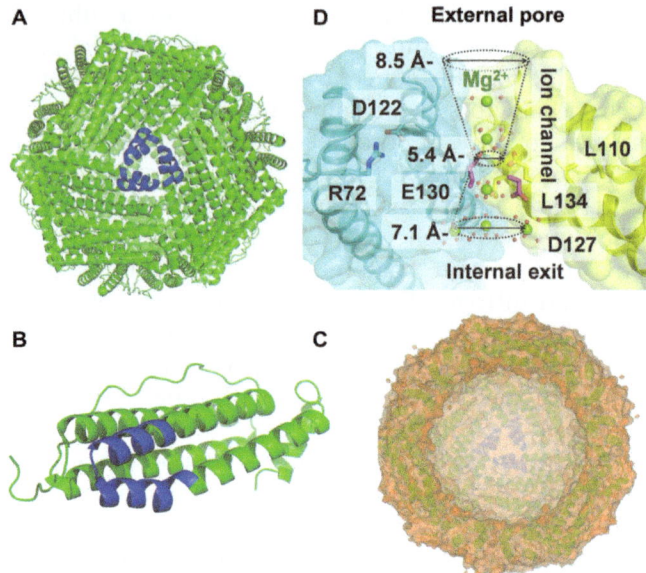

Figure 13.1 Ferritin protein nano-cage structure.
Figure taken from the author's own work in ref. 1.

C. Transport and nucleation of $[Fe^{III}\text{-}O\text{-}Fe^{III}]^{+}$ units;

D. Surface-limited reduction of the $Fe_2O_3 \cdot H_2O$ biomineral and exit of Fe^{2+} (*via* the ion channels).

In general, ferritin cages are composed of many polypeptide subunits (12 in mini-ferritins or 24 in maxi-ferritins) folded in 4α-helix bundles (Figure 13.1A,B), with multiple ion channels. Consequently, there are multiple internal sites past which Fe^{2+} ions are moving and in some cases reacting. A single ion entry channel at one of the eight three-fold symmetry axes transports Fe^{2+} ions to multiple catalytic sites (up to 3) in most ferritins since there is one active site in each H subunit. An exception might be the ferritin from horse spleen, which, unlike all other known animal, plant or microbial ferritins, has only about 5 H subunits/24-subunit cage. The stoichiometry of the ferroxidase catalysis that occurs at these sites may be represented simply by the following equation:

$$2\,Fe^{2+} + O_2 + 2\,H^+ \rightarrow 2\,Fe^{3+} + H_2O_2$$

The reaction in eukaryotic ferritins generates a peroxo-diferric intermediate, formulated as $[Fe^{III}\text{-}O\text{-}O\text{-}Fe^{III}]^{4+}$. Subsequent diferric species feature oxo/hydroxo ligands and enter nucleation channels that connect the catalytic sites to the large, central, mineral growth cavity. The water-based ligands of these products promote hydrolysis inside the nucleation channels to form larger multimers and small mineral nuclei. These emerge from the protein cage *via* four subunit channels and enter the mineral growth cavity around the four-fold symmetry axes of the cage. Thus several ferric-oxo/hydroxo mineral nuclei will emerge into the growth cavity and, oriented by the protein nucleation channels, are able to form ordered aggregates. This order is conferred on the mineral precursors by the juxtaposition of nucleation channels, catalytic centres and the protein cage.

Ferritin proteins from different sources are heteropolymers of similar subunits, which impose differences in catalytic rates and in the ordering of the mineral structure.[19] In animal tissues, differences in the mineral order coincide with different ratios of the H and L subunits. This follows from the fact that the former are catalytically active while the latter are catalytically inactive. In tissue such as heart, high H : L ratios lead to minerals that are more highly ordered than those in tissue such as liver whose ferritin has a low H : L ratio.

All ferritin protein cages exhibit an overall symmetry that was, until recently, only an object of crystallographic wonder. However, recent discoveries have correlated ferritin iron chemistry with the local *pseudo-*symmetry associated with the three- and four-fold cage axes. Entry of Fe^{2+} occurs *via* ion channels at the N-terminal ends of the three subunits that define the three-fold axes (Figure 13.1A). This flux of ions fills the di-iron catalytic sites cooperatively (Hill coefficient $= 3$). The ferric oxo/hydroxo mineral nuclei exit from the nucleation channels defined by each subunit; they proceed towards the C-terminal ends of four subunits that converge around the four-fold cage axes. They emerge into the central cavity and co-alesce to form the ferritin biomineral of stoichiometry $Fe_2O_3 \cdot H_2O$.

13.2 Ion Entry, Catalysis, Nucleation and Mineralization in Eukaryotic and Bacterial Ferritins

All ferritins share the properties of transporting ferrous ions through ion channels (Figure 13.1D). In analogy to crossing cell membranes, these ions cross the protein cage from the exterior (cytoplasm) to the internal mineral growth cavity. They are then transferred either to buried, di-iron catalytic centres (maxi-ferritins and some mini-ferritins) or to catalytic centres on the inner surface of the protein cage (mini-ferritins).

13.2.1 Maxi-ferritins: Eukaryotes and Bacteria

The eight Fe^{2+} ion entry channels have been recognized since the first low-resolution crystal structures.[20,21] However, recognition of their participation in mineral dissolution and Fe^{2+} exit lagged by 20 years.[22] The amino acid residues that form the walls of the channels are highly conserved. They are contributed by helical segments from three separate subunits. Residues in the helix 3–4 loop participate in the pore gate[23] and those in helix 3 stabilize the channel.[13] The cytoplasmic regulators of pore folding are still to be identified. They are likely to be peptides or proteins, based on the effects of physiological concentrations of urea or synthetic, binding peptides (see refs 47, 53). The eight ion channels embedded in maxi-ferritin protein cages share many properties with ion channels embedded in cell membranes. In ferritin, each channel supplies the Fe^{2+} substrate to a maximum of three H subunit di-iron sites.

Ferrous ions traverse the ion channels and reach the di-iron catalytic centres. Two Fe^{2+} ions react with dioxygen in less than 10 ms, an estimate based on formation rates of the peroxo-diferric catalytic intermediate detected in eukaryotic ferritins. Reaction rates are comparable in bacterial ferritins (BFR) that contain haem, but the active site ligands bind iron more strongly (Table 13.1). After the first catalytic turnover event at each of the 24 catalytic sites in BFR, two iron atoms are retained and act as cofactors for subsequent reactions. Turnover in BFR appears to involve the protein-bound haem and a third iron binding site (designated C).[15]

Bacterial and eukaryotic ferritins share only the substrates Fe^{2+} and O_2 and the final product $Fe_2O_3 \cdot H_2O$, together with the ion channels and overall cage structure. Structural and functional divergence in catalytic mechanisms is greater among Archaeal maxi-ferritins and in all mini-ferritins (Dps proteins). Much less is known about these systems and this sometimes results in premature attempts to develop unifying mechanisms. Even among the more extensively studied bacterial and eukaryotic ferritins, marked differences in properties reflect differences in primary structure: amino acid sequences can vary by as much as 80%.[24]

During ferritin protein catalysis (ferroxidase activity), stoichiometric formation of hydrogen peroxide can be detected as the peroxo-diferric intermediate decays (2 Fe/active site; 48 Fe/cage, 24 H_2O_2/cage). The stoichiometric equation below fits this aspect:

$$2\,Fe^{2+} + O_2 + H_2O \rightarrow [Fe^{III}\text{-}O\text{-}O\text{-}Fe^{III}]^{4+} + H_2O \rightarrow [Fe^{III}\text{-}O\text{-}Fe^{III}]^{4+} + H_2O_2$$

However, the intermediates in multiple turnover events (>48 Fe/protein cage) and the overall mineralization reaction in eukaryotic ferritins are very complex.[25] In addition, a detailed mechanism that includes nucleation and

Table 13.1   Di-iron(II) ligand sites Fe-1 and Fe-2 in maxi-ferritins.

Source	Fe-1	Fe-2	Reference
Eukaryotic[a]	E,ExxH	E,QxxA/S/D	1, 2, 4, 10
E. coli Heme-BFR[b]	E,ExxH	E,ExxH	14
E. coli FTN[b]	E,ExxH	E,E xxE	17
P. furiosus	E,xxH	E,ExxE	19
	E,ExxH,Q[c]	Y,QxxE, E	22

[a]Determined by mutagenesis and X-ray diffraction analyses of metal-protein crystals.
[b]*E. coli* and other bacteria can express 3–4 maxi-ferritins and mini-ferritins in the same cells.
[c]Both Q and E are bridging Fe-1 and Fe-2 in this structure,[22] whereas only E is bridging in eukaryotic and bacterial ferritins and in another structure of the same archaeal protein.[19]

growth of large caged hydrated ferric oxide minerals ($N \sim 2$–4000 Fe; written as $Fe_2O_3 \cdot H_2O$) is beyond current knowledge.

The peroxo-diferric complex has been studied in vertebrate ferritins (human and frog)[26-28] but has been difficult to detect in plant and bacterial ferritins. It has been detected in recombinant soy ferritin but detection was buffer-dependent (E. C. Theil and B. K. Fuqua, unpublished observations). In an invertebrate ferritin from the *Pseudo-nitzschia* multiseries, dioxygen binding was dependent on the presence of only a single Fe^{2+} at the active site,[29] contrasting with the active sites in vertebrates. The Fe: H_2O_2 stoichiometry after the first turnover is not known. Also unknown are the sources of coordinated water in the mineral (ligand of entering Fe^{2+}?; ordered water on the protein?). Other undefined components include the entry pathways of reactant dioxygen, the exit pathways of product H^+ and the details of protein-based nucleation processes occurring in the catalytically active H subunits of animals.[19]

Animal ferritin minerals are much more ordered than plant and bacterial ferritin minerals. The latter contain phosphorus (Fe : P $\sim 1 : 1$) and are largely amorphous.[30-32] Current hypotheses about the influence of high percentages of H subunits in animal ferritins on the order in ferritin minerals (as in heart tissue[33]) and of the tetrameric arrangement of nucleation channel exits around the four-fold symmetry axes of maxi-ferritin protein cages have been discussed above and in ref. 34. The genetic control of H and L subunit ratios in different animal tissues indicates physiological relevance. A possible explanation is that slower dissolution of hard/ordered ferritin minerals is desirable in the high dioxygen environment of the heart.[35]

Catalytic ferroxidase sites in ferritins have similar ligands to di-iron sites in oxygenases; in haem-containing BFRs, they are identical (Figure 13.2, Table 13.1).[15,36] Another similarity is the detection in eukaryotic ferritins of the peroxo-diferric intermediate $[Fe^{III}\text{-}O\text{-}O\text{-}Fe^{III}]^{4+}$. A third similarity may be the binding of a specific reductant such as the ferredoxin bound by one of the BFR proteins in *P aeruginosa*.[36] However, the peroxo-diferric intermediate has not been detected in those ferritins that have active site ligand structures most similar to those of the oxygenases (BFRs; Table 13.1). However, BFRs have a third site (labelled C) that participates in catalysis. Note that dissociation of dioxygen has never been observed among ferritins.

Variations in ferritin active site structure (Table 13.1) and in the detectable intermediates (*e.g.*, refs. 28, 29, 37) indicate that "one size fits all" models for ferritin structures found in organisms ranging from hyperthermophilic Archaea to humans[18] are premature at best and possibly misleading. Convergent evolution is consistent with the fact that variations of up to 80% in primary sequence among ferritins still lead to very similar secondary, tertiary and quaternary structures. It can also rationalize the fact that different active site ligands in different (compensatory?) protein backgrounds have different mechanistic intermediates but the same product, *i.e.* the caged mineral $Fe_2O_3 \cdot H_2O$.

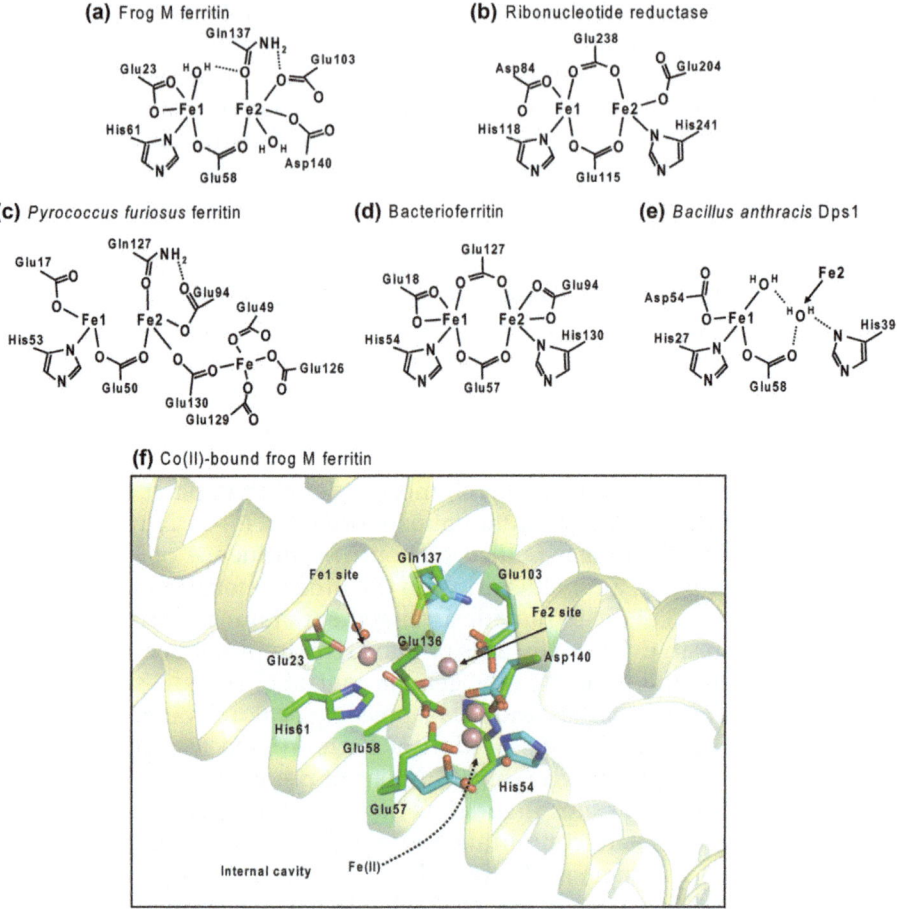

Figure 13.2 Di-iron sites at catalytic centres in (a, c–e) ferritins and (b) a di-iron oxygenase (ribonucleotide reductase). (a,b) show ligand structure in ferric-peroxo intermediates.
Figure from the author's own work in ref. 1.

13.2.2 Mini-ferritins: Bacteria and Archaea (Dps Proteins)

Mini-ferritin protein cages self-assemble from 12 subunits and feature four ion channels. They are also called Dps proteins (DNA protection during stress) and were not recognized as members of the ferritin family until an X-ray crystal structure was obtained.[38] They share the antioxidant role of maxi-ferritins but whether mini-ferritins participate in concentrating metabolic iron remains to be revealed.

Mini-ferritin ion channels are composed of segments from subunit helix 4, contributed by each of three subunits (Figure 13.1).[39] Each of the four ion channels form around the three-fold symmetry axes of the protein cages, as seen in maxi-ferritins. However, unlike maxi-ferritins, the same set of channel amino acids that control Fe^{2+} entry is used to control Fe^{2+} exit from the pores after mineral dissolution.[40] The forces that propel Fe^{2+} ions into and out of mini-ferritin protein cages can be attributed exclusively to the carboxylate ion gradient.[41] This functional behaviour contrasts with that of maxi-ferritins (see below).[3,23] Gates to the channels of mini-ferritins, if they exist, have not been identified.

The catalytic centres in mini-ferritins differ dramatically from those in maxi-ferritins in location, substrates and ligands. In most mini-ferritins, the catalytic centres are located between two subunits facing the mineral growth cavity, contrasting with those in eukaryotic ferritins where the catalytic centres are buried in the subunit helices. While the predominant oxidizing substrate in mini-ferritins is H_2O_2,[39] O_2 can be a substrate in one of the two types of Dps proteins in *B. anthracis*.[42] The physiological roles of these different Dps proteins are unclear but they are often associated with virulence because they can protect the invader from the antibacterial oxidants released by cells of the host.[43,44] Binding of Fe^{2+} at mini-ferritin sites appears to be weaker than in maxi-ferritins, especially at the Fe-2 site where Fe^{2+} binding appears to follow oxidant binding.[39]

The ligands at each of the Dps catalytic sites are H, DxxE in Fe-1 and HxxD in Fe-2; there are 12 active sites/cage. Protein-bound free radicals of tyrosine have been observed during the reaction[45] but whether this represents protein damage or is part of the mechanism remains unclear. In general, the mechanism in hydrogen peroxide-dependent mini-ferritins has been little studied; even less is known about the Dps protein reaction between Fe^{2+} and O_2. Current knowledge of the latter indicates that ferroxidase catalysis with H_2O_2 as oxidant is most reasonably represented by the simple stoichiometric reaction:

$$2\,Fe^{2+} + H_2O_2 + 2\,H^+ \rightarrow 2\,Fe^{3+} + 2\,H_2O$$

Transfer of the ferric products of catalysis to the mineral growth cavity is potentially simpler in mini-ferritins than in maxi-ferritins, since the mini-ferritin active sites are usually on the inner surface of the protein cage. However, very little is known at present about the growth, physical properties and biological function of the iron minerals in Dps proteins.

13.3 Iron Biomineral Dissolution and Exit of Ferrous Ions

Once formed, the caged mineral $Fe_2O_3 \cdot H_2O$ is unusually stable on the time-scale of normal physiology. Explanations of this stability include: 1. Inorganic iron-oxo minerals are very insoluble under neutral conditions; the solubility of the hydrated ion $Fe^{3+}{}_{aq}$ is $\sim 10^{-18}$ M, in sharp contrast to that of $Fe^{2+}{}_{aq}$; 2. Dissolution of all minerals is surface-limited, so only a small fraction of the mineral is able to interact with solvent at any one time; 3. The protein cages protect the mineral from reductants in the cytoplasm that might react with the mineral surface and cause leakage of pro-oxidant Fe^{2+}.

Two models for Fe^{2+} recovery from ferritin protein-caged biominerals have been proposed:

Model 1. After autophagocytosis from the cytoplasm to lysosomes, the protein shell is degraded proteolytically and the iron mineral exposed. It is then dissolved under the more acidic conditions by reductants and the ferrous iron is transported back into the cytoplasm.[46] This model originates from studies by pathologists who observed accumulation of ferritin in lysosomes under iron excess conditions (toxicity).

Model 2. Unfolding/opening of gated pores at the cytoplasmic side of the ion channels exposes the biomineral to cytoplasmic reductants, allowing dissolution and release of Fe^{2+}, which is bound by external chelators). This model derives from two sets of observations. Firstly, changes in the rates of Fe^{2+} reduction and chelation have been observed recently, both in solution and in cultured cells. Secondly, the dynamics of ferritin pore unfolding is changed by substitutions of conserved amino acids lining the ion channels (Figure 13.3)[23,47,48] and could be affected by cytoplasmic peptides or proteins with similar behaviour. In addition, increased proteasomal degradation of empty protein cages occurs during cellular iron deficiency[49] (Figures 13.1D, 13.3).

The idea that ferritin is degraded to recover the iron, as in Model 1, is counterintuitive upon consideration of the extraordinarily complex methods required to regulate ferritin protein biosynthesis (RNA and DNA targets and repressor turnover; see below) and the energy cost of synthesizing such a

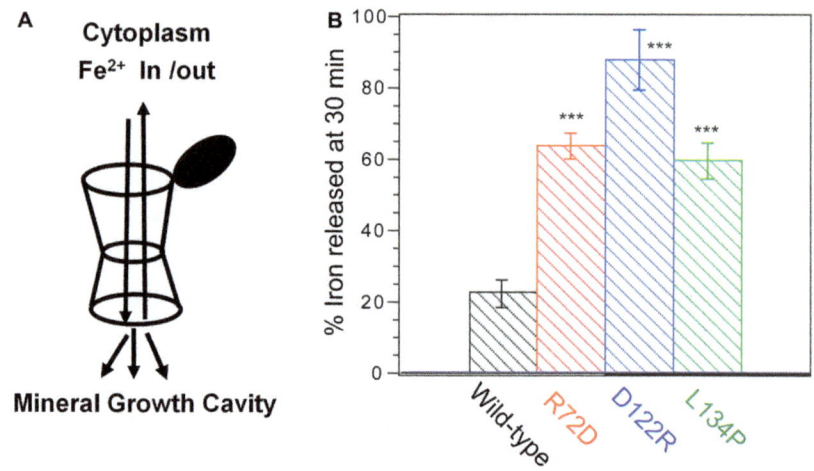

Figure 13.3   Effects of changing pore folding on caged mineral dissolution. Replacing conserved amino acids in ferritin entry/exit ion (Fe^{2+}) channels increases rates of mineral dissolution and chelation when an external reductant (NADH/ FMNH$_2$) is added.
Taken from the author's own work.[3]

large protein (molar mass >470 000 Da; 4200 amino acids per maxi-ferritin cage). Moreover, when ferritin protein is damaged under pathological conditions with formation of hemosiderin, the exposed mineral is toxic.

In contrast, changing the rate of pore opening/closing to change access of cellular reductants to the mineral is only a small drain on cellular energy budgets. It is likely that the accumulation/degradation of ferritin in lysosomes (Model 1) is related to the experimental use of large amounts of added iron and may reflect a toxic state. On the other hand, regulated opening and closing of ferritin pores to maintain normal cell iron levels (Model 2) seems to be a more likely situation for normal homeostasis.

Fe^{3+} chelators are able to remove only small amounts of iron from ferritin.[50,51] The entry of the reductant $FMNH_2$ to initiate ferric mineral reduction has been proposed since the earliest crystal structures[20] and solution studies.[52] However, it remains unresolved whether a reductant must interact directly with the mineral or whether there is an electron transfer pathway through the protein to the mineral core.[51]

Rates of recovery of iron from ferritin wild-type protein in solution in the presence of reductant and chelator are increased when the environment is changed (1 mM urea; 0.1 mM guanidine; certain peptides).[47,53] Comparable changes are achieved in solution by substitution of highly conserved ion channel amino acids.[1,3,23] The elusive physiological mechanism for changing rates of iron recovery may involve binding of a cellular regulatory molecule at external pores of the protein cages.

Model peptides that increase or decrease the rate of Fe^{2+} removal in solution were identified in a combinatorial search.[53] However, identification of the putative *in vivo* pore regulator has not been achieved. The ion channel amino acids that control the exit of Fe^{2+} from ferritin are highly conserved in eukaryotic maxi-ferritins and differ from those for Fe^{2+} entry (Figures 13.1D, 13.3)[3]. Entry is controlled by rings of three aspartate residues at the external pore and three glutamate residues in the middle of the pore at a channel constriction. They have no effect on Fe^{2+} exit.[54] By contrast, the carboxylates that control Fe^{2+} entry in mini-ferritins also control Fe^{2+} exit.[40] The N-terminal extensions in mini-ferritins are of variable length, structure and function, a situation that contrasts sharply with eukaryotic maxi-ferritins. Here, the N-terminal peptide sequence is highly conserved and functions as a gate for Fe^{2+} exit. The N-terminal peptide is held in place by a conserved H-bond network involving amino acids R72 and D122 in the subunit 4-α-helix bundle, as illustrated in Figure 13.1D.[3] Manipulation of ferritin pores to increase rates of iron chelation from ferritin minerals holds promise as a novel way to improve removal of excess iron in conditions such as iron overload from hypertransfusion treatment of genetic anaemias (thalassaemia and sickle-cell disease).

13.4 Fe^{2+}-RNA Control of Ferritin Biosynthesis

Ferritin protein synthesis increases when concentrations of iron increase inside cells (caused by environmental increases administered experimentally as ferrous or ferric salts, haem, *etc.*) or *in vivo* when iron absorption barriers are breached by injected red cells to treat genetic anaemias. There are two mechanisms to increase ferritin protein synthesis: increased rates of DNA transcription or increased rates of ferritin mRNA translation (protein biosynthesis). Under extreme oxidant stress, when denatured iron proteins increase the levels of iron in the cytoplasmic pool, both mechanisms are activated.

Increased oxidant stress increases ferritin DNA transcription by activating ARE (<u>A</u>ntioxidant <u>R</u>esponse <u>E</u>lement) sequences in multiple genes that encode several antioxidant response proteins. These include thioredoxin reductase I, NADPH-quinone reductase I and the genes for both ferritin H and L subunits.[55] ARE elements are part of the maf regulatory pathway and are inactivated when bound to the repressor protein Bach 1. Haem/Bach 1 protein interactions occur when oxidant stress is signalled by haem itself or other stressors such as *t*-butylhydroquinone. Such binding decreases the affinity of Bach 1 for its DNA sequence, allowing DNA transcription. Fe^{2+} itself is a relatively poor inducer of transcription,[56] possibly because the main transcriptional activator Bach 1 is a haem-binding protein.[55] After transcription and processing, mature ferritin mRNA is relatively stable.

Increased concentrations of iron in the cytoplasm activate the stored ferritin mRNA for translation. The mechanisms of activation depend on binding of Fe^{2+} to a non-coding, ribo-regulatory structure in the 5'-UTR of the mRNA named IRE-RNA (<u>I</u>ron <u>R</u>esponsive <u>E</u>lement-RNA). The IRE ribo-regulatory structure is present in mRNA, and encodes a group of iron homeostatic proteins that includes ferroportin (iron export), DMT1 (intestinal iron absorption), transferrin receptor (serum iron transport) and both the ferritin H and L subunits. Binding of protein repressors (IRP 1 or 2) to IRE-mRNA near the ribosome binding site in the 5'-UTR inhibits ribosome binding and translation. Binding of Fe^{2+} to IRE-RNA destabilizes the mRNA/IRP complex. While transcriptional regulation in general has been studied intensively, translational regulation is much less studied. IRE-mRNA is the most extensively characterized ribo-regulator (NMR, crystallography of the RNA/IRP complex, chemical probing and site-directed mutagenesis) and one of the few identified in eukaryotic mRNAs.[55]

13.4.1   Cytoplasmic "Labile Iron Pool"

The signal for increasing ferritin protein biosynthesis (translation of ferritin mRNA or any other IRE-mRNA) is the size of the cytoplasmic "labile iron pool". A proposal that Fe^{2+} bound by glutathione has the properties of the "labile iron pool" has been published recently;[12] Fe^{2+}-glutathione can serve as a reservoir when more specific complexes (metallochaperones, enzyme cofactors, regulatory proteins) are saturated. In iron-deficient cells, this "iron pool" will be essentially empty but will be full when iron is in excess. In conditions where the intracellular Fe^{2+}-glutathione concentration is elevated above normal, Fe^{2+} will be available to bind to "sensor" sites that increase synthesis of cell protectants. Examples of such sensor proteins are cytoplasmic ferritin encoded in an IRE-mRNA, which protects cells by consuming the excess of substrates (Fe^{2+} and O_2).[34,57] Ferroportin, a membrane protein, also encoded in an IRE-mRNA, lowers intracellular iron concentrations by transporting Fe^{2+} out of the cell.

13.4.2 mRNA Riboregulator (IRE-RNA)

A group of metabolically related mRNAs in animals contain a non-coding, three-dimensional structure of ~30 nt in the mRNA leader sequence (5′-UTR). This is the IRE ribo-regulator that senses iron and binds two regulatory proteins (Figure 13.4). Since the cellular concentrations of the proteins encoded in the mRNAs increase when body and cell iron concentrations increase, the regulatory sequence was named IRE (Iron Responsive Element). Because there is potential for confusion among IRE (Insulin Responsive Element), a transcription regulator in DNA, and IRES (Internal Ribosome Entry Sites) in viral and some cellular mRNAs, the non-coding, iron-sensitive mRNA regulatory element is designated "IRE-RNA". IRE-RNA sequences are highly conserved phylogenetically with >90% sequence identity for each mRNA in which they appear.[58] However, among the mRNAs encoding multiple iron-regulated proteins in the same animal, the IRE sequence can vary by as much as 35%.[58] Differences in IRE-RNA sequence and/or folded structure change the stability of the RNA complexes with

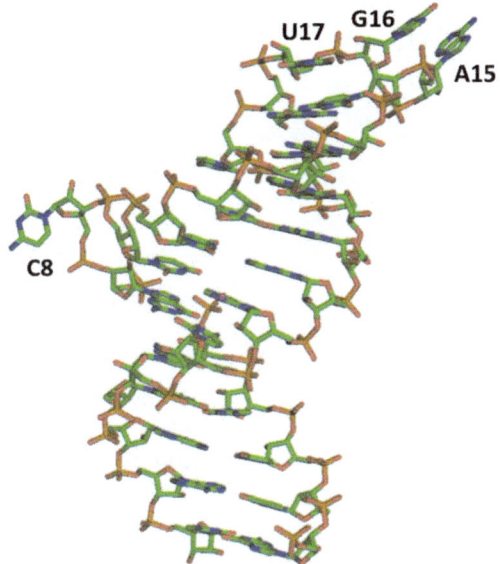

Figure 13.4 Ferritin IRE-RNA structure when bound to repressor protein (X-ray diffraction; PDB 2IPY). Note the distortion of the RNA helix.[5] In the free IRE-RNA in solution (NMR) C^8, A^{15} and G^{16} are disordered, although the C14-G18 base pair that spans the terminal loop of IRE-RNAs is present.[7,8]

regulatory proteins such as the IRP. The result is a hierarchy of RNA-repressor binding affinities and thus of mRNA sensitivities to intracellular Fe^{2+} concentrations.[59–61] Plant mRNAs do not contain the IRE riboregulators; IRE-RNA from animals functions only weakly when inserted into plant ferritin mRNA. The most primitive ferritin IRE-RNA is in sponges and sea anemones, the evolutionarily earliest animal mRNA with an IRE riboregulator. Other IRE-RNAs appeared at different times in more advanced animals. The complex, multiple IRE-RNA regulatory structure in the 3′-UTR (after the coding region) of transferrin receptor mRNA is one of the most recent IRE-RNAs to evolve and is found only in vertebrates.[58]

The IRE-RNA secondary structure is a terminal loop with the sequence CAGUGX that has a loop-spanning C-G base pair, a short helix of 5 base pairs, a C bulge and a lower helix of 4–5 base pairs (Figure 13.4), Distortions in the helix around the C^8 bulge cause a twist that allows that bulge to insert into the IRP protein at one of the protein-RNA contact sites (Figure 13.5A). Similarly the C-G pair distorts the terminal loop, effectively creating a tri-loop $A^{15}G^{16}U^{17}$ so that $A^{15}$and G^{16} can be inserted into the IRP protein at the second of the two protein-RNA contact sites.[5] These two protein-RNA contacts C^8 and A^{15}, G^{16} are present in all IRE-RNAs. The remaining bases and base pairs in the helical stem of IRE-RNAs vary for the different IRE-RNAs that encode different iron-regulated proteins in the same species. Thus, in human ferritin mRNA, the base pairs encoding the IRE helix are different

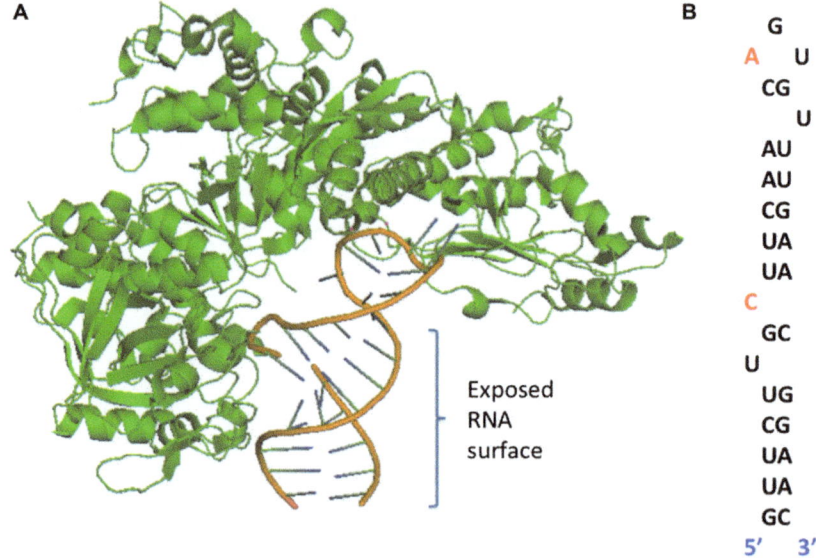

Figure 13.5   Structure of the ferritin H IRE-RNA complexed with protein (IRP). (A) Drawn with PYMOL using data from PDB ID 3SNP. (B) Predicted secondary structure of ferritin H IRE; red bases are IRP contact sites and also sites where the conformation is changed by Fe^{2+} binding, based on fluorescence changes in 2-aminopurine substituted for C^8 and A^{15}.[9]

from those in human mitochondrial (mt-)aconitase mRNA that encodes the IRE helix. As a result, IRE-mRNAs in different mRNAs of the same animal have different IRP/IRE-RNA binding stabilities[60,61] and different kinetics of binding/dissociation. Thus, under the same conditions, the fraction of mRNA dissociated from the IRP repressor will be different for each type of IRE-mRNA. For example, the IRE sequence in ferritin mRNA binds IRP more tightly than does the IRE-RNA in mt-aconitase mRNA. The differences in IRP binding stabilities means that dissociation of the IRP from the RNA will have different effects for each IRE-mRNA. Because of the different IRE-RNA/IRP binding interactions, the effect of Fe^{2+} binding to ferritin mRNA is greater than for mt-aconitase and the rate of ferritin protein synthesis increases more than for mt-aconitase. Such differences in protein synthesis rates have been observed both *in vitro* and *in vivo*.[9,60,61]

Large surface areas of the IRE-RNA remain exposed in the complex with the IRP proteins (Figure 13.5A). Superposition of contact sites between IRE-RNA reagents[5,59] indicated that metal binding sites on the IRE-RNA surfaces remain accessible in the RNA-IRP complex.[60] Fe^{2+} binds readily to IRE-RNA. There is no evidence of Fe^{2+} binding to the IRP protein.[60] Based on changes in 2-aminopurine (AP) fluorescence, Fe^{2+} changed the conformation of IRE-RNA when the base replaced one of the two bases in contact with IRP (C^8 or A^{15}). The AP fluorescence intensity decreased at position 8 upon Fe^{2+} addition but increased at position 15[9] (Figure 13.6). This is indicative of the

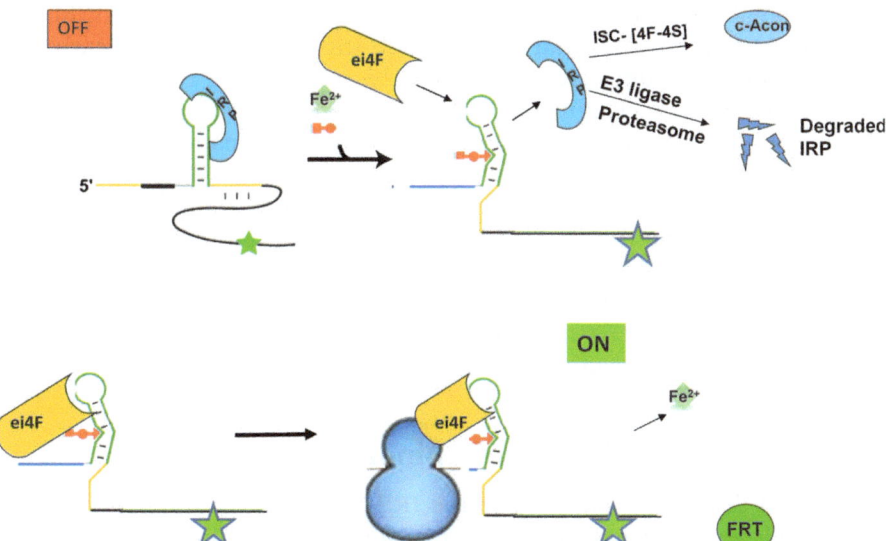

Figure 13.6   Cartoon showing how direct Fe^{2+} binding to the IRE ribo-regulator in IRE-mRNA controls the binding of a negative regulatory protein (IRP) and a positive regulatory protein (eIF-4F) that is a translation initiation (ribosome binding) factor.
From the author's own work in ref. 9.

selectivity of the changes in different parts of the ferritin H IRE-RNA structure and the large magnitude of the conformational changes that are induced by Fe^{2+} binding.

13.4.3   IRE-RNA Protein Complexes

Ferritin IRE-RNA can bind two types of regulatory proteins, IRP1 and IRP2.[‡] The binding of either prevents ribosome binding and mRNA translation (protein biosynthesis).[9] Ferritin IRE-RNA can also bind the translation initiator eIF-4F, which enhances mRNA translation. EIF-4F is a trimeric protein that facilitates protein biosynthesis by binding the mRNA cap (4E subunit), unwinding structured mRNA (4A subunit) and binding complexes of the ribosome with mRNA and other proteins (4G subunit) (Figure 13.6). Because ferritin mRNA is a relatively stable mRNA, its interactions with regulatory proteins have a large effect on rates of ferritin protein biosynthesis beyond transcription (mRNA synthesis).

Many current models of IRE-IRP regulation by intracellular iron focus on degradation (turnover) of the repressor proteins IRP1 and IRP2. In these models, the "iron signal" is the insertion of an Fe_4S_4 cluster, which converts IRP1 into cytosolic aconitase.[62] Alternatively, it is the dependence upon iron of an E3 ligase FBXL5 that catalyses ubiquitinylation of IRP2 as a signal for IRP2 degradation by proteasomes; FBXL5 has a haemerythrin-like binuclear iron site.[63] It seems unlikely that iron cluster insertion or ubiquitinylation/proteosomal degradation occurs on the IRP1–RNA complex itself. Many of the amino acid residues that interact with the Fe-S cluster are nearby or at the RNA-IRP1 binding site[5] and would be inaccessible except in the free protein. To date no evidence has been obtained that Fe^{2+} binds directly to IRP1.[60] While IRP1 can accept an Fe_4S_4 cluster, IRP2 cannot. The iron sensitivity of IRE-RNA regulation by both IRP1 and IRP2 likely relates to the recent observation that Fe^{2+} binds to IRE-RNA itself and changes IRE-RNA conformation.[9] The role of Fe_4S_4 cluster insertion can then be assigned to the inhibition of IRP1 binding to IRE-RNA. Less is known about the nature of the IRP2 ligands that bind to IRE-RNA. Isolation of pure, stable IRP2 is more difficult than for IRP1; X-ray diffraction studies of IRE-RNA complexed with IRP2 have yet to be accomplished.

13.4.4   The Iron Signal: Fe^{2+} Binding to IRE-RNA

Testing of direct Fe^{2+} binding to the IRE-RNA as the initial iron signal in living cells was initiated for several reasons. Firstly, while such binding destabilizes the mRNA-IRP complex,[62,63] this is unlikely to be the case for the large complex comprised of ferritin mRNA (>500 kDa) plus an IRP (90 kDa).

[‡]See also Chapters 11 and 12 for aspects of the action of IRP1 and IRP2.

Secondly, a large surface area of the RNA is exposed in the IRP1-IRE-mRNA complex (Figure 13.5A) and includes sites previously known to bind metal ions.[5,59] After initial studies of Fe^{2+}-IRE-RNA binding,[60] Fe^{2+} has recently been shown to share specific roles in RNA binding and in folding of ribozymes that transcend simple electrostatic interactions with divalent cations.[64] As an analogy, Mg^{2+} binds specifically to sites on the bases of a variety of RNAs and, for example, controls catalytic RNA activity.[65]

The ferrous ion has had a major role in metabolism since early evolutionary times and that continues today. While the appearance of dioxygen has increased the efficiency of many metabolic reactions in contemporary organisms, the stability of Fe^{2+} inside cells is well established (*e.g.* ref. 12) and the segregation of Fe^{2+} and O_2 inside cells is likely. Fe^{2+} selectively weakens the affinity of IRP for IRE-RNA under anaerobic conditions; the K_d increases 17-fold and, in electrophoretic mobility shift assays, the fraction of the RNA/protein complex observed decreases.[60] Electrostatics as the main effect is ruled out by the fact that two orders of magnitude higher Mg^{2+} concentration is required to have a similar effect.[60] Mn^{2+} has similar effects in solution but, *in vivo*, Mn^{2+}-glutathione and Fe^{2+}-glutathione will behave very differently.[12]

The observations that indicate that the biological signal for regulating IRE-mRNA function is Fe^{2+} can be summarized as: 1. Fe^{2+} is the "free" or "labile iron" in the cell cytoplasm because in healthy cells the redox potential is negative; 2. Cellular concentrations of Fe^{2+} available to bind to IRE-RNA will increase as total cellular iron concentrations increase; 3. When Fe^{2+} binds to IRE-RNA the affinity for inhibitors of translation (IRP proteins) decreases and the affinity of enhancers of translation (eIF-4F) increases.

13.5 Ferritin Absorption and Physiology

Iron is the most abundant "trace" element in life, but is required at vastly lower levels than metals such as sodium, potassium and calcium. The human body contains 3–5 g of iron ($\sim$2.5 g is in haemoglobin), contrasting with calcium, which is present at $\sim$1000 g ($\sim$100 g in the skeleton). Calcium is the main cationic constituent of tooth and bone minerals while iron is present mainly in protein cofactors (haem, non-haem and iron-sulfur centres); only in ferritin does iron contribute to a mineral. The chemical properties of iron in solution under physiological conditions are incompatible with biological needs since Fe^{2+} is unstable in air and Fe^{3+} is insoluble at neutral pH [$K_{sp} \sim 10^{-39}$ M, as hydrated ferric oxide (rust): $Fe_2O_3 \cdot H_2O/Fe(OH)_3$. The apparent mismatch between iron requirements and solubility is solved by binding iron ions to a stabilizing ligand or protein and by concentrating them as the ferritin mineral after oxidizing Fe^{2+} and forming ferric oxo/hydroxo bridges (with the release of H^+) inside the soluble ferritin protein cage. Iron deficiency is the most common nutrient deficiency among humans.[66] Iron is central to the major electron chains in biology, as exemplified by the haem proteins cytochrome *c* and cytochrome *c*

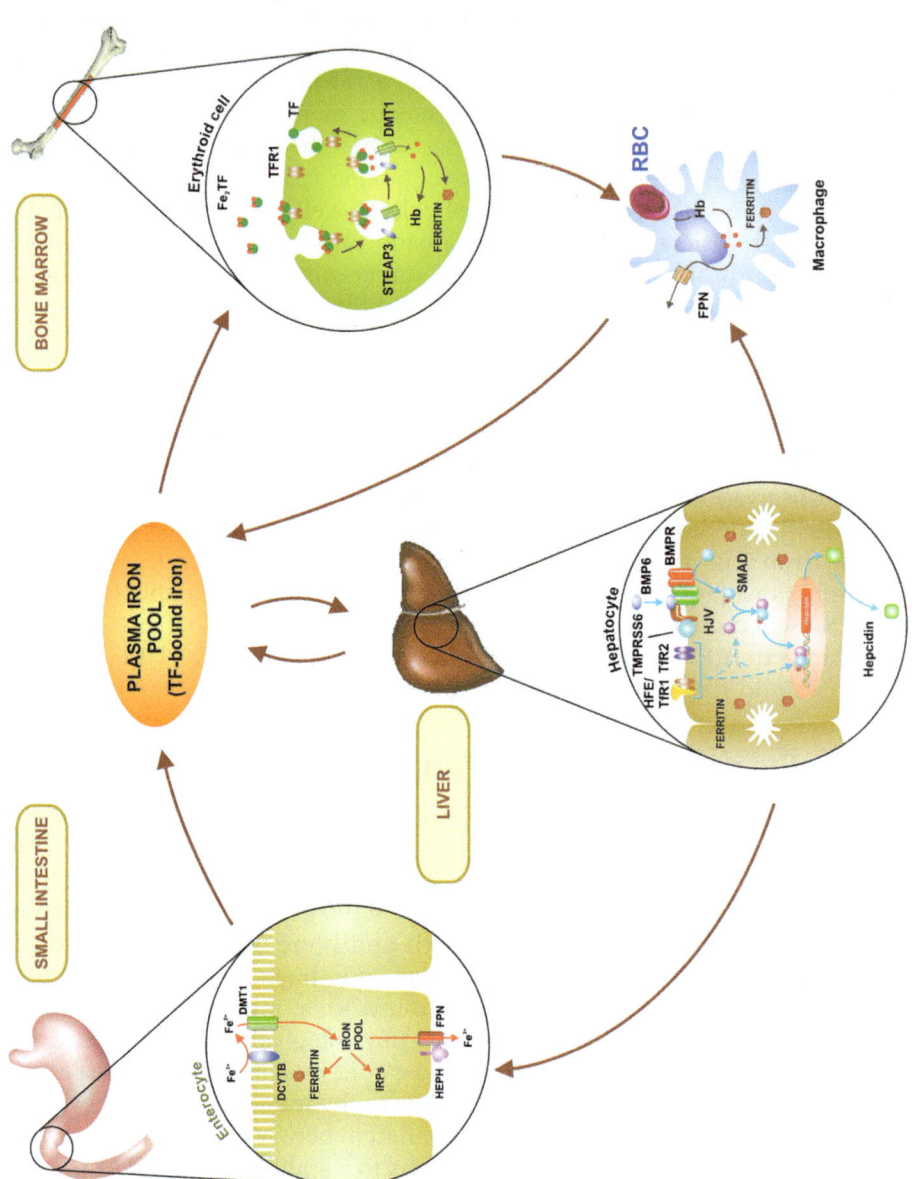

oxidase for respiration and iron-sulfur clusters (ISC) such as Fe_4S_4 present in proteins for photosynthesis. Iron absorption is the process that controls the entry of environmental iron into living organisms; animals excrete iron poorly, in general, making control of iron absorption particularly important. The major target for scientific study of iron absorption in animals is the small intestine (Figure 13.7).[§]

All organisms have evolved multiple mechanisms for iron absorption and the human intestine is no exception; both ion transport (Fe^{2+} or haem) and receptor-mediated endocytosis (Fe-protein complexes) participate in absorption of iron from food.[67,68] The iron, in various chemical forms, passes through the stomach where much is converted to free haem (from myoglobin in meat), inorganic iron (Fe^{3+}) and non-absorbable, chelated iron such as ferric phytate, Fe^{III}_{1-6} (inositol hexaphosphate). Ferritin, as the protein-coated nano-mineral, survives gastric digestion and is absorbed by a mechanism independent of the haem and non-haem species of low molar mass.[21] The long-held idea that after digestion all food iron is converted to "non-haem" (Fe^{3+} or Fe^{2+} ions) or haem is an over-simplification. Recent work revealed that much of the ferritin eaten by humans survives gastric digestion and arrives in the intestine intact.[21] There it is recognized by a receptor, yet to be identified; the ferritin is absorbed intact by receptor-mediated endocytosis, observed by electron microscopy to detect intact ferritin inside cells and by specific inhibitors of receptor-mediated endocytosis.[11] Ferritin gene deletion targeted to the intestine showed that digestion was impeded. Therefore, intestinal cell ferritin contributes to iron absorption.[69] Inside the intestinal cells, the ferritin absorbed from food is degraded and the iron incorporated into intestinal cell ferritin or exported into the blood (Figure 13.7).[11,21,70]

When there is sufficient iron in the body for normal metabolism, the levels of the iron uptake pumps, including the unidentified ferritin receptor,

---

Figure 13.7   Vertebrate iron homeostasis. Iron enters the intestinal enterocytes on the apical side (facing the food). Fe^{2+} is transported by the H^+-coupled transporter DMT1, haem iron by the folate transporter HCP1[6] and intact ferritin by an unidentified endocytic receptor.[11] All types of absorbed iron are proposed to be converted to Fe^{2+}-glutathione[12,13] and then move into the circulation *via* the FPN transporter and then, after oxidation to Fe(III), *via* the serum transporter protein transferrin. Iron is delivered to cells when transferrin binds to the endocytic transferrin receptor TfR followed by internalization and separation of Fe^{3+} from transferrin. The TfR receptor returns to the cell surface for another cycle of iron binding/delivery. The iron is released to the intracellular iron pool, thought to be Fe^{2+}-glutathione.[12] Hepcidin, a peptide hormone synthesized in the liver, controls iron export from the intestine (food iron) and the spleen (old red cell iron). Most of the transferrin iron is used for haemoglobin synthesis in bone marrow erythroid cells or for cellular iron reserves in ferritin. Figure from ref. 16.

[§]A detailed account of iron uptake is given in Chapter 11.

decrease on the cell surface, as judged by the decreased levels of iron and ferritin iron absorbed.[21,16] Iron is unique among absorbed trace elements because regulation of its uptake is highly restrictive, apparently because excretion of iron is low.

Iron absorption and physiology (Figure 13.7) is complex for many reasons. Firstly, high concentrations of iron are needed by living organisms for multiple physiological functions. Secondly, the uncontrolled chemistry of iron with dioxygen, a chemical element also required by many living organisms, leads to accumulations of products that are toxic to life. Examination of most physiological schemes for iron homeostasis (Figure 13.7) reveals a heavy emphasis on the proteins controlling cellular iron entry/exit and utilizing or transporting iron extracellularly.[71,72] With regard to dioxygen, haemoglobin and myoglobin are well-studied extracellular transporters and the peptide hepcidin is an oxygen-responsive serum hormone. However, little is known about intracellular fluxes of dioxygen or iron.

13.6   Perspective

The idea of hydrated Fe^{2+} as a key to biological activity is underrated by the heavy emphasis, in most definitions, on its charge and the distinction of ferrous iron from ferric iron, Fe^{3+}. The concept of transition metal ions forming covalent bonds that induce conformational changes in large biological molecules is relatively rare outside of inorganic chemistry. A further complicating factor is the well-known instability of Fe^{2+} and the insolubility of Fe^{3+} under "physiological conditions" (neutral solutions in air) that make them difficult to study under typical biochemical or cell biological experimental conditions. Nevertheless, iron is abundant in biological systems and maintaining an iron balance compatible with life is never ending. Iron homeostasis is the biological range of normal iron balance and the maintenance mechanisms. Multiple genes and pathways control the introduction of iron into living organisms from the environment. Since efflux is usually low relative to uptake, iron homeostasis depends more on control of iron intake from the environment and on the intracellular management of absorbed iron, an immense difference from other transition metals in biology. Zinc, for example, almost as abundant as iron in biology, can be readily excreted if too much is ingested. The acquisition from the environment is also often difficult and in many cases, growth limiting.

Ferritins play a major role in intracellular iron homeostasis by maintaining intracellular iron concentrations depending on the cellular iron concentrations. As a result, cellular ferritin concentrations vary and are tightly regulated in preparation for future developments. Consequently, intracellular ferritin concentrations are high in leaves before photosynthesis becomes active, decrease in the chlorophyll-rich parts of leaves and then increase again in senescent leaves; as the leaves fall, they create an iron-rich bed in which seeds may germinate. Similarly, during rapid growth in vertebrate larvae or embryos, red blood cells carry an extra reservoir of ferritin iron in addition to the liver to ensure an immediate supply of iron.

Fe^{2+} is the key to both ferritin function and regulation. It functions as (a) the ion transported into and out of the ferritin molecule; (b) the substrate for the ferritin protein ferroxidase redox catalysis; (c) the product of ferritin mineral dissolution and release; (d) the regulatory signal for ferritin biosynthesis. The ion appears to bind selectively to sites in the non-coding riboregulator (IRE-RNA) to change IRE-mRNA conformations and increase biosynthesis rates of ferritin protein and other iron homeostatic proteins encoded in different IRE-RNAs. The direct effects of Fe^{2+}-RNA binding are unique, at this time, in facilitating exchange of a specific protein repressor (IRP) for a generic translation initiation protein (eIf-4F). The triple functions of eIF-4F-mRNA cap binding, mRNA helix unwinding and ribosome binding lead to efficient, iron-dependent translation of IRE-mRNAs to produce ferritin protein subunits and IRE-RNA-encoded proteins required for normal iron balance. Participation of Fe^{2+} in regulating ferritin protein biosynthesis and as a ferritin protein substrate creates a biochemical feedback loop with inorganic Fe^{2+} as an effector. The emerging awareness of Fe^{2+} bioinorganic chemistry to mRNA function and protein biosynthesis complements the well-known roles of Fe^{2+} in protein function and illustrates a richness of bioinorganic chemistry that has only begun to be appreciated.

References

1. E. C. Theil, R. K. Behera and T. Tosha, *Coord. Chem. Rev.*, 2013, **257**, 579–586.
2. P. Arosio and S. Levi, *Biochim. Biophys. Acta*, 2010, **1800**, 783–792.
3. T. Tosha, R. K. Behera and E. C. Theil, *Inorg. Chem.*, 2012, **51**, 11406–11411.
4. P. D. Hempstead, S. J. Yewdall, A. R. Fernie, D. M. Lawson, P. J. Artymiuk, D. W. Rice, G. C. Ford and P. M. Harrison, *J. Mol. Biol.*, 1997, **268**, 424–448.
5. W. E. Walden, A. I. Selezneva, J. Dupuy, A. Volbeda, J. C. Fontecilla-Camps, E. C. Theil and K. Volz, *Science*, 2006, **314**, 1903–1908.
6. S. Le Blanc, M. D. Garrick and M. Arredondo, *Am. J. Physiol. Cell Physiol.*, 2012, **302**, C1780–1785.
7. Z. Gdaniec, H. Sierzputowska-Gracz and E. C. Theil, *Biochemistry*, 1998, **37**, 1505–1512.
8. L. Toussaint, L. Bertrand, L. Hue, R. R. Crichton and J. P. Declercq, *J. Mol. Biol.*, 2007, **365**, 440–452.
9. J. Ma, S. Haldar, M. A. Khan, S. D. Sharma, W. C. Merrick, E. C. Theil and D. J. Goss, *Proc. Natl Acad. Sci. USA*, 2012, **109**, 8417–8422.
10. J. K. Schwartz, X. S. Liu, T. Tosha, E. C. Theil and E. I. Solomon, *J. Am. Chem. Soc.*, 2008, **130**, 9441–9450.
11. C. D. San Martin, C. Garri, F. Pizarro, T. Walter, E. C. Theil and M. T. Nunez, *J. Nutr.*, 2008, **138**, 659–666.
12. R. C. Hider and X. Kong, *Dalton Trans.*, 2013, **42**, 3220–3229.

13. T. Tosha, H. L. Ng, O. Bhattasali, T. Alber and E. C. Theil, *J. Am. Chem. Soc.*, 2010, **132**, 14562–14569.

14. A. Crow, T. L. Lawson, A. Lewin, G. R. Moore and N. E. Le Brun, *J. Am. Chem. Soc.*, 2009, **131**, 6808–6813.

15. N. E. Le Brun, A. Crow, M. E. Murphy, A. G. Mauk and G. R. Moore, *Biochim. Biophys. Acta*, 2010, **1810**, 732–744.

16. G. Anderson, in *Iron Section of the Encyclopedia of Metalloproteins*, eds. V. N. Uversky, R. H. Kretsinger and E. A. Permyakov, Springer, 2013.

17. J. Tatur, W. R. Hagen and P. M. Matias, *J. Biol. Inorg. Chem.*, 2007, **12**, 615–630.

18. K. Honarmand Ebrahimi, E. Bill, P. L. Hagedoorn and W. R. Hagen, *Nat. Chem. Biol.*, 2012, **8**, 941–948.

19. P. Turano, D. Lalli, I. C. Felli, E. C. Theil and I. Bertini, *Proc. Natl Acad. Sci. USA*, 2010, **107**, 545–550.

20. S. H. Banyard, D. K. Stammers and P. M. Harrison, *Nature*, 1978, **271**, 282–284.

21. E. C. Theil, H. Chen, C. Miranda, H. Janser, B. Elsenhans, M. T. Nunez, F. Pizarro and K. Schurmann, *J. Nutr.*, 2012, **142**, 478–483.

22. H. Takagi, D. Shi, Y. Ha, N. M. Allewell and E. C. Theil, *J. Biol. Chem.*, 1998, **273**, 18685–18688.

23. T. Tosha, R. K. Behera, H.-L. Ng, O. Bhattasali, T. Alber and E. C. Theil, *J. Biol. Chem.*, 2012, **287**, 13016–13025.

24. L. E. Bevers and E. C. Theil, *Prog. Mol. Subcell. Biol.*, 2011, **52**, 29–47.

25. G. N. L. Jameson, W. Jin, C. Krebs, A. S. Perreira, P. Tavares, X. Liu, E. C. Theil and B. H. Huynh, *Biochemistry*, 2002, **41**, 13435–13443.

26. A. S. Pereira, W. Small, C. Krebs, P. Tavares, D. E. Edmondson, E. C. Theil and B. H. Huynh, *Biochemistry*, 1998, **37**, 9871–9876.

27. F. Bou-Abdallah, G. C. Papaefthymiou, D. M. Scheswohl, S. D. Stanga, P. Arosio and N. D. Chasteen, *Biochem. J.*, 2002, **364**, 57–63.

28. X. Liu and E. C. Theil, *Proc. Natl Acad. Sci. USA*, 2004, **101**, 8557–8562.

29. S. Pfaffen, R. Abdulqadir, N. E. Le Brun and M. E. Murphy, *J. Biol. Chem.*, 2013, **288**, 14917–14925.

30. S. Mann, J. M. Williams, A. Treffrey and P. M. Harrison, *J. Mol. Biol.*, 1987, **198**, 405–416.

31. J. S. Rohrer, Q. T. Islam, G. D. Watt, D. E. Sayers and E. C. Theil, *Biochemistry*, 1990, **29**, 259–264.

32. V. J. Wade, A. Treffry, J.-P. Laulhere, E. R. Bauminger, M. I. Cleton, S. Mann, J.-F. Briat and P. M. Harrison, *Biochim. Biophys. Acta*, 1993, **1161**, 91–96.

33. T. G. St Pierre, K. C. Tran, J. Webb, D. J. Macey, B. R. Heywood, N. H. Sparks, V. J. Wade, S. Mann and P. Pootrakul, *Biol. Met.*, 1991, **4**, 162–165.

34. E. C. Theil, *Curr. Opin. Chem. Biol.*, 2011, **15**, 304–311.

35. E. C. Theil and E. Westhof, *Acc. Chem. Res.*, 2011, **44**, 1255–1256.

36. H. Yao, Y. Wang, S. Lovell, R. Kumar, A. M. Ruvinsky, K. P. Battaile, I. A. Vakser and M. Rivera, *J. Am. Chem. Soc.*, 2012, **134**, 13470–13481.

37. A. Treffry, Z. Zhao, M. A. Quail, J. R. Guest and P. M. Harrison, *Biochemistry*, 1995, **34**, 15204–15213.
38. R. A. Grant, D. J. Filman, S. E. Finkel, R. Kolter and J. M. Hogle, *Nat. Struct. Biol.*, 1998, **5**, 294–303.
39. E. Chiancone and P. Ceci, *Biochim. Biophys. Acta*, 2010, **1800**, 798–805.
40. G. Bellapadrona, S. Stefanini, C. Zamparelli, E. C. Theil and E. Chiancone, *J. Biol. Chem.*, 2009, **284**, 19101–19109.
41. P. Ceci, G. Di Cecca, M. Falconi, F. Oteri, C. Zamparelli and E. Chiancone, *J. Biol. Inorg. Chem.*, 2011, **16**, 869–880.
42. X. Liu, K. Kim, T. Leighton and E. C. Theil, *J. Biol. Chem.*, 2006, **281**, 27827–27835.
43. G. H. Gauss, M. A. Reott, E. R. Rocha, M. J. Young, T. Douglas, C. J. Smith and C. M. Lawrence, *J. Bacteriol.*, 2012, **194**, 15–27.
44. J. M. Colburn-Clifford, J. M. Scherf and C. Allen, *Appl. Environ. Microbiol.*, 2012, **76**, 7392–7399.
45. G. Bellapadrona, M. Ardini, P. Ceci, S. Stefanini and E. Chiancone, *Free Radic. Biol. Med.*, 2010, **48**, 292–297.
46. T. Z. Kidane, E. Sauble and M. C. Linder, *Am. J. Physiol. Cell Physiol.*, 2006, **291**, C445–455.
47. X. Liu, W. Jin and E. C. Theil, *Proc. Natl Acad. Sci. USA*, 2003, **100**, 3653–3658 (issue cover).
48. M. R. Hasan, T. Tosha and E. C. Theil, *J. Biol. Chem.*, 2008, **283**, 31394–31400.
49. I. De Domenico, M. B. Vaughn, L. Li, D. Bagley, G. Musci, D. M. Ward and J. Kaplan, *EMBO J.*, 2006, **25**, 5396–5404.
50. R. R. Crichton, F. Roman and F. Roland, *J. Inorg. Biochem.*, 1980, **13**, 305–316.
51. J. Johnson, J. Kenealey, R. J. Hilton, D. Brosnahan, R. K. Watt and G. D. Watt, *J. Inorg. Biochem.*, 2011, **105**, 202–207.
52. T. Jones, R. Spencer and C. Walsh, *Biochemistry*, 1978, **17**, 4011–4017.
53. X. S. Liu, L. D. Patterson, M. J. Miller and E. C. Theil, *J. Biol. Chem.*, 2007, **282**, 31821–31825.
54. S. Haldar, L. E. Bevers, T. Tosha and E. C. Theil, *J. Biol. Chem.*, 2011, **286**, 25620–25627.
55. E. C. Theil and D. J. Goss, *Chem. Rev.*, 2009, **109**, 4568–4579.
56. K. J. Hintze and E. C. Theil, *Proc. Natl Acad. Sci. USA*, 2005, **102**, 15048–15052.
57. D. M. Ward and J. Kaplan, *Biochim. Biophys. Acta*, 2012, **1823**, 1426–1433.
58. P. Piccinelli and T. Samuelsson, *RNA*, 2007, **13**, 952–966.
59. E. C. Theil and R. S. Eisenstein, *J. Biol. Chem.*, 2000, **275**, 40659–40662.
60. M. A. Khan, W. E. Walden, D. J. Goss and E. C. Theil, *J. Biol. Chem.*, 2009, **284**, 30122–30128.
61. J. B. Goforth, S. A. Anderson, C. P. Nizzi and R. S. Eisenstein, *RNA*, 2010, **16**, 154–169.
62. R. Lill, B. Hoffmann, S. Molik, A. J. Pierik, N. Rietzschel, O. Stehling, M. A. Uzarska, H. Webert, C. Wilbrecht and U. Muhlenhoff, *Biochim. Biophys. Acta*, 2012, **1823**, 1491–1508.

63. A. A. Vashisht, K. B. Zumbrennen, X. Huang, D. N. Powers, A. Durazo, D. Sun, N. Bhaskaran, A. Persson, M. Uhlen, O. Sangfelt, C. Spruck, E. A. Leibold and J. A. Wohlschlegel, *Science*, 2009, **326**, 718–721.

64. S. S. Athavale, A. S. Petrov, C. Hsiao, D. Watkins, C. D. Prickett, J. J. Gossett, L. Lie, J. C. Bowman, E. O'Neill, C. R. Bernier, N. V. Hud, R. M. Wartell, S. C. Harvey and L. D. Williams, *PLoS One*, 2012, 7, e38024.

65. P. Auffinger, N. Grover and E. Westhof, *Met. Ions Life Sci.*, 2011, **9**, 1–35.

66. WHO, *World Health Organization*, 2013, http://www.who.int/nutrition/topics/ida/en/.

67. G. W. Anderson, ed., *Iron Metabolism in Disease*, Springer, Heidelberg, 2013.

68. A. Shawki, P. B. Knight, B. D. Maliken, E. J. Niespodzany and B. Mackenzie, *Curr. Top. Membr.*, 2012, **70**, 169–214.

69. L. Vanoaica, D. Darshan, L. Richman, K. Schumann and L. C. Kuhn, *Cell Metab.*, 2010, **12**, 273–282.

70. E. Antileo, C. Garri, V. Tapia, J. P. Munoz, M. Chiong, F. Nualart, S. Lavandero, J. Fernandez and M. T. Nunez, *Am. J. Physiol. Gastrointest. Liver Physiol.*, 2013, **304**, G655–661.

71. B. E. Eckenroth, A. N. Steere, N. D. Chasteen, S. J. Everse and A. B. Mason, *Proc. Natl Acad. Sci. USA*, 2012, **108**, 13089–13094.

72. J. Kaplan, D. M. Ward and I. De Domenico, *Int. J. Hematol.*, 2011, **93**, 14–20.

Cobalt and Nickel

PETER T. CHIVERS

Department of Chemistry, School of Biological and Biomedical Sciences, and Biophysical Sciences Institute, Durham University, Durham, UK
Email: peter.chivers@durham.ac.uk

14.1 Overview

Nickel and cobalt are required by a limited number of enzyme catalysts. This restricted use has led to the declaration that nickel- and cobalt-containing enzymes are "remnants from early life".[1] The limited but entrenched use of these two transition metals likely reflects biological evolution over billions of years in response to geochemical change, some 2.4 billion years ago.[2,3] Changes in metal availability occurred then due to the loss of reduced sulfur ligands *via* increasing levels of dioxygen in the atmosphere. This would have restricted the growth niches of microbes like the methanogenic *Archaea* that require higher levels of nickel for their existence.[4] Bioinformatics analyses of a wide variety of sequenced genomes support the emergence of new copper- and zinc-binding proteins in more recently evolved species,[5,6] a shift in metal availability that agrees with the classification of nickel and cobalt as components of essential but ancient enzymes.

The differing requirements of nickel and cobalt across the domains of life reflect the environments in which nickel- and cobalt-dependent organisms are found. In some instances, either or both metals are essential for growth, or can limit growth because of a role in responses to stress.[7] Water-soluble nickel and cobalt compounds are present on average at low (nM) levels in most locations.[8–13] Both Ni and Co exhibit a nutrient-like profile in open water environments, with depletion close to the surface where microbial

RSC Metallobiology Series No. 2
Binding, Transport and Storage of Metal Ions in Biological Cells
Edited by Wolfgang Maret and Anthony Wedd
© The Royal Society of Chemistry 2014
Published by the Royal Society of Chemistry, www.rsc.org

growth is more abundant, and somewhat higher concentrations at lower depths.[11,13] At least some proportion of the total level of each metal ion in aqueous environments is coordinated by non-aqua ligands and this may reduce bioavailability.[9,13,14]

Nickel and cobalt are most commonly found as divalent ions, ($E°$ −0.25 and −0.28 V for M(II) + 2e⁻ ⇆ M°, respectively). In aqueous solution under ambient conditions, the higher trivalent state of each metal ion is easily reduced to the divalent state (Co(III) + e⁻ ⇆ Co(II), $E°$ 1.92 V and Ni(III) + e⁻ ⇆ Ni(II), $E°$ 1.09 V). Consequently, these species are not found in most growth niches, in contrast to the case of iron where oxidation of the ferrous to ferric ion ($E°$ 0.77 V) is favoured. On the other hand, the mono- and trivalent forms of nickel and cobalt are known to be important catalytically and can readily occur in the protective environment of enzyme active sites within cells. These more accessible reduction potentials result from coordination to amino acid side-chains. Hard ligands, such as carboxylates, favour higher oxidation states while soft ligands, such as the cysteine thiolate, favour lower oxidation states.[15] There are no examples known of any oxidation state other than M(II) when Co or Ni ions are bound to non-catalytic proteins.

The scarcity of these metals has given rise to the presence of protein systems to specifically acquire them from the environment and efficiently utilize them once internalized.[1,2] The importance of controlling metal ion targeting has been described elsewhere.[16] The total intracellular concentrations of nickel and cobalt enzymes typically exceed the average extracellular availability. For example, in a growth environment containing 1 nM nickel, a microbe with a cell volume of 0.5 to 1 fL with 1000 copies of a nickel enzyme would require access to intracellular Ni concentrations of roughly 1–3 μM. This chapter focuses on the transport, trafficking and regulation of Ni(II) and Co(II) ions by various protein systems. These proteins are often essential for growth under normal (*i.e.* low) metal ion availability because of their role in metallo-enzyme assembly. Detailed aspects of the structure, function and biochemistry of these Ni(II), Co(II) and cobalamin-binding proteins important in metal acquisition and metal ion assembly have been reviewed extensively very recently.[17–21] This chapter emphasizes the structural features of the metal-binding sites of proteins with different biological roles to identify relationships between metal site properties and protein function.

14.1.1 Nickel and Cobalt Utilization in Biological Systems

Nickel and cobalt enzymes are found almost exclusively in microbes,[22,23] reflective of the reduced availability of these metals in the environment as multicellular life began to appear on Earth. The distribution of nickel enzymes, however, is generally widespread. Both metals can participate in enzyme-catalysed hydrolytic and redox reactions. Cobalt is required by many organisms in the form of corrinoids, the tetrapyrroles, most familiarly known as cofactor B_{12}. Despite the importance of this cofactor, only a subset of microbes are capable of synthesizing B_{12} *de novo*.[23] Apart from cobalamin-dependent reactions, cobalt enzymes are rare. There is no evidence yet for any nickel enzyme expressed in mammalian cells and the cobalt requirement is restricted to the microbially synthesized cobalamin (B_{12}) cofactor. The role of nickel in biological systems and its interactions with various types of biomolecules has been covered in great detail.[24]

14.1.1.1 Nickel-dependent Enzymes

Nickel-dependent enzymes (Table 14.1) are restricted almost exclusively to bacteria and Archaea.[23,25] The exception is urease, which is also found in plants and single-celled eukaryotes. Urease and hydrogenase are the most

Table 14.1 Examples of nickel-, cobalt- and cobalamin-utilizing enzymes.

	Distribution	Reaction
Nickel		
[Ni-Fe] hydrogenase (H_2ase)	Bacteria/ Archaea	$2H^+ + 2e^- \leftrightarrows H_2$
Urease	Bacteria/ Archaea/ Eukarya	$(NH_2)_2CO + H_2O \rightarrow 2NH_3 + CO_2$
MethylCoM reductase (F_{430})	Archaea	$MethylCoM + CoB \rightarrow CoM\text{-}S\text{-}S\text{-}CoB + CH_4$
Glyoxalase I	Bacteria/ Eukarya	$GSH + methylglyoxal \leftrightarrows (R)\text{-}S\text{-lactoylGSH}$
Acireductone dioxygenase	Bacteria	1,2-dihydroxy-5-(methylthio)pent-1-en-3-one $+ O_2 \leftrightarrows$ 3-(methylthio)propanoate $+$ formate $+ CO$
Superoxide dismutase	Bacteria	$2O_2^- + 2H^+ \leftrightarrows O_2 + H_2O_2$
AcetylCoA synthase[a]	Bacteria	$CO + MeCbl + CoASH \leftrightarrows AcetylCoA + Cbl$
CO dehydrogenase	Bacteria	$CO + H_2O + A \leftrightarrows CO_2 + AH_2$
Alanyl-tRNA hydrolase	Archaea	Unknown
Cupin	Archaea	Unknown
Cobalt		
(Aceto)nitrile hydratase	Bacteria	$R\text{-}C\equiv N + H_2O \rightarrow R\text{-}C(O)NH_2$
Prolidase	Bacteria	$Xaa\text{-}Pro\text{-}COO^- \rightarrow Xaa\text{-}COO^- + Pro$
Vitamin B_{12}		
Methionine synthase	Eukarya	Homocysteine $\rightarrow$ Methionine
MethylmalonylCoA mutase	Eukarya	$MethylmalonylCoA \leftrightarrows SuccinylCoA$
Ribonucleotide reductase	Bacteria	$NDP + NADPH + H^+ \leftrightarrows dNDP + NADP^+$
Glutamate mutase	Bacteria	L-threo-3-methylaspartate $\leftrightarrows$ L-glutamate
Diol dehydratase	Bacteria	Propane-1,2-diol $\leftrightarrows$ propanal $+ H_2O$
D-lysine 5,6 aminomutase	Bacteria	D-lysine $\leftrightarrows$ 2,5-diaminohexanoate
Ethanolamine ammonia-lyase	Bacteria	Ethanolamine $\leftrightarrows$ acetaldehyde $+ NH_3$

[a]The only reaction known to utilize both Ni and Co atoms.

widespread of the nickel enzymes. Interestingly, the known enzymes participate overwhelmingly as catalysts in reactions involving gas molecules.[26,27]

Despite the restriction of nickel enzymes to microbes, they are intimately connected with human health and disease. The restriction of nickel from the diet of rats yields a range of physiological and metabolic effects, suggesting that it is an important nutrient.[28-30] In contrast, the requirement for nickel in the diet of ruminants to support microbial growth is well established. Most studies of nickel in the human diet have not considered the considerable contribution of nickel enzymes to the metabolism of the normal intestinal microflora, which is well known to affect physiology.[31-33] The daily intake of nickel and cobalt in humans, typically 100–200 μg,[34,35] will depend upon diet. Nickel is present in plants (leafy green vegetables) and plant-derived foods (chocolate), so the proportion of vegetables consumed will affect dietary nickel content. Stable isotope mass spectrometry studies indicate that roughly two-thirds of ingested nickel passes through the digestive tract without being absorbed.[36] Of the remainder, $\sim 20\%$ is excreted in the urine after absorption by the body. Of the total nickel ingested, $\sim 10\%$ remained unaccounted for after five days. These ions may remain in the body or be part of the microbial consortium of the digestive tract that comprises the primary nickel-utilizers of the human ecosystem. Nickel concentrations in urine are in the low nM range.[37]

Ni-Fe hydrogenases are typically associated with anaerobic metabolism.[27,38] Uptake hydrogenases (H_2 oxidation) have been shown to be essential in colonization of humans by the pathogens *Helicobacter pylori* and *Salmonella typhimurium*.[39-41] The dihydrogen is supplied by microbial fermentation reactions.[31,42,43] H_2 is also oxidized by hydrogenase to provide reducing equivalents in methanogenesis.[4] Urea hydrolysis catalysed by urease is used to supply nitrogen, as NH_4^+. Microbes also use urease to raise the pH of acidic growth environments. Examples include *Helicobacter pylori* in the stomach and *Proteus mirabilis* in the bladder.

A nickel superoxide dismutase was discovered relatively recently in *Streptomyces*.[44,45] This type of enzyme is known to be essential in some marine cyanobacteria, notably *Synechococcus* and *Prochlorococcus* species.[7] Lesser-studied nickel-enzymes include the metal-dependent glyoxalase I, which is typically a zinc-enzyme. However, it is selective for nickel in some bacteria and single-celled eukaryotes because of sequence and length variations in the N-terminal portion of the sequence that change the metal selectivity.[46,47]

14.1.1.2   *Cobalt-dependent Enzymes*

Cobalt-dependent enzymes are less prevalent than has been suggested.[48] This overestimation arises because Co(II) readily substitutes for Zn(II) in many enzymes that participate in hydrolytic reactions. This has led to several declarations of cobalt-dependence based only on activity assays with purified enzyme. The strict cobalt-dependence of an enzyme requires demonstration of *in vivo* loading with cobalt ions or of the dependence of enzyme activity on cobalt availability in the growth medium linked to the presence of a cobalt-specific transporter protein. Cobalt-requiring enzymes are thus limited to one or two *bona fide* examples (Table 14.1). Cobalt-dependent nitrile and acetonitrile hydratases (NHases) are known. Careful restriction of cobalt in the growth medium as well as genetic studies demonstrated the essential requirement of cobalt for growth on acetonitrile by *Rhodococcus jostii* RHA1.[49] As with glyoxalase I, amino acid sequence differences match metal selectivity in the NHase family.[50]

14.1.1.3 Nickel- and Cobalt-containing Tetrapyrrole Cofactors

Both nickel and cobalt are found in tetrapyrrole complexes (Figure 14.1). Cobalt is best known as part of the cobalamin factor (B_{12}) that is essential for many organisms in methyl transfer reactions and fatty acid catabolism. However, only a few microbes can synthesize this complex molecule. B_{12}-dependent enzymes are more abundant in microbial species, and more are likely yet to be discovered.[51] There are two B_{12}-dependent enzymes in humans (Table 14.1) that require mechanisms for uptake and transport of B_{12} from the digestive tract.

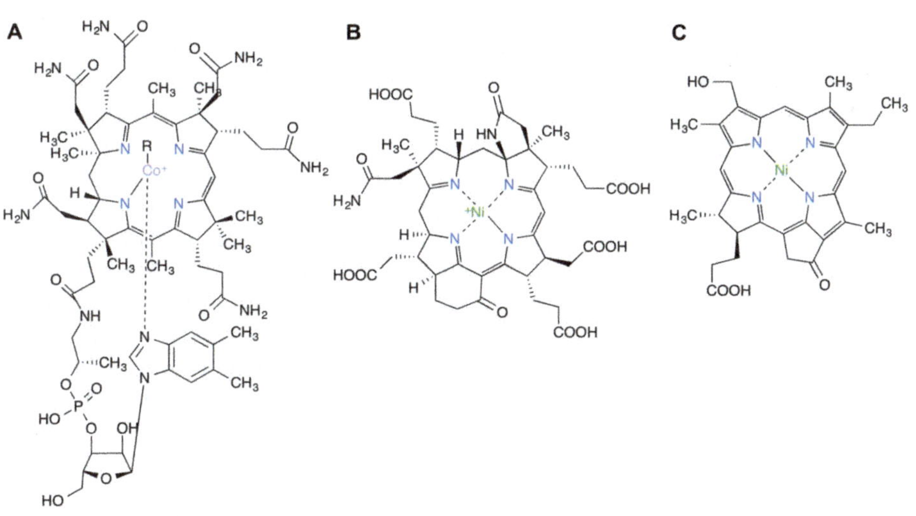

Figure 14.1   The structures of cobalt- and nickel-containing tetrapyrrole cofactors. A. Cobalamin (Cbl), the R group in enzymes is either $-CH_3$ (MeCbl) or adenosine (AdoCbl). In the industrially produced version R = $-$ CN, (CNCbl), which does not occur naturally. aquaCbl, R = $-$ OH, occurs environmentally as a result of UV-dependent cleavage of the Co–C bond. B. Cofactor F_{430} from methanogenic archaea required by methyl CoM reductase. C. Tunichlorin, a nickel tetrapyrrole of unknown function isolated from a marine microorganism.

Nickel is found in Factor 430 (F_{430}), a key component of the methyl CoM reductase enzyme required for methanogenesis by the Archaea.[52] Methanogens are always found in ruminant stomachs and are sometimes present in the human digestive tract.[42] A nickel-tetrapyrrole of unknown function, named tunichlorin, has been previously isolated from tunicates, *e.g.* sea squirts.[53] Tunichlorin (Figure 14.1) is proposed to be a nickelated derivative of an algal-derived chlorophyll molecule.[54]

14.1.1.4 Emerging Roles for Nickel and Cobalt in Biological Systems

The list of Ni- and Co-dependent enzymes may grow slightly as advances in combined proteome/metallome analyses are directed at less well-studied microbes.[55] Two new nickel enzymes have been identified (tRNA hydrolase and cupin, Table 14.1) based on analysis of the metalloproteome from *Pyrococcus furiosus*.[56] The discovery of new proteins and enzymes undoubtedly depends upon the growth conditions used. Thus, even the comprehensive and revealing examination of *E. coli*, *P. furiosus* and *Sulfolobus solfataricus* metallomes are only snapshots under specific growth conditions in monoculture.

The proliferation of sequenced genomes has allowed bioinformatic analyses of genome sequences that have further elaborated our knowledge of nickel and cobalt utilization in biological systems, most often prokaryotes. The identification of regulatory elements of known function can identify genes and operons of putative function. This approach was used, in conjunction with proteomics, to identify a key protein required for cobalamin uptake in marine diatoms.[57]

These newer, more comprehensive experimental approaches will undoubtedly expand the repertoire of nickel- and cobalt-requiring proteins. Following this type of study, the use of classical genetic and biochemical approaches, in combination with studies of cell growth and physiology, combined with an understanding of bioinorganic chemistry will help to fully elucidate the roles of known and yet to be discovered nickel- and cobalt-binding biomolecules. It is now well established that cobalt and nickel ions outside the cell bind to several selective sites before ending up in the final enzyme active site target.

14.1.2 Coordination Chemistry of Ni(II) and Co(II)

An understanding of the mechanisms of transition metal acquisition and intracellular targeting and regulation requires knowledge of the physical and chemical properties of the ions when complexed with various types of ligands. Although transition metal ions are often depicted as point charges, their electronic structures, in particular the distribution of electrons in the 3d orbitals, are critical to understanding both their functional roles and the mechanisms by which biomolecules distinguish them. General properties of biologically important transition metal ions relevant to their function are discussed comprehensively elsewhere.[1,2,15,58,59]

14.1.2.1 Coordination Preferences

In acidic aqueous solution, the bivalent metal ions favour the aqua ions $[M(II) \cdot (H_2O)_6]^{2+}$ species. Ni(II) $(3d^8)$ most commonly adopts four (square planar) or six (octahedral) coordinate geometries when bound to biomolecules.[17] It is the only first row transition metal that readily adopts square planar geometry as its low spin electron configuration is favoured in this coordination environment. Ni(II) also forms octahedral complexes with the favoured high-spin configuration. Co(II) $(3d^7)$ also adopts four (tetrahedral) and six coordinate geometries in biomolecules. The two environments are not strongly discriminated because different $3d^7$ electron configurations are not especially favoured by a particular ligand distribution. Co(II) has long been used as a probe of metalloprotein binding sites.[60]

Coordination geometry affects electronic structure, leading to slight variations in the atomic radius of each ion (0.69 Å for square planar Ni(II) and 0.72 Å for tetrahedral Co(II)). These values are slightly larger in six-coordinate environments. This similarity in size means that proteins usually discriminate between transition metal ions on the basis of coordination geometry, driven by the electronic structure of the ion. This is coupled to conformational changes in the protein. Thus, proteins generally bind first-row divalent metals with affinities consistent with the Irving–Williams series. However, only the right coordination number and geometry leads to a functional response.

Divalent nickel and cobalt ions have slower ligand dissociation rates than many other first-row transition metals (10^4 and $10^{5.5}$ s^{-1} for H_2O exchange from Ni(II) and Co(II), respectively, compared to 10^6 to 10^8 s^{-1} and higher for divalent Fe, Cu and Zn). This difference in rates can impose kinetic constraints on mechanisms of metal acquisition and selectivity.[61,62] Coordination geometry will also influence the kinetic mechanisms of ligand exchange. Reactions of coordinately saturated octahedral complexes will generally proceed *via* a dissociative mechanism. In contrast, those for square planar or tetrahedral geometries can proceed *via* an associative mechanism as the electronic structure of the metal ion can accommodate an additional incoming ligand. Changes in protein structure can also affect ligand exchange rate by altering the stability of a particular coordination environment.

The importance of electronic structure also rationalizes why Co^{III}($3d^6$) has limited utility in biological systems. This electron configuration is particularly stable in low-spin, octahedral coordination environments. Consequently, Co(III) complexes are kinetically inert on biological time-scales owing to very slow ligand exchange rates ($\sim 10^{-5}$ s^{-1}). Thus, any protein or enzyme that binds Co(III), such as the nitrile hydratases, will need to

influence the ligand field stabilization energy sufficiently to overcome slow
ligand exchange kinetics.

14.1.2.2 Biological Ligands

Ni(II) and Co(II) are both considered intermediate on the hard-soft Lewis
acid scale and thus tend to preferentially coordinate with ligands containing
N and S atoms (His and Cys in proteins, for example). However, these metals
are not always found bound to proteins. The nature of the coordination
complexes formed by Ni(II) and Co(II) in extracellular environments is important to an understanding of the forms that must be recognized and
transported into the cell. Additionally, stable (kinetically non-labile) complexes of low molar mass can provide an energetically efficient way to store
excess metal ions. This scenario is not always ideal, however, as it may lead
to problems with accessing the store (slow off rates).

Nickel complexes. A number of biologically relevant Ni(II)-ligand complexes have been identified, which promote transport or storage of excess
nickel in biological systems. Aside from $[Ni^{II}(H_2O)_6]^{2+}$, the most widespread simple Ni complex may be octahedral $Ni^{II}(\text{L-His})_2$, which has no net
charge at neutral pH. The amino acid acts as a tridentate ligand *via* the -
$\underline{N}H_2$, COO^- and imidazole nitrogen atoms (Figure 14.2A). It has functional
roles in nickel transport in human plasma,[63] in nickel detoxification in
plants and yeast[64,65,66] and in nickel uptake in bacteria.[67] Stepwise formation
constants have been determined by several methods, most recently by
isothermal titration calorimetry ($K_1 = 6.42 \times 10^7$ M^{-1}, $K_2 = 4.65 \times 10^6$ M^{-1}).[68]
Its abundance depends strongly upon L-His levels, which vary in different
environments (50–200 μM in the lower intestine,[69] 100 μM in the *E. coli*
cytosol[70] and low mM in some plant tissues).[64] Nickel binding to other
amino acids and His-containing peptides has been reviewed.[71,72]

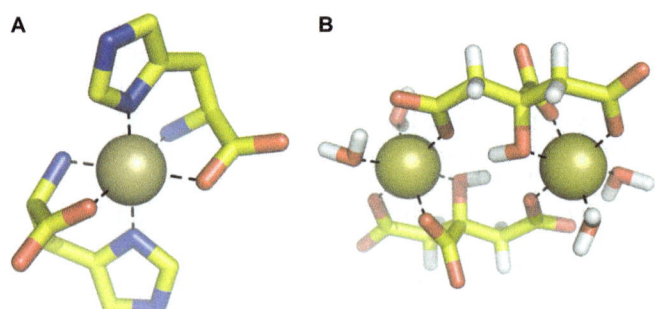

Figure 14.2   The structures of nickel in complex with small biomolecule chelators.
A. $Ni^{II}(\text{L-His})_2$ and $[Ni^{II}_2(\text{Citrate})_2(OH_2)_2]_2^{2-}$. Atom colouring – C,
yellow; H, pale grey; N, blue; Ni, gold; O, red. Black dashes indicate
the primary coordination sphere.

Plants are unique in coping with metal stress because they are non-motile. They often use small molecules to accumulate the metal ions and so protect against potentially toxic effects.[100,109,110] Plants also compose a dietary source of nickel for microbes that inhabit the digestive tract of mammals. The pH-dependent stability of these complexes will impact their lifetime once they enter the stomach, with the coordination geometry influencing the mechanism of release. Thus, transport and storage of Ni(II) in plants is relevant to understanding Ni(II)-uptake in other organisms.[64,73,74] Potential ligands include the acids malate and citrate, the amines histidine and nicotianamine, and the inositol phosphates, *e.g.* phytate. The affinity of Ni(II) for citrate is relatively weak ($K_d = 4 \times 10^{-6}$ M; Figure 14.2B) but citrate can be produced in abundance and so drive complex formation. The regulation of citrate levels can therefore affect the availability of nickel. The recently identified nicotianamine binds several transition metal ions with high affinity ($K_d = 8 \times 10^{-17}$ M for Ni(II)).[73] The structure of the 1 : 1 Ni(II) complex is unknown, but the coordination environment is expected to be octahedral. There is evidence for a bound, non-labile pool of nickel in some marine environments, but the identity of the coordinating molecule is unknown.[75]

Small thiol-containing ligands such as cysteine and glutathione do not appear to play a significant role in Ni(II) chemistry, either inside or outside the cell. The levels of these molecules may be elevated in organisms that accumulate nickel, but evidence suggests that this increase provides a protective effect against the oxidative stress resulting from the high concentration of nickel.[76,77]

Cobalt complexes. Cobalt is considered to be a beneficial element in plants, meaning that it may be essential or promote growth only in particular taxa.[78] Oxalic and malic acids are important for long distance transport.[79] The plant tissue location of the excess metal deposition is also different for Ni(II) and Co(II).[80] Cysteine has been implicated as a ligand in some hyperaccumulators.[81]

There is a high-affinity cobalt-containing complex found in the ocean.[13] It may be aquacobalamin, formed from the light-induced breakdown of adenosylcobalamin (AdoCbl; Figure 14.1).[82,83] The source of this cobalt pool is undoubtedly B_{12}-synthesizing microbes, but the origin of the cobalt that is originally acquired by the microbes for *de novo* B_{12} biosynthesis is not known.

14.2 Transport through the Membrane

Metal acquisition from the extracellular environment generally requires energy-dependent transporter proteins and protein complexes that exhibit saturable behaviour. Transport requires energy because extracellular cobalt and nickel ion concentrations are at an order of magnitude less than is required intracellularly.[84] Cobalt and nickel transporters identified in one bacterial species usually function in others, a feature that has enabled the discovery and study of nickel transporters by heterologous expression.[85,86]

However, quantitative studies of transport are limited. The determination of a K_m value of a specific transporter can be useful because it does not vary with transporter expression level and also provides an indication of metal ion availability in the native growth environment. An additional challenge in studies of metal transport is the speciation of the metal in the growth medium, which can be complex and poorly defined. Organic ligands may enhance or inhibit transport by affecting metal availability or even recognition.

14.2.1 Nickel Uptake

There are now three main classes of nickel transporters known. There are two types of ATP-dependent multiprotein complexes – the ABC transporters and the ECF-type transporters. Different classes of single polypeptide transporters have also been identified but are not structurally characterized. Some microbes possess one of each type of transporter.

14.2.1.1 NixA/HoxN/HupN

Homologous single component proteins have been identified in bacteria.[85,87,88] Nickel uptake is driven by the proton gradient. There is little information about the structural basis for recognition, although mutagenesis has identified key residues required for Ni recognition in these proteins from different organisms.[89–91] The K_m values measured in transport assays range between 10^{-9} and 10^{-5} M. These values can be difficult to obtain due to the low capacity of these transporters. This means that metal adsorption to the cell surface provides significant background signal.

14.2.1.2 NikABCDE

NikABCDE from *E. coli* was the first nickel-selective ABC transporter to be identified.[92,93] This gene cluster is found primarily in the *Enterobacteria*.[23] Other structurally related NikABCDE-like transporters have been identified in both Gram-positive and -negative microbes that are associated with other mammalian growth niches.[23] In all cases, the NikA proteins ($\sim$520 amino acids) are structurally homologous to di- and oligopeptide binding proteins.[94]

In *E. coli*, Ni(L-His)$_2$ has been identified as the substrate for NikABCDE ($K_m = 5 \times 10^{-8}$ M).[67] It binds to purified NikA with mid-nanomolar affinity ($K_d = 7.5 \times 10^{-8}$ M).[67] D-His cannot substitute in either assay. The structure of the NikA:Ni(L-His)$_2$ complex was reported recently, and shows several interesting features (Figure 14.3A and B).[95] The Ni(II) atom is coordinated by five ligand atoms from the two L-His molecules but one of the COO$^-$ ligands of free Ni(L-His)$_2$ has been displaced by the His$_{416}$ side-chain of NikA. Conserved Arg residues play critical roles in orienting the Ni(L-His)$_2$ complex by interacting with the COO$^-$ groups of L-His (Figure 14.3B). These interactions serve to select for L-His over D-His and stabilize the displaced COO$^-$ ligand. Additionally, the imidazole ligands from the two His molecules are in the *cis* rather than in the *trans* configuration seen in the free

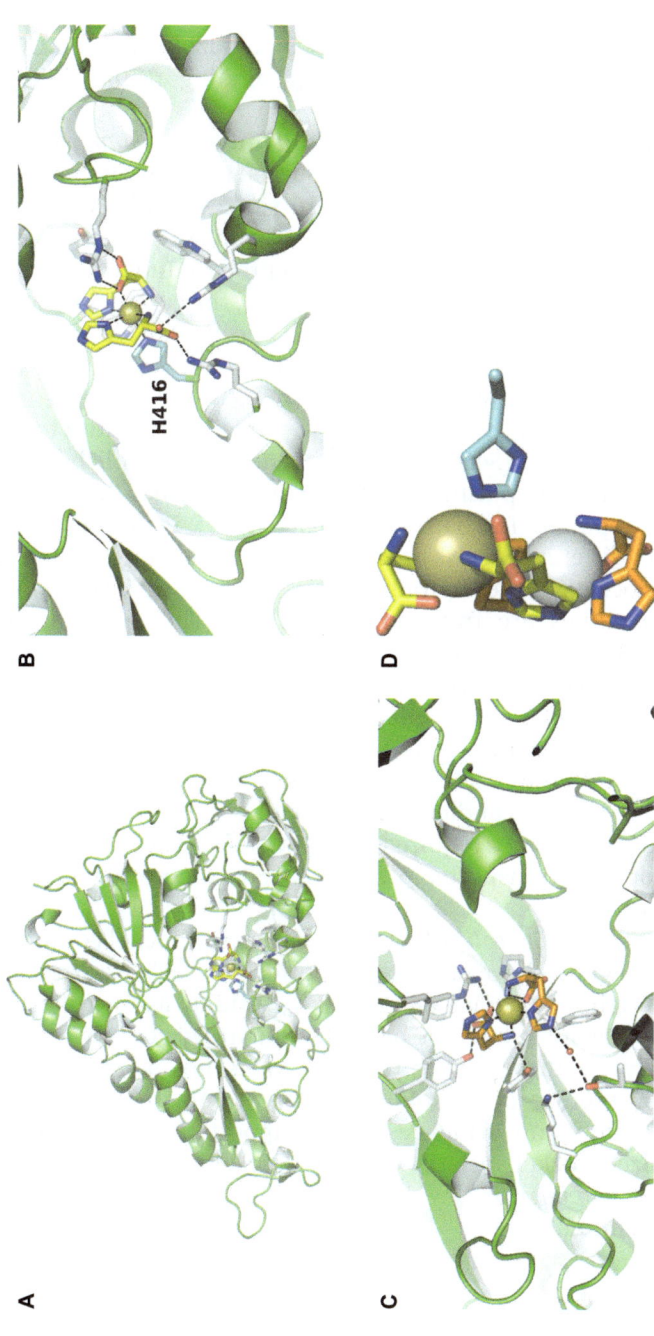

Figure 14.3 The structures of nickel-binding sites in microbial uptake proteins. A. *E. coli* NikA in complex with Ni(L-His)$_2$ showing the location of the binding pocket. B. Close-up view of Ni(L-His)$_2$ in the binding pocket. This view is rotated roughly 120° from panel A. The C atoms of free L-His groups are yellow, while the C atoms of the protein are grey. His$_{416}$ is labelled and coloured cyan. C. *S. aureus* NikA-like protein in complex with Ni(L-His)$_2$. The small red sphere corresponds to the O atom of an H$_2$O molecule. Black dashes show the contacts between atoms that coordinate Ni directly as well as those that coordinate atoms of the L-His molecules. D. Superposition of the Ni(II) · (L-His)$_2$ complexes based on the orientation of NikA shown in panel A. The *S. aureus* Ni(II) · (L-His)$_2$ complex is in grey and orange.

Ni(L-His)$_2$ structure (Figure 14.2A).[96] Co(II) is known to form different geometric isomers with L-His and some of these can be trapped by oxidation to Co(III) and subsequently separated.[97,98] It is not known whether the Ni(L-His)$_2$ complex binds initially to NikA in a *cis* imidazole and one COO$^-$ ligand is then displaced by His$_{416}$ or if His$_{416}$ binding results in a rearrangement of a *trans*-imidazole form.

A Ni(L-His)$_2$ complex has also been observed bound to the *Staphylococcus aureus* NikA-like protein (Figure 14.3C), an unpublished structure available in the Protein Data Bank.[99] However, there are some notable differences compared to the *E. coli* NikA structure (Figure 14.3C and D). First, there is no direct coordination of the Ni atom by the *S. aureus* protein. Indeed, there are no His side-chains within the binding pocket. Second, the complex is bound as the *trans*-NH$_2$-*cis*-imidazole-stereoisomer. Like *E. coli* NikA, Arg residues in the *S. aureus* protein also recognize the Ni(L-His)$_2$ complex *via* interactions with the COO$^-$ moieties of the two L-His ligands.

L-histidine is present in the lower intestine in sufficient quantity to support Ni(II) uptake by *E. coli* (Section 14.1.2). It is also established as a transporter of Ni(II) in human plasma.[100] Initial studies of NikA-like proteins have recently been reported for pathogenic bacteria,[101] including *S. aureus*,[102,103] that infect different organs or tissues, besides the intestinal tract. While the binding pockets of these proteins have sequence differences from *E. coli* NikA, Arg residues are frequently conserved. The apparent convergence upon Ni(L-His)$_2$ recognition by the *E. coli* NikA and the *S. aureus* NikA-like proteins suggests that there are multiple structural solutions to recognition of the Ni(L-His)$_2$ complex. It will be interesting to see whether other NikA-like proteins recognize Ni(L-His)$_2$ in a different conformation or recognize an entirely different nickel complex. Recently, experiments with *H. pylori* have identified both a candidate outer membrane receptor protein[104] and an ABC-type transporter[105] required for Ni-uptake. These both share homology to siderophore transporter proteins ($\sim$330 amino acids) rather than to the NikA family.

Several other structures of *E. coli* NikA in the presence of nickel complexed with polycarboxylate ligands have also been reported.[106,107] These structures revealed a single protein-Ni contact *via* the His$_{416}$ residue, as for Ni(L-His)$_2$. However, robust Ni transport has not been demonstrated for any of these ligands. Additionally, in these structures NikA does not adopt the fully closed conformation seen in other solute-binding proteins bound to their preferred solute. There is no evidence yet to support the secretion of a bacterially synthesized nickel-specific chelator (nickelophore) that would complex Ni ions and be recognized by the transport machinery by analogy to microbial iron acquisition.

14.2.2   Cobalt Uptake

In contrast to studies of B$_{12}$ import, structural studies of cobalt transporters are not well advanced. This is likely due to the smaller number of organisms that need Co(II) directly. Two distinct multiprotein cobalt importers have been identified and limited quantitative information is available about their transport properties. The CbiMNQO complex is found in B_{12}-synthesizing bacteria. ATP is used to drive transport. Cobalt ion affinity is submicromolar ($K_m = 3 \times 10^{-7}$ M).[86] There are no structural data available for Co(II) binding to any protein of this complex. The CbtJKL transporter was identified in *S. meliloti*, a B_{12}-synthesing microbe, as essential for growth in defined medium. The Co(II) affinity for transport is likely similar to that for CbiMNQO based on published data.[108]

A single polypeptide transporter NhlF that is homologous to the Nix-A/HoxN/HupN family (Section 14.2.1.1) is required for the activity of cobalt-dependent nitrile hydratases. Transport assays suggest a K_m of 10^{-7} to 10^{-8} M.[109] Mutagenesis studies have identified residues that contribute to selectivity and capacity.[110]

The CorA membrane protein is a member of the divalent metal ion transporter family (DMT or DMIT). It functions as an allosterically regulated channel for magnesium ($K_m = 1.5 \times 10^{-5}$ M). While Co(II) can enter ($K_m = 2 \times 10^{-6}$ M) the cell *via* this route, it is easily out-competed by Mg(II) ions and the Co(II) concentrations required are non-physiological. Studies of *Thermotoga maritima* CorA suggest a selectivity for Co(II) over Mg(II). However, there is no evidence yet of an essential role for CorA in *T. maritima* cobalt physiology. Neither has the affinity of this CorA protein for Co(II) been shown to be in the range of environmental cobalt levels. These could be high at the thermal vent locations where *T. maritima* can grow.

14.2.3 Cobalamin Uptake

A large number of organisms must acquire cobalamin from dietary sources or the environment, a particular challenge in the ocean. The exceptions are the limited number of microbes that can synthesize cobalamin *de novo* and those organisms that do not use B_{12}-dependent enzymes. The mechanism of B_{12} internalization in mammals is mechanistically more complex than for microbes because of the need to circulate the nutrient and, therefore, there is a requirement for B_{12} to bind in different environments and be transported across multiple membranes. Cobalamin uptake across membranes in mammals occurs *via* recognition of cobalamin trafficking proteins (Section 14.3.3).

14.2.3.1 Cobalamin Uptake in Prokaryotes – BtuB and BtuF

Cobalamin (Cbl) uptake relies on an ABC-transporter, using the solute binding protein to initiate the process. In *E. coli*, B_{12} is too large to diffuse through the porins and so must first traverse the outer membrane before it encounters the solute binding protein.[111] The BtuB receptor binds B_{12} molecules and the cofactor is subsequently internalized with energetic assistance from the TonB/ExbBD system. There is no basis for selectivity between different cobalamin forms (cyano, Ado, aqua) because this portion of the molecule is exposed to solvent (Figure 14.4A).[112]

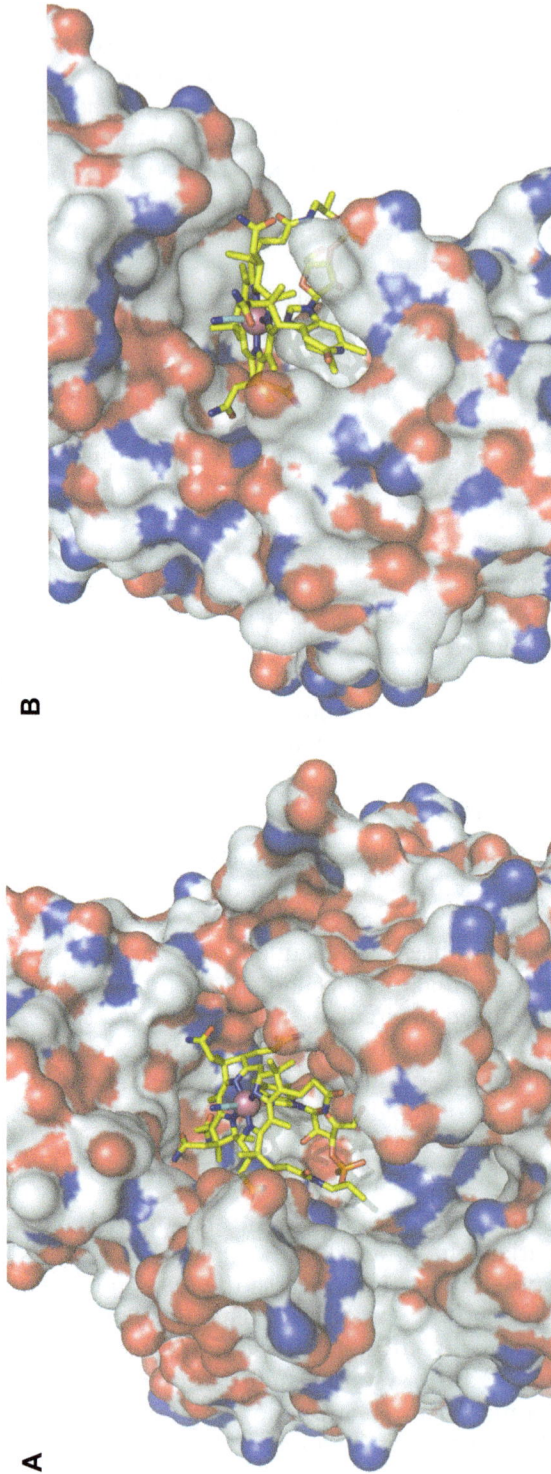

Figure 14.4 The microbial cobalamin uptake proteins. A. *E. coli* BtuB, an outer membrane protein receptor with CNCbl bound. The –CN group (cyan and blue) is exposed and solvent accessible. B. *E. coli* BtuF, the periplasmic solute binding protein with CNCbl bound. The –CN group (cyan and blue) protrudes into a large cavity. Proteins are represented as surfaces – C, grey; N, blue; O, red. The cobalamin molecules are shown as sticks – C, yellow; Co, pink; N, blue; O, red; P, orange.

Binding to the periplasmic protein BtuF is the first step in moving cobalamins across the inner membrane. B_{12} analogues bound to BtuF can be delivered to the BtuCD complex for ATP-driven import. As with BtuB, there is considerable flexibility in binding different cobalamins (Figure 14.4B). Both BtuB and BtuF bind to cobalamins with high affinity ($K_d = 10^{-10}$ to 10^{-8} M; Table 14.2).[113,114]

Marine diatoms were recently shown to express a protein in response to cobalamin limitation.[57] The protein, called CBA1, was not identified but a structural homology search using PHYRE[115] predicts high similarity to BtuB-like solute binding proteins.

14.3 Protein-mediated Transport within the Cell

Metal ions and metal-ion complexes are frequently escorted through the cell by proteins that more specifically direct the metal to its intracellular targets. These chaperone proteins, in principle, provide an escort from the cytoplasmic face of the uptake protein complex to the apo-enzyme active site. These molecular systems ensure efficient use of limiting quantities of the metal while preventing inappropriate interactions with other proteins and biomolecules and subsequent mis-metalation. In essence, they create a pathway for the flow of metal ions from the importer protein to the enzyme active site.

Nickel enzymes have long been known to utilize chaperone proteins for metal insertion. The complexity of the metal sites of some Ni-containing enzymes also demands an ordered, rather than random, assembly of metallocluster components, thereby providing an additional mechanistic requirement for chaperones. While the discovery of these proteins was made some time ago, it is only recently that their specific roles in metal cluster assembly have begun to be elucidated. This in part is due to the difficulty of reconstituting these systems *in vitro* with all the components present as well as generating stable, soluble versions of individual proteins. Cobalamin trafficking proteins were identified because of the direct link to cobalamin-deficiency in some individuals.

14.3.1 Nickel Trafficking and Enzyme Assembly

The nature of the accessory proteins required to efficiently assemble a functional metalloenzyme active site has been studied particularly well for two nickel enzymes – hydrogenase and urease.[116,117] Both enzymes require proteins of similar function – a Ni(II)-binding protein, an NTP-hydrolysing protein and additional proteins that help to expose the apo-enzyme active site. The recent availability of purified components for the assembly of both enzymes is significantly improving our understanding of the molecular steps required for Ni-insertion. Notably, there appears to be at least one handoff between these proteins before delivery to the enzyme active site and NTP hydrolysis is coupled to one of the metal transfer steps. It is likely that some nickel trafficking proteins remain to be discovered. For example, a nickel

Table 14.2 Ni(II)- and Co(II)-binding site features of proteins required for transport, trafficking and storage.

	Organism	Affinity (K_d, M)	Metal-site features[a]		Method[b]
Ni(II)					
Uptake					
NikA	E. coli	7.5×10^{-8} for Ni•(L-His)$_2$	Six-coordinate		4I8C
			His$_{416}$ Nε	2.4	
			His$_A$ NH$_2$	2.2	
			His$_A$ Nδ	2.2	
			His$_A$ COO$^-$	2.0	
			His$_B$ NH$_2$	2.4	
			His$_B$ Nδ	2.5	
NikA	S. aureus	nr[c]	Six-coordinate		3RQT
			His$_A$ NH$_2$	2.1	
			His$_A$ Nδ	2.1	
			His$_A$ COO$^-$	2.1	
			His$_B$ NH$_2$	2.1	
			His$_B$ Nδ	2.1	
			His$_B$ COO$^-$	2.1	
H$_2$ase assembly					
HypA/HybF	H. pylori	1.1×10^{-6}	Six-coordinate		XAS
	E. coli	7.5×10^{-8}	1 Imid (His$_2$)	2.04	
			2 N/O	2.04	
			3 N/O	2.15	
HypB	Site 1	5.0×10^{-5}	Planar		2HF8
			Cys	2.17	
			1 Imid	2.07	
			4 N/O	2.07	
	Site 2 (E. coli)	1.3×10^{-13}	Planar		XAS
			-NH$_2$	1.87	
			Cys$_2$ S	2.17	
			Cys$_5$ S	2.17	
			Cys$_7$ S	2.17	
SlyD	T. thermophilus	nr	Square pyramidal		3CGM
			His$_{145}$ Nδ	2.2	
			His$_{147}$ Nε	2.3	
	E. coli	$\leq 1 \times 10^{-10}$	His$_{149}$ Nε	2.2	
			His$_{155}$ Nε	2.3	
			His$_{157}$ Nε	2.0	
HybD (HycI)	E. coli	nr	Trigonal bipyramidal		1CFZ
			Glu$_{16}$ Oε$_1$	2.2	
			Glu$_{16}$ Oε$_2$	2.5	
			Asp$_{62}$ Oδ$_1$	2.5	
			Asp$_{62}$ Oδ$_2$	2.8	
			His$_{93}$ Nε	2.3	
Urease assembly					
UreE	H. pylori	1.5×10^{-7}	Trigonal bipyramidal		3TJ8
			His$_{102}$ Nε	2.1	
			His$_{152}$ Nε	1.9	
			His$_{102'}$ Nε	2.2	
			HOH	2.1	
			Glu$_{4''}$ Oε$_2$	2.1	
CODH					
CooC	C. hydrogenoformans	nr	Tetrahedral		3KJI (ADP bound)
			Cys$_{112}$ Sγ	2.1	
			Cys$_{114}$ Sγ	2.0	
			Cys$_{112'}$ Sγ	2.1	
			Cys$_{114'}$ Sγ	2.0	

Table 14.2 (*Continued*)

	Organism	Affinity (K_d, M)	Metal-site features[a]		Method[b]
			Cys_{112} Sγ	2.3	3KJH (no
			Cys_{114} Sγ	2.5	ADP)
			$Cys_{112'}$ Sγ	2.2	
			$Cys_{114'}$ Sγ	2.6	
Co(II)					
CbiK	*D. vulgaris*	nr	Four-coordinate		2XVZ
			His_{154}Nε	2.0	
			His_{216}Nε-Cε	2.0	
			HOH	2.2	
			Peroxo O_1	1.8	
			(Peroxo O_2	2.6)	
AnhE	*R. jostii*	$K_1 = 1.2 \times 10^{-10}$	Octahedral		*UV-vis*
		$K_2 = 1.1 \times 10^{-7}$			

[a]Ligand atom identity is listed for crystal structures. The likely amino acid identity for XAS studies also includes sequence homology and mutagenesis results when reported. All distances reported in Å.
[b]PDBID numbers from www.rcsb.org, where appropriate; XAS, X-ray absorption spectroscopy; UV-vis, UV-visible spectroscopy.
[c]Not reported.

chelatase has been proposed for the F_{430} biosynthetic pathway and candidates have been suggested based on gene conservation between methanogenic Archaea.[118] Quantitative and structural features of the Ni(II)-binding sites are listed in Table 14.2.

14.3.1.1 *[Ni-Fe] Hydrogenase*

Microbes express one or more hydrogenase isoenzymes. There are both general and specific isoenzyme assembly proteins that are expressed from operons.[119] In *E. coli*, which has multiple hydrogenases (Hya/Hyb/Hyc), the HypBCDEF proteins are general to all isoenzymes. HypCDEF have roles in insertion of the essential iron atom, and in providing diatomic ligands (CO and CN^-) that coordinate the Fe atom.[116] HypB is required for the subsequent Ni-insertion process, along with the isoenzyme specific HypA or HybF proteins. More recently, the SlyD protein was demonstrated to increase the efficiency of hydrogenase assembly. Finally, an isoenzyme-specific C-terminal endoprotease, HyaD/HybD/HycI, engages in a nickel-specific cleavage reaction that allows formation of the correct nickel-coordination site in the enzyme.

HypA (HybF) is a small, dimeric protein (~110 amino acids). All HypA proteins contain a Zn(II) ion that is coordinated by four conserved Cys residues.[120] *E. coli* HypA associates with the large structural subunit of the hydrogenase enzyme (HycE), and is thus thought to aid in nickel-transfer to the large subunit from the HypB and SlyD proteins. Consistent with this model,[121] nickel-free HypA can accept Ni ions from HypB in its GDP-bound form.[122] Purified *H. pylori* HypA binds a single Ni(II) atom in an octahedral environment.[123] The affinity of this site shows some species-to-species variation ($K_d = 7.5 \times 10^{-8}$ M and $K_d = 1 \times 10^{-6}$ M for *E. coli* and *H. pylori*, respectively),[122,124] which may be linked to functional roles in different cytosol environments. An NMR structure of *H. pylori* HypA proposed a four-coordinate, planar Ni-coordination geometry. However, the protein employed contained two additional non-native N-terminal residues (Gly-Ser) that resulted in a sequence reminiscent of the amino-terminal Cu(II) and N(II) (ATCUN) binding motif of serum albumin proteins,[125] known to provide a four-coordinate binding site. Consequently, the observed planar geometry is unlikely to be physiologically relevant.[126]

Interestingly, *H. pylori* HypA also participates in nickel insertion into the urease enzyme. HypA is known to form heterodimers with the UreE assembly proteins.[127,128] Experimental data suggest this HypA protein may utilize a pH-dependent conformational switch to enable the protein to participate in urease activation in order to combat acid stress. Under the acidic pH 6.3, the Ni:HypA stoichiometry decreases to 1 : 2, with small changes in the Ni(II) coordination sphere. The Zn(II) site undergoes more substantial changes and gains two imidazole (His) ligands at the expense of two Cys residues. Both lower pH and Cys to His mutations to mimic the Zn(II)-coordination sphere at lower pH decreased the affinity for Ni(II), $K_d = 1.7$–2.0×10^{-5} M. This value is weak and suggests that in the cell low pH may liberate Ni from HypA.

HypB was initially identified as a GTPase based on sequence homology and activity assays.[129] HypB possesses a low-affinity Ni binding site ($K_d = 5.0 \times 10^{-5}$ M) near the GTP-binding site (Figure 14.5A and B) at the interface between two monomers. This site is four-coordinate with two Cys residues coming from each subunit, likely similar to CooC (Section 14.3.1.4 and Figure 14.5E). The crystal structure contains bound Zn(II) at this site. However, experimental studies with the *E. coli* HypA-HypB pair show that Ni(II), but not Zn(II), can be transferred from pre-loaded HypB to apo-HypA.[122]

E. coli HypB contains a second, high-affinity Ni(II)-binding site at its N-terminus ($K_d = 1.3 \times 10^{-13}$ M) that is not conserved in many HypB proteins.[130] Experimental data suggest this site plays a regulatory role in HypB activity with Ni-occupancy required for full function. Nickel binding to this motif (NH_2-CysXaaXaaCysXaaCys) can be replicated in a small peptide and also fused to other protein domains,[131,132] indicating that its intrinsic metal-binding properties do not depend on context.

SlyD was first identified as an *E. coli* protein required for proline isomerization in a bacteriophage protein.[133] Its participation in hydrogenase assembly emerged after it was found to associate with HypB in a screen for *E. coli* protein-protein interactions.[134,135] The association of SlyD with HypB results in the transfer of Ni(II) from SlyD to HypB and the subsequent hydrolysis of GTP.[136] This suggests that SlyD primes HypB for nickel transfer to HypA. This activity requires a His- and Cys-rich C-terminal domain, despite the ability of SlyD to bind Ni(II) when lacking this domain ($K_d = 6.5 \times 10^{-7}$ M; octahedral coordination).[137] This implicates the C-terminal domain in the metal transfer reaction between the two proteins. Full-length *E. coli* SlyD

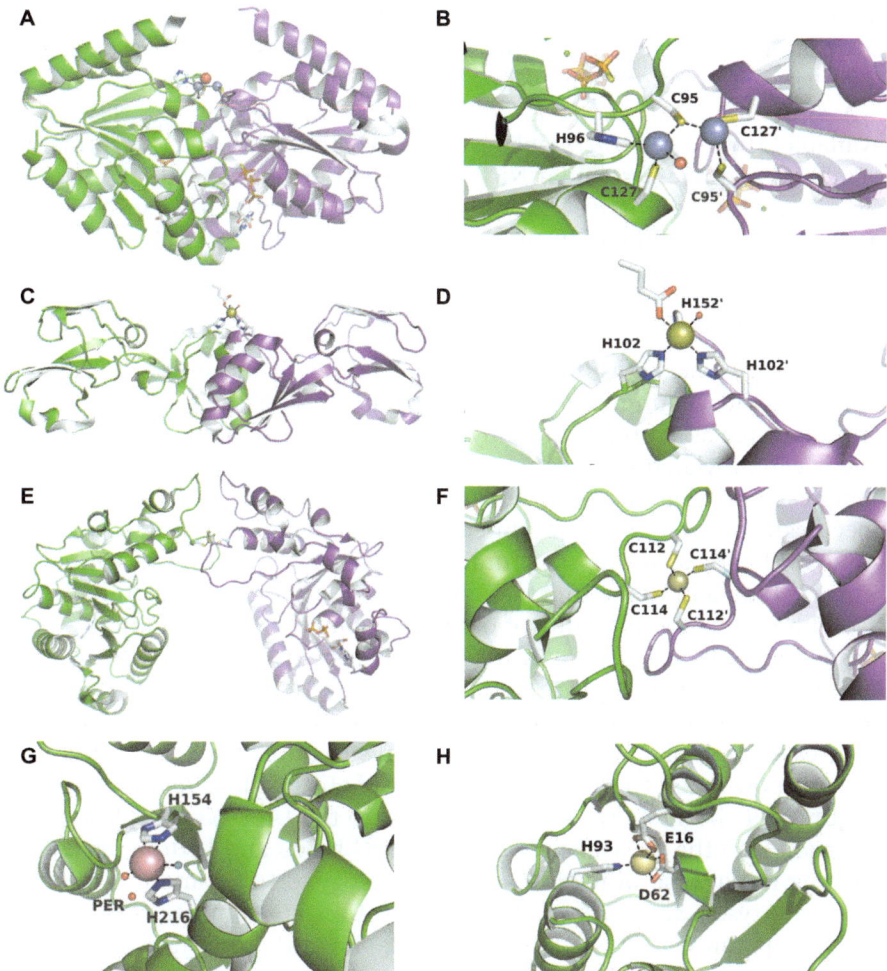

Figure 14.5 The structures of Ni(II)- and Co(II)-binding sites in metal trafficking and metalloenzyme assembly. Binding site residues are shown as sticks and balls – C, grey; Co, pink; N, blue; Ni, gold; O, red; P, orange. A. The structure of the *M. jannaschii* HypB dimer. The bound ATPγS is shown in stick representation. B. Close-up of the metal binding site at the dimer interface. Bound Zn(II) ions are shown in purple. C. The *H. pylori* UreE dimer. The glutamate residue from the upper left comes from a monomer chain that is nearby as a result of crystal packing. D. Close-up view of the UreE Ni(II) binding site at the interface between two dimers. E. The *C. hydrogenoformans* CooC dimer, the bound ADP is visible as a stick in the lower right and upper left. F. Close-up view of the proposed Ni(II)-binding site in CooC. G. Structure of the Co(II)-binding site in the *Desulfovibrio vulgaris* CbiK cobaltochelatase. The cyan sphere corresponds to the O of a water molecule, the red spheres correspond to the O atoms of a peroxy group. H. The structure of the proposed Ni(II) coordination environment of the *E. coli* HybD hydrogenase-specific endoproteinase.

binds multiple Ni(II) ions per polypeptide with a variety of coordination environments and competition experiments suggest affinities of some of these sites are tighter than 10^{-10} M.[138]

A structure of an intact *E. coli* SlyD protein with bound Ni(II) is not available. However, *T. thermophilus* SlyD with an introduced C-terminal His$_6$ tag, shows square-pyramidal Ni(II)-coordination (PDBID 3CGM). It is possible that this non-native coordination may resemble a Ni-species in the native protein at some point in the Ni(II)-transfer process.[139]

14.3.1.2   Urease

Like hydrogenase, active urease contains a divalent nickel site but the additional ligand requirements are less complex. Four genes (*ureDEFG*) are important for assembling the fully active enzyme. An active site lysine residue is carbamylated by CO_2 prior to metal insertion to provide a negatively charged carboxylate that bridges the two metal ions. Recent reviews focusing specifically on urease have outlined the assembly process in great detail,[140] along with advances in structural characterization of the assembly proteins.[141] Elucidating the functions of these latter proteins has until recently been limited by the difficulty of obtaining soluble, pure protein.

UreE was identified as a Ni(II)-binding protein ($K_d = 1.5 \times 10^{-7}$ M and $K_d = 5 \times 10^{-6}$ M for *H. pylori* and *K. aerogenes*, respectively). The *Klebsiella aerogenes* UreE has a long His-rich C-terminal extension that is not essential for urease assembly.[142] Several UreE crystal structures have been determined and all show metal coordination by a pair of His residues (Figure 14.5C and D). UreE has been shown to deliver nickel to the UreDFG complex, prior to insertion into the urease active site.[143] UreG binds and hydrolyses GTP, indicating it may also bind Ni(II) ions by analogy to HypB and CooC (below). A recent crystal structure of the *H. pylori* UreDFG complex (PDBID 4HI0) reveals a potential 2 Cys/2 His Ni-binding site at the interface between two UreG monomers (Cys-Pro-His motif in each monomer) that are buried in the complex.[144] This site is reminiscent of those observed in HypB and CooC (Figures 14.5B and G). Substitution of Cys or His by Ala abolishes Ni-dependent effects on UreG dimerization, which is required for urease activation. It seems unlikely that any nickel transfer between UreE and UreG uses an intermediate not yet observed structurally. The recent availability of expressed versions of the individual chaperone proteins[145] will aid in understanding the individual Ni transfer reactions that occur before final insertion into the apo-enzyme active site.

14.3.1.3   Additional Nickel-binding Proteins in H. pylori

The unique acidic growth niche of *H. pylori* in the stomach has resulted in the emergence or modification of several additional proteins with nickel-specific functions based on observed effects of gene deletions on urease or hydrogenase activity. The *H. pylori* Hpn proteins are a specific class of His-rich proteins that also have a high proportion of glutamine residues. Their

Ni-binding properties are complex. They bind multiple nickel atoms that promote oligomerization. The strong coupling of these two activities makes it difficult to assess stepwise metal binding constants. An average dissociation constant for Ni(II) (7×10^{-6} M) has been reported for a mixture of oligomeric states up to 20 subunits.[146] Recent studies with an Hpn peptide suggest that increasing glutamine content increases the affinity for Ni.[147] HspA is a homologue of the well-known GroES protein, a folding chaperone. HspA contains a C-terminal His/Cys sequence and the Cys residues are critical for Ni binding. Interestingly, it is important for Ni-Fe hydrogenase activity in *H. pylori* but not for urease activity.[148] *H. pylori* Mua (HP0868) was recently shown to weakly bind two equivalents of nickel ($K_d \sim 2 \times 10^{-5}$ M).[149] The presence of Mua correlates with reduced urease expression, suggesting a role for modulating urea hydrolysis under conditions where it could be deleterious to the cell. The mechanism by which Mua acts has not been elucidated, but it appears likely to require Ni binding.

14.3.1.4 Carbon Monoxide Dehydrogenase (CODH)

Several proteins (CooCTJ) have a role in providing Ni(II) for CODH cluster assembly in *Rhodospirillum rubrum*.[150] CooC has a P-loop nucleotide-binding motif, analogous to HypB and UreG. CooJ (12.5 kDa) has a His-rich motif in the C-terminal portion of its sequence (15 of the last 30 residues). The protein binds four Ni(II) per monomer ($K_d = 4.3 \times 10^{-6}$ M).[151]

Carboxydothermus hydrogenoformans CooC has been studied both structurally and functionally. Purified CooC forms dimers in the presence of Ni(II), with a stoichiometry of 2 CooC : 1 Ni(II) ($K_d = 4.1 \times 10^{-7}$ M).[152] Sequence analysis and UV-visible spectral features were consistent with Cys-Ni coordination *via* a CXC motif (one from each dimer). Mutation of the Cys residues eliminated the spectral change observed upon Ni binding. The structure of a Zn-bound form of CooC has been determined and is consistent with the nickel coordination observed in solution, as well as the role of Ni(II) in CooC dimerization (Figure 14.5F and G).[153]

Ni-insertion into acetyl CoA synthase is little studied. A putative HypB homologue, AcsF from *Clostridium thermoaceticum*, has been identified owing to the presence of an NTP binding domain.[154] However, a specific role in Ni binding or insertion could not be demonstrated.

14.3.1.5 Proteolytic Processing and Nickel-binding Site Formation

Two nickel enzymes, hydrogenase and NiSOD, undergo proteolytic processing to form the coordination environment required for enzyme activity.

The HypA-HypB-SlyD protein complement is necessary to target nickel ions to hydrogenase but one further nickel-dependent step is required to form the active enzyme. The peptidase HycI recognizes the Ni(II)-loaded form of the large subunit and subsequently removes the last 15 residues.[155]

HycI is specific for the H_2-evolving hydrogenase and two other orthologues are present in *E. coli* to deal with the H_2-oxidizing isoenzymes. A crystal structure of HybD bound to Cd(II) has been determined (Figure 14.5H),[156] and on this basis it has been inferred that the protease uses Asp_{62} to interrogate the metal identity and thus be selective for Ni(II) insertion. HycI may also use the Ni(II) to activate water for hydrolysis, although Asp_{62} has been shown to be essential for this process.[157] It is also possible that the position of the Cd ion does not represent a normal part of the HycI-large subunit interaction during hydrolysis.[158]

NiSOD is not known to require a chaperone for metal-ion delivery but requires N-terminal processing to be active. A second gene in the NiSOD operon, encoding a protein denoted by SodX, was identified as essential for normal processing of the enzyme in the native microbial species.[159] Little is known about the structural basis for processing and, in particular, whether SodX would directly contact the nickel ion, as has been proposed for HycI in hydrogenase processing. Structurally, the two endopeptidases are unrelated. A SodX structure has not been reported, but threading prediction using the PHYRE web server[115] suggests that SodX is similar in structure to the signal peptidase and the bacterial proteins, LexA and UmuD, which undergo self-catalysed site-specific hydrolysis in response to cellular stress.

14.3.2 Cobalt Trafficking

There are far fewer examples of microbial Co(II) trafficking proteins than are known for Ni(II), reflecting the lower number of Co(II)-dependent enzymes.[19]

14.3.2.1 Cobalamin Biosynthesis – CbiK Cobalt Chelatase

There are two mechanistically distinct pathways used for cobalamin biosynthesis.[160,161] These pathways differ in the chemical steps used for tetrapyrrole synthesis, and also differ in the point at which Co(II) is inserted. The Cbi pathway inserts Co(II) with additional modifications of the heterocycle structure still required. The Cob pathway inserts Co(II) much later when only the two axial ligands are left to be added. The Cbi and Cob pathways are often referred to as anaerobic and aerobic, respectively, as a result of the oxygen sensitivity of the organisms in which they were first identified.

In either case, Co(II) insertion requires the action of a chelatase enzyme, CbiK or CobNST. CbiK has been characterized structurally, using proteins from different organisms and with different substrates and products bound.[162] Co(II) is coordinated by two His side-chains, and an unusual sideways coordination by an imidazole N-C bond of one of these was proposed (Figure 14.5G). It was suggested that this coordination mode facilitates proton transfer during cobalt insertion.

CobNST is heterotrimeric, and requires ATP-hydrolysis for Co(II) insertion. No structures have been reported for this Co-bound protein, but a Michaelis constant for Co(II) has been reported ($K_m = 5$ μM).[163] Ni(II), Cu(II) and Zn(II) inhibited transfer of Co(II) to CobNST at the 50% level at concentrations ranging from 0.5 to 3 µM. However, none of the metals was inserted. This observation indicates that conformational changes linked to the electronic structure of the bound metal are required for successful insertion of the metal, rather than selectivity associated with metal binding to the chelatase.

14.3.2.2 Nitrile Hydratase Assembly

Cobalt-dependent nitrile hydratases (NHases) are found in a limited number of microbes. Accessory proteins are known for each of nitrile hydratase and acetonitrile hydratase (ANHase). They use different mechanisms for Co(II) insertion.

NhlE is required for cobalt insertion into the low-molecular-weight NHase from *Rhodococcus rhodochrous*.[164] Interestingly, NhlE forms a heterotrimeric, cobalt-containing complex with one subunit of the enzyme (NhlA:NhlE$_2$) and then undergoes exchange with the other catalytic subunit (NhlB) to form the active enzyme.[164] There is no information available on the affinity for cobalt(II) or the nature of the coordinating residues in the precursor complex. A similar protein, NhhG, is required for cobalt insertion into the high molecular weight NHase from *Rhodococcus rhodochrous*.[165] The oxidation of Co(II) to Co(III) occurs after insertion and appears to be O$_2$-dependent.[166]

AnhE is required for Co(II) insertion into *R. jostii* ANHase.[49] Purified AnhE (11.1 kDa polypeptide) exists in a monomer–dimer equilibrium (100–500 µM protein concentration). Cobalt binding drives dimerization and appears to be negatively cooperative ($K_{d1} = 1.3 \times 10^{-10}$ M and $K_{d2} = 1.4 \times 10^{-7}$ M). The visible spectrum is consistent with a six-coordinate site. The coordinating residues are unknown, although the lone His residue in the protein does not appear to be important.

14.3.3 Mammalian Cobalamin-trafficking Proteins

Cobalamin trafficking in mammals requires passage through several physiological environments and cellular locations to reach the target enzymes. Three B$_{12}$-binding proteins play roles in trafficking in mammals – haptocorrin (HC), intrinsic factor (IF) and transcobalamin (TC).[20,167] The proteins are similar in size (~45 kDa) and show primary sequence and 3D structural homology. They are, however, differently decorated by carbohydrates, a feature that affects their biological stability and also aids in receptor recognition. The proteins all exhibit very tight binding to Cbl molecules (Table 14.3), which means that Cbl release is a rate-limiting step and can be controlled by changes in physiological conditions (pH) or *via* regulated proteolysis.

14.3.3.1 Haptocorrin/Intrinsic Factor/Transcobalamin

HC binds to cobalamins ($K_d = 1 \times 10^{-14}$ M for MeCbl) as they become available in the digestive tract. It is present in the saliva, but likely encounters few molecules of Cbl liberated from food. The interaction between HC and

Table 14.3　Cobalamin-binding molecules in uptake, trafficking and regulation.

Macromolecule	Origin	Selectivity^a	K_d (M)	PDBID
Uptake				
BtuB	Bacteria	R-Cbl	3×10^{-10}	1NQH
BtuF	Bacteria	R-Cbl	1.5×10^{-8}	1N4A
Trafficking				
Haptocorrin	Mammal	R-Cbl	1×10^{-14}	–
Intrinsic factor	Mammal	AquaCbl	1×10^{-12}	3KQ4
Transcobalamin	Mammal	AquaCbl	5×10^{-15}	2BB5
MMAHCC (cobalamin processing)	Mammal	R-Cbl	–	3SC0
Regulation				
btuB riboswitch	Bacteria	AdoCbl	2.5×10^{-7}	4GMA
btuB riboswitch	Bacteria	AdoCbl	–	4GXY
*env4*AqCbl riboswitch	Metagenome	AquaCbl	1.9×10^{-8}	4FRG
*env8*AqCbl riboswitch	Metagenome	AquaCbl	7.5×10^{-9}	4FRN

^{a}Abbreviations are listed in Figure 14.1. R-Cbl, no apparent selectivity.

cobalamins is stable under acidic conditions, indicating a protective role for the HC in the stomach when Cbl liberation is maximized. The biological stability of HC decreases as the pH rises in the duodenum, and results in the release of Cbl molecules after proteolysis. IF (Figure 14.6A) binds to Cbl molecules $(K_d = 1 \times 10^{-12}$ M) in the intestine after their release from HC. The IF-Cbl complex is recognized by a cell-surface receptor called cubulin for uptake into the epithelial cells lining the digestive tract. TC is largely responsible for distribution of Cbl to tissues *via* the circulatory system. It coordinates the cobalt ion directly *via* a His residue (Figure 14.6B). This requirement excludes binding to any species of Cbl other than AqCbl $(K_d = 5 \times 10^{-15}$ M) due to the energetic barrier to displacing ligands with a C-Co linkage (Ado and CN).

14.3.3.2　Cobalamin Processing Enzyme (MMAHCC)

Cobalamins often require processing (*e.g.* demethylation or decyanation) in order to be converted into their active forms (*e.g.* AdoCbl). In humans, this activity is carried out by the MMAHCC protein. The structure of MMAHCC with bound cobalamin has been determined (Figure 14.6C), and shows the displacement of the dimethylbenzimidazole ligand in the bound cofactor.[168] This conformation activates the cobalt atom for the reductive decyanation reaction.

14.4　Sensing, Buffering and other Control Mechanisms

The presence of transcriptional regulatory mechanisms indicates that at some point in their lifecycle, microbes face metal ion stress even when growing under metal-limiting conditions.[169] A switch from aerobic to an-aerobic metabolism or a change in available energy sources will affect the

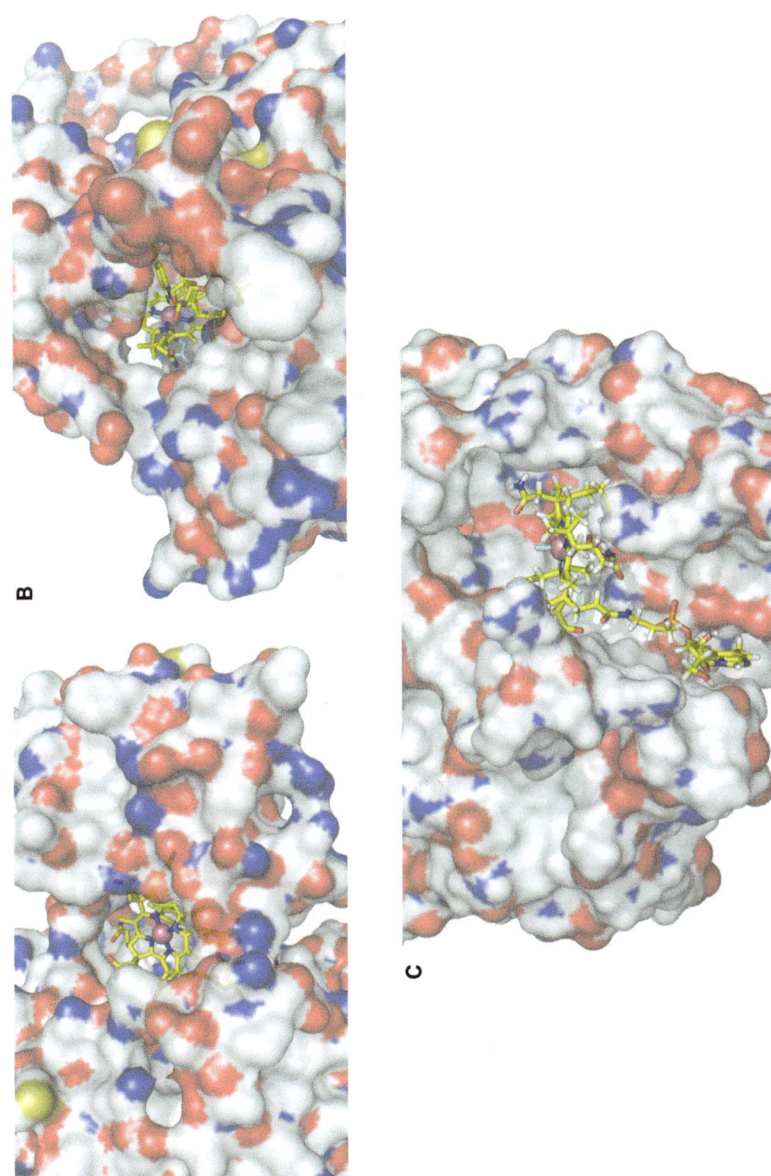

Figure 14.6 The structures of cobalamin binding proteins required for transport. A. View of cobalamin bound to *H. sapiens* Intrinsic Factor. An axial R group was not present in this structure. B. View of cobalamin bound *H. sapiens* Haptocorrin, the C atoms of the His residue from haptocorrin that coordinates the Co atom are coloured cyan. C. View of cobalamin bound to *H. sapiens* Methylmalonic Aciduria and Homocystinuria Type C Protein (MMACHC). The cyan stick represents a C atom from a –CH₃ group. Proteins are represented as surfaces – C, grey; N, blue; O, red; S, yellow. The cobalamin molecules are shown as sticks – C, yellow; Co, pink; N, blue; O, red; P, orange.

proteome and thus the metallome. Transcriptional regulators are a well-studied class of metalloproteins from a structural perspective. Prokaryotic transcriptional metalloregulatory proteins fall within several structural classes.[170] DNA binding affinity is controlled allosterically by metal binding so that the metal does not directly mediate a protein-DNA interaction.[171] For metals with higher affinities in the Irving–Williams series (Co, Ni, Cu, Zn), the correct metal coordination geometry is coupled to an allosteric response so that, even if other metals can bind with higher affinity, they are not capable of eliciting the correct protein conformational change to regulate DNA-binding affinity. Studies of nickel-responsive regulators, in particular NikR and RcnR from *E. coli*,[172,173] have contributed to these advances in understanding molecular function. Features of the metal-binding sites of these proteins are described in Table 14.4. Additional regulators for which structural information has not yet been reported are reviewed elsewhere.[17]

Table 14.4   Ni(II) and Co(II) binding site features in proteins involved in transcriptional regulation.

	Organism	K_d (M)	Metal-site features[a]		Method[b]
Ni(II)					
NikR	*E. coli*	5.0×10^{-13c}	Planar		2HZV
	H. pylori		$His_{76'} N\varepsilon_2$	1.91	2Y3Y
	P. furiosus		$His_{87} N\varepsilon_2$	1.91	2BJ1
	(high-affinity)		$His_{89} N\delta_1$	1.91	
			$Cys_{95} S\gamma$	2.13	
	E. coli	3.0×10^{-8}	Six-coordinate		XAS
	(low-affinity)		2His	1.99	
			4 N/O	2.11	
	H. pylori	nr[d]	Six-coordinate		2CAD
	(alternate site)		$His_{74} N\varepsilon_2$	2.1	3QSI
			$His_{88'} N\varepsilon_2$	2.3	
			$His_{101} N\varepsilon_2$	2.4	
			HOH	2.0	
			HOH	2.1	
			HOH	2.3	
	H. pylori	nr	Six-coordinate		2CAD
	(alternate site)		$His_{74} N\varepsilon_2$	2.1	
			$His_{75'} N\varepsilon_2$	2.3	
			$Glu_{104} O$	2.4	
			Citrate O	2.0	
			Citrate O	2.1	
			Citrate O	2.3	
	P. horikoshii	nr	Square pyramidal		2BJ8
	(alternate site)		$His_{65} N\varepsilon_2$	2.4	
			$Asp_{65} O\delta_1$	2.1	
			$Asp_{65} O\delta_2$	2.6	
			$Asp_{75} O\delta$	2.3	
			HOH	2.2	
Nur	*S. coelicolor*	nr	Planar		3EYY
			$His_{33} N\varepsilon_2$	2.2	
			$His_{86} N\delta_1$	2.6	
			$His_{88} N\varepsilon_2$	2.4	
			$His_{90} N\varepsilon_2$	2.2	

Table 14.4 (*Continued*)

	Organism	K_d (M)	Metal-site features[a]		Method[b]
RcnR	*E. coli*	2.5×10^{-8}	Six-coordinate		XAS
			Cys_{35} $S\gamma$	2.54	
			His_{60}	1.99	
			His_{64}	1.99	
			NH_2	2.11	
			2 N/O	2.11	
CnrX	*C. metallidurans*	nr	Six-coordinate		2Y39
			His_{42} $N\varepsilon_2$	2.10	
			His_{46} $N\varepsilon_2$	2.12	
			Glu_{63} $O\varepsilon_1$	2.20	
			Glu_{63} $O\varepsilon_2$	2.23	
			His_{119} $N\varepsilon_2$	2.16	
			Met_{123} $S\delta$	2.45	
NmtR	*M. tuberculosis*	9.1×10^{-11}	Six-coordinate		XAS
			Ni-N (3 His, 3 N/O)	2.08	
			Gly-NH_2		
			His_3		
			Asp_{91}		
			His_{93}		
			His_{104}		
			His_{107}		
Co(II)					
RcnR	*E. coli*	5×10^{-9}	Six-coordinate		XAS
			Cys_{35}	2.24	
			His_{60}	1.94	
			His_{64}	1.94	
			NH_2	2.11	
			2 N/O	2.11	
CnrX	*C. metallidurans*	nr	Six-coordinate		2Y3B
			His_{42} $N\varepsilon_2$	2.11	
			His_{46} $N\varepsilon_2$	2.11	
			Glu_{63} $O\varepsilon_1$	2.13	
			Glu_{63} $O\varepsilon_2$	2.16	
			His_{119} $N\varepsilon_2$	2.12	
			Met_{123} $S\delta$	2.54	

[a] ligand atom identity is listed for crystal structures. The likely amino acid identity for XAS studies also includes sequence homology and mutagenesis results when reported. All distances reported in Å.
[b] PDBID numbers from www.rcsb.org, where appropriate; XAS, x-ray absorption spectroscopy.
[c] value for *E. coli* NikR.
[d] not reported.

14.4.1 Nickel-responsive Transcriptional Regulation

14.4.1.1 NikR

NikR was the first nickel-dependent regulator of nickel uptake to be identified.[174,175] NikR orthologues are found in both bacteria and Archaea.[23] In some cases, NikR regulates Ni-enzyme expression in addition to controlling Ni-transporter expression.[176–180]

NikR is a modular, tetrameric protein composed of a dimeric N-terminal ribbon-helix-helix DNA-binding domain and a tetrameric C-terminal Ni(II)-binding domain (Figure 14.7A).[181,182] When expressed separately, each domain retains its basic function,[175,183] but they must be covalently linked to function properly. The primary Ni(II)-binding site (four per tetramer) is located at the interface between C-domain monomers.[181] The four-coordinate, planar site[184] is composed of His_{87}, His_{89} and Cys_{95} from one subunit and His_{76} from the subunit across the interface (Figure 14.7B). Metal-binding to this site is required for DNA-binding. Competition experiments in the presence of EGTA demonstrated a tight affinity for this site ($K_d = 5.0 \times 10^{-13}$ M).[183,185]

A second, lower affinity Ni-binding site in *E. coli* NikR was inferred from DNA-binding studies.[183] Titrating Ni(II) against NikR-DNA complexes indicated a lower affinity for this site ($K_d = 3.0 \times 10^{-8}$ M).[186] A combination of crystallographic, mutational and spectroscopic analyses has indicated that the lower affinity site is most likely octahedral, and uses the non-conserved His_{48} and His_{110} residues from NikR as well as solvent molecules.[187–189]

Additional nickel-binding sites have been observed in different NikR proteins. In particular, Ni(II) atoms have been observed binding at sites very near to the planar site (Figure 14.7C to E). These sites have higher coordination numbers and could represent intermediates in which Ni(II) atoms have been captured either entering or exiting the planar site. Alternatively, these sites may reflect differences in Ni(II) loading that enable selective binding to different DNA promoters. This is supported by mutation of residue His_{74} in HpNikR (equivalent to His_{62} in EcNikR). This residue is not part of the planar four-coordinate primary coordination sphere, but DNA-binding to one promoter region is substantially weakened by this mutation.[190]

14.4.1.2 Nur

Nur is a member of the Fur family of metalloregulators. It was discovered in *Streptomyces* as a regulator of superoxide dismutase (SOD) expression and nickel uptake.[191] A crystal structure revealed two distinct metal-binding sites (Figure 14.7G and H).[192] The first is a four-coordinate, planar site comprised of His ligands while the second is a six-coordinate site composed of protein ligands and an anion from the crystallization buffer. The positions of these two sites are consistent with metal site locations in other Fur family members. This homology indicates that the four-coordinate site is likely the primary nickel-sensing site while the six-coordinate site may provide an additional mechanism for nickel-sensing, analogous to that seen for EcNikR when bound to DNA.

14.4.1.3 NmtR/KmtR

Two distinct Ni sensors have been identified in *Mycobacterium tuberculosis*.[193,194] NmtR senses lower nickel concentrations than does KmtR. Both are members of the ArsR/SmtB family of metalloregulators. The biological

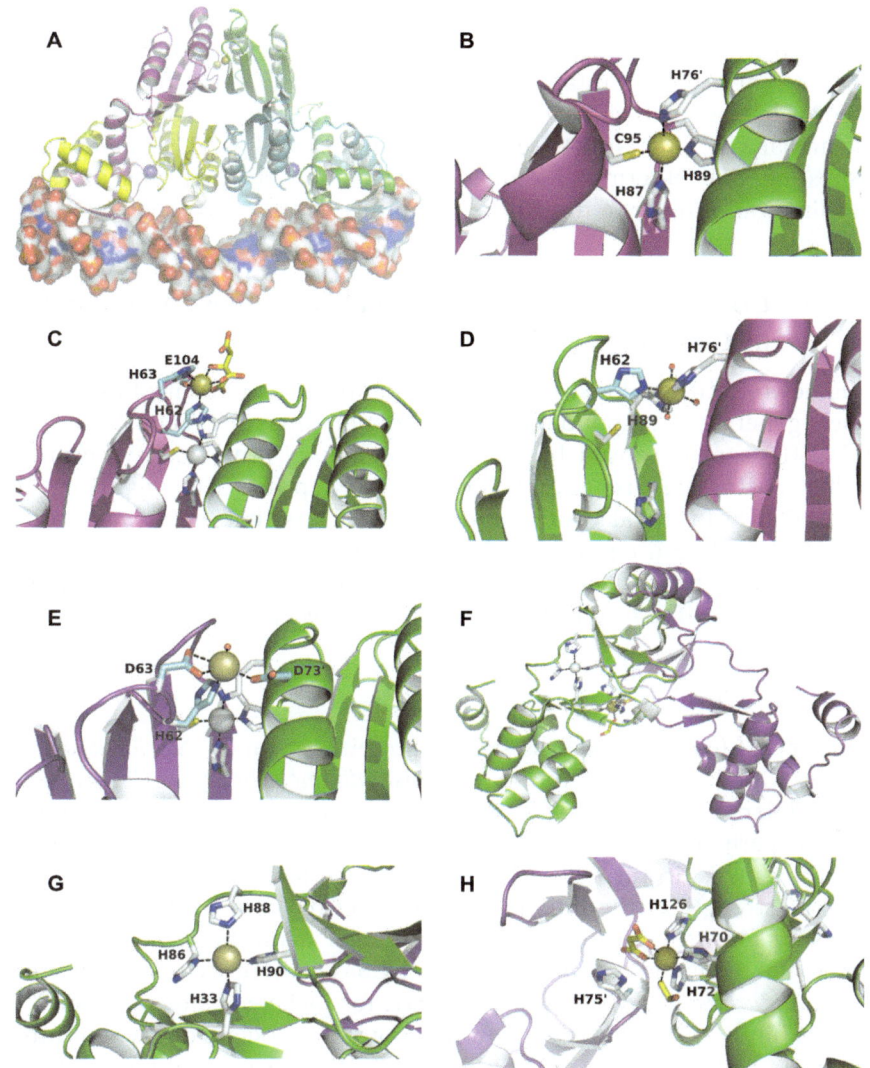

Figure 14.7 The structures of Ni(II) binding sites in transcriptional regulation used to control metal uptake and utilization. A. The structure of *E. coli* NikR tetramer bound to operator DNA. The Ni(II) atoms are visible as gold spheres. B. Close-up view of the square planar NikR binding site observed in all NikR proteins. C and D. Six-coordinate Ni(II)-binding sites observed in *H. pylori* NikR structures in the absence of DNA. The high-affinity square planar site can be seen in C. The Ni(II) site in D uses one of the His ligands from the square planar site, so that the site is no longer occupied. E. A five-coordinate Ni(II)-binding site observed in *P. horikoshii* NikR. This view is similar to that in panel C. Residue numbers are all normalized to EcNikR for ease of comparison. F. The structure of the *S. coelicolor* Nur dimer in the absence of DNA. G. View of the square planar Ni(II)-binding site in Nur. H. View of the six-coordinate bound Ni(II) atom in Nur.

necessity for having two regulators in the same cytosol is unclear but interesting, as it suggests a complex physiology centred over a wide range of intracellular total nickel concentrations. NmtR has been more extensively characterized. This six-coordinate site has high affinity for Ni(II) (picomolar K_d).[195,196] NmtR uses the N-terminus as a ligand, similar to RcnR, suggesting that N-terminal sequence extensions are a means of varying coordination number to discriminate between Ni(II)/Co(II) and Cu(I)/Zn(II).

14.4.2 Combined Cobalt- and Nickel-responsive Regulation of Gene Expression

The known Co(II)-responsive regulators that function to control metal efflux appear generally to respond biologically to more than one metal. Thus, Ni(II) and Co(II) can be sensed by the same regulatory proteins. Two examples of this sensing are discussed below. The overlap in selectivity arises from the similarity in metal coordination mechanisms that yields small differences in metal binding affinity. One protein, CoaR, appears to be selective towards Co(II), but little is known about its metal-binding site except that a Cys-His-Cys motif is implicated in metal responsiveness.[197]

14.4.2.1 CnrX

The CnrYXH (cobalt nickel resistance) protein system of *Cupriavidus metallidurans* CH34, formerly known as *Ralstonia metallidurans* CH34, senses extracellular nickel and cobalt ions.[198,199] CnrX contains two domains, a globular periplasmic portion and a transmembrane region that associates with CnrY. This, in turn, associates with CnrH. Metal binding to CnrX dictates the strength of the interaction between CnrY and CnrH. Free CnrH is necessary for RNA polymerase recognition of a promoter controlling expression of the metal resistance genes (*cnrCBAT*). There is a related system in *C. metallidurans* 31A that is known as *ncc*, for nickel-cobalt-cadmium.[200]

CnrX binds Ni(II) or Co(II) in a six-coordinate, octahedral environment (Figure 14.8 and Table 14.4). In contrast, Zn(II) is five-coordinate, inferring that the missing ligand (MetSδ) is a critical metal-contact for initiating the transcriptional response. Ni(II) or Co(II)-MetSδ coordination has not been observed in other proteins. Because CnrX is a periplasmic protein, the use of CysSγ as a coordinating ligand may be precluded because of the more oxidizing conditions of the periplasm.

14.4.2.2 RcnR

RcnR (resistance to cobalt and nickel) represses transcription of the gene encoding the RcnA efflux protein in the absence of Ni(II) or Co(II).[201] RcnR is structurally homologous to the Cu(I)-responsive CsoR protein (PDB code: 2HH7).[170,202]

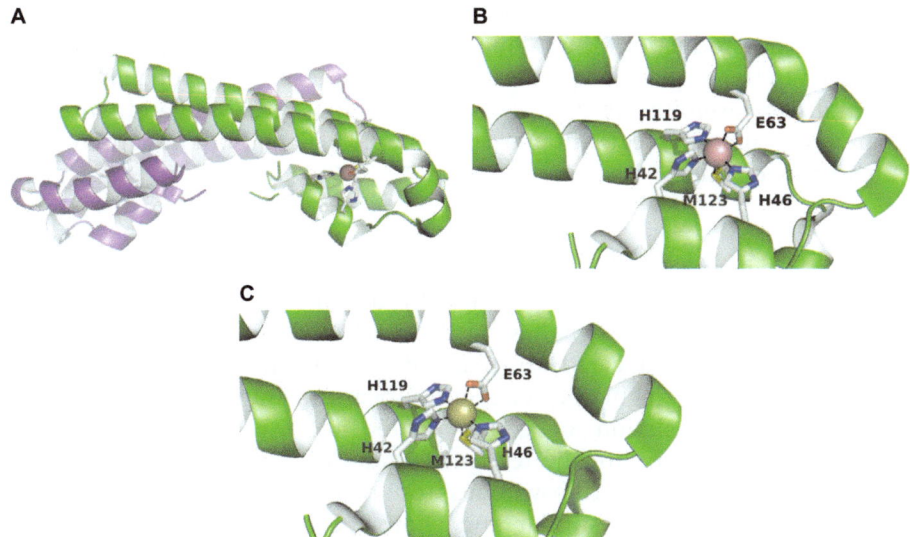

Figure 14.8 The structures of Ni(II) and Co(II) binding sites in regulation of the response to excess metal ions. A. The *C. metallidurans* CnrX periplasmic domain homodimer. There is one metal site in each monomer with no interchain coordination. B. The Co(II)/Ni(II)-binding site of CnrX. The binding site residues are represented by sticks – C, grey; N, blue; O, red; S, yellow. The metal shown here (pink sphere) is a Co atom.

Cobalt and nickel binding to RcnR have been studied extensively *via* mutational and spectroscopic methods.[203–205] Each metal binds with six-coordinate geometry (Table 14.4), but the two sites are not composed of an identical set of ligands, suggesting some flexibility in coordination. RcnR uses a conserved Cys residue to coordinate both Ni(II) and Co(II), but the Cys is not necessary for Ni-responsive regulation. Both metals utilize an N-terminal coordination motif that includes the N-terminal amine. Non-responsive metals [Cu(I/II), Zn(II)] are not able to coordinate *via* the N-terminal motif,[204] making it likely that the allosteric network is initiated by co-ordination to this portion of the protein. Competition experiments indicate that Co(II) and Ni(II) affinities are in the low nanomolar range (Table 14.4), consistent with a function in regulating efflux protein expression in response to a higher level of excess free metal ions when compared to NikR.

InrS is a structural homologue of RcnR reported in cyanobacteria and is involved in regulating nickel efflux. Based on UV-visible spectra, Ni(II)-coordination is likely to be four-coordinate planar. The affinity for Ni(II) ($K_d = 2.0 \times 10^{-14}$ M) is much tighter than RcnR.[206]

14.4.3 Cobalamin-responsive Regulation of Gene Expression

Cobalamin-dependent regulation occurs in two forms – the cobalamin-binding riboswitches in microbes and the photosensitive regulation of carotenoid biosynthesis in *Myxococcus xanthus*. In the first case, cobalamins regulate the production of full-length RNA transcripts that express proteins involved in the acquisition (*e.g.* BtuB) or biosynthesis of the cobalamins. The second case exploits the light sensitivity of AdoCbl as a ligand-switch for expression of gene regulation by the classical ligand-binding protein repressor mechanism.

14.4.3.1   B$_{12}$-responsive Riboswitches

Cobalamins regulate gene expression by binding to upstream mRNA sequences, known as riboswitches.[207] The *btuB* riboswitch is a ~200 nucleotide RNA sequence located upstream of the translational start site of the *btuB* gene.[208] This regulatory element appears in numerous bacteria and controls the translation of genes related to B$_{12}$ acquisition or biosynthesis.[209,210] It specifically recognizes AdoCbl, based on recently determined structures of *btuB* riboswitches from two thermophilic bacteria.[211,212] The structures are very similar and show specific base-base interactions between the RNA and the adenosyl moiety of AdoCbl to enhance selectivity (Figure 14.9). There are few intermolecular contacts between the RNA and the rest of the corrinoid molecules, suggesting that the remainder of ligand-riboswitch interaction is largely shape-driven.

A second type of cobalamin-responsive riboswitch has also been reported. The sequence was identified in a metagenomic sample from the Pacific Ocean. Interestingly, these switches are selective for aquaCbl and bind the molecule in a way that excludes AdoCbl interaction as the aqua group is buried at an interface between the switch and the ligand. It is not surprising that such a second type of cobalamin-binding riboswitch exists, as AdoCbl is light sensitive and would be in short supply in the surface marine environment due to conversion to aquaCbl.[82,83]

The riboswitch-cobalamin ligand interactions are tight (Table 14.3), with K_d values in the low to mid nanomolar range. The use of equilibrium binding affinities as a measure for understanding metabolite sensing and transcription control by riboswitches is debated.[207] Affinities acquired in this way frequently underestimate the concentration of metabolite required to block transcription in an *in vitro* assay. The selection of RNA conformation in sequences containing riboswitches is likely under kinetic control due to the competition between metabolite binding and continued transcription by RNA-polymerase. Thus, the riboswitch conformation is likely less abundant during active transcription, thus requiring higher cobalamin concentrations to repress RNA polymerase activity.

14.4.3.2   Light-dependent Gene Regulation Using Ado-Cbl

Cobalamins are known only to use the riboswitch mechanism for homeostasis. However, an AdoCbl-dependent protein-mediated mechanism of gene regulation unrelated to cobalamin biosynthesis was recently identified in

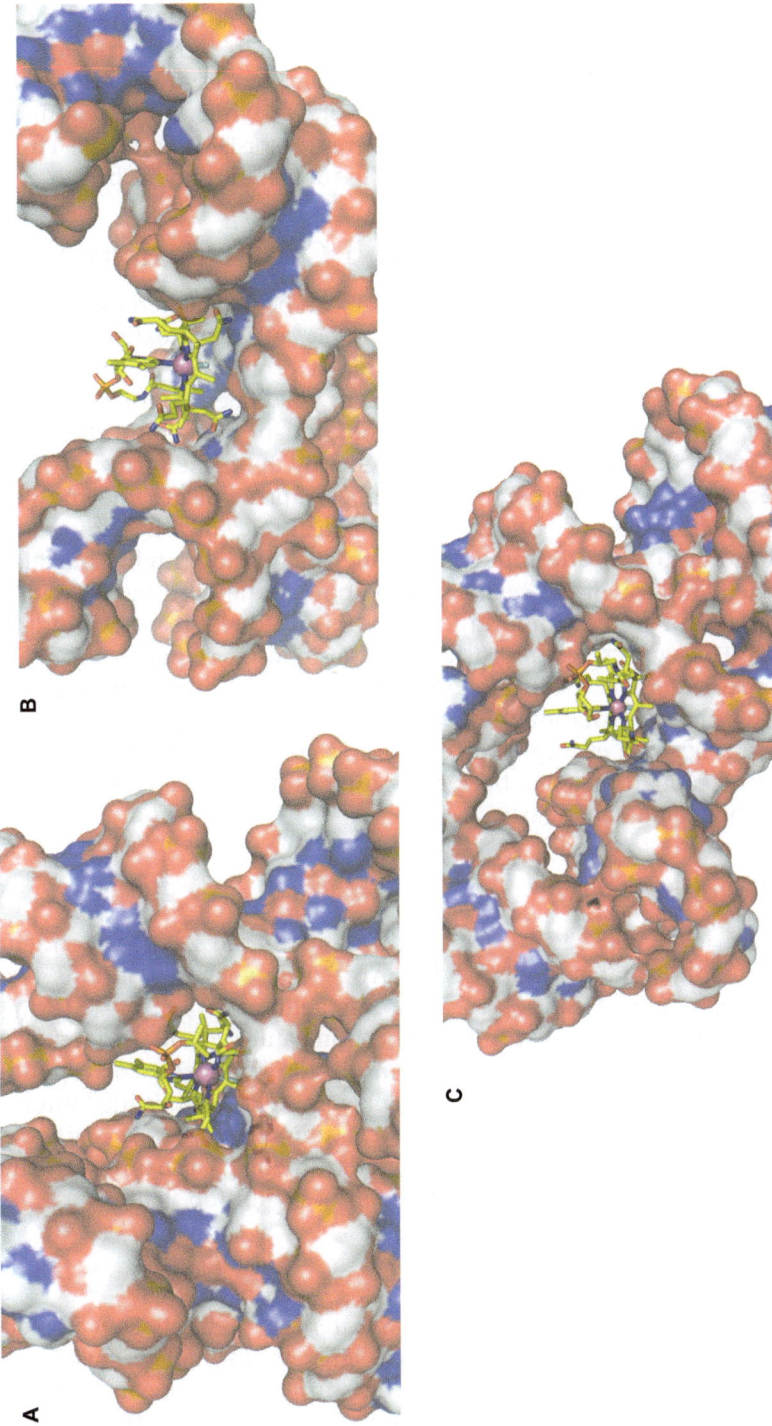

Figure 14.9 The structures of cobalamin-responsive riboswitches. A. The *btuB*-AdoCbl complex with the Ado group buried in a pocket. B and C. The *env4*- and *env8*-aquaCbl complexes; the shallow pockets in each riboswitch prevent AdoCbl binding. RNA portions are represented as surfaces – C, grey; N, blue; O, red; P, orange. The cobalamin molecules are shown as sticks – C, yellow; Co, pink; N, blue; O, red; P, orange. Ado and aqua moieties are in cyan.

Myxococcus xanthus.[213] CarH, a MerR-type transcriptional regulator, re-presses expression of carotenoid (Car) biosynthetic genes in the dark. AdoCbl binds to CarH ($K_d \sim 8 \times 10^{-7}$ M) and promotes protein tetramerization,[214] which leads to increased DNA-binding affinity. The sensitivity of AdoCbl to visible light results in its conversion to aquaCbl,[82,83] which has low affinity for CarH and results in destabilization of the repressor-DNA complex. Thus, the CarH protein must be able to distinguish AdoCbl and aquaCbl.

14.5 Mechanisms of Nickel and Cobalt Toxicity

Nickel and cobalt, like other transition metals, can be toxic when their levels exceed the normal binding capacity within the cell. The toxicity derives primarily from displacing native metals from a protein or enzyme.[215,216] Thus, the mechanism of Co(II) and Ni(II) toxicity will depend upon the target proteins that are present in the cell or microorganism, and could therefore vary with growth conditions. Toxicity also depends upon the form in which the ion enters the cell or organism.[217] A number of proteins are known to be inhibited by Ni(II) *in vitro*,[215] but the physiological relevance is seldom tested. Ni(II)-binding to peptide sequences from eukaryotic proteins has also been examined, and much has been learned about metal-binding to peptide motifs, in particular the importance of His residues in the binding motif.[71,72,218] However, the removal of a metal-binding sequence from its protein context can affect its affinity and the coupling of metal binding to a change in protein conformation in ways that are not physiologically relevant. In this section, specific examples of nickel and cobalt toxicity are given in which the molecular target has been identified and the mechanism of inhibition can be elucidated clearly.

14.5.1 Nickel in *E. coli*

Nickel toxicity in *E. coli* results from inhibition of the Zn(II)-dependent fructose 1,6 bisphosphatase aldolase (FbaA).[219] The enzyme contains two zinc-sites, ~ 6.8 Å apart. Mutation of the amino acids (Asp_{144}, Glu_{174}, Glu_{181} and the nearby Cys_{177}) that coordinate the non-catalytic Zn(II) in a tetrahedral binding site resulted in decreased inhibition by nickel. Correspondingly, cell growth was less sensitive to nickel in cells containing mutant versions of the enzyme, or when FbaA is not required for growth. While the structure of the Ni(II) site has not been determined, its proximity to the active site Zn(II) suggests that the mechanism of inhibition is allosteric.

14.5.2 Cobalt in *E. coli*

Cobalt stress in enterobacteria (*E. coli* and *S. enterica*)[216] targets Fe-S cluster-containing enzymes. Co(II) inserts into Fe-S clusters during the cluster assembly process,[220] rather than displacing iron from intact enzymes. When inserted into Fe-S cluster dependent enzymes (*e.g.* aconitase, MiaB, FhuF), the mixed Co/Fe-S clusters are deleterious to function. UV-visible spectroscopy indicates that Co(II) binds directly to the Fe-S cluster assembly proteins IscU and SufA.[†]

14.5.3 Contact Dermatitis – TLR4-like Receptor

Environmental sources of nickel and cobalt can cause an allergic response in humans known as contact dermatitis.[221] Recent studies have identified the TLR4 receptor protein (toll-like receptor 4) as the molecular target for nickel and cobalt binding in the dermatological response.[221,222] TLR4 normally dimerizes in the presence of bacterial lipopolysaccharide and subsequently initiates a specific signalling cascade. Nickel and cobalt mediate TLR4 dimerization *via* a pair of His residues (His$_{456}$-Xaa-His$_{458}$) from each subunit (PDB CODE: 3FXI). The geometry of the site is not known but likely relies heavily on His-coordination, favouring four-coordinate planar geometry. However, a six-coordinate site could also occur. Murine TLR4 lacks the His-Xaa-His motif, consequently mice are not susceptible to contact dermatitis due to nickel/cobalt exposure.

14.5.4 Nickel and Cancer – Displacing Iron

Nickel is a known carcinogen in humans,[223] especially when particulate forms are present in the body, usually *via* occupational inhalation. Cobalt is a less potent carcinogen.[224] Both metals act on proteins involved in DNA or histone modification that can lead to changes in gene expression due to effects on chromatin structure.

14.5.4.1 Inhibition of Non-haem Iron Oxygenases

Ni(II) binds to non-haem Fe(II)-dependent enzymes that are involved in oxidative demethylation of methylated DNA and methylated histones. Ni(II) displaces Fe(II) from the H3K9 demethylase JMJD1A, resulting in increased methylation levels (Lys9 of histone 3; hence H3K9), and from the DNA demethylase ABH2 that repairs endogenously formed 1-methyladenine and 3-methylcytosine bases in duplex DNA. Purified ABH2 has a marginally higher affinity for Ni(II) than Fe(II) ($K_d = 1.7 \times 10^{-6}$ M *versus* 4.5×10^{-6} M, respectively).[225,226] Studies of ABH2 in different binding states (metal-bound, metal + α-ketoglutarate, and metal + α-ketoglutarate + substrate) reveal that Ni(II) binds to the same site as Fe(II) in the metallated enzyme with and without α-ketoglutarate.[225] Coordination is *via* two His residues, one Asp residue and α-ketoglutarate in the cofactor-bound enzyme. The situation is similar to that in the crystal structure of the inactive, six-coordinate Mn(II)-substituted form of the enzyme bound to cofactor and substrate. Only in the

---

[†]The properties and structures of Fe-S clusters are discussed in Chapter 12.

presence of the methylated DNA substrate is the inhibitory mechanism revealed: Fe(II) retains a five-coordinate geometry that allows O_2 coordination, whereas Ni(II) adopts a six-coordinate geometry that excludes O_2 coordination.[225] An unpublished crystal structure of JMJD2A with Ni(II) and inhibitor present (PDB code: 3RVH) is similar to the X-ray absorption spectroscopy data in the absence of DNA.

14.5.4.2 *Modification of Histone Structure*

Ni(II) binds to the N- and C-terminal regions of histone subunits *via* the His residues present in these regions.[227–229] Such binding could affect protein conformation and therefore function in a way that changes gene expression due to changes in chromatin structure. A second, more drastic modification is the Ni(II)-induced cleavage of specific histone subunits (H2A and H2B). In the former case, cleavage releases a peptide $(NH_3^+\text{-SHHKAKGK-CO}_2^-)$ that binds Ni(II) with planar coordination. This modification occurs both *in vitro* with the purified histone H2A subunit and also in living cells.[230,231] The four-coordinate planar Ni-peptide complex is a potent activator of reactive species such as H_2O_2,[125,232] suggesting a possible mechanism for causing DNA-strand breaks. The cleavage of H2B *in vivo* appears to depend both upon Ni(II) binding as well as a cellular protease, as the purified H2B subunit was not susceptible to cleavage *in vitro* in the presence of added Ni(II).[233]

14.6   Conclusions and Open Questions

The last decade has seen a substantial advance in our understanding of specific recognition of Ni(II) and Co(II) by proteins of varied biochemical activity and biological function. The structural characterization of regulatory proteins and metalloenzyme accessory proteins has flourished. From these studies have emerged some recurrent features of nickel- and cobalt-binding sites.

14.6.1   Structural Features

The nickel- and cobalt-binding sites reported thus far show some clear trends. Four-coordinate planar sites for Ni(II) in NikR, Nur and HypB act as Ni sensors at low concentration and serve as activation signals for function. Six-coordinate sites are used to sense Ni and Co in RcnR and CnrX. These sites have lower affinities and are linked to metal efflux processes. The nickel-enzyme assembly proteins UreE, HypB and CooC all show Ni coordination at a monomer/monomer interface in a homodimer assembly. This location is likely important for function in an assembly pathway that requires metal transfer between different protein subunits.

  Several of the Ni(II)-binding sites described in this chapter use a coordination motif that requires the N-terminus. The participation of the amine function as well as an amino acid side-chain close to the N-terminus has been observed for HypA and the *E. coli* HypB high-affinity site, as well as for the regulators NmtR and RcnR. In the case of NmtR and RcnR, this motif is not present in homologous metalloregulators that are selective for metals with lower coordination number preferences (*e.g.* Zn(II) and Cu(I)). Thus, N-terminal coordination motifs may enhance selectivity for Ni(II) and Co(II), which prefer higher coordination numbers. Whether this relationship between coordination number and selectivity also holds for HypA and HypB remains to be determined. Furthermore, an N-terminal coordination motif is likely to strongly impact the kinetic mechanisms of metal loading and unloading. These aspects are relatively unstudied for Ni proteins.

Several proteins involved in nickel homeostasis have histidine-rich regions, *e.g.* SlyD. It has been argued recently that His coordination is unstable due to changes in pH due that affect His protonation.[234] This feature, in fact, may be important for their recurrence in nickel trafficking and regulation as the reversibility of these binding sites makes them useful in responding to changes in available Ni, in cell physiology (*e.g.* pH) or even in protein-microenvironment upon hetero-oligomer formation. Additionally, His-rich motifs could serve to increase rates of nickel transfer during enzyme assembly by providing a localized high concentration of nickel ions that could be transferred in a controlled, stepwise mechanism.

14.6.2 Protein-protein Metal Transfer Reactions

The progress that has been made in understanding the metal-binding properties of individual protein components and the associated proteins involved in nickel- and cobalt-enzyme assembly has been illuminating. The next step will be to understand how these metal sites change when hetero-oligomeric protein complexes are formed. Certainly, the locations of metal sites in UreE, HypB and CooC are well suited to metal-transfer reactions upon formation of a new protein:protein interface. Heterooligomeric formation could alter metal-site stability *via* protein conformational change, thus providing a kinetic basis for controlling the fidelity of metalloenzyme assembly. Nucleotide hydrolysis will also contribute to these processes. Structural studies of heterooligomer complexes, with both wild-type and mutant proteins, will help to visualize steps in this process. The ability to obtain good quantitative kinetic parameters from these processes will depend upon the presence of resolvable spectroscopic signals, either metal or protein, in the different steps of the transfer process. To date, only Ni(II) transfer between Ni-(L-His)$_2$ and serum albumin has been studied from a kinetic perspective.[235]

14.6.3 Dynamic Changes in Metal Homeostasis

Experimental studies of nickel and cobalt utilization and regulation typically focus on equilibrium binding conditions. Enzyme assembly is a dynamic process that requires the flow of metal ions from one protein to another.

Similarly, changes in growth conditions can result in increases or decreases in metal importer expression, metalloenzyme expression and turnover, pH and extracellular metal availability. All of these changes will result in changes in metal availability and localization within the cell. Unlike the direct metal transfer reactions in metalloenzyme assembly, control of metal ion levels requires a shift between proteins of different functional types, *e.g.* metalloenzyme assembly *versus* metalloregulation or metal efflux, and may not involve protein-protein interactions. Knowledge of the species of Ni(II) and Co(II) that first enter the cell will be critical to understanding how these metal ions are partitioned into assembly pathways instead of being detected by metal sensors, from both thermodynamic and kinetic perspectives. Studies of cellular responses to these small but physiologically relevant surges should provide a context with which to understand the kinetic and thermodynamic features of nickel and cobalt binding sites that control metal targeting under different metal availabilities and metal requirements.

Note Added in Proof

A structure of H. pylori HypB (PDBID 4LPS) with Ni(II) bound shows four coordinate, planar tetrathiolate coordination, with two Cys coming from each monomer in the HypB dimer (see Figure 14.5B and F for related structures). (Ref.: Sydor, A. M., Lebrette, H., Ariyakumaran, R., Cavazza, C., and Zamble, D. B., *J. Biol. Chem.* 2014, **289**, 3828–3841).

A structure of the periplasmic H. pylori CeuE protein (PDBID 4LS3) with Ni-(L-His)2 bound shows recognition features similar to those depicted in Figure 14.3, e.g., Arg coordination of L-His COO- groups. However, the Ni-(L-His)2 stereoisomer is distinct from the two structures published previously. (Ref: Shaik, M. M., Cendron, L., Salamina, M., Ruzzene, M., and Zanotti, G., *Mol. Microbiol.* 2014, **91**, 724–735).

References

1. J. J. R. Fraústo da Silva and R. J. P. Williams, *The Biological Chemistry of the Elements: The Inorganic Chemistry of Life*, Oxford University Press, Oxford, 1991.
2. R. J. P. Williams and J. J. R. Fraústo da Silva, *The Natural Selection of the Chemical Elements: The Environment and Life's Chemistry*, Oxford University Press, Oxford, 1996.
3. M. A. Saito, *Nature*, 2009, **458**, 714–715.
4. R. K. Thauer, *Microbiology*, 1998, **144**, 2377–2406.
5. C. L. Dupont, A. Butcher, R. E. Valas, P. E. Bourne and G. Caetano-Anolles, *Proc. Natl Acad. Sci. USA*, 2010, **107**, 10567–10572.
6. C. L. Dupont, S. Yang, B. Palenik and P. E. Bourne, *Proc. Natl Acad. Sci. USA*, 2006, **103**, 17822–17827.

7. C. L. Dupont, K. Barbeau and B. Palenik, *Appl. Environ. Microbiol.*, 2008, **74**, 23–31.

8. F. R. Sclater, E. Boyle and J. M. Edmond, *Earth Planet. Sci. Lett.*, 1976, **31**, 119–128.

9. C. L. Dupont, K. N. Buck, B. Palenik and K. Barbeau, *Deep-Sea Res. Pt I*, 2010, **57**, 553–566.

10. M. A. Saito, J. W. Moffett and G. R. DiTullio, *Geochim. Cosmochim. Acta*, 2003, **67**, A409–A409.

11. K. W. Bruland, *Earth Planet. Sci. Lett.*, 1980, **47**, 176–198.

12. M. A. Saito, J. W. Moffett and G. R. DiTullio, *Global Biogeochem. Cycles*, 2004, **18**.

13. J. M. Vraspir and A. Butler, *Ann. Rev. Mar. Sci.*, 2009, **1**, 43–63.

14. R. N. Collins and A. S. Kinsela, *Chemosphere*, 2010, **79**, 763–771.

15. J. A. Cowan, *Inorganic Biochemistry: An Introduction*, Wiley-VCH, New York, 2nd edn, 1997, pp. 1–24.

16. R. J. P. Williams, in *Concepts and Models in Bioinorganic Chemistry*, ed. H.-B. Kraatz and N. Metzler-Nolte, Wiley-VCH, Weinheim, 2006, pp. 1–24.

17. K. A. Higgins, C. E. Carr and M. J. Maroney, *Biochemistry*, 2012, **51**, 7816–7832.

18. A. M. Sydor and D. B. Zamble, in *Metallomics and the Cell*, ed. L. Banci, Springer, Dordrecht, 2013, pp. 375–416.

19. S. Okamoto and L. D. Eltis, *Metallomics*, 2011, **3**, 963–970.

20. S. N. Fedosov, *Subcell. Biochem.*, 2012, **56**, 347–367.

21. V. Cracan and R. Banerjee, in *Metallomics and the Cell*, ed. L. Banci, Springer, Dordrecht, 2013, pp. 333–374.

22. Y. Zhang and V. N. Gladyshev, *J. Biol. Chem.*, 2010, **285**, 3393–3405.

23. Y. Zhang, D. A. Rodionov, M. S. Gelfand and V. N. Gladyshev, *BMC Genomics*, 2009, **10**, 78.

24. A. Sigel, H. Sigel and R. K. O. Sigel, *Nickel and Its Surprising Impact in Nature*, Wiley, Chichester, England, 2007.

25. U. Ermler, W. Grabarse, S. Shima, M. Goubeaud and R. K. Thauer, *Curr. Opin. Struct. Biol.*, 1998, **8**, 749–758.

26. S. W. Ragsdale, *J. Biol. Chem.*, 2009, **284**, 18571–18575.

27. J. C. Fontecilla-Camps, P. Amara, C. Cavazza, Y. Nicolet and A. Volbeda, *Nature*, 2009, **460**, 814–822.

28. F. H. Nielsen, T. R. Shuler, T. G. McLeod and T. J. Zimmerman, *J. Nutr.*, 1984, **114**, 1280–1288.

29. F. H. Nielsen, *FASEB J.*, 1991, **5**, 2661–2667.

30. F. H. Nielsen, in *Handbook of Metal-Ligand Interactions in Biological Fluids – Bioinorganic Medicine*, ed. G. Berthon, Marcel Dekker, Inc., New York, 1995, vol. 1, pp. 257–260.

31. J. H. Cummings and G. T. Macfarlane, *J. Appl. Bacteriol.*, 1991, **70**, 443–459.

32. S. R. Gill, M. Pop, R. T. Deboy, P. B. Eckburg, P. J. Turnbaugh, B. S. Samuel, J. I. Gordon, D. A. Relman, C. M. Fraser-Liggett and K. E. Nelson, *Science*, 2006, **312**, 1355–1359.

33. B. S. Samuel and J. I. Gordon, *Proc. Natl Acad. Sci. USA*, 2006, **103**, 10011–10016.

34. M. A. Flyvholm, G. D. Nielsen and A. Andersen, *Z. Lebensm. Unters. Forsch.*, 1984, **179**, 427–431.

35. P. Trumbo, A. A. Yates, S. Schlicker and M. Poos, *J. Am. Diet. Assoc.*, 2001, **101**, 294–301.

36. M. Patriarca, T. D. Lyon and G. S. Fell, *Am. J. Clin. Nutr.*, 1997, **66**, 616–621.

37. J. Wang, E. Harald Hansen and B. Gammelgaard, *Talanta*, 2001, **55**, 117–126.

38. G. Unden and J. Bongaerts, *Biochim. Biophys. Acta*, 1997, **1320**, 217–234.

39. J. Olson and R. Maier, *Science*, 2002, **298**, 1788–1790.

40. R. J. Maier, A. Olczak, S. Maier, S. Soni and J. Gunn, *Infect. Immun.*, 2004, **72**, 6294–6299.

41. R. J. Maier, *Biochem. Soc. Trans.*, 2005, **33**, 83–85.

42. G. R. Gibson, J. H. Cummings, G. T. Macfarlane, C. Allison, I. Segal, H. H. Vorster and A. R. Walker, *Gut*, 1990, **31**, 679–683.

43. G. R. Gibson, G. T. Macfarlane and J. H. Cummings, *Gut*, 1993, **34**, 437–439.

44. H. D. Youn, E. J. Kim, J. H. Roe, Y. C. Hah and S. O. Kang, *Biochem. J.*, 1996, **318**, 889–896.

45. H. D. Youn, H. Youn, J. W. Lee, Y. I. Yim, J. K. Lee, Y. C. Hah and S. O. Kang, *Arch. Biochem. Biophys.*, 1996, **334**, 341–348.

46. S. L. Clugston and J. F. Honek, *J. Mol. Evol.*, 2000, **50**, 491–495.

47. N. Sukdeo, S. L. Clugston, E. Daub and J. F. Honek, *Biochem. J.*, 2004, **384**, 111–117.

48. M. Kobayashi and S. Shimizu, *Eur. J. Biochem.*, 1999, **261**, 1–9.

49. S. Okamoto, F. Van Petegem, M. A. Patrauchan and L. D. Eltis, *J. Biol. Chem.*, 2010, **285**, 25126–25133.

50. E. T. Yukl and C. M. Wilmot, *Curr. Opin. Chem. Biol.*, 2012, **16**, 54–59.

51. V. Cracan and R. Banerjee, *Biochemistry*, 2012, **51**, 6039–6046.

52. R. K. Thauer and L. G. Bonacker, *Ciba Found. Symp.*, 1994, **180**, 210–222.

53. K. C. Bible, M. Buytendorp, P. D. Zierath and K. L. Rinehart, *Proc. Natl Acad. Sci. USA*, 1988, **85**, 4582–4586.

54. H. L. Sings, K. C. Bible and K. L. Rinehart, *Proc. Natl Acad. Sci. USA*, 1996, **93**, 10560–10565.

55. S. M. Yannone, S. Hartung, A. L. Menon, M. W. Adams and J. A. Tainer, *Curr. Opin. Biotech.*, 2012, **23**, 89–95.

56. A. Cvetkovic, A. L. Menon, M. P. Thorgersen, J. W. Scott, F. L. Poole, 2nd, F. E. Jenney, Jr., W. A. Lancaster, J. L. Praissman, S. Shanmukh, B. J. Vaccaro, S. A. Trauger, E. Kalisiak, J. V. Apon, G. Siuzdak, S. M. Yannone, J. A. Tainer and M. W. Adams, *Nature*, 2010, **466**, 779–782.

57. E. M. Bertrand, A. E. Allen, C. L. Dupont, T. M. Norden-Krichmar, J. Bai, R. E. Valas and M. A. Saito, *Proc. Natl Acad. Sci. USA*, 2012, **109**, E1762–1771.

58. R. J. P. Williams and J. J. R. Fraústo da Silva, *The Chemistry of Evolution: The Development of Our Ecosystem*, Elsevier, Amsterdam, Boston, 1st edn, 2006.

59. S. J. Lippard and J. M. Berg, *Principles of Bioinorganic Chemistry*, University Science Books, Mill Valley, CA, 1994.

60. W. Maret and B. L. Vallee, *Methods Enzymol.*, 1993, **226**, 52–71.

61. F. M. M. Morel, R. J. M. Hudson and N. M. Price, *Limnol. Oceanogr.*, 1991, **36**, 1742–1755.

62. R. J. M. Hudson and F. M. M. Morel, *Deep-Sea Res. Pt I*, 1993, **40**, 129–150.

63. M. Lucassen and B. Sarkar, *J. Toxicol. Environ. Health*, 1979, **5**, 897–905.

64. D. L. Callahan, A. J. Baker, S. D. Kolev and A. G. Wedd, *J. Biol. Inorg. Chem.*, 2006, **11**, 2–12.

65. I. C. Farcasanu, M. Mizunuma, F. Nishiyama and T. Miyakawa, *Biosci. Biotechnol. Biochem.*, 2005, **69**, 2343–2348.

66. D. A. Pearce and F. Sherman, *J. Bacteriol.*, 1999, **181**, 4774–4779.

67. P. T. Chivers, E. L. Benanti, V. Heil-Chapdelaine, J. S. Iwig and J. L. Rowe, *Metallomics*, 2012, **4**, 1043–1050.

68. Y. Zhang, S. Akilesh and D. E. Wilcox, *Inorg. Chem.*, 2000, **39**, 3057–3064.

69. S. A. Adibi and D. W. Mercer, *J. Clin. Invest.*, 1973, **52**, 1586–1594.

70. B. D. Bennett, E. H. Kimball, M. Gao, R. Osterhout, S. J. Van Dien and J. D. Rabinowitz, *Nat. Chem. Biol.*, 2009, **5**, 593–599.

71. I. Sovago and K. Osz, *Dalton Trans.*, 2006, 3841–3854.

72. H. Kozlowski, W. Bal, M. Dyba and T. Kowalik-Jankowska, *Coord. Chem. Rev.*, 1999, **184**, 319–346.

73. M. J. Haydon and C. S. Cobbett, *New Phytol.*, 2007, **174**, 499–506.

74. E. Montarges-Pelletier, V. Chardot, G. Echevarria, L. J. Michot, A. Bauer and J. L. Morel, *Phytochemistry*, 2008, **69**, 1695–1709.

75. J. R. Donat, K. A. Lao and K. W. Bruland, *Anal. Chim. Acta*, 1994, **284**, 547–571.

76. J. L. Freeman, M. W. Persans, K. Nieman and D. E. Salt, *Appl. Environ. Microbiol.*, 2005, **71**, 8627–8633.

77. J. L. Freeman, M. W. Persans, K. Nieman, C. Albrecht, W. Peer, I. J. Pickering and D. E. Salt, *Plant Cell*, 2004, **16**, 2176–2191.

78. E. A. Pilon-Smits, C. F. Quinn, W. Tapken, M. Malagoli and M. Schiavon, *Curr. Opin. Plant Biol.*, 2009, **12**, 267–274.

79. W. Wei, Y. Wang, Z. G. Wei, H. Y. Zhao, H. X. Li and F. Hu, *Biol. Trace Elem. Res.*, 2009, **131**, 165–176.

80. R. Tappero, E. Peltier, M. Grafe, K. Heidel, M. Ginder-Vogel, K. J. Livi, M. L. Rivers, M. A. Marcus, R. L. Chaney and D. L. Sparks, *New Phytol.*, 2007, **175**, 641–654.

81. M. Oven, E. Grill, A. Golan-Goldhirsh, T. M. Kutchan and M. H. Zenk, *Phytochemistry*, 2002, **60**, 467–474.

82. R. T. Taylor, L. Smucker, M. L. Hanna and J. Gill, *Arch. Biochem. Biophys.*, 1973, **156**, 521–533.

83. W. H. Pailes and H. P. Hogenkamp, *Biochemistry*, 1968, 7, 4160–4166.

84. C. E. Outten and T. V. O'Halloran, *Science*, 2001, **292**, 2488–2492.

85. H. L. Mobley, R. M. Garner and P. Bauerfeind, *Mol. Microbiol.*, 1995, **16**, 97–109.

86. D. A. Rodionov, P. Hebbeln, M. S. Gelfand and T. Eitinger, *J. Bacteriol.*, 2006, **188**, 317–327.

87. C. Fu, S. Javedan, F. Moshiri and R. J. Maier, *Proc. Natl Acad. Sci. USA*, 1994, **91**, 5099–5103.

88. T. Eitinger and B. Friedrich, *J. Biol. Chem.*, 1991, **266**, 3222–3227.

89. J. F. Fulkerson, Jr., R. M. Garner and H. L. Mobley, *J. Biol. Chem.*, 1998, **273**, 235–241.

90. T. Eitinger, L. Wolfram, O. Degen and C. Anthon, *J. Biol. Chem.*, 1997, **272**, 17139–17144.

91. L. Wolfram and P. Bauerfeind, *J. Bacteriol.*, 2002, **184**, 1438–1443.

92. L. F. Wu, M. A. Mandrand-Berthelot, R. Waugh, C. J. Edmonds, S. E. Holt and D. H. Boxer, *Mol. Microbiol.*, 1989, **3**, 1709–1718.

93. L.-F. Wu and M.-A. Mandrand-Berthelot, *Biochimie*, 1986, **68**, 167–179.

94. M. Sandy and A. Butler, *Chem. Rev.*, 2009, **109**, 4580–4595.

95. H. Lebrette, M. Iannello, J. C. Fontecilla-Camps and C. Cavazza, *J. Inorg. Biochem.*, 2012, **121**, 16–18.

96. K. A. Fraser and M. M. Harding, *J. Chem. Soc. (A)*, 1967, 415–420.

97. L. J. Zompa, *Acta Crystallogr. E*, 2005, **61**, M849–M851.

98. L. J. Zompa, *J. Chem. Soc. D Chem. Commun.*, 1969, 783.

99. G. Minasov, A. Halavaty, L. Shuvalova, I. Dubrovska, J. Winsor, O. Kiryukhina, F. Falugi, M. Bottomly and W. F. Anderson, 10.2210/pdb3rqt/pdb.

100. C. Harford and B. Sarkar, in *Handbook of Metal-Ligand Interactions in Biological Fluids*, ed. G. Berthon, Marcel Dekker, Inc., New York, 1995, vol. 1, pp. 62–70.

101. R. M. Howlett, B. M. Hughes, A. Hitchcock and D. J. Kelly, *Microbiology*, 2012, **158**, 1645–1655.

102. L. Remy, M. Carriere, A. Derre-Bobillot, C. Martini, M. Sanguinetti and E. Borezee-Durant, *Mol. Microbiol.*, 2013, **87**, 730–743.

103. A. Hiron, B. Posteraro, M. Carriere, L. Remy, C. Delporte, M. La Sorda, M. Sanguinetti, V. Juillard and E. Borezee-Durant, *Mol. Microbiol.*, 2010, 77, 1246–1260.

104. K. Schauer, B. Gouget, M. Carriere, A. Labigne and H. de Reuse, *Mol. Microbiol.*, 2007, **63**, 1054–1068.

105. J. Stoof, E. J. Kuipers, G. Klaver and A. H. van Vliet, *Infect. Immun.*, 2010, **78**, 4261–4267.

106. M. V. Cherrier, C. Cavazza, C. Bochot, D. Lemaire and J. C. Fontecilla-Camps, *Biochemistry*, 2008, **47**, 9937–9943.
107. C. Cavazza, L. Martin, E. Laffly, H. Lebrette, M. V. Cherrier, L. Zeppieri, P. Richaud, M. Carriere and J. C. Fontecilla-Camps, *FEBS Lett.*, 2011, **585**, 711–715.
108. J. Cheng, B. Poduska, R. A. Morton and T. M. Finan, *J. Bacteriol.*, 2011, **193**, 4405–4416.
109. H. Komeda, M. Kobayashi and S. Shimizu, *Proc. Natl Acad. Sci. USA*, 1997, **94**, 36–41.
110. O. Degen and T. Eitinger, *J. Bacteriol.*, 2002, **184**, 3569–3577.
111. H. Nikaido, *J. Biol. Chem.*, 1994, **269**, 3905–3908.
112. D. P. Chimento, R. J. Kadner and M. C. Wiener, *J. Mol. Biol.*, 2003, **332**, 999–1014.
113. D. P. Chimento, A. K. Mohanty, R. J. Kadner and M. C. Wiener, *Nat. Struct. Biol.*, 2003, **10**, 394–401.
114. N. Cadieux, C. Bradbeer, E. Reeger-Schneider, W. Koster, A. K. Mohanty, M. C. Wiener and R. J. Kadner, *J. Bacteriol.*, 2002, **184**, 706–717.
115. L. A. Kelley and M. J. Sternberg, *Nat. Protoc.*, 2009, **4**, 363–371.
116. M. Blokesch, A. Paschos, E. Theodoratou, A. Bauer, M. Hube, S. Huth and A. Bock, *Biochem. Soc. Trans.*, 2002, **30**, 674–680.
117. S. B. Mulrooney and R. P. Hausinger, *J. Bacteriol.*, 1990, **172**, 5837–5843.
118. A. K. Kaster, M. Goenrich, H. Seedorf, H. Liesegang, A. Wollherr, G. Gottschalk and R. K. Thauer, *Archaea*, 2011, **2011**, 973848.
119. L. Casalot and M. Rousset, *Trends Microbiol.*, 2001, **9**, 228–237.
120. A. Atanassova and D. B. Zamble, *J. Bacteriol.*, 2005, **187**, 4689–4697.
121. K. C. Chan Chung and D. B. Zamble, *J. Biol. Chem.*, 2011, **286**, 43081–43090.
122. C. D. Douglas, T. T. Ngu, H. Kaluarachchi and D. B. Zamble, *Biochemistry*, 2013, **52**, 1788–1801.
123. D. C. Kennedy, R. W. Herbst, J. S. Iwig, P. T. Chivers and M. J. Maroney, *J. Am. Chem. Soc.*, 2007, **129**, 16–17.
124. R. W. Herbst, I. Perovic, V. Martin-Diaconescu, K. O'Brien, P. T. Chivers, S. S. Pochapsky, T. C. Pochapsky and M. J. Maroney, *J. Am. Chem. Soc.*, 2010, **132**, 10338–10351.
125. C. Harford and B. Sarkar, *Acc. Chem. Res.*, 1997, **30**, 123–130.
126. W. Xia, H. Li, K. H. Sze and H. Sun, *J. Am. Chem. Soc.*, 2009, **131**, 10031–10040.
127. J. W. Olson, N. S. Mehta and R. J. Maier, *Mol. Microbiol.*, 2001, **39**, 176–182.
128. S. L. Benoit, J. L. McMurry, S. A. Hill and R. J. Maier, *Biochim. Biophys. Acta*, 2012, **1820**, 1519–1525.
129. T. Maier, A. Jacobi, M. Sauter and A. Bock, *J. Bacteriol.*, 1993, **175**, 630–635.
130. M. R. Leach, S. Sandal, H. Sun and D. B. Zamble, *Biochemistry*, 2005, **44**, 12229–12238.

131. C. D. Douglas, A. V. Dias and D. B. Zamble, *Dalton Trans.*, 2012, 7876–7878.
132. K. C. Chung, L. Cao, A. V. Dias, I. J. Pickering, G. N. George and D. B. Zamble, *J. Am. Chem. Soc.*, 2008, **130**, 14056–14057.
133. W. D. Roof, S. M. Horne, K. D. Young and R. Young, *J. Biol. Chem.*, 1994, **269**, 2902–2910.
134. J. W. Zhang, G. Butland, J. F. Greenblatt, A. Emili and D. B. Zamble, *J. Biol. Chem.*, 2005, **280**, 4360–4366.
135. G. Butland, J. M. Peregrin-Alvarez, J. Li, W. Yang, X. Yang, V. Canadien, A. Starostine, D. Richards, B. Beattie, N. Krogan, M. Davey, J. Parkinson, J. Greenblatt and A. Emili, *Nature*, 2005, **433**, 531–537.
136. H. Kaluarachchi, J. W. Zhang and D. B. Zamble, *Biochemistry*, 2011, **50**, 10761–10763.
137. H. Kaluarachchi, M. Altenstein, S. R. Sugumar, J. Balbach, D. B. Zamble and C. Haupt, *J. Mol. Biol.*, 2012, **417**, 28–35.
138. H. Kaluarachchi, D. E. Sutherland, A. Young, I. J. Pickering, M. J. Stillman and D. B. Zamble, *J. Am. Chem. Soc.*, 2009, **131**, 18489–18500.
139. C. Low, P. Neumann, H. Tidow, U. Weininger, C. Haupt, B. Friedrich-Epler, C. Scholz, M. T. Stubbs and J. Balbach, *J. Mol. Biol.*, 2010, **398**, 375–390.
140. E. L. Carter, N. Flugga, J. L. Boer, S. B. Mulrooney and R. P. Hausinger, *Metallomics*, 2009, **1**, 207–221.
141. B. Zambelli, F. Musiani, S. Benini and S. Ciurli, *Acc. Chem. Res.*, 2011, **44**, 520–530.
142. T. G. Brayman and R. P. Hausinger, *J. Bacteriol.*, 1996, **178**, 5410–5416.
143. A. Soriano, G. J. Colpas and R. P. Hausinger, *Biochemistry*, 2000, **39**, 12435–12440.
144. Y. H. Fong, H. C. Wong, M. H. Yuen, P. H. Lau, Y. W. Chen and K. B. Wong, *PLoS Biol.*, 2013, **11**, e1001678.
145. J. K. Kim, S. B. Mulrooney and R. P. Hausinger, *J. Bacteriol.*, 2006, **188**, 8413–8420.
146. R. Ge, Y. Zhang, X. Sun, R. M. Watt, Q. Y. He, J. D. Huang, D. E. Wilcox and H. Sun, *J. Am. Chem. Soc.*, 2006, **128**, 11330–11331.
147. N. M. Chiera, M. Rowinska-Zyrek, R. Wieczorek, R. Guerrini, D. Witkowska, M. Remelli and H. Kozlowski, *Metallomics*, 2013, **5**, 214–221.
148. K. Schauer, C. Muller, M. Carriere, A. Labigne, C. Cavazza and H. De Reuse, *J. Bacteriol.*, 2010, **192**, 1231–1237.
149. S. L. Benoit and R. J. Maier, *mBio*, 2011, **2**, e00039–00011.
150. R. L. Kerby, P. W. Ludden and G. P. Roberts, *J. Bacteriol.*, 1997, **179**, 2259–2266.
151. R. K. Watt and P. W. Ludden, *J. Biol. Chem.*, 1998, **273**, 10019–10025.
152. J. H. Jeoung, T. Giese, M. Grunwald and H. Dobbek, *Biochemistry*, 2009, **48**, 11505–11513.
153. J. H. Jeoung, T. Giese, M. Grunwald and H. Dobbek, *J. Mol. Biol.*, 2010, **396**, 1165–1179.

154. H. K. Loke and P. A. Lindahl, *J .Inorg. Biochem.*, 2003, **93**, 33–40.
155. E. Theodoratou, A. Paschos, W. Mintz and A. Bock, *Arch. Microbiol.*, 2000, **173**, 110–116.
156. E. Fritsche, A. Paschos, H. G. Beisel, A. Bock and R. Huber, *J. Mol. Biol.*, 1999, **288**, 989–998.
157. E. Theodoratou, A. Paschos, A. Magalon, E. Fritsche, R. Huber and A. Böck, *Eur. J. Biochem.*, 2000, **267**, 1995–1999.
158. J. Pei and N. V. Grishin, *Protein Sci.*, 2002, **11**, 691–697.
159. T. Eitinger, *J. Bacteriol.*, 2004, **186**, 7821–7825.
160. E. Raux, H. L. Schubert and M. J. Warren, *Cell. Mol. Life Sci.*, 2000, **57**, 1880–1893.
161. M. J. Warren, E. Raux, H. L. Schubert and J. C. Escalante-Semerena, *Nat. Prod. Rep.*, 2002, **19**, 390–412.
162. E. Raux, C. Thermes, P. Heathcote, A. Rambach and M. J. Warren, *J. Bacteriol.*, 1997, **179**, 3202–3212.
163. L. Debussche, M. Couder, D. Thibaut, B. Cameron, J. Crouzet and F. Blanche, *J. Bacteriol.*, 1992, **174**, 7445–7451.
164. Z. Zhou, Y. Hashimoto, K. Shiraki and M. Kobayashi, *Proc. Natl Acad. Sci. USA*, 2008, **105**, 14849–14854.
165. Z. Zhou, Y. Hashimoto, T. Cui, Y. Washizawa, H. Mino and M. Kobayashi, *Biochemistry*, 2010, **49**, 9638–9648.
166. Z. Zhou, Y. Hashimoto and M. Kobayashi, *J. Biol. Chem.*, 2009, **284**, 14930–14938.
167. R. Banerjee, C. Gherasim and D. Padovani, *Curr. Opin. Chem. Biol.*, 2009, **13**, 484–491.
168. M. Koutmos, C. Gherasim, J. L. Smith and R. Banerjee, *J. Biol. Chem.*, 2011, **286**, 29780–29787.
169. M. D. Rolfe, C. J. Rice, S. Lucchini, C. Pin, A. Thompson, A. D. Cameron, M. Alston, M. F. Stringer, R. P. Betts, J. Baranyi, M. W. Peck and J. C. Hinton, *J. Bacteriol.*, 2012, **194**, 686–701.
170. Z. Ma, F. E. Jacobsen and D. P. Giedroc, *Chem. Rev.*, 2009, **109**, 4644–4681.
171. A. J. Guerra and D. P. Giedroc, *Arch. Biochem. Biophys.*, 2012, **519**, 210–222.
172. J. S. Iwig and P. T. Chivers, *Nat. Prod. Rep.*, 2010, **27**, 658–667.
173. S. C. Wang, A. V. Dias and D. B. Zamble, *Dalton Trans.*, 2009, 2459–2466.
174. K. De Pina, V. Desjardin, M. A. Mandrand-Berthelot, G. Giordano and L. F. Wu, *J. Bacteriol.*, 1999, **181**, 670–674.
175. P. T. Chivers and R. T. Sauer, *Protein Sci.*, 1999, **8**, 2494–2500.
176. E. L. Benanti and P. T. Chivers, *J. Bacteriol.*, 2010, **192**, 4327–4336.
177. A. H. van Vliet, F. D. Ernst and J. G. Kusters, *Trends Microbiol.*, 2004, **12**, 489–494.
178. F. D. Ernst, E. J. Kuipers, A. Heijens, R. Sarwari, J. Stoof, C. W. Penn, J. G. Kusters and A. H. van Vliet, *Infect. Immun.*, 2005, **73**, 7252–7258.
179. B. Zambelli, A. Danielli, S. Romagnoli, P. Neyroz, S. Ciurli and V. Scarlato, *J. Mol. Biol.*, 2008, **383**, 1129–1143.

180. N. S. Dosanjh, A. L. West and S. L. Michel, *Biochemistry*, 2009, **48**, 527–536.
181. E. R. Schreiter, M. D. Sintchak, Y. Guo, P. T. Chivers, R. T. Sauer and C. L. Drennan, *Nat. Struct. Biol.*, 2003, **10**, 794–799.
182. E. R. Schreiter and C. L. Drennan, *Nat. Rev. Microbiol.*, 2007, **5**, 710–720.
183. P. T. Chivers and R. T. Sauer, *Chem. Biol.*, 2002, **9**, 1141–1148.
184. P. Carrington, P. Chivers, F. al-Mjeni, R. Sauer and M. Maroney, *Nat. Struct. Biol.*, 2003, **10**, 126–130.
185. S. C. Wang, A. V. Dias, S. L. Bloom and D. B. Zamble, *Biochemistry*, 2004, **43**, 10018–10028.
186. S. L. Bloom and D. B. Zamble, *Biochemistry*, 2004, **43**, 10029–10038.
187. C. M. Phillips, E. R. Schreiter, C. M. Stultz and C. L. Drennan, *Biochemistry*, 2010, **49**, 7830–7838.
188. S. C. Wang, Y. Li, M. Ho, M. E. Bernal, A. M. Sydor, W. R. Kagzi and D. B. Zamble, *Biochemistry*, 2010, **49**, 6635–6645.
189. S. Leitch, M. J. Bradley, J. L. Rowe, P. T. Chivers and M. J. Maroney, *J. Am. Chem. Soc.*, 2007, **129**, 5085–5095.
190. A. L. West, S. E. Evans, J. M. Gonzalez, L. G. Carter, H. Tsuruta, E. Pozharski and S. L. Michel, *Proc. Natl Acad. Sci. USA*, 2012, **109**, 5633–5638.
191. B. E. Ahn, J. Cha, E. J. Lee, A. R. Han, C. J. Thompson and J. H. Roe, *Mol. Microbiol.*, 2006, **59**, 1848–1858.
192. Y. J. An, B. E. Ahn, A. R. Han, H. M. Kim, K. M. Chung, J. H. Shin, Y. B. Cho, J. H. Roe and S. S. Cha, *Nucleic Acids Res.*, 2009, **37**, 3442–3451.
193. D. R. Campbell, K. E. Chapman, K. J. Waldron, S. Tottey, S. Kendall, G. Cavallaro, C. Andreini, J. Hinds, N. G. Stoker, N. J. Robinson and J. S. Cavet, *J. Biol. Chem.*, 2007, **282**, 32298–32310.
194. J. S. Cavet, W. Meng, M. A. Pennella, R. J. Appelhoff, D. P. Giedroc and N. J. Robinson, *J. Biol. Chem.*, 2002, **277**, 38441–38448.
195. H. Reyes-Caballero, C. W. Lee and D. P. Giedroc, *Biochemistry*, 2011, **50**, 7941–7952.
196. M. A. Pennella, J. E. Shokes, N. J. Cosper, R. A. Scott and D. P. Giedroc, *Proc. Natl Acad. Sci. USA*, 2003, **100**, 3713–3718.
197. J. C. Rutherford, J. S. Cavet and N. J. Robinson, *J. Biol. Chem.*, 1999, **274**, 25827–25832.
198. G. Grass, C. Grosse and D. H. Nies, *J. Bacteriol.*, 2000, **182**, 1390–1398.
199. C. Tibazarwa, S. Wuertz, M. Mergeay, L. Wyns and D. van Der Lelie, *J. Bacteriol.*, 2000, **182**, 1399–1409.
200. T. Schmidt and H. G. Schlegel, *J. Bacteriol.*, 1994, **176**, 7045–7054.
201. J. S. Iwig, J. L. Rowe and P. T. Chivers, *Mol. Microbiol.*, 2006, **62**, 252–262.
202. T. Liu, A. Ramesh, Z. Ma, S. K. Ward, L. Zhang, G. N. George, A. M. Talaat, J. C. Sacchettini and D. P. Giedroc, *Nat. Chem. Biol.*, 2007, **3**, 60–68.
203. J. S. Iwig, S. Leitch, R. W. Herbst, M. J. Maroney and P. T. Chivers, *J. Am. Chem. Soc.*, 2008, **130**, 7592–7606.
204. K. A. Higgins, P. T. Chivers and M. J. Maroney, *J. Am. Chem. Soc.*, 2012, **134**, 7081–7093.

205. K. A. Higgins, H. Q. Hu, P. T. Chivers and M. J. Maroney, *Biochemistry*, 2013, **52**, 84–97.
206. A. W. Foster, C. J. Patterson, R. Pernil, C. R. Hess and N. J. Robinson, *J. Biol. Chem.*, 2012, **287**, 12142–12151.
207. R. R. Breaker, *Cold Spring Harbor Perspect. Biol.*, 2012, **4**, a003556.
208. M. D. Lundrigan, W. Koster and R. J. Kadner, *Proc. Natl Acad. Sci. USA*, 1991, **88**, 1479–1483.
209. A. G. Vitreschak, D. A. Rodionov, A. A. Mironov and M. S. Gelfand, *RNA*, 2003, **9**, 1084–1097.
210. D. A. Rodionov, A. G. Vitreschak, A. A. Mironov and M. S. Gelfand, *J. Biol. Chem.*, 2003, **278**, 41148–41159.
211. J. E. Johnson, Jr., F. E. Reyes, J. T. Polaski and R. T. Batey, *Nature*, 2012, **492**, 133–137.
212. A. Peselis and A. Serganov, *Nat. Struct. Mol. Biol.*, 2012, **19**, 1182–1184.
213. M. C. Perez-Marin, S. Padmanabhan, M. C. Polanco, F. J. Murillo and M. Elias-Arnanz, *Mol. Microbiol.*, 2008, **67**, 804–819.
214. J. M. Ortiz-Guerrero, M. C. Polanco, F. J. Murillo, S. Padmanabhan and M. Elias-Arnanz, *Proc. Natl Acad. Sci. USA*, 2011, **108**, 7565–7570.
215. L. Macomber and R. P. Hausinger, *Metallomics*, 2011, **3**, 1153–1162.
216. F. Barras and M. Fontecave, *Metallomics*, 2011, **3**, 1130–1134.
217. D. Schaumloffel, *J. Trace Elem. Med. Biol.*, 2012, **26**, 1–6.
218. I. Sovago, C. Kallay and K. Varnagy, *Coord. Chem. Rev.*, 2012, **256**, 2225–2233.
219. L. Macomber, S. P. Elsey and R. P. Hausinger, *Mol. Microbiol.*, 2011, **82**, 1291–1300.
220. C. Ranquet, S. Ollagnier-de-Choudens, L. Loiseau, F. Barras and M. Fontecave, *J. Biol. Chem.*, 2007, **282**, 30442–30451.
221. M. E. Rothenberg, *Nat. Immunol.*, 2010, **11**, 781–782.
222. M. Schmidt, B. Raghavan, V. Muller, T. Vogl, G. Fejer, S. Tchaptchet, S. Keck, C. Kalis, P. J. Nielsen, C. Galanos, J. Roth, A. Skerra, S. F. Martin, M. A. Freudenberg and M. Goebeler, *Nat. Immunol.*, 2010, **11**, 814–819.
223. T. F. Cheng, S. Choudhuri and K. Muldoon-Jacobs, *J. Appl. Toxicol.*, 2012, **32**, 643–653.
224. D. Beyersmann and A. Hartwig, *Arch. Toxicol.*, 2008, **82**, 493–512.
225. N. C. Giri, H. Sun, H. Chen, M. Costa and M. J. Maroney, *Biochemistry*, 2011, **50**, 5067–5076.
226. H. Chen, N. C. Giri, R. Zhang, K. Yamane, Y. Zhang, M. Maroney and M. Costa, *J. Biol. Chem.*, 2010, **285**, 7374–7383.
227. M. A. Zoroddu, T. Kowalik-Jankowska, H. Kozlowski, H. Molinari, K. Salnikow, L. Broday and M. Costa, *Biochim. Biophys. Acta*, 2000, **1475**, 163–168.
228. M. A. Zoroddu, M. Peana, T. Kowalik-Jankowska, H. Kozlowski and M. Costa, *J. Chem. Soc. Dalton Trans.*, 2002, 458–465.
229. W. Bal, H. Kozlowski and K. S. Kasprzak, *J. Inorg. Biochem.*, 2000, **79**, 213–218.

230. A. A. Karaczyn, W. Bal, S. L. North, R. M. Bare, V. M. Hoang, R. J. Fisher and K. S. Kasprzak, *Chem. Res. Toxicol.*, 2003, **16**, 1555–1559.
231. W. Bal, R. Liang, J. Lukszo, S. H. Lee, M. Dizdaroglu and K. S. Kasprzak, *Chem. Res. Toxicol.*, 2000, **13**, 616–624.
232. Y. Jin, M. A. Lewis, N. H. Gokhale, E. C. Long and J. A. Cowan, *J. Am. Chem. Soc.*, 2007, **129**, 8353–8361.
233. A. A. Karaczyn, F. Golebiowski and K. S. Kasprzak, *Chem. Res. Toxicol.*, 2005, **18**, 1934–1942.
234. M. Rowinska-Zyrek, S. Potocki, D. Witkowska, D. Valensin and H. Kozlowski, *Dalton Trans.*, 2013, **45**, 93–104.
235. M. Tabata and B. Sarkar, *J. Inorg. Biochem.*, 1992, **45**, 93–104.

CHAPTER 15

Platinum

FABIO ARNESANO,* MAURIZIO LOSACCO AND
GIOVANNI NATILE

Department of Chemistry, University of Bari "Aldo Moro", Bari, Italy
*Email: fabio.arnesano@uniba.it

15.1 Introduction

The name platinum is derived from the Spanish term *platina*, which is literally translated into *little silver*. It is one of the rarest elements on Earth's crust (the average abundance is approximately 5 µg kg^{-1}) and the least reactive (the noblest among metals). It occurs in some nickel and copper ores along with some native deposits. Native platinum was first used by pre-Columbian South American natives to produce artefacts. The first European reference to platinum appeared in 1557 in the writings of the Italian humanist Julius Caesar Scaliger as a description of an unknown noble metal found in Mexico, "which no fire nor any Spanish artifice has yet been able to liquefy", but it was not until Antonio de Ulloa published a report on a new metal of Colombian origin in 1748 that it became investigated by scientists.

15.1.1 Compounds

Elemental platinum is used in catalytic converters, laboratory equipment, electrical contacts and electrodes, dentistry equipment and jewellery.

It is generally unreactive, but dissolves in hot *aqua regia* to give aqueous chloroplatinic acid (H$_2$[PtCl$_6$], Figure 15.1), as in the following reaction:

$$Pt + 4\,HNO_3 + 6\,HCl \rightarrow H_2[PtCl_6] + 4\,NO_2 + 4\,H_2O$$

RSC Metallobiology Series No. 2
Binding, Transport and Storage of Metal Ions in Biological Cells
Edited by Wolfgang Maret and Anthony Wedd
© The Royal Society of Chemistry 2014
Published by the Royal Society of Chemistry, www.rsc.org

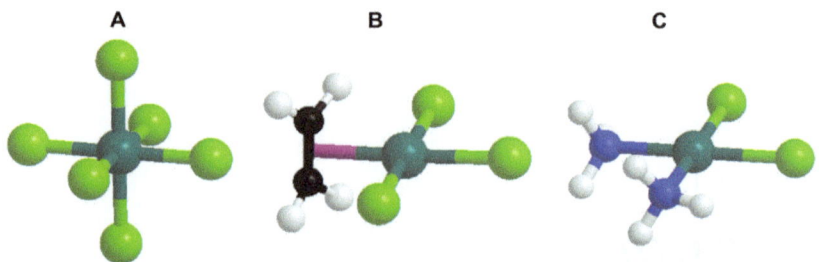

Figure 15.1   Ball and stick structures of hexa-chloroplatinate (A), Zeise's salt anion
(B) and cisplatin (C). Chlorine atoms are shown as green spheres,
carbon atoms as black spheres, nitrogen atoms as blue spheres and
hydrogen atoms as grey spheres. Platinum is displayed in pale blue.

The most common oxidation states of platinum are +2 and +4. The +1 and
+3 oxidation states are less common, and are often stabilized by metal
bonding in bimetallic (or polymetallic) species. $[PtCl_4]^{2-}$ can be obtained by
reduction of $[PtCl_6]^{2-}$ with a reductant such as hydrazine or glutathione and
ascorbic acid (to mention two biologically relevant species). While $[PtCl_6]^{2-}$,
as other six-coordinate Pt(IV) compounds, has the most common octahedral
geometry, $[PtCl_4]^{2-}$, as other four-coordinate Pt(II) compounds, adopts the
16-electron square planar geometry.

In the case of chloro species, the electric potentials for reduction of Pt(IV)
to Pt(II) and of Pt(II) to Pt(0) are: +0.68 V, $E°([PtCl_6]^{2-}+2e^- \rightarrow$
$[PtCl_4]^{2-}+2Cl^-$), and +0.73 V, $E°([PtCl_4]^{2-}+2e^- \rightarrow Pt^0+4Cl^-$).

These values are lower than $E°(Cl_2+2e^- \rightarrow 2Cl^-)$ (+1.36 V), $E°(Br_2+2e^- \rightarrow$
$2Br^-)$ (+1.07 V), and $E°(H_2O_2+2e^- \rightarrow 2OH^-)$ (+0.95 V), but higher than
$E°(I_2+2e^- \rightarrow 2I^-)$ (+0.54 V). Thus Pt(IV) species can be obtained from their
Pt(II) counterparts by oxidation with Cl_2, Br_2 or H_2O_2.

As a soft acid, platinum has a great affinity for nitrogen, sulfur and
phosphorous donor ligands, as well as other soft ligands such as ethylene
(C_2H_4). Zeise's salt anion $[PtCl_3(C_2H_4)]^-$, containing an ethylene ligand, was
one of the first discovered organometallic compounds (Figure 15.1). Its in-
ventor, W. C. Zeise, a professor at the University of Copenhagen, prepared
this compound in 1825 while investigating the reaction of $PtCl_2$ with boiling
ethanol, and proposed that the resulting compound contained ethylene.
Justus von Liebig, an influential chemist of that era, often criticized Zeise's
proposal, but Zeise's theories were decisively supported in 1868 when Karl
Birnbaum prepared the complex using ethylene.[1]

15.1.2   Ligand Substitution In Square-planar Complexes

Just as the inert $Co^{3+}L_6$ complexes have been widely studied as a model for
octahedral substitution, also the four-coordinate, square-planar, complexes
of d^8 Pt^{2+} have been used as model compounds for ligand substitution in
that geometry (substitution rate constants close to those of Co^{3+} complexes).

Square-planar complexes undergo substitution predominantly by an associative mechanism. Since an associative mechanism brings the two reactants together in the transition state without significant bond breaking, the negative $\Delta S^{\ddagger}$ is a good diagnostic of the increased order in such a mechanism. Moreover, an associative mechanism is much more reasonable for square-planar complexes than for octahedral complexes. The coordination number is lower, so that there is potential coordinative unsaturation. Furthermore, the planar arrangement leaves access open even for an entering ligand with fairly large cone angle unless the planar ligands themselves are quite bulky. Substitution in Pt^{2+} square-planar complexes occurs with retention of configuration, thus the first structure formed (the entering ligand E at the apex of a square-pyramid) must shift to a trigonal bipyramid in which the entering and leaving groups and the ligand *trans* to the leaving group are in the trigonal plane. This latter will then switch to a square-pyramid in which the leaving-group is now in the apical position (Scheme 15.1).

When comparing entering group (E) effect (the initial L_3Pt moiety and the leaving group D held constant) or leaving group (D) effect (the initial L_3Pt moiety and the entering group E held constant), it can be seen that there is a rate change of five-orders of magnitude or more. In the series H_2O, Cl^-, Br^-, I^- and NCS^-, H_2O is the worst entering and the best leaving group while NCS^- is the best entering and the worst leaving group.[2]

Besides the observable effects of entering and leaving groups on the kinetics of substitution in square-planar complexes, there is a strong and theoretically interesting effect due to the non-reacting ligand *trans* to the leaving ligand. The effect on the rate of substitution can be quite large (up to six orders of magnitude). The general order of *trans*-labilizing effect is:

$$CN^- \approx C_2H_4 \approx CO \approx NO > PR_3 \approx H^- \approx SC(NH_2)_2 > CH_3^- > C_6H_5^-$$
$$> SCN^- > NO_2^- > I^- > Br^- > Cl^- > py \approx NH_3 > OH^- > H_2O.$$

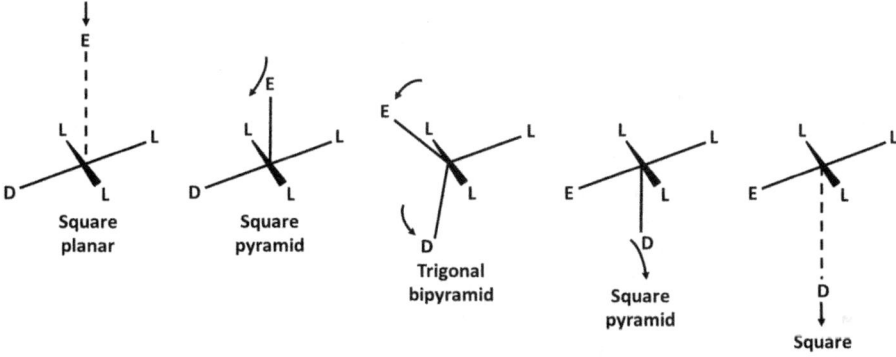

Scheme 15.1 Substitution mechanism for square-planar complexes with retention of configuration.

15.1.3 Catalysis

The most common use of platinum is as a catalyst in chemical reactions, often as platinum black (finely powdered metal). It has been employed in this application since the early nineteenth century, when platinum powder was used to catalyse the ignition of hydrogen. Its most important application is in automobiles as catalytic converters (nearly 50% of the 245 tons of Pt sold in 2010 were used for vehicle emission control devices), which allows the complete combustion of low concentrations of unburned hydrocarbons from the exhaust into carbon dioxide and water vapour (in 2007 Gerhard Ertl won the Nobel Prize in Chemistry for determining the detailed molecular mechanisms of the catalytic oxidation of carbon monoxide over platinum). Platinum is also used in the petroleum industry as a catalyst in a number of processes, but especially in catalytic reforming of straight run naphtha into higher-octane gasoline, which becomes rich in aromatic compounds. PtO_2, also known as Adams' catalyst, is used as a hydrogenation catalyst, specifically for vegetable oils. Being a heavy metal, platinum leads to health issues upon exposure to its salts but, due to its corrosion resistance, it is not as toxic as some other metals.

15.1.4 Cisplatin and Derivatives

Cisplatin, *cis*-diamminedichloridoplatinum(II) [*cis*-$PtCl_2(NH_3)_2$], was synthesized nearly two centuries ago by Michele Peyrone, an Italian doctor who, while visiting von Liebig's laboratory in Giessen, synthesized a new platinum compound containing two ammines and two chlorines per platinum (Figure 15.1).[3]

However, it was only in 1965 that Barnett Rosenberg, a biophysicist at Michigan State University, published his serendipitous discovery of the inhibition of microbial cell division by Pt complexes. This turned out to be a discovery of tremendous importance in the struggle against cancer. While investigating the effect of an electric field on division of *Escherichia coli* cells, by accident he generated electrochemically, from the platinum electrodes and the components of the culture medium, some Peyrone-like compounds.[4–6] Rosenberg noticed that bacteria stopped dividing normally and produced long filaments (some 300 times the normal length) rather than typical short rods. After a detailed chemical analysis, one component of this biological effect was identified as the neutral *cis*-isomer [$PtCl_4(NH_3)_2$]. Soon afterwards the reduced species [$PtCl_2(NH_3)_2$] was found to be more active and was later named *cisplatin*. A key finding was the early observation that cisplatin completely inhibited the development of Sarcoma-180 solid tumour in mice.[7–8]

After some years of clinical trials, the final approval for clinical use was granted from the United States Food and Drug Administration (US FDA) in 1978.[9] This first success triggered a renaissance in medicinal inorganic chemistry in the field of cancer treatment and fostered the synthesis and biological testing of several thousand platinum derivatives.

The investigation has been expanded to neighbouring elements of the periodic table, such as palladium, ruthenium and gold.

Since its approval by FDA, cisplatin has become one of the best-selling anti-cancer drugs; moreover many other Pt compounds display good or at least promising anti-cancer properties (Figure 15.2).

Cisplatin is totally curative for testicular cancer and is one of the most effective agents against melanoma and non-small-cell lung carcinoma. In combination therapy it is also noticeably active against breast, ovarian and bladder cancer.[10] Cisplatin and related drugs are regularly used for the treatment of both slowly growing and rapidly growing tumours, thanks to the ligand-exchange kinetics, which are in the same order of magnitude as the cell cycle progression of tumour cells.

Traditionally, anti-cancer therapy is based on drugs that target the DNA of rapidly dividing malignant cells. A severe drawback of this strategy is that healthy cells belonging to tissues with fast recycling and turnover are affected too, causing harmful and uncontrolled side effects. Many efforts have been aimed at the elimination, or at least reduction, of nephrotoxicity, ototoxicity and hepatotoxicity (the most common side effects of cisplatin) and to the enhancement of drug oral bioavailability (to avoid demanding intravenous infusions).

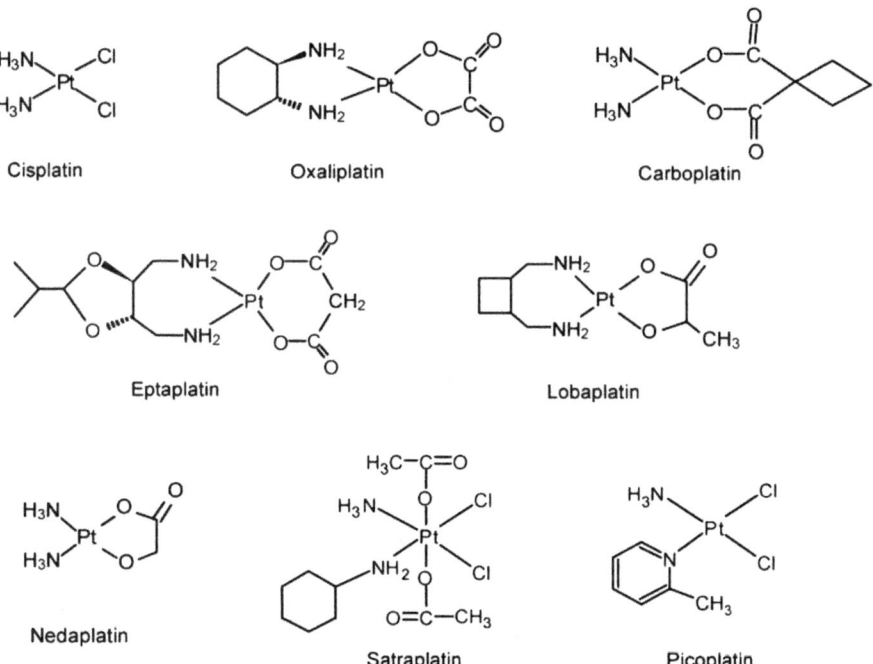

Figure 15.2 Structures of the clinically active Pt-drugs in use worldwide (first row) or regionally, plus satraplatin and picoplatin, both in advanced clinical trials (second and third row).

A rational design of novel platinum compounds directly targeted to specific cellular pathways of cancer cells would limit side effects and render cisplatin-based therapy better tolerated and safer for patients. Another major issue is tumour resistance either intrinsic (as for prostate, lung or breast cancer) or acquired during prolonged cycles of therapy (as occurs with ovarian cancer). Such resistance reduces the cure rate. Furthermore, the cellular response conferring resistance to cisplatin is multifactorial and not well understood.

Some of these safety questions have been solved, at least to some extent, by the introduction of second- and third-generation drugs, carboplatin and oxaliplatin, respectively.

One aspect of cisplatin toxicity was ascribed to the relatively fast *aquation* rate that favours interaction of the drug with many biological ligands on its way to nuclear DNA. In response, the less labile compound carboplatin was developed. It contains the same carrier ligands as cisplatin (two ammines) but uses the less labile chelating cyclobutanedicarboxylate (CBDCA) ligand as leaving group (Figure 15.2).

Carboplatin displays reduced nephrotoxicity, which largely compensates for its reduced cytotoxicity (approximately ten-fold less than that of cisplatin).[11] It was approved for clinical use by the US FDA in 1989. The mechanism of action of carboplatin is similar to that of cisplatin, leading to extensive cross resistance.

In an attempt to overcome this cross resistance, a third-generation platinum drug has been developed: oxaliplatin harbours *R,R*-1,2-diaminocyclohexane (DACH) as carrier ligand and oxalate as leaving group (Figure 15.2). It is significantly less nephrotoxic than cisplatin and carboplatin and, as its cytotoxic DNA-adducts are processed differently from those formed by the other two compounds, it is effective against cisplatin/carboplatin-insensitive malignancies and, in particular, against colorectal cancer.[12] Oxaliplatin was authorized for clinical use in 2002, after US FDA endorsement. The main side effects of carboplatin and oxaliplatin are myelosuppression and neuropathy, respectively.

During three decades, several other platinum derivatives sharing the same *cis*-PtX$_2$(amine)$_2$ structure of cisplatin, where X is a leaving group and amine is any primary or secondary amine, have displayed equivalent or even boosted cytotoxic activity towards various cancer cell lines. Nevertheless, only 10–15 complexes have moved to advanced clinical trials and only three have been approved for clinical practice, but not with worldwide extension (nedaplatin in Japan, lobaplatin in China and eptaplatin in South Korea: see Figure 15.2).

Among complexes in advanced clinical trials, there are water soluble Pt(IV) compounds, such as satraplatin (JM216; Figure 15.2), which could be suitable for oral administration. Specifically, satraplatin has been investigated for the treatment of hormone-resistant prostate cancer.[10] Another promising drug appears to be picoplatin (Figure 15.2), which has entered Phase III clinical trials for small-cell lung cancer.

New-generation Pt-based drugs that do not follow the traditional structure-activity rules established for platinum cytoxicity are *trans*-compounds (JM335) or even polynuclear complexes (BBR3464).[13]

More recently, aspects of drug targeting and delivery are receiving increasing consideration, owing to their potential for improving effectiveness and reducing side effects. A variety of biological targets can be exploited, particularly those that are expressed differently in normal and cancerous tissues and cells. From the point of view of delivery, the use of liposomes for carrying anti-cancer drugs appears to be very promising. These vectors can greatly increase the amount of drug available to a malignant cell. Several liposomal formulations of platinum drugs have been developed. Lipoplatin is currently the most favourable as it is taken up by tumour cells in preference to cisplatin and displays much reduced nephrotoxicity and related side effects while retaining comparable cytotoxic efficacy.[14] It is now undergoing various Phase III clinical trials.

Oestrogen receptors are overexpressed in several cancers including breast, uterine and ovarian cancers; therefore, oestrogens have been linked to cisplatin derivatives or to Pt(IV) compounds. Some of these compounds turned out to be much more cytotoxic than cisplatin, indicating that Pt compounds with high receptor binding affinity are very promising therapeutic tools.[15] Additional interest in using oestrogenic steroids to deliver platinum-based compounds to tumour cells arises from the fact that oestrogen can induce overexpression of HMGB1 (high mobility group binding protein 1), a protein that protects cisplatin-DNA adducts from nucleotide excision repair, thereby conferring higher sensitivity to the drug.

15.2 Coordination Chemistry of the Platinum Core

15.2.1 Inside Cells

Cisplatin is a neutral inorganic compound with square planar geometry, containing a Pt(II) core bound to two inert ammine ligands and two labile chlorido ligands in *cis*-configuration (see Figure 15.1). It is administered intravenously, dissolved in a large volume preparation that includes 154 mM NaCl. Since cisplatin is inert, it is assumed that its mechanism of action involves drug activation by hydrolysis (*i.e. aquation*) inside cells, where the Cl^- concentration is considerably lower than in the extracellular environment (3–20 mM *versus* >100 mM, Figure 15.3).[16–17] Outside cells, the hydrolysis of cisplatin appears to be suppressed by the high chloride concentration present in the blood plasma and interstitial fluids allowing the drug to reach the outer surface of cells as a neutral molecule. Inside cells, the strategic process of hydrolysis can take place.

Substitution of a chloride ligand by water in the first *aquation* step leads to the cation *cis*-$[PtCl(OH_2)(NH_3)_2]^+$, which undergoes proton dissociation to form the neutral chlorido-hydroxido complex *cis*-$[PtCl(OH)(NH_3)_2]$. The hydrolysis is generally considered a prerequisite for reaction with the ultimate target, nuclear DNA.[18] The *aquated* species can also be involved in reactions

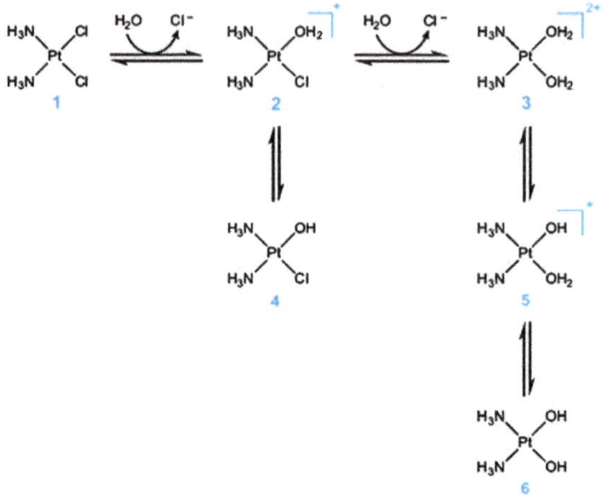

compound	100mM	4mM
1	68	3
2	7	5
3	<1	<1
4	24	30
5	<1	28
6	<1	35

Figure 15.3   Scheme showing the stepwise *aquation* and hydrolysis of cisplatin in aqueous solution. The percentage of each species present at equilibrium (pH 7.4) in high (100 mM) and low (4 mM) chloride concentrations is reported in the table.
Adapted with permission from ref. 21. Copyright 2008 Annual Reviews.

with other biomolecules that can contribute to the overall cancer cells toxicity. The monoaqua cation $[PtCl(OH_2)(NH_3)_2]^+$ (formation half-time $t_{1/2} \approx 2$ h) is much more reactive than the parent cisplatin and can covalently bind to N7 of guanine or adenine bases in DNA, to form a monofunctional adduct ($t_{1/2} \approx 0.1$ h). Hydrolysis of the second chloride followed by coordination of a second purine base can then lead to the formation of a cross-link. These cross-links (either inter- or intrastrand) induce changes to the secondary structure of DNA, including remarkable bending and unwinding. In turn, these inhibit fundamental processes that require DNA-strand separation, such as transcription and replication, thereby promoting cell killing.[19–20] Although the ultimate outcome is usually apoptotic cell death, the complex pathway intervening between binding to DNA and apoptosis has still to be unravelled in detail.

In order to modify nuclear DNA, cisplatin must enter the nucleus. How platinum drugs can cross the nuclear membrane is not yet clear. Given the slow ligand exchange characteristic of the platinum core, it is generally assumed that the reaction of cisplatin with cellular constituents is under kinetic rather than thermodynamic control.

This premise can rationalize why cisplatin can bind to DNA in the nucleus, instead of exclusively reacting with other abundant biological substrates, particularly those containing the soft sulfur-donor amino acids cysteine and methionine that have a high affinity for platinum.[22] What is more, platinum can migrate from S-donor ligands to guanine bases.[23] Analyses carried out on DNA extracted from cisplatin-treated patients revealed the presence of about 65% 1,2-d(GpG), 25% 1,2-d(ApG) and 5–10%

1,3-d(GpNpG) intrastrand cross-links.[24] The residual part comprises inter-strand cross-links and monofunctional adducts. In comparison to cisplatin, carboplatin and oxaliplatin have different leaving groups and so exhibit different reaction rates. However, they do not appear to generate different adducts with DNA.

Intrastrand cross-links cause a significant alteration of the DNA structure. The most frequent adduct, the 1,2-intrastrand cross-link, unwinds the DNA duplex near the platination site, bends the double helix by 32–35° toward the major groove and causes the minor groove to widen (Figure 15.4). Conversely, the interstrand DNA cross-link bends the duplex DNA by 20–40° toward the minor groove, as revealed by the crystal structure of a double-stranded DNA decamer complexed with cisplatin.[25]

It is conceivable that each DNA alteration is unique and may evoke distinctive recognition and processing pathways mediated by cellular proteins. These differences very likely result in specific anti-cancer properties for each of the Pt compounds analysed.

Once the Pt-GG adduct is recognized by proteins belonging to the various DNA repair systems, nucleotide excision (NER) and mismatch repair (MMR)

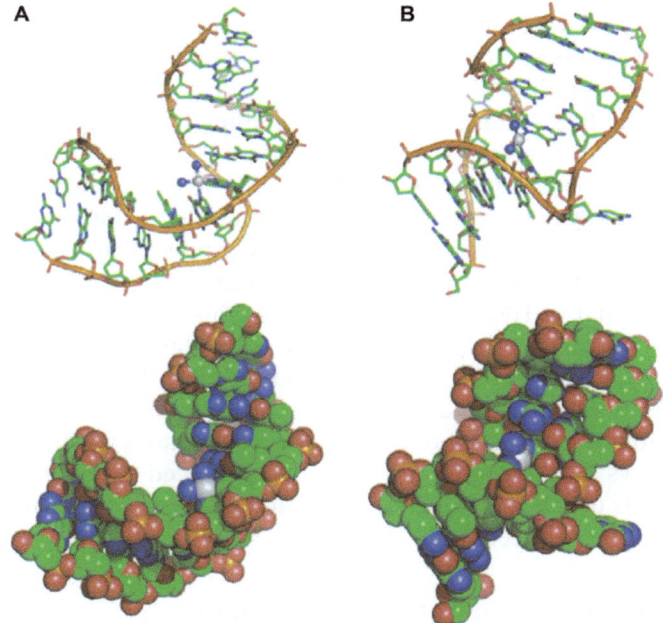

Figure 15.4 Structure of the platinum-DNA adducts containing (A) cisplatin 1,2-d(GpG) intrastrand and (B) interstrand cross-links, generated by PyMol. The DNA sequences are d(CCTCTG*G*TCTCC)d(GGAGACCA-GAGG) and d(CCTCG*CTCTC)d(GAGAG*CGAGG), respectively. Guanine residues linked by cisplatin at the N7 position are indicated by asterisks in the DNA sequences.
Adapted with permission from ref. 27. Copyright 2007 American Chemical Society.

being the most operative, the fate of the kinked DNA structure could be either its stabilization or its removal.[26]

The biological consequences of the protein binding at platinated purine sites are hence a crucial point to clarify in order to develop dedicated drugs. There are different views regarding which lesion is responsible for greater cytotoxicity. At present, it is believed that the most abundant, 1,2-d(GpG) intrastrand cross-link is also the most important. Apart from its high frequency, this lesion is specific for cisplatin (the inactive isomer of cisplatin, *transplatin*, is only able to form 1,3-intra- and interstrand adducts) and the therapeutic outcome generally correlates with the concentration of the 1,2-d(GpG)Pt(NH$_3$)$_2$ metabolite in the biological fluids.[28] Furthermore, high mobility group binding proteins (HMGB) recognize and bind to DNA at the site of the 1,2-d(GpG) intrastrand cross-link. HMG domains are approximately 80 amino acid-long motifs that are closely associated with the curvature of chromatin. Their presence is crucial for rendering cancer cells sensitive to Pt-drugs. Members of the HMGB family, including HMGB1, able to bind to DNA modified by platinum 1,2-d(GpG) cross-links, can prevent the translesion synthesis.[29]

Moreover, some HMGB proteins can block NER machinery by shielding the cisplatin-induced DNA-lesion.[30] The cisplatin-DNA-HMGB1 ternary complex has also been shown to inhibit the activity of transcription factors, consequently stopping both transcription and replication. As a result of hampering the latter vital functions, a signalling cascade, prompted by the DNA damage, can eventually initiate apoptosis.[31] The testicular tissue overexpresses several HMGB proteins, and is thus extremely sensitive to cisplatin. Recently, a new member of the mammalian HMGB protein family, HMGB4, has been found to be preferentially expressed in the testis and to display a stronger binding affinity ($K_D = 4.35$ nM) for the platinated adduct compared to that of HMGB1 ($K_D = 120$ nM). This higher DNA-repair inhibition ability mirrors the hypersensitivity of testicular tumours to cisplatin.[32] Interestingly, the same authors lately reported that the redox cellular state can strongly influence the affinity of HMGB1 protein for platinated DNA, hence modulating the activity of cisplatin as an anti-cancer drug. *In vitro* analysis shows that the protein with the two reduced cysteine residues C22 and C44 binds to the 1,2-d(GpG) intrastrand cross-link with a 10-fold greater affinity than the oxidized one.[33]

15.2.2   In the Extracellular Environment

After cisplatin reaches the bloodstream (by intravenous injection or infusion), the drug is transported through all body districts. The exchange of the chlorido ligands with solvent molecules takes place on a time-scale of a few hours, although the excess of chloride ions in the blood favours the dichlorido species. Nevertheless, the small fraction of the compound undergoing *aquation* can interact with proteins present in high concentration in the blood plasma.

Human serum albumin (HSA, 66 kDa) is the most abundant plasma protein (0.6 mM) and any anti-cancer metallodrug is expected to interact in some way with this macromolecule, with significant consequences on its bioavailability and systemic toxicity. Without exception, platinum drugs bind extensively to HSA.[34]

One day after intravenous administration of cisplatin, most of the drug (about 65–98%) is found irreversibly bound to HSA and consequently kept in circulation. The major platinum binding sites in HSA appear to be methionine (Met) residues rather than cysteine. Met298 has the highest surface accessibility of the six Met residues present in HSA and is therefore considered to be the main cisplatin-binding site. The Met residue binds to cisplatin forming monofunctional adducts as well as *S,N*-chelates.[35]

Oxaliplatin also binds to human plasma proteins (albumin and γ-globulins).[36] However, other platinum-based drugs in clinical use in some countries, such as nedaplatin and lobaplatin, show poorer association with plasma proteins. Although the quasi-irreversible albumin-cisplatin binding would not support the idea of a drug reservoir with therapeutic potential, a number of studies have shown that the drug-protein interaction can have a positive clinical effect. For example, both free and protein-bound cisplatin appear to exhibit similar effects in seven tumour models and administration of cisplatin-HSA seems to increase the concentration of Pt in tumour cells.[37–38]

In addition, some reports have pointed out that decreased levels of plasma albumin can increase the major marrow-, nephro-, hepato- and ototoxicity of cisplatin.

Human serum transferrin (HST, 80 kDa) is another protein that actively binds platinum agents. Iron-loaded transferrin enters cells *via* a receptor-mediated endocytotic process. It is conceivable that HST can help delivering Pt to tumour cells, on the basis that some tumours can overexpress such a receptor. The use of transferrin with isotopically labelled $^{13}CH_3$-Met groups allowed the identification of specific methionines that are platinated.[39] These studies suggest again, as shown for HSA, that the preferred binding sites for platinum are the monodentate solvent-exposed Met sulfur atoms. Since S-binding of monodentate Met to Pt(II) is reversible, HST could act as a delivery tool for Pt anti-cancer complexes.

15.3 Transport through the Cytoplasmic Membrane

With respect to entry of platinum drugs into the cell, it is mostly accepted that the neutral intact drug crosses the plasma membrane by passive diffusion.[40] In the case of cisplatin, this assumption is supported by the observed linear cellular uptake that occurs without saturation up to 1 mM concentration. Furthermore, cisplatin analogues fail to inhibit its absorption, demonstrating the absence of specific cellular transporters.[41]

However, accumulated evidence suggests that, together with passive diffusion, active mechanisms can contribute to cisplatin uptake and efflux.[42–44]

In this scenario, the primary actors appear to be the organic cation transporters (OCTs), belonging to the SoLute Carrier superfamily (SLC), and the major copper influx transporter CTR1. Moreover, other non-saturable systems, such as fluid-phase *endocytosis* mediated by membrane invagination, have been proposed.[45]

15.3.1 Organic Cation Transporters

The role of the SLC superfamily in the facilitated transport of anti-cancer platinum drugs has emerged recently. The subgroup of OCTs comprises three members: SLC22A1 (OCT1), SLC22A2 (OCT2) and SLC22A3 (OCT3), all characterized by 12 transmembrane domains.

They are involved in drug absorption, distribution and excretion. An indirect indication that they may have a role in platinum transport came from the finding that tissues overexpressing OCTs, such as the kidney, are frequently affected by platinum-derived side effects. Unlike SCL22A1, for which there is no clear evidence, SLC22A2 has been convincingly demonstrated to be involved in the uptake and cytotoxicity modulation of various platinum compounds.[46] Oxaliplatin is definitely an excellent SLC22A2 substrate. Several studies have demonstrated that the uptake of oxaliplatin in transfected hOCT2 cells is significantly higher than that observed in control cells and also significantly reduced in the presence of specific transporter inhibitors, such as cimetidine. Furthermore, the increased uptake was accompanied by augmented cytotoxicity.[47]

15.3.2 Copper Transporters

The proteins that mediate Cu influx and efflux were first associated with Pt-drugs when it was noticed that cells selected for resistance to cisplatin were cross-resistant to various other metals, including copper. In a complementary way, cells selected for resistance to Cu were found to be cross-resistant to cisplatin. It was hypothesized that Pt-drugs could utilize the cellular copper transport machinery for entry into and export from cells (Figure 15.5).

Experimental evidence confirms that alterations in the expression levels of the influx transporters CTR1 and CTR2, the soluble metallochaperone ATOX1 and the efflux transporters ATP7A and ATP7B modify the sensitivity of the cells to the toxic effect of Pt-drugs.[43,48–50] Additional indications that transporters of the CTR family were involved in the uptake of cisplatin came from experiments performed on baker's yeast cells. When the CTR1 gene was deleted or silenced, there was an increased resistance to cisplatin that correlated with reduced uptake of the drug and reduced formation of the cell-killing platinum-DNA adducts. What renders all copper transporter proteins very likely to interact with the platinum drugs is the high affinity for platinum of the copper binding motifs, as supported by crystallographic and mass spectrometry experiments.[51–53] Although these observations suggest

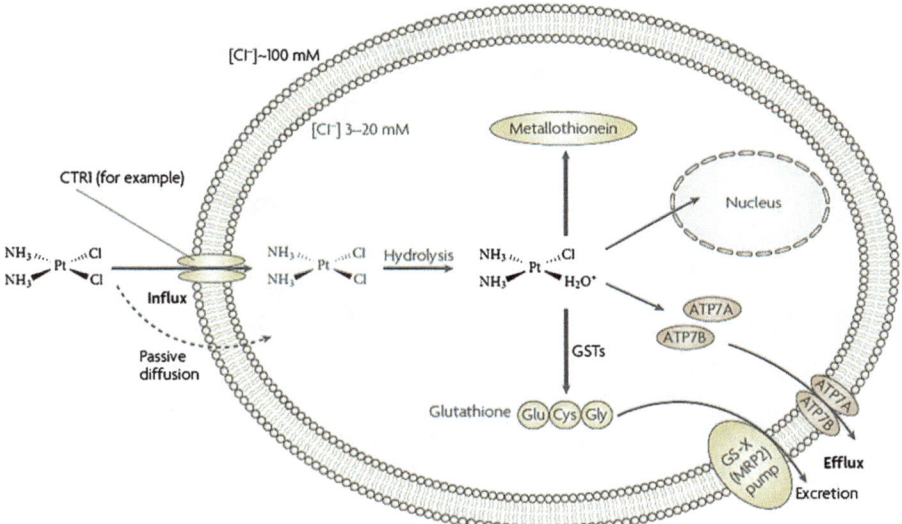

Figure 15.5 Schematic diagram depicting the path of cisplatin once entered in the cell, through CTR1 or passive diffusion, along its way to the nucleus. Reprinted with permission from ref. 10. Copyright 2007 Nature Publishing Group.

that cisplatin uptake and efflux is influenced by Cu transporters, the mechanism involved is poorly understood.

15.3.2.1 CTR1

Copper transporter 1 (*SLC31A1*) is an evolutionarily conserved copper influx transporter present in plants, yeast and mammals, and is the main copper importer in mammalian cells. Human CTR1 is expressed in all tissues and is crucial for the finely tuned homeostatic regulation of intracellular copper levels to guarantee nutritional delivery of copper to enzymes, such as cytosolic Cu,Zn-superoxide dismutase and mitochondrial cytochrome c oxidase, while impeding copper accumulation, which can be harmful.[54]

Human CTR1 contains 190 amino acids organized into i) three transmembrane (TM) domains, ii) an N-terminal extracellular domain rich in methionines and histidines, iii) a large intracellular loop and iv) a short intracellular C-terminal domain. Conserved methionine-rich motifs and individual methionines, histidines and cysteines, essential for Cu transport, are located within the extracellular domain, within the second and third transmembrane domains and in the C-terminal tail (Figure 15.6).

All the most recent reports indicate that CTR1 forms a homotrimer in the membrane, and structural studies suggest that this trimer features an inverted, cone-shaped pore through which Cu(I) is conducted from one side to the other side of the membrane.[55]

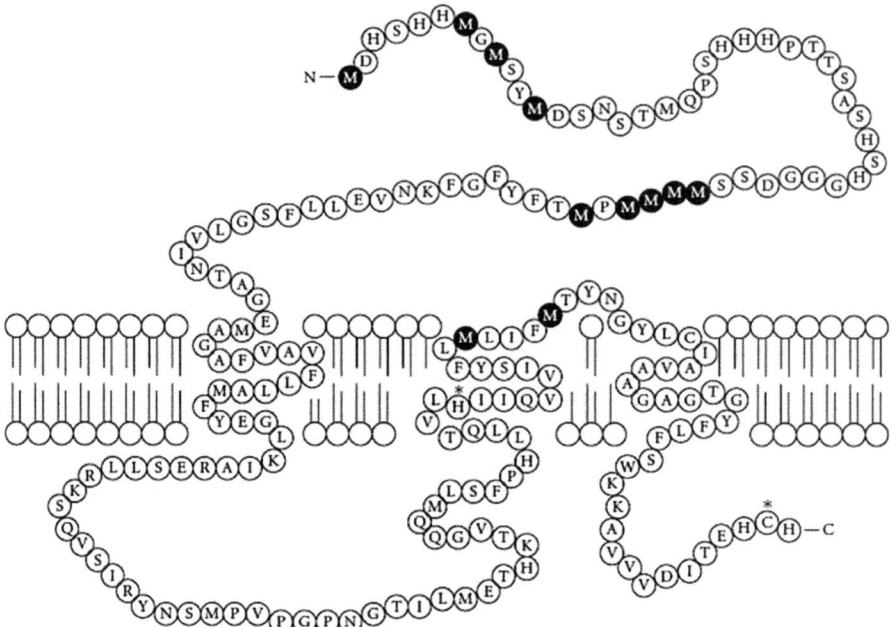

Figure 15.6 Schematic diagram of human CTR1 showing membrane topology and amino acidic sequence. Important methionine residues within the metal binding sequences and transmembrane regions are shown as darkened residues. The asterisks denote other amino acids (H139 and C189) crucial for Cu transport.
Reprinted with permission from ref. 57.

Recent computational studies provide further support for the relevance to the mechanism of transport of several of these conserved residues, particularly those on the second transmembrane domain.[56] Early studies have recognized the importance of Met150, Met154 and His139 for the transport of copper. Structural studies have shown how these amino acids are placed in the narrowest part of the trimeric CTR1 pore, hence in the most crucial region where they could favour copper transit by coordination (methionines and histidines are frequently involved in copper binding to various proteins).

To investigate the mechanism by which hCTR1 transports both copper and cisplatin, cancer cell lines were transfected with an hCTR1 mutant in which both Met150 and Met154 were converted to isoleucines, or in which His139 was converted to alanine. Both variants stopped copper transport and increased resistance to the cytotoxicity mediated by copper. Surprisingly, such modifications increased cisplatin accumulation and the cell killing effect, which was greater than that of wild-type hCTR1. Neither variant impaired the capacity of cisplatin to elicit hCTR1 degradation. These results support the pivotal role of hCTR1 in cisplatin transport and identify Met150 and Met154 as potential pharmacological targets for improving the cellular uptake of cisplatin.

After passing into the cell through the CTR1 pore, Cu is delivered to chaperones. In all these steps, the *free* Cu(I) ion appears to be present. Relatively little is known about the fate of a platinum drug using the same transport pathway: does Pt retain the ammine ligands or enter in a ligand-free form?

In a spectroscopic study, a seven-residue peptide having the amino acidic sequence of one extracellular methionine-rich motif of yeast CTR1 (called Mets7) was reacted with a selection of platinum complexes, showing that cisplatin loses both ammine ligands and appears to be finally bound to the three thioether ligands supplied by the Met residues of Mets7 (Figure 15.7). In contrast, transplatin retains its two ammine ligands.[51]

Both circular dichroism (CD) and theoretical studies indicated that the peptide conformation changed from that typical of a random coil for apo-Mets7 to that typical of a β-turn after reaction with cisplatin.[58]

These results underline a perfect parallelism between copper and platinum, both being metal ions taken up by Met-rich motifs in the naked form. But if this is physiological for copper, which is then taken up by soluble chaperones, it is not acceptable for platinum, which must keep the ammine ligands to remain pharmacologically active.[60–61]

A *stripped* platinum ion would be useless in stimulating anti-tumour activity. An attractive hypothesis, still to be confirmed, is that platinum binding to CTR1 could induce formation of endocytic vesicles incorporating a portion of the extracellular solution, which may contain active cisplatin (Figure 15.8). Incorporated in this way, the drug may be protected from attack by cytosolic platinophiles (*e.g.* metallothioneins and glutathione) until it is delivered to subcellular targets, which might include the nucleus.[51] Existing data suggest that, at least in human ovarian cancer cells, cisplatin noticeably increases the degree of CTR1 internalization through macropinocytosis, followed by proteasomal degradation. A probable explanation, yet to be proven, is that drug binding to the extracellular N-terminal domain

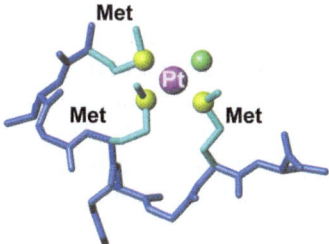

Figure 15.7   Structural model of the adduct between cisplatin and Mets7 peptide. As an example, a chloride ion was placed at the fourth coordination position but it could be a solvent molecule as well. Sulfur atoms of methionines are shown as yellow spheres, the chloride ion as a green sphere and platinum in magenta.
Reprinted from ref. 59. Copyright 2013 WILEY-VCH.

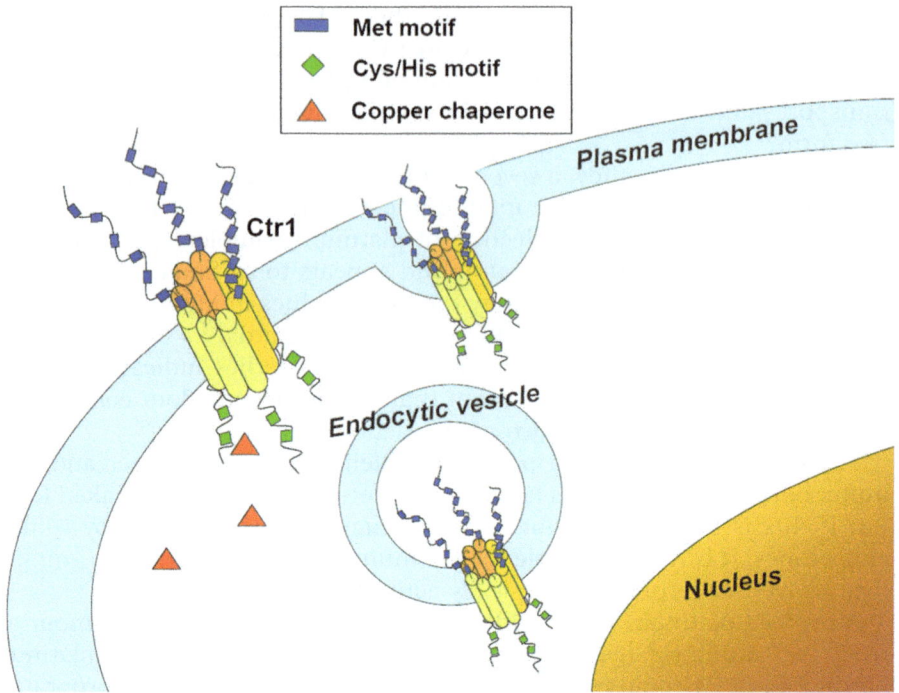

Figure 15.8   Model of cisplatin trafficking mediated by CTR1. The formation of endocytic vesicles incorporating extracellular solution containing active drug entities could prevent drug inactivation by metallothioneins and other platinophiles.
Reprinted with permission from ref. 51. Copyright 2007 WILEY-VCH.

of CTR1 induces a conformational modification, tagging the protein to macropinosomes.[62]

Current data also indicate that copper and cisplatin interact with CTR1 differently, as indicated by FRET measurements carried out on yCTR1.[63] In HEK293 cells, treatment with cisplatin results in the formation of a reduction-resistant trimeric form of endogenous CTR1, which can be visualized by gel electrophoresis. Formation of such stable trimers is not observed following copper treatment of the same cells. Furthermore, Met to Ala substitution in both Mets motifs of hCTR1 (A^7GASYA^{12} and $A^{40}AAAPA^{45}$) prevents the cisplatin-induced stabilization of CTR1 trimers, presumably due to loss of binding. By contrast, copper still binds and induces CTR1 endocytosis.[64] Altogether, these observations suggest that cisplatin is likely to bind to the N-terminus of CTR1 and cross-link the CTR1 monomers without blocking copper binding and copper-induced endocytosis.[65] In this case, cisplatin might be taken into cells as a *passenger* during the normal course of copper-dependent endocytosis of CTR1.

The C-terminus is required as well for the cellular uptake of both copper and cisplatin. A recent work has addressed different C-terminal mutated variants of CTR1, showing that only a deletion encompassing the last 12 residues at the C-terminus (^{179}KAVVVDITEHCH^{190}) totally abolishes the cellular uptake of cisplatin.[66] A single mutation at Cys189, a site that is involved in the stabilization of the CTR1 multimer, affects cisplatin uptake only partially.[66] The C-terminal portion of CTR1 is expected to be similarly important for cisplatin-mediated endocytosis, as observed for copper-dependent endocytosis and degradation in yeasts, where two critical lysine residues (Lys340 and Lys345) in the C-terminal cytoplasmic tail of CTR1 function as acceptor sites for ubiquitination by the ubiquitin ligase RSP5.[67] This hypothesis is also supported by the observation that mouse CTR1, which has two lysine residues at the C-terminal end (Lys185 and Lys186), is ubiquitinated *via* a process requiring ATOX1.[68]

On this basis, impairment of copper and cisplatin influx due to truncation of the C-terminal tail of human CTR1, which contains two neighbouring lysine residues as well, could be attributed to the lack of the interaction site for ubiquitination enzymes, needed to favour CTR1 endocytosis, instead of the block of metal delivery to ATOX1, which has been proposed to proceed *via* transchelation reactions ultimately involving Cys189 at the C-terminus of CTR1.

The fate of cisplatin taken up by the proposed CTR1 endocytosis has yet to be defined. While not being detected at significant levels in the cytoplasm of ovarian carcinoma cells, fluorescein-labelled cisplatin (F-DDP) accumulates in the vesicular structures belonging to the lysosomes, *trans*-Golgi network (TGN) and secretory pathways expressing the copper efflux protein ATP7A. In addition, redistribution of F-DDP to other subcellular compartments occurs mainly *via* vesicle trafficking.[69]

A recent study reported the co-localization of CTR1 with a fluorescent cisplatin analogue in vesicular structures inside both A2780 and A2780 cisplatin-resistant ovarian tumour cells. This finding suggests that cisplatin may be rapidly endocytosed with CTR1 during the continuous recycling of the transporter between the intracellular compartment and the plasma membrane, but it is still not clear whether or not the drug is bound to CTR1 during internalization.[70]

Hence, cisplatin would require release from the endocytic vesicles into the cytosol and the nucleus. What molecular machinery might mediate this step remains to be determined.

More recently, the correlation between CTR1 overexpression and increased Pt-drug uptake in human ovarian cisplatin-sensitive tumour cells was confirmed; however, it was proposed that hCTR1 is not the major entry route of platinum drugs and that the copper transporter is not internalized in response to extracellular drug.[71]

15.3.2.2 CTR2

The intricacy of cellular uptake of cisplatin was further highlighted by experiments exploring the role of a CTR1 homologue, CTR2. This pump is mainly distributed in the late endosome and lysosome/vacuole compartments of the cell and is structurally similar to CTR1, notwithstanding only 22% amino acid sequence identity.[72] Lacking an extended N-terminal domain, CTR2 contains the three putative transmembrane domains with a MXM motif at the N-terminus and a conserved MXXXM motif in TM2. Similarly to its yeast homologue, human CTR2 releases copper from intracellular copper stores. However, there is also evidence in mammalian systems that CTR2 can function as a copper influx transporter inside the subcellular compartments in case of excess concentration of extracellular copper.[73–74]

CTR2 has a significant influence on the accumulation of both cisplatin and carboplatin.[49] However, instead of limiting import, knockdown of CTR2 increases uptake of both drugs by 2–3 times, an effect opposite to that of knocking down CTR1. In addition, copper chelators that are able to decrease the expression level of CTR2 also increase Pt-drug uptake. This increase translates into increased Pt-DNA adduct formation and cytotoxicity. The issue of CTR2 expression being affected by exposure to Cu and cisplatin has been addressed in various model systems but with ambiguous results. Copper depletion leads to a decrease of CTR2 mRNA and protein half-life, suggesting both transcriptional and post-transcriptional regulation.[75] For instance, the decrease in CTR2 half-life following copper depletion is, at least partly, attributable to increased proteasomal degradation: the proteasome inhibitor bortezomib is able to hamper this effect.[76] This post-transcriptional regulation of CTR2 seems to depend on the copper chaperone ATOX1. ATOX1 binds both Cu(I) and cisplatin, using the two cysteine residues of its CXXC motif, and may function as a metal sensor signalling the switch between up- and down-regulation of CTR1 and CTR2. However, the molecular details of such a mechanism are yet to be clarified.[57]

It is assumed that CTR2 also forms a trimeric pore in the plasma membranes. It limits cisplatin uptake and this appears to be the result of reduced influx rather than greater efflux. This effect is independent of CTR1 and it seems that these two pumps mediate cisplatin accumulation by distinct mechanisms. Based on its role in *S. cerevisiae*, the primary function of CTR2 seems to be the delivery of Cu- and Pt-drugs to intracellular compartments. However, the observation that Pt accumulation in the microsomal fraction is unaffected in cells in which the expression of CTR2 is abolished questions the role of CTR2 in the sequestration and efflux of cisplatin from vesicles.[49] Therefore, how CTR1 increases and CTR2 limits Pt-drug uptake remains to be settled.

15.3.2.3   *Other Transporters*

In recent studies it was proposed that the multidrug resistance-associated protein MRP2 (belonging to the family of ABC-transporters) could be involved in the transport of cisplatin. Higher levels of MRP2 in human carcinoma cell lines confer elevated cisplatin resistance, poorer intracellular accumulation of cisplatin and decreased DNA adduct formation and cytotoxicity.[77–78]

Likewise, overexpression of antisense MRP2 in human hepatic cancer cells leads to an increased sensitivity to cisplatin.[79] MRP2 functions as an ATP-dependent efflux pump that, very likely, eliminates platinum drugs through the formation of Pt-GSH conjugates.[80]

Most recently, it was shown that the endocytic-lysosomal recycling pathway is altered in cisplatin-resistant cell lines. Using the epidermal growth factor (EGF) as a marker of the endocytic pathway, it was reported that while the kinetics of EGF uptake are similar for cisplatin-sensitive and cisplatin-resistant lines, the latter cells have an impaired internalization of the protein, due to failure of the endosomal recycling compartment.[81] These results depict cisplatin as an endosomal poison, probably working as a protein cross-linker that damages the physiology of membrane-associated processes, including endocytosis. CTR1 would then be an expected target.

Some recent studies also suggest the involvement of transporters belonging to the multidrug and toxin extrusion (MATE) family in the active efflux of various anti-cancer platinum agents. Human MATE-type transporters, hMATE1 and hMATE2-K, are able to traffic both cisplatin and oxaliplatin, but not carboplatin.[82] It has been proposed that MATEs are implicated in the renal elimination of platinum compounds and are critical factors in platinum-induced nephrotoxicity.[83]

15.4 Transport within the Cell (Metallochaperones *vs.* other Binders)

As previously mentioned, cisplatin does not react directly with DNA. It must first undergo a rate-limiting *aquation* yielding *cis*-$[PtCl(H_2O)(NH_3)_2]^+$ or a related form. Thus, it is of striking importance to know the half-life for the formation of *cis*-$[PtCl(H_2O)(NH_3)_2]^+$ in the cytoplasm, assuming that it is not associated with a carrier molecule. The half-life for formation of *cis*-$[PtCl(H_2O)(NH_3)_2]^+$ in water (or buffer) is around 120 minutes.[84] Sulfur-containing nucleophiles in the cytoplasm are likely to react with cisplatin at a considerably faster rate, reducing the cytoplasmic concentration of *cis*-$[PtCl(H_2O)(NH_3)_2]^+$.

Even today, almost 50 years after the discovery of its anti-cancer activity, the half-life of cisplatin in cancer cells is not yet known with certainty. Unfortunately, this information cannot be easily retrieved, even with the cutting-edge experimental techniques available today.

15.4.1 Metallothioneins

The reaction of metallothioneins with several platinum complexes has revealed that inactive transplatin preserves both ammines notwithstanding the fact that these proteins are highly rich in sulfhydryl groups. On the other hand, cisplatin readily loses its ammine ligands.[85] Cisplatin loses immediately one ammine ligand even reacting with cytochrome *c*, which contains a single sulfur ligand (a methionine) on its surface.[86]

Several authors have stressed the fact that, given the strong thermodynamic preference of Pt for S-donor ligands and the presence of so many cellular platinophiles in the cytosol, cisplatin itself should never reach the nucleus.[87]

Cellular thiols represent a major reservoir for metal ion binding, including platinum agents. Metallothioneins (MTs) are the major thiol-containing proteins that bind to Pt-based drugs and other metal ions. MTs are a group of proteins with 61–62 amino acids, 20 of which are conserved Cys. MTs bind many metal ions including Zn, Cu, Cd, Hg, Ag and Ni. Among these, Cu is the most tightly bound and can displace the other ions. An *apo*-protein (thionein) is able to load up to 12 Cu(I) ions. Increased MTs levels have been observed in cisplatin resistant murine and human tumour cells and over-expression of human MT-IIA by transfection confers resistance to cisplatin.[88–89] Moreover, double knock-out mice carrying MT-I/II deletions exhibit elevated sensitivity to cisplatin-induced hepatocellular carcinoma.[90]

15.4.2 Glutathione

GSH, a tripeptide with the amino acid sequence γ-glutamyl-cysteinylglycine, is another important cellular thiol that interacts with Cu and Pt-based agents. It is common to find elevated GSH levels (up to millimolar) in cisplatin-resistant tumour models.[91] The biosynthesis of GSH is controlled by the rate-limited enzyme, γ-glutamyl-cysteine synthetase (γ-GCS). Over-expression of γ-GCS is induced in cultured cells treated with cisplatin and, in many cases where the enzyme content was measured, the levels of γ-GCS mRNA correlate with cisplatin resistance. GSH reacts with cisplatin to form a platinum complex of apparent stoichiometry $Pt(GS)_2$ with the peptide acting as a chelating ligand (coordinating *via* cysteinyl sulfur and nitrogen atoms). The complex is stable in acidic as well as in neutral solutions. Complex formation is also observed in cultured cells treated with high concentrations of cisplatin.[92]

While binding to GSH would reduce the amount of intracellular Pt that can exert cytotoxic effects, it remains to be determined whether competition of platinum for binding to GSH would alter the expression of copper transporters. Unlike the slow formation of $Pt(GS)_2$, the formation of a copper complex is almost instantaneous.

15.4.3 Metallochaperones

The participation of many actors in metal ion homeostasis could imply the existence of a complex network of cross-regulation, a mechanism that ultimately results in cross-interference between copper and Pt-drugs, without the direct involvement of CTR1 in platinum transport.

Metallochaperones represent a group of important proteins involved in the intracellular distribution of various metal ions. As far as copper is concerned, in addition to ATOX1 that delivers Cu(I) to ATP7A and ATP7B for injection into the TGN, at least two other pathways are involved in the distribution of copper: COX17, which transfers copper to the subunits of cytochrome *c* oxidase in the mitochondria, and CCS, which incorporates Cu into superoxide dismutase 1 (SOD1).[93–94]

In fact, very little is known about the intracellular distribution of cisplatin. By attaching to cisplatin a fluorescent probe, such as carboxyfluorescein-diacetate (CFDA, developed by Reedjik and co-workers), it has been demonstrated that platinum rapidly enters tumour cells and is initially localized throughout the whole cytoplasm. After 2–3 h of exposure, the probe accumulates also in the nucleus. Co-localization of CFDA-Pt with TGN vesicles and other cell organelles is observed after 6–8 h and even longer incubation times.[95] TGN-localized Pt is likely to associate with ATP7A and ATP7B transporters which are able to export Pt.[69] However, as discussed above, it remains to be proved whether translocation of Pt ions to the TGN requires the intervention of the metallochaperone ATOX1.

ATOX1 is a small protein with a ferredoxin-like fold ($\beta_1\alpha_1\beta_2\beta_3\alpha_2\beta_4$) and a metal-binding motif, CXXC, located in the solvent-exposed $\beta1$-$\alpha1$ loop, which binds a single Cu(I) ion. This structure is highly conserved among metallochaperones and soluble domains of Cu-transporting ATPases.[96] Moreover, a lysine-rich region (KKTGK), located at the C-terminus, could represent a nuclear localization signal needed for ATOX1 translocation to the nucleus. Evidence now exists that ATOX1 is a Cu-dependent nuclear transcription factor implicated in cell proliferation (overexpression of cyclin-D1) and oxidative stress modulation in the cardiovascular system (overexpression of extracellular SOD3).[97–98]

The single metal binding domain of ATOX1 is structurally very similar to the six N-terminal metal binding domains of ATP7A and ATP7B, and is expected to bind cisplatin and other Pt-based drugs and have an influence on their accumulation. Both *Drosophila* ATOX1$^{-/-}$ mutants and mouse ATOX1$^{-/-}$ fibroblast cell lines display lower sensitivity to cisplatin than the wild-type controls. In fact, loss of ATOX1 reduces cisplatin uptake and accumulation in vesicular compartments and in DNA, impairing the process of cisplatin-induced CTR1 internalization and degradation needed to mediate drug uptake by macropinocytosis.[99–100]

It is also noteworthy that ATOX1-deficient cells possess higher levels of proteasome activity, which could be linked to its transcriptional activity. In fact, many of the subunits constituting the proteasome contain the putative ATOX1 binding site GAAAGA within their promoter regions.[68]

Two crystal structures of ATOX1 bound to cisplatin are available.[53] In one, Pt was stripped of all of its endogenous ligands and bound to Cys12 (*N,S*) and Cys15 (*S*) of ATOX1 with tris(2-carboxyethyl)phosphine (TCEP, a reductant used in the preparation of the *apo* protein) completing the square planar environment. The other structure contained molecules of stoichiometry *cis*-Pt(NH$_3$)$_2$(ATOX1)$_2$ featuring the Cys15 sulfur atoms from each protein molecule as ligands.

It is of interest to note that the crystal structure of Hg(ATOX1)$_2$ involves metal ion coordination by the two CXXC motifs.[101] Thus, like Cu and Hg, Pt can be accommodated within the CXXC motif.

Furthermore, it has been shown that cisplatin can bind to the ATOX1 metal-binding site even in the presence of Cu. Remarkably, addition of cisplatin to preformed Cu(I)-ATOX1, or *vice versa*, results in changes of CD spectra resembling those of metal clusters in MTs. This similarity indicates that cisplatin is close to the Cu(I) ion when bound to ATOX1, as if both copper and cisplatin occupy the same metal-binding site.[102]

In-cell NMR spectroscopy has been used to monitor the interaction of cisplatin with ATOX1 in a physiological environment (*Escherichia coli* cells), avoiding the employment of reducing agents, which may be competitor ligands for soft metal ions such as Pt(II). ESI-MS, inductively coupled plasma mass spectrometry (ICP-MS) and solution NMR experiments on purified ATOX1 in the presence of DTT were used to complement the in-cell analysis.

Cisplatin administration to *E. coli* cells overexpressing ^{15}N-ATOX1 caused a significant modification of the signals belonging to residues of the CXXC metal binding domain (Figure 15.9). The major spectral changes involved residues Cys12, Cys15 and Ala16, whose signals shifted in the same direction as in the corresponding *in vitro* experiment. Elemental analysis *via* ICP-MS on ATOX1 extracted from cisplatin-treated *E. coli* cells showed the presence of Pt in over 70% of ATOX1 molecules, proving that cisplatin crosses the cell membrane and actively binds to the CXXC motif of ATOX1 in living *E. coli* cells.[103] The same study reported that in *E. coli* cells, where there is no compartmentalization of DNA, overexpression of ATOX1 reduces DNA platination and ameliorates cell survival. However, the situation occurring in eukaryotic cells, where DNA is constrained in the nucleus, could be different. Both X-ray and NMR studies indicate that cisplatin promotes ATOX1 dimerization, hypothesizing that such a species can cross the nuclear membrane and interact with a consensus sequence of DNA, as observed for Cu-loaded ATOX1.[104]

We have highlighted the fact that the Cu(I)-binding site of ATOX1 is similar to those of the soluble domains of copper ATPases and of MTs, all characterized by CXXC sequences. In an elegant approach, an ATP7B recombinant variant, containing four out of six Cu(I)-binding domains, was recently shown to sequester Pt in *E. coli* cells, thus reducing the abnormal filamentous growth promoted by cisplatin.[105]

This behaviour suggests that resistance to cisplatin chemotherapy could be associated with overexpression of Cu transporters, which sequester cisplatin, with undesired beneficial effect on cancer cell survival. Interestingly, some cancer cell lines with higher resistance to cisplatin have correspondingly higher levels of ATOX1, up to 3.3-fold, than the cells with lower cisplatin resistance. By binding to ATOX1 in the cytoplasm, the drug could

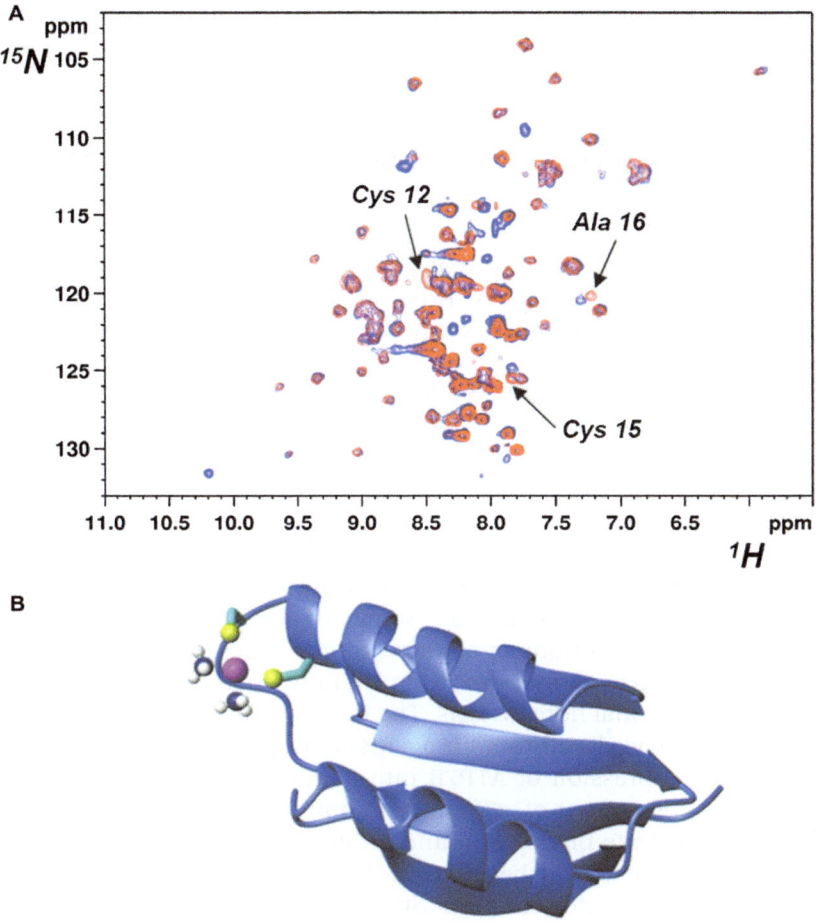

Figure 15.9 (A) Overlay of ^{1}H,^{15}N-SOFAST-HMQC spectra of *E. coli* cells expressing ^{15}N-ATOX1 in the presence (red) or absence (blue) of cisplatin. Cross-peaks of residues Cys12, Cys15 and Ala16 are indicated with arrows. (B) Structural model of the adduct between ATOX1 and cisplatin, solved by solution NMR. The Pt(II) ion is shown as a magenta sphere. Side-chains of Cys12 and Cys15 are shown as cyan sticks. S = yellow, N = blue, H = white.
Adapted with permission from ref. 103. Copyright 2011 American Chemical Society.

be easily targeted for ATP7B-mediated cell export, or simply stored such as in a sink.[102]

In summary, stable binding of cisplatin to ATOX1 could interfere with the key cytotoxic activity of the drug, namely DNA binding and cross-links formation. Furthermore, cisplatin bound to ATOX1 may alter Cu homeostasis and cellular defence against reactive oxygen species, hence providing an alternative route leading to cell killing. On the contrary, it is also conceivable that, rather than deactivating the drug, cisplatin-induced ATOX1 monomers or dimers could facilitate platinum translocation to the nucleus, hampering DNA replication and transcription.

15.5 Transport through Subcellular Membranes

15.5.1 ATPases

Recent studies have suggested that the copper export system also functions for cisplatin and its analogues. Excess copper is eliminated by two P-type ATPases, acting as efflux pumps. These proteins are functionally conserved from yeast to mammals.

In humans, elimination of excess copper is handled by ATP7A and ATP7B and mediated by ATOX1.[106–107] The two copper efflux transporters are located in the TGN, as occurs in yeasts, and undergo copper-stimulated trafficking.[108–110]

ATP7A and ATP7B share about 54% amino acid sequence identity and contain eight highly conserved putative transmembrane helices, a long N-terminal region containing six putative metal binding domains (MBD), each containing the typical CXXC sequence, with about 100 amino acid intervals. Under conditions of elevated copper levels, ATP7A migrates from the TGN to the plasma membrane and pumps the extra copper out of the cell, restoring normal homeostasis. The involvement of these ATPases in the transport of Pt-based anti-tumour agents derives from several experimental findings. Overexpression of ATP7B into human epidermal carcinoma cells resulted in development of resistance to both cisplatin and copper with an enhanced efflux rate of cisplatin and reduced accumulation of cisplatin in the transfected cells. Furthermore, ATP7B is frequently overexpressed in various cancers refractory to cisplatin therapy.[111–112] For instance, patients affected by ATP7B-positive ovarian cancer have significantly poorer responses to cisplatin-based chemotherapy than ATP7B-negative counterparts.[113] These results suggest that ATP7B may also be involved in the efflux of Pt-based antitumour agents. ATP7A is also likely to play a role. Human ovarian carcinoma cells transfected with an ATP7A overexpression construct also show higher resistance to cisplatin, carboplatin and oxaliplatin. Moreover, it has been reported that elevated expression of ATP7A is associated with reduced clinical outcome for patients with ovarian cancer treated with Pt-based drugs.[113]

Resistance is not necessarily associated to a decrease in total Pt drug accumulation, suggesting that the overexpressed transporters may alter intracellular Pt distribution with platinum accumulation in non-toxic compartments. Therefore, more work is needed to fully demonstrate that these copper efflux transporters are also involved in outward transport of cisplatin.

Although it has been suggested that the stoichiometry of copper binding to both ATP7A and ATP7B is one atom per metal binding motif, only mutations on MBD5 and MBD6 can rule out the metal-induced translocation of

ATP7A from TGN to the plasma membrane.[114] It has been demonstrated that initial contact between Cu(I)-ATOX1 and ATP7A occurs at MBD4 or MBD2, from which copper is transferred to MBD6/MBD5 before the ATP-driven transport of the Cu(I) ion across the vesicular membrane can take place.[115–116]

The mechanism of copper transfer between ATOX1 and the ATP7A and ATP7B transporters is not completely clarified, but certainly involves the conserved CXXC motifs. The crystal structure of ATOX1 in the presence of Cu(I) revealed that a copper ion is coordinated by two adjacent Cys residues from two ATOX1 molecules.[101] It has been proposed that such a coordination enables the formation of metal-bridged intermediates, a halfway step to Cu(I) transfer.

Likewise, our knowledge of how Pt is delivered to copper ATPases and in which way the transport machinery operates downstream is continuously growing. Similarly to copper, it is now documented that cisplatin also uses the CXXC metal binding domains and that, at least for ATP7B, such interaction is essential for its trafficking ability and to mediate drug resistance.[117] However, the mechanism through which Cu-ATPases affect cell resistance to platinum drugs remains unclear, with ATP7B seemingly more involved than ATP7A. Several studies have suggested that cisplatin is effluxed from cells by ATPases, although the increase in ATP7B expression does not always mirror a reduced intracellular accumulation of cisplatin. It has also been proposed that ATP7B modulates resistance through drug sequestration within intracellular vesicles, rather than through active outward transport as is the case for copper.[118]

A fluorescence study showed that ATP7B appears to co-localize with cisplatin in vesicles, likely suggesting binding at the vesicle surface.[119] Cisplatin transport *via* ATP7B was assessed in vesicles isolated from Sf9 cells expressing either the wild-type or an inactive ATP7B mutant.[120] Vesicles with wild-type ATP7B or with ATP7B mutant unable to transport copper contained more cisplatin than those lacking ATP7B, even in the absence of ATP, indicating that cisplatin only binds to the transporter at the vesicle surface. Moreover, when the pH was lowered, cisplatin accumulated inside vesicles expressing wild-type ATP7B by requiring ATP, but the rate was considerably slower than for copper. Unlike Sf9 cells, experiments carried out at physiological pH in hepatocytes showed that cisplatin does not induce trafficking of ATP7B nor competes with copper for transport. Furthermore, no correlation was observed between cisplatin resistance and endogenous ATP7B levels.[121] Taken together, these experiments support a model in which ATP7B controls drug availability in the cytosol not necessarily through direct transport but perhaps by sequestering the drug within the multiple metal-binding sites present at the N-terminus. This possibility is further supported by studies in ovarian cell cultures, in which the overexpression of the ATP7B N-terminal soluble domain alone confers resistance to cisplatin, although at a lower extent than does the full-size protein.[122]

To our knowledge, the first convincing experimental proof that ATPases actively transport Pt drugs, similarly to Cu, is described in very recent work that measured ATP-dependent electrical current movements on microsomes from COS-1 cells enriched with ATP7A and ATP7B. The outcome shows that cisplatin is pumped by both ATPases upon ATP addition and that co-administration of Pt drug and Cu inhibits transport of both metal ions.[123]

15.6   Conclusions and Open Questions

Platinum drugs remain an essential class of anti-cancer agents, with cisplatin and carboplatin extensively used in handling testicular, ovarian, head and neck and lung malignancies, and oxaliplatin becoming the basic treatment for colorectal cancer. These agents are characterized by the capacity to generate lesions on DNA that induce the majority of cytotoxic effects.

However, clinical complications arising from tumour resistance and side effects severely limit the pharmacological potential of platinum drugs. The latest findings have also stressed the role of active transporters in the uptake/efflux of Pt-anti-tumour drugs, including the copper transporters CTR1, ATOX1, ATP7A and ATP7B. While the molecular mechanisms by which these proteins mediate the transport of Pt-drugs remain elusive, it is clear that they deeply influence drug uptake, cytotoxicity and resistance.

While it is apparent that copper transporters play a critical role in the uptake of platinum drugs, it is still debated whether they can be considered clinically relevant biomarkers of sensitivity of human carcinomas to cisplatin, carboplatin and oxaliplatin. In the future, it is possible that regulation of their expression may play an essential role in enhancing the responsiveness to platinum-based chemotherapy.

Furthermore, given that, before reaching nuclear DNA, Pt-drugs can encounter several cytoplasmic binders, particularly those containing sulfur atoms, the design of new pharmaceuticals should aim to limit such detrimental interactions in order to enhance the anti-cancer effect.

A better knowledge of all transport phenomena associated with platinum drugs could be the basis for elaborating a strategy aiming to overcome the platinum resistance of tumours as well as to reduce adverse side effects. The impact on the large population of cancer patients currently receiving these drugs would be enormous.

The effectiveness of platinum complexes against cancer represents the most outstanding success of inorganic drugs and has triggered many investigations aiming to identify new chemotherapeutic agents employing transition metals other than Pt. The significant similarity between the coordination chemistry of Pt(II) and Pd(II) has promoted studies on Pd(II) complexes to be used as anti-tumour drugs. Compared to platinum analogues, the hydrolysis of palladium complexes is much faster (around 10^5 times), leading to very reactive species that could never reach their pharmacological targets. However, some palladium complexes, with encouraging activity against tumour cell lines, have been synthesized employing mainly the same designing strategy used for platinum drugs.[124] Different types of monodentate and bidentate ligands have been used, the latter meeting the need to prevent unfavourable *cis-trans* isomerization. There appears to be no palladium drug in advanced clinical trials.

On the other hand, at least two ruthenium complexes and a titanium complex have completed Phase-II clinical testing.[125–126] It is interesting to note that Ru(II) has an intrinsic ligand substitution rate comparable to that of Pt(II). It has been shown that transferrin can accommodate a ruthenium complex in the pocket where Fe(III) is hosted, suggesting that transferrin could represent a means of drug delivery to tumour cells.[127]

Beyond its employment in the clinic, platinum (and palladium as well) is extensively used in automotive catalytic converters with the consequence that these metals have become environmental pollutants. Indeed, exposure of humans and animals to platinum group elements may represent a pernicious health risk. Therefore, while contact with environmental concentrations of these elements has been neglected, there is increasing concern that they may exert subtle toxic effects, especially at a chronic, subclinical, level.[128]

References

1. D. Seyferth, *Organometallics*, 2001, **20**, 2.
2. W. W. Porterfield, *Inorganic Chemistry – A Unified Approach*, Academic Press Inc., San Diego, CA, 1993.
3. M. Peyrone, *Ann. Chem. Pharm.*, 1845, **51**, 1.
4. B. Rosenberg, L. VanCamp and T. Krigas, *Nature*, 1965, **205**, 698.
5. B. Rosenberg, E. Renshaw, L. VanCamp, J. Hartwick and J. Drobnik, *J. Bacteriol.*, 1967, **93**, 716.
6. B. Rosenberg, L. VanCamp, E. B. Grimley and A. J. Thomson, *J. Biol. Chem.*, 1967, **242**, 1347.
7. B. Rosenberg, L. VanCamp, J. E. Trosko and V. H. Mansour, *Nature*, 1969, **222**, 385.
8. B. Rosenberg and L. VanCamp, *Cancer Res.*, 1970, **301**, 799.
9. T. W. Hambley, *Dalton Trans.*, 2007, **43**, 4929.
10. L. Kelland, *Nat. Rev. Cancer*, 2007, 7, 573.
11. A. Ardizzoni, L. Boni, M. Tiseo, F. V. Fossella, J. H. Schiller, M. Paesmans, D. Radosavljevic, A. Paccagnella, P. Zatloukal, P. Mazzanti, D. Bisset and R. Rosell, *J. Natl Cancer Inst.*, 2007, **99**, 847.
12. M. Mishima, G. Samimi, A. Kondo, X. Lin and S. B. Howell, *Eur. J. Cancer*, 2002, **38**, 1405.
13. M. Coluccia and G. Natile, *Anticancer Agents Med. Chem.*, 2007, 7, 111.
14. G. P. Stathopoulos, T. Boulikas, M. Vougiouka, G. Deliconstantinos, S. Rigatos, E. Darli, V. Viliotoy and J. G. Stathopoulos, *Oncol. Rep.*, 2005, **13**, 589.
15. V. Gagnon, M. E. St-Germain, C. Descteaux, J. Provencher-Mandeville, S. Parent, S. K. Mandal, E. Asselin and G. Bèrubè, *Bioorg. Med. Chem. Lett.*, 2004, **14**, 5919.
16. B. Rosenberg, *Met. Ions Biol. Syst.*, 1980, **11**, 127.
17. M. Howe-Grant and S. J. Lippard, *Met. Ions Biol. Syst.*, 1980, **11**, 63.

18. T. W. Hambley, *Dalton Trans.*, 2001, **19**, 2711.
19. S. L. Bruhn, J. H. Toney and S. J. Lippard, *Prog. Inorg. Chem.*, 1990, **38**, 477.
20. A. Eastman, *Cancer Cells*, 1990, **2**, 275.
21. M. D. Hall, M. Okabe, D. W. Shen, X. J. Liang and M. M. Gottesman, *Annu. Rev. Pharmacol. Toxicol.*, 2008, **48**, 495.
22. J. Reedijk, *Proc. Natl Acad. Sci. USA*, 2003, **100**, 3611.
23. J. Reedijk, *Chem. Rev.*, 1999, **99**, 2499.
24. M. Kartalou and J. M. Essigmann, *Mutat. Res.*, 2001, **478**, 1.
25. F. Coste, J. M. Malinge, L. Serre, W. Shepard, M. Roth, M. Leng and C. Zelwer, *Nucleic Acids Res.*, 1999, **27**, 1837.
26. V. Brabec, *Progr. Nucleic Acid Res. Mol. Biol.*, 2002, **71**, 1.
27. Y. Jung and S. J. Lippard, *Chem. Rev.*, 2007, **107**, 1387.
28. A. Eastman and M. A. Barry, *Biochemistry*, 1987, **26**, 3303.
29. A. Vaisman, S. E. Lim, S. M. Patrick, W. C. Copeland, D. C. Hinkle, J. J. Turchi and S. G. Chaney, *Biochemistry*, 1999, **38**, 11026.
30. R. Reeves and J. E. Adair, *DNA Repair*, 2005, **4**, 926.
31. Z. H. Siddik, *Oncogene*, 2003, **22**, 7265.
32. S. Park and S. J. Lippard, *Biochemistry*, 2012, **51**, 6728.
33. S. Park and S. J. Lippard, *Biochemistry*, 2011, **50**, 2567.
34. A. R. Timerbaev, S. S. Aleksenko, K. Polec-Pawlak, R. Ruzik, O. Semenova, C. G. Hartinger, S. Oszwaldowski, M. Galanski, M. Jarosz and B. K. Keppler, *Electrophoresis*, 2004, **25**, 1988.
35. A. I. Ivanov, J. Christodoulou, J. A. Parkinson, K. J. Barnham, A. Tucker, J. Woodrow and P. J. Sadler, *J. Biol. Chem.*, 1998, **273**, 14721.
36. M. A. Graham, G. F. Lockwood, D. Greenslade, S. Brienza, M. Bayssas and E. Gamelin, *Clin. Cancer Res.*, 2000, **6**, 1205.
37. P. A. de Simone, L. Brennan, M. L. Cattaneo and E. Zukka, *Proc. Am. Soc. Clin. Oncol.*, 1987, **6**, 33.
38. J. D. Holding, W. E. Lindup, C. van Laer, G. C. M. Vreeburg, V. Schiling, J. A. Wilson and P. M. Stell, *Br. J. Clin. Pharmacol.*, 1992, **33**, 75.
39. E. J. Beatty, M. C. Cox, T. A. Frenkiel, B. M. Tam, A. B. Mason, R. T. A. Macgillivray, P. J. Sadler and R. C. Woodworth, *Biochemistry*, 1996, **35**, 7635.
40. S. P. Binks and M. Dobrota, *Biochem. Pharmacol.*, 1990, **40**, 1329.
41. D. P. Gately and S. B. Howell, *Br. J. Cancer*, 1993, **67**, 1171.
42. A. K. Holzer, G. Samimi, K. Katano, W. Naerdemann, X. Lin, R. Safaei and S. B. Howell, *Mol. Pharmacol.*, 2004, **66**, 817.
43. R. Safaei and S. B. Howell, *Crit. Rev. Oncol. Hematol.*, 2005, **53**, 13.
44. R. Safaei, *Cancer Lett.*, 2006, **234**, 34.
45. X. J. Liang, D. W. Shen, K. G. Chen, S. M. Wincovitch, S. H. Garfield and M. M. Gottesman, *J. Cell Physiol.*, 2005, **202**, 635.
46. H. Burger, A. Zoumaro-Djayoon, A. W. Boersma, J. Helleman, E. M. Berns, R. H. J. Mathijssen, W. J. Loos and E. A. C. Wiemer, *Br. J. Pharmacol.*, 2010, **159**, 898.

47. S. Zhang, K. S. Lovejoy, J. E. Shima, L. L. Lagpacan, Y. Shu, A. Lapuk, Y. Chen, T. Komori, J. W. Gray, X. Chen, S. J. Lippard and K. M. Giacomini, *Cancer Res.*, 2006, **66**, 8847.
48. S. Ishida, J. Lee, D. J. Thiele and I. Herskowitz, *Proc. Natl Acad. Sci. USA*, 2002, **99**, 14298.
49. B. G. Blair, C. Larson, R. Safaei and S. B. Howell, *Clin. Cancer Res.*, 2002, **15**, 4312.
50. C. A. Larson, B. G. Blair, R. Safaei and S. B. Howell, *Mol. Pharm.*, 2009, **75**, 324.
51. F. Arnesano, S. Scintilla and G. Natile, *Angew. Chem. Int. Ed.*, 2007, **46**, 9062.
52. C. M. Sze, G. N. Khairallah, Z. Xiao, P. S. Donnelly, R. A. J. O'Hair and A. G. Wedd, *J. Biol. Inorg. Chem.*, 2009, **14**, 163.
53. A. K. Boal and A. C. Rosenzweig, *J. Am. Chem. Soc.*, 2009, **131**, 14196.
54. E. Luk, L. T. Jensen and V. C. Culotta, *J. Biol. Inorg. Chem.*, 2003, **8**, 803.
55. C. J. De Feo, S. G. Aller, G. S. Siluvai, N. J. Blackburn and V. M. Unger, *Proc. Natl Acad. Sci. USA*, 2009, **106**, 4237.
56. M. Schushan, Y. Barkan, T. Haliloglu and N. Ben-Tal, *Proc. Natl Acad. Sci. USA*, 2010, **107**, 10908.
57. P. Abada and S. B. Howell, *Met. Based Drugs*, 2010, 317581.
58. T. H. Nguyen, F. Arnesano, S. Scintilla, G. Rossetti, E. Ippoliti, P. Carloni and G. Natile, *J. Chem. Theory Comp.*, 2012, **8**, 2912.
59. F. Arnesano, M. Losacco and G. Natile, *Eur. J. Inorg. Chem.*, 2013, **2013**, 2701.
60. V. Marchán, V. Moreno, E. Pedroso and A. Grandas, *Chemistry*, 2001, **7**, 808.
61. E. R. Jamieson and S. J. Lippard, *Chem. Rev.*, 1999, **99**, 2467.
62. A. K. Holzer and S. B. Howell, *Cancer Res.*, 2006, **66**, 10944.
63. D. Sinani, D. J. Adle, H. Kim and J. Lee, *J. Biol. Chem.*, 2007, **282**, 26775.
64. Y. Guo, K. Smith and M. J. Petris, *J. Biol. Chem.*, 2004, **279**, 46393.
65. Y. Guo, K. Smith, J. Lee, D. J. Thiele and M. J. Petris, *J. Biol. Chem.*, 2004, **279**, 17428.
66. X. Du, X. Wang, H. Lia and H. Sun, *Metallomics*, 2012, **4**, 679.
67. J. Liu, A. Sitaram and C. G. Burd, *Traffic*, 2007, **8**, 1375.
68. R. Safaei, M. H. Maktabi, B. G. Blair, C. A. Larson and S. B. Howell, *J. Inorg. Biochem.*, 2009, **103**, 333.
69. R. Safaei, K. Katano, B. J. Larson, G. Samimi, A. K. Holzer, W. Naerdemann, M. Tomioka, M. Goodman and S. B. Howell, *Clin. Cancer Res.*, 2005, **11**, 756.
70. G. V. Kalayda, C. H. Wagner and U. Jaehde, *J. Inorg. Biochem.*, 2012, **116**, 1.
71. K. D. Ivy and J. H. Kaplan, *Mol. Pharmacol.*, 2013, **83**, 1237.
72. B. Zhou and J. Gitschier, *Proc. Natl Acad. Sci. USA*, 1997, **94**, 7481.
73. J. Bertinato and M. R. L'Abbe, *J. Biol. Chem.*, 2003, **278**, 35071.
74. P. V. van den Berghe, D. E. Folmer, H. E. Malingre, B. E. van Beurden, A. E. Klomp, S. B. van De Sluis, M. Merkx, R. Berger and L. W. Klomp, *Biochem. J.*, 2007, **407**, 49.

75. B. G. Blair, C. A. Larson, P. L. Adams, P. B. Abada, R. Safaei and S. B. Howell, *Mol. Pharmacol.*, 2010, 77, 912.
76. D. D. Jandial, S. Farshchi-Heydari, C. A. Larson, G. I. Elliott, W. J. Wrasidlo and S. B. Howell, *Clin. Cancer Res.*, 2009, 15, 553.
77. K. Taniguchi, M. Wada, K. Kohno, T. Nakamura, T. Kawabe and M. Kawakami, *Cancer Res.*, 1996, 56, 4124.
78. B. Liedert, V. Materna, G. L. Schadendorf, J. Thomale and H. Lage, *J. Invest. Derm.*, 2003, 121, 172.
79. K. Koike, T. Kawabe, T. Tanaka, S. Toh, T. Uchiumi, M. Wada, S. Akiyama, M. Ono and M. Kuwano, *Cancer Res.*, 1997, 57, 5475.
80. Y. Toyoda, Y. Hagiya, T. Adachi, K. Hoshijima, M. T. Kuo and T. Ishikawa, *Xenobiotica*, 2008, 38, 833.
81. X. J. Liang, S. Mukherjee, D. W. Shen, F. R. Maxfield and M. M. Gottesman, *Cancer Res.*, 2006, 66, 2346.
82. A. Yonezawa, S. Masuda, S. Yokoo, T. Katsura and K. Inui, *J. Pharmacol. Exp. Ther.*, 2006, 319, 879.
83. T. Terada and K. Inui, *Biochem. Pharmacol.*, 2008, 75, 1689.
84. D. P. Bancroft, C. A. Lepre and S. J. Lippard, *J. Am. Chem. Soc.*, 1990, 112, 6860.
85. M. Knipp, A. V. Karotki, S. Chesnov, G. Natile, P. J. Sadler, V. Brabec and M. Vasak, *J. Med. Chem.*, 2007, 50, 4075.
86. A. Casini, C. Gabbiani, G. Mastrobuoni, R. Z. Pellicani, F. P. Intini, F. Arnesano, G. Natile, G. Moneti, S. Francese and L. Messori, *Biochemistry*, 2007, 46, 12220.
87. J. Reedijk, *Proc. Natl Acad. Sci. USA*, 2003, 100, 3611.
88. K. Kasahara, Y. Fujiwara, K. Nishio, T. Ohmori, Y. Sugimoto, K. Komiya, T. Matsuda and N. Saijo, *Cancer Res.*, 1991, 51, 3237.
89. S. L. Kelley, A. Basu, B. A. Teicher, M. P. Hacker, D. H. Hamer and J. S. Lazo, *Science*, 1988, 241, 1813.
90. M. P. Waalkes, J. Liu, K. S. Kasprzak and B. A. Diwan, *Int. J. Cancer*, 2006, 119, 28.
91. Z. H. Siddik, *Cancer Treat. Res.*, 2002, 112, 263.
92. T. Ishikawa and F. Ali-Osman, *J. Biol. Chem.*, 1993, 268, 20116.
93. N. J. Robinson and D. R. Winge, *Annu. Rev. Biochem.*, 2010, 79, 537.
94. V. C. Culotta, M. Yang and T. V. O'Halloran, *Biochim. Biophys. Acta*, 2006, 1763, 747.
95. C. Molenaar, J. M. Teuben, R. J. Heetebrij, H. J. Tanke and J. Reedijk, *J. Biol. Inorg. Chem.*, 2000, 5, 655.
96. F. Arnesano, L. Banci, I. Bertini, S. Ciofi-Baffoni, E. Molteni, D. L. Huffman and T. V. O'Halloran, *Genome Res.*, 2002, 12, 255.
97. S. Itoh, K. Ozumi, H. W. Kim, O. Nakagawa, R. D. McKinney, R. J. Folz, I. N. Zelko, M. Ushio-Fukai and T. Fukai, *Free Radic. Biol. Med.*, 2008, 46, 95.
98. S. Itoh, H. W. Kim, O. Nakagawa, K. Ozumi, S. M. Lessner, H. Aoki, K. Akram, R. D. McKinney, M. Ushio-Fukai and T. Fukai, *J. Biol. Chem.*, 2008, 283, 9157.

99. H. Hua, V. Gunther, O. Georgiev and W. Schaffner, *Biometals*, 2011, **24**, 445.
100. Y. Itoh, M. Tamai, K. Yokogawa, M. Nomura, S. Moritani, H. Suzuki, Y. Sugiyama and K. Miyamoto, *Anticancer Res.*, 2002, **22**, 1649.
101. A. K. Wernimont, D. L. Huffman, A. L. Lamb, T. V. O'Halloran and A. C. Rosenzweig, *Nat. Struct. Biol.*, 2000, **7**, 766.
102. M. E. Palm, C. F. Weise, C. Lundin, G. Wingsle, Y. Nygren, E. Bjorn, P. Naredi, M. Wolf-Watz and P. Wittung-Stafshede, *Proc. Natl Acad. Sci. USA*, 2011, **108**, 6951.
103. F. Arnesano, L. Banci, I. Bertini, I. C. Felli, M. Losacco and G. Natile, *J. Am. Chem. Soc.*, 2011, **133**, 18361.
104. P. A. Muller and L. W. Klomp, *Int. J. Biochem. Cell. Biol.*, 2009, **41**, 1233.
105. N. V. Dolgova, D. Olson, S. Lutsenko and O. Y. Dmitriev, *Biochem. J.*, 2009, **419**, 51.
106. C. Vulpe, B. Levinson, S. Whitney, S. Packman and J. Gitschier, *Nat. Gen.*, 1993, **3**, 7.
107. P. C. Bull, G. R. Thomas, J. M. Rommens, J. M. Forbes, J. R. Forbes and D. W. Cox, *Nat. Genet.*, 1993, **5**, 327.
108. J. Camakaris, I. Voskoboinik and J. F. Mercer, *Biochem. Biophys. Res. Comm.*, 1999, **261**, 225.
109. M. DiDonato and B. Sarkar, *Biochem. Biophys. Acta*, 1997, **1360**, 3.
110. J. R. Prohaska and A. A. Gybina, *J. Nutr.*, 2004, **134**, 1003.
111. M. Komatsu, T. Sumizawa, M. Mutoh, Z. S. Chen, K. Terada, T. Furukawa, X. L. Yang, H. Gao, N. Miura, T. Sugiyama and S. Akiyama, *Cancer Res.*, 2000, **60**, 1312.
112. K. Nakayama, A. Kanzaki, K. Terada, M. Mutoh, K. Ogawa, T. Sugiyama, S. Takenoshita, K. Itoh, N. Yaegashi, K. Miyazaki, N. Neamati and Y. Takebayashi, *Clin. Cancer Res.*, 2004, **10**, 2804.
113. G. Samimi, N. M. Varki, S. Wilczynski, R. Safaei, D. S. Alberts and S. B. Howell, *Clin. Cancer Res.*, 2003, **9**, 5853.
114. D. Strausak, S. La Fontaine, J. Hill, S. D. Firth, P. J. Lockhart and J. F. R. Mercer, *J. Biol. Chem.*, 1999, **274**, 11170.
115. D. Achila, L. Banci, I. Bertini, J. Bunce, S. Ciofi-Baffoni and D. L. Huffman, *Proc. Natl Acad. Sci. USA*, 2006, **103**, 5729.
116. L. Banci, I. Bertini, F. Cantini, C. T. Chasapis, N. Hadjiliadis and A. Rosato, *J. Biol. Chem.*, 2005, **280**, 38259.
117. R. Safaei, P. L. Adams, M. H. Maktabi, R. A. Mathews and S. B. Howell, *J. Inorg. Biochem.*, 2012, **110**, 8.
118. A. Gupta and S. Lutsenko, *Future Med. Chem.*, 2009, **1**, 1125.
119. K. Katano, R. Safaei, G. Samimi, A. Holzer, M. Tomioka, M. Goodman and S. B. Howell, *Clin. Cancer Res.*, 2004, **10**, 4578.
120. R. Safaei, S. Otani, B. J. Larson, M. L. Rasmussen and S. B. Howell, *Mol. Pharmacol.*, 2008, **73**, 461.
121. K. Leonhardt, R. Gebhardt, J. Mossner, S. Lutsenko and D. Huster, *J. Biol. Chem.*, 2009, **284**, 7793.

122. L. S. Mangala, V. Zuzel, R. Schmandt, E. S. Leshane, J. B. Halder, G. N. Armaiz-Pena, W. A. Spannuth, T. Tanaka, M. M. Shahzad, Y. G. Lin, A. M. Nick, C. G. Danes, J. W. Lee, N. B. Jennings, P. E. Vivas-Mejia, J. K. Wolf, R. L. Coleman, Z. H. Siddik, G. Lopez-Berestein, S. Lutsenko and A. K. Sood, *Clin. Cancer Res.*, 2009, **15**, 3770.

123. F. Tadini-Buoninsegni, G. Bartolommei, M. R. Moncelli, G. Inesi, A. Galliani, M. Sinisi, M. Losacco, G. Natile and F. Arnesano, *Angew. Chem. Int. Ed.*, 2014, **53**, 1297.

124. E. Gao, C. Liu, M. Zhu, H. Lin, Q. Wu and L. Liu, *Anticancer Agents Med. Chem.*, 2009, **9**, 356.

125. A. Bergamo, C. Gaiddon, J. H. M. Schellens, J. H. Beijnen and G. Sava, *J. Inorg. Biochem.*, 2012, **106**, 90.

126. U. Olszewski and G. Hamilton, *Anticancer Agents Med. Chem*, 2010, **10**, 302.

127. O. Mazuryk, K. Kurpiewska, K. Lewinski, G. Stochel and M. Brindell, *J. Inorg. Biochem.*, 2012, **116**, 11.

128. C. L. Wiseman and F. Zereini, *Sci. Total Environ.*, 2009, **407**, 2493.

Copper in Prokaryotes

NICK E. LE BRUN

Centre for Molecular and Structural Biochemistry, School of Chemistry, University of East Anglia, Norwich Research Park, Norwich, NR4 7TJ, UK
Email: n.le-brun@uea.ac.uk

16.1 Introduction

16.1.1 Fundamentals of Copper Chemistry

The transition metal element copper is the first member of Group 11 of the periodic table, which also includes silver and gold. Although it shares some properties with these precious metals, copper is far more abundant in the Earth's crust (though approximately 1000-fold less abundant than iron). It can be found as deposits of its native, elemental form, which were exploited by man in the first applications some 10 000 years ago. The development of smelting techniques enabled recovery of the metal from ores such as chalcopyrite ($CuFeS_2$, the most commonly found), chalcocite (Cu_2S), cuprite (Cu_2O) and malachite ($Cu_2CO_3(OH)_2$). The name "copper" is derived from Cyprium, the Latin name for Cyprus, where much of the copper was mined in Roman times.[1] The corrosion-resistant properties of the metal made it useful for many applications, including fixing materials in ship building and in water pipes. Although an advance on what was previously available, copper is not sufficiently hard to be an ideal material for tools and weapons. This problem was overcome with the development of alloying techniques, in which copper was mixed with other metals, most notably tin to form bronze. These alloys are hard and durable and these advances dominated life until ways to smelt iron were developed, leading to a transition

RSC Metallobiology Series No. 2
Binding, Transport and Storage of Metal Ions in Biological Cells
Edited by Wolfgang Maret and Anthony Wedd
Published by the Royal Society of Chemistry, www.rsc.org

from the bronze age to the iron age. Copper and its alloys (including brass, an alloy of copper and zinc) continued to find use in coins and as decoration of buildings, furniture and art, but it was not until the industrial revolution that copper again took centre stage.[1] Its high electrical conductivity (second only to silver) and relative abundance meant that it became the dominant material for electrical wiring, motors and devices, which remains its major application today. It is also still used in plumbing, roofing and cladding.

Copper has an electronic configuration of $1s^2 2s^2 2p^6 3s^2 3p^6 4s^1 3d^{10}$, and can lose an electron from the 4s orbital first and subsequently from the 3d orbitals. The most common oxidation states of copper are $+1$ and $+2$, though $+3$ is also known. The chemistry of copper is dominated by its ability to form coordination complexes with a range of Lewis bases. In terms of the Hard-Soft Acid-Base (HSAB) classification system, Cu(I), being a relatively large ion of low charge, is a soft metal (Lewis acid) and therefore forms its most stable coordination complexes with soft ligands (Lewis bases) such as those containing polarizable coordinating atoms such as sulfur, for example, thiols (R-SH) and thioethers (R-S-R). This simple classification system alone can account for much of the biological chemistry of copper. Simple Cu(I) compounds have low coordination numbers, being commonly linear, although trigonal and tetrahedral geometries are also common.

Cu(II), being more highly charged and somewhat smaller, is a borderline metal in the HSAB classification system and its ligand binding preferences are wider and less easily predicted. As a d^9 ion, Cu(II) has an incomplete 3d shell and so its complexes are subject to crystal field stabilization effects, leading to the lifting of d-orbital degeneracy and the stabilization of some orbitals depending on the coordination geometry. Cu(II) typically forms complexes with coordination numbers four, five and six, with square planar, pyramidal or tetragonally distorted octahedral arrangement. The absence of pure octahedral complexes is associated with the Jahn–Teller effect that leads to a distortion that lifts the degeneracy of the octahedral d^9 e_g orbital subset and provides additional stabilization. This effect is extremely important because it results in Cu(II) forming the most stable complexes of the divalent first row transition metals, as described by the Irving–Williams series, with important consequences for the biological chemistry of copper.

The d^9 configuration means that Cu(II) contains a single unpaired electron and is therefore paramagnetic (with total spin $S = \frac{1}{2}$), in contrast to d^{10} Cu(I), which is diamagnetic ($S = 0$). Thus, Cu(II) is amenable to study by magnetic techniques such as electron paramagnetic resonance (EPR) spectroscopy, which can provide detailed information about the paramagnetic centre. The difference in electronic configuration also has consequences for the colours of copper complexes. With their filled electronic shell, Cu(I) complexes are typically colourless, unless ligand-based transitions or charge transfer transitions (between metal ion and ligands) occur in the visible region of the spectrum. For example, Cu(I) complexes of phenanthroline and related ligands (that have low-lying unoccupied π^* orbitals) exhibit d→π^* metal to ligand charge transfer transitions (reviewed in ref. 2) and are intensely coloured. On the other hand, Cu(II) complexes have uniform but less intense colours (typically blue to blue-green) as the d^9 configuration results in (formally forbidden) d→d transitions. Ligand to metal charge transfer bands can result in complexes with intense colours (for example, Cys thiolate to Cu(II) in blue copper proteins).

Cu(I) salts are generally sparingly soluble in water, but can dissolve in the presence of suitable ligands. Cu(II) salts are of varying solubility (generally following the rules of solubility); some, such as copper sulfate, are highly soluble, while others, such as copper sulfide are highly insoluble.

Most of the useful properties of copper in biological cells depend on the capability of the metal to cycle between the +2 and +1 oxidation states. The half reaction and standard reduction potential for the couple connecting the aqua ions is given in Equation 16.1.[3]

$$Cu^{II}_{(aq)} + e^- \rightarrow Cu^I_{(aq)} \qquad E° = +0.153 \text{ V} \qquad (16.1)$$

This relatively low potential means that, although Cu(I) is the more stable form thermodynamically, it is highly susceptible to oxidation in the presence of O_2 and so in an aerobic atmosphere, Cu(II) is the predominant form.

The Cu(II)/Cu(I) potential is highly dependent on the nature of the metal environment; soft ligands that favour Cu(I) will result in a more oxidizing (higher) potential, while harder ligands that favour Cu(II) will tend to decrease the potential. This allows tight control of electron transfer processes. For example, the blue copper proteins (see Section 16.1.2) exhibit a very wide range of redox potentials. The coordination environment, consisting of a trigonal core of one Cys and two His residues, favours Cu(I) so the intrinsic redox potential is relatively high. Additional, longer-range interactions with one (*e.g.* Met in plastocyanins) or two (*e.g.* Met and a peptide carbonyl in azurins) ligands fine tune the potential. The nature of the Cu-O=C interaction in the latter case is also important, with a shorter bond length leading to a lower potential (favouring Cu(II)). Solvent accessibility is also important: secondary interactions with water favour Cu(II) and will decrease the potential, but exclusion of water has the effect of driving the potential up (as high as +680 mV in rusticyanin from *Thiobacillus ferrooxdians*).[4]

The oxidation of copper metal does not occur readily (see Equation 16.2).[3] This explains why elemental copper can be found naturally, and why copper metal is resistant to corrosion.

$$Cu^I_{(aq)} + e^- \rightarrow Cu^0_{(s)} \qquad E° = +0.52 \text{ V} \qquad (16.2)$$

Combination of redox couples 1 and 2 above demonstrates that Cu(I) in aqueous solution is unstable and undergoes disproportionation, ie, Cu(I) is both oxidized and reduced (Equation 16.3).

$$2\,Cu(I)_{(aq)} \rightarrow Cu(II)_{(aq)} + Cu_{(s)} \qquad E° = +0.37 \text{ V} \qquad (16.3)$$

This aspect and sensitivity to aerial oxidation means that it can be difficult to maintain stable solutions of Cu(I) ions. However, the addition of suitable ligands can lead to stable Cu(I) complexes.

16.1.2 Evolutionary Aspects of Copper Biochemistry and Roles of Copper Proteins in Prokaryotes

When life first evolved on Earth, it did so in an atmosphere lacking in O_2. Under these conditions, elemental copper and Cu(I) ores were the most stable forms and, because of their insolubility, copper was present at very low concentrations in water. It was therefore not bioavailable and early life did not utilize copper. The evolution of photosynthetic organisms began the long process of oxygenation of the atmosphere to the approx. 20% levels of today. As explained in Section 16.1.1, Cu(I) is readily oxidized in the presence of O_2 and Cu(II) compounds are generally much more soluble. Sulfide also underwent oxidation with the result that concentrations of copper (as Cu(II)) in water increased, having been released from Cu^I_2S. As the metal became bioavailable, it was incorporated into life processes[5] and, for this reason, copper is sometimes referred to as a "modern" bio-element.

The oxygenation of the atmosphere facilitated the evolution of more complex organisms through the increased energy that could be obtained from more efficient, O_2-based respiratory processes. Copper played a central role: as increased O_2 levels released previously unavailable copper, the versatile coordination chemistry and redox properties of copper were ideally suited to harness the oxidative power of O_2. Thus, the major roles of copper in life are associated with metabolism of O_2. While some anaerobic prokaryotes contain no copper proteins at all, a greater number of copper enzymes are found in more complex organisms.[6]

Oxidases of various types employ a copper centre for the reduction of O_2 to water or hydrogen peroxide. Amongst these are the haem-copper respiratory oxidases that couple the reduction of O_2 to water with the translocation of protons across a membrane to generate a proton motive force that drives unfavourable processes, including ATP synthesis. The active site copper ion (Cu_B) forms a binuclear centre with a high-spin haem. In addition, some haem-copper oxidases contain a second copper centre consisting of a pair of copper ions coordinated by Cys and His residues and referred to as a Cu_A centre. This unit functions in transferring electrons from cytochrome c to the binuclear centre. Other oxidases that contain multiple copper ions (generally referred to as multicopper oxidases) catalyse the oxidation of a range of specific substrates, such as phenols in laccase, with reduction of O_2 to water. The haem-copper and multicopper oxidases are the key catalytic centres for safe four-electron reduction of O_2 to water. In addition, amine oxidases contain a single copper ion and catalyse oxidation of amines to aldehydes with two-electron reduction of O_2 to H_2O_2. Oxygenases also react with O_2, but catalyse the insertion of oxygen into their substrates. An interesting example is tyrosinase, which catalyses tyrosine oxidation in pigment formation.

Superoxide dismutase is another copper-dependent enzyme that is associated with O_2 metabolism, though not in the same way as oxidases or oxygenases. This enzyme plays a crucial role in defence against oxidative stress. It catalyses the disproportionation of superoxide anion (O_2^-) *via* its sequential oxidation to O_2 (with reduction of the Cu(II) centre to Cu(I)) and reduction to H_2O_2 (with re-oxidation to Cu(II)). Not all copper proteins are associated with O_2 or reactive oxygen species (ROS). Blue copper proteins function as electron transfer proteins in, for example, photosynthesis. Two copper enzymes are also associated with the nitrogen cycle: copper-dependent nitrite reductase catalyses the reduction of NO_2^- to NO, and nitrous oxide reductase is the final enzyme of denitrification, reducing N_2O to N_2.

Copper proteins can also be classified according to the type of copper centre they contain.[7] Thus, the Type I copper centre of blue copper proteins features coordination by a Cys and two His residues, augmented by other longer-range interactions. The Cu_B site of haem-copper oxidases and the copper of superoxide dismutase are Type II centres, where the copper is coordinated by at least three His residues. The dinuclear copper centre of tyrosinase is an example of a Type III centre. Multicopper oxidases can contain Types I, II and III copper centres. Some copper sites do not fall into these categories. For example, the copper centre of nitrous oxide reductase contains both an oxidase-like Cu_A centre and a unique Cu_Z centre, consisting of four copper ions bridged by two sulfide anions.[8]

16.1.3 Cytotoxicity of Copper

The cytotoxic properties of copper have been exploited through the ages, for example to prevent fouling of drinking water (first discovered by the ancient Egyptians and Greeks) and to treat various infectious diseases.[9] It was used to protect the hulls of wooden ships from rotting and weed growth as early as 3000 years ago (by Mediterranean civilizations) and more recently in the eighteenth century by the British navy.[10] The phrase "copper-bottomed", meaning a safe plan or investment, is believed to derive from the reliability of ships with copper-sheathed hulls. It is also effective as a fungicide and copper sulfate is the active ingredient of the most effective fungicidal sprays that were first developed in the vineyards of France in the late nineteenth century. The antimicrobial properties of copper are today increasingly exploited in hospitals, where copper alloys are used as a covering for frequently touched surfaces to help control and prevent the spread of pathogenic microorganisms. Studies have shown that many bacteria are killed within minutes of contact with a copper alloy surface.[11] Recent evidence indicates that copper is concentrated in the macrophage phagosome of the human host during bacterial infection and therefore may constitute an important component of the innate immune system response.[12]

Molecular understanding of the effectiveness of copper as a biocide, which is by no means complete, is a much more recent development. In general terms, it is the very properties that make copper essential for a range of life processes that also make it potentially extremely toxic. The capacity of copper to shuttle between $+1$ and $+2$ oxidation states with a reduction potential in a physiologically relevant range means that it can catalyse the formation of the highly reactive, and therefore toxic, hydroxyl radical, $OH^\bullet$. This occurs through the reaction of Cu(I) with H_2O_2, in a reaction analogous to the Fenton reaction in which Fe(II) is oxidized, with one-electron reduction of peroxide to $OH^\bullet$. The resulting Cu(II) can be reduced by superoxide (or other cellular components, such as low-molecular-weight (M_w) thiol compounds), resulting in a Haber–Weiss-like cycle (Equations 16.4–16.6).

$$Cu^{II} + O_2^{-\bullet} \rightarrow Cu^I + O_2 \qquad (16.4)$$

$$Cu^I + H_2O_2 \rightarrow Cu^{II} + OH^- + OH^\bullet \qquad (16.5)$$

$$\text{Overall: } O_2^{-\bullet} + H_2O_2 \rightarrow O_2 + OH^- + OH^\bullet \qquad (16.6)$$

Disease states, in which copper accumulates, demonstrate the toxicity of the metal when in excess. The best known is Wilson's disease, an autosomal recessive genetic disorder in which copper accumulates in the liver and brain. The condition is fatal unless levels of "free" copper are reduced through chelation therapy. The ability of copper to form high-affinity complexes in both $+1$ and $+2$ states means that another principal route of copper toxicity is through the displacement of native metal ions. For example, studies of *E. coli* revealed that copper exerted toxicity through displacement of iron from the iron-sulfur clusters in dehydratase enzymes that are required for biosynthesis of essential branched chain amino acids.[13]

16.1.4 Overview of Chapter

The mechanisms by which copper is transported into, around and out of the cell are considered in some detail, including uptake for incorporation into copper-requiring proteins and export from the cytoplasm/periplasm. The emphasis is on recent developments in our understanding of these complex metabolic pathways. By necessity, moving copper around the cell requires transport across membranes and a number of different types of transmembrane transporters are now known. Although characterized principally as central components of detoxification systems, it is now becoming clear that such transporters are also centrally involved in copper trafficking for biosynthesis purposes. Uptake of copper into the cell is also considered and the focus here is on the methanotrophs, organisms that have an unusually high demand for copper. Transport of copper in the reducing environment of the cytoplasm and the more oxidizing periplasm are discussed in separate sections, with the focus on metallochaperone proteins that interact specifically with substrate proteins to effect transfer of copper ions. The variety of regulatory mechanisms that enable prokaryotes to sense their copper status and respond to it accordingly are examined subsequently. Finally, overall conclusions and major outstanding questions are considered.

16.2 Transport of Copper through the Membrane

Little is currently known about how copper is taken up into prokaryotic cells, and there are very few, if any, *bona fide* proteins or enzymes located in the cytoplasm that require copper. However, it is clear for some systems that trafficking of copper to membrane-located copper proteins/enzymes involves a pathway *via* the cytoplasm. Given the detailed information about copper exporters, it remains a puzzle as to why so little is known about copper import. Some P-type ATPases have been proposed to function as importers, but it was recently proposed that all such proteins transport copper out of the cytoplasm (see Section 16.2.1). A family of copper import proteins may have been recently identified (see Section 16.2.4) but this remains to be verified through structure/function studies. One possibility is that copper is taken up into the cytoplasm *via* a number of pathways, making individual routes difficult to identify.

A principal function of the copper transport systems characterized to date is detoxification. There are a number of different membrane transporters that efflux excess copper across membranes. The occurrence of these across prokaryotes points to a kind of "mix and match" evolutionary process that has resulted in significant diversity of copper metabolism proteins. P-type ATPases, which pump copper across the cytoplasmic membrane, are the most widespread amongst prokaryotes. On the other hand, copper-proton anti-porter tripartite transporters, found in Gram-negative bacteria, pump copper across both the cytoplasmic and outer membranes.

16.2.1 Copper Transporting P-type ATPases

Members of the P-type ATPase superfamily of transporters catalyse the transfer of substrates (mostly cations) across biological membranes by coupling the transport process to the hydrolysis of ATP. Their name is derived from the phosphorylation of an Asp residue in a conserved motif during the reaction cycle. Phosphorylation, actuator and nucleotide-binding domains are common to all P-type ATPases (Figure 16.1A) but structural variations mean that P-type ATPases constitute 11 distinct classes. The P_{IB} class is amongst the largest and most widespread and contains transporters of soft metal ions, including Cu(I), Cd(II), Pb(II) and Ag(I), and borderline metal ions such as Zn(II) and Co(II). The IB class can be further divided into five subgroups on the basis of specificity, with the P_{IB-1} subgroup representing Cu(I)-transporters.[14] The membrane-spanning region of these transporters consists of six to eight transmembrane helices, including a CPX motif that plays a key role in determining the metal ion substrate specificity.

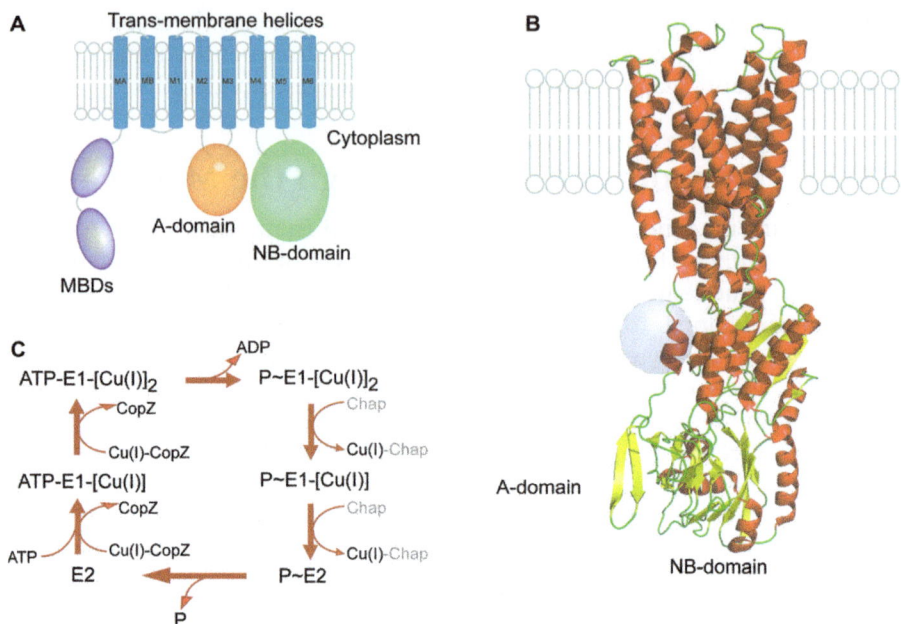

Figure 16.1   Copper-transporting P_{IB}-type ATPase CopA. (A) A schematic view of a
typical copper-transporting ATPase showing the N-terminal metal-
binding domains (MBD), eight transmembrane helices, the actuator
(A-) domain and the nucleotide-binding (NB-) domain. The figure was
generated using PyMOL and pdb 3RFU. (B) Cartoon representation of
the X-ray structure of CopA from *Legionella pneumophilia*. The MBD of
the protein was not resolved in the structure but its approximate
location is indicated by a grey sphere. (C) Scheme of the catalytic
transport cycle illustrating the resting E2 state and Cu(I)-bound E1
states. The putative interaction with metallochaperones (Chap) is
indicated by the grey text.

Also unique to this class are a variable number of usually N-terminal metal
binding domains (Figure 16.1A and Table 16.1).

The first copper-transporting P-type ATPase was discovered in the Gram-
positive bacterium *Enterococcus hirae* (reviewed in ref. 15): CopB exports
Cu(I) under conditions of high copper concentration. Similar copper-
detoxification transporters were subsequently identified in a wide range of
organisms, and it seems that most organisms have this type of detoxification system. Increased copper in the growth medium leads to up-
regulation of the genes encoding these transporters. Inactivation of these
genes leads to a significant increase in sensitivity to copper toxicity and
increased levels of copper in the cell. In the case of pathogenic organisms,
increases in environmental copper levels are due to the recently discovered
active concentration of copper in the host phagosome as part of the innate
immune response; consistent with this, inactivation of detoxification

Table 16.1 Principal bacterial copper metabolism proteins.

Protein	Description/function	Copper site(s)	Ref.
Transmembrane transporters			
CopA	Cu(I)-transporting P-type ATPase. Involved in Cu(I) export for detoxification or delivery for cofactor insertion	Two trigonal membrane transport sites: 1) Tyr, 2×Cys (CPC motif); 2) Asn, Met, Ser N-terminal ferredoxin-like domain distorted digonal site(s): 2×Cys (MXCXXC motif) + low-M_w thiol/Cys? Some transportertain a cuperdoxin N-terminal domain with a dinuclear Cu(I) centre	14, 23, 29
CusABC	Tripartite efflux system involved in Cu(I) detoxification (particularly under anaerobic conditions)	CusA: Met-rich Cu(I)-transport channel CusB: trigonal 3×Met	33
PcoD (CopD)	Possible copper (Cu(II)?) importer for PcoA (CopA) assembly	Unknown	46
Cu(I)-detoxifying oxidases			
CueO	Multi-copper oxidase (MCO). Functions as O_2-dependent Cu(I) oxidase for detoxification of Cu(I) in the periplasm	In addition to the type I, II and III sites of the MCO, contains tetrahedral substrate (Cu(I)) oxidation site: 2×Met, 2×Asp. Also contains two Cu(I) binding site(s): 1) 2×Met + 1 unclear ligand, probably Met or His; 2) 3×Met	56
PcoA/CopA	MCOs encoded by plasmid-located copper resistance system. CueO homologues		46
Cytoplasmic metallochaperones			
CopZ (Atx1)	Small soluble protein involved in tight coordination of Cu(I) in the cytoplasm, providing buffering/storage capacity, preventing deleterious reactions and delivering of Cu(I) to cognate P-type ATPase. Homologue of yeast Atx1 and human Hah1	Distorted digonal site: 2×Cys (MXCXXC motif) + low-M_w thiol/Cys? $Cu_4(CopZ)_2$ forms have bridging Cys and terminal Cys/His coordination	27, 53
CupA	Small membrane-tethered protein with cupredoxin-like domain that functions similarly to CopZ. Interacts with P-type ATPase containing a similar cupredoxin-like MBD	Dinuclear Cu(I) centre High-affinity site has distorted digonal coordination by 2×Cys. Low-affinity site has distorted trigonal planar coordination from Cys, 2×Met + 1 exogenous anion	29
Periplasmic metallochaperones			
CueP	Metallochaperone required for copper insertion into the periplasmic CuZn superoxide dismutase SodCII. Also important for copper detoxification under anaerobic conditions. Not yet structurally characterized	Unknown	19, 57

Table 16.1 *(Continued)*

Protein	Description/function	Copper site(s)	Ref.
CusF	Soluble protein with OB fold required for full copper resistance through the Cus system. Likely delivers Cu(I) from periplasm to CusB	Distorted trigonal Cu(I) site: His, 2×Met (MXM motif) with π-cation interaction with nearby Trp	59, 60
PcoC/CopC	Copper resistance plasmid-encoded seven-stranded β-barrel proteins required for full resistance	Trigonal/tetrahedral Cu(I) site: 2–3×Met, His Square planar Cu(II) site: 2×His, NH$_2$-terminus, H$_2$O Cu(I) and Cu(II) sites separated by ∼30 Å	47, 63
CopK	Copper resistance plasmid-encoded dimeric β-sandwich protein	Binds Cu(I) and Cu(II) cooperatively Two alternative Cu(I) sites: 1) 3×Met, H$_2$O; 2) 4×Met Cu(II) site: His70 + N- and C-terminal residues	67, 68
Sco	Thioredoxin-like fold protein required for the assembly of cytochrome *c* oxidase (CcO)	Binds Cu(II) or Cu(I) at the same red copper site: 2×Cys (CXXXC motif), His + unidentified ligand	69, 72
Cox11	Immunoglobulin-like fold protein required for the assembly of Cu$_B$ in *aa*$_3$-type CcOs	Trigonal Cu(I) site: 3×Cys (CFCF motif) located between subunits of dimer	76
PCu$_A$C	Cupredoxin-like fold protein involved in the delivery of copper (Cu$_A$ or Cu$_B$) to a range of CcOs	Tetrahedral Cu(I) site: 2×His, 2×Met, or His, 3×Met (motif H(M)X$_{10}$MX$_{21}$HXM)	74, 77
Regulatory proteins			
CueR	Member of MerR family of dimeric winged-helix regulators, required for copper-responsive regulation in a wide range of bacteria	Buried digonal Cu(I) site: 2×Cys	80, 82
CsoR	Tetrameric four-helix bundle protein required for copper-responsive regulation in (mostly) Gram-positive bacteria	Trigonal Cu(I) site: 2×Cys, His located between subunits	83,84
CopY	Dimeric helix-turn-helix superfamily member, required for copper-responsive regulation in a narrow range of bacteria (mostly within the order Lactobacillales)	Binds two Cu(I) ions through a C-terminal CXCX$_6$CXC motif in each monomer generating a luminescent dinuclear centre	88
CusS	Periplasmic sensor of Cu(I) that, together with the regulator CusR, forms a two-component sensor-regulator system. Copper resistance plasmid-encoded PcoRS and CopRS are homologues	Unknown	90, 91

transporters significantly affects virulence.[12] Some prokaryotes contain more than one gene encoding Cu(I)-transporting P-type ATPases, indicating that these transporters are involved in additional functions. For example, PacS from the cyanobacterium *Synechocystis* PPC 6803 is located in the thylakoid membrane where it supplies copper for the synthesis of the electron transfer protein plastocyanin. Phylogenetic analyses revealed that many of these additional transporters belong to a distinct subgroup, referred to as the FixI/CopA2-like ATPases,[16] which in many cases are involved in the assembly of one or more cytochrome *c* oxidase enzymes (*e.g.* ref. 17). Thus inactivation of the genes encoding such transporters does not affect sensitivity to copper toxicity, but does result in a deficiency in oxidase activity. In some cases, transporters have been proposed to be involved in the import of copper into the cytoplasm, but recent studies suggest that all P-type ATPases transport copper in the same direction (*i.e.* export from the cytoplasm). The activities of FixI/CopA2 transporters were found to be much lower than the archetypal detoxification transporters, and it was argued that this is consistent with a role in the assembly of copper-requiring proteins outside of the cytoplasm.[16,18] Recently, it was shown that, in *Salmonella enterica* serovar *Typhimurium*, either of two transporters, CopA or GolT, is needed to activate a periplasmic Cu,Zn superoxide dismutase (SodCII),[19] indicating that proteins other than oxidases are dependent on the transport of copper from the cytoplasm. In some cases, uncertainty remains and further work is needed to confirm direction of transport and the function of such atypical transporters.

Progress towards structural characterization of bacterial Cu(I)-transporting ATPase was first made through high-resolution structures of N-terminal domains (*e.g.* ref. 20), the actuator domain[21] and the nucleotide binding domain.[22] For the last two domains, these revealed significant similarity to P-type ATPases belonging to other classes. In 2011 the first full structure (at 3.2 Å resolution) of a class 1B Cu(I)-transporter, CopA from the bacterium *Legionella pneumophilia*, was reported.[23] The protein was in a copper-free form but revealed the key features of the transporter and, in particular, the interactions between membrane and soluble domains (Figure 16.1B).

Although smaller than in other characterized P-type ATPases and despite low sequence conservation, the soluble domains were found to retain a core structure. The membrane spanning region consists of eight transmembrane helices: six core helices (M1–M6) and two helices, MA and MB, that are unique to the P_{IB} class (Figure 16.1A). Helix MB is kinked at the membrane-cytoplasm interface, with the C-terminal half being amphipathic in nature sitting across the membrane surface. The kink is a result of two conserved Gly residues, suggesting that this is a common structural feature of CopA transporters. Conversely, M1 is not kinked as it is in other characterized P_{IB} transporters. MA and kinked MB are proposed to be part of an entry pathway for Cu(I) into the membrane spanning region, and may represent a docking site for delivery of Cu(I) from a soluble chaperone (see Section 16.3.1) or the N-terminal domains (see below). There is no Cu(I) bound in the structure, but biochemical and spectroscopic studies have shown that two, likely trigonal, Cu(I) ions bind at the transport site in the membrane-spanning region.[24] One Cu(I) is coordinated by Cys382 and Cys384 from the CPC motif of M4 and Tyr688 from M5 (*L. pneumophilia* CopA numbering), while the other is coordinated by Asn689 from M5 and Met717 and Ser721 from M6. The metal coordination is strikingly different from that typically associated with Cu(I). Binding affinities have been estimated for CopA from *Archaeoglobus fulgidus* and were found to be non-cooperative and in the femtomolar range, consistent with very high-affinity binding in preparation for transport.[24] Studies of *A. fulgidus* CopA showed that Cu(I) accesses the site bound to CopZ, an Atx1-like copper metallo-chaperone (Section 16.3.1).[25,26]

The soluble metal binding domains (MBDs) of P_{IB} transporters often have significant primary sequence similarity to Atx1-like metallochaperones. Several have been structurally characterized and shown to adopt the same $\beta\alpha\beta\beta\alpha\beta$ ferredoxin-like fold (see Figure 16.2 and ref. 27 for a review), and to bind Cu(I) at an MXCXXC motif. The coordination geometry of the bound Cu(I) is typically distorted digonal (*i.e.* the bond angle is <180°), suggesting the *in vivo* interaction of a third ligand, derived from either the protein or an exogenous source. In prokaryotes there are one (*e.g.* E. coli CopA) or two (*e.g.* B. subtilis CopA) of these domains, located usually at the N-terminus of the protein. Up to six are found in higher eukaryotes. The N-terminal soluble domain of *L. pneumophilia* CopA is not well defined in the crystal structure. However, some electron density was observed, and an approximate location was modelled, indicating proximity to the actuator and nucleotide binding domains, consistent with electron microscopy studies (Figure 16.1B).[28] Recently, a novel type of MBD was discovered in *Streptococcus pneumonia* CopA. Rather than a ferredoxin-like fold, the MBD has a cupredoxin-like fold consisting of an eight stranded β-barrel. The domain contains a dinuclear Cu(I)-binding centre with one site coordinated by two Cys residues and the other by one Cys and two Met residues, where one of the Cys residues bridged the two sites.[29] This unusual P-type ATPase is paired with a novel type of membrane-bound chaperone that closely resembles the MBD of the transporter (see Section 16.3.1).

The function of the N-terminal MBDs has not yet been demonstrated unequivocally. It was initially proposed that they might supply Cu(I) to the membrane transport site. However, they are not essential for Cu(I) transport[30,31] and the transporter can be activated by the Cu(I)-bound cognate metallochaperone in the absence of the MBDs.[25] Furthermore, the available structural information indicates that the soluble domains may be too far away from the proposed Cu(I)-entry site such that a significant structural rearrangement would be required for the domains to interact directly with the membrane transport site. Proximity to the actuator and nucleotide-binding domains, along with protein-protein interaction studies, has led to the proposal that the metal binding domain(s) fulfil a regulatory function through local interactions. Such a regulatory role would still require a

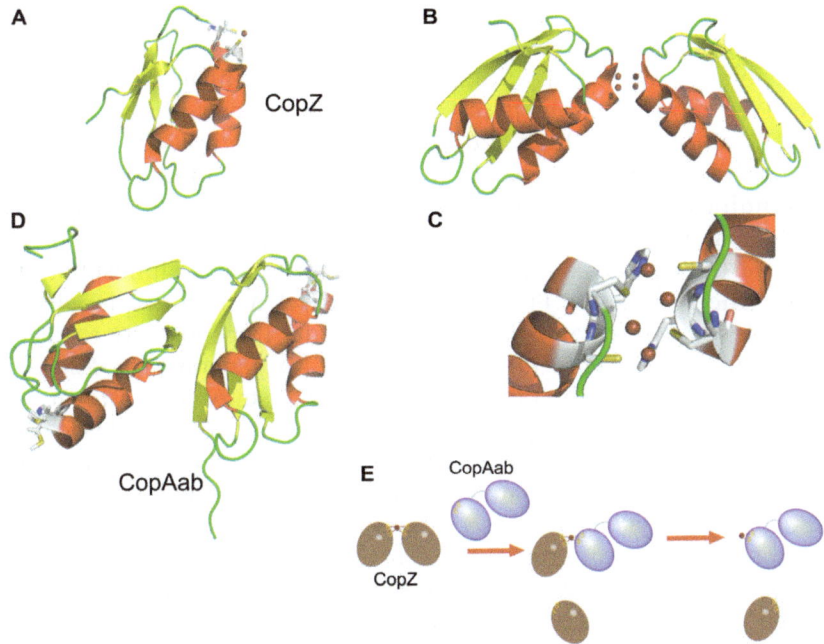

Figure 16.2 CopZ and CopA, cytoplasmic copper trafficking proteins. (A) Cartoon structure of *B. subtilis* CopZ generated from pdb 1K0V. Cu(I) is coordinated by two Cys residues and by a likely third, exogenous thiol (not shown). (B) Dimeric CopZ containing a tetranuclear Cu(I) cluster at the protomer interface (pdb 2QIF). (C) The Cu(I)-binding interface region showing coordinating Cys and His residues. (D) Cartoon representation of the two N-terminal soluble domains of *B. subtilis* CopA (CopAab) (pdb 2RML). Note the structural similarity of each CopA domain to CopZ. (E) Scheme for the transfer of Cu(I) from CopZ to CopAab, *via* a process of stepwise ligand exchange that is likely generally applicable to chaperone-target exchange reactions. CopZ is a dimer in the presence of Cu(I) and the first step involves the exchange of one CopZ molecule for a CopAab molecule.

conformational change of some sort, and it has been shown that the interaction between the N-terminal MBDs and the nucleotide binding domain of *A. fulgidus* CopA is sensitive to the presence of Cu(I) and ATP.[32] This led to the proposal that, in the absence of Cu(I) bound to the MBDs, interactions with the nucleotide-binding domain would inhibit the conformational changes needed for efficient Cu(I) transfer. On the other hand, as Cu(I) levels increase, so would the occupancy of the MBDs, leading to reduced interaction between domains and consequent activation of the transporter.[18] In the context of conformational changes, it is noteworthy that studies of the two N-terminal MBDs of *B. subtilis* CopA revealed that binding of more than one Cu(I) per two domains led to the formation of a luminescent multinuclear Cu(I) centre together with significant changes in secondary structure.[20] Thus, Cu(I)-dependent conformational changes are, in some cases, relatively complex and might point to the functional significance of more than one N-terminal metal binding domain.

In general, P-type ATPases function *via* a mechanism described by the Albers–Post model, in which the enzyme adopts two principal conformations, E1 and E2, with high and low affinity, respectively, for the extruded substrate (Figure 16.1C). It has been shown that ATP binding at the nucleotide-binding domain is required for binding of Cu(I) at both transport sites. Hydrolysis of ATP results in transfer of the γ-phosphate to the invariant Asp residue of the conserved DKTG motif, generating an acylphosphate intermediate (the E1-P state). Phosphorylation drives a conformational change that results in exposure of the now low-affinity transport site to the periplasmic side and release of the Cu(I) ions (to generate the E2-P state). Dephosphorylation regenerates the resting E2 state. The rate at which transport occurs is likely regulated by the N-terminal MBDs. Release of Cu(I) on the outside of the cytoplasmic membrane might involve a metallochaperone in Gram-negative organisms. CueP, a periplasmic metallochaperone in *S. enterica* sv *Typhimurium* (see Section 16.3.2.1), is required to transfer Cu(I) from P-type ATPase transporters to a superoxide dismutase (SodCII).[19] In detoxification pathways, CueO (see Section 16.3.2.1) would be a candidate in some organisms. Metallochaperone-mediated release is perhaps less likely in Gram-positive organisms that lack a periplasmic space and outer membrane. Here, another possibility is that release might require oxidation to the more labile Cu(II).

16.2.2 The Cus Copper (and Silver) Resistance System of *E. coli* and other Gram-negative Bacteria

In Gram-negative bacteria, copper pumped out of the cytoplasm into the periplasm remains toxic, particularly if it accumulates as Cu(I), and so additional detoxification pathways are commonly found. Under aerobic conditions, Cu(I) can be oxidized to the much less toxic Cu(II) and this is a strategy employed by some organisms. Under anaerobic conditions, the lack of a suitable oxidant means that a different approach is required, and principal amongst these is the efflux of Cu(I) out of the cell entirely, driven by the Cus system.

The Cus system, comprising CusCBA, belongs to a superfamily (called the resistance-nodulation-cell division or RND superfamily) of tripartite efflux systems that plays crucial roles in intrinsic and acquired resistance to a range of antibacterial compounds. Many members of this family are multidrug efflux pumps, while Cus functions as a heavy metal efflux pump. The tripartite system is composed of a cytoplasmic membrane transporter (CusA, the pump), a periplasmic protein that contacts both the cytoplasmic and the outer membrane components (CusB, the adaptor), and an outer membrane channel protein CusC belonging to the outer membrane factor family (Figure 16.3A).[33] Export of copper (and silver) through CusA is driven by the proton motive force (*i.e.* proton import drives copper export). CusF is an

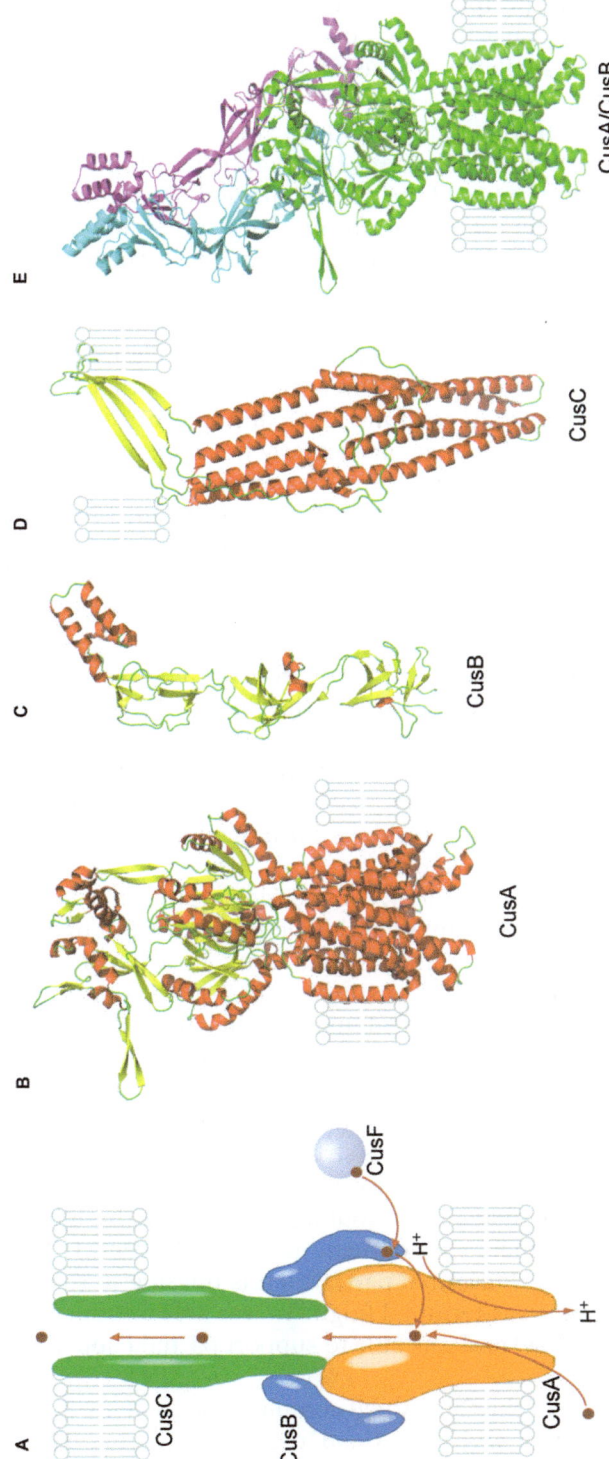

Figure 16.3 The Cus copper detoxification system. (A) Schematic view of the CusCBA complex that straddles both outer and cytoplasmic membranes. Cu(I) is transported principally from the periplasm, delivered by the chaperone CusF *via* CusB to the channel formed by CusA and CusC. Transport across the cytoplasmic membrane is driven by exchange with H⁺ from outside the membrane. Cartoon representations of the X-ray structures of (B) *E. coli* CusA (generated from pdb 3K07), (C) *E. coli* CusB (pdb 3OOC), (D) *E. coli* CusC (pdb 3PIK) and (E) a CusAB complex (pdb 3NE5).

additional protein in the periplasm that functions as a metallochaperone for the delivery of copper to CusCBA (see Section 16.3.2.2). X-ray structures of all Cus proteins are now available,[34–36] as is that of a complex of CusA and CusB.[37]

CusA is a homo-trimer with each subunit consisting of 12 transmembrane helices with two large periplasmic loops between helices 1 and 2, and 7 and 8. Helices 4–10 form the centre of the membrane domain, while helices 4, 5, 10 and 11 extend into the cytoplasm, forming a soluble domain. Helices 2 and 8 extend into the periplasm, forming part of the periplasmic domain. The relative positions of helices 1–6 are symmetrically related to those of 7–12 (Figure 16.3B). The periplasmic domain stabilizes the trimeric arrangement and also interacts with the N-terminus of CusB (Figure 16.3E). At the base of the periplasmic domain are three Met residues that form a Cu(I) binding site that is important for function.[38] Binding of Cu(I) to this site results in a conformational change, particularly in the periplasmic domain, leading to the opening of a channel for presumed entry of Cu(I) from the periplasm. Changes also occur in the membrane domain, where all of the helices except helix 2 undergo significant lateral or vertical movement. In particular, helix 8 shifts substantially away from the core, perhaps initiating the translocation of a proton that must accompany efflux of Cu(I) from the cytoplasm.

CusA contains 34 Met residues, with 18 of these in the membrane domain, including three pairs, which, together with the cluster of three Met residues (described above) and an additional pair in the periplasmic domain, line a channel stretching right across the transmembrane domain up to and into the periplasmic domain (Figure 16.3). Site-directed mutagenesis studies support the proposal that these residues form a relay network of binding sites important for copper transport.[33]

The structure of CusB (solved at 3.4 Å and lacking the first 88 residues), the first of a metal effluxing member of the membrane fusion protein (MFP) family, revealed four distinct domains that result in a flexible, elongated shape (Figure 16.3C). The first, consisting of six β-strands arranged into a barrel, interacts with the periplasmic domain of the pump CusA.[34] Domains 2 and 3 are mainly composed of β-strands, while domain 4 is composed of a three α-helical bundle, a structural feature that is not present in other characterized members of the MFP family, which normally contain a helical hairpin in this location. In addition to a structural role in connecting the two integral membrane proteins, CusB also plays an important role in copper (and silver) binding. One metal ion is bound per protein with high affinity in an all-sulfur three-coordinate environment generated by Met21, Met36 and Met38, all located in the N-terminal part, which is missing in the structure. A protein consisting of the N-terminal fragment was able to transfer metal to and from CusF (see below) and was able to support partial copper resistance *in vivo*.[39]

The crystal structure of the CusA-CusB complex, solved at 2.9 Å,[37] revealed that each CusA interacts with two CusB molecules (Figure 16.3E), with specific interactions between domains 1–3 of the two CusB molecules. Thus, the overall complex is $(CusA)_3(CusB)_6$ with the six CusB molecules tilted to form a ring at the top of the complex, with a channel (~ 62 Å long and, on average, 37 Å wide) running through the ring. The interior surface of the channel is negatively charged and, in addition to bridging the cytoplasmic and outer membrane proteins, the adaptor is likely to play an active role in extruding copper.

The crystal structure of CusC was solved recently (to 2.3 Å) and shown to be a trimer, similar to previously characterized OMF proteins such as TolC (Figure 16.3D).[36] Each of the monomers contributes four β-strands to form a 12-stranded β-barrel in the outer membrane, and four α-helices to form a periplasmic α-barrel. Two of the four helices are actually made up of two smaller helices that interact end to end. On the inside of the trimeric structure is a hydrophilic cylindrical shaped channel. It also appears that the protein is triacylated at the N-terminal Cys residue with *N*-acyl and *S*-diacylglycerol functions, the first example of an OMF protein with a lipid anchor. CusC has no particular features that mark it out as a copper pore, and so its specificity for copper is most likely determined by its specific interactions with CusA and B. The structure of CusC does not reveal how it interacts with CusA and B, though modelling studies suggest that the trimeric CusC α-barrel would fit well within the hexameric CusB channel[33,37] (Figure 16.3A). Once Cu(I) ions reach the CusC channel, they most likely complete their exit by diffusion.

16.2.3 Methanobactin

Methanotrophs are proteobacteria that can utilize methane as a carbon source. Oxidation of methane to methanol, the first step of the incorporation process, is catalysed in many cases by the copper-dependent integral membrane enzyme methane monooxygenase (MMO), usually referred to as particulate MMO or pMMO. There is considerable interest in this transformation because of the greenhouse gas properties of methane in the atmosphere. Because pMMO is highly abundant (it accounts for 20% of the total cellular proteome), methanotrophs have a considerable requirement for copper, which they satisfy by secreting a low-M_w (~ 1200 Da) high-affinity Cu(I) chelator called methanobactin.[40] This strategy is analogous to that utilized by many microorganisms to satisfy their requirement for iron. They synthesize and secrete low-M_w Fe(III) chelators called siderophores, of which there are many different examples (*e.g.* see Chapter 11). Copper chelators, which by analogy with iron might be referred to as chalkophores (greek for copper-bearers), are thus far confined to the methanotrophs, reflecting that most organisms can obtain sufficient copper (which is much more bio-available than iron) without needing to employ a high-affinity chelator strategy. The unusual metabolic requirement for copper in methanotrophs is an exception to this general situation.

Methanobactin accumulates in the growth media of methanotrophs under copper-limited conditions and is rapidly taken up by cells as the Cu(I) complex. The structure of methanobactin from *Methylosinus trichosporium* OB3b consists of 7 amino acid residues (Gly1, Ser2, Cys3, Tyr4, Ser5, Cys6 and Met7), 2 oxazolone rings, each with a neighbouring thioamide group (generating two 4-hydroxy-5-thionyl imidazoles), a 3-methylbutanoyl group and a pyrrolidinyl group (Figure 16.4A). Coordination of copper occurs through the thionyl imidazole moieties, which provide a distorted tetra-hedral geometry around the metal. Overall, the molecule has a compact pyramid-like structure that is stabilized by a disulfide bridge between the two Cys residues.

Recent structures of methanobactins from three *Methylocystis* strains re-vealed major structural variation.[41] Methanobactins from these strains had fewer amino acid residues (5 or 4), which are also different (Ala1, Ser2, Ala3

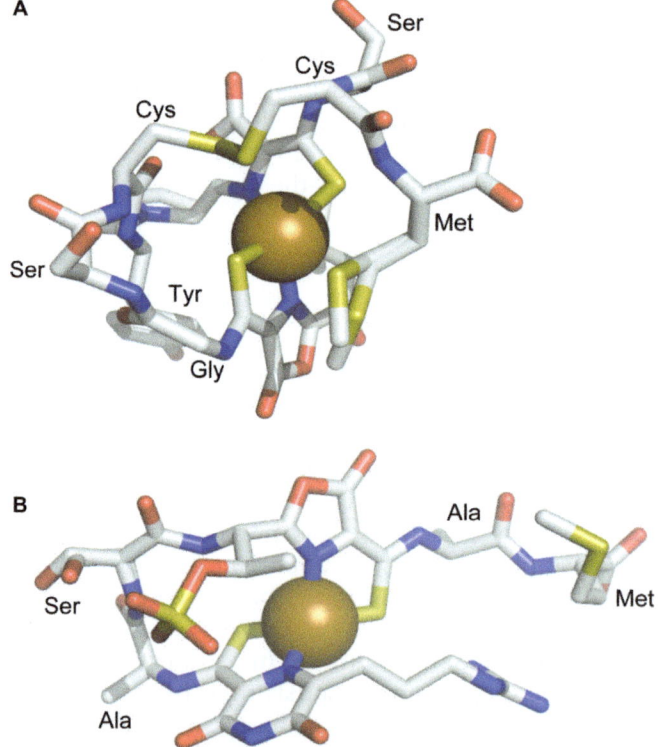

Figure 16.4 Structures of two methanobactins, the first examples of small-molecule copper chelators (chalkophores). Structures of methanobactin from (A) *Methylosinus trichosporium* OB3b (pdb 2XJH) and (B) *Methylocystis* strain M (pdb 2YGJ). The latter is one of two principal methanobactins synthesized by this organism. Amino acid residues are indicated. Note that the S_2N_2 coordination is conserved in both methanobactins.

and either Met4 or Ala4; in one form, there is an additional Thr residue at the C-terminal end), and the shorter backbone results in a hairpin-like structure (Figure 16.4B). Furthermore, one of the coordinating oxazolone rings is replaced by an unusual pyrazinedione ring. However, the overall S_2N_2 distorted coordination sphere is very similar.[41]

The Cu(I) affinities of these methanobactins are similar (in the region of $K \sim 10^{21}$ M^{-1}), but their Cu(II) affinities are quite different (in the range $K \sim 10^{11}$–10^{14} M^{-1}), influencing their relative abilities to sequester Cu(II), the predominant environmental form. Differences in binding affinities result in variable midpoint reduction potentials for the Cu(II)/Cu(I) couple, in the range $+480$–$+750$ mV (*versus* SHE at pH 7.5). Note that these potentials are consistent with a much higher affinity for Cu(I) and facile reduction of Cu(II) to Cu(I).

Little is known about the synthesis of methanobactins. Recent evidence from studies of *M. trichosporium* OB3b indicated that they are chromosomally encoded as a precursor peptide;[42] this work also showed that methanobactin plays a role in regulating expression of MMO. Little is also known about how the Cu(I)-methanobactin complex is transported into the cell. While unchelated Cu(II) might gain entry through the outer membrane *via* porins, Cu(I)-methanobactin is too large to enter *via* this route. Recently, it was demonstrated that Cu(I)-methanobactin is specifically taken up into the cytoplasm of methanotrophs in a process that is dependent on the proton motive force.[43] Thus, again by analogy to Fe(III)-siderophore uptake, the pathway could involve a TonB-dependent transport system, in which TonB functions to couple energy from the proton motive force to active transport across the outer membrane. Once inside the cell, Cu(I) must be released from its complex with methanobactin. The very high stabilities of such complexes suggest that this is unlikely to occur *via* a simple dissociation mechanism. The variation in reduction potentials measured thus far suggests that, in some cases at least, oxidation to Cu(II), which is much less tightly bound, and generally much more labile, would be a reasonable mechanism.[41] However, in methanobactins with very high potentials, an alternative mechanism(s) is likely to exist.

16.2.4 Other Transmembrane Copper Transporting Systems Including Possible Importers

Some bacteria have the ability to grow in environments containing high amounts of copper that would normally overwhelm chromosomally encoded detoxification systems. This is due to the presence of an additional plasmid-borne copper detoxification system. Examples of these are the homologous *pco* and *cop* systems found, respectively, in copper-resistant *E. coli* and *Pseudomonas syringae*. The *pco* and *cop* gene clusters are arranged in two operons (*pco/copABCD* and *pco/copRS*) with the *pco* system containing an additional gene, *pcoE*, further downstream.[44,45] While the *RS* genes encode for a two-component regulator (see Section 16.4.4), the *ABCD* genes encode for inner and outer membrane and periplasmic proteins. PcoA and CopA (not to be confused with the P-type ATPase discussed in Section 16.2.1) are periplasmic multicopper oxidases and therefore homologues of CueO (see Section 16.3.2.1). Like CueO, they contain an N-terminal twin arginine signal suggesting that their copper sites are loaded in the cytoplasm and the mature protein is transported across the cytoplasmic membrane *via* the TAT pathway. PcoB/CopB are predicted to be outer membrane proteins, while PcoC and CopC are small periplasmic proteins that bind copper (and are discussed in more detail in Section 16.3.2.3). PcoD and CopD are predicted integral cytoplasmic membrane proteins with eight putative transmembrane segments, which likely transport copper across the cyto-plasmic membrane.

It appears that the Pco/CopAB proteins are the key elements of the de-toxification system and that they likely work in concert.[46] Pco/CopCD (and PcoE) are required for full detoxification activity but are not essential com-ponents. It is proposed that PcoA functions as a copper oxidase, converting Cu(I) to the less toxic Cu(II) in the periplasm[47] (see Section 16.3.2.1). The obvious role of a detoxification transporter would be in copper efflux, but the role of PcoD may be in the import of copper into the cytoplasm for synthesis of PcoA. Interestingly, *Bacillus subtilis* encodes for homologues of PcoC and D in a single fusion protein known as YcnJ. This protein has recently been implicated in the uptake of copper into the cell,[48] consistent with an importer function. Although PcoC has been shown to interact with PcoA, it is possible that it also interacts with PcoD.

It is known that some cytoplasmic membrane P-type ATPases function in the specific delivery of Cu(I) to membrane- or periplasm-located enzymes. There are now also indications that other types of transporters are involved in copper supply. For example, in *Rhodobacter capsulatus*, supply of copper to the Cu_B site of the cbb_3 cytochrome c oxidase requires a transporter, CcoA, with homology to the family of major facilitator superfamily (MFS)-type transporter proteins.[49]

16.3   Transport of Copper within the Cell

To minimize the toxicity of copper, cells have evolved often complex traf-ficking pathways involving metallochaperones and their specific substrate (or target) proteins. These invariably have high affinity for copper, but the characteristics of the copper site vary with cellular compartment. Cyto-plasmic copper chaperones exclusively bind Cu(I) at sites dominated by Cys thiolate coordination. On the other hand, periplasmic chaperones bind Cu(I) or Cu(II) and, in some cases, have distinct sites for each oxidation state. In the latter case, the Cu(I) sites tend to be dominated by Met thioether and His imidazole ligands, while the Cu(II) sites are usually dominated by His. The relative redox characteristics of the cytoplasmic and periplasmic compart-ments suggest an explanation for these differences. Cu(II) cannot readily persist in the reducing environment of the cytoplasm (which has millimolar concentrations of low-M_w thiols such as glutathione). In the more oxidizing periplasm, the Met thioether is less susceptible to oxidation than the Cys thiol, consistent with its dominance in this compartment. Coordinating Met residues are often found in Met-rich motifs. Cu(II) is stable in this compartment and is much less toxic than Cu(I) and is thus the "preferred" oxidation state in terms of controlling toxicity.

16.3.1 Copper Transport and Detoxification in the Cytoplasm

As discussed in Section 16.2, there are few, if any, cytoplasmic copper proteins besides those that function in trafficking or cellular regulation of the metal. The major cytoplasmic copper transport protein is CopZ (sometimes known as Atx1), which is homologous to yeast Atx1 and human Hah1 (Atox1) and functions principally in the transport of Cu(I) to a Cu(I)-transporting P-type ATPase. CopZ is present in a very wide range of bacteria and the proteins from a number of these, including *Enterococcus hirae*, *Bacillus subtilis* and *Synechocystis*, have been extensively characterized (reviewed in ref. 27). CopZ is a small ($\sim$70 residues) protein that adopts a characteristic ferredoxin $\beta\alpha\beta\beta\alpha\beta$-fold in which the antiparallel β strands form a β-sheet on which the two α-helices are superimposed (known as an open-faced β-sandwich; Figure 16.2A). A MXCXXC Cu(I)-binding motif is situated on a flexible solvent-exposed loop at the beginning of the first α-helix. The Cu(I) ion is bound with trigonal geometry derived from the two motif Cys residues and a third sulfur ligand (from an exogenous low-M_w thiol molecule or another CopZ molecule, see below), yielding an extremely high-affinity site ($K\sim10^{17}$–10^{18} M^{-1}). The affinity of thiolates for Cu(I) together with the high concentration of low-M_w thiols in the cytoplasm suggests that such molecules could play an important role in Cu(I) trafficking in the cytoplasm. More work will be required in order to answer this, but studies in *Synechocystis* showed that glutathione is not required for copper trafficking from cytoplasm to the thylakoid membrane.[50] Furthermore, a number of CopZ proteins have been shown to undergo dimerization upon Cu(I)-binding, even in the presence of glutathione or Cys.[27]

CopZ functions in the delivery of copper to its cognate P-type ATPase, both to the N-terminal regulatory domain(s) and to the integral membrane transporter site (see Section 16.2.1). The N-terminal soluble, cytoplasmic MBDs are highly related structurally to CopZ (Figure 16.2D) and bind Cu(I) with affinities that are similar to those of CopZ.[20] Because binding affinities are in effect a ratio of on and off rate constants, such high affinities imply extremely low rates of Cu(I) dissociation. This is a favourable characteristic when it comes to protecting Cu(I) from participating in deleterious reactions, but does not assist the movement of the metal around the cell. This is facilitated by specific interactions between CopZ/Atx1 and the N-terminal MBDs, which have been demonstrated both in cells, through two-hybrid assays (*e.g.* ref. 51), and *in vitro* through functional assays[25] and the detection of Cu(I)-dependent hetero-complexes by NMR chemical shift mapping experiments.[52] Because of similar tight binding, there is no large thermodynamic driving force for transfer of Cu(I) from CopZ/Atx1 to the MBDs. The formation of transient hetero-complexes, in which Cu(I) is believed to be transferred through a ligand exchange reaction (Figure 16.2E), appears to facilitate transfer by lowering the kinetic barrier to dissociation of the original Cu(I)-protein complex (thereby increasing its rate).

It is proposed that CopZ/Atx1 can also deliver Cu(I) to the membrane transport domain of the ATPase and this has been shown for *A. fulgidus* CopZ/CopA. It was also shown that Cu(I) is not delivered from the soluble domains to the membrane transport domain.[25] Recently, it was shown that electrostatically complementary regions of *A. fulgidus* CopA (a positively charged region close to a proposed entry pathway for Cu(I) into the membrane) and CopZ (negatively charged region close to the CXXC motif) are important for the transfer of Cu(I), consistent with this being the docking site for the Cu(I)-bound chaperone.[26] In *Synechocystis* inactivation of the metallochaperone did not significantly affect the delivery of Cu(I) to plastocyanin, a thylakoid-located electron transfer protein involved in photosynthesis.[50] Thus, in some cases at least, CopZ is not essential for the delivery of Cu(I) to the transporter, and it is noteworthy that some bacteria that have a copper-transporting P-type ATPase do not possess a *copZ* gene. If the chaperone is not essential for copper delivery then why is it so ubiquitous in life?

The answer to this question has at least three interconnected parts. Firstly, as described above, delivery of Cu(I) to the N-terminal domains is most likely not part of the transport pathway and, therefore, serves an alternative function. As discussed in Section 16.2.1, this is likely to be a regulatory one, in which the N-terminal domain(s) modulate transport activity *via* Cu(I)-dependent interactions with the actuator and nucleotide-binding domains of the transporter. It is noteworthy that some CopZ dimers are capable of accommodating up to four Cu(I) ions, generating tetranuclear clusters at the monomer interface, *e.g.* see Figure 16.2B and C.[53] Transfer of Cu(I) from *B. subtilis* CopZ containing multiple Cu(I) ions to the N-terminal domains of CopA (CopAab) is thermodynamically highly favourable and binding of more than one Cu(I) per CopAab leads to its dimerization and the formation of a luminescent Cu(I) cluster.[20] Thus, the structure of the N-terminal domains of CopA is sensitive to Cu(I)-loading, suggesting a mechanism for regulation of activity. Secondly, deletion of *copZ* in *B. subtilis* resulted in a significant decrease in cellular copper content.[51] This implies that CopZ may act as a cytoplasmic copper buffer (or store) under normal conditions, and the ability of CopZ to accommodate multiple Cu(I) ions is perhaps relevant here also. Thirdly, by tightly binding Cu(I) and directing it along specific transport pathways governed by protein-protein interactions, CopZ/Atx1 prevents Cu(I) from participating in deleterious reactions. This has been assumed for some time but direct evidence was lacking until recently. Deletion of the chaperone in *Synechocistis* does not affect delivery of copper but does lead to increased toxicity, most likely due to the mislocation of Cu(I) to other metal-binding sites.[50] Another function of CopZ is the transfer of Cu(I) to transcriptional regulatory proteins (see Section 16.4).

Amongst the organisms that do not have a CopZ protein is *S. pneumoniae*, which contains a Cu(I)-transporting P-type ATPase with a novel kind of N-terminal MBD (see Section 16.2.1).[29] Instead of CopZ, this organism, and many other related bacteria, contains a copper chaperone called CupA, which, remarkably, is tethered to the membrane *via* a single transmembrane helix. The soluble domain of the protein has a cupredoxin-like fold with a dinuclear Cu(I) centre that is highly similar to the MBD of *S. pneumoniae* CopA, thus preserving the structural similarity between chaperone and transporter MBD that is a feature of CopZ/ATPase systems.[29] The two Cu(I) sites of CupA (and the MBD of CopA) have inequivalent affinities for Cu(I), with the bis-thiolate site having a significantly higher affinity (by three orders of magnitude). Delivery of Cu(I) was found to occur from the low-affinity site of the chaperone to the high-affinity site of the MBD (down a thermodynamic gradient). In this case, CupA was shown to be essential for copper resistance in *S. pneumoniae*, leading to the conclusion that CupA is required to deliver Cu(I) to the transporter. As for other transporters, the MBD was not essential for resistance.[29]

Metallothioneins, which are small, Cys-rich proteins, are ubiquitous and well characterized heavy metal-binding proteins in eukaryotic cells. However, they are not widely distributed in prokaryotes and do not usually function in copper metabolism. An exception is MymT from *Mycobacterium tuberculosis*, which appears to function in the sequestration of (mainly) surplus copper, to alleviate copper toxicity, particularly in the presence of reactive oxygen and nitrogen species.[54] The protein binds six Cu(I) ions, generating a luminescent cluster core in which the Cu(I) ions are protected from participating in deleterious reactions. Since enhanced copper resistance appears to be crucial for pathogens, it may be that this and related metallothioneins are important for virulence.

16.3.2 Copper Transport and Detoxification in the Periplasm

16.3.2.1 The Cue System

In Gram-negative bacteria, transport of Cu(I) out of the cytoplasm to the periplasm alleviates toxicity in the former, but contributes to it in the latter and so additional detoxification pathways are commonly found. Principal amongst these is CueO, a multi-copper oxidase (MCO) that, together with the co-regulated CopA, constitutes the Cue system. Like other MCOs (Section 16.1.2), CueO couples the four-electron reduction of O_2 to water to four one-electron substrate oxidation steps. The substrate for CueO appears to be Cu(I), resulting in its oxidation to the much less toxic Cu(II) (see Section

16.1.3). The crystal structure of *E. coli* CueO, initially determined at 1.4 Å resolution,[55] revealed an overall structure similar to those of previously characterized MCOs, with a type 1 copper site along with a trinuclear copper centre consisting of a type 2 site associated with a binuclear type 3 site. In addition, the protein contained a Met-rich 45 residue insert. Of the 14 Met residues in the insert, nine lie on a helix that covers the entrance to the type 1 copper site and contributes ligands, and restricts access, to the substrate Cu(I) binding/oxidation site, which is formed by two Met and two Asp residues. Subsequent structures revealed additional Cu(I)-binding sites in the Met-rich region, which appear to be important for activity and it seems likely that these are the initial sites of Cu(I) binding prior to transfer to the Cu(I) substrate site.[56] The electron resulting from oxidation of Cu(I) here is released to the Type 1 copper site and then on to the trinuclear site where it is combined with O_2 to generate water.

Clearly, CueO is dependent on O_2 and so is only active under aerobic conditions. Thus, under anaerobic conditions, or when the Cue system is overwhelmed by high copper concentrations, additional components or an alternative system are required. In some Gram-negative bacteria, particularly pathogens (*e.g. Salmonella* and *Yersinia*), the Cue system contains an additional periplasmic protein called CueP that appears to play a crucial role in copper detoxification under anaerobic conditions,[57] and in *S. enterica* sv Typhimurium, at least, is the predominant periplasmic copper-binding protein.[58] Very recently, it was shown that CueP is required for copper insertion into the periplasmic CuZn superoxide dismutase SodCII, which is associated with virulence,[19] and thus appears to be a *bona fide* periplasmic metallochaperone. It is also possible that CueP detoxifies periplasmic copper through complexation. Interestingly, genes encoding CueP-like proteins are also found in Gram-positive bacteria, with the common feature that most are pathogens. Thus, CueP appears to represent a detoxification pathway that may be adapted to resisting the chemical assault that invading organisms face in the macrophage.

16.3.2.2   CusF

More widespread than CueP as an anaerobic detoxification system is the Cus system (see Section 16.2.2), which, along with the tripartite transporter, contains the periplasmic protein CusF, which is required for full resistance against copper (and silver). The crystal structure of CusF (solved at 1.5 Å) revealed a five-stranded β-barrel, classified as an OB (oligonucleotide binding) fold not previously observed for a copper metallochaperone.[59] The protein binds Cu(I) *via* His36, Met47 and Met49 located in β-strands 2 and 3 (Figure 16.5A).[38,59] There is also a novel π-cation interaction between the Cu(I) and Trp44, which appears to play an important role in controlling the redox and binding properties of Cu(I).[60] CusF has been shown to exchange copper with CusB,[61] indicating that it plays a role in the efflux of periplasmic copper *via* CusCBA (Figure 16.3A).

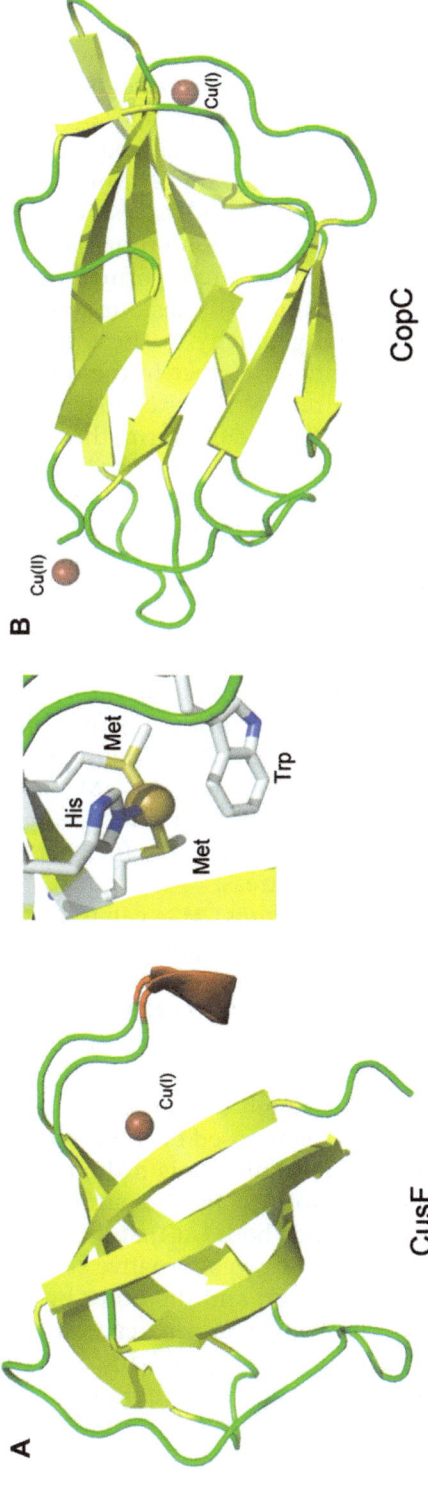

Figure 16.5 CusF and CopC, periplasmic copper chaperones. (A) Cartoon structure of *E. coli* CusF (pdb 2VB2). To the right is a close-up view showing the Cu(I)-binding environment involving one His and two Met residues. The interacting Trp residue is also shown. (B) Cartoon representation of *Pseudomonas syringae* CopC generated from pdb 2C9Q, showing the two remote binding sites for Cu(I) and Cu(II), as indicated.

16.3.2.3   *Periplasmic Pco and Cop Proteins*

The plasmid-borne *pco* and *cop* copper resistance systems of *E. coli* and *P. syringae*, respectively, include periplasmic proteins that are required for full resistance. PcoC and CopC are homologous seven-stranded β-barrel proteins with a Met-rich motif lying on a solvent exposed loop.[62] Both proteins can bind Cu(I) and Cu(II) at separate high-affinity sites (Figure 16.5B). The Cu(II) site features coordination by two His residues, the amino terminus and a water molecule (note these are borderline/hard Lewis bases) in a square planar geometry, while the Cu(I) site is formed by two or three Met residues from the Met-rich motif and a His residue (soft/borderline Lewis bases) in a trigonal or tetrahedral geometry.[63] These sites, which are separated by ~30 Å, have little affinity for copper in the other oxidation state, such that oxidation or reduction resulted in exchange of copper from one site to the other.[62,63] In the Cu(I)/Cu(II)-bound form, the Cu(I) site is stable to O_2-mediated oxidation, indicating that a high-affinity receiving site is required to promote facile oxidation. However, in the presence of the MCO PcoA, the Cu(I) ion bound in Cu(I)/Cu(II)-PcoC was readily oxidized to Cu(II), which dissociated from the protein, yielding Cu(II)-PcoC.[47] This form of the protein inhibited further Cu(I) oxidase activity of PcoA, consistent with the interaction of PcoC and PcoA, possibly through interactions of Met-rich regions of both proteins. Thus, it is proposed that PcoA catalyses the O_2-driven oxidation of Cu(I) bound to PcoC to the less toxic Cu(II).

CusRS-regulated PcoE is a periplasmic protein with a high proportion of metal-coordinating amino acid residues and a high capacity for binding Cu(I) and Ag(I) ions with high affinity ($K = \sim 10^{12}$ M^{-1}). It can also bind Cu(II) ions, though these are fewer in number and less tightly bound than Cu(I). These properties are consistent with a role for PcoE as a periplasmic copper "sponge" in times of copper stress.[64]

The bacterium *Cupriavidus metallidurans* CH34 can tolerate the presence of many heavy metals, including copper. Compared to other copper-resistant bacteria, it contains a greatly expanded *cop* cluster containing, in addition to the *cop/pco* genes discussed above, a P-type ATPase (CopF), a number of predicted cytoplasmic Cys-rich proteins (CopJ, CopG and CopL) and several predicted periplasmic proteins.[65] One of these, CopK, is highly abundant in the periplasm of cells grown under high copper levels. The protein is Met-rich and adopts a β-sandwich fold, composed of two perpendicularly aligned β-sheets and an unstructured C-terminus. It is a dimer in the absence of copper, with the dimer interface formed by the surface of the C-terminal β-sheet.[66] Like Cop/PcoC, CopK binds both Cu(I) and Cu(II) ions at separate sites. In fact, Cu(I) can bind at one of two alternative sites, one coordinated by three Met residues (3S) and a water, and the other by four Met residues (4S).[67] The two sites share two residues and so cannot be simultaneously occupied. Both sites are high affinity ($K = \sim 10^{10}$–10^{12} M^{-1}), with the 4S site somewhat more stable. Cu(II)-binding to apo-CopK, on the other hand, occurs with relatively low affinity ($K = < \sim 10^{6}$ M^{-1}). However, binding of Cu(I) has a remarkable effect on the affinity of the Cu(II) site, increasing it by six orders of magnitude.[67] The Cu(I) affinity also increases (by two orders of magnitude) when the Cu(II) site is occupied. Such strongly cooperative Cu(I)/Cu(II) binding has not been observed previously and structural studies revealed the basis for this behaviour. Whereas binding of Cu(I) at the 3S site does not result in major structural change, binding at the 4S site, which is preferred in solution, does, resulting in the release of the C-terminal β-strand from the β-sandwich, disrupting the dimer interface and causing dissociation of the dimer into monomers. This also releases the N-terminus, which forms a largely disordered loop, generating the high-affinity Cu(II) site, which involves coordination from His70 and residues in the N- and C-terminal regions.[68] The flexibility of Cu(I) coordination is likely to be important for Cu(I) transfer along trafficking pathways (*e.g.* to transporters). Furthermore, these properties mean that, in the presence of Cu(I), CopK promotes the oxidation of a single Cu(I) ion to generate the Cu(I)/Cu(II)-form, which is then rather stable to further oxidation.

16.3.2.4 Proteins Involved in the Assembly of Cu Centres in Cytochrome c Oxidases

16.3.2.4.1 Sco Proteins. Sco proteins, which are widely distributed in all types of cells, have an N-terminal membrane anchored segment connected to a soluble domain that adopts a thioredoxin-like fold. This contains a CXXXC motif, rather than the more usual CXXC motif of disulfide exchange proteins, but can still cycle between di-thiol and disulfide forms. They also have the capacity to bind copper and the precise nature of their functional role(s) is the subject of some controversy: Sco has been proposed to function in the delivery of copper to the Cu_A centre of cytochrome c oxidase (CcO; *e.g.* ref. 69) or, alternatively, to be involved in thiol-disulfide redox processes that likely precede copper insertion[70] or that function more generally in redox signalling.[71] Some prokaryotes contain Sco but not a CcO, while others contain multiple Sco proteins, and it is likely that some Sco proteins do not function directly in copper transport.

Copper binding to Sco generates a red copper site coordinated by two Cys and one His, and a fourth, unidentified ligand. This site is related to those of blue copper proteins (see Section 16.1.2), sharing intense, albeit shifted, Cu(II)–Cys charge transfer bands, but with a low covalency of the Cu–S bond and overall distinct coordination and geometry.[72] This site can accommodate both Cu(II) and Cu(I), with evidence indicating that, unlike blue copper proteins, there are significant differences in the site on cycling between oxidation states. Studies of *B. subtilis* and *Streptomyces lividans* Sco proteins indicate that the ability of the protein to coordinate Cu(II) is functionally important, and mutations which disrupt this coordination result in loss of function without necessarily abolishing copper binding.[69,72,73] Whether Sco directly interacts with the Cu_A centre of CcO is not entirely clear. In some cases, Sco also appears to play a role in the assembly of the Cu_B centre of cbb_3-type cytochrome oxidases (Cox) and appears to interact directly with Cox for the delivery of copper (*e.g.* ref. 74).

16.3.2.4.2 Cox11. Cox11 consists of a soluble domain with an N-terminal membrane anchor and is known to be an assembly factor for the Cu_B centre of eukaryotic CcOs. Homologues of this protein are also found in several Gram-negative α-proteobacteria, where it appears to function similarly in the assembly of aa_3-type CcOs.[75] Interestingly, it does not appear to be involved in Cu_B assembly in cbb_3-type oxidases, where Sco (see above) and PCu_AC (see below) functionally replace it. The structure of Cox11 (also referred to as CoxG and CtaG) from *Sinorhizobium meliloti* revealed an immunoglobulin-like fold with unusual β-strand organization. The protein binds two Cu(I) at the dimer interface of two Cox11 subunits, with trigonal coordination from three Cys residues: two from one subunit (in a conserved CFCF motif) and one from the other (juxtaposed) subunit, such that two of the total of four Cys residues bridge Cu(I) ions.[76]

16.3.2.4.3 PCu_AC. In *Thermus thermophilus*, the delivery of Cu into the Cu_A site of the ba_3 oxidase has been shown to be dependent on the periplasmic protein PCu_AC.[70] The protein, which adopts a cupredoxin-like fold, belongs to a previously recognized family of proteins that possess a high-affinity Cu(I)-binding motif $H(M)X_{10}MX_{21}HXM$ in which the metal is coordinated by two His and two Met residues, or by one His and three Met residues, in a tetrahedral site.[77] It is noteworthy that the PCu_AC fold is similar to that of other bacterial extracytoplasmic copper proteins like CopC, though the metal-binding sites are different. The often close proximity of genes encoding these proteins with *sco* genes suggested a potential role in copper-binding and Cu_A assembly. It was shown *in vitro* that, while Sco could reduce disulfide bonds at the Cu_A site prior to copper insertion, it could not deliver copper while PCu_AC could.[70] However, in *R. sphaeroides*, Sco seems to be more important for copper insertion and appears to function downstream of PCu_AC in the process of copper delivery.[78] Both Sco and PCu_AC were also found to be important for Cu_B insertion into the cbb_3 oxidase.

A homologue of PCu_AC, PcuC from *Bradyrhizobium japonicum*, was recently reported to be important for the assembly of both aa_3 and cbb_3 CcOs. It binds Cu(I) with high affinity, apparently at a $His_2 Met_2$ site, consistent with previous studies of the PCu_AC family.[74] Thus, it seems that, in α-proteobacteria at least, Sco and PCu_AC function collectively in the copper insertion into both the aa_3- and cbb_3-type CcOs. Their precise roles in delivery of copper to Cu_A and Cu_B sites remains to be clarified, though it is suggested that PCu_AC may deliver Cu(I) to Sco (and perhaps to other metallochaperones, such as Cox11).[74]

16.4 Sensing, Buffering and other Control Mechanisms (Cellular Homeostasis)

Regulatory systems for the sensing and control of responses to copper broadly fall into two categories: one-component regulators that consist of a single, cytoplasmically located protein and two-component regulators that consist of a sensor kinase located in the cytoplasmic membrane and a response regulator in the cytoplasm. Single component regulators are most prevalent in bacteria. It is now also becoming apparent that, in some cases, post-transcriptional/translational regulation also plays a part.

16.4.1 CueR

CueR is widely found in proteobacteria, in which it commonly regulates transcription of genes encoding P-type ATPases and CueO, and may also regulate a wider range of genes involved in energy metabolism and stress response.[79] The crystal structure of dimeric CueR revealed that it belongs to the MerR family of winged-helix regulators (Figure 16.6A).[80] Three N-terminal helices, which form a core that includes the helix-turn-helix motif is followed by a β-hairpin (the "wing"), which is followed by a second helix-turn-helix motif.[81] In addition to the DNA-binding domain, there is a coiled-coil dimerization domain and a Cu(I)-binding sensory domain. MerR family regulators are involved in sensing a range of metals. The specificity of CueR for Cu(I), which is bound with extraordinarily high affinity ($K = 10^{21}$ M^{-1}), appears to be due to an unusually buried Cys-coordinated site that restricts the metal ion to linear geometry, which is relatively rare in copper proteins.[80,82] CueR binds DNA at a specific inverted repeat sequence in both copper-free and bound forms, but only the copper-bound form activates transcription through distortion of the DNA. It is the only known copper regulator that acts positively (*i.e.* it is an activator rather than a repressor, see Figure 16.6A).

16.4.2 CsoR

The most recently discovered family of copper regulators is the CsoR family, first identified in *Mycobacterium tuberculosis*.[83] Examples are also present in a wide range of Gram-positive bacteria, including *B. subtilis*, *Staphyloccocus aureus* and *Streptomyces lividans*, where they are involved in regulation of operons that include genes for P-type ATPases. Homologues are also present in proteobacteria and cyanobacteria but it is not known whether these function in copper regulation. The crystal structure of *M. tuberculosis* CsoR revealed a DNA-binding domain unlike that of any other regulator family, consisting of a homodimeric antiparallel four-helix bundle (Figure 16.6B). The protein binds Cu(I) with extremely high affinity ($K = \sim 10^{18}$–10^{21} M^{-1})[84,85] at bridging intermonomer sites, with trigonal coordination by two Cys (one from monomer 1 and one from monomer 2) and one His (from

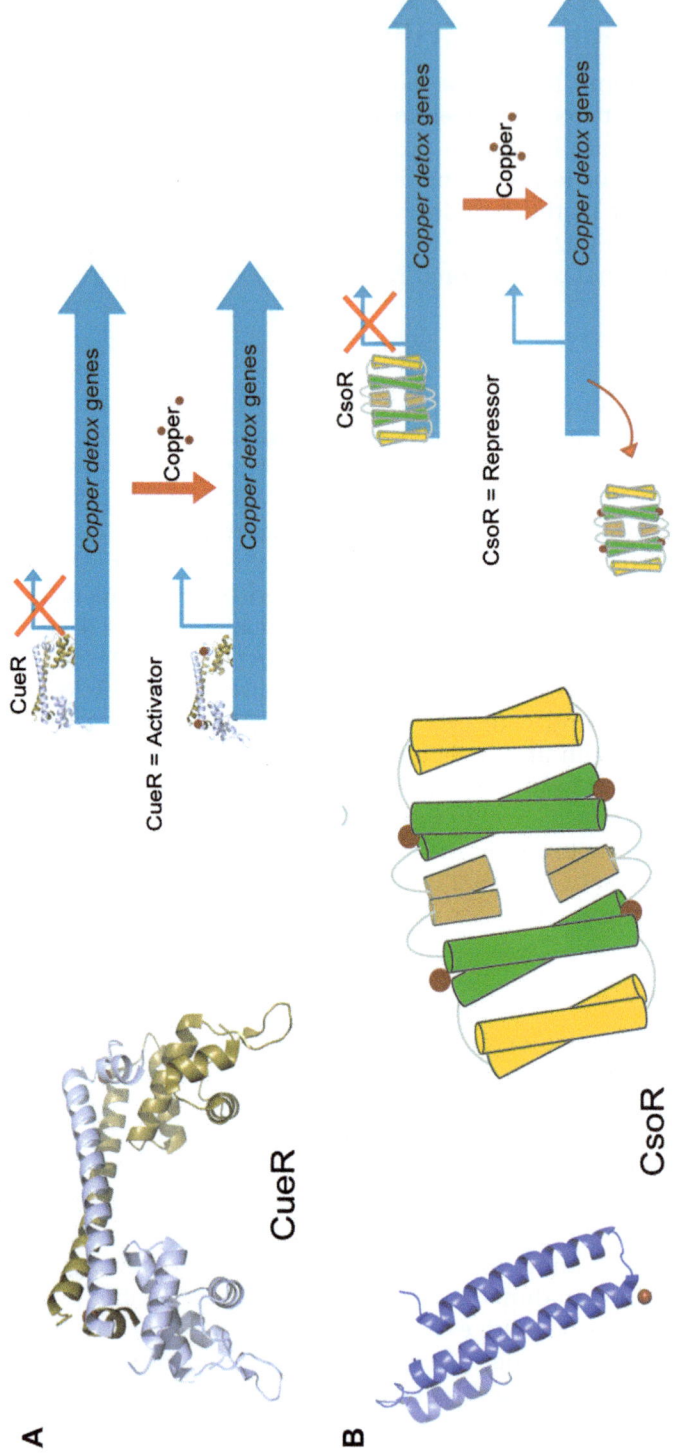

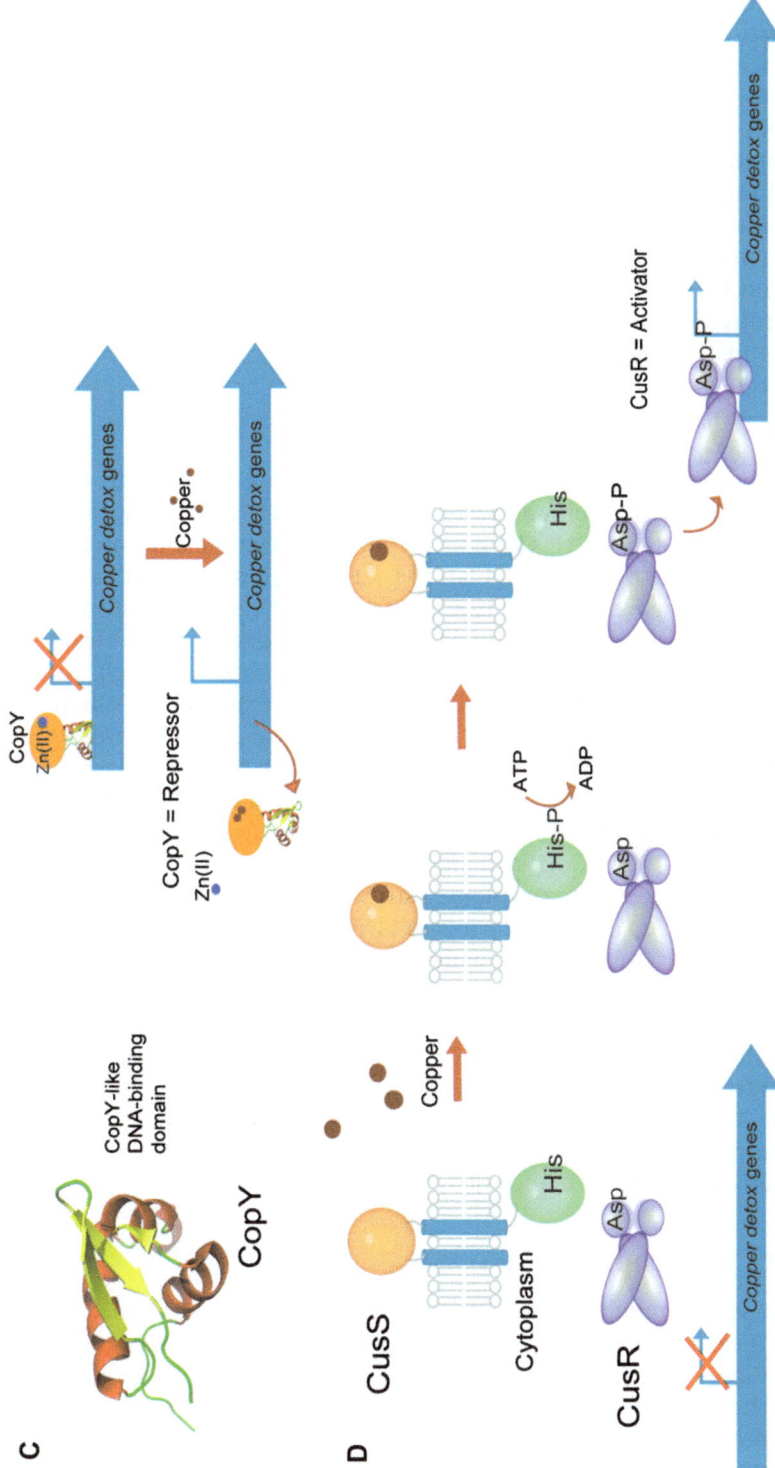

Figure 16.6 Copper-responsive regulatory proteins. (A) A cartoon of *E. coli* CueR functions as a Cu(I)-responsive transcriptional activator (right). (B) A cartoon of how CueR functions as a Cu(I)-responsive transcriptional activator (right). (B) A cartoon of *M. tuberculosis* CsoR generated from pdb file 2HH7 (left) and a representation of the CsoR solution tetramer, containing four ions bound at sites located between subunits (centre). A scheme of how CsoR functions as a transcriptional repressor is shown on the right. (C) A cartoon of the DNA-binding domain of *Lactococcus lactis* CopR (a CopY homologue) generated from pdb 2K4B (left) and a scheme of how CopY functions as a transcriptional repressor (right). CopZ, the cytoplasmic chaperone, has been shown to deliver Cu(I) to CopY and it is likely that Cu(I) is delivered to other regulators described here in a similar way. (D) A scheme of how copper-responsive two-component regulators such as CusRS function as transcriptional activators.

monomer 2) residues. While the *M. tuberculosis* protein structure revealed a dimeric structure, it and other CsoR proteins are tetramers in solution.[84,85] CsoR recognizes a pseudo-palindromic DNA sequence in the operator-promoter region only in its copper-free form. Cu(I)-binding results in loss of high-affinity DNA binding and dissociation of the protein from DNA, alleviating transcriptional repression (Figure 16.6B). Some insight into the allosteric switching mechanism was gained from the incorporation of the non-natural residue Nε2-methyl-histidine in place of the Cu(I) coordinating His61 residue. This suggested that a formation of a H-bonding network involving Tyr35 and Glu81 is initiated by the Nε2 atom of the imidazole ring upon Cu(I)-coordination to the Nδ1 atom of the ring, and that this is a key for driving the allosteric switch.[86]

16.4.3 CopY

Copper trafficking in the Gram-positive bacterium *E. hirae* involves an operon encoding four proteins: CopA and CopB are P-type ATPases involved in copper transport (Section 16.2.1); CopZ is an Atx1-like cytoplasmic metallochaperone (Section 16.3.1); CopY is a small ($\sim$16 kDa) protein that functions as a copper-responsive transcriptional regulator of the *cop* operon.[87] Under low copper conditions, CopY binds a Zn^{2+} ion, is dimeric and binds DNA at a conserved inverted repeat (known as a *cop* box) through its N-terminal domain, thereby repressing expression of the cop operon. When copper levels increase, two Cu(I) ions are delivered to the C-terminal $CXCX_4CXC$ motif of each CopY monomer by the chaperone CopZ, displacing the Zn^{2+} ion and forming a luminescent dinuclear Cu(I) cluster.[88] This form of the protein no longer binds DNA with high affinity and repression is alleviated (Figure 16.6C). The Cu(I)-transferring interaction with CopZ is specific, based on electrostatic complementarity between interacting domains. Homologues of CopY are found in the Firmicutes phylum and most are within the order Lactobacillales, which includes *Enterococcus*, *Lactobacillus* and *Streptococcus*.[89] Structural NMR studies of the DNA-binding domain of the CopY homologue CopR from *L. lactis* showed that the protein belongs to the winged helix family that in turn belongs to the helix-turn-helix superfamily (Figure 16.6C). The N-terminal helix-turn-helix motif is followed by helix α3 (the likely DNA-recognition helix) and this is followed by an antiparallel β-hairpin (the "wing"). It is distinct, however, from CueR, another winged helix regulator (see above).

16.4.4 CusRS and CopRS/PcoRS

Two-component sensor-regulators, which are extremely widespread in bacteria, consist of a sensor kinase located in the cytoplasmic membrane and a soluble response regulator in the cytoplasm. On detection of their signal in the periplasm, the sensor undergoes auto-phosphorylation at a conserved His residue located on the cytoplasmic side. The phospho-group is then spontaneously transferred to a conserved aspartate residue of the response regulator, causing a conformational change that activates transcription of the regulated genes (Figure 16.6D). In *E. coli*, the Cus copper detoxification system (see Section 16.2.2) that is crucial under anaerobic conditions is regulated by the two-component system (TCS) CusRS. CusS is the sensor that responds to periplasmic levels of Cu(I)[90,91] and controls activation of regulator CusR, which binds to a palindromic sequence between the divergently transcribed *cusCFBA* and *cusRS* operons. Although the Cus system appears to be able to pump Cu(I) from the cytoplasm across both the inner and outer membranes, it seems that removing periplasmic copper is its major function, consistent with the role of CusS in sensing periplasmic Cu(I). Similar TCS regulators are found in the plasmid-borne *pcoABCDRS* and *copABCDRS* operons that are found in copper-resistant strains of *E. coli* and *P. syringae*, respectively, and which confer resistance principally through the oxidation of Cu(I) to the less toxic Cu(II) *via* a multicopper oxidase (see Section 16.2.4). The precise mode(s) of copper binding and how this stimulates phosphorylation is currently not known.

16.4.5 Other Copper-responsive Regulators

A number of other copper-responsive regulators exist, which are either confined to a few bacteria or for which there is currently relatively little information. The *comR* gene from *E. coli* encodes a TetR-like transcriptional regulator that regulates the transcription of *comC* in a copper-dependent fashion. ComC is a predicted outer membrane protein that appears to function in lowering the permeability of the outer membrane to copper, providing an important first clue about how these (and perhaps many other) bacteria might regulate their copper uptake.[92] Copper-resistant strains of the plant pathogen *Xanthomonas axonopodis* contain plasmid-borne copper-resistance-genes similar to the *cop* and *pco* operons discussed in Section 16.2.4. However, these are not regulated by a TCS. Instead, a soluble protein, CopL, rich in His and Cys residues, has been shown to be required for regulation.[93] Whether this is a single-component regulatory system has not yet been demonstrated.

The ArsR/SmtB family transcriptional repressor BxmR from the cyanobacterium *Oscillatoria brevis* regulates the expression of *bxa1* (encoding a metal-transporting P-type ATPase (see Section 16.2.1)) and *bmtA* (encoding an intracellular metallothionein) in response to Cu(I)/Ag(I) and Zn(II)/Cd(II) ions,[94] and is the first ArsR protein reported to sense both mono- and divalent ions. The dimeric protein contains two metal sites (termed the α3N and α5C sites) per subunit, which have different specificities and, unusually, can accommodate a Cu(I)$_2$S$_4$ cluster at the α3N site,[94] analogous to that proposed for *E. hirae* CopY (see above). The stepwise affinity for each Cu(I) is estimated to be $K \sim 10^7$ M^{-1}, much lower than those measured for other regulators, and consistent with a role in copper detoxification only under severe copper stress.

Though a common means of regulating expression of genes in bacteria, sigma factors are not normally associated with metal regulation. However, in the δ-proteobacterium *Myxococcus xanthus* an extracytoplasmic function (ECF) sigma factor named CorE was shown to activate transcription of genes encoding two P_{1B}-type ATPases and a multi-copper oxidase. No evidence could be found for the existence of an anti-sigma factor (that normally controls the availability of the sigma factor for DNA binding). Instead it appears that copper, which most likely binds in the Cys-rich C-terminal domain, is required as a co-activator for DNA binding by CorE.[95] The mechanism of CorE-activation of transcription is not yet clear, however.

16.4.6 Post-transcriptional/-translational Regulation

Transcriptional control is the major mechanism by which bacteria regulate their cellular proteome. However, examples of post-transcriptional control are now emerging. In *R. capsulatus*, levels of the multi-copper oxidase CueO are regulated in response to copper, but this does not occur through the type of transcriptional regulation described above, because the promoter for transcription of the tricistronic operon containing *cueO* is constitutively "on" and unresponsive to copper.[96] Regulation is achieved through modulation of the stability of the mRNA and is dependent on a stem-loop structure formed upstream of the *cueO* message. It is proposed that an as yet unidentified cellular factor (protein) binds the stem-loop under conditions of high copper, thus stabilizing the mRNA for increased levels of translation.

Post-translational regulation, through the modulation of protein stability, is another mechanism by which the cellular proteome is controlled. *E. hirae* CopZ (see Section 16.3.1) appears to be under post-translational control, as levels of the protein were found to diminish at very high copper levels and to be mediated by specific proteolysis.[97] The reason why the protein, which functions in copper detoxification, would be controlled in this way is not yet clear. Though information about post-transcriptional/-translational control mechanisms is currently somewhat sparse, such mechanisms may be widespread in bacteria.

16.5 Conclusions with Open Questions

Since copper trafficking proteins were discovered during the early/mid 1990s, progress in understanding such systems and in discovering new ones has been spectacular. The wealth of genome sequence information, together with increasingly tractable genetics in a range of prokaryotes, has facilitated functional understanding, while recent progress in gaining high-resolution X-ray structures, in particular, of membrane transporters has paved the way for detailed molecular understanding. A wealth of information on Cu(I) and Cu(II) coordination by metallochaperones in the cytoplasm (Cu(I)) and periplasm (Cu(I) and Cu(II)) has revealed general properties associated with proteins operating in the different cellular compartments.

Despite the recent progress, it is clear that, in most cases, much remains to be learned about mechanisms of copper metabolism and their regulation. With one or two exceptions, very little is known about how copper enters bacterial cells, and it is possible that under normal conditions, high-affinity uptake systems are not required because copper is sufficiently abundant and available in the majority of environments. The exceptions are where there is a heavy cellular demand for copper, or where copper availability is limited.

Another emerging theme is that, despite the fact that copper does not appear to be required *per se* in the cytoplasm, cytoplasmic membrane transport systems are essential in some cases for the assembly of extra-cytoplasmic copper proteins, and thus biosynthetic trafficking systems are apparently more complex than first appeared. It is likely that much more information about these pathways will emerge in the near future. The recent demonstration that the toxicity of copper is employed by the immune system to defeat invading pathogens has begun a new chapter in the story, familiar in connection to iron, of the war between pathogens and their human host for control of metal ions. How pathogens respond to this copper challenge is already beginning to emerge.

Acknowledgements

The author thanks the UK's Biotechnology and Biological Sciences Research Council for supporting his work on metals in biology.

References

1. B. Webster Smith, *Sixty Centuries of Copper*, Hutchinson, London, 1965.
2. D. V. Scaltrito, D. W. Thompson, J. A. O'Callaghan and G. J. Meyer, *Coord. Chem. Rev.*, 2000, **208**, 243.
3. F. A. Cotton, G. Wilkinson, C. A. Murillo and M. Bochmann, *Advanced Inorganic Chemistry: A Comprehensive Text*, Wiley-Blackwell, New York, 6th edn, 1999, pp. 854–876.
4. B. G. Malmstrom and J. Leckner, *Curr. Opin. Chem. Biol.*, 1998, **2**, 286.
5. L. Decaria, I. Bertini and R. J. Williams, *Metallomics*, 2011, **3**, 56.
6. N. J. Robinson and D. R. Winge, *Annu. Rev. Biochem.*, 2010, **79**, 537.
7. J. T. Rubino and K. J. Franz, *J. Inorg. Biochem.*, 2012, **107**, 129.
8. A. Pomowski, W. G. Zumft, P. M. Kroneck and O. Einsle, *Nature*, 2011, **477**, 234.
9. W. Richardson, *Handbook of Copper Compounds and Applications*, Marcel Dekker, New York, 1997.
10. J. M. Bingeman, J. P. Bethell, P. Goodwin and A. T. Mack, *Int. J. Naut. Archaeol.*, 2000, **29**, 218.
11. C. Espirito Santo, E. W. Lam, C. G. Elowsky, D. Quaranta, D. W. Domaille, C. J. Chang and G. Grass, *Appl. Environ. Microbiol.*, 2011, **77**, 794.
12. C. White, J. Lee, T. Kambe, K. Fritsche and M. J. Petris, *J. Biol. Chem.*, 2009, **284**, 33949.

13. L. Macomber and J. A. Imlay, *Proc. Natl Acad. Sci. USA*, 2009, **106**, 8344.
14. J. M. Arguello, M. Gonzalez-Guerrero and D. Raimunda, *Biochemistry*, 2011, **50**, 9940.
15. M. Solioz, H. K. Abicht, M. Mermod and S. Mancini, *J. Biol. Inorg. Chem.*, 2010, **15**, 3.
16. M. Gonzalez-Guerrero, D. Raimunda, X. Cheng and J. M. Arguello, *Mol. Microbiol.*, 2010, **78**, 1246.
17. B. K. Hassani, C. Astier, W. Nitschke and S. Ouchane, *J. Biol. Chem.*, 2010, **285**, 19330.
18. D. Raimunda, M. Gonzalez-Guerrero, B. W. Leeber, 3rd, and J. M. Arguello, *Biometals*, 2011, **24**, 467.
19. D. Osman, C. J. Patterson, K. Bailey, K. Fisher, N. J. Robinson, S. E. Rigby and J. S. Cavet, *Mol. Microbiol.*, 2012, **87**, 466.
20. C. Singleton, L. Banci, S. Ciofi-Baffoni, L. Tenori, M. A. Kihlkenl, R. Boetzel and N. E. Le Brun, *Biochem. J.*, 2008, **411**, 571.
21. M. H. Sazinsky, S. Agarwal, J. M. Arguello and A. C. Rosenzweig, *Biochemistry*, 2006, **45**, 9949.
22. M. H. Sazinsky, A. K. Mandal, J. M. Arguello and A. C. Rosenzweig, *J. Biol. Chem.*, 2006, **281**, 11161.
23. P. Gourdon, X. Y. Liu, T. Skjorringe, J. P. Morth, L. B. Moller, B. P. Pedersen and P. Nissen, *Nature*, 2011, **475**, 59.
24. M. Gonzalez-Guerrero, E. Eren, S. Rawat, T. L. Stemmler and J. M. Arguello, *J. Biol. Chem.*, 2008, **283**, 29753.
25. M. Gonzalez-Guerrero and J. M. Arguello, *Proc. Natl Acad. Sci. USA*, 2008, **105**, 5992.
26. T. Padilla-Benavides, C. J. McCann and J. M. Arguello, *J. Biol. Chem.*, 2013, **288**, 69.
27. C. Singleton and N. E. Le Brun, *Biometals*, 2007, **20**, 275.
28. C. C. Wu, W. J. Rice and D. L. Stokes, *Structure*, 2008, **16**, 976.
29. Y. Fu, H. C. Tsui, K. E. Bruce, L. T. Sham, K. A. Higgins, J. P. Lisher, K. M. Kazmierczak, M. J. Maroney, C. E. Dann, 3rd, M. E. Winkler and D. P. Giedroc, *Nat. Chem. Biol.*, 2013, **9**, 177.
30. B. Fan, G. Grass, C. Rensing and B. P. Rosen, *Biochem. Biophys. Res. Commun.*, 2001, **286**, 414.
31. A. K. Mandal and J. M. Arguello, *Biochemistry*, 2003, **42**, 11040.
32. M. Gonzalez-Guerrero, D. Hong and J. M. Arguello, *J. Biol. Chem.*, 2009, **284**, 20804.
33. F. Long, C. C. Su, H. T. Lei, J. R. Bolla, S. V. Do and E. W. Yu, *Philos. Trans. R. Soc. Lond. B Biol. Sci.*, 2012, **367**, 1047.
34. C. C. Su, F. Yang, F. Long, D. Reyon, M. D. Routh, D. W. Kuo, A. K. Mokhtari, J. D. Van Ornam, K. L. Rabe, J. A. Hoy, Y. J. Lee, K. R. Rajashankar and E. W. Yu, *J. Mol. Biol.*, 2009, **393**, 342.
35. F. Long, C. C. Su, M. T. Zimmermann, S. E. Boyken, K. R. Rajashankar, R. L. Jernigan and E. W. Yu, *Nature*, 2010, **467**, 484.
36. R. Kulathila, M. Indic and B. van den Berg, *PLoS One*, 2011, **6**, e15610.

37. C. C. Su, F. Long, M. T. Zimmermann, K. R. Rajashankar, R. L. Jernigan and E. W. Yu, *Nature*, 2011, **470**, 558–562.
38. S. Franke, G. Grass, C. Rensing and D. H. Nies, *J. Bacteriol.*, 2003, **185**, 3804.
39. T. D. Mealman, M. Zhou, T. Affandi, K. N. Chacon, M. E. Aranguren, N. J. Blackburn, V. H. Wysocki and M. M. McEvoy, *Biochemistry*, 2012, **51**, 6767.
40. H. J. Kim, D. W. Graham, A. A. DiSpirito, M. A. Alterman, N. Galeva, C. K. Larive, D. Asunskis and P. M. Sherwood, *Science*, 2004, **305**, 1612.
41. A. El Ghazouani, A. Basle, J. Gray, D. W. Graham, S. J. Firbank and C. Dennison, *Proc. Natl Acad. Sci. USA*, 2012, **109**, 8400.
42. J. D. Semrau, S. Jagadevan, A. A. Dispirito, A. Khalifa, J. Scanlan, B. H. Bergman, B. C. Freemeier, B. S. Baral, N. L. Bandow, A. Vorobev, D. H. Haft, S. Vuilleumier and J. C. Murrell, *Environ. Microbiol.*, 2013, **15**, 3077.
43. R. Balasubramanian, G. E. Kenney and A. C. Rosenzweig, *J. Biol. Chem.*, 2011, **286**, 37313.
44. N. L. Brown, S. R. Barrett, J. Camakaris, B. T. Lee and D. A. Rouch, *Mol. Microbiol.*, 1995, **17**, 1153.
45. M. A. Mellano and D. A. Cooksey, *J. Bacteriol.*, 1988, **170**, 2879.
46. C. Rensing and G. Grass, *FEMS Microbiol. Rev.*, 2003, **27**, 197.
47. K. Y. Djoko, Z. Xiao and A. G. Wedd, *ChemBioChem*, 2008, **9**, 1579.
48. S. Chillappagari, M. Miethke, H. Trip, O. P. Kuipers and M. A. Marahiel, *J. Bacteriol.*, 2009, **191**, 2362.
49. S. Ekici, H. Yang, H. G. Koch and F. Daldal, *MBio*, 2012, **3**, 1–11.
50. S. Tottey, C. J. Patterson, L. Banci, I. Bertini, I. C. Felli, A. Pavelkova, S. J. Dainty, R. Pernil, K. J. Waldron, A. W. Foster and N. J. Robinson, *Proc. Natl Acad. Sci. USA*, 2012, **109**, 95.
51. D. S. Radford, M. A. Kihlken, G. P. Borrelly, C. R. Harwood, N. E. Le Brun and J. S. Cavet, *FEMS Microbiol. Lett.*, 2003, **220**, 105.
52. L. Banci, I. Bertini, S. Ciofi-Baffoni, N. G. Kandias, N. J. Robinson, G. A. Spyroulias, X. C. Su, S. Tottey and M. Vanarotti, *Proc. Natl Acad. Sci. USA*, 2006, **103**, 8320.
53. S. Hearnshaw, C. West, C. Singleton, L. Zhou, M. A. Kihlken, R. W. Strange, N. E. Le Brun and A. M. Hemmings, *Biochemistry*, 2009, **48**, 9324.
54. B. Gold, H. Deng, R. Bryk, D. Vargas, D. Eliezer, J. Roberts, X. Jiang and C. Nathan, *Nat. Chem. Biol.*, 2008, **4**, 609.
55. S. A. Roberts, A. Weichsel, G. Grass, K. Thakali, J. T. Hazzard, G. Tollin, C. Rensing and W. R. Montfort, *Proc. Natl Acad. Sci. USA*, 2002, **99**, 2766.
56. S. K. Singh, S. A. Roberts, S. F. McDevitt, A. Weichsel, G. F. Wildner, G. B. Grass, C. Rensing and W. R. Montfort, *J. Biol. Chem.*, 2011, **286**, 37849.
57. L. B. Pontel and F. C. Soncini, *Mol. Microbiol.*, 2009, **73**, 212.
58. D. Osman, K. J. Waldron, H. Denton, C. M. Taylor, A. J. Grant, P. Mastroeni, N. J. Robinson and J. S. Cavet, *J. Biol. Chem.*, 2010, **285**, 25259.

59. I. R. Loftin, S. Franke, S. A. Roberts, A. Weichsel, A. Heroux, W. R. Montfort, C. Rensing and M. M. McEvoy, *Biochemistry*, 2005, **44**, 10533.

60. Y. Xue, A. V. Davis, G. Balakrishnan, J. P. Stasser, B. M. Staehlin, P. Focia, T. G. Spiro, J. E. Penner-Hahn and T. V. O'Halloran, *Nat. Chem. Biol.*, 2008, **4**, 107.

61. I. Bagai, C. Rensing, N. J. Blackburn and M. M. McEvoy, *Biochemistry*, 2008, **47**, 11408.

62. F. Arnesano, L. Banci, I. Bertini, S. Mangani and A. R. Thompsett, *Proc. Natl Acad. Sci. USA*, 2003, **100**, 3814.

63. L. Zhang, M. Koay, M. J. Maher, Z. Xiao and A. G. Wedd, *J. Am. Chem. Soc.*, 2006, **128**, 5834.

64. M. Zimmermann, S. R. Udagedara, C. M. Sze, T. M. Ryan, G. J. Howlett, Z. Xiao and A. G. Wedd, *J. Inorg. Biochem.*, 2012, **115**, 186.

65. S. Monchy, M. A. Benotmane, R. Wattiez, S. van Aelst, V. Auquier, B. Borremans, M. Mergeay, S. Taghavi, D. van der Lelie and T. Vallaeys, *Microbiology*, 2006, **152**, 1765.

66. B. Bersch, A. Favier, P. Schanda, S. van Aelst, T. Vallaeys, J. Coves, M. Mergeay and R. Wattiez, *J. Mol. Biol.*, 2008, **380**, 386.

67. L. X. Chong, M. R. Ash, M. J. Maher, M. G. Hinds, Z. Xiao and A. G. Wedd, *J. Am. Chem. Soc.*, 2009, **131**, 3549.

68. M. R. Ash, L. X. Chong, M. J. Maher, M. G. Hinds, Z. Xiao and A. G. Wedd, *Biochemistry*, 2011, **50**, 9237.

69. N. R. Mattatall, J. Jazairi and B. C. Hill, *J. Biol. Chem.*, 2000, **275**, 28802.

70. L. A. Abriata, L. Banci, I. Bertini, S. Ciofi-Baffoni, P. Gkazonis, G. A. Spyroulias, A. J. Vila and S. Wang, *Nat. Chem. Biol.*, 2008, **4**, 599.

71. A. C. Badrick, A. J. Hamilton, P. V. Bernhardt, C. E. Jones, U. Kappler, M. P. Jennings and A. G. McEwan, *FEBS Lett.*, 2007, **581**, 4663.

72. G. S. Siluvai, M. Mayfield, M. J. Nilges, S. Debeer George and N. J. Blackburn, *J. Am. Chem. Soc.*, 2010, **132**, 5215.

73. K. L. Blundell, M. T. Wilson, D. A. Svistunenko, E. Vijgenboom and J. A. Worrall, *Open Biol.*, 2013, **3**, 120163.

74. F. Serventi, Z. A. Youard, V. Murset, S. Huwiler, D. Buhler, M. Richter, R. Luchsinger, H. M. Fischer, R. Brogioli, M. Niederer and H. Hennecke, *J. Biol. Chem.*, 2012, **287**, 38812.

75. L. Hiser, M. Di Valentin, A. G. Hamer and J. P. Hosler, *J. Biol. Chem.*, 2000, **275**, 619.

76. L. Banci, I. Bertini, F. Cantini, S. Ciofi-Baffoni, L. Gonnelli and S. Mangani, *J. Biol. Chem.*, 2004, **279**, 34833.

77. L. Banci, I. Bertini, S. Ciofi-Baffoni, E. Katsari, N. Katsaros, K. Kubicek and S. Mangani, *Proc. Natl Acad. Sci. USA*, 2005, **102**, 3994.

78. A. K. Thompson, J. Gray, A. Liu and J. P. Hosler, *Biochim. Biophys. Acta*, 2012, **1817**, 955.

79. C. J. Kershaw, N. L. Brown, C. Constantinidou, M. D. Patel and J. L. Hobman, *Microbiology*, 2005, **151**, 1187.

80. A. Changela, K. Chen, Y. Xue, J. Holschen, C. E. Outten, T. V. O'Halloran and A. Mondragon, *Science*, 2003, **301**, 1383.

81. F. W. Outten, C. E. Outten, J. Hale and T. V. O'Halloran, *J. Biol. Chem.*, 2000, **275**, 31024.
82. K. Chen, S. Yuldasheva, J. E. Penner-Hahn and T. V. O'Halloran, *J. Am. Chem. Soc.*, 2003, **125**, 12088.
83. T. Liu, A. Ramesh, Z. Ma, S. K. Ward, L. Zhang, G. N. George, A. M. Talaat, J. C. Sacchettini and D. P. Giedroc, *Nat. Chem. Biol.*, 2007, **3**, 60.
84. Z. Ma, D. M. Cowart, R. A. Scott and D. P. Giedroc, *Biochemistry*, 2009, **48**, 3325.
85. S. Dwarakanath, A. K. Chaplin, M. A. Hough, S. Rigali, E. Vijgenboom and J. A. Worrall, *J. Biol. Chem.*, 2012, **287**, 17833.
86. Z. Ma, D. M. Cowart, B. P. Ward, R. J. Arnold, R. D. DiMarchi, L. Zhang, G. N. George, R. A. Scott and D. P. Giedroc, *J. Am. Chem. Soc.*, 2009, **131**, 18044.
87. M. Solioz and J. V. Stoyanov, *FEMS Microbiol. Rev.*, 2003, **27**, 183.
88. P. A. Cobine, G. N. George, C. E. Jones, W. A. Wickramasinghe, M. Solioz and C. T. Dameron, *Biochemistry*, 2002, **41**, 5822.
89. C. Rademacher and B. Masepohl, *Microbiology*, 2012, **158**, 2451.
90. G. P. Munson, D. L. Lam, F. W. Outten and T. V. O'Halloran, *J. Bacteriol.*, 2000, **182**, 5864.
91. S. A. Gudipaty, A. S. Larsen, C. Rensing and M. M. McEvoy, *FEMS Microbiol. Lett.*, 2012, **330**, 30.
92. M. Mermod, D. Magnani, M. Solioz and J. V. Stoyanov, *Biometals*, 2012, **25**, 33.
93. A. E. Voloudakis, T. M. Reignier and D. A. Cooksey, *Appl. Environ. Microbiol.*, 2005, **71**, 782.
94. T. Liu, X. Chen, Z. Ma, J. Shokes, L. Hemmingsen, R. A. Scott and D. P. Giedroc, *Biochemistry*, 2008, **47**, 10564.
95. N. Gomez-Santos, J. Perez, M. C. Sanchez-Sutil, A. Moraleda-Munoz and J. Munoz-Dorado, *PLoS Genet.*, 2011, 7, e1002106.
96. C. Rademacher, R. Moser, J. W. Lackmann, B. Klinkert, F. Narberhaus and B. Masepohl, *J. Bacteriol.*, 2012, **194**, 1849.
97. Z. H. Lu and M. Solioz, *J. Biol. Chem.*, 2001, **276**, 47822.

CHAPTER 17

Copper in Mitochondria

KATHERINE E. VEST AND PAUL A. COBINE*

Department of Biological Sciences, Auburn University, 101 Rouse Life
Sciences Building, Auburn, AL, 36849, USA
*Email: paul.cobine@auburn.edu

17.1   Introduction to Mitochondria

17.1.1   Mitochondria: Cofactors and Protein Import

The mitochondrion is well known as the major site of energy metabolism in
eukaryotes.[1] It also serves as the site of haem and iron-sulfur cluster biosynthesis. Mitochondria have a porous outer membrane that allows for gated
diffusion of metabolites and an impermeable inner membrane that is folded
into cristae. These two membranes define three separate spaces, each with
its own unique proteome. The mitochondrial matrix, which is enclosed by
the inner membrane (IM), contains the mitochondrial genome and ribosomes as well as the enzymes of the tricarboxylic acid cycle and those required for cofactor biosynthesis. This compartment accounts for most of the
volume of the organelle. The IM, which contains the highest ratio of proteins
to lipids in eukaryotic membranes, houses the complexes of oxidative
phosphorylation: the electron transport chain (ETC) and ATP synthase. This
membrane is sealed under most conditions to maintain the proton motive
force used for ATP production. The area between the two membranes is
known as the intermembrane space (IMS) and contains proteins required for
energy production, assembly of the embedded ETC complexes and transport
and insertion of membrane proteins.

RSC Metallobiology Series No. 2
Binding, Transport and Storage of Metal Ions in Biological Cells
Edited by Wolfgang Maret and Anthony Wedd
© The Royal Society of Chemistry 2014
Published by the Royal Society of Chemistry, www.rsc.org

The majority of the ~1000 mitochondrial proteins are encoded by the nuclear genome. These proteins must be imported into the organelle and properly inserted into the membranes when required. Import and insertion is mediated by a set of tightly regulated protein complexes that mediate transit of unfolded polypeptides across the IM or insertion of these proteins into the IM without disrupting the membrane potential required for ATP synthesis (Figure 17.1).[2] To enter the mitochondrion, all proteins pass through the translocase of the outer membrane (TOM), a multi-subunit complex that interacts with cytosolic molecular chaperones that hold the imported cargo in an unfolded state. Proteins that are destined for the matrix enter *via* the translocase of the inner membrane (TIM) complex, specifically the complex containing Tim23. While there are no known chaperones associated with this TIM complex, it is closely associated with the TOM complex to allow proteins to efficiently enter the matrix. Two different TIM complexes, either Tim22-or Tim23-containing complexes, are required for insertion of inner membrane proteins, while the β-barrel

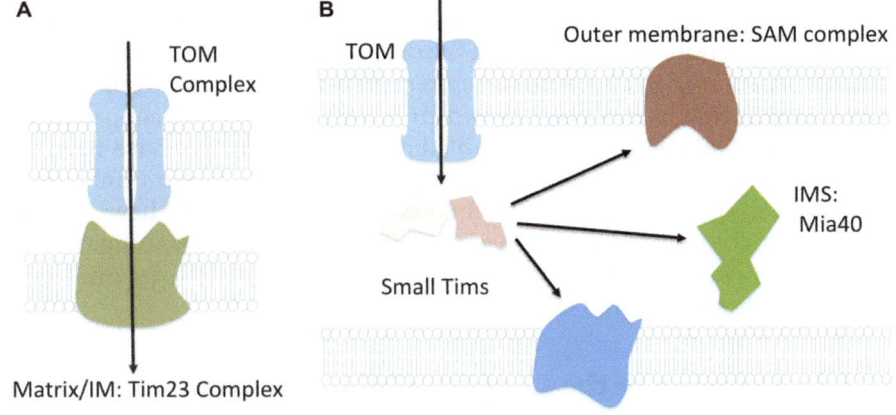

Figure 17.1 Mitochondrial import pathways. The mitochondrial proteome is made up of proteins that must be imported in an unfolded state. A series of protein complexes mediate the transit through the lipid bilayers; targeting to each complex is achieved by characteristic protein sequences. A) Proteins with matrix localization or inner membrane (IM) tethering domains are imported by the coupled translocase of the outer membrane (TOM) and translocase of the inner membrane (TIM) complexes. Tim23 is the major component of the TIM complex that directs this insertion. B) Proteins destined for the outer membrane (OM), intermembrane space or proteins with multiple IM transmembrane domains are inserted after interaction with small TIM proteins. Outer membrane proteins are inserted by the sorting and assembly machinery (SAM) complex. A TIM complex containing Tim22 inserts a subset of IM proteins. Cysteine-rich IMS proteins (including the small TIMs) are retained in the IMS *via* the Mia40 pathway.

sorting and assembly machinery (SAM) directs OM proteins to their proper location. A series of small protein chaperones, known as the small TIMs, direct OM and IM proteins to either the SAM complex or select TIM complexes. Tim22 is required for insertion of the integral membrane proteins including those of the mitochondrial carrier family that are required for transport of metabolites across the inner membrane. Cysteine-containing IMS localized proteins have a dedicated pathway mediated by Mia40, a soluble protein that participates in import *via* transient disulfide bond formation with the incoming targets and then, in the final stages of the import process, facilitates the formation of disulfide bonds within the IMS protein. The most important aspect of all these import systems is that the majority of mitochondrial proteins enter in an unfolded state. Consequently, structural modifications including cofactor insertion must occur after import of the apo-protein.

The most important mitochondrial copper enzyme for aerobic eukaryotes is the terminal ETC enzyme cytochrome *c* oxidase (CcO) (Figure 17.2). A series of copper-binding chaperones have been identified that participate in the assembly and insertion of the CcO cofactors within the IM and IMS. Because all of these copper-binding proteins must be unfolded for import, the mitochondrion also contains a storage pool of copper to provide the cofactor when required. A soluble, non-proteinaceous ligand found in the matrix compartment has been shown to facilitate storage of copper in mitochondria.[3] The details of the established assembly factors have been uncovered using genetic strategies in combination with *in vitro* analysis of copper binding and copper transfer assays. These studies have demonstrated that the copper chaperones are required to partition copper to assemble the two different copper sites in CcO *via* a series of sequential steps as the complex is matured to its final state.

17.1.2 Overview of the Copper in Mitochondria

Copper is found within each of the mitochondrial compartments. In the IM, CcO is the multi-subunit complex that catalyses the final steps of the ETC. It accounts for about 25% of mitochondrial copper and is the major cuproenzyme present in this organelle.[4,5] In addition, a small percentage of cellular Cu, Zn superoxide dismutase (Sod1) and its copper chaperone, Ccs1, is localized to the IMS. However, these proteins do not account for a significant proportion of the total mitochondrial Cu. The majority ($\sim$70%) is found in the matrix and is bound to a biochemically characterized but structurally undefined complex known as the copper ligand (CuL).[6] The presence of accumulated copper in the matrix suggests that mitochondrial copper transporters exist. The first candidate identified to contribute to matrix copper import is a mitochondrial carrier family protein capable of transporting CuL as a substrate. As is the case with the metallochaperones, this protein is not expected to account for significant binding or storage of copper in the organelle.

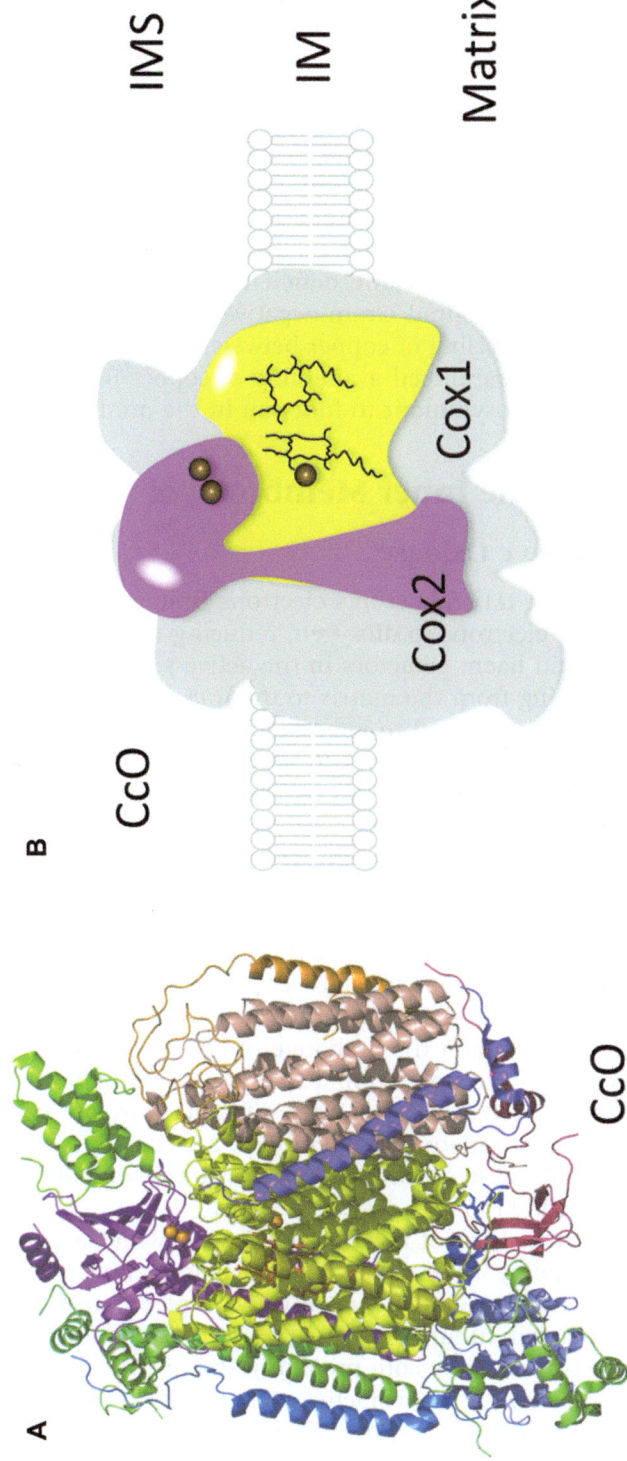

Figure 17.2 Cytochrome *c* oxidase: multi-subunit terminal electron transport chain complex. A) Monomeric bovine cytochrome *c* oxidase (PDB: 2OCC) structure. B) Cartoon representation highlighting Cox1 and Cox2 as the mitochondrially encoded, cofactor-binding core of the complex.

The current model of mitochondrial copper homeostasis is reminiscent of the copper-handling strategies utilized by Gram-negative bacteria, which typically import copper to the cytosol before redistribution to the periplasm.[7] The endosymbiotic theory of eukaryotic origin states that a Gram-negative α-proteobacterium was engulfed by another cell and became the mitochondrion in early eukaryotes.[8] While the biosynthesis of iron-sulfur clusters is the sole essential function of modern mitochondria, initial stages of selection for retaining the endosymbiont could have depended on ATP production. It therefore would behoove the bacterium to maintain a stable copper pool even if the host became deficient or attempted to withhold copper. Maintaining the original Gram-negative strategy appears to have generated a complicated cycling of copper between the matrix and the IMS. This strategy may have guaranteed availability of copper for CcO assembly allowing the original endosymbiont to function in the production of ATP.

17.2   Copper in the Inner Membrane of Mitochondria

17.2.1   Cytochrome *c* Oxidase

CcO is embedded in the IM and accepts electrons from cytochrome *c* in the IMS. It shuttles these electrons to dioxygen, reducing it to water *via* a combination of copper and haem cofactors in the active site.[9] This reaction results in proton pumping from the matrix to the IMS and contributes to the membrane potential that is used to generate ATP. Cells use the concerted action of a number of protein factors to assemble CcO.[1] This process is critical for the growth of *Saccharomyces cerevisiae* on non-fermentable carbon sources, a phenotype that has been used to probe CcO assembly. A large-scale genetic screen in *S. cerevisiae* identified a wide of array of mutations that resulted in loss of CcO activity.[10] In fact, 20 of the 30 genetic complementation groups related to CcO deficiency in yeast are due to mutations in genes encoding assembly factors.[11] CcO appears to assemble *via* the formation of separate individual modules containing both structural subunits and chaperones.[12] Protein-protein interactions control the formation of transient subcomplexes and the exchange of certain members into and out of these en route to the final complex allows for regulation of the assembly process.[13] In *S. cerevisiae*, it is very difficult to isolate partially assembled CcO as inappropriate subassemblies are rapidly degraded. In humans, stable partially assembled intermediates have been identified and have been used to define the CcO assembly steps.[1]

Mammalian CcO is made up of 13 subunits, three of which are encoded by the mitochondrial genome while the remaining subunits are encoded by the nuclear genome. The hydrophobic core of the complex is made up of the mitochondrially encoded subunits Cox1–3 while the ten nuclear-encoded subunits surround this catalytic core providing stability, sites of regulation and potentially protection for the cofactors. The complex is assembled in a modular fashion within the IM.[9] The assembled intermediates provide a platform to coordinate insertion of mitochondrially encoded subunits with import of nuclear-encoded subunits and the required cofactors.[12] The latter consist of three copper ions and two specialized haem groups plus single zinc, sodium and magnesium ions.

Specific activators coordinate the translation of Cox 1, 2 and 3 from their spliced transcripts. Subsequently, a subset of these proteins act as chaperones to stabilize the nascent protein before insertion into the IM.[11] The copper and haem cofactors are bound by mitochondrially encoded core subunits (Cox1 and 2) and therefore must be inserted early in the assembly process and concomitant with membrane insertion. The Cox1 subunit is made up of helical bundles and is located in the centre of the complex, buried in the inner membrane (Figure 17.3). Cox1 binds two specialized haem *a* cofactors, which are modified by the addition of farnesyl tails to enhance binding and modulate protein packing. There are additional modifications of methyl groups to vinyl groups to modulate the redox potential of these haems.[14] The heterometallic cofactor that consists of a copper and a haem a_3 that binds the substrate oxygen for conversion to water is known as the Cu_B site (Figure 17.3 and Table 17.1). This site is buried in the IM, limiting its accessibility.[9] Three histidine residues coordinate the Cu in the Cu_B site with one histidine cross-linked to an adjacent tyrosine residue (in the bovine structures). Protons are transported through the membrane *via* the D- and K-pathways that allow for the reduction of water and also contribute to the membrane potential.

Cox2 has two transmembrane helices that anchor it in the complex and an IMS-exposed globular domain at the top of the complex that binds the binuclear copper site known as the Cu_A site (Figure 17.2B; Table 17.1). Cu_A is an unusual mixed valence site that accepts electrons from cytochrome *c* before transfer to the cofactors in Cox1. Two cysteine residues provide sulfurs that bridge the two copper atoms. Coordination is completed by a histidine residue for each copper atom plus a single methionine for one atom and a single aspartic acid for the other (Figure 17.3B). The site is positioned approximately 8 Å above the inner membrane and roughly 20 Å above the Cu_B site. The transfer of electrons from cytochrome *c* reduces the site from the delocalized mixed valence state $[Cu_2]^{3+}$ ($[Cu^{II}Cu^{I}]$ in the localized oxidation state formalism) to a fully reduced state $[Cu^{I}_2]^{2+}$. Electrons are subsequently transferred to the Cox1 cofactors and to dioxygen at the Cu_B site. The details and structural insights into the nature of the proton translocation and electron transfer continue to be refined/defined, particularly using the homologous enzymes from *Thermus thermophilus*, *Paracoccus denitrificans* and *Rhodobacter sphaeroides*.[15]

17.2.2 Sco Proteins and Assembly of the Cu_A Site

Many informative studies of the function of Sco proteins have used bacterial homologues. Results from these studies have provided the platform for determining the presumptive mechanisms of copper delivery and

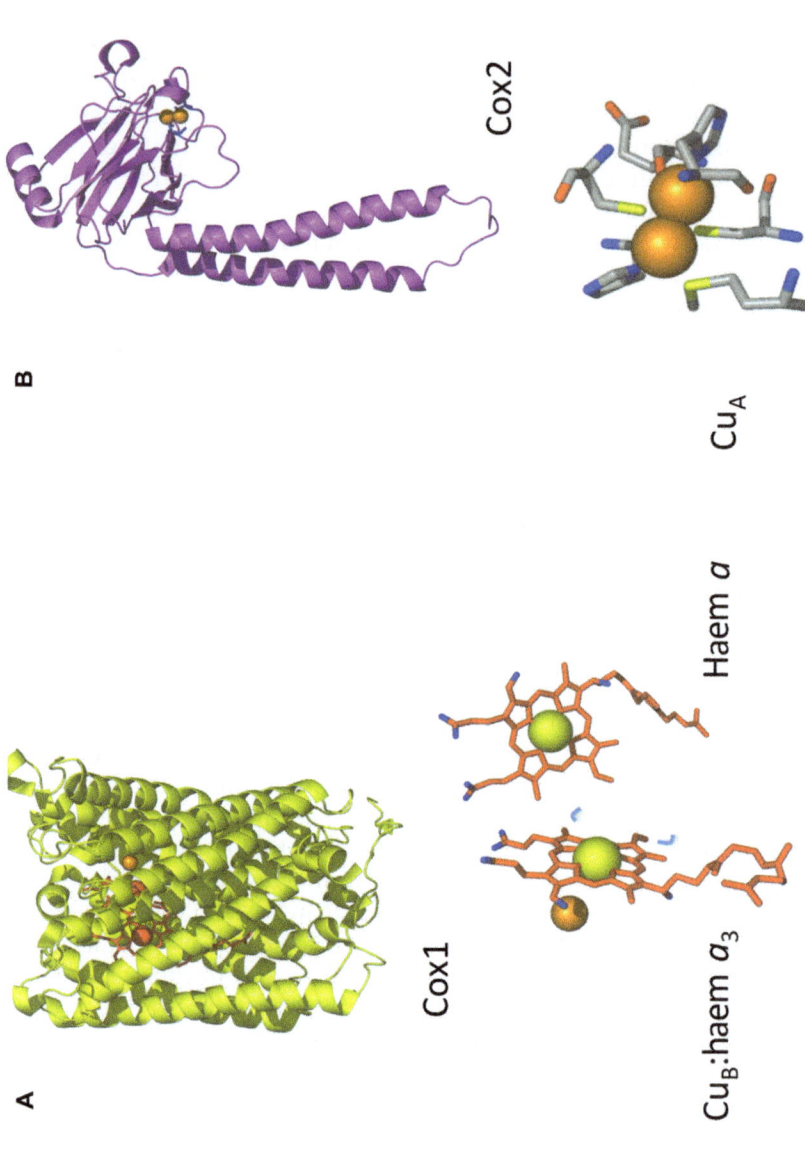

Figure 17.3 Copper coordinating subunits of CcO. A) Cox1 is an all-helical structure that forms the core of the enzyme complex and binds two haem *a* cofactors and a single Cu. The Cu$_B$:haem a_3 centre is the site of oxygen reduction. B) Cox2 has a globular beta-sheet domain that extends above the inner membrane and binds two copper ions in the Cu$_A$ site that accepts electrons from cytochrome *c* and then donates them to the haem *a* of Cox1.

Table 17.1 Copper-binding components redox state and coordination.

Protein/complex	Compartment	Cu redox state	Ligand donors
COX17	IMS	+1	Cys, Cys
SCO1	IM	+1	Cys, Cys, His
SCO1	IM	+2	Cys, Cys, His, Asp
SCO2	IM	+1	Cys, Cys, His
COX11	IM	+1	Cys, Cys
COX1	IM	+2	His, His, His
COX2	IM	+1.5	His, His, Met, backbone carbonyl, 2 Cys (bridging)
Matrix pool	Matrix	+1	Sulfur and nitrogen/oxygen
CuL	Matrix	+1	–
SOD1	IMS	+1	His, His, His, His
CCS	IMS	+1	Cys, Cys
CMC1	IMS	+1	Cys, Cys
COX19	IMS	+1	Cys, Cys
Plastocyanin	Chloroplast	+2	Cys, Met, His, His

thiol-disulfide oxido-reductase properties of the eukaryotic versions of these proteins.[16–25]

SCO1 was identified in yeast as being required for post-translational stability of Cox2.[26] An early biochemical experiment showed that Sco1 could interact with Cox2 in a pull-down assay, suggesting a physical interaction may be involved.[27] Yeast Sco1 is an integral IM protein with a large globular domain facing the IMS (Figure 17.4). The transmembrane domain is required for function. The large globular domain of Sco1 has a peroxiredoxin-like fold. Peroxiredoxins have active site cysteine residues on an exposed loop to allow reduction of a variety of peroxide substrates.[28] This fold in Sco1 presents a surface exposed CxxxC motif (Figure 17.4)[28] This motif acts as a copper-binding site and as a redox active site in the assembly of CcO in different organisms.[29] Soluble, truncated variants of Sco proteins are capable of binding either Cu(I) or Cu(II) (Table 17.1). The trigonal Cu(I) site features the Cys thiolate functions of the CxxxC as ligands as well as an imidazole nitrogen ligand from a conserved His residue (Figure 17.4).[30]

The same site can also bind Cu(II), presumably in a square planar stereochemistry using the same three ligands plus an additional oxygen ligand from the carboxyl group of a conserved Asp residue. The relevance of the ability to bind either Cu(I) or Cu(II) is unknown, as the Cu_A site should be capable of accepting either Cu(I) or Cu(II) in the assembly process. Mutation of the Asp and His residues results in an *in vivo* CcO defect, strongly suggesting that copper binding is absolutely required for the function of Sco proteins.[30] The copper binding function is supported by the fact that *SCO1* is a multicopy suppressor of a non-fermentable growth defect of a *COX17* mutant *S. cerevisiae*. Cox17 is the copper metallochaperone localized to the IMS (Section 17.3.1) and the ability to bypass by Sco1 confirms that these two proteins are in the same pathway of assembly. In the absence of Cox17, either Sco1 could be holding Cox2 in a conformation that allows spontaneous

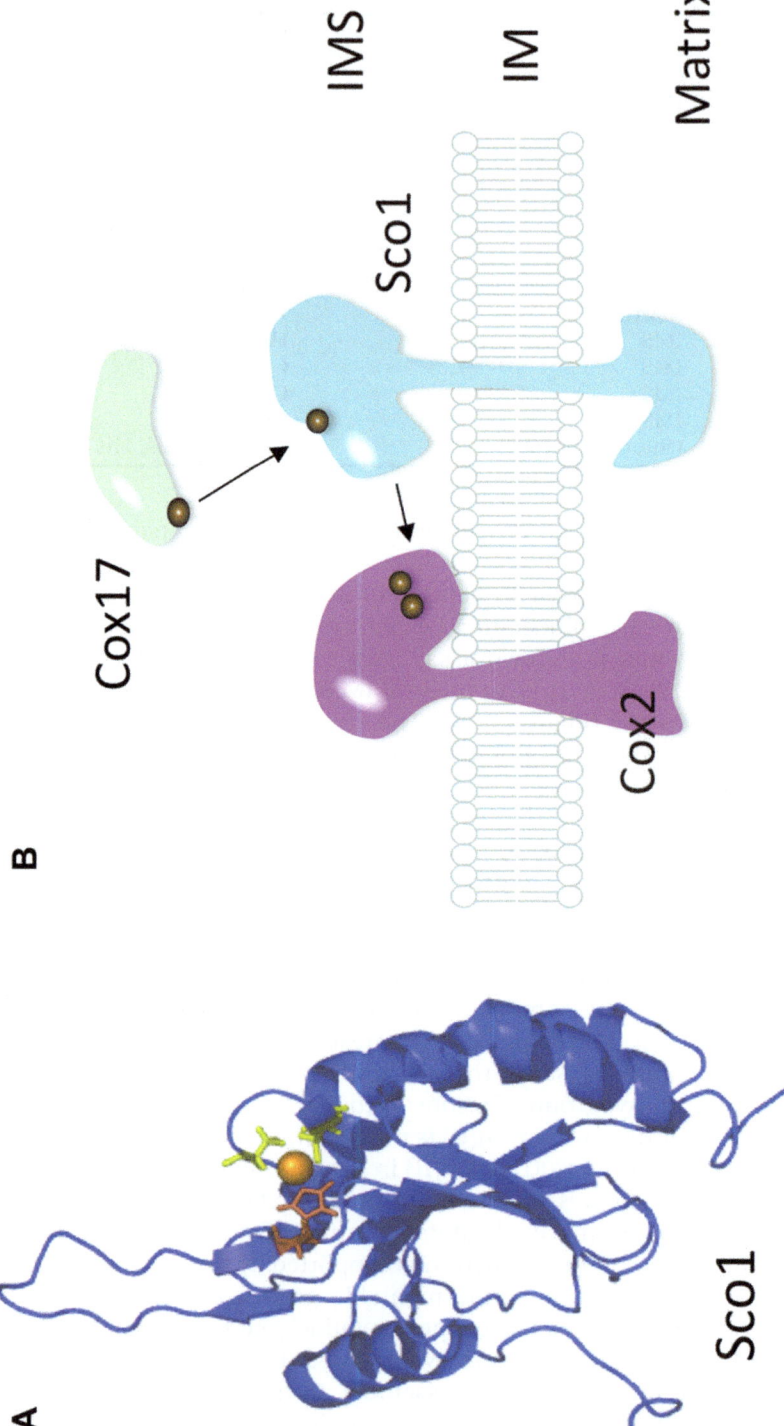

Figure 17.4 Sco1 assembly of the Cu$_A$ site of CcO. A) Solution structure of the globular IMS domain of Sco1 with the two cysteine residues and one histidine residue that provide the sulfur and nitrogen copper binding ligands highlighted in ball-and-stick format. B) Cartoon representation of Cu transfer reaction with Cox17 delivering Cu to Sco1 for subsequent insertion into Cox2 to form the Cu$_A$ site.

formation of Cu_A or the increased Sco1 levels may help facilitate scavenging of limited amounts of Cu in the IMS for direct delivery to Cox2.[31] It should be noted that no suppressor of a *SCO1* deletion (even under high copper conditions) has been identified in yeast, reflecting its crucial role in Cu_A assembly. The *in vitro* truncation studies of copper binding have been complemented with one study of purified intact Sco1 from yeast mitochondria that also demonstrated copper binding.[32]

A role for SCO in redox chemistry was initially proposed due to its structural similarity to the peroxiredoxin family.[28] The peroxiredoxin proteins participate in antioxidant signalling pathways *via* modification of active site cysteine residues. Observations of hydrogen peroxide sensitivity in yeast lacking *SCO1* suggested a role in general oxidative stress in mitochondria.[33] However, subsequent studies have shown that this sensitivity is due to a misassembled CcO intermediate that has the haem *a* moiety bound and acts as a pro-oxidant.[34] This sensitivity to hydrogen peroxide could be suppressed with alleles of *SCO1* lacking the cysteine motif and therefore any redox activity. The proposed mechanism to suppress hydrogen peroxide sensitivity relates to Sco1 binding to Cox2 that is required for Cox1 interactions to block formation of the pro-oxidant. Based on this mechanism the hydrogen peroxide growth assay was combined with an *in vitro* copper transfer assay to determine that a flexible loop adjacent to the CxxxC motif in the SCO proteins is required for interaction with Cox2 but dispensable for interaction with Cox17.[34,35]

Humans have two *SCO* genes and despite their ubiquitous tissue expression these proteins have non-redundant functions in CcO assembly, and mutation in either gene results in distinct clinical presentations.[36] A mutation in *SCO1* (P174L) causes neonatal liver failure and ketoacidotic coma while a pathogenic allele of human *SCO2* (E140K) causes fatal cardioencephalomyopathy. Both pathologies are due to a CcO deficiency.[37] Identification of these mutant pedigrees has been crucial to uncovering the details of the role for SCO proteins in assembly of Cu_A in Cox2.[38,39] The differences in disease phenotypes has led to a series of structural and protein chemistry studies to try and uncover the difference in function of the Sco proteins. No obvious defining characteristics have been observed regarding *in vitro* copper binding or structure but differences have been observed in redox activities and the ability to be copper-loaded by Cox17.[30,40] The yeast genome also contains a second *SCO* gene, but CcO assembly requires only Sco1 under all conditions reported. Copper transfer from COX17 to SCO has been studied extensively using the human proteins. In addition to demonstrating transfer, these NMR studies show the reorganization of the Cu(I) coordinating cysteine and histidine residues and have revealed numerous aspects of the electron transfer within SCO and COX17 (Figure 17.5). Solution structures of human SCO in multiple conformations provided evidence that the protein can exist in an oxidized state, even while maintaining metal binding, at least in a nickel-bound form.[41] When considering the redox function of SCO, the most detailed data demonstrating a role as a thiol-disulfide

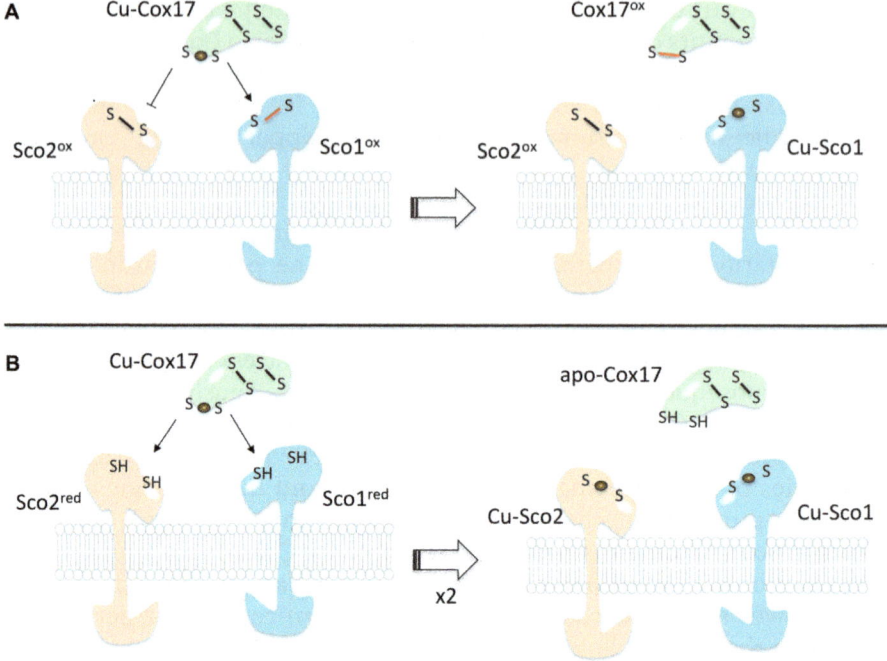

Figure 17.5 Schematic of the transfer of copper to oxidized or reduced target Sco
proteins by Cox17. A) Cu-Cox17 donates copper to oxidized Sco1,
reducing the Sco1 disulfide transferring copper and then becoming
oxidized itself. Cu-Cox17 cannot interact, reduce or deliver copper to
oxidized Sco2. B) Cu-Cox17 is able to donate copper to either Sco1 or
Sco2 when the proteins are reduced. This transfer does not result in
oxidation of Cox17.

oxidoreductase arose from experiments in human cell culture.[42] Mutations
in *SCO2* or changes in its expression level affected the ratio of oxidized and
reduced cysteine residues in SCO1. This result supports a model where SCO2
can act on SCO1 to modulate its redox state.[43] The role of SCO2 as a modifier
of SCO1 explains in part the non-redundant but overlapping roles of these
proteins in CcO assembly in humans.

Perhaps the most surprising result related to SCO proteins is the role these
proteins have in regulating mammalian cellular copper status.[44] Immortalized fibroblasts from patients with mutations in SCO1 or SCO2 have a severe
copper deficiency as a result of increased export of copper from the cell. The
mutant SCO protein present in these patient cells triggers a signal that results in inappropriate export of copper from the cell. This signal can be
activated in other CcO assembly mutants without mutation in SCO1 or
SCO2, suggesting that the signal plays a role in normal cellular copper
homeostasis. Importantly, the copper export signalling in fibroblasts with
mutations in genes other than *SCO* can be reversed by expression of SCO
proteins (or chimeric SCO proteins). The reversal of this phenotype by

increased expression of SCO demonstrates its specific role in this pathway.[44] This signalling cascade from SCO in mitochondria to the copper export pathway is dependent on levels and localization of the dual localized (IMS and cytosol) Cx_9C protein COX19 (Section 17.3.2).[45]

17.2.3 Cox11 and Assembly of the Cu_B Site

The Cu_B required for binding and reduction of dioxygen is buried in the mitochondrial inner membrane. Insertion of copper and haem a_3 must occur at an early step in the CcO assembly process due to the buried position of these cofactors in the final complex. The role of Cox11 in assembly of the heterometallic haem a_3-linked Cu_B site in Cox1 was established by both phenotypic analysis in yeast and biochemical analysis in yeast and bacterial homologues.[46] *In vitro* experiments showed that yeast Cox11 could bind copper *via* cysteine residues in a CxC motif and that these ligands were required for CcO assembly *in vivo* (Table 17.1).[47,48] The close proximity to the transmembrane domains of this copper-binding site as visualized from the bacterial homologue[49] appears to be optimal for the insertion of copper into the buried Cu_B site (Figure 17.6). The most striking demonstration to date of the role of Cox11 in providing the Cu to CcO remains the purification of the partially assembled enzyme in Cox11 mutants from *Rhodobacter sphaeroides*.[47]

Consistent with the coordinated insertion of haem a_3 and Cu, Cox11 transiently interacts with Shy1 (SURF1 homologue in yeast) that is required for haem a_3 insertion into Cox1.[50] The importance of this interaction was uncovered by a peroxide sensitivity that is induced by deletion of *COX11* (or *SCO1*) by formation of a pro-oxidant intermediate of Cox1 that contains the haem a but not Cu.[34] The critical role of incorporating Cu to block formation of this intermediate shows the degree of coordination required to form the

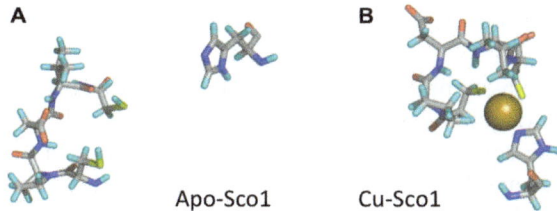

Apo-Sco1   Cu-Sco1

Figure 17.6   Cox11 mediates the assembly of Cu_B in Cox1. A) Solution structure of the bacterial homologue of Cox11 shows an immunoglobulin like fold; the protein is anchored in the membrane by a transmembrane domain that was truncated in the solution structure. Cysteine residues responsible for binding copper are highlighted. Copper binding stoichiometry of Cox11 has been reported as both 1Cu:monomer and 1Cu:dimer. B) Cartoon representation of copper transfer from Cox17 to Cox11 and then to Cox1.

early core modules of CcO. The potential risks of assembling these sites appear to be balanced by proteases that are poised to remove subcomplexes that fail to progress to the final active complex.[51,52]

17.3   Copper in the IMS

17.3.1   Cox17 and Cu Delivery to SCO and Cox11

Copper availability in the IMS appears to be limited. This is an extension of the situation in the eukaryotic cytosol where transport and sequestration limit the available copper, leading to a demand for copper chaperones.[3] In the IMS, Cox17 presents bioavailable copper to Sco1 and Cox11 (Figure 17.7). Cox17 is a relatively small soluble protein with a characteristic twin Cx_9C motif. This protein adopts a coiled coil-helix-coiled coil-helix fold stabilized by two disulfide bonds that are separate from those in the Cx_9C motif that binds a single copper (Figure 17.7).[53–56] Mutational analyses have shown that the two disulfide bridges provide stability but are not essential for function. In fact, multiple conformations of this protein have been isolated from heterologous expression systems, but the single copper conformer seems to be the biologically relevant species (Table 17.1).[55,56]

The Cox17-Sco1 reaction is the most thoroughly studied copper transfer of the IMS. Cox17 donates Cu(I) to the exposed CxxxC site on Sco1 through an interaction face that is perturbed by the pathogenic human mutation *SCO1* P174L.[38,39] *In vitro* studies with Cox17 isolated from heterologous expression systems support this interaction.[57] In addition, MS and NMR studies have shown that the reactions proceed from the Cu-loaded, partially oxidized conformer of human COX17 to SCO1 *via* transient interactions.[40,55] In particular, the Cu(I)-COX17 conformer featuring Cu(I) bound at the Cx_9C motif and the two additional disulfide bonds (Figure 17.7A) reacted with oxidized apo-SCO1 (featuring a disulfide bond) to yield fully oxidized COX17 (with three disulfide bonds) and Cu(I)-SCO1. In other words, the reaction involved transfer of one Cu(I) and two electrons (Figure 17.5).[40] This reaction is specific, as it did not occur with a highly homologous SCO2 as the recipient protein.[40]

This copper delivery pathway is also supported by a combination of *in vitro* observations and genetic suppression experiments. The phenotype of the $COX17^{C57Y}$ allele can be suppressed by overexpression of *SCO1*.[31] Additionally, defects observed in a complete *COX17* deletion are reversed by overexpression of *SCO1* in the presence of additional copper. These studies were further complemented by data showing that the $COX17^{C57Y}$ allele could not be rescued by yeast *SCO1* with a P153L mutation (equivalent to the human pathogenic SCO1 P174L mutation).[38] In this case, the combination of two crippled alleles that disturbed the interaction interfaces of both proteins prevented assembly of CcO in yeast. These results support the idea that Cox17 delivers copper to Sco1 for the subsequent insertion into Cox2 to allow assembly of CcO.

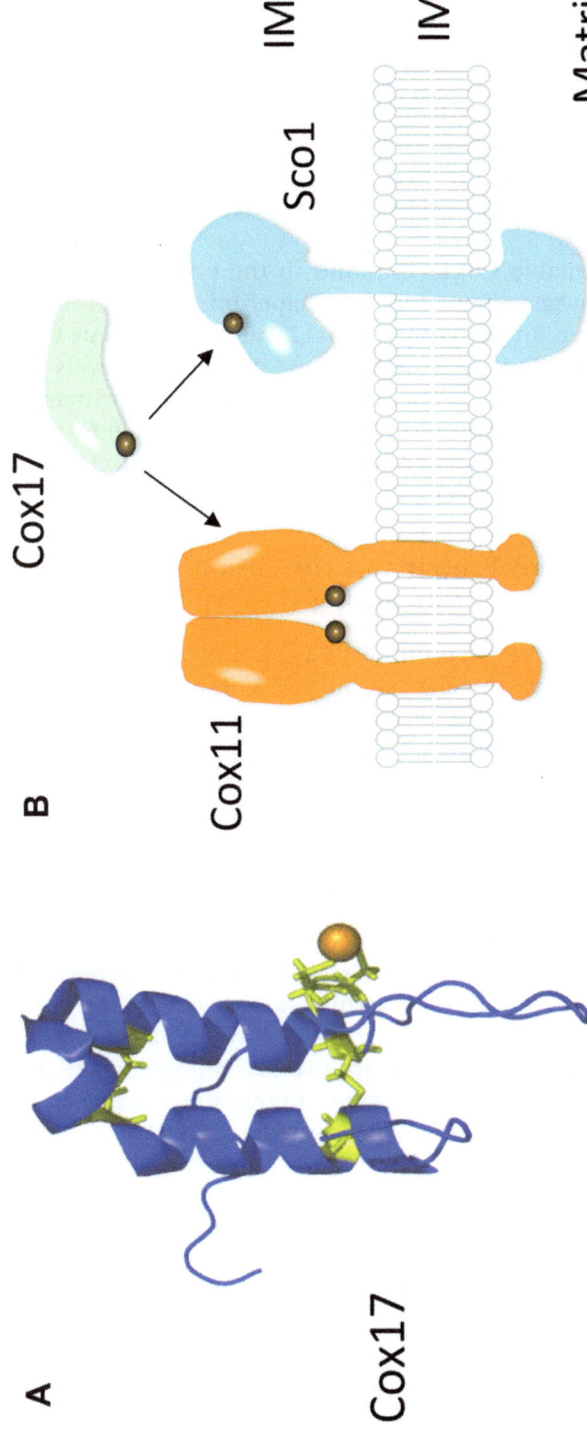

Figure 17.7 Cox17 mediates of copper delivery to both Sco1 and Cox11. A) Twin Cx₉C motifs in Cox17 form disulfide bonds to stabilize the two-helix structure. Additional cysteine residues provide the sulfur ligands for a single copper to bind. B) Cartoon representation of Cox17 as the Cu donor for both Sco1 and Cox11 in the inner membrane. *In vivo* and *in vitro* experiments suggest independent surfaces of Cox17 mediate the different interactions.

Cox11 is also a recipient of copper from Cox17. Copper transfer occurs *via* an interaction interface that is distinct from that used for interaction with Sco1. This was demonstrated by the fact that copper transfer could proceed between Cox11 and a C57Y variant of Cox17 in two independent assays.[57] The C57Y mutation is sufficient to disrupt the copper transfer from Cox17 to Sco1.[57] The requirement for Cox17 in the IMS to facilitate copper loading reinforces the idea that limited copper availability and specific protein-protein interactions are required for the correct metallation of cuproenzymes.

About 40% of available Cox17 is found in the cytosol, which led to the suggestion that it was responsible for chaperoning copper to the mitochondrion.[53] However, an IM-tethered Cox17 was able to rescue COX activity in a cell lacking an endogenous *COX17* gene and copper levels are not affected by deletion of *COX17*.[6,58] Therefore Cox17 is not the primary means of copper delivery to the mitochondrion. Cox17 is responsible for the delivery of copper in the IMS to the globular domains of Sco1 and Cox11 for assembly of the Cu_A and Cu_B sites of CcO.

17.3.2 Other Dithiol Proteins of the Inner Membrane Space Involved in CcO Biogenesis

Several Cx_9C proteins have been implicated as copper-binding proteins with loosely defined functions in CcO assembly.[54] These include Cox19, Cmc1, Cmc2 and Cox23. While these proteins are required under different conditions in various models, further investigation is needed to place them definitively in the pathway of copper delivery to CcO.[59–61] One feature they all share is that their import into the IMS is dependent on the Mia40 oxidative import pathway, indicating that they are imported and subsequently oxidized to fold into their active conformations.

CMC1 encodes a Cx_9C IMS protein required for CcO assembly; deletion of this gene in yeast and mammalian cell culture results in changes in the distribution of copper in the IMS with increased Sod1 and decreased CcO.[59,62] These data perhaps suggest a role for this Cx_9C protein in the regulation of Cox17 copper loading or general copper availability within the IMS.[59,62] A role in cellular copper regulation has been observed for human COX19. It has been implicated in regulation of cellular copper homeostasis acting in the signal cascade from mitochondria to the copper exporter ATP7A.[45]

17.3.3 SOD1

The IMS also houses 1–5% of the total cellular superoxide dismutase (Sod1).[63] The copper chaperone for superoxide dismutase Ccs1 is also localized in the IMS. Mitochondrial localization of Ccs1 is dependent on the Mia40 import pathway that uses a disulfide relay exchange system to fold and trap proteins in the IMS.[64] Sod1 accumulation in the IMS is dependent on the presence of Ccs1 in this compartment.

17.4 Copper in the Matrix

The most abundant and dynamic pool of copper in mitochondria is located in the mitochondrial matrix. Quantification of total copper in mitochondria from fermenting and respiring yeast have shown that between 60% and 80% of mitochondrial copper is not associated with CcO or Sod1.[6,65,66] In fact mitochondrial copper levels do not change when these major cuproenzymes are absent. The levels of matrix copper are responsive to increases in cellular copper: increased copper availability leads to increases in the matrix pool.[6,65] X-ray fluorescence techniques and metal-responsive chelators have been able to detect Cu(I) in mitochondria and independent spectroscopic assays measuring total copper *versus* EPR-detectable copper (Cu_A and Cu_B) support the existence of this pool (Table 17.1). It should be noted that activity that is indicative of a copper amine oxidase has been identified in the rat liver mitochondrial matrix.[67,68] However, no detailed analyses of the copper-binding characteristics of the protein have been described.

17.4.1 The Matrix Copper Ligand

Biochemical and analytical experiments have been able to uncover certain characteristics of this copper pool but identification of the molecules involved has proven to be difficult. The current data are consistent with the presence of a stable, non-proteinaceous copper complex known as the copper ligand (CuL).[69] Copper can be extracted from the mitochondrial matrix by anion exchange and reverse phase chromatography. These characteristics are inconsistent with a free ionic form of copper. Cu(I) in the isolated CuL complex is highly stable and can be treated with proteases, nucleases and organic solvents without disrupting any of its physical characteristics. The ligand has been partially characterized by fluorescence spectroscopy and thin layer chromatography. However, its exact identity and the pathways involved in its biogenesis are not yet known.

The current model of non-proteinaceous recruitment to mitochondria is consistent with the strategy that methanotrophs use to recruit copper for the enzyme particulate methane monooxygenase (pMMO).[70] When required, these bacteria produce the ligand methanobactin. It is exported and its copper complex is taken up by the cell for extraction of the copper to be inserted into pMMO.[71] pMMO is loaded in convoluted membrane invaginations, reminiscent of the mitochondrial IM cristae. In addition, a non-proteinaceous ligand recruitment model involving an intracellular siderophore for recruitment of iron to mitochondria has been proposed in eukaryotes.[72] This intracellular siderophore appears to be required at least indirectly for mitochondrial iron import as yeast mutated unable to synthesize the intracellular siderophore exhibit decreased haem synthesis.[72] Synthesis of intracellular siderophores has been described in a number of fungi and has been shown to be responsible for storage and protection against reactive oxygen stress as well as fertility. A role for the CuL complex in *S. cerevisiae* in storage in cytoplasm or protection against stress is yet to be demonstrated.

17.4.2   The Matrix Pool Provides Cu to CcO and SOD1

The most thorough examination of the matrix copper pool has involved genetic analysis of *S. cerevisiae.* Mutants of *S. cerevisiae* that lack the Mn-utilizing *SOD2* cannot propagate in high-dioxygen conditions. However, this growth defect can be rescued by targeting the copper-dependent human Sod1 to the matrix, indicating that bioavailable copper exists in the matrix.[6] This accessibility of the matrix pool observed in *sod2Δ* cells suggested that the copper pool could be modified *via* expression of heterologous copper-binding competitor molecules. In fact, targeting human Sod1 or yeast metallothionein (Crs5) to the matrix can exert a dominant negative effect on the assembly of CcO and result in partial depletion of Sod1 from the IMS.[69] These phenotypes are reversed by supplementation of the medium with copper that returns CuL levels in the matrix to normal levels, suggesting that Cox17 and Ccs1 were receiving copper from this pool for assembly (Figure 17.8).

The only example of depletion of the matrix copper pool without heterologous expression of competitors relates to assembly factors for Cox1.[73] The assembly factors Coa1 and Shy1 cooperate in early stages of the Cox1 assembly.[73] Mitochondria isolated from *S. cerevisiae* lacking either *COA1, SHY1* or both are copper deficient. While the relevance of this copper deficiency is unknown, copper supplementation to the medium can reverse the growth defect on non-fermentable carbon sources in strains lacking *COA1* or *SHY1.* These data hint at a coordination of copper availability with the translation of a copper target (Cu$_B$ in Cox1). However, despite extensive proteomic and genetic experiments with tagged or mutant *COA1* or *SHY1*, no connection has been identified with a candidate transport protein.

17.4.3   Maintenance of Mitochondrial Copper

Localization of a copper store inside the mitochondrial matrix provides the cell with a mechanism to protect this copper from export.[42,65] Homeostatic mechanisms normally maintain cytosolic Cu levels but in certain pathological states copper export can be inappropriately activated.

Mutation in *SCO1* (P174L) appears to activate a signal that stimulates copper export.[74] This signal is transduced by the Cx$_9$C protein COX19 that regulates the activity of ATP7A by an unidentified mechanism.[45] The SCO1 P174L mutation results in profoundly copper deficient cells, yet the mitochondrial matrix pool is maintained.[65] This pathological state of inappropriate copper export suggests that the matrix storage strategy may have evolved to protect copper in mitochondria and, in turn, protect energy generation.

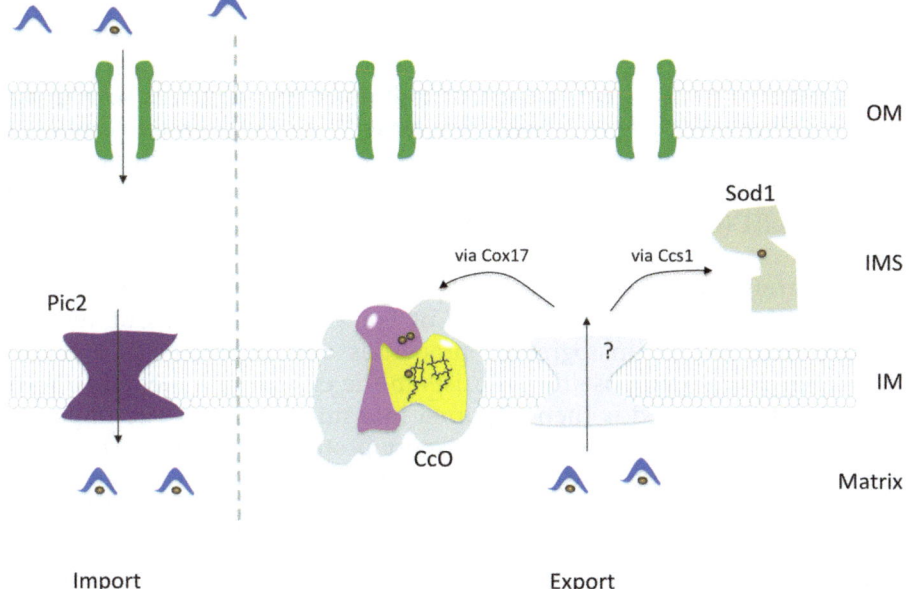

Figure 17.8 Model for copper homeostasis in mitochondria. CuL appears to be present as an apo molecule in the cytopplasm where it is available to bind copper. The CuL molecule that carries a net negative charge is transported into the mitochondria presumably *via* the voltage-dependent anion channels (porins) into the IMS. It is then transported into the matrix by at least one mitochondrial carrier family protein (Pic2). Pic2 is capable of transporting both the intact complex and ionic copper. Once in the matrix it is stored until release to the IMS for Cox17 binding. The mechanism of transport and the transporter used are unknown. But biochemical depletion of the CuL complex in the matrix results in a loss of activity of CcO and Sod1 in the IMS.

A second observation in the same setting also supported the maintenance of the matrix pool and a role for matrix copper as the source of copper for the assembly of CcO. The P174L allele of *SCO1* acts as a dominant signal, inducing copper deficiency even in the presence of wild-type *SCO1*. In this cell line (SCO1 P174L expressing WT SCO1), the CcO deficiency is completely reversed, indicating adequate copper stores are available for this enzyme. This means that the pool of copper required to assemble CcO is intact in *SCO1* patient fibroblasts, in spite of the global copper deficit.[74] This observation was subsequently confirmed by using a mitochondrially targeted fluorescent copper sensor that showed that in spite of global cellular deficiency mitochondrial copper was retained.[65]

17.5 Copper Transport into the Mitochondrion

The mechanism of copper delivery through the cytosol to the mitochondrion has not been determined. Initially Cox17 was suggested to act as a copper shuttle to mitochondria. This hypothesis was based on dual localization, copper-binding characteristics and the fact that copper could suppress CcO-related defect in yeast lacking *COX17*.[53] However, Cox17 is fully functional when tethered in the inner membrane and no copper deficit is seen in mitochondria when *COX17* is deleted from yeast.[6,58] No other protein candidate has emerged and therefore it is assumed that the mitochondrion delivery pathway may be distinct from the well-described protein-mediated pathways used by Sod1 (delivered by Ccs1) and the multi-copper oxidases (delivered *via* Atx1).[3]

The current model is that copper is recruited to mitochondria by the small molecule ligand found in the matrix CuL complex. While the CuL remains unidentified, its chromatographic properties and fluorescence signature have been used to propose that it exists in the cytosol in a metal-free form that binds copper and is then imported into mitochondria. The import of CuL to the matrix must involve transport proteins to facilitate transport across the impermeable IM. Metabolites are transported routinely across the IM by the mitochondrial carrier family (MCF) proteins, ATP binding cassette-transporters and multimeric associations of single transmembrane domain proteins.[75–77] MCF proteins appear to be ideally suited to transport a metabolite such as CuL, and this family has been implicated in mitochondrial iron homeostasis.[78–82]

In vitro transport of CuL into the mitochondrial matrix is saturable and temperature dependent but independent of membrane potential.[78] These characteristics are consistent with MCF-mediated transport. Recently, an MCF protein Pic2 was found to be involved in copper import into the matrix in *S. cerevisiae*.[78] Strains with a deletion of *PIC2* show copper-dependent growth defects and mitochondria from these cells have lower total mitochondrial copper. Importantly, Pic2 expressed in heterologous systems was able to transport copper when provided with either CuL or ionic copper.[78] Interestingly, the mitochondrial iron transporters, Mrs3 and Mrs4, were also reported to transport ionic forms of Fe and Cu, even though our current understanding of eukaryotic metal homeostasis suggests these are not normal substrates.[79] However, depletion of cytoplasmic metallothionein in yeast, which allows for accumulation of "free" copper, does result in significant expansion of the mitochondrial matrix copper pool.[6]

The importance of maintaining mitochondrial copper and the deleterious effects of overload are observed in the rat model of Wilson disease.[83] High levels of copper accumulate within cells due to a block in export. As copper accumulates, it can be found in mitochondria where it appears to form insoluble particulates that damage the membranes, resulting in aberrant mitochondrial morphology.[83] This damage may be a consequence of overwhelming the capacity of metallothionein and even the mitochondrial ligand pool in the cytosol. The transport of ionic copper into yeast mitochondria suggests copper could enter the matrix under these conditions and could induce oxidative stress or form inappropriate complexes that lead to pathology. This model is not unlike that proposed for iron as a contributing factor in the pathology of Friedreich's ataxia.[84]

17.6 Copper in Chloroplasts

Like mitochondria, chloroplasts originated from a bacterial symbiont.[85] The outer membrane of these organelles surrounds a series of stacked inner thylakoid membranes, which house the copper-containing enzymes plastocyanin and Cu, Zn SOD.[86] Plastocyanin is a type 1, blue copper protein that carries electrons from cytochrome *f* in the quinol oxidase complex to photosystem I.[87] The crystal structure of plastocyanin shows a single Cu(II) bound in a tetrahedral geometry using thiol groups from cysteine and methionine residues along with two imidazole nitrogens from histidine residues.[88] The copper loading of plastocyanin appears to date to be independent of any known metallochaperone.

Higher plants, including *Arabidopsis thaliana* also contain a Cu, Zn SOD within the chloroplast. A homologue of yeast Ccs1 exists in *A. thaliana* that contains a putative chloroplast targeting sequence.[89] Recent evidence also suggests that this Cu, Zn SOD may be activated in the absence of the chaperone.[90] Protein import machinery into chloroplasts favours the import of unfolded proteins, so cofactors must be assembled within the organelle.[91] The copper requirement necessitates transport into chloroplasts, which is mediated by the P-type ATPases in the outer envelope as well as the thylakoids.[86,92]

17.7 Conclusions and Open Questions

Even though the major cuproenzymes of the mitochondrion have been identified, many unanswered questions regarding copper in this organelle still exist. The number of mitochondrial proteins with copper-binding capacity continues to grow, but the requirement for Cu-binding in many of these proteins remains unexplained. One aspect of the existence of these proteins may be that eukaryotes have built in a level of redundancy or limited specificity to maintain delivery and to allow for CcO assembly under different conditions of stress.

Pathological states have been critical to uncovering the mechanism of copper homeostasis in mitochondria. This is highlighted by the important role that partially functional alleles of *SCO* in humans have played in defining the nature of the proteins that maintain mitochondrial copper homeostasis. They have also provided information regarding preservation of the matrix copper pool. When export is blocked (*e.g.* Wilson disease[83]) the mechanisms used to ensure mitochondrial copper is maintained under deficiency could become responsible for accumulation and subsequent damage. Further identification of transporters and the biosynthesis of CuL will remove one of the rate-limiting steps in our understanding of the regulation and mechanisms of distributions of copper in the mitochondria.

References

1. E. A. Shoubridge, *Am. J. Med. Genet.*, 2001, **106**, 46–52.
2. J. Dudek, P. Rehling and M. van der Laan, *Biochim. Biophys. Acta*, 2013, **1833**, 274–285.
3. N. J. Robinson and D. R. Winge, *Annu. Rev. Biochem.*, 2010, **79**, 537–562.
4. P. A. Cobine, F. Pierrel and D. R. Winge, *Biochim. Biophys. Acta*, 2006, **1763**, 759–772.
5. S. C. Leary, D. R. Winge and P. A. Cobine, *Biochim. Biophys. Acta*, 2009, **1793**, 146–153.
6. P. A. Cobine, L. D. Ojeda, K. M. Rigby and D. R. Winge, *J. Biol. Chem.*, 2004, **279**, 14447–14455.
7. C. L. Dupont, G. Grass and C. Rensing, *Metallomics*, 2011, **3**, 1109–1118.
8. T. Lithgow and A. Schneider, *Philos. Trans. R. Soc. Lond. B Biol. Sci.*, 2010, **365**, 799–817.
9. T. Tsukihara, H. Aoyama, E. Yamashita, T. Tomizaki, H. Yamaguchi, K. Shinzawa-Itoh, R. Nakashima, R. Yaono and S. Yoshikawa, *Science*, 1996, **272**, 1136–1144.
10. A. Tzagoloff, A. Akai and R. B. Needleman, *J. Bacteriol.*, 1975, **122**, 826–831.
11. I. C. Soto, F. Fontanesi, J. Liu and A. Barrientos, *Biochim. Biophys. Acta*, 2012, **1817**, 883–897.
12. G. P. McStay, C. H. Su and A. Tzagoloff, *Mol. Biol. Cell*, 2013, **24**, 440–452.
13. D. U. Mick, T. D. Fox and P. Rehling, *Nat. Rev. Mol. Cell Biol.*, 2011, **12**, 14–20.
14. H. S. Carr and D. R. Winge, *Acc. Chem. Res.*, 2003, **36**, 309–316.
15. J. A. Lyons, D. Aragao, O. Slattery, A. V. Pisliakov, T. Soulimane and M. Caffrey, *Nature*, 2012, **487**, 514–518.
16. A. C. Badrick, A. J. Hamilton, P. V. Bernhardt, C. E. Jones, U. Kappler, M. P. Jennings and A. G. McEwan, *FEBS Lett.*, 2007, **581**, 4663–4667.
17. L. Banci, I. Bertini, G. Cavallaro and S. Ciofi-Baffoni, *FEBS J.*, 2011, **278**, 2244–2262.
18. L. Banci, I. Bertini, G. Cavallaro and A. Rosato, *J. Proteome Res.*, 2007, **6**, 1568–1579.
19. B. Bennett and B. C. Hill, *FEBS Lett.*, 2011, **585**, 861–864.
20. D. Buhler, R. Rossmann, S. Landolt, S. Balsiger, H. M. Fischer and H. Hennecke, *J. Biol. Chem.*, 2010, **285**, 15704–15713.
21. I. Imriskova-Sosova, D. Andrews, K. Yam, D. Davidson, B. Yachnin and B. C. Hill, *Biochemistry*, 2005, **44**, 16949–16956.
22. E. Lohmeyer, S. Schroder, G. Pawlik, P. I. Trasnea, A. Peters, F. Daldal and H. G. Koch, *Biochim. Biophys. Acta*, 2012, **1817**, 2005–2015.
23. A. G. McEwan, A. Lewin, S. L. Davy, R. Boetzel, A. Leech, D. Walker, T. Wood and G. R. Moore, *FEBS Lett.*, 2002, **518**, 10–16.
24. P. Saenkham, P. Vattanaviboon and S. Mongkolsuk, *FEMS Microbiol. Lett.*, 2009, **293**, 122–129.

25. A. K. Thompson, J. Gray, A. Liu and J. P. Hosler, *Biochim. Biophys. Acta*, 2012, **1817**, 955–964.
26. M. Schulze and G. Rodel, *Mol. Gen. Genet.*, 1988, **211**, 492–498.
27. A. Lode, M. Kuschel, C. Paret and G. Rodel, *FEBS Lett.*, 2000, **485**, 19–24.
28. Y. V. Chinenov, *J. Mol. Med. (Berl.)*, 2000, **78**, 239–242.
29. E. Balatri, L. Banci, I. Bertini, F. Cantini and S. Ciofi-Baffoni, *Structure*, 2003, **11**, 1431–1443.
30. Y. C. Horng, S. C. Leary, P. A. Cobine, F. B. Young, G. N. George, E. A. Shoubridge and D. R. Winge, *J. Biol. Chem.*, 2005, **280**, 34113–34122.
31. D. M. Glerum, A. Shtanko and A. Tzagoloff, *J. Biol. Chem.*, 1996, **271**, 20531–20535.
32. J. Beers, D. M. Glerum and A. Tzagoloff, *J. Biol. Chem.*, 2002, **277**, 22185–22190.
33. J. C. Williams, C. Sue, G. S. Banting, H. Yang, D. M. Glerum, W. A. Hendrickson and E. A. Schon, *J. Biol. Chem.*, 2005, **280**, 15202–15211.
34. O. Khalimonchuk, A. Bird and D. R. Winge, *J. Biol. Chem.*, 2007, **282**, 17442–17449.
35. K. Rigby, P. A. Cobine, O. Khalimonchuk and D. R. Winge, *J. Biol. Chem.*, 2008, **283**, 15015–15022.
36. S. C. Leary, B. A. Kaufman, G. Pellecchia, G. H. Guercin, A. Mattman, M. Jaksch and E. A. Shoubridge, *Hum. Mol. Genet.*, 2004, **13**, 1839–1848.
37. L. C. Papadopoulou, C. M. Sue, M. M. Davidson, K. Tanji, I. Nishino, J. E. Sadlock, S. Krishna, W. Walker, J. Selby, D. M. Glerum, R. V. Coster, G. Lyon, E. Scalais, R. Lebel, P. Kaplan, S. Shanske, D. C. De Vivo, E. Bonilla, M. Hirano, S. DiMauro and E. A. Schon, *Nat. Genet.*, 1999, **23**, 333–337.
38. P. A. Cobine, F. Pierrel, S. C. Leary, F. Sasarman, Y. C. Horng, E. A. Shoubridge and D. R. Winge, *J. Biol. Chem.*, 2006, **281**, 12270–12276.
39. L. Banci, I. Bertini, S. Ciofi-Baffoni, I. Leontari, M. Martinelli, P. Palumaa, R. Sillard and S. Wang, *Proc. Natl Acad. Sci. USA*, 2007, **104**, 15–20.
40. L. Banci, I. Bertini, S. Ciofi-Baffoni, T. Hadjiloi, M. Martinelli and P. Palumaa, *Proc. Natl Acad. Sci. USA*, 2008, **105**, 6803–6808.
41. L. Banci, I. Bertini, V. Calderone, S. Ciofi-Baffoni, S. Mangani, M. Martinelli, P. Palumaa and S. Wang, *Proc. Natl Acad. Sci. USA*, 2006, **103**, 8595–8600.
42. S. C. Leary, *Antioxid. Redox Signal.*, 2010, **13**, 1403–1416.
43. S. C. Leary, F. Sasarman, T. Nishimura and E. A. Shoubridge, *Hum. Mol. Genet.*, 2009, **18**, 2230–2240.
44. S. C. Leary, P. A. Cobine, B. A. Kaufman, G. H. Guercin, A. Mattman, J. Palaty, G. Lockitch, D. R. Winge, P. Rustin, R. Horvath and E. A. Shoubridge, *Cell Metab.*, 2007, **5**, 9–20.
45. S. C. Leary, P. A. Cobine, T. Nishimura, R. M. Verdijk, R. de Krijger, R. de Coo, M. A. Tarnopolsky, D. R. Winge and E. A. Shoubridge, *Mol. Biol. Cell*, 2013, **24**, 683–691.
46. A. Tzagoloff, N. Capitanio, M. P. Nobrega and D. Gatti, *EMBO J.*, 1990, **9**, 2759–2764.

47. L. Hiser, M. Di Valentin, A. G. Hamer and J. P. Hosler, *J. Biol. Chem.*, 2000, **275**, 619–623.

48. H. S. Carr, G. N. George and D. R. Winge, *J. Biol. Chem.*, 2002, **277**, 31237–31242.

49. L. Banci, I. Bertini, F. Cantini, S. Ciofi-Baffoni, L. Gonnelli and S. Mangani, *J. Biol. Chem.*, 2004, **279**, 34833–34839.

50. O. Khalimonchuk, M. Bestwick, B. Meunier, T. C. Watts and D. R. Winge, *Mol. Cell. Biol.*, 2010, **30**, 1004–1017.

51. O. Khalimonchuk, M. Y. Jeong, T. Watts, E. Ferris and D. R. Winge, *J. Biol. Chem.*, 2012, **287**, 7289–7300.

52. F. Pierrel, O. Khalimonchuk, P. A. Cobine, M. Bestwick and D. R. Winge, *Mol. Cell. Biol.*, 2008, **28**, 4927–4939.

53. D. M. Glerum, A. Shtanko and A. Tzagoloff, *J. Biol. Chem.*, 1996, **271**, 14504–14509.

54. G. Cavallaro, *Mol. Biosyst.*, 2010, **6**, 2459–2470.

55. L. Banci, I. Bertini, C. Cefaro, S. Ciofi-Baffoni and A. Gallo, *J. Biol. Chem.*, 2011, **286**, 34382–34390.

56. L. Banci, I. Bertini, S. Ciofi-Baffoni, A. Janicka, M. Martinelli, H. Kozlowski and P. Palumaa, *J. Biol. Chem.*, 2008, **283**, 7912–7920.

57. Y. C. Horng, P. A. Cobine, A. B. Maxfield, H. S. Carr and D. R. Winge, *J. Biol. Chem.*, 2004, **279**, 35334–35340.

58. A. B. Maxfield, D. N. Heaton and D. R. Winge, *J. Biol. Chem.*, 2004, **279**, 5072–5080.

59. D. Horn, H. Al-Ali and A. Barrientos, *Mol. Cell. Biol.*, 2008, **28**, 4354–4364.

60. M. H. Barros, A. Johnson and A. Tzagoloff, *J. Biol. Chem.*, 2004, **279**, 31943–31947.

61. K. Rigby, L. Zhang, P. A. Cobine, G. N. George and D. R. Winge, *J. Biol. Chem.*, 2007, **282**, 10233–10242.

62. D. Horn, W. Zhou, E. Trevisson, H. Al-Ali, T. K. Harris, L. Salviati and A. Barrientos, *J. Biol. Chem.*, 2010, **285**, 15088–15099.

63. V. C. Culotta, L. W. Klomp, J. Strain, R. L. Casareno, B. Krems and J. D. Gitlin, *J. Biol. Chem.*, 1997, **272**, 23469–23472.

64. D. P. Gross, C. A. Burgard, S. Reddehase, J. M. Leitch, V. C. Culotta and K. Hell, *Mol. Biol. Cell*, 2011, **22**, 3758–3767.

65. S. C. Dodani, S. C. Leary, P. A. Cobine, D. R. Winge and C. J. Chang, *J. Am. Chem. Soc.*, 2011, **133**, 8606–8616.

66. J. Garber Morales, G. P. Holmes-Hampton, R. Miao, Y. Guo, E. Munck and P. A. Lindahl, *Biochemistry*, 2010, **49**, 5436–5444.

67. R. Stevanato, S. Cardillo, M. Braga, A. De Iuliis, V. Battaglia, A. Toninello, E. Agostinelli and F. Vianello, *Amino Acids*, 2011, **40**, 713–720.

68. S. Cardillo, A. D. Iuliis, V. Battaglia, A. Toninello, R. Stevanato and F. Vianello, *Arch. Biochem. Biophys.*, 2009, **485**, 97–101.

69. P. A. Cobine, F. Pierrel, M. L. Bestwick and D. R. Winge, *J. Biol. Chem.*, 2006, **281**, 36552–36559.

70. G. E. Kenney and A. C. Rosenzweig, *ACS Chem. Biol.*, 2012, 7, 260–268.

71. R. Balasubramanian, G. E. Kenney and A. C. Rosenzweig, *J. Biol. Chem.*, 2011, **286**, 37313–37319.
72. L. R. Devireddy, D. O. Hart, D. H. Goetz and M. R. Green, *Cell*, 2010, **141**, 1006–1017.
73. F. Pierrel, M. L. Bestwick, P. A. Cobine, O. Khalimonchuk, J. A. Cricco and D. R. Winge, *EMBO J.*, 2007, **26**, 4335–4346.
74. S. C. Leary, P. A. Cobine, B. A. Kaufman, G. H. Guercin, A. Mattman, J. Palaty, G. Lockitch, D. R. Winge, P. Rustin, R. Horvath and E. A. Shoubridge, *Cell Metab.*, 2007, **5**, 403–403.
75. J. M. Baughman, F. Perocchi, H. S. Girgis, M. Plovanich, C. A. Belcher-Timme, Y. Sancak, X. R. Bao, L. Strittmatter, O. Goldberger, R. L. Bogorad, V. Koteliansky and V. K. Mootha, *Nature*, 2011, **476**, 341–345.
76. D. K. Bricker, E. B. Taylor, J. C. Schell, T. Orsak, A. Boutron, Y. C. Chen, J. E. Cox, C. M. Cardon, J. G. Van Vranken, N. Dephoure, C. Redin, S. Boudina, S. P. Gygi, M. Brivet, C. S. Thummel and J. Rutter, *Science*, 2012, **337**, 96–100.
77. E. R. Kunji, *FEBS Lett.*, 2004, **564**, 239–244.
78. K. E. Vest, S. C. Leary, D. R. Winge and P. A. Cobine, *J. Biol. Chem.*, 2013, **288**, 23884–23892.
79. E. M. Froschauer, R. J. Schweyen and G. Wiesenberger, *Biochim. Biophys. Acta*, 2009, **1788**, 1044–1050.
80. U. Muhlenhoff, J. A. Stadler, N. Richhardt, A. Seubert, T. Eickhorst, R. J. Schweyen, R. Lill and G. Wiesenberger, *J. Biol. Chem.*, 2003, **278**, 40612–40620.
81. M. Yang, P. A. Cobine, S. Molik, A. Naranuntarat, R. Lill, D. R. Winge and V. C. Culotta, *EMBO J.*, 2006, **25**, 1775–1783.
82. D. M. Gordon, E. R. Lyver, E. Lesuisse, A. Dancis and D. Pain, *Biochem. J.*, 2006, **400**, 163–168.
83. H. Zischka, J. Lichtmannegger, S. Schmitt, N. Jagemann, S. Schulz, D. Wartini, L. Jennen, C. Rust, N. Larochette, L. Galluzzi, V. Chajes, N. Bandow, V. S. Gilles, A. A. DiSpirito, I. Esposito, M. Goettlicher, K. H. Summer and G. Kroemer, *J. Clin. Invest.*, 2011, **121**, 1508–1518.
84. A. Bayot, R. Santos, J. M. Camadro and P. Rustin, *BMC Med.*, 2011, **9**, 112.
85. K. W. Osteryoung and J. Nunnari, *Science*, 2003, **302**, 1698–1704.
86. K. E. Vest, H. F. Hashemi and P. A. Cobine, *Met. Ions Life Sci.*, 2013, **12**, 451–478.
87. J. D. Rochaix, *Biochim. Biophys. Acta*, 2011, **1807**, 375–383.
88. J. M. Guss and H. C. Freeman, *J. Mol. Biol.*, 1983, **169**, 521–563.
89. S. E. Abdel-Ghany, J. L. Burkhead, K. A. Gogolin, N. Andres-Colas, J. R. Bodecker, S. Puig, L. Penarrubia and M. Pilon, *FEBS Lett.*, 2005, **579**, 2307–2312.
90. C. H. Huang, W. Y. Kuo, C. Weiss and T. L. Jinn, *Plant Physiol.*, 2012, **158**, 737–746.
91. P. Jarvis, *New Phytol.*, 2008, **179**, 257–285.
92. S. Puig, N. Andres-Colas, A. Garcia-Molina and L. Penarrubia, *Plant Cell Environ.*, 2007, **30**, 271–290.

CHAPTER 18

Copper in Eukaryotes

NINIAN J. BLACKBURN,*[a] NAN YAN[b] AND
SVETLANA LUTSENKO[b]

[a] Institute of Environmental Health, Oregon Health and Sciences
University, Portland, OR, 97239, USA; [b] Department of Physiology, The
Johns Hopkins University School of Medicine, Baltimore, MD, 21205, USA
*Email: blackbni@ohsu.edu

18.1 Introduction

Eukaryotic cells utilize copper to power the chemical reactivity of an important group of enzymes whose activity spans a wide variety of physiological processes. Most important to the metabolic health of the cell is cytochrome c oxidase (CCO), a component of the mitochondrial energy-generating machine.[1] CCO contains copper in two distinct sites: the CuA centre, which shuttles electrons from cytochrome c, and the CuB centre which, in concert with haem-$a3$, is the site of dioxygen binding and reduction to water. The reaction catalysed by CCO converts the energy of oxygen reduction ($E^0 = 0.86$ V) into a transmembrane proton concentration gradient that in turn drives ATP synthesis, the major fuel for cellular metabolism. Other copper-dependent enzymes include peptidylglycine monooxygenase (biosynthesis of neuropeptide hormones essential for proper endocrine function[2]), dopamine β-monoxygenase (biosynthesis of catecholamines, the major neurotransmitters of the sympathetic nervous system[3]), monoamine oxidases (metabolism of biogenic amines and connective tissue formation[4]), superoxide dismutases 1 and 3 (removal of harmful superoxide radicals[5]) and tyrosinase, which uses copper and dioxygen to produce pigments such as melanin and is important in protection from the carcinogenic effects of

RSC Metallobiology Series No. 2
Binding, Transport and Storage of Metal Ions in Biological Cells
Edited by Wolfgang Maret and Anthony Wedd
© The Royal Society of Chemistry 2014
Published by the Royal Society of Chemistry, www.rsc.org

excess sunlight.[6] Finally, ceruloplasmin and hephaestin are members of the multicopper oxidase superfamily catalysing the oxidation of Fe(II) to Fe(III), a critical step in cellular iron storage and transport.[7,8]

The copper centres of these enzymes exhibit a spectrum of very different structures (Figure 18.1), and catalyse widely different chemical transformations. A common theme, however, is their role in activation and reduction of dioxygen, which occurs *via* a mechanism involving cycling of the copper between the +2 and +1 oxidation states. This cycling requires the presence of a copper-binding ligand set that is capable of stabilizing both oxidation states. As discussed in Chapter 16, cupric centres prefer electrostatic bonding to nitrogen and/or oxygen donor ligands, whereas cuprous centres prefer the more covalent environment of S donors. Because of these differences, the ligand sets chosen by copper-dependent redox enzymes tend to be a mix of these two classes, with histidine, methionine and cysteine dominating. The challenge for the cell is therefore to produce a robust set of copper importers, trafficking molecules and exporters that are capable of the selective and vectorial delivery of copper to the critical enzymes located in distinct cell compartments.

Mammalian cells differ from bacterial systems in their mechanisms of copper uptake and utilization. Serum copper levels in humans typically range between 10 and 20 μM, while levels within the cell range between 10 and 100 μM. It is still unclear (for either extracellular or intracellular copper) which fraction of this total copper is bioavailable and can be exchanged, and which is tightly bound to copper-utilizing enzymes. In addition, copper levels may vary significantly depending on cell type, environment and developmental stage. For example, while macrophages exhibit basal copper concentrations in the 10–20 μM range, exposure to immune stimulants that mimic the host-defence response to microbial infection results in a 10-fold increase. Massive copper redistribution between liver (copper decreases) and mammary gland (copper increases) occurs prior to and during lactation. Likewise numerous disease states such as Parkinson's disease, Menkes disease, Wilson's disease, Alzheimer's disease and cancer are associated with changes in copper levels or distribution.

Eukaryotic cells have a greater number of critical copper-dependent enzymes as well as additional mechanisms for copper sequestration and storage. Bioavailable copper is less abundant in food and in biologic fluids than, for example, in soils or in other components of the biosphere. Consequently, eukaryotic organisms, and mammals in particular, need high-affinity importers (seemingly absent in most bacteria), together with the mechanisms for selective partitioning of copper between various intracellular compartments. The general features of the eukaryotic transport machinery are now well understood,[9,10] as shown schematically in Figure 18.2. Copper enters the cell *via* the high-affinity transporter CTR1 and is then partitioned into at least three independent pathways *via* binding to metallochaperones, small proteins that bind the metal tightly in the Cu(I) state.

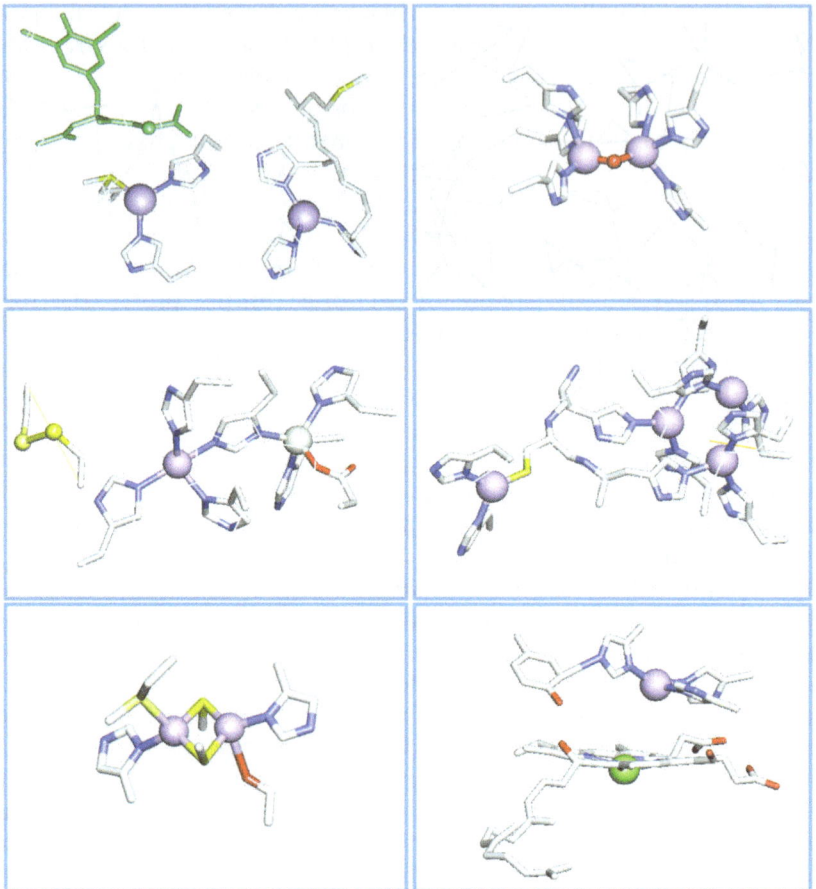

Figure 18.1   Active sites of some essential cuproproteins in eukaryotic cells (PDB file entries in parentheses). Top left, peptidylglycine monooxygenase with bound substrate (1OPM); top right, tyrosinase with bridging OH (3NTM); middle left, Cu,Zn superoxide dismutase (1HL5); middle right, the trinuclear copper site of the multicopper oxidase Fet3p (1ZPU); bottom left, CuA of cytochrome oxidase (2CUA); bottom right, binuclear Fe-Cu site of cytochrome oxidase (1QLE). Copper atoms are violet, zinc is metallic grey, iron is green.

A generally accepted paradigm supposes that metallochaperones can interact directly with CTR1[11,12] but such interactions are not obligatory for copper uptake, since variation in chaperone levels (down-regulation or over-expression) has little influence on the initial rates of uptake.[13]

The chaperone CCS (copper chaperone for superoxide dismutase) shuttles copper to the cytosolic radical scavenging enzyme SOD1. CCS is essential for SOD1 metallation in *Saccharomyces cerevisiae*, but in mammalian cells this reaction can be facilitated by glutathione in the absence of CCS.[14] A second

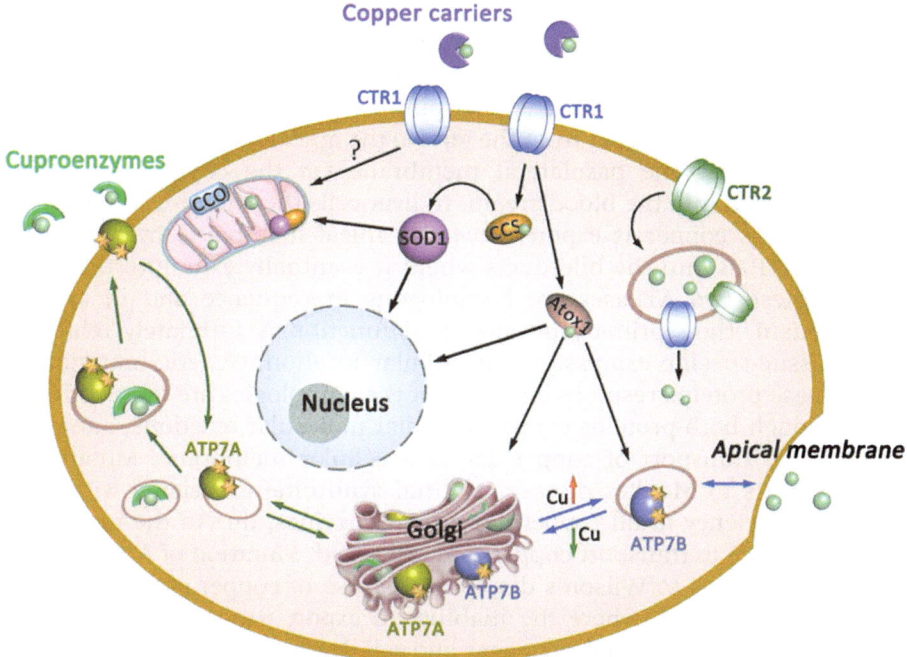

Figure 18.2 Eukaryotic pathways for copper transport. High-affinity copper uptake permease CTR1 accepts copper from the extracellular copper carriers and transfers copper into cytosol. The low-affinity copper transporter CTR2 may regulate copper uptake and intracellular copper stores by modulating CTR1 or (in yeast) directly release copper from the intracellular stores. Upon copper entry, copper chaperones distribute copper to different cellular targets. CCS transfers copper to SOD1 in cytosol and mitochondria, whereas Atox1 transfers copper to the secretory pathway and, possibly, nucleus. An ensemble of proteins regulates copper delivery to cytochrome *c* oxidase (CCO) in mitochondria. Cu-ATPases transport copper to the secretory pathways for metallation of cuproenzymes and mediate copper excretion by sequestering excess copper in vesicles. The distribution of Cu-ATPases between these two locations is controlled by cellular copper level and is associated with phosphorylation by a kinase (indicated by yellow stars), which increases in response to copper elevation and decreases when copper is low.

shuttle system involves the chaperone Atox1 delivering copper to the ATP7A or ATP7B copper pumps, which then incorporate the metal into cuproproteins in the *trans*-Golgi network and/or vesicles of the secretory pathway. A third system incorporates copper into mitochondrial Cu pools for the metallation of CCO.[15] Molecular details of this latter pathway are not well defined and the state-of-play is discussed in Chapter 17.

Whereas bacteria require robust export systems primarily to combat copper toxicity from environmental onslaught, mammalian cells use exporters to maintain homeostasis, to metallate copper-dependent enzymes and to transfer copper from one cell/tissue to another. Both copper entry into the bloodstream following ingestion of food and eventual removal of copper from the body require copper pumps.[16] Intestinal cells (enterocytes) function to import copper from the gut *via* the apical membrane and then to export it across the basolateral membrane *via* the copper-transporting ATPase ATP7A into the bloodstream. In liver cells (hepatocytes) on the other hand, excess copper is exported *via* the apical membrane by the related ATPase ATP7B into the bile ducts where it eventually exits into the colon. While these two ATPases are homologous in sequence and in catalytic mechanism, their primary physiological function is intimately related to their tissue-specific expression and cellular location. Genetic mutations in both these proteins result in disease, but the pathologies are quite different, even though both proteins carry out similar molecular reactions, namely an ATP-driven transport of copper across a cellular membrane. Mutation in ATP7A leads to Menkes disease, a lethal syndrome associated with acute copper deficiency in all copper-dependent enzymes, due to the inability of the enterocyte to transport copper into the blood. Mutation of ATP7B on the other hand leads to Wilson's disease, a disease of copper excess mainly in the liver and brain, where the inability to export copper leads to copper overload, distinct metabolic changes and cell death. Patients with neurologic symptoms often have visible copper deposits in their eyes at the cornea-sclera border known as Kayser–Fleischer rings. These examples illustrate a number of important principles regarding transport processes in eukaryotic cells. Firstly, metallation mechanisms require *selectivity* to ensure that out of the many potential copper binding sites present in the cell, copper is targeted only to the cuproproteins critical for function. This can be achieved by either of two mechanisms. One is selective binding of copper to metallochaperones and handing off to partner proteins *via* molecular recognition coupled to favourable thermodynamic and kinetic factors. Alternatively, selectivity can be achieved by spatial colocation of partner proteins in specific organelles or vesicles. Cells use both of these methods either separately or in concert. Secondly, maintaining cellular homeostasis requires *regulation via* systems that sense cellular copper levels and activate molecular mechanisms either to increase (import) or decrease (export) copper levels in the cell. These mechanisms will be discussed in the following sections, emphasizing the current state of knowledge and areas where additional information and research are required.

18.2 Factors Affecting Selectivity in Eukaryotic Copper Transport

Organisms must develop robust and reliable systems to ensure that copper is delivered only to proteins and enzymes where it is required for function. At first glance, one might argue that this would be determined purely by thermodynamics and, if the cell behaved like a dilute homogeneous solution, the distribution of copper ions would be set by the affinity constants of all of the potential binding sites, provided the system had enough time to reach equilibrium. A version of such a model has been suggested in the recent study that investigated the copper-binding affinities of individual components of the copper handing machinery.[17] Closer examination of this hypothesis reveals a caveat, namely that copper would become trapped in the sites of highest thermodynamic stability. In reality, enzymes vary greatly in their affinity for copper, such that the more weakly binding cuproteins could never out-compete those with higher affinity, and would only become metallated under conditions of copper excess. Secondly, the cell deviates significantly from an ideal solution, and more closely resembles an ensemble of permeable compartments (organelles and vesicles) embedded in a fairly concentrated solution of proteins and small molecules (the cytosol). The distribution and concentration of copper in the cellular milieu is therefore extremely heterogeneous, which provides a greater opportunity for dynamic partitioning into different pathways where kinetic effects probably can dominate. For example, if a copper ion enters the cell *via* an importer, its fate will depend more on the proximity of a metallochaperone or small molecule chelator, even though the affinity constant for the chaperone may be lower than that of more spatially distant cellular components, simply because the complexity of the cellular compartment lowers the rate of diffusion. These considerations suggest that thermodynamic stability is important but not necessarily definitive in determining the mechanisms of selectivity.

Nonetheless, thermodynamic stability provides an excellent starting point for understanding selectivity. As discussed in Chapter 16, Cu(II) is generally classified as a Lewis acid of intermediate "hardness" while Cu(I) is classified as a "soft" acid. The unfilled d-shell of the d^9 Cu(II) ion provides incomplete shielding of the nuclear charge, such that complex formation occurs primarily *via* electrostatic (non-covalent) interactions, with ligands containing more electronegative donors (O, N) providing greater electrostatic stabilization than those with S, P or Se. In addition, as the Cu^{II}_{aq} ion is hydrated by six water molecules, these waters are released when it is sequestered by a protein binding site generating a large increase in entropy. Cu(I) on the other hand contains a filled d-shell, which effectively shields the nuclear charge and allows the empty 4s and 4p orbitals to overlap with ligand donor orbitals to form stable bonds with much greater covalent character. Cu(I) therefore forms strongest complexes with thiols, thioethers, phosphines, selenols and selenoethers, and much weaker complexes with the more electronegative N- and O-donor ligands. As "free" Cu(I) is only weakly hydrated, little or no entropy stabilization accrues upon binding to a protein site. A further difference is found between the coordination number and/or coordinate geometry. Cu(II) forms mainly four- or five-coordinate complexes (which maximizes the electrostatic interaction) whereas Cu(I) prefers low-coordination, with two- and three-coordinate systems predominating (in order to optimize orbital overlap). An exception to these rules is the imidazole group of the histidine side-chain, which is able to form strong complexes with both Cu(I) and Cu(II), albeit with the geometric preferences of each oxidation state still expressed.

The large differences in preferred ligand-type and coordinate geometry provide the cell with multiple options for designing selective processes, particularly in the choice of oxidation state. Cupric ions are far less selective with respect to other biogenic metals (Fe, Mn, Co, Zn) as all of these metals form similar complexes as divalent ions for the same reasons as discussed for Cu(II) above (although the relative affinities vary). On the other hand, Cu(I) has a unique coordination chemistry, comparable only to its d^{10} congener Ag(I). Mammalian cells are able to leverage these inorganic properties, since the redox potential of the cytosol is highly reducing due to high concentrations of glutathione. Thus, provided that copper import is selective for the Cu(I) state, the cell can utilize the unique coordination properties of the Cu(I) ion to build selective trafficking pathways.

There is, however, a price to pay for choosing Cu(I) as the primary oxidation state for eukaryotic transport. Firstly, in the absence of any coordinating ligand, the greater solvation of the cupric state in aqueous medium leads to disproportionation of aqueous Cu(I) *via* the following reaction:

$$2Cu(I) \rightarrow Cu(II)(aq) + Cu(0) \qquad (18.1)$$

where Cu(0) represents metallic copper, which generally precipitates out as crystalline nanoparticles, and drives the reaction to completion. Secondly in the presence of hydrogen peroxide, a ubiquitous by-product of oxygen metabolism, Cu(I) catalyses the formation of hydroxyl radicals *via* the so-called Fenton reaction:

$$Cu(I) + H_2O_2 \rightarrow Cu(II) + OH^- + OH^\bullet \qquad (18.2)$$

Hydroxyl radicals are extremely harmful to the cell and lead to oxidative damage of protein and DNA molecules *via* H-atom abstraction. Taken together, these two properties imply that the concentration of "free" (*i.e.* uncomplexed) Cu(I) must be minimized if serious cellular damage is to be avoided. For yeast it has indeed been calculated that the concentration of free copper (in either oxidation state) is many times less than an average of one atom per cell.[18] This estimate was based on the observation that whereas the yeast superoxide dismutase (SOD1) can be metallated by either free Cu(I) or free Cu(II) ions in the cell when copper levels are replete, under normal conditions enzyme activation by copper requires the presence of the copper chaperone for superoxide dismutase (yCCS). Using a yeast strain lacking yCCS, the maximum number of active SOD1 molecules (metallated by free copper) was estimated at 200, relative to an estimate of the total SOD1 population of 50 000 with the ratio f of active to total of 4×10^{-3}. The maximum free copper in the ΔCCS cell could then be calculated from equations for equilibrium binding to SOD1 and mass balance:

$$Cu_{free} + apoSOD1 \rightarrow CuSOD1 \quad K_{Cu(II)} \approx 10^{15}M^{-1}; K_{cu(I)} \approx 10^{20}M^{-1}$$

(18.3)

$$SOD1_{total} = apoSOD1 + Cu(II)SOD1 \tag{18.4}$$

which could be then be solved for the concentration of free Cu(II) in terms of the equilibrium constant K, and the ratio f of metallated (CuSOD1) to total SOD1 (SOD1$_{Total}$):

$$Cu_{free} = \frac{f}{K(1-f)} \tag{18.5}$$

This calculation leads to an upper limit for the free copper concentration of 10^{-18} M for Cu(II) and 10^{-23} M for Cu(I). Since one free copper atom in a cell volume of 10^{-14} L corresponds to a concentration of 10^{-10} M, it is clear that the number of free copper ions of either oxidation state in the cytosol is essentially zero.

The ability of the molecular components of the cell to effectively co-ordinate every free copper ion protects against Fenton chemistry and dis-proportionation reactions but introduces challenges in dynamic Cu(I) transport, since the binding constants are high enough to restrict "off rates" to values far less than the half-life of the transporter molecules. Thus, copper is seldom free to find its proper cellular target by passive diffusion. Instead the copper-binding metallochaperones and their partner proteins are be-lieved to find each other by molecular recognition, driven by electrostatic complementarity of protein surfaces, and exchange their copper cargo within these protein-protein complexes. To understand the mechanism and selectivity of this transport mechanism it has been necessary to develop accurate methods to measure the binding constants of Cu(I) to each member of a transport pathway. Historically, attempts to measure Cu(I) affinities led to a large range of values spanning ten or more orders of magnitude, due to use of different conditions and methodologies. Recently, standardization of experimental approaches has led to unification of these affinities, such that the relative binding strength of chaperone-target pairs can be accurately assessed.[19] The consensus approach utilizes the Cu(I)-specific ligands BCA (bicinchoninic acid) and BCS (bathocuproine sulfonate) as competitive binding agents for Cu(I) (Figure 18.3). Both BCA and BCS form 2 : 1 com-plexes with Cu(I) (CuL$_2$) and give rise to intense ligand-to-metal charge transfer bands in the visible region of the spectrum, where the extinction coefficient and β_2 value for the formation of the CuL$_2$ complex have been determined to a high degree of accuracy. The experiment is conducted by titrating apo-protein into a solution of the CuL$_2$ and measuring the decrease in absorbance of CuL$_2$ as the apo-protein competes with the ligand for the available Cu(I). Alternatively, metal-loaded protein can be titrated with

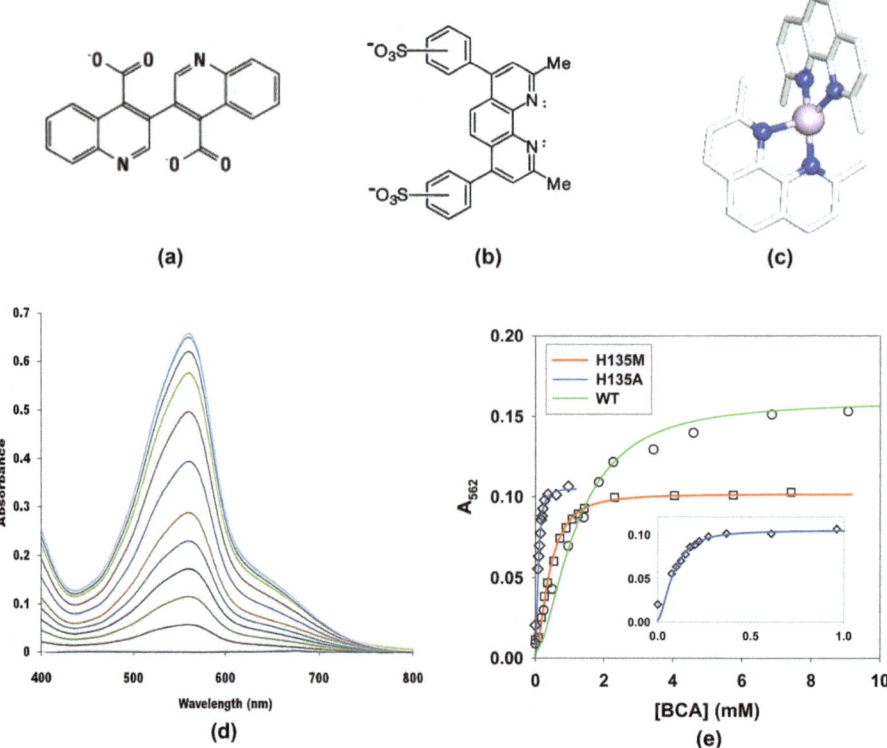

(a) **(b)** **(c)**

(d) **(e)**

Figure 18.3 Determination of Cu(I) binding affinities using competitive binding of chromophoric Cu(I)-binding ligands. (a) Bicinchoninic acid (BCA), (b) bathocuproin sulfonate (BCS), (c) structure of the [Cu(I)-(BCS)$_2$] complex showing the steric hindrance of the 9,10-methyl groups, which leads to a preferred tetrahedral coordination favouring the Cu(I) oxidation state. (d) Titration of a 90 μM solution of Cu(I)-loaded protein with increasing concentrations of BCA. (e) Plots of A$_{562}$ *versus* [BCA] for the WT *B. subtilis* Sco and two mutants H135A and H135M. The plots demonstrate the weaker binding by the H135A variant as evidenced by the lower concentration of competitor ligand required to elicit full colour response due to removal of Cu(I) from the protein. The inset shows an expanded view of the low [BCA] range for the alanine variant. Solid lines are fits to the data using DYNAFIT as described in the text.

increasing amounts of free BCA or BCS, and the increase in the absorbance of CuL$_2$ determined at each titration point (Figure 18.3). Because the equilibrium constant for the formation of the coloured CuL$_2$ complex is accurately known, the free Cu(I) concentration can be calculated for all points in the titration from the relationship:

$$[Cu_F] = \frac{\dfrac{A}{\varepsilon}}{\beta_2 \times \left\{ [L]_T - 2\dfrac{A}{\varepsilon} \right\}^2} \tag{18.6}$$

where A is the absorbance at point i, ε is the molar extinction coefficient of the coloured CuL_2 ligand complex and β_2 is its overall stability constant. Assuming the concentration of free Cu(I) is vanishingly small, the degree of formation of the protein complex Y can be calculated from the known total concentrations of copper (Cu_T) using the expression:

$$Y = \frac{[CuP]}{Cu_T} = \frac{[Cu_T] - [Cu(BCS)_2]}{Cu_T} = \frac{[Cu_T] - \dfrac{A}{\varepsilon}}{Cu_T} \qquad (18.7)$$

Finally, the association constant K_M for Cu(I) binding to the protein is determined from the relationship:

$$Y = \frac{K_M[Cu_F]}{1 + K_M[Cu_F]} \qquad (18.8)$$

Typically, these equations are solved using non-linear regression methods. One such example is the program DYNAFIT available online as freeware.[20] Inputs include equations defining the equilibrium of copper-binding to protein and competitor ligand, extinction coefficients of all known coloured species and starting concentrations of all reagents. The program fits the experimental absorbance *versus* competitor ligand concentration to a theoretical curve defined by the values of ligand β_2 and K_D for Cu(I) binding to the protein.

18.3 Copper Import – Structure and Function

Copper enters the eukaryotic cell *via* the high-affinity CTR1 transporter,[21] a membrane-bound permease located in the plasma membrane[22] (Figure 18.2). While the speciation of copper in serum is uncertain, CTR1 almost certainly transports copper in the Cu(I) oxidation state and in the yeast *S. cerevisiae* the plasma membrane metalloreductase FRE1p has been implicated in the reduction of both Fe(III) and Cu(II) prior to uptake into the cell.[23] FRE proteins belong to a family of NADH dehydrogenases that couple the oxidation of cytoplasmic NADH to Fe(III) or Cu(II) reduction, and contain a cytoplasmic NADH oxidoreductase and a transmembrane haem domain. Whether these or similar proteins are involved in Cu(II) reduction in mammalian cells is still an open question, and little evidence exists either way. For example, the STEAP (6-transmembrane epithelial antigen of prostate) family of proteins, which share domain homology with FRE1p and are overexpressed in metastatic tumours, have been suggested as candidates for such a role, although existing data are conflicting.

The CTR1 monomer consists of four domains, which include an extracellular N-terminal domain, three transmembrane (TM) helices, a TM1-TM2-connecting loop exposed into cytosol and a short C-terminal tail (Figure 18.4). The protein architecture built with three TM helices means that N-terminal and C-terminal domains lie on opposite sides of the plasma membrane, and identification of N-glycosylation sites has firmly identified the membrane topology with an extracellular N-terminus and an

(a)   (b)

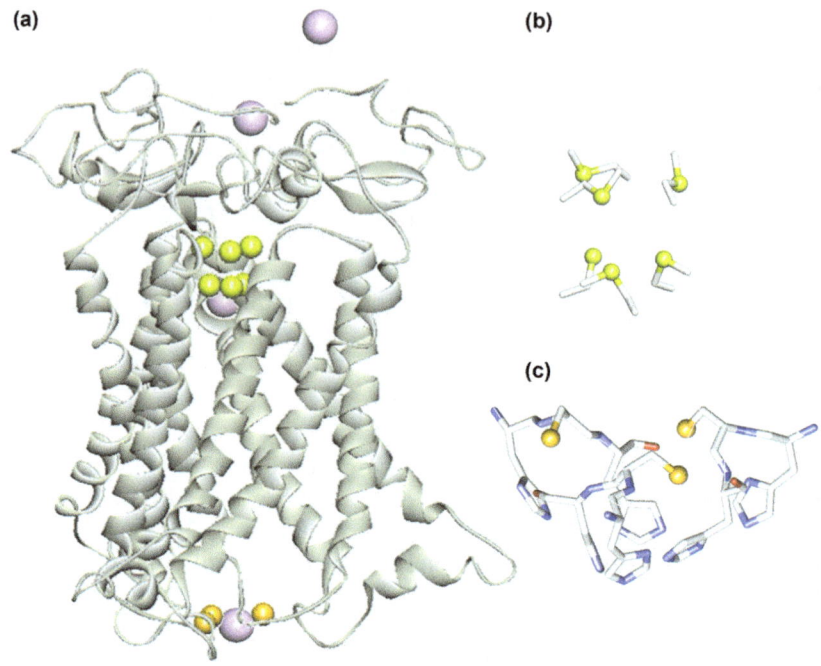

(c)

Figure 18.4   Visualization of a suggested mechanism for copper transport through the CTR1 transporter. Cu(I) enters the pore through the N-terminal funnel and is drawn downwards to the double triad of Met residues in the vestibule (shown in pale yellow). This is followed by a conformational change coupled to organization of the Cys triad (orange) on the cytoplasmic side so as to provide a driving force for transport through the pore. (b) Expanded view of vestibular double triad of Met residues. (c) Expanded view of the triad of C-terminal HCH motifs, showing the cluster of His residues at the exit of the pore. Taken from the *in silico* representation of the CTR1 structure reported in reference 34.

intracellular C-terminus.[24,25] The N-terminal domain contains a number of His-rich and Met-rich sequences where the predominance of methionine is suggestive of a role as a Cu(I)-selective filter. However, mutational analysis within these sequences has led to conflicting conclusions regarding their role in transport. Thiele's laboratory measured the effect of site-directed mutations on the ability of CTR1 to complement a yeast CTR1 knock-out or to stimulate ^{64}Cu uptake in HEK293 cells.[26] Their results suggested that the Met-rich N-terminal motifs were not essential since a single Met residue in the Met-motif closest to TM1 (M43 or M45) was sufficient to induce Cu uptake. On the other hand, the pair of Met residues in TM2 (M150, 154) appeared critical for growth under copper-limiting conditions. Their observation that Cys could replace Met at these positions provided evidence that the pair of methionines might be important in Cu(I) binding. In a more quantitative study, Kaplan's laboratory was able to dissect the contributions to K_M and V_{max} produced by a similar series of mutants.[27] While conclusions regarding the role of the N-terminal Met clusters were generally similar, CTR1 stably transfected in sf9 cells was capable of withstanding the effect of a double M150/M154A substitution, retaining 30% of WT ^{64}Cu transport activity. Further, a critical tyrosine (Y156) was discovered whose position close to Met150 and 154 suggested a role in correctly orienting the Met residues. Truncation of the putative Cu(I)-binding HCH motif in the C-terminal tail likewise did not abolish function, but slowed the rate of intramembrane transport, perhaps through an effect on the release step. These studies collectively led to an early model for CTR1 function[28] in which the protein was proposed to assemble as a trimer, thereby creating an intramembrane pore at the centre of the nine TM helices. The N-terminal domain served to sequester and concentrate Cu(I) out of the extracellular milieu, and to guide it into the pore.

The low-resolution structure determined by electron crystallography has shed additional light on the architecture and possible transport mechanism of hCTR1.[29–31] As suggested by the earlier studies, the protein indeed forms a trimer in the membrane. The three protomers interact to form a funnel-shaped pore, narrower at the upper (extracellular) end (8 Å), and wider at the lower (intracellular) end (21 Å). Transmembrane helix 2 lines the inner section of the pore, which places the M150xxxM154 motif at the upper end of the pore forming a "vestibule" lined with potential Cu(I)-binding thioether ligands (Figure 18.4). Although the resolution of the structure is too low to locate the side-chains of these residues, the variant M154C can cross-link between two subunits, suggesting that the M154 Met residues of at least two subunits would be close enough to coordinate a Cu(I) ion. The short C-terminus of the human protein contains another potential Cu(I) binding motif HCH, suggesting a second site at which Cu(I) might bind. Consistent with this prediction, a detergent-solubilized form of hCTR1 was found to bind 2 Cu(I) ions tightly (binding constants not determined). Subsequent X-ray absorption spectroscopic (XAS) studies identified these as most likely arising from one three-coordinate $Cu^I(Met)_3$ site and a second three-coordinate $Cu^I(Cys)_3$ site corresponding to the three Met residues at the entrance to the pore, and to the three Cys residues at the C-terminal pore exit. Consistent with this model, mutation of the C-terminal Cys residue generated a protein with a mixed Cu-N(His) and Cu-S(Met) coordination, suggesting that in the absence of the C-terminal Cys residue, Cu(I) bound to one or both of the His residues of the HCH motif.[30]

It is not known at present whether this solubilized form represents an on-pathway intermediate or whether during transport only one of these sites is occupied at any one time. Binding to a high-affinity site coupled to a conformational change is certainly one mechanism that could impart selectivity by opening a pore or channel allowing access and transport of the metal ion. With the channel in the open conformation, the selected metal ion could be transported, perhaps assisted by weaker (transient) binding sites along the channel pathway. Although there is as yet no evidence either way, the CTR1 transport mechanism can be visualized within these general principles.

The transporter is selective for Cu(I) and Ag(I), presumably due to a binding site that discriminates against other metals, and we envisage that this function is provided by the MxxxM site at the entrance to the vestibule. Binding to this site may be assisted by the His and Met rich sequences in the N-terminal domain, but mutagenesis studies have shown that these are not critical. What is critical, however, is to establish a driving force that favours unidirectional transport of Cu(I) into the cell. The Cys triad at the intracellular pore exit could potentially provide a thermodynamic gradient since a $Cu^I(Cys)_3$ site is likely to have a much larger association constant than $Cu^I(Met)_3$. This follows since the latter has been measured in a peptide model system to have a $\log K_M$ (assoc) ~ 6,[32] while the former is estimated to be many orders of magnitude larger based on measured values for $Cu(Cys)_2$ ($\log K_M$ (assoc) $= 16–21$).[19,33] On the other hand unidirectional diffusion of copper within the channel and out of the pore can be facilitated by intracellular components such as glutathione,[13] or the apo form of a metallochaperone (*vide infra*).[13] Cuprous ions would thus selectively enter the pore and bind to the Met triad. If the Cys triad is empty, Cu(I) would be driven through the pore by the thermodynamic gradient, where it would bind in the $Cu(Cys)_3$ site, to be finally handed off to an intracellular metallochaperone or small molecule carrier.

A recent *in silico* study based on the low-resolution structure of CTR1 has generated a set of coordinates for the main-chain and all side-chain residues.[34] While further experimental evidence is needed for this model, it provides a compelling visualization of the important structural elements of the transporter. The two triads of Met residues locate to the top of the pore where they appear to form potential Cu(I) binding sites (Figure 18.4). The HCH motif localizes to the cytoplasmic side of the pore, and creates a high-affinity triad of Cys residues together with a cluster of His residues, which may assist in handing off the Cu(I) to its initial partner in the cytoplasm. Further, the N-terminal domain is replete with S(Met) and N(His) coordinating donors, which may serve to initially capture the copper from the extracellular milieu and draw it down towards the Met triads in the vestibule (Figure 18.5). Using this model, it is possible to visualize the transport of copper first to the vestibule and then onwards to the cytoplasmic gate. Although speculative, this mechanism incorporates all of the structural and thermodynamic elements of the CTR1 system. Specific elements of the mechanism such as the role of each of the two Met triads formed by the MxxxM motif or whether Cu(I) binding at one site affects the affinity for Cu(I) at the other are yet to be determined. For example, it is possible that Cu(I) binding at the exit site might change the configuration of Met residues in the vestibule so as to prevent more copper entering the pore until successful hand-off to the cytoplasmic acceptor. Alternatively, loading of the N-terminus with Cu under copper replete conditions might act as a signal for CTR1 down-regulation *via* endocytosis from the plasma membrane.[35]

While the mechanism for mammalian CTR1 transport is compelling, the existence of CTR proteins with different structural elements suggests that

(a) **(b)**

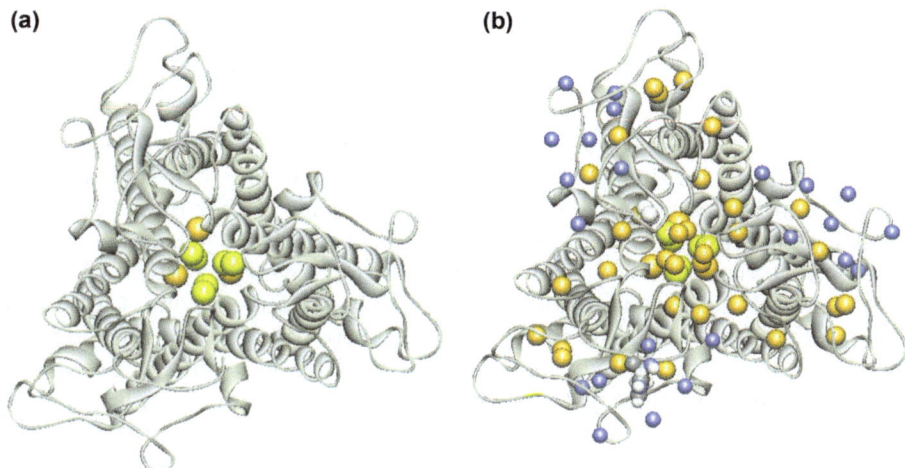

Figure 18.5 Top views of the proposed CTR1 Structure. (a) View looking down the N-terminal funnel towards the vestibular Met triads (yellow) and the C-terminal HCH motifs (orange). (b) Same view but with N-terminal S atoms from Met clusters and Nε atoms from His clusters added. Note the large density of S(Met) and N(His) that line the funnel.

the details of transport mechanisms, such as copper delivery and retrieval, may vary. For example, the yeast *Saccharomyces* contains two high-affinity transporters CTR1 and CTR3 whose roles are not clearly defined but are redundant with respect to function. The C-terminus of yeast CTR1 is longer than the mammalian homologue, and contains a cluster of Cys residues that have been shown to form a multinuclear Cu(I) cluster[36] which exchanges Cu(I) with the Atx1 metallochaperone.[37] However, mutants, in which the C-terminal domain has been cleaved, are competent in copper import but have increased sensitivity to copper, suggesting a regulatory role in protecting the transporter against toxic copper levels.[38] An additional low-affinity transporter CTR2, present in yeast and in humans, was shown to be involved in mobilization of Cu(I) from a vacuole.

18.4   Transport within the Cell and the Requirement for Metallochaperones

The fate of copper as it exits the CTR pore and enters the intracellular milieu is one of the remaining uncertainties of eukaryotic transport. While evidence exists that Cu(I) is sequestered by cytoplasmic metallochaperones for selective transport to specific organelles and/or cuproprotein targets, it is unclear whether the chaperones interact directly with the CTR C-terminus or whether some intermediate Cu(I) carrier is involved. As mentioned above, down-regulation of metallochaperones has little impact on copper uptake, suggesting that other entities such as glutathione may accept copper as it exits the membrane, or may substitute for metallochaperones in their absence.[13] In this section we pick up the story as the Cu(I) binds to each of the metallochaperones and transports it onwards for eventual loading into cuproteins such as the ATPases ATP7A and 7B, and Cu/Zn superoxide dismutase.

Metallochaperones are small proteins with high affinity for Cu(I) and almost no affinity for Cu(II). There are a number of reasons why such systems are chosen by eukaryotes. Firstly, transport of Cu(I) rather than Cu(II) can leverage the unique coordination chemistry of Cu(I) and Ag(I), and eliminate competitive reactions with other divalent cations (Mn(II), Fe(II), Co(II) and Zn(II)), which have similar binding preferences to Cu(II). Second, Cu(I) must be tightly bound inside the cell to avoid disproportionation or formation of oxygen radicals due to Fenton chemistry. This leaves open the possibility that Cu(I) could be transported by a low-molecular-weight complex such as $Cu(I)(GSH)_2$, especially since the cytoplasmic concentration of glutathione is ~ 5 mM. However, this scenario would also cause difficulties for the cell since Cu(I) binding to GSH is relatively weak (estimates of log K_D are between -13 and -15), resulting in the possibility of unselective incorporation of Cu(I) into sites that require other metals such as FeS centres or Zn-finger motifs, especially under copper replete conditions. In addition, $Cu(I)(GSH)_2$ is known to react with molecular oxygen to generate superoxide radicals, which would be harmful to the cell. As we will describe in further detail in subsequent sections, use of metallochaperones circumvents all these issues (a) by providing tight selective binding (log $K_D \sim -17$) with low off rates for the isolated Cu(I)-loaded chaperone, (b) by ensuring selectivity through complementary molecular surfaces that recognize partner proteins with high fidelity and (c) by catalysing metal exchange within an intermediate involving shared thiolate ligands. In this way, copper transport can occur in a rapid and highly selective fashion, with low probability of copper incorporation into sites designed for other metals.

18.5 The ATOX1 Pathway

Secreted cuproteins enter the TGN as folded apo-proteins that still require metallation prior to being shipped to the cell surface or exported into the extracellular milieu. This function is performed by copper transporting ATPases,[39–42] which are members of the P1B family of heavy metal transporters found in all forms of life from bacteria to mammals. The mammalian ATP7A & 7B reside in the membranes of TGN or post-Golgi vesicles and pump copper from the cytosolic to the lumenal side of the membrane. The proteins have a multidomain structure with an N-terminal regulatory domain, a cytosolic ATP binding domain and a transmembrane domain, which is comprised of eight transmembrane helices.[43,44] The bacterial CopA from *Legionella pneumophila*[45] is currently the only copper-transporting ATPase for which a high-resolution crystal structure is available. This and two low-resolution EM structures,[46,47] together with homology to the well-studied Ca ATPase SERCA, have provided a template for understanding the mechanism of transmembrane copper transport. However, the mammalian homologues ATP7A and 7B are twice as large as CopA due to the presence of various regulatory elements. These include a far more complex N-terminal regulatory domain comprised of six metal binding subdomains called MBDs, each of which binds Cu(I); a small domain between TMs 1 and 2, which extends outwards from the lumenal side of the membrane, and a long C-terminal tail involved in regulation of intracellular trafficking and protein stability.

Cuproprotein loading is a catalytic process beginning with a high-affinity conformer of the ATPase, E1, which is stabilized when copper binds to a transmembrane site from the cytosolic site. This binding activates ATP-hydrolysis and couples to a transient phosphorylation of the protein at the invariant aspartate (E1-P), which in turn drives a conformational change leading to the occlusion of copper from the cytosolic side (E2-P). Copper release on the lumenal side is accompanied by dephosphorylation, and the generation of the low-affinity E2 form, ready to rebind copper and restart the catalytic cycle.[48,49] As discussed above, eukaryotic cells provide a selective pathway for copper from its entry point *via* CTR1 to its binding site on the ATPase. The selective binding and cytosolic transport of copper ions destined for the TGN and post-TGN secretory vesicles is handled by the metallochaperone Atox1, also known by the organism-specific acronyms such as Atx1 (yeast) and HAH1 (human).

Atx1 and other Atx-like chaperones have been extensively studied by NMR and crystallography, and their structures are known in great detail.[9,12,50,51] The proteins adopt the "ferredoxin fold" comprised of a βαββαβ secondary structure (Figure 18.6). All members of the family contain the MxCxxC motif, which is capable of binding Cu(I) as well as a number of other class-B heavy metals such as Hg(II) and Cd(II). These metal ions bind to the two Cys residues to form a two-coordinate structure, which is linear in the Hg derivative but deviates from linearity ($120 \pm 40°$) in the Cu(I) derivative. Important differences exist between the structure of apo and Cu(I) bound forms in solution, most of which map to the metal binding region.[52] In the apo forms, both Cys residues of the metal binding loop (C15 & C18 in yeast, C12 & C15 in human) are exposed to solvent, whereas when Cu(I) binds one of the Cys residues (yeast C18) flips so as to become buried. A conserved lysine (yeast K65, human K64) is located close to the metal binding centre and is believed to stabilize the Cu(I)-bound form through interaction with C15/12, thereby neutralizing the negative charge associated with the Cu(I)-bis-thiolate structure. Evidence in favour of this assertion comes from the Hg(II)-Atx1 and apo-structures where the lysine side-chain is displaced away from the metal centre: the Hg(II)-bis-thiolate is neutral while in the apo-protein both thiolates are protonated so that no additional stability is gained by an interaction with a positive lysine group. Comparison of the structures of the Cu(I) and apo forms of Atx1 provides clues to the mechanism of copper release. The two forms have different conformations, determined largely by the position of helix2 and the following loop 5 which contains K65. Movement of K65 away from the copper centre would

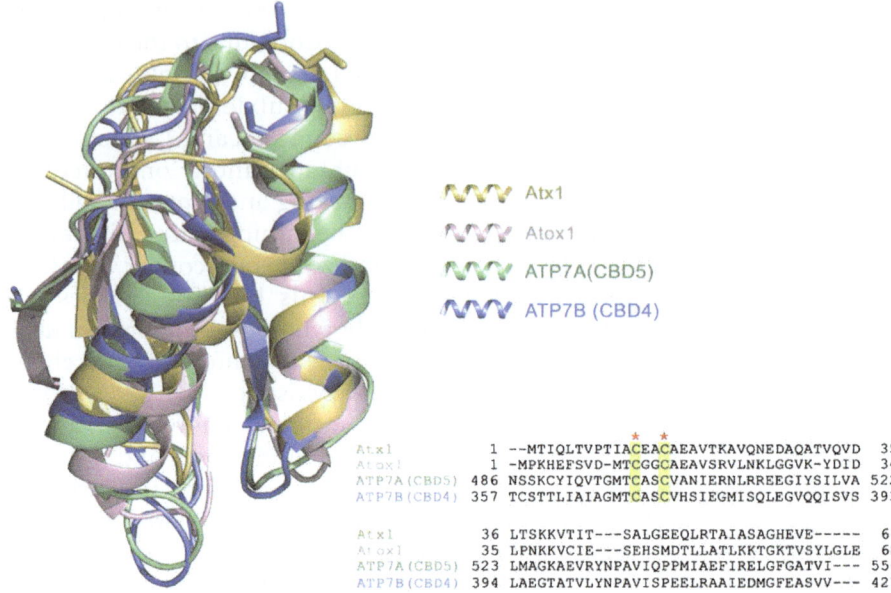

```
              ~~~  Atx1
              ~~~  Atox1
              ~~~  ATP7A(CBD5)
              ~~~  ATP7B (CBD4)

                                *    *
Atx1             1  --MTIQLTVPTIACEACAEAVTKAVQNEDAQATVQVD  35
Atox1            1  -MPKHEFSVD-MTCGGCAEAVSRVLNKLGGVK-YDID  34
ATP7A(CBD5)    486  NSSKCYIQVTGMTCASCVANIERNLRREEGIYSILVA  522
ATP7B(CBD4)    357  TCSTTLIAIAGMTCASCVHSIEGMISQLEGVQQISVS  393

Atx1            36  LTSKKVTIT---SALGEEQLRTAIASAGHEVE-----  64
Atox1           35  LPNKKVCIE---SEHSMDTLLATLKKTGKTVSYLGLE  68
ATP7A(CBD5)    523  LMAGKAEVRYNPAVIQPPMIAEFIRELGFGATVI---  556
ATP7B(CBD4)    394  LAEGTATVLYNPAVISPEELRAAIEDMGFEASVV---  427
```

Figure 18.6   Sequence and structural alignment of Atx1, Atox1, ATP7A(MBD5) and ATP7B(MBD4). Colour code: wheat, Atx1 (PDB ID: 2XMT); pink, Atox1 (PDB ID: 1FEE); pale-green, ATP7A(CBD5) (PDB ID: 1Y3J) and blue, ATP7B(CBD4) (PDB ID: 2ROP). The potential copper binding cysteines are highlighted in stick model. (Domains are labelled CBD in the figure to denote copper binding domain).

destabilize the structure and cause the copper and its cysteine ligands to move from its buried position to a more solvent-exposed location, while interaction of the chaperone with its partner would be a logical trigger for this event.

To further understand the mechanism of copper transfer, NMR was used to study the interaction of Atx1 with the metal-binding domain of its partner Ccc2.[52,53] In yeast, the important function of Ccc2 is to metallate the multicopper oxidase Fet3, an important component of iron homeostasis. The N-terminal metal binding domains (MBDs) of Ccc2 are homologous to Atx1, contain the MxCxxC motif and could be metallated by the Cu-loaded chaperone *in vitro*, suggesting that they serve as acceptors for copper on the ATPase. Early NMR structures for the first MBD of yeast Ccc2 and for the Ag-bound form of MBD4 of human ATP7A both exhibited the Atx1 βαββαβ fold, and could be modelled to bind Cu(I) *via* the two Cys residues. However, the MBD structures differed from Atx1 in that the apo- and Cu(I)-bound forms were similar, with no helical movement on Cu-binding.

The nature of the protein-protein interaction between Atx1 and its Ccc2 partner was investigated by forming 1 : 1 adducts of Cu-Atx1 with apo-Ccc2 or *vice versa*. These studies showed conclusively that positively charged lysine residues on the surface of the chaperone interacted with corresponding negatively charged residues on Ccc2 leading to a highly specific molecular interaction between the two proteins. Modelling the structure of the adduct suggested that the Ccc2 component was not greatly perturbed from its isolated structure implying that its metal binding site was "preformed". For the Atx1 component, however, residues in the vicinity of the copper binding site including both Cys residues and K65 were significantly perturbed, adopting a conformation intermediate between metal-bound and metal-free states of the chaperone.

While this model supported the idea that protein-protein complex formation triggered copper release, the Cu(I) coordination in the adduct could not be determined since the single Cu(I) ion was in fast exchange with both metal binding sites on the NMR time-scale. In a subsequent study, it was shown that in addition to complementary electrostatic surfaces, adduct formation is driven *via* formation of a three-coordinate intermediate.[54] Adduct formation was observed for C18A-Atx1 + WT-Ccc2 and WT-Atx1 + C15A-Ccc2 but not for Cys to Ala mutations of the other Cys residues. Therefore, complex formation must proceed through three-coordinate intermediates involving C15-Atx1 and both Cys residues on Ccc2 or C13-Ccc2 and both Cys residues on Atx1. This finding provided the basis for the proposed mechanism for metal transfer between Atx-like chaperones and their partner ATPase MBDs. The chaperone and target proteins first interact *via* complementary electrostatic molecular recognition of their surface charges forming an initial protein-protein complex organized such that the two copper-binding elements are in proximity. Thiolate sharing within this complex further stabilizes the interaction, and leads to a three-coordinate intermediate involving Cu(I) and three cysteine thiolates, two from Atx1 and one from the MBD. Metal transfer within this complex evolves, probably *via* further cysteine swapping, to form a secondary three-coordinate intermediate involving both cysteines of the MBD and only one from the chaperone. Finally, this complex dissociates to generate the transfer products. The proposed interaction between chaperone and partner MBD is illustrated by the structure of Atox1 with ATP7A MBD1[55] in Figure 18.7.

There is much additional evidence in support of the three-coordinate activated complex. The Cu(I) derivative of human Atox1 chaperone crystallizes as a dimer, held together by a similar three-coordinate interaction at the copper centre, with two Cu(I)-thiolate interactions from one protomer and a third from the second protomer.[51] The fourth cysteine is only weakly associated with the Cu(I). It has also been shown by XAS that while Cu(I)-Atox1 is a classical two-coordinate linear bis-cysteinate complex when endogenous cuprophilic ligands (thiols, phosphines) are rigorously excluded, weak three-coordinate adducts form when the complex is exposed to potential ligands such as DDT or the phosphine TCEP.[56] Perhaps of greatest significance, very careful determination of the binding affinities of Cu(I) for chaperone and MBD pairs has revealed almost identical binding constants (Table 18.1) implying that there is no thermodynamic gradient for metal transfer.[19] Therefore, transfer must be a kinetically driven process, where

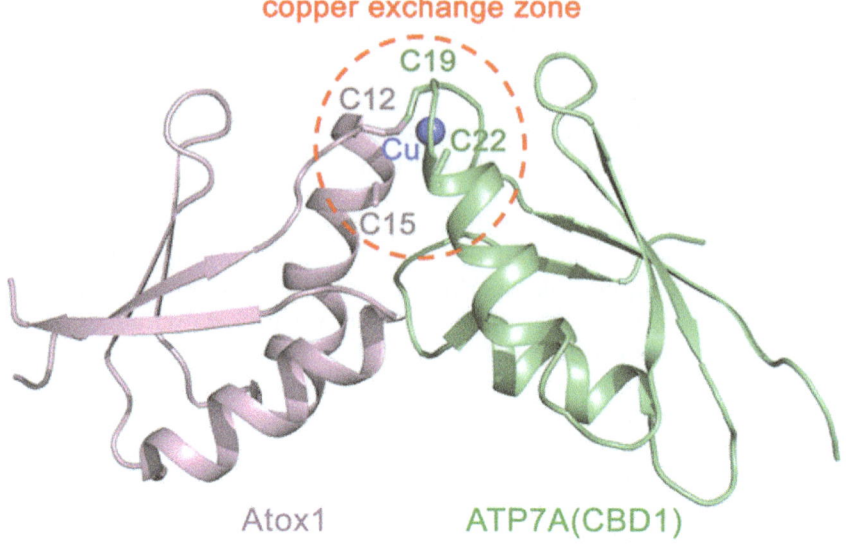

Figure 18.7   Solution structure of Atox1-Cu-CBD1 complex. Copper chaperone Atox1 and MBD1 of ATP7A are represented in pink and pale-green, respectively. The key cysteine residues (C12 and C15 of Atox1, C19 and C22 of MBD1) are shown in stick model. (The ATP7A is labelled CBD in the figure to denote <u>c</u>opper <u>b</u>inding <u>d</u>omain.)

Table 18.1   Representative affinity constants (log K_D) for some copper transporters and chaperones for copper(I). All constants are determined at pH 7.5, and are relative to the β_2 values for the competing ligand as listed in the table. For Sco the decrease in affinity on mutation of the coordinating H135 ligand to alanine is clearly seen in the data. For CusF, the W44 ligand which caps the Cu(I) site and forms a weak π-interaction with Cu(I) can be mutated to alanine with little effect on affinity. However, mutation to Met causes a large increase in affinity (decrease in K_D) due to replacement by an additional coordinating Met residue.

Protein	Complexing agent	Log K_D	Reference
	BCA	19.8 (log β_2)	19
	BCS	17.3 (log β_2)	19
	DTT	−14.6	19
ATOX1	BCS	−17.7	19
WND-MBD5 − 6	BCS	−17.6	19
CCC2	BCS	−18.8	19
CCS-D1	BCS	−17.7	84
CCS-D3	BCA	−16.4	84
Sco1 WT	BCA	−15.6	97
Sco H135H	BCA	−13.5	97
Sco H135M	BCA	−15.4	97
CusF WT	BCA	−13.9	98
CusF W44A	BCA	−14.2	98
CusF W44M	BCA	−16.9	98

the protein-protein interaction generates an activated complex, which accelerates metal exchange *via* formation of the three-coordinate intermediate. In this respect, the process is akin to an enzyme-catalysed reaction where the rate is enhanced *via* formation of the ES complex that optimizes the orientation of substrates for reaction.

Whereas copper exchange between MBDs and Atx1/Atox1 is well understood, the process of copper delivery to the transmembrane sites and the mechanism of copper release from the membrane into the lumen of TGN/vesicles are much less clear. It has been recently proposed (based on studies of the truncated copper chaperone CopZ and an archaeal copper ATPase CopA) that the chaperone delivers copper directly to the transmembrane site by docking to the positively charged platform in the vicinity of a conserved Met, Asp and Glu triad of residues that are believed to form the "entrance" of the Cu(I) translocation pathway.[57,58] This hypothesis has some experimental support[59] since mutations of these residues and/or mutations in the putative negative patch on the chaperone eliminates the chaperone-based transfer of copper *in vitro*. However, the model needs stronger evidence (such as using the full-length copper ATPase in a native membrane environment along with a full length chaperone). Furthermore, when applied to eukaryotic cells, the model of copper delivery needs to explain the structural and mechanistic basis for the proposed dual role for the copper chaperone, *i.e.* docking and transfer of copper to structurally unrelated and dissimilar N-terminal and transmembrane sites.

An alternative model, based on a two-step process of copper delivery, may better accommodate all available data. The N-terminal MBDs of mammalian Cu-ATPases (ATP7A and ATP7B) have very distinct surface charges and are packed into the structure with the pairs of CxxC motifs located in proximity to each other. Although isolated recombinant MBD4 exhibits the strongest protein-protein interaction with Atox1 in solution,[60] in the context of the full-length N-terminal domain MBD2 of ATP7B is the preferential site to accept copper from holo-Atox1.[61] Mutations of the Cu-coordination residues (C154 and C157) in MBD2 of the intact ATP7B completely abolished the Cu-dependent catalytic phosphorylation when holo-Atox1 was the only source of copper.[62] The folding of the intact N-terminus of Cu-ATPases may conceal the copper-binding motif of MBD4, which is exposed to solvent in its isolated state.

The process of copper transfer from holo-Atox1 to the MBD2 of the N-terminal domain of copper-ATPases can be proposed as the first step of metal delivery into the transporter. Metal loading into MBD2 leads to conformational changes in the N-terminal domain of Cu-ATPases,[62] and results in the possible access of copper to the other MBDs, such as 5 and 6, which are closest to the transmembrane segment and are important for the functional activity of ATP7A and ATP7B. Structural changes that occur upon copper binding to MBDs may open an access to the transmembrane transport site (for direct copper delivery) or allow preferential binding of Atox1 at the positively charged surface near the predicted entry site (Figure 18.8). As discussed above, supporting experimental evidence exists for both models,

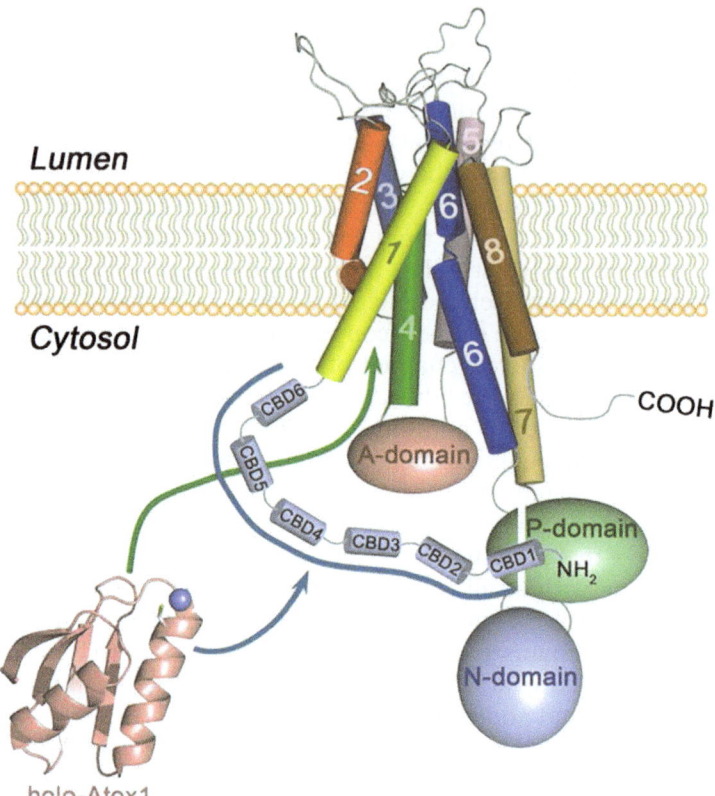

Figure 18.8   Proposed Atox1-dependent copper delivery pathways to human Cu-ATPases. The green arrow indicates the direct docking of holo-Atox1 at the platform near the transmembrane segments (TM1 and TM2) of Cu-ATPases and the transfer of copper into the copper exit site. The blue arrow indicates the indirect delivery of copper from holo-Atox1 to Cu-ATPases, *i.e.* holo-Atox1 transfers copper to the N-terminal metal binding domain (including MBD1, 2, 3, 4, 5 and 6) at first, and then copper is further delivered to the copper exit site in the trans-membrane segment of Cu-ATPases. (Note: N-terminal domains are labelled CBD in the figure to denote copper binding domain.)

yet neither fully explains the communication and functional inter-dependence between the copper-binding sites in the N-terminal domain and the copper-binding sites of the intramembrane region.

The N-terminal domain of ATP7A and ATP7B is likely to be partially metallated under low copper concentrations with only a fraction of MBDs occupied by copper. However, as the copper concentration increases, all MBDs are likely to become fully saturated. Compared to its apo-form, the fully saturated N-terminal domain of ATP7B exhibits a significant con-formational change.[63] This structural change within the regulatory domain has important consequences: it disrupts the interaction of the fully saturated

N-terminal domain with the ATP-binding domain, and speeds up catalytic phosphorylation,[64] which in turn presumably increases ATPase activity and copper transport. Reciprocally, the decrease in the intracellular copper and increased fraction of apo-Atox1 may result in the reverse sequence of events, a reaction that has been demonstrated *in vitro*. Therefore, Atox1 may act not only as a cytosolic chaperone for copper delivery, but also as an essential copper sensor required for both the copper-dependent up-regulation of ATP7A&7B activity and for "resetting" the transporters when the intracellular copper concentration decreases.

The ability of Atox1 to stimulate copper transport by ATP7A and 7B may resemble the activity of another metal-binding protein, calmodulin, which is essential for activation of various calcium-transport systems. Plasma membrane Ca^{2+}-ATPase exists in an auto-inhibited state under basal conditions. Upon calcium binding, calmodulin disrupts the auto-inhibitory interdomain interactions within the Ca^{2+}-ATPase[65] leading to calcium binding at the transport sites and further conformation transitions required for calcium export. An auto-inhibitory interdomain interaction also exists between the N-terminal domain and other functional domains in copper-ATPases. Atox1 may act like calmodulin disrupting this auto-inhibitory interaction (by displacing structurally similar MBDs) and stimulating the copper transport activity when cytosolic copper concentrations are increased.

In addition to its role as a chaperone for the copper export pathway, Atox1 may function as an antioxidant. Recent studies indicate that the genetic ablation of Atox1 sensitizes cells to the fluctuation of glutathione (GSH) levels.[66] When compared to the wild-type mouse embryonic fibroblasts, the $Atox1^{-/-}$ fibroblasts are much more sensitive to buthioninesulfoximine (BSO, GSH synthesis inhibitor). Relatively low concentrations of BSO (42 μM) inhibit growth of fibroblasts lacking Atox1 and trigger cell death. Since thiol-containing biomolecules/metabolites, including GSH, play an important role in maintaining the redox balance of the cytosol, sensitivity of $Atox1^{-/-}$ cells to GSH depletion points to the essential role of Atox1 in buffering cell redox equilibrium. Moreover, the increased sensitivity of $Atox1^{-/-}$ cells to GSH depletion is Cu-independent, suggesting that the antioxidant and copper chaperone functions of Atox1 may not be directly correlated. Detailed characterization of the redox potential of the CxxC site of Atox1 revealed a value (−229 mV) similar to that of the GSH:GSSG pair (−240 mV), indicating that fluctuations in a cellular redox balance may influence the oxidation state of Atox1 and thus alter its copper binding capacity as well as delivery of copper to the secretory pathway.

18.6 Intramembrane Transport and Release

Within the transmembrane portion, the CPC motif is essential for copper transport and is presumed to be directly involved in copper coordination, by analogy with archaeal CopA. In yeast, evidence for distinct roles of these two cysteines was reported.[67] Mutation of both cysteines of the (583)CPC(585) motif in yeast copper ATPase Ccc2 to (583)SPS(585) resulted in a non-functional protein, whereas mutation of individual Cys to Ser variants had different functional consequences. Catalytic phosphorylation by ATP was still observed for the C583S variant, indicative that copper binding to the transport site is not destroyed by this mutation. In contrast, no catalytic activity was detected for the C585S variant. The authors proposed that Cys(583) was necessary for copper dissociation and/or enzyme dephosphorylation and Cys(585) was required for copper binding and catalytic phosphorylation by ATP.

The stoichiometry of copper transport and the number of copper binding sites within the copper translocation pathway of any eukaryotic copper-ATPases are still not known, but assumed to be two by analogy with the bacterial orthologues. In mammalian transporters, rates of copper transport are stimulated at low pH, presumably due to facilitation of copper release from the transport sites.[68] Recent studies of the human ATP7A have also identified the role of the lumenal loop located between TM1 and TM2 in binding of the released copper and further transfer of this copper to the acceptor enzyme.[69,70] The TM1,2 loop is rich in Met and His residues and may accommodate more than one copper atom. Binding to this loop stabilizes Cu(I), although Cu(II) can also be bound.

Given that the entry to the transmembrane portion appears to be formed by cytosolic ends of TM1,2, the copper transport pathway can be visualized as a system with two coupled gates. The opening of the cytosolic gate (driven by binding of ATP in the ATP-binding domain) would allow transfer of copper to the transport site and subsequent occlusion would close this gate, whereas phosphorylation and E1–E2 transition may open the luminal gate allowing copper release and subsequent dephosphorylation of the transporter. Further studies are necessary to test this model and to firmly establish the stoichiometry of copper transport, as well as the mechanism of copper entry, binding and release.

18.7 Superoxide Dismutase and Its Metallochaperone CCS

Superoxide dismutase (SOD1) is a homodimeric protein that binds one Cu and one Zn atom per monomer. In the oxidized form (Cu(II) state), the Zn(II) ion is coordinated by a monodentate carboxylate from Asp and the Nε of the imidazole groups of three His residues (Figure 18.1). One of these His ligands bridges to the Cu(II) ion, which is further coordinated by three terminal His residues. In the reduced form (Cu(I) state) the bond from copper to the bridging histidine is broken, resulting in a three-coordinate Cu(I) centre. The enzyme catalyses the dismutation of superoxide to oxygen and hydrogen peroxide *via* a mechanism that involves both oxidation and reduction of the $O_2^{\bullet-}$ anion (disproportionation):

$$O_2^{\bullet-} + Cu^{2+} \rightarrow O_2 + Cu^+ \tag{18.9}$$

$$O_2^{\bullet-} + Cu^+ + 2H^+ \rightarrow H_2O_2 + Cu^{2+} \qquad (18.10)$$

From these equations it can be seen that the enzyme cycles between cupric and cuprous oxidation states, but is probably metallated in its Cu(I) state. Little is known about how the apo-protein acquires zinc, but this step is believed to occur before addition of copper to form the EZn-SOD1, which is a monomer. The active enzyme requires the formation of a critical disulfide bond between C57 and C146, which helps to pre-form the active site for the binding of Cu, and stabilizes the active dimeric molecule. Activation of apo-SOD1 therefore requires three post-translational modifications, which are believed to follow the order of Zn binding, then Cu(I) incorporation and finally disulfide bond formation.

The activation of SOD1 *in vivo* was first studied in *S. cerevisiae* where it was determined that formation of the mature enzyme was 100% dependent on the presence of an accessory protein LYS7, later identified as the copper chaperone for superoxide dismutase yCCS. Since much of the early work on the SOD1-CCS interaction was conducted in yeast or with yeast recombinant proteins, the absolute dependence on CCS for SOD maturation became the accepted paradigm. More recent studies have shown that in fact *S. cerevisiae* is a rare case of an organism that shows this absolute dependence, and that most organisms including mammals have both CCS-dependent and -independent pathways for activating SOD1.[71] An important conclusion from the comparison of the CCS-dependent and independent pathways is that differing requirements for Cu insertion and disulfide formation may determine which pathway is utilized, and may depend on the reduction potential of the cytoplasmic environment. SOD1 is believed to be one of very few cytoplasmic proteins that has an oxidized disulfide, and the dependence of protein maturation on CCS may relate to the extent to which the thiol-disulfide equilibrium can be driven towards the disulfide state at the different cytoplasmic reduction potentials operating in different organisms.

CCS is a 27 kDa three-domain protein that has been extensively characterized by both biochemical[72,73] and crystallographic[74–76] studies. Domain 1 (D1) (human, residues 1–77, Figure 18.9) has an Atx1-like structure and was predicted to bind a single Cu(I) at the MxCxxC motif. The early studies in yeast suggested that this domain is only functionally important when cellular copper was depleted,[73] although more recent studies in mammalian species report that it is essential.[77] Domain 2 (D2) (residues 78–234) is a SOD1-like polypeptide, which in the human protein retains its Zn binding site and a vestigial Cu site lacking one of the copper-binding histidine residues. Domain 3 (D3) (residues 235–272) contains a CXC motif critical for SOD1 maturation activity.

The reaction chemistry of yCCS with its partner SOD has been shown to proceed *via* the intermediacy of a heterodimeric protein-protein interaction, which was first observed in solution and subsequently characterized by crystallography. When yCCS was allowed to interact with either WT

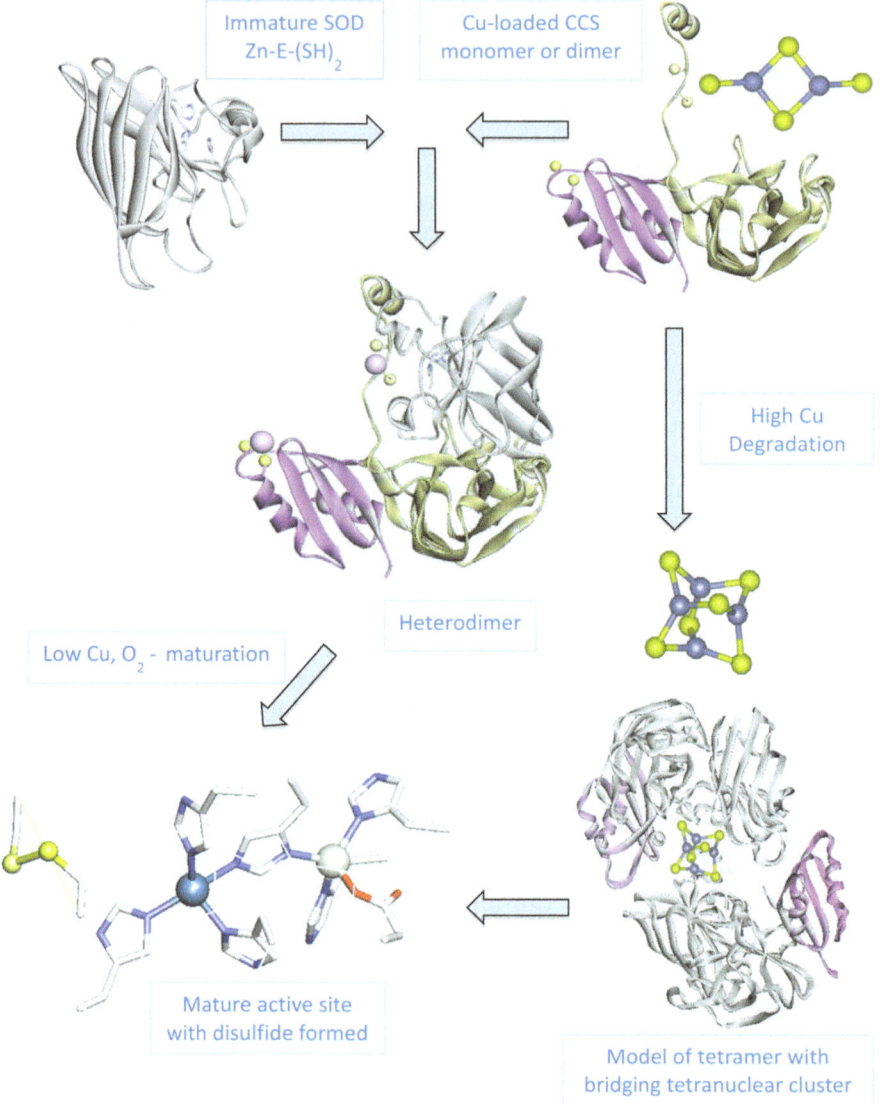

Figure 18.9 Elements of the SOD-CCS molecular interaction, showing the maturation process occurring *via* the intermediacy of the heterodimer.
Possible structures for dinuclear and tetranuclear copper clusters
that may form in CCS at high copper, and may act as signals for
degradation are also shown (right). CCS domains are colour coded
pink (D1) and gold (D2) (except in the model tetrameric structure
where CCS D2 is grey).

ySOD1 or a mutant (H46F) designed to be unable to bind copper,[78,79] a 42
kDa band was detected by gel-filtration and SDS PAGE, which stained with
antibodies to both proteins. A crystal structure of the yCCS-H46F SOD

heterodimer[80] revealed that the heterocomplex contained an intermolecular disulfide between copper ligand C229 of yCCS and C57 of SOD, and suggested a possible role for CCS in forming the SOD disulfide bond. This role in forming the essential disulfide was firmly established in later studies,[81,82] which demonstrated that EZnSOD^{S-H} but not EZnSOD^{S-S} was the substrate for CCS, reacting with the chaperone in a Cu- and O$_2$-dependent manner to generate fully active CuZnSOD^{S-S}. It was proposed that the process was driven at least in part by differences in the relative stability of SOD monomers and dimers: immature EZnSOD^{S-H} was more stable in the monomeric form facilitating docking with the chaperone, whereas CuZnSOD^{S-S} was more stable as the dimer facilitating product release.

A major unresolved question is how copper is transferred and whether the transfer chemistry is linked to the thiol oxidase/disulfide isomerase activity of CCS. Studies on copper binding in the presence of imidazole-, DTT- and glutathione-containing buffers reported copper to protein ratios between 1 and 1.5 for both yeast[18,73,79] and human[83] while more recent reconstitution studies in the presence of the Cu(I) chelator BCA have shown tight binding of 2 Cu(I) per CCS monomer.[84] Reconstitution using [Cu(MeCN)$_4$]PF$_6$ in a non-coordinating buffer resulted in slightly higher copper to protein ratios in the range 2–2.5.[85–87] Given the presence of well-defined Cu(I)-binding motifs such as CxxC in domain 1 and CXC in domain 3, the expectation was that Cu(I) would bind to one or both of these motifs, but despite attempts to metallate the crystalline forms, none of the structures retained copper at either locus. X-ray absorption (XAS) was therefore used to determine the coordination and nuclearity of copper ligation in human CCS[85,86] and showed that Cu(I) forms a monomeric three-coordinate CuS$_2$X (X = exogenous thiol, or solvent) complex at the MTCQSC site in D1 but forms a cluster at the interface of two D3 polypeptides within a hCCS dimer or tetramer. Using advanced spectroscopic analysis tools together with selective substitution of copper-binding cysteines with the Se analogue selenocysteine, the molecular structure at the D3 CXC site was formulated as a Cu$_4$S$_6$ adamantane-type multinuclear cluster.[88,89] When combined with data on the dependence of oligomerization on copper loading, the results suggested a model in which Cu(I) binding converted the apo-dimer with a D2-D2 interface to a new "holo" dimer connected by cluster formation at two D3 CXC motifs. The predominance of a dimer over a tetramer in the cluster-containing species further suggested that the D2 interface was open in the holo dimer and available for sequestering a SOD1 monomer. The XAS studies provided a molecular framework for understanding how copper binding could trigger heterodimer formation, but fell short of providing any mechanistic detail on the reactivity of the cluster in the redox-driven copper/thiol-disulfide exchange between hCCS and its SOD1 partner.

A complication arises from the fact that the dependence of SOD maturation on copper loading of CCS domains relied in these earlier experiments on measuring SOD activity, which requires both copper loading and disulfide formation. In more recent work the determinants of copper transfer were dissected from those of disulfide formation by using a combination of ESI mass spectrometry and NMR.[33] ESI-MS on copper-loaded domain fragments was used to monitor copper transfer and heterocomplex formation, while NMR of mixtures of unlabelled CCS and ^{15}N-labelled SOD detected copper loading, disulfide formation and the oligomerization state of the SOD product. By monitoring the effects of mixtures of Cu(I)-loaded full-length hCCS, its C244AC246A variant, and the domain fragments D1, D1-D2 and D2-D3, a modified paradigm for SOD activation emerged: that CCS D1 was required for copper transfer, D2 was required for heterocomplex formation and D3 appeared to function as the thiol-disulfide isomerase. Competition between CCS D1 and Cu(I)-loaded DTT monitored by ESI-MS led to a value for the K_D of Cu(I) binding to the D1 CxxC motif of 2.3×10^{-18} M, close to that determined for the corresponding domains in Atox1, Atx1 and the ATP7A and ATP7B N-terminal MBDs and significantly weaker than the reported K_D for Cu(I) binding to SOD of 10^{-20} M. Thus, even though the heterodimeric complex with a D2/SOD dimer interface would prevent the close approach of the CCS D1 site with the SOD catalytic cleft, copper could transfer down a relatively strong thermodynamic gradient. Cu(I) binding to D3 in the presence of BCA resulted in a value for the D3-Cu(I) affinity about one order of magnitude weaker than D1.[84] A consensus mechanism for SOD activation by CCS is depicted in Figure 18.9.

These studies illustrate a classic dilemma, namely that biophysical determination of structure and metal coordination of necessity requires high concentrations of both metal and protein, which are often far above the physiological concentrations in the cell. Although the XAS and MS/NMR studies were both carried out using hCCS prepared by incubation with excess Cu(I) as its acetonitrile complex, the role of the multinuclear cluster in copper transfer and/or disulfide bond formation is unclear. The lower affinity of Cu(I) for D3 may signal that under normal physiological copper levels, only D1 is populated, allowing D3 to undergo thiol-disulfide isomerase activity *via* a mixed disulfide intermediate without the involvement of a coordinated metal ion. Higher copper levels on the other hand could lead to formation of the cluster, and the formation of the open cluster-bridged dimer. It is known that hCCS is regulated by copper, *i.e.* that excess copper levels lead to degradation of the chaperone. This process is dependent on the presence of the Cys residues in the D3 CXC motif, which are required for cluster formation.[77] We may speculate that the formation of the cluster at high copper acts as a signal for degradation, perhaps *via* exposure of residues at the D2/D2 dimer interface for proteolysis or ubiquitination. Future studies hopefully will shed light on these unresolved issues.

18.8   Concluding Remarks and Future Directions

The last decade has seen a revolution in our understanding of the genetic and biochemical processes that mediate the transport of copper in eukaryotic cells. It is now possible to track the fate of the metal from entry into the cell, partitioning between essential pathways, and export when cellular levels are replete. Attention will now turn to understanding copper distribution between different cell types and organs, and utilization of this knowledge to combat the growing number of diseases that are believed to be associated with aberrant copper transport. For example, what is the function of the large pool of copper in brain, and is altered copper chemistry in the brain associated with Alzheimer's, Parkinson's or Huntington's diseases?[90] Why do many cancer cells have altered levels of copper?[91] What is the mechanism for the influx of copper into macrophages to combat bacterial infection, and how can the toxicity of copper to microbial invaders be leveraged as a new antibacterial weapon?[91] What is the fate of copper after chelation by drugs such as tetrathiomolybdate (TM) used to treat Wilson's disease?[92] The latter question is of significance due to the long recognized prevalence of copper deficiency in ruminants feeding on soils high in molybdenum. Indeed a recent crystal structure of the complex between Atox1 and TM shows the formation of a stable multinuclear copper-molybdenum cluster, which may sequester copper into an unavailable form.[93] Approaches to solving these and other questions relating to cellular copper speciation and localization will come in part from advances in technologies that can determine the molecular identity and localization within eukaryotic cells at subcellular resolution. These include mapping the subcellular distribution of copper using novel Cu(I)-specific fluorescent imaging reagents[94] or X-ray fluorescence microprobes,[95] and emerging techniques that allow the visualization of single molecules of fluorescently labelled transporters as they interact with their partners in real time.[96] While these techniques are beyond the scope of the present chapter, they signal an exciting new era for copper transport chemistry that is likely to have significant biomedical impact in the years ahead.

References

1. H. S. Carr and D. R. Winge, *Acc. Chem. Res.*, 2003, **36**, 309–316.
2. S. T. Prigge, R. E. Mains, B. A. Eipper and L. M. Amzel, *Cell. Mol. Life Sci.*, 2000, **57**, 1236–1259.
3. J. P. Klinman, *J. Biol. Chem.*, 2006, **281**, 3013–3016.
4. B. J. Brazeau, B. J. Johnson and C. M. Wilmot, *Arch. Biochem. Biophys.*, 2004, **428**, 22–31.
5. J. A. Tainer, E. D. Getzoff, J. S. Richardson and D. C. Richardson, *Nature*, 1983, **306**, 284–287.
6. A. C. Rosenzweig and M. H. Sazinsky, *Curr. Opin. Struct. Biol.*, 2006, **16**, 729–735.
7. D. J. Kosman, *J. Biol. Inorg. Chem.*, 2010, **15**, 15–28.
8. P. J. Sargent, S. Farnaud and R. W. Evans, *Curr. Med. Chem.*, 2005, **12**, 2683–2693.

9. L. Banci, I. Bertini, F. Cantini and S. Ciofi-Baffoni, *Cell. Mol. Life Sci*, 2010, **67**, 2563–2589.

10. S. Lutsenko, *Curr. Opin. Chem. Biol.*, 2010, **14**, 211–217.

11. T. V. O'Halloran and V. C. Culotta, *J. Biol. Chem.*, 2000, **275**, 25057–25060.

12. A. C. Rosenzweig, *Acc. Chem. Res.*, 2001, **34**, 119–128.

13. E. B. Maryon, S. A. Molloy and J. H. Kaplan, *Am. J. Physiol. Cell. Physiol.*, 2013, **304**, C768–779.

14. M. C. Carroll, J. B. Girouard, J. L. Ulloa, J. R. Subramaniam, P. C. Wong, J. S. Valentine and V. C. Culotta, *Proc. Natl Acad. Sci. USA*, 2004, **101**, 5964–5969.

15. Y. C. Horng, P. A. Cobine, A. B. Maxfield, H. S. Carr and D. R. Winge, *J. Biol. Chem.*, 2004, **279**, 35334–35340.

16. J. H. Kaplan and S. Lutsenko, *J. Biol. Chem.*, 2009, **284**, 25461–25465.

17. L. Banci, I. Bertini, S. Ciofi-Baffoni, T. Kozyreva, K. Zovo and P. Palumaa, *Nature*, 2010, **465**, 645–648.

18. P. J. Rae, P. J. Schmidt, R. A. Pufahl, V. C. Culotta and T. V. O'Halloran, *Science*, 1999, **284**, 805–808.

19. Z. Xiao, J. Brose, S. Schimo, S. M. Ackland, S. La Fontaine and A. G. Wedd, *J. Biol. Chem.*, 2011, **286**, 11047–11055.

20. P. Kuzmic, *Anal. Biochem.*, 1996, **237**, 260–273.

21. C. R. Pope, A. G. Flores, J. H. Kaplan and V. M. Unger, *Curr. Top. Membr.*, 2012, **69**, 97–112.

22. J. F. Eisses, Y. Chi and J. H. Kaplan, *J. Biol. Chem.*, 2005, **280**, 9635–9639.

23. R. Hassett and D. J. Kosman, *J. Biol. Chem.*, 1995, **270**, 128–134.

24. J. F. Eisses and J. H. Kaplan, *J. Biol. Chem.*, 2002, 277, 29162–29171.

25. J. Lee, M. M. Pena, Y. Nose and D. J. Thiele, *J. Biol. Chem.*, 2002, **277**, 4380–4387.

26. S. Puig, J. Lee, M. Lau and D. J. Thiele, *J. Biol. Chem.*, 2002, **277**, 26021–26030.

27. J. F. Eisses and J. H. Kaplan, *J. Biol. Chem.*, 2005, **280**, 37159–37168.

28. E. B. Maryon, S. A. Molloy, A. M. Zimnicka and J. H. Kaplan, *Biometals*, 2007, **20**, 355–364.

29. S. G. Aller and V. M. Unger, *Proc. Natl Acad. Sci. USA*, 2006, **103**, 3627–3632.

30. C. J. De Feo, S. G. Aller, G. S. Siluvai, N. J. Blackburn and V. M. Unger, *Proc. Natl Acad. Sci. USA*, 2009, **106**, 4237–4242.

31. C. J. De Feo, S. G. Aller and V. M. Unger, *Biometals*, 2007, **20**, 705–716.

32. J. Jiang, I. A. Nadas, M. A. Kim and K. J. Franz, *Inorg. Chem.*, 2005, **44**, 9787–9794.

33. L. Banci, I. Bertini, F. Cantini, T. Kozyreva, C. Massagni, P. Palumaa, J. T. Rubino and K. Zovo, *Proc. Natl Acad. Sci. USA*, 2012, **109**, 13555–13560.

34. I. F. Tsigelny, Y. Sharikov, J. P. Greenberg, M. A. Miller, V. L. Kouznetsova, C. A. Larson and S. B. Howell, *Cell Biochem. Biophys.*, 2012, **63**, 223–234.

35. Y. Guo, K. Smith, J. Lee, D. J. Thiele and M. J. Petris, *J. Biol. Chem.*, 2004, **279**, 17428–17433.
36. Z. Xiao, F. Loughlin, G. N. George, G. J. Howlett and A. G. Wedd, *J. Am. Chem. Soc.*, 2004, **126**, 3081–3090.
37. Z. Xiao and A. G. Wedd, *Chem. Commun.*, 2002, 588–589.
38. X. Wu, D. Sinani, H. Kim and J. Lee, *J. Biol. Chem.*, 2009, **284**, 4112–4122.
39. T. C. Steveson, G. D. Ciccotosto, X.-M. Ma, G. P. Mueller, R. E. Mains and B. A. Eipper, *Endocrinology*, 2003, **144**, 188–200.
40. M. J. Petris, D. Strausak and J. F. Mercer, *Hum. Mol. Genet.*, 2000, **9**, 2845–2851.
41. S. R. Setty, D. Tenza, E. V. Sviderskaya, D. C. Bennett, G. Raposo and M. S. Marks, *Nature*, 2008, **454**, 1142–1146.
42. Z. Qin, S. Itoh, V. Jeney, M. Ushio-Fukai and T. Fukai, *FASEB J.*, 2006, **20**, 334–336.
43. S. Lutsenko, N. L. Barnes, M. Y. Bartee and O. Y. Dmitriev, *Physiol. Rev.*, 2007, **87**, 1011–1046.
44. M. Schushan, A. Bhattacharjee, N. Ben-Tal and S. Lutsenko, *Metallomics*, 2012, **4**, 669–678.
45. P. Gourdon, X. Y. Liu, T. Skjorringe, J. P. Morth, L. B. Moller, B. P. Pedersen and P. Nissen, *Nature*, 2011, **475**, 59–64.
46. S. Chintalapati, R. Al Kurdi, A. C. van Scheltinga and W. Kuhlbrandt, *J. Mol. Biol.*, 2008, **378**, 581–595.
47. C. C. Wu, W. J. Rice and D. L. Stokes, *Structure*, 2008, **16**, 976–985.
48. S. Lutsenko, E. S. LeShane and U. Shinde, *Arch. Biochem. Biophys.*, 2007, **463**, 134–148.
49. M. Y. Bartee and S. Lutsenko, *Biometals*, 2007, **20**, 627–637.
50. A. K. Boal and A. C. Rosenzweig, *Chem. Rev.*, 2009, **109**, 4760–4779.
51. A. K. Wernimont, D. L. Huffman, A. L. Lamb, T. V. O'Halloran and A. C. Rosenzweig, *Nat. Struct. Biol.*, 2000, 7, 766–771.
52. F. Arnesano, L. Banci, I. Bertini, F. Cantini, S. Ciofi-Baffoni, D. L. Huffman and T. V. O'Halloran, *J. Biol. Chem.*, 2001, **276**, 41365–41376.
53. L. Banci, I. Bertini, S. Ciofi-Baffoni, D. L. Huffman and T. V. O'Halloran, *J. Biol. Chem.*, 2001, **276**, 8415–8426.
54. L. Banci, I. Bertini, F. Cantini, I. C. Felli, L. Gonnelli, N. Hadjiliadis, R. Pierattelli, A. Rosato and P. Voulgaris, *Nat. Chem. Biol.*, 2006, 2, 367–368.
55. L. Banci, I. Bertini, V. Calderone, N. Della-Malva, I. C. Felli, S. Neri, A. Pavelkova and A. Rosato, *Biochem. J.*, 2009, **422**, 37–42.
56. M. Ralle, S. Lutsenko and N. J. Blackburn, *J. Biol. Chem.*, 2003, **278**, 23163–23170.
57. M. Gonzalez-Guerrero and J. M. Arguello, *Proc. Natl Acad. Sci. USA*, 2008, **105**, 5992–5997.
58. A. C. Rosenzweig and J. M. Arguello, *Curr. Top. Membr.*, 2012, **69**, 113–136.
59. T. Padilla-Benavides, C. J. McCann and J. M. Arguello, *J. Biol. Chem.*, 2013, **288**, 69–78.

60. D. Achila, L. Banci, I. Bertini, J. Bunce, S. Ciofi-Baffoni and D. L. Huffman, *Proc. Natl Acad. Sci. USA*, 2006, **103**, 5729–5734.
61. J. M. Walker, D. Huster, M. Ralle, C. T. Morgan, N. J. Blackburn and S. Lutsenko, *J. Biol. Chem.*, 2004, **279**, 15376–15384.
62. J. M. Walker, R. Tsivkovskii and S. Lutsenko, *J. Biol. Chem.*, 2002, **277**, 27953–27959.
63. B. Sarkar, *J. Inorg. Biochem.*, 2000, **79**, 187–191.
64. D. Huster and S. Lutsenko, *J. Biol. Chem.*, 2003, **278**, 32212–32218.
65. E. Carafoli, E. Garcia-Martin and D. Guerini, *Experientia*, 1996, **52**, 1091–1100.
66. Y. Hatori, S. Clasen, N. M. Hasan, A. N. Barry and S. Lutsenko, *J. Biol. Chem.*, 2012, **287**, 26678–26687.
67. J. Lowe, A. Vieyra, P. Catty, F. Guillain, E. Mintz and M. Cuillel, *J. Biol. Chem.*, 2004, **279**, 25986–25994.
68. R. Safaei, S. Otani, B. J. Larson, M. L. Rasmussen and S. B. Howell, *Mol. Pharmacol.*, 2008, **73**, 461–468.
69. A. N. Barry, A. Otoikhian, S. Bhatt, U. Shinde, R. Tsivkovskii, N. J. Blackburn and S. Lutsenko, *J. Biol. Chem.*, 2011, **286**, 26585–26594.
70. A. Otoikhian, A. N. Barry, M. Mayfield, M. Nilges, Y. Huang, S. Lutsenko and N. J. Blackburn, *J. Am. Chem. Soc.*, 2012, **134**, 10458–10468.
71. J. M. Leitch, P. J. Yick and V. C. Culotta, *J. Biol. Chem.*, 2009, **284**, 24679–24683.
72. P. J. Schmidt, C. Kunst and V. C. Culotta, *J. Biol. Chem.*, 2000, **275**, 33771–33776.
73. P. J. Schmidt, T. D. Rae, R. A. Pufahl, T. Hamma, J. Strain, T. V. O'Halloran and V. C. Culotta, *J. Biol. Chem.*, 1999, **274**, 23719–23725.
74. L. T. Hall, R. L. Sanchez, S. P. Holloway, H. Zhu, J. E. Stine, T. J. Lyons, B. Demeler, V. Schirf, J. C. Hansen, A. M. Nersissian, J. S. Valentine and P. J. Hart, *Biochemistry*, 2000, **39**, 3611–3623.
75. A. L. Lamb, A. K. Wernimont, R. A. Pufahl, V. C. Culotta, T. V. O'Halloran and A. C. Rosenzweig, *Nat. Struct. Biol.*, 1999, **6**, 724–729.
76. A. L. Lamb, A. K. Wernimont, R. A. Pufahl, T. V. O'Halloran and A. C. Rosenzweig, *Biochemistry*, 2000, **39**, 1589–1595.
77. A. L. Caruano-Yzermans, T. B. Bartnikas and J. D. Gitlin, *J. Biol. Chem.*, 2006, **281**, 13581–13587.
78. A. S. Torres, V. Petri, T. D. Rae and T. V. O'Halloran, *J. Biol. Chem.*, 2001, **276**, 38410–38416.
79. A. L. Lamb, A. S. Torres, T. V. O'Halloran and A. C. Rosenzweig, *Biochemistry*, 2000, **39**, 14720–14727.
80. A. L. Lamb, A. S. Torres, T. V. O'Halloran and A. C. Rosenzweig, *Nat. Struct. Biol.*, 2001, **8**, 751–755.
81. Y. Furukawa, A. S. Torres and T. V. O'Halloran, *EMBO J.*, 2004, **23**, 2872–2881.
82. N. M. Brown, A. S. Torres, P. E. Doan and T. V. O'Halloran, *Proc. Natl Acad. Sci. USA*, 2004, **101**, 5518–5523.

83. T. D. Rae, A. S. Torres, R. A. Pufahl and T. V. O'Halloran, *J. Biol. Chem.*, 2001, **276**, 5166–5176.
84. S. Allen, A. Badarau and C. Dennison, *Biochemistry*, 2012, **51**, 1439–1448.
85. J. F. Eisses, J. P. Stasser, M. Ralle, J. Kaplan and N. J. Blackburn, *Biochemistry*, 2000, **39**, 7337–7342.
86. J. P. Stasser, J. F. Eisses, A. N. Barry, J. H. Kaplan and N. J. Blackburn, *Biochemistry*, 2005, **44**, 3143–3152.
87. J. Stasser, G. S. Siluvai, A. N. Barry and N. J. Blackburn, *Biochemistry*, 2007, **46**, 11845–11856.
88. A. N. Barry and N. J. Blackburn, *Biochemistry*, 2008, **49**, 4916–4928.
89. A. N. Barry, K. M. Clark, A. Otoikhian, W. A. van der Donk and N. J. Blackburn, *Biochemistry*, 2008, **47**, 13074–13083.
90. K. M. Davies, D. J. Hare, V. Cottam, N. Chen, L. Hilgers, G. Halliday, J. F. Mercer and K. L. Double, *Metallomics*, 2013, **5**, 43–51.
91. B. B. Cheung and G. M. Marshall, *Curr. Cancer Drug Targets*, 2011, **11**, 826–836.
92. L. Zhang, J. Lichtmannegger, K. H. Summer, S. Webb, I. J. Pickering and G. N. George, *Biochemistry*, 2009, **48**, 891–897.
93. H. M. Alvarez, Y. Xue, C. D. Robinson, M. A. Canalizo-Hernandez, R. G. Marvin, R. A. Kelly, A. Mondragon, J. E. Penner-Hahn and T. V. O'Halloran, *Science*, 2010, **327**, 331–334.
94. T. Hirayama, G. C. Van de Bittner, L. W. Gray, S. Lutsenko and C. J. Chang, *Proc. Natl Acad. Sci. USA*, 2012, **109**, 2228–2233.
95. S. Vogt and M. Ralle, *Anal. Bioanal. Chem.*, 2013, **405**, 1809–1820.
96. P. Chen, N. M. Andoy, J. J. Benitez, A. M. Keller, D. Panda and F. Gao, *Nat. Prod. Rep.*, 2010, **27**, 757–767.
97. G. S. Siluvai, M. Nakano, M. Mayfield and N. J. Blackburn, *J. Biol. Inorg. Chem.*, 2011, **16**, 285–287.
98. I. R. Loftin, N. J. Blackburn and M. M. McEvoy, *J. Biol. Inorg. Chem.*, 2009, **14**, 905–912.

CHAPTER 19

Silver

NIC. R. BURY

King's College London, Division of Diabetes and Nutritional Sciences, Franklin Wilkins Building, 150 Stamford Street, London, SE1 9NH, UK
Email: nic.bury@kcl.ac.uk

19.1 Introduction

Silver (Ag) is a member of Group 11 of the periodic table and of the second transition series, being situated below copper and above gold and roentgenium. It has an atomic mass of 107.87 and is the 67th most abundant element in the Earth's crust.[1] Silver can be found in a number of oxidation states but is more often encountered as elemental Ag(0) or monovalent Ag(I). Unlike copper (Cu), Ag is not known to support enzymatic redox functions and has no confirmed biological use; thus, it is classified as a non-essential metal.

Silver has been mined for millennia being valued for its use in jewellery and, more recently, the photo-sensitivity of Ag halides saw it used in the development of Polaroid photographs. However, the advent of digital cameras has seen a decline in the use of Ag in the photographic industry. Its antibacterial properties have been known for centuries.[2] In antiquity, it was often added to water vessels to ensure a reliable supply of drinking water during military conflicts and applied to dressings to prevent infection.[3] The phrase "born with a silver spoon in his/her mouth" first appeared in print in English as early as 1719, in Peter Anthony Motteux's translation of the novel *Don Quixote* (quoted in ref. 3). It is often used to refer to someone who is fortunate to be from a family with high status in society. It may reflect the ability of this socio-economic class to purchase silver cutlery but another

RSC Metallobiology Series No. 2
Binding, Transport and Storage of Metal Ions in Biological Cells
Edited by Wolfgang Maret and Anthony Wedd
© The Royal Society of Chemistry 2014
Published by the Royal Society of Chemistry, www.rsc.org

positive effect of using these utensils would have been to reduce exposure during infancy to infectious disease that was prevalent in the eighteenth century. It was only the advent of modern antibiotics that saw a decline in the use of Ag as an anti-microbial agent in medicine, but it is precisely this property that has seen an exponential increase in the production of colloidal Ag nanoparticles (Ag-NP) in recent years. The uses of Ag-NP are widespread and include water purification, washing powders, deodorants, coatings of food containers, catheters and as a component of wound dressings. The majority of colloidal Ag is relatively insoluble in aqueous environments but, due to the large surface area of these nanoparticles, there is significant dissociation of ionic Ag^+. This ion, which binds water weakly, is extremely toxic to bacteria and aquatic organisms[1,2,4] and, potentially, to mammalian cells.[5]

This chapter will provide a brief overview of the uptake process of Ag(I) across epithelia and its intracellular metabolism. Its uptake is linked inextricably to Cu(I) and thus the review provides an overview of characteristics of the copper-transporting proteins and chaperones, providing examples of where Ag(I) has been used as a Cu(I) surrogate in these studies.[†] The chapter will briefly assess Ag toxicity and its mode-of-action in bacteria, mammalian cells and aquatic organisms and will show how these data, along with geochemical speciation models, have been used to assist environmental regulators develop predictive tools for acute Ag toxicity. More extensive reviews on Ag toxicity, especially of Ag-NP, have been published.[5–8] Finally, the chapter will present an overview of the mechanisms of Ag resistance in bacteria and its significance in medicine.

19.2 Silver Uptake across Epithelial Membranes and Release into the Blood

19.2.1 Silver(I) Uptake

With no known specific uptake process for non-essential metals, these compounds enter cells by mimicking essential metals in their uptake pathways[9] or by binding to other nutrients (*e.g.* amino acids) and so being taken up by their pathways. The transport across membranes into or out of cells often occurs against large electrochemical gradients. It can occur *via* an array of different transport proteins belonging to the RND, ABC, CDF and P-type ATPase super-families, as well as *via* other metal-specific membrane channels or pores. Integral to this process are intracellular metallochaperones that ensure metal ions, once they have entered the cell, are trafficked to specific intracellular sites. In the case of silver, passive diffusion of circumneutral complexes (such as $AgCl_{aq}$) has also been implicated in uptake in fish.[10,11]

[†]Chapters 16 and 17 provide extended discussion of the Cu(I) uptake pathways.

Experimental evidence of uptake, intracellular transport, export from cells and systemic handling of Ag is linked to the control of cellular and systemic Cu homeostasis. Ag^+ is isoelectronic with the cuprous ion Cu^+, and has the same d-shell electron configuration but a slightly larger ionic radius. Both Cu^+ and Ag^+ are soft acids and thus preferentially bind to amino acids with side-chains that provide soft ligands, such as the heterocyclic nitrogen of His and the sulfur donors of Cys and Met. Cu(I) and Ag(I) appear to form coordination complexes of similar structures but with Ag^+ preferring a digonal two-coordinate geometry.[12]

The strong link between Cu and Ag metabolism suggests that intestinal cellular entry of Ag is *via* the high-affinity copper transporter CTR1, SLC31A1.[13–15] This protein has been localized to the apical membrane of intestinal cells and is involved in uptake of Cu(I) and Ag(I) from the diet.[13] There is also evidence that it is present on the plasma membrane of other cells and involved in cellular uptake of both metals from the circulation.[14,16] CTR1 pumps are homotrimers with three transmembrane domains (TMD) per monomer. The structure forms a pore across the membrane.[17] Though the precise details of transfer have not been elucidated, the MXXM motif of TM2 is conserved within CTR1s,[18,19] and may play an integral part in channelling Cu(I) for transfer across the membrane.[17] In the less reducing environment of the extracellular matrix, MXXM motifs appear to provide trigonal $Cu^I(S\text{-Met})_3$ sites for Cu(I) and Ag (I). Compared to cysteine thiolates, Met has a longer side-chain, has no ionizable groups and is less polar. So it may provide a mechanism of ion selectivity *via* geometrical preference and enough flexibility to allow for exchange of Cu(I) and Ag(I) from site to site as it passes through the membrane.[20] Indeed, Met residues in the model peptide $MX_{1\text{-}2}MX_{1\text{-}2}M$ specifically bind Ag(I) and Cu(I) rather than divalent metal ions, supporting the hypothesis that the arrays of these three methionines of CTR1 determine metal specificity.[18]

The C-terminal HCH motifs of trimeric CTR1 avidly bind Cu(I) and Ag(I) and facilitate the transfer of the metal to the chaperones. It is the reducing environment of the cytosol (-190 to -250 mV) and the high concentrations of glutathione that prevent disulfide bridge formation between adjacent Cys residues. This would favour Cu(I)/Ag(I) trigonal or tetrahedral coordination with Cys/His ligands.[18] This binding motif, along with the other cytosolic metal-binding proteins maintains an exceptionally low internal concentration of free Cu^+ ions and, presumably, of free Ag^+ ions, although this has not been confirmed for silver.[21] CTR1 appears to make use of this concentration gradient to allow Cu(I) and Ag(I) to pass from its external to internal compartments without the need for ATP.[17] Recombinant hCTR1 transports Ag at low μM concentrations.[15] However, Ctr1-deficient mouse embryonic fibroblasts (Ctr1 $-/-$) still accumulate a significant amount of silver when exposed to the metal ion,[15] suggesting an alternative uptake pathway.

Pharmaceutical inhibitor and monovalent cation competition studies in freshwater fish suggest that Ag(I) may enter the cells of the gills *via* a Na^+-dependent pathway.[22] Epithelial sodium channels (ENaC) appear to be involved, as are V-type ATPases, which are proton pumps that excrete H^+ to generate the electrochemical gradient necessary for Na^+ passage through the ENaC. Their respective inhibitors, phenamil and bafilomycin A1, both reduce uptake of Na^+ and Ag^+ by 50%.[22] The pathway that accounts for the other 50% has not been identified but has been hypothesized to involve a copper uptake mechanism similar to that of CTR1.[23] The overall situation is uncertain as the difference in size between Na^+ and Ag^+ ions and the known selectivity of vertebrate ENaCs[24,25] seem to suggest that Ag^+ is unlikely to be entering *via* this channel. Indeed, no ENaC proteins have been identified in the genomes of fish.[26]

In an attempt at clarification, an oocyte system overexpressing the ENaC was examined.[27] The study looked for inward induced current generated by Ag^+, but was unable to distinguish between currents generated by Ag^+ passage through the channel and the currents induced by Ag^+ in the native oocyte. The precise channels that Ag^+ stimulates to induce these currents are difficult to decipher but both mechano-sensitive and non-selective cation channels were implicated.[27] The reversible nature of the current and its reduction in the presence of dithiothreitol, a compound that reduces disulfide bridges, suggest it is due to interaction of Ag(I) with external sulfur-containing moieties. Similarly, Ag^+-induced inward short circuit currents have also been observed in frog skin preparations,[28,29] and generated by both Na^+ and K^+, with Ag(I) also crossing this epithelia.

P-type ATPase metal transporters are named because they form a phosphoenzyme intermediate during transfer of the gamma-phosphate from ATP to a highly conserved DKTGT motif.[30] The P-type ATPases that catalyse the translocation of metals also possess a Cys-Pro-Xaa (where Xaa = Ser, Cys or His) motif and are referred to as P_{1B}ATPases.[31] Most of these transporters are involved in export from the cell of monovalent metal (Cu^+ or Ag^+) or divalent metal (Zn^{2+} or Cd^{2+}) ions and are integral in cellular metal detoxification in prokaryotes. In humans, the Cu^+-ATPases ATP7a and ATP7b are linked to two diseases characterized by copper deficiency (Menkes disease) and copper accumulation in the liver (Wilson's disease).[32,33] ATP7a is expressed predominantly in the intestine and ATP7b in the liver, but both are found in other tissues, and located to the *trans*-Golgi network (TGN) where they transport Cu(I) into the TGN for insertion into cuproenzymes. A rise in intracellular copper causes the trafficking of ATP7b from the TGN into vesicles that fuse with the apical membrane of liver to facilitate the excretion of Cu(I) (Figure 19.1).[34] Similarly, ATP7a traffics to the enterocyte basolateral membranes for passage of Cu(I) into circulation.[32,35]

Evidence for monovalent Ag(I) ions being transported by Cu^+-ATPases is available for both prokaryotes and eukaryotes.[36–38] The transport of Ag(I) *via* a Cu^+-ATPase is dependent on delivery of the ion to the metal binding site of the protein. Ag(I) is able to bind to the human Cu(I) metallochaperone Atox1 (or Hah1) and its yeast homologue Atx1.[39] Its delivery to the transporter is determined by the complementary geometries of the interacting surfaces of the metallochaperone and the metal-accepting sites of the pump. In the

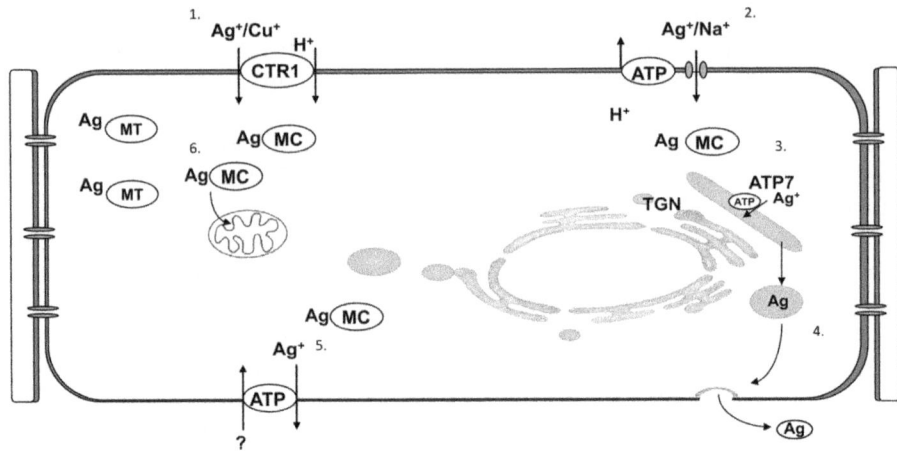

Figure 19.1   Silver cellular uptake routes and metabolism in vertebrates. Silver enters cells *via* copper transporter 1 (CTR1) (1) or a Na^+ uptake pathway in freshwater fish gills (2). On entry Ag(I) is bound to the metallochaperones (MC) Atox1 (or equivalent homologue in other species) where it is trafficked to Cu-ATPases ATP7a or b located on the *Trans*-Golgi Network (TGN) membrane for internalization (3). TGN vesicles laden with Ag will either pass to membrane for excretion of silver (4) into circulation if taken up from the food, or into the bile for excretion from the liver or back into the airway of the lungs if inhaled and taken up by bronchial cells. Alternatively, there is evidence for a basolateral membrane P-type ATPase capable of transporting Ag(I) in fish gills (5). Ag(I) may also bind to the copper chaperone for superoxide dismutase (SOD), CCS, and replace Cu in the Cu, Zn SOD and is also known to displace Zn from metallothionein (MT). Ag has been observed to accumulate in mitochondria, but the transport process is unknown (6).

bacterium *Legionella pneumophila*, the docking and transfer of Cu(I), and presumably Ag(I), from metallochaperone to the pump CopA is achieved by favourable electrostatic forces, the negatively charged chaperone docking with the positively charged pump.[40] The process is metal-dependent as the apo-chaperone does not interact. This imposes a unidirectional transfer of metal ion from metallochaperone to transporter.[41] The apparent site of chaperone docking allows for the interaction of metal ions with conserved Met, Asp and Glu residues that may form a trigonal binding site that is the initial step in transfer across the membrane.[42] Interestingly, in the absence of metallochaperone, the metal binding sites of the extremophile *Archaeoglobus fulgidis* Cu^+-ATPase show selectivity for monovalent metal ions Cu^+ and Ag^+ over divalent Cu^{2+}, Zn^{2+} or Ni^{2+}.[43]

The binding of metal ion induces conformational changes to the enzyme that enable the catalytic hydrolysis of ATP and subsequent metal transport *via* different configurations E1 and E2 of the P-type ATPase. According to the Albers–Post model, phosphorylation of the E1 pump results in the metal integration into the transmembrane region with a conformational change to E2 that results in the release of Ag(I) with subsequent dephosphorylation and return of the enzyme to the E1 conformation. The cytoplasmic N-terminal metal binding sites of the Cu^+-ATPases consist of numerous CXXC repeats. Despite evidence that these sites bind Cu(I) with high affinity, they are not involved directly in transport of Cu(I) or Ag(I) and are thought to play a role in regulating activity.[35] The turnover rates for metal P-type ATPases are slow (<10 ions s^{-1}).[31]

In fish, gill basolateral membrane vesicles (BLMV) accumulate silver in a process that follows saturable kinetics and is inhibited by orthovanadate.[44] In addition, in the presence of Ag(I) and ATP, the BLMV form acylphosphate intermediates. These properties are consistent with the involvement of a P-type ATPase in vesicle uptake.[44] The candidates for Ag^+ transport in the fish gill are ATP7a, b, which have been cloned and identified in a number of fish species.[19] However, whether these pumps from fish are capable of transporting Ag(I) has not been confirmed and other P-type ATPases have been implicated.[44] The BLMV preparation excludes Golgi network membranes and thus the unusual aspect of this fish branchial Ag(I) transport system is that it demonstrates an active ATPase at the basolateral membrane capable of pumping this ion from the cytoplasm into the blood. This property differs markedly from the mammalian model of trafficking of TGN vesicles laden with copper or silver ions for extrusion from the cell.

Rainbow trout respond to sublethal exposure to silver by increasing BLMV Ag(I) transport capacity (V_{max}) compared to controls.[45] This is hypothesized to be part of the defence mechanism against Ag exposure whereby Ag is exported from the gills into the circulation for detoxification in other tissues (*e.g.* liver). Transport of a toxic metal to the internal organs appears to be counterintuitive, but the gills are vital organs for ion homeostasis in freshwater fish. They actively take up major cations (Na^+ and Ca^{2+}) to compensate for ion loss against a large concentration gradient.[46] On exposure to metals, fish have to protect the gills against metal toxicity or they pose a risk from death due to perturbed ion homeostasis.[47] Maintenance of low intracellular Ag concentrations may result from a reduction in apical uptake, from intracellular chelation or from export. The export of Ag across the basolateral membrane away from the Na^+/K^+ ATPase, the site of toxicity, appears to be one of these protective mechanisms.[45] Ag detoxification in mouse ciliated lung epithelial cells has been linked to the Cu(I) P-type ATPase.[38] In this model, in contrast to the fish, Ag(I) did not induce the translocation of ATP7a to the basolateral membrane for Ag(I) export into the circulation; ATP7b was trafficked instead to the apical membrane and Ag(I) transported back into the airway.

An interesting observation in rainbow trout is that increasing chloride concentrations in water do not affect silver accumulation, but ameliorate Ag(I) toxicity in these salmonid fish.[48,49] As chloride concentrations in water increase, circum-neutral $AgCl_{aq}$ forms, which is then followed by other $AgCl_{(n-)}$ species and finally insoluble ceragyrite (see Figure 1 of ref. 11). The protection by chloride in trout occurs at concentrations where $AgCl_{aq}$ is formed, but $AgCl_0$ is bioavailable to these fish.[11] Chloride may be protective because sodium ion uptake, one target for Ag(I) toxicity, may occur only in one subset of cells in the trout gill (the mitochondria-rich cells that constitute 10% of the cell number[46]). On the other hand, $AgCl_{aq}$ exhibits a higher log K_{ow} (octanol/water partition coefficient) value compared to Ag^+ (0.09 *versus* 0.03[50]) and indiscriminately enters all cells. This effectively dilutes the concentration of Ag^+ at the site of toxicity.[11] However, it is difficult to understand how $AgCl_{aq}$ protects because this ion pair exhibits a relatively low conditional equilibrium binding constant (log K 3.3–5.0) compared to the biological ligand (log K 10).[51] Consequently, whether entry into a cell is in the form of $AgCl_{aq}$ or Ag^+, the ion is expected to bind to intracellular sulfur sites.

19.2.2 Silver Nanoparticle Uptake

One mechanism of silver uptake from Ag-NP may follow the Ag^+ entry route, as described above, following the dissociation of Ag^+ from the nanoparticles. However, uptake of the latter has been observed directly in aquatic species, such as *Daphnia magna* and carp (*Cyprinus carpio*). Such uptake has also been detected in cell lines derived from human intestinal epithelial cells.[52] In addition, Ag-NP uptake is reduced by inhibitors of different endocytotic pathways in *in vitro* studies.[53–57] For, example, in macrophages Ag-NP internalization is inhibited by compounds that affect phagocytosis and macropinocytosis and, more specifically, by dextran sulfate, indicating the involvement of scavenger receptor-mediated phagocytosis.[55,56] Nystatin and genistein are inhibitors, respectively of caveolin- and clathrin-dependent endocytosis but had little effect in these studies. Similarly, nystatin had little effect on Ag-NP uptake in human mesenchymal stem cells, but the uptake is inhibited by both chloropromazine (an alternative inhibitor of clathrin-dependent endocytosis) and wortmannin (an inhibitor of macropinocytosis).[57] Reduced Ag-NP uptake, measured as cellular Ag content, was observed with chloroquine (an inhibitor of macropinocytosis) but not with wortmannin.[53] Interestingly, amiloride also reduced uptake of silver. While this reagent inhibits Na^+/H^+ exchangers (NHE) that are involved in the process of macropinocytosis,[58] it also inhibits ENaCs function. Whether dissociation of Ag^+ from Ag-NP prior to uptake *via* a Na^+-uptake pathway, similar to that in fish,[22] is a component has not been evaluated.

19.3 Silver Transport in the Blood

Once Ag(I) has entered the bloodstream, it circulates bound to alpha 1-macroglobulin[59] and ceruloplasmin (CP).[60] CP is a multicopper oxidase, and incorporation of Ag into CP causes misfolding, loss of enzyme activity and an inability to bind Cu.[60,61] Treatment of rats or mice with a diet high in Ag(I) (50 mg AgCl per kg body weight) reduces serum Cu concentrations, as well as CP oxidase and ferrous oxidase activity, presumably also altering plasma transferrin bound iron concentration (though not measured).[62] The reduction of serum *holo*-ceruloplasmin concentrations did not affect cellular copper transporter expression and cuproenzyme activity. Ag(I) accumulated in liver, lung and adrenal glands in these rodents and was excreted into the bile. This suggests that Ag(I), once in the body, follows a Cu excretory pathway *via* the liver and bile (Figure 19.1).

19.4 Transport and Binding of Silver within the Cell

Ctr1 acts as a Cu(I) and Ag(I) permease to move the metal from the extra-cellular compartment into the cell. Once Ag(I) or Cu(I) enters the cells, the Cu(I) chaperone Atox1 trafficks the metal ion to the TGN where it interacts with the CXXC motif of ATP7a,b to enable entry to the organelle (Figure 19.1). Alternatively, Cu(I) is bound to another copper chaperone, the copper chaperone for superoxide dismutase (CCS) and is transferred for insertion into the cytosolic Cu,Zn enzyme superoxide dismutase SOD1.[63] The N-terminal domain of Atox1 that binds the metal is homologous with that of CCS having a MXCXXC binding motif. Yeast grown in the presence of Ag^+ shows an increase in Cu,Zn SOD1 levels, demonstrating that Ag can regulate expression. However, the activity of the enzyme is lost as Ag replaces Cu because Ag does not support the enzymatic redox function of Cu.[64] *In vitro* studies with purified bovine Cu,Zn SOD1 have also shown that Ag can replace Cu at its binding site.[65] The propensity for Ag(I) to mimic Cu(I) in its trafficking in cells and evidence for inclusion into SOD1 could be a potential mechanism of toxicity, but this has seldom been considered.

On exposure of the bronchial cells BEAS-2b to silver, the intracellular Ag(I) content was measured at 1.2 mM, *i.e.* 12 000 times the concentration (0.1 μM) required to inhibit complexes I-IV of the mitochondrial electron transport chain. The result suggests that there must be effective buffering of cytosolic Ag(I).[66] The cysteine-rich cytoplasmic protein metallothionein (MT) plays a key role in buffering internal metal concentrations.[‡] The affinity of MT for metal ions indicates that it preferentially binds Ag over most other metals: $Hg(II) > Ag(I) > Cu(I) > Cd(II) > Pb(II) > Zn(II)$.[67–70] Twelve mol eq of Ag^+ are needed to completely displace the usual complement of 7 mol eq Zn^{2+},[67] or Cd^{2+} atoms.[68] Under certain conditions, MTs have been shown to bind up to 16–18 mol eq Ag^+.[69,71]

The susceptibility of cells to Ag(I) is likely to depend on MT content or an ability to readily synthesize apo-MT in response to a Ag(I) challenge. For example, in the BEAS-2b cells, a rapid increase in Ag(I)-MT was observed, followed by a decrease as cellular partition of the Ag^+ ions altered from the MT pool to an insoluble fraction that is presumably bio-inactive.[66] In a number of aquatic invertebrate species, the buffering of cytoplasmic metal concentration is likely to depend initially on metallothionein-like proteins

---

[‡]Further discussion of the structure and function of metallothioneins is provided in Chapter 21.

(MTLP). However, the main long-term detoxification strategy is the formation of metal-rich granules such as insoluble Ag-sulfides.[72] On the other hand, there is a dynamic exchange between intracellular pools that, in aquatic organisms at least, is dependent on exposure route (water or diet) and water chemistry.[73] There are also species differences. For example, the gills of the marine mussel *Pernaverdis* immediately partition silver into an insoluble fraction (organelles, cellular debris, membranes, inorganic metal-rich granules) and only a relatively small proportion is found in the MTLP (<30%).[74]

The uptake of Ag-NP *via* different endocytotic pathways results in Ag-NP-laden intracellular vesicles.[53–57] A property of these NPs is a large surface area from which Ag^+ dissociates. It is unclear whether there is a mechanism of ionic Ag export from the vesicle or indeed a process by which Ag-NPs may be released.[75]

Rats fed a diet contaminated with silver chloride exhibited an accumulation of silver in the liver and, more specifically, in the mitochondrial fraction. This result indicates that Ag enters this organelle *via* the Cu(I) pathway that imports copper for insertion into mitochondrial cuproenzymes. The mechanism of this transport process is unknown. The existence of a mitochondrial chalcophore has been suggested.[62,76] This might be a homologue of methanobactin, a Cu(I)/Ag(I) binding peptide identified in methanotrophic bacteria.[77] The latter acts in a similar way to siderophores but scavenges for Cu instead of Fe in the environment. They are also proposed to play a role in intracellular Cu(I) regulation.[77] The incorporation of Ag(I) into mitochondrial cuproenzymes has not been studied in depth, but because replacing Cu(I) with Ag(I) inhibits enzyme activity (*e.g.* ceruloplasmin and Cu,Zn SOD1), it has the potential to be toxic.

19.5   Mechanisms of Silver Toxicity

Silver is not known to be overtly toxic to humans. However, excess consumption leads to deposition, firstly in the eye and then the skin. Photochemical reduction of accumulated Ag(I) leads to discolouration that gives the eye or skin a pale grey/blue appearance, termed agyria.[78] It has been difficult to identify the body burden of Ag that results in argyria and total body content estimates range from 0.6–6.4 g Ag.[79,80]

The World Health Organization suggests a human No Observable Adverse Effect Level (NOAEL) of 10 g over a lifetime, based on epidemiological and pharmacokinetic studies.[81] Drinking water typically contains trace levels of Ag (0–0.5 μg L^{-1}), but this may increase substantially to 50 μg L^{-1} if the water has been treated with Ag for its disinfection. As silver binds avidly to thiol groups, it is retained in all food in the range of 10–100 μg kg^{-1},[82] providing an estimated dietary intake of 4.5 μg day^{-1}.[83] If the average human consumes 2L of water per day, then 2.7 g Ag will be ingested from the most contaminated drinking water and food over a period of 70 years. It is highly unlikely that the average person reaches the NOAEL. In fact, accumulation in the body is likely to be much less as the estimated release of Ag from food, such as oysters, during digestion has been measured at between 35 and 63%[84] and the absorption by the gastrointestinal tract of rats, mice and monkeys is estimated to be relatively poor, being approximately 10% of the ingested dose.[85] In fish, similar assimilation efficiencies (AE) of Ag have been measured from natural diets (1–6% in zebrafish),[86] with an AE of 10% for brown trout where the diet of fish was spiked with radiolabelled silver cyanide.[87] Humans are exposed to Ag throughout their lives and this is reflected in tissue Ag accumulation (Table 19.1).

Acute toxicity is highly unlikely in humans as it is rare to receive an unintentional dose at a concentration that is lethal. The estimated lethal dose for humans is 10 g (cited in ref. 88 from Hill and Pillsbury, 1939). There is concern over the potential for non-lethal toxic effects given the prevalent use of Ag in wound dressings and the utilization of its antimicrobial properties in catheters. Both practices expedite direct entry into the circulation. Modern Western lifestyle inadvertently leads to exposure to Ag-NPs that are present in a plethora of consumer products and this has reignited the debate concerning the potential cytotoxic effects of Ag (*e.g.* see refs. 89–98).

Like Cu^+, Ag^+ has high and indiscriminate affinities for sulfur compounds. Consequently, it may interact and be detrimental at a number of intracellular sites and exhibit a plethora of modes of toxicity. It is known to affect cell membrane integrity and morphology,[99] as well as DNA synthesis.[100] It also inhibits cuproenzyme,[63] lactate dehydrogenase[101] and Na^+/K^+ ATPase activity[1] and has been found to affect a variety of tissues and cell types.[52,57]

Table 19.1 Adult tissue silver content.[159]

Tissue	Ag content
Adrenal gland	0.05–0.8 $\mu g \, g^{-1}$
Bile	408 $\mu g \, g^{-1}$
Blood (total)	0.0034–0.12 $mg \, L^{-1}$
Bone	<0.1–1.1 $\mu g \, g^{-1}$
Brain	<0.1–1.7 $\mu g \, g^{-1}$
Gastrointestinal tract	0.02–0.08 $\mu g \, g^{-1}$
Hair	0.025–3.8 $\mu g \, g^{-1}$
Heart	<0.1–0.26 $\mu g \, g^{-1}$
Kidney	<0.1–0.36 $\mu g \, g^{-1}$
Liver	<0.1–1.7 $\mu g \, g^{-1}$
Lung	0.002–0.1 $\mu g \, g^{-1}$
Muscle	0.002–0.06 $\mu g \, g^{-1}$
Ovary	0.002–0.5 $\mu g \, g^{-1}$
Pancreas	<0.1–0.5 $\mu g \, g^{-1}$
Skin	0.0004–4 $\mu g \, g^{-1}$
Spleen	<0.1–2.68 $\mu g \, g^{-1}$
Testis	<0.1–1.3 $\mu g \, g^{-1}$
Thyroid	0.03 $\mu g \, g^{-1}$
Tooth	0.004–2.2 $\mu g \, g^{-1}$
Urinary bladder	0.07 $\mu g \, g^{-1}$
Urine	0–10 $\mu g \, L^{-1}$

On the other hand, Ag^+, unlike Cu^+, is not a direct source of Fenton–Haber–Weiss chemistry, which generates toxic radicals and other reactive species, generically termed reactive oxygen species (ROS). However, use of soluble Ag salts in cytotoxicity studies does evoke oxidative damage, due to its ability to replace copper in SOD that results in an inability of this enzyme to take care of superoxide.[70,102,103] Such treatment also induces a reduction in the levels of intracellular glutathione, a major source of thiols that are proposed to mop up ROS.[70,99] Exposure of primary human skin fibroblast cells to Ag^+ causes a number of different intracellular effects: it targets the mitochondria with the production of superoxide anion radicals; it lowers glutathione concentrations; it reduces the expression of transcription factor Nrf2 that regulates the expression of antioxidant genes.[70] In a human bronchial cell line, $AgNO_3$ inhibited the electron transport chain complexes I–IV of mitochondria by inducing increased ROS production.[66] In addition, $AgNO_3$ induced a transient rise in intracellular zinc concentrations.[70] The source of this Zn is unknown but may involve exchange of Zn bound to MT and other Zn containing proteins[70] or from intracellular stores.[102] The rise in intracellular Zn will affect the expression of metal-responsive genes *via* binding to metal transcription factor 1 (MTF1) and interaction of MTF1 with metal-response elements upstream of target genes.[104] But, an increase in Zn can also affect cell-signalling cascades *via* the inhibition of protein-tyrosine phosphatase activity.[105]

Subcutaneous administration of Ag-NPs, defined as having one dimension <100 nm, into the systemic circulation of rats caused their accumulation in many organs (liver, spleen, kidney and brain), inducing toxicity.[106] The precise mode-of-action of Ag-NP toxicity has received much attention recently (*e.g.* refs 5,52,89–98). Although the debate has not been fully resolved, the toxicity is likely to be due to the release of highly reactive Ag ions from the nanoparticle surface rather than to a direct effect of Ag-NPs. They have been shown to increase respiration rates in cells, inducing ROS production and mitochondrial-dependent apoptosis in NIH3T3 fibroblast cells.[107] In contrast, in isolated liver mitochondria, Ag-NPs depressed state 3 respiration but increased state 4 respiration, resulting in uncoupling of the oxidative phosphorylation system and subsequent proton leakage into the mitochondria.[108] However, there was no observable effect on ROS production. The uncoupling of oxidative phosphorylation results in an ineffective transfer of electrons to oxygen at the terminal oxidase of the electron chain and, whether ROS increases or not, apoptosis is likely to occur because of mitochondrial malfunction.[107,109] In addition, in human Chang liver cells and Chinese hamster lung fibroblasts, Ag-NPs induced endoplasmic reticulum stress-mediated apoptosis; this was documented by increased expression of genes that are markers of ER stress, by a release of Ca^{2+} and overloading of Ca^{2+} in mitochondria and by an increase in CCAAT/enhancer-binding protein-homologous protein (CHOP), a protein that leads to growth arrest and apoptosis.[110]

The antimicrobial activity of Ag impregnated into medical dressing is directly related to the release of Ag^+.[111] The precise mechanisms by which Ag is toxic to bacteria is also unclear, but the same mode of actions (cell wall damage, DNA interactions, inhibition of enzymes and ROS production) as those that cause cytotoxicity in eukaryotes have been reported.[112] A difference is that bacterial cells are more sensitive (minimal inhibitory concentration (MIC) ~ 3 µM; *e.g.* see ref. 112) than some eukaryotic cells (EC50 ~ 40 µM; *e.g.* see ref. 70) and it is precisely the reason why Ag^+ can be used as a therapeutic agent. Ag(I) disrupts bacterial respiratory chain activity, generating ROS.[112,113] In *E. coli*, the ion binds to the flavin groups of complex I on the cytoplasmic membrane associated with NADH binding.[114] The dehydrogenases of *E. coli* contain flavin cofactors that are bound by cysteine residues and it is here that Ag^+ presumably binds to affect the uncoupling of oxidative phosphorylation.

19.5.1 Silver Toxicity to Aquatic Organisms and the Biotic Ligand Model

Ag acute toxicity in freshwater fish is associated with a disruption of gill physiology. The fish gill is an important multifunctional organ that provides for aquatic gases exchange, osmotic and ionic regulation, acid-base regulation and excretion of nitrogenous wastes.[46] Exceptionally high concentrations of Ag, only observed in practice in industrial waste effluents, cause severe gill irritation that results in epithelium oedemas and subsequent excess mucus production. The effect is to increase the distance for oxygen to diffuse from the water into the bloodstream. The disruption to gas exchange leads to asphyxiation. At lower concentrations (nM–µM range), Ag interferes with ion homeostasis.[1] The internal milieu of freshwater fish are hyperosmotic to their surroundings and they constantly lose ions across their integument to the environment. Ag(I) has been shown to interfere with both Na^+ and Cl^- uptake.[49,115,116] Ag^+ competes with Na^+ entry at the apical membrane[22] and inhibits the basolateral Na^+/K^+-ATPase activity[117] *via* competition with the Mg^{2+}-ATP binding site on the α subunit. The conditional equilibrium binding constant log K was reported to be 7.3–8.0.[1] Another mechanism of disruption of Na^+ balance, and more specifically Cl^- uptake, is the inhibition of branchial carbonic anhydrase (CA) activity.[116] CA catalyses the hydration of CO_2 and the products H^+ and HCO_3^- are counterions in the exchange of Cl^- and Na^+, respectively, as they enter the cell across the apical surface.[46] The progressive loss of Na^+ and Cl^- from the plasma results in fluid volume disturbances resulting in circulatory failure that leads to death.[1,47] Silver toxicity in freshwater is closely linked to metal speciation and the effective concentration of Ag^+.[48,49] Thus, in water polluted with Ag where concentrations of total metal may reach 10 µg L^{-1} (9.2 nM), the concentration of Ag^+ is considerably lower due to the presence of chelating agents such as dissolved organic matter and sulfur compounds.[1,49,118]

The strong correlation between aquatic $[Ag^+]$ and toxicity has resulted in the use of geochemical speciation models to develop computer-based tools for prediction of acute metal toxicity.[119] These tools use empirical data to ascertain the binding affinity of Ag to the gill, termed the biotic ligand (BL).[51] The assumption is that the Ag-BL complex is in equilibrium and, once inserted into the model, the gill is simply seen as a static ligand that enables an estimation of the concentration of Ag that will accumulate on the gill (Figure 19.2). It is also assumed that all water chemical parameters are known. A value of gill Ag accumulation that causes mortality must be ascertained from toxicity studies for the model to be predictive of toxicity.[119] However, the initial binding affinity constants to be derived[51] were found to be too high and over-predictive and it was suggested that this was because they were not representative of the association between intracellular Ag and the site of toxicity (*e.g.* Na^+/K^+ ATPase).[47] There have been a number of re-iterations of the model and the binding affinity has been refined based on toxicity data and inhibition of Na^+/K^+ATPase.[1] The chemistry behind these adjustments has been questioned, but the model does produce a useful tool to predict acute freshwater toxicity[120] and a silver biotic ligand model (BLM) is currently used by environmental regulators in the USA.

The challenge for toxicologists and epidemiologists is to identify cause and effect relationships in chronic exposure scenarios. However, given current data, it is highly unlikely that any links between effects on human health and Ag(I) or Ag-NP exposure will be revealed. The data for such analysis will take decades to accumulate, at which stage exposure will already have occurred. Lessons may be learnt from environmental epidemiology. A summary of 25 years of data from a US Geological Survey long-term monitoring program in San Francisco Bay found a relationship between a

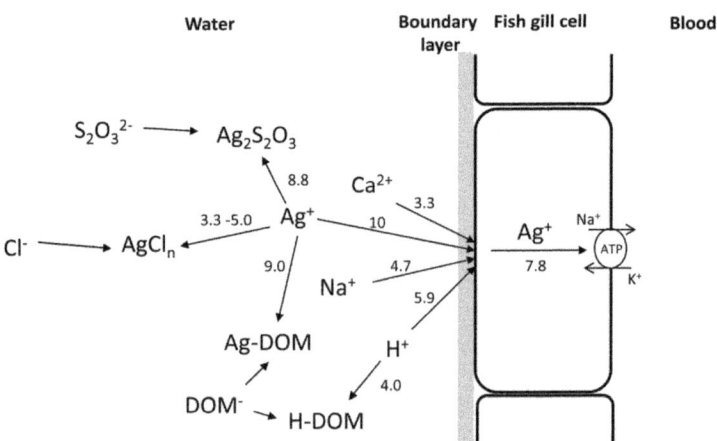

Figure 19.2   Conceptualization of the silver Biotic Ligand Model (BLM) modified from ref. 120. Values represent the conditional equilibrium constants for silver binding to various ligands in the water, including dissolved organic matter (DOM), as well as the conditional equilibrium constants for other cations binding to the gill and silver binding to the internal site of toxicity in fish gill cells; the Na^+/K^+-ATPase.

decline in reproductive output of the clam (*Macoma baltica*) and the sediment Ag(I) concentrations.[121] This analysis suggests Ag may be a metallic endocrine disruptor, and a laboratory study conducted at around the same time indicated that dietary Ag also interfered with reproduction in another invertebrate, *Daphnia magna*.[122] Similar metal-induced reproductive disturbances, which are not considered in regulatory risk assessment,[123] have been observed for cadmium[124,125] and arsenic exposure.[126,127] Interestingly, one of these studies suggested that, if reproductive success is used as a measure of toxicity, the food chain transfer of Ag may be more significant than water-borne exposure.[122] This conclusion was based on the fact that daphnids raised in clean water but fed an algal diet that was cultured in media containing 1 nM Ag had a significant reduction in reproductive output compared to organisms raised in media containing 100 nM Ag but fed a non-contaminated diet.[122]

19.6 Silver Resistance in Bacteria

Silver-resistant bacteria have been identified in hospitals, both human and veterinary, and the environment. The resistance is due to either expression of genes encoded for Ag efflux pumps, often in conjunction with Ag-binding proteins[128] (Figure 19.3) or the ability to reduce external Ag^+ to form elemental Ag.[129,130]

pMG101 is a 180 kb IncH1 resistant plasmid identified in a *Salmonella* spp from a hospital burns ward that caused the death of three people.[128] The plasmid contains nine open reading frames, seven of which have been defined and are collectively referred to as Sil genes.[128] In order of expression, the genes encode for a metal-binding protein, SilE; a responder and metal sensor; SilS and SilR; three proteins that constitute the Ag^+ efflux transporter SilCBA and a P-type ATPase, SilP.[128,131] This cassette also confers resistance to mercury, tellurite and antibiotics (*e.g.* ampicillin and streptomycin), and is found on a number of other plasmids of the IncHI-2 incompatibility group that enables silver resistance in a number of other bacterial species.[132] The products of these genes are closely related to those involved in copper detoxification in bacteria, such as *Bacillus subtilis* (a P-type ATPase CopA and a chaperone CopZ),[133] *Archeoglobus fulgidus* (CopA and Z)[41] and *Cupriavidus metallidurans* (Czc CBA)[134] and silver resistance in *C. metallidurans* (SilABC).[135] In *E. coli*, there are two systems that confer Cu resistance. The first Cue consists of a P-type ATPase (CopA),[136] which pumps Cu(I) as well as Ag(I) into the periplasm where a multicopper oxidase (CueO) converts Cu^+ to Cu^{2+}.[137] Under aerobic conditions on exposure to Ag and Cu, the Cue system is redundant as, despite the propensity for CopA to transport Ag(I), the CueO enzyme is inhibited by Ag(I).[138] Ag(I) resistance is conferred by a second system, termed Cus, which consists of a CusCFBA efflux system.[139,140] The expression of the Ag- and Cu-resistant genes are under the control of the two-component system RS (CueRS, CsoR, SilRS) that serves as silver responsive (SilR) transcription factors.[131] Very few

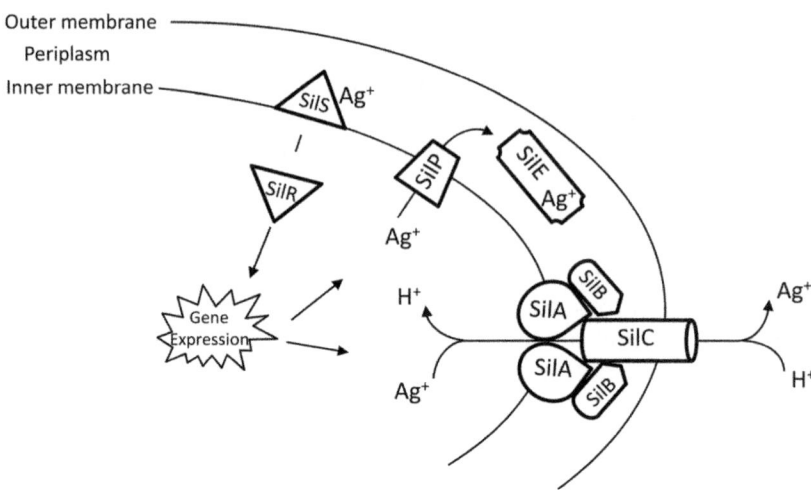

Figure 19.3   Mechanism of bacterial resistance to silver. Proteins involved in silver
resistance identified in *Salmonella* are termed Sil. SilS acts as a
detector of periplasmic Ag(I) and binding of Ag(I) to SilS induces the
response regulator SilR to mediate the expression of genes associated
with Ag resistance. SilP is a P-type ATPase that pumps Ag(I) from the
cytoplasm to the periplasm where it is bound by SilE. SilCBA forms a
tri-polypeptide transmembrane potential dependent resistance nodulation division (RND) Ag(I)/protein antiporter that connects the inner
and outer membranes forming a mechanism by which Ag(I) is transported out of the cells across the periplasm.
Modified from ref. 131.

of the Sil genes have been fully characterized and, where they have not, their
function has been proposed based on sequence homology and analysis of
Ag(I) interaction with members of the Cus system, CopA and CopZ.

19.6.1   SilE and Silver-binding Polypeptides

SilE is a periplasmic Ag-binding protein believed to be the first line of defence against Ag toxicity. The gene encodes for a 143 amino acid protein with
47% similarity to the PcoE metal-binding protein of *E. coli* but there is
a >50% and 100% homology in methionine and histidine residues, respectively, between the two proteins, suggesting that these are significant for
metal binding.[141] The purification of SilE from the periplasm and subsequent N-terminal sequencing revealed that the active protein lacks the first
20 amino acids. In aerobic conditions, the SilE protein binds ∼five Ag(I), but
less than one Cu(II) and Cd(II). In contrast to other metal binding proteins,
SilE lacks cysteine residues. However, it does possess 10 His and, on exposure to Ag, the 10 His form cross-links with the Ag ions ordering the
structure of the apo-silE protein into one of 19 α-helices and 2 β-sheets. NMR
analysis confirmed binding to the nitrogen donor of imidazole. The initial

four Ag^+ ions bind to the four pairs of HisX6His and modelling indicates that these are located on either side of α-helices and fold to form the secondary structure. The final Ag^+ ions bind to the last two N-terminal His residues located in the β-sheets and on α-helices.[131]

SilE on its own does not confer Ag resistance and has to be expressed with other components of the Sil cassette. In contrast, *Pseudomonas stutzeri* AG259, isolated from a silver mine with the ability to withstand 50 mM Ag, employs an entirely different strategy. An array of small peptides (7–14 amino acids) are expressed that exhibit an affinity for Ag^+. They induce precipitation of crystals of metallic silver, approximately 200 nm in diameter.[142] One of these putative peptides, AgBP2, was linked to the *E. coli* maltose-binding protein that was expressed in either the cytoplasm or the periplasm.[142] Those strains that expressed the AgBP2 protein in the periplasm tolerated Ag^+ at concentrations of 28 μM in the growth media, whereas those that expressed the peptide in the cytoplasm did not. This suggests that prevention of entry of silver into the cell is the most effective defence.

Other polypeptides that bind Ag(I) have been identified. Their potential in the production of Ag-NPs is being explored *via* a biotechnology termed "molecular bio-mimetics". The amino acid sequences of AgBP2 and Ag4 polypeptide and the two other sequences reported in[143] are dissimilar, but they do possess a number of hydrophobic and hydroxyl polar amino acids that appear to be important. Most of the Ag is deposited within the periplasm in the elemental form,[144] but the precise mechanism by which these polypeptides induce formation of elemental Ag is unclear.

19.6.2 SilCBA/CusCFBA – Chemiosmotic Cation/Proton Antiporters

The silCBA genes produce a tri-polypeptide pump that is a member of the transmembrane potential-dependent resistance nodulation division (RND) class of cation/proton antiporters. This class can be divided into the hydrophobe/amphiphile efflux family (HAE-RND) that confer resistance to drugs and the heavy-metal efflux family (HME-RND).[145] SilCBA are homologous to the metal export proteins involved in Cu(I) and Ag(I) resistance in *E. coli* (CusCFBA).[139,140,§] The three proteins are arranged so that SilA/CusA is located in the inner membrane and SilC/CusC transcends the periplasmic space linking the inner and outer membranes, whereas SilB/CusB is located in the periplasmic space, and collectively they form a pore for the passage of Ag(I) out of the cell.[131] CusF, the fourth protein to confer resistance in *E. coli*, is a periplasmic metallochaperone that facilitates transfer of Ag(I) to CusB.[147,148] It exhibits a higher binding affinity for Ag(I) ($K_D = 38$ nM) than for Cu(I) ($K_D = 495$ nM).[146]

---

§The molecular structures of the CusCBA pump and chaperone CusF are displayed in Chapter 16.

Analysis of the crystal structure of CusA on binding Ag(I) and Cu(I) has provided evidence of how these proteins operate to pump metals out of the cell.[149] CusA has 12 transmembrane (TM) helices with two periplasmic loops situated between TM1 and 2 and TM7 and 8, and TM4, 5, 10 and 11 extend into the cytoplasm. Subdomains of the periplasmic region (PN1, PN2, PC1 and PC2) form a pore and two other subdomains (DN and DC) are proposed to interact with CusC. The apo-CusA periplasmic region contains an α-helix and three β-sheets close to the opening of the metal binding cleft. Three methionines (M573, M623 and M672) of PC1 are presumed to make specific trigonal binding sites for Ag(I). Once the metal is bound to these Met ligands, a conformational change of the protein occurs, noticeably PC2 moves out of the way to form a cleft for metal entry and in doing so tilts to allow M672 to move towards M573 and M623. TM8 also moves allowing the cleft region to interact with its N-terminal region; at the same time the other TM bonds also shift. The conformational changes of the protein that take place are identical in the presence of Cu(I) or Ag(I) (see Figure 3 of ref. 149). Two distinct processes allow the passage of Ag(I) through the pore. First, there are three methionine pairs (M391–M1009; M486–M403; M410–M501) located in the pore that transcend the inner membrane, which allow for the Ag^+ to be passed along the pore. Mutation studies whereby these Met are replaced by Ile increases *E. coli* sensitivity to Ag^+. Similarly the D405 residue of TM4 is also important in conferring metal resistance[150] and interacts with E939 and K984 to form a mechanism by which protons are passed along the proteins in exchange for the metal (Figure 19.3).

The CusB proteins have recently been shown to play an integral part in Ag(I) resistance and, unusually for the proteins of the HME-RND family, which are associated with the periplasm, show evidence of substrate binding. Isothermal titration calorimetry was used to determine enthalpy changes in the apo-CusB protein and the binding affinity ($K_d = 24$ nM). Extended X-ray absorbance fine structure (EXAFS) analysis suggests a three-coordinate binding environment and the parsimonious model would be three Ag/Cu-S bonds.[151] CusB lacks cysteine and the probable interactions occur with three methionine residues,[146] similar to that reported for other Cu/Ag transporters. This suggests that potentially in addition to acting to stabilize the CBA complex, CusB also plays a more active role in metal transfer across the periplasm.[148,151] Two mechanisms by which this may occur have been proposed; either CusB acts as a periplasmic chaperone passing Ag(I) to CusC, or the binding of Ag(I) to CusB alters the conformation of the CusCB complex that opens the pore within CusC, allowing passage of Ag(I) from the cell.

In metal-resistant *Cupriavidus metallidurans* CH34, the SilABC system confers Ag resistance.[135] The metal binding characteristics of the C-terminus of SilB (SilB$_{440-521}$) has similar metal coordination sites to those of the periplasmic copper chaperone CusF, and also preferentially binds monovalent Ag and Cu compared to other divalent cations, including Cu^{2+}.[135] NMR spectroscopy revealed a slow motion detected by ^{15}N relaxation in the metal binding loop of SilB$_{440-521}$ that is comparable to CusF[135] and has been seen in other metallochaperones.[152] The metallochaperone properties of the C-terminus of SilB suggest that this region may coordinate the transfer of silver from the periplasm to the N-terminal region before transfer to the transporter, in this instance SilA, for export.[135] Thus, the SilB protein of *C. metallidurans* CH34 combines the role of CusF and CusB of *E. coli*.[135]

19.6.3 SilP and CopA

SilP encodes a protein of 824 aa that possesses the conserved features of other metal P-type ATPases, including the conserved aspartyl residues of the phosphoenzyme cycle, the thr-gly-glu-ser phosphate binding region and the metal binding cys-pro-cys motif in the predicted sixth transmembrane domain.[128] However, SilP lacks the CXXC motif characteristic of the N-terminal regions of other Cu(I) transporters, and the cysteines appear to be replaced by histidine. Details of transport of Ag by P-type ATPases can be found in the section above.

19.6.4 SilSR/CurRS – Metal Response Regulators

The Ag response regulators consist of an inner membrane bound SilS/CueS and a cytoplasmic SilR/CueR.[131,153] SilS/CueS is a histidine kinase sensor that acts as an outer-cytoplasmic detector of increased periplasmic Ag concentrations. The signal from Ag(I) binding to SilS is mediated *via* the response regulator SilRl/CueR.[154,155] CueR activates the *CopA* promoter in the presence of Ag, Cu and Au, but not Ni, Zn or Cd.[153,154] The precise mechanism by which this occurs has yet to be determined.

19.6.5 Alternate Mechanism of Silver Resistance in Bacteria

Three bacterial isolates, *Pseudomonas* spp., *Bacillus* spp. and *Shewanella* spp., from Thames estuary sediment were found to withstand up to 10 mM AgNO$_3$. The isolates were cultured in aerobic conditions and, despite this, the bacteria produced elemental Ag deposits in the culture vessels when cultured in the presence of AgNO$_3$.[129] The mechanisms by which this occurred were not elucidated, but the mechanism of Ag(I) detoxification *via* the reduction of Ag(I) to Ag(0) has been proposed.[156] *Shewanella oneidensis* can use a wide range of electron acceptors for growth including Ag(I)[130] and the Fe(III)-reducing bacterium *Geobacter sulfreducens* precipitates Ag particles around the cell surface if grown in a medium containing Ag(I).[157]

19.7 Conclusions and Open Questions

The uptake and intracellular metabolism of Ag(I) is inextricably linked to that of Cu(I), sharing the same uptake pathways *via* CTR1, binding to Cu-chaperones for trafficking to ATP7A and B for transport into the TGN. There is evidence that Ag(I) can also replace Cu(I) in a number of cuproenzymes, thereby severely impairing their activity. This is a potential mode of toxic action but, because Ag(I) avidly binds to sulfur centres, there are a number of toxic mechanisms, including perturbation of membrane morphology and DNA synthesis and inducement of ROS. Ag(I) inhibits the activity of purified mammalian Na^+/K^+-ATPase[158] and this enzyme has been identified as a target for Ag(I) toxicity in fish gill cells. However, the activity of this important enzyme has not been considered in many Ag(I) cytotoxicity studies. The contrast between the extreme toxicity of Ag(I) to bacteria and its relative tolerance in humans is remarkable and difficult to reconcile. The difference is likely to be due to the ability of eukaryotic cells to affect intracellular Ag(I) concentrations *via* metal-binding proteins such as metallothioneins that detoxify the metal ion.

The use of Ag(I) as an antimicrobial agent has increased as drug-resistant strains of bacteria evolve. However, silver-resistant bacterial strains have evolved that employ a combination of Ag(I) efflux pumps and periplasmic metal-binding proteins.[128] As Ag(I) is often used as an antimicrobial agent of last resort in hospitals, the evolution of silver-resistant bacteria will have serious implications for medicine.

The growing use of Ag-NPs has re-ignited the debate on silver cytotoxicity as they deliver a localized package of concentrated metal. The possible risk to human health is being debated. There is growing evidence that Ag-NPs can cause cytotoxicity in a number of different cell types *via* a number of different mechanisms. However, the key question is whether humans and wildlife are being exposed to the nanoparticles at concentrations that are harmful. Currently, this is felt to be unlikely but identification of long-term chronic effects remains important for assessment of the risk to humans.

References

1. C. M. Wood, *Homeostasis and Toxicology of Non-Essential Metals*, ed. C. M. Wood, A. P. Farrell and C. J. Brauner, Academic Press, Amsterdam, 2012, Fish Physiology vol. 31B, ch. 1, pp. 1–65.
2. A. D. Russell and W. B. Hugo, *Prog. Med. Chem.*, 1994, **31**, 351–370.
3. J. W. Alexander, *Surg. Infect.*, 2009, **10**, 289–292.
4. A. Bianchini and C. M. Wood, *Environ. Toxicol. Chem.*, 2003, **22**, 1361–1367.
5. T. De Lima, A. B. Seabra and N. Durán, *J. Appl. Toxicol.*, 2012, **32**, 867–879.
6. M. C. Stensberg, Q. Wei, E. S. McLamore, D. M. Porterfield, A. Wei and M. S. Sepúlveda, *Nanomedicine*, 2011, **6**, 879–898.
7. H. S. Sharma and A. Sharma, *CNS Neurol. Disord. Drug Targets*, 2012, **11**, 65–80.
8. P. L. Drake and K. J. Hazelwood, *Ann. Occup. Hyg.*, 2005, **49**, 575–585.
9. C. C. Bridges and R. K. Zalups, *Toxicol. Appl. Pharmacol.*, 2005, **204**, 274–308.

10. C. Hogstrand, F. Galvez and C. M. Wood, *Environ. Toxicol. Chem.*, 1995, **15**, 1102–1108.
11. N. R. Bury and C. Hogstrand, *Environ. Sci. Technol.*, 2002, **36**, 2884–2888.
12. M. A. Kihlken, C. Singleton and N. E. Le Brun, *J. Biol. Inorg. Chem.*, 2008, **13**, 1011–1023.
13. J. Lee, M. J. Petris and D. J. Thiele, *J. Biol. Chem.*, 2002, **277**, 40253–40259.
14. A. M. Zimnicka, K. Ivy and J. H. Kaplan, *Am. J. Physiol. Cell Physiol.*, 2011, **300**, C588–C599.
15. J. Bertinato, L. Chang, R. Hogue and L. J. Plouffe, *J. Trace Elem. Med. Biol.*, 2010, **24**, 178–184.
16. J. F. Eisses, Y. Chi and J. H. Kaplan, *J. Biol. Chem.*, 2005, **280**, 9635–9639.
17. C. J. De Feo, S. G. Aller, G. S. Siluvai, N. J. Blackburn and V. M. Unger, *Proc. Natl Acad. Sci. USA*, 2009, **106**, 4237–4242.
18. J. T. Rubino and K. J. Franz, *J. Inorg. Biochem.*, 2012, **107**, 129–143.
19. M. Minghetti, M. J. Leaver, E. Carpenè and S. J. George, *Comp. Biochem. Physiol. C Toxicol. Pharmacol.*, 2008, **147**, 450–459.
20. A. V. Davis and T. V. O'Halloran, *Nat. Chem. Biol.*, 2008, **4**, 148–151.
21. T. D. Rae, P. J. Schmidt, R. A. Pufahl, V. C. Culotta and T. V. O'Halloran, *Science*, 1999, **284**, 805–808.
22. N. R. Bury and C. M. Wood, *Am. J. Physiol. – Reg. I*, 1999, **46**, R1385–R1391.
23. M. Grosell and C. M. Wood, *J. Exp. Biol.*, 2002, **205**, 1179–1188.
24. S. Kellenberger, I. Gautschi and L. Schild, *Proc. Natl Acad. Sci. USA*, 1999, **96**, 4170–4175.
25. A. N. Takeda, I. Gautschi, M. X. van Bemmelen and L. Schild, *J. Biol. Chem.*, 2007, **282**, 31928–31936.
26. N. R. Bury and J. C. McGeer, *Osmoregulation and Ion Transport: Integrating Physiology, Molecular and Environmental Aspects*, ed. R. D. Handy, N. R. Bury and G. Flik, Society for Experimental Biology, London, Essential Review in Experimental Biology, vol. 1, ch. 6, pp. 181–203.
27. M. K. Schnizler, R. Bogdan, A. Bennert, N. R. Bury, M. Fronius and W. Clauss, *Biochim. Biophys. Acta*, 2007, **1768**, 317–323.
28. P. F. Curran, *Biochim. Biophys. Acta*, 1972, **288**, 90–97.
29. G. A. Gersencer, K. M. Cornetet, S. Y. Loo and S. K. Hong, *Life Sci.*, 1977, **20**, 1883–1889.
30. J. M. Argüello, *J. Membr. Biol.*, 2003, **195**, 93–108.
31. J. M. Argüello, D. Raimunda and M. González-Guerrero, *J. Biol. Chem.*, 2012, **287**, 13510–13517.
32. C. Vulpe, B. Levinson, S. Whitney, S. Packman and J. Gitschier, *Nat. Genet.*, 1993, **3**, 7–13.
33. R. E. Tanzi, K. Petrukhin, I. Chernov, J. L. Pellequer, W. Wasco, B. Ross, D. M. Romano, E. Parano, L. Pavone, L. M. Brzustowicz, M. Devoto, J. Peppercorn, A. I. Bush, I. Sternlieb, M. Pirastu, J. F. Gusella, O. Evgrafov, G. K. Penchaszadeh, B. Honig, I. S. Edelman, M. B. Soares, I. H. Scheinberg and T. C. Gilliam, *Nat. Genet.*, 1993, **5**, 344–350.

34. N. M. Hasan, A. Gupta, E. Polishchuk, C. H. Yu, R. Polishchuk, O. Y. Dmitrev and S. Lutsenko, *J. Biol. Chem.*, 2012, **287**, 36041–36050.
35. S. Lutsenko, E. S. LeShane and U. Shinde, *Arch. Biochem. Biophys.*, 2007, **463**, 134–148.
36. M. Solioz and A. Odermatt, *J. Biol. Chem.*, 1995, **270**, 9217–9221.
37. A. K. Mandal and J. M. Argüello, *Biochemistry*, 2003, **42**, 11040–11047.
38. A. Ibricevic, S. L. Brody, W. J. Youngs and C. L. Cannon, *Toxicol. Appl. Pharm.*, 2010, **243**, 315–322.
39. S. Narindrasorasak, X. Zhang, E. A. Roberts and B. Sarkar, *Bioinorg. Chem. Appl.*, 2004, **2**, 105–123.
40. T. Padilla-Benavides, C. J. McCann and J. M. Argüello, *J. Biol. Chem.*, 2013, **288**, 69–78.
41. M. González-Guerrero and J. M. Argüello, *Proc. Natl Acad. Sci. USA*, 2008, **105**, 5992–5997.
42. P. Gourdon, X. Y. Liu, T. Skjørringe, J. P. Morth, L. B. Møller, B. P. Pedersen and P. Nissen, *Nature*, 2011, **475**, 59–64.
43. A. K. Mandal, W. D. Cheung and J. M. Argüello, *J. Biol. Chem.*, 2002, **277**, 7201–7208.
44. N. R. Bury, M. Grosell, C. M. Wood and A. K. Grover, *Toxicol. Appl. Pharmacol.*, 1999, **159**, 1–8.
45. N. R. Bury, *Aquat. Toxicol.*, 2005, **72**, 135–145.
46. D. H. Evans, P. M. Piermarini and K. P. Choe, *Physiol. Rev.*, 2005, **85**, 97–177.
47. C. M. Wood, R. C. Playle and C. Hogstrand, *Environ. Toxicol. Chem.*, 1999, **18**, 71–83.
48. J. C. McGeer and C. M. Wood, *Can. J. Fish. Aquat. Sci.*, 1998, **55**, 2447–2454.
49. N. R. Bury, J. C. McGeer and C. M. Wood, *Environ. Toxicol. Chem.*, 1999, **18**, 49–55.
50. J. R. Reinfelder and S. I. Chang, *Environ. Sci. Technol.*, 1999, **33**, 1860–1863.
51. N. Janes and R. C. Playle, *Environ. Toxicol. Chem.*, 1995, **14**, 1847–1858.
52. B. K. Gaiser, T. F. Fernandes, M. Jepson, J. R. Lead, C. R. Tyler and V. Stone, *Environ. Health*, 2004, **8**(I), S2.
53. E. M. Luther, Y. Koehler, J. Diendorf, M. Epple and R. Dringen, *Nanotechnology*, 2011, **22**, 375101.
54. A. Haase, S. Rott, A. Mantion, P. Graf, J. Plendl, A. F. Thünemann, W. P. Meier, A. Taubert, A. Luch and G. Reiser, *Toxicol. Sci.*, 2012, **126**, 457–468.
55. R. P. Singh and P. Romero, *Toxicol Lett.*, 2012, **213**, 249–259.
56. H. Wang, L. Wu and B. M. Reinhard, *ACS Nano*, 2012, **6**, 7122–7132.
57. C. Greulich, J. Diendorf, T. Simon, G. Eggeler, M. Epple and M. Köller, *Acta Biomater.*, 2001, **7**, 347–354.
58. M. Koivusalo, C. Welch, H. Hayashi, C. C. Scott, M. Kim, T. Alexander, N. Youret, K. M. Hahn and S. Grinstein, *J. Cell Biol*, 2010, **188**, 547–563.

59. S. R. Hanson, S. A. Donley and M. C. Linder, *J. Trace Elem. Med. Biol.*, 2001, **15**, 243–253.
60. E. Ilyechova, A. Skvortsov, E. Zatulovsky, N. Tsymbalenko, M. Shavlovsky, M. Broggini and L. Puckova, *J. Trace Elem. Med. Biol.*, 2011, **25**, 27–35.
61. F. Hirasawa, Y. Kawarda, M. Sato, S. Suzuki, K. Terada, N. Miura, M. Fuji, K. Kato, Y. Takizawa and T. Sugiyama, *Biochim. Biophys. Acta*, 1997, **1336**, 195–201.
62. E. A. Zatulovskiy, A. N. Skvortsov, P. Rusconi, E. Y. Ilyechova, P. S. Babich, N. V. Tsymbalenko, M. Broggini and L. V. Pushkova, *J. Inorg. Biochem.*, 2012, **116**, 88–96.
63. L. Banci, I. Bertini, F. Cantini, T. Kozyreva, C. Massagni, P. Palumaa, J. T. Rubino and K. Zovo, *Proc. Natl Acad. Sci. USA*, 2012, **109**, 13555–13560.
64. M. R. Ciriolo, P. Civitareale, M. T. Carri, A. De Martino, F. Galiazzo and G. Rotilio, *J. Biol. Chem.*, 1994, **269**, 25783–25787.
65. J. A. Roe, R. Peoples, D. M. Scholler and J. S. Valentine, *J. Am. Chem. Soc.*, 1990, **112**, 1538–1545.
66. T. Miyayama, Y. Arai, N. Suzuki and S. Hirano, *Toxicology*, 2013, **305**, 20–29.
67. K. B. Nielson, C. L. Atkin and D. R. Winge, *J. Biol. Chem.*, 1985, **260**, 5342–5350.
68. H. Li and J. D. Otvos, *Biochemistry*, 1996, **35**, 13929–13936.
69. Ó. Palacios, K. Polec-Pawlak, R. Lobinski, M. Capdevila and P. González-Duarte, *J. Biol. Inorg. Chem.*, 2003, **8**, 831–842.
70. M. M. Cortese-Krott, M. Munchow, E. Pirev, F. Hessner, A. Bozkurt, P. Uciechowsi, N. Pallua, K.-D. Kronke and C. V. Suschek, *Free Radic. Biol. Med.*, 2009, **47**, 1570–1577.
71. A. J. Zelazowski, Z. Gasyna and M. J. Stillman, *J. Biol. Chem.*, 1989, **264**, 17091–17099.
72. C. Amiard-Triquet, B. Berthet and R. Martoja, *Biol. Met.*, 1991, **4**, 144–150.
73. S. N. Luoma and P. S. Rainbow, *Metal Contamination in Aquatic Environments*, Cambridge University Press, United Kingdom, 2008.
74. T. N.-T. Ng and W.-X. Wang, *Environ. Toxicol. Chem.*, 2005, **24**, 2365–2372.
75. M. C. Hohnholt, M. Geppert, E. M. Luther, C. Petters, F. Bulcke and R. Drihgen, *Neurochem. Res.*, 2013, **38**, 227–239.
76. P. A. Cobine, L. D. Ojeda, K. M. Rigby and D. R. Winge, *J. Biol. Chem.*, 2006, **281**, 36552–36559.
77. R. Balasubramanian and A. C. Rosenzweig, *Curr. Opin. Chem. Biol.*, 2008, **12**, 245–249.
78. Y. Kim, H. S. Suh, H. J. Cha, S. H. Kim, K. S. Jeong and D. H. Kim, *Am. J. Ind. Med.*, 2009, **52**, 246–250.
79. L. E. Gaul and A. H. Staud, *J. Am. Med. Assoc.*, 1935, **104**, 1387–1390.
80. B. W. East, K. Boddy, E. D. Williams, D. Macintyre and A. L. Mclay, *Clin. Exp. Dermatol.*, 1980, **5**, 305–311.

81. World Health Organization, Guidelines for drinking water quality, 2nd edn, vol. 2, Health criteria and other supporting information (WHO/SDE/WSH/03.04/14), 1996.

82. R. S. Gibson and C. A. Scythes, *Biol. Trace Elem. Res.*, 1984, **6**, 105–116.

83. G. V. Iyenger, J. T. Tanner, W. R. Wolf and R. Zeisler, *Sci. Total Environ.*, 1987, **61**, 235–252.

84. V. Bragigand, B. Berthet, J. C. Amiard and P. S. Rainbow, *Food Chem. Toxicol.*, 2004, **42**, 1893–1902.

85. J. E. Furchner, C. R. Richmond and G. A. Drake, *Health Phys.*, 1968, **15**, 505–514.

86. D. Boyle, C. Hogstrand and N. R. Bury, *Aquat. Toxicol.*, 2010, **105**, 21–28.

87. J. Garnier and J. P. Boudin, *Water Air Soil Pollut.*, 1990, **50**, 409–421.

88. US Environmental Protection Agency, Ambient water quality criteria for silver, Washington DC, 1980 (EPA 440/5-80-071).

89. H. G. Petering, *Pharmacol. Ther. A*, 1976, **1**, 127–130.

90. M. A. Hollinger, *Crit. Rev. Toxicol.*, 1996, **26**, 255–260.

91. H. A. Tran and S. Song, *Pathology*, 2007, **39**, 456–458.

92. J. H. Sung, J. H. Ji, J. D. Park, J. U. Yoon, D. S. Kim, K. S. Jeon, M. Y. Song, J. Jeong, B. S. Han, J. H. Han, Y. H. Chung, H. K. Chang, J. H. Lee, M. H. Cho, B. J. Kelman and I. J. Yu, *Toxicol. Sci.*, 2009, **108**, 452–461.

93. S. Kim, J. E. Choi, J. Choi, K. H. Chung, K. Park, J. Yi and D. Y. Ryu, *Toxicol. In Vitro*, 2009, **23**, 1076–1084.

94. B. K. Gaiser, T. F. Fernandes, M. A. Jepson, J. R. Lead, C. R. Tyler, M. Baalousha, A. Biswas, G. J. Britton, P. A. Cole, B. D. Jonston, Y. Ju-Nam, P. Rosenkranz, T. M. Scown and V. Stone, *Environ. Toxicol. Chem.*, 2012, **31**, 144–154.

95. S. Hachenberg, A. Scherzed, M. Kessler, S. Hummerl, A. Technau, K. Froelich, C. Ginzkey, C. Koehler, R. Hagen and N. Kleinsasser, *Toxicol. Lett.*, 2011, **201**, 27–33.

96. J. S. Hyun, B. S. Lee, H. Y. Ryu, J. H. Sung, K. H. Chung and I. J. Yu, *Toxicol. Lett.*, 2008, **182**, 24–28.

97. H. J. Johnston, G. Hutchinson, F. M. Christensen, S. Peters, S. Hankin and V. Stone, *Crit. Rev. Toxicol.*, 2010, **40**, 328–346.

98. S. W. P. Wijnhoven, W. J. G. M. Peijnenburg, C. A. Herberts, W. I. Hagens, A. G. Oomen, E. H. W. Heugens, B. Roszek, J. Bisschops, I. Gosens, D. Van de Meent, S. Dekkers, W. H. De Jong, M. Van Zijverden, A. J. A. M. Sips and R. E. Geersma, *Nanotoxicology*, 2009, **3**, 109–138.

99. C. Baldi, C. Minoia, A. Di Nucci, E. Capodaglio and L. Manzo, *Toxicol. Lett.*, 1998, **41**, 261–268.

100. E. Hidalgo and C. Domíngues, *Toxicol. Lett.*, 1998, **98**, 169–179.

101. K. S. Rogers, *Biochim. Biophys. Acta*, 1972, **263**, 309–314.

102. T. Yoshimaru, Y. Suzuki, T. Inoue, O. Niide and C. Ra, *Free Radic. Biol. Med.*, 2006, **40**, 1949–1959.

103. B. Hultberg, A. Andersson and A. Isaksson, *Toxicology*, 1997, **117**, 89–97.

104. R. A. Colvin, W. R. Holmes, C. P. Fontaine and W. Maret, *Metallomics*, 2010, **2**, 306–317.
105. M. Wilson, C. Hogstrand and W. Maret, *J. Biol. Chem.*, 2012, **287**, 9322–9326.
106. J. Tang, L. Xiong, S. Wang, J. Y. Wang, L. Liu, J. G. Li, F. Q. Yuan and T. F. Xi, *J. Nanosci. Nanotechnol.*, 2009, **9**, 4924–4932.
107. Y. H. Hsin, C. F. Chena, S. Huang, T. S. Shih, P. S. Lai and P. J. Chueh, *Toxicol. Lett.*, 2008, **179**, 130–139.
108. J. S. Teodoro, A. M. Simões, F. V. Duarte, A. P. Rolo, R. C. Murdocj, S. M. Hussain and C. M. Palmeira, *Toxicol. In Vitro*, 2011, **25**, 664–670.
109. R. Foldberg, D. A. Dand and H. Autrup, *Arch. Toxicol.*, 2011, **85**, 743–750.
110. R. Zhang, M. J. Piao, K. C. Kim, A. D. Kim, J.-Y. Choi, J. Choi and J. W. Hyun, *Int. J. Biochem. Cell B*, 2012, **44**, 224–232.
111. P. L. Taylor, A. L. Ussher and R. E. Burrell, *Biomater.*, 2005, **26**, 7221–7229.
112. S. Echhardt, P. S. Brunetto, J. Gagnon, M. Priebe, B. Giese and K. M. Fromm, *Chem. Rev.*, 2013, **113**, 4708–4754.
113. K. B. Holt and A. Bard, *Biochemistry*, 2005, **44**, 13214–13223.
114. S. Y. Liau, D. C. Read, W. J. Pugh, J. R. Furr and A. D. Russell, *Lett. Appl. Microbiol.*, 1997, **25**, 279–283.
115. C. M. Wood, C. Hogstrand, F. Galvez and R. S. Munger, *Aquat. Toxicol.*, 1996, **35**, 93–109.
116. I. J. Morgan, R. P. Henry and C. M. Wood, *Aquat. Toxicol.*, 1997, **38**, 145–163.
117. E. A. Ferguson, D. A. Leach and C. Hogstrand. *4th International Conference on Transport, Fate and Effects of Silver in the Environment*, eds. A. W. Andren and T. W. Bober, University of Wisconsin Grant Institute, Madison, USA, pp 191–196.
118. C. Hogstrand, F. Galvez and C. M. Wood, *Environ. Toxicol. Chem.*, 1996, **15**, 1102–1108.
119. P. R. Paquin, J. W. Gorsuch, S. Apte, G. E. Batley, K. C. Bowles, P. G. C. Campbell, C. G. Delos, D. M. Di Toro, R. L. Dwyer, F. Galvez, R. W. Gensemer, G. G. Goss, C. Hogstrand, C. R. Janssen, J. C. McGeer, R. B. Naddy, R. C. Playle, R. C. Santore, U. Schneider, W. A. Stubblefield, C. M. Wood and K. B. Wu, *Comp. Biochem. Physiol. Toxicol. Pharmacol.*, 2002, **133C**, 3–35.
120. J. C. McGeer, R. C. Playle, C. M. Wood and F. Galvez, *Environ. Sci. Technol.*, 2000, **34**, 4199–4207.
121. M. I. Hornberger, S. N. Luoma, D. J. Cain, F. Parchaso, C. L. Brown, R. M. Rouse, C. Wellise and J. K. Thompson, *Environ. Sci. Technol.*, 2000, **34**, 2401–2409.
122. S. E. Hook and N. S. Fisher, *Environ. Toxicol. Chem.*, 2001, **20**, 568–574.
123. J. S. Meyer, W. J. Adams, K. V. Brix, S. N. Luoma, S. R. Mount, W. A. Stubblefield and C. M. Wood, *Toxicity of Dietborne Metals to Aquatic Organisms*, Society of Environmental Toxicology and Chemistry (SETAC) Press, 2005.

124. H. M. Lizardo-Daudt, O. S. Bains, C. R. Singh and C. J. Kenedy, *Arch. Environ. Contam. Toxicol.*, 2008, **55**, 103–110.

125. M. K. Sellin and A. S. Kolok, *Arch. Environ. Contam. Toxicol.*, 2006, **51**, 594–599.

126. J. C. Davey, J. E. Bodwell, J. A. Gosse and J. W. Hamilton, *Toxicol. Sci.*, 2007, **98**, 75–86.

127. D. Boyle, K. V. Brix, H. Amlund, A. K. Lundebye, C. Hogstrand and N. R. Bury, *Environ. Sci. Technol.*, 2008, **42**, 5354–5360.

128. A. Gupta, M. Matsui, J.-F. Lo and S. Silver, *Nat. Med.*, 1999, **5**, 183–188.

129. A. Otitoloju, G. B. Rogers, N. R. Bury, C. Hogstrand and K. D. Bruce, *Environmentalist*, 2009, **29**, 85–92.

130. H. Wang, N. Law, G. Pearson, B. E. van Dongen, R. M. Jarvis, R. Goodacre and J. R. Lloyd, *J. Bacteriol.*, 2010, **192**, 1143–1150.

131. S. Silver, *FEMS Microbiol Rev.*, 2003, **27**, 341–353.

132. A. Gupta, L. T. Phung, D. E. Taylor and S. Silver, *Microbiology*, 2001, **147**, 3393–3402.

133. C. Singleton, S. Hearnshaw, L. Zhou, N. E. Le Brun and A. M. Hemmings, *Biochem. J.*, 2009, **424**, 347–356.

134. D. Munkelt, G. Grass and D. H. Nies, *J. Bacteriol.*, 2004, **186**, 8036–8043.

135. B. Bersch, K.-M. Derfoufi, F. De Angelis, V. Auquier, E. Ngonlong Ekendé, M. Mergeay, J.-M. Ruysschaert and G. Vadenbussche, *Biochemistry*, 2011, **50**, 2194–2204.

136. C. Rensing, B. Fan, R. Sharma, B. Mitra and B. P. Rosen, *Proc. Natl Acad. Sci. USA*, 2000, **97**, 652–656.

137. F. W. Outten, D. L. Huffman, J. A. Hale and T. V. O'Halloran, *J. Biol. Chem.*, 2001, **276**, 30670–30677.

138. S. K. Singh, S. A. Roberts, S. F. McDevitt, A. Weichsel, G. F. Wildner, G. B. Grass, C. Rensing and W. R. Montfort, *J. Biol. Chem.*, 2011, **286**, 37849–37857.

139. C. N. Lok, C. M. Ho, R. Chen, P. K. Tam, J. F. Chiu and C. M Che, *J. Proteome Res.*, 2008, 7, 2351–2356.

140. B.-Y. Yun, Y. Xu, S. Piao, N. Kim, J.-H. Yoon, H.-S. Cho, L. Lee and N.-C. Ha, *J. Microbiol.*, 2010, **48**, 829–835.

141. M. Zimmerman, S. R. Udagedara, C. M. Sze, T. M. Ryan, G. J. Howlett, Z. Xiao and A. G. Wedd, *J. Inorg. Biochem.*, 2012, **115**, 186–197.

142. R. H. Sedlak, M. Hnilova, C. Grosh, H. Fong, F. Baneyx, D. Schwartz, M. Sarikaya, C. Tamerler and B. Traxler, *Appl. Environ. Microbiol.*, 2012, **78**, 2289–2296.

143. M. Sarikaya, C. Tamerler, A. K.-Y. Jen, K. Schulten and F. Baneyx, *Nat. Mater.*, 2003, 2, 577–585.

144. T. Klaus, R. Joerger, E. Olsson and G. C. Granqvist, *Proc. Natl Acad. Sci. USA*, 1999, **96**, 13611–136114.

145. D. H. Nies, *FEMS Microbiol. Rev.*, 2003, **27**, 313–339.

146. J. T. Kittleson, I. R. Loftin, A. C. Haisrath, K. P. Engelhardt, C. Rensing and M. M. McEvoy, *Biochemistry*, 2006, **45**, 11096–11102.

147. T. D. Mealmann, I. Bagai, P. Singh, D. R. Goodett, C. Rensing, H. Zhou, V. H. Wysocki and M. M. McEvoy, *Biochemistry*, 2011, **50**, 2559–2566.

148. T. D. Mealman, M. Zhou, T. Affandi, K. N. Chacón, M. E. Aranguren, N. J. Blackburn, V. H. Wysocki and M. M. McEvoy, *Biochemistry*, 2012, **51**, 6767–6775.

149. F. Long, C.-C. Su, M. T. Zimmermann, S. E. Boyken, K. R. Rajashankar, R. L. Jernigan and E. W. Yu, *Nature*, 2010, **467**, 484–490.

150. S. Franke, G. Grass, C. Rensing and D. H. Nies, *J. Bacteriol.*, 2003, **185**, 3804–3812.

151. I. Bagai, W. Liu, C. Rensing, N. J. Blackburn and M. M. McEvoy, *J. Biol. Chem.*, 2007, **282**, 35695–35702.

152. I. Anastassopoulou, L. Banci, I. Bertini, F. Cantini, E. Katsari and A. Rosato, *Biochemistry*, 2004, **43**, 13046–13053.

153. F. W. Outten, C. E. Outten, J. Hale and T. V. O'Halloran, *J. Biol. Chem.*, 2000, **275**, 31024–31029.

154. J. V. Stoyanov and N. L. Brown, *J. Biol. Chem.*, 2003, **278**, 1407–1410.

155. J. V. Stoyanov, J. L. Hobman and N. L. Brown, *Mol. Microbiol.*, 2001, **39**, 502–511.

156. S. L. Percival, P. G. Bowler and D. Russell, *J. Hosp. Infect.*, 2005, **60**, 1–7.

157. N. Law, S. Ansari, F. R. Livens, J. C. Renshaw and J. R. Lloyd, *Appl. Environ. Microbiol.*, 2008, **74**, 7090–7093.

158. S. Hussain, R. M. Anner and B. M. Anner, *Biochem. Biophys. Res. Comm.*, 1992, **189**, 1444–1449.

159. G. V. Iyengar, W. E. Kollmer and H. J. M. Bowen, *The Elemental Composition of Human Tissues and Body Fluids*, Verlag Chemie, New York, 1978.

CHAPTER 20

Gold

FERNANDO C. SONCINI* AND SUSANA K. CHECA*

Instituto de Biología Molecular y Celular de Rosario, Departamento de Microbiología, Facultad de Ciencias Bioquímicas y Farmacéuticas, Universidad Nacional de Rosario, Consejo Nacional de Investigaciones Científicas y Técnicas. Ocampo y Esmeralda, Predio CONICET-Rosario, 2000 – Rosario, Argentina
*Email: soncini@ibr-conicet.gov.ar; checa@ibr-conicet.gov.ar

20.1   Introduction

The chemical symbol for gold is Au, derived from the latin aurum or "glow of sunrise". It must have been one of the first metals known to man. Because of its rarity, beauty and durability, it has been fashioned into a variety of objects and jewellery and stored as a valuable treasure. This follows from the fact that it does not tarnish or corrode as it does not easily react with non-metallic elements such as dioxygen. It is also malleable, ductile and a good conductor of heat and electricity. These properties mean that it has been used extensively in engineering, aerospace and catalysis as well as in medicine. This includes its use in dentistry (filling dental cavities and replacing teeth) and lately in high-tech industries to make electronic components for products such as computers and mobile phones.[1,2]

The continuous increase in the market price of gold demanded the development of economically viable and eco-friendly technologies to improve its detection and recovery.[3] One alternative is the application of naturally occurring or genetically modified organisms that, in turn, must be linked to a deep knowledge of the biogeochemical cycle of Au. Such knowledge has advanced considerably over the last few years.[4] In addition, the medical

RSC Metallobiology Series No. 2
Binding, Transport and Storage of Metal Ions in Biological Cells
Edited by Wolfgang Maret and Anthony Wedd
© The Royal Society of Chemistry 2014
Published by the Royal Society of Chemistry, www.rsc.org

application of Au nanoparticles for diagnosis or as drug carriers for the treatment of a number of diseases has greatly enhanced research in these topics.[5,6]

Gold is among the elements considered to be non-essential for life. However, an ability to metabolize it seems to have been advantageous for survival of certain microorganisms in environments such as metalliferous soils.[4,7] The toxic effects of Au and its antibacterial capacity have been documented since the nineteenth century, although the mechanism of toxicity is far from understood.[8,9]

20.1.1 Essential Chemistry of Gold

Gold is a transition metal of Group 11 of the periodic table of elements along with its lighter congeners copper (Cu) and silver (Ag). Consequently, Au atoms feature a ground state electronic configuration of $[Xe]4f^{14}6s^{1}5d^{10}$, *i.e.* a half-filled 6s orbital among filled shells. This particular electronic structure and its large size make the Au atom highly polarizable and so it can form weakly covalent interactions with ligands.[10] Au can be found at its unreactive reduced metallic form [formally Au(0)] or in six oxidation states (+1, +2, +3, +4, +5 and +7), although only the +1 [Au(I)] and +3 [Au(III)] states are commonly found in nature.

The existence of aqua ions is thermodynamically unfavourable as the reduction potentials of both Au(I) (d^{10}) and Au(III) (d^{8}) ions are more positive than that of water.[11] Under oxidizing conditions at ambient temperature, gold is found as Au(III), which tends to form four-coordinate, square planar complexes with anionic ligands.[12] Under reducing conditions, gold is found as Au(I), which favours a two-coordinate linear stereochemistry. According to the hard–soft classification of Pearson,[13] this oxidation state is classified as a soft Lewis acid because the ion is large, is of low charge and features a highly polarizable filled d shell. In consequence, the monovalent ion forms stable complexes with soft Lewis bases, such as S-, C- or N-bearing ligands.[10] In general, Au(III) compounds tend to be more stable than those of Au(I), although most decompose easily in aqueous solution, even at neutral pH, releasing the metal ion, which is reduced to the metallic form.[11]

Solubilization and dissemination of gold in the environment has been attributed historically to geochemical reactions occurring in both surface and deep-surface conditions.[14] However, the existence of an active, microbiologically driven biogeochemical cycle for Au that operates mainly under surface conditions has been proposed recently.[4] Its existence is supported by the recent discovery of bacterial biofilms over gold nuggets[15–18] and the identification of Au-sensing and -resistant bacterial systems.[9,19] These observations seem also attractive for the development of biotechnological tools to improve Au detection and recovery.[3] This chapter will review knowledge on the influence of biota in Au cycling and *vice versa*. The focus is primarily on the current understanding of the bacterial regulatory proteins involved in Au detection, on the mechanisms employed to alleviate toxicity caused by Au and how these systems might be manipulated for biotechnological uses.

20.2 Geo-chemistry, Geo-biology and Toxicity of Gold

Gold is scarce and not distributed homogeneously in nature. The average abundance of Au in the Earth's mantle and crust is about 1.3 $\mu g \ kg^{-1}$, and 0.005–0.01 $\mu g \ L^{-1}$ in seawater.[14] These levels increase at mineable deposits.[17]

The environment as well as its chemical properties determine the forms in which this element is found in nature.[14] The native metal can be found mainly as microscopic particles embedded in rock or as grains in alluvial deposits. More rarely, it appears as dendrites, wires, plates, sheets and nuggets. A number of classification systems of Au deposits have been developed by geologists. The most basic one is to group them into the two broad categories of primary and secondary deposits.[14,20] Primary deposits form by precipitation of Au during chemical reactions between metal-rich hydrothermal fluids and rocks in the Earth's crust. These deposition systems can occur either over preformed rocks or at the same time as that at which the surrounding rocks are formed. In contrast, secondary Au deposits, also commonly referred to as "placer or supergene deposits", are formed from the weathering and erosion of primary deposits followed by concentration and re-precipitation under surface conditions into sedimentary deposits. Because of its origin, native primary Au is commonly found as a variety of alloys with other metals, mainly Ag (from a few percent up to 50–60 wt% of Ag), and much more rarely with Cu, Hg, Al, Fe, Pd, Pt, Pb, Sn, Sb or Bi, amongst others.[20] The associated minerals are commonly quartz, pyrite, arsenopyrite, tellurides, selenides, silicates and oxides.[14] By contrast, gold is nearly pure (*i.e.* of "high fineness") in secondary deposits and usually contains less than 1 wt% of Ag.[20]

From the geochemical point of view, mobilization of gold in nature begins with the weathering of Au-bearing rocks and minerals (Figure 20.1A), resulting in the release of the metal.[14] The particular gold compounds formed depends on environmental factors such as the characteristics of the mineralization (geomorphology, mineralogical composition and permeability), the physicochemical characteristics of the groundwater, the presence of organic matter and the climatic conditions. From sulfide-containing minerals, the metal is mobilized as the thiosulfate complex anion $[Au^I(S_2O_3)_2]^{3-}$ when pH conditions are neutral to weakly alkaline and the environment is not too oxidizing. Under more reducing conditions, sulfido/hydrosulfido complexes form, such as $Au^I(SH)(OH_2)$, $[Au^I(HS)_2]^-$ and $[Au^I_2S_2]^{2-}$.[11] In saline or acidic groundwater from arid or semi-arid environments, gold is mobilized mainly as complexes such as $[Au^{III}Cl_4]^-$, $[Au^ICl_2]^-$ or $Au^I(OH)(OH_2)$.[11] Complexation with organic ligands occurs in organic matter-rich sediments from tropical and equatorial climates.[17] Under these

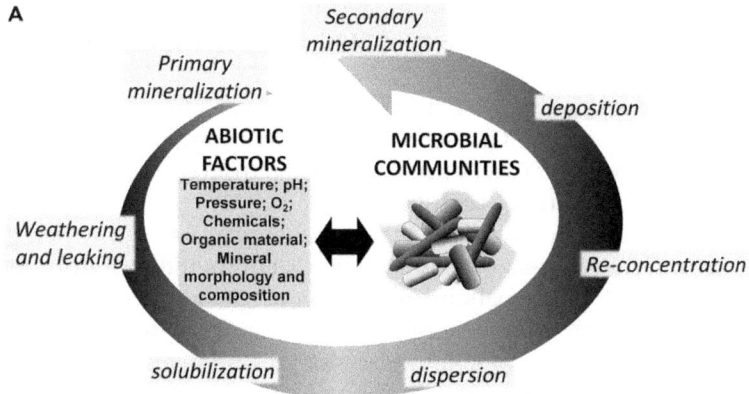

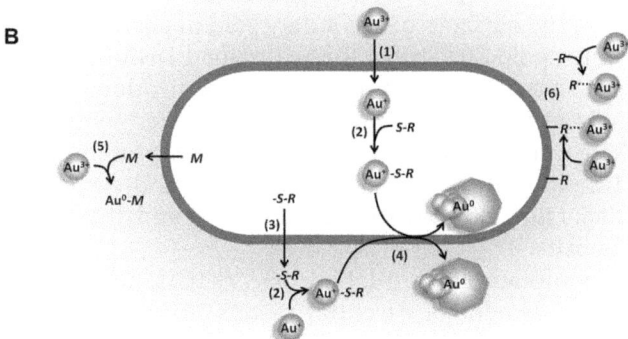

Figure 20.1 The biogeochemical cycle of Au in nature. (A) Gold is mobilized from primary mineralization sites by weathering and leaking that leads to solubilization and dispersion into the environment. Soluble Au can concentrate under appropriate conditions resulting in precipitation and deposition as secondary deposits that usually exhibit much higher purity than do the primary deposits. The local abiotic geological factors and the biota (mainly prokarya living in contact with the mineralization) each participate in every step of Au cycling. While those abiotic factors and the presence of bioavailable toxic metals influence the species predominating in the microbial community, the microorganisms themselves alter the environment as a consequence of their metabolic activities. (B) Mechanisms employed by microorganisms to mobilize Au. The predominant extracellular species is Au(III) that enters the cell and is rapidly reduced (1) to Au^I in the reducing environment of the cytoplasm. Complexes may then form (2) with the S-groups of cellular metabolites or polypeptides (S-R). Other S-containing ligands can be secreted to the extracellular milieu and interact with Au^I (3). Both intracellular and extracellular Au(I)-complexes may undergo reduction mediated by the cell membrane (4). This may generate Au^0 that precipitates, depending on the microorganism, in the cytoplasm, the cell wall or the extracellular milieu. Some microorganisms can secrete metallophores (M) that bind and/or induce precipitation of the metal (5). Ionic Au can be absorbed by the cell wall or extracellular matrix components (R) as well as by organic debris from the milieu (6).

conditions, the most abundant species is the stable complex anion $[Au^I(CN)_2]^-$, although it can also be mobilized by ligands such as amines or organic acids.

Metallic Au can also be dispersed into the environment in the form of colloids. When the metal is present as Au^0 in solution, its high polarizability allows association in microscopically disperse clusters with charged surfaces (mainly negative) that repel each other.[11]

It was assumed for a long time that the chemical form in which gold occurs in the environment as well as the kinetics of its redistribution depends exclusively on the physicochemical and geological processes mentioned above (Figure 20.1A). New evidence, however, points to the existence of biologically driven cycling of gold under surface and near-surface conditions and, in particular, to the key contribution of microbiota to this process.[4,15,18] The chemical structure of the minerals produced by living organisms cannot be distinguished from those produced by geochemical mechanisms. However, visualization on the surfaces of secondary gold deposits of bacteria-like forms or bacterioform gold structures that contained DNA has fostered the investigation of the role of microbes in gold mineralization.[21] Recent molecular profiling of the bacterial communities associated with gold grains indicates that their composition differs from those present in the surrounding areas. In addition to bacteria, fungi, algae and even plants have been linked to dispersion of gold in the biosphere and the formation of secondary gold deposits.[4,22]

All oxidized Au compounds are highly toxic.[4,9] It is predicted that the toxicity involves displacement of essential metals from the active sites of enzymes as well as ectopic interactions with different cellular components. In the reduced environment of the cytoplasm, Au(III) should be converted rapidly to Au(I), probably by membrane-associated reductases.[9] This may result in the induction of oxidative stress, which, in turn, would up-regulate oxidative stress response genes.[23] Au(I) has a high affinity for protein thiol groups, but also for the softer N- and C-based functional groups from other macromolecules.[10] This will affect metabolic processes and membrane permeability and may finally result in cell death. For several bacterial species including *Cupriavidus metallidurans*, a metal-resistant bacterium living in Au-rich environments, the minimal inhibitory concentration of the acid $HAu^{III}Cl_4$ is in the range 1–100 µM,[23,24] although its toxic effects probably start at much lower concentrations. Even though not essential, gold complexes can provide metabolic energy to some bacteria, offering a survival advantage in Au-rich ecological niches. For example, the Gram-positive actinomycete *Micrococcus luteus* uses a Au(III)-containing enzyme to oxidize methane produced anaerobically in sediments around gold mines.[25] Iron sulfide-reducing bacteria such as *Acidithiobacillus* sp. can efficiently use Au-containing minerals to obtain energy during respiration.[26]

In order to survive, bacteria have developed different strategies to detoxify Au. These include (i) active efflux of Au ions from the cytoplasm (*S. enterica*[24,27] and *C. metallidurans*[19,23]); (ii) intracellular complexation of the toxic ion[23,24] and (iii) the production of potential ligands such as thiosulfate, amino acids, cyanide and metallophores to complex and solubilize Au in forms that can be reduced and precipitated as inert gold nanoparticles either in the cytoplasm or on the cell surface[7,16] (Figure 20.1B; see also Figure 20.3). Some species such as the photosynthetic cyanobacterium *Plectonema boryanum* use membrane vesicles to shed Au from solution.[28] On the other hand, *Burkholderia cepacia* was shown to complex Au into polyhydroxybutyrate granules and to secrete a low-molar-mass protein that binds Au when the cells are grown in the presence of Au ions.[29]

The ability to develop detoxifying mechanisms would certainly be advantageous for microorganisms living near Au deposits where total and bioavailable levels of gold increase substantially.[7] Interestingly, the same mechanisms used for the defence against Au can also contribute to dispersion of the metal and to formation of biominerals.

20.3 Microbes in Gold Cycling and Mineral Formation

The presence of toxic metals in the biosphere certainly influences the composition of microbiota.[17] On the other hand, microorganisms can alter the environment as a consequence of their metabolic activities, also affecting metal speciation (Figure 20.1A). A wide range of chemolithotrophic and heterotrophic microorganisms, including bacteria and fungi, alter the redox conditions and/or pH of the medium as a consequence of their metabolic activities and/or through the release of metabolites.[4,30] These processes may disturb the integrity of both inorganic and organic Au-containing compounds, increasing or decreasing dispersion of gold in the environment and/or driving its non-specific precipitation. An outline of the commonly used microbial pathways that affect Au mobilization in the environment is given in Figure 20.1B. The production of ligands and the precipitation of gold metal in the cytoplasm, the cell wall or the extracellular milieu is usually included.[4,15,16,31] In addition, gold can be adsorbed by cell wall components such as the carboxylate or phosphate groups of lipopolysaccharides, peptidoglycans and teichuronic and teichoic acids, by capsules or exopolymers present in biofilms as well as by organic debris from the medium[4,32] (Figure 20.1B). The metal initially forms small nanoparticles that can evolve to crystals and later to more complex structures that in some cases can be visualized as bacterioform Au.[15,18] As these structures are also found in sites considered free from microorganisms, the real contribution of microorganisms in their development and in Au cycling in general is still controversial.[4,20]

One of the first groups of microorganisms linked to Au-cycling were ferrous- and sulfide-oxidizing bacteria, such as *A. ferrooxidans* and *A. thiooxidans*.[33] These bacteria live in communities over primary sulfide minerals (pyrite and arsenopyrite) mainly in arid, surficial environments poor in organic matter. These lithoautotrophic γ-proteobacteria obtain energy from the oxidation of primary sulfide minerals, releasing Au mainly as $[Au(S_2O_3)_2]^{3-}$ in the process (Figure 20.1B).[33] In the presence of oxygen, further oxidation of thiosulfate results in the reduction and precipitation of Au associated with the cytoplasmic membrane. Au-bioleaching can be achieved also by other groups of microorganisms such as different strains of antinomyces that produce thiosulfate as the result of different metabolic pathways.[7]

Bacteria and archaea that carry out dissimilatory sulfate- and Fe^{3+}-reduction are known to reduce gold thiosulfato- or chloro-complexes to obtain energy, precipitating metallic gold in the process (Figure 20.1B). This includes Gram-negative bacteria from the *Geobacter*, *Pseudomonas*, *Shewannella* and *Desulfovibrio* genera and even *Escherichia coli*, as well as several cyanobacterial and some hyperthermophilic bacterial and archaeal species (*e.g. Pyrobaculum islandicum*, *Pyrococcus furiosis*, *Thermotoga maritime*).[7,34-37] Thermophilic microorganisms that live in deep anoxic subsurface environments, as well as in hydrothermal vents and hot springs, have been proposed to be involved in the formation of primary Au-bearing sulfide minerals. They also may mediate the formation of secondary deposits.[4]

In arid environments, intracellular Au precipitation can arise from soluble $[AuCl_4]^-$. This was observed for many microorganisms including the alkalotolerant actinomycete *Rhodococcus* sp.,[38] amongst others. Au(III) reduction in these microorganisms involves the formation of an intermediate Au(I) species probably bound by sulfur ligands from cellular polypeptides (Figure 20.1B). By contrast, in auriferous soils rich in organic matter, gold mobilization appears to be linked to the activity of heterotrophic bacteria and more specifically to the excretion of cyanide or amino acids (such as aspartic and glutamic acids) to the medium (Figure 20.1B).[7] Cyanide is produced by microorganisms at the end of their growth phase as a by-product of the oxidative decarboxylation of glycine.[39,40] Several HCN-forming bacteria, such as species of the *Bacillus* and *Pseudomonas* genera and other organisms such as *Chromobacterium violaceum*, were shown to solubilize and mobilize Au as the stable cyanide complex $[Au^I(CN)_2]^-$.[39,41,42]

Besides bacteria and archaea, eukaryotic organisms like fungi are also proposed to participate in biologically driven gold cycling.[4,30] *Rhizopus oryzae*, a fungus that lives in dead organic matter, is able to reduce $[Au^{III}Cl_4]^-$ in a process that also involves a Au(I) intermediate. This leads to generation of Au-nanoparticles that accumulate both in the cell wall and in the cytoplasm.[31] Other species, including yeast, also accumulate Au as nanoparticles.[43] A range of microorganisms, including bacteria, archaea, fungi or even microbial consortia, are employed currently in the extraction of Au from mine tailings or to obtain Au nanoparticles.[3]

Until recently, the role of microorganisms in Au-cycling was studied mostly in laboratory experiments using pure cultures.[4] In consequence, it was difficult to extrapolate these results to real events occurring with complex microbial communities living in natural environments. Recent advances in instrumentation are allowing the characterization of the Au-associated microbial communities.[15–17,23] These advances include high-resolution microscopies (transmission electron microscopy (TEM), high-resolution scanning electron microscopy (SEM), confocal scanning laser microscopy (CSLM)), spectroscopic techniques (X-ray absorption spectroscopy (XAS), inductively coupled plasma mass spectrometry (ICP-MS)) and molecular biology techniques (16S rDNA sequencing and PCR-DGGE fingerprinting; three-domain multiplex-terminal restriction fragment length polymorphism (M-TRFLP), shotgun cloning and sequencing and microarrays).

Two phylogenetically related β-proteobacteria, *C. metallidurans* and *Delftia acidovorans*, were found to dominate biofilms associated with Au grains in secondary gold deposits from temperate and tropical Australian sites.[23] These microorganisms share several heavy-metal resistant genes suggesting that they probably underwent one or more horizontal gene transfer events during evolution.[18,44] In recent reports,[15,16,21] *C. metallidurans* or *D. acidovarans* were cultured in the presence of Au ions, simulating the conditions found in nature. It was demonstrated that *C. metallidurans* accumulates Au from solution and intracellularly precipitates pure gold particles[15,21] while *D. acidovarans* creates gold particles outside its cell wall.[16] Synchrotron X-ray analysis revealed that, at least for *C. metallidurans*, Au accumulation is coupled to the formation of intracellular Au(I)-sulfido complexes.[23] Resident microbial communities in Au-rich soils from different regions of Australia were analysed by molecular fingerprints and phylogenetic and functional microarrays combined with field mapping and geochemical analyses to obtain a realistic picture of the microcosm.[17] These results indicate that Au-rich samples were dominated by a great diversity of α-proteobacteria (especially *Sphingomonadales*), *Firmicutes* and *Actinobacteria*. Functional gene array analysis showed an increased presence of metal-resistance genes in the samples from auriferous soils and, in particular, of the genes *chrA*, *czcD*, *copA* and *zntA* involved, respectively, in Cr, Cd/Zn/Co, Cu/Ag and Zn transport and detoxification. These mixed microbial communities forming biofilms in auriferous soils created not only cores of microbial activity but also physicochemical gradients that may affect the rates of mineral weathering and metal mobilization. The formation of secondary gold deposits may also be affected by dispersion of soluble Au complexes and nanoparticles. In these conditions, bacterial lysis or biofilm disruption can also enhance the mobility of Au nanoparticles with the formation of pseudo-crystalline gold, which could evolve to more complex structures and Au deposition.[18]

20.4 Accumulation of Gold in Plants and Animals

Anthropological activities, especially mining, have increased Au concentrations at numerous locations throughout the world. For example, they have reached concentrations as high as 19 μg L^{-1} in a freshwater stream near a gold mining site, 843 µg kg^{-1} in alluvial soil near a gold mine and 256.0 mg kg^{-1} in freshwater sediments near a gold mine tailings pile.[45] As a consequence, most organisms living in these environments accumulate the metal as a defensive strategy against its toxic effects. There have been reports of fish and aquatic invertebrates with gold concentration up to 38 µg kg^{-1} of dry weight (DW). Gold concentrates mainly in the hair of humans. While the normal range is between 6 and 880 µg kg^{-1} DW, hair of goldsmiths accumulates up to 1.4 mg kg^{-1} DW. Also, about 1 µg L^{-1} Au was determined in the urine of dental technicians and up to 2 µg L^{-1} in breast milk (attributed to dental fillings and jewellery).[46] Exposure to gold also results in its accumulation in several organs such as liver, kidney and spleen, as well as in adrenal and adipose tissue. For example, patients treated with Au-compounds contained up to 2.4 mg L^{-1} of Au in blood.[47]

Several studies have indicated that Au is taken up passively by plant roots and is then complexed, transported and immobilized into different tissues.[22] Although the background levels detected in plants are usually very low and rarely exceed 10 µg kg^{-1} of DW tissue, they increase substantially in aquatic and terrestrial plants, particularly in trees growing close to gold exploitation.[45] For example, more than 1 mg kg^{-1} DW of Au were reported in algal mats and up to 100 mg kg^{-1} DW in gold accumulator plants. The latter are able not only to grow on metalliferous soils but also to accumulate high amounts of heavy metals in aerial organs without suffering phytotoxic effects. The ability of plants living on mineralized zones to acquire gold from soil and to accumulate the metal in leaves and rootlets (tissues that would be discarded by the plant) suggests that plants also have a role in the biogeochemical cycling of gold.[22]

Besides abiotic factors, different metabolic processes taking place at the plant roots influence Au bioavailability and uptake. Some cyanogenic plants produce and secrete cyanide that increases Au mobility by forming stable [Au(CN)$_2$]$^-$ that can be easily taken up by roots.[48] Cyanide can be released from plants as the result of tissue damage as well as from leaf litter decomposing on the soil surface.[49] Indeed, the microbial communities living in the root zone (the rhizosphere) or as endophytes in plant tissues appear to secrete compounds that help to mobilize the metal, so enhancing plant bioaccumulation.[50]

Gold accumulates in different tissues of plants and algae as Au complexes or as nanoparticles.[22] Nanoparticles of diameter 5–50 nm were observed in *Brassica juncea* by TEM,[51] and gold particles (diameter < 20 nm) were found both inside as well as on the cell surface of *Lyngbya majuscula* exposed to HAuCl$_4$ solution.[52] These observations led to consideration of the use of accumulator plants for phytoremediation of areas near gold exploitations or to serve as biogeochemical indicators of local gold deposits. The production of Au-nanoparticles or even phyto-mining have been canvassed. In this latter case, improvement in the recovery of the precious metal was achieved by adding to the soil complexing agents such as ammonium or sodium thiocyanate, ammonium thiosulfate, thiourea, sodium or potassium cyanide, *etc.* [22] The aim was to mobilize and promote plant uptake and accumulation of the residual gold from tailings, waste rock or ores. The process was termed "induced hyper-accumulation". With this technique, up to 57 mg kg[-1] DW of gold was reported to have been recovered from *B. juncea*, the mustard plant. A recent report indicates that *Chilopsis linearis* plants grown in hydroponics with medium containing 10 mg kg[-1] Au accumulated up to 1000, 200 and 30 mg kg[-1] DW of Au in roots, stems and leaves, respectively.[53]

20.5 Molecular Bases for Bacterial Gold Sensing

In natural environments, gold is usually accompanied by other metals such as Ag and Cu.[20] These also belong to Group 11 of the periodic table and so share similar physicochemical properties. This compromises their individual recognition, even though certain bacterial species harbour sensor proteins that can achieve selective detection of Au.[9] Upon detecting the metal ion in the cytoplasm (probably as Au(I)), these proteins activate expression of resistance determinants that initiate elimination of the toxic metal. A selective and sensitive gold sensor was discovered initially in *S. enterica*[24] (Figure 20.2A) and was followed by a xenologue in *C. metallidurans*.[19] These protein sensors, respectively named GolS and CupR, belong to the MerR family of bacterial transcriptional regulators that control resistance to heavy metals, xenobiotics and other types of stress signals.[54,†] The closest homologues are the copper sensors present in *E. coli* and other Gram-negative bacteria.[9]

Dimeric MerR metalloregulators activate expression of their target genes, which code for metal efflux or detoxification systems.[54] Each monomer can be divided into three functional domains: the N-terminal DNA-binding region containing a characteristic winged helix-turn-helix structure, a long anti-parallel coiled-coil dimerization helix and the C-terminal inducer binding domain. Binding of the metal ion at a solvent-inaccessible site at the dimer interface provokes an allosteric change at the N-terminal DNA binding region of the protein, which, in turn, induces changes in the promoter structure resulting in transcriptional activation of the target genes.[54,55] Typically, these metalloregulators recognize inverted repeat sequences spanning σ^{70}-dependent promoters with an unusual 19 or 20 bp spacer between the -35 and -10 elements.[54]

The ability of the metal sensor to detect an ion depends on the intrinsic properties of the metal ion itself (charge, size, coordination chemistry, intracellular availability) and on the protein environment, that is, the array of ligands forming the first and second coordination spheres within the folded protein.[55] In consequence, many metal-ion sensors are poorly selective as they cannot distinguish between ions with similar physicochemical properties and coordination chemistry. Of the essential metal ions, Cu(I)

---

†The prototypical MerR protein was discovered as a regulator of mercury resistance (see Chapter 24). Other aspects are discussed in Chapters 16, 23 and 28.

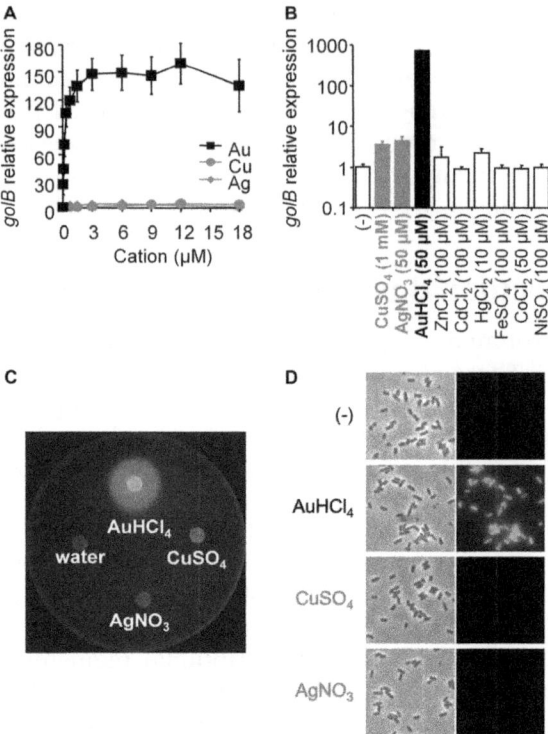

Figure 20.2   The Au-selective response of the *Salmonella gol* regulon. (A) Relative
expression of the GolS-dependent *golB::lacZ* transcriptional fusion from
overnight-grown cells in SM9 minimal media containing the indicated
concentrations of AuHCl$_4$ (Au), CuSO$_4$ (Cu) or AgNO$_3$ (Ag). A 148-fold
induction was achieved with 3 µM AuHCl$_4$ in comparison with an 8-fold
increase at 18 µM CuSO$_4$ and a 9-fold increase at 6 µM AgNO$_3$. (B)
Relative expression from a *golB::lacZ* transcriptional fusion *S. typhimur-*
ium cells grown overnight in LB broth with the addition of the indicated
concentrations of different heavy-metal salts. Circa 650-fold *golB* induction was detected by addition of 50 µM AuHCl$_4$ to the culture
medium, while addition of 50 µM AgNO$_3$, 1 mM CuSO$_4$, 10 µM HgCl$_2$ or
100 µM ZnCl$_2$ resulted in a 9-fold, a 3-fold, a 0.9-fold or a 0.6-fold
increased expression of the reported gene, respectively. No appreciable
effect was detected with the rest of the metal salts. GolS-controlled
promoters can also be used to develop highly sensitive and selective
bacterial biosensors for the detection of soluble Au compounds in the
environment. Either wild-type *Salmonella* or an engineered *Escherichia*
coli strain with the gene coding for GolS ectopically inserted into the
chromosome under its own promoter, carrying a *gfp*-reporter gene
under the control of the GolS-controlled *golB* promoter in a multicopy plasmid, were generated to selectively detect bioavailable gold.
These biosensors were able to produce a detectable 2.31-fold increase in
fluorescence with the presence of as low as 33.23 nM Au(I), equivalent
to 6.54 mg L^{-1} (or ppm). Au-selective induced fluorescence can be
visualized in diffusion assays on agar plates (C), by fluorescence microscopy (D) or quantified in a fluorescence spectrometer (not shown).
Adapted from refs. 24, 60.

(3d^{10}) is uniquely able to sustain linear two-coordination. Consequently, Cu(I) sensors such *E. coli* CueR provide two conserved cysteine residues (Cys112 and Cys120) that define a metal-binding loop in the C-terminal domain.[56,57] However, CueR cannot differentiate between the Group 11 ions Cu(I), Ag(I) or Au (I).[56,58] A serine residue (Ser77) at the beginning of the dimerization helix of the other monomer approaches the metal-binding region in the folded dimer providing a shielded hydrophobic environment that may disfavour binding of metal ions such as Zn(II) and Hg(II), which typically require a higher coordination number.[55,56]

A common ancestor of these three sensors GolS, CupR and CueR is predicted to have evolved the binding motifs that characterize the group of monovalent metal ion sensors, *i.e.* the C-terminal C-X-G-X4-D-C-P metal-binding loop containing the conserved Cys112 and Cys120 residues of CueR plus the shielding serine residue (Ser77).[9] However, the Au sensors are able to distinguish Au(I) from Cu(I) and Ag(I) and serve as paradigms for the molecular ability of metalloregulators of the MerR family to discriminate between such similar metal ions. The ability of GolS to sense Au(I) was used in the construction of a highly sensitive and selective *E. coli* biosensor for the detection of soluble Au compounds in the environment (Figure 20.2B–D).[59,60]

It appears that the selectivity of GolS for Au(I) over Cu(I) or Ag(I) compared to CueR is based on subtle differences in the detailed interaction of the metal ion with its binding site. There is also an efficient cellular Cu detection and detoxification system in *Salmonella*, the *cue* regulon.[9] CueR-like copper responsiveness was observed in a GolS variant whose residues within the Cys112-Cys120 metal-binding loop were replaced by their counterpart in the copper sensor.[24] Conversely, a lower Cu response was observed in a CueR variant with the metal binding loop of GolS.[61] Interestingly, GolS and its homologues have alanine and proline at positions 113 and 118, respectively, while CueR-like proteins have proline and alanine at these respective positions. It appears that the metal-selectivity relies mainly on these residues.[61]

X-ray fluorescence studies performed on the Cu(I)-CupR complex were best fitted to a three-coordinate structure composed of two short Cu-S bonds (to conserved C112 and C120) and one longer Cu-S interaction, assumed to involve Cys137 from the "CHH" motif at the C-terminus of the protein.[19] CueR features an analogous Cys/His rich "CCHH" motif (C129-H132) at its C-terminus. However, both structural and spectroscopic studies were consistent with a linear two-coordinate geometry for the binding of Cu(I), Ag(I) or Au(I) to CueR.[56,57] Truncated versions of CueR and CupR with these Cys/His rich motifs missing still recognize Au(I) and Cu(I), albeit with altered affinities.[19,58] It is possible that these motifs may modulate but not determine metal selectivity. Interestingly, the *Salmonella* GolS sensor features neither a similar motif nor a Cys residue in its C-terminal region. In fact, the presence of such a motif adjacent to the metal binding loop is the exception rather than the rule for MerR monovalent metal sensors.[24,62] Moreover, replacement of the C-terminal region of GolS (after the metal-binding loop) with that of *S. typhimurium* CueR did not affect its metal selectivity.[24] Until structural studies for the Au sensors are available, the current data indicate that metal selectivity depends on the nature of the metal-binding loop in these sensors.[61]

All MerR monovalent metal ion sensors characterized to date recognize very similar operator sequences, two inverted repeat sequences placed in between and overlapping the -35 and -10 promoter elements.[9,54] Several bacterial species such as *Salmonella* harbour more than one gene coding for a MerR sensor in their genomes and it is evident that each of these gene products is simultaneously expressed in a single bacterial cell and also controls the expression of different sets of target genes. This then posed a conundrum about how these regulators recognize their target operators to prevent ectopic activation of heterologous resistance factors. Recent work shows that, at least in *Salmonella*, *in vivo* cross-activation of the target genes controlled by GolS or by CueR is prevented by i) the presence of signature nucleotide bases in the centre of the operators,[62] which is correlated with particular amino acid residues in the α2-helix of the N-terminal DNA-binding motif in the sensor,[63] and by ii) tight control of the intracellular concentration of each of these transcription factors.[62]

Remarkably, the presence of these signature nucleotide base/amino acid residue pairs that differentiate *gol versus cue* regulons in *Salmonella* is not restricted to this species but is conserved in all bacterial GolS-like regulators and their target genes, including *C. metallidurans* CupR. This clearly differentiates them from CueR-like sensors and their predicted operators.[62] All GolS xenologues are predicted to control the expression of their target genes by recognizing GolS-like target binding sites, *i.e.* sequences harbouring A and T at positions 3′ and 3, respectively, from the centre of the operators dyad. Conversely, CueR-like regulators are predicted to control genes with operators harbouring either a C or a G at the same positions. Accordingly, GolS-like proteins harbour a conserved Met at position 16 while Thr, Ala or Ser but not Met are found in CueR homologues. Recent results indicate that the replacement of GolS Met16 (in the α2-helix of the winged helix-turn-helix motif that is predicted to contact the DNA[64]) by Ala reduces the recognition of the native *gol* promoters and increases its affinity for *cue*-like operators.[63] Conversely, a Met residue in place of the Ala16 in CueR switches its promoter selectivity. These studies not only confirmed that GolS and CueR can be distinguished as paralogues that evolved from a common ancestor but also provide the demonstration of phylogenetic differences between GolS xenologues and the CueR cluster to avoid cross-activation.[62]

20.6 Bacterial Gold Resistance by Active Efflux and Sequestration

One of the more effective strategies of defence against heavy-metal toxicity developed by prokaryota is the rapid elimination of the toxic ions from the cell by directing active efflux.[65,66] This task is most commonly carried out by metal-exporting proteins such as P-type ATPases, CBA-efflux pumps and cation diffusion facilitators that transport the metal ion from the cytoplasm or the periplasm to the outside milieu (Figure 20.3).

The Au sensing/response locus from *Salmonella* includes genes coding for a P-type ATPase GolT, a metal-binding protein GolB, and a CBA-type efflux system GesABC, which is encoded in an operon forming a divergon with *golTS*[27] (Figure 20.3). Deletion of any of the *gol* genes affects Au resistance, although the most severe phenotype is attained by either the inactivation of either the whole *gol* locus (deletion of all the GolS target genes *gesABC*, *golT* and *golB*) or deletion of *golS*, the gene coding for the transcriptional regulator.[24,27] Either GolS exerts tight regulation on its target genes or, more probably, there is sequestration of the metal ion by this sensor protein that is, in essence, a metal-binding protein. In fact, the intracellular concentration of GolS increases by up to 400-fold after exposure to Au.[62]

GolT and GolB are expressed rapidly after exposure to Au(I) and may comprise the first line of defence against toxicity, serving as export pump and metallochaperone, respectively (Figure 20.3).[24] However, the precise function of GolB is under investigation. GolB may sequester the toxic metal using two conserved Cys residues in the N-terminal MXCXXC motif (characteristic of Cu(I) metallochaperones) or it may be required for the delivery of the metal ion to the P-type transporter GolT. Recently, an *E. coli* strain was constructed in which GolB was fused to the N-terminal region of OmpA, an *E. coli* outer membrane protein.[67] The engineered strain could effectively sequester Au ions from the medium, a biotechnological tool that has potential for exploitation or bioremediation purposes to recover or remove gold from the environment.

The *Salmonella* GesABC system is unusual in several aspects. Firstly, its transcription is delayed and requires higher GolS concentrations than those of *golTS* and *golB*.[27] Moreover, it is induced only by the presence of Au in a GolS-dependent manner, even in a variant strain deleted of the copper-resistance *cue* regulon.[27,62] Secondly, it has higher homology with members of the CBA/RND family of transporters that export organic molecules (such as AcrAB and MexEF-OprN systems) than those members of this family responsible for metal-ion resistance (such as CusCFBA, SilCBA or CzcCBA). Because of its delayed induction, it can be considered to be a second line of defence, required either under prolonged metal stress or at higher gold levels.

The GesABC complex is likely to span the cell envelope from the cytoplasm to the outer space, directing its substrate from either the periplasm or the cytoplasm to the outside milieu[68] (Figure 20.3). Elimination of periplasmic Au(I) ions may serve to avoid the damage caused by the displacement of essential Cu(I) ions from their binding sites in this compartment. In cells lacking the main drug transporter AcrAB, GesABC can also confer resistance to different antimicrobial agents and chemical compounds either by ectopic overexpression[69,70] or, physiologically, by its Au-activated GolS-mediated induction.[27] In view of these observations and the similarity of GesB (the

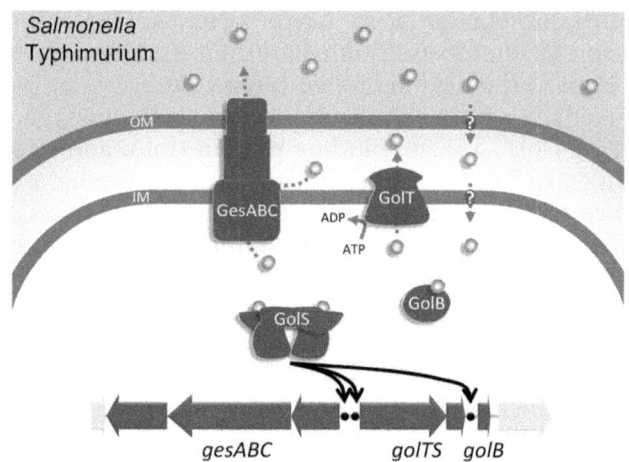

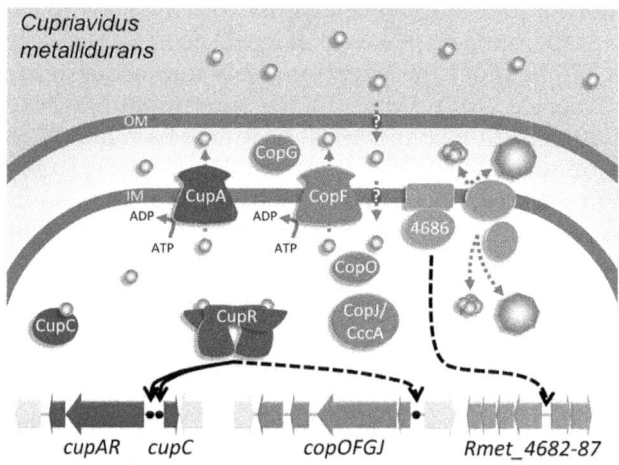

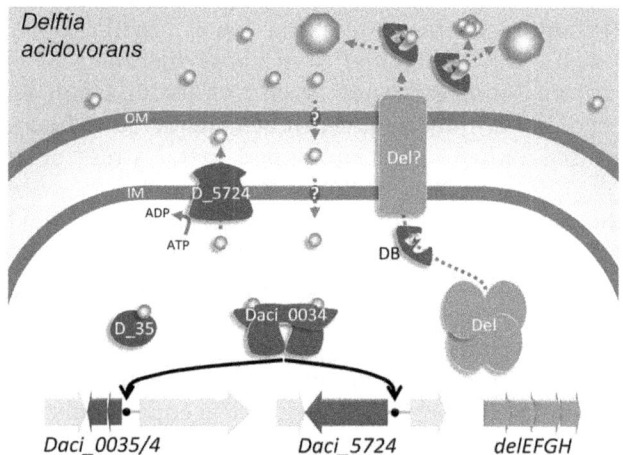

substrate-specificity component) to xenobiotics detoxification proteins such as AcrB and MexF, it is tempting to hypothesize that this system is involved in the elimination of a gold compound or cellular metabolites damaged by the toxic metal ion, rather than free Au(I) ions.

The presence of genes linked to Au resistance (such as those coding for homologues to the *C. metallidurans* CupA P-type transporter[17]) in bacterial communities living in auriferous soils suggests possible horizontal gene transfer events in these environments. In fact, genes encoding GolS/CupR-like proteins, as well as P-type ATPases and metallochaperones are predicted in the genomes of *D. acidovorans*, *Stenotrophomonas maltophilia* and *Burkholderia cenocepacia*, all found associated with biofilms on gold nuggets.[9,23,62]

The *C. metallidurans* Au sensor CupR activates expression of the P_{1B}-type ATPase CupA, encoded in the same operon as the transcriptional regulator, and the small metal-binding protein CupC (Figure 20.3).[19,23,71] Transcriptomic analysis conducted in *C. metallidurans* CH43 exposed to HAuCl$_4$ for 10 and 30 min revealed the presence of several gold-inducible genes, including the Au-sensing/response *cup* locus and the Rmet_4682-Rmet_4687 cluster that codes for a group of predicted proteins of unknown function and a sigma/anti-sigma transcription system that may control their expression.[23] Other genes from the pMOL30 plasmid predicted to be involved in copper resistance[72] were reported to be up-regulated by gold, including the *copOFGJ* operon (Figure 20.3). This finding fits with the *in silico* prediction of a putative CupR-operator upstream of this transcriptional unit.[24,62]

Figure 20.3 Gold-responsive/resistance systems present in the enteropathogen *Salmonella* Typhimurium and in the gold-nuggets biofilm forming species *Cupriavidus metallidurans* and *Delftia acidovorans*. In the presence of submicromolar concentration of Au, the *Salmonella* GolS transcription factor induces transcription of genes encoding the P-type ATPase GolT, the metal binding protein GolB and the GesABC efflux system. These act to reduce the free metal ion levels in the cytoplasm or the periplasm.[9] Similarly, CupR from *C. metallidurans* induces the expression of the GolT- and GolB-homologues CupA and CupC, respectively. A CupR-operator was predicted in the promoter region of the operon *copOFGJ*, coding for putative Cu-resistance factors. The Rmet_4682-4687 locus is among the highest Au-induced genes detected in a microarray analysis.[23] This is composed by the Rmet_4285-82 operon coding for a group of predicted membrane and transmembrane proteins of unknown function and an extracytoplasmic function sigma factor (Rmet_4686) and its putative anti-sigma factor (Rmet_4287), which are transcribed as an operon and are expected to regulate the expression of the Rmet_4285-82 operon. These genes could be involved in the reduction and the formation of nanoparticulate Au⁰. Finally, the presence of a *gol*-like regulon is predicted in *D. acidovorans* that includes the sensor, the membrane transporter and the cytoplasmic metal-binding protein. *D. acidovorans* synthesizes and secretes a linear polyketide-nonribosomal peptide, named delftibactin, that sequesters and precipitates Au.[16]

A recent report disclosed a new strategy of defence against toxic gold levels operating in these bacterial communities.[16] *D. acidovorans* synthetizes and secretes a linear polyketide-nonribosomal peptide delftibactin that sequesters Au in the bacterial surroundings (Figure 20.3). This secondary metabolite is predicted not only to interact with Au(III) but also to favour the formation of colloidal gold nanoparticles and small aggregates similar to those found in natural deposits.[7,20] The presence of delftibactin can effectively lower the bioavailable gold levels from toxic to non-toxic for *D. acidovorans* and probably for other microorganisms forming a mixed biofilm.[16] This observation represents the first example of an extracellular tool that can be linked to biomineralization to defend against toxic gold levels produced by a microorganism living in an Au-rich environment.

20.7 Gold in Human Hands: Therapeutic Use of Gold Compounds and Nanoparticles

Gold-based preparations have been used in medicine since ancient times and the practice has been designated as "chrysotherapy" or "aurotherapy". However, the biological properties of gold compounds were not supported by experimental laboratory data until the last century.[5,8] In 1890, Robert Koch had observed that gold cyanide was effective against *Mycobacterium tuberculosis* and, at the beginning of the twentith century, different Au-containing compounds were applied to the treatment of a number of human diseases. For example, krysolgan (sodium 4-amino-2-auromercaptobenzoate) was reported as an effective treatment of lupus erythematosus, and sanochrysine ($Na_3[Au^I(S_2O_3)_2]$), myochrysine (aurothiomalate) and allochrysine (sodium gold(I) thiopropanolsulfonate) were each described as treatments of skin and pulmonary tuberculosis (Figure 20.4).[5] This last finding led to the application of these and other Au(I)thiolate drugs such as solganol (aurothioglucose) and auronofin [(2,3,4,6-tetra-*O*-acetyl-1-(thio-κS)-β-D-glucopyranosato)(triethylphosphine)gold(I)] for the treatment of rheumatoid arthritis. One of the major problems with these compounds is their elevated toxicity, which in most cases appears at concentrations close to their therapeutic doses.

Auranofin was also shown to have an antiproliferative action on HeLa and other carcinoma-derived cell lines.[73,74] Subsequent research led to the development of new and improved gold-based anticancer complexes that show antiproliferative properties against selected human tumour cell lines, including those resistant to the classic platinum drugs.[5,75] Some of these compounds were remarkably effective, even in tumour models *in vivo*.

Both Au(I) and Au(III) compounds are currently subject to clinical investigation, not only for the treatment of different types of cancer but also for HIV/AIDS therapy, in which they showed promising results.[76] Au(III) is isoelectric with platinum(II) and forms square planar compounds (Figure 20.4) similar to cisplatin *cis*-Pt(NH$_3$)$_2$Cl$_2$, the successful anticancer drug.[75]

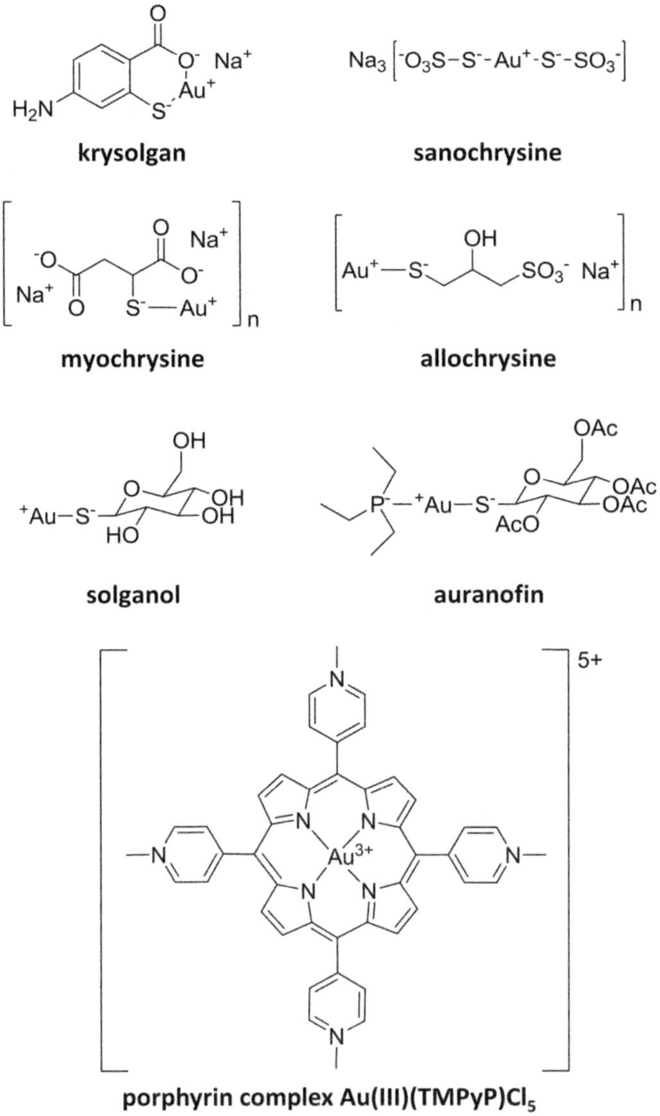

Figure 20.4 Structures of gold complexes with therapeutic uses.

However, Au(III) complexes are highly unstable under physiological conditions as the metal is rapidly reduced to Au(I). The stabilization of the redox-active Au(III) ion with porphyrin resulted in complexes with promising *in vivo* antitumour properties towards nasopharyngeal carcinoma, hepatocellular carcinoma, colon cancer, melanoma and neuroblastoma.[77]

Despite numerous biochemical studies, the mechanisms through which gold compounds produce their biological effects are not understood. In addition, whether they are subject to processing before the active moiety reaches its cellular target is unknown. Extensive lines of evidence show that, in contrast to cisplatin, the cytotoxicity of gold compounds is not due to targeting DNA.[75] Au(I) compounds were found to form complexes with glutathione, with albumin and with haemoglobin in red blood cells.[78,79] They are also reported to inhibit a number of essential enzymes, including the zinc finger protein PARP1[80,81] and proteases[82,83] involved in progression of both cancer and rheumatoid arthritis.

Au(I) and Au(III) compounds were shown to inhibit cytoplasmic and mitochondrial selenocysteine-containing thioredoxin reductases.[84,85] These selenoproteins catalyse the NADPH-dependent reduction of thioredoxin, their native substrate. They contain a redox-active selenocysteine residue in their active sites proposed to be the main target for the metal ion. Recent results demonstrated that Au(I) and, to a lesser extent, the Au(I)-triethyl-phosphine fragment from auranofin (Figure 20.4), forms adducts with the Cu(I) metallochaperone Atox-1 and with the glutathione-S-transferase GST P1, involved in cellular detoxification processes of toxic compounds of both endogenous and xenobiotic origin.[86,87] Stable porphyrin-Au(III) compounds have been shown to induce apoptosis by the activation of the p38 mitogen-activated protein kinase.[88] Whether these actions involve the stable complexes or whether reduction and ligand dissociation occurs prior to apoptosis induction is still under analysis. Overall, the unique chemistry of Au and, in particular, its affinity for selenoproteins and/or thiol-containing proteins is of great interest in the development of new gold-based therapeutics.

Gold has also been used in the preparation of nanoparticles that offer an interesting combination of chemical and physical properties with potential for biomedical applications.[6] They can be adapted to bind different ligands at their surfaces and to shield unstable drugs or poorly soluble substrates, facilitating their efficient delivery to otherwise inaccessible regions of the body. In some cases Au nanoparticles can even act as intrinsic drug agents.[89] Besides, these particles not only enter into cells but also accumulate at sites of tumour growth/inflammation. This, in addition to their ability to efficiently convert light into heat, make them useful for specific thermal ablation of tumours or infected tissues.[90] Because of their ability to absorb X-rays, they can enhance cancer radiation therapy or increase imaging contrast.[91] Their intense photophysical properties also allow their use in bio-diagnostic assays as well as in the detection of several molecules at low concentrations. Detailed descriptions of these and other applications of Au nanoparticles can be found in recent reviews.[6,92]

20.8 Conclusions and Perspectives

The role of organisms, in particular bacteria, in the biogeochemical cycle of gold has been the target of much recent work. The accumulated evidence indicates that several species not only interact with gold-containing minerals but are also directly responsible for the formation of biogenic gold deposits

(Figure 20.1). The characterization of the mechanism engaged by micro-organisms to counteract the toxic effects of the metal in its bioavailable forms allowed identification of Au-activated transcriptional regulators that direct the expression of metal transporters and/or metal-binding proteins involved in Au resistance (Figures 20.2, 20.3). This signal-responsive regulatory circuit has been employed as a platform for the design of whole-cell bacterial Au biosensors. Furthermore, the resistance pathways that different organisms employ and that increase Au mobility and favour re-precipitation can be manipulated to develop eco-friendly and economical technologies to improve recovery, recycling and even remediation. Industrial-scale bio-leaching operations for gold recovery or bio-mining employing chemo-lithoautotrophic microbes have been established in several countries including South Africa, Australia, Brazil and Chile. Another attractive alternative would be the use of genetically manipulated accumulator plants.

Promising experimental results are boosting the development of novel gold compounds for medical purposes and for the application of innovative technologies. The development of biological methods for the synthesis of Au nanomaterials appears as a viable alternative to traditional physicochemical techniques applied in microelectronics and nanotechnology.

Acknowledgements

We thank L. I. Llarrull for her help in the generation of ChemDraw structures. Our work is supported by grants from Agencia Nacional de Promoción Científica y Tecnológica and from the National Research Council (CONICET). F.C.S. and S.K.C. are career investigators of the same institution. F.C.S. is also a career investigator of the Rosario National University Research Council (CIUNR).

References

1. C. R. M. Butt and R. M. Hough, *Elements*, 2009, **5**, 277–280.
2. C. M. Cobley and Y. Xia, *Elements*, 2009, **5**, 309–313.
3. C. M. Zammit, N. Cook, J. Brugger, C. L. Ciobanu and F. Reith, *Miner. Eng.*, 2012, **32**, 45–53.
4. G. Southam, M. F. Lengke, L. Fairbrother and F. Reith, *Elements*, 2009, **5**, 303–307.
5. S. J. Berners-Price and A. Filipovska, *Metallomics*, 2011, **3**, 863–873.
6. E. C. Dreaden, A. M. Alkilany, X. Huang, C. J. Murphy and M. A. El-Sayed, *Chem. Soc. Rev.*, 2012, **41**, 2740–2779.
7. F. Reith, M. F. Lengke, D. Falconer, D. Craw and G. Southam, *ISME J.*, 2007, **1**, 567–584.
8. T. G. Benedek, *J. Hist. Med. Allied Sci.*, 2004, **59**, 50–89.
9. S. K. Checa and F. C. Soncini, *Biometals*, 2011, **24**, 419–427.
10. C. E. Housecroft and E. C. Constable, *Chemistry*, Pearson Education, Essex, UK, 3rd edn, 2006.

11. A. E. Williams-Jones, R. J. Bowell and A. A. Migdisov, *Elements*, 2009, **5**, 281–287.

12. A. Usher, D. C. McPhail and J. Brugger, *Geochim. Cosmochim. Acta*, 2009, **73**, 3359–3380.

13. R. G. Pearson, *J. Am. Chem. Soc.*, 1963, **85**, 3533–3539.

14. J. L. Walshe and J. S. Cleverley, *Elements*, 2009, **5**, 288–296.

15. L. Fairbrother, B. Etschmann, J. Brugger, J. Shapter, G. Southam and F. Reith, *Environ. Sci. Technol.*, 2013, **47**, 2628–2635.

16. C. W. Johnston, M. A. Wyatt, X. Li, A. Ibrahim, J. Shuster, G. Southam and N. A. Magarvey, *Nat. Chem. Biol.*, 2013, **9**, 241–243.

17. F. Reith, J. Brugger, C. M. Zammit, A. L. Gregg, K. C. Goldfarb, G. L. Andersen, T. Z. DeSantis, Y. M. Piceno, E. L. Brodie, Z. Lu, Z. He, J. Zhou and S. A. Wakelin, *ISME J.*, 2012, **6**, 2107–2118.

18. F. Reith, L. Fairbrother, G. Nolze, O. Wilhelmi, P. L. Clode, A. Gregg, J. E. Parsons, S. A. Wakelin, A. Pring, R. Hough, G. Southam and J. Brugger, *Geology*, 2010, **38**, 843–846.

19. X. Jian, E. C. Wasinger, J. V. Lockard, L. X. Chen and C. He, *J. Am. Chem. Soc.*, 2009, **131**, 10869–10871.

20. R. M. Hough, C. R. M. Butt and J. Fischer-Bühner, *Elements*, 2009, **5**, 297–302.

21. F. Reith, S. L. Rogers, D. C. McPhail and D. Webb, *Science*, 2006, **313**, 233–236.

22. V. Wilson-Corral, C. W. N. Anderson and M. Rodriguez-Lopez, *J. Environ. Manage.*, 2012, **111**, 249–257.

23. F. Reith, B. Etschmann, C. Grosse, H. Moors, M. A. Benotmane, P. Monsieurs, G. Grass, C. Doonan, S. Vogt, B. Lai, G. Martinez-Criado, G. N. George, D. H. Nies, M. Mergeay, A. Pring, G. Southam and J. Brugger, *Proc. Natl Acad. Sci.*, 2009, **106**, 17757–17762.

24. S. K. Checa, M. Espariz, M. E. Audero, P. E. Botta, S. V. Spinelli and F. C. Soncini, *Mol. Microbiol.*, 2007, **63**, 1307–1318.

25. L. A. Levchenko, A. P. Sadkov, N. V. Lariontseva, E. M. Koldasheva, A. K. Shilova and A. E. Shilov, *J. Inorg. Biochem.*, 2002, **88**, 251–253.

26. L. X. Sun, X. Zhang, W. S. Tan and M. L. Zhu, *J. Biosci. Bioeng.*, 2012, **114**, 531–536.

27. L. B. Pontel, M. E. Audero, M. Espariz, S. K. Checa and F. C. Soncini, *Mol. Microbiol.*, 2007, **66**, 814–825.

28. M. Lengke and G. Southam, *Geochim. Cosmochim. Acta*, 2006, **70**, 3646–3661.

29. D. P. Higham, P. J. Sadler and M. D. Scawen, *J. Inorg. Biochem.*, 1986, **28**, 253–261.

30. G. M. Gadd, *Mycol. Res.*, 2007, **111**, 3–49.

31. S. K. Das, J. Liang, M. Schmidt, F. Laffir and E. Marsili, *ACS Nano*, 2012, **6**, 6165–6173.

32. N. Das, *Hydrometallurgy*, 2010, **103**, 180–189.

33. M. G. Aylmore and D. M. Muir, *Miner. Eng.*, 2001, **14**, 135–174.

34. S. Karthikeyan and T. J. Beveridge, *Environ. Microbiol.*, 2002, **4**, 667–675.

35. Y. Konishi, T. Tsukiyama, N. Saitoh, T. Nomura, S. Nagamine, Y. Takahashi and T. Uruga, *J. Biosci. Bioeng.*, 2007, **103**, 568–571.
36. R. Brayner, H. Barberousse, M. Hemadi, C. Djedjat, C. Yéprémian, T. Coradin, J. Livage, F. Fiévet and A. Couté, *J. Nanosci. Nanotechnol.*, 2007, 7, 2696–2708.
37. K. Deplanche and L. E. Macaskie, *Biotechnol. Bioeng.*, 2008, **99**, 1055–1064.
38. A. Ahmad, S. Senapati, M. I. Khan, R. Kumar, R. Ramani, V. Srinivas and M. Sastry, *Nanotechnology*, 2003, **14**, 824–828.
39. L. Fairbrother, J. Shapter, J. Brugger, G. Southam, A. Pring and F. Reith, *Chem. Geol.*, 2009, **265**, 313–320.
40. C. Blumer and D. Haas, *Arch. Microbiol.*, 2000, **173**, 170–177.
41. M. A. Faramarzi, M. Stagars, E. Pensini, W. Krebs and H. Brandl, *J. Biotechnol.*, 2004, **113**, 321–326.
42. M. A. Faramarzi and H. Brandl, *FEMS Microbiol. Lett.*, 2006, **259**, 47–52.
43. C. Arun, Z. Swaleha, T. Saba, S. Asif, S. Mohammad, C. R. Suri, A. Amir and O. Mohammad, *Int. J. Nanomed.*, 2011, **6**, 2305–2319.
44. R. Houdt, S. Monchy, N. Leys and M. Mergeay, *Antonie Van Leeuwenhoek*, 2009, **96**, 205–226.
45. R. Eisler, *Environ. Monit. Assess.*, 2004, **90**, 73–88.
46. M. Krachler, T. Prohaska, G. Koellensperger, E. Rossipal and G. Stingeder, *Biol. Trace Elem. Res.*, 2000, **76**, 97–112.
47. S. Hirohata, *Clin. Immunol. Immunopathol.*, 1996, **81**, 175–181.
48. M. Zagrobelny, S. Bak and B. L. Møller, *Phytochemistry*, 2008, **69**, 1457–1468.
49. D. Kadow, K. Voß, D. Selmar and R. Lieberei, *Ann. Bot.*, 2012, **109**, 1253–1262.
50. A. G. Khan, *J. Trace Elem. Med. Biol.*, 2005, **18**, 355–364.
51. A. T. Marshall, R. G. Haverkamp, C. E. Davies, J. G. Parsons, J. L. Gardea-Torresdey and D. van Agterveld, *Int. J. Phytorem.*, 2007, **9**, 197–206.
52. N. Chakraborty, A. Banerjee, S. Lahiri, A. Panda, A. Ghosh and R. Pal, *J. Appl. Phycol.*, 2009, **21**, 145–152.
53. J. Gardea-Torresdey, E. Rodriguez, J. Parsons, J. Peralta-Videa, G. Meitzner and G. Cruz-Jimenez, *Anal. Bioanal. Chem.*, 2005, **382**, 347–352.
54. N. L. Brown, J. V. Stoyanov, S. P. Kidd and J. L. Hobman, *FEMS Microbiol. Rev.*, 2003, **27**, 145–163.
55. Z. Ma, F. E. Jacobsen and D. P. Giedroc, *Chem. Rev.*, 2009, **109**, 4644–4681.
56. A. Changela, K. Chen, Y. Xue, J. Holschen, C. E. Outten, T. V. O'Halloran and A. Mondragón, *Science*, 2003, **301**, 1383–1387.
57. K. Chen, S. Yuldasheva, J. E. Penner-Hahn and T. V. O'Halloran, *J. Am. Chem. Soc.*, 2003, **125**, 12088–12089.
58. J. V. Stoyanov and N. L. Brown, *J. Biol. Chem.*, 2003, **278**, 1407–1410.
59. S. K. Checa, M. D. Zurbriggen and F. C. Soncini, *Curr. Opin. Biotechnol.*, 2012, **23**, 766–772.

60. S. Cerminati, F. C. Soncini and S. K. Checa, *Biotechnol. Bioeng.*, 2011, **108**, 2553–2560.

61. M. M. Ibañez, S. Cerminati, S. K. Checa and F. C. Soncini, *J. Bacteriol.*, 2013, **195**, 3084–3092.

62. M. E. Pérez Audero, B. M. Podoroska, M. M. Ibáñez, A. Cauerhff, S. K. Checa and F. C. Soncini, *Mol. Microbiol.*, 2010, **78**, 853–865.

63. M. V. Humbert, R. M. Rasia, S. K. Checa and F. C. Soncini, *J. Biol. Chem.*, 2013, **288**, 20510–20519.

64. K. J. Newberry and R. G. Brennan, *J. Biol. Chem.*, 2004, **279**, 20356–20362.

65. D. Raimunda, M. González-Guerrero, B. Leeber, III and J. Argüello, *Biometals*, 2011, **24**, 467–475.

66. A. O. Summers, *Curr. Opin. Microbiol.*, 2009, **12**, 138–144.

67. W. Wei, T. Zhu, Y. Wang, H. Yang, Z. Hao, P. R. Chen and J. Zhao, *Chem. Sci.*, 2012, **3**, 1780–1784.

68. H. Nikaido and Y. Takatsuka, *Biochim. Biophys. Acta Proteins Proteomics*, 2009, **1794**, 769–781.

69. O. Conroy, E.-H. Kim, M. M. McEvoy and C. Rensing, *FEMS Microbiol. Lett.*, 2010, **308**, 115–122.

70. K. Nishino, T. Latifi and E. A. Groisman, *Mol. Microbiol.*, 2006, **59**, 126–141.

71. D. Julian, C. Kershaw, N. Brown and J. Hobman, *Antonie Van Leeuwenhoek*, 2009, **96**, 149–159.

72. S. Monchy, M. A. Benotmane, R. Wattiez, S. van Aelst, V. Auquier, B. Borremans, M. Mergeay, S. Taghavi, D. van der Lelie and T. Vallaeys, *Microbiology*, 2006, **152**, 1765–1776.

73. T. M. Simon, D. H. Kunishima, G. J. Vibert and A. Lorber, *J. Rheumatol. Suppl.*, 1979, **5**, 91–97.

74. T. M. Simon, D. H. Kunishima, G. J. Vibert and A. Lorber, *Cancer*, 1979, **44**, 1965–1975.

75. C.-M. Che and R. W.-Y. Sun, *Chem. Commun.*, 2011, **47**, 9554–9560.

76. P. Fonteh, F. Keter and D. Meyer, *Biometals*, 2010, **23**, 185–196.

77. C. T. Lum, A. S.-T. Wong, M. C. M. Lin, C.-M. Che and R. W.-Y. Sun, *Chem. Commun.*, 2013, **49**, 4364–4366.

78. Y. Zhang, E. V. Hess, K. G. Pryhuber, J. G. Dorsey, K. Tepperman and R. C. Elder, *Inorg. Chim. Acta*, 1995, **229**, 271–280.

79. C. F. Shaw, III, A. A. Isab, M. T. Coffer and C. K. Mirabelli, *Biochem. Pharmacol.*, 1990, **40**, 1227–1234.

80. F. Mendes, M. Groessl, A. A. Nazarov, Y. O. Tsybin, G. Sava, I. Santos, P. J. Dyson and A. Casini, *J. Med. Chem.*, 2011, **54**, 2196–2206.

81. M. Serratrice, F. Edafe, F. Mendes, R. Scopelliti, S. M. Zakeeruddin, M. Gratzel, I. Santos, M. A. Cinellu and A. Casini, *Dalton Trans.*, 2012, **41**, 3287–3293.

82. A. Chircorian and A. M. Barrios, *Bioorg. Med. Chem. Lett.*, 2004, **14**, 5113–5116.

83. S. P. Fricker, *Metallomics*, 2010, **2**, 366–377.

84. R. Rubbiani, I. Kitanovic, H. Alborzinia, S. Can, A. Kitanovic, L. A. Onambele, M. Stefanopoulou, Y. Geldmacher, W. S. Sheldrick,

G. Wolber, A. Prokop, S. Wölfl and I. Ott, *J. Med. Chem.*, 2010, **53**, 8608–8618.

85. J. L. Hickey, R. A. Ruhayel, P. J. Barnard, M. V. Baker, S. J. Berners-Price and A. Filipovska, *J. Am. Chem. Soc.*, 2008, **130**, 12570–12571.
86. C. Gabbiani, F. Scaletti, L. Massai, E. Michelucci, M. A. Cinellu and L. Messori, *Chem. Commun.*, 2012, **48**, 11623–11625.
87. A. De Luca, C. G. Hartinger, P. J. Dyson, M. Lo Bello and A. Casini, *J. Inorg. Biochem.*, 2013, **119**, 38–42.
88. Y. Wang, Q.-Y. He, C.-M. Che, S. W. Tsao, R. W.-Y. Sun and J.-F. Chiu, *Biochem. Pharmacol.*, 2008, **75**, 1282–1291.
89. A. M. Alkilany and C. J. Murphy, *J. Nanopart. Res.*, 2010, **12**, 2313–2333.
90. G. von Maltzahn, J.-H. Park, A. Agrawal, N. K. Bandaru, S. K. Das, M. J. Sailor and S. N. Bhatia, *Cancer Res.*, 2009, **69**, 3892–3900.
91. J. F. Hainfeld, H. M. Smilowitz, M. J. O'Connor, F. A. Dilmanian and D. N. Slatkin, *Nanomedicine*, 2013, **8**, 1601–1609.
92. R. R. Arvizo, S. Bhattacharyya, R. A. Kudgus, K. Giri, R. Bhattacharya and P. Mukherjee, *Chem. Soc. Rev.*, 2012, **41**, 2943–2970.

Metallothioneins

CLAUDIA A. BLINDAUER

Department of Chemistry, University of Warwick, Coventry, CV4 7AL, UK
Email: c.blindauer@warwick.ac.uk

21.1 Introduction and Overview

Metallothioneins (MTs) constitute a large group of metal-binding proteins. Although they are often described as forming a protein superfamily, this notion should be interpreted with caution as MTs from different phyla are not related by either structural similarity or physiological function. Those from different phyla are thought not to be evolutionarily related but instead are characterized by their highly unusual primary structures: they are all relatively small (usually up to *ca.* 10 kDa), have a high cysteine content (*ca.* 15–30%) and have a low abundance of aromatic amino acids. The cysteine residues engage in the formation of metal-thiolate clusters, typically including both bridging and non-bridging thiolates (Figure 21.1). MTs are in principle capable of binding a variety of metal ions, but physiologically, the d^{10} ions Zn(II), Cd(II) and Cu(I) are the most important binding partners.[1] Because of this lack of a clear association with a particular metal plus their functional diversity, MTs are discussed in this separate chapter.

21.1.1 Discovery and Early Work on Metallothioneins

The first metallothionein was discovered in horse kidney cortex by Margoshes and Vallee in 1957, whilst they were searching for cadmium-binding proteins.[2] The new protein corresponded to 1–2% of total protein in this tissue, and hence was quite abundant. In further work by Kägi and Vallee,

RSC Metallobiology Series No. 2
Binding, Transport and Storage of Metal Ions in Biological Cells
Edited by Wolfgang Maret and Anthony Wedd

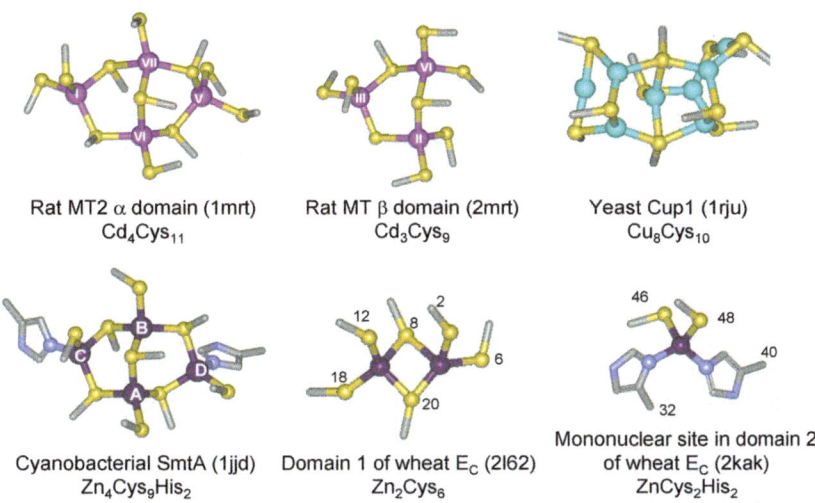

Rat MT2 α domain (1mrt)
Cd_4Cys_{11}

Rat MT β domain (2mrt)
Cd_3Cys_9

Yeast Cup1 (1rju)
Cu_8Cys_{10}

Cyanobacterial SmtA (1jjd)
$Zn_4Cys_9His_2$

Domain 1 of wheat E_C (2l62)
Zn_2Cys_6

Mononuclear site in domain 2 of wheat E_C (2kak)
$ZnCys_2His_2$

Figure 21.1 Representative metal-thiolate clusters and binding motifs in metallo-thioneins. Zn(II) (purple) and Cd(II) (magenta) are coordinated tetra-hedrally while Cu(I) (cyan) typically forms trigonal planar and digonal sites with thiolate sulfurs (yellow). Histidine residues (coordinating nitrogens as blue spheres) can also participate in metal coordination, *e.g.* in cyanobacterial SmtA and wheat E_C. The cluster topology of SmtA is otherwise closely similar to that in the α-domain of mamma-lian MTs. The structure determination of wheat E_C has added two coordination motifs unprecedented in MTs, namely a 2-metal cluster in domain 1 (γ-domain), similar to that found in GAL4 fungal transcription factors, and a mononuclear site with the most common coordination motif found in zinc fingers. The roman numerals for rat MT-2 refer to [113]Cd NMR resonances and are also reflected in Figure 21.5. The letters for SmtA refer to [111]Cd NMR resonances, and are also reflected in Figures 21.4 and 21.5. The small numbers near the ligands in the two last structures (both from wheat E_C) refer to their location in the E_C sequence (Table 21.1), starting with Gly1 (alternative labelling starting with the initiation Met residue can also be found in the literature). The structure shown for the γ-domain is one of several possibilities; instead of Cys20, Cys2 could also be one of the bridging residues. Both possibilities are compatible with NMR-derived restraints.[201]

the name "metallothionein" was coined, since the isolated protein con-tained 2.9% cadmium, 0.6% zinc and 4.1% sulfur (all per weight).[3] Further purification steps yielded a preparation with 5.9% cadmium, 2.2% zinc and 8.5% sulfur, giving a sulfur : metal molar ratio of *ca.* 3 : 1.[4] The metal-free *apo*-protein was named "thionein". These early works laid the foundations for defining some of the characteristic properties of MTs: (i) a scarcity or absence of aromatic amino acids, as judged from the low absorbance at 280 nm and (ii) the high thermodynamic stability of the metal-protein complexes with Zn(II) and Cd(II). The latter property was reflected in the low pH values that were required to remove all metal ions from the protein.

This property allowed the initial estimation of pH-independent binding constants of log $K_{Cd} \approx 17$ and log $K_{Zn} \approx 14$ (see Section 21.2.2 for refined values). The reactivity of the metals bound to the isolated protein was also probed in various ways, and it became quickly apparent that, despite the high thermodynamic stability of the complexes, the metal ions were bound in a kinetically rather labile fashion. Bound Zn(II) could be readily and stoichiometrically displaced by Cd(II), but not *vice versa*. All metal ions bound in the protein as isolated could be displaced by Ag(I) and reaction with *p*-chloromercuribenzoate also led to the complete and rapid release of all bound Zn(II) and Cd(II). In contrast, reaction of the metal-loaded protein with the cysteine probe *N*-ethylmaleimide was relatively slow while removal of the metal ions led to much more reactive thiols.[4]

Additional proteins with similar characteristics were soon isolated from the livers of rabbits that had been exposed to cadmium[5] and from human kidney cortex.[6] The refined purification methodologies used in these studies had indicated that more than one protein was obtained, leading to an in-depth study of rabbit MT using isoelectric focusing.[7] At this stage it was not clear whether the proteins merely differed by their metal content or in amino acid composition, or both. That the microheterogeneity may be due to the presence of forms with different amino-acid composition was suggested by non-integer values for several amino acids,[8] but it was not possible to resolve these forms at the time. Two years later, individual MT variants were separated by anion exchange chromatography and the first full amino-acid sequence (pertaining to equine renal MT-1B) was reported.[9] Besides confirming the presence of 20 Cys residues and the absence of aromatic amino acids, the study explicitly acknowledged the existence of MT variants.

The second half of the 1970s were a particularly exciting time for MT researchers: they brought indications for the existence of MTs from organisms other than mammals, namely from fish,[10] invertebrates,[11-13] fungi[14-16] and cyanobacteria.[17] They also featured the first International Meeting on Metallothioneins in 1978, a forum to share current knowledge and find consensus on nomenclature and good practice. The occurrence of "true" metallothioneins (see Section 21.1.2) in a plant (the grass *Agrostis gigantea*) was first demonstrated in 1980,[18] followed by further discoveries in wheat germ[19] and garden pea.[20] Their existence was hotly debated for a number of years (*e.g.* ref. 21) before becoming widely accepted.[22-24]

As of May 2013, about 13 000 articles dealing with metallothioneins have been published. A range of monographs[25-28] and journal special issues[29] have been compiled after several international meetings[30] were dedicated to metallothioneins, with further edited books,[31,32] journal issues[33] and reviews (too numerous to be listed) being published over the years. A dedicated website created by Binz (http://www.bioc.unizh.ch/mtpage/MT.html) is available but has not been updated recently. Biophysical and biochemical studies have been compiled in a volume of *Methods in Enzymology*.[34] Many of the techniques and approaches described therein are still state-of-the-art. It is, however, noteworthy that the early work by Vallee and Kägi essentially introduced nearly all major experimental and chemical concepts,[4] with the exception of those that have only become possible by advances in bio-technological and instrumental techniques. The most salient findings from the past three decades of MT biochemistry research will be summarized in the following sections.

21.1.2 Definition, Occurrence, Primary Structures and Classification

As the amount of biochemical information increased, the community recognized that new criteria were needed to define and classify MTs. In 1985, when most work occurred at the protein level, the main goal was to distinguish MTs from other metal-binding proteins. It was proposed that "polypeptides resembling equine renal metallothionein in several of their features can be designated as 'metallothionein'",[35] and this definition is still in use today.[36] The "features" cited were (i) low molar mass, (ii) high metal content, (iii) characteristic amino acid composition (high cysteine content, low content of aromatic amino acids), (iv) unique amino acid sequence with characteristic distribution of Cys residues, *e.g.* CxC and CC motifs, (v) spectroscopic features characteristic of metal thiolates and (vi) presence of metal thiolate clusters. Further potentially useful criteria were discussed:[37] (vii) inducibility (see Section 21.1.3), and operational characteristics such as (viii) 50% loss of Zn at pH 4.5 and 50% loss of Cd at pH 3.0, and (ix) ultraviolet absorption spectra of the Cd complexes characterized by a high 250 nm/280 nm ratio indicative of cadmium-thiolate bonds. Clearly, criterion (vi) requires structural work, and criterion (vii) requires *in vivo* studies that monitor induction of gene transcription at either mRNA or protein level. It is still true today that both chemical and biological properties need to be studied to adequately understand and describe the functions of MTs. One of the reasons for this ongoing requirement is the fact that confirmed metallothioneins, *i.e.* proteins that fulfil several of the above criteria, from different phyla display no discernible evolutionary relationship. This issue became ever more obvious and pertinent once work on the level of genes, or even genomes, became more common.

The first attempt to develop a classification system was based on sequence similarity with the first MT discovered, *i.e.* equine renal MT. Three classes were defined: class I comprising vertebrate and crustacean MTs as well as an MT from the mould *Neurospora crassa* (later, some mollusc MTs were added to this class),[38] class II comprising MTs that did not display any clear sequence similarity to vertebrate MTs, and class III comprising biosynthetic polypeptides such as phytochelatins and cadcystins, which due to their roles in cadmium detoxification were considered functionally and phenotypically related to MTs.[35] For a number of reasons, this classification system is no longer appropriate. Besides the fact that phytochelatins and cadcystins, due to their fundamentally different biosynthesis, should definitely not be called MTs, pooling all MTs that are not from two or three small branches of the tree of life into one big group does not do justice to the enormous diversity of primary and 3D structures and biological functions of these proteins.

Accordingly, a refined classification system was developed (http://www.bioc.unizh.ch/mtpage/classif.html). This is based on sequence similarity, including intron/exon structure of the gene and analysis of untranslated 5′ regions upstream of the genes that include the presence of metal-response elements (MREs; see Section 21.1.3.1).[39] These latter criteria require the availability of genomic DNA sequence information. The MT sequences known in 1997 were divided into 15 families, and these were further divided into subfamilies, subgroups and isoforms, some of which are shown in Table 21.1.

For example, vertebrate MTs form family 1, and contain the subfamilies m1, m2, m3, m4, m, a, a1, a2, b, ba, t – the letters standing for mammalian, avian, batracian (amphibians), anura (frogs) and teleost (fish), respectively. With few exceptions, all known sequences (60–68 amino acids) contain 20 Cys residues. Mammals harbour four distinct subtypes. Mice have one of each (MT-1 to MT-4), whilst humans have at least 10 functional MT genes, as besides MT-2A (often referred to as MT-2), MT-3 and MT-4, there are seven different expressed isoforms of MT-1, which are labelled MT-1A, -B, -E, -F, -G, -H and -X. There are also six additional genes for MT-1 (*MT-1c, -d, -i, -j, -k, -l*) and one classified as *MT-2*, which are thought not to be expressed, *i.e.* pseudogenes,[40] although this conclusion may not be justified in all cases.[41] Inconsistencies in the nomenclature of human MT-1 proteins and genes have been discussed.[41] Other organisms also harbour multiple MT genes; the fruit fly has at least five isoforms, *Caenorhabditis elegans* has two, bacteria between none and five and *Arabidopsis thaliana* has seven functional MT genes with four defined subgroups.

Implicitly, the classification system suggests that "metallothioneins" are a polyphyletic group, as evolutionary relationships between families are unclear – hence, MTs from different phyla may not have evolved from a common ancestor. This means that there is a strong likelihood for "metallothionein" having evolved more than once, and that it is not appropriate to infer biological function(s) from the mere fact that a protein has several MT-like features or has been named a metallothionein. Biological functions as well as biophysical/biochemical properties will be addressed in subsequent sections but, clearly, there is an immense diversity in primary sequences – sometimes even within single species. For example, the two MTs from yeast, Cup1 and Crs5 (Table 21.1), show no discernible sequence similarity.

The advent of widespread availability of genome sequencing and cDNA analyses, such as expressed sequence tag (EST) analyses, has brought new challenges to MT research. In particular, because of the very characteristics of their primary sequences, namely their shortness and their low complexity, they may easily be overlooked during genome annotation. The fact that MTs from different phyla show no recognizable evolutionary relationship

Table 21.1 Diversity of primary structures of metallothioneins from different species. The table contains MT sequences for which structures are available (Table 21.3), plus sequences for which biophysical data exist and further examples from different phyla. The eukaryotic sequences are grouped according to the scheme in Figure 21.2, and the Kägi/Binz classification symbols are given in the final column. Results on metal stoichiometry from *in vitro* studies are summarized in the penultimate column. Empty fields indicate that no corresponding information is available.

Species	Name/ in vivo metal	Sequence	No. of Cys residues	Experimentally confirmed metal stoichiometry	Classification
Animals					
Bilateria					
Homo sapiens	MT-1A	MDPNCSCATGGSCTCTGSCKCKECKCTSCKKSCCSCCPMSCAKCAQGCICKGASEKCSCCA	20	7 M(II), or 12 M(I)	m1
Homo sapiens	MT-2A	MDPNCSCAAGDSCTCAGSCKCKECKCTSCKKSCCSCCPVGCAKCAQGCICKGASDKCSCCA	20	7 M(II), or 12 M(I)	m2
Homo sapiens	MT-3	MDPETCPCPSGGSCTCADSCKCEGCKCTSCKKSCCSCCPAECEKCAKDCVCKGGEAAEAEAEKCSCCQ	20	7 M(II), Zn$_4$Cu$_4$, Zn$_3$Cu$_4$	m3
Homo sapiens	MT-4	MDPRECVCMSGGICMGGDNCKCTTCNCKTCRKSCCPCPPGCAKCARGCICKGGSDKCSCCP	20	7 M(II)	m4
Fish (*N. coriiceps*)	MTA	MDPCECESKSGTCNCGGSCTCTNCSCKSCKKSCCPCPSGCTKCASGCVCKGKTCDTSCCQ	20	7 M(II)	t
Reptile (*C. chalcides*)		MDPQDCSCNTGGTCTCAGSCKCNCKCTSCKKSCCSCCPAGCDNCAKGCVCKEPLSGKCSCCH	20		n.a.
Bird (chicken)		MDPQDCTCAAGDSCSCAGSCKCKNCRCRSCRKSCCSCCPAGCNNCAKGCVCKEPASSKCSCCH	20	7 M(II), 10-12 Cu(I)	a1

Table 21.1 (Continued)

Species	Name/in vivo metal	Sequence	No. of Cys residues	Experimentally confirmed metal stoichiometry	Classification
Lancelet (B. floridae)	MT-1	MPDPCCSACEGCSSTCNKCSCDCCKCCASCKACGPAADCNCGCACCKCCASC-DGCKSSCTSC- SCDCCK	27	7-8 M(II); 12 M(I)	n.a.
	MT-2	MPDPCCESCKACGPT---------VGCNCGCACCKCCSSCVQTCKPGCTNCPGCDCCK	18	6 M(II); 10 M(I)	n.a.
Sea urchin (S. purpureus)	MTA	MPDVKCVCCKEGKECACFGQDCCKTGECCKDGTCCGICTNAACKCANGCKCGSGCSCTEGNCAC	20	7 M(II)	n.a.
Blue crab	CdMT-1	MPGPCCNDKCVCQEGGCKAGCQCTSCRCSPCQKCTSGCKCATKEECSKTCTKPCSCCPK	18	6 M(II)	
Blue crab	CuMT-2	MPCGCGTSCKCGSGKCCCGSTCNCTTCPSKQSCSCNDGACGSACQCKTSCCGADCKCSPCPMK	21		n.a.
Insect (O. cincta)	OcMT (Cd)	MSSTQGSASEAIRNCLCCGENCKCGGAEGKSPTCKCEKKCCGGGATQTASCCTCCGPDCVCKDGASLP CCANKTCCK	19	7 or 8 Cd(II)	n.a.
Insect (Drosophila melanogaster)	MtnA (Cu)	MPCPCGSGCKCASQATKGSCNCGSDCKCGGDKKSACGCSE	10	4-5 Zn(II), 8-9 Cu(I)	d1
	MtnB (Cd)	MVCKGCGTNCQCSAQKCGDNCACNKDCQCVCKNGPKDQCCSNK	12	3-4 Zn(II), 8 Cu(I)	d2
	MtnC	MVCKGCGTNCKCQDTKCGDNCACNQDCKCVCKNGPKDQCCKSK	12	4 M(II), 8-9 Cu(I)	d2
	MtnD	MGCKACGTNCQCSATKCGDNCACSQQCQCSKNGPKDKCCSTKN	12	4 M(II), 5 Cu(I)	d2
	MtnE	MPCKGCGNNCQCSAGKCGGNCAGNSQCQCAAKTGAK--CCQAK	10	4 or 5 M(II), 5 Cu(I)	n.a.

Organism	MT	Sequence	Length	Metal	Code
Mussel (*Mytilus galloprovincialis*)	MT-10	MPAPCNCIESNVCICGTGCSGEGCRCGDACKCSGADCKCSGCKVVCKCSGSCACEAGCTGPSTCRCAPG	21	7 Cd(II)	mo1
	(Cd)	CSCK			
	MT-20-IV	MAGPCNCIATNVCICGTGCSEKCCQCGDACKCE-SGCGCSGCKVVCRCSGTCACGCGCTGPTNCKCESGCSCN	23	7–8 M(II)?	mo2
Snail (*Helix pomatia*)	CdMT	SGKGKGEKCTSACRSEPCQCGSKCQCGEGCTCAACKTCNCTSDGCKCGKECTGPDSCKCGSSCSCK	18	6 Cd(II) (or 6 Zn(II))	mog
	CuMT	--SGRGKNCGGACNSNPCSCGNDCKCGAGCNCDRCSSCHCSNDDCKCGSQCTGSGSCKCGSACGCK	18	12 Cu(I)	mog
	CdCuMT	--SGKGSNCAGSCNSNPCSCGDDCKCGAGCSCVQCHSCQCNNDTCKCGNQCSASGSCKCGS-CGCK	18	mixture	mog
Earthworm (*E.foetida*)	MT-Cd	DTQCCGKSTCAREGSTCCCTNCRCLKSECLPGCKKLCCADAEKGKCGNAGCKCSAGSCAAG	21	domain 1: 4 Cd(II)	n.a.
		CKKGCCGD			
Nematode (*C. elegans*)	MTL-1	MACKCDCKNKQCKCGDKCECSGDKCCEKYCCEEASEKKCCPAGCKGDCKCANCHCAEQKQCGDKT	19	7 M(II)	n1
		HQHQGTAAAH			
	MTL-2	MVCKCDCKNQNCSCNTGTKDCDCSDAKCCEQYCCPTASEKKCCKSGCAGGCKCANCEAQAAH	18	6 M(II)	n2
Flatworm (*Schistosoma mansonii*)	Cd-MT	MPSIGPELRCRYQCNSLTQPHCADYFDNRTFPLVACPNDGRNYSRCVKMIQEMYLDGKWTRRYYRDCAVT	34		n.a.
		GVIGAEDGRWCIDRLGTYRVKVRYCNCNNKNGCACKEYGESCMKTVFNKCCGSNVCQLIGPFNGKCVR			
		CIQTGYSCLHNGECCSRNCWLFRCCRALGEKCSKTVFDRCCGDTVCHLSSPFHGKCVKCLKEGTLCVSD			
		KNCCSHKCNIGKCTKEKHHY			
Radiata					
Coral (*Acropora cervicornis*)	Cu	SPCNCIEIAPCNCIELHSVEKCPGRKNVPTFILVRGIFALCEPATCKCCKYQRSTKCDSSC	10		n.a.
Placozoa					
Trichoplax adhaerens	n.a.	MDPPCNCAETGDCCQCPSNCACTNCKCAPNCRLCGGKTCKCAENISSCVCTTCICTDCKCPKGCKN	19		n.a.

Table 21.1 (Continued)

Species	Name/ in vivo metal	Sequence	No. of Cys residues	Experimentally confirmed metal stoichiometry	Classification
Fungi					
Yeast	Cup1	QNEGHECQCGSCKNNEQCQKSCSCPTGCNSDDKCPCGN	10	8 Cu(I)	f5
	Crs5	MTVKICDCEGECCKDSCHCGSTCLPSCSGGEKCKCDHSTGSPQCKSCGEKCKCETTCTCIEKSKCNCEKC	19	5–7 Zn(II), 10–12 Cu(I), 6–8 Cd(II)	f6
Candida glabrata	MT-1	MANDCKCPNGCSCPNCANGGCQCGDKCECKKQSCHGCGEQCKGSHGSSCHGSCGCGDKCECK	18	11–12 Cu(I)	f2
	MT-2	MPEQVNCQYDCHCSNCACENTCNCCAKPACACTNSASNECSCQTCKCQTCKC	16	10 Cu(I)	f3
Candida albicans		MACSAAQCVCAQKSTCSCGKQPALKCNCSKASVENVVPSSNDACACGKRNKSSCTCGANAICDGTRDG	12		n.a.
		ETDFTNLKMSKFELVNYASGCSCGADCKCASETECKCASKK	6		n.a.
Fission yeast	Zym1 (Zn)	MEHTTQCKSKQGKPCDCQSKCGCQDKESCGCKSSAVDNCKCSSCKCASK	12		n.a.
Yarrowia lipolytica	MT-1	MEFTTAMFGTSLIFTTSTQSKHNLVNNCCCSSSTSESSMPASCACTKCGCKTCKC	9		f4
Neurospora crassa		GDCGCSGASSCNCGSGCSCSNCGSK	7		f1
Magnaporthe grisea	MTMT1	MCGDNCTGASCSCSSCGTHGK	6	2 Zn(II)	n.a.
Heliscus lugdunensis	Cd	SPCTCSTCNCAGACBSCSCTSCSH	8		n.a.

Green Plants

Arabidopsis	MT-1A	MADSNCGCGSSCKCGDSCSCEKNYNKECDNCSCGSNCSCGSNCNC	13		p1
thaliana	MT-1C	MAGSNCGCGSSCKCGDSCSCEKNYNKECDNCSCGSNCSCGSSCNC	13		p1
	MT-2A	MSCCGGNCGCGSGCKCGNGCGGCKMYPDLGFSGETTTETFVLGVAPAMKNQYEASGESNNAENDAC KCGSDCKCDPCTCK	14		p2
	MT-2B	MSCCGGSCGCGSACKCGNGCGGCKRYPDLENTATETLVLGVAPAMNSQYEASGETFVAENDACKCGSD CKCNPCTCK	14		p2
	MT-3	MADSNCGCGSSCKCGDSCSCEKNYNKECDNCSCGSNCSCGSNCNC	10		p3
	MT-4A	MADTGKGSSVAGCNDSCGCPSPCPGGNSCRCRMREASAGDQCHMVCPCGEHCGCNPCNCPKTQTQTS AKGCTCGEGCTCASCAT	17		pec
	MT-4B	MADTGKGSASASCNDRCGCPSPCPGGESCRCKMMSEASGGDQCEHNTCPCGEHCGCNPCNCPKTQTQTS AKGCTCGEGCTCATCAA	17		pec
Wheat	E_C–I/II	GCDDKCGCAVPCPGGTGCRCTSARSGAAAGEHTTCGCGEHCGCNPCACGREGTPSGRANRRANCSG AACNCASCGSATA	17	6 Zn(II)	pec
Xerophyta humilis	n.a.	MADVKKCDETCGCPVPCSLDTACKCSVESGNAASRHATCTCGEHCSCNPCSCGKVPIGVQAGKGSCSC GSGCNCEKCSC	18		pec
Douglas Fir	n.a.	MASTCTRGAECGCGETCACVDKCGAGSTATPSDQTSGGNAYCKCGENCSCNPCNCSKSDQTASGKSY CKCGENCACETCSCSRAQM	19		pec
Selaginella moellendorfii	n.a.	MARTEGCGCATPCPGDACKCGMDKEKSSAPGAETSSSFCSCGEKCSCDPCSCSKVSASGDGFCKCGSE CKCDKCSCAKSVVARA	18		pec
Azolla filiculoides	AzMT2	MSCCGGNCGCGSGCRCGKRSFDETAFDAPMEVEVTVFGNEGSNCGFDSSCGFESDAQATSQNGCSCGSN CTCNPCRC	14		n.a.

Table 21.1 (Continued)

Species	Name/ in vivo metal	Sequence	No. of Cys residues	Experimentally confirmed metal stoichiometry	Classification
Chromalveolata					
Fucus vesiculosus		MAGTGCKIWEDCKGAAACSCGDSCTCGTVKKGTTSRAGAGCPCGPKCKCTGQGSCNCVKDDCCGCGK	16	6 Cd(II), 5 As(III)	n.a.
Thalassiosira pseudonana CCMP1335		MNANANSESGKKRSINENDTVPTTTNNNPTNTTCKCSKSRCIKLYCDCFHGGNLCNSLCNCTDCKNTTE FREEREWKMKEVLKLNPKAFSEDSDKFNTKRQRMSRGNGCACPSSHCLKKYCSCFGADAGCTDKCSCN DCE	18		n.a.
Tetrahymena pyriformis	MT1	MDKVNNNCCCGENAKPCCTDPNSGCCCVSETNNCCKSDKKECCTGTGEGCKCTGCKCCEPAKSGCCC GDKAKACCTDPNSGCCCSSKTNKCCDSTNKTECKTCCCK	31	9–11 Zn(II), 11 Cd(II)	ci
Tetrahymena thermophila	MTT1 (Cd)	MDKVNSCCCGVNAKPCCTDPNSGCCCVSKTDNCCKSDTKECCTGTGEGCKCVNCKCCKPQANCCCG VNAKPCCFDPNSGCCCVSKTNNCCKSDTKECCTGTGEGCKCTSCQCCKPVQQGCCGDKAKACCTDP NSGCCCSNKANKCCDATSKQECQTCQCCK	48		ci
Tetrahymena thermophila	MTT2 (Cu)	MDTQTQTKVTVGCSCNPCKCQPLCKCGTTAACNCQPCENCDPCSCNPCKCGVTESCCGNPCKCAECKC GSHTEKTSACKCNPCACNPCNCGSTSNCKCNPCKCAECKC	32		ci
Excavata					
Trichomonas vaginalis	n.a.	MSSQKTCDCNKQNAGKSQWFKCQCCPGCKCGSNCHCTKDNKCSPDCHCGEGCNCNEGCYCNEGCKC GSNCHCTKDNKCSPDCHCGEGCNCNEGCYCNEGCKCGSNCHCTKDNKCSPDCHCGEGCNCNEGCKC GSNCHCTKDNKCSPDCHCGEGCNCNEGCYCNEGCKCGSNOHCTKDNKCSPDCHCGEGCNCNEGCKC NEGCKCGSNCHCTKDSKCSPDCHCGEGCNCNEGCKCGADCHCNN	91		n.a.

Bacteria

Organism	Protein	Sequence			
Synechococcus PCC7942	SmtA	MTSTTLVKCACEPCLCNVDPSKAIDRNGLYYCSEACADGHTGGSKGCGHTGCNCHG	9	4 M(II)	pr
Synechococcus	Sync_1081	MTVTVVKCACSSCTCEVSSSSAISRNGHSYCSDACASGHRNNEPCHDAAGACGCNCGS	10		pr
CC9311	Sync_2379	MATSNQVCACDPCSCAVSVESAVQKDGKVYCSQPCADGHSGSDECCKSCDCC	11		pr
	Sync_2426	MTTNLVRCDCPPCTCSIEEATAAMYGNKLFCSEACATAHINQEPSNSAEHTECSCGC	9		pr
	Sync_0853	MNEVLLLCDCSLCKRSVEESRSIRIGGQHFCSESCAKGHPNMEPCDGERDGCNCGIAELELLLAAAD	8		pr
Pseudomonas aeruginosa		MNSETCACPKCTCQPGADAVER- DGQHYCCAACAGGHPQGEPCRDADCPCGGTTRPQVAEDRQLD DALKETFPASDPISP	10	3–4 M(II)	pr
Mycobacterium tuberculosis	MymT	MRVIRMTNYEAGTLLTCSHEGCGCRVRIEVPCHCAGAGDAYRCTCGDELAPVK	7	6 Cu(I)?	n.a.
Streptomyces lividans		MRTASVAHGEGARLAAAVSLVLASAKVPDHVADGDGLLARRPVALAARRIAGGWDAAGARGGGAVGF DTAVLVDAVDRQGGIEALAGPGTPLLTVTEPTETATAAAFAHTAHLAGRPHNAVPLAEAGRLFGRLAHL LDAVEDREADAASGAWNPLTATGTPLAEARRLADDAVHGVRLALREAEFTDGRLVHRLLVHELGSSVD RAFGTVSCAHGSHPYAPPGAPGPGGTPPEPPRRDRRGLLAGCAVWLGLACTCQMCCGTFNDPWSGQ RREGLCSQCDCGNCDCDCDCCSNCCGDDGCGCGDCGCGSC	24		n.a.

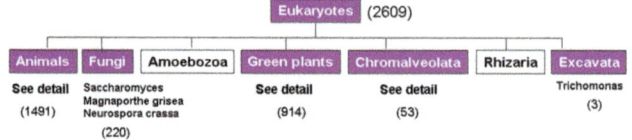

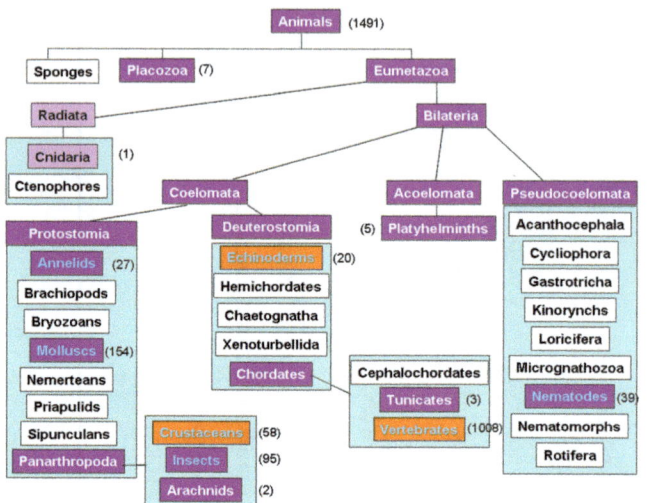

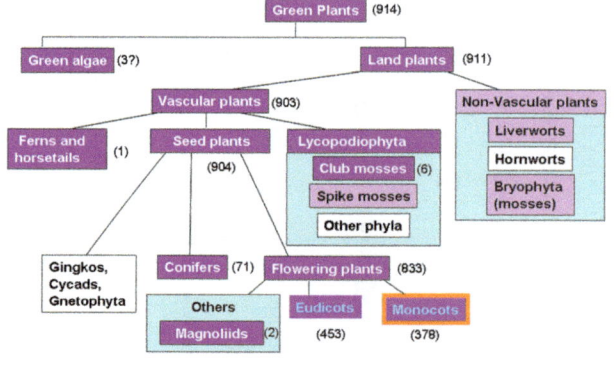

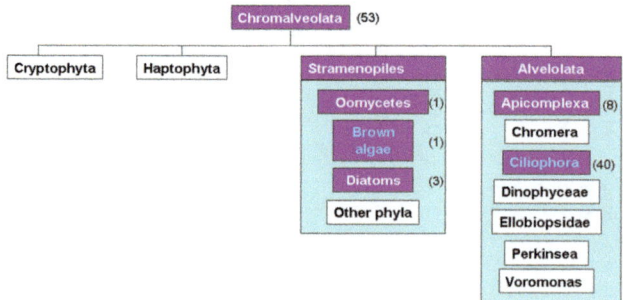

(Table 21.1) exacerbates this problem. It is therefore not surprising that new metallothioneins are still being discovered; examples include the recent addition of MtnE from *Drosophila melanogaster*,[42] MymT from *Mycobacterium tuberculosis*[43] and two MTs from lancelets (*Branchiostoma floridae* and *B. lanceolatum*). The latter animals are regarded to be most closely related to vertebrates,[44] but even their MTs show no clear sequence similarity to vertebrate MTs. A recent study on actinobacterial genomes[45] has approached the problem of overlooked MTs by developing a dedicated search method for small cysteine-rich proteins. This has delivered several new candidate MTs, but since MTs are not the only proteins with such a high cysteine content, their metal-binding ability will need to be tested *in vivo* and *in vitro*.

The NCBI "Nucleotide" database (accessed January 2013) returned 3714 hits for the search "metallothionein", with 2609 classified entries for eukaryotes (Figure 21.2). Spot-checks of these entries indicated that most (but not all) indeed fulfil the criteria for metallothioneins as far as they pertain to primary sequence; however, this is not the case for the 934 entries retrieved for bacteria. The majority of these refer to S-adenosyl-L-methionine-dependent methyltransferase, which shares the abbreviation "smta" with the first characterized bacterial metallothionein, SmtA from *Synechococcus* sp. PCC7942. At some point in time, this has led to the erroneous assignment of an S-adenosyl-L-methionine-dependent methyltransferase as a metallothionein, and this mis-assignment has been perpetuated since. Careful genome mining[46–48] has revealed that true SmtA homologues are present in many but not all cyanobacterial genomes (a search in November 2012 found true BmtAs (abbreviation for bacterial metallothioneins homologous to SmtA) in 71 out of 116 genomes), in a small number of α- and γ-proteobacteria (methylobacteria, pseudomonads, *Granulibacter bethesdensis*, *Nitroscococcus* sp., *Magnetospirillum magentotacticum*) and one firmicute (*Staphylococcus epidermidis*). Recent bioinformatic analysis has also suggested that related genes are present in actinobacteria.[48] The literature also contains further indications for the presence of MTs in other species, *e.g.* sponges,[49] but since no molecular or sequence information is available, these are not reflected in Figure 21.2.

In summary, metallothioneins are diverse, small proteins with strong metal-binding abilities mediated predominantly through cysteine residues.

Figure 21.2   Sequence and structural data for eukaryotic metallothioneins. Boxes coloured in magenta signify availability of sequence data at either the nucleotide or protein level, the number in brackets refers to the number of sequences deposited in the "nucleotide" database at NCBI. Boxes shaded in pink refer to availability of additional sequence data in dbEST, which contains 5066 mRNA sequences annotated as metallothioneins (January 2013). Orange boxes highlight availability of 3D structural data. Blue lettering indicates that biophysical data for MTs from this phylum are available.

This statement should immediately make clear that it is inappropriate to assume that biological function or *in vitro* properties of two MTs from different phyla are similar enough to allow extrapolation from one to the other, and the inadmissibility of such attempts has been demonstrated by recent biochemical studies.[50] Similarly, phrases such as "metallothionein is a protein..." without specifying which particular MT (or which MT family) is meant should be avoided. Such formulations ignore the staggering diversity in structure, biochemistry and functions, and subsequently may lead to inappropriate conclusions.

21.1.3   Considerations Regarding Physiological Roles

The shared molecular (*i.e.* chemical) functions of metallothioneins are encoded in fundamental chemistry: because of the presence of thiol(ate)s, they all have the ability to bind metal ions, and are in principle redox-active. Both of these chemical properties can be useful in a variety of biological contexts, and it is therefore not surprising that metallothioneins are not only ubiquitous, but also involved in a diverse range of physiological processes. It is important to recognize that demonstrating *in vitro* that a particular MT can bind a particular metal ion and/or is redox-active is insufficient to define a physiological role for this protein – only cell and molecular biology approaches can potentially furnish this information. Since different organisms live in different environments and since their requirements, utilization and sensitivities for metal ions and hence metal metabolism differ significantly,[51] it can also be expected that metallothioneins may have different predominant biological functions in different organisms, and this indeed seems to be the case.

In general, MTs have roles in normal physiological processes but also respond to all kinds of stresses, be they a result of normal physiology, disease or environmental influences. In multicellular organisms, they are present mostly in the cytosol but can translocate to the nucleus[52] and the intermembrane space of mitochondria.[53] In addition, some vertebrate MTs can be detected in extracellular fluids,[54] including blood plasma and cerebrospinal fluid.[55] The latter study also demonstrated that astrocytes actively secrete MT. At least three different states of MT, namely oxidized thionin, reduced thionein and metallothionein could be distinguished in rat liver and cultured cells.[56,57] Like other proteins, MTs are continuously synthesized and degraded. It is thought that degradation occurs in lysosomes,[58] and, intriguingly, the degradation rate may depend on the metallation state. Whilst the metal-free thionein can be rapidly degraded by lysosomal proteases at pH 5.5, fully metallated MTs were resistant towards degradation under the same conditions.[59] It is of note that, *in vitro*, the identity of the bound metal ion also influenced degradation rates, with Zn-MT being turned over more slowly than Cd-MT. More recent work quantifying free MT thiols from tissue extracts suggested that, *in vivo*, partially metallated MTs are more stable and more common than previously thought.[60]

21.1.3.1 Regulation of MT Gene Expression

Metallothioneins are expressed at basal levels in most organisms in which they have been investigated, but the synthesis of the majority of MTs is inducible by Cd(II), Zn(II) and/or Cu(I) (and often other metal ions), no matter from which organism. There are some notable exceptions. For example, gene expression of mammalian brain-specific MT-3 and the seed-specific MT-4 from plants is not metal-responsive. In the case of mammalian MT-1 and -2, a range of additional stimuli, including oxidative stress, hormones and cytokines, also induce MT synthesis (Figure 21.3).[61–63] Predominantly, MT expression is regulated on the transcriptional level[61,64] although, more recently, an epigenetic mechanism has been demonstrated in mice.[65] In some cases, for example mammalian MT-1 and -2, it is known that transcriptional regulation is mediated by so-called metal-response elements (MREs), specific DNA sequences that occur in the promoter region of MTs[66] and other genes. MREs have been studied in vertebrates[64] including fish,[67] and in invertebrates such as insects[64] and snails.[68] Their presence in other phyla is less clear.

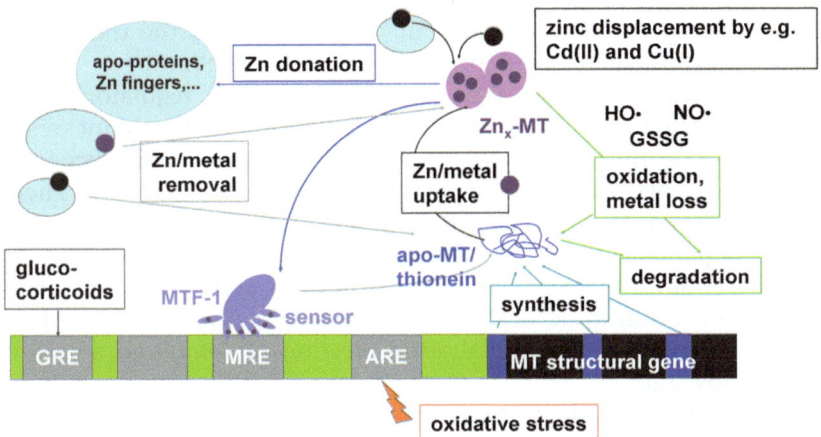

Figure 21.3 Simplified schematic summarizing regulation and potential reactions of MTs,[61,62] using mammalian MTs as an example. The horizontal bar symbolizes genomic DNA. MT synthesis is regulated by the metal-response element (MRE)-binding transcription factor (MTF-1), which contains six zinc fingers and activates MT transcription upon binding to the MRE region of DNA. Zinc binding to the fingers is necessary for activation, and may originate from cellular stores including Zn_xMT, which may release Zn(II) either by displacement by other metals (Cd(II), Cu(I)), or triggered by thiol oxidation by ROS, NOS or biological disulfides. Other factors inducing MT synthesis not shown in the schematic include hormones that interact through Ca^{2+}-dependent signalling cascades, and cytokines such as interleukin-6 and -11, which interact through STAT (signal transducers and activator of transcription) proteins.

The transcription factor that recognizes MREs in vertebrates and *Drosophila* is called MTF-1 (for metal-regulatory transcription factor 1),[64] and a related protein has been identified in snails and termed MTF-2.[68] MTF-1 from either vertebrates or *Drosophila* contains six zinc fingers with zinc-dependent DNA-binding ability and is involved in the response of cells to heavy metal stress, oxidative stress and also hypoxia.[64] At least some of these responses are thought to be mediated *via* the intracellular release of Zn(II), which subsequently binds to and activates MTF-1 (Figure 21.3). Importantly, in mammals, MT-1/MT-2 are central to mediating Zn(II) release elicited by both toxic levels of heavy metals as well as reactive oxygen species including peroxide (Sections 21.1.3.3, 21.1.3.5 and 21.2.7.4).[69] Drosophila MTF-1 also contains an additional copper-sensing element distinct from the zinc-sensing moieties.[64] Putative MREs are also present in the genomes of the nematode *C. elegans*,[70] the earthworm *Lumbricus rubellus*[71] and plants, *e.g.* tomato,[72] but no homologue of MTF-1 has been identified in any of these organisms and it is not clear whether these putative MREs are functional. Fungal and bacterial MT genes also harbour specific recognition sequences in their promoter sequences, although these differ from those for higher eukaryotes and also bind different transcription factors. In the yeast *Saccharomyces cerevisae*, the copper-responsive transcription factor Ace1 mediates transcription of Cup1 and Crs5,[73] and in some, but not all,[47] cyanobacteria, the transcriptional repressor SmtB mediates Zn-dependent expression of SmtA.[74]

For vertebrate MT-1 and MT-2 genes, several other regulatory elements and transcription factors have been identified.[61] Antioxidant response elements (AREs) mediate response to oxidative stress, and glucocorticoid response elements (GREs) are involved in other stress responses. There are additional recognition sequences that link MT genes to cytokine- and hormone-dependent signalling cascades.[61,62] The latter finding may explain their differential expression under a plethora of conditions, and their involvement in a multitude of cellular processes. It also suggests a primary function in vertebrate zinc metabolism (Section 21.1.3.5) and a link to the cellular redox state.[75]

The two other isoforms in mammals, MT-3 and MT-4, do not respond to these various stress conditions but their expression is highly tissue-specific. MT-3[76] is particularly abundant in certain hippocampal neurons and MT-4[77] is restricted to cornified and stratified squamous epithelia. In addition, the expression of MT-3[78] and MT-4[79] underlies tight developmental regulation. The expression of brain MT-3, previously named GIF (for growth-inhibitory factor), has been reported to be down-regulated in Alzheimer's disease,[76] although this notion is still under debate.[80] It is the only MT that was isolated on the basis of a biological function and it is one of the few MTs for which biological functions have been defined: it inhibits survival of cortical neurons and is thought to prevent neuronal sprouting and the formation of neurofibrillary tangles.[81] In addition, *via* its interactions with copper, it may play a role in preventing the formation of amyloid plaques and in inhibiting Cu(I)/(II) redox cycling that may lead to generation of reactive oxygen species

(ROS) (Section 21.1.3.3). Both tangles and plaques are hallmarks of Alzheimer's disease. Overall, MT (isoforms 1–3) expression in the brain is also affected by a range of other neurological diseases as well as by brain inflammation and injury. MTs in the central nervous system are thought to have neuroprotective functions.[81]

21.1.3.2 Tolerance towards Toxic Metal Ions

For many MTs, it has been demonstrated that they confer higher tolerance to excess metal ions, be they essential Zn(II) or Cu(I), or toxic Cd(II). It has been argued that the up-regulation of MT expression in response to exposure to Cd(II), at least in mammals, is not related to an evolutionarily conserved biological function but rather is a "property" of MTs.[82,83] This contention has been based on the fact that higher animals are unlikely to have evolved a protein in response to a toxin that has only recently been mobilized through human activity. The case may, however, be different for such organisms that have been and still are constantly exposed to low levels of cadmium in water and soil, where this metal ion is typically only two to three orders of magnitude less abundant than zinc. These organisms include plants, terrestrial[71,84] and aquatic[51] invertebrates, fungi and bacteria. One or more MTs in many of these organisms, for example earthworms[71] and snails,[85] are most strongly up-regulated in response to Cd(II), and this fact forms the basis of a large body of work in ecotoxicology.[86] There is some evidence that high levels of Cd(II) also lead to differential MT gene regulation in plants,[87] although it is thought that in plants phytochelatins are more relevant for cadmium detoxification.[24] Interestingly, this is also the case for the soil-dwelling nematode *C. elegans*,[88] even though the expression of its two MT genes is also more or less strongly up-regulated by cadmium. Nevertheless, constitutive elevated gene expression of certain MTs has also been observed in metal hyperaccumulating plants, *e.g.* the Zn/Cd hyperaccumulator *Arabidopsis halleri*.[89] The utilization of such natural or genetically engineered[90] MT-overexpressing plant species or genotypes is an attractive approach to the bioremediation of contaminated soils.

Whether a "real" function or not, expression of MTs in mammals does protect from effects of acute cadmium toxicity.[91] Similarly, (over-)expression of MT enhances tolerance against mercury,[81] arsenic[92] and platinum in mammals.[93] However, MTs are also responsible for prolonged retention of cadmium in the body, especially in liver and kidneys, with detrimental long-term effects.[94] This is a major reason why the cadmium detoxifying function of vertebrate MTs has been questioned.

Increased metal accumulation has frequently been observed as a consequence of MT overexpression in a variety of organisms – either in the natural host or in a heterologous/recombinant system. For example, the "*in vivo*" metal-binding properties of plant MTs are often studied in genetically modified yeast, following an approach developed by Goldsbrough.[95] The accumulation of particular metal ions as a result of MT overexpression, and the restoration of tolerance towards a metal ion that the yeast has been artificially sensitized against (usually by knocking out the respective yeast gene), are often taken as indicators for the *in planta* function. It should, however, be considered that neither observation necessarily reflects the physiological function in the natural host – especially bearing in mind that the two hosts are from different phyla. Conversely, it has been observed recently that MT knockout mutants of *C. elegans* also accumulate elevated levels of zinc and cadmium, suggesting that, in this organism, MTs are involved in the excretion of these metal ions.[96] This observation, which contrasts with observations in mammals and fungi, re-emphasizes that caution must be applied when comparing functional information on MTs from different phyla.

Besides response to and protection against cadmium and copper toxicity (Section 21.1.3.4), a range of other xenobiotic metal ions (Ag(I), Hg(II), Pb(II), Pt(II), As(III), Bi(III) and others) may influence MT gene expression and bind MT, at least *in vitro* (Section 21.2.1).

21.1.3.3 Protection against Oxidative Stress

Their large number of cysteine residues means that, in principle, MTs are powerful antioxidants (Section 21.2.5). This is borne out by the fact that MTs in a variety of organisms are inducible by oxidative stress.[97–100]

Studies in primary cultures of MT-1/2-null cardiomyocytes from the mouse have demonstrated increased cellular ROS accumulation induced by the intercalator doxorubicin.[101] Even under normal conditions, cells have to deal constantly with ROS and although a series of specialized systems supplies the major line of defence, the presence of AREs (antioxidant response elements) in the promoter regions of mammalian MTs suggests that at least these MTs form part of the mechanisms that protect from the deleterious effects of ROS. Cultured embryonic mouse cells not expressing MT-1 and -2 were hypersensitive towards external peroxide and the superoxide-generating compound paraquat.[102] Even basal levels of peroxide were markedly elevated in single and double MT knock-out mutants in the nematode *C. elegans*.[98]

Two major ways in which MTs can have a protective effect may be considered: (i) the redox-activity of the thiols can disarm ROS and other oxidants including radical species (either directly or indirectly by, *e.g.*, thiol-disulfide interchange reactions with cellular disulfides such as glutathione disulfide GSSG) and (ii) MTs can bind copper ions and render them innocuous. This may be of particular importance in extracellular fluids, in particular in the mammalian brain, where Cu(II) ions may occur and participate in the generation of ROS. Zinc-loaded MT-3 protects cultured neurons from the toxicity of the Aβ peptide,[103] plaques of which are found in the brains of Alzheimer's patients. Neurotoxicity of Aβ is associated with copper binding and ROS production. It has been shown that Zn_7MT-3 "swaps" bound metals with Cu(II)-Aβ, reducing Cu(II) to Cu(I) in the process and abolishing ROS production by sequestering Cu(I).[104] Similar reactions also occur with α-synuclein[105] and prion protein,[106] and other MT isoforms have shown the same capability and protection of neurons.[107] The potential for MTs inhibiting copper redox cycling was previously demonstrated in cell cultures.[108]

In the case of mammalian MTs, it has been suggested that their protective action in oxidative stress conditions is intimately linked to the cellular zinc status. This link goes some way towards explaining the action of redox-inactive zinc as a "pro-antioxidant" as well as the oxidative stress elicited by thiophilic metals such as cadmium.[109] Both oxidative stress[110] and exogenous NO[111,112] have been shown to lead to release of intracellular zinc, and it is thought that this increase in free zinc originates from MTs. Although, *in vitro*, MTs have been shown to directly react with ROS[98,113] and NO, it is not clear whether the effects observed *in vivo* involve direct interaction between MT and the reactive species or are mediated by interactions with other components of the cellular redox system, *e.g.* the glutathione redox couple.[114] The link between redox balance and cellular zinc homeostasis is further discussed in Sections 21.1.3.5 and 21.2.7.

21.1.3.4   *Copper Homeostasis*

The first indication that MTs could also be involved in biological copper binding was the finding that the first preparations of MTs also contained certain quantities of this metal ion. Based on this fact and the discovery of a "Neonatal Hepatic Mitochondrocuprein" rich in cysteine,[115] Weser and Rupp prepared fully Cu-exchanged chicken liver MT, and suggested that the "cuprein" may indeed be identical to MT.[116] Copper-containing MT was also isolated from the livers of calves and sheep,[117] and MTs isolated from foetal and neonatal livers also have higher copper contents (up to 2.5 mol eq)[118] than those isolated from adult livers or kidneys. Current opinion holds that mammalian MT-1 and -2 have no major role in copper homeostasis,[61] even though elevated copper levels result in increased Cu-MT complexation. There is, however, a strong possibility that MT-3 plays a role in detoxifying extracellular copper in the brain (Section 21.1.3.3).[119] This conclusion is also supported by the fact that MT-3 isolated from brain contains significant quantities of not only Zn but also Cu.[120,121] Currently available evidence suggests that MT-4 might deal with both Zn and Cu,[77,122,123] but further work is required.

In contrast, specific roles in copper homeostasis and/or detoxification have been demonstrated for metallothioneins from other phyla, for example fungi,[124] crustaceans,[125] insects[126] and at least some molluscs, *e.g.* gastropods, where a role in copper supply for hemocyanin biosynthesis was suggested.[85,127] Some plant MTs also appear to have roles in copper handling[128] and the recently discovered MymT[43] from *Mycobacterium tuberculosis* has a clear role in protecting this pathogen against phagosomal copper stress.[129]

21.1.3.5   *Zinc Homeostasis*

A large body of evidence attests to a role of many MTs in zinc homeostasis. It should be noted that this section distinguishes "normal homeostasis" from response to extreme toxic levels.

The cellular functions of mammalian MTs (isoforms 1 to 3) have been studied most extensively.[61,62,83,130,131] About 10% of all Zn(II) in hepatocytes is bound to MT[130] but virtually all mammalian cell types express one or more isoform. The two main MTs in the majority of mammalian cells are MT-1 and MT-2,[79] and these are *in vivo* associated mostly with Zn(II).[132] Therefore, it is thought that their main function concerns normal physiological zinc homeostasis and they have been suggested to play a key role in intracellular Zn(II) buffering and muffling.[133] Although the viability of MT-1/MT-2 knock-out mice suggested that these MTs are not the sole players for handling intracellular Zn(II), the fact that these mice cannot cope well with either zinc excess or zinc deficiency supports this idea.[134] These proteins have also been shown to be necessary for zinc re-distribution from plasma to liver during acute phase response to inflammation.[135] Other examples of MTs with likely roles in zinc homeostasis include the cyanobacterial BmtAs,[46–48] type 4 MTs from plants[19] and MTL-1 from *C. elegans*.[96]

In comparison to metal tolerance, a less explored physiological function for MTs concerns protection against zinc deficiency. Nevertheless, some studies have indicated that the (over-)expression of MTs can also help to cope with deficiencies of this essential metal ion. This is particularly relevant for Zn(II) in mammals and has been demonstrated in mice:[136] transgenic mice overexpressing MT-1 were not only able to accumulate more Zn(II) in their cells, but also resisted the negative effects of Zn(II) deficiency, including teratogenic effects. There are indications that such a function may be more widespread: the promoter region of the BmtA gene in the marine cyanobacterium *Synechococcus* sp. WH8102 contains a recognition sequence for the zinc-uptake regulator protein Zur. This suggests that this MT is overexpressed under zinc depletion conditions.[46]

Intimately linked to a role in zinc homeostasis is the concept that MTs, at least in vertebrates, provide a link between the redox state of the cell and zinc homeostasis/zinc signalling.[75,137,138] As indicated in Section 21.1.3.3, biological oxidants including ROS, NO and GSSG trigger the intracellular release of Zn(II) from MT. Hence, the cellular redox balance regulates the concentration of intracellular free Zn(II). The redox-balance dependent changes in Zn(II) levels may have a range of downstream effects, including the population/de-population of zinc-binding sites in zinc fingers and enzymes, which may be either activated or inhibited by binding Zn(II). Sections 21.2.5–21.2.7 provide further information on this topic.

21.1.3.6   General Conclusions Regarding MT Functions

Ultimately, the most reliable methods to establish the physiological function(s) of MTs are studies in the natural host and, in particular, study of the effects of knock-out mutations or of gene silencing, overexpressing mutants and transcriptomics. Even then, "the" function of a particular MT may not necessarily be determined. It is now clear that many MTs serve more than one function and many organisms contain several isoforms that may partially substitute for each other. Furthermore, since Zn(II) itself impacts on virtually every conceivable physiological process, mammalian MTs with their central impact on cellular zinc levels have been linked to a wide variety of pathological conditions, including cancer,[139,140] neurodegenerative diseases[81] and diabetes[141] and its associated cardiovascular complications.[142] In many of these instances, the effect of MTs on zinc buffering and redistribution is inseparable from their redox activity.[137] Whilst their absolute indispensability has not been demonstrated in any organism or cell type,[62] it is evident that the presence of functional MTs correlates with a healthy organism, and it is therefore not surprising that MT overexpression in mice increases their lifespan,[143] that knock-out deletions in *C. elegans* affect growth, lifespan and fertility[144] and that MTs have been suggested as biomarkers for ageing.[145]

21.2 Properties and Reactions

The following sections will summarize studies of *in vitro* properties (21.2.1–21.2.5) and reactions (21.2.6–21.2.8) of metallothioneins. Where appropriate, these will be put in the context of potential *in vivo* functions. Pertinent reactions are also shown in the schematic of Figure 21.3.

21.2.1 Metal Content

The identity and number of metal ions bound to a metallothionein is probably one of the most important parameters that needs to be determined in any *in vitro* study involving an MT protein. This is true for studies concerning structure and reactivity of isolated, purified proteins, but also of pivotal importance for understanding the role of MTs *in vivo*. The determination of *in vitro* metal:protein stoichiometries (Table 21.1) is comparably straightforward nowadays, providing due care is taken during sample preparation. More challenging is the endeavour to study the *in vivo* metallation state of MTs, but a few studies are also available on this topic.

The metal-free protein is called either *apo*-MT or thionein. More recently, these terms have also been used to refer to the portion of free thiols in partially metallated MTs. Thionein (T) in rat tissues was distinguished from the fully metallated form (MT) either by differential reactivity of the thiol(ates) (Section 21.2.5) with suitable electrophilic dyes[60] or, alternatively, free thiols have been probed by titrations with Cd(II).[146,147] The most important finding of such studies was the realization that T can form a significant portion of cellular MT. It is important to note that both approaches determine the ratio of free thiols to total thiol $(T/(MT + T))$, without the possibility to specify whether free and Zn-bound thiols are present in the same MT molecule or whether the ratio reflects pure, completely demetallated T. There are, however, various indications that partially metallated MT constitutes a significant proportion of cellular MT. The prominence of under-metallated MT is also supported by the observation of corresponding species in ESI-MS spectra for substoichiometric Zn : MT ratios.[148] In contrast, co-existence of pure T and Zn_7MT has never been observed at neutral pH.

Obtaining reliable speciation data on isolated MTs depends on a range of conditions, and also the objective of the study. Two pivotal questions concern (i) the stoichiometry of the fully metallated protein and (ii) the likely metal ion(s) bound *in vivo*. These parameters were first determined in the original study on horse kidney MT,[2] in the form of weight percentages, and further refined subsequently (5.9% Cd, 2.2% Zn, 8.5% S).[4] Once the total number of Cys residues was known, it became clear that seven divalent metal ions (Zn(II) or Cd(II)) were bound by 20 Cys residues. Subsequent work revealed that vertebrate MTs can also bind Cu(I) with a preferred stoichiometry of 12 Cu(I) per protein.[149] It is possible to force the binding of additional copper ions. Table 21.1 includes information on experimentally determined metal stoichiometries.[50,150] In general, the ratio of cysteine to M(II) in fully metal-loaded MTs, no matter from which species, is around 3 : 1. It is much lower for Cu(I) owing to its lower coordination number preference (Sections 21.2.3 and 21.2.4). The stoichiometries for many of the more recently studied MTs were determined by Electrospray Ionization Mass Spectrometry (ESI-MS) in combination with Inductively Coupled Plasma Optical Emission Spectroscopy (ICP-OES).[87,151–157]

In vitro, MTs have been shown to bind other metal ions in addition to Zn(II), Cd(II) and Cu(I). These include Fe(II)[158] and Co(II)[159,160] (both requiring exclusion of air), Ni(II) (preparations with limited stability irrespective of the presence of dioxygen),[159] Bi(III),[161] As(III),[162] Pt(II),[163,164] Hg(II),[165] Ag(I),[149] Au(I)[166] and a range of other metal ions.[149] Whether and which of these additional interactions plays any role *in vivo* has not been comprehensively studied and will, of course, also depend on the organism and its environment. The interaction with Pt(II) is thought to play a role in reducing the cytotoxicity of Pt-based drugs or the development of resistance against them.[164] Bi(III)-containing drugs are known not only to induce MT synthesis in the liver, but also to reduce the toxic side effects of Pt-based drugs by inducing MT synthesis.[161] It is not known whether these effects are due to direct interactions with MT or a result of zinc displacement from other sites.

21.2.2 Stability Constants

Limited quantitative information on the metal ion affinity of metallothioneins is available, but this is perhaps true for metalloproteins in general. Some data on mammalian MTs are summarized in ref. 1 and several additional studies that include affinity data have appeared since (Table 21.2). Three approaches have been used: (i) spectrophotometric pH titrations, which gave the first estimates for Zn(II) and Cd(II) affinities,[3] (ii) competition reactions of the fully metallated proteins with a chelator and, (iii) more

Table 21.2 Experimental stability constants for MTs. Constants for which a specific pH value is given are conditional constants valid for these conditions.

	Log K_{Zn} (low I)[a]	Log K_{Zn} (high I)[b]	Log K_{Cd}	Method/conditions	Reference
Horse liver MT	14		17	pH-independent, estimated from pH titrations (5 mM phosphate/succinate)	3
mammalian MT	18		22	pH-independent, calc. from pH titrations	167
Rabbit liver MT-2	11.5–12.1		14.5–15.1	Various methods and conditions	1
Rabbit liver MT-1		11.3		Competition with NTA	174
Rabbit liver MT-2		11.2		or KTSM; 5 mM Tris, 0.1 M KCl, pH 7.4, 25 °C	174
Human recombinant MT-2		11.8 10.45 9.95 7.7	–	Titration of thionein with Zn(II) in presence of competing chelator; $I = 0.1$ M (50 mM HEPES and KNO$_3$); pH 7.4	173
MT-2	11.5		14.8[c]	Competition with 5F-BAPTA (Zn only)	171
MT-3	10.8		14.3[c]	Competition with 5F-BAPTA (Zn only)	171
Isolated β-domain (human MT-2)		11.3	15.2	pH titrations; values for pH 7.0, $I = 0.1$ M	169
SmtA	~13	10.9		Competition with 5F-BAPTA	151
H40C SmtA	~13	10.5		Competition with 5F-BAPTA	151
H49C SmtA	~13	9.7		Competition with 5F-BAPTA	151
H40C/H49C SmtA	~13	9.6		Competition with 5F-BAPTA	151
Wheat E$_C$	10.6	8.6	12–13[a]	Competition with 5F-BAPTA	170
C. elegans MTL-1	12.0		13.1[a]	Competition with 5F-BAPTA	96,172
Cd$_6$Zn$_1$MTL-1	11.6		14.2[a]	Competition with 5F-BAPTA	96,172
C. elegans MTL-2	12.0		15.0[a]	Competition with 5F-BAPTA	96,172

[a] $I = 4$ mM, pH 8.0–8.1, 10 mM Tris, unless stated otherwise.
[b] $I = 100 \pm 8$ mM, pH 7.4, 10 or 50 mM Tris, unless stated otherwise.
[c] Determined by spectrophotometric pH titration and extrapolated to pH 7.0.

recently, titrations of T with the respective metal ion, either in the presence or absence of a competing chelator. It is of note that the first two approaches use the fully metallated MT as a starting point, whilst the latter starts with T.

The most straightforward parameter to be derived from a pH titration is the pH of half-displacement. Relative metal loading can be monitored either by electronic absorption spectrophotometry using the LMCT (ligand-to-metal charge transfer) bands[165,167] or by measuring bound metal by atomic spectroscopy after removal of unbound metal. The first method is easier, quicker and tends to give smaller errors, but since it is based on the assumption that bound metal correlates linearly with the intensity of the LMCT band, it may not necessarily be more accurate than the last method. In theory, ESI-MS could also be employed but, since it is not clear whether the ionization efficiencies of different metal species are the same, such approaches need to be benchmarked with a complementary technique. An overview of representative pH of half-displacement values can be found in ref. 168. The pH titration curves can also be used to calculate average stability constants; in this way, the pH-independent stoichiometric stability constants for mammalian MT were estimated to be log $K_{Zn} = 18$ and log $K_{Cd} = 22$.[167] From these, the conditional stability constants at pH 7 were estimated as log $K_{Zn}' = 12$ and log $K_{Cd}' = 16$. Apparent stability constants for the β domain at pH 7 have been derived as 11.3 (Zn) and 15.2 (Cd), whereas the titration curves for the isolated α domains were more complicated and did not allow a straightforward derivation of a stability constant.[169] Furthermore, although no corresponding stability constant values for Cu(I) are provided in the literature, the pH of half-displacement values for Cu(I) are generally even lower than those for either Zn(II) or Cd(II) and, accordingly, the respective stability constants will be higher, leading to the overall trend Zn(II) < Cd(II) < Cu(I). These trends and values are in broad agreement with corresponding values for complexes with synthetic thiolate ligands, suggesting that the strengths of the M-S bonds govern the overall *in vitro* affinity. Section 21.2.4 discusses the issue of "metal specificity" in more detail.

Stability constants of similar orders of magnitude have been obtained by other techniques and for other MTs (Table 21.2). It should be noted that, like all stability constants, those for metal-MT complexes are dependent upon pH and on the ionic strength of the medium. A decrease in stability is expected for an increase in ionic strength and this has been demonstrated for bacterial SmtAs[151] and wheat E_C.[170] The difference between values determined at low and medium ionic strength is particularly dramatic for wheat E_C, and it may be pertinent to consider that the ionic strength in cells is even higher than those used in the reported study.

A frequently used chelator for competition studies, suitable for Zn(II) and Cd(II), is 5F-BAPTA (1,2-bis(2-amino-5-fluorophenoxy)ethane N,N,N',N'-tetraacetic acid), first employed to compare mammalian MT-2 and MT-3.[171] The proportion of metal transferred from the fully loaded MT to 5F-BAPTA is measured *via* ^{19}F NMR spectroscopy, and then, with knowledge of the conditional stability constant of the M(II)-5F-BAPTA complex, is used to calculate an overall conditional stability constant for the MT under study. This method has also been used to determine stability constants for bacterial wild-type and mutant SmtAs,[151] wheat E_C[170] and MTL-1 and -2 from *C. elegans*.[96,172] The drawback of this method is that it implies that all binding sites are of equal strength, which is likely not the case. Nevertheless, these values are useful for comparisons of the equilibrium behaviour of metallothioneins. The values compiled in Table 21.2 testify to the considerable differences in affinity for MTs from different phyla.

The third approach is illustrated by a titration of human T with Zn(II) followed by spectrophotometry (LMCT region), which indicated that the seventh Zn(II) bound with lower affinity. Further titration experiments in the presence of the metallochromic dyes FluoZin-3 and RhodZin-3 indicated that at least three classes of binding sites could be distinguished (Table 21.2).[173] In contrast, direct competition of FluoZin-3 and other chelating agents with fully metallated rabbit liver Zn_7MT-2 did not reveal a weaker site.[174] Thus, it is possible that the species formed during metallation of the *apo*-protein are not identical to those formed during demetallation by a chelator and, hence, these species may display different affinities.

21.2.3 3D Structures

As of 2013, the protein databank contains 35 entries from 20 structural studies on metallothioneins (Table 21.3).[175–180] Nine of the studies pertain to

Table 21.3 Summary of 3D structures of metallothioneins. Entries are given in chronological order.

Organism and isoform	pdb entries	Metal ions, domains	Technique	Reference
Rabbit liver MT-2A	1mrb, 2mrb	$Cd_4\alpha$ and $Cd_3\beta$	NMR	190
Rat liver MT-2	1mrt, 2mrt	$Cd_4\alpha$ and $Cd_3\beta$	NMR	191
Human MT-2	1mhu, 2mhu	$Cd_4\alpha$ and $Cd_3\beta$	NMR	192
Rat liver MT-2	4mt2	$Cd_5Zn_2\alpha\beta$	X-ray	194,195
Blue crab	1dmc, 1dmd, 1dme, 1dmf	Cd_6	NMR	175
Yeast Cup1	1aoo, 1aqq	Ag_7	NMR	211
Yeast Cup1	1aqr, 1aqs	Cu_7	NMR	211
Sea urchin	1qik,1qil	$Cd_4\alpha$ and $Cd_3\beta$	NMR	176
Mouse MT-1	1dft, 1dfs	$Cd_4\alpha$ and $Cd_3\beta$	NMR	177
Yeast Cup1	1fmy	Protein part only	NMR	208
Synechococcus SmtA	1jjd	Zn_4	NMR	200
Mouse MT-3	1ji9	$Cd_4\alpha$	NMR	205
Lobster	1j5m, 1j5l	β-N, synthetic peptide and isolated β-C	NMR	178
Lobster			NMR	178
Notothenia coriiceps	1m0j,1m0g	$Cd_4\alpha$ and $Cd_3\beta$	NMR	179
Neurospora crassa	1t2y	Protein part only	NMR	209
Yeast Cup1	1rju	Cu_8; truncated	X-ray	210
Human MT-3	2fj5, 2f5h	$Cd_4\alpha$	NMR	180
Wheat E_C	2kak	Zn_4; β-E domain	NMR	203
Wheat E_C	2l61, 2l62	γ-domain	NMR	201
Wheat E_C	2mfp	Circularized γ-domain	NMR	202

vertebrate MTs, three to yeast Cup1, two to crustacean MTs, three to the two domains of wheat E_C, one to the two domains of sea urchin MT and one each for the MTs from the fungus *Neurospora crassa* and the cyanobacterium *Synechococcus* PCC 7942.

With two exceptions, all of these structures have been determined by solution state NMR spectroscopy rather than X-ray crystallography. There are two main reasons for this heavy bias towards NMR spectroscopy: firstly, the protein backbones of MTs tend to be quite dynamic, a feature that renders crystallization difficult and, secondly, their small molecular size makes MTs attractive targets for structure determination by NMR. In addition, their high affinity for Cd(II) permits the utilization of one of the two spin $I = \frac{1}{2}$ nuclei of Cd, ^{111}Cd or ^{113}Cd, enabling the development of heteronuclear approaches to determine the connectivities between Cd(II) ions and amino acid side-chains of MTs (Figure 21.4).[181] However, despite instrumental and methodological progress, the determination of an NMR structure of an MT is still not trivial and is possibly the main reason why so few structures have been published. Crucially, it is the metal-to-sulfur bonds that define and stabilize the structure and so direct information on these connectivities is central not

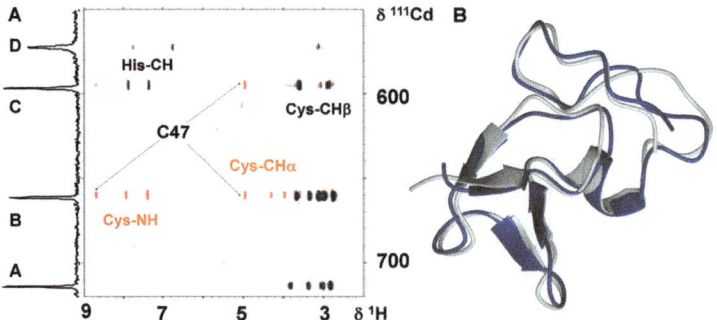

Figure 21.4 ^{111}Cd and ^{113}Cd NMR spectroscopic experiments have been instrumental in determining the structures of MTs. (A) Overlay of a [^{1}H,^{111}Cd] HSQC spectrum (black; $^3J_{1H,111Cd} = 25$ Hz) with a [^{1}H,^{111}Cd] HSQC-TOCSY spectrum (red; $^3J = 45$ Hz) of Cd$_4$SmtA. The corresponding 1D ^{111}Cd spectrum is shown on the left. Typically, CdCys$_4$ sites (*e.g.* sites A and B in SmtA) display resonances above 650 ppm, whilst CdCys$_3$His sites (*e.g.* sites C and D) display shifts below this value. The two 2D spectra give complementary information. At long mixing times (small ^3J), 3-bond couplings between ^{111}Cd and the Hε1 and Hδ2 protons of His, as well as several couplings to Hβ protons of Cys residues, can be detected. The HSQC-TOCSY spectrum correlates individual ^{111}Cd resonances with not only β protons, but also α and NH protons, allowing for better distinction between individual residues. (B) For natively Zn-binding MTs, the approach relies on isostructural replacement of Zn(II) by Cd(II); for SmtA, this has been demonstrated by separate structure determinations for Zn$_4$SmtA (grey) and Cd$_4$SmtA (blue) – only the protein backbones are shown. See also Figure 21.7 for a comparison of [^{1}H,^{1}H] TOCSY spectra of the two metalloforms.

only to understanding the cluster structures but also to defining the overall protein structure with sufficient resolution. The most recent work on MT structures has highlighted that this information is not necessarily straightforward to obtain, as a prerequisite for this approach is the isostructural replacement of the native metal by Cd(II). This outcome is harder to achieve than previously thought (see Section 21.2.4).

The first ^{1}H NMR studies of an MT were reported in 1974,[182] and the first ^{113}Cd NMR spectra were presented in 1978.[183,184] Proton-decoupled ^{113}Cd NMR spectra of rabbit liver MT-2, in which all native metal ions had been replaced by ^{113}Cd(II), permitted the observation of ^{113}Cd-^{113}Cd 2-bond couplings. These demonstrated the presence of two metal-thiolate clusters, one with three and one with four metal ions, and that the ^{113}Cd(II) ions were directly connected by bridging thiolates.[185] The tetrahedral coordination of Zn(II) and Cd(II) was inferred from the ^{113}Cd chemical shifts as well as from the spectroscopic characteristics of Co(II)-substituted horse kidney MT-1A.[186] The detailed structure of the clusters remained elusive for a little while longer but inspired research on synthetic metal-thiolate model compounds.[187,188]

The first 3D structure of an MT determined by X-ray crystallography was published in 1986. However, this study turned out to be not only fundamentally flawed but also had a severe negative impact on initially accepting NMR spectroscopy as an alternative method for structure determination.[189] In 1985, Kurt Wüthrich and his team had also solved the structure of rat liver MT-2, but this differed considerably from the crystal structure that was about to be published. Although both X-ray and NMR studies reported a two-domain structure with an M_3Cys_9 cluster in the β-domain and an M_4Cys_{11} cluster in the α-domain, neither the protein fold nor the metal-to-sulfur bond assignments were in agreement. To exclude the possibility that MTs from different mammalian species had different structures, the Wüthrich team determined the structures of rat, rabbit and human MTs,[190–193] the latter with either Cd(II) or Zn(II) bound. In each case the same fold and metal-to-sulfur connectivities were obtained (Figures 21.5 and 21.6A). Eventually, the X-ray data for rat liver Cd_5Zn_2-MT-2 were re-analysed and the previous interpretation was found to be erroneous; correct analysis resulted in identical NMR solution and X-ray crystallographic structures (Figures 21.6A and B).[194,195] One piece of information that only the X-ray structure could furnish concerns the mutual orientation of the two domains, as the NMR experiments detected no interactions between the two domains. The characteristic dumbbell shape was confirmed for rabbit MT-2A by scanning tunnelling microscopy, and the observed shape was consistent with the corrected X-ray structure of rat MT-2.[196]

The clusters in the available structures of vertebrate MTs show the same overall features (Figure 21.1): the N-terminal β-domain holds an M_3Cys_9 cluster, in which three divalent metal ions and three sulfur atoms form a 6-membered ring in boat conformation; the C-terminal α-domain contains an M_4Cys_{11} cluster, in which the metal ions and five sulfur atoms form two

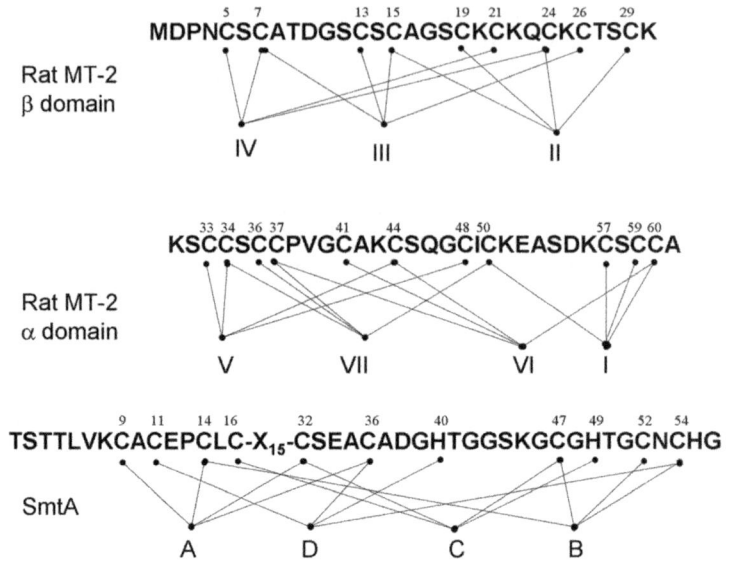

Figure 21.5 Metal-ligand connectivity patterns derived from heteronuclear [^{1}H,$^{111/113}$Cd] NMR spectroscopic experiments for mammalian and bacterial MTs. There is no discernible similarity between the patterns for the two M$_4$L$_{11}$ clusters in the MT-2 α domain and SmtA; the location and order of terminal and bridging Cys residues is completely different, as is the order of metal sites (A and B structurally correspond to VI and VII, and C and D to I and V); yet the 3D structures of these clusters are closely similar (Figure 21.1).

fused 6-membered rings, again in boat conformations giving a bicyclo[3.3.1]nonane arrangement. Besides the bridging thiolates, both clusters contain six terminal thiolate ligands. The four-metal clusters contain two types of metal sites; sites I and V have two terminal thiolates while sites VI and VII have a single terminal thiolate only. These structures are not analogous to those seen in synthetic clusters; the latter predominantly form highly symmetrical adamantane-type cages with all six-membered rings in a chair conformation, corresponding to the zinc blende structure of the cubic form of ZnS. In contrast, the clusters found so far in MTs resemble the packing in hexagonal Wurtzite (ZnS) or Greenockite (CdS).[197]

M$_3$Cys$_9$ clusters similar to those in mammalian MTs (Figures 21.1 and 21.6A) are present in the MTs from crustaceans (Figure 21.6D) and the C-terminal β-domain of sea urchin MTA (Figure 21.6C). More recently, similar clusters have been discovered in the SET domains of certain histone lysine methyltransferases,[198] and the male-specific lethal 2 protein from *Drosophila melanogaster*.[199] The N-terminal α-domain cluster in sea urchin MTA (Figure 21.6C) and the M$_4$Cys$_9$His$_2$ cluster in the bacterial MT SmtA (Figures 21.1 and 21.6F) closely resemble the M$_4$Cys$_{11}$ cluster of mammalian MTs (Figures 21.1 and 21.6A).[200] However, in the latter case, this structural similarity extends only to the metal cores: the protein folds are entirely

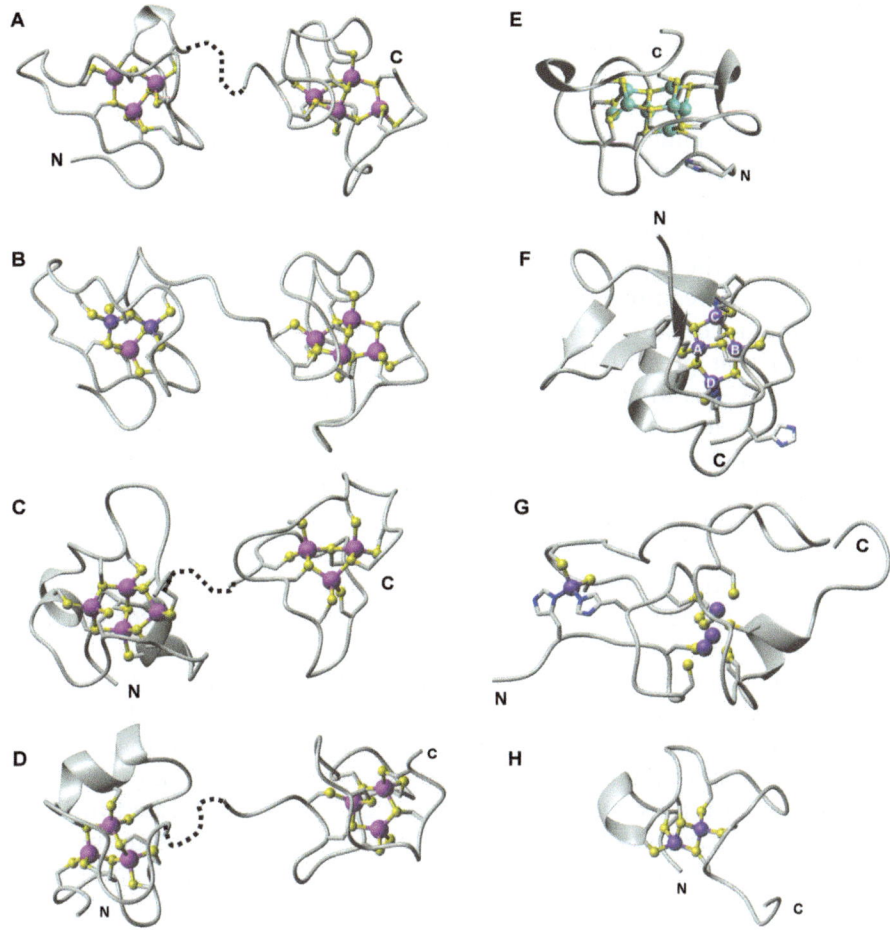

Figure 21.6 Selected 3D structures of MTs. (A) Solution NMR structures for the α and β domains of rat liver ^{113}Cd$_7$MT-2 (pdb 1mrt and 2mrt[191]). (B) Corrected X-ray crystal structure for Cd$_5$Zn$_2$MT-2 (pdb 4mt2[194,195]). (C) NMR structure of sea urchin MTA (pdb 1qik and 1qil[176]). (D) NMR structure of *Callinectes sapidus* Cd$_6$MT (pdb 1dmc, 1dmd, 1dme, 1dmf[175]). (E) X-ray structure of truncated yeast Cu$_8$Cup1 (pdb 1rju[210]). (F) NMR structure of bacterial Zn$_4$SmtA from *Synechococcus elongatus* sp. PCC7942 (pdb 1jjd[200]). (G) Domain 2 (also called β-E domain) of wheat E$_C$ (pdb 2kak[203]) with four bound Zn(II) ions. (H) Domain 1 (also called γ-domain) of wheat E$_C$ (pdb 2l62[201]) with 2 bound Zn(II) ions. All protein backbones are coloured in grey, sulfurs in yellow and metal ions as in Figure 21.1. The letters N and C refer to N- and C-terminus, respectively.

unrelated and, in addition to the replacement of two terminal Cys residues with two His residues, all other connectivities are completely different (Figure 21.5). This diversity in protein folds and connectivity patterns is the rule, not the exception; Figure 21.6 gives an impression of the structural diversity of MTs from different species.

In general, regular secondary structure elements such as α-helices and β-strands are rare in MTs and this is one reason why structural elucidation of MTs remains challenging. The bacterial MT SmtA is a notable exception as the first 38 residues, together with one of the cluster zinc ions, form a so-called treble-clef zinc finger fold. Section 21.2.4 will discuss selected examples of how protein folding influences reactivity and metal selectivity.

Studies on the type 4 metallothionein E_C from wheat have added two structural features not previously observed in MTs. The 20-residue N-terminal domain contains a Zn_2Cys_6 cluster (Figures 21.1 and 21.6H)[201,202] similar to that present in fungal transcription factors of the GAL4 type. The larger C-terminal domain contains a mononuclear $ZnCys_2His_2$ site (Figures 21.1 and 21.6G),[203] as well as three additional zinc ions bound by nine Cys residues in a cluster whose structure has remained elusive. The replacement of Zn(II) with Cd(II) did not yield a well-folded protein – an example of an MT in which Zn(II) and Cd(II) are not bound isostructurally.[156,203] Unfavourable structural dynamics were also encountered with Cd-substituted β-domains of several mammalian MT-3s,[204] and, for this reason, only models based on previously resolved structures for other vertebrate MTs have been published.[205,206]

Whilst both Zn(II) and Cd(II) are tetrahedrally coordinated in MTs, Cu(I) is coordinated in a trigonal planar or even digonal (two sulfur ligands) fashion. As for the divalent metal ions, the dominant coordination mode closely resembles that found in the respective natural mineral. In the case of Cu(I), this is Cu_2S as chalcocite, in which Cu(I) ions are located in the trigonal voids between three adjacent sulfur atoms. Structure determinations for Cu(I)-MTs (the issue of metal selectivity is discussed in Section 21.2.4) have met with even greater challenges than those for the M(II)-MTs. Like zinc, copper has no practically useful NMR-active isotopes. In principle, Ag(I), with the two NMR active nuclei ^{107}Ag and ^{109}Ag, may provide a probe for Cu(I), but the coordination chemistries of these two ions differ considerably more than those of Zn(II) and Cd(II) and, therefore, non-isostructural replacement, resulting in either no or incorrect/incomplete structural information, is even more likely.[207] It is therefore not surprising that two entries in the Protein Data Bank pertaining to Cu(I)-MTs refrain from showing any metal connectivities,[208,209] and that there is some disagreement between the latest X-ray structure[210] and the earliest NMR structure[211] of yeast Cup1.

NMR spectroscopy does not yield direct information on the M-S or M-N bond lengths and so other techniques have been employed. They include Extended X-ray Absorption Fine Structure Spectroscopy (EXAFS)[168,212,213] and theoretical approaches such as molecular dynamics simulations[206,214] or Density Functional Theory (DFT) calculations.[215] The latter approaches, especially, have confirmed the expectation that the lengths of terminal M-S bonds are shorter than those of bridging ones. EXAFS has given some insight into the structure of the Cu_4-β domain cluster in mammalian MT-3[213] and supports the hypothesis that His residues participate in metal-binding in the type 4 plant MT E_C from wheat.[168]

No structures of partially metallated or even metal-free thioneins are available, and it is thought that the latter are intrinsically disordered.[75] Molecular dynamics simulations on mammalian T made different predictions, depending on whether the starting point was a random coil or a metallated conformation.[216] Fluorescence Resonance Energy Transfer (FRET) studies on human MT-2 suggested that metal removal from the two domains had little effect on the distance between the respective fluorescent labels,[217] and a more recent molecular dynamics simulation study suggested that T proteins may be more organized than previously thought.[218] In the case of bacterial SmtA, partially metallated Zn_1SmtA still retains the zinc finger fold whereas the 17 C-terminal residues are disordered, according to ^1H NMR spectroscopic results.[219]

Even though the folding of MTs is dominated by metal coordination, a number of protein-derived weak interactions are also important. NH-O hydrogen bonds are present, manifesting in slow H/D exchange rates.[220] In addition, the presence of NH-S hydrogen bonds reduces dynamic structural disorder.[221] These bonds may also increase the overall stability of clusters and decrease their reactivity, as they can reduce negative charge density – the formal charge of an M_3Cys_9 cluster is −3. Furthermore, lysine residues are relatively abundant in many MTs (up to 14%),[222] and their positive charge may further stabilize the negative charges on metal-thiolate clusters. The X-ray structure of rat liver MT-2 (pdb 4mt2) displays an H-bond from the ε-amino group of Lys31 to the sulfur of Cys19 in the β-domain.

21.2.4 The Interplay between Protein Structure and Metal Specificity

ESI-MS speciation studies of metal-saturated MTs have revealed repeatedly that a single species tends to dominate the speciation if the "correct" metal ion is bound. On the other hand, the observation of multiple metallospecies is more likely if the protein is either expressed or reconstituted with a non-native metal ion.[153,223] This is not surprising for Cu(I) *vs.* Zn(II)/Cd(II) binding as Cu(I) prefers trigonal planar or linear/digonal coordination geometry whereas Zn(II) and Cd(II) prefer tetrahedral coordination with thiolates. It follows that the protein backbone needs to adopt different conformations depending on which metal ion is bound, which, depending on the overall flexibility, may be facile or not. The need for changes in the protein fold has indeed been demonstrated experimentally: the displacement of Zn(II) by Cu(I) in the two isolated α and β domains of mouse MT-1 yielded mixed metal species $Zn_xCu_3(\alpha)$ and $Zn_xCu_4(\beta)$, whose protein folds were substantially different from those for $Zn_4(\alpha)$ and $Zn_3(\beta)$. Higher Cu(I):Zn(II) ratios led to structural disorder.[224] Different protein folds are associated with different energetics – even if the protein is highly flexible and largely unfolded in the absence of metal ions. Hence, whilst a predominant, optimal fold and metal-loaded species may exist for the "native" metal ion, there may be several energetically similar folds and compositions for non-native metal ions, none of them optimal. This may be the underlying cause for the observation of several different metallospecies. It is very important to note, however, that this does not mean that some MTs have an *absolute* preference for M(II) ions over Cu(I), as the superior strength of the Cu(I)-S bonds dominates the overall energetics, even if that results in a structurally disordered protein.

To obtain a complete picture, it is essential to consider to which metal ions the respective MT is likely to be exposed *in vivo.*[225] There is an emerging concept that MT protein folds (and hence their sequences) have evolved to best accommodate the metal ion that is to be dealt with *in vivo.*[153] The most striking illustration of this concept at work are probably the MTs from snails. The two major isoforms of MT from the snail *Helix pomatia* (see Table 21.1; entries CdMT and CuMT) contain the same Cys residue patterns within 38 fully and 22 semi-conserved residues over 64 and 66 amino acids, respectively (*i.e.* >90% similarity). Although both MTs were simultaneously present in the tissue extracts from Cd-exposed snails, one contained copper only whereas the other was associated overwhelmingly with cadmium.[85,226] Whilst it is clear that both MTs are specific to different tissues and cells, it has recently been demonstrated that their *in vivo* metal specificities are also reflected in their *in vitro* behaviour, *i.e.* yielding clean homo-metallic preparations only when recombinantly expressed in the presence of the "correct" metal ion (*i.e.* the one found bound *in vivo*), but giving multiple species when the non-native metal was supplied in the expression media.[226]

The multiple species observed in the presence of the "wrong" metal ion may include both under- and over-metallated species; for example the Zn-MT E_C from wheat binds six Zn(II) ions in an orderly fashion but can bind up to seven Cd(II) ions with significant affinity[212] but with loss of ordered structure.[156] Over-metallation *in vitro* has also been observed for human MT-1A reacted with Cd(II),[227] and for mammalian MT-3 reacted with Zn(II).[228–230] Mixed-metal species are also common.[153] It needs to be noted that the concept described above heavily relies on the metallation of MTs during heterologous expression in *E. coli*, which is targeted at the production of fully metal-saturated preparations. Hence, the isolated products, even when the correct metal is supplied, may not necessarily reflect the species prevalent in the natural host, which, as we have seen in Section 21.2.2., may comprise partially metallated forms.

Previously, site-selectivity in mammalian MTs had been suggested when reconstituting thionein with mixtures of divalent metal ions[231] and when investigating metal-exchange reactions (Section 21.2.8). The observed preferential formation of homo-metallic clusters was attributed to the influence of the protein structure,[231] and has led to suggestions that a single MT may function simultaneously in the metabolism of different metal ions. Mammalian MT-3 is perhaps the most intriguing example. Bovine MT-3 was isolated from the native host with both Cu(I) (4–5 mol eq) and Zn(II) (2–2.5 mol eq) bound.[232] Spectroscopic analysis of such preparations, together with *in vitro* data of recombinant MT-3's suggested that domain-specific metallation is likely, with four Cu(I) ions in the β-domain and three or four Zn(II) ions in the α domain.[120,121] Remarkably, the Cu(I)$_4$ cluster is air-stable, an uncommon feature for Cu(I)-thiolate complexes. In contrast, the Zn$_4$ cluster in the α-domain is redox-labile and prone to air oxidation.[121] Hence, mammalian MT-3 is a particularly interesting case regarding correlations between protein structure and dynamics on one hand and metal specificity and biological activity on the other. The growth inhibitory activity – although not known to be dependent on the metalloform – of MT-3 has been associated with particular amino acids in the N-terminal β-domain[171,233] but this domain so far has withstood all efforts aimed at a 3D structure determination. Specifically, biological activity, unusual cluster dynamics in Cd-substituted MT-3[204] and reduced zinc-binding affinity[171] hinge to a large extent on the CPCP motif in the β-domain. Whilst prolines undoubtedly impart considerably different backbone structure and dynamics, it may be worth exploring whether the structure of Cu(I)$_4$-β-MT-3 is more ordered than those with M(II) metals.

More recently, several MTs from non-vertebrate eukaryotes also showed a degree of discrimination between Zn(II) and Cd(II) based on affinities. Co-ordination geometry cannot be the cause for differential behaviour towards these two closely related metal ions, which differ mainly by their ionic radii (0.74 *vs.* 0.92 Å) and their Pearson classification: Cd(II) is much softer (and hence thiophilic) than Zn(II).[234] The two MTs from *C. elegans* display almost identical Zn(II) affinities, but their average Cd(II) affinities differ by two orders of magnitude (Table 21.2).[96,172] This could only partially be attributed to the participation of histidine residues in the binding of one M(II) ion; the remaining six Cd(II) ions in MTL-1 are still bound with a *ca.* 10-fold lower affinity than those in MTL-2 (Table 21.2). The origin of this difference is not yet clear, but expected to reside in subtle structural differences. Once more, the *in vitro* behaviour correlates with biological data: the MTL-2 knock-out mutant accumulated much more Cd(II) than the wild-type and the MTL-1 knock-out mutant,[96] whilst the latter accumulated much more Zn(II) than either wild-type of MTL-2 knock-out mutant worms.

Plant type 4 MTs, exemplified by wheat E$_C$, present an intriguing case of histidine residues modulating the behaviour of Zn(II) *vs.* Cd(II). As expected, the overall affinity for Cd(II) is higher by several orders of magnitude[156] (Table 21.2), and titration of Zn$_6$E$_C$ with Cd(II) led to the complete and stoichiometric replacement of Zn(II) by Cd(II).[235] However, the replacement is not isostructural: whilst E$_C$ is a well-folded protein in the presence of Zn(II), its domain 2 is completely disordered in the presence of Cd(II), as indicated by the complete absence of resolved backbone NH signals for this domain in 2D ^{1}H spectra (see Figure 21.7, which also compares an MT in

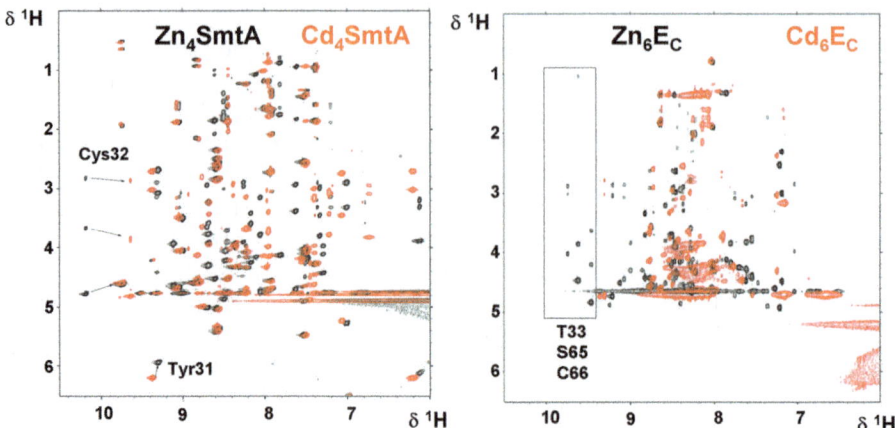

Figure 21.7 Isostructural and non-isostructural replacement of zinc by cadmium. Overlays of 2D [^{1}H,^{1}H] TOCSY NMR spectra for MTs bound to either zinc (black) or cadmium (red). Each vertical line of cross-peaks corresponds to one residue. SmtA (left hand panel) is an example for which isostructural replacement of Zn(II) by Cd(II) occurs (Cd$_4$SmtA can be produced by reconstitution of apo-SmtA with Cd(II)). This is reflected in the good dispersion of resonances in both cases, indicating well-folded proteins, and the close similarity in chemical shifts for most residues. The arrows indicate two residues whose NH or CH(α) resonances shift considerably; for the two high-lighted residues this can be related to the structural effect that the larger Cd(II) has on an H-bond between the NH of Cys32 and the S of Cys9 in site A. Wheat E$_C$ (right hand panel) is a spectacular example for non-isostructural replacement: whilst the spectrum for Zn-loaded E$_C$ (black), whether native or recombinant, shows good dispersion of resonances, Cd-loaded E$_C$ (red) shows much unresolved intensity between 8 and 8.5 ppm, indicating the presence of unfolded protein sections. Sequence assignment of the remaining resolved resonances indicates that they all refer to domain 1.[154] A few low-field shifted domain 2 residues of Zn$_6$E$_C$ with no counterparts in Cd$_6$E$_C$ are picked out in the box.

which isostructural replacement does take place).[156] Similarly, either no or very small and broad[168] signals were observed for the Cd(II) ions bound to domain 2 in ^{111}Cd or ^{113}Cd NMR spectra. This can be rationalized by understanding the central importance of the mononuclear Cys$_2$His$_2$ site for an ordered protein fold of domain 2, and considering the difference in affinities of Cys$_2$His$_2$ sites *vs.* Cys$_4$ sites for Zn(II) and Cd(II): whilst their affinity at neutral pH for Zn(II) is almost the same, that for Cd(II) differs by four orders of magnitude, as judged from measurements with mononuclear model peptides.[236] As a consequence, the MCys$_2$His$_2$ + M$_3$Cys$_9$ arrangement associated with an ordered protein fold is favourable for Zn(II), but alternative clusters, *e.g.* M$_4$Cys$_{11}$, are likely to be formed with Cd(II) – and the associated dynamics of (multiple) misfolded species precludes the observation of well-defined NMR signals for either $^{111/113}$Cd or ^{1}H. The latter hypothesis was also supported by studying His-to-Ala and His-to-Cys mutants.[156] Hence, and in stark contrast to many other MTs, wheat E_C "discriminates" between Zn(II) and Cd(II) *via* metal-specific protein folding. Whether the folding state of E_C is correlated with cellular fate, *e.g. via* enhanced degradation of misfolded species, and concomitant metal remobilization, remains to be explored.

21.2.5 Redox Properties

Even though Zn(II) and Cd(II) do not exhibit redox activity, the cysteine residues of MTs mean that they are redox-active under physiological conditions.[114,237] At pH 7.3, free thiols in MTs exhibit a redox potential of *ca.* −0.6 V *vs.* Ag/AgCl (*ca.* −0.4 V *vs.* NHE; normal hydrogen electrode). Data for MTs from a cyanobacterium, the crab *Scylla serrata* (MT-2) and human and mouse liver are available.[238,239] Bound metal ions decreased the reducing character of thiols, with the potential shifting to −0.38 V for Zn-MT and to around −0.30 V for Cd-MT (*vs.* Ag/AgCl). The reduction of bound and free redox-active metal ions could also be measured, and although these reductions are physiologically not relevant, it is clear that MT-bound metals are harder to reduce than the free ions. It should be noted that the numbers quoted above are single average values, but that it is likely that the redox potentials of different thiolates in the folded, metallated forms may vary over a certain range. It is also clear that redox and metal binding equilibria are coupled, and difficult to dissect.

The behaviour of Cu-MTs appears to be even less clear cut, as far as electrochemical data are concerned. However, it is well known that MTs can only co-exist with bound Cu(I), as Cu(II) leads to oxidation of the thiols, Cu(II) being more oxidizing than either metal-bound or free thiols. This has been demonstrated most recently by the reaction of protein-bound Cu(II) with human Zn$_7$MT-3 (Section 21.2.8).[119] Compared to the free ion, Cu(I) bound to MT is considerably less prone to oxidation; the extraordinary air-stability of the Cu(I)$_4$-β cluster in mammalian MT-3 was mentioned in the previous section.

The redox potentials for the thiolates in Zn-MTs permit in principle redox reactions with physiologically relevant oxidizing agents, including superoxide, peroxide, nitric oxide, dehydroascorbate, cytochrome *c*, selenate and biological disulfides including glutathione disulfide (GSSG).[237] Selected studies regarding reactions with some of these oxidants will be described in Sections 21.2.7.3 and 21.2.7.4.

21.2.6 Metallation Reactions

Experimentally, metallation reactions are possibly the most challenging to study, as they require the presence of thioneins that are generally extremely air-sensitive. In addition, metallation reactions with Zn(II) and Cd(II) are too fast to be studied directly.[240] However, analysis of the products of reactions with substoichiometric amounts suggested that both Zn(II) and Cd(II) metallate human MT-1A in a non-cooperative fashion,[148,241,242] as was suggested previously for MT-3.[228] In contrast, earlier studies involving titrations of rabbit T with substoichiometric quantities of [113]Cd(II) suggested that α domain formation is completed before full population of the β cluster. This observation suggests cooperative cluster formation.[243] The issue of cooperativity is difficult, as the start and end points of the various possible steps for a given MT do not necessarily have the same structures. The picture is further complicated by the possibility that partially and differentially loaded species can exchange metal ions. Metallation reactions with non-native metal ions tend to be non-cooperative.[159,162] Different speciation during metallation and demetallation may also be partly responsible for conflicting results regarding the affinities of MTs for Zn(II) (see Section 21.2.2).

Thioneins can also acquire metal ions from other proteins *in vitro*. Examples include the zinc finger transcription factors Sp1[244] and TF-IIIa,[245,246] Cd(II)-loaded carbonic anhydrase[247] and the Zn(II)-bound insulin hexamer.[248] Zn(II) abstraction from the transcription factors inhibited their normal interactions with their cognate DNA, raising the idea that MTs may have an indirect impact on gene regulation. More recently, it has been shown that incubation of the complete Zn(II) proteome of U-87 MG human glioblastoma-astrocytoma cells with T led to less than 10% of transferred Zn(II) from this pool, and a separate experiment demonstrated that Zn(II) could not be abstracted from Zn_3Sp1 when bound to DNA.[249]

The domain specificity of mammalian MTs was studied by reconstitution of *apo*-MT-2A (rabbit liver) with mixtures of Co(II)/Cd(II), Zn(II)/Cd(II) or Co(II)/Zn(II).[231] The most important finding was that the distribution of these metal ions was not random, leading to the suggestion that protein structural elements favoured particular re-constitutions.

Finally, thioneins can in principle be fully metallated by Zn(II) bound to Zn-inhibited enzymes, *e.g.* glyceraldehyde 3-phosphate dehydrogenase.[250] In contrast, zinc transfer to T from enzymes that require Zn(II) for activity was not significant.[250]

21.2.7 Demetallation Reactions Including Metal Transfer to other Proteins

Such reactions can be elicited either by direct interaction with the metal ions (by ligand exchange; 21.2.7.1), or by attack on the thiolate sulfurs, usually by electrophilic agents.[251] The most simple "electrophile" is the proton; the utility of the pH of half-displacement to characterize the thermodynamics of metal binding was discussed in Section 21.2.2. Most other reactions involving the thiol(ate) sulfur atoms are redox reactions that result in their oxidation. Notably, metal-bound thiolates are much less reactive than free thiol(ate)s. This follows from the reduced nucleophilicity of the ligands due to the Lewis acidity of, *e.g.*, bound Zn(II) that leads not only to greatly depressed pK_a values but also to decreased reactivity with any electrophile. Strong electrophiles can nevertheless react with the metal-bound thiolates, ultimately leading to the ejection of the metal ion. Depending on the electrophilic agent, various products may result, including oxidized thionin, in which adjacent sulfur atoms have formed disulfide bridges. Such species have been detected in MT-overexpressing mouse heart[110] and other tissues and cell lines.[57] This section will briefly discuss reactions with compounds used to probe the reactivity of MTs as well as reactions with biologically relevant molecules.

21.2.7.1 Ligand Exchange with Small Molecule Chelators

MTs are renowned for their unique combination of binding metal ions with kinetic lability and high thermodynamic stability. In part, kinetic lability may be promoted by cluster dynamics and it has been suggested that metal-sulfur bonds are continuously made and broken, thus enabling the interaction of both metal ion and thiolate ligand with exogenous reactants.[252] The lability of metal ions can be probed either by reactions with high-affinity small-molecule chelators, such as EDTA,[155,248,253–256] nitrilotriacetic acid (NTA)[173,257] 3-ethoxy-2-oxobutyraldehyde-bis(thiosemicarbazone) (KTS)[174] and 4-(2-pyridylazo)resorcinol (PAR),[237] or by metal-exchange reactions (discussed in Section 21.2.8). Reactions of mammalian MTs, whether loaded with Zn(II) or Cd(II), with EDTA and NTA display biphasic kinetics that were attributed to differential reactivity of the two clusters. Rabbit MT-2 was used to demonstrate that Zn(II)-MT reacted faster with EDTA than did Cd(II)-MT.[256] In principle, stoichiometric quantities of EDTA are sufficient to remove all metals bound to MTs whereas chelators with lower affinities have a more limited ability to abstract metals from MTs.[173] For example, PAR appears to react with only a small subpopulation of metals bound in either mammalian or plant MTs.[155,169] Low-affinity dyes such as Zincon (2-[5-(2-hydroxy-5-sulfophenyl)-3-phenyl-1-formazyl]benzoic acid) can be used to monitor zinc release triggered exclusively by other mechanisms, *e.g.* through thiolate oxidation (Sections 21.2.7.3 and 21.2.7.4).

To illustrate the complexity of such studies, three examples will be discussed briefly. The reaction of full-length rabbit liver Cd$_7$MT (reconstituted from Zn-MT) with EDTA showed biphasic kinetics, and by studying isolated α- and β-domains, it was possible to allocate the faster phase to the three Cd(II) ions in the β-domain and the slower phase to four Cd(II) in the α-domain.[253] Both of these processes depended on the EDTA concentration, suggesting a mechanism that involves the formation of EDTA-MT adducts. In the isolated α-domain, a further and faster step was observed corresponding to the loss of one Cd(II), independent of EDTA concentration. This suggested that one Cd(II) is relatively labile and dissociates from the protein before binding to EDTA – at least in the isolated α-domain. In addition, the slowest step was even slower at higher protein concentrations, suggesting that dimer formation may decrease the rate of metal loss from the α-domain.

The same reaction was also studied by ^{111}Cd NMR spectroscopy, ultra-filtration and gel filtration chromatography. Interestingly, results from ultrafiltration indicated slower loss of Cd(II) from MT than that observed by the other three techniques, again supporting the idea of EDTA-MT adduct formation.

Reactions with EDTA were also studied for the single-domain bacterial Zn$_4$SmtA.[254] A partially folded intermediate corresponding to Zn$_1$SmtA was observed, and metal loss occurred in three phases: the first rate-determining step corresponds to direct removal of the metal ion in site C (Figure 21.6F), followed by rapid loss of metal ions B and D, leading to the formation of the site A intermediate in which the zinc finger fold is still intact. Eventually, this last zinc ion is also lost, leading to unfolded *apo*-SmtA. Overall, the reaction was relatively slow, permitting its observation by ^{1}H NMR. This, in contrast, was not possible in a similar study on the type 4 plant MT E$_C$ from wheat.[155] Here, stoichiometric amounts of EDTA at the concentrations required for NMR studies led to rapid demetallation and complete unfolding of the protein within 5 min. Titrations with substoichiometric amounts of EDTA allowed the conclusion that domain 2 unfolded first. Faster methods allowed the study of the time-dependence of metal loss, and confirmed that the zinc release rates of Zn$_6$E$_C$ are much higher than those of other MTs.

These three separate studies highlight that the metal transfer behaviour of metallothioneins towards the chelating agent EDTA can differ quite dramatically.

21.2.7.2   Alkylating Agents, Thiol-reactive Probes and Reagents

The reactivity of thiol(ate)s in MTs have been probed with alkylating agents such as iodoacetamide[240,258] and *N*-ethylmaleimide (NEM).[259] As for other electrophilic reactions, the reactivity of free thiols is much higher than that of metal-bound thiolates and this can be used to differentiate between the two forms. Iodoacetamide is a particularly useful reagent for this purpose.[240] Nevertheless, metallated MTs can react slowly with alkylating agents, either with (iodoacetamide) or without (NEM) concomitant metal loss. In general, both Zn- and Cd-MTs can be alkylated.

The thiol-specific fluorescent dye 7-fluorobenz-2-oxa-1,3-diazole-4-sulfonamide (ABD-F) reacts very rapidly and quantitatively with free thiols but not with Zn-bound thiolates. This has been exploited for differential labelling that allows distinction between these two pools in tissue homogenates and cell extracts, clearly indicating the presence of reduced, unmetallated thiols *in vivo*.[60] ABD-F can also be used to quantify oxidized thionin after reduction by *tris*(2-carboxyethyl)phosphine (TCEP), a reducing agent that, in contrast to thiol-containing reductants (Section 21.2.7.3), does not react with metallated MT.[56]

Furthermore, aldehydes have been shown to release Zn(II) from MTs, likely through formation of hemithioacetals.[260] Acrolein was particularly effective and it was suggested that this may provide a link between oxidative stress (including lipid peroxidation) and zinc signalling. The increase in intracellular free Zn(II) upon administration of exogenous aldehydes was also demonstrated in HepG2 liver cells.

21.2.7.3 Disulfides, Thiols and Selenium Compounds

The most frequently used reagent to probe the redox state of protein thiols is Ellman's reagent, 5,5'-dithiobis-(2-nitrobenzoic acid) (DTNB). The reduction potential of the TNB/DTNB redox couple is -0.35 V (*vs.* Ag/AgCl), close to the redox potentials measured for metallated thiol groups in MT (see Section 21.2.5). Accordingly, DTNB is capable of being reduced by MT thiolates.[169,251,261] The relatively slow reaction with DTNB proceeds *via* thiol/disulfide interchange and results in the oxidation of the MT sulfur atoms to form intra- and interprotein disulfide bridges, as well as mixed disulfides.[262] This also leads to slow metal release.[262] Biphasic kinetics were observed for full-length rabbit liver MT-2, and study of the isolated α-domain revealed that the two phases reflect differential reactivity of the two domains: the α-domain reacts faster.[263] The rate laws for both steps contain DTNB-dependent and -independent components. The reaction with native MT isolated from rabbit liver was initially found to be independent of metal composition,[261] but further studies with the disulfide GSSG revealed that the release of Cd(II) was less efficient than that of Zn(II).[262] Interestingly, even though very little Cd(II) was removed by substoichiometric ratios of DTNB, no signals were observed in [111]Cd NMR spectra, suggesting that the reaction products are structurally highly dynamic.[262] More recently, the existence of partially metallated MTs *in vivo* (mammalian systems) became apparent and, therefore, the reactivity of partially metallated MT was studied *in vitro*.[173] MT with four or more bound Zn(II) displayed dramatically decreased reactivity for the stoichiometry of one DTNB per Cys. In addition, the reactivity of fully metallated MT with DTNB was studied in the presence of a range of small-molecule chelators (citrate, ATP, L-histidine, bipyridine, NTA, EDDA [ethylenediamine diacetic acid] and EDTA; each at 1000-fold excess over MT). The pseudo-first order rates were shown to correlate with the stability constants of the Zn(II) complexes of the small chelators. Since this relationship also held for the chelators with affinities much weaker than the average Zn(II) affinity of MT (*e.g.* citrate: log $K' \approx 5$), it was concluded that site(s) with much weaker affinities must be present in the MT under study. Chelators with even weaker affinity (*e.g.* glycine: log $K' \approx 2.9$) did not affect the rate and the data provide a lower limit for the potential low-affinity site(s) in MT.

The oxidized form of glutathione, GSSG, has also been shown to elicit Zn(II) release *in vitro*.[114,262] GSSG promotes the transfer of Zn(II) to *apo*-enzymes (Section 21.2.7.5) whereas reduced GSH inhibits metal release.[264] Most efficient transfer was achieved with mixtures of GSSG and GSH, a condition similar to that encountered *in vivo*. Hence, the rate and amount of released Zn(II) depend on the GSSG concentration and the GSSG/GSH ratio.

It is therefore thought that MTs, together with the GSSG/GSH redox couple ($E_0' = -0.24$ V *vs.* NHE at pH 7.0),[237] provide a link between cellular redox state and zinc homeostasis.[75,138] Furthermore, it is of note that selenium compounds in a variety of oxidation states catalytically enhance the reactivity of MTs with GSSG/GSH.[265] It is also of note that the bacterial protein disulfide isomerase DsbA was able to elicit Zn(II) release from MT-2.[237]

21.2.7.4 *Reactive Oxygen and Nitrogen Species*

Cysteine thiols are notoriously air-sensitive and often react with molecular oxygen, usually leading to the formation of disulfides. Other oxidants also react readily with thiols but, as for other electrophilic reactions, the reactivity of zinc-bound thiolates is greatly reduced. Biologically relevant oxidants include superoxide, peroxide and hypochlorite which are, for example, produced by neutrophils during inflammation.[266] Superoxide and hydroxyl radicals can also be generated enzymatically by xanthine oxidase and were also shown to lead to the oxidation of thiolates with concomitant loss of Zn(II) from rabbit liver MT-1.[113]

These typical reactive oxygen species (ROS) have the capacity to mobilize Zn(II) from MTs *in vitro*, although it is not always clear whether this occurs *in vivo* to a significant extent, and whether this involves direct interaction between ROS and MT, or is mediated by other biological agents, *e.g.* the GSH/GSSG redox couple (Section 21.2.7.3).[267] Peroxide elicited Zn(II) release from mammalian Zn$_7$MT-3.[268] Since MT-3 is also present in the more oxidizing extracellular milieu, and peroxide and other ROS are generated in Alzheimer's disease brains, this reaction may be of pathophysiological importance. A recent study by ESI-MS on the two MTs from *C. elegans* has demonstrated that peroxide-induced Zn(II) loss is accompanied by the formation of disulfide bonds. In addition, a single Cys residue in MTL-1 formed a sulfinic acid.[98]

Thiols also react with nitric oxide or NO donors; common products are *S*-nitrosothiols, which readily undergo further reactions. For the reaction of *S*-nitrosocysteine or NO gas with mammalian Zn- or Cd-loaded MT, *S*-nitrosylation and Zn(II) release was demonstrated *in vitro*[269] and *in cellulo*.[111] However, more recent work indicated that Zn-bound thiolates do not react with NO under strictly anaerobic conditions, although they do in the presence of dioxygen.[270] Similarly, *S*-nitrosothiols were not formed upon reaction of Zn-MT (from human foetal liver) with NO: cysteines were oxidized to cystine with concomitant metal release.[271] Reactions of mammalian zinc-loaded MT-1, -2 and -3 with the NO donor *S*-nitroso-cysteine led predominantly to trans-nitrosation (formal transfer of NO$^+$) and also Zn(II) release.[272] MT-3 was by far the most reactive isoform, consistent with the presence of a redox-labile Zn(II) site in the α-domain.

The weaker NO donors SNAP (*S*-nitrosopenicillamine) and GSNO (*S*-nitrosoglutathione) did not react with Zn-MT or Cd-MT under anaerobic conditions in the dark.[273] Incubation studies with mammalian cells (derived from a medullo-blastoma tumour and kidney tubular cells) suggested that the NO donor DEA/NO (Diethylamine NONOate) reacted with free cellular thiols including those in T. However, no NO-induced Zn(II) mobilization was observed. In contrast, in vascular endothelial cells, an MT FRET sensor[274] indicated that the provision of 10 µM NO led to the loss of Zn(II) from the MT portion of the MT-GFP fusion protein.

21.2.7.5 Metal Transfer to Proteins

The ability of thioneins to abstract metal ions from proteins was described in Section 21.2.6. Conversely, *apo*-proteins can also acquire Zn(II) from metallated MTs, a conclusion based on *in vivo* metal speciation and numerous *in vitro* experiments. MTs have been proposed to act as reservoirs for readily mobilizable zinc (Figure 21.3). Activation of a number of zinc-requiring *apo*-enzymes (alkaline phosphatase, aldolase, thermolysin, carbonic anhydrase) was demonstrated in 1980, giving one of the earliest experimental indications for a potential role of MTs in zinc homeostasis.[275]

Apo zinc finger peptides rapidly acquire a Zn(II) ion from Zn_7MT, if the two proteins are in the same solution, but negligible transfer was observed when the proteins were separated by a dialysis membrane.[276] This observation has led to suggestions that the Zn(II) transfer occurs *via* an associative mechanism, *i.e.* involves a protein-protein interaction. Interestingly, only about three out of the seven Zn(II) ions were "transferable", and the predominant intermediate MT species observed was Zn_4MT, besides the fully metallated Zn_7MT. Similarly, only one equivalent of Zn(II) was transferable to sorbitol dehydrogenase,[264] with the β-domain the more likely donor.[169] Reactivation of alkaline phosphatase and carboxypeptidase A also required an equimolar amount of Zn_7MT.[277] The activation of carbonic anhydrase by Zn-MT was studied by several groups, but the stoichiometry of transfer was not established.[248,255]

The mechanism of acquisition of Zn(II) by cytosolic proteins including enzymes is still unclear. Once the extremely low levels of "free" cytosolic zinc became evident, Zn(II) transfer involving direct protein-protein interactions became an even more popular proposition.[278] The *in vivo* transfer of zinc from MT to mitochondrial aconitase (a zinc-inhibited enzyme) has been demonstrated in mouse heart and was proposed to involve a direct protein-protein interaction.[279] This may be a possibility for transfer to and from surface-exposed Zn(II) sites or sites that are only formed upon Zn(II) incorporation (such as in zinc fingers). However, Zn(II) supply from MTs to large, globular, largely folded metalloenzymes with deeply buried active sites is likely to involve a different mechanism, probably dissociative. For example, it has been shown that a mixture of GSSG/GSH promotes Zn(II) release and subsequent activation of the zinc-dependent enzyme sorbitol dehydrogenase.[264] The observation is consistent with GSSG-promoted Zn(II) release from MT followed by subsequent Zn(II) uptake by the enzyme. As mentioned in Section 21.1.3.3, a number of studies have shown that various physiological agents may elicit the intracellular release of MT-bound zinc and that this may subsequently be redistributed to other protein binding sites. Hence, it is still possible that the metallation of cytosolic zinc-dependent enzymes occurs *via* a small but not negligible, and above all dynamic, pool of free ionic Zn(II) and that the concentration of this pool is controlled by the degree of Zn(II) saturation of MT.[280] This idea also permits the activation and deactivation of various enzymes by regulation of available Zn(II): enzymes that are activated by Zn(II) have zinc affinities around $\log K = 10$–12.[281] On the other hand, inhibitory sites often have lower affinities, although this is not a strict rule – recent work on a protein tyrosine phosphatase demonstrated that picomolar Zn(II) is inhibitory to this enzyme.[282]

Finally, although it does not involve transfer of metal from MT to another protein, a recent study reported that in mixtures of Zn-loaded MT and Fe(III)-loaded ferritin, both Fe(II) and Zn(II) become mobilized from their respective protein.[283] This involves reduction of Fe(III) to Fe(II), and the oxidation of Cys thiols that, in turn, mobilizes the MT-bound Zn(II). It is currently unclear how the electron transfer from thiolate to Fe(III) proceeds and whether and how this reaction may impact biological systems. It can be speculated that it may play a role under conditions that are characterized by metal misbalance, such as for various neurodegenerative diseases and ageing.

21.2.8 Metal Exchange Reactions, Including Metal Exchange with other Proteins

Studies of metal exchange reactions of MTs serve mainly two purposes: (i) they provide a facile handle on the kinetics and thermodynamics of metal binding and (ii) they may support studies regarding roles in protection against metal toxicity, where appropriate.

The earliest studies established that all bound metal ions could be displaced by Ag(I).[3] Reaction of radioactive rat liver ^{65}Zn-MT-2 with a range of metal ions established that Cd(II) had the greatest ability to displace bound Zn(II), closely followed by Cu(II),[284] although the reaction products were not further analysed. It is, however, clear that Cu(II) is reduced by MTs,[119] resulting in the generation and binding of Cu(I), which has very high affinity (Section 21.2.2) and can even displace Cd(II).[285]

Titrations of various Zn-loaded MTs with Cd(II) revealed that slight excesses of Cd(II) will normally displace Zn(II) completely, and the same tends to be true of titrations with Cu(I) or Ag(I). Hence, the exchange/displacement behaviour usually follows the order of stability, and the metal exchange reactions are fast and usually complete within seconds. A notable exception to these rules is the bacterial MT SmtA. Here, the Cys$_4$ zinc finger site A (Figures 21.1 and 21.6), which is expected to have an absolute preference for Cd(II), is kinetically inert towards metal exchange.[200,286] Notably, the

Cys$_4$ site B was substituted preferentially, followed by the two Cys$_3$His sites C and D, clearly due to the higher relative affinity of site B for Cd(II). As in this relatively simple case, ^{111}Cd/^{113}Cd NMR spectroscopy has previously provided a wealth of information on exchange behaviour in mammalian MTs.[285,287] The particular strength of this approach is its ability to obtain site-specific information. The $^{111/113}$Cd chemical shifts are so sensitive to their environment that all seven cadmium ions can be assigned and, in favourable cases, different mixed-metal species can be identified. This feature has been useful in studying whether the two domains of mammalian MTs display differential specificity (Section 21.2.4). Titrations of Zn$_7$MT-2 (rabbit liver) with ^{113}Cd(II) revealed that mixed clusters were preferentially formed.[287] However, at substoichiometric levels, much fewer species than statistically possible were observed, indicating a certain level of site specificity. This finding is important because it clearly demonstrates that the metal sites in MTs are not all equivalent – in agreement with the zinc affinity data for human MT (Section 21.2.2) that had indicated three different binding site classes. Furthermore, the spectrum generated at a 2 : 1 Cd : Zn ratio did not resemble that of Cd,Zn-MT-2 from the livers of Cd-exposed animals, initially suggesting that the respective species are not formed by displacement of Zn(II) from Zn$_7$MT-2. In contrast, titrations of ^{113}Cd$_7$MT-2 with Zn$_7$MT-2 led, at the appropriate ratio, to a spectrum that was indistinguishable from the "native" spectra.[287] Further work using CD and MCD (magnetic circular dichroism) spectroscopy established that the spectral fingerprint of "native" Cd,Zn-MT could be generated from the species obtained by titrations with free Cd(II) by heating them to 50 °C.[288] This suggested that titrations of fully metallated Zn-MT with free Cd(II) lead to the formation of kinetically rather than thermodynamically controlled products. Furthermore, a recent detailed kinetic study of the reaction of Zn$_7$MT-2 itself, the isolated α- and β-domains and also the α-domain partially substituted with Cd(II), suggested that the α-domain reaction (at least) involves an associative mechanism, followed by metal exchange.[289] Some site-specificity within the isolated α-domain was observed at low concentrations, with sites V and VI preferentially occupied by Cd(II) after addition of 1 mol eq of Cd(II).

Conversely, titrations of ^{111}Cd$_7$MT-2 with Ag(I) or Cu(I)[290] demonstrated that substoichiometric amounts of Ag(I) bound preferentially in the β-domain, with the dominant formation of Cd$_4$(α)Ag$_6$(β)MT-2. The fully substituted form was Ag$_6$(α)Ag$_6$(β)MT-2. The selective formation of homometallic clusters suggested that the metal displacement was to some extent cooperative in this case, with the domains acting independently from one another. Cu(I) behaved in a broadly similar way, but mixed clusters were also formed. Whilst many of the metalloforms discussed above are not likely to occur *in vivo*, they are relevant to aid our understanding of the concept of metal specificity discussed in Section 21.2.4 – if there can be individual sites with higher relative affinities towards a particular metal ion within a single MT, it is also conceivable that different MTs have different relative affinities towards different metal ions. This concept was illustrated by a titration of a mixture of the two Zn-loaded MTs from *C. elegans* with Cd(II).[172] The preferential partitioning of Cd(II) into MTL-2 was observed, consistent with the respective affinity data (Table 21.2) and with functional data (Section 21.2.4).[96] In addition, MTL-1 contains a CysHis$_3$ site that displays an absolute preference for Zn(II) over Cd(II) and hence this site cannot be populated by Cd(II) through metal-exchange reactions.

Interprotein metal exchange was first demonstrated by mixing ^{113}Cd$_7$MT-2 with Zn$_7$MT-2.[287,291] The resulting ^{113}Cd NMR spectra suggested that Cd(II) was preferentially transferred from the 3-metal cluster to the 4-metal cluster, resulting in a preferential location of Zn(II) in the 3-metal cluster, as indeed was observed in the crystal structure of rat MT.[195] Metal exchange occurred on a time-scale of minutes. The fact that MT-bound Cd(II) and free Cd(II) led to different reaction products not only suggested the importance of kinetic control, but also strongly supported the idea that metal exchange involved direct protein-protein interactions.

Mixtures of homo-metallated ^{111}Cd$_7$MT-2 and Ag$_{12}$MT-2 or Cu$_{12}$MT-2 were also studied;[290] in the case of Ag(I), the same products as in titrations with free Ag(I) were observed. In the case of Cu(I), partitioning of Cu(I) into the β domain and Cd(II) into the α domain was observed.

Most studies described above concern competition between two different metal ions, and benefited from exploiting the spectroscopic properties of Cd(II). Comparably fewer studies are available on metal self-exchange, in particular that of Zn(II). Either radioactive ^{65}Zn or stable isotopes such as ^{67}Zn can be employed for this purpose. The lability of Zn(II) in mammalian MTs was demonstrated elegantly by mixing ^{65}Zn$_7$MT-1 and Zn$_7$MT-2 (from rabbit liver); two kinetic steps were observed with rates of 5000 min^{-1} M^{-1} and 200 min^{-1} M^{-1}, thought to correspond to differential reactivity of the α- and β-domains.[292]

The same study also demonstrated metal exchange at a rate of 800 min^{-1} M^{-1} between ^{65}Zn$_7$MT-2 and Zn$_2$GAL4, a fungal transcription factor. Metal exchange between Zn-MTs and Cd(II)-metallated carbonic anhydrase was studied spectrophotometrically; quantitative exchange was established after 12 h.[247] Slow metal exchange was also observed for the Cd-substituted zinc finger transcription factor TFIIIA and Zn-MT, but not the reverse couple.[246] Both latter studies illustrate how Zn-MTs may perform a short-term protective role in acute cadmium poisoning. Potentially, the most exciting metal exchange reactions involving a zinc-loaded MT concern the brain-specific mammalian MT-3. The Zn$_7$MT-3 form has been shown to "swap" metals with the amyloid-beta peptide,[104,293] as well as with other Cu(II)-binding peptides and proteins involved in neurodegeneration (Section 21.1.3.3).[105,106] The *in vitro* reaction resulted in the generation of Cu(I)$_4$(β)Zn(II)$_4$(α)MT-3 with two disulfide bonds. It is thought that this particular metal exchange reaction may also occur *in vivo*, specifically in the extracellular matrix in brains affected by Alzheimer's disease. The ability of MT-3 to protect against oxidative stress mediated by peptide-bound copper is also highlighted in Section 21.1.3.3.

21.2.9 General Observations Regarding Reactivity and Dynamic Behaviour

For two-domain MTs, differential reactivity of the two clusters in metal up-take, release and exchange, as well as their potential interactions, is of interest. The most extensive data are available for mammalian MT-2. There is no clear trend to judge which of the two domains is the more reactive; this depends on the nature of the reaction and the reactant, and this will be true for reactions both *in vitro* and *in vivo*. An early study established that the β-domain reacted preferentially with iodoacetamide[258] but no preferential or cooperative behaviour was observed with NEM.[259] The α-domain is more reactive in thiol/disulfide interchange with DTNB, reactions with some chelators and with aurothiomalate.[166] The β-domain reacts faster with a range of oxidizing reagents, as well as with PAR, even though the α-domain transfers a larger proportion of Zn(II) to PAR.[169] The β-domain is also characterized by faster structural dynamics, and NMR saturation transfer experiments have suggested that the three β-domain ^{113}Cd ions undergo intersite exchange on a time-scale of seconds, whereas those in the α-domain exchange on a time-scale of hours.[291] It is also clear that the presence of the other domain may influence the reactivity of the domain under study.[169] In addition, the few studies available for MTs from other phyla have demon-strated that although they may undergo similar reactions as governed by their chemical similarity, neither their thermodynamic nor kinetic be-haviour is presently predictable from the wealth of data available for vertebrate MTs.

An accurate description of the reactions that may occur *in vivo* involves both mechanistic insight as well as quantitative parameters such as ther-modynamic and kinetic constants. Such data, obtained by the various *in vitro* biophysical approaches described above, are vital ingredients for systems biology-based approaches, *e.g.* involving the modelling of biological zinc fluxes.[133]

21.3 Concluding Remarks

The chemical properties and reactivity of MTs prime them for their in-volvement in a wide range of biological processes linked to metal homeo-stasis and stress response. Whilst structural and chemical studies have been and still are[294] of utmost importance to understand the molecular mech-anisms underlying the modes of action of MTs, work on their functions in cells and whole organisms is at least of equal importance – as for any other protein. Extensive work on the cellular and organismal level has been carried out for vertebrate MTs,[40] but also for plants[24,295] and terrestrial in-vertebrates[84,127] including insects.[64] Fewer studies are available for aquatic invertebrates[51] and for microbial MTs.[296] Overall, the investigations have revealed the complexity of MT gene regulation and the involvement of MTs in a staggering range of fundamental cellular processes including cell cycle and (programmed) cell death.[297] In the case of vertebrate MTs, work on the cellular and organismal level has demonstrated roles in development and growth, immune function and ageing,[298] as well as in major diseases, in particular those related to the central nervous system,[81] diabetes[299] and cancer.[139,140] MTs in plants may prove important in food security and nutrition,[295] and MTs in microbial pathogens may play roles in virulence.[130] Apart from these direct impacts on human health, MT in many other organisms may have important functions in the context of environmental toxicology. Aside from their demonstrated importance in the areas mentioned and beyond, MTs remain deeply captivating objects of study.

Abbreviations

5F-BAPTA:	1,2-bis(2-amino-5-fluorophenoxy)ethane N,N,N',N'-tetraacetic acid
ABD-F:	7-fluorobenz-2-oxa-1,3-diazole-4-sulfonamide
AES:	Atomic Emission Spectroscopy
ATP:	adenosine triphosphate
BmtA:	Bacterial metallothionein homologous to SmtA
CD:	Circular Dichroism
CE:	Capillary Electrophoresis
DEA/NO:	Diethylamine NONOate
DTNB:	5,5'-dithiobis(2-nitrobenzoic acid)
EDDA:	ethylenediamine-N,N'-diacetic acid
EDTA:	ethylenediamine-N,N,N',N'-tetraacetic acid
ESI-MS:	Electrospray Ionization Mass Spectrometry
EXAFS:	Extended X-ray Absorption Fine Structure
FRET:	Fluorescence Resonance Energy Transfer
FT-ICR-MS:	Fourier-Transform-Ion Cyclotron Resonance Mass Spectrometry
GFP:	green fluorescent protein
GSH:	glutathione
GSNO:	S-nitroso-glutathione
GSSG:	glutathione disulfide
GST:	glutathione S-transferase
HPLC:	High Performance Liquid Chromatography
HSQC:	Heteronuclear Single Quantum Coherence
ICP-OES:	Inductively Coupled Plasma Optical Emission Spectroscopy
KTS:	3-ethoxy-2-oxobutyraldehyde-bis(thiosemicarbazone)
LMCT:	ligand-to-metal-charge-transfer
MCD:	Magnetic Circular Dichroism
MS:	Mass Spectrometry
MT:	metallothionein
NEM:	N-ethylmaleimide
NHE:	Normal hydrogen electrode
NMR:	Nuclear Magnetic Resonance

PAR: 4-(2-pyridylazo)resorcinol
SDS-PAGE: sodium dodecyl sulfate polyacrylamide gel electrophoresis
SNAP: *S*-nitroso-penicillamine
T: thionein
TCEP: *tris*(2-carboxyethyl)phosphine
TOCSY: Total Correlation Spectroscopy
UV-Vis: Ultraviolet-Visible Spectroscopy

References

1. N. Romero-Isart and M. Vašák, *J. Inorg. Biochem.*, 2002, **88**, 388–396.
2. M. Margoshes and B. L. Vallee, *J. Am. Chem. Soc.*, 1957, **79**, 4813–4814.
3. J. H. R. Kägi and B. L. Vallee, *J. Biol. Chem.*, 1960, **235**, 3460–3465.
4. J. H. R. Kägi and B. L. Vallee, *J. Biol. Chem.*, 1961, **236**, 2435–2442.
5. M. Piscator, *Nordisk hygienisk tidskrift*, 1964, **45**, 76–82.
6. P. Pulido, J. H. R. Kägi and B. L. Vallee, *Biochemistry*, 1966, **5**, 1768–1777.
7. G. F. Nordberg, M. Nordberg, O. Vesterbe and M. Piscator, *Biochem. J.*, 1972, **126**, 491–498.
8. J. H. R. Kägi, S. R. Himmelhoch, P. D. Whanger, J. L. Bethune and B. L. Vallee, *J. Biol. Chem.*, 1974, **249**, 3537–3542.
9. Y. Kojima, C. Berger, B. L. Vallee and J. H. R. Kägi, *Proc. Natl Acad. Sci. USA*, 1976, **73**, 3413–3417.
10. E. Marafante, *Experientia*, 1976, **32**, 149–150.
11. A. G. Howard and G. Nickless, *Chem. Biol. Interact.*, 1977, **16**, 107–114.
12. V. Talbot and R. J. Magee, *Arch. Environ. Contam. Toxicol.*, 1978, 7, 73–81.
13. R. W. Olafson, R. G. Sim and A. Kearns, *Experientia Supplementum 34*, Birkhäuser, Basel, Boston, Stuttgart, 1979, pp. 197–204.
14. M. L. Failla, C. D. Benedict and E. D. Weinberg, *J. Gen. Microbiol.*, 1976, **94**, 23–36.
15. R. Prinz and U. Weser, *Hoppe-Seylers Z. Physiol. Chem.*, 1975, **356**, 767–776.
16. K. Lerch, in Metallothionein, eds. J. H. R. Kägi and M. Nordberg, *Experientia Supplementum 34*, Birkhäuser, Basel, Boston, Stuttgart, 1979, pp. 173–180.
17. R. W. Olafson, K. Abel and R. G. Sim, *Biochem. Biophys. Res. Commun.*, 1979, **89**, 36–43.
18. W. E. Rauser and N. R. Curvetto, *Nature*, 1980, **287**, 563–564.
19. B. Lane, R. Kajioka and T. Kennedy, *Biochem. Cell Biol.*, 1987, **65**, 1001–1005.
20. I. M. Evans, L. N. Gatehouse, J. A. Gatehouse, N. J. Robinson and R. R. D. Croy, *FEBS Lett.*, 1990, **262**, 29–32.
21. J. C. Steffens, *Annu. Rev. Plant Biol.*, 1990, **141**, 553–575.
22. N. J. Robinson, A. M. Tommey, C. Kuske and P. J. Jackson, *Biochem. J.*, 1993, **295**, 1–10.

23. W. E. Rauser, *Cell Biochem. Biophys.*, 1999, **31**, 19–48.
24. C. Cobbett and P. Goldsbrough, *Annu. Rev. Plant Biol.*, 2002, **53**, 159–182.
25. J. H. R. Kägi and M. Nordberg (eds.), *Metallothionein*, Birkhäuser Verlag, Basel, Boston, Stuttgart, 1979, pp. 1–378.
26. J. H. R. Kägi and Y. Kojima (eds.), *Metallothionein II*, Birkhäuser Verlag, Basel, Boston, Stuttgart, 1987, pp. 1–755.
27. K. T. Suzuki, N. Imura and M. Kimura (eds.), *Metallothionein III: Biological Roles and Medical Implications*, Birkhäuser Verlag, Basel, Boston, Stuttgart, 1993, pp. 1–479.
28. C. D. Klaassen (ed.), *Metallothionein IV*, Birkhäuser Verlag, Basel, Boston, Stuttgart, 1999, pp. 1–645.
29. M. J. Stillman, K. T. Suzuki and C. F. Shaw, III (eds.), *Metallothioneins, Proceedings of the Pacifichem 2000 Metallothionein Symposium*, Honolulu, Hawaii, 14–19 December 2000, *J. Inorg. Biochem.* 2002, **88**, 119–240.
30. M. Nordberg and G. F. Nordberg, in *Metal Ions in Life Sciences*, eds. A. Sigel, H. Sigel and R. K. O. Sigel, RSC Publishing, Cambridge, UK, 2009, vol. 5, pp. 1–29.
31. A. Sigel, H. Sigel and R. K. O. Sigel (eds.), *Metallothioneins and Related Chelators, vol. 5 of Metal Ions Life Sci.*, RSC Publishing, Cambridge, UK, 2009.
32. C. D. Klaassen and K. T. Suzuki (eds.), *Metallothionein in Biology and Medicine*, CRC Press, Boca Raton, 1991, pp. 1–432.
33. *Metallothioneins: Chemical and Biological Challenges, J. Biol. Inorg. Chem.*, 2011, **16**, 975–1134.
34. B. L. Vallee and J. F. Riordan (eds.), *Metallobiochemistry Part B: Metallothionein and Related Molecules, vol. 205 of Methods in Enzymology*, Academic Press, 1991.
35. B. A. Fowler, C. E. Hildebrand, Y. Kojima and M. Webb, *Experientia Suppl.*, 1987, **52**, 19–22.
36. Y. Kojima, P.-A. Binz and J. H. R. Kägi, in *Metallothionein IV*, ed. C. D. Klaassen, Birkhäuser Verlag, Basel, Boston, Berlin, 1999, pp. 1–6.
37. M. Vašák and I. Armitage, *Environ. Health Perspect.*, 1986, **65**, 215–216.
38. J. H. R. Kägi, M. Vašák, K. Lerch, D. E. O. Gilg, P. Hunziker, W. R. Bernhard and M. Good, *Environ. Health Perspect.*, 1984, **54**, 93–103.
39. P.-A. Binz and J. H. R. Kägi, in *Metallothionein IV*, ed. C. D. Klaassen, Birkhäuser Verlag, Basel, Boston, Berlin, 1999, pp. 7–13.
40. J. Hidalgo, R. Chung, M. Penkowa and M. Vašák, in *Metal Ions in Life Sciences*, eds. A. Sigel, H. Sigel and R. K. O. Sigel, RSC Publishing, Cambridge, UK, 2009, vol. 5, pp. 279–317.
41. Y. Li and W. Maret, *J. Anal. At. Spectrom.*, 2008, **23**, 1055–1062.
42. L. Atanesyan, V. Günther, S. E. Celniker, O. Georgiev and W. Schaffner, *J. Biol. Inorg. Chem.*, 2011, **16**, 1047–1056.
43. B. Gold, H. Deng, R. Bryk, D. Vargas, D. Eliezer, J. Roberts, X. Jiang and C. Nathan, *Nat. Chem. Biol.*, 2008, **4**, 609–616.
44. M. Guirola, S. Perez-Rafael, M. Capdevila, O. Palacios and S. Atrian, *PLoS One*, 2012, 7, e43299.

45. A. Schmidt, M. Hagen, E. Schutze and E. Kothe, *J. Basic Microbiol.*, 2010, **50**, 562–569.
46. J. P. Barnett, A. Millard, A. Z. Ksibe, D. J. Scanlan, R. Schmid and C. A. Blindauer, *Front. Microbiol.*, 2012, **3**, 142.
47. C. A. Blindauer, *Chem. Biodiv.*, 2008, **5**, 1990–2013.
48. C. A. Blindauer, *J. Biol. Inorg. Chem.*, 2011, **16**, 1011–1024.
49. B. Berthet, C. Mouneyrac, T. Perez and C. Amiard-Triquet, *Comp. Biochem. Physiol. C Toxicol. Pharmacol.*, 2005, **141**, 306–313.
50. C. A. Blindauer and O. I. Leszczyszyn, *Nat. Prod. Rep.*, 2010, **27**, 720–741.
51. J. C. Amiard, C. Amiard-Triquet, S. Barka, J. Pellerin and P. S. Rainbow, *Aquat. Toxicol.*, 2006, **76**, 160–202.
52. C. Schmidt and D. Beyersmann, *Arch. Biochem. Biophys.*, 1999, **364**, 91–98.
53. B. Ye, W. Maret and B. L. Vallee, *Proc. Natl Acad. Sci. USA*, 2001, **98**, 2317–2322.
54. M. A. Lynes, K. Zaffuto, D. W. Unfricht, G. Marusov, J. S. Samson and X. Y. Yin, *Exp. Biol. Med.*, 2006, **231**, 1548–1554.
55. R. S. Chung, M. Penkowa, J. Dittmann, C. E. King, C. Bartlett, J. W. Asmussen, J. Hidalgo, J. Carrasco, Y. K. J. Leung, A. K. Walker, S. J. Fung, S. A. Dunlop, M. Fitzgerald, L. D. Beazley, M. I. Chuah, J. C. Vickers and A. K. West, *J. Biol. Chem.*, 2008, **283**, 15349–15358.
56. H. Haase and W. Maret, *Anal. Biochem.*, 2004, **333**, 19–26.
57. A. Krezel and W. Maret, *Biochem. J.*, 2007, **402**, 551–558.
58. A. T. Miles, G. M. Hawksworth, J. H. Beattie and V. Rodilla, *Crit. Rev. Biochem. Mol. Biol.*, 2000, **35**, 35–70.
59. C. D. Klaassen, S. Choudhuri, J. M. McKim, L. D. Lehman-McKeeman and W. C. Kershaw, *Environ. Health Perspect.*, 1994, **102**, 141–146.
60. Y. Yang, W. Maret and B. L. Vallee, *Proc. Natl Acad. Sci. USA*, 2001, **98**, 5556–5559.
61. S. R. Davis and R. J. Cousins, *J. Nutr.*, 2000, **130**, 1085–1088.
62. P. Coyle, J. C. Philcox, L. C. Carey and A. M. Rofe, *Cell. Mol. Life Sci.*, 2002, **59**, 627–647.
63. F. Haq, M. Mahoney and J. Koropatnick, *Mutat. Res. Fundam. Mol. Mech. Mutagen.*, 2003, **533**, 211–226.
64. V. Günther, U. Lindert and W. Schaffner, *Biochim. Biophys. Acta Mol. Cell Res.*, 2012, **1823**, 1416–1425.
65. F. Okumura, Y. Li, N. Itoh, T. Nakanishi, M. Isobe, G. K. Andrews and T. Kimura, *Biochim. Biophys. Acta Gene Regul. Mech.*, 2011, **1809**, 56–62.
66. M. Karin, A. Haslinger, H. Holtgreve, R. I. Richards, P. Krauter, H. M. Westphal and M. Beato, *Nature*, 1984, **308**, 513–519.
67. C. Hogstrand, D. Zheng, G. Feeney, P. Cunningham and P. Kille, *Biochem. Soc. Trans.*, 2008, **36**, 1252–1257.
68. M. Höckner, K. Stefanon, D. Schuler, R. Fantur, A. De Vaufleury and R. Dallinger, *J. Exp. Zool.*, 2009, **311A**, 776–787.
69. B. Zhang, O. Georgiev, M. Hagmann, C. Günes, M. Cramer, P. Faller, M. Vašák and W. Schaffner, *Mol. Cell. Biol.*, 2003, **23**, 8471–8485.

70. J. H. Freedman, L. W. Slice, D. Dixon, A. Fire and C. S. Rubin, *J. Biol. Chem.*, 1993, **268**, 2554–2564.

71. S. R. Stürzenbaum, O. Georgiev, A. J. Morgan and P. Kille, *Environ. Sci. Technol.*, 2004, **38**, 6283–6289.

72. C. A. Whitelaw, J. A. LeHuquet, D. A. Thurman and A. B. Tomsett, *Plant Mol. Biol.*, 1997, **33**, 503–511.

73. C. Gross, M. Kelleher, V. R. Iyer, P. O. Brown and D. R. Winge, *J. Biol. Chem.*, 2000, **275**, 32310–32316.

74. J. W. Huckle, A. P. Morby, J. S. Turner and N. J. Robinson, *Mol. Microbiol.*, 1993, 7, 177–187.

75. W. Maret, *J. Biol. Inorg. Chem.*, 2011, **16**, 1079–1086.

76. Y. Uchida, K. Takio, K. Titani, Y. Ihara and M. Tomonaga, *Neuron*, 1991, 7, 337–347.

77. C. J. Quaife, S. D. Findley, J. C. Erickson, G. J. Froelick, E. J. Kelly, B. P. Zambrowicz and R. D. Palmiter, *Biochemistry*, 1994, **33**, 7250–7259.

78. H. Kobayashi, Y. Uchida, Y. Ihara, K. Nakajima, S. Kohsaka, T. Miyatake and S. Tsuji, *Mol. Brain Res.*, 1993, **19**, 188–194.

79. M. Vašák and G. Meloni, *J. Biol. Inorg. Chem.*, 2011, **16**, 1067–1078.

80. Y. Manso, J. Carrascao, G. Comes, G. Meloni, P. A. Adlard, A. I. Bush, M. Vašák and J. Hidalgo, *Cell. Mol. Life Sci.*, 2012, **69**, 3683–3700.

81. A. K. West, J. Hidalgo, D. Eddins, E. D. Levin and M. Aschner, *Neuro-Toxicology*, 2008, **29**, 489–503.

82. B. L. Vallee, *Experientia Suppl.*, 1987, **52**, 5–16.

83. R. D. Palmiter, *Proc. Natl Acad. Sci. USA*, 1998, **95**, 8428–8430.

84. R. Dallinger, B. Berger, C. Gruber, P. Hunziker and S. Stürzenbaum, *Cell. Mol. Biol.*, 2000, **46**, 331–346.

85. R. Dallinger, B. Berger, P. Hunziker and J. H. R. Kägi, *Nature*, 1997, **388**, 237–238.

86. A. Sarkar, D. Ray, A. N. Shrivastava and S. Sarker, *Ecotoxicology*, 2006, **15**, 333–340.

87. M. A. Pagani, M. Tomas, J. Carrillo, R. Bofill, M. Capdevila, S. Atrian and C. S. Andreo, *J. Inorg. Biochem.*, 2012, **117**, 306–315.

88. S. L. Hughes, J. G. Bundy, E. J. Want, P. Kille and S. R. Sturzenbaum, *J. Proteome Res.*, 2009, **8**, 3512–3519.

89. H. C. Chiang, J. C. Lo and K. C. Yeh, *Environ. Sci. Technol.*, 2006, **40**, 6792–6798.

90. G. DalCorso, S. Farinati, S. Maistri and A. Furini, *J. Integr. Plant. Biol.*, 2008, **50**, 1268–1280.

91. C. D. Klaassen, J. Liu and B. A. Diwan, *Toxicol. Appl. Pharmacol.*, 2009, **238**, 215–220.

92. J. Liu, M. L. Cheng, Q. Yang, K. R. Shan, J. Shen, Y. S. Zhou, X. J. Zhang, A. L. Dill and M. P. Waalkes, *Environ. Health Perspect.*, 2007, **115**, 1101–1106.

93. M. Satoh, Y. Aoki and C. Tohyama, *Cancer Chemother. Pharmacol.*, 1997, **40**, 358–362.

94. N. A. Wolff, M. Abouhamed, P. J. Verroust and F. Thevenod, *J. Pharmacol. Exp. Ther.*, 2006, **318**, 782–791.
95. J. M. Zhou and P. B. Goldsbrough, *Plant Cell*, 1994, **6**, 875–884.
96. S. Zeitoun-Ghandour, J. M. Charnock, M. E. Hodson, O. I. Leszczyszyn, C. A. Blindauer and S. R. Stürzenbaum, *FEBS J.*, 2010, **277**, 2531–2542.
97. J. Hidalgo, M. Penkowa, M. Giralt, J. Carrasco and A. Molinero, *Methods Enzymol.*, 2002, **348**, 238–249.
98. S. Zeitoun-Ghandour, O. I. Leszczyszyn, C. A. Blindauer, F. M. Geier, J. G. Bundy and S. R. Stürzenbaum, *Mol. BioSyst.*, 2011, 7, 2397–2406.
99. T. Dalton, R. D. Palmiter and G. K. Andrews, *Nucleic Acids Res.*, 1994, **22**, 5016–5023.
100. V. H. Hassinen, A. I. Tervahauta, H. Schat and S. O. Karenlampi, *Plant Biol.*, 2011, **13**, 225–232.
101. Y. J. Kang, Y. Chen, A. D. Yu, M. Voss-McCowan and P. N. Epstein, *J. Clin. Invest.*, 1997, **100**, 1501–1506.
102. J. S. Lazo, Y. Kondo, D. Dellapiazza, A. E. Michalska, K. H. A. Choo and B. R. Pitt, *J. Biol. Chem.*, 1995, **270**, 5506–5510.
103. Y. Irie and W. M. Keung, *Biochem. Biophys. Res. Commun.*, 2001, **282**, 416–420.
104. G. Meloni, V. Sonois, T. Delaine, L. Guilloreau, A. Gillet, J. Teissie, P. Faller and M. Vašák, *Nature Chem. Biol.*, 2008, **4**, 366–372.
105. G. Meloni and M. Vašák, *Free Radic. Biol. Med.*, 2011, **50**, 1471–1479.
106. G. Meloni, A. Crameri, G. Fritz, P. Davies, D. R. Brown, P. M. H. Kroneck and M. Vašák, *ChemBioChem*, 2012, **13**, 1261–1265.
107. R. S. Chung, C. Howells, E. D. Eaton, L. Shabala, K. Zovo, P. Palumaa, R. Sillard, A. Woodhouse, W. R. Bennett, S. Ray, J. C. Vickers and A. K. West, *PLoS One*, 2010, **5**, e12030.
108. J. P. Fabisiak, L. L. Pearce, G. G. Borisenko, Y. Y. Tyurina, V. A. Tyurin, J. Razzack, J. S. Lazo, B. R. Pitt and V. E. Kagan, *Antioxid. Redox Signal.*, 1999, **1**, 349–360.
109. Q. Hao and W. Maret, *J. Alzheimers Dis.*, 2005, **8**, 161–170.
110. W. K. Feng, F. W. Benz, J. Cai, W. M. Pierce and Y. J. Kang, *J. Biol. Chem.*, 2006, **281**, 681–687.
111. D. Berendji, V. Kolb-Bachofen, K. L. Meyer, O. Grapenthin, H. Weber, V. Wahn and K. D. Kröncke, *FEBS Lett.*, 1997, **405**, 37–41.
112. W. Lin, B. Mohandas, C. P. Fontaine and R. A. Colvin, *Biometals*, 2007, **20**, 891–901.
113. P. J. Thornalley and M. Vašák, *Biochim. Biophys. Acta*, 1985, **827**, 36–44.
114. W. Maret, *Proc. Natl Acad. Sci. USA*, 1994, **91**, 237–241.
115. H. Porter, *Biochim. Biophys. Acta*, 1971, **229**, 143–154.
116. H. Rupp and U. Weser, *FEBS Lett.*, 1974, **44**, 293–297.
117. I. Bremner and R. B. Marshall, *Br. J. Nutr.*, 1974, **32**, 293–300.
118. J. R. Riordan and V. Richards, *J. Biol. Chem.*, 1980, **255**, 5380–5383.
119. G. Meloni, P. Faller and M. Vašák, *J. Biol. Chem.*, 2007, **282**, 16068–16078.
120. B. Roschitzki and M. Vašák, *J. Biol. Inorg. Chem.*, 2002, 7, 611–616.

121. B. Roschitzki and M. Vašák, *Biochemistry*, 2003, **42**, 9822–9828.
122. G. Meloni, K. Zovo, J. Kazantseva, P. Palumaa and M. Vašák, *J. Biol. Chem.*, 2006, **281**, 14588–14595.
123. L. Tio, L. Villarreal, S. Atrian and M. Capdevila, *J. Biol. Chem.*, 2004, **279**, 24403–24413.
124. B. Dolderer, H.-J. Hartmann and U. Weser, in *Metal Ions in Life Sciences*, eds. A. Sigel, H. Sigel and R. K. O. Sigel, RSC Publishing, Cambridge, UK, 2009, vol. 5, pp. 83–105.
125. L. Vergani, in *Metal Ions in Life Sciences*, eds. A. Sigel, H. Sigel and R. K. O. Sigel, RSC Publishing, Cambridge, UK, 2009, vol. 5, pp. 199–237.
126. D. Egli, J. Domenech, A. Selvaraj, K. Balamurugan, H. Q. Hua, M. Capdevila, O. Georgiev, W. Schaffner and S. Atrian, *Genes Cells*, 2006, **11**, 647–658.
127. M. Höckner, R. Dallinger and S. R. Stürzenbaum, *J. Biol. Inorg. Chem.*, 2011, **16**, 1057–1065.
128. W. J. Guo, M. Meetam and P. B. Goldsbrough, *Plant Physiol.*, 2008, **146**, 1697–1706.
129. J. L. Rowland and M. Niederweis, *Tuberculosis*, 2012, **92**, 202–210.
130. W. Maret, *Biometals*, 2009, **22**, 149–157.
131. J. Hidalgo, R. Chung, M. Penkowa and M. Vašák, in *Metal Ions in Life Sciences*, eds. A. Sigel, H. Sigel and R. K. O. Sigel, RSC Publishing, Cambridge, UK, 2009, vol. 5, pp. 279–317.
132. J. H. Kägi, *Methods Enzymol.*, 1991, **205**, 613–626.
133. R. A. Colvin, W. R. Holmes, C. P. Fontaine and W. Maret, *Metallomics*, 2010, **2**, 306–317.
134. E. J. Kelly, C. J. Quaife, G. J. Froelick and R. D. Palmiter, *J. Nutr.*, 1996, **126**, 1782–1790.
135. A. M. Rofe, J. C. Philcox and P. Coyte, *Biochem. J.*, 1996, **314**, 793–797.
136. T. Dalton, K. Fu, R. D. Palmiter and G. K. Andrews, *J. Nutr.*, 1996, **126**, 825–833.
137. W. Maret, *Exp. Geront.*, 2008, **43**, 363–369.
138. S. G. Bell and B. L. Vallee, *ChemBioChem*, 2009, **10**, 55–62.
139. M. G. Cherian, A. Jayasurya and B. H. Bay, *Mutat. Res. Fundam. Mol. Mech. Mutagen.*, 2003, **533**, 201–209.
140. M. O. Pedersen, A. Larsen, M. Stoltenberg and M. Penkowa, *Prog. Histochem. Cytochem.*, 2009, **44**, 29–64.
141. M. S. Islam and D. T. Loots, *Biofactors*, 2007, **29**, 203–212.
142. L. Cai, *Curr. Med. Chem.*, 2007, **14**, 2193–2203.
143. W. R. Swindell, *Ageing Res. Rev.*, 2011, **10**, 132–145.
144. S. Hughes and S. R. Stürzenbaum, *Environ. Pollut.*, 2007, **145**, 395–400.
145. E. Mocchegiani, R. Giacconi, C. Cipriano, M. Muzzioli, P. Fattoretti, C. Bertoni-Freddari, G. Isani, P. Zambenedetti and P. Zatta, *Brain Res. Bull.*, 2001, **55**, 147–153.

146. A. Pattanaik, C. F. Shaw, D. H. Petering, J. Garvey and A. J. Kraker, *J. Inorg. Biochem.*, 1994, **54**, 91–105.

147. D. H. Petering, J. Zhu, S. Krezoski, J. Meeusen, C. Kiekenbush, S. Krull, T. Specher and M. Dughish, *Exp. Biol. Med.*, 2006, **231**, 1528–1534.

148. D. E. K. Sutherland, K. L. Summers and M. J. Stillman, *Biochemistry*, 2012, **51**, 6690–6700.

149. K. B. Nielson, C. L. Atkin and D. R. Winge, *J. Biol. Chem.*, 1985, **260**, 5342–5350.

150. E. Freisinger, *Dalton Trans.*, 2008, 6663–6675.

151. C. A. Blindauer, M. T. Razi, D. J. Campopiano and P. J. Sadler, *J. Biol. Inorg. Chem.*, 2007, **12**, 393–405.

152. C. A. Blindauer, M. D. Harrison, A. K. Robinson, J. A. Parkinson, P. W. Bowness, P. J. Sadler and N. J. Robinson, *Mol. Microbiol.*, 2002, **45**, 1421–1432.

153. R. Bofill, M. Capdevila and S. Atrian, *Metallomics*, 2009, **1**, 229–234.

154. M. Capdevila, J. Domenech, A. Pagani, L. Tio, L. Villarreal and S. Atrian, *Angew. Chem. Intern. Ed.*, 2005, **44**, 4618–4622.

155. O. I. Leszczyszyn and C. A. Blindauer, *Phys. Chem. Chem. Phys.*, 2010, **12**, 13408–13418.

156. O. I. Leszczyszyn, C. R. J. White and C. A. Blindauer, *Mol. BioSyst.*, 2010, **6**, 1592–1603.

157. P. M. Gehrig, C. H. You, R. Dallinger, C. Gruber, M. Brouwer, J. H. R. Kägi and P. E. Hunziker, *Protein Sci.*, 2000, **9**, 395–402.

158. X. Q. Ding, C. Butzlaff, E. Bill, D. L. Pountney, G. Henkel, H. Winkler, M. Vašák and A. X. Trautwein, *Eur. J. Biochem.*, 1994, **220**, 827–837.

159. M. Vašák, J. H. R. Kägi, B. Holmquist and B. L. Vallee, *Biochemistry*, 1981, **20**, 6659–6664.

160. I. Bertini, C. Luchinat, L. Messori and M. Vašák, *J. Am. Chem. Soc.*, 1989, **111**, 7296–7300.

161. H. Z. Sun, H. Y. Li, I. Harvey and P. J. Sadler, *J. Biol. Chem.*, 1999, **274**, 29094–29101.

162. T. T. Ngu, A. Easton and M. J. Stillman, *J. Am. Chem. Soc.*, 2008, **130**, 17016–17028.

163. A. Pattanaik, G. Bachowski, J. Laib, D. Lemkuil, C. F. Shaw, D. H. Petering, A. Hitchcock and L. Saryan, *J. Biol. Chem.*, 1992, **267**, 16121–16128.

164. A. V. Karotki and M. Vašák, *J. Biol. Inorg. Chem.*, 2009, **14**, 1129–1138.

165. M. Vašák, J. H. R. Kägi and H. A. O. Hill, *Biochemistry*, 1981, **20**, 2852–2856.

166. J. E. Laib, C. F. Shaw, D. H. Petering, M. K. Eidsness, R. C. Elder and J. S. Garvey, *Biochemistry*, 1985, **24**, 1977–1986.

167. M. Vašák and J. H. R. Kägi, in *Metal Ions in Biological Systems: Zinc and Its Role in Biology and Nutrition*, ed. H. Sigel, Marcel Dekker, New York, 1983, vol. 15, pp. 213–273.

168. E. A. Peroza, A. Al Kaabi, W. Meyer-Klaucke, G. Wellenreuther and E. Freisinger, *J. Inorg. Biochem.*, 2009, **103**, 342–353.

169. L. J. Jiang, M. Vašák, B. L. Vallee and W. Maret, *Proc. Natl Acad. Sci. USA*, 2000, **97**, 2503–2508.

170. O. I. Leszczyszyn, R. Schmid and C. A. Blindauer, *Proteins: Struct. Funct. Bioinf.*, 2007, **68**, 922–935.

171. D. W. Hasler, L. T. Jensen, O. Zerbe, D. R. Winge and M. Vašák, *Biochemistry*, 2000, **39**, 14567–14575.

172. O. I. Leszczyszyn, S. Zeitoun-Ghandour, S. R. Stürzenbaum and C. A. Blindauer, *Chem. Commun.*, 2011, **47**, 448–450.

173. A. Krezel and W. Maret, *J. Am. Chem. Soc.*, 2007, **129**, 10911–10921.

174. M. A. Namdarghanbari, J. Meeusen, G. Bachowski, N. Giebel, J. Johnson and D. H. Petering, *J. Inorg. Biochem.*, 2010, **104**, 224–231.

175. S. S. Narula, M. Brouwer, Y. X. Hua and I. M. Armitage, *Biochemistry*, 1995, **34**, 620–631.

176. R. Riek, B. Precheur, Y. Y. Wang, E. A. Mackay, G. Wider, P. Güntert, A. Z. Liu, J. H. R. Kägi and K. Wüthrich, *J. Mol. Biol.*, 1999, **291**, 417–428.

177. K. Zangger, G. Öz, J. D. Otvos and I. M. Armitage, *Protein Sci.*, 1999, **8**, 2630–2638.

178. A. Munoz, F. H. Försterling, C. F. Shaw and D. H. Petering, *J. Biol. Inorg. Chem.*, 2002, **7**, 713–724.

179. C. Capasso, V. Carginale, O. Crescenzi, D. Di Maro, E. Parisi, R. Spadaccini and P. A. Temussi, *Structure*, 2003, **11**, 435–443.

180. H. Wang, Q. Zhang, B. Cai, H. Y. Li, K. H. Sze, Z. X. Huang, H. M. Wu and H. Z. Sun, *FEBS Lett.*, 2006, **580**, 795–800.

181. M. H. Frey, G. Wagner, M. Vašák, O. W. Sorensen, D. Neuhaus, E. Wörgötter, J. H. R. Kägi, R. R. Ernst and K. Wüthrich, *J. Am. Chem. Soc.*, 1985, **107**, 6847–6851.

182. H. Rupp, W. Voelter and U. Weser, *FEBS Lett.*, 1974, **40**, 176–179.

183. P. J. Sadler, A. Bakka and P. J. Beynon, *FEBS Lett.*, 1978, **94**, 315–318.

184. K. T. Suzuki and T. Maitani, *Experientia*, 1978, **34**, 1449–1450.

185. J. D. Otvos and I. M. Armitage, *Proc. Natl Acad. Sci. USA*, 1980, **77**, 7094–7098.

186. M. Vašák, *J. Am. Chem. Soc.*, 1980, **102**, 3953–3955.

187. I. G. Dance, *J. Am. Chem. Soc.*, 1980, **102**, 3445–3451.

188. G. Henkel and B. Krebs, *Chem. Rev.*, 2004, **104**, 801–824.

189. K. Wüthrich, *Nature Struct. Biol.*, 2001, **8**, 923–925.

190. A. Arseniev, P. Schultze, E. Wörgötter, W. Braun, G. Wagner, M. Vašák, J. H. R. Kägi and K. Wüthrich, *J. Mol. Biol.*, 1988, **201**, 637–657.

191. P. Schultze, E. Wörgötter, W. Braun, G. Wagner, M. Vašák, J. H. R. Kägi and K. Wüthrich, *J. Mol. Biol.*, 1988, **203**, 251–268.

192. B. A. Messerle, A. Schaffer, M. Vašák, J. H. R. Kägi and K. Wüthrich, *J. Mol. Biol.*, 1990, **214**, 765–779.

193. B. A. Messerle, A. Schaffer, M. Vašák, J. H. R. Kägi and K. Wüthrich, *J. Mol. Biol.*, 1992, **225**, 433–443.

194. A. H. Robbins, D. E. McRee, M. Williamson, S. A. Collett, N. H. Xuong, W. F. Furey, B. C. Wang and C. D. Stout, *J. Mol. Biol.*, 1991, **221**, 1269–1293.

195. W. Braun, M. Vašák, A. H. Robbins, C. D. Stout, G. Wagner, J. H. R. Kägi and K. Wüthrich, *Proc. Natl Acad. Sci. USA*, 1992, **89**, 10124–10128.
196. J. J. Davis, H. A. O. Hill, A. Kurz, C. Jacob, W. Maret and B. L. Vallee, *PhysChemComm*, 1998, **1**, 12–22.
197. C. A. Blindauer and P. J. Sadler, *Acc. Chem. Res.*, 2005, **38**, 62–69.
198. X. Zhang, H. Tamaru, S. I. Khan, J. R. Horton, L. J. Keefe, E. U. Selker and X. D. Cheng, *Cell*, 2002, **111**, 117–127.
199. S. D. Zheng, J. Wang, Y. G. Feng, J. F. Wang and K. Q. Ye, *PLoS One*, 2012, 7, e45437.
200. C. A. Blindauer, M. D. Harrison, J. A. Parkinson, A. K. Robinson, J. S. Cavet, N. J. Robinson and P. J. Sadler, *Proc. Natl Acad. Sci. USA*, 2001, **98**, 9593–9598.
201. J. Loebus, E. A. Peroza, N. Blüthgen, T. Fox, W. Meyer-Klaucke, O. Zerbe and E. Freisinger, *J. Biol. Inorg. Chem.*, 2011, **16**, 683–694.
202. K. Tarasava, S. Johannsen and E. Freisinger, *Molecules*, 2013, **18**, 14414–14429.
203. E. A. Peroza, R. Schmucki, P. Güntert, E. Freisinger and O. Zerbe, *J. Mol. Biol.*, 2009, **387**, 207–218.
204. P. Faller, D. W. Hasler, O. Zerbe, S. Klauser, D. R. Winge and M. Vašák, *Biochemistry*, 1999, **38**, 10158–10167.
205. G. Öz, K. Zangger and I. M. Armitage, *Biochemistry*, 2001, **40**, 11433–11441.
206. F.-Y. Ni, B. Cai, Z.-C. Ding, F. Zheng, M.-J. Zhang, H.-M. Wu, H.-Z. Sun and Z.-X. Huang, *Proteins Struct. Funct. Bioinf.*, 2007, **68**, 255–266.
207. O. Palacios, K. Polec-Pawlak, R. Lobinski, M. Capdevila and P. Gonzalez-Duarte, *J. Biol. Inorg. Chem.*, 2003, **8**, 831–842.
208. I. Bertini, H. J. Hartmann, T. Klein, G. H. Liu, C. Luchinat and U. Weser, *Eur. J. Biochem.*, 2000, **267**, 1008–1018.
209. P. A. Cobine, R. T. McKay, K. Zangger, C. T. Dameron and I. M. Armitage, *Eur. J. Biochem.*, 2004, **271**, 4213–4221.
210. V. Calderone, B. Dolderer, H. J. Hartmann, H. Echner, C. Luchinat, C. Del Bianco, S. Mangani and U. Weser, *Proc. Natl Acad. Sci. USA*, 2005, **102**, 51–56.
211. C. W. Peterson, S. S. Narula and I. M. Armitage, *FEBS Lett.*, 1996, **379**, 85–93.
212. I. L. Abrahams, I. Bremner, G. P. Diakun, C. D. Garner, S. S. Hasnain, I. Ross and M. Vašák, *Biochem. J.*, 1986, **236**, 585–589.
213. R. Bogumil, P. Faller, P. A. Binz, M. Vašák, J. M. Charnock and C. D. Garner, *Eur. J. Biochem.*, 1998, **255**, 172–177.
214. C. D. Berweger, W. Thiel and W. F. van Gunsteren, *Proteins*, 2000, **41**, 299–315.
215. K. P. Kepp, *J. Inorg. Biochem.*, 2012, **107**, 15–24.
216. K. E. Rigby and M. J. Stillman, *Biochem. Biophys. Res. Commun.*, 2004, **325**, 1271–1278.
217. S. H. Hong, Q. Hao and W. Maret, *Protein Eng. Des. Sel.*, 2005, **18**, 255–263.

218. K. L. Summers, A. K. Mahrok, M. D. M. Dryden and M. J. Stillman, *Biochem. Biophys. Res. Commun.*, 2012, **425**, 485–492.

219. O. I. Leszczyszyn, C. D. Evans, S. E. Keiper, G. Z. L. Warren and C. A. Blindauer, *Inorg. Chim. Acta*, 2007, **360**, 3–13.

220. B. A. Messerle, M. Bos, A. Schaffer, M. Vašák, J. H. R. Kägi and K. Wüthrich, *J. Mol. Biol.*, 1990, **214**, 781–786.

221. N. Romero-Isart, B. Oliva and M. Vašák, *J. Mol. Model.*, 2010, **16**, 387–394.

222. M. Vašák, C. E. McClelland, H. A. O. Hill and J. H. R. Kägi, *Experientia*, 1985, **41**, 30–34.

223. O. Palacios, S. Atrian and M. Capdevila, *J. Biol. Inorg. Chem.*, 2011, **16**, 991–1009.

224. B. Dolderer, H. Echner, A. Beck, H. J. Hartmann, U. Weser, C. Luchinat and C. Del Bianco, *FEBS J.*, 2007, **274**, 2349–2362.

225. A. W. Foster and N. J. Robinson, *BMC Biol.*, 2011, **9**, 25.

226. O. Palacios, A. Pagani, S. Perez-Rafael, M. Egg, M. Höckner, A. Brandstätter, M. Capdevila, S. Atrian and R. Dallinger, *BMC Biol.*, 2011, **9**, 4.

227. D. E. K. Sutherland, M. J. Willans and M. J. Stillman, *J. Am. Chem. Soc.*, 2012, **134**, 3290–3299.

228. P. Palumaa, E. Eriste, O. Njunkova, L. Pokras, H. Jörnvall and R. Sillard, *Biochemistry*, 2002, **41**, 6158–6163.

229. P. Palumaa, I. Tammiste, K. Kruusel, L. Kangur, H. Jörnvall and R. Sillard, *Biochim. Biophys. Acta Proteins Proteomics*, 2005, **1747**, 205–211.

230. G. Meloni, T. Polanski, O. Braun and M. Vašák, *Biochemistry*, 2009, **48**, 5700–5707.

231. M. Good, R. Hollenstein and M. Vašák, *Eur. J. Biochem.*, 1991, **197**, 655–659.

232. R. Bogumil, P. Faller, D. L. Pountney and M. Vasak, *Eur. J. Biochem.*, 1996, **238**, 698–705.

233. A. K. Sewell, L. T. Jensen, J. C. Erickson, R. D. Palmiter and D. R. Winge, *Biochemistry*, 1995, **34**, 4740–4747.

234. R. G. Pearson, *J. Am. Chem. Soc.*, 1963, **85**, 3533–3539.

235. E. A. Peroza and E. Freisinger, *J. Biol. Inorg. Chem.*, 2007, **12**, 377–391.

236. B. A. Krizek, D. L. Merkle and J. M. Berg, *Inorg. Chem.*, 1993, **32**, 937–940.

237. W. Maret and B. L. Vallee, *Proc. Natl Acad. Sci. USA*, 1998, **95**, 3478–3482.

238. R. W. Olafson, *Bioelectrochem. Bioenerg.*, 1988, **19**, 111–125.

239. M. Dabrio and A. R. Rodriguez, *Anal. Chim. Acta*, 2000, **424**, 77–90.

240. J. Ejnik, J. Robinson, J. Y. Zhu, H. Försterling, C. F. Shaw and D. H. Petering, *J. Inorg. Biochem.*, 2002, **88**, 144–152.

241. D. E. K. Sutherland and M. J. Stillman, *Biochem. Biophys. Res. Commun.*, 2008, **372**, 840–844.

242. D. E. K. Sutherland and M. J. Stillman, *Metallomics*, 2011, **3**, 444–463.

243. M. Good, R. Hollenstein, P. J. Sadler and M. Vašák, *Biochemistry*, 1988, **27**, 7163–7166.

244. J. Zeng, R. Heuchel, W. Schaffner and J. H. R. Kägi, *FEBS Lett.*, 1991, **279**, 310–312.

245. J. Zeng, B. L. Vallee and J. H. R. Kägi, *Proc. Natl Acad. Sci. USA*, 1991, **88**, 9984–9988.

246. M. Huang, C. F. Shaw and D. H. Petering, *J. Inorg. Biochem.*, 2004, **98**, 639–648.

247. J. Ejnik, A. Munoz, T. Gan, C. F. Shaw and D. H. Petering, *J. Biol. Inorg. Chem.*, 1999, **4**, 784–790.

248. J. Zaia, D. Fabris, D. Wei, R. L. Karpel and C. Fenselau, *Protein Sci.*, 1998, 7, 2398–2404.

249. U. Rana, R. Kothinti, J. Meeusen, N. M. Tabatabai, S. Krezoski and D. H. Petering, *J. Inorg. Biochem.*, 2008, **102**, 489–499.

250. W. Maret, C. Jacob, B. L. Vallee and E. H. Fischer, *Proc. Natl Acad. Sci. USA*, 1999, **96**, 1936–1940.

251. A. Munoz, D. H. Petering and C. F. Shaw, *Inorg. Chem.*, 1999, **38**, 5655–5659.

252. M. Vašák, *Environ. Health Perspect.*, 1986, **65**, 193–197.

253. T. Gan, A. Munoz, C. F. Shaw and D. H. Petering, *J. Biol. Chem.*, 1995, **270**, 5339–5345.

254. O. I. Leszczyszyn, C. D. Evans, S. E. Keiper, G. Z. L. Warren and C. A. Blindauer, *Inorg. Chim. Acta*, 2007, **360**, 3–13.

255. T. Y. Li, A. J. Kraker, C. F. Shaw and D. H. Petering, *Proc. Natl Acad. Sci. USA*, 1980, 77, 6334–6338.

256. S. Yue, W. Zhong, B. Zhang, L. Zhu and W. Tang, *J. Inorg. Biochem.*, 1996, **62**, 243–251.

257. H. Li and J. D. Otvos, *J. Inorg. Biochem.*, 1998, **70**, 187–194.

258. W. R. Bernhard, M. Vašák and J. H. R. Kägi, *Biochemistry*, 1986, **25**, 1975–1980.

259. C. F. Shaw, L. B. He, A. Munoz, M. M. Savas, S. Chi, C. L. Fink, T. Gan and D. H. Petering, *J. Biol. Inorg. Chem.*, 1997, **2**, 65–73.

260. Q. Hao and W. Maret, *FEBS J.*, 2006, **273**, 4300–4310.

261. T. Y. Li, D. T. Minkel, C. F. Shaw and D. H. Petering, *Biochem. J.*, 1981, **193**, 441–446.

262. M. M. Savas, C. F. Shaw and D. H. Petering, *J. Inorg. Biochem.*, 1993, **52**, 235–249.

263. M. M. Savas, D. H. Petering and C. F. Shaw, *Inorg. Chem.*, 1991, **30**, 581–583.

264. L. J. Jiang, W. Maret and B. L. Vallee, *Proc. Natl Acad. Sci. USA*, 1998, **95**, 3483–3488.

265. C. Jacob, W. Maret and B. L. Vallee, *Proc. Natl Acad. Sci. USA*, 1999, **96**, 1910–1914.

266. H. Fliss and M. Menard, *Arch. Biochem. Biophys.*, 1992, **293**, 195–199.

267. B. Zhang, O. Georgiev, M. Hagmann, C. Gunes, M. Cramer, P. Faller, M. Vašák and W. Schaffner, *Mol. Cell. Biol.*, 2003, **23**, 8471–8485.

268. J. Durand, G. Meloni, C. Talmard, M. Vašák and P. Faller, *Metallomics*, 2010, **2**, 741–744.
269. K. D. Kröncke, K. Fehsel, T. Schmidt, F. T. Zenke, I. Dasting, J. R. Wesener, H. Bettermann, K. D. Breunig and V. Kolb-Bachofen, *Biochem. Biophys. Res. Commun.*, 1994, **200**, 1105–1110.
270. J. Kozhukh and S. J. Lippard, *Inorg. Chem.*, 2012, **51**, 7346–7353.
271. C. T. Aravindakumar, J. Ceulemans and M. De Ley, *Biochem. J.*, 1999, **344**, 253–258.
272. Y. Chen, Y. Irie, W. M. Keung and W. Maret, *Biochemistry*, 2002, **41**, 8360–8367.
273. J. Y. Zhu, J. Meeusen, S. Krezoski and D. H. Petering, *Chem. Res. Toxicol.*, 2010, **23**, 422–431.
274. L. L. Pearce, R. E. Gandley, W. P. Han, K. Wasserloos, M. Stitt, A. J. Kanai, M. K. McLaughlin, B. R. Pitt and E. S. Levitan, *Proc. Natl Acad. Sci. USA*, 2000, **97**, 477–482.
275. A. O. Udom and F. O. Brady, *Biochem. J.*, 1980, **187**, 329–335.
276. Y. Hathout, D. Fabris and C. Fenselau, *Int. J. Mass Spectrom.*, 2001, **204**, 1–6.
277. C. Jacob, W. Maret and B. L. Vallee, *Proc. Natl Acad. Sci. USA*, 1998, **95**, 3489–3494.
278. L. C. Costello, C. C. Fenselau and R. B. Franklin, *J. Inorg. Biochem.*, 2011, **105**, 589–599.
279. W. Feng, J. Cai, W. M. Pierce, R. B. Franklin, W. Maret, F. W. Benz and Y. J. Kang, *Biochem. Biophys. Res. Commun.*, 2005, **332**, 853–858.
280. A. Krezel and W. Maret, *J. Biol. Inorg. Chem.*, 2008, **13**, 401–409.
281. W. Maret and Y. Li, *Chem. Rev.*, 2009, **109**, 4682–4707.
282. M. Wilson, C. Hogstrand and W. Maret, *J. Biol. Chem.*, 2012, **287**, 9322–9326.
283. R. Orihuela, B. Fernandez, O. Palacios, E. Valero, S. Atrian, R. K. Watt, J. M. Dominguez-Vera and M. Capdevila, *Chem. Commun.*, 2011, **47**, 12155–12157.
284. M. P. Waalkes, M. J. Harvey and C. D. Klaassen, *Toxicol. Lett.*, 1984, **20**, 33–39.
285. H. Li and J. D. Otvos, *Biochemistry*, 1996, **35**, 13937–13945.
286. C. A. Blindauer, N. C. Polfer, S. E. Keiper, M. D. Harrison, N. J. Robinson, P. R. R. Langridge-Smith and P. J. Sadler, *J. Am. Chem. Soc.*, 2003, **125**, 3226–3227.
287. D. G. Nettesheim, H. R. Engeseth and J. D. Otvos, *Biochemistry*, 1985, **24**, 6744–6751.
288. M. J. Stillman, W. Cai and A. J. Zelazowski, *J. Biol. Chem.*, 1987, **262**, 4538–4548.
289. J. Ejnik, C. F. Shaw and D. H. Petering, *Inorg. Chem.*, 2010, **49**, 6525–6534.
290. H. Li and J. D. Otvos, *Biochemistry*, 1996, **35**, 13929–13936.
291. J. D. Otvos, H. R. Engeseth, D. G. Nettesheim and C. R. Hilt, *Experientia Suppl.*, 1987, **52**, 171–178.

292. W. Maret, K. S. Larsen and B. L. Vallee, *Proc. Natl Acad. Sci. USA*, 1997, **94**, 2233–2237.
293. J. T. Pedersen, C. Hureau, L. Hemmingsen, N. H. H. Heegaard, J. Ostergaard, M. Vašák and P. Faller, *Biochemistry*, 2012, **51**, 1697–1706.
294. M. Capdevila, R. Bofill, O. Palacios and S. Atrian, *Coord. Chem. Rev.*, 2012, **256**, 46–62.
295. O. I. Leszczyszyn, H. T. Imam and C. A. Blindauer, *Metallomics*, 2013, **5**, 1146–1169.
296. N. J. Robinson, S. K. Whitehall and J. S. Cavet, *Adv. Microb. Physiol.*, 2001, **44**, 183–213.
297. D. Beyersmann and H. Haase, *Biometals*, 2001, **14**, 331–341.
298. E. Mocchegiani, R. Giacconi, C. Cipriano, L. Costarelli, E. Muti, S. Tesei, C. Giuli, R. Papa, F. Marcellini, E. Mariani, L. Rink, G. Herbein, A. Varin, T. Fulop, D. Monti, J. Jajte, G. Dedoussis, E. S. Gonos, I. P. Trougakos and M. Malavolta, *Ann. NY Acad. Sci.*, 2007, **1119**, 129–146.
299. J. Jansen, W. Karges and L. Rink, *J. Nutr. Biochem.*, 2009, **20**, 399–417.

CHAPTER 22

Zinc

CHRISTER HOGSTRAND*[a] AND DAX FU[b]

[a] King's College London, Metal Metabolism Group, Diabetes and Nutritional Sciences, School of Medicine, Franklin-Wilkins Building, 150 Stamford Street, London, SE1 9NH, UK; [b] The Johns Hopkins University School of Medicine, Department of Physiology, 202 Physiology, 725 North Wolfe Street, Baltimore, MD, 21205, USA
*Email: christer.hogstrand@kcl.ac.uk

22.1 Introduction

Zinc is essential to all cells in all known organisms and it is the second most abundant trace element, after iron, in most vertebrates. Whilst iron is used in relatively few but abundant proteins, such as haemoglobin, the body uses zinc much more diversely and often in minute quantities in less abundant proteins.[1] Zinc is required for a variety of basic biological processes including metabolism of proteins, nucleic acids, carbohydrates and lipids and is also involved in more advanced functions, such as the immune system, neurotransmission, hormone secretion and signalling. It has been estimated that about 10% of all proteins in eukaryotic cells bind zinc and that there are approximately 3000 zinc proteins in humans.[2] Almost all of the zinc in cells is bound to proteins, peptides and amino acids, but there is a minute fluctuating pool of free cytosolic zinc(II) which is involved in cell signalling. With respect to the latter, zinc is unique among transition metals in that the Zn^{2+} ion functions as an intracellular messenger. Thus, much akin to Ca^{2+} transients, cellular Zn^{2+} waves regulate numerous biological processes. In animals, compartmentalization of zinc between tissues and within cells is managed principally by two evolutionary conserved families of zinc

RSC Metallobiology Series No. 2
Binding, Transport and Storage of Metal Ions in Biological Cells
Edited by Wolfgang Maret and Anthony Wedd
© The Royal Society of Chemistry 2014
Published by the Royal Society of Chemistry, www.rsc.org

transporters, the Zinc Transporter (ZnT; SLC30A) family and the Zrt-, Irt-like proteins (ZIP; SLC39A) family, which between them have 24 paralogues in human. Bacteria and plants also have zinc-transporting ATPases, such as zinc-transporting P-type ATPases in bacteria and the cadmium/zinc transporting Heavy Metal ATPases (HMA) in plants and fungi. Distinct distributions and activities of these transporters determine the distribution of zinc among organelles of cells and tissues of organisms.

22.2 Transport through the Cytoplasmic Membrane and Subcellular Membranes

22.2.1 The ZNT (SLC30 Family) of Zinc Transporters

ZNT1 (SLC30a1) was the first eukaryotic zinc transporter to be discovered.[3] It was identified as a protein that conferred resistance to zinc in cell culture and it is homologous to the "cation diffusion facilitator" (CDF) family of proteins, whose members efflux zinc and other transition metals from bacteria[4] and is ubiquitous among organisms. Ten SLC30 paralogues have been identified in mammalian genomes (Figure 22.1). Among the human SLC30 proteins, sequence homology is indicative of biological function (Figures 22.1, 22.2). Of all mammalian ZNT proteins only the closely related

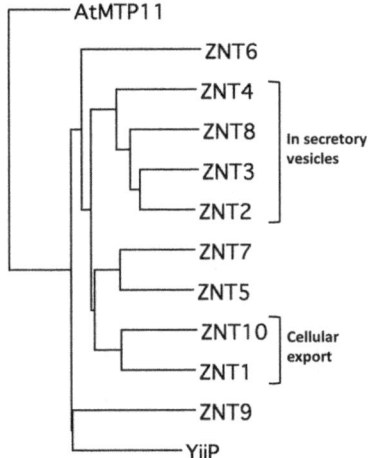

Figure 22.1 Phylogenetic analysis of SLC30 proteins. A multiple sequence alignment was generated with ClustalW and optimized "by eye" of Cation Diffusion Facilitator (CDF) protein amino acid sequences, including the 10 human SLC30 (ZNT) sequences, *Escherichia coli* YiiP and manganese transporter MTP11 from *Arabidopsis thaliana*. The output phylogenetic tree file was used to generate a phylogram built *via* neighbour joining with systematic tie-breaking (MacVector) and rooted with *A. thaliana* MTP11. SLC30 proteins involved in zinc transport into secretory vesicles/granules are structurally closely related as are ZNT1 and ZNT10, which export zinc out of the cell.

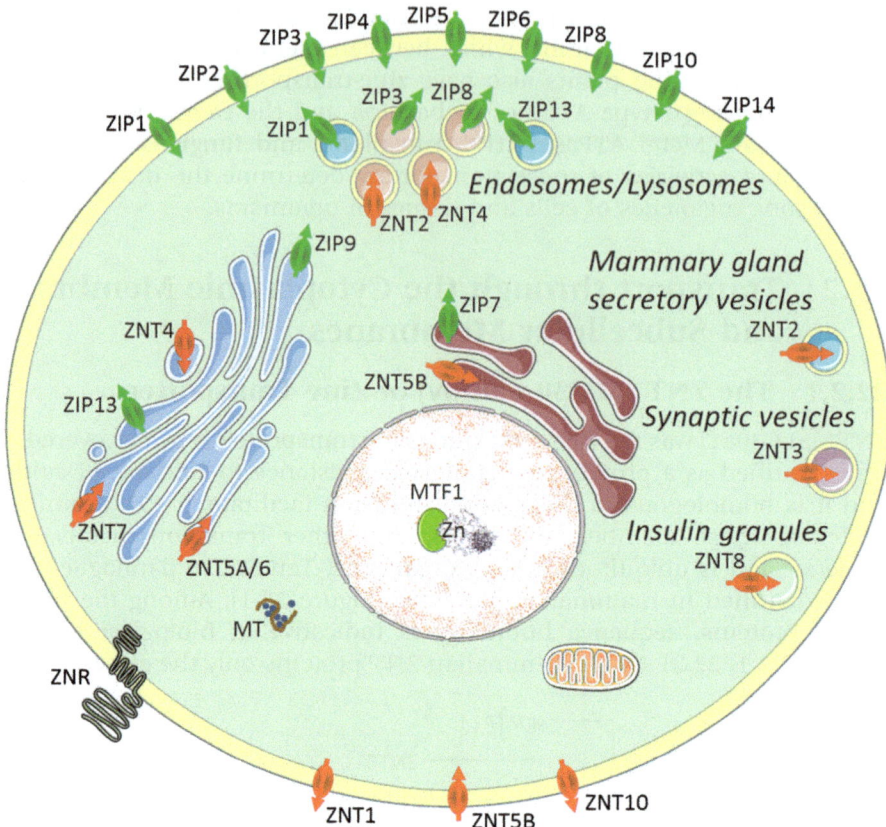

Figure 22.2   Inventory of zinc regulatory proteins in human cells. The zinc content of human cells and their organelles is maintained primarily by two families of zinc transporters, ZNT (SLC30) and ZIP (SLC39), shown in red and green, respectively. The likely direction of transport during physiological conditions is indicated with arrows. Metallothionein (MT) is the principal zinc buffering protein in cells and humans have multiple MT isoforms. Metal-responsive transcription factor 1 (MTF1) is an intracellular zinc sensor, mediating transcriptional activation and repression of genes in response to an elevated cytosolic concentration of free zinc. The zinc-sensing receptor (ZNR) is a G-protein coupled receptor (GPR39), which when activated by extracellular zinc triggers cytosolic release of Ca^{2+} from intracellular stores *via* the inositol 1,4,5-trisphosphate (IP3) pathway. The endoplasmic reticulum (ER) is depicted in violet-red and the Golgi apparatus in blue.

ZNT1 and ZNT10 operate as a zinc extrusion system from the cell while ZNT2–4 and 6–8 are located in the membranes of intracellular organelles (Figure 22.2). ZNT2, -3, -4 and -8 are close neighbours in the phylogenetic tree of ZNT proteins and they are all involved in transport of zinc into secretory vesicles/granules of different cell types (Figure 22.2). ZNT5 occurs as two splice variants, of which ZnT5B is located at the plasma membrane and ER whilst ZnT5A appears to transport zinc into the Golgi apparatus. All of the SLC30 paralogues, except ZNT5A, have six transmembrane domains, with N- and C-termini being located on the cytoplasmic side. ZNT5A is much larger with 12 transmembrane domains and it forms a heterodimer with ZNT6, which by itself does not appear to have transporter function.[5] This is different from other SLC30 proteins, which appear to form homodimers. The shorter of the ZNT5 splice variants (ZNT5B) may also be unique among SLC30 proteins in that apparently it can transport zinc into the cytosol.[6] The longer ZNT5 splice variant (ZNT5A) along with all other SCL30 isoforms examined transport zinc out of the cytosol and either out of the cell (ZNT1, ZNT10) or into different organelles. ZNT9 is structurally the most dissimilar of the human SLC30 proteins and clusters phylogenetically together with zinc exporter YiiP from *Escherichia coli* (Figure 22.1). Curiously, ZNT9 was first described as HUEL, a nuclear receptor co-activator.[7] Whether ZNT9 is also a zinc transporter remains to be established.

Although there appears to be redundancy in terms of zinc transporter function, many of the described zinc transporters have distinct biological roles relating to their unique subcellular or tissue distribution. For example, human ZNT2 transports zinc into secretory vesicles of the mammary glands, ZNT3 into synaptic vesicles and ZNT8 into insulin granules of pancreatic β-cells (Figure 22.2).[8] Interestingly, fish do not seem to have ZNT3, but its function in the brain may be carried out by ZNT2, which in fish is more or less exclusive to neural tissue.[9]

22.2.2 The ZIP (SLC39 Family) of Zinc Transporters

The ZIP family, SLC39, of zinc transporters has 14 members in mammalian genomes (Figure 22.3). This family also has plant, fungi and prokaryotic zinc-transporting orthologues and was first identified as a family of proteins that mediate zinc transport in *Arabidopsis thaliana*.[10] All mammalian ZIP paralogues appear to transport zinc into the cytosol, at least during normal physiological conditions (Figure 22.2). Also in contrast to the ZNT family, most human ZIP proteins, except ZIP7, -9 and -13, are functional in the plasma membrane and mediate tissue-specific zinc uptake. ZIP7 releases zinc from the endoplasmic reticulum (ER), and ZIP9 and -13 from the *trans*-Golgi network. ZIP8 may transport zinc into some cells, but has been found in lysosomes in other cells. It should be noted that shuttling of zinc transporters between organelles and the plasma membrane is one of the mechanisms involved in regulation of their activities, so finding a zinc transporter in an organelle does not necessarily mean that this is the locus of its zinc-transporting function.

ZIP proteins have eight transmembrane spanning domains with both N- and C-termini located away from the cytosol. They all have a long cytoplasmic loop (between transmembrane domains III and IV), which typically has several clustered histidine residues. These might act as temporary binding sites for zinc as it transverses the protein and allow donation of zinc

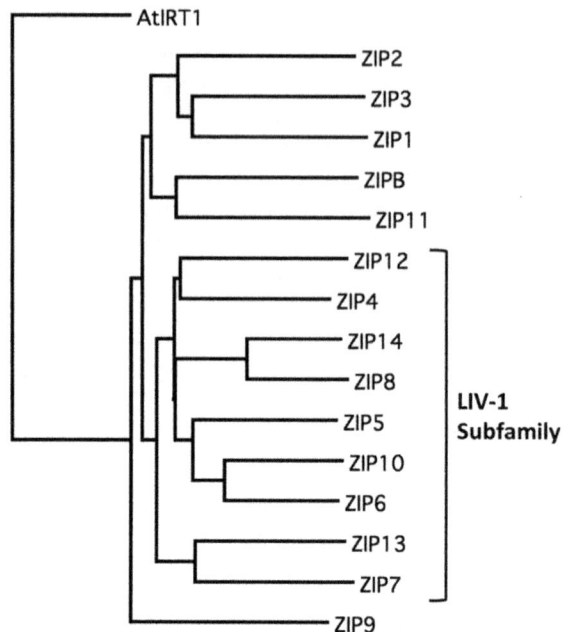

Figure 22.3 Phylogenetic analysis of SLC39 proteins. A multiple sequence alignment was generated with ClustalW and optimized "by eye" of phylogetically related protein amino acid sequences of metal transporters, including the 14 human SLC39 (ZIP), *Bordetella bronchiseptica*, ZIPB and Fe^{2+} transport protein 1 (IRT1) from *Arabidopsis thaliana*. The output phylogenetic tree file was used to generate a phylogram built *via* neighbour joining with systematic tie-breaking (MacVector) and rooted with *A. thaliana* IRT1. SLC39 proteins of the LIV-1 subfamily are indicated.

to cytosolic proteins through protein-protein interaction. Proteins in a subfamily of the vertebrate ZIP proteins, the LIV-1 subfamily (Figure 22.3), have a long extracellular N-terminal stretch, which is also rich in histidine residues, and in some members has a proteolytic cleavage site that is of importance for post-translational processing.[11] ZIP7 is the only zinc release transporter in the ER, and ZIP12 is expressed in fenestrated epithelia.

22.2.3 Zinc-transporting ATPases

Prokaryotes, yeast and plants express P-type ATPases capable of extruding soft metals, including zinc. The P-type ATPases occur in all forms of life and the P-type refers to a phosphorylated enzyme intermediate as the enzyme hydrolyses ATP to extract the energy required to move a charged solute across a biomembrane against its electrochemical gradient. The phylogenetically related CPx-type ATPase family contains several groups of soft-metal transporting proteins.[12] These transporters are distinguished by having eight transmembrane domains, in contrast to the soft-metal P-type ATPases, which have ten. Among these CPx-type ATPases is ZntA, which has orthologues in numerous bacteria.[13] ZntA is induced by high zinc concentrations and confers tolerance through zinc efflux.

Plants and yeast express several groups of Heavy Metal ATPases (HMA), which share substantial homology with the bacterial CPx-type ATPases.[14] A subgroup of the HMA, known as the P1B-type ATPases, has members that mediate zinc export from the cytoplasm. In Arabidopsis *hma2*, *hma3* and *hma4* provide zinc and cadmium tolerance and are believed to translocate these elements across membranes. Arabidopsis double mutants of hma2 and hma4 are zinc deficient in the shoots and show chlorosis, indicative of impaired chlorophyll synthesis, stunted growth, increased shoot branching and infertility. Both hma2 and hma4 appear to be required for the loading of zinc into the xylem by pumping zinc out of the adjacent cells in the root. While hma3 is expressed in the vacuole, both hma2 and hma4 are believed to reside in the plasma membrane. Thus, as for the ZNT and ZIP family proteins, existence of in-species HMA paralogues may be explained by their differential distribution in tissues and cells. At present, zinc transporting P-type ATPases have yet to be identified in mammals.

22.2.4   Zinc Transport through Calcium Channels

Already in the 1970s it was shown that voltage gated calcium channels in insect muscle are permeable to zinc.[15] It was later shown that zinc can enter mammalian neurons through VGCC and α-amino-3-hydroxy-5-methyl-4-isoxazole propionic acid receptors (AMPAR). Similarly, several members of the TRP super-family of Ca^{2+} channels can mediate Zn^{2+} movement into cells. For fish, this is of particular significance as the competition of Zn^{2+} with Ca^{2+} for the epithelial calcium channel (Ecac/Trpv6), located on the apical membrane of the gill, is a key mechanism of toxicity to elevated zinc concentrations in the water.

22.2.5   Uptake in the Intestinal Cells and Release into the Circulation

Intestinal uptake of zinc is highest in the ileum, followed by jejunum and duodenum. Uptake of zinc in the gut starts with the diffusion into the unstirred layer followed by binding to the mucus of the intestinal epithelium. The metal is then transported as the Zn^{2+} ion into the epithelial cells. Several zinc transporters are expressed in the brush border membrane of human enterocytes, including ZNT5 and members of the ZIP family. The most important of these is arguably ZIP4. Mutations in ZIP4 lead to the disease acrodermatitis enteropathica, in which individuals die from zinc deficiency at a young age unless their diet is supplemented with high levels of zinc.[16] Similarly, the homozygote mouse ZIP4 knock-out dies during embryonic development and the heterozygote is very sensitive to zinc deficiency.[17,18] Inside the intestinal enterocyte, zinc is bound to metallothioneins, which are a family of zinc-binding proteins that buffer the intracellular free Zn^{2+} ion concentration.[†] At the basolateral membrane of the enterocyte, zinc is extruded into the circulation by ZNT1.[19] There is also a retrograde transport of zinc into the enterocyte at the basolateral membrane through ZIP5.[20] Expression of the zinc transporters in the enterocyte is regulated by the intestinal zinc content at the transcriptional, translational and post-translational levels. There is little evidence for systemic homeostatic regulation according to whole body zinc status. Instead it appears that zinc uptake in the intestine is regulated locally by ingested zinc and the zinc concentration in the intestinal tissue.[21]

22.3 Biochemical Principles of Zinc Transporters (ZIPs and ZNTs)

The control over zinc flow among various cellular membrane compartments are principally achieved by selective binding of the zinc ion to the transport site and its energetic coupling to the transmembrane electrochemical potential that drives zinc entry from one side of the membrane and releases it at the opposite membrane face. Additional control over zinc transport are possible with extra-membranous regulatory sites. Cell signalling events, such as phosphorylation, may modify zinc transporters post-translationally.[22] Phosphorylation of surface residues may be transmitted *via* a conformational change to activate the membrane-embedded zinc transport site, switching on zinc transport activity as a part of a cell signalling cascade. In a feedback loop of zinc homeostatic controls, cellular zinc fluctuation may be sensed through a change in the binding occupancy of a regulatory zinc binding site.[23] Again, a protein conformational change would be required to propagate the zinc binding event at an extra-membrane binding site to the distant zinc transport site.

The coupling of structure dynamics and metal selectivity is a unique feature of the zinc transport site in ZNT proteins. The zinc transport reaction occurs on a millisecond to second time-scale.[24] The kinetics of zinc turnover is vastly faster than zinc metabolism in zinc enzymes, where a catalytic zinc ion, once inserted into the reaction centre, will stay there throughout the lifespan of the enzyme.[25] By comparison, the zinc transport site is capable of flipping its membrane facing to capture a zinc ion from one side of the membrane and release the bound zinc ion to the opposite side. The vacant transport site then rapidly recoils to its original membrane facing to complete a transport reaction cycle. The rapid binding and releasing of zinc is highly coordinated to ensure a vectorial zinc ion movement across the membrane barrier while maintaining zinc selectivity against similar metal

---

[†]Chapter 21 provides an extended discussion of metallothioneins.

ions that may be several orders of magnitude more abundant in the local environments. The integration of metal selectivity, energetic transmembrane movement and functional regulation in the zinc transport site provides an opportunity to learn how metal coordination chemistry is built into the protein structure, and how the protein structure dynamics enable selective zinc transport across the membrane barrier.

22.4 Structure of Zinc Transporters

Although structures have been resolved for single domains of several zinc transporting proteins, the atomic-level structural information for intact zinc transporters comes from the crystal structure of YiiP, a CDF zinc transporter from *Escherichia coli*. The first YiiP structure was determined at 3.8 Å resolution in complex with zinc ions.[26] A later refined structure at 2.9 Å resolution defines the atomic detail of the transport site in a putative zinc translocation pathway across the membrane.[23] Overall, YiiP adopts a Y-shaped homodimeric architecture arranged around a two-fold axis oriented perpendicular to the membrane plane (Figure 22.4). Each monomer comprises an N-terminal transmembrane domain (TMD), followed by a C-terminal domain (CTD) that protrudes into the cytoplasm. Two CTDs of a YiiP homodimer juxtapose each other in parallel to form a major dimerization contact. At the CTD and TMD juncture, highly conserved intersubunit salt bridges and hydrophobic contacts form another dimerization contact, from which two TMDs swing outward into the membrane. Each TMD contains six transmembrane helices (TM1-6) with both the N- and C-terminus in the cytoplasm. TM1 to TM6 are linked together by three extracellular loops and two intracellular loops. An additional intracellular loop connects TM6 to the CTD. The intracellular loop connecting TM4 and TM5 is only four residues long from position 141 to 144. Many CDF members contain an extended TM4-TM5 loop that harbours metal binding motifs with multiple histidine residues, known as the His-loop. On the extracellular membrane surface, all three extracellular loops adopt extended conformations with full exposure to solvent.

The TMD consists of two relatively independent subdomains with limited interactions, namely TM1, TM2, TM4 and TM5 form a compact four-helix bundle, whereas the remaining TM3 and TM6 cross over in an antiparallel configuration outside the bundle. The orientation of the TM3–TM6 pairs is stabilized by four interlocked salt bridges formed between Lys77 of TM3 and Asp207 from a short loop that connects TM6 to the CTD. These highly conserved charged residues are arranged in a circular fashion to form a (Lys77-Asp207)$_2$ interlock that bundles together the cytoplasmic ends of two TM3-TM6 pairs at the dimer interface.[23] Among the six TMs, TM5 is conspicuously short in length, accounting for four helical turns from residue Gln145 to Met159. This short helix is largely sequestered within the hydrophobic core of the TMD, thereby giving rise to one extracellular and one intracellular cavity on either side of the membrane (Figure 22.5). The extracellular cavity is accessible from the bulk solvent and exposed to the

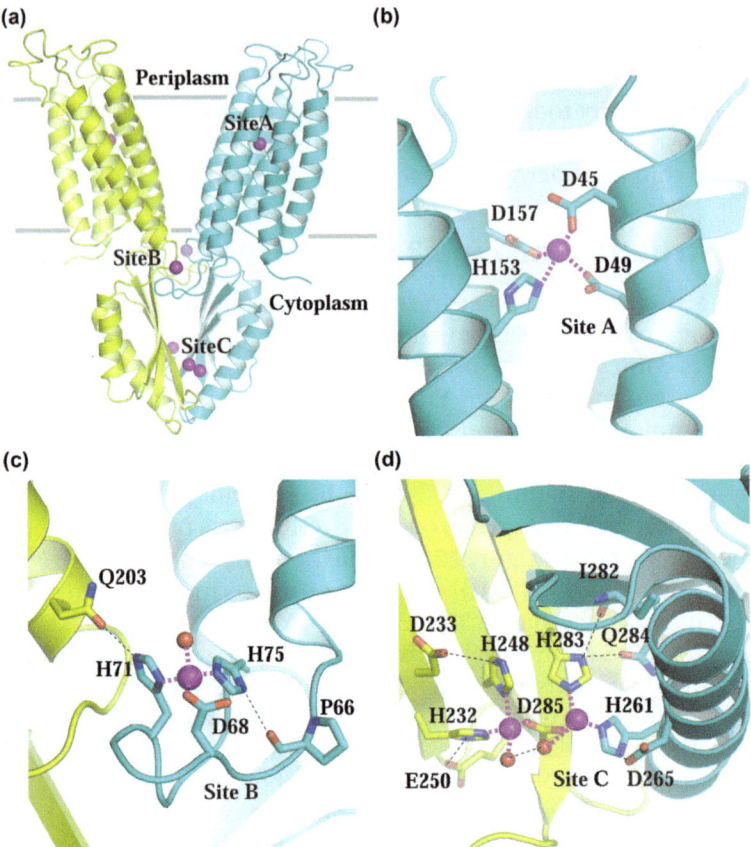

Figure 22.4   YiiP structure and zinc binding sites. (a) Ribbon representation of YiiP homodimer (yellow and cyan for each protomer) viewed from the membrane plane. Magenta spheres represent bound zinc ions in three zinc binding-sites as indicated. (b) The transport site (Site A) with a bound zinc ion and four coordination residues (sticks) arranged in a tetrahedral configuration. (c) Site B with a coordinated water molecule (red sphere). (d) Site C at the CTD-CTD dimer interface.

membrane outer leaflet. This cavity spans nearly half of the membrane thickness, and is lined with negatively charged residues that provide a favourable electrostatic environment for a bound zinc ion near the bottom of the cavity. On the other hand, the intracellular cavity is relatively shallow. The two cavities from different sides of the membrane define an occluded zinc translocation pathway with hydrophobic residues between the two cavities to seal off intercavity zinc diffusion.[26]

The CTD of YiiP adopts an open α-β fold, exhibiting an overall structural similarity with the copper metallochaperone Hah1, although there is no sequence homology between YiiP and Hah1.[27] This structural similarity to metallochaperone proteins places CTD into the same structural category as a

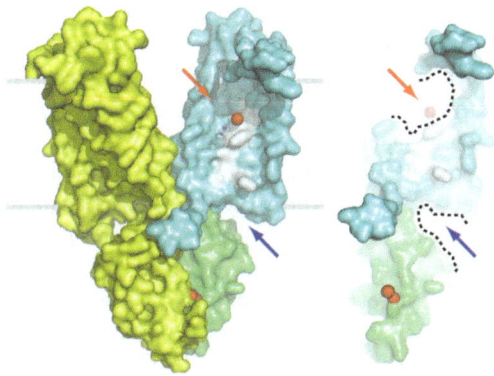

Figure 22.5 Extracellular and intracellular cavities. YiiP dimer in the molecular surface representation is viewed from the membrane plane. The front portions of TM1, TM2 and CTD of one protomer (cyan) are removed to reveal the embedded TM5 (grey patch) and cavities as indicated by arrows. TMD and CTD are coloured cyan and green, respectively.

possible Zn^{2+} receiving domain. The second important feature of CDT is represented by a charge-rich CTD-CTD interface stretching from the (Lys77-Asp207)$_2$ interlock at the TMD-CTD juncture to the tip of the CTD. The electrostatic repulsion between two CTDs is neutralized by zinc binding at the CTD-CTD interface.[23]

The crystal structure reveals four coordinated Zn^{2+} (Z1–Z4) in each monomer (Figure 22.4). Z1–Z4, together with protein ligands and water molecules that participate in their first coordination shells, constitute three distinct Zn^{2+} binding sites, termed SA, SB and SC, respectively.[23] Z1 in SA is tetra-coordinated by highly conserved Asp45 and Asp49 from TM2, and His153 and Asp157 from TM5. They form a coordination system with four-ligand groups (Asp45 Oδ1, Asp49 Oδ2, His153 Nε2 and Asp157 Oδ2) positioned at the vertices of a nearly ideal tetrahedron. Z2 in SB is five-coordinated by two His-imidazoles, a bidentate Asp-carboxylate and a water molecule that takes up the fifth position. All Z2 coordinating residues are located within an intracellular loop that connects TM2 and TM3. Z3 and Z4 are 3.8 Å apart in the binuclear SC at the CTD-CTD interface. The carboxylate group of a conserved Asp285 bridge Z3 and Z4. Additional coordinating residues to Z3 are His232 and His248, and those to Z4 are His283 and His261 from the neighbouring subunit. These residues are tucked into the CTD interface in a cleft capped by two coordinated water molecules. The positive charges on the two Zn^{2+} ions in SC balance the strong electronegativity at the CTD-CTD interface, thereby stabilizing dimeric association.

22.5 Metal Selectivity

Phenotype analysis of deletion or overexpression of the *yiiP* gene suggested that YiiP may play a physiological role in protecting *Escherichia coli* against toxic iron overexposure. Thus YiiP was originally named Fief for ferrous ion efflux, implying that YiiP is a Fe^{2+} selective transporter.[28] However, attempts to measure transport of radioactive Fe^{2+} were not successful using membrane vesicles derived from a mutant *E. coli* strain with Fief overexpression. Rather, overexpression of Fief was found to enhance vesicular Zn^{2+} accumulation.[28] The discrepancy between the iron tolerance phenotype and the lack of Fe^{2+} transport activity necessitated a more rigorous biochemical characterization of metal selectivity with purified YiiP protein in detergent micelles and reconstituted proteoliposomes. Direct measurements of metal uptake by inductively coupled plasma mass spectroscopy (ICP-MS) showed that the purified YiiP transported Zn^{2+} and Cd^{2+}, but rejected all other transition metal ions in the fourth period.[29] Further, direct measurement of metal binding to the transport site (SA) by isothermal titration calorimetry showed no evidence of Fe^{2+} binding, although Fe^{2+} binding to SB and/or SC was detected.[30] The spectrum of metal binding selectivity was characterized by introducing a reporter cysteine residue (D157C) to the transport site. Binding of Zn^{2+} and Cd^{2+}, but not Fe^{2+}, Hg^{2+}, Co^{2+}, Ni^{2+}, Mn^{2+}, Ca^{2+} and Mg^{2+}, protected D157C from thiol-specific labelling.[30] Thus, the metal binding selectivity revealed by binding competition analysis is fully consistent with the metal transport selectivity as determined by ICP-MS. It is clear that that the transport site of YiiP is indeed selective to Zn^{2+} and Cd^{2+} against Fe^{2+} and other transition metal ions, although the cellular mechanism of Fe^{2+} tolerance has yet to be elucidated.[31]

Zinc, cadmium, mercury and copernicium of Group 12 are unique in having a completely filled d-shell (d^{10}). Mercury further differs from zinc and cadmium due to the presence of a filled 4f shell. As such, cadmium is the most closely related metal to zinc on the entire periodic table. Their full d-orbital electronic configurations determine a single oxidation state of $+2$ by losing two outermost s-orbital electrons. The resulting vacant $4s^1 4p^3$ or $5s^1 5p^3$ orbitals enable an $s^1 p^3$-type orbital hybridization with four lone electron pairs donated by four ligand groups resulting in a certain preference for tetrahedral geometry. By comparison, other d-block metal ions, with additional vacant orbitals of a partially filled d-shell, prefer $d^2 s^1 p^3$-type orbital hybridization that gives rise to octahedral coordination involving six ligand groups. A mercury ion, with a filled inner f shell, prefers linear coordination *via* $s^1 p^1$-type orbital hybridization. Therefore, zinc and cadmium ions, by virtues of their similar electronic structures, are distinct from other transition metals in terms of coordination number and geometry. They coordinate to only four ligands mainly because of their electronic structures that accommodate four pairs of electrons in their principal quantum shells ($4s4p^3$ or $5s5p^3$).[32] Zinc ions are often observed forming five- and six-coordinate complexes in proteins. These higher coordination numbers may be viewed as tetrahedral complexes with one or two pairs of alternate coordinates.[33] When a Zn^{2+} encounters an extra number of coordinates, alternation occurs, namely, one of the $4s4p^3$ hybrid orbitals of zinc alternately accommodates one of two coordinates that appear as bidentate coordinates in the protein crystal structures.[34]

A correlation of the unique coordination chemistry of zinc and cadmium ions with the crystal structure of YiiP sheds light on the chemical and structural basis of metal selectivity. The transport site is characterized by an electron-rich environment in the hydrophobic core of the membrane. The availability of lone electron pairs from the ligand groups can satisfy the need to fill in vacant orbitals of an incoming metal ion. It is the exact match of the coordination number and geometry that dictates the metal selection for Zn^{2+} and Cd^{2+}. In the entire hydrophobic core of YiiP, only four potential coordination residues are available, providing the exact number of lone electron pairs required for tetrahedral coordination. The positioning of co-ordination ligand groups at the vertices of a tetrahedron fits the tetrahedral arrangement of the $4s4p^3$ hybrid orbitals, satisfying the geometric requirement for Zn^{2+} and Cd^{2+} coordination.

The metal selectivity appears conserved from bacterial to human homologues. This conservation arises from sequence conservation of the transport site, and is further strengthened by a unique positioning of this site in the hydrophobic core of the protein. The lack of additional hydrophilic residues in the vicinity of the transport site deprives the opportunity to form hydrogen bonds with the remaining part of the protein, thereby insulating the influence of sequence variation outside of the transport site.[23] As such, the four coordination residues are expected to play a dominant role in determining the metal selectivity. The tetrahedral transport site of YiiP has a DD-HD residue composition, differing by a single residue from the HD-HD transport site of mammalian ZnTs. While YiiP transports both Zn^{2+} and Cd^{2+} with equal specificity, human homologues exhibit a refined Zn^{2+} selectivity over Cd^{2+}.[29] Intriguingly, a single D-to-H substitution is sufficient to generate preference for Zn^{2+} over Cd^{2+} in the bacterial YiiP. The Cd^{2+} binding affinity to the mutated transport site was weakened by more than 30-fold whereas Zn^{2+} affinity remains unchanged. This differential effect of a histidine substitution gives rise to the observed Zn^{2+}-over-Cd^{2+} selectivity. As a result, the mutant YiiP exhibited significantly reduced Cd^{2+} transport activity at a physiological concentration, although the loss of Cd^{2+} transport activity can be overridden at higher Cd^{2+} concentrations. Likewise, a single H-to-D mutation in human ZnTs completely abolishes the Zn^{2+} over Cd^{2+} selectivity, converting the ZnT transport site to a YiiP-like one. The reciprocal conversion of transport specificity by a single residue substitution to both bacterial and human transport site demonstrates remarkable evolutionary fidelity of the selectivity determinant in the tetrahedral transport site.[29]

The strict Zn^{2+} and Cd^{2+} selectivity of YiiP is determined by the unique structural features of the transporter site. It is not clear whether this canonical tetrahedral site can be remodelled structurally to accept other transition metal ions. In many CDF homologues, hydrophilic residues in addition to the four known coordination residues are predicted near the transport site.[35] The participation of additional ligand groups may adapt the tetrahedral site to octahedral coordination. Indeed, an extensive body of genetic complementation data exists in support of a broad spectrum of metal selectivity. Overexpression of various CDF homologues was found to confer metal-resistant phenotypes on host cells that were exposed to toxic levels of Zn, Cd, Co, Ni, Fe and Mn.[36] Despite the lack of direct biochemical evidence, the genetic data from different laboratories consistently indicate that the spectrum of metal selectivity is far beyond Zn^{2+} and Cd^{2+}. Notably, single or cluster mutations, each to a distinct location in the protein structure, could narrow or widen metal selectivity, or even switch metal selectivity from one metal ion to another.[37] Understanding the structural and chemical basis of metal selectivity for a broad spectrum of transition metal ions remains a focus at the forefront of CDF research.

22.6 Mechanisms of Zinc Transport

22.6.1 SLC30 Family

Zinc mobility makes the fundamental distinction between zinc transporters and zinc metalloenzymes. The YiiP crystal structure is characterized by a pair of cavities that extend into the lipid bilayer from opposite membrane surfaces (Figure 22.5). The unique molecular architecture of zinc transport proteins provides the accessibility of a zinc ion from an aqueous membrane surface to the membrane-embedded transport site. Water within the cavities may significantly reduce the free energy barrier to Zn^{2+} diffusion compared to the low dielectric environment of the inner membrane, thereby defining a putative zinc translocation pathway. Once a Zn^{2+} ion enters the transport site from one side of the membrane, the binding energy is expected to trigger a protein conformational change that flips the accessibility of the transport site to the opposite face of the membrane. The change of membrane facing is thought to be linked to a switch in tetrahedral coordination from a role of zinc acquisition to zinc release. Thus, a tight coupling of zinc binding with protein dynamics is essential to zinc transport through a conformational change between two alternating states of a single transport site with distinct membrane facing and zinc binding affinities. The crystal structure of YiiP in complex with coordinated zinc ions revealed an outward facing conformation while an electron crystallographic structure of a YiiP homologue may portray an inward facing conformation (Figure 22.6).[38] These two structures suggest how a conformational change may reorient the transport site to enable directional zinc movement from the intracellular to the extracellular cavity.[26,39]

In the outward facing conformation of YiiP, the transport site is situated at the bottom of the extracellular cavity, anchored to a hydrophobic seal that separates the extracellular cavity from the intracellular one (Figure 22.5).[26] The partially buried tetrahedral site is open to both the membrane outer leaflet and the periplasm. Although the zinc exit pathway is evident in the crystal structure, the release of the coordinated zinc ion may need to overcome a significant energy barrier because of high electron affinity of a densely changed zinc ion in a low dielectric environment. It is particularly

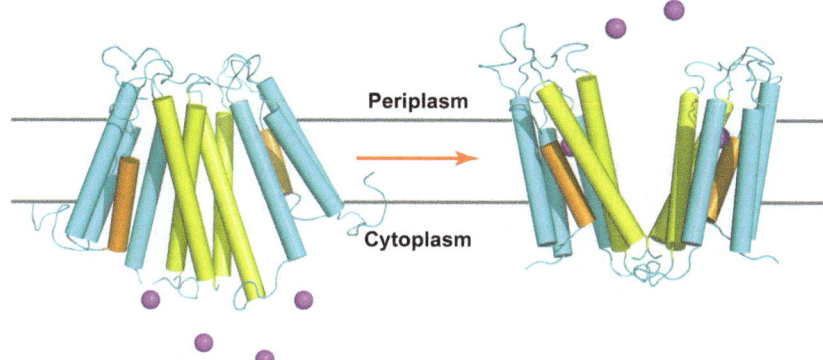

Figure 22.6   Alternate access of the transport site to either side of the membrane. Transmembrane helixes in cylindrical representations are reoriented to give an inward facing (left) and outward facing conformation (right) based on an electron and X-ray crystallographic structure, respectively. The conformational change allows the transport site to bind a zinc ion (magenta sphere) from the cytoplasm, and then release the bound zinc ion to the periplasm. The CTD is omitted for clarity.

challenging for the release of the coordinated zinc ion on a 10–100 ms time-scale as expected for a zinc transporter.[24] This rate of zinc dissociation is many orders of magnitudes faster than the Zn^{2+} dissociation rates for typical zinc metalloproteins. The transport site of YiiP exhibits three features well suited to rapid on-off switching of zinc coordination. First, all coordinating residues in the transport site are completely unconstrained, and thus no making or breaking of outer-shell interactions is needed for Zn^{2+} binding and release during a transport cycle. Second, each of the three Asp-carboxylates in the transport site is mono-coordinated, leaving an uncoordinated carboxylate oxygen ligand free of outer-shell interaction. These carboxylate oxygens can potentially be rearranged to enter the coordination shell, resulting in a highly adaptable inner-shell arrangement with varied coordination geometries. The resultant conformational flexibility may contribute to a rapid Zn^{2+} turnover rate. Third, the transport site is confined exclusively between TM2 and TM5. A small interhelix shift between TM2 and TM5 can lead to a large readjustment of the zinc coordination geometry in favour of zinc binding or release, allowing for rapid binding mode change through protein conformational changes.

In a putative inward facing conformation of a YiiP homologue observed by cryo-EM in a zinc-free state, the absence of zinc binding to the transport site caused a scissoring movement of the Y-shaped homodimer that partially closed up the 30-degree wedge between two transmembrane domains (Figure 22.6).[38] Concomitant to the overall intersubunit movement, each protomer undergoes an intrasubunit subdomain shifting. As described earlier, the transmembrane domain consists of two relatively independent components, namely the M3-M6 helix pair and the M1-M2-M4-M5 four-helix bundle, which harbours the transport site between TM2 and TM5.[23] The EM structure suggests that the cytoplasmic part of the four helix bundle may move away from the helix pair. The tilting of the four-helix bundle relative to the M3-M6 pair collapsed the periplasmic cavity while expanding the cyto-plasmic cavity. Although no atomic detail is revealed in the low-resolution EM-structure, the expanded cytoplasmic cavity may become sufficiently deep to include the transport site, making it accessible to Zn^{2+} entry from the cytoplasm. As a result, the conformational change appears to switch the membrane facing of the transport site from the periplasm to cytoplasm.

The conformational change provides structural insights into the access of a zinc ion to the transport site from either side of the membrane surface. A critical energy barrier for zinc transport, however, is to move a zinc ion across the hydrophobic seal between the two cavities. The transport activity of YiiP in everted membrane vesicles requires a NADH-energized trans-membrane pH gradient.[28] A more detailed kinetic analysis of the energetic process was performed on a purified YiiP homologue, ZitB in reconstituted proteoliposomes.[24] In the absence of a proton gradient, zinc transport is arrested despite an imposing zinc gradient across the membrane. Applying proton motive forces across the membrane drove Zn^{2+} fluxes both into and out of proteoliposomes against the imposed H^+ gradients. The proton coupled Zn^{2+} transport is electrogenic, indicating that a charge imbalance resulted from Zn^{2+}-for-proton exchange. Since Cd^{2+} binding to YiiP was shown to displace protons with a $\sim 1 : 1$ exchange stoichiometry, the ex-change stoichiometry of the proton-coupled transport reaction is likely to be $1 : 1$ with one net positive charge across the membrane per transport cycle. Among the four coordination residue in the DD-HD transport site of YiiP, the histidine residue is a likely candidate for proton exchange. Protonation of the histidine may facilitate zinc release into the extracellular cavity while deprotonation may be required for zinc entry into the transport site from the intracellular cavity. Given the sequence conservation of the transport site from bacteria to humans, it is likely that the proton coupled zinc antiport is a general energetic mechanism.[40] Indeed, all the characterized CDF proteins so far seem to utilize the transmembrane proton gradient to drive zinc transport reactions.

The proton-coupled transport reaction suggests an active role of water molecules as proton donors and acceptors. Proton antiport in YiiP requires water access to the transport site, which is completely sequestered from proton exchange with neighbouring residues. The $1 : 1$ zinc-for-proton stoichiometric exchange further necessitates a highly coordinated access of both water and the zinc ion to the transport site. The extracellular water access is evident in the existing crystal structure, but water access from the cytoplasm is blocked by the intercavity hydrophobic seal (Figure 22.5). A transient opening of the seal is expected to increase water interactions. Thus, change in water accessibility could provide dynamic information re-garding the transition between the outward and inward facing conformation associated with the intercavity crossing of a zinc ion. The conformational transition was probed with the hydroxyl radical, which is an ideal water accessibility probe attributed to its similar van der Waals area and solvent properties to those of the water molecule.[41] Hydroxyl radicals can be generated by isotropic radiolysis wherever water is present. As a water molecule approaches the transport site, the residues in direct contact with the water molecule become targets for hydroxyl radical-mediated oxidation. The modified residue can be identified by high-resolution mass spectrometry, and the extent of modification can be quantified in terms of water accessibility.[42] Time-resolved experiments based on the production of hydroxyl radicals by millisecond pulses of focused synchrotron X-rays revealed only two conformational hotspots in the entire YiiP protein sequence. Both sites are mapped to the vicinity of the transport site and exhibit more than a 1000-fold change in water accessibility as a function of zinc binding. In addition to this remarkably localized conformational change, analysis of moderate changes in water accessibility revealed a reciprocal pattern of conformational change along two opposite faces of TM5, indicating that zinc binding triggers a rotational motion in TM5. Since the shortened helix length of TM5 in the hydrophobic core of TMD is largely responsible for the formation of two water-filled cavities from either side of the membrane (Figure 22.5), It is conceivable that a small TM5 rotation relative to its surrounding TMs suffices to break the intercavity seal, creating a breaching point for the crossing of zinc and water. Furthermore, any TM5 motion relative to TM2 is expected to distort the tetrahedral geometry of the transport site, switching on and off zinc coordination.[23] Therefore, zinc binding, release and intercavity movements are highly coordinated. Further structural analysis and computational modelling are needed to elucidate the atomic details of the proton-coupled zinc transport reaction based on the established experimental framework.

22.6.2 SLC39 Family

There is presently no three-dimensional structure resolved for proteins of the SLC39 (ZIP) family and available data on transport mechanisms come from empirical studies. Based on electrographic and chromatographic data it appears that ZIP transporters form homodimers in lipid bilayers.[43,44] In experiments on purified ZIPB from *Bordetella bronchiseptica* that was functionally reconstituted into proteoliposomes, zinc flux through ZIPB was non-saturable and electrogenic, generating membrane potentials that could be predicted by the Nernst equation.[45] Conversely, membrane potentials drove zinc fluxes with a linear voltage-flux relationship. These results suggest that this prokaryotic protein operates as an electro-diffusional channel. ZIPB is selective for two Group 12 transition metal ions, Zn^{2+} and Cd^{2+}, whereas rejecting transition metal ions in other groups.

ZIPB from *Bordetella bronchiseptica* shows considerable homology with the human SLC39 sequences and clusters phylogenetically together with ZIP11 (Figure 22.3). However, it is not clear if all vertebrate ZIP proteins have retained the channel properties of ZIPB. Transport activity of human ZIP2, ZIP8 and ZIP14 is stimulated by external HCO_3^- and shows maximum transport activity at pH 7.5,[46–48] which would suggest transport of zinc as a bicarbonate complex or possibly a Zn^{2+} HCO_3^- symport mechanism. In contrast, zinc transport by the pufferfish (*Takifugu rubripes*) Zip3 orthologue (FrZip2) was stimulated by a slightly acidic medium (pH 5.5–6.5) and inhibited by HCO_3^-.[49] Through manipulation of medium chemistry and modelling of zinc speciation it was determined that for pufferfish Zip3 the Zn^{2+} ion is likely the transported species. Collectively, these results are indicative of a Zn^{2+}-H^+ symport function for pufferfish Zip3. The activities of all functionally characterized vertebrate ZIP transporters are all inhibited by other bivalent metals, such as Fe^{2+}, Co^{2+}, Mn^{2+}, Cu^{2+} and Cd^{2+}. In addition, Cd^{2+} and Mn^{2+} can be transported by ZIP8 whilst ZIP14 can mediate Fe^{2+} uptake into cells.[48,50,51]

22.7 Allosteric Regulation

A rapid and precise control over cellular and subcellular zinc concentrations is critically important for zinc to have regulatory roles in cellular signalling. Zinc-related signalling pathways involve functional modulation of neurotransmission and hormone secretion, transient stabilization of zinc fingers in transcription factors, inhibition of phosphatases and assembly of multiprotein complexes. In these roles, transient increases in the cytosolic free zinc concentration can occur through the activation of zinc transporters that promote the influx of extracellular Zn^{2+} or the release of free Zn^{2+} from intracellular stores. For example, phosphorylation of ZIP7 in the ER by protein kinase CK2 results in zinc release from the ER and the subsequent activation of multiple downstream pathways, leading to enhanced cell proliferation and migration.[43] Since the fluctuation of the intracellular zinc level can influence the outcomes of cell signalling, expression, assembly, activity and stability of the zinc transporters are highly regulated by changes in zinc status, and often subjected to elaborate feedback controls around zinc homeostatic set points. At present, the physiological events linked to the activation of CDF zinc transporters have yet to be identified. Structural models based on the YiiP crystal structure and emerging experimental evidence suggest that two modular domains, the CTD and the His-rich loop are potential regulatory components. They are both zinc-binding domains located on the cytoplasmic membrane surface, probably serving as zinc sensors to receive the cytoplasmic zinc signal and responding with conformational changes, which can be transmitted to the remote transport site to enable a tuneable or even gated zinc transport activity.

The C-terminal cytoplasmic domain (CTD) in YiiP is the zinc-sensing module with a high-affinity zinc binding site. This TMD-CTD modular architecture has emerged as a common structural theme for diverse divalent cation transporters.[52,53] While the clustering of negative charges surrounding metal site C (SC) provides a favourable electrostatic environment for binuclear zinc binding to the CTD interface, zinc release from SC would produce unbalanced electronegativity at the CTD-CTD interface, causing two juxtaposed CTDs to swing apart by electrostatic repulsion in a hinge-like motion pivoting around the (Lys77–Asp207)$_2$ interlock. As the CTD portion of the YiiP structure is essentially identical to the structure of a soluble fragment of a CDF homologue, CzrB,[54] in either the presence or the absence of zinc binding, the zinc-induced CTD conformational changes may be largely restricted to *en bloc* movements.[23] In the transmembrane region, the TM3-TM6 pair embraces TM5 at one corner of the rectangle-shaped four-helix bundle while TM2 is anchored in the four-helix bundle only. Since the charge-interlock at the CTD-TMD juncture stabilizes the orientation of the TM3-TM6 pair, the inter-CTD motion may be transmitted through the charge-interlock to the rigid helix shafts of the TM3-TM6 pair, causing a rotational motion in TM5 as described above (Figure 22.7).[23] The TM5 motion with respect to TM2 and other surrounding TMs may ultimately lead to a transient breaching of the intercavity seal and distortion of the tetrahedral coordination geometry of the transport site, thereby switching on zinc binding from the intracellular cavity and its subsequent release into the extracellular cavity. The trigger of this sequence of protein conformational changes is the zinc binding to SC that stabilizes the CTD-CTD dimer association. By comparison with the transport site, SC is highly constrained through extensive outer shell interactions (Figure 22.4). Each of the outer-shell residues can potentially accept additional hydrogen bonds from neighbouring protein donors or water molecules to further expand the interactions throughout the entire CTD-CTD interface, resulting in an allosteric pathway that propagates zinc-mediated local conformational change to the transport site in the TMD. As such, zinc binding to SC may trigger an autoregulatory mechanism by which the *en bloc* movements of CTDs convert increasing cytoplasmic zinc concentrations to elevated efflux activities of YiiP.[23] A single R325W mutation in human ZnT8 has been

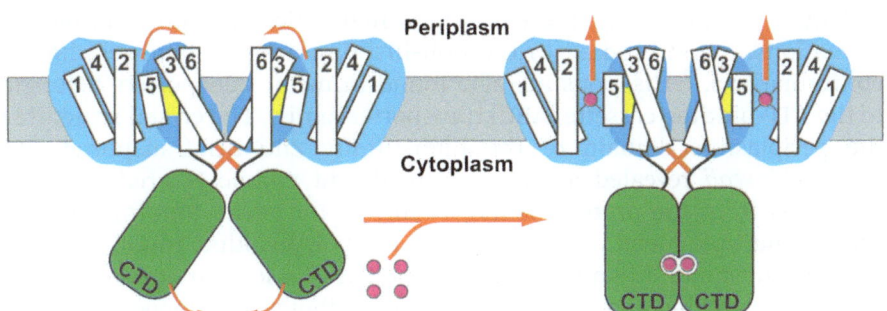

Figure 22.7 Schematic drawing of allosteric regulation of the transport site. Cytosolic zinc binding to site C triggers an inter-CTD motion, propagating into TMD where the tetrahedral transport-site is switched on and off by a shifting of the TM2-TM5 orientation. The red cross at the TMD-CTD juncture represents the conserved charge-interlock.

linked to the risk of type 2 diabetes.[55] Homology modelling of ZnT8 localizes the R325W mutation to the CTD interface. This mutation is unlikely to have a direct effect on the distant transport site. Rather, R325W may impair the allosteric pathway that transmits the activation signal of cytoplasmic zinc binding to the transport site in ZnT8.

Intriguingly, the CTD adopts a metallochaperone-like fold despite a lack of sequence homology.[56] The apparent structural similarity may have arisen independently through convergent evolution at the structural level. Metallochaperones are cytoplasmic metal carrier proteins that deliver metal ions to various protein targets, including the metallochaperone-like domains of metal-transporting P-type ATPases.[57] Metal transfer is mediated by the docking of a metal-loaded metallochaperone protein to its targeted acceptor, followed by a swapping of metal coordination at the docking interface.[27] While the metallochaperone module has been the focus of extensive studies of intracellular metal trafficking, its functional role in metal transport is still poorly understood. The metallochaperone-like CTD structure raises the possibility that CTD may act as a tethered metallochaperone that binds Zn^{2+} and transfers it to the transport sites. A model of metallochaperone-YiiP interactions is proposed on the basis of the crystal structures of YiiP and the metallochaperone Hah1 in a homodimer form. Superposition of a Hah1 monomer and a CTD allows placement of a second Hah1 structure onto a putative docking position. The docked metallochaperone fits snugly between TMD and CDT with a putative Zn^{2+} transfer site positioned right at the entrance to the intracellular cavity. In this model, a zinc-loaded metallochaperone would directly deliver its cargo into the intracellular cavity to initiate zinc transport.

The second potential cytoplasmic zinc sensing module is a histidine-rich loop between TM4 and TM5 in many eukaryotic CDF homologues. For example, the His-loop in *Arabidopsis thaliana* AtMTP1 consists of 81 residues including 25 histidine residues. Deletion of this His-loop seems to increase vacuolar zinc accumulation attributed to an enhanced AtMTP1 transport activity.[58] Biochemically, the His-loop is highly flexible with multiple conformations. Metal binding to the poly-histidine motifs stabilizes the loop conformations.[59] His-loops are also found in metal transporters belonging to the ATP-binding cassette (ABC) transporter family. The crystal structure of the periplasmic domain of the high-affinity ZnuABC transporter from *Escherichia coli* revealed a highly charged and mobile His-rich loop that protrudes from the protein in the vicinity of the metal binding site.[60] Although multiple metal ions can bind to the poly-histidine motif, only Zn^{2+} binding substantially stabilizes ZnuA. The His-rich loop in ZnuA may play a role in zinc sensing and acquisition, and contribute to the recognition of the transport module ZnuB, which is responsible for subsequent Zn^{2+} transport. At present, structural information on the His-loop is entirely lacking for any CDF protein. TM4 and TM5 of YiiP are connected by a tight peptide turn. Despite the missing His-loop, the positioning of the TM4-TM5 turn in the YiiP crystal structure suggests that the His-loop is localized to the opening of the intracellular cavity. Given the loop conformational flexibility, the poly-histidine binding motifs may extend into the cytoplasm to capture Zn^{2+} and increase the local zinc concentration *via* their multiple zinc binding sites. Alternatively, the flexible loop may enter the intracellular cavity, establishing a potential interaction with the transport site in the inward-facing conformation. As such, the His-rich loop may facilitate zinc delivery to the transport site. A direct interaction between the histidine binding motif and the transport site would bring in additional coordination residues to the tetrahedral site, broadening the metal selectivity to octahedral coordination. Therefore, residues within the cytoplasmic His-loop may contribute to metal selectivity by interaction with residues near the transport site. It has been shown that mutations in the His-loop can change metal selectivity in yeast and plant CDF homologues as indicated by a selective loss of transport for some metals and increased activity for others under various metal-selective growth conditions.[61] The structural basis underlying the potential regulatory role of the His-loop and the molecular mechanism of its actions remain to be elucidated.

22.8   Sensing, Buffering and Zinc Signalling

Zinc regulates transcription of genes through metal-responsive transcription factor-1 (MTF1), which is conserved through evolution and is present in diverse animal taxa including insects and fish.[62,63] This protein has about 590 amino acids including six zinc finger domains, two of which bind zinc with lower affinity than the others. If there is a rise in the cytosolic free zinc(II) concentration, the low-affinity zinc binding sites will be filled enabling MTF1 to associate with its cognate DNA motif (5'-TGCRCNC-3') known as a Metal Response Element (MRE) and thereby regulating transcription. There are other mechanisms, such as phosphorylation, that modulate MTF1 activity,[62] but the zinc fingers are sufficient in regulating DNA binding in response to zinc because a chimeric construct consisting of the DNA binding domain from MTF1 and the transactivation domain of Gal4 mediates Zn-dependent transcription when transfected into cells.[64] Several metals, including Cd, Hg, Ag and Cu can induce MTF1-mediated gene expression, but zinc appears to be the only metal activating the protein to any significant extent. Thus the effects of other metals on MTF1 activity must be either by alternative mechanisms, such as phosphorylation, or by increasing the cytosolic free zinc(II) concentration. The former could be achieved through activation of kinases or inhibition of protein phosphatases and the latter by displacement of zinc from proteins. MTF1 was first identified in mouse as the transcription factor responsible for induction of metallothionein (MT) expression in response to metal exposure.[65] MT is a family of metal binding proteins that can protect cells against metal insult and function as a redox switch to release Zn^{2+} ions for zinc signalling events.[66,‡]

---

‡Chapter 21 provides a discussion of the redox biochemistry of metallothioneins.

There have been several searches for MTF1 target genes and the list is now considerably expanded.[67–71] Using microarray technology with a multifactorial experimental design, including RNAi knockdown of Mtf1 and zinc treatments in zebrafish ZF4 cells, it was shown that regulation of over 1000 genes was Mtf1 dependent.[69] However, sequence analysis of these genes showed that only 43 of them contained MRE in configurations and locations compatible with Mtf1 responsiveness and it was concluded that remaining genes were likely induced as downstream ripples of the signalling cascade. Of the 43 putative Mtf1 targets, 19 were genes involved in development. There was also a staggering over-representation of transcription factors, which made up almost half (48%) of the identified genes. The results strongly suggest that MTF1 has roles beyond stress responses and that genomic zinc signalling *via* MTF1 is of particular importance during embryonic development. These findings are in keeping with earlier findings that MTF1 knock-out mice die during development from failed organogenesis while excision of MTF1 in liver and bone marrow with Cre recombinase after birth is non-lethal but causes leukopenia and sensitivity to metal stress.[72] Specific functions of MTF1 during organogenesis open up another hypothetical mechanism of metal toxicity. Exposure of embryos to metals would likely cause inappropriate activation of MTF1 and it is therefore tempting to speculate that metal toxicity during embryogenesis involves disruption of zinc signalling through MTF1.

Zinc activation of MTF1 is also of importance in the defence against free radical stress. MT is an important antioxidant and it reduces free radicals while metal thiolate bonds are oxidized and zinc is released.[73,74] This results in a transient increase in the free zinc concentration of the cell, activation of MTF1 and consequential expression of several key antioxidant genes.

The existence of non-genomic zinc signalling was proposed by R. J. P Williams at the University of Oxford, who correctly predicted that zinc may interact with calcium metabolism and act on cell signalling by inhibition of proteins.[75] Until relatively recently, it was difficult to separate a biological function of an elevated free zinc concentration from a pharmacological response. However, with the demonstration of sudden changes in cytosolic free zinc concentrations in the absence of zinc treatment, and consequential changes in cell signalling pathways, the involvement of free zinc in various signalling pathways is now evident. One such pathway is the epithelial-to-mesenchyme transition (EMT), which requires the presence of Zip6 in the gastrula organizer cells of the zebrafish embryo.[76] During EMT, cells downregulate expression of the cell-cell adhesion protein, E-cadherin (CDH1), allowing them to migrate into new positions. The involvement of zinc and ZIP6 (or its orthologues) in EMT of cells is evolutionarily conserved between fish, human and insects. EMT is pathologically activated during progression and metastasis of cancers; and human ZIP6 was originally discovered as one of the most up-regulated genes in oestrogen-dependent breast cancers.[77] EMT is initiated by STAT3-mediated transcription of ZIP6, which when present at the plasma membrane allows Zn^{2+} influx leading to a phosphorylation cascade starting with AKT, which in turn phosphorylates GSK3β.[78] The phosphorylated GSK3β phosphorylates the transcriptional repressor, SNAIL, causing it to be exported from the nucleus and degraded. In the absence of Zn^{2+} and ZIP6, un-phosphorylated SNAIL binds to the promoter of and represses expression of *CDH1*, which codes for the cell-cell adhesion protein E-cadherin. Thus, ZIP6 mediated Zn^{2+} signalling results in reduced E-cadherin expression, which allows the cells to migrate and invade tissues.

Zinc regulates the exit from meiosis in mouse oocytes and chelation of zinc blocks meiosis past telophase I.[79] Zinc is also involved in regulation of cell proliferation and in mammalian somatic cells an increase in cytosolic $[Zn^{2+}]$ appears to be required for activation of proliferative protein tyrosine kinases, such as v-src sarcoma (Schmidt-Ruppin A-2) viral oncogene homologue (avian) (SrcSRC), epidermal growth factor receptor (EGFR), insulin receptor substrate 1 and 2 (IRS1/2) and the insulin-like growth factor 1 receptor (IGF1R).[43] Zinc is probably not directly activating these kinases, but rather acts through inhibition of specific protein tyrosine phosphatases, resulting in an increased phosphorylation status of protein tyrosine kinases.[80] In some human cancer cell types that rely on protein tyrosine kinase pathways for aggressive growth, the zinc signal for activation seems to come from ZIP7, which is located in the ER and releases Zn^{2+} into the cytosol upon phosphorylation by CK2 (see Section 22.7 above). Release of Zn^{2+} from the ER is also a part of cell signalling in mast cells during the response to antigens. Cross-linking of the high-affinity immunoglobin E receptor (FcεR) triggers release of Zn^{2+} into the cytosol from the ER and perinuclear area within minutes of the stimulus.[81] The wave of Zn^{2+} was dependent on a prior Ca^{2+} release and activation of mitogen-activated protein kinase (MAPK). Again, phosphatases were indicated as the targets of Zn^{2+} signalling, resulting in phosphorylation of ERK1/2 and JNK1/2. Impairment of the immune system is one of the first symptoms of zinc deficiency and zinc signals have been observed also in monocytes and dendritic cells after stimulation with lipopolysaccharide and in T-cells after treatment with phorbol esters.[82] Zinc signals have now been shown to be involved in cytokine production in monocytes, maturation in dendritic cells, degranulation of mast cells and apoptosis in lymphocytes.

There is strong evidence for involvement of zinc signalling in bone and cartilage formation. Zinc deficiency suppresses matrix mineralization and delays osteogenic activity in mouse osteoblastic MC3T3-E1 cells through down-regulation of RUNX2 (runt-related transcription factor 2).[83] Furthermore, knock-out of any of several zinc transporters in mice, including *Znt1*, *Znt5*, *Zip4*, *Zip13* and *Zip14* results in skeletal deformities.[84] *Zip13* −/− mice have a phenotype typified by defects in bone, teeth and connective tissue and these malformations are linked to changes in the BMP/TGF-β signalling pathway.[85] Mice deficient in the ZIP14 protein (*Zip14* −/−) have retarded growth, accompanying the shortened proliferative zone and the extended hypertrophic zone in their growth plates.[86] ZIP14 is present in the plasma membrane and is

expressed in the growth zone of bone. Zinc influx through ZIP14 in proliferative chondrocytes inhibits phosphodiesterase (PDE). This inhibition reduces cAMP breakdown, aiding parathyroid hormone-related peptide 1
receptor (PTH1R) signalling upon parathyroid hormone-related peptide
(PTHrP) binding. PTH1R signalling is critical for the inhibition of premature
hypertrophic differentiation of proliferative chondrocytes during the endo-
chondral ossification process. In *Zip14* $-/-$ mice, decreased cAMP accelerates
the conversion of proliferative chondrocytes to hypertrophic chondrocytes resulting in stunted growth. Taken together an abundance of data suggest that
zinc deficiency, caused by reduced zinc uptake or disruption of zinc transporters, leads to dysregulation of mineralizing cells in a process that involves
effects on chondrocyte and osteoblast differentiation.

The growth-promoting effect of zinc is well established as illustrated by the
retardation in growth during zinc deficiency.[87] Growth hormone release from
the anterior pituitary is under neuroendocrine control of growth hormone-
releasing hormone, which is produced in the hypothalamus. *Zip14* $-/-$ mice
also show impairment in this signalling pathway, through reduced cAMP
concentrations, resulting in decreased GH release and consequential systemic
growth retardation.[86] Another mechanism behind the growth-promoting effect
of zinc is that zinc has insulin mimetic properties and stimulates cellular
glucose uptake.[88–90] Zinc causes an increase in phosphorylation of the insulin
receptor through inhibition of PTP1B, leading to activation of downstream
signalling pathways and recruitment of insulin-responsive glucose transporter
4 to the plasma membrane.[80,89] Thus, there is mounting evidence involving
diverse pathways that Zn^{2+} acts as an intracellular signalling ion, similarly to
Ca^{2+}. The principal difference between Zn^{2+} and Ca^{2+} signalling is that
whereas Ca^{2+} signals are in the micromolar range Zn^{2+} signals operate at
picomolar to low nanomolar concentrations.

Zinc is accumulated in specific areas of the mammalian brain, such as
the mossy fibres of the CA3 area of the hippocampus.[91] Zinc is specifically
found in synaptic vesicles of certain glutamatergic neurons and is released
together with glutamate.[92] In mammals, zinc is loaded into synaptic vesicles of
Zn-enriched neurons by the zinc transporter ZNT3 (SLC30A3).[94] Synaptically
released Zn^{2+} modulates the activity of *N*-methyl-D-aspartate (NMDA) and γ-
aminobutyric acid type A (GABAA) receptors. In addition, the zinc receptor,
GRP39, is expressed in "zincergic" neurons and may respond to presynaptic
zinc release.[93] Thus, zinc influences both excitatory and inhibitory synapses in
the brain and it could be argued that zinc is a neurotransmitter in its own right.

22.9   Cytosolic Handling

Because of its ability to influence the activities of a great number of molecules in the cell, the cytosolic Zn^{2+} concentration has to be kept very low.
How low has been a matter of a long debate that really has been limited to
extrapolations following a series of assumptions. With the advent of specific
fluorescent probes for zinc it is now possible to measure the concentration

of free zinc(II) within a cell and such analyses have shown that (1) the free zinc(II) concentration fluctuates during different phases of the cell cycle, (2) the resting zinc(II) concentration in eukaryotic cells is probably in the pM range and (3) during instances of zinc signalling the zinc(II) concentration may reach mid nM levels.[95–97] Although resting zinc(II) concentrations likely vary between cell types, these measurements fit well with the set-point for activation of the cell's own zinc sensor, MTF1, which has an activation set-point in the low nM range.[98]

MT and glutathione are major zinc-binding molecules in the cytosol. MT is a family of cysteine-rich proteins that can bind up to seven zinc atoms and when isolated from tissues it is saturated with zinc, or zinc in combination with other metals. Recent detailed analysis has shown that the different zinc binding sites differ by four orders of magnitude in their affinities for zinc and that unsaturated MT with up to three available zinc binding sites exists in the cell.[74]

The classic view of zinc handling by the cell has been that zinc first binds to glutathione and that an overload of the glutathione pool will activate MTF1, which will then stimulate *de novo* synthesis of ZNT1 and apo-metallothionein (thionein), which will then increase the capacity for zinc extrusion and sequester the zinc excess. We now know that because MT in the cell is not normally saturated with metal, it can contribute to the initial buffering of zinc entering the cell.[74,95] However, when experimental data of labile zinc concentrations in the cytosol of cells exposed to zinc were fitted to quantitative mathematical models, it was clear that the buffering capacity of MT and glutathione could not account for observed data.[99] Instead, the model that best fit the data was the rapid translocation of the entering zinc to a "deep store" by a vehicle that also had buffering capacity before the zinc reappeared in the cytosol at a later stage. This vehicle was termed a "muffler" to distinguish from a pure buffer and the "deep store" was predicted to perhaps be one or several organelles. This mathematical model of cellular zinc kinetics would explain observations made in Tamoxifen-resistant breast cancer cells (TamR).[43,100] Adding exogenous zinc to these cells causes a massive activation of growth factor receptors, such as ErB, IGF-1R, EGFR and c-SRC, through Zn-stimulated tyrosine phosphorylation, followed by increased growth and invasive behaviour. Interestingly, in Zn-exposed cells where ZIP7 was silenced by siRNA, there was no free Zn(II) in the cytosol, no growth factor receptor activation and no effect on growth or invasiveness. The conclusion may be that before zinc has effects on proteins in the cytosol it has to emerge from the ER and it must get there within seconds or minutes following zinc treatment of the cells. Thus, zinc that enters the cell is muffled, perhaps by MT in combination with glutathione, and then immediately moved into organelles, such as the ER or vesicles, before it may re-enter the cytosol through ZIP7. With continuous zinc exposure MTF1-mediated gene expression is activated, resulting in increased zinc muffling by MT and export by ZNT1, but as opposed to effects on cell signalling, which happen within minutes, gene expression responses take hours to be manifested.

The fraction of total tissue zinc bound to MT varies enormously amongst tissues, zinc content and organismal species, and may be as little as 6% in some tissues or as much as 74% in the liver of female squirrelfish and soldierfish.[101,102] In females of these species of fish, zinc is stored in copious amounts in the liver (up to 70 mmoles kg^{-1}) bound to MT before it is translocated to the ovaries where zinc is taken up by the developing oocytes.[103] It appears that in this family of fish extraordinarily high amounts of zinc are required for viability of embryos. In addition to MT, small molecules, such as glutathione, cysteine and histidine, are probably important zinc ligands in cells but the quantitative roles of these zinc ligands is not well researched.

References

1. A. Cvetkovic, A. L. Menon, M. P. Thorgersen, J. W. Scott, F. L. Poole, 2nd, F. E. Jenney, Jr., W. A. Lancaster, J. L. Praissman, S. Shanmukh, B. J. Vaccaro, S. A. Trauger, E. Kalisiak, J. V. Apon, G. Siuzdak, S. M. Yannone, J. A. Tainer and M. W. Adams, *Nature*, 2010, **466**, 779–782.
2. C. Andreini, L. Banci, I. Bertini and A. Rosato, *J. Proteome Res.*, 2005, **5**, 196–201.
3. R. D. Palmiter and S. D. Findley, *EMBO J.*, 1995, **14**, 639–649.
4. D. H. Nies and S. Silver, *J. Ind. Microbiol.*, 1995, **14**, 186–199.
5. A. Fukunaka, T. Suzuki, Y. Kurokawa, T. Yamazaki, N. Fujiwara, K. Ishihara, H. Migaki, K. Okumura, S. Masuda, Y. Yamaguchi-Iwai, M. Nagao and T. Kambe, *J. Biol. Chem.*, 2009, **284**, 30798–30806.
6. R. A. Valentine, K. A. Jackson, G. R. Christie, J. C. Mathers, P. M. Taylor and D. Ford, *J. Biol. Chem.*, 2007, **282**, 14389–14393.
7. D. L. C. Sim, W. M. Yeo and V. T. K. Chow, *Int. J. Biochem. Cell Biol.*, 2002, **34**, 487–504.
8. T. Kambe, *Biosci. Biotechnol. Biochem.*, 2011, **75**, 1036–1043.
9. G. P. Feeney, D. Zheng, P. Kille and C. Hogstrand, *Biochim. Biophys. Acta*, 2005, **1732**, 88–95.
10. N. Grotz, T. Fox, E. Connolly, W. Park, M. L. Guerinot and D. Eide, *Proc. Natl Acad. Sci. USA*, 1998, **95**, 7220–7224.
11. K. M. Taylor and R. I. Nicholson, *Biochim. Biophys. Acta*, 2003, **1611**, 16–30.
12. C. Rensing, M. Ghosh and B. P. Rosen, *J. Bacteriol.*, 1999, **181**, 5891–5897.
13. C. Rensing, B. Mitra and B. P. Rosen, *Proc. Natl Acad. Sci. USA*, 1997, **94**, 14326–14331.
14. S. A. Sinclair and U. Krämer, *Biochim. Biophys. Acta*, 2012, **1823**, 1553–1567.
15. J. Fukuda and K. Kawa, *Science*, 1977, **196**, 309–311.
16. S. Kury, B. Dreno, S. Bezieau, S. Giraudet, M. Kharfi, R. Kamoun and J. P. Moisan, *Nat. Genet.*, 2002, **31**, 239–240.

17. J. Dufner-Beattie, B. P. Weaver, J. Geiser, M. Bilgen, M. Larson, W. H. Xu and G. K. Andrews, *Hum. Mol. Genet.*, 2007, **16**, 1391–1399.
18. J. Geiser, K. J. T. Venken, R. C. De Lisle and G. K. Andrews, *PloS Genet.*, 2012, **8**, e1002766.
19. R. J. McMahon and R. J. Cousins, *FASEB J.*, 1997, **11**, 81–81.
20. J. Dufner-Beattie, Y. M. Kuo, J. Gitschier and G. K. Andrews, *J. Biol. Chem.*, 2004, **279**, 49082–49090.
21. D. Zheng, G. P. Feeney, R. D. Handy, C. Hogstrand and P. Kille, *Metallomics*, 2014, **6**, 154–165.
22. K. M. Taylor, S. Hiscox, R. I. Nicholson, C. Hogstrand and P. Kille, *Sci. Signal.*, 2012, **5**, ra11.
23. M. Lu, J. Chai and D. Fu, *Nat. Struct. Mol. Biol.*, 2009, **16**, 1063–1067.
24. Y. Chao and D. Fu, *J. Biol. Chem.*, 2004, **279**, 12043–12050.
25. J. Frausto da Silva and R. Williams, *The Biological Chemistry of the Elements: The Inorganic Chemistry of Life*, Oxford University Press, 2001.
26. M. Lu and D. Fu, *Science*, 2007, **317**, 1746–1748.
27. A. K. Wernimont, D. L. Huffman, A. L. Lamb, T. V. O'Halloran and A. C. Rosenzweig, *Nat. Struct. Biol.*, 2000, **7**, 766–771.
28. G. Grass, M. Otto, B. Fricke, C. J. Haney, C. Rensing, D. H. Nies and D. Munkelt, *Arch. Microbiol.*, 2005, **183**, 9–18.
29. E. Hoch, W. Lin, J. Chai, M. Hershfinkel, D. Fu and I. Sekler, *Proc. Natl Acad. Sci. USA*, 2012, **109**, 7202–7207.
30. Y. Wei and D. Fu, *J. Biol. Chem.*, 2005, **280**, 33716–33724.
31. D. H. Nies, *Science*, 2007, **317**, 1695–1696.
32. Y. P. Pang, K. Xu, J. E. Yazal and F. G. Prendergas, *Protein Sci.*, 2000, **9**, 1857–1865.
33. B. Tamames, S. F. Sousa, J. Tamames, P. A. Fernandes and M. J. Ramos, *Proteins*, 2007, **69**, 466–475.
34. S. F. Sousa, P. A. Fernandes and M. J. Ramos, *J. Am. Chem. Soc.*, 2007, **129**, 1378–1385.
35. B. Montanini, D. Blaudez, S. Jeandroz, D. Sanders and M. Chalot, *BMC Genomics*, 2007, **8**, 107.
36. D. H. Nies, *FEMS Microbiol. Rev.*, 2003, **27**, 313–339.
37. H. Lin, A. Kumanovics, J. M. Nelson, D. E. Warner, D. M. Ward and J. Kaplan, *J. Biol. Chem.*, 2008, **283**, 33865–33873.
38. N. Coudray, S. Valvo, M. Hu, R. Lasala, C. Kim, M. Vink, M. Zhou, D. Provasi, M. Filizola, J. Tao, J. Fang, P. A. Penczek, I. Ubarretxena-Belandia and D. L. Stokes, *Proc. Natl Acad. Sci. USA*, 2013, **110**, 2140–2145.
39. M. Hattori, Y. Tanaka, R. Ishitani and O. Nureki, *Acta Crystallogr. Sect. F Struct. Biol. Cryst. Communi.*, 2007, **63**, 771–773.
40. E. Ohana, E. Hoch, C. Keasar, T. Kambe, O. Yifrach, M. Hershfinkel and I. Sekler, *J. Biol. Chem.*, 2009, **284**, 17677–17686.
41. G. Xu and M. R. Chance, *Chem. Rev.*, 2007, **107**, 3514–3543.
42. S. D. Maleknia, M. Brenowitz and M. R. Chance, *Anal. Chem.*, 1999, **71**, 3965–3973.

43. K. M. Taylor, S. Hiscox, R. I. Nicholson, C. Hogstrand and P. Kille, *Sci. Signaling*, 2012, 5(210): ra11.

44. B.-H. Bin, T. Fukada, T. Hosaka, S. Yamasaki, W. Ohashi, S. Hojyo, T. Miyai, K. Nishida, S. Yokoyama and T. Hirano, *J. Biol. Chem.*, 2011, **286**, 40255–40265.

45. W. Lin, J. Chai, J. Love and D. Fu, *J. Biol. Chem.*, 2010, **285**, 39013–39020.

46. L. A. Gaither and D. J. Eide, *J. Biol. Chem.*, 2000, **275**, 5560–5564.

47. K. Girijashanker, L. He, M. Soleimani, J. M. Reed, H. Li, Z. W. Liu, B. Wang, T. P. Dalton and D. W. Nebert, *Mol. Pharmacol.*, 2008, **73**, 1413–1423.

48. L. He, K. Girijashanker, T. P. Dalton, J. Reed, H. Li, M. Soleimani and D. W. Nebert, *Mol. Pharmacol.*, 2006, **70**, 171–180.

49. A. D. Qiu and C. Hogstrand, *Biochem. J.*, 2005, **390**, 777–786.

50. D. W. Nebert, M. Galvez-Peralta, E. Ben Hay, H. Li, E. Johansson, C. Yin, B. Wang, L. He and M. Soleimani, *Metallomics*, 2012, **4**, 1218–1225.

51. J. P. Liuzzi, F. Aydemir, H. Nam, M. D. Knutson and R. J. Cousins, *Proc. Natl Acad. Sci. USA*, 2006, **103**, 13612–13617.

52. M. Hattori, Y. Tanaka, S. Fukai, R. Ishitani and O. Nureki, *Nature*, 2007, **448**, 1072–1075.

53. V. V. Lunin, E. Dobrovetsky, G. Khutoreskaya, R. Zhang, A. Joachimiak, D. A. Doyle, A. Bochkarev, M. E. Maguire, A. M. Edwards and C. M. Koth, *Nature*, 2006, **440**, 833–837.

54. V. Cherezov, N. Hofer, D. M. Szebenyi, O. Kolaj, J. G. Wall, R. Gillilan, V. Srinivasan, C. P. Jaroniec and M. Caffrey, *Structure*, 2008, **16**, 1378–1388.

55. R. Sladek, G. Rocheleau, J. Rung, C. Dina, L. Shen, D. Serre, P. Boutin, D. Vincent, A. Belisle, S. Hadjadj, B. Balkau, B. Heude, G. Charpentier, T. J. Hudson, A. Montpetit, A. V. Pshezhetsky, M. Prentki, B. I. Posner, D. J. Balding, D. Meyre, C. Polychronakos and P. Froguel, *Nature*, 2007, **445**, 881–885.

56. A. C. Rosenzweig and T. V. O'Halloran, *Curr. Opin. Chem. Biol.*, 2000, **4**, 140–147.

57. M. Gonzalez-Guerrero and J. M. Arguello, *Proc. Natl Acad. Sci. USA*, 2008, **105**, 5992–5997.

58. M. Kawachi, Y. Kobae, T. Mimura and M. Maeshima, *J. Biol. Chem.*, 2008, **283**, 8374–8383.

59. N. Tanaka, M. Kawachi, T. Fujiwara and M. Maeshima, *FEBS Open Bio*, 2013, **3**, 218–224.

60. B. Wei, A. M. Randich, M. Bhattacharyya-Pakrasi, H. B. Pakrasi and T. J. Smith, *Biochemistry*, 2007, **46**, 8734–8743.

61. M. W. Persans, K. Nieman and D. E. Salt, *Proc. Natl Acad. Sci. USA*, 2001, **98**, 9995–10000.

62. V. Günther, U. Lindert and W. Schaffner, *Biochim. Biophys. Acta Mol. Cell Res.*, 2012, **1823**, 1416–1425.

63. T. Kimura, N. Itoh and G. K. Andrews, *J. Health Sci.*, 2009, **55**, 484–494.

64. P. J. Daniels, D. Bittel, I. V. Smirnova, D. R. Winge and G. K. Andrews, *Nucleic Acids Res.*, 2002, **30**, 3130–3140.
65. G. Westin and W. Schaffner, *EMBO J.*, 1988, **7**, 3763–3770.
66. W. Maret, *Neurochem. Int.*, 1995, **27**, 111–117.
67. C. Gunes, R. Heuchel, O. Georgiev, K. H. Muller, P. Lichtlen, H. Bluthmann, S. Marino, A. Aguzzi and W. Schaffner, *EMBO J.*, 1998, **17**, 2846–2854.
68. P. Lichtlen and W. Schaffner, *Bioessays*, 2001, **23**, 1010–1017.
69. C. Hogstrand, D. Zheng, G. Feeney, P. Cunningham and P. Kille, *Biochem. Soc. Trans.*, 2008, **36**, 1252–1257.
70. P. A. Walker, P. Kille, A. Hurley, N. R. Bury and C. Hogstrand, *Toxicol. Appl. Pharmacol.*, 2008, **230**, 67–77.
71. M. Wang, F. Yang, X. Zhang, H. Zhao, Q. Wang and Y. Pan, *Mamm. Genome*, 2010, **21**, 287–298.
72. Y. Wang, U. Wimmer, P. Lichtlen, D. Inderbitzin, B. Stieger, P. J. Meier, L. Hunziker, T. Stallmach, R. Forrer, T. Rulicke, O. Georgiev and W. Schaffner, *FASEB J.*, 2004, **18**, 1071–1079.
73. W. Maret, *Proc. Natl Acad. Sci. USA*, 1994, **91**, 237–241.
74. A. Krezel and W. Maret, *J. Am. Chem. Soc.*, 2007, **129**, 10911–10921.
75. R. J. Williams, *Endeavour*, 1984, **8**, 65–70.
76. S. Yamashita, C. Miyagi, T. Fukada, N. Kagara, Y. S. Che and T. Hirano, *Nature*, 2004, **429**, 298–302.
77. M. A. Dressman, T. M. Walz, C. Lavedan, L. Barnes, S. Buchholtz, I. Kwon, M. J. Ellis and M. H. Polymeropoulos, *Pharmacogenomics J.*, 2001, **1**, 135–141.
78. C. Hogstrand, P. Kille, M. L. Ackland, S. Hiscox and K. M. Taylor, *Biochem. J.*, 2013, **455**, 229–237.
79. A. M. Kim, S. Vogt, T. V. O'Halloran and T. K. Woodruff, *Nat. Chem. Biol.*, 2010, **6**, 674–681.
80. H. Haase and W. Maret, *Biometals*, 2005, **18**, 333–338.
81. S. Yamasaki, K. Sakata-Sogawa, A. Hasegawa, T. Suzuki, K. Kabu, E. Sato, T. Kurosaki, S. Yamashita, M. Tokunaga, K. Nishida and T. Hirano, *J. Cell Biol.*, 2007, **177**, 637–645.
82. H. Haase and L. Rink, *Annu. Rev. Nutr.*, 2009, **29**, 133–152.
83. I.-S. Kwun, Y.-E. Cho, R.-A. R. Lomeda, H.-I. Shin, J.-Y. Choi, Y.-H. Kang and J. H. Beattie, *Bone*, 2010, **46**, 732–741.
84. T. Fukada, S. Hojyo and T. Furuichi, *J. Bone Miner. Metab.*, 2013, **31**, 129–135.
85. T. Fukada, N. Civic, T. Furuichi, S. Shimoda, K. Mishima, H. Higashiyama, Y. Idaira, Y. Asada, H. Kitamura, S. Yamasaki, S. Hojyo, M. Nakayama, O. Ohara, H. Koseki, H. G. dos Santos, L. Bonafe, R. Ha-Vinh, A. Zankl, S. Unger, M. E. Kraenzlin, J. S. Beckmann, I. Saito, C. Rivolta, S. Ikegawa, A. Superti-Furga and T. Hirano, *PLoS One*, 2008, **3**, e3642.
86. S. Hojyo, T. Fukada, S. Shimoda, W. Ohashi, B.-H. Bin, H. Koseki and T. Hirano, *PLoS One*, 2011, **6**(3), e18059.

87. A. S. Prasad, *J. Am. Coll. Nutr.*, 2009, **28**, 257–265.
88. L. Coulston and P. Dandona, *Diabetes*, 1980, **29**, 665–667.
89. X.-h. Tang and N. F. Shay, *J. Nutr.*, 2001, **131**, 1414–1420.
90. V. V. T. Wong, P. M. Nissom, S.-L. Sim, J. H. M. Yeo, S.-H. Chuah and M. G. S. Yap, *Biotechnol. Bioeng.*, 2006, **93**, 553–563.
91. M. Kalinowski, G. Wolf and M. Markefski, *Acta Histochem.*, 1983, **73**, 33–40.
92. S. L. Sensi, P. Paoletti, A. I. Bush and I. Sekler, *Nat. Rev. Neurosci.*, 2009, **10**, 780–791.
93. L. Besser, E. Chorin, I. Sekler, W. F. Silverman, S. Atkin, J. T. Russell and M. Hershfinkel, *J. Neurosci.*, 2009, **29**, 2890–2901.
94. R. D. Palmiter, T. B. Cole, C. J. Quaife and S. D. Findley, *Proc. Natl Acad. Sci. USA*, 1996, **93**, 14934–14939.
95. R. A. Colvin, W. R. Holmes, C. P. Fontaine and W. Maret, *Metallomics*, 2010, **2**, 306–317.
96. Y. Li and W. Maret, *Exp. Cell Res.*, 2009, **315**, 2463–2470.
97. E. A. Bellomo, G. Meur and G. A. Rutter, *J. Biol. Chem.*, 2011, **286**, 25778–25789.
98. J. H. Laity and G. K. Andrews, *Arch. Biochem. Biophys.*, 2007, **463**, 201–210.
99. R. A. Colvin, A. I. Bush, I. Volitakis, C. P. Fontaine, D. Thomas, K. Kikuchi and W. R. Holmes, *Am. J. Physiol. Cell Physiol.*, 2008, **294**, C726–C742.
100. C. Hogstrand, P. Kille, R. I. Nicholson and K. M. Taylor, *Trends Mol. Med.*, 2009, **15**, 101–111.
101. C. Hogstrand and C. Haux, *Mar. Biol.*, 1996, **125**, 23–31.
102. C. Hogstrand, N. J. Gassman, B. Popova, C. M. Wood and P. J. Walsh, *J. Exp. Biol.*, 1996, **199**, 2543–2554.
103. E. D. Thompson, G. D. Mayer, C. N. Glover, T. Capo, P. J. Walsh and C. Hogstrand, *PLoS One*, 2012, 7, e46127.

CHAPTER 23

Cadmium

JEAN-MARC MOULIS,*[a,b,c] JACQUES BOURGUIGNON[d,e,f,g] AND
PATRICE CATTY[a,b,c]

[a] CEA, Institut de Recherches en Technologies et Sciences pour le Vivant,
Laboratoire Chimie et Biologie des Métaux, 17 rue des Martyrs, F-38054,
Grenoble, France; [b] CNRS, UMR5249, F-38054, Grenoble, France;
[c] Université Joseph Fourier-Grenoble I, UMR5249, F-38041, Grenoble,
France; [d] CEA, Institut de Recherches en Technologies et Sciences pour le
Vivant, Laboratoire Physiologie Cellulaire et Végétale, F-38054, Grenoble,
France; [e] CNRS, UMR5168, F-38054, Grenoble, France; [f] Université Joseph
Fourier-Grenoble I, UMR5168, F-38041, Grenoble, France; [g] INRA,
USC1359, F-38054, Grenoble, France
*Email: jean-marc.moulis@cea.fr

23.1 Introduction

The impact of cadmium on health has been considered for some time,
particularly since links between renal dysfunction or lung cancer and environmental exposure to the metal were established.[1–3] Most of the mobilization of cadmium in the environment is due to anthropogenic activities
and, despite attempts at limiting uncontrolled cadmium release, part of that
already released is available for interaction with living species.

23.1.1 Geological Origin

Cadmium occurs in the lithosphere with zinc or, to a lesser extent, with other
metals such as lead and copper.[4] Its ores do not generally occur as pure
crystals. The major exception is as cadmium sulfide in greenockite or

RSC Metallobiology Series No. 2
Binding, Transport and Storage of Metal Ions in Biological Cells
Edited by Wolfgang Maret and Anthony Wedd
© The Royal Society of Chemistry 2014
Published by the Royal Society of Chemistry, www.rsc.org

hawleyite. It is often found at the surface of zinc ores such as sphalerite (ZnS). In more oxidized zones, cadmium may be found in calcite ($CaCO_3$) and in the isostructural cadmium mineral otavite ($CdCO_3$). It also occurs as an impurity in smithsonite ($ZnCO_3$) and hemimorphite ($Zn_4Si_2O_7(OH)_2 \cdot (H_2O)$), which both received the alternate name of calamine. Lead (cerussite, $PbCO_3$, or pyromorphite, $Pb_5(PO_4)_3Cl$) and copper (azurite, $Cu_3(CO_3)_2(OH)_2$, or malachite, $Cu_2(CO_3)(OH)_2$) crystals may contain cadmium. Natural release of cadmium into the environment is due to the weathering of the above rocks, to volcanic activity and, since cadmium interacts with living species, to forest fires.

23.1.2   Circumstances of Discovery

As cadmium was discovered in the second half of the 1810s, the detailed circumstances of its identification are now somewhat muddled.[5] Hermann, a German manufacturer of the healing compound zinc oxide, was oxidizing some calamine ($ZnCO_3$) from Silesia by heating. Once, a yellow compound was produced instead of the whitish zinc oxide. The yellow compound was first proposed to contain harmless iron but then suspected to be orpiment (As_2S_3) by the Medicinal Counselor Roloff. The controversy was tackled in 1817 by the German chemist Friedrich Stromeyer (1776–1835) of the University of Göttingen. He isolated a new metal, to which he gave the name of cadmium, originating from calamine, *i.e.* cadmean earth near Thebes (Greece), a city founded by Cadmus (from Greek Καδμος). The isolation by Stromeyer consisted of successive steps of acid and base dissolution interspersed with sulfide precipitations and oxidation-reductions to separate the then unknown cadmium from zinc and copper.[5] Cadmium may have been independently discovered by other German chemists, Meissner and Karsten. The initial controversy about cadmium between business and public regulatory bodies with the involvement of academia still underlies cadmium studies, now with the additional involvement of biologists and toxicologists.

23.1.3   Historical and Contemporary Uses

The occurrence of cadmium in small proportions in zinc ores makes it an inescapable by-product of the metallurgy of its lighter congener. It may be estimated that ~ 3 kg of cadmium can be recovered per tonne of zinc produced. The world mine production of cadmium has been at *ca.* 20 000 tons *per annum* for several years.

   The first use of cadmium was as pigment in paintings soon after its discovery. The colourful cadmium chalcogenides later spread into plastics, ceramics, glasses and enamels. The rich colour-palette of cadmium compounds can now be safely substituted in conventional paints, but its replacement in materials requiring harsh processing conditions, such as high temperatures and strong light intensities, is less easy. Thus, cadmium organic salts have been used to stabilize polyvinyl chloride polymers.

Cadmium melts at 320 °C, a property used in detection systems of fire sprinklers, and its addition to solder alloys eases brazing. Cadmium also improves the resistance of alloys to fatigue, corrosion and mechanical stress. The limited reactivity of cadmium in alkaline environments and in other harsh conditions was used for most of the twentieth century for coating materials, mainly by electroplating, and to protect copper, brass and other metal pieces.

More recently, the largest use of cadmium (70% or more) has been in nickel-cadmium (nicad) batteries, in which the cadmium cathode oxidizes to the hydroxide upon discharging. These types of batteries were the first to be marketed as rechargeable accumulators, and, despite their replacement by other systems in small size electronics, they remain widely used in heavy duty applications needing large capacitance.

The light-driven generation of hole-electron pairs in some cadmium crystals is used to convert light energy into electricity in solar cells and in many light detectors for imaging purposes. In photovoltaics, cadmium-based thin films are cheap options compared to other materials. Cadmium chalcogenides are now used in a series of nanomaterials, including quantum dots. Independently, cadmium-containing alloys are among the best neutron capturers and they have been used as neutron absorbers in nuclear energy reactors.

The above man-made cadmium-containing compounds have contributed to cadmium dispersal since they have generally not been disposed of properly after use, except for efforts with nicad batteries and rods of nuclear power plants.

23.1.4 Cadmium Coordination Chemistry

23.1.4.1 General Properties of Cadmium

Cadmium can be solubilized in biologically relevant environments, such as water or lipids, after coordination by biologically compatible ligands. This largely depends on its electronic characteristics: cadmium has an outer electronic configuration $4d^{10}5s^2$, with filled shells like zinc and mercury. Only the 5s electrons can be readily exchanged, which leaves Cd^{2+} as the only relevant cation in biology. The standard reduction potential of Cd^{2+} is –0.4 V, whereas that of Zn^{2+} is -0.76 V, making elemental zinc a better reductant than cadmium. The 4d shell of Cd^{2+} is filled, which is the main feature that distinguishes the Group 12 elements from the other transition metals. Incomplete d orbitals can readily be engaged in bonds, but their absence in cadmium (and zinc) does not impair formation of complexes with anions and Lewis bases. Electrons of the filled 4d shell are easily polarized. Neither the electro-neutral cadmium hydroxide nor chalcogenides (formed with S^{2-}, Se^{2-}, Te^{2-}) are soluble in water, but halides display solubilities above 2 M. Oxo salts (nitrates, sulfates, *etc.*) are soluble, which contributes to cadmium mobilization in the environment.

The ionic radius of Cd^{2+} in hexa-coordinated complexes is within 5% of those of silver and calcium ions in similar environments. The similar sizes of cadmium and calcium ions are of biological importance. The ionic radii of hexa-coordinated divalent 3d transition metals are about 20% less than that of Cd^{2+}. This accounts for the preference of Cd^{2+} for higher coordination numbers than Zn^{2+} (which favours tetra-coordination). No crystal field stabilization is expected for Group 12 dications, but significant contributions of the electron-rich d shell to the metal-ligand bonds determine their relative affinities for ligands.[6] In other words, Cd^{2+} and Zn^{2+} can be seen through the lenses of the hard-soft acid-base (HSAB) concept: the softer cadmium ion is expected to more easily combine with soft ligands, such as thiolates and thioethers, than the harder (actually borderline) Zn^{2+}. This is illustrated by the relative affinities of both ions for zinc finger structures: Zn^{2+} is a preferred binder to His_2Cys_2 sites, whereas Cd^{2+} is more tightly bound by Cys_4 sites.

In biological systems, many aspects of the speciation of ions are under thermodynamic control, congruent with good agreement between the above HSAB description and the observed behaviour. But important exceptions exist for ions regulating biological processes. Calcium is a major biological messenger, which dynamically, and often transiently, binds to specific targets to transmit signals. Cadmium interference in calcium regulatory processes changes the kinetics of metal exchange, and this should perturb signal transmission.

23.1.4.2 *Chemical Comparison with Zinc and Mercury*

The chemical differences between cadmium and zinc (ionic radii, favoured coordination numbers, polarizability) do not always allow biological systems to discriminate between the essential (Zn) and the toxic (Cd) elements. This makes chemical mimicry a valid concept to analyse the biological behaviour of cadmium, as illustrated by the only known naturally occurring cadmium active site of *Thalassiosira weissflogii* ζ-carbonic anhydrase[7] (Figure 23.1).

In contrast, the properties of elemental mercury (*e.g.* liquid at biologically relevant temperatures, high volatility) and of its ionic forms (*e.g.* two accessible valence states, propensity to form molecular species) significantly affect the way mercury interacts with living species as compared to zinc and cadmium. Hg^{2+} prefers lower coordination numbers (≤ 4) and it forms organomercuric compounds far more readily than Zn^{2+} and Cd^{2+}. The methylmercury ion is of particular environmental and toxicological importance, whereas organo-zinc and organo-cadmium complexes can be safely neglected in biology.[†]

[†]Chapter 24 provides an extended discussion of the environmental and toxicological significance of mercury.

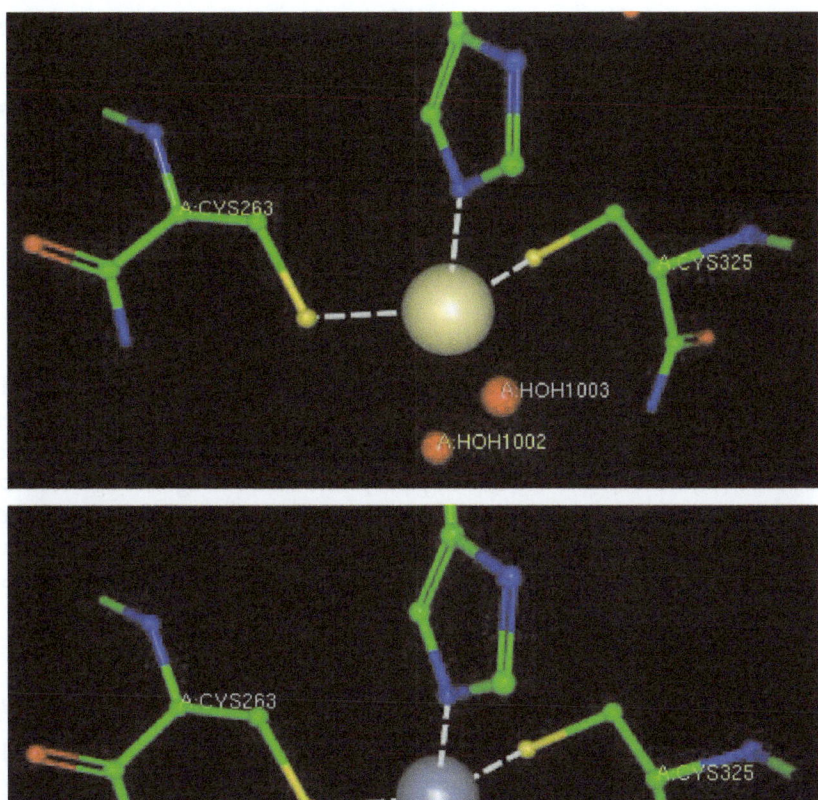

Figure 23.1 Structure of the metal binding site of the ζ carbonic anhydrase from *Thalassiosira weissflogii.*[7] Top: site occupied by Cd^{2+} (PDB ID 3BOB); bottom: Zn^{2+} form (3BOC). The two forms are almost isostructural with two cysteine residues, one histidine and two water molecules in the coordination sphere. The catalytic efficiency (k_{cat}/K_m) for CO_2 hydration of the Cd-form is less than 10-fold lower than that of the Zn-form under similar conditions, both being very high among carbonic anhydrases. The functional replacement of zinc by cadmium at this active site illustrates the adaptability of the phytoplankton to varying but generally low concentrations of metals as found in some oceanic regions.

23.2   General Properties of Cadmium Transporters

As stated above, most chemical forms of cadmium are ionic, which prevents passive diffusion through biological membranes. A variety of transmembrane molecules have been proposed to mediate cadmium cellular uptake, export and intracellular traffic, as detailed below. The main properties of the major families of these molecules are summarized in Table 23.1.

23.3   Cadmium Entry into Organisms

23.3.1   Occurrence in Nature

As a result of its use in various applications (Section 23.1.3), cadmium turnover in the environment occurs through the smelting of other metals, the burning of fossil fuels, the incineration of waste materials, including electronic (e-) waste, and the use of phosphate and sewage sludge fertilizers. For e-waste, some ends up in landfill and some is incinerated, sometimes in places that are remote from the sites of intensive use. This economy of materials and waste is responsible for environmental cadmium dispersal.

23.3.2   Assimilation of Cadmium by Living Species

23.3.2.1   *Passive Biosorption of Cadmium on Bacteria and Yeast*

The first barrier encountered by metal ions is the cell wall of unicellular microorganisms and their interaction is called biosorption,[8] passive biosorption[9] or passive uptake.[10] This process is independent of cellular metabolism and it can occur with dead cells. In contrast, bioaccumulation (also called active uptake or active biosorption) requires metabolic energy to transport metals into the cell. It is not known whether passive biosorption is a prerequisite for the bioaccumulation of metals. Passive biosorption involves interactions between metal ions and functional groups present at the cell surface. Biosorption capacity depends on the chemical composition of the cell wall and therefore varies from one family of microorganisms to another.

A ubiquitous component of all microbial surfaces is a polysaccharide, be it simple (cellulose, chitin, β-glucan or mannan as in yeast cell walls) or conjugated (peptidoglycan in bacteria or lipopolysaccharide as found predominantly in Gram-negative bacteria). Metal ions interact with the sugars through hydroxyl, carboxyl or phosphoryl groups. However, they also bind to other components of cell walls, such as sulfhydryl groups of glycoproteins and lipoproteins, phospholipids of Gram-negative bacteria and teichoic acids present only in Gram-positive bacteria.[11,12] Displacement of ions such as Ca^{2+}, Mg^{2+}, K^+, or H^+ at the surface may induce morphological changes of the cells.

Cadmium biosorption by microorganisms has been studied with non-growing or dead biological materials for bioremediation purposes, but

Table 23.1 Main characteristics of transmembrane transport systems of relevance for cadmium.

Common name/aliases or examples	TMD/TMh	Energy source	Transported molecules (besides Cd^{2+})	Co-transport
TRANSPORT INTO THE CYTOPLASM				
ZIP/SLC39; IRT; ZRT	1/8	2	Zn^{2+}; Fe^{2+}; Mn^{2+}	H^+; HCO_3^- (symport)
NRAMP/SLC11; Smf; MntH; DMT1; DCT1	1/11–12	2	Fe^{2+}; Ni^{2+}; Mn^{2+}; Zn^{2+}; Cu^{2+}; Co^{2+}; *etc.*	H^+ (symport)
T-type Ca channels/Ca_v3	–/>20	2	Ca^{2+}	–
TRANSPORT FROM THE CYTOPLASM				
P-type ATPases	1/6–10	1	Ca^{2+}; Na^+; K^+; H^+; Cu^+; Cu^{2+}; Mn^{2+}; Zn^{2+}; Pb^{2+}; amino-phospholipids	–
RND proteins[a] (Resistance-Nodulation-Division)/AcrABC; CusABC	12 (×3: trimer)	2	Many, often lipophilic, drugs (*e.g.* antibiotics); Zn^{2+}; Co^{2+}; $Cu^{+/2+}$	H^+ (antiport)
CDF (Cation Diffusion Facilitator)/SLC30; ZnT; MTP; Yiip; ZitB	6 (×2)	2	Zn^{2+}; Co^{2+}; Fe^{2+}; Ni^{2+}; Mn^{2+}	H^+; K^+ (antiport)
TRANSPORT TO AND FROM THE CYTOPLASM				
ABC	2–3/10–20	1	Sugars; lipids; pigments; xenobiotics; antibiotics; metal ions	–

TMD: number of transmembrane domains; TMh: number of transmembrane helices.
Energy source: 1: primary = the transporter generates its own energy source, 2: secondary = the transporter depends on an external energy source, which may be provided by the dissipation of the ionic gradient formed by the co-substrate.
[a] In association with an outer membrane factor and a membrane fusion protein.

studies with growing organisms[10] are scant. Yet the latter provide information not only on the capacity and mechanisms of cadmium fixation at the surface but also on putative bioaccumulation with insight into metal resistance. Soil microorganisms resistant to cadmium such as *Pseudomonas* strain I1a fix cadmium outside the cell as bound to the exopolysaccharide layer. Alternatively, *Pseudomonas* strain H1 and *Bacillus sp.* strain H9 accumulate cadmium inside the cell, and they display a 10-fold higher resistance than *Pseudomonas* strain I1a.[13] Comprehensive analytical studies showed that cadmium triggered significant cell morphology changes in the Gram-positive bacterium *Streptomyces zincresistens*.[14] Cadmium distribution at the cell surface, within the cell wall and inside the cell was 10%, 60% and 30%, respectively, after eight days of exposure. Infrared spectroscopy identified $C = O$, NH and CN groups as involved in metal binding. For the Gram-negative bacterium *Acidiphilum symbioticum H8*, the amide, carboxyl and hydroxyl groups of the cell wall accounted for, respectively, 42, 26 and 18% of the metal binding.[15] Growing cells of *S. cerevisiae* accumulated 3–4-fold more cadmium than dry or wet pressed cells.[16] The difference was probably due to the accumulation of cadmium as cadmium phosphate in the vacuole.

23.3.2.2 *Passive Biosorption of Cadmium by other Cells*

The plant cell wall is one of the main locations where metal cations accumulate since it is rich in compounds that are capable of binding divalent and trivalent metal cations *via* functional groups such as -COOH, -OH and -SH. Homogalacturonans, one of the four major polysaccharide domains of pectin, is the main domain responsible for this binding because of the high density of carboxyl groups it provides.[17] Cd^{2+} may substitute Ca^{2+} and it may cross-link these polymers of low methyl-esterified pectin present in the cell wall. Cd^{2+} may also provoke a remodelling of the methyl-esterification pattern of these homogalacturonans by increasing pectin methyl-esterase and peroxidase activities leading to a higher accumulation of metals in the cell wall.[17,18] This could be considered as a defence mechanism adopted by plants to limit uptake of trace metals, since this modification leads to thickening of the cell wall, which then may act as a physical barrier, lowering penetration into the plant cell. Novel cysteine-rich peptides from *Digitaria ciliaris* and *Oryza sativae* localize at the plasma membrane where their increased levels enhance tolerance to cadmium, probably by preventing its entry and accumulation in the cells.[19]

In mammals only a fraction of ingested cadmium is retained by the body. Retention by the mucosal cells is an initial step of the absorption process.[20] The trapped proportion of ingested cadmium depends on the administered dose and, to a lesser extent, on the composition of the diet. Thus, interaction of cadmium salts with the outside of mammalian cells plays a role in the uptake reaction.

23.3.2.3 Active Uptake

23.3.2.3.1 Microorganisms

23.3.2.3.1.1 Bacterial NRAMP. In bacteria, cadmium uptake has been studied in *Staphylococcus aureus*[21] and *Bacillus subtilis*[22] by direct measurement of accumulation of radioactive metal tracer and competition with the endogenous transport of manganese. *S. aureus* strain 17810S accumulates 17.1 nmoles of ^{115m}Cd mg^{-1} of dry mass when exposed to 10 μM ^{115m}CdCl$_2$ for 40 min ($K_m = 5.4$ μM). The amount of accumulated cadmium is reduced by a factor of four when the experiment is performed in the presence of MnCl$_2$ (100 μM) but is insensitive to the presence of the same concentration of ZnCl$_2$. Cadmium uptake is inhibited by the protonophore carbonyl cyanide *m*-chlorophenylhydrazone (CCCP), in agreement with the involvement of NRAMP (Table 23.1).[21] In *B. subtilis*, cadmium transport ($K_m = 1.77 \pm 0.4$ μM; $V_{max} = 1.53 \pm 0.11$ μmoles of ^{109}Cd (g of dry mass.min)$^{-1}$) is CCCP-sensitive.[22] Bacterial NRAMP transporters, also called MntH for proton-dependent manganese transporter, originated early in evolution and are present in both Gram-positive and -negative bacteria and in some archaea. But *Cupriavidus metallidurans CH34*, which tolerates high concentrations of heavy metals, does not contain any NRAMP transporters, suggesting that the absence of MntH explains the resistance to cadmium.[23] MntH in whole *E. coli* cells transports cadmium ($K_m = 2.6 \pm 0.5$ μM; $V_{max} = 13.4 \pm 2.1$ nmol.min^{-1} at pH 6.5). These values increase by a factor of 3 at pH 7.5. Cadmium transport is associated with an intracellular acidification as measured by a pH-dependent ratiometric probe pHluorin. From mutational analysis, Asp34 in TM1 is essential for coupling H$^+$ and Cd^{2+} uptake with a possible direct role in metal binding, whereas His211 in TM6 is required for proton binding to the transporter.[24] TM7 and 11 also participate to the mechanism of transport.

23.3.2.3.1.2 Bacterial ZIP. In the *E. coli* strain K-12, cadmium uptake is saturable and follows Michaelis–Menten kinetics ($K_m = 2.10 \pm 0.27$ μM; $V_{max} = 0.830 \pm 0.095$ μmol (g of dry mass.min)$^{-1}$). Cadmium uptake is diminished by zinc and inhibited by the energy uncoupler CCCP.[25] The ability of ZupT (ygiE) to transport cadmium was demonstrated in overexpressing *E. coli* ECA580 cells.[26] This strain lacks MntH and ZupT, the cadmium binding protein YodA and the cadmium exporters ZntA and RcnA. Complementation by ZupT increased the sensitivity to cadmium by a factor of 5 and the amount of accumulated cadmium by a factor of about 4.[26] Purified ZIPB, the ZupT homologue from *Bordetella bronchiseptica*, also transfers Cd^{2+} into proteoliposomes.[27] Glutamate and histidine residues belonging to three transmembrane domains and one loop of ZupT are mandatory for transport.[28]

23.3.2.3.1.3 Other Bacterial Cadmium Uptake Systems. In the *Cupriavidus metallidurans* metal-sensitive strain AE104, a derivative of the CH34 strain which lacks the two resistance-carrying plasmids pMOL28 and pMOL30, cadmium uptake follows Michaelis–Menten kinetics ($K_m = 135 \pm 22$ µM; $V_{max} = 3.4 \pm 1.2$ µmoles (g of dry mass.min)$^{-1}$). On the basis of competition experiments, cadmium was proposed to enter the cell *via* the magnesium uptake system.[29]

In *Lactobacillus plantarum*, cadmium accumulation in manganese-starved cells is characterized by $K_m = 0.44 \pm 0.07$ µM and $V_{max} = 3.57 \pm 0.13$ µmol of Cd (g of dry mass.min)$^{-1}$. Cadmium uptake is strongly reduced in manganese-sufficient cells and its energy requirement is witnessed by its inhibition by uncouplers.[30] MntA, the protein responsible for this high-affinity cadmium uptake, belongs to the P-type ATPase family,[31] whose members are generally involved in metal efflux (Table 23.1). A second, low-affinity cadmium transporter called CdtB also exists in *L. plantarum* and both MntA and CdtB were successfully used in combination with a metallothionein in metal bioremediation experiments.[32,33]

23.3.2.3.1.4 Yeast NRAMP.

The yeast *S. cerevisiae* contains three paralogue NRAMP transporters, Smf1p, Smf2p and Smf3p.[34,35] Based on resistance tests and uptake measurements, Smf1p has been described as a manganese transporter located at the plasma membrane.[36] The role of Smf1p in the uptake of cadmium has been demonstrated indirectly in a strain where the BSD2 gene regulating metal homeostasis had been deleted. This Δ*bsd2* strain is particularly sensitive to cadmium and accumulates over twice the levels of the wild-type strain upon exposure to CdCl$_2$ (7.5 µM) for 30 min. If the SMF1 gene (also called SBS1 for suppressor of bsd2) is also deleted, normal uptake of intracellular cadmium occurs with significant improvement in resistance to cadmium. The Δ*bsd2*Δ*smf2* strain is slightly more resistant than the Δ*bsd2* strain, but much less so than the Δ*bsd2*Δ*smf1* strain. This result indicates different properties for the two yeast NRAMP pumps with respect to cadmium,[37] as confirmed by gene dosage of *SMF1* and *SMF2* vs. cadmium toxicity.[38]

When cells are exposed to an excess of manganese, Smf1p is internalized into the Golgi apparatus where it undergoes ubiquitinylation by the Rsp5p/Bsd2p complex and is eventually degraded in the vacuole.[39] Like manganese, cadmium also affects the residence time of Smf1p at the plasma membrane. However, the cadmium-dependent regulation of Smf1p does not involve the Golgi apparatus and it specifically requires two arrestin-like proteins.[40]

23.3.2.3.1.5 Other Cadmium Uptake Systems in S. cerevisiae.

Other putative cadmium entry pathways may contribute to intracellular accumulation. A strain deleted in the ZRT1 gene was unable to take up cadmium from a CdSO$_4$ solution (47.8 µM) for 25 h, as compared to the wild-type strain, which accumulated 23% of the cadmium present in the extracellular medium under the same conditions.[41] This suggested that Zrt1p, a member of the ZIP transporter family, was a cadmium importer. More recently, experiments of ^{109}Cd accumulation on whole cells indicated that the calcium channel Mid1p/Cch1p may also contribute to cadmium uptake.[42]

23.3.2.3.2 Plants

23.3.2.3.2.1 ZIP. Although some trace metals may enter the plant *via* endocytosis,[17] this does not seem to have been reported for cadmium. Cd^{2+} ions present in the mobile fraction of the soil are apparently taken up by root cells in plants mainly *via* Fe^{2+}, Zn^{2+} and Mn^{2+} transporters and probably by Ca^{2+} channels. In general, these results were obtained from indirect evidence, and biochemical characteristics of the metal transporters are usually not available.

IRT1, the founding member of the ZIP transporter family, represents the main Fe^{2+} uptake system for plants like *Arabidopsis thaliana*.[43,44] An Arabidopsis *IRT1* knock-out mutant failed to accumulate iron. Nor did it accumulate manganese, zinc and cadmium under low-iron conditions.[44] However, plants in which *IRT1* was overexpressed did accumulate higher levels of cadmium and zinc in their roots compared to wild-type plants.[45] Furthermore, iron deficiency (which induces *IRT1* expression) increased the capacity of plants to absorb other metals. These plants included peas (cadmium),[46] the hyperaccumulator *Noccaea caerulescens* (formerly *Thlaspi caerulescens*) ecotype Ganges (zinc/cadmium)[47] and also rice.[48] Functional analyses of the *Arabidopsis* transporter expressed in *S. cerevisiae* showed that IRT1 transports Fe^{2+} $(K_{m(app)} = 6 \pm 1$ $\mu M)$[43] and also manganese $(K_{m(app)} = 9 \pm 1$ $\mu M)$,[49] with uptake of both metal ions being inhibited by cadmium. Heterologous expression in yeast revealed that RIT1 (the IRT orthologue in the pea *Pisum sativum*) had a high affinity for Fe^{2+} $(K_m$ in the range 54–93 nM) but a much lower affinity for zinc and cadmium $(K_m$ around 4 and 100 μM, respectively).[50] These data suggest that RIT1 could be an entry port for cadmium in peas only in the case of massive pollution.

In addition to IRT-type proteins, cadmium was also shown to compete with zinc uptake by plant ZIP transporters such as Arabidopsis ZIP1, 2 and 3[51] and rice OsZIP1.[52] The *Medicago truncatula* MtZIP1 expressed in yeast is capable of transporting zinc $(K_m$ *ca.* 1 μM). MtZIP1 zinc uptake is inhibited by 55% in the presence of cadmium (3 μM). On the other hand, MtZIP5 and MtZIP6, which have higher affinities for zinc $(K_m = 0.4$ and 0.3 μM, respectively) are inhibited by 20–30% only. MtZIP6 was shown to transport cadmium at a higher rate than zinc but with a lower apparent affinity $(K_m = 57$ $\mu M)$,[53] whereas the ZIP transporter ZNT1 from *Noccaea caerulescens* (ecotype Prayon) mediated uptake of zinc and cadmium with high affinity when expressed in yeast.[54] NcZNT1 is a plasma membrane transporter that plays a role in zinc, but surprisingly not cadmium, uptake from the soil.[55]

23.3.2.3.2.2 NRAMP. Plant NRAMP pumps are metal transporters for a broad range of ionic substrates (Table 23.1). OsNRAMP5, located in the plasma membrane and constitutively expressed in roots, seems to be the major transporter responsible for manganese and cadmium uptake in rice.[56] Low expression of *NRAMP5* correlated with lower concentrations of manganese and cadmium in roots and shoots.[56] The substrate specificity of plant NRAMPs has not been established. Nevertheless, from cadmium short-term uptake experiments *in planta*,[56] an apparent K_m of *ca.* 0.4 µM was estimated for OsNRAMP5. OsNRAMP1 was also shown to be involved in cadmium uptake in rice and overexpression of OsNRAMP1 in rice increased accumulation in leaves.[57]

23.3.2.3.2.3 Calcium Channels. Ca^{2+} uptake systems could also be involved in cadmium entry into plants. Patch-clamp studies performed on guard cell protoplasts showed that Ca^{2+} channels are permeable to cadmium and that cadmium disturbs plant water status by affecting the guard cell regulation in an abscissic acid-independent manner.[58] Nevertheless, no Ca^{2+} channels present in the plasma membrane of root epidermis cells were identified as actors in cadmium uptake by plant roots, although the low-affinity cation transporter LCT1 from wheat functionally complemented the yeast Mid1p Ca^{2+}-channel and LCT1 was proposed to mediate cadmium transport.[59]

23.3.2.3.2.4 Aerial Uptake. Cadmium dust deposition on leaves may be another source of cadmium for plants and would be expected to occur near industrial activities such as smelters and cement factories.[60,61] Analyses of plants contaminated with airborne metals are scant but, in plants cultivated under controlled conditions, uptake of cadmium from the leaves reached almost 50% in some cases.[62] In real field conditions in an industrial environment, cadmium deposition on vegetable crops was too low to be determined.[63]

23.3.2.3.3 Animals. Throughout this chapter, "animals" should be mainly understood as "mammals" and even more restrictively as "humans". This is not to say that cadmium handling by other animals is of little importance, particularly in the study of contaminated ecosystems, but simply that it cannot be considered here for lack of space.

23.3.2.3.3.1 Inhalation. Cadmium in aerosols is generally diluted but inhalation can occur in two particular situations. The first is occupational for workers using heated cadmium as occurs in isolation of the metal from minerals and in processing or recycling it. The second is recreational tobacco smoking. In both cases, cadmium is mainly included in dust and smokes. The forms of airborne metal reaching the lung are likely to be mainly particles of different sizes. Hence, cadmium from these particles may be assimilated by endocytosis, or any liberated cadmium cations in the alveolar fluid may be transported through the plasma membrane. In any case, lungs incorporate inhaled cadmium with subsequent health problems.[64] The size of the particles determines the site of deposit in the respiratory tract, with smaller particles penetrating deeper into the broncho-alveolar region. Particulate material in the lungs induces inflammation,[65] which influences the organ sensitivity to their content, and the surface activity of the particles impacts the air-blood barrier and the uptake mechanism of cadmium.

Specific uptake of cadmium chloride by rat alveolar type II monolayers is characterized by a Michaelis constant of the order of 6 μM.[66] A contribution of paracellular permeation of cadmium was also observed. *ZIP8* (SCL39A8) is abundantly expressed in lungs and it can transport Cd^{2+}.[67] When mouse *ZIP8* mRNA was transcribed in *Xenopus* oocytes, uptake of both Cd^{2+} and Zn^{2+} occurred from a synthetic medium with similar transport parameters ($K_m \sim 0.5$ μM; $V_{max} \sim 1.8$ pmol.(oocyte.h)$^{-1}$ and $K_m \sim 0.25$ μM; $V_{max} \sim 1$ pmol.(oocyte.h)$^{-1}$, respectively).[68] Pertaining to the interaction between particles and the lung epithelia, *ZIP8* is modulated by inflammation,[67] but several regulating pathways are involved. Consequently, the up- or downregulation of *ZIP8* depends intricately on the conditions. In addition, metal cations also regulate ZIP8 and influence its ability to carry Cd^{2+} across membranes. For instance, cadmium-induced lung disorders from cigarette smoke appear to be more prevalent in zinc-deficient individuals as compared to those with sufficient zinc intake.[69] In addition, zinc inhibition of cadmium transport in human lung cell lines agrees with competition of both cations for ZIP8.[70] ZIP8 is not selective, except for Cu^{2+}, which is not transported. The cell surface concentration of ZIP8 seems to follow iron availability by a post-transcriptional mechanism.[71] The actual involvement of ZIP8 in the physiological trafficking of transition metals and its precise role in cadmium transport remain unclear, partly because of its complex regulation.[72]

Intratracheal instillation of cadmium chloride decreased the amount of cadherins (calcium-dependent adhesion proteins) present in epithelial alveolar cells and in vascular endothelial cells.[73] Subsequent disruption of cell-cell interactions may facilitate permeation of the metal through the lung epithelium.

23.3.2.3.3.2 Ingestion. The other assimilatory pathway for cadmium by animals is through the diet. Studies using a range of biological samples from cell cultures to whole animals have indicated an inverse relationship between intestinal cadmium absorption and iron availability.[74] This pointed to the main intestinal ferrous ion transporter DMT1 (NRAMP2; SLC11A2) as the mediator of intestinal cadmium uptake. Since DMT1 is a proton symporter, its transport activity is enhanced at low pH, as prevails at the apical side of enterocytes where iron absorption occurs. Any loosely bound Cd^{2+} present in the diet would be assimilated *via* DMT1. Heterologous production of human DMT1 in *Xenopus* oocytes indicated Cd^{2+} transport at pH 6.0 with $K_m \sim 1$ μM,[75] and a higher V_{max} than for ZIP8 at neutral pH. Similar concentrations of active transporters per oocyte are assumed for this comparison. Divalent iron and lead and, to a lesser extent, manganese were the most potent inhibitors of cadmium transport, where-as zinc had no effect. It appears that Cd^{2+} mainly uses the major iron up-take system to cross the intestinal barrier.

However, the characteristic features of cadmium transport by a human intestinal cell line are not compatible with the involvement of DMT1 only.[76] For instance, zinc interferes at neutral pH. On this basis, members of the ZIP family, such as ZIP8 and one form of ZIP14 which are present in enterocytes, are candidates for cadmium transport.[77] The presence of the (initially called) hZLT1 zinc transporter in the luminal membrane of enterocytes was also considered as a possible intestinal absorption route for cadmium.[78] This protein turned out to be ZnT5, a member of the SLC30 group which can bi-directionally transport Zn^{2+} including at the plasma membrane.[79] Yet, in view of the recently rationalized selectivity of human ZnT transporters against Cd^{2+},[80] it is highly unlikely that SLC30 proteins play a significant role in Cd^{2+} uptake and trafficking. Mice fed diets de-ficient in one among the elements calcium, copper, magnesium, zinc and iron[81] had increased levels of orally provided hepatic cadmium compared to mice fed a mineral-replete diet. These high levels of hepatic cadmium did not correlate with increased intestinal expression of DMT1, except for iron-starved animals. Also, mice expressing an inactive mutated form of DMT1 accumulated Cd^{2+} from drinking water as efficiently as wild-type animals.[82] The uptake of divalent cations can be complex due to a mixture of regulatory cross-talk and molecular mimicry among essential and toxic cations, as is more readily shown in cases of resistance to diverse metal ions.[72,83] It is thus premature to consider, for instance, the calcium transporter CaT1 as an efficient pathway for cadmium absorption, even in calcium-deficient animals.[84] Thus, DMT1 and those members of the ZIP family that are lo-cated at the plasma membrane of enterocytes (ZIP8, ZIP14 and possibly ZIP4) remain the most likely candidates to mediate the crossing of Cd^{2+} through the intestinal barrier.

Intestinal cadmium absorption by animals depends on cadmium speci-ation in food. Human food mainly consists of plants and meat. As described below (Section 23.4), a contaminated plant-derived diet contains cadmium bound to phytochelatins and probably to other thiol-containing compounds. Cadmium-metallothionein (Cd-MT) is a component of a contaminated diet rich in meat, particularly in offal (liver and kidney). Cadmium-phytochela-tins (Cd-PC) are less readily absorbed from the gastro-intestinal tract than $CdCl_2$, but the proportion ending up in the kidneys is higher than that in the liver.[85] Different transport systems for $CdCl_2$ and Cd-PC are probably in-volved at the intestinal level,[86] but the actual mechanism of Cd-PC uptake and the effects on intestinal cells do not seem to have been studied in detail. Similarly, ingestion of Cd-MT by animals leads to absorption and distri-bution with different characteristics than those for $CdCl_2$.[87] Precise mo-lecular information is not available. The transporters involved in cadmium uptake in the organisms considered in this section are collected in Table 23.2.

Table 23.2 Cellular cadmium uptake systems.

	Transporter type	Protein name	K_m (μM)	Evidence	Comments
BACTERIA					
S. aureus	NRAMP[a]		5.4	2, 4	Mn transport
B. subtilis	NRAMP[a]		1.8	2, 4	At pH 6.5
E. coli	NRAMP	MntH	2.5	2	
	ZIP	ZupT	2.1	1, 2	Zn transport
B. bronchiseptica	ZIP	ZIPB		5, 6	
C. metallidurans CH34	Mg²⁺ uptake system[a]	MntA	135	2	Mn transport
L. plantarum	P-type ATPase	CdtB	0.45	2	
YEAST					
S. cerevisiae	NRAMP	Smf1p		1, 2	Mn transport
	ZIP	Zrt1p		3	
	Calcium channel	Mid1p/Cch1p		2	
PLANT					
A. thaliana	ZIP	IRT1		2, 4	Mainly Fe transport
	ZIP	ZIP1, 2, 3		4, 6	Zn transport
P. sativum	ZIP	RIT1	100	5, 6	Fe transport
N. caerulescens	ZIP	IRT1		2	
O. sativa	ZIP	NcZNT1		5, 6	Zn transport
	ZIP	OsZIP1		4, 6	Zn transport
	NRAMP	NRAMP1		2	Fe transport
	NRAMP	NRAMP5	0.4	2	Mn transport

Table 23.2 *(Continued)*

M. truncatula	ZIP	MtZIP1	1	4, 6	Zn transport
	ZIP	MtZIP5, 6		6	Zn transport
	ZIP	MtZIP6	57	5, 6	Zn transport
MAMMALS					
Lung	ZIP	ZIP8	0.5	2, 4, 6	Mn, Fe, Zn transport
Intestine	NRAMP	Nramp2	1	2, 4, 6	at pH 6, Fe transport
Liver, kidney proximal tubules, heart, pancreas, *etc.*	ZIP	ZIP8, 14		2, 5, 6	Mn, Fe, Zn transport
	Voltage gated calcium channel	CaV3.1 (α1G)		2, 5	Transfected cells
	TRP channels[a]	TRPM7, TRPV6	low μM	2	Transfected cells
Kidney	Endocytosis	Megalin/cubilin			Cd-MT and other proteins
		Lipocalin-2			Cd-MT and other proteins

1: Growth sensitivity to cadmium; 2: Cadmium content in cells or tissues; 3: Extracellular cadmium depletion; 4: Metal uptake competition with cadmium; 5: Cadmium-dependent activity or cadmium binding; 6: In heterologous systems.
[a] Putative

23.4   Intracellular Binding and Trafficking

Even though loosely bound Cd^{2+} may be present in the environment and should be considered for metal uptake (Section 23.3), once inside cells Cd^{2+} is expected to encounter many strong ligands.

23.4.1   Glutathione

As it is present at mM levels in the cytosol, glutathione (L-γ-glutamyl-L-cysteinylglycine, GSH) is expected to be a significant ligand of Cd^{2+} in most living cells. It is synthesized in two steps by two different enzymes, γ-glutamyl cysteine synthase and glutathione synthetase.[88,89] GSH can also be imported from the extracellular medium, for instance by the ABC transporter yilABCD in *E. coli*[90] and Hgt1p, a carrier belonging to the class of oligopeptide transporters, in yeast.[91] The solution structure of the complex between cadmium and GSH indicates predominantly a 1 : 2 stoichiometry with a dynamic cadmium coordination sphere containing sulfur and water oxygen atoms under physiological conditions.[92]

23.4.1.1   *Microorganisms*

23.4.1.1.1   Bacteria.   In *E. coli*, the role of GSH in the tolerance to cadmium was assessed by measuring the growth of strains genetically altered in either the GSH biosynthetic pathway or in its efflux systems. GSH-deprived cells were more sensitive to Cd^{2+} toxicity than the reference strain, and the removal of the exporter ZntA ATPase (Section 23.6) increased the sensitivity by more than two orders of magnitude.[93] Furthermore, exposure of *E. coli* cells to Cd^{2+} increased transcription of the GSH biosynthetic genes, besides other changes in different pathways.[94] The involvement of bacterial GSH in the handling of Cd^{2+} appears to be widespread.[95]

23.4.1.1.2   Yeast.   In *S. cerevisiae*, transcription of the γ-glutamyl cysteine synthase gene depends on the transcription factor Yap1p and other transcription factors involved in sulfur utilization (Section 23.8). The induction of the γ-glutamyl cysteine synthase gene is fast and efficient at high cadmium concentrations.[96,97] Consistently, the intracellular concentration of GSH significantly increases from 2 to 15 mM upon exposure to 50 μM $CdSO_4$ for six hours.[98] Even though a threshold effect may not occur, induction is much reduced at lower cadmium concentrations,[99] which questions the need for enhanced GSH synthesis to cope with moderate cadmium levels in yeast. GSH, whose lifetime strongly decreases in the presence of cadmium, is degraded in yeast by the cytosolic complex composed of the proteins Dug1p, Dug2p and Dug3p, the transcription of the corresponding last two genes being controlled by Met4p (Section 23.8.1.2).[100]

The situation described for *S. cerevisiae* is likely to apply to other species, such as the fission yeast *Schizosaccharomyces pombe* since deletion of the genes encoding γ-glutamyl cysteine synthase and YAP1 orthologues led to qualitatively similar sensitivity to cadmium.[101]

23.4.1.2   Mammalian Cells

The contribution of GSH to cadmium handling in mammalian cells is beyond doubt. One line of evidence comes from the identification of resistance mechanisms against cadmium exposure that increase the production of GSH by, for instance, enhancing the activity of the limiting biosynthetic enzyme γ-glutamyl-cysteine synthase. Accordingly, the intracellular level of GSH is at least transiently perturbed in the presence of cadmium.

23.4.1.3   Plant Cells

Hypersensitivity to cadmium was observed in *Arabidopsis* plants lacking enzymes of GSH synthesis.[102] Plant exposure to cadmium enhances the biosynthesis of glutamate, glycine and cysteine, including enzymes of the sulfate assimilation pathway which are required for GSH biosynthesis.[103] However, and in contrast with other organisms (see above), the enzymes involved in GSH synthesis remain stable at both the transcriptional and protein levels in the presence of cadmium.[104-106] This strongly suggests that amino acid availability, rather than enzyme activity, is limiting the production of GSH in plants, with consequences for both conjugation of toxic compounds and production of phytochelatins.

23.4.2   Phytochelatins

23.4.2.1   Schizosaccharomyces pombe

In addition to GSH, the fission yeast *S. pombe* also contains cysteine-rich peptides of the general formula $(\gamma\text{Glu-Cys})_n\text{-Gly}$ (n = 2–11) called phytochelatins (PC). PC are synthesized from GSH by the phytochelatin synthase PCS1. PCS1-deleted strains are sensitive to cadmium. Heterologous expression of the PC from *A. thaliana*, *Triticum aestivum* (wheat) and *S. pombe* significantly increase the tolerance of *S. cerevisiae* to cadmium.[107] When *S. pombe* was exposed to Cd^{2+} (200 μM), the production of GSH and PC was induced with maximal amounts reached first for GSH and later for PC.[108,109] Gel-filtration analysis revealed the formation of low molar mass PC-Cd complexes and high molar mass ones containing sulfide as well.

23.4.2.2   Higher Plants

As discussed above for *S. pombe*, GSH and PC are both present in plant cells. PC synthase is post-translationally activated by the $Cd(GS)_2$ complex[110] and Arabidopsis plants lacking enzymes required for GSH (Cad2) or PC synthesis (Cad1) are hypersensitive to cadmium.[102,111] In plants exposed to cadmium, amino acid availability, which limits GSH synthesis (Section 23.4.1.3), also impacts PC production. Consequently, different isoforms of PC with Ser, Ala or Glu replacing the C-terminal Gly are produced in this same organism with different kinetics of synthesis for each of them.[112,113]

23.4.3 Thioneins Involved in Cadmium Handling

Metallothioneins (MT) are low-molecular-weight proteins rich in cysteine which were discovered in mammals as cadmium-binding proteins.[114] This emphasizes their importance in cellular cadmium handling.[‡] They also bind other metals, predominantly zinc and copper, and they have been found also to be widespread in plants and yeasts, and, more recently, in some bacteria.[115,116]

23.4.3.1 *Prokaryotes*

There are only a few examples of bacterial metallothioneins. The most studied is SmtA from *Synechococcus PCC7942*. Deletion of the *smtA* gene slightly increased the sensitivity to cadmium.[117] The protein SmtA binds cadmium *in vitro* at several sites, including some containing His residues,[118] which sets apart this prokaryotic MT from its counterparts from higher organisms (see below). In *Mycobacterium tuberculosis*, expression of the MymT gene, encoding a Cu-binding thionein, was induced by a factor of 100 after exposure to cadmium (50 μM).[119]

23.4.3.2 *Saccharomyces cerevisiae*

The expression of the *S. cerevisiae* MT genes is under the control of Ace1p, which triggers the transcriptional response to copper in yeast. *CUP*1, which encloses the two tandem genes *CUP*1-1 and *CUP*1-2, is not induced by cadmium.[97,99] Its deletion increases sensitivity to copper but not to cadmium.[120] However, CUP1 overexpression confers increased resistance to cadmium,[121] which is probably induced by the increased cadmium trapping properties of overexpressing cells. Biochemical characterization of Cup1p, with 12 Cys among its 53 amino acids, indicated binding of up to 4 Cd^{2+},[122] with a contribution of exogenous sulfide ions in the coordination sphere.[123] Crs5p (19 Cys in 69 amino acids) is another yeast MT that is an orthologue of those of higher organisms. It can bind 6–8 cadmium ions per molecule when purified or produced heterologously in a cadmium-overloaded medium in *E. coli*.[124] However, as observed for CUP1, CRS5 deletion does not increase sensitivity to cadmium and CRS5 transcription is not induced by cadmium.[125] These observations cast doubt on the role of Crs5p in cadmium handling.

[‡]See Chapter 21 for a detailed discussion of the structures and properties of metallothioneins.

23.4.3.3 Plants

Plant MTs are classified into four types based on the conserved positions of cysteine residues,[126] each type seeming to have its tissue-specific pattern of expression.[127] They are involved in metal tolerance and homeostasis but also in the scavenging of reactive oxygen species, being able to bind various transition metal ions including Zn^{2+}, Cd^{2+} and Cu^+.[128] *Noccaea caerulescens* MT3 binds up to five equivalents of Cd^{2+},[129] while *Quercus suber* MT2 binds up to six, with the coordination sphere being filled with four sulfide ligands.[130]

Several reports link MT with tolerance to cadmium. Down-regulation by RNAi of the Arabidopsis MT1a, b and c isoforms is associated with hypersensitivity to cadmium. In addition, accumulation of cadmium (but also zinc and arsenic from arsenite) was several-fold lower in the MT1-depleted plants than in the wild-type, while Cu and Fe levels remained unaffected.[131] The double Arabidopsis mutant $\Delta(mt1a-2\ mt2b-1)$ had normal metal tolerance, suggesting overlapping roles of the MT1 family members. However, when MT deficiency was combined with PC deficiency (in the absence of the PC synthase gene *cad1*), growth of the *mt1a-2 mt2b-1 cad1-3* triple mutant was more sensitive to copper and cadmium compared to the *cad1-3* mutant alone.[127] The expression of *Brassica juncea* MT2 increases copper and cadmium tolerance in *Arabidopsis thaliana*.[132]

23.4.3.4 Animals

After the initial discovery of MT in animal cells,[114] more than a dozen genes and pseudo-genes were found in the human genome. Although metal-specific forms appear to occur in some animal species,[133] MT isolated from mammals usually contain a mixture of metals including zinc, copper and cadmium. Interestingly, the cadmium content of mammalian MT seems to be correlated with the amount of the metal being accumulated in the organ examined. For instance, kidney MT contains significant concentrations of cadmium (Section 23.7).[114] Mammalian MT display dynamic structures and interactions with metals, so it may be unrealistic to associate binding constants to the binding of metals to thionein. However, early work has clearly established that Cd^{2+} binds very tightly to mammalian MT, and its rate of release is more than two orders of magnitude slower than that for zinc under similar conditions.[134] It follows that the zinc exchange properties of MT are different from those with cadmium and that the presence of Cd^{2+} in MT is likely to impair proper homeostasis of the essential metals interacting with this family of proteins.

23.4.4 Other Thiols of Low Molar Mass

GSH is the most abundant intracellular thiol-containing compound in proteobacteria and cyanobacteria. In other prokaryotes, other thiol-containing compounds are synthesized such as co-enzyme-M in methanogenic micro-organisms, trypanothiones in some protozoa, mycothiols in mycobacteria and Streptomyces, or ergothioneine (which binds cadmium *in vitro*[135]) in some actinomycetales. All these metabolites, and others, could interact with metals *via* their thiol groups and thereby they could have similar roles to GSH. However, there is so far no clear evidence that any of them actually plays a role in the resistance to cadmium. The main reason might be that the thiolate groups of these molecules must be readily accessible *in cellulo* and sufficiently abundant to bind metal ions according to their specific dissociation constants. Except for GSH, these two conditions are unlikely to be met under normal physiological conditions.

23.4.5 Other Cadmium-binding Molecules

Proteomic approaches, combined with genetic and biochemical analysis, have been applied to candidate proteins that accumulate in *Arabidopsis* cells treated with cadmium. The results highlighted the role of a putative Selenium Binding Protein (SBP) in a new heavy metal detoxification mechanism.[113,136] SBP1 overexpression protects *Arabidopsis* from the toxic effects of cadmium. Biochemical studies indicated a binding stoichiometry of approximately 3 Cd^{2+} : SBP1.[136] *SBP1* expression was induced in response to cadmium stress and also in response to increased needs for sulfur. In addition, SBP1 overexpression enhanced tolerance to stresses that require GSH for detoxification.[137]

Beyond the binding molecules described in this paragraph, cadmium may interact with other metal binding molecules by mimicry or with normally vacant cation sites. This may occur in cases of massive cadmium overload, but such occurrences are rare and poorly characterized.

23.5 Subcellular Membrane Transport

23.5.1 Unicellular eukaryotes

23.5.1.1 *S. cerevisiae*

The most important transporter involved in the compartmentalization of cadmium in *S. cerevisiae* is the ABC transporter Ycf1p, a 1515 amino acids protein of the vacuolar membrane belonging to the ABC-C family.[138,139] Experiments with model substrates GS-X, such as *S*-(2,4-dinitrophenyl)glutathione (DNP-GS) and glutathione-monochlorobimane, applied to whole cell or vacuolar membrane vesicles showed that Ycf1p is a MgATP-dependent GS-X pump.[140] The uptake of DNP-GS by the vacuolar fraction is saturable (K_m for MgATP and DNP-GS of $\sim$80 and $\sim$15 µM, respectively).[140,141] The uptake of cadmium by Ycf1p was measured by ^{109}Cd fluxes with purified vacuolar membrane vesicles, which confirmed that it follows Michaelis–Menten kinetics ($K_m \sim$40 µM; $V_{max} \sim 3 \times V_{max}$(DNP-GS)).[142]

Transport is also MgATP-dependent and the substrate was identified as $Cd(GS)_2$. Within the vacuole, that complex may be degraded by peptidases.[143] Two other ABC-C transporters Bpt1p[144,145] and Vmr1p[146], described as GS-X vacuolar pumps, might play a role in the intracellular localization of cadmium in *S. cerevisiae* but their involvement seems minor compared to that of Ycf1p.

S. cerevisiae contains five genes coding for CDF transporters: MSC2, COT1, ZRC1, MMT1 and MMT2. Among them, ZRC1 codes for a vacuolar zinc transporter that transports cytosolic zinc to the vacuole. Zrt3p reverses the zinc flux from the vacuole and both transporters control zinc homeostasis in yeast.[147] ZRC1 transcription is not induced by cadmium,[148] but it affects the resistance to cadmium in a dose-dependent way. Strains containing zero, one or several copies of ZRC1 stopped growing at 50, 100 and 200 μM $CdCl_2$, respectively.[149] This ZRC1-dependent resistance to cadmium and the inhibition of Zrc1p-mediated zinc uptake in vacuolar vesicles by cadmium[150] suggested that Zrc1p was responsible for cadmium accumulation in the vacuole.

23.5.1.2   *S. pombe*

The *hmt1* gene was first identified by its capability to restore cadmium tolerance and normal accumulation of the phytochelatin-cadmium-sulfide complex to a cadmium-sensitive *S. pombe* variant.[151] It was later detected in a genome-wide screen of genes involved in cadmium resistance.[101] Accordingly, *hmt1*-deletion significantly increased *S. pombe* sensitivity to cadmium.[152,153] The HMT1 protein is an ABC-B transporter located in the vacuolar membrane.[151,154] It is not induced by cadmium[151] but it transports *apo*-phytochelatin and Cd-phytochelatin complexes through purified vacuolar membranes in an ATP-dependent reaction showing a marked preference for the smaller Cd-PC complexes.[154] However, cadmium transport by HMT1 is not strictly dependent on the presence of PC, as cadmium translocation occurs in organisms that do not have PC, such as *S. cerevisiae* and *E. coli*. In such cases, the substrate of HMT1 is likely to be the $Cd(GS)_2$ complex.[153]

23.5.2   Plants

Like yeasts, plant cells sequester cadmium in the vacuole using different types of transporters. The *Arabidopsis* CAtion/H^+ eXchangers CAX2 and CAX4 are thought to mediate cadmium compartmentalization in vacuoles. Their expression in tobacco leads to enhanced cadmium selective transport into root tonoplast vesicles and to higher tolerance of the plant to the toxic metal.[155–157] The *Hordeum vulgare* L. (barley) CAX1a isoform was found to be up-regulated in a differential proteomic study of vacuoles purified from plants grown on moderate cadmium levels (20 μM).[158]

P_{1B}-type ATPases, like the *Arabidopsis thaliana* (Wassilewskija ecotype) HMA3 highly expressed in guard cells, hydathodes, vascular tissues and root apex, also participate in vacuolar sequestration of cadmium.[159,160] Whereas

HMA3 is a pseudogene in the Columbia ecotype of *A. thaliana*,[161] the *HMA3* orthologue is highly expressed in the hyperaccumulators *Arabidopsis halleri*[162,163] and *Noccaea caerulescens*.[164] NcHMA3 expression was seven-fold higher in the higher Cd-accumulating Ganges ecotype of *N. caerulescens* compared to the lower Cd-accumulating ecotype (Prayon).[164] NcHMA3 also localizes in the tonoplast, *i.e.* the vacuolar membrane, and it is more specific for cadmium than its *A. thaliana* (WS) orthologue. Its presence in large amounts in leaves likely explains the hypertolerance to cadmium of the Ganges ecotype.[164] The rice OsHMA3 sequesters cadmium in the root vacuoles limiting its translocation to the above-ground tissues.[165] Consequently, HMA3-deficient rice plants strongly accumulate cadmium in their shoots.[166,167]

Arabidopsis ABC transporters MRP1 (ABC-C1) and MRP2 (ABC-C2) are abundant in the tonoplast[168] and they are able to transport a broad range of substrates, including glutathione and glucuronate conjugates and chlorophyll catabolites.[169,170] They are also the long-sought-after transporters of metal-PC complexes:[171,172] they carry As(III)-, Cd(II)- and Hg(II)-PC complexes into the vacuole with low affinity, since the As(III)-PC$_2$ concentration for half-maximal uptake by isolated vacuoles (~ 350 μM) is ~ 5 times higher than for herbicide-GS conjugates.[169]

AtNRAMP3 and AtNRAMP4 at the tonoplast are essential for the remobilization of vacuolar iron for seed germination under low iron conditions and of vacuolar manganese for optimal photosynthesis and growth under manganese deficiency.[173,174] Their absence strongly increases *A. thaliana* sensitivity to cadmium and zinc.[175] NRAMP3 and NRAMP4 from *N. caerulescens* are efficiently expressed in yeast, where they transport iron, manganese and cadmium.[175]

23.5.3 Animals

23.5.3.1 *Caenorhabditis elegans and Drosophila melanogaster*

Cadmium storage into acidic vesicles has also been described in these higher organisms. In *C. elegans*, the *ce-hmt-1* gene (for *C. elegans* heavy metal tolerance 1, alias haf5) encodes an ABC transporter with 51% similarity to *S. pombe* HMT1.[176] Complementation experiments have shown that expression of *ce-hmt-1* was capable of restoring cadmium tolerance in the *hmt-1⁻ S. pombe* LK100 strain. In *C. elegans*, injection of *ce-hmt-1* RNAi causes growth arrest of the progeny at an early larval stage and cellular necrosis upon exposure to 5 μM cadmium.[176] A growth defect was also observed in the presence of cadmium in two mutant alleles of *ce-hmt-1*.[177] In *C. elegans*, *ce-hmt-1* and *pcs-1* (coding for the phytochelatin synthase) are co-expressed in the coelomocytes.[177] However, the role of *hmt-1* does not depend exclusively on *pcs-1*.[176]

DmHMT1 from *D. melanogaster* possesses 55% and 57% sequence similarity with SpHMT1 and CeHMT1, respectively, and it can complement the function of SpHMT1 in *S. pombe*. In *Drosophila*, which is devoid of phytochelatin synthase, the substrate of DmHMT-1 is probably the Cd(GS)$_2$ complex.[178]

23.5.3.2 *Mammals*

After long-term intoxication of laboratory animals, a large majority of cadmium resides in the cytosol with 10 to 15% found in organelles.[179] Bound forms of cadmium, such as Cd-MT, can release Cd^{2+} in endosomes of cells such as those of the proximal tubules of the kidney. Cd^{2+} crosses the endosomal membrane through the constitutive and non-iron-regulated form(s) of DMT1.[180] Apoptosis, autophagy and endoplasmic reticulum stress are triggered by cadmium exposure. This strongly suggests interaction of the metal ion with organelles.

23.5.3.2.1 Mitochondria. Many reports describe the effects of cadmium on mitochondrial function, associated oxidative damage and apoptotic consequences, but few have attempted to unravel the relevant mechanisms and actors. A Michaelis constant of *ca.* 3×10^{-8} M for the mitochondrial uptake of cadmium was estimated for gill cells of trout.[181] The increased cadmium concentrations in mitochondria upon exposure of whole animals or of isolated organelles indicate that inward transport is more efficient than export, if there is any. Consequently, cadmium-induced damage has been repeatedly reported in mitochondria with contributions from reactive oxygen species and inactivation of mitochondrial activities (Krebs cycle – mainly isocitrate dehydrogenases and aconitase – respiratory chain, *etc.*).

Morphological mitochondrial alterations (swelling) follow cadmium exposure due to activation of water channels. The triggering event could be Cd^{2+} mitochondrial import by the Ca^{2+} uniporter (MCU), as witnessed by the sensitivity of the observed effects to its inhibitors in proximal tubule cells and models.[182] This channel transports divalent cations with less efficiency than it transports calcium but it may be capable of supporting entry of Cd^{2+} into mitochondria. However, permeation of cadmium has not been measured.

The above data were obtained at μM concentrations of cadmium, which are above most toxicological conditions. However, cadmium interference with activities involved in the intracellular traffic of cations can occur at far lower concentrations, in the nM range.[183] The detailed succession of events after cadmium exposure may also depend on the cell type and the complex form of cadmium in the cytosol that interacts with mitochondria.[184] But it seems that, in most cells, caspase-independent apoptosis occurs more readily than processes that follow opening of the permeability transition pore and inhibition of the respiratory chain. Necrosis is mainly observed at high cadmium concentrations.[182,185]

23.5.3.2.2 Endoplasmic Reticulum. Among the many effects of cadmium on the different functions of eukaryotic cells, endoplasmic reticulum (ER) stress has been investigated recently.[42,186] This particular phenomenon is characterized by the accumulation of unfolded and misfolded proteins, which triggers activation of a specific response named the Unfolded Protein Response (UPR). This response makes use of parallel pathways targeting transcription (of chaperone proteins, in particular), translation and protein degradation, initially to decrease the impact of the stress. In cases of sustained effects of the toxicant, apoptosis ensues. It appears that ER stress is induced by a redox imbalance, by increased superoxide concentrations in particular. It is not clear whether cadmium has to penetrate into the ER to activate the UPR, as cadmium-induced oxidative events, such as calcium release from intracellular stores, may trigger the UPR.

23.5.3.2.3 Other Locations. There is no definitive evidence for cadmium accumulation in the nucleus. Yet, a wealth of transcription factors (TF) is sensitive to cadmium but the effects may be indirectly imposed by redox-dependent activation. The nuclear action of TF may originate from reactions occurring in the cytosol such as loss of Nrf2 degradation by Keap1 action. Furthermore, inhibition of DNA repair enzymes by Cd^{2+} has been assigned to replacement of zinc at the four-cysteine finger site of XPA (Xeroderma Pigmentosum complementation group A). The nucleus is the site of action of this protein initiating recruitment of the nucleotide excision repair complex, as are the Cys-rich domains of base excision repair enzymes, such as 8-oxoguanine glycosylase and poly(ADP-ribose) polymerase 1. Upstream events, occurring in the cytosol and modulating DNA repair enzymes after translocation of the signal, cannot be ruled out, as in the case of transcription factors discussed above.[187] In yeast, cadmium inhibits mismatch repair by inactivating the ATPase activity of the Msh2p-Msh6p subunits of the DNA-bound complex.[188] Overall, the presence of cadmium in the nucleus needs to be demonstrated under pathophysiological conditions, and its most sensitive nuclear targets must be identified.

Another likely location for cadmium is the lysosomal system, which appears as a key organelle in the processing of cadmium, in particular Cd-MT.[184,189] This acidic compartment facilitates Cd^{2+} release from MT, which contributes to intracellular cadmium turnover and inhibition of lysosomal metabolic activity.

The transporters involved in cadmium intracellular traffic are listed in Table 23.3.

23.6 Export of Cadmium from Cells

23.6.1 Forms of Expelled Cadmium

It can be concluded from the number of Cd^{2+} ligands available inside cells, including the ones of high affinity (Section 23.4), that the most likely chemical forms of expelled cadmium are bound ones. Indeed, there is no clear evidence that cadmium is expelled as the isolated ("free") cation

Table 23.3 Intracellular transport.

	Transporter type	Protein name	Evidence	Intracellular localization	Comments
YEAST					
S. cerevisiae	ABC-C	Ycf1p	1, 5	Vacuole	GS-complex $K_m = 40\ \mu M$
	ABC-C	Bpt1p	1	Vacuole	GS-complex
	ABC-C	Vmr1p	1	Vacuole	GS-complex
	CDF	Zrc1p	1, 4	Vacuole	
S. pombe	ABC-B	SpHMT1	1	Vacuole	PC- or GS-complexes
PLANT					
A. thaliana	Cation/proton exchanger	CAX2, CAX4	2, 5	Vacuole	
	P_{1B}-type ATPase	HMA3	1, 1a, 2, 5	Vacuole	
	ABC-C	AtABCC1, AtABCC2	2, 5	Vacuole	PC-complexes
	NRAMP	AtNRAMP3, AtNRAMP4	1, 1a	Vacuole	Fe and Mn transport
H. vulgare	Cation/proton exchanger	CAX1a	6	Vacuole	
A. halleri	P_{1B}-type ATPase	HMA3	1, 2	Vacuole	
N. caerulescens	P_{1B}-type ATPase	HMA3	1, 2	Vacuole	
	NRAMP	NcNRAMP3, NcNRAMP4	1, 1a	Vacuole	
ANIMALS					
C. elegans	ABC-B	CeHMT1	1, 7	Vacuole	
D. melanogaster	ABC-B	DmHMT1	1	Vacuole	
mammals	NRAMP	DMT1	7	Endosomes	Fe transport
Most	Calcium uniporter	MCU	7	Mitochondria	

1: Growth sensitivity to cadmium, 1a: in heterologous systems. 2: Changes of cadmium content. 4: Metal uptake competition with cadmium. 5: Cadmium-dependent activity or cadmium binding. 6: Cadmium induced mRNA or protein production. 7: Sensitive to inhibitors (siRNA).

Cd_{aq}^{2+}. But, due to this lack of information, both possibilities, *i.e.* outward transport of Cd^{2+} or of complexes, have been considered in various studies.

23.6.2   Unicellular Organisms

23.6.2.1   Bacteria

23.6.2.1.1 P_{IB}-type ATPases. Cadmium export by P_{IB}-type ATPases was suggested decades ago for numerous bacterial species (*e.g. S. aureus*,[190] *H. pylori*,[191] *R. metallidurans* CH34[192]). This conclusion followed from growth measurements using ATPase-deleted strains in media containing cadmium, or by complementation assays of the $Cd^{2+}/Zn^{2+}/Pb^{2+}$-ATPase ZntA from *E. coli*.[193] Cadmium induces P_{IB}-type ATPases in several species (*e.g. L. monocytogenes*,[194] *S. aureus*,[195] *E. coli*[196]). Biochemical characterization of *E. coli* ZntA and *L. monocytogenes* CadA has been carried out in detail.§ The amino terminal metal binding domain is not mandatory for the transport activity but its removal affects enzymatic parameters.[197,198] The structure of this domain in ZntA[199] and CadA[200] revealed the involvement of conserved carboxylate residues in the binding of divalent metal ions. Micromolar concentrations of cadmium trigger the formation of phosphorylated intermediates[198,201] and reverse phosphorylation from phosphate provides a way to estimate an apparent affinity constant of 0.5 μM for Cd^{2+} in CadA.[202] ATPase activity measurements on purified ZntA revealed that the apparent affinity constant for cadmium varied from 5.5 μM (for metal provided as cadmium acetate) to 252 μM with $Cd(GS)_2$, the presumed genuine substrate of ZntA. The value of V_{max} for the complex was 20-fold higher than that for the dissociated ion.[203] ZntA and CadA differ by the number of cadmium ions bound at the transport site, one for ZntA[204] and probably two for CadA.[202]

23.6.2.1.2   RND Proteins. Heavy metal transporting RND-dependent systems (HME-RND) are largely restricted to Gram-negative bacteria[205] with the most studied systems being CzcCBA, involved in resistance to cobalt, zinc and cadmium in *Ralstonia metallidurans* CH34, and CusCBA (an important component of copper homeostasis in *E. coli*).¶

In *R. metallidurans* CH34, the resistance to cadmium was identified on the plasmid pMOL30, whose deletion reduced the minimal inhibitory concentration (MIC) of the metal from 2.5 mM in the CH34 strain to 0.6 mM (with increased accumulation of the metal) in the deleted AE104 strain.[206,207] Selective deletions in the *czc* operon showed that the absence of CzcA strongly decreased the resistance to cadmium, cobalt and zinc, whereas deletion of the CDF transporter CzcD had minor effects.[208,209] Expression in *E. coli* and measurements of metal transport in inverted vesicles led to the

---

§The molecular properties of zinc transporters are discussed in Chapter 22.
¶The molecular properties of copper transporters are discussed in Chapter 16.

determination of $K_m = 266$ µM for cadmium transport by the CzcCBA system.[210] This value compares with the concentration of ~95 µM required for half maximal induction of the *czc* operon[209] but it is far lower than the estimate of 8 mM for the purified CzcA protein reconstituted in proteoliposomes.[211] These experiments indicate that the CzcCBA system cannot be reduced to its unique RND component when discussing cadmium transport and that significant efflux of cadmium from *R. metallidurans* CH34 *via* CzcCBA occurs only at relatively large concentrations.

23.6.2.1.3 Interplay between RND Proteins and P$_{1B}$-type ATPases. There are few studies comparing the respective contributions of P-type ATPases and RND-dependent transport systems to the resistance to cadmium. The plasmid-free AE104 strain of *R. metallidurans* CH34 displays a minimal inhibitory concentration for cadmium of 350 µM, which collapses to *ca.* 1 µM when the chromosomal genes encoding the two P-type ATPases CadA and ZntA are deleted. Expression of the CzcCBA system in the AE104 Δ*zntAcadA* strain does not restore the tolerance to cadmium. Furthermore, the two P-type ATPases of the AE104 strain do not provide resistance to cadmium matching that of the wild-type CH34 strain (MIC of *ca.* 3 mM), which contains both RND and P$_{1B}$-type ATPases.[192] Qualitatively similar results were observed with the *P. putida* KT2440 strain,[212] suggesting that P$_{1B}$-type ATPases and RND-dependent systems work in synergy to protect bacteria from cadmium toxicity.

23.6.2.2 *S. cerevisiae*

23.6.2.2.1 P$_{1B}$-type ATPases. The CAD2 gene was first identified in the cadmium-resistant mutated strain X3382-3A[213] and later by systematic sequencing of the *S. cerevisiae* genome (then called PCA1).[214] CAD2 and PCA1 are isogenes encoding active and inactive forms of the same P-type ATPase, respectively. Among the 27 amino acid differences between Cad2p and Pca1p, the R→G substitution at position 970 is critical for the activity of the transporter.[215] The CAD2 gene dosage correlated with the yeast resistance to cadmium and it was inversely proportional to the intracellular cadmium content.[216] Cad2p appears to be protected from ubiquitination, and thereby degradation, by co-translational cadmium binding to a cysteine residue of its N-terminal domain in the endoplasmic reticulum. This increases the amount of the transporter at the plasma membrane when *S. cerevisiae* is exposed to cadmium.[217,218]

23.6.2.2.2 The ABC-C Transporter Yor1p. Among the six ABC-C transporters of the *S. cerevisiae* genome, five (including Ycf1) are located at the vacuolar membrane[219] while the sixth, Yor1p, is resident in the plasma membrane.[220] Yor1p is structurally different in lacking the N-terminal transmembrane domain. Although less pronounced than for Ycf1p (which is 31% sequence identical in a 1259 residues overlap), YOR1 deletion increases the sensitivity to cadmium and overexpression increases toler-ance.[221,222] Cadmium transport in whole cells by Yor1p is ATP-dependent and the $Cd(GS)_2$ complex is the substrate.[221] In contrast to YCF1 and the gene of glutamyl cysteine synthase (Section 23.4.1.1.2), YOR1 is regulated not by Yap1p but by transcription factors involved in the pleiotropic drug resistance (PDR).[223]

23.6.3 Plants

23.6.3.1 P_{IB}-type ATPases

In *A. thaliana*, the P_{1B}-type ATPases AtHMA2 and AtHMA4 are localized at the plasma membrane and are essential for zinc homeostasis and zinc trans-location from root to shoot tissue.[161,224] HMA4 deletion lowers cadmium translocation from root to shoot, whereas its overexpression enhances metal tolerance.[224] The double *hma2hma4* mutant displays a near-complete abo-lition of cadmium translocation, identifying these transporters as key components of cadmium export from roots.[225] Similar results have been obtained with the rice P_{1B}-type ATPase OsHMA2.[226–228] HMA4 is expressed at a very high level in the metal hyperaccumulators *N. caerulescens*[229] and *A. halleri* where it maps within the major quantitative trait locus for cadmium tolerance.[230] This high constitutive expression is attributable to a combin-ation of modified *cis*-regulatory sequences and HMA4 triplication.[231] HMA4 purified from *N. caerulescens* roots disclosed a cadmium-dependent ATPase activity with a maximum turnover rate of 300 s^{-1} at concentrations of Cd^{2+} in the range 0.03–1.0 μM.[232,233]

23.6.3.2 ABC Transporters

AtPDR8 is a member of the pleiotropic drug resistance (PDR) subfamily of ABC transporters present in the plasma membrane and especially in those of the root hair and epidermal cells. It is likely to transport a very broad range of substrates and might also be involved in extrusion of cadmium or cad-mium conjugates.[234]

23.6.4 Animals

A seemingly natural exporter for Cd^{2+} could be the only characterized fer-rous ion outward transporter ferroportin (SLC40A1 or MTP1) since, in add-ition to iron, it expels zinc from cells. It is transcriptionally up-regulated by cadmium through Metal Responsive Elements (MRE).[235] However, there is no experimental evidence for transport of cadmium by ferroportin. The possible failure of ferroportin to use cadmium as substrate may be due to the absence of available unbound Cd^{2+} within cells and to the likely pre-dominance of more easily mobilized forms of cadmium, such as $Cd(GS)_2$. Likely candidates for export of the latter are members of the Multidrug

Resistance Protein (MRP) family. At least some of these transporters can be induced in a human myelocytic leukaemia cell line by Cd^{2+} salts, with concomitant increase in transport activity and glutathione synthesis.[236] Export of Cd^{2+} from cells should thus occur through transport of a complex unless the metal ion dissociates before crossing the membrane. Cadmium up-regulates MRP2/ABC-C2, mainly through the MRE of the promoter, in most examined cells of excretory organs (intestine, liver, kidney). An indirect measure of intracellular cadmium suggests that it decreases in proportion of MRP2 expression.[237] However, such data cannot be considered as an irrefutable proof of cadmium export by MRP2 since similar characteristics for ABC-B1/PgP affect resistance against cadmium-induced apoptosis by means other than transporting cadmium out of cells.[238] CFTR/ABC-C7 expression is also sensitive to cadmium and its activity is modulated by Cd^{2+}. However, these effects seem to depend on the exact experimental conditions including the cell type, and no direct measurement of cadmium export by this transporter is available.[239,240] Overall, it has not been unequivocally proven that any ABC transporter exports cadmium in animal cells, although some members of this family are clearly sensitive to the presence of cadmium.

A similar statement holds for other suggested exporters of cadmium complexes. They include the H^+ antiporter of the organic cation transport system[241] and the bidirectional organic anion transporters, although the latter do transport metal chelators and mercury complexes.[242] Cadmium complexes with amino acids, other small molecules and oligopeptides may also be expelled from cells by other transporters, but the expected low affinities of these ligands for Cd^{2+} makes such events of marginal importance at best.

An outstanding candidate for cadmium export from cells has not been identified after more than 40 years of investigation. This statement is in line with the low rate of removal of cadmium from the body after its absorption. Cadmium is partly excreted into the biliary tract, possibly as a complex with GSH,[243] and the liver is a major site of the expression of thioneins upon cadmium induction (Section 23.4). Yet, Cd-MT appears to decrease biliary excretion of cadmium.[244] In contrast, Cd-MT seems to be released from the liver into the bloodstream before reaching its eventual target organ, the kidney. However, the mechanism of Cd-MT cellular release remains unclear. It can only be suggested that some sort of exocytosis occurs or that cadmium-induced cellular damage contributes to cadmium exit from hepatocytes and other cells.

The putative transporters for cadmium export are listed in Table 23.4.

23.7 Distribution of Cadmium Ions in Multicellular Organisms

23.7.1 Traffic in Plants

The usual fate of cadmium entering roots is to be sequestered into the vacuoles. In the Indian mustard *Brassica juncea* and in *A. thaliana* roots, this

Table 23.4 Export of cadmium from cells.

	Transporter type	Protein name	K_m (μM)	Evidence	Comments
BACTERIA					
E. coli	P_{1B}-type ATPase	ZntA	5.5 [Cd^{2+}] 252 [$Cd(GS)_2$]	1, 3, 4, 5	
L. monocytogenes	P_{1B}-type ATPase	CadA	0.5 [Cd^{2+}]	1, 3, 4, 5	
C. metallidurans CH34	RND	CzcCBA	266	1, 2, 4, 5	
YEAST					
S. cerevisiae	P_{1B}-type ATPase	Cad2p/Pca1p		1, 2	
	ABC-C	Yor1p		1, 4	Cd-glutathione complex
PLANT					
A. thaliana	P_{1B}-type ATPase	AtHMA2, AtHMA4		1, 2	
	Pleiotropic drug resistance (PDR) subfamily of ABC transporters	AtPDR8		1, 2	
A. halleri	P_{1B}-type ATPase	HMA4		1, 2	
N. caerulescens	P_{1B}-type ATPase	HMA4		1, 1a, 2, 5	
O. sativa	P_{1B}-type ATPase	OsHMA2		1, 1a, 2	
	Low-affinity cation transporter	OsLCT1		2	
MAMMALS					
	ABC-C	ABC-C2, ABC-C7		2, 5	

1: Growth dependence on cadmium, 1a: in heterologous systems. 2: Cadmium content in cells or tissues. 3: Metal binding competition with cadmium. 4: Cadmium-dependent activity or cadmium binding. 5: Cadmium-induced mRNA or protein production.

seems to occur as complexes of cadmium with sulfur ligands, most likely GSH or PC.[245,246] In the *A. thaliana* root cortex, Cd is also associated with phosphorus whereas, in the endodermis, sequestration of cadmium associated with sulfur-donor ligands in fine granular deposits is observed in the vacuole and in the cytosol.

Part of the cadmium present in plants is loaded into the xylem for root to shoot translocation driven by transpiration in an abscissic acid dependent process.[245] Efficient translocation is one of the most important characteristics of the metal hyperaccumulator *A. halleri*, in which cadmium concentrations can reach up to 0.3% of the mass of the shoot without growth inhibition.[231,247,248] In *A. halleri*, cadmium is translocated in the xylem as hydrated Cd^{2+}, with cysteine and GSH being undetected in xylem sap.[248] In *Brassica juncea*, the majority of the cadmium in the xylem was coordinated with O/N ligands.[245]

In leaves, trichomes (epidermal hairs) may represent the site of cadmium accumulation in several plants species.[245,246,249] Tobacco plants exposed to toxic levels of cadmium increase the number of trichomes and excrete cadmium as Cd/Ca crystals through their head cells.[250] In the hyper-accumulators *A. halleri* and *N. caerulescens*, cadmium is mostly found in the mesophyll tissues,[247,251] which implies efficient root to shoot transfer.

OsLCT1, a rice low-affinity cation transporter, exports cadmium from yeast: it is a plasma membrane transporter involved in cadmium loading to the phloem. Down-regulation of Os*LCT*1 by RNAi reduces cadmium accumulation two-fold in rice grains.[252] Real-time dynamics of ^{107}Cd distribution was studied by positron-emitting tracer imaging (PETIS) in different rice cultivars. Implementation of this new method allowed discrimination between low-cadmium accumulating cultivars (*japonica* type) from high-cadmium accumulating ones (*indica* type). The latter exhibited an increased translocation of the metal to shoots and panicles.[253] Cadmium may be re-allocated to seeds through the phloem where substantial cadmium precipitates were observed in *A. thaliana*.[254] The cadmium concentration was more than four-fold higher in the phloem than in the xylem sap of *Brassica napus* together with large amounts of PC and GSH, suggesting that these ligands function as long-distance carriers of cadmium.[255] In intact seeds and isolated embryos of *Thlaspi praecox*, almost two-thirds of the cadmium ligands are thiol groups and one-third are oxygen- and phosphate-containing ligands.[256]

23.7.2 Traffic in Animals

Cadmium distribution in animal bodies indicates that the liver and kidneys are the target organs accumulating the toxic metal after exposure.[257] This distribution is independent of the route of absorption and beyond the as-similatory organs, namely, the lungs and the intestine (Section 23.3). Compared to other organs, the kidneys appear to be the tissue accumulating more cadmium over time. They can be viewed as some sort of dead-end for cadmium trafficking. On the other hand, the liver is more involved in processing absorbed cadmium, with thionein synthesis and binding of the ion.

Two main aspects determine cadmium traffic: speciation in the circulation and uptake by the different tissues.

23.7.2.1 Speciation in Blood

The distribution of cadmium in the blood is not fully characterized, but a fraction of the metal is associated with reticulocytes. From the available data,[258] high-affinity uptake systems of red blood cells display Michaelis constants for Cd^{2+} in the low μM range. Inhibition of a Cl^-/HCO_3^- anion exchanger in the plasma membrane of erythrocytes limits Cd^{2+} transport.[259] The other fraction of circulating cadmium can be bound to proteins such as albumin, α_2-macroglobulin, immunoglobulins, transferrin[260] and metallothionein.[261] Low-molar-mass species, likely involving thiol-containing molecules such as GSH and cysteine, may co-exist with protein bound cadmium in the circulation. However, cadmium speciation in blood, especially under chronic low level exposure, does not seem to have been comprehensively probed with modern analytical methods.

23.7.2.2 Entry into Different Cells

Circulating cadmium can be taken up by a variety of mechanisms depending on the cell type. Assuming that some of the circulating cadmium complexes can readily dissociate, Cd^{2+} itself may penetrate into cells by alternative routes to those described for absorption in Section 23.3.

23.7.2.2.1 Calcium Channels[||]. Cd^{2+} may have a dual role with respect to calcium channels, one of potential substrate and another of channel blocker. The latter action has been largely employed to inhibit cation permeation in electrophysiological studies, usually at high concentrations (>100 μM). This effect will not be further considered here. Among the range of voltage-gated calcium channels, the T-type channel CaV3.1 (α1G) is a strong candidate for cadmium entry into a variety of animal cells, mainly excitable ones (*e.g.* neuron and heart).[262] Although the estimated rate of cadmium transport ($1\ s^{-1}$ $channel^{-1}$) is very modest, it may contribute to Cd^{2+} accumulation in the long run due to the large proportion of opened channels upon activation. L-type calcium channels have also been proposed to transport Cd^{2+} in a process regulated by the zinc exporter ZnT-1.[263,264] Cd^{2+} permeation also most likely occurs through the less selective divalent cation channels. TRP-melastatin-related 7 (TRPM7) channels containing cells, such as osteoblasts, accumulate cadmium, preferably at low pH (<8) or under magnesium deficiency, with an apparent Michaelis constant of a few μM.[265,266] Cells transfected with the vanilloid transient receptor potential channel TRPV6 show increased fluorescence of different cation probes in the presence of Cd^{2+}.[267] These are the most

---

[||]Molecular details of calcium channels are discussed in Chapter 5.

substantiated examples of cadmium permeation through ionic channels and it cannot be excluded that, in specific cells, additional molecules participate in Cd^{2+} transport through the plasma membrane. This is strongly suggested for neurons.[268]

23.7.2.2.2 Cellular Entry of Bound Forms of Cadmium. Release of protein-bound Cd^{2+} before transfer across membranes is not documented and, because of the size of these putative complexes, endocytosis is the most relevant mechanism. The receptor may change with the type of protein binding cadmium. The very large membrane protein cubilin is mainly present in the epithelium of intestine and kidney and it binds a range of proteins of different sizes. Its intracellular trafficking is modulated by other proteins, among which is another very large receptor binding many ligands: megalin or low-density lipoprotein-related protein 2.[269] Endocytosis of transferrin, albumin or ferritin may contribute to cadmium entry into cells, and megalin-cubilin is likely involved in some of these cases at least.[239] Cd-MT also binds to megalin and it can reach intracellular locations by this route. The receptor for lipocalin-2 (SLC22A17 or 24p3R), a protein binding the iron chelator dihydroxybenzoate to deprive infecting microorganisms of iron, participates in the uptake of the same cadmium-binding proteins (MT, albumin, transferrin).[270] Receptor-mediated endocytosis is thus an important uptake mechanism for protein-bound cadmium in different cells.[78]

Cadmium accumulates independently of the absorption route, from less than 5 µg l^{-1} in blood to more than 200 µg g^{-1} in the kidney cortex of individuals with renal damage. Cd-MT may cross biological barriers without being trapped and destroyed[87] but it is easily filtered at the glomerulus and internalized at the renal tubules. These cells have the capability of liberating the cadmium cation in acidic compartments. This aspect may explain why the renal tubules are the major site for the health effects of cadmium exposure.

23.8 Sensing, Buffering and other Control Mechanisms

23.8.1 Cadmium Cellular Sensing?

23.8.1.1 *In bacteria*

Various examples show that cadmium is sensed in bacteria by the three main types of metal-sensing transcriptional regulators: the metal de-repressors of the ArsR-SmtB family, the metal co-repressors of the DtxR/MntR family and the metal activators of the MerR family.[271]

In their dimeric *apo*-form ArsR-SmtB-type metalloregulators (CadC, CmtR and BxmR) inhibit transcription by association with the promoter region of their target genes. Metal binding disassembles this association and consequently induces transcription. In this family, CadC from *S. aureus* or

L. monocytogenes is involved in the cadmium-dependent induction of the P$_{1B}$-ATPase encoded by the CadA genes.[194,272] CadC from *S. aureus* forms a homodimer containing two pairs of metal binding sites at the interface between the two monomers: a type 1 thiol-rich cadmium- and lead-responsive site (affinity in the picomolar range) and a type 2 oxygen/nitrogen zinc binding site.[273] CmtR metalloregulators occur in several bacteria and regulate P$_{1B}$-ATPases or CDF encoding genes.[274–276] In *M. tuberculosis*, CmtR has one pair of metal-binding sites, whereas two pairs are found in the version from *Streptomyces coelicolor*. In the latter, the affinities for cadmium of the two sites are in the picomolar range, but only the type 2 site differentiates between cadmium and lead. While CadC and CmtR exclusively respond to divalent metals, BxmR from the cyanobacterium *Oscillatoria brevis* derepresses the *bxa1*, *bxmR* and *bmtA* genes (coding for a P$_{1B}$-ATPase and two metallothioneins, respectively) in the presence of cadmium and zinc, but also copper and silver.[277]

The DtxR/MntR metalloregulators repress transcription of target genes in their dimeric metal bound form. In *B. subtilis*, MntR binds to the promoter region of the NRAMP MntH coding gene in the presence of manganese and cadmium. Mutation of the *mntR* gene increases the sensitivity to cadmium.[278] In *Streptococcus gordonii*, ScaR, an orthologue of DtxR, represses the transcription of an ABC-type manganese importer under high manganese concentrations. It binds other transition metal ions including cadmium.[279]

The MerR metalloregulators activate transcription of target genes in their metal bound form.[280] ZntR in *E. coli* and CadR in *Pseudomonas putida* both mediate cadmium-dependent activation of P$_{1B}$-ATPase transcription.[196,281]

It thus appears that bacterial transcription factors are far more sensitive to cadmium (K_m in the pM or nM range) than are transport systems. The latter generally display Michaelis constants in the μM range. This should trigger an early cellular response upon exposure to even moderate amounts of cadmium. However, no metalloregulator exclusively dedicated to cadmium sensing has been identified in any prokaryote.

23.8.1.2 *In Yeast*

In *S. cerevisiae*, Yap transcription factors, members of the bZIP (basic-leucine Zipper) protein family,[282] are sensitive to cadmium and bind to a sequence called YRE (for Yap Responsive Element) in the promoter region of the regulated genes (Section 23.4.1.1.2). Mutations of YRE increase the sensitivity to cadmium[283,284] and the YAP:YRE stoichiometry determines cadmium tolerance.[284,285] Activation occurs by increased residence time in the nucleus of Yap1p and its paralogue Yap2p, which depends on the nuclear exporter Crm1p. Association of Cd^{2+} with cysteines 356 and 391 of Yap2p and with cysteines in the C-terminal region of Yap1p may contribute to nuclear retention of the active YRE-binding forms.[286] Yap1 also regulates the transcription of Ycf1p, the major protein involved in cadmium vacuolar sequestering.[138]

Other transcription factors involved in sulfur utilization, such as in methionine biosynthesis, also regulate γ-glutamylcysteine synthase.[287,288] Their activities influence the resistance of yeast to cadmium.[289] The transcriptional activator Met4p of yeast sulfur homeostasis is mainly post-translationally regulated by proteasomal degradation. Cadmium activates Met4p by promoting the dissociation of the F-box protein Met30p from the ubiquitin ligase complex SCF^{Met30} thus impairing ubiquitination.[290,291]

23.8.1.3   In Plants

The modification of the ultrastructure of the cell wall when plants are exposed to cadmium suggests that plants detect external cadmium. When entering into the cells, toxic metals trigger, like other plant stressors, a series of signal transduction events.[292] Cadmium, *via* induction of NADPH oxidase activity at the plasma membrane, provokes a first burst of oxidative response, witnessed by the accumulation of H_2O_2, as a result of an increased cytosolic calcium concentration.[293,294] The increase of oxygen derivatives (ROS) was proposed as a rapid, long-distance, auto-propagating signal through the plant.[295] Analysis of protein phosphorylation revealed the activation of the four mitogen-activated protein kinases (MAPK): SIMK, MMK$_2$, MMK$_3$ and SAMK in alfalfa (*Medicago sativa*) by cadmium, and also by copper. However, distinct signal transduction pathways are induced for the two metals and different kinetics are observed.[296] In *Arabidopsis*, cadmium is capable of activating MAPK, MPK3 and MPK6 in a dose-dependent manner. The accumulation of ROS seems to be necessary for this activation since a pre-treatment with the ROS scavenger GSH drastically inhibits it.[297,298] Recently, microRNAs have emerged as a class of regulators participating in the cellular response to stress, including that from cadmium.[292,299] As an example, miRNA398 induces the two Cu/Zn superoxide dismutases (named CSD1/2 in plants) under the oxidative stress conditions generated by large copper or cadmium concentrations.[300,301] Other miRNAs target a multitude of genes under cadmium stress,[301-303] and this constitutes a major challenge for deciphering the regulatory networks involved in metal homeostasis and toxicology. Nevertheless, an emerging feature is that miRNAs may regulate expression of transcription factors that control plant growth and development under stress conditions, particularly the miRNAs that regulate auxin perception and signalling.[304]

23.8.1.4   In Mammals

Exposure of epithelial cultures to µM concentrations of cadmium salts disrupts cell-cell contacts with simultaneous morphological changes. Cd^{2+} binding to fragments of the cell-adhesion protein E-cadherin occurs with lower dissociation constants (higher affinity) than for calcium.[305] If intact cadherins bind cadmium present in usual toxicological situations, the subsequent release of β-catenin and its signalling function should transmit the presence of cadmium and participate in chronic cadmium poisoning.[306] Although interaction of external cadmium with cell surface receptors has been discussed regularly,[307] specific effects of cadmium on signalling initiated by the estrogen receptor (ER) has received particular attention. Cd^{2+} binds strongly to the ERα with a dissociation constant of *ca.* 4.5×10^{-10} M, only 1.5 times that of estradiol itself and compatible with the observation that low level exposure to cadmium has estrogenic effects. Cd^{2+} perturbs 17β-estradiol binding[308] and, whereas the cadmium activation of the receptor seems less efficient than that induced by estrogens, qualitatively it has similar consequences.[309] The molecular interaction between Cd^{2+} and ERα is often considered to be the triggering event of some cases of breast cancer. The apparent tight binding of Cd^{2+} to ERα is a feature that has no reason to be an exception, given the large number of receptors present at the surface of the different animal cell types and the even larger range of their agonists. Thus, some of the biological effects of cadmium may be mediated by interaction at the cell surface as exemplified by that with ERα.

Intracellularly, cells respond to cadmium by activating the Metal Transcription Factor (MTF1) that up-regulates a range of genes, among which are those encoding MT and γ-glutamylcysteine synthetase.[310] The DNA sequences recognized by MTF1 change with the inducing metal,[311] and the effect of cadmium on transcription is not exclusively mediated by MTF1.[312] These observations suggest that MTF1 cannot be considered as a genuine cadmium sensor, but that it displays some specificity in its sensitivity to cadmium. Another DNA motif, called the cadmium response element, triggers haem oxygenase-1 induction by cadmium with the involvement of the Heat Shock Factor 1[313] or the nucleolar protein Pescadillo,[314] whereas other less specific DNA regulatory sequences, such as Maf responsive elements (MARE), are also sensitive to cadmium.

These occurrences may be considered sensing mechanisms, although the primary function of the responsive molecules is not the detection of cadmium. There is no evidence that any specific sensing system monitors the presence of this toxic metal in cells. Yet cellular adaptation may occur after some time using various means to fight against the cadmium insult.[315]

23.8.2 Buffering Systems

Once inside cells, cadmium becomes bound to GSH and MT, simultaneously or in succession. This process constitutes a significant cadmium scavenging capacity. The tighter binding of cadmium to MT as compared to zinc (see above) should maintain a concentration of easily mobilized ("free") cadmium below that of zinc, which was estimated to be 2×10^{-9} M at most under different conditions.[316] These estimates may vary with the cell type and with environmental conditions as, for instance, the degradation of the Cd-MT complex is more efficient in renal cells than in intestinal ones (Section 23.5). Thus, the scavenging capacity of a given cell, regulated mainly by GSH and MT synthesis, should be able to meet the cadmium challenge, up to a point. Oxidative conditions trigger zinc release from MT[317] and the same reaction may occur with cadmium, although this has not been assessed.

The large amount of vacuolar cadmium[251] is mainly bound to malate in the leaves of *N. caerulescens*,[318] but it is surrounded by O/N ligands in the shoots of *A. thaliana*[246] and by O ligands provided by the cell walls and by organic acids stored in vacuoles in the hyperaccumulator *Thlaspi praecox*.[256] Metal speciation depends also on the age of the tissues as shown in *N. caerulescens* where 25% of cadmium is bound to thiol-containing ligands in young leaves whereas it is almost exclusively bound to oxygen-containing ligands in senescent ones.[319] These observations indicate that plants implement several strategies with different ligands to shield themselves from deleterious cadmium.

Overall, eukaryotic cells can trap cadmium in high-affinity ligand molecules but there is no evidence that this can be done in a regulated way nor that a buffering role for cadmium can be assigned to any cellular component.

23.8.3 Cellular Response through Endogenous Regulatory Mechanisms

The regulatory mechanisms impacted by cadmium in cells of different kinds can be divided into three main classes with shared elements: maintenance of the redox status, calcium signalling and homeostasis of essential metals. The inhibitory properties of cadmium toward a range of activities supporting fundamental functions, such as photosynthesis and respiration, lead to cellular redox imbalance. The cell response is induction of anti-oxidant activities, including those involved in ROS scavenging, and defence against other forms of stress, such as misfolding, by synthesis of heat shock proteins and the UPR (Section 23.5.3.2.2). Conserving cellular sulfur to promote the synthesis of GSH and PC is part of this response. Cadmium fluxes influence the calcium-dependent cellular behaviour through perturbation of calcium traffic and signalling.[306] In the guard cells of plants, Ca^{2+} signalling and osmoregulation are disturbed with consequences on the plant-water relationship.[58] Considering the number of calcium-dependent functions and the importance of calcium as a second messenger, cadmium is thus a significant troublemaker in a variety of cell types. Signalling pathways relying on other messengers, such as the brassinosteroid plant hormone, are involved in the response to cadmium, since addition of this hormone increases the toxicity of cadmium and its depletion increases tolerance,[320] and such effects occur in many biological contexts.[306] Zinc is another important secondary messenger like calcium[321] and MT is a major point of convergence between cellular cadmium effects and zinc homeostasis. Zinc release, MT up-regulation and MTF1 activation are some consequences of the presence of cadmium, even at low concentrations. Besides zinc, the homeostasis of other metal cations could be affected by cadmium at different steps, but mechanistic details are often missing.[72] Yet, cadmium traffic depends on the homeostatic status of other metals, as in the case of iron deficiency in animals, which enhances cadmium uptake by DMT1 (Section 23.3). Reciprocally, plant exposure to cadmium influences the concentration of other metals such as zinc,[106] iron,[255] calcium and potassium.[322,323]

The central position of the cadmium-sensitive pathways and their intimate interrelationships in the regulation of the fate of the cell illustrate the main features of cadmium effects: proliferation or death, mainly by apoptosis, depending on the relative importance of these regulatory networks in each cellular system considered.

23.9 Conclusions with Open Questions

The literature dealing with all conceivable aspects of the interaction of cadmium with living cells is very large. Yet, it contains minimal quantitative delineation of the induced events and consequences. This applies to transport, for which candidate molecules abound (Tables 23.2–23.4), but for which biochemical features often remain of dubious significance in the environmental conditions pertaining to natural cadmium exposure. The work and effort needed to clarify these questions, including ruling out inappropriate suggestions,[238] are rarely addressed. They usually do not receive the attention they deserve. For instance, the often-discussed similarity between zinc and cadmium should not be taken for granted as a toxicity mechanism for cadmium. This has been recently demonstrated with the ZnT family of human transporters.[80] There is no doubt about ion mimicry among metals, including Cd^{2+}, but this cannot be a generally satisfactory explanation for the effects of cadmium. Too often, speculation dominates the (important) biological questions associated with cadmium and it is hoped that future investigations will focus on quantitative measurements in a relevant biological context. This will require significant technical improvements since functionally monitoring selected molecules within cells or whole multicellular organisms remains a challenging task. Monitored changes can be minute under low-level chronic exposure to toxic compounds such as cadmium.

The available dissociation constants for biologically relevant cadmium complexes bring forth a general picture in which competition of cysteine-rich sites for Cd^{2+} dominates its interactions with cellular molecules. This should occur at low concentrations, in the nanomolar range and below. However, the specificity of these interactions, if any, is virtually unknown. The phenotypical consequences of cadmium binding to thiolate-containing molecules vary with the biological framework, for reasons that remain unclear.

The biochemistry of cadmium is characterized by massive pleiotropic biological effects. This statement questions any proposed mechanism for the biological action of cadmium. Progress in the area will come from studies that sort out the relevant cadmium-sensitive pathways from the dubious.

References

1. K. Aoshima, *Tohoku J. Exp. Med.*, 1987, **152**, 151.
2. T. Nawrot, M. Plusquin, J. Hogervorst, H. A. Roels, H. Celis, L. Thijs, J. Vangronsveld, E. Van Hecke and J. A. Staessen, *Lancet Oncol.*, 2006, 7, 119.
3. G. F. Nordberg, *Lancet Oncol.*, 2006, 7, 99.
4. J. W. Anthony, R. A. Bideaux, K. W. Bladh and M. C. Nichols, *Mineralogical Society of America*, Chantilly, VA 20151-1110, USA, 2003, vol. 2012.
5. M. E. Weeks, *J. Chem. Educ.*, 1932, **9**, 1046.
6. W. Maret and J. M. Moulis, in *Cadmium: From Toxicity to Essentiality*, ed. A. Sigel, H. Sigel and R. K. O. Sigel, Springer Science + Business Media, Dordrecht, 2013, Metal Ions in Life Sciences, 11, p 1.
7. Y. Xu, L. Feng, P. D. Jeffrey, Y. Shi and F. M. Morel, *Nature*, 2008, **452**, 56.
8. K. Chojnacka, *Environ. Int.*, 2010, **36**, 299.
9. J. Wang and C. Chen, *Biotechnol. Adv.*, 2006, **24**, 427.
10. A. Malik, *Environ. Int.*, 2004, **30**, 261.
11. J. Wang and C. Chen, *Biotechnol. Adv.*, 2009, **27**, 195.
12. B. Volesky, *Water Res.*, 2007, **41**, 4017.
13. T. M. Roane, K. L. Josephson and I. L. Pepper, *Appl. Environ. Microbiol.*, 2001, **67**, 3208.
14. Y. Lin, X. Wang, B. Wang, O. Mohamad and G. Wei, *Ecotoxicol. Environ. Saf.*, 2012, 77, 7.
15. R. Chakravarty and P. C. Banerjee, *Bioresour. Technol.*, 2012, **108**, 176.
16. B. Volesky, H. May and Z. R. Holan, *Biotechnol. Bioeng.*, 1993, **41**, 826.
17. M. Krzeslowska, *Acta Physiol. Plant.*, 2011, **33**, 35.
18. F. Paynel, A. Schaumann, M. Arkoun, O. Douchiche and C. Morvan, *Ann. Bot.*, 2009, **104**, 1363.
19. M. Kuramata, S. Masuya, Y. Takahashi, E. Kitagawa, C. Inoue, S. Ishikawa, S. Youssefian and T. Kusano, *Plant Cell Physiol.*, 2009, **50**, 106.
20. L. Järup, M. Berglund, C. G. Elinder, G. Nordberg and M. Vahter, *Scandinav. J. Work Environ. Health*, 1998, **24**, 1.
21. Z. Tynecka, Z. Gos and J. Zajac, *J. Bacteriol.*, 1981, **147**, 305.
22. R. A. Laddaga, R. Bessen and S. Silver, *J. Bacteriol.*, 1985, **162**, 1106.
23. A. Kirsten, M. Herzberg, A. Voigt, J. Seravalli, G. Grass, J. Scherer and D. H. Nies, *J. Bacteriol.*, 2011, **193**, 4652.
24. P. Courville, E. Urbankova, C. Rensing, R. Chaloupka, M. Quick and M. F. Cellier, *J. Biol. Chem.*, 2008, **283**, 9651.
25. R. A. Laddaga and S. Silver, *J. Bacteriol.*, 1985, **162**, 1100.
26. G. Grass, M. D. Wong, B. P. Rosen, R. L. Smith and C. Rensing, *J. Bacteriol.*, 2002, **184**, 864.
27. W. Lin, J. Chai, J. Love and D. Fu, *J. Biol. Chem.*, 2010, **285**, 39013.
28. N. Taudte and G. Grass, *Biometals*, 2010, **23**, 643.
29. D. H. Nies and S. Silver, *J. Bacteriol.*, 1989, **171**, 4073.

30. Z. Hao, H. R. Reiske and D. B. Wilson, *Appl. Environ. Microbiol.*, 1999, **65**, 4741.

31. Z. Hao, S. Chen and D. B. Wilson, *Appl. Environ. Microbiol.*, 1999, **65**, 4746.

32. S. K. Kim, B. S. Lee, D. B. Wilson and E. K. Kim, *J. Biosci. Bioeng.*, 2005, **99**, 109.

33. X. Deng, X. E. Yi and G. Liu, *J. Hazard. Mater.*, 2007, **139**, 340.

34. M. E. Portnoy, X. F. Liu and V. C. Culotta, *Mol. Cell. Biol.*, 2000, **20**, 7893.

35. A. Cohen, H. Nelson and N. Nelson, *J. Biol. Chem.*, 2000, **275**, 33388.

36. F. Supek, L. Supekova, H. Nelson and N. Nelson, *Proc. Natl Acad. Sci. USA*, 1996, **93**, 5105.

37. X. F. Liu, F. Supek, N. Nelson and V. C. Culotta, *J. Biol. Chem.*, 1997, **272**, 11763.

38. R. Ruotolo, G. Marchini and S. Ottonello, *Genome Biol.*, 2008, **9**, R67.

39. X. F. Liu and V. C. Culotta, *J. Mol. Biol.*, 1999, **289**, 885.

40. H. E. Stimpson, M. J. Lewis and H. R. Pelham, *EMBO J.*, 2006, **25**, 662.

41. D. S. Gomes, L. C. Fragoso, C. J. Riger, A. D. Panek and E. C. Eleutherio, *Biochim. Biophys. Acta*, 2002, **1573**, 21.

42. A. Gardarin, S. Chedin, G. Lagniel, J. C. Aude, E. Godat, P. Catty and J. Labarre, *Mol. Microbiol.*, 2010, **76**, 1034.

43. D. Eide, M. Broderius, J. Fett and M. L. Guerinot, *Proc. Natl Acad. Sci. USA*, 1996, **93**, 5624.

44. G. Vert, N. Grotz, F. Dedaldechamp, F. Gaymard, M. L. Guerinot, J. F. Briat and C. Curie, *Plant Cell*, 2002, **14**, 1223.

45. E. L. Connolly, J. P. Fett and M. L. Guerinot, *Plant Cell*, 2002, **14**, 1347.

46. C. K. Cohen, T. C. Fox, D. F. Garvin and L. V. Kochian, *Plant Physiol.*, 1998, **116**, 1063.

47. E. Lombi, K. L. Tearall, J. R. Howarth, F. J. Zhao, M. J. Hawkesford and S. P. McGrath, *Plant Physiol.*, 2002, **128**, 1359.

48. H. Nakanishi, I. Ogawa, Y. Ishimaru, S. Mori and N. K. Nishizawa, *Soil Sci. Plant Nutr.*, 2006, **52**, 464.

49. Y. O. Korshunova, D. Eide, W. G. Clark, M. L. Guerinot and H. B. Pakrasi, *Plant Mol. Biol.*, 1999, **40**, 37.

50. C. K. Cohen, D. F. Garvin and L. V. Kochian, *Planta*, 2004, **218**, 784.

51. N. Grotz, T. Fox, E. Connolly, W. Park, M. L. Guerinot and D. Eide, *Proc. Natl Acad. Sci. USA*, 1998, **95**, 7220.

52. S. A. Ramesh, R. Shin, D. J. Eide and D. P. Schachtman, *Plant Physiol.*, 2003, **133**, 126.

53. B. W. Stephens, D. R. Cook and M. A. Grusak, *BioMetals*, 2011, **24**, 51.

54. N. S. Pence, P. B. Larsen, S. D. Ebbs, D. L. D. Letham, M. M. Lasat, D. F. Garvin, D. Eide and L. V. Kochian, *Proc. Natl Acad. Sci. USA*, 2000, **97**, 4956.

55. M. J. Milner, E. Craft, N. Yamaji, E. Koyama, J. F. Ma and L. V. Kochian, *New Phytol.*, 2012, **195**, 113.

56. A. Sasaki, N. Yamaji, K. Yokosho and J. F. Ma, *Plant Cell*, 2012, **24**, 2155.

57. R. Takahashi, Y. Ishimaru, T. Senoura, H. Shimo, S. Ishikawa, T. Arao, H. Nakanishi and N. K. Nishizawa, *J. Exp. Bot.*, 2011, **62**, 4843.

58. L. Perfus-Barbeoch, N. Leonhardt, A. Vavasseur and C. Forestier, *Plant J.*, 2002, **32**, 539.

59. S. Clemens, D. M. Antosiewicz, J. M. Ward, D. P. Schachtman and J. I. Schroeder, *Proc. Natl Acad. Sci. USA*, 1998, **95**, 12043.

60. C. Pruvot, F. Douay, F. Herve and C. Waterlot, *J. Soils Sed.*, 2006, **6**, 215.

61. B. Isikli, T. A. Demir, T. Akar, A. Berber, S. M. Urer, C. Kalyoncu and M. Canbek, *Chemosphere*, 2006, **63**, 1546.

62. R. M. Harrison and M. B. Chirgawi, *Sci. Total Environ.*, 1989, **83**, 13.

63. L. De Temmerman and M. Hoenig, *J. Atmosph. Chem.*, 2004, **49**, 121.

64. A. G. Davison, P. M. Fayers, A. J. Taylor, K. M. Venables, J. Darbyshire, C. A. Pickering, D. R. Chettle, D. Franklin, C. J. Guthrie, M. C. Scott, H. Holden, A. L. Wright and D. Gompertz, *Lancet*, 1988, **331**, 663.

65. I. M. McKenna, M. P. Waalkes, L. C. Chen and T. Gordon, *Toxicol. Appl. Pharmacol.*, 1997, **146**, 196.

66. C. Jumarie, *Biochim. Biophys. Acta*, 2002, **1564**, 487.

67. S. Jenkitkasemwong, C. Y. Wang, B. Mackenzie and M. D. Knutson, *BioMetals*, 2012, **25**, 643.

68. Z. Liu, H. Li, M. Soleimani, K. Girijashanker, J. M. Reed, L. He, T. P. Dalton and D. W. Nebert, *Biochem. Biophys. Res. Commun.*, 2008, **365**, 814.

69. Y. S. Lin, J. L. Caffrey, M. H. Chang, N. Dowling and J. W. Lin, *Respir. Res.*, 2010, **11**, 53.

70. M. Mantha, L. El Idrissi, T. Leclerc-Beaulieu and C. Jumarie, *Toxicol. In Vitro*, 2011, **25**, 1701.

71. C. Y. Wang, S. Jenkitkasemwong, S. Duarte, B. K. Sparkman, A. Shawki, B. Mackenzie and M. D. Knutson, *J. Biol. Chem.*, 2012, **287**, 34032.

72. J. M. Moulis, *BioMetals*, 2010, **23**, 877.

73. C. A. Pearson, P. C. Lamar and W. C. Prozialeck, *Life Sci.*, 2003, **72**, 1303.

74. J. P. Bressler, L. Olivi, J. H. Cheong, Y. Kim and D. Bannona, *Ann. NY Acad. Sci.*, 2004, **1012**, 142.

75. M. Okubo, K. Yamada, M. Hosoyamada, T. Shibasaki and H. Endou, *Toxicol. Appl. Pharmacol.*, 2003, **187**, 162.

76. F. Elisma and C. Jumarie, *Biochem. Biophys. Res. Commun.*, 2001, **285**, 662.

77. D. W. Nebert, M. Galvez-Peralta, E. B. Hay, H. Li, E. Johansson, C. Yin, B. Wang, L. He and M. Soleimani, *Metallomics*, 2012, **4**, 1218.

78. R. K. Zalups and S. Ahmad, *Toxicol. Appl. Pharmacol.*, 2003, **186**, 163.

79. R. A. Valentine, K. A. Jackson, G. R. Christie, J. C. Mathers, P. M. Taylor and D. Ford, *J. Biol. Chem.*, 2007, **282**, 14389.

80. E. Hoch, W. Lin, J. Chai, M. Hershfinkel, D. Fu and I. Sekler, *Proc. Natl Acad. Sci. USA*, 2012, **109**, 7202.

81. K. S. Min, H. Ueda, T. Kihara and K. Tanaka, *Toxicol. Sci.*, 2008, **106**, 284.

82. T. Suzuki, K. Momoi, M. Hosoyamada, M. Kimura and T. Shibasaki, *Toxicol. Appl. Pharmacol.*, 2008, **227**, 462.

83. E. Rousselet, P. Richaud, T. Douki, J. G. Chantegrel, A. Favier, A. Bouron and J. M. Moulis, *Toxicol. Appl. Pharmacol.*, 2008, **230**, 312.
84. K. S. Min, H. Ueda and K. Tanaka, *Toxicol. Lett.*, 2008, **176**, 85.
85. Y. Fujita, H. I. el Belbasi, K. S. Min, S. Onosaka, Y. Okada, Y. Matsumoto, N. Mutoh and K. Tanaka, *Res. Commun. Chem. Pathol. Pharmacol.*, 1993, **82**, 357.
86. C. Jumarie, C. Fortin, M. Houde, P. G. Campbell and F. Denizeau, *Toxicol. Appl. Pharmacol.*, 2001, **170**, 29.
87. M. G. Cherian, *Environ. Health Perspect.*, 1979, **28**, 127.
88. I. Pocsi, R. A. Prade and M. J. Penninckx, *Adv. Microb. Physiol.*, 2004, **49**, 1.
89. M. J. Penninckx, *FEMS Yeast Res.*, 2002, **2**, 295.
90. H. Suzuki, T. Koyanagi, S. Izuka, A. Onishi and H. Kumagai, *J. Bacteriol.*, 2005, **187**, 5861.
91. A. Bourbouloux, P. Shahi, A. Chakladar, S. Delrot and A. K. Bachhawat, *J. Biol. Chem.*, 2000, **275**, 13259.
92. O. Delalande, H. Desvaux, E. Godat, A. Valleix, C. Junot, J. Labarre and Y. Boulard, *FEBS J.*, 2010, **277**, 5086.
93. K. Helbig, C. Bleuel, G. J. Krauss and D. H. Nies, *J. Bacteriol.*, 2008, **190**, 5431.
94. K. Helbig, C. Grosse and D. H. Nies, *J. Bacteriol.*, 2008, **190**, 5439.
95. E. Bianucci, A. Fabra and S. Castro, *BioMetals*, 2012, **25**, 23.
96. D. W. Stephen and D. J. Jamieson, *Mol. Microbiol.*, 1997, **23**, 203.
97. K. Vido, D. Spector, G. Lagniel, S. Lopez, M. B. Toledano and J. Labarre, *J. Biol. Chem.*, 2001, **276**, 8469.
98. A. Lafaye, C. Junot, Y. Pereira, G. Lagniel, J. C. Tabet, E. Ezan and J. Labarre, *J. Biol. Chem.*, 2005, **280**, 24723.
99. Y. H. Jin, P. E. Dunlap, S. J. McBride, H. Al-Refai, P. R. Bushel and J. H. Freedman, *PloS Genet.*, 2008, **4**, e1000053.
100. P. Baudouin-Cornu, G. Lagniel, C. Kumar, M. E. Huang and J. Labarre, *J. Biol. Chem.*, 2012, **287**, 4552.
101. P. J. Kennedy, A. A. Vashisht, K. L. Hoe, D. U. Kim, H. O. Park, J. Hayles and P. Russell, *Toxicol. Sci.*, 2008, **106**, 124.
102. R. Howden, C. R. Andersen, P. B. Goldsbrough and C. S. Cobbett, *Plant Physiol.*, 1995, **107**, 1067.
103. F. Villiers, C. Ducruix, V. Hugouvieux, N. Jarno, E. Ezan, J. Garin, C. Junot and J. Bourguignon, *Proteomics*, 2011, **11**, 1650.
104. M. J. May, T. Vernoux, C. Leaver, M. Van Montagu and D. Inze, *J. Exp. Bot.*, 1998, **49**, 649.
105. S. Herbette, L. Taconnat, V. Hugouvieux, L. Piette, M. L. M. Magniette, S. Cuine, P. Auroy, P. Richaud, C. Forestier, J. Bourguignon, J. P. Renou, A. Vavasseur and N. Leonhardt, *Biochimie*, 2006, **88**, 1751.
106. M. Weber, A. Trampczynska and S. Clemens, *Plant Cell Environ.*, 2006, **29**, 950.
107. S. Clemens, E. J. Kim, D. Neumann and J. I. Schroeder, *EMBO J.*, 1999, **18**, 3325.

108. J. G. Vande Weghe and D. W. Ow, *Mol. Microbiol.*, 2001, **42**, 29.
109. W. Bae and X. Chen, *Mol. Cell. Proteomics*, 2004, **3**, 596.
110. O. K. Vatamaniuk, S. Mari, Y. P. Lu and P. A. Rea, *J. Biol. Chem.*, 2000, **275**, 31451.
111. R. Howden, P. B. Goldsbrough, C. R. Andersen and C. S. Cobbett, *Plant Physiol.*, 1995, **107**, 1059.
112. C. Ducruix, C. Junot, J. B. Fievet, F. Villiers, E. Ezan and J. Bourguignon, *Biochimie*, 2006, **88**, 1733.
113. J. E. Sarry, L. Kuhn, C. Ducruix, A. Lafaye, C. Junot, V. Hugouvieux, A. Jourdain, O. Bastien, J. B. Fievet, D. Vailhen, B. Amekraz, C. Moulin, E. Ezan, J. Garin and J. Bourguignon, *Proteomics*, 2006, **6**, 2180.
114. M. Margoshes and B. L. Vallee, *J. Am. Chem. Soc.*, 1957, **79**, 4813.
115. C. A. Blindauer, *J. Biol. Inorg. Chem.*, 2011, **16**, 1011.
116. O. Palacios, S. Atrian and M. Capdevila, *J. Biol. Inorg. Chem.*, 2011, **16**, 991.
117. J. S. Turner, A. P. Morby, B. A. Whitton, A. Gupta and N. J. Robinson, *J. Biol. Chem.*, 1993, **268**, 4494.
118. M. J. Daniels, J. S. Turner-Cavet, R. Selkirk, H. Sun, J. A. Parkinson, P. J. Sadler and N. J. Robinson, *J. Biol. Chem.*, 1998, **273**, 22957.
119. B. Gold, H. Deng, R. Bryk, D. Vargas, D. Eliezer, J. Roberts, X. Jiang and C. Nathan, *Nat. Chem. Biol.*, 2008, **4**, 609.
120. M. Thorsen, G. G. Perrone, E. Kristiansson, M. Traini, T. Ye, I. W. Dawes, O. Nerman and M. J. Tamas, *BMC Genomics*, 2009, **10**, 105.
121. D. J. Ecker, T. R. Butt, E. J. Sternberg, M. P. Neeper, C. Debouck, J. A. Gorman and S. T. Crooke, *J. Biol. Chem.*, 1986, **261**, 16895.
122. D. R. Winge, K. B. Nielson, W. R. Gray and D. H. Hamer, *J. Biol. Chem.*, 1985, **260**, 14464.
123. R. Orihuela, F. Monteiro, A. Pagani, M. Capdevila and S. Atrian, *Chemistry*, 2010, **16**, 12363.
124. A. Pagani, L. Villarreal, M. Capdevila and S. Atrian, *Mol. Microbiol.*, 2007, **63**, 256.
125. V. C. Culotta, W. R. Howard and X. F. Liu, *J. Biol. Chem.*, 1994, **269**, 25295.
126. C. Cobbett and P. Goldsbrough, *Annu. Rev. Plant Biol.*, 2002, **53**, 159.
127. W. J. Guo, M. Meetam and P. B. Goldsbrough, *Plant Physiol.*, 2008, **146**, 1697.
128. E. Freisinger, *J. Biol. Inorg. Chem.*, 2011, **16**, 1035.
129. L. R. Fernandez, G. Vandenbussche, N. Roosens, C. Govaerts, E. Goormaghtigh and N. Verbruggen, *Biochim. Biophys. Acta Proteins Proteomics*, 2012, **1824**, 1016.
130. J. Domenech, A. Tinti, M. Capdevila, S. Atrian and A. Torreggiani, *Biopolymers*, 2007, **86**, 240.
131. A. M. Zimeri, O. P. Dhankher, B. McCaig and R. B. Meagher, *Plant Mol. Biol.*, 2005, **58**, 839.
132. A. Zhigang, C. J. Li, Y. G. Zu, Y. J. Du, A. Wachter, R. Gromes and T. Rausch, *J. Exp. Bot.*, 2006, **57**, 3575.

133. M. Höckner, R. Dallinger and S. R. Sturzenbaum, *J. Biol. Inorg. Chem.*, 2011, **16**, 1057.
134. D. H. Petering and C. F. Shaw, 3rd, *Methods Enzymol.*, 1991, **205**, 475.
135. C. E. Hand and J. F. Honek, *J. Nat. Prod.*, 2005, **68**, 293.
136. C. Dutilleul, A. Jourdain, J. Bourguignon and V. Hugouvieux, *Plant Physiol.*, 2008, **147**, 239.
137. V. Hugouvieux, C. Dutilleul, A. Jourdain, F. Reynaud, V. Lopez and J. Bourguignon, *Plant Physiol.*, 2009, **151**, 768.
138. J. A. Wemmie, M. S. Szczypka, D. J. Thiele and W. S. Moye-Rowley, *J. Biol. Chem.*, 1994, **269**, 32592.
139. M. S. Szczypka, J. A. Wemmie, W. S. Moye-Rowley and D. J. Thiele, *J. Biol. Chem.*, 1994, **269**, 22853.
140. Z. S. Li, M. Szczypka, Y. P. Lu, D. J. Thiele and P. A. Rea, *J. Biol. Chem.*, 1996, **271**, 6509.
141. J. F. Rebbeor, G. C. Connolly, M. E. Dumont and N. Ballatori, *Biochem. J.*, 1998, **334**, 723.
142. Z. S. Li, Y. P. Lu, R. G. Zhen, M. Szczypka, D. J. Thiele and P. A. Rea, *Proc. Natl Acad. Sci. USA*, 1997, **94**, 42.
143. P. D. Adamis, A. D. Panek and E. C. Eleutherio, *Toxicol. Lett.*, 2007, **173**, 1.
144. M. Klein, Y. M. Mamnun, T. Eggmann, C. Schuller, H. Wolfger, E. Martinoia and K. Kuchler, *FEBS Lett.*, 2002, **520**, 63.
145. K. G. Sharma, D. L. Mason, G. Liu, P. A. Rea, A. K. Bachhawat and S. Michaelis, *Eukaryot. Cell*, 2002, **1**, 391.
146. D. Wawrzycka, I. Sobczak, G. Bartosz, T. Bocer, S. Ulaszewski and A. Goffeau, *FEMS Yeast Res.*, 2010, **10**, 828.
147. S. Miyabe, S. Izawa and Y. Inoue, *Biochem. Biophys. Res. Commun.*, 2001, **282**, 79.
148. S. Miyabe, S. Izawa and Y. Inoue, *Biochem. Biophys. Res. Commun.*, 2000, **276**, 879.
149. A. Kamizono, M. Nishizawa, Y. Teranishi, K. Murata and A. Kimura, *Mol. Gen. Genet.*, 1989, **219**, 161.
150. C. W. MacDiarmid, M. A. Milanick and D. J. Eide, *J. Biol. Chem.*, 2002, 277, 39187.
151. D. F. Ortiz, L. Kreppel, D. M. Speiser, G. Scheel, G. McDonald and D. W. Ow, *EMBO J.*, 1992, **11**, 3491.
152. P. Perego, J. Vande Weghe, D. W. Ow and S. B. Howell, *Mol. Pharmacol.*, 1997, **51**, 12.
153. S. Preveral, L. Gayet, C. Moldes, J. Hoffmann, S. Mounicou, A. Gruet, F. Reynaud, R. Lobinski, J. M. Verbavatz, A. Vavasseur and C. Forestier, *J. Biol. Chem.*, 2009, **284**, 4936.
154. D. F. Ortiz, T. Ruscitti, K. F. McCue and D. W. Ow, *J. Biol. Chem.*, 1995, **270**, 4721.
155. K. D. Hirschi, V. D. Korenkov, N. L. Wilganowski and G. J. Wagner, *Plant Physiol.*, 2000, **124**, 125.
156. V. Koren'kov, S. Park, N. H. Cheng, C. Sreevidya, J. Lachmansingh, J. Morris, K. Hirschi and G. J. Wagner, *Planta*, 2007, **225**, 403.

157. V. Korenkov, K. Hirschi, J. D. Crutchfield and G. J. Wagner, *Planta*, 2007, **226**, 1379.
158. T. Schneider, M. Schellenberg, S. Meyer, F. Keller, P. Gehrig, K. Riedel, Y. Lee, L. Eberl and E. Martinoia, *Proteomics*, 2009, **9**, 2668.
159. A. Gravot, A. Lieutaud, F. Verret, P. Auroy, A. Vavasseur and P. Richaud, *FEBS Lett.*, 2004, **561**, 22.
160. M. Morel, J. Crouzet, A. Gravot, P. Auroy, N. Leonhardt, A. Vavasseur and P. Richaud, *Plant Physiol.*, 2009, **149**, 894.
161. D. Hussain, M. J. Haydon, Y. Wang, E. Wong, S. M. Sherson, J. Young, J. Camakaris, J. F. Harper and C. S. Cobbett, *Plant Cell*, 2004, **16**, 1327.
162. I. N. Talke, M. Hanikenne and U. Kramer, *Plant Physiol.*, 2006, **142**, 148.
163. M. Becher, I. N. Talke, L. Krall and U. Kramer, *Plant J.*, 2004, **37**, 251.
164. D. Ueno, M. J. Milner, N. Yamaji, K. Yokosho, E. Koyama, M. C. Zambrano, M. Kaskie, S. Ebbs, L. V. Kochian and J. F. Ma, *Plant J.*, 2011, **66**, 852.
165. D. Ueno, N. Yamaji, I. Kono, C. F. Huang, T. Ando, M. Yano and J. F. Ma, *Proc. Natl Acad. Sci. USA*, 2010, **107**, 16500.
166. H. Miyadate, S. Adachi, A. Hiraizumi, K. Tezuka, N. Nakazawa, T. Kawamoto, K. Katou, I. Kodama, K. Sakurai, H. Takahashi, N. Satoh-Nagasawa, A. Watanabe, T. Fujimura and H. Akagi, *New Phytol.*, 2011, **189**, 190.
167. D. Ueno, E. Koyama, N. Yamaji and J. F. Ma, *J. Exp. Bot.*, 2011, **62**, 2265.
168. M. Jaquinod, F. Villiers, S. Kieffer-Jaquinod, V. Hugouvieux, C. Bruley, J. Garin and J. Bourguignon, *Mol. Cell. Proteomics*, 2007, **6**, 394.
169. Y. P. Lu, Z. S. Li, Y. M. Drozdowicz, S. Hortensteiner, E. Martinoia and P. A. Rea, *Plant Cell*, 1998, **10**, 267.
170. G. S. Liu, R. Sanchez-Fernandez, Z. S. Li and P. A. Rea, *J. Biol. Chem.*, 2001, **276**, 8648.
171. W. Y. Song, J. Park, D. G. Mendoza-Cozatl, M. Suter-Grotemeyer, D. Shim, S. Hortensteiner, M. Geisler, B. Weder, P. A. Rea, D. Rentsch, J. I. Schroeder, Y. Lee and E. Martinoia, *Proc. Natl Acad. Sci. USA*, 2010, **107**, 21187.
172. J. Park, W. Y. Song, D. Ko, Y. Eom, T. H. Hansen, M. Schiller, T. G. Lee, E. Martinoia and Y. Lee, *Plant J.*, 2012, **69**, 278.
173. V. Lanquar, F. Lelievre, S. Bolte, C. Hames, C. Alcon, D. Neumann, G. Vansuyt, C. Curie, A. Schroder, U. Kramer, H. Barbier-Brygoo and S. Thomine, *EMBO J.*, 2005, **24**, 4041.
174. V. Lanquar, M. S. Ramos, F. Lelievre, H. Barbier-Brygoo, A. Krieger-Liszkay, U. Kramer and S. Thomine, *Plant Physiol.*, 2010, **152**, 1986.
175. R. Oomen, J. Wu, F. Lelievre, S. Blanchet, P. Richaud, H. Barbier-Brygoo, M. G. M. Aarts and S. Thomine, *New Phytol.*, 2009, **181**, 637.
176. O. K. Vatamaniuk, E. A. Bucher, M. V. Sundaram and P. A. Rea, *J. Biol. Chem.*, 2005, **280**, 23684.
177. M. S. Schwartz, J. L. Benci, D. S. Selote, A. K. Sharma, A. G. Chen, H. Dang, H. Fares and O. K. Vatamaniuk, *PLoS One*, 5, e9564.

178. T. Sooksa-Nguan, B. Yakubov, V. I. Kozlovskyy, C. M. Barkume, K. J. Howe, T. W. Thannhauser, M. A. Rutzke, J. J. Hart, L. V. Kochian, P. A. Rea and O. K. Vatamaniuk, *J. Biol. Chem.*, 2009, **284**, 354.

179. K. Waku, *Environ. Health Perspect.*, 1984, **54**, 37.

180. M. Abouhamed, N. A. Wolff, W. K. Lee, C. P. Smith and F. Thévenod, *Am. J. Physiol. Renal Physiol.*, 2007, **293**, F705.

181. F. Galvez, D. Wong and C. M. Wood, *Am. J. Physiol. Regul. Integr. Comp. Physiol.*, 2006, **291**, R170.

182. W. K. Lee, U. Bork, F. Gholamrezaei and F. Thevenod, *Am. J. Physiol. Renal Physiol.*, 2005, **288**, F27.

183. P. M. Verbost, M. H. Senden and C. H. van Os, *Biochim. Biophys. Acta*, 1987, **902**, 247.

184. I. Sabolic, D. Breljak, M. Skarica and C. M. Herak-Kramberger, *BioMetals*, 2010, **23**, 897.

185. Y. Liu and D. M. Templeton, *J. Cell. Physiol.*, 2008, **217**, 307.

186. M. Kitamura and N. Hiramatsu, *BioMetals*, 2010, **23**, 941.

187. A. Hartwig, *BioMetals*, 2010, **23**, 951.

188. S. Banerjee and H. Flores-Rozas, *Nucleic Acids Res.*, 2005, **33**, 1410.

189. B. A. Fowler, *Environ. Health Perspect.*, 1978, **22**, 37.

190. G. Nucifora, L. Chu, T. K. Misra and S. Silver, *Proc. Natl Acad. Sci. USA*, 1989, **86**, 3544.

191. L. Herrmann, D. Schwan, R. Garner, H. L. Mobley, R. Haas, K. P. Schafer and K. Melchers, *Mol. Microbiol.*, 1999, **33**, 524.

192. A. Legatzki, G. Grass, A. Anton, C. Rensing and D. H. Nies, *J. Bacteriol.*, 2003, **185**, 4354.

193. C. Rensing, B. Mitra and B. P. Rosen, *Proc. Natl Acad. Sci. USA*, 1997, **94**, 14326.

194. M. Lebrun, A. Audurier and P. Cossart, *J. Bacteriol.*, 1994, **176**, 3040.

195. K. P. Yoon, T. K. Misra and S. Silver, *J. Bacteriol.*, 1991, **173**, 7643.

196. M. R. Binet and R. K. Poole, *FEBS Lett.*, 2000, **473**, 67.

197. B. Mitra and R. Sharma, *Biochemistry*, 2001, **40**, 7694.

198. N. Bal, E. Mintz, F. Guillain and P. Catty, *FEBS Lett.*, 2001, **506**, 249.

199. L. Banci, I. Bertini, S. Ciofi-Baffoni, L. A. Finney, C. E. Outten and T. V. O'Halloran, *J. Mol. Biol.*, 2002, **323**, 883.

200. L. Banci, I. Bertini, S. Ciofi-Baffoni, X. C. Su, R. Miras, N. Bal, E. Mintz, P. Catty, J. E. Shokes and R. A. Scott, *J. Mol. Biol.*, 2006, **356**, 638.

201. Z. Hou and B. Mitra, *J. Biol. Chem.*, 2003, **278**, 28455.

202. C. C. Wu, A. Gardarin, A. Martel, E. Mintz, F. Guillain and P. Catty, *J. Biol. Chem.*, 2006, **281**, 29533.

203. R. Sharma, C. Rensing, B. P. Rosen and B. Mitra, *J. Biol. Chem.*, 2000, **275**, 3873.

204. J. Liu, S. J. Dutta, A. J. Stemmler and B. Mitra, *Biochemistry*, 2006, **45**, 763.

205. D. H. Nies, *FEMS Microbiol. Rev.*, 2003, **27**, 313.

206. M. Mergeay, D. Nies, H. G. Schlegel, J. Gerits, P. Charles and F. Van Gijsegem, *J. Bacteriol.*, 1985, **162**, 328.

207. D. H. Nies and S. Silver, *J. Bacteriol.*, 1989, **171**, 896.
208. D. H. Nies, A. Nies, L. Chu and S. Silver, *Proc. Natl Acad. Sci. USA*, 1989, **86**, 7351.
209. A. Legatzki, S. Franke, S. Lucke, T. Hoffmann, A. Anton, D. Neumann and D. H. Nies, *Biodegradation*, 2003, **14**, 153.
210. D. H. Nies, *J. Bacteriol.*, 1995, **177**, 2707.
211. M. Goldberg, T. Pribyl, S. Juhnke and D. H. Nies, *J. Biol. Chem.*, 1999, **274**, 26065.
212. A. Leedjarv, A. Ivask and M. Virta, *J. Bacteriol.*, 2008, **190**, 2680.
213. H. Tohoyama, M. Inouhe, M. Joho and T. Murayama, *Curr. Genet.*, 1990, **18**, 181.
214. M. R. Rad, L. Kirchrath and C. P. Hollenberg, *Yeast*, 1994, **10**, 1217.
215. D. J. Adle, D. Sinani, H. Kim and J. Lee, *J. Biol. Chem.*, 2007, **282**, 947.
216. E. Shiraishi, M. Inouhe, M. Joho and H. Tohoyama, *Curr. Genet.*, 2000, **37**, 79.
217. D. J. Adle, W. Wei, N. Smith, J. J. Bies and J. Lee, *Proc. Natl Acad. Sci. USA*, 2009, **106**, 10189.
218. D. J. Adle and J. Lee, *J. Biol. Chem.*, 2008, **283**, 31460.
219. C. M. Paumi, M. Chuk, J. Snider, I. Stagljar and S. Michaelis, *Microbiol. Mol. Biol. Rev.*, 2009, **73**, 577.
220. D. J. Katzmann, E. A. Epping and W. S. Moye-Rowley, *Mol. Cell. Biol.*, 1999, **19**, 2998.
221. Z. Nagy, C. Montigny, P. Leverrier, S. Yeh, A. Goffeau, M. Garrigos and P. Falson, *Biochimie*, 2006, **88**, 1665.
222. Z. Cui, D. Hirata, E. Tsuchiya, H. Osada and T. Miyakawa, *J. Biol. Chem.*, 1996, **271**, 14712.
223. B. Rogers, A. Decottignies, M. Kolaczkowski, E. Carvajal, E. Balzi and A. Goffeau, *J. Mol. Microbiol. Biotechnol.*, 2001, **3**, 207.
224. F. Verret, A. Gravot, P. Auroy, N. Leonhardt, P. David, L. Nussaume, A. Vavasseur and P. Richaud, *FEBS Lett.*, 2004, **576**, 306.
225. C. K. E. Wong and C. S. Cobbett, *New Phytol.*, 2009, **181**, 71.
226. R. Takahashi, Y. Ishimaru, H. Shimo, Y. Ogo, T. Senoura, N. K. Nishizawa and H. Nakanishi, *Plant Cell Environ.*, 2012, **35**, 1948.
227. F. F. Nocito, C. Lancilli, B. Dendena, G. Lucchini and G. A. Sacchi, *Plant Cell Environ.*, 2011, **34**, 994.
228. N. Satoh-Nagasawa, M. Mori, N. Nakazawa, T. Kawamoto, Y. Nagato, K. Sakurai, H. Takahashi, A. Watanabe and H. Akagi, *Plant Cell Physiol.*, 2012, **53**, 213.
229. C. Bernard, N. Roosens, P. Czernic, M. Lebrun and N. Verbruggen, *FEBS Lett.*, 2004, **569**, 140.
230. M. Courbot, G. Willems, P. Motte, S. Arvidsson, N. Roosens, P. Saumitou-Laprade and N. Verbruggen, *Plant Physiol.*, 2007, **144**, 1052.
231. M. Hanikenne, I. N. Talke, M. J. Haydon, C. Lanz, A. Nolte, P. Motte, J. Kroymann, D. Weigel and U. Kramer, *Nature*, 2008, **453**, 391.

232. A. Parameswaran, B. Leitenmaier, M. Yang, P. M. H. Kroneck, W. Welte, G. Lutz, A. Papoyan, L. V. Kochian and H. Kupper, *Biochem. Biophys. Res. Commun.*, 2007, **363**, 51.

233. B. Leitenmaier, A. Witt, A. Witzke, A. Stemke, W. Meyer-Klaucke, P. M. H. Kroneck and H. Kupper, *Biochim. Biophys. Acta Biomembranes*, 2011, **1808**, 2591.

234. D. Y. Kim, L. Bovet, M. Maeshima, E. Martinoia and Y. Lee, *Plant J.*, 2007, **50**, 207.

235. M. B. Troadec, D. M. Ward, E. Lo, J. Kaplan and I. De Domenico, *Blood*, 2010, **116**, 4657.

236. T. Ishikawa, J. J. Bao, Y. Yamane, K. Akimaru, K. Frindrich, C. D. Wright and M. T. Kuo, *J. Biol. Chem.*, 1996, **271**, 14981.

237. Y. Long, Q. Li, S. Zhong, Y. Wang and Z. Cui, *Comp. Biochem. Physiol. C Toxicol. Pharmacol.*, 2011, **153**, 381.

238. W. K. Lee, B. Torchalski, N. Kohistani and F. Thevenod, *Toxicol. Sci.*, 2011, **121**, 343.

239. F. Thévenod, *BioMetals*, 2010, **23**, 857.

240. J. Rennolds, S. Butler, K. Maloney, P. N. Boyaka, I. C. Davis, D. L. Knoell, N. L. Parinandi and E. Cormet-Boyaka, *Toxicol. Sci.*, 2010, **116**, 349.

241. T. Endo, *Comp. Biochem. Physiol. C Toxicol. Pharmacol.*, 2002, **131**, 223.

242. D. H. Sweet, *Toxicol. Appl. Pharmacol.*, 2005, **204**, 198.

243. N. Sugawara, Y. R. Lai, K. Arizono and T. Ariyoshi, *Toxicology*, 1996, **112**, 87.

244. C. D. Klaassen, J. Liu and B. A. Diwan, *Toxicol. Appl. Pharmacol.*, 2009, **238**, 215.

245. D. E. Salt, R. C. Prince, I. J. Pickering and I. Raskin, *Plant Physiol.*, 1995, **109**, 1427.

246. M. P. Isaure, B. Fayard, G. Saffet, S. Pairis and J. Bourguignon, *Spectrochim. Acta Part B Atomic Spectroscopy*, 2006, **61**, 1242.

247. H. Kupper, E. Lombi, F. J. Zhao and S. P. McGrath, *Planta*, 2000, **212**, 75.

248. D. Ueno, T. Iwashita, F. J. Zhao and J. F. Ma, *Plant Cell Physiol.*, 2008, **49**, 540.

249. S. Huguet, V. Bert, A. Laboudigue, V. Barthes, M. P. Isaure, I. Llorens, H. Schat and G. Sarret, *Environ. Exp. Bot.*, 2012, **82**, 54.

250. Y. E. Choi, E. Harada, M. Wada, H. Tsuboi, Y. Morita, T. Kusano and H. Sano, *Planta*, 2001, **213**, 45.

251. J. F. Ma, D. Ueno, F. J. Zhao and S. P. McGrath, *Planta*, 2005, **220**, 731.

252. S. Uraguchi, T. Kamiya, T. Sakamoto, K. Kasai, Y. Sato, Y. Nagamura, A. Yoshida, J. Kyozuka, S. Ishikawa and T. Fujiwara, *Proc. Natl Acad. Sci. USA*, 2011, **108**, 20959.

253. S. Ishikawa, N. Suzui, S. Ito-Tanabata, S. Ishii, M. Igura, T. Abe, M. Kuramata, N. Kawachi and S. Fujimaki, *BMC Plant Biol.*, 2011, 11.

254. F. Van Belleghem, A. Cuypers, B. Semane, K. Smeets, J. Vangronsveld, J. d'Haen and R. Valcke, *New Phytol.*, 2007, **173**, 495.

255. D. G. Mendoza-Cozatl, E. Butko, F. Springer, J. W. Torpey, E. A. Komives, J. Kehr and J. I. Schroeder, *Plant J.*, 2008, **54**, 249.
256. K. Vogel-Mikus, I. Arcon and A. Kodre, *Plant Soil*, 2010, **331**, 439.
257. T. Kjellstrom, *Environ. Health Perspect.*, 1979, **28**, 169.
258. D. L. Savigni and E. H. Morgan, *J. Physiol.*, 1998, **508**, 837.
259. M. Lou, R. Garay and J. O. Alda, *J. Physiol.*, 1991, **443**, 123.
260. B. J. Scott and A. R. Bradwell, *Clin. Chem.*, 1983, **29**, 629.
261. G. F. Nordberg, K. Nogawa, M. Nordberg and L. T. Friberg, in *Handbook on the Toxicology of Metals*, ed. G. F. Nordberg, B. A. Fowler, M. Nordberg and L. T. Friberg, Academic Press – Elsevier, Burlington, MA, USA, 3rd edn, 2007.
262. K. V. Lopin, F. Thevenod, J. C. Page and S. W. Jones, *Mol. Pharmacol.*, 2012, **82**, 1183.
263. E. Ohana, I. Sekler, T. Kaisman, N. Kahn, J. Cove, W. F. Silverman, A. Amsterdam and M. Hershfinkel, *J. Mol. Med.*, 2006, **84**, 753.
264. S. Levy, O. Beharier, Y. Etzion, M. Mor, L. Buzaglo, L. Shaltiel, L. A. Gheber, J. Kahn, A. J. Muslin, A. Katz, D. Gitler and A. Moran, *J. Biol. Chem.*, 2009, **284**, 32434.
265. M. K. Monteilh-Zoller, M. C. Hermosura, M. J. Nadler, A. M. Scharenberg, R. Penner and A. Fleig, *J. Gen. Physiol.*, 2003, **121**, 49.
266. C. Martineau, E. Abed, G. Medina, L. A. Jomphe, M. Mantha, C. Jumarie and R. Moreau, *Toxicol. Lett.*, 2010, **199**, 357.
267. G. Kovacs, T. Danko, M. J. Bergeron, B. Balazs, Y. Suzuki, A. Zsembery and M. A. Hediger, *Cell Calcium*, 2011, **49**, 43.
268. C. Usai, A. Barberis, L. Moccagatta and C. Marchetti, *J. Neurochem.*, 1999, **72**, 2154.
269. E. I. Christensen, P. J. Verroust and R. Nielsen, *Pflugers Arch.*, 2009, **458**, 1039.
270. C. Langelueddecke, E. Roussa, R. A. Fenton, N. A. Wolff, W. K. Lee and F. Thevenod, *J. Biol. Chem.*, 2012, **287**, 159.
271. K. J. Waldron, J. C. Rutherford, D. Ford and N. J. Robinson, *Nature*, 2009, **460**, 823.
272. G. Endo and S. Silver, *J. Bacteriol.*, 1995, **177**, 4437.
273. J. Ye, A. Kandegedara, P. Martin and B. P. Rosen, *J. Bacteriol.*, 2005, **187**, 4214.
274. C. Zheng, Y. Li, L. Nie, L. Qian, L. Cai and J. Liu, *Curr. Microbiol.*, 2012, **65**, 117.
275. Y. Wang, L. Hemmingsen and D. P. Giedroc, *Biochemistry*, 2005, **44**, 8976.
276. Y. Wang, J. Kendall, J. S. Cavet and D. P. Giedroc, *Biochemistry*, 2011, **49**, 6617.
277. T. Liu, S. Nakashima, K. Hirose, M. Shibasaka, M. Katsuhara, B. Ezaki, D. P. Giedroc and K. Kasamo, *J. Biol. Chem.*, 2004, **279**, 17810.
278. Q. Que and J. D. Helmann, *Mol. Microbiol.*, 2000, **35**, 1454.
279. K. E. Stoll, W. E. Draper, J. I. Kliegman, M. V. Golynskiy, R. A. Brew-Appiah, R. K. Phillips, H. K. Brown, W. A. Breyer, N. S. Jakubovics,

H. F. Jenkinson, R. G. Brennan, S. M. Cohen and A. Glasfeld, *Biochemistry*, 2009, **48**, 10308.

280. J. L. Hobman, *Mol. Microbiol.*, 2007, **63**, 1275.

281. S. W. Lee, E. Glickmann and D. A. Cooksey, *Appl. Environ. Microbiol.*, 2001, **67**, 1437.

282. C. Rodrigues-Pousada, R. A. Menezes and C. Pimentel, *Yeast*, 2010, **27**, 245.

283. A. L. Wu and W. S. Moye-Rowley, *Mol. Cell. Biol.*, 1994, **14**, 5832.

284. A. Serero, J. Lopes, A. Nicolas and S. Boiteux, *DNA Repair (Amst.)*, 2008, **7**, 1262.

285. A. Wu, J. A. Wemmie, N. P. Edgington, M. Goebl, J. L. Guevara and W. S. Moye-Rowley, *J. Biol. Chem.*, 1993, **268**, 18850.

286. D. Azevedo, L. Nascimento, J. Labarre, M. B. Toledano and C. Rodrigues-Pousada, *FEBS Lett.*, 2007, **581**, 187.

287. U. H. Dormer, J. Westwater, N. F. McLaren, N. A. Kent, J. Mellor and D. J. Jamieson, *J. Biol. Chem.*, 2000, **275**, 32611.

288. G. L. Wheeler, E. W. Trotter, I. W. Dawes and C. M. Grant, *J. Biol. Chem.*, 2003, **278**, 49920.

289. T. A. Lee, P. Jorgensen, A. L. Bognar, C. Peyraud, D. Thomas and M. Tyers, *Mol. Biol. Cell*, 2010, **21**, 456.

290. J. L. Yen, N. Y. Su and P. Kaiser, *Mol. Biol. Cell*, 2005, **16**, 1872.

291. R. Barbey, P. Baudouin-Cornu, T. A. Lee, A. Rouillon, P. Zarzov, M. Tyers and D. Thomas, *EMBO J.*, 2005, **24**, 521.

292. Y. F. Lin and M. G. M. Aarts, *Cell. Mol. Life Sci.*, 2012, **69**, 3187.

293. L. Garnier, F. Simon-Plas, P. Thuleau, J. P. Agnel, J. P. Blein, R. Ranjeva and J. L. Montillet, *Plant Cell Environ.*, 2006, **29**, 1956.

294. K. Smeets, K. Opdenakker, T. Remans, S. Van Sanden, F. Van Belleghem, B. Semane, N. Horemans, Y. Guisez, J. Vangronsveld and A. Cuypers, *J. Plant Physiol.*, 2009, **166**, 1982.

295. R. Mittler, S. Vanderauwera, N. Suzuki, G. Miller, V. B. Tognetti, K. Vandepoele, M. Gollery, V. Shulaev and F. Van Breusegem, *Trends Plant Sci.*, 2011, **16**, 300.

296. C. Jonak, H. Nakagami and H. Hirt, *Plant Physiol.*, 2004, **136**, 3276.

297. X. M. Liu, K. E. Kim, K. C. Kim, X. C. Nguyen, H. J. Han, M. S. Jung, H. S. Kim, S. H. Kim, H. C. Park, D. J. Yun and W. S. Chung, *Phytochemistry*, 2010, **71**, 614.

298. K. Opdenakker, T. Remans, E. Keunen, J. Vangronsveld and A. Cuypers, *Environ. Exp. Bot.*, 2012, **83**, 53.

299. Y. F. Ding and C. Zhu, *Biochem. Biophys. Res. Commun.*, 2009, **386**, 6.

300. R. Sunkar, A. Kapoor and J. K. Zhu, *Plant Cell*, 2006, **18**, 2051.

301. Z. S. Zhou, S. Q. Huang and Z. M. Yang, *Biochem. Biophys. Res. Commun.*, 2008, **374**, 538.

302. F. L. Xie, S. Q. Huang, K. Guo, A. L. Xiang, Y. Y. Zhu, L. Nie and Z. M. Yang, *FEBS Lett.*, 2007, **581**, 1464.

303. Z. S. Zhou, J. B. Song and Z. M. Yang, *J. Exp. Bot.*, 2012, **63**, 4597.

304. R. Sunkar, Y. F. Li and G. Jagadeeswaran, *Trends Plant Sci.*, 2012, **17**, 196.

305. W. C. Prozialeck, *Toxicol. Appl. Pharmacol.*, 2000, **164**, 231.
306. F. Thévenod, *Toxicol. Appl. Pharmacol.*, 2009, **238**, 221.
307. B. Faurskov and H. F. Bjerregaard, *Pflugers Arch.*, 2002, **445**, 40.
308. A. Stoica, B. S. Katzenellenbogen and M. B. Martin, *Mol. Endocrinol.*, 2000, **14**, 545.
309. C. L. Siewit, B. Gengler, E. Vegas, R. Puckett and M. C. Louie, *Mol. Endocrinol.*, 2010, **24**, 981.
310. P. Lichtlen, Y. Wang, T. Belser, O. Georgiev, U. Certa, R. Sack and W. Schaffner, *Nucleic Acids Res.*, 2001, **29**, 1514.
311. H. I. Sims, G. W. Chirn and M. T. Marr, 2nd, *Proc. Natl Acad. Sci. USA*, 2012, **109**, 16516.
312. W. A. Chu, J. D. Moehlenkamp, D. Bittel, G. K. Andrews and J. A. Johnson, *J. Biol. Chem.*, 1999, **274**, 5279.
313. S. Koizumi, P. Gong, K. Suzuki and M. Murata, *J. Biol. Chem.*, 2007, **282**, 8715.
314. E. M. Sikorski, T. Uo, R. S. Morrison and A. Agarwal, *J. Biol. Chem.*, 2006, **281**, 24423.
315. J. M. Moulis, in *Encyclopedia of Metalloproteins*, ed. V.N. Uversky, R. H. Kretsinger and E. A. Permyakov, Springer Science + Business Media, LLC, 2013, http://www.springerreference.com/docs/html/chapterdbid/372757.html.
316. A. Krezel and W. Maret, *J. Biol. Inorg. Chem.*, 2006, **11**, 1049.
317. W. Maret, *Antioxid. Redox Signal.*, 2006, **8**, 1419.
318. D. Ueno, J. F. Ma, T. Iwashita, F. J. Zhao and S. P. McGrath, *Planta*, 2005, **221**, 928.
319. H. Kupper, A. Mijovilovich, W. Meyer-Klaucke and P. M. H. Kroneck, *Plant Physiol.*, 2004, **134**, 748.
320. F. Villiers, A. Jourdain, O. Bastien, N. Leonhardt, S. Fujioka, G. Tichtincky, F. Parcy, J. Bourguignon and V. Hugouvieux, *J. Exp. Bot.*, 2012, **63**, 1185.
321. T. Fukada, S. Yamasaki, K. Nishida, M. Murakami and T. Hirano, *J. Biol. Inorg. Chem.*, 2011, **16**, 1123.
322. M. Rodriguez-Serrano, M. C. Romero-Puertas, D. M. Pazmino, P. S. Testillano, M. C. Risueno, L. A. del Rio and L. M. Sandalio, *Plant Physiol.*, 2009, **150**, 229.
323. S. Li, J. L. Yu, M. J. Zhu, F. G. Zhao and S. Luan, *Plant Cell Environ.*, 2012, **35**, 1998.

CHAPTER 24

Mercury

STEPHANIE J. B. FRETHAM[a] AND MICHAEL ASCHNER*[b]

[a] Department of Biology, Luther College; 700 College Drive, Decorah, IA 52101, USA; [b] Department of Molecular Pharmacology, Albert Einstein College of Medicine; Forchheimer 209, 1300 Morris Park Avenue, Bronx, NY 10461, USA
*Email: Michael.Aschner@einstein.yu.edu

24.1 Introduction

Mercury (Hg) is a naturally occurring element present in small quantities in the environment and atmosphere. However, release of Hg into the atmosphere has increased significantly recently, with as much as 75% resulting from anthropogenic activity.[1] Due to these rising levels and its toxic nature, there has been much interest in determining the mechanisms through which both prokaryotic and eukaryotic organisms handle this metal. This chapter will provide an overview of the biogeochemical cycle of Hg and a detailed description of prokaryotic and eukaryotic transport and cellular handling of both the metal itself and its compounds.

24.1.1 Forms of Mercury

There are three primary forms: metallic Hg (elemental mercury), and inorganic and organic Hg compounds (Table 24.1). Metallic Hg (Hg^0) is a silvery liquid at room temperature and readily forms vapours at ambient temperatures due to its high vapour pressure. Elemental Hg is rarely found in nature but can be extracted from inorganic mineral ores (*e g.* cinnabar) by heating. Its unique liquid nature and high vapour pressure have made

RSC Metallobiology Series No. 2
Binding, Transport and Storage of Metal Ions in Biological Cells
Edited by Wolfgang Maret and Anthony Wedd
© The Royal Society of Chemistry 2014
Published by the Royal Society of Chemistry, www.rsc.org

Table 24.1 Mercury compounds.

	Oxidation state	Chemical formula	Compound name	Sources	Uses
Elemental	0	Hg	Mercury	Cinnabar ore HgS, atmospheric vapour, bacterial reduction of Hg^{2+}	Chlorine gas production, precious metal refinement, electrical components
Inorganic	I	$[Hg_2^I]^{2+}$	Mercurous ion	Abiotic atmospheric transformation, mineral deposits	
		Hg_2Cl_2	Mercury(I)chloride		Diuretic, disinfectant, laxative
		$Hg_6Cl_3O(OH)$	Eglestonite		
	II	Hg^{2+}	Mercuric ion	Mineral deposits, cinnabar, industrial processes	Fungicide, cosmetics, commercial processes
		HgS	Mercury(II)sulfide		
		$HgCl_2$	Mercury(II)chloride		
Organic	II	$[CH_3Hg]^+$	Methylmercury ion	Biotic transformation of Hg^{2+}, industrial processes	Fungicide, vaccine preservatives, commercial processes
		$(CH_3)_2Hg$	Dimethylmercury		
		$[C_2H_5Hg]^+$	Ethylmercury ion		
		$[C_6H_5Hg]^+$	Phenylmercury ion		

metallic Hg a valuable tool for many industrial processes, such as the production of chlorine gas and caustic soda, extraction of gold and silver from ores and as a component of thermometers, barometers, batteries, electrical switches and dental amalgams.

Non-metallic Hg can exist in either the mercurous Hg^I or mercuric Hg^{II} oxidation state (Table 24.1). In the atmosphere, small amounts of mercurous Hg are present as the mercurous ion dimer ($[Hg^I_2]^{2+}$). In the environment, Hg^I forms stable compounds with metal-metal bonds, including mercury(I)chloride (known as calomel, Hg_2Cl_2), which has a linear geometry: $Cl-Hg^I-Hg^I-Cl$. At one time, calomel was used extensively as a medical diuretic, disinfectant and laxative. Hg^I is also found in deposits of eglestonite ($Hg_6Cl_3O(OH)$), a mineral often found near deposits of mercuric minerals, such as cinnabar. The mercurous ion is susceptible to disproportionation to form Hg^0 and Hg^{II}.[2]

Mercuric Hg is the most commonly found oxidation state and is observed in both inorganic and organic compounds. The former are the most abundant and are found in large quantities in cinnabar deposits. Cinnabar (mercury(II)sulfide, HgS) is deposited by ascending aqueous (Hg^{2+}) solutions released by volcanic activity or alkaline hot springs. As these solutions rise to the Earth's surface and cool, Hg^{2+} reacts with H_2S to form insoluble HgS, deposited as red or black cinnabar. Red cinnabar (α-HgS) is the most stable form while black cinnabar, also known as metacinnabar (β-HgS), adopts a less stable zinc blende structure which oxidizes rapidly in air. The mercuric ion (Hg^{2+}) readily forms highly corrosive inorganic salts including mercuric chloride, mercuric sulfide and mercuric acetate.

Organomercurial species such as methylmercury $[CH_3Hg]^+$, dimethylmercury $(CH_3)_2Hg$, ethylmercury $[C_2H_5Hg]^+$ and phenylmercury $[C_6H_5Hg]^+$ are mercuric compounds. These organometallic compounds are generated from inorganic forms including cinnabar and Hg^{2+} by aquatic microorganisms and to a lesser extent through abiotic processes. Organomercurials and especially $[CH_3Hg]^+$ are much more bioavailable than the metal itself and the inorganic compounds. This results in uptake and bioaccumulation of organic Hg in the food chain. Like metal and inorganic mercury compounds, organomercurials have been previously used in fungicide treatments and in limited commercial processes. It is now being phased out from use as a preservative (thimerosal) in children's vaccines.

24.1.2 Biogeochemical Cycle

Like many elements, Hg cycles through land, sea and air in a complex biogeochemical cycle, undergoing several biotic and abiotic transformations (Figure 24.1). The global Hg cycle is characterized by large differences in Hg levels with the highest concentrations found in soil followed, in decreasing order, by fish and other food sources, water and the atmosphere.[2]

As described above, inorganic forms are the most abundant and, although their bioavailability is very low, they can be released into the atmosphere and

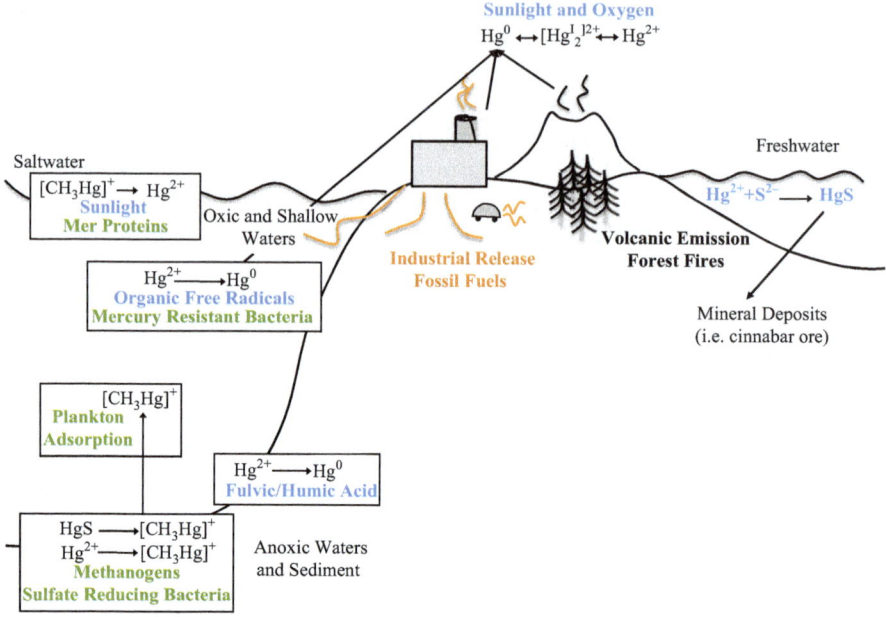

Figure 24.1 The mercury biogeochemical cycle. Abiotic transformations are indicated in blue, biotic transformations in green and anthropogenic Hg release in orange.

be transformed into forms with higher bioavailability. There are a number of natural and anthropogenic sources of environmental Hg. The most significant natural sources include release of elemental Hg vapour from volcanoes and forest fires and the release of inorganic Hg by weathering and movement of water.[3] Anthropogenic sources include the burning of coal and fossil fuels, mining, precious metal refinement and industrial uses including electrical and automotive part manufacture and chemical processing. To these must be added intentional release through waste incineration, landfills and the release of industrial contamination into water systems.

The major route of global Hg distribution is through the atmosphere where the majority of Hg (90%) is elemental. Atmospheric Hg^0 is relatively stable and can remain in the atmosphere for approximately one year.[2] The remaining 10% of atmospheric Hg is found as $[Hg^I_2]^{2+}$, Hg^{2+} and $[CH_3Hg]^+$ ions. Once in the atmosphere, sunlight and dioxygen act to oxidize elemental mercury to the water-soluble mercuric ion,[1,2] which binds water and is re-deposited on the Earth's surface in the form of contaminated rain.

Sources of aquatic Hg include the contaminated rain described above as well as anthropogenic contamination from industrial and chemical waste. Several biotic and abiotic processes contribute to the transformation of aquatic Hg. Abiotic transformations include formation of salts and thiol complexes. In salt water, high levels of chloride lead to production of highly corrosive mercury(II) chloride. In freshwater environments with lower chloride levels, Hg^{2+} binds to thiols found in dissolved organic matter and is ultimately converted to HgS, which precipitates and settles into the sediment, decreasing bioavailability.[4,5] In shallow and surface waters, aquatic Hg^{2+} can be reduced to Hg^0 by organic free radicals generated by photolysis of dissolved organic compounds in the presence of dioxygen.[6,7] Sunlight can also promote conversion of $[CH_3Hg]^+$ to Hg^{2+}.[1] In deeper and anoxic waters and sediments, fulvic and humic acid radicals can reduce Hg^{2+} to Hg^0.[8–10]

Hg is also transformed *via* biotic reactions in aquatic microorganisms. In anoxic sediments, $[CH_3Hg]^+$ is formed from Hg^{2+} and HgS by methanogens and sulfate-reducing bacteria. The cation is highly lipid soluble and so can diffuse out of the cell where it accumulates in anoxic waters. $[CH_3Hg]^+$ may then diffuse into cells of other microorganisms and is adsorbed by exoskeletons of small plankton. In oxic waters, mercury-resistant bacteria can break down organomercurials to form Hg^{2+} and reduce it to Hg^0, which then vaporizes. Dissolved Hg^0 can enter the atmosphere or be oxidized through catalase reactions to form Hg^{2+}.

These abiotic and biotic transformations result in an overall increase in mercury bioavailability, particularly by formation of $[CH_3Hg]^+$. The latter is adsorbed by zooplankton and other small organisms that form the base of both the freshwater and ocean food chains. These organisms are consumed by small fish, which are then eaten by larger fish, and so on up the food chain. In this way, $[CH_3Hg]^+$ and to a lesser extent the other alkylated forms of mercury are concentrated by as much as a million-fold in large, fatty, predatory fish (relative to the surrounding water).[2]

24.2 Prokaryotes

Prokaryotes including both archaea and bacteria are central actors in the biogeochemical cycle of Hg as agents of methylation and reduction. This section will provide a detailed description of the mechanisms through which prokaryotes bind, transport and transform Hg.

24.2.1 Biomethylation

Organic mercury is generated by the formation of a mercury-carbon covalent bond. The first step in the biotic formation of organomercurials is methylation to form $[CH_3Hg]^+$. While small, although potentially significant, amounts of $[CH_3Hg]^+$ can be formed through abiotic mechanisms, the largest source is the activity of methanogens and sulfate-reducing bacteria, which generate $[CH_3Hg]^+$ as a by-product of normal metabolic processes.

The biotic nature of $[CH_3Hg]^+$ formation was first demonstrated by Jensen and Jernelöv in 1969.[11] At that time, the source of methylmercury in fish from waters contaminated with inorganic Hg and phenylmercury industrial waste was unknown. These investigators found that incubation of bottom sediments from freshwater aquaria containing $HgCl_2$ produced $[CH_3Hg]^+$ in a concentration-dependent manner. Sterilization of the sediments *via* autoclaving eliminated $[CH_3Hg]^+$ production and indicated bioactivity as the source.

The first organisms identified as capable of generating $[CH_3Hg]^+$ were methanogens (anaerobic archaea that use CO_2 and/or acetate $[CH_3COO^-]$ to generate ATP)[12]. These organisms (which normally produce methane gas as a metabolic by-product) are found nearly everywhere on Earth including the human gastrointestinal tract and extreme environments such as ocean vents. In the presence of Hg^{2+}, methanogenic extracts produce $[CH_3Hg]^+$ and $(CH_3)_2Hg$ rather than methane.[12]

Methanogens were thought to be the only organisms capable of methylating Hg^{2+} until $[CH_3Hg]^+$ production was observed in sulfate-reducing bacteria (SRB).[13] $[CH_3Hg]^+$ production nearly doubled in saltmarsh sediment spiked with $HgCl_2$ and treated with 2-bromoethane sulfonate to inhibit methanogen activity. However, inhibition of SRB with sodium molybdate decreased Hg^{2+} methylation by more than 95% with a similar increase in methane production. An individual strain of *Desulfovibrio desulfuricans* was isolated from the saltmarsh sediment and was capable of methylating Hg^{2+} in culture and previously sterilized sediment.[13]

Hg methylation by SRB occurs through two distinct mechanisms dependent on genus and metabolic conditions. *Desulfovibrio* genus members use lactate, pyruvate and alcohols as substrates for sulfate reduction, producing acetate as the end product. $[CH_3Hg]^+$ formation occurs in *Desulfovibrio* during fermentative conditions when sulfate is limiting and organic substrates such as pyruvate and lactate are available for fermentation.[13] Support for $[CH_3Hg]^+$ generation during fermentative growth rather than during sulfate reduction is provided by reduced $[CH_3Hg]^+$ generation by *Desulfovibrio* in high sulfate conditions that reduce fermentation.[14] Approximately half of the *Desulfovibrio* strains examined were capable of Hg^{2+} methylation.[15]

However, this fermentative methylation does not occur in other genera of SRB. Strains of the *Desulfobacterium* genus are capable of completely oxidizing acetate to form CO_2 and are incapable of fermentation. Nevertheless, *Desulfobacterium* methylates Hg^{2+} at a high rate dependent on sulfate availability and metabolic rate.[16] Furthermore, $[CH_3Hg]^+$ production by another complete oxidizer *Desulfococcus multivorans* is inhibited by limitation of the methyltransferase cofactor B_{12} while methylation by the incomplete oxidizer *Desulfovibrio africanus* is not affected by limited B_{12} availability.[17] Methanogens and SRB are also capable of degrading $[CH_3Hg]^+$ to form Hg^{2+}, thereby creating a cycle of methylation and demethylation that results in an equilibrium between $[CH_3Hg]^+$ and Hg^{2+}.[18]

24.2.2 Reduction (Mer Operon)

Prokaryotes have developed resistance to many toxic environmental factors. Resistance to mercury conferred by the prokaryotic mer operon is one of the most widely studied resistance systems. Unlike most other resistance mechanisms, which provide protection by preventing uptake and promoting efflux of the toxic substance, proteins under control of the mer operon protect cells by reducing both organic and inorganic forms Hg. The resultant Hg^0 is relatively unreactive and can diffuse out of the cell. Hg resistance has been found in every bacterial genus that has been examined and is even present in some archaea.

The first studies of this bacterial resistance found co-occurrence of Hg resistance and resistance to other factors such as penicillin. This co-occurrence was the result of a single plasmid containing the genetic elements that mediate resistance to both Hg and penicillin.[19,20] The mer operon contains several *mer* genes that encode proteins that bind, transport and reduce organic and/or inorganic forms of Hg^{II} (Table 24.2).

In the course of evolution, many mer operon variants have developed and they contain varying numbers and types of *mer* genes. Despite these variations, all Hg-resistant bacteria are capable of sequestering and enzymatically reducing the mercuric ion to the level of elemental Hg. Hg^0, like $[CH_3Hg]^+$, is lipid soluble and can diffuse through the cell membrane. However, unlike $[CH_3Hg]^+$, which can accumulate locally, Hg^0 readily vaporizes. In this way, the transformation not only decreases the cellular toxicity of Hg^{2+} but also prevents Hg^{2+} from accumulating in nearby waters or precipitating into anoxic sediments where it could be methylated by methanogens and SRB. In addition to the elements necessary to reduce Hg^{2+}, some mer operons contain structural genes encoding proteins that degrade organic mercury and confer broad-spectrum resistance to both inorganic and organic compounds. Organisms without these additional elements are not resistant to organomercurials and are referred to as narrow-spectrum resistant bacteria.

Table 24.2 Functions, coordination and binding affinity of mer proteins.

	Function	K_d μM	Coordination	Ref.
MerP	Hg^{2+} binding in periplasm	3.7 ± 1.3	Bicoordinate	103,104
MerT	Hg^{2+} transport across cytoplasmic membrane			
MerC	Hg^{2+} transport across cytoplasmic membrane			28
MerF	Hg^{2+} transport across cytoplasmic membrane			30
MerE	Hg^{2+} and $[CH_3Hg]^+$ import			32
MerG	Periplasmic inhibition of $[C_6H_5Hg]^+$ uptake			33
MerA	Hg^{2+} reductase	2.1 ± 1.9	Bicoordinate and Tricoordinate	105,106
MerB	Organomercurial lyase		Bicoordinate	44
MerR	Transcriptional regulation	0.1 ± 0.3	Tricoordinate metal-bridged dimer	47,107
MerD	MerR antagonist			48,49

Table 24.3   Coordination chemistry and binding affinities of Hg compounds with cellular amino acids and proteins.

Ligand	Mercury species	Log K	Coordination chemistry	Ref.
Cysteine	Hg^{2+}	20.1	Bicoordinate	108
	$[CH_3Hg]^+$	15.7	Bicoordinate	108
Selenocysteine	Hg^{2+}		Tetrahedral	75
	$[CH_3Hg]^+$	17.4		71,75,80
GSH	Hg^{2+}	16.6	Bicoordinate (tricoordinate with excess GSH)	86,87,108
	$[CH_3Hg]^+$	16.5	Bicoordinate	88,108
Albumin	Hg^{2+}	10.1 ± 0.2	Bicoordinate	109
	$[CH_3Hg]^+$	12.9 ± 0.4	Bicoordinate	109,110
Haemoglobin (human)	Hg^{2+}		Mixed tricoordinate coordination with GSH	67
	$[CH_3Hg]^+$	6.69		111
Metallothionein	Hg^{2+}		Tetrahedral	83
	$[CH_3Hg]^+$		Does not bind to MT	112

Each mer protein contains key cysteine residues essential for interaction with Hg^{II} compounds. Hg^{II} is a soft electrophile and has an extremely high affinity for thiol (-SH) and selenol (-SeH) groups, the only biological soft nucleophiles (Table 24.3). Hg compounds react specifically with these ligands forming stable complexes of defined stoichiometry. In biological media, $[CH_3Hg]^+$ is always bound to ligands such as glutathione, cysteine and selenocysteine to form two-coordinate digonal complexes.[21] In both prokaryotic and eukaryotic systems, thiol and selenol groups determine the biological consequences of protein-Hg interactions. In most instances Hg compounds bind thiol or selenol groups with high affinity, impairing protein function; however, Hg does not alter the Cys residues of mer proteins.

24.2.2.1   Cellular Uptake

One unique aspect of bacterial Hg resistance is the specific transport of Hg^{2+} into cells by the mer proteins MerP and MerT. MerP is the most abundant mer protein and is found in the periplasm where it acts as an "Hg sponge". MerP is not absolutely necessary for cellular Hg^{2+} import, as its deletion does not prevent Hg^{2+} uptake;[22] however, monomeric MerP acts to sequester Hg^{2+} and prevent damage to other proteins. It binds a single Hg^{2+} ion employing the thiol groups of conserved cysteine residues 14 and 17[23,24] (Figure 24.2). These residues are key elements in the binding motif CMxCxxC. This or related motifs (such as CxxC) are common features of binding sites for transition metal cations in prokaryotic and eukaryotic transport proteins and chaperones.[25]

Once bound to MerP, Hg^{2+} is transferred to MerT, a membrane protein with three transmembrane helices, which facilitates movement of Hg^{2+}

from the periplasm into the cytosol. Hg^{2+} is transferred from MerP to a periplasmic cysteine pair on MerT and is then transported to another MerT cysteine pair on the cytoplasmic face (Figure 24.2).[26] The transfer from MerP to MerT is achieved through three-coordinate intermediates involving one of the two MerP cysteines and the two periplasmic cysteines of MerT.[27] The latter are located in the first transmembrane domain, and the cytoplasmic pair is located between transmembrane domains 2 and 3.

MerT is the most common transmembrane mer protein; however, some mer operons contain MerC and MerF that are also capable of transporting

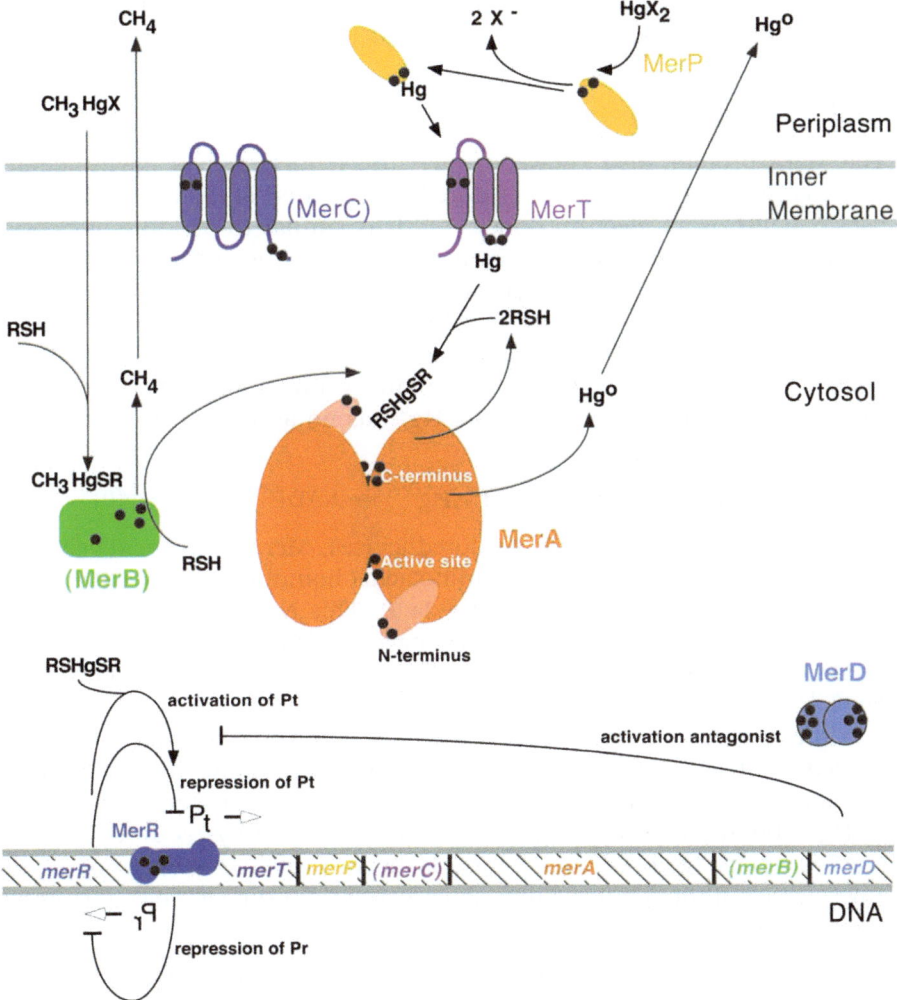

Figure 24.2 Model of a typical mer operon. Cysteine residues are indicated by black dots and X refers to a generic solvent nucleophile. RSH is a low-molecular-mass, cytosolic thiol such as glutathione.
Adapted from Barkay *et al.*, 2003 (ref. 1).

Hg^{2+} into the cell. Like MerT, they both contain two pairs of cysteine residues critical for Hg^{2+} transport. MerC is the largest mer protein with four transmembrane domains. It facilitates Hg^{2+} import in the absence of MerP (Figure 24.2).[28,29] MerF is also capable of transporting Hg^{2+} in the absence of MerP and contains two transmembrane domains with two pairs of cysteine residues required for Hg^{2+} import.[30]

There is evidence for the involvement of MerP and MerT in the uptake of arylmercurials, such as phenylmercury.[31] There is also recent evidence suggesting that MerE, a small transmembrane protein, facilitates import of both Hg^{2+} and $[CH_3Hg]^+$.[32] Many broad-spectrum mer operons express MerG in the periplasm, which prevents phenylmercury uptake without altering Hg^{2+} resistance, perhaps by altering membrane permeability.[33]

24.2.2.2   Detoxification

The MerA protein, also known as mercuric reductase, is a flavin adenine dinucleotide (FAD)-dependent oxidoreductase found in the cytoplasm[34] that uses NADPH as an electron donor.[35–38] It is similar to pyridine nucleotide thiol/disulfide oxidoreductases including glutathione reductase and lipoamide dehydrogenase.[39] MerA has an active site similar to other members of this family with two essential cysteine residues. However, in MerA these cysteines are not involved in oxidoreduction, but rather form high-affinity complexes with Hg^{2+} to enable Hg^{2+} reduction by the transfer of electrons from NADPH (eqn (1)). The product Hg^0 has no affinity for the site and is released.

$$\text{MerA Reduction: } NADPH + Hg^{2+} \rightarrow NADP^+ + H^+ + Hg(0) \tag{1}$$

Unlike other thiol/disulfide oxidoreductases, MerA has a relatively large N-terminal domain including 70 amino acids homologous to MerP and also including a pair of cysteines (Figure 24.2). This N-terminal domain is not essential for Hg^{2+} reduction, but may have a role in delivering Hg^{2+} to the catalytic core through a mechanism similar to transfer of Hg^{2+} from MerP to MerT.[1,27]

While mercury resistance is essentially provided by MerA mediated reduction of Hg^{2+}, the lyase MerB is required to confer broad-spectrum resistance to organomercurials. MerB is a lyase that catalyses protonolysis of Hg-C bonds, releasing Hg^{2+}, which is subsequently reduced by MerA. MerB is a small (22 kDa), monomeric, cytosolic protein that does not require any cofactors although it requires an excess of thiol such as GSH or cysteine. It can protonate most organomercurials.[40]

$$\text{Protonation: } [R\text{-}Hg]^+ + H^+ \rightarrow Hg^{2+} + R\text{-}H \text{ (R} = \text{alkyl, aryl ligand)} \tag{2}$$

There are no known eukaryotic MerB homologues although there are several prokaryotic MerB variants. MerB activity depends on three highly conserved cysteines: 96, 117 and 159. Cysteines 96 and 159 are essential for enzymatic activity while cysteine 117 contributes to protein conformation and protein solubility.[41] Organomercurials are stabilized in the MerB active site by digonal coordination with cysteine residues 96 and 159, aspartate residue 99 and a water molecule.[42] The precise protonation mechanism is unknown; however, once a covalent adduct is formed between the organomercurial and cysteines 96 and 159, the alkyl leaving group is protonated by either cysteine 159 or aspartate 99.[42]

24.2.2.3 Sensing and Regulation

MerR is a 144 amino acid protein with three highly conserved cysteines that form a binding site for Hg^{2+}. It regulates transcriptional expression of mer genes. As a homodimer, MerR binds the DNA sequence of the mer operator upstream of the *merT* gene (Figure 24.2). When MerR is bound to the operator, it attracts RNA polymerase and forms a stable, non-transcribing pre-initiation complex. In this state, transcription of all the mer operon genes, including *merR*, is inhibited. Very low, nanomolar concentrations of Hg^{2+} bind MerR with a stoichiometry of one Hg^{2+} per homodimer.[43] Hg^{2+} binding induces a conformational change in the MerR-DNA-RNA polymerase complex enabling a nearly instantaneous transcriptional response to cytosolic Hg^{2+}.[44] Like all other mer proteins, MerR cysteine residues mediate coordination with Hg^{2+}; however, MerR forms a high affinity planar trigonal coordination complex between three cysteine residues rather than the digonal coordination observed between cysteine pairs in other mer proteins (Table 24.2).[45] This high-affinity MerR trigonal coordination retains Hg^{2+} even after cytosolic Hg^{2+} is practically eliminated. This prolonged activation of MerR results in continued expression of mer components including *merA*.

Some mer operons contain MerD, a small, Cys-rich protein with an N-terminal domain similar to MerR. MerD is expressed in small amounts and antagonizes MerR and represses *merA* transcription very quickly once Hg^{2+} is eliminated (Figure 24.2). MerD binds to the mer operator with lower affinity than does MerR and promotes its turnover. Deletion of MerD results in continued expression of *merA* and other structural genes after Hg^{2+} levels have decreased.[46,47]

24.3 Eukaryotes

24.3.1 Sources of Exposure and Systemic Uptake Mechanisms

24.3.1.1 Yeast, Algae, Plants

Similarly to prokaryotes, eukaryotes including yeast, algae and plants are primarily exposed to inorganic and organic mercurial compounds in the soil and water. In yeast exposed to $HgCl_2$, several electron microscopic and radiographic studies have demonstrated that the majority of Hg accumulated by yeast is bound to the cell wall and the cytoplasmic membrane in the form of insoluble HgS compounds.[48,49] This binding generates insoluble mercury complexes and acts as an adsorption filter to limit the amount of Hg^{2+} that enters the cytoplasm in yeast and plants.[50] Hg^{2+} from the soil is readily taken up and accumulates in root systems with up to 80% bound to cell walls.[51] In addition to uptake from the soil, elemental Hg^0 can be absorbed through leaves. Total mercury content of corn and wheat plants exposed to Hg^0 vapours reflected total Hg concentrations in the air rather than the soil.[52] The exact mechanism of Hg^0 uptake is not known, but likely involves gas exchange through stomata. Hg that does enter the cytosol, roots or leaves can be transferred into shoots through xylem-uploading or phytochelatin binding.[50]

The bioavailability of Hg is affected by certain compounds present in soil and water. For instance, dissolved organic matter is a heterogeneous mixture of organic compounds including amino acids, carbohydrates and humic substances generated by decomposition of plants and animals. It diminishes Hg bioavailability through coordination with reduced sulfur residues. These large organic Hg complexes are much less likely to undergo biotic methylation compared to Hg^{2+}, HgS and $HgCl_2$. As a result, high levels of dissolved organic matter increase the likelihood that Hg will remain bound in large organic complexes and undergo photolytic reduction and volatilization, thereby ultimately reducing the concentration of Hg in the water.[4] Natural and applied chemicals can also alter Hg uptake. For example, the presence of arsenic (an environmental toxicant often found in water used for drinking and irrigation) increased uptake of Hg^{2+} in cultured rice plants.[53] Thiosulfate treatment of contaminated soil near a large mercury mine in China increased the bioavailability and uptake of Hg^{2+} by plants.[54]

24.3.1.2 Fish

The primary source of Hg exposure in larger eukaryotic organisms including fish is consumption of contaminated plankton and small animals. Hg, particularly $[CH_3Hg]^+$, can reach concentrations as much as 10^6-fold higher than surrounding water because it is excreted at a much slower rate than it is consumed.[55] There is a tendency for animals to accumulate inorganic and organic Hg species differentially in tissues. X-ray fluorescence imaging of zebrafish larvae demonstrated that Hg uptake and tissue distribution varies with exposure to different Hg species. Exposure to $[CH_3Hg]^+$ led to higher larval Hg burden compared to inorganic exposure. The highest levels were found in the eye while the highest levels of inorganic compounds were found in the kidney.[56] In sea bass exposed to varying levels of environmental contamination, Hg is deposited hierarchically in different tissues with the highest levels observed in the liver followed, in decreasing order, by kidney, muscle, brain and gills and, finally, blood.[57] At high levels of exposure, liver concentrations increased disproportionately to other tissues examined, indicating that, in fish, Hg may be sequestered in the liver to reduce Hg damage in other tissues. This sequestration of Hg at the tissue may reduce the toxicity of Hg and contribute to the high concentrations of Hg present in some fish.

24.3.1.3 Humans

Humans take in $[CH_3Hg]^+$ and Hg^{2+} through ingestion of contaminated fish and plants. However, occupational exposure and other anthropogenic sources of Hg can also be significant sources of Hg. Exposure to metallic Hg can occur through ingestion, skin contact and inhalation of vapours. Occupational exposure is the most common route, particularly in certain occupations in manufacturing, dentistry, house painting and waste disposal. Non-occupational exposure can occur through Hg spills from broken thermometers or light bulbs and from dental fillings. Due to its high density, less than 0.01% of ingested liquid Hg is absorbed. However, elemental Hg vapours are highly lipid soluble and are rapidly generated from metallic Hg at ambient temperatures and during industrial processes. As a result, inhalation of Hg vapours is the most significant source of metallic Hg absorption, with as much as 80% of the inhaled vapours directly entering the bloodstream. Upon absorption, elemental Hg enters all tissues and accumulates in the central nervous system and kidneys.

Absorption of inorganic Hg occurs largely through occupational exposure, especially in electrical and auto parts manufacturing, chemical processing, metal processing and construction. However, significant levels of non-occupational exposure can occur through consistent use of cosmetic and medical products containing inorganic mercury (such as skin lightening creams and laxatives). It can also occur through consumption of contaminated fish because as much as 20% of the Hg present in fish may be inorganic. In contrast to metallic Hg, inorganic Hg compounds do not easily vaporize. As a result, ingestion is the most common route of exposure with up to 10% of ingested Hg^{2+} being absorbed in the gastrointestinal tract. Inorganic Hg has low lipid solubility and does not readily cross cell membranes. Once absorbed, the majority of Hg^{2+} accumulates in the kidney and liver.

Human exposure to organic Hg takes place chiefly through consumption of contaminated fish and marine mammals with $[CH_3Hg]^+$ being the predominant organomercurial. As much as 90% of ingested $[CH_3Hg]^+$ is absorbed through the intestine and forms thiol complexes with proteins and amino acids (such as glutathione and cysteine) in the liver. Some $[CH_3Hg]^+$ enters the general circulation where it is distributed throughout all tissues including the brain. However, most of the absorbed $[CH_3Hg]^+$ is incorporated into bile, secreted into the intestine and reabsorbed. This enterohepatic circulation extends the time during which $[CH_3Hg]^+$ can be absorbed. Other organomercurials such as ethylmercury are metabolized to Hg^{2+} more rapidly than $[CH_3Hg]^+$, resulting in a shorter half-life and different patterns of tissue accumulation. As for Hg^{2+}, the highest levels of $[C_2H_5Hg]^+$ are found in the kidney while very little Hg accumulation is observed in neural tissues.[58,59]

24.3.2 Coordination Chemistry in Blood

Recall that Hg^{2+} is a soft electrophile with high affinity for soft nucleophile thiol (-SH) and selenol (-SeH) groups (Table 24.3). As with mer proteins, the interaction of Hg^{II} compounds with these groups forms the foundation for the transport and activity of Hg species within eukaryotic organisms. The coordination chemistry of Hg compounds in blood is determined largely by the key cysteine residues of serum albumin, glutathione (γ-glutamyl-cysteinyl-glycine; GSH) and haemoglobin and by interactions with the selenocysteine residues of selenoprotein P.

The primary Hg species found in plasma is the mercuric ion Hg^{2+}. Most plasma Hg^{2+} (>90%) is bound to serum albumin. This protein contains a single reactive cysteine residue that interacts with both Hg^{2+} and $[CH_3Hg]^+$. Studies in mice, whole human blood and plasma surrogate have demonstrated that Hg^{2+} forms a digonal complex with the cysteines of two albumin molecules while monovalent $[CH_3Hg]^+$ interacts similarly with a single serum albumin molecule.[60–62] The remaining fraction of serum Hg^{2+} is bound to globins and other plasma proteins including selenoprotein P, as discussed below. In contrast, the majority of $[CH_3Hg]^+$ in blood is not found in plasma but is bound to GSH as CH_3Hg-SG in erythrocytes.[63,64]

Selenium (Se) is a heavier congener of sulfur and is an essential trace element that is incorporated into several prokaryotic and eukaryotic proteins *via* the amino acid selenocysteine. The interaction between Hg and Se was first identified in humans by Kosta and colleagues in 1975, who observed a 1 : 1 ratio of Hg : Se in *post-mortem* tissue from mine workers exposed to high levels of inorganic Hg.[65] Subsequent studies revealed that the affinity of inorganic and organic Hg compounds is higher for selenol groups than for thiol groups (Log $K = 38.9$ and 45.0 for HgS and HgSe, respectively).[66–68] The first selenoprotein identified was glutathione peroxidase (GPx);[69] other selenoproteins include iodothyronine deiodinase, thioredoxin reductase and selenoprotein P. The latter is abundant in plasma and contributes to whole body Se metabolism. It contains ten selenocysteine residues and is the only selenoprotein to contain more than one selenocysteine. Hg compounds inhibit the function of selenoproteins in general, including the reduction of glutathione peroxidase activity by both Hg^{2+} and CH_3Hg^+ in red blood cell lysates.[70] However, inorganic and organic Hg compounds can also form 1 : 1 stoichiometric compounds with the selenide anion Se^{2-} that are not toxic to animals.[71–75] These compounds form crystal-like zinc blende structures that can be conjugated to GSH to form aggregates that can be represented by $(GSH)_5(HgSe)_{100}$.[76] As many as 35 of these aggregates can bind a single selenoprotein P.[77] In this way, selenoprotein P effectively sequesters Hg.

24.3.3 Cellular Transport

As discussed in the previous section, the high affinity of Hg for thiol and selenol groups forms the basis of its interactions with biological molecules and its transport through cells and entire organisms. The metal itself and Hg compounds have high affinities for thiol groups and bind to thiol-rich intracellular proteins, peptides and amino acids, including metallothionein and GSH. There are no known dedicated eukaryotic Hg transport systems; however, these Hg-thiolate complexes enable transport of cellular Hg *via* several distinct transport systems.

24.3.3.1 *Mercury Binding Proteins*

Metallothioneins (MTs) are small, cysteine-rich proteins that bind both essential (Zn, Cu) and non-essential metals (Cd, Hg and Ag).[†] They contribute to regulation of metal homeostasis and protect against metal toxicity and oxidative stress. The MT metal-binding domains contain 20 conserved cysteine residues located in two distinct regions of the metal-binding domains (α and β) that are separated by a cysteine free spacer region. Under normal conditions, MTs are found in the cytoplasm and mitochondria and each individual protein can bind up to 7 divalent cations and 12 monovalent cations.[78,79] Most metals are bound in the metal binding domains by tetrahedral coordination with the cysteine residues. Zinc is the most common metal associated with metallothioneins; however, Hg^{2+} can displace Zn and result in Hg sequestration. Although there is limited evidence from EXAFS studies demonstrating that MTs can bind Hg^{2+} with tetrahedral or distorted tetrahedral coordination, the most likely coordination geometry is digonal.[80,81]

The importance of MTs in tissue binding and accumulation of Hg has been observed in three different species of arctic seal where tissue concentrations of Hg correlated positively with increased levels of MT expression.[113] This is consistent with the well-characterized induction of MT expression by stressors such as heavy metals. Hg regulation of MT expression may also occur in plants. A cDNA screen for Hg exposure-induced gene expression changes identified two putative MTs.[82]

GSH is a highly abundant tripeptide that serves as a reducing agent for reactive oxygen species and other unstable molecules, a reaction catalysed by the selenoprotein GPx. GSH can also be conjugated to xenobiotics including Hg^{2+} and $[CH_3Hg]^+$ to facilitate excretion during detoxification. Hg^{2+} forms linear complexes GS-Hg-SG and Cys-Hg-Cys with the thiol groups of both GSH and cysteine while $[CH_3Hg]^+$ forms complexes GS-HgCH$_3$ and Cys-HgCH$_3$.[83–86] The GSH complexes facilitate systemic excretion through the liver and kidneys as well as cellular efflux of $[CH_3Hg]^+$ from astrocytes.[87]

[†]See Chapter 21 for a comprehensive account of metallothionein proteins.

24.3.3.2   Amino Acid Transporters

The structure of Hg-thiolate complexes can be recognized by certain amino acid transport systems, specifically the $B^{0,+}$ and L systems that transport neutral amino acids such as methionine.[88] $B^{0,+}$ and L systems are hetero-dimeric amino acid exchangers comprised of a heavy chain and a light chain.[89]

The L system is nearly ubiquitous in non-epithelial cells and transports large neutral (zwitterionic) amino acids such as leucine, isoleucine, tyrosine, valine, tryptophan, phenylalanine, threonine, methionine and cysteine.[90] There are two subtypes: LAT-1 is a high-affinity transporter while LAT-2 is a lower affinity transporter that can also transport other smaller neutral amino acids including alanine, glycine and serine but also glutamate and cysteine.[89] The L system is sodium-independent and operates as a stoichiometric 1 : 1 exchanger.[91] The first indication that $[CH_3Hg]^+$ Cys-S conjugates may be transported across the blood-brain barrier by a selective transport mechanism such as the L system came from the observation of competitive inhibition of methylmercury uptake in astrocytes by large, neutral amino acids.[92,93] When expressed in CHO cells, LAT-1 causes uptake of $[CH_3Hg]^+$ Cys-S conjugates.[94]

The $B^{0,+}$ system is found in the kidney, intestine and brain and transports neutral and dibasic amino acids in a sodium-dependent manner. In the kidney, it is localized on the luminal plasma membrane of proximal tubular epithelial cells and facilitates reabsorption of essential amino acids such as methionine. *Xenopus* oocytes transfected with a cDNA encoding system $B^{0,+}$ were capable of taking up $[CH_3Hg]^+$ Cys-S conjugates. Uptake is inhibited by competition with methionine, suggesting that, as in the L system, $[CH_3Hg]^+$ Cys-S is transported as a mimic of methionine.[95] The involvement of $B^{0,+}$ in methylmercury transport was further confirmed in proximal tubule segments isolated from rabbit, which were capable of taking up $[CH_3Hg]^+$ GS and $[CH_3Hg]^+$ Cys-S conjugates. This uptake was inhibited by competition with methionine, cysteine and acivicin (a specific γ-glutamyltransferase inhibitor). Treatment with chelating agents decreases this reuptake and facilitates urinary excretion of $[CH_3Hg]^+$ thiol conjugates.[96]

24.3.3.3   Multi-drug Resistance Transporters

In addition to amino acid transporters, multidrug resistance proteins (MRPs) also transport Hg-thiolate complexes. MRP genes encode transporters that export GSH-protein complexes and xenobiotics. They, or their homologues, have been found in many different types of animals and facilitate cellular export and excretion of both inorganic and organic Hg-thiolate complexes.[97,98] $HgCl_2$ sensitivity and accumulation was reduced in a metal-sensitive strain of *E. coli* transfected with human MDR-1 and the bacterial homologues LMR and OmrA. Consistent with the *E. coli* study, deletion of *ycf1p*, a yeast MRP homologue that transports glutathione *S*-conjugates into the vacuole, increased sensitivity of *S. cerevisiae* to Hg and demonstrated that $Hg(GS)_2$ is a substrate for ycf1p.[99] The role of MRP in

Hg-thiolate conjugate efflux is further supported by studies in primary cortical cultures where inhibition of MRP1 increased sensitivity and cellular accumulation of $[CH_3Hg]^+$.[100] Furthermore, rats lacking MRP2 have increased accumulation of Hg in liver and kidney concomitant with decreased faecal and urinary excretion of both Hg^{2+} and $[CH_3Hg]^+$ Cys-S conjugates.[101]

Organic anion transporters (OATs) are broad-spectrum, plasma membrane exporters found in liver, brain, kidney and intestine. Together with MRPs, they provide a mechanism for cellular export of many endogenous compounds including bile and xenobiotics such as prescription drugs and GSH conjugated proteins.[102] In hyperbilirubinemic rats that lack both OAT and MRP expression, excretion of inorganic Hg and GSH-Hg conjugates was greatly reduced compared to control rats, suggesting that OAT facilitates biliary excretion of inorganic Hg.[97]

24.4 Conclusions

Hg is a non-essential, naturally occurring, highly toxic metal. Prokaryotes can enzymatically transform Hg as a by-product of normal metabolic pathways as well as through specific cellular uptake and reduction by well-characterized proteins controlled by the mer operon/transcriptional regulator. There is convincing evidence that eukaryotes take up Hg compounds through molecular mimicry by amino acid transporters and MRPs. However, very little is known about the specific mechanisms of Hg transport and sensing within cells. Looking forward, recent advances in genetic and molecular technology are enabling more detailed exploration of Hg transport and leading to a better understanding of how Hg cycles in the global environment.

Acknowledgements

This chapter was supported in part by grants from the National Institute of Environmental Health Sciences ESR01-10563, R01-07331, ES T32-007028 and the Molecular Toxicology Center ES P30-000267.

References

1. T. Barkay, S. M. Miller and A. O. Summers, *FEMS Microbiol. Rev.*, 2003, **27**, 355–384.
2. T. W. Clarkson, *Crit. Rev. Clin. Lab. Sci.*, 1997, **34**, 369–403.
3. S. Bose-O'Reilly, K. M. McCarty, N. Steckling and B. Lettmeier, *Curr. Probl. Pediatr. Adolesc. Health Care*, 2010, **40**, 186–215.
4. M. Ravichandran, *Chemosphere*, 2004, **55**, 319–331.
5. S. Haverstock, T. Sizmur, J. Murimboh and N. J. O'Driscoll, *Chemosphere*, 2012, **88**, 1220–1226.
6. J. Nriagu and C. Becker, *Sci. Total Environ.*, 2003, **304**, 3–12.
7. H. Zhang and S. E. Lindberg, *Environ. Sci. Technol.*, 2001, **35**, 928–935.

8. R. K. Skogerboe and S. A. Wilson, *Anal. Chem.*, 1981, **53**, 228–232.

9. B. Allard and I. Arsenie, *Water Air Soil Pollut.*, 1991, **56**, 457–464.

10. J. H. Weber, *Chemosphere*, 1993, **26**, 2063–2077.

11. S. Jensen and A. Jernelov, *Nature*, 1969, **223**, 753–754.

12. J. M. Wood, F. S. Kennedy and C. G. Rosen, *Nature*, 1968, **220**, 173–174.

13. G. C. Compeau and R. Bartha, *Appl. Environ. Microbiol.*, 1985, **50**, 498–502.

14. K. R. Pak and R. Bartha, *Appl. Environ. Microbiol.*, 1998, **64**, 1013–1017.

15. C. C. Gilmour, D. A. Elias, A. M. Kucken, S. D. Brown, A. V. Palumbo, C. W. Schadt and J. D. Wall, *Appl. Environ. Microbiol.*, 2011, 77, 3938–3951.

16. J. K. King, J. E. Kostka, M. E. Frischer and F. M. Saunders, *Appl. Environ. Microbiol.*, 2000, **66**, 2430–2437.

17. E. B. Ekstrom and F. M. Morel, *Environ. Sci. Technol.*, 2008, **42**, 93–99.

18. K. Pak and R. Bartha, *Bull. Environ. Contam. Toxicol.*, 1998, **61**, 690–694.

19. M. H. Richmond and M. John, *Nature*, 1964, **202**, 1360–1361.

20. R. P. Novick and C. Roth, *J. Bacteriol.*, 1968, **95**, 1335–1342.

21. S. Fretham, S. Caito, E. Martinez-Finley and M. Aschner, *Toxicol. Res.*, 2012, **1**, 32–28.

22. A. P. Morby, J. L. Hobman and N. L. Brown, *Mol. Microbiol.*, 1995, **17**, 25–35.

23. L. Sahlman and E. G. Skarfstad, *Biochem. Biophys. Res. Commun.*, 1993, **196**, 583–588.

24. T. M. DeSilva, G. Veglia, F. Porcelli, A. M. Prantner and S. J. Opella, *Biopolymers*, 2002, **64**, 189–197.

25. P. C. Bull and D. W. Cox, *Trends Genet.*, 1994, **10**, 246–252.

26. S. Silver and L. T. Phung, *J. Ind. Microbiol. Biotechnol.*, 2005, **32**, 587–605.

27. N. L. Brown, Y. C. Shih, C. Leang, K. J. Glendinning, J. L. Hobman and J. R. Wilson, *Biochem. Soc. Trans.*, 2002, **30**, 715–718.

28. L. Sahlman, W. Wong and J. Powlowski, *J. Biol. Chem.*, 1997, **272**, 29518–29526.

29. L. Sahlman, E. M. Hagglof and J. Powlowski, *Biochem. Biophys. Res. Commun.*, 1999, **255**, 307–311.

30. J. R. Wilson, C. Leang, A. P. Morby, J. L. Hobman and N. L. Brown, *FEBS Lett.*, 2000, **472**, 78–82.

31. Y. Uno, M. Kiyono, T. Tezuka and H. Pan-Hou, *Biol. Pharm. Bull.*, 1997, **20**, 107–109.

32. M. Kiyono, Y. Sone, R. Nakamura, H. Pan-Hou and K. Sakabe, *FEBS Lett.*, 2009, **583**, 1127–1131.

33. M. Kiyono and H. Pan-Hou, *J. Bacteriol.*, 1999, **181**, 726–730.

34. A. O. Summers and L. I. Sugarman, *J. Bacteriol.*, 1974, **119**, 242–249.

35. T. Barkay and I. Wagner-Dobler, *Adv. Appl. Microbiol.*, 2005, **57**, 1–52.

36. K. Izaki, Y. Tashiro and T. Funaba, *J. Biochem.*, 1974, **75**, 591–599.

37. J. L. Schottel, *J. Biol. Chem.*, 1978, **253**, 4341–4349.

38. K. Furakawa and K. Tonomura, *Agr. Biol. Chem.*, 1972, **36**, 2441–2448.

39. B. Fox and C. T. Walsh, *J. Biol. Chem.*, 1982, **257**, 2498–2503.

40. T. P. Begley, A. E. Walts and C. T. Walsh, *Biochemistry*, 1986, **25**, 7192–7200.
41. K. E. Pitts and A. O. Summers, *Biochemistry*, 2002, **41**, 10287–10296.
42. J. M. Parks, H. Guo, C. Momany, L. Liang, S. M. Miller, A. O. Summers and J. C. Smith, *J. Am. Chem. Soc.*, 2009, **131**, 13278–13285.
43. T. V. O'Halloran, *Science*, 1993, **261**, 715–725.
44. T. O'Halloran and C. Walsh, *Science*, 1987, **235**, 211–214.
45. J. G. Wright, H. T. Tsang, J. E. Penner-Hahn and T. O'Halloran, *J. Am. Chem. Soc.*, 1990, **112**, 2434–2434.
46. D. Mukhopadhyay, H. R. Yu, G. Nucifora and T. K. Misra, *J. Biol. Chem.*, 1991, **266**, 18538–18542.
47. G. Nucifora, S. Silver and T. K. Misra, *Mol. Gen. Genet.*, 1990, **220**, 69–72.
48. A. D. Murray and D. K. Kidby, *J. Gen. Microbiol.*, 1975, **86**, 66–74.
49. S. G. Whittaker, D. G. Smith, J. R. Foster and I. R. Rowland, *J. Histochem. Cytochem.*, 1990, **38**, 823–827.
50. J. Chen and Z. M. Yang, *Biometals*, 2012, **25**, 847–857.
51. Y. Wang and M. Greger, *J. Environ. Qual.*, 2004, **33**, 1779–1785.
52. Z. Niu, X. Zhang, Z. Wang and Z. Ci, *Environ. Pollut.*, 2011, **159**, 2684–2689.
53. X. Du, Y. G. Zhu, W. J. Liu and X. S. Zhao, *Environ. Exp. Bot.*, 2005, **54**, 1–7.
54. J. Wang, X. Feng, C. W. Anderson, H. Wang, L. Zheng and T. Hu, *Environ. Sci. Technol.*, 2012, **46**, 5361–5368.
55. N. Ballatori and J. L. Boyer, *Toxicol. Appl. Pharmacol.*, 1986, **85**, 407–415.
56. M. Korbas, T. C. Macdonald, I. J. Pickering, G. N. George and P. H. Krone, *ACS Chem. Biol.*, 2012, **7**, 411–420.
57. C. L. Mieiro, M. Pacheco, M. E. Pereira and A. C. Duarte, *Arch. Environ. Contam. Toxicol.*, 2011, **61**, 135–143.
58. J. G. Dorea, M. Farina and J. B. Rocha, *J. Appl. Toxicol.*, 2013, **38**, 1–8.
59. T. M. Burbacher, D. D. Shen, N. Liberato, K. S. Grant, E. Cernichiari and T. Clarkson, *Environ. Health Perspect.*, 2005, **113**, 1015–1021.
60. J. Sundberg, B. Ersson, B. Lonnerdal and A. Oskarsson, *Toxicology*, 1999, **137**, 169–184.
61. S. Trumpler, S. Nowak, B. Meermann, G. A. Wiesmuller, W. Buscher, M. Sperling and U. Karst, *Anal. Bioanal. Chem.*, 2009, **395**, 1929–1935.
62. A. Yasutake, K. Hirayama and M. Inoue, *Arch. Toxicol.*, 1989, **63**, 479–483.
63. T. W. Clarkson, J. B. Vyas and N. Ballatori, *Am. J. Ind. Med.*, 2007, **50**, 757–764.
64. D. L. Rabenstein and A. A. Isab, *Biochim. Biophys. Acta*, 1982, **721**, 374–384.
65. L. Kosta, A. R. Byrne and V. Zelenko, *Nature*, 1975, **254**, 238–239.
66. D. Dryssen and M. Wedborg, *Water Air Soil Pollut.*, 1991, **56**, 507–519.
67. C. Sasakura and K. T. Suzuki, *J. Inorg. Biochem.*, 1998, **71**, 159–162.
68. M. A. Khan and F. Wang, *Environ. Toxicol. Chem.*, 2009, **28**, 1567–1577.
69. J. T. Rotruck, A. L. Pope, H. E. Ganther, A. B. Swanson, D. G. Hafeman and W. G. Hoekstra, *Science*, 1973, **179**, 588–590.
70. H. M. Mykkanen and H. E. Ganther, *Bull. Environ. Contam. Toxicol.*, 1974, **12**, 10–16.

71. Y. Sugiura, Y. Hojo, Y. Tamai and H. Tanaka, *J. Am. Chem. Soc.*, 1976, **98**, 2339–2341.
72. Y. Sugiura, Y. Tamai and H. Tanaka, *Bioinorg. Chem.*, 1978, **9**, 167–180.
73. A. Naganuma and N. Imura, *Pharmacol. Biochem. Behav.*, 1981, **15**, 449–454.
74. A. Naganuma, J. Tabata and N. Imura, *Res. Commun. Chem. Pathol. Pharmacol.*, 1982, **38**, 291–299.
75. A. Naganuma, K. Kosugi and N. Imura, *Toxicol. Lett.*, 1981, **8**, 43–48.
76. T. Garcia-Barrera, J. L. Gomez-Ariza, M. Gonzalez-Fernandez, F. Moreno, M. A. Garcia-Sevillano and V. Gomez-Jacinto, *Anal. Bioanal. Chem.*, 2012, **403**, 2237–2253.
77. J. Gailer, G. N. George, I. J. Pickering, S. Madden, R. C. Prince, E. Y. Yu, M. B. Denton, H. S. Younis and H. V. Aposhian, *Chem. Res. Toxicol.*, 2000, **13**, 1135–1142.
78. G. Henkel and B. Krebs, *Chem. Rev.*, 2004, **104**, 801–824.
79. P. Babula, M. Masarik, V. Adam, T. Eckschlager, M. Stiborova, L. Trnkova, H. Skutkova, I. Provaznik, J. Hubalek and R. Kizek, *Metallomics*, 2012, **4**, 739–750.
80. M. Vasak, J. H. Kagi and H. A. Hill, *Biochemistry*, 1981, **20**, 2852–2856.
81. D. E. Sutherland and M. J. Stillman, *Metallomics*, 2011, **3**, 444–463.
82. P. Venkatachalam, A. K. Srivastava, K. G. Raghothama and S. V. Sahi, *Environ. Sci. Technol.*, 2009, **43**, 843–850.
83. V. Mah and F. Jalilehvand, *J. Biol. Inorg. Chem.*, 2008, **13**, 541–553.
84. V. Mah and F. Jalilehvand, *Chem. Res. Toxicol.*, 2010, **23**, 1815–1823.
85. K. L. Pei, M. Sooriyaarachchi, D. A. Sherrell, G. N. George and J. Gailer, *J. Inorg. Biochem.*, 2011, **105**, 375–381.
86. R. K. Mehra, J. Miclat, V. R. Kodati, R. Abdullah, T. C. Hunter and P. Mulchandani, *Biochem. J.*, 1996, **314**, 73–82.
87. J. Fujiyama, K. Hirayama and A. Yasutake, *Biochem. Pharmacol.*, 1994, **47**, 1525–1530.
88. N. Ballatori, *Environ. Health Perspect.*, 2002, **110**, 689–694.
89. C. A. Wagner, F. Lang and S. Broer, *Am. J. Physiol. Cell Physiol.*, 2001, **281**, C1077–1093.
90. C. Valdovinos-Flores and M. E. Gonsebatt, *Neurochem. Int.*, 2012, **61**, 405–414.
91. C. Meier, Z. Ristic, S. Klauser and F. Verrey, *EMBO J.*, 2002, **21**, 580–589.
92. M. Aschner, N. B. Eberle, S. Goderie and H. K. Kimelberg, *Brain Res.*, 1990, **521**, 221–228.
93. M. Aschner, N. B. Eberle and H. K. Kimelberg, *Brain Res.*, 1991, **554**, 10–14.
94. Z. Yin, H. Jiang, T. Syversen, J. B. Rocha, M. Farina and M. Aschner, *J. Neurochem.*, 2008, **107**, 1083–1090.
95. C. C. Bridges and R. K. Zalups, *J. Pharmacol. Exp. Ther.*, 2006, **319**, 948–956.
96. Y. Wang, R. K. Zalups and D. W. Barfuss, *Toxicol. Lett.*, 2012, **213**, 203–210.

97. N. Sugawara, Y. R. Lai, C. Sugaware and K. Arizono, *Toxicology*, 1998, **126**, 23–31.
98. I. Bosnjak, K. R. Uhlinger, W. Heim, T. Smital, J. Franekic-Colic, K. Coale, D. Epel and A. Hamdoun, *Environ. Sci. Technol.*, 2009, **43**, 8374–8380.
99. O. Gueldry, M. Lazard, F. Delort, M. Dauplais, I. Grigoras, S. Blanquet and P. Plateau, *Eur. J. Biochem.*, 2003, **270**, 2486–2496.
100. M. Achard-Joris and J. P. Bourdineaud, *Biometals*, 2006, **19**, 695–704.
101. C. C. Bridges, L. Joshee and R. K. Zalups, *Toxicol. Appl. Pharmacol.*, 2011, **251**, 50–58.
102. J. Konig, *Handb. Exp. Pharmacol.*, 2011, **201**, 1–28.
103. L. Sahlman and B. H. Jonsson, *Eur. J. Biochem.*, 1992, **205**, 375–381.
104. R. A. Steele and S. J. Opella, *Biochemistry*, 1997, **36**, 6885–6895.
105. N. Schiering, W. Kabsch, M. J. Moore, M. D. Distefano, C. T. Walsh and E. F. Pai, *Nature*, 1991, **352**, 168–172.
106. E. Rossy, L. Champier, B. Bersch, B. Brutscher, M. Blackledge and J. Coves, *J. Biol. Inorg. Chem.*, 2004, **9**, 49–58.
107. J. D. Helmann, B. T. Ballard and C. T. Walsh, *Science*, 1990, **247**, 946–948.
108. R. B. Simpson, *J. Am. Chem. Soc.*, 1961, **83**, 4711–4717.
109. Y. Li, X. P. Yan, C. Chen, Y. L. Xia and Y. Jiang, *J. Proteome. Res.*, 2007, **6**, 2277–2286.
110. A. Yasutake and K. Hirayama, *Bull. Environ. Contam. Toxicol.*, 1990, **45**, 662–666.
111. R. Doi and M. Tagawa, *Toxicol. Appl. Pharmacol.*, 1983, **69**, 407–416.
112. R. W. Chen, H. E. Ganther and K. G. Hoekstra, *Biochem. Biophys. Res. Commun.*, 1973, **51**, 383–390.
113. C. Sonne, O. Aspholm, R. Dietz, S. Andersen, M. H. Berntssen and K. Hylland, *Sci. Total Environ.*, 2009, **407**(24), 6166–6172.

CHAPTER 25

Antimony and Bismuth

TIANFAN CHENG AND HONGZHE SUN*

Department of Chemistry, The University of Hong Kong, Pokfulam Road, Hong Kong, China
*Email: hsun@hku.hk

25.1 General Properties of Antimony and Bismuth

Antimony (Stibium, Sb) and bismuth (Bisemutum, Bi) are the heaviest elements in Group 15 of the periodic table (pnictogens), sharing the outer shell electronic configuration ns^2np^3 with the lighter members of the group. Their most common oxidation states are +3 and +5, respectively. Derivatives of antimony are often labelled as "antimonites" and "antimonates", respectively. Antimony is considered a metalloid while bismuth displays more pronounced metallic properties. Their general physical and chemical properties have been reviewed in detail.[1]

The elements are Lewis acids in the pentavalent state, *e.g.* pentahalides such as SbX_5 (X = halogen). In the trivalent state, they are generally considered as Lewis bases due to the presence of an influential lone electron pair. In triorganyl compounds, however, basicity decreases upon descending in the group, especially for bismuth, leading to only a few triorganobismuthine compounds having been prepared. When bonded with significantly more electronegative atoms or groups, antimony and bismuth show substantial Lewis acidity, as in SbX_3, BiX_3 and the phenyl dihalides.[2,3] The common coordination numbers are four, five and six for Sb(V) and Bi(V), corresponding to tetrahedral, trigonal bipyramidal or octahedral geometries. For the +3 oxidation states, the coordination numbers are three, four, five and six with trigonal pyramidal, trigonal bipyramidal, square-based

RSC Metallobiology Series No. 2
Binding, Transport and Storage of Metal Ions in Biological Cells
Edited by Wolfgang Maret and Anthony Wedd

pyramidal and octahedral geometries. A more detailed compilation of geometries is available.[1] Furthermore, chalcogeno-bridged antimony or bismuth complexes (chalcogen = O, S, Se) are often observed. They form oligomers, polymers or even extended clusters or cages.[1]

25.2 Antimony and Bismuth in Medicine

25.2.1 Antimony: Anti-parasite and Anti-cancer Activities

Antimony has a long history for medical application as antimonials. Since the sixteenth century, antimony-based drugs have been employed for the treatment of parasitic, protozoa-caused diseases.[4] Antimony (III) potassium tartrate (also known as potassium antimonyl tartrate), for example, had been applied for the treatment of trypanosomiasis and leishmaniasis since the beginning of the twentieth century,[5] and lately even of schistosomiasis, before its replacement by praziquantel.[6] Two major antimony-based drugs, sodium stibogluconate (Pentostam®) and meglumine antimonate (Glucantim®) (Figure 25.1) have been used for the treatment of cutaneous and visceral leishmaniasis, though increasing resistance and side effects are observed.[1,7] Formulations based on liposomes as carriers of pentavalent antimonial drugs in the therapy of leishmaniasis are applied widely.[7–9] In spite of extensive clinical applications, the molecular structures, metabolism and mechanisms of action of the drugs are still unclear. It was assumed that pentavalent antimony acts as a pro-drug that is reduced to more toxic trivalent antimony at or near the site of action;[10] however, other investigations support the direct involvement of pentavalent antimony in drug action.[7] Peptides, proteins and ribonucleosides may mediate the action of the drugs.[7]

Besides their well-known anti-trypanosomal use in clinical practices, the potential anti-cancer activity of antimonials was also explored.[1,11–13] An antimony-containing technetium-99m (^{99m}Tc) agent, ^{99m}Tc antimony (tri)-sulfide colloid, is applied in medical radiography for tumour diagnosis.[14] Anti-viral activity was also investigated, *e.g.* in the case of hepatitis C virus (HCV) and human immunodeficiency virus (HIV).[1]

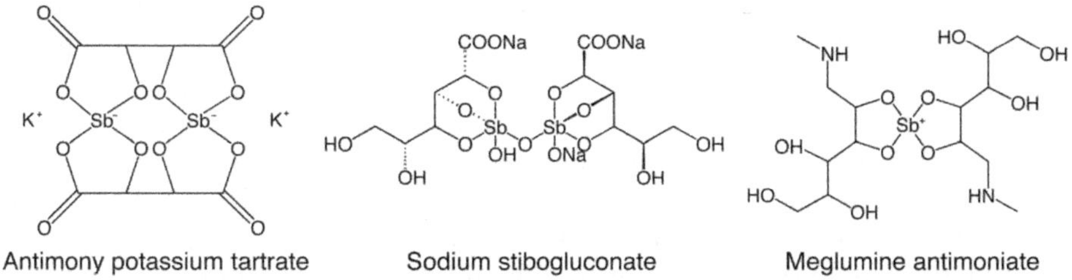

Antimony potassium tartrate   Sodium stibogluconate   Meglumine antimoniate

Figure 25.1 Formulae of three antimonials: antimony potassium tartrate, sodium stibogluconate (Pentostam®) and meglumine antimonate (Glucantim®).

25.2.2   Bismuth: Anti-*Helicobacter pylori* and Anti-cancer Activities

Bismuth compounds have long been used to treat various infections caused by microorganisms,[15] especially gastric diseases (peptic ulcers by bismuth subnitrate, subcarbonate and subcitrate).[16] Nowadays, since the discovery of the relationship between *H. pylori* infection and gastric ulcers, bismuth subsalicylate (BSS, Pepto-Bismol®), colloidal bismuth subcitrate (CBS, De-Nol® and ranitidine bismuth citrate (RBC, Pylorid®) are widely used to treat *H. pylori*-infection-related peptic diseases.[15,17] The bacterium apparently does not develop resistance to bismuth compounds.[1] Furthermore, for over a century, BSS has been used to treat minor digestive diseases, especially traveller's diarrhoea, due to its antibacterial action.[18,19,20]

The detailed molecular mechanism of how bismuth affects *H. pylori* is not well understood. It is generally believed that bismuth drugs, once administered, form a protective "coating" over stomach ulcers by forming polymers.[21,22] The soluble fraction enters the pathogen and then binds to zinc, nickel and iron sites of some enzymes and proteins in the bacterium, disrupting their functions.[23] Using a comparative and metalloproteomics approach, several potential protein targets of Bi(III) were proposed based on how CBS affects expression levels.[24,25] These included heat shock proteins (HspA and HspB) and the oxidative stress-related proteins NapA and TsaA. Enzymes identified as interacting with bismuth drugs include fumarase, the urease subunit UreB, a translational factor (Ef-Tu), phospholipases A and C, pepsin and alcohol dehydrogenase.[24,26–28]

25.3   Biomethylation of Antimony and Bismuth

While biomethylation processes involving arsenic compounds have been investigated intensively, the studies on antimony and bismuth compounds are still limited. The present knowledge of biomethylated derivatives of antimony and bismuth has been reviewed.[1,29] Numerous species of microorganisms were found to be capable of methylating different antimony substrates. The fungi *Scopulariopsis brevicaulis*, *Phaeolus schweinitzii* and *Asterotremella humicola* (synonym: *Cryptococcus humicola*) as well as the bacterium *Flavobacterium sp.* were identified as capable of aerobic and/or biphasic methylation of antimony. In addition, several bacterial species of *Clostridium* and a number of methanoarchaea (*e.g. Methanobacterium formicicum*) promote anaerobic methylation. Methanoarchaea are also major players in methylation of bismuth. Some bacteria (*Desulfovibrio piger*, *Eubacterium eligens*, *Lactobacillus acidophilus* and *Clostridium collagenovorans*) were also found to be able to methylate bismuth. Some species can methylate a range of arsenic, antimony and bismuth compounds.

Enzymatic methylation of antimony was reported by the As(III) *S*-adenosylmethionine methyltransferase from the thermoacidophilic eukaryotic red algae *Cyanidioschyzon merolae* (CmArsM). This enzyme can bind both inorganic Sb(III) and Sb(GS)$_3$ (GS = anion of glutathione GSH) and methylate Sb(III) to form trimethylantimony bromide Me$_3$Sb^{V}Br$_2$.[30] The study did not define a mechanism or pathway for these reactions. It did propose a mechanism of As(III) methylation to Me$_3$As^{III} by CmArsM using *S*-adenosylmethionine (SAM) as methyl donor and glutathionylated As(III) as substrates. It involved three steps of metalloid methylation, GSH oxidation and SAM reduction, during which arsenic maintains its trivalent oxidation state. This proposal is different from the classic Challenger mechanism,[31] which requires oxidation and reduction of metalloid in each step. The methylation of antimony by CmArsM may involve a different pathway since only the Me$_3$Sb^V centre was identified. However, *S. brevicaulis* was demonstrated to methylate Sb(III) to trimethylstibine Me$_3$Sb^{III},[32] and it was suggested that SAM acts as methyl donor in a process similar to the Challenger mechanism.[33] Methylation of antimony was enhanced in the presence of inorganic arsenic (III) and (V) species during culture of the aerobic yeast *C. humicolus*,[34] indicating the existence of inter-relationships between biomethylation of arsenic and antimony species.

In addition to SAM, methylcobalamin (CH$_3$Cob(III)) also plays a role in biomethylating antimony, in cooperation with GSH[35] or cob(I)alamin (Cob(I)).[36] Further investigation found that the methanol-utilizing methanogenic pathway of the methanoarchaeon *Methanosarcina mazei* promotes the methylation of the metal(loid)s (As, Se, Sb, Te and Bi).[37] The system involved the methanol-co-enzyme M-methyltransferase complex, consisting of three subunits: MtaA, MtaB and MtaC.[38] MtaA is a zinc-containing corrinoid : co-enzyme M methyltransferase. In the methanol-utilization, the MtaB-MtaC complex cleaves methanol (bound to MtaB) and transfers the methyl group to the cobalt of Cob(I) (bound to MtaC), forming CH$_3$Cob(III), while the MtaA-MtaC complex transfers the methyl group from CH$_3$Cob(III) (bound to MtaC) to 2-mercaptoethanesulfonate (co-enzyme M (CoM), bound to MtaA) in methanogenesis.[37] For the metal(loid) methylation, MtaA catalyses methyl transfer from CH$_3$Cob(III) to the metal(loid)s. In addition, in the absence of MtaA, the methyl group can also be transferred from CH$_3$Cob(III) to the metal(loid)s in the presence of Cob(I),[37] whose standard reduction potential $E_0 = -610$ mV.[36] Methylcobalamin-based methylation may be important for corrinoid-containing microorganisms.

Hitherto, the reports on biomethylation of antimony and bismuth are still focused on microorganisms rather than mammalian systems. Recently, CBS and bismuth cysteine but not bismuth glutathione were found to be methylated by the human hepatocellular carcinoma cell line HepG2.[39]

25.4 Cellular Transport of Antimony and Bismuth

25.4.1 Uptake of Antimony

Although sodium stibogluconate and meglumine antimonate (Figure 25.1) have been widely used in the clinic as anti-leishmania drugs, the mechanism and pathway for the influx of Sb(V) into cells remains unclear. The most stable form of Sb(V) (antimonate) in aqueous solutions at neutral pH is a six-coordinate species $[Sb(OH)_6]^-$. This anion is not isoelectronic with aqueous arsenate $[H_2AsO_4]^-$ and phosphate $[H_2PO_4]^-$,[40] and most probably is not taken up by phosphate transporters that also transport arsenate.[40,41] The intracellular uptake of Sb(V) (as sodium stibogluconate) in leishmania is partially inhibited by gluconate but not by arsenate or phosphate.[41]

Aqueous Sb(V) is reduced to Sb(III) in cells, a reaction that occurs in both *Leishmania* amastigotes[42,43] and host macrophages.[44] Sb(III) is more toxic towards parasites than Sb(V).[45] Trypanothione (N^1, N^8-bis(glutathionyl)-spermidine, $T(SH)_2$, a conjugate of glutathione (GSH) with spermidine and a major low-molecular-mass thiol present in the *Leishmania* parasite in addition to glutathione and cysteine, Figure 25.2) reduces Sb(V) compounds non-enzymatically. Sb(III) binds trypanothione to form a binary or ternary complex, which is considered to be important for the antileishmanial properties of the drugs.[46,47] Sb(V) can also be reduced by glutathione.[48] It has also been suggested that Sb(V) is reduced by Cys or Cys-Gly within the acidic compartments of mammalian cells, *e.g.* in macrophages.[49]

Enzymatic reduction of aqueous Sb(V) may be a major contributor. LmACR2, an arsenate reductase from *Leishmania major*, is a physiological antimonate reductase.[50] The expression of the *LmACR2* gene in *L. infantum* enhances the parasite's sensitivity to stibogluconate. It is also able to restore arsenate resistance in either a $\Delta arsC$ *E. coli* strain or a $\Delta ScACR2$ *Saccharomyces cerevisiae* strain.[50] The protein belongs to the <u>r</u>hodanese <u>h</u>omology <u>d</u>omain (RHOD) superfamily, which also includes Cdc25 phosphatases. LmACR2 represents a type of As/Sb reductase that also exhibits phosphatase activity. Several putative candidates were proposed by phylogenetic analysis, including those from Trypanosomatid protozoa and some yeasts, *e.g.* *Kluyveromyces lactis*, *S. cerevisiae* (ScACR2 or ScAcr2p).[51]

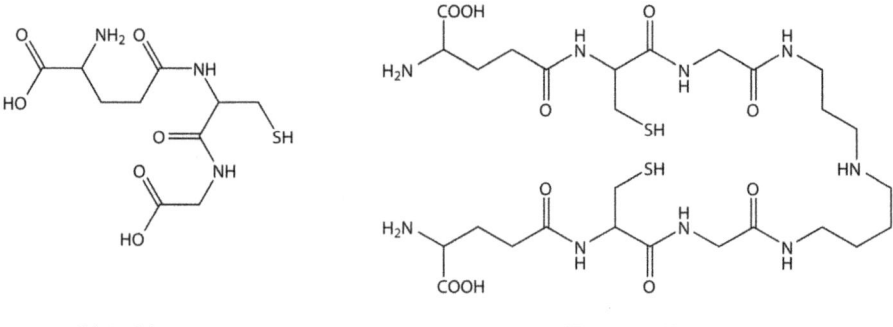

Glutathione   Trypanothione

Figure 25.2   Structures of two major low-molecular-mass thiol donors: glutathione and trypanothione.

The active site of LmACR2 includes the sequence 75-**CAQSLVR**-81. Cys75 is the catalytic residue (*vide infra*) and the other six form an active-site loop that includes Ser78 and Arg81, which are conserved throughout the eukaryotic RHOD superfamily.[51] The potential for ScACR2 to reduce Sb(V) was not ruled out.[52] Further work is required to investigate the reduction in other eukaryotes. Based on the phylogenetic analysis, As/Sb reductases from trypanosomatids and yeasts do not cluster in the same node of the phylogenetic tree and are distinct from other proteins in the superfamily.

The real physiological function of LmACR2 is proposed to be that of a tyrosine phosphatase, potentially important for host-parasite interactions.[53] Its X-ray crystal structure reveals an overall fold similar to rhodanese: a central five-stranded parallel sheet (βA, βE, βD, βB and βC) in counterclockwise twist with two groups of helices (αA, αB, αE; αC, αD) packed around the sheet (Figure 25.3a).[51] A significant rearrangement in secondary structure is apparent, in comparison to other Cdc25 phosphatases. The catalytic thiolate of Cys75 is located at the geometric centre of the active-site pocket, a shallow circular depression formed by residues 76–81 at the protein surface (Figure 25.3b). It is noteworthy that, in the crystal structure that contains bound sulfate, the side-chain of Arg81 is hydrogen-bonded with the carboxyl group of Asp34, leading to an extended conformation that is roughly planar with the active-site loop (Figure 25.3c). The detailed molecular mechanism of LmACR2 for the reduction of antimonate to antimonite is still unclear, though the reduction of arsenate by prokaryotic arsenate reductase (ArsC) from *Staphylococcus aureus* plasmid pI258 has

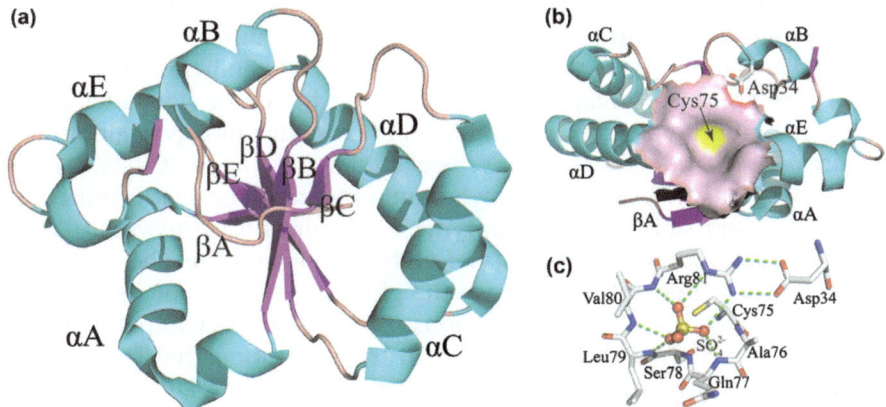

Figure 25.3 X-ray crystal structure of Sb/As reductase LmACR2 from *Leishmania major* (PDBID: 2J6P). (a) The overall structure of LmACR2 with α helices in cyan and β-sheets in magenta. (b) Surface conformation of the active-site pocket highlighted in light pink. The surface of Cys75 is in yellow and Asp34 in sticks. (c) The active-site loop and Asp34 are shown by sticks. Hydrogen bonds are shown as dashed lines. The bound sulfate in the pocket is shown by balls and sticks. The figures were generated by PyMol.

been proposed to include an intramolecular dynamic disulfide cascade, which involves sequential forming of disulfide bonds between Cys10 and Cys82 and between Cys82 and Cys89 of ArsC.[54,55]

TDR1 is another enzyme in *L. major* that is capable of reducing Sb(V) species. It possesses both thiol transferase and dehydroascorbate reductase activities and belongs to the glutathione *S*-transferase (GST) superfamily.[56] Its homologue in *Trypanosoma cruzi* is Tc52, which is crucial for the survival and virulence of that parasite.[57] However, *Leishmania* and other trypanosomatids possess the trypanothione-based redox system (not the glutathione-based one) and this is critical for safeguarding the organisms against oxidative stress and toxic heavy metals.[58] TDR1 uses glutathione as a reductant to convert Sb(V) species into trivalent ones, although it has stronger arsenate reductase activity. TDR1 is present in both promastigote and axenic amastigote forms of *L. infantum*, with the expression level being higher in amastigotes.[57] It is localized in the cytosol and exists as a trimer in its native state. Unlike Tc52, TDR1 cannot modulate effectively the host immune response.

TDR1 is in a homotrimer of 150 kDa and forms a triangular prism (Figure 25.4a).[59] Each subunit consists of two GST-like domains (Figure 25.4b), which are connected by a short linker and are each composed of N-terminal glutaredoxin-like and C-terminal α-helical subdomains (Figures 25.4c and 4d). The two domains are structurally similar though sequence identity is only 30%. The α-helical subdomains diverge more than the glutaredoxin-like subdomains. The structures of TDR1 domains are more closely related to the omega- and tau-family of GSTs and the glutaredoxin-like subdomains are more similar to GSTs than to glutaredoxin. The active site is formed by two domains *via* intersubunit interactions contributed by different subunits (Figure 25.4a). Each subunit contains two glutathione-binding sites (G-sites), resulting in six GST catalytic centres on the TDR1 assembly. The backbone of the valine of the G-site (Val55/Val280) in each domain forms a hydrogen bond with the cysteinyl moiety of glutathione (Figures 25.4e and 4f). The conserved glutamate (Glu709/Glu293) forms hydrogen bonds with the γ-glutamyl group of glutathione that is critical for the activity and stability of the enzyme.[60] Cysteines (Cys14/Cys240) at the active site interact with the thiols of glutathiones and the phenylalanines (Phe16/Phe242) interact weakly with the glutamyl moieties of the two GSH molecules *via* van der Waals interactions. The presence of comparable levels of GSH and T(SH)$_2$ in *Leishmania* led to the suggestion that GSH plays a regulatory role in trypanosomatids that is dependent on TDR1.[61]

In contrast to the unknown Sb(V) transporter, Sb(III) transporters have been widely identified in bacteria, yeasts, mammals and even in plants.[62] They belong to the aquaglyceroporin subfamily of major intrinsic proteins (MIPs). The first of this family was identified in *E. coli* as the glycerol facilitator GlpF by screening of Sb(III)-tolerant mutants.[63] A function in the uptake of both Sb(III) and As(III) was demonstrated.[64] At neutral pH, the major form derived from antimonous acid is its neutral monomeric species Sb(OH)$_3$, due to its pK_a of 11.8. This suggests that this inorganic equivalent

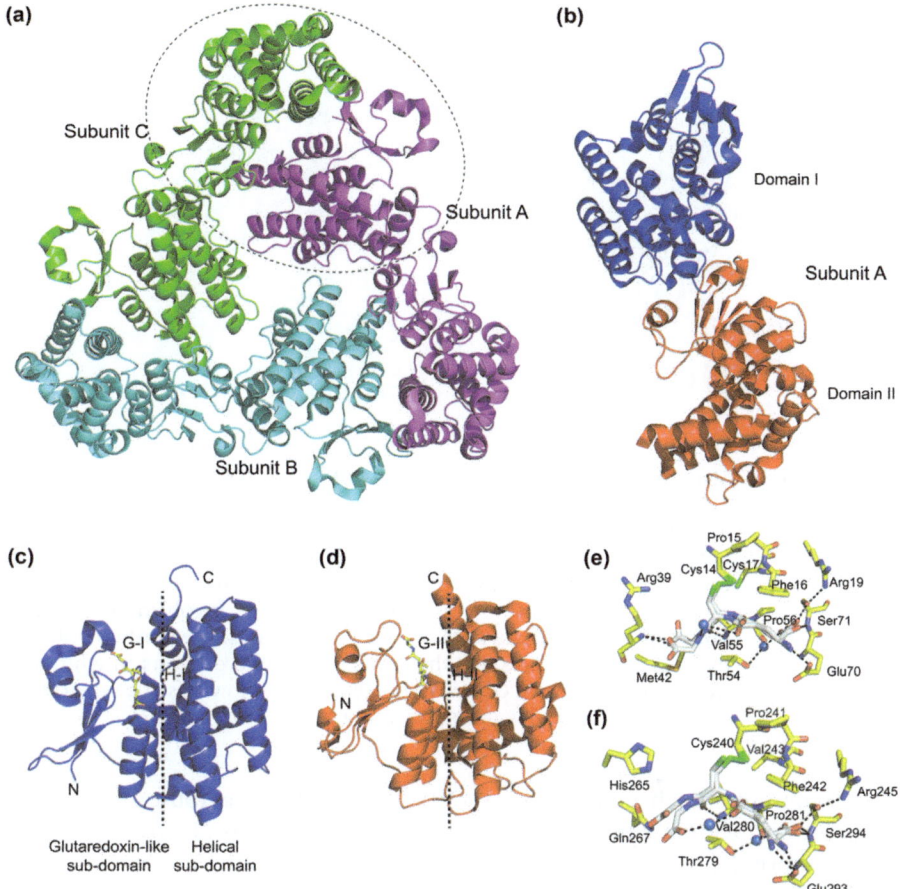

Figure 25.4 Structure of thiol-dependent reductase I (TDR1) from *Leishmania infantum* (PDBID: 4AGS). (a) Overall structure of the TDR1 trimer. One of the GST-like dimers formed by two different domains from two different subunits is highlighted by the circle. (b) Domain organization of each subunit, exemplified by Subunit A. (c, d) Domain I (blue) and Domain II (red) presented separately in the same orientation. Glutathione in each domain is shown as sticks (carbon in yellow). The G- and H-sites are labelled. (e, f) Glutathione binding sites G-I and G-II within Subunit A shown by sticks. Colour code: carbons in proteins and GSH in yellow and metallic white, respectively; sulfurs in green. Water molecules are shown as marine blue spheres. Hydrogen bonds are shown by dashed lines. Note that GSH is in a mixture of both reduced and oxidized forms.

of a polyol (*e.g.* glycerol) is the substrate of GlpF.[63,64] The crystal structure revealed the presence of a glycerol-conducting channel[65] (Figure 25.5a). Each monomer is formed by six transmembrane and two half-membrane-spanning α helices (M1 to M8), resulting in a right-handed helical bundle around each amphipathic channel. The whole GlpF protein is a tetramer, leading to

(a)

(b)

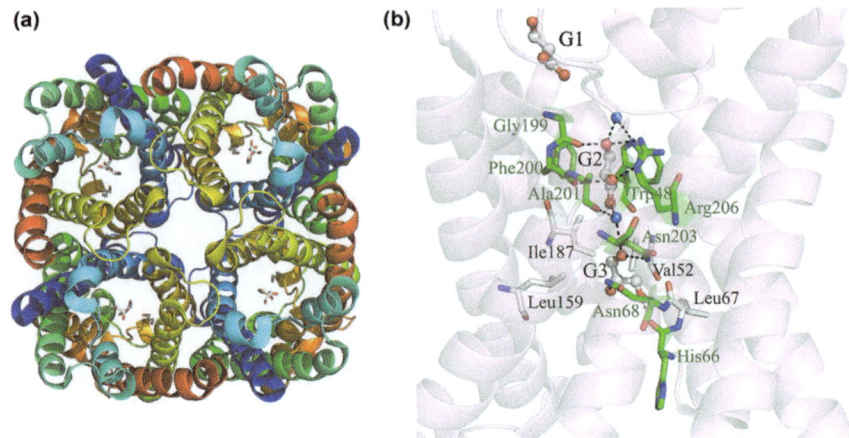

Figure 25.5 Structure of glycerol facilitator (GlpF) from *Escherichia coli* (PDBID: 1FX8). (a) Ribbon representation of the overall structure of the GlpF tetramer viewed from the periplasmic side. Glycerol molecules in the channel of each monomer are shown as sticks. (b) The hydrogen bonding network (dashed lines). The first glycerol molecule (G1) does not participate in the interactions. Residues involved in hydrogen bonding are shown as sticks with carbons. Residues only involved in hydrophobic contacts are shown as sticks with carbons in white (and numbered in black). Glycerol molecules are shown as sticks and balls. Two water molecules are shown as marine blue balls.

four channels. Each channel is ~ 15 Å wide with a vestibule on the periplasmic surface. This features a constriction of ~ 3.8 Å by 3.4 Å at a distance of 8 Å above the quasi two-fold axis. The glycerol molecules engage in a series of hydrogen-bonding networks and hydrophobic contacts in what constitutes the "selectivity filter" of the channel (Figure 25.5b).

Similar glycerol channels were also found in, for example, Fps1p from *S. cerevisiae*,[66] mammalian AQP9 (aquaporin, AQP),[67] AQP1 from *Leishmania*,[68] AQP from *Schistosoma mansoni*[69] and NIP$_{1;1}$ (nodulin 26-like intrinsic protein, NIP) from *Arabidopsis thaliana*.[70] Although structure-based studies on the selectivity of water and glycerol were reported,[65,71–74] the molecular mechanism of selective transport of Sb(III) instead of As(III) by certain aquaglyceroporin homologues remains unclear.

25.4.2 Efflux of Antimony

The overall resistance response of cells to antimony also relies on the extrusion of intracellular antimony, *i.e.* of Sb(III). The *E. coli* arsenite translocation system is coded by the *ars* operon of plasmid R773. The expressed pump ArsAB provides resistance not only to arsenite but also to antimonite.[75] It consists of a soluble ATPase ArsA and a membrane channel ArsB that combine to extrude both As(III) and Sb(III) from the cytosol.[76]

ArsA undergoes a number of conformational transitions during the catalytic cycle, from an opened *apo*-form to the closed nucleotide- and metalloid-bound form(s).[77] The crystal structure of Sb(III)-bound ArsA of *E. coli* shows that the protein consists of two structurally homologous L-shaped domains A1 and A2, with a β-sheet core (one antiparallel, seven parallel strands) with flanking helices (Figure 25.6).[78] The two domains overlap at each arm to form a diamond-shaped molecule with a central cavity (Figure 25.6a). As it is an ATPase, there are two nucleotide binding sites A1NBS and A2NBS located at the domain interfaces. The former is composed

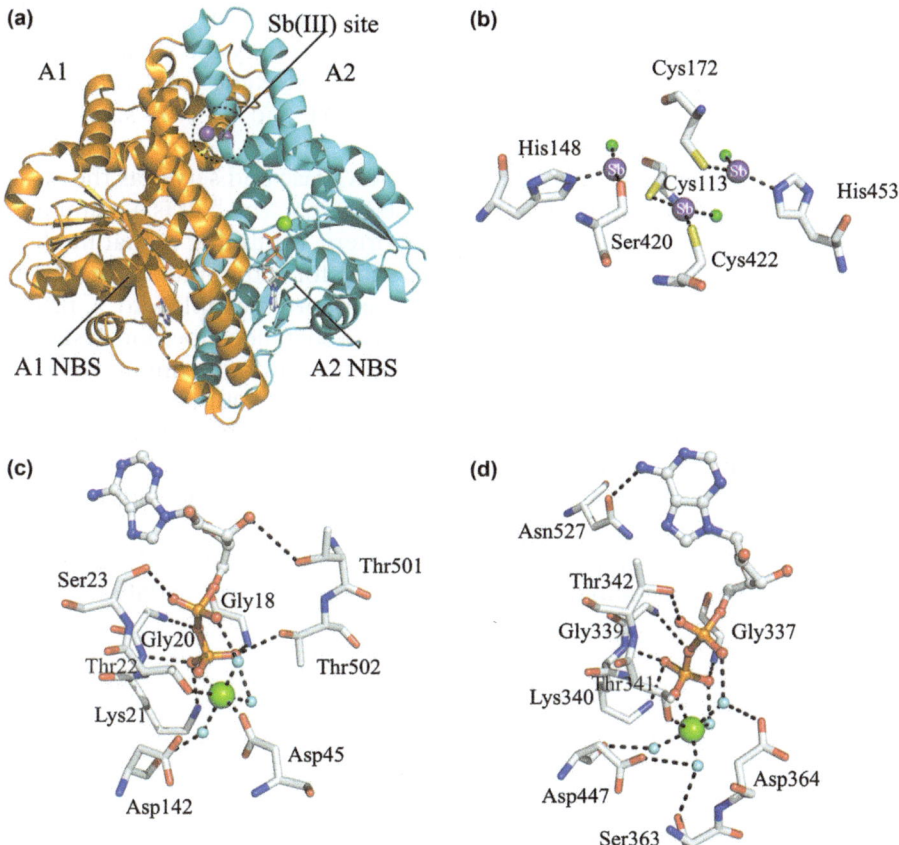

Figure 25.6 Structure of ArsA ATPase from *Escherichia coli* (PDBID: 1F48). (a) Overall structure of ArsA dimer (A1 in orange and A2 in cyan, respectively). Sb(III) ions are shown as violet purple balls and Mg(II) ions as chartreuse balls. ADP molecules are shown as sticks. Two nucleotide-binding sites (NBS) are labelled. (b) The allosteric site with a bound trinuclear Sb(III) cluster. Chlorides (Cl$^-$) are shown as green balls. (c, d) Polar contacts (dashed lines) of Mg-ADP at the A1 and A2 NBS. Mg^{2+} octahedrally coordinates the protein. ADP molecules are shown as sticks and balls. Water molecules are shown as cyan balls.

of residues mostly from A1 whereas the latter is composed of residues mostly from A2. The terminal phosphate group of ADP binds to the conserved GKGGVGKT sequence of the P-loop in both domains with Mg^{2+} coordinated octahedrally at both sites. Each domain binds three Sb(III) ions at the allosteric metal-binding site, forming a trinuclear cluster (Figure 25.5b). Each Sb(III) centre is bound to two residues from the protein and a Cl^- ion. The first coordinates His148 (A1) and Ser420 (A2), the second Cys113 (A1) and Cys422 (A2) and the third Cys172 (A1) and His453 (A2). Only the second is a high-affinity metalloid binding site.[79] Both the low coordination number and the involvement of an exogenous ligand Cl^- may facilitate metal trafficking to downstream receptors. The most intriguing characteristic of the structure is the inference of coordinating the functions of the catalytic and allosteric sites to provide a transduction path to convey the signal of metal binding to the ATPase active site. Two identical stretches of seven-residue sequences $D_{142/447}TAPTGH_{148/453}$ connect the A1 and A2 NBS sites to the metal-binding site (Figures 25.6c and 6d).

ArsB is the membrane anchor for the ArsA ATPase and serves as the anion channel[80] that allows efflux of As(III) and Sb(III) in *E. coli*.[64] It is a member of the major facilitator superfamily of transporters, exhibiting a dual mode of energy coupling. It both interacts with ArsA to form the functioning ArsAB ATPase and acts as an $As(III)/Sb(III)/H^+$ antiporter coupled to a proton electrochemical gradient.[64,80,81] There is also a metallochaperone ArsD that transfers As(III) or Sb(III) to ArsA[82,83] and, in addition, exhibits weak repressor activity.[83,84] A crystal structure (residues 1–109) shows that each monomer in a dimer has a core consisting of four β strands flanked by four α-helices adopting, overall, a thioredoxin fold (Figure 25.7).[85] Three conserved cysteine residues Cys12, Cys13 and Cys18 form a metalloid binding site.[85] The protein interacts with ArsA through its helix α1.[86] There are two additional vicinal cysteine pairs, Cys112-Cys113 and Cys119-Cys120, forming two independent metalloid binding sites.[87] Although no metalloid-bound

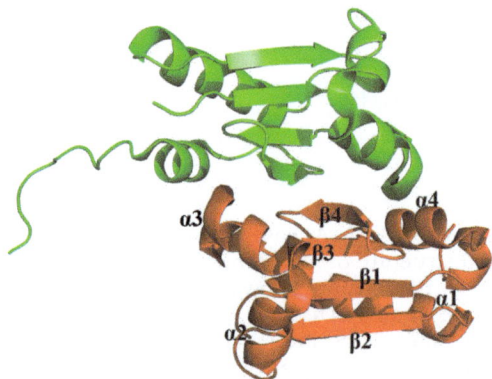

Figure 25.7 X-ray crystal structure of ArsD (PDBID: 3MWH), showing an extended interface between monomers.

structure of ArsD is available, the transfer of As(III) from ArsD to ArsA has been examined.[83] In the cytosol, As(OH)$_3$ reacts with 3 GSHs to form As(GS)$_3$ that then transfers As(III) to the binding site (Cys12, Cys13, Cys18) of ArsD. The loaded chaperone binds to an open form of ArsA and, subsequently, As(III) is transferred to the thiols of the binding site (Cys113, Cys172, Cys422) of ArsA and induces dissociation of the ArsD-ArsA complex. Loaded ArsA is now in a closed conformation competent for activation of ATP hydrolysis. Based on the structural similarity of Sb(OH)$_3$ and As(OH)$_3$, a similar mechanism can be considered for Sb(III) transport.

ArsA homologues occur extensively in many organisms, including eukaryotes.[88,89] The first eukaryotic version was found in humans.[90] A study of the homologue ASNA1 in *Caenorhabditis elegans* demonstrated that the *asna-1*-mutated nematode exhibited an increase in sensitivity towards both As(III) and Sb(III). In addition, the ATPase activity of the enzyme ASNA1 was stimulated by As(III) and Sb(III)[91] and appears to enhance insulin secretion in the nematode as well as in mammalian cells.[92] Similarly, ASNA1 is involved in resistance to the metalloids and cross-resistance to cisplatin in mammalian cells.[93–96] It displays the so-called RASP phenotype (cross-resistance between arsenite, antimony and cisplatin). In contrast, the yeast homologue Arr4p is not resistant to arsenic but it may be involved in tolerance to metal- and heat-induced stress.[97]

No eukaryotic ArsB homologues have been found to date. In *S. cerevisiae*, Acr3p was demonstrated to be a metalloid/H$^+$ antiporter that mediates the efflux of both As(III) and Sb(III) and confers tolerance.[98,99] The process depends on the pH gradient generated by the H$^+$-ATPase in the plasma membrane. The transport of As(III) is three times faster than that of Sb(III) despite similar affinities. Acr3p belongs to the bile/arsenite/riboflavin transporter (BART) superfamily. The majority of Acr3 members are specific for As(III) transport[99] although Acr3p may have a minor role in Sb(III) tolerance in yeast. A vacuolar ABC transporter Ycf1p is more important in detoxification of Sb(III) than As(III).[100] It catalyses the ATP-driven uptake of GSH conjugated Sb(III) (Sb(GS)$_3$) into the vacuole. *S. cerevisiae ycf1Δ* cells exhibit a higher sensitivity to Sb(III) than to As(III) while the *acr3Δ ycf1Δ* double mutant is sensitive to both metalloid ions.[100,101] Vmr1 is a close homologue of Ycf1p and was also reported to participate in the transport of As(III) but not of Sb(III).[102]

Some ABC transporters are involved in efflux of arsenite or antimonite. For example, while multidrug resistance-associated protein1 (MRP1/ABCC1) of the MRP subfamily C of the ABC transporter superfamily[103] is frequently overexpressed in cancer cells, its expression also leads to cross-resistance to antimony potassium tartrate.[104] MRP1-mediated effluxes of arsenic and antimony do not require the formation of the relevant M(GS)$_3$ species, which cannot be formed at short term, suggesting that As- and Sb-containing species can be co-transported with GSH, considering no active efflux in GSH-depleted cells.[105] Human MRP1 overexpressed in *A. thaliana* protoplasts can also extrude antimony.[106] The conserved Trp1246 residue is critical for the selectivity of drug resistance: its mutation leads to partial loss of resistance to sodium arsenite but not to antimony tartrate, as well as loss of resistance to natural product chemotherapeutic agents. This observation suggests that structural determinants in the protein are necessary for recognition and transport of heavy metal oxyanions and that these are different from those required for recognition of the chemotherapeutic agents.[107] Antimony in pentavalent antimonial drugs (*e.g.* meglumine antimonate) does not show significant antileishmanial activity. Instead, it is responsible for cytotoxic activity against mammalian cells and the MRP1-mediated resistance to these drugs.[108] MRP1/Mrp1 can decrease the cytotoxicity of As(III), As(V) and Sb(III) (and Sb(V) to a lesser extent) in cultured human and murine cells due, possibly, to GSH-dependent extrusion.[109,110] Mature MRP1 is a 190 kDa glycoprotein consisting of a 171 kDa polypeptide arranged in the domain order of TMD0-TMD1-NBD1-TMD2-NBD2, with *N*-linked glycosylation at Asn19 and Asn23 of an extracytosolic N-terminus and at Asn1006 of a C-proximal membrane-spanning domain (MSD3).[111,112]

An ABC transporter in *Leishmania*, labelled PGPA or MRPA, localizes in membrane vesicles of the parasites, close to the flagellar pocket where endo- and exo-cytosis occur.[113–115] Amplification of the gene is only observable in metalloid-resistant cells. When grown in the absence of the metalloid compounds, cells lost the amplicon of the *pgpA* gene and some of their partial resistance.[116] PGPA confers resistance to metals and metalloids, including Sb(III), Sb(V) and As(III), in the transfectants due to the amplification of its gene[116–118] that is accompanied by trypanothione in the forms of metal/metalloid-trypanothione conjugates.[119] However, it has been shown that PGPA is not responsible for the efflux of these conjugates across the plasma membrane. Instead, its overexpression promotes a decrease of influx of antimony (at least),[120] a property distinct from those of its mammalian counterpart MRP1. Consequently, it was suggested that PGPA could be an intracellular transporter for antimony rather than an efflux transporter.[121] Its negative effect on the influx of antimony is likely due to interaction with other membrane proteins.[122] The PGPA-mediated antimony resistance is reversed by sitamaquine,[123] an orally active 8-aminoquinoline analogue used in the treatment of visceral leishmaniasis.[124] A second ABC transporter PRP1 (pentamidine resistance protein 1) was shown to confer cross-resistance to trivalent but not to pentavalent antimony.[125] This protein exhibits less sequence identity to other members of the ABC families and no homologues were found in trypanosomatid species other than *Leishmania*.[126,127] It is localized intracellularly next to the kinetoplast and at the tubulovesicular element responsible for the exocytic and endocytic pathways.[126,128] Recently, overexpression of a novel protein of the leucine-rich repeats (LRRs) super-family has been found to confer resistance to Sb(III) in axenic amastigotes and to Sb(V) in intracellular parasites.[129] Figure 25.8 summarizes our present knowledge of the uptake and efflux of antimony in three representative systems.

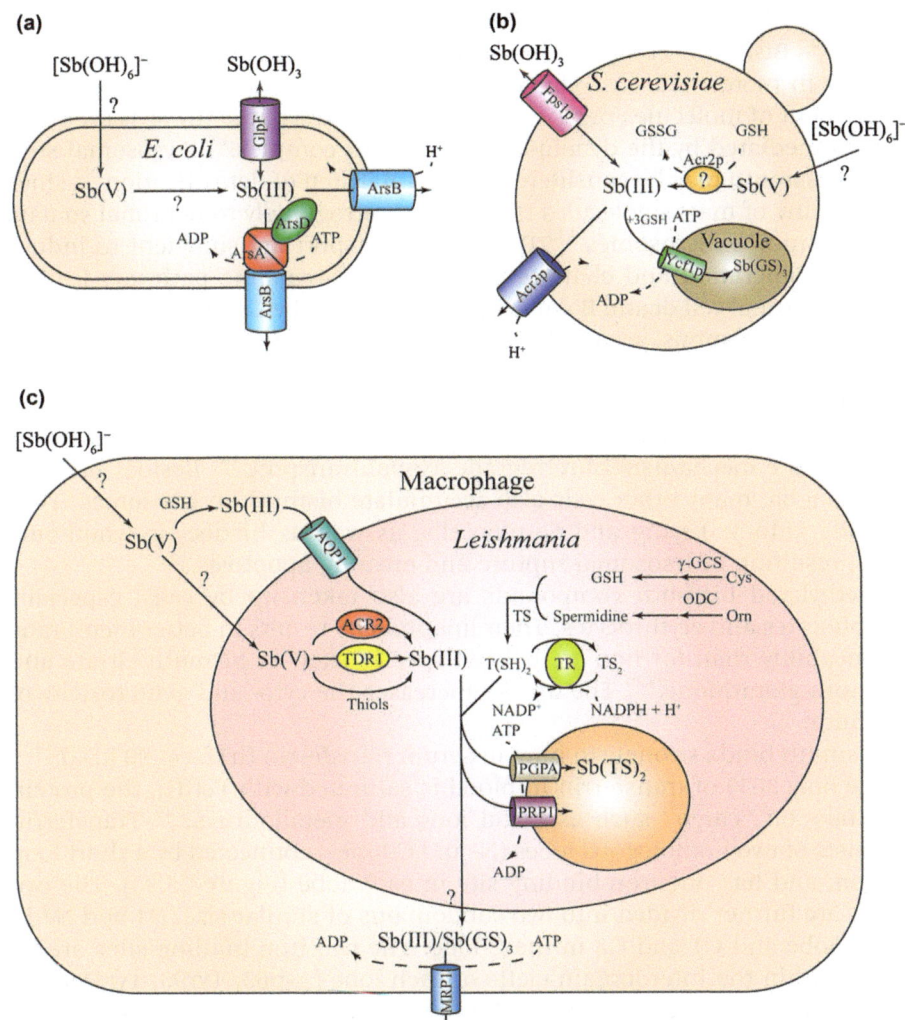

Figure 25.8 The uptake and efflux of antimony in three representative systems: (a) *E. coli*; (b) *S. cerevisiae*; (c) macrophage host cell and intracellular parasite *Leishmania* (Sb-resistant). Abbreviations: γ-GCS: γ-glutamyl-cysteine synthetase; ODC: ornithine decarboxylase; Orn: L-ornithine; TR: trypanothione reductase; TS: trypanothione synthetase; T(SH)$_2$: reduced trypanothione; TS$_2$: oxidized trypanothione; Sb(TS)$_2$: conjugate of Sb(III) with trypanothione.

25.4.3 Transport of Bismuth

Bismuth can be found in macrophages at the base and margins of gastric ulcers.[130] Limited studies only have been performed on the intracellular transport of bismuth compounds. Although they show little toxicity, bismuth can penetrate the blood-brain barrier and lead to encephalopathy[131] and it can be taken up and accumulated by spinal cord motor neurons of mice.[132] Bismuth compounds accumulate in lysosome-like organelles, mainly in motor neurons. This occurs *via* retrograde axonal transport,[133] a movement of molecules/organelles away from the synapse towards the soma that is mediated by the dynein-dynactin motor complex.[134] Lysosomal storage of bismuth may be considered as the final step of detoxification[133] since the destiny of macromolecules transported retrogradely to neuronal somata is deposition in lysosomes.[135] Defects in transport are sufficient to induce neurodegeneration and changes in retrograde signalling pathways lead to rapid neuronal cell death. Bismuth is also taken up by CA1 pyramidal cells in the rat hippocampus, accumulating and localizing subcellularly in lysosome-like organelles.[136] High concentrations of bismuth citrate destroy the cyto-architecture of the hippocampus completely. Therefore, neurons are more prone to bismuth accumulation in lysosomes than glia and endothelial cells through the mechanism of retrograde axonal transport.[137] Besides neurons and ganglia, many other cells also accumulate bismuth in lysosomes. They include kidney, Leydig and Sertoli cells, as well as histiocytic lymphoma cells, resulting in lysosomal rupture and ensuing apoptosis.[138]

Methylated bismuth compounds are also taken up by cells, especially lymphocytes and erythrocytes. Their lipophilicity results in better membrane permeability than for non-methylated species such as bismuth citrate and bismuth glutathione.[139] The uptake increases the cyto- and geno-toxicity of bismuth.

Bismuth binds strongly to human serum transferrin (hTf, *ca.* 80 kDa).[140],† Since only 30% of transferrin in blood is saturated with Fe(III), the protein can take on "cargo" such as metal ions and metallodrugs.[141] Transferrin consists of two homologous lobes (N- and C-lobes) connected by a short loop region, and has one iron-binding site in each lobe (Figure 25.9a). The two lobes are further divided into two subdomains of similar size, N1 and N2 in the N-lobe and C1 and C2 in the C-lobe. The two iron-binding sites are located within the interdomain clefts of each lobe (Asp63, Tyr95, Tyr188 and His249 in the N-lobe and Asp392, Tyr426, Tyr517 and His585 in the C-lobe). Fe(III) coordinates to the two sets of four residues together with a bidentate carbonate (synergistic anion) in a distorted octahedral geometry (Figures 25.9a, b and c). Crystal structures of transferrins and related proteins from various species have been characterized extensively.[142] *Apo*-hTF is normally in a "fully opened" conformation whereas the *holo* form adopts a "fully closed" conformation in both lobes. Bi(III) binds strongly to the sites in both the N- and C-lobes with binding constants (log K^*) of 19.4 and 18.6, respectively.[140] Selective labelling of transferrin (ε-[^{13}C]Met-hTF) allowed detection of binding of the Bi(III) anti-ulcer drug (ranitidine bismuth citrate) to transferrin in preference to albumin, even in blood plasma. This suggests transferrin as a potential target for Bi(III) drugs.[143] Bismuth binds firstly to

---

†Aspects of the action of transferrin in Fe(III) transport are discussed in Chapter 10.

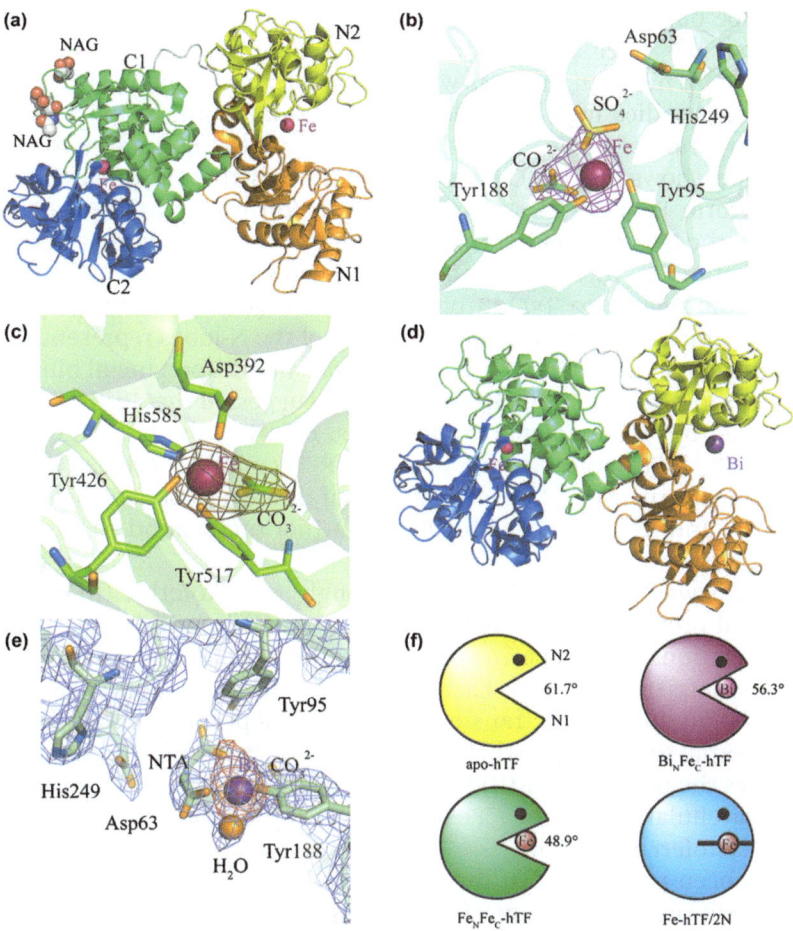

Figure 25.9 Structure of human serum transferrin (hTf). (a) Ribbon representation of the overall structure of $Fe_N Fe_C$-hTf (PDBID: 3QYT) with subdomains N1 in orange, N2 in yellow, C1 in lime green, C2 in marine blue and the peptide linker in grey. The two *N*-acetylglucosamine moieties (NAG) are represented as sphere models. Fe(III) ions are shown as pink balls. (b) Coordination of Fe(III) in the *N*-lobe of $Fe_N Fe_C$-hTf. Putative binding residues are shown as sticks. The F_{obs}-F_{calc} map (computed before the Fe(III), CO_3^{2-} and SO_4^{2-} were modelled) is contoured at 3.0 σ. Note that the carboxyl group of Asp63 and the imidazole group of His249 are ca. 7 and 10 Å away from the Fe(III). (c) Coordination of the Fe(III) in the C-lobe of $Fe_N Fe_C$-hTf. Metal-binding residues and the bound carbonate are shown as sticks. The F_{obs}-F_{calc} map is contoured at 3.0 σ. (d) Ribbon representation of the overall structure of $Bi_N Fe_C$-hTf (PDBID: 4H0W). Bi(III) is shown as a purple ball. (e) Coordination of Bi(III) in the N-lobe of $Bi_N Fe_C$-hTF. The $2F_{obs}$-F_{calc} map (blue) is contoured at 1.0 σ and the anomalous electron map (red) is contoured at 10.0 σ. Nitrilotriacetate (NTA) is represented as sticks and the water molecule as an orange ball. (f) A schematic diagram shows the extent of opening of the N-lobe (as a Pac-Man) in the structures of apo-hTf (yellow), $Bi_N Fe_C$-hTF (purple), $Fe_N Fe_C$-hTF (green) relative to the Fe-hTF/2N (cyan, PDBID: 1N84).

the C-lobe binding site (in less than 100 ms) and more slowly to the N-lobe (less than 0.5 s for the second conformational change and 50 s for the third).[144,145] This form interacts with the transferrin receptor (TfR) in a single kinetic step, indicating a possible difference from Fe(III)-bound hTF.[146]

The structure of Bi_NFe_C-hTf, with Bi(III) bound to the N-lobe and Fe(III) to the C-lobe, shows that the C-lobe is in a "fully closed" conformation as in the diferric form Fe_NFe_C-hTf and in other transferrins reported previously. In contrast, the N-lobe in either protein (Figures 25.9a, d) is largely in the open conformation.[147] In addition to anions such as carbonate, Bi(III) in the N-lobe coordinates to Tyr188 only whereas Fe(III) binds both Tyr95 and Tyr188 (Figure 25.9b, e). The half-openness of the N-lobes represents a series of snapshots of the dynamic motion of subdomains upon metal binding and dissociation (Figure 25.9f). Therefore, it was suggested that Bi(III) is taken up into cells by TfR-mediated endocytosis of hTf, similar to Fe(III) uptake.[145,148–150] Structural studies of the hTfR-Tf complex enriched knowledge of the transferrin cycle.[145,149] It would be of interest to obtain the structures of Bi_2-hTf and its complex with hTfR to further understand hTFR-mediated Bi(III) transport.

Bismuth was also found to bind strongly to lactoferrin (hLf), a multi-functional protein of the transferrin family found in milk, tears, saliva and other secretions, at the specific iron sites. Either carbonate or oxalate can act as the synergistic anion.[151] Lactoferrin binds iron about 100 times more strongly than does serum transferrin. It acts by depriving microorganisms of iron essential for their growth, thus exerting a bacteriostatic function.[151] ATP facilitates the release of Bi(III) from Bi_2-hLf at lower pH and Bi_2-hLf blocked the intracellular uptake of ^{59}Fe-hLf, suggesting possible uptake of Bi_2-hLf *via* the lactoferrin receptor.

It has been proposed that the uptake of bismuth in prokaryotes strongly depends on iron transport mechanisms and that the antimicrobial action of bismuth is largely non-specific and competitive with iron transport.[152] Resistance to bismuth correlates with the production of siderophores, which is relevant to the virulence of many Gram-negative bacteria, *e.g. E. coli*, *Aeromonas hydrophila* or *Pseudomonas aeruginosa*. Using time-resolved ICP-MS, a competitive transport pathway involving iron and bismuth for individual *H. pylori* cells was observed, indicating a protective effect of ferric ions against colloidal bismuth subcitrate (CBS, De-Nol®) accumulation in *H. pylori*,[153] in agreement with a previous report.[154]

25.5 Potential Targets of Antimony and Bismuth

25.5.1 Binding of Antimony to Proteins and Peptides

Similar to As(III), Sb(III) forms stable complexes with a thiolate, such as cysteine and glutathione and trypanothione (Figure 25.2).[46,155] *In vitro*, Sb(III) (as $Sb(GS)_3$) can substitute Zn(II) in one of the two zinc finger motifs ($CX_2CX_4HX_4C$, CCHC zinc finger) of nucleocapsid protein p7 (NCp7) of human immunodeficiency virus type 1 (HIV-1).[156] Moreover, a bound Sb-GSH species was observed, suggesting a function for glutathione conjugates beyond simple chemical reduction. Interestingly, Sb(III) exhibits high affinity for a CCCH zinc finger motif (essentially $CX_7CX_5CX_3H$ or $CX_8CX_5CX_3H$), which is present mainly in RNA binding proteins with regulatory functions at all stages of mRNA metabolism. The affinity is higher than that for a CCHC motif present in one of the kinetoplastid proteins in *Leishmania*.[157]

Reports of the binding of antimony to proteins are limited, other than to those transporters mentioned above. The most important structure is that of trypanothione reductase (TR) from *L. infantum*.[158] This enzyme has a higher affinity for Sb(III) than for As(III), making it a putative target of antimonials as binding results in inhibition of activity.[159] Together with trypanothione synthetase (TS), TR is an important component in the unique thiol-based metabolism of all parasites of the Trypanosomatidae family including *Leishmania*, while TS synthesizes trypanothione. TR maintains the reduced state of trypanothione (Figure 25.8c). This trypanothione/TR system is unique and is absent in all other organisms, which possess, instead, the glutathione/glutathione reductase (GR) and thioredoxin/thioredoxin reductase systems. *L. infantum* TR exists as a dimer with each monomer consisting of three different domains (Figure 25.10a), namely, the FAD-binding domain, the NADPH-binding domain and the interface domain, all of which are conserved in other TRs.[160] The dimeric interface contains two symmetric catalytic clefts, where the oxidized trypanothione is reduced. The FAD-binding domain adopts a Rossmann fold as do other GR family member proteins. This is formed by a three-stranded antiparallel β-sheet, a five-stranded parallel β-sheet and four α-helices. The NADPH-binding domain undergoes significant conformational changes between the oxidized, Sb(III)-bound and NADPH forms. The most significant changes occur at Tyr198 and Arg222. The phenolic ring of Tyr198 is rotated by about 120° around the Cα-Cβ bond to accommodate the nicotinamide ring of NADPH (Figure 25.10b) while Arg222 is rotated by about 30° around the Cα-Cβ bond for the adenine ring. The two Sb(III) ions bind in tetrahedral sites at the two catalytic clefts. The ligands are Cys 52, Cys57 and Thr335 from one monomer and His461' of another (Figure 25.10c). In the native TR structure in an oxidized state, the two sulfur atoms of Cys52 and Cys57 are present with a disulfide bond. However, Sb(III) binds upon reduction. Sb(III) interrupts the direct communication between His461', Cys52, Cys57 and the sulfur atoms of trypanothione, leading to inhibition of the activity of TR.

A structure is also available for another antimony-binding protein, the *Gloeobacter violaceous* ligand-gated ion channel (GLIC),[161] a homopentameric transmembrane pump that opens and closes in response to the binding of a ligand. The protein is selective for cations and is inhibited by a diverse number of positively charged molecules such as the quaternary ammonium ions tetrabutylammonium (TBA) and tetraethylammonium (TEA). It consists of an extracellular ligand-binding domain and a transmembrane domain forming a narrow channel with four membrane-spanning helices

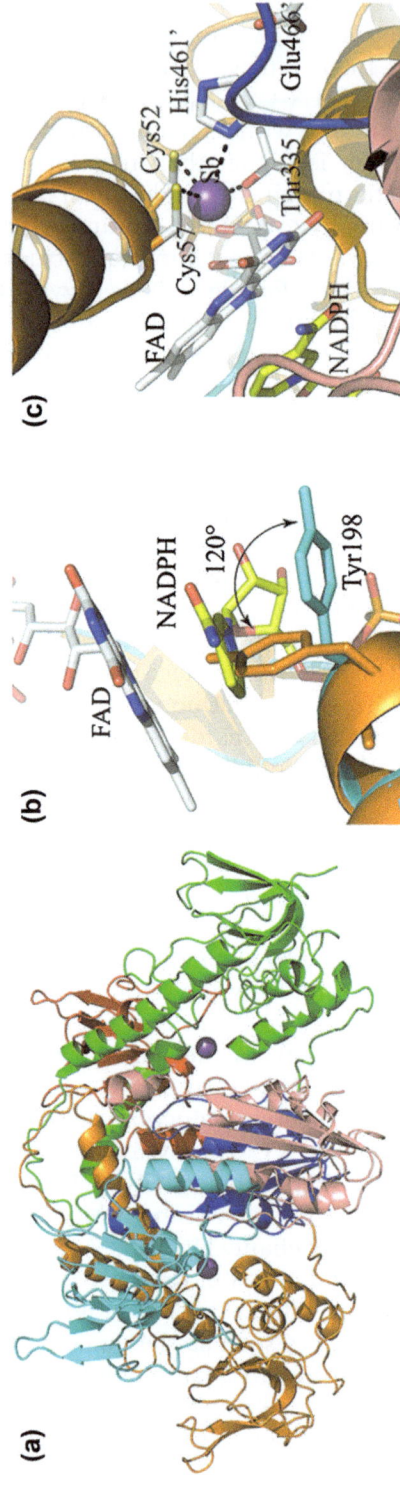

Figure 25.10 Structure of trypanothione reductase (TR) from *Leishmania infantum* (PDBID: 2W0H). (a) Overall structure of reduced TR with Sb(III) and NADPH bound. FAD binding domains (residues 1–160 and 289–360): orange for subunit A and green for B; NADPH binding domains (residues 161–289): cyan for subunit A and red for B; interface domain (residues 361–488): salmon for subunit A and blue for B; Sb(III) as purple balls. (b) Superposition between the NADPH binding domain of native oxidized TR (in orange, PDBID: 2JK6) and the one of Sb(III)- and NADPH-bound TR (in cyan). Note that the phenolic ring of Tyr198 undergoes a 120° rotation upon NADPH binding. (c) Sb(III) binding site. Ligand residues (Cys52, Cys57, Thr 335 and His461') are shown as sticks, as are Glu466', FAD and NADPH.

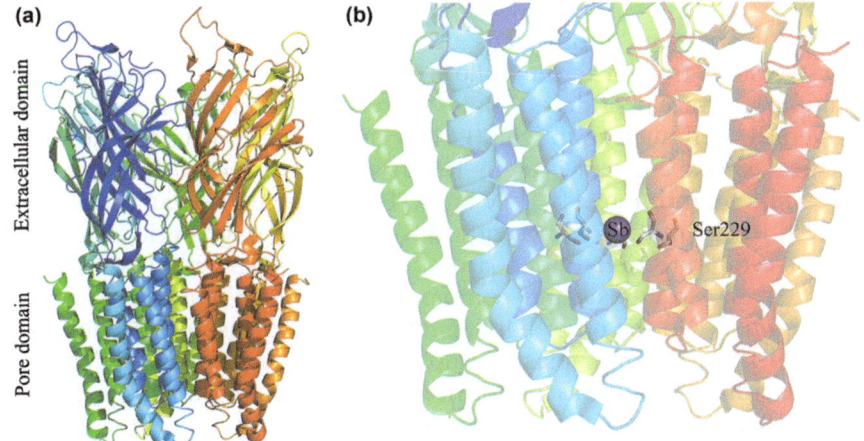

Figure 25.11 Structure of *Gloeobacter violaceous* pentameric ligand-gated ion channel (GLIC, PDBID: 2XQA) in complex with tetrabutylantimony. (a) Overall structure of the GLIC pentamer. The protein consists of an extracellular domain and a pore domain. (b) Zoomed-in view of the pore domain. Sb(III) of TBSb is highlighted as a purple ball that is located in a cavity close to five residues of Ser229 shown as sticks.

(Figure 25.11a). Tetrabutylantimony (TBSb) and tetraethylarsonium (TEAs) cations are analogues of TBA and TEA by replacement of the central nitrogen atom with antimony and arsenic, respectively. They show a voltage-dependent inhibition similar to those induced by the ammonium ions and block the channel in the same way. TBSb is located in a cavity close to Ser229 at the boundary between the hydrophobic and hydrophilic parts of the pore (Figure 25.11b).

Proteomic approaches were applied to investigate the direct or indirect influences of antimonials on eukaryotic cells, mostly *Leishmania*.[162–166] In a proteome mapping of *L. donovani*, several proteins in both membrane-enriched and cytosolic fractions were identified for their overexpression in clinical isolate promastigotes resistant to sodium antimony gluconate (SAG). ABC transporters, the cysteine-leucine rich protein, HSP-83, GPI protein transamidase and 60S ribosomal protein L23a were found to be the major proteins in the membrane fraction. Considering that ABC transporters are involved in drug resistance, including antimony resistance, the three-fold overexpression of these proteins is not unexpected[163] and they are probably the direct binding proteins/targets. Two-fold-enhancement was observed for the cysteine-leucine rich protein, considered to be involved in protein-protein interactions, transcription, RNA processing and drug resistance. Strikingly, HSP83 expression was up by a factor of five and it was suggested to be involved in drug resistance and related activation of programmed cell death after the disruption of the mitochondrial membrane.[167] A putative drug target protein, GPI protein transamidase,[168] also exhibited a three-fold increase. Unlike for ABC transporters, there are no reports on the direct interactions between antimonials and the other overexpressed proteins.

In cytosolic fractions, the following proteins were identified: HSP70, β-tubulin, carboxypeptidase, enolase, fructose-1,6-bisphosphate aldolase, proliferative cell nuclear antigen (PCNA) and proteasome complexes. HSP70 and heat shock cognate protein HSC70 are involved in the tolerance of *Leishmania* to antimony.[169] They are ATP-dependent molecular chaperones involved in assembly, folding and translocation of oligomeric proteins and provide cyto-protection under stress.[170] Both HSP83 and HSP70 are up-regulated in Sb(V)-resistant compared with Sb(V)-sensitive isolates of *L. donovani*.[167] On the other hand, enolase, 20S proteasome α5 subunit and SKCRP14.1 (small kinetoplastid calpain related protein) were down-regulated. A similar study was performed on *L. infantum* Sb(III)-sensitive and resistant axenic amastigotes. In contrast, β-tubulin was down- rather than up-regulated. The LACK receptor, proteasome activator protein pa26 subunit, pyruvate kinase and kinetoplastid membrane protein 11 (KMP-11) and argininosuccinate synthase (ARGG; linked to MRPA) were overexpressed.[162] The overexpression of ARGG may enhance the efflux of antimony by MRPA. Due to possible membrane association and localization at both the flagellar pocket and the flagella themselves, the reduction of KMP-11 expression was suggested to alter the interaction of either MRPA or AQP1, leading to increased export of Sb(III). Further differential proteome analysis revealed that *S*-adenosylmethionine synthetase (SAMS) and *S*-adenosylhomocysteine hydrolase (SAHH), the two sulfur amino acid-metabolizing enzymes, were overexpressed in *L. panamensis* Sb(III)-resistant lines and clinical isolates. These enzymes are critical in synthesizing precursors of trypanothione.[166] Moreover, down-regulation of mitogen-activated protein kinase 1 (MAPK1) in *L. donovani* is associated with antimony resistance.[171]

A quantitative proteomic approach was applied to demonstrate more detailed protein expression profiles of antimony-susceptible/-resistant *L. donovani* strains.[164] Glycolytic enzymes were up-regulated (*e.g.* glycosomal glyceraldehyde 3-phosphate dehydrogenase) enhancing glycolysis in resistant isolates and perturbing energy metabolism. In agreement with previous reports,[167,169] Hsp70 was overexpressed in resistance isolates, together with several other chaperones and stress proteins. So were β-tubulin, 60S ribosomal protein L23a, aldolase, ATP-dependent RNA helicase and pyruvate kinase. Several ribosomal proteins were found to be overexpressed in relation to antimony resistance. For example, elongation factor 1α had higher expression levels in resistance isolates. Elongation factor 1β, which was shown to associate with trypanothione *S*-transferase and trypanothione-peroxidase activities,[172] was also overexpressed in one resistant isolate.

No antimony-binding proteins were identified from these studies. Nevertheless, due to similarities between antimony and arsenic, arsenic-binding proteins may provide some hints.[173–175] Dozens of arsenic-binding proteins were identified in the human breast cancer cell line MCF-7 cells, in three major functional categories, metabolic enzymes, structural proteins and stress response proteins.[173] Previously identified arsenic-binding proteins (such as glyceraldehyde 3-phosphate dehydrogenase (GAPDH), malate dehydrogenase (MDH), pyruvate kinase, glucose-6-phosphate dehydrogenase, T-complex protein 1 (tcp1), elongation factor 1 and 2, eukaryotic translation initiation factor, 14-3-3 protein, β-tubulin and hsp70) were also present in the comparative proteomic studies on antimony-resistant *Leishmania* isolates. ATP-dependent RNA helicase, fructose-1,6-bisphosphate aldolase, HSC70 and mitochondrial hsp60 were found to bind arsenic in A549 human lung carcinoma cells.[174] Furthermore, certain proteins, whose expression levels do not vary between antimony-resistant and -sensitive *Leishmania* isolates, were also found to bind arsenic. Enolase and pyruvate carboxylase, present in the modelled pathway of antimony resistance in *L. donovani*,[173] were also included as arsenic-binding proteins. Arsenic can react with closely spaced cysteine residues of cysteine-rich proteins, leading to modulation or disruption of their function. For example, arsenic inhibited the polymerization of tubulin by blocking the active site, where GTP has access.[173] This is also the case for *L. donovani*.[176] These proteins are potential binders of antimony. Additional (metallo)proteomic studies are needed to elucidate the antimony-protein interaction networks and to determine the molecular mechanism of antimony resistance, not only in *Leishmania* cells but also in bacterial and other eukaryotic cells.

25.5.2 Binding of Bismuth to Proteins

Besides transferrin and lactoferrin, human serum albumin (HSA) can also bind Bi(III) with a binding constant (log K_a) of 11.2.[177] Although HSA is 13 times more abundant in human blood serum, it was shown that 70% Bi(III) binds to *apo*-transferrin and 30% to HSA, suggesting HSA as the second major binding target for Bi(III) in the blood serum.

Metallothioneins (MT) are low-molecular-mass (~7 kDa), cysteine-rich proteins, which contain 61 amino acids, of which 20 are cysteine residues.[178,‡] Bi(III) was found to bind to MT-II with higher affinity than Zn(II), with a molar ratio [Bi]/[MT] of 7. Moreover, Bi(III) can replace both Zn(II) and Cd(II) from MT-II in a biphasic process. Importantly, bismuth can enhance the expression of the MT gene(s) in macrophage cells as defence and/or repair mechanism(s) under different forms of stress.[179]

As bismuth-containing complexes are used as anti-*H. pylori* drugs, it is of interest to know their targets in prokaryotes. Bi(III) binds to the *Staphylococcus aureus* metal sensor protein CadC, which regulates the expression of the metal-ion-resistance *cad* operon on the pI258 plasmid.[180] Bi(III) binds to CadC *via* the four cysteine residues Cys7, Cys11, Cys58 and Cys60. Bismuth drugs suppressed production of acetaldehyde by *H. pylori* due to inhibition

---

‡Metallothioneins are discussed in detail in Chapter 21.

of alcohol dehydrogenase, a zinc metalloenzyme.[28] Binding of Bi(III) disrupts the quaternary structure of ADH from its native tetramer into a dimer.

Binding of Bi(III) to CadC reduced its affinity for DNA significantly, enabling access of RNA polymerase to the promoter. This led to de-repression of transcription and the expression of *cadA*, which encodes a calcium- and lead-specific ATPase efflux pump CadA. Bismuth, especially in the form of bismuth-thiol conjugates, also inhibits other proteins in cells (*e.g.* the rho protein, a transcription termination factor in *E. coli*).[181] Ranitidine bismuth citrate inhibits the phospholipase A2 of *H. pylori* and also that in the venom from the Indian cobra *Naja naja*. This may contribute to the anti-*H. pylori* action of this drug.[182] Bismuth also inhibits the activity of urease from *Klebsiella aerogenes via* binding to the conserved Cys319, a residue at the entrance of the urease active site.[183]

In *H. pylori*, several proteins were found to interact with bismuth compounds, indicative of a mechanism of action for bismuth drugs.[15,23,45] A unique histidine-rich protein Hpn, which is exclusively found in *H. pylori* and was considered crucial for intracellular nickel storage, binds about four Bi(III) per monomer ($K_d \sim 11$ μM).[184] Knock-out of the gene *hpn* leads to four times higher susceptibility of the bacteria to bismuth drugs.[185] To find bismuth-binding proteins and targets in *H. pylori* in high-throughput screens, a metalloproteomic analysis was performed.[24,153,186,187] Seven proteins were identified to bind Bi(III): the heat shock proteins HspA and HspB (homologues of chaperonins GroES and GroEL), the urease subunit UreB, fumarase, the translational factor Ef-Tu, the alkyl hydroperoxide reductase TsaA and the neutrophil-activating protein NapA.[24] HspA was subsequently found to bind two Bi(III) per monomer.[188] One binding site consisted of His45, Cys51 and Cys53 in the apical domain[189] with another one in the histidine-rich domain at the C-terminus. The binding converted the quaternary structure from a heptamer into a dimer, compromising function.

In addition to antimicrobial activity, Bi(III) exhibits antiviral activity.[190,191] It binds to the cysteine-rich metal binding domain of NTPase/helicase of the severe acute respiratory syndrome coronavirus (SCV), inducing a conformational change of the protein that results in the inhibition of its DNA unwinding activity, both *in vitro* and *in vivo*.

25.6 Perspectives

The therapeutic application of metal-containing drugs is drawing increasing attention. In spite of significant progress in the biochemistry and cell biology of antimony and bismuth, further investigation is necessary to uncover molecular mechanisms of action including toxicity in living systems. Proteomic analyses are anticipated to elucidate the effects of antimony and bismuth on pathogens, host cells and cancer cells. The effects of metals on the pathogen-host interaction proteomes, which are important for drug development and application, should also be emphasized. Based on proteomics and metallomics investigations, potential targets can be characterized by an integrative approach. As ancient medical agents, antimony- and bismuth-containing complexes will continue to have great potential for medical and healthcare purposes.

Acknowledgements

We thank the RGC of Hong Kong, RGC/NSFC joint research scheme, Croucher Foundation, Livzon Pharmaceutical Group and the University of Hong Kong (an emerging Strategic Research Theme (e-SRT) on *Integrative Biology*) for their support of our recent research on medicinal inorganic chemistry and metallomics of antimony and bismuth.

References

1. H. Sun, ed., *Biological Chemistry of Arsenic, Antimony and Bismuth*, John Wiley & Sons Ltd, Chichester West Sussex, UK, 2011.
2. C. J. Carmalt, A. H. Cowley, A. Decken and N. C. Norman, *J. Organomet. Chem.*, 1995, **496**, 59–67.
3. C. J. Carmalt, W. Clegg, M. R. J. Elsegood, R. J. Errington, J. Havelock, P. Lightfoot, N. C. Norman and A. J. Scott, *Inorg. Chem.*, 1996, **35**, 3709–3712.
4. R. I. McCallum, *Proc. R. Soc. Med.*, 1977, **70**, 756–763.
5. G. C. Low, *Trans. R. Soc. Trop. Med. Hyg.*, 1916, **10**, 37–42.
6. A. Harder, *Parasitol. Res.*, 2002, **88**, 395–397.
7. F. Frézard, C. Demicheli and R. R. Ribeiro, *Molecules*, 2009, **14**, 2317–2336.
8. C. R. Alving, E. A. Steck, W. L. Chapman, Jr., V. B. Waits, L. D. Hendricks, G. M. Swartz, Jr. and W. L. Hanson, *Proc. Natl Acad. Sci. USA*, 1978, **75**, 2959–2963.
9. R. R. C. New, M. L. Chance, S. C. Thomas and W. Peters, *Nature*, 1978, **272**, 55–56.
10. W. L. Roberts, J. D. Berman and P. M. Rainey, *Antimicrob. Agents Chemother.*, 1995, **39**, 1234–1239.
11. D. C. Reis, M. C. Pinto, E. M. Souza-Fagundes, L. F. Rocha, V. R. Pereira, C. M. Melo and H. Beraldo, *Biometals*, 2011, **24**, 595–601.
12. B. Xu, J. Ding, K.-X. Chen, Z.-H. Miao, H. Huang, H. Liu and X.-M. Luo, in *Recent Advances in Cancer Research and Therapy*, eds. X.-Y. Liu, S. Pestka and Y.-F. Shi, Elsevier, London, 2012, pp. 287–350.
13. K. S. O. Ferraz, N. F. Silva, J. G. da Silva, L. F. de Miranda, C. F. D. Romeiro, E. M. Souza-Fagundes, I. C. Mendes and H. Beraldo, *Eur. J. Med. Chem.*, 2012, **53**, 98–106.
14. C. Tsopelas, *Appl. Radiat. Isot.*, 2003, **59**, 321–328.
15. N. Yang and H. Sun, *Coord. Chem. Rev.*, 2007, **251**, 2354–2366.
16. G. G. Briand and N. Burford, *Chem. Rev.*, 1999, **99**, 2601–2658.
17. B. J. Marshall and J. R. Warren, *Lancet*, 1984, **1**, 1311–1315.
18. H. L. DuPont and C. D. Ericsson, *N. Engl. J. Med.*, 1993, **328**, 1821–1827.

19. T. E. Sox and C. A. Olson, *Antimicrob. Agents Chemother.*, 1989, **33**, 2075–2082.
20. H. L. DuPont, *Nat. Clin. Pract. Gastroenterol. Hepatol.*, 2005, **2**, 191–198.
21. W. Li, L. Jin, N. Y. Zhu, X. M. Hou, F. Deng and H. Sun, *J. Am. Chem. Soc.*, 2003, **125**, 12408–12409.
22. H. Sun, L. Zhang and K. Y. Szeto, *Metal Ions Biol. Syst.*, 2004, **41**, 333–378.
23. H. Y. Li and H. Sun, *Curr. Opin. Chem. Biol.*, 2012, **16**, 74–83.
24. R. G. Ge, X. S. Sun, Q. Gu, R. M. Watt, J. A. Tanner, B. C. Y. Wong, H. H. X. Xia, J. D. Huang, Q. Y. He and H. Sun, *J. Biol. Inorg. Chem.*, 2007, **12**, 831–842.
25. X. S. Sun, C. N. Tsang and H. Sun, *Metallomics*, 2009, **1**, 25–31.
26. A. Ottlecz, J. J. Romero, S. L. Hazell, D. Y. Graham and L. M. Lichtenberger, *Dig. Dis. Sci.*, 1993, **38**, 2071–2080.
27. R. Stables, C. J. Campbell, N. M. Clayton, J. W. Clitherow, C. J. Grinham, A. A. McColm, A. McLaren and M. A. Trevethick, *Aliment. Pharmacol. Ther.*, 1993, 7, 237–246.
28. L. Jin, K. Y. Szeto, L. Zhang, W. H. Du and H. Sun, *J. Inorg. Biochem.*, 2004, **98**, 1331–1337.
29. R. Bentley and T. G. Chasteen, *Microbiol. Mol. Biol. Rev.*, 2002, **66**, 250–271.
30. K. Marapakala, J. Qin and B. P. Rosen, *Biochemistry*, 2012, **51**, 944–951.
31. F. Challenger, *Adv. Enzymol. Relat. Areas Mol. Biol.*, 1951, **12**, 429–491.
32. P. Andrewes, W. R. Cullen and E. Polishchuk, *Appl. Organomet. Chem.*, 1999, **13**, 659–664.
33. P. Andrewes, W. R. Cullen and E. Polishchuk, *Chemosphere*, 2000, **41**, 1717–1725.
34. L. M. Hartmann, P. J. Craig and R. O. Jenkins, *Arch. Microbiol.*, 2003, **180**, 347–352.
35. S. Wehmeier, A. Raab and J. Feldmann, *Appl. Organomet. Chem.*, 2004, **18**, 631–639.
36. O. Wuerfel, F. Thomas, M. S. Schulte, R. Hensel and R. A. Diaz-Bone, *Appl. Organomet. Chem.*, 2012, **26**, 94–101.
37. F. Thomas, R. A. Diaz-Bone, O. Wuerfel, B. Huber, K. Weidenbach, R. A. Schmitz and R. Hensel, *Appl. Environ. Microbiol.*, 2011, 77, 8669–8675.
38. A. Hoeppner, F. Thomas, A. Rueppel, R. Hensel, W. Blankenfeldt, P. Bayer and A. Faust, *Acta Crystallogr. D*, 2012, **68**, 1549–1557.
39. M. Hollmann, J. Boertz, E. Dopp, J. Hippler and A. V. Hirner, *Metallomics*, 2010, **2**, 52–56.
40. R. Zangi and M. Filella, *Chem-Biol. Interact.*, 2012, **197**, 47–57.
41. C. Brochu, J. Y. Wang, G. Roy, N. Messier, X. Y. Wang, N. G. Saravia and M. Ouellette, *Antimicrob. Agents Chemother.*, 2003, 47, 3073–3079.
42. D. Sereno, M. Cavaleyra, K. Zemzoumi, S. Maquaire, A. Ouaissi and J. L. Lemesre, *Antimicrob. Agents Chemother.*, 1998, **42**, 3097–3102.
43. P. Shaked-Mishan, N. Ulrich, M. Ephros and D. Zilberstein, *J. Biol. Chem.*, 2001, **276**, 3971–3976.

44. C. Hansen, E. W. Hansen, H. R. Hansen, B. Gammelgaard and S. Stürup, *Biol. Trace Elem. Res.*, 2011, **144**, 234–243.

45. R. Ge and H. Sun, *Acc. Chem. Res.*, 2007, **40**, 267–274.

46. S. C. Yan, F. Li, K. Y. Ding and H. Sun, *J. Biol. Inorg. Chem.*, 2003, **8**, 689–697.

47. S. C. Yan, I. L. K. Wang, L. M. C. Chow and H. Sun, *Chem. Comm.*, 2003, 266–267.

48. F. Frézard, C. DeMicheli, C. S. Ferreira and M. A. P. Costa, *Antimicrob. Agents Chemother.*, 2001, **45**, 913–916.

49. C. D. Ferreira, P. S. Martins, C. Demicheli, C. Brochu, M. Ouellette and F. Frézard, *BioMetals*, 2003, **16**, 441–446.

50. Y. Zhou, N. Messier, M. Ouellette, B. P. Rosen and R. Mukhopadhyay, *J. Biol. Chem.*, 2004, **279**, 37445–37451.

51. R. Mukhopadhyay, D. Bisacchi, Y. Zhou, A. Armirotti and D. Bordo, *J. Mol. Biol.*, 2009, **386**, 1229–1239.

52. R. Mukhopadhyay, J. Shi and B. P. Rosen, *J. Biol. Chem.*, 2000, **275**, 21149–21157.

53. Y. Zhou, H. Bhattacharjee and R. Mukhopadhyay, *Mol. Biochem. Parasitol.*, 2006, **148**, 161–168.

54. I. Zegers, J. C. Martins, R. Willem, L. Wyns and J. Messens, *Nat. Struct. Biol.*, 2001, **8**, 843–847.

55. J. Messens, J. C. Martins, K. Van Belle, E. Brosens, A. Desmyter, M. De Gieter, J.-M. Wieruszeski, R. Willem, L. Wyns and I. Zegers, *Proc. Natl Acad. Sci. USA*, 2002, **99**, 8506–8511.

56. H. Denton, J. C. McGregor and G. H. Coombs, *Biochem. J.*, 2004, **381**, 405–412.

57. A. M. Silva, J. Tavares, R. Silvestre, A. Ouaissi, G. H. Coombs and A. Cordeiro-da-Silva, *Parasite Immunol.*, 2012, **34**, 345–350.

58. S. Müller, E. Liebau, R. D. Walter and R. L. Krauth-Siegel, *Trends Parasitol.*, 2003, **19**, 320–328.

59. P. K. Fyfe, G. D. Westrop, A. M. Silva, G. H. Coombs and W. N. Hunter, *Proc. Natl Acad. Sci. USA*, 2012, **109**, 11693–11698.

60. N. Allocati, M. Masulli, E. Casalone, S. Santucci, B. Favaloro, M. W. Parker and C. Di Ilio, *Biochem. J.*, 2002, **363**, 189–193.

61. R. L. Krauth-Siegel and M. A. Comini, *Biochim. Biophys. Acta Gen. Subj.*, 2008, **1780**, 1236–1248.

62. M. Filella, N. Belzile and M.-C. Lett, *Earth-Sci. Rev.*, 2007, **80**, 195–217.

63. O. I. Sanders, C. Rensing, M. Kuroda, B. Mitra and B. P. Rosen, *J. Bacteriol.*, 1997, **179**, 3365–3367.

64. Y. L. Meng, Z. J. Liu and B. P. Rosen, *J. Biol. Chem.*, 2004, **279**, 18334–18341.

65. D. X. Fu, A. Libson, L. J. W. Miercke, C. Weitzman, P. Nollert, J. Krucinski and R. M. Stroud, *Science*, 2000, **290**, 481–486.

66. R. Wysocki, C. C. Chéry, D. Wawrzycka, M. Van Hulle, R. Cornelis, J. M. Thevelein and M. J. Tamás, *Mol. Microbiol.*, 2001, **40**, 1391–1401.

67. Z. J. Liu, J. Shen, J. M. Carbrey, R. Mukhopadhyay, P. Agre and B. P. Rosen, *Proc. Natl Acad. Sci. USA*, 2002, **99**, 6053–6058.
68. K. Figarella, N. L. Uzcategui, Y. Zhou, A. LeFurgey, M. Ouellette, H. Bhattacharjee and R. Mukhopadhyay, *Mol. Microbiol.*, 2007, **65**, 1006–1017.
69. Z. Faghiri and P. J. Skelly, *FASEB J.*, 2009, **23**, 2780–2789.
70. T. Kamiya and T. Fujiwara, *Plant Cell Physiol.*, 2009, **50**, 1977–1981.
71. E. Tajkhorshid, P. Nollert, M. Ø. Jensen, L. J. W. Miercke, J. O'Connell, R. M. Stroud and K. Schulten, *Science*, 2002, **296**, 525–530.
72. Y. Wang, K. Schulten and E. Tajkhorshid, *Structure*, 2005, **13**, 1107–1118.
73. Z. E. R. Newby, J. O'Connell, III, Y. Robles-Colmenares, S. Khademi, L. J. Miercke and R. M. Stroud, *Nat. Struct. Mol. Biol.*, 2008, **15**, 619–625.
74. J. D. Ho, R. Yeh, A. Sandstrom, I. Chorny, W. E. C. Harries, R. A. Robbins, L. J. W. Miercke and R. M. Stroud, *Proc. Natl Acad. Sci. USA*, 2009, **106**, 7437–7442.
75. S. Silver, K. Budd, K. M. Leahy, W. V. Shaw, D. Hammond, R. P. Novick, G. R. Willsky, M. H. Malamy and H. Rosenberg, *J. Bacteriol.*, 1981, **146**, 983–996.
76. C. M. Chen, T. K. Misra, S. Silver and B. P. Rosen, *J. Biol. Chem.*, 1986, **261**, 5030–5038.
77. T. Q. Zhou, S. Radaev, B. P. Rosen and D. L. Gatti, *J. Biol. Chem.*, 2001, **276**, 30414–30422.
78. T. Zhou, S. Radaev, B. P. Rosen and D. L. Gatti, *EMBO J.*, 2000, **19**, 4838–4845.
79. X. Ruan, H. Bhattacharjee and B. P. Rosen, *Mol. Microbiol.*, 2008, **67**, 392–402.
80. L. S. Tisa and B. P. Rosen, *J. Biol. Chem.*, 1990, **265**, 190–194.
81. S. Dey and B. P. Rosen, *J. Bacteriol.*, 1995, **177**, 385–389.
82. Y. F. Lin, A. R. Walmsley and B. P. Rosen, *Proc. Natl Acad. Sci. USA*, 2006, **103**, 15617–15622.
83. A. A. Ajees, J. B. Yang and B. P. Rosen, *Biometals*, 2011, **24**, 391–399.
84. Y. X. Chen and B. P. Rosen, *J. Biol. Chem.*, 1997, **272**, 14257–14262.
85. J. Ye, A. A. Ajees, J. B. Yang and B. P. Rosen, *Biochemistry*, 2010, **49**, 5206–5212.
86. J. B. Yang, A. A. A. Salam and B. P. Rosen, *Mol. Microbiol.*, 2011, **79**, 872–881.
87. Y. F. Lin, J. B. Yang and B. P. Rosen, *J. Biol. Chem.*, 2007, **282**, 16783–16791.
88. B. P. Rosen, *Trends Microbiol.*, 1999, **7**, 207–212.
89. H. Bhattacharjee, Y. S. Ho and B. P. Rosen, *Gene*, 2001, **272**, 291–299.
90. B. Kurdi-Haidar, S. Aebi, D. Heath, R. E. Enns, P. Naredi, D. K. Hom and S. B. Howell, *Genomics*, 1996, **36**, 486–491.
91. Y. Y. Tseng, C. W. Yu and V. H. C. Liao, *FEBS J.*, 2007, **274**, 2566–2572.
92. G. Kao, C. Nordenson, M. Still, A. Rönnlund, S. Tuck and P. Naredi, *Cell*, 2007, **128**, 577–587.

93. P. Naredi, D. D. Heath, R. E. Enns and S. B. Howell, *Cancer Res.*, 1994, **54**, 6464–6468.

94. P. Naredi, D. D. Heath, R. E. Enns and S. B. Howell, *J. Clin. Invest.*, 1995, **95**, 1193–1198.

95. O. Hemmingsson, Y. Zhang, M. Still and P. Naredi, *Cancer Chemother. Pharmacol.*, 2009, **63**, 491–499.

96. O. Hemmingsson, G. Kao, M. Still and P. Naredi, *Cancer Res.*, 2010, **70**, 10321–10328.

97. J. Shen, C. M. Hsu, B. K. Kang, B. P. Rosen and H. Bhattacharjee, *Biometals*, 2003, **16**, 369–378.

98. E. Maciaszczyk-Dziubinska, D. Wawrzycka, E. Sloma, M. Migocka and R. Wysocki, *Biochim. Biophys. Acta Biomembr.*, 2010, **1798**, 2170–2175.

99. E. Maciaszczyk-Dziubinska, D. Wawrzycka and R. Wysocki, *Int. J. Mol. Sci.*, 2012, **13**, 3527–3548.

100. M. Ghosh, J. Shen and B. P. Rosen, *Proc. Natl Acad. Sci. USA*, 1999, **96**, 5001–5006.

101. R. Wysocki, P.-K. Fortier, E. Maciaszczyk, M. Thorsen, A. Leduc, Å. Odhagen, G. Owsianik, S. Ulaszewski, D. Ramotar and M. J. Tamás, *Mol. Biol. Cell*, 2004, **15**, 2049–2060.

102. D. Wawrzycka, I. Sobczak, G. Bartosz, T. Bocer, S. Ułaszewski and A. Goffeau, *FEMS Yeast Res.*, 2010, **10**, 828–838.

103. S. P. Cole, G. Bhardwaj, J. H. Gerlach, J. E. Mackie, C. E. Grant, K. C. Almquist, A. J. Stewart, E. U. Kurz, A. M. Duncan and R. G. Deeley, *Science*, 1992, **258**, 1650–1654.

104. L. Vernhet, A. Courtois, N. Allain, L. Payen, J. P. Anger, A. Guillouzo and O. Fardel, *FEBS Lett.*, 1999, **443**, 321–325.

105. M. Salerno, M. Petroutsa and A. Garnier-Suillerot, *J. Bioenerg. Biomembr.*, 2002, **34**, 135–145.

106. L. Gayet, N. Picault, A.-C. Cazalé, A. Beyly, P. Lucas, H. Jacquet, H.-P. Suso, A. Vavasseur, G. Peltier and C. Forestier, *FEBS Lett.*, 2006, **580**, 6891–6897.

107. K. Ito, S. L. Olsen, W. Qiu, R. G. Deeley and S. P. C. Cole, *J. Biol. Chem.*, 2001, **276**, 15616–15624.

108. S. A. Dzamitika, C. A. B. Falcão, F. B. de Oliveira, C. Marbeuf, A. Garnier-Suillerot, C. Demicheli, B. Rossi-Bergmann and F. Frézard, *Chem-Biol. Interact.*, 2006, **160**, 217–224.

109. S. P. C. Cole, K. E. Sparks, K. Fraser, D. W. Loe, C. E. Grant, G. M. Wilson and R. G. Deeley, *Cancer Res.*, 1994, **54**, 5902–5910.

110. B. D. Stride, C. E. Grant, D. W. Loe, D. R. Hipfner, S. P. C. Cole and R. G. Deeley, *Mol. Pharmacol.*, 1997, **52**, 344–353.

111. D. R. Hipfner, K. C. Almquist, E. M. Leslie, J. H. Gerlach, C. E. Grant, R. G. Deeley and S. P. C. Cole, *J. Biol. Chem.*, 1997, **272**, 23623–23630.

112. D. R. Hipfner, R. G. Deeley and S. P. C. Cole, *Biochim. Biophys. Acta Biomembr.*, 1999, **1461**, 359–376.

113. M. Ouellette, A. Haimeur, K. Grondin, D. Légaré and B. Papadopoulou, *Methods Enzymol.*, 1998, **292**, 182–193.

114. D. Légaré, D. Richard, R. Mukhopadhyay, Y.-D. Stierhof, B. P. Rosen, A. Haimeur, B. Papadopoulou and M. Ouellette, *J. Biol. Chem.*, 2001, **276**, 26301–26307.

115. A. Parodi-Talice, J. M. Araújo, C. Torres, J. M. Pérez-Victoria, F. Gamarro and S. Castanys, *Biochim. Biophys. Acta Biomembr.*, 2003, **1612**, 195–207.

116. B. Papadopoulou, G. Roy, S. Dey, B. P. Rosen and M. Ouellette, *J. Biol. Chem.*, 1994, **269**, 11980–11986.

117. H. L. Callahan and S. M. Beverley, *J. Biol. Chem.*, 1991, **266**, 18427–18430.

118. A. Mukherjee, P. K. Padmanabhan, S. Singh, G. Roy, I. Girard, M. Chatterjee, M. Ouellette and R. Madhubala, *J. Antimicrob. Chemother.*, 2007, **59**, 204–211.

119. A. Haimeur, C. Brochu, P. A. Genest, B. Papadopoulou and M. Ouellette, *Mol. Biochem. Parasitol.*, 2000, **108**, 131–135.

120. H. L. Callahan, W. L. Roberts, P. M. Rainey and S. M. Beverley, *Mol. Biochem. Parasitol.*, 1994, **68**, 145–149.

121. F. Weise, Y. D. Stierhof, C. Kühn, M. Wiese and P. Overath, *J. Cell Sci.*, 2000, **113**, 4587–4603.

122. Ashutosh, S. Sundar and N. Goyal, *J. Med. Microbiol.*, 2007, **56**, 143–153.

123. J. M. Pérez-Victoria, B. I. Bavchvarov, I. R. Torrecillas, M. Martínez-García, C. López-Martín, M. Campillo, S. Castanys and F. Gamarro, *Antimicrob. Agents Chemother.*, 2011, **55**, 3838–3844.

124. C. Yeates, *Curr. Opin. Investig. Drugs*, 2002, **3**, 1446–1452.

125. A. C. Coelho, S. M. Beverley and P. C. Cotrim, *Mol. Biochem. Parasitol.*, 2003, **130**, 83–90.

126. A. C. Coelho, E. H. Yamashiro-Kanashiro, S. F. Bastos, R. A. Mortara and P. C. Cotrim, *Mol. Biochem. Parasitol.*, 2006, **150**, 378–383.

127. P. Leprohon, D. Légaré, I. Girard, B. Papadopoulou and M. Ouellette, *Eukaryot. Cell*, 2006, **5**, 1713–1725.

128. A. C. Coelho, N. Messier, M. Ouellette and P. C. Cotrim, *Antimicrob. Agents Chemother.*, 2007, **51**, 3030–3032.

129. P. A. Genest, A. Haimeur, D. Légaré, D. Sereno, G. Roy, N. Messier, B. Papadopoulou and M. Ouellette, *Mol. Biochem. Parasitol.*, 2008, **158**, 95–99.

130. R. L. Soutar and S. B. Coghill, *Gastroenterology*, 1986, **91**, 84–93.

131. J. F. Ross, R. D. Broadwell, M. R. Poston and G. T. Lawhorn, *Toxicol. Appl. Pharmacol.*, 1994, **124**, 191–200.

132. R. Pamphlett, M. Stoltenberg, J. Rungby and G. Danscher, *Neurotoxicol. Teratol.*, 2000, **22**, 559–563.

133. M. Stoltenberg, J. D. Schiønning and G. Danscher, *Acta Neuropathol.*, 2001, **101**, 123–128.

134. E. Perlson, S. Maday, M. M. Fu, A. J. Moughamian and E. L. F. Holzbaur, *Trends Neurosci.*, 2010, **33**, 335–344.

135. K. Kristensson, *Annu. Rev. Pharmacol. Toxicol.*, 1978, **18**, 97–110.

136. L. J. Locht, L. Munkøe and M. Stoltenberg, *J. Neurosci. Methods*, 2002, **115**, 77–83.

137. A. Larsen, M. Stoltenberg, C. Søndergaard, M. Bruhn and G. Danscher, *Basic Clin. Pharm. Toxicol.*, 2005, **97**, 188–196.

138. M. Stoltenberg, A. Larsen, M. Zhao, G. Danscher and U. T. Brunk, *Acta PMIS*, 2002, **110**, 396–402.

139. U. von Recklinghausen, L. M. Hartmann, S. Rabieh, J. Hippler, A. V. Hirner, A. W. Rettenmeier and E. Dopp, *Chem. Res. Toxicol.*, 2008, **21**, 1219–1228.

140. H. Y. Li, P. J. Sadler and H. Z. Sun, *J. Biol. Chem.*, 1996, **271**, 9483–9489.

141. H. Sun, H. Y. Li and P. J. Sadler, *Chem. Rev.*, 1999, **99**, 2817–2842.

142. K. Mizutani, M. Toyoda and B. Mikami, *Biochim. Biophys. Acta Gen. Subj.*, 2012, **1820**, 203–211.

143. H. Sun, H. Y. Li, A. B. Mason, R. C. Woodworth and P. J. Sadler, *J. Biol. Chem.*, 2001, **276**, 8829–8835.

144. G. Miquel, T. Nekaa, P. H. Kahn, M. Hémadi and J.-M. E. H. Chahine, *Biochemistry*, 2004, **43**, 14722–14731.

145. J.-M. E. H. Chahine, M. Hémadi and N.-T. Ha-Duong, *Biochim. Biophys. Acta Gen. Subj.*, 2012, **1820**, 334–347.

146. N.-T. Ha-Duong, M. Hémadi, Z. Chikh and J.-M. E. H. Chahine, *Biochem. Soc. Trans.*, 2008, **36**, 1422–1426.

147. N. Yang, H. M. Zhang, M. J. Wang, Q. Hao and H. Sun, *Sci. Rep.*, 2012, **2**, DOI: 10.1038/srep00999.

148. Y. Cheng, O. Zak, P. Alsen, S. C. Harrison and T. Walz, *Cell*, 2004, **116**, 565–576.

149. H. Drakesmith and A. Prentice, *Nat. Rev. Microbiol.*, 2008, **6**, 541–552.

150. B. E. Eckenroth, A. N. Steere, N. D. Chasteen, S. J. Everse and A. B. Mason, *Proc. Natl Acad. Sci. USA*, 2011, **108**, 13089–13094.

151. L. Zhang, K. Y. Szeto, W. B. Wong, T. T. Loh, P. J. Sadler and H. Sun, *Biochemistry*, 2001, **40**, 13281–13287.

152. P. Domenico, J. Reich, W. Madonia and B. A. Cunha, *J. Antimicrob. Chemother.*, 1996, **38**, 1031–1040.

153. C. N. Tsang, K. S. Ho, H. Sun and W. T. Chan, *J. Am. Chem. Soc.*, 2011, **133**, 7355–7357.

154. M. V. Bland, S. Ismail, J. A. Heinemann and J. I. Keenan, *Antimicrob. Agents Chemother.*, 2004, **48**, 1983–1988.

155. H. Sun, S. C. Yan and W. S. Cheng, *Eur. J. Biochem.*, 2000, **267**, 5450–5457.

156. C. Demicheli, F. Frézard, J. B. Mangrum and N. P. Farrell, *Chem. Commun.*, 2008, 4828–4830.

157. F. Frézard, H. Silva, A. M. D. Pimenta, N. Farrell and C. Demicheli, *Metallomics*, 2012, **4**, 433–440.

158. P. Baiocco, G. Colotti, S. Franceschini and A. Ilari, *J. Med. Chem.*, 2009, **52**, 2603–2612.

159. S. Wyllie, M. L. Cunningham and A. H. Fairlamb, *J. Biol. Chem.*, 2004, **279**, 39925–39932.

160. Y. H. Zhang, C. S. Bond, S. Bailey, M. L. Cunningham, A. H. Fairlamb and W. N. Hunter, *Protein Sci.*, 1996, **5**, 52–61.

161. R. J. C. Hilf, C. Bertozzi, I. Zimmermann, A. Reiter, D. Trauner and R. Dutzler, *Nat. Struct. Mol. Biol.*, 2010, **17**, 1330–1336.

162. K. El Fadili, J. Drummelsmith, G. Roy, A. Jardim and M. Ouellette, *Exp. Parasitol.*, 2009, **123**, 51–57.

163. A. Kumar, B. Sisodia, P. Misra, S. Sundar, A. K. Shasany and A. Dube, *Br. J. Clin. Pharmacol.*, 2010, **70**, 609–617.

164. N. Biyani, A. K. Singh, S. Mandal, B. Chawla and R. Madhubala, *Mol. Biochem. Parasitol.*, 2011, **179**, 91–99.

165. D. Paape and T. Aebischer, *J. Proteomics*, 2011, **74**, 1614–1624.

166. J. Walker, R. Gongora, J. J. Vasquez, J. Drummelsmith, R. Burchmore, G. Roy, M. Ouellette, M. A. Gomez and N. G. Saravia, *Mol. Biochem. Parasitol.*, 2012, **183**, 166–176.

167. B. Vergnes, B. Gourbal, I. Girard, S. Sundar, J. Drummelsmith and M. Ouellette, *Mol. Cell. Proteomics*, 2007, **6**, 88–101.

168. K. Nagamune, K. Ohishi, H. Ashida, Y. C. Hong, J. Hino, K. Kangawa, N. Inoue, Y. Maeda and T. Kinoshita, *Proc. Natl Acad. Sci. USA*, 2003, **100**, 10682–10687.

169. C. Brochu, A. Haimeur and M. Ouellette, *Cell Stress Chaperon.*, 2004, **9**, 294–303.

170. T. Liu, C. K. Daniels and S. S. Cao, *Pharmacol. Ther.*, 2012, **136**, 354–374.

171. Ashutosh, M. Garg, S. Sundar, R. Duncan, H. L. Nakhasi and N. Goyal, *Antimicrob. Agents Chemother.*, 2012, **56**, 518–525.

172. T. J. Vickers, S. Wyllie and A. H. Fairlamb, *J. Biol. Chem.*, 2004, **279**, 49003–49009.

173. X. Y. Zhang, F. Yang, J. Y. Shim, K. L. Kirk, D. E. Anderson and X. X. Chen, *Cancer Lett.*, 2007, **255**, 95–106.

174. H. M. Yan, N. Wang, M. Weinfeld, W. R. Cullen and X. C. Le, *Anal. Chem.*, 2009, **81**, 4144–4152.

175. A. Mizumura, T. Watanabe, Y. Kobayashi and S. Hirano, *Toxicol. Appl. Pharmacol.*, 2010, **242**, 119–125.

176. K. G. Jayanarayan and C. S. Dey, *Int. J. Parasitol.*, 2004, **34**, 915–925.

177. H. Sun and K. Y. Szeto, *J. Inorg. Biochem.*, 2003, **94**, 114–120.

178. H. Wang, Q. Zhang, B. Cai, H. Li, K. H. Sze, Z. X. Huang, H. M. Wu and H. Sun, *FEBS Lett.*, 2006, **580**, 795–800.

179. N. E. Magnusson, A. Larsen, J. Rungby, M. Kruhøffer, T. F. Ørntoft and M. Stoltenberg, *Cell Tissue Res.*, 2005, **321**, 195–210.

180. L. S. Busenlehner, J. L. Apuy and D. P. Giedroc, *J. Biol. Inorg. Chem.*, 2002, 7, 551–559.

181. A. P. Brogan, J. Verghese, W. R. Widger and H. Kohn, *J. Inorg. Biochem.*, 2005, **99**, 841–851.

182. A. Ottlecz, J. J. Romero and L. M. Lichtenberger, *Aliment. Pharmacol. Ther.*, 1999, **13**, 875–881.

183. L. Zhang, S. B. Mulrooney, A. F. K. Leung, Y. B. Zeng, B. B. C. Ko, R. P. Hausinger and H. Z. Sun, *BioMetals*, 2006, **19**, 503–511.

184. R. G. Ge, Y. Zhang, X. S. Sun, R. M. Watt, Q. Y. He, J. D. Huang, D. E. Wilcox and H. Sun, *J. Am. Chem. Soc.*, 2006, **128**, 11330–11331.

185. H. L. T. Mobley, R. M. Garner, G. R. Chippendale, J. V. Gilbert, A. V. Kane and A. G. Plaut, *Helicobacter*, 1999, **4**, 162–169.

186. C. N. Tsang, J. Bianga, H. Sun, J. Szpunar and R. Lobinski, *Metallomics*, 2012, **4**, 277–283.

187. L. Hu, T. Cheng, B. He, L. Li, Y. Wang, Y. T. Lai, G. Jiang and H. Sun, *Angew. Chem. Int. Ed.*, 2013, **52**, 4916–4920.

188. S. J. Cun, H. Y. Li, R. G. Ge, M. C. M. Lin and H. Sun, *J. Biol. Chem.*, 2008, **283**, 15142–15151.

189. S. J. Cun and H. Sun, *Proc. Natl Acad. Sci. USA*, 2010, **107**, 4943–4948.

190. N. Yang, J. A. Tanner, B. J. Zheng, R. M. Watt, M. L. He, L. Y. Lu, J. Q. Jiang, K. T. Shum, Y. P. Lin, K. L. Wong, M. C. M. Lin, H. F. Kung, J. D. Huang and H. Sun, *Angew. Chem. Int. Ed.*, 2007, **46**, 6464–6468.

191. N. Yang, J. A. Tanner, Z. Wang, J. D. Huang, B. J. Zheng, N. Zhu and H. Sun, *Chem. Commun.*, 2007, 4413–4415.

CHAPTER 26

Actinides in Biological Systems

GERHARD GEIPEL*[a] AND KATRIN VIEHWEGER[b]

[a] Institute of Resource Ecology, Helmholtz Center Dresden Rossendorf, Germany; [b] Institute of Radiopharmaceutical Cancer Research, Helmholtz Center Dresden Rossendorf, Germany
*Email: g.geipel@hzdr.de

26.1 Introduction

Elements with atomic numbers >89 are named actinides or, in the case of atomic numbers >103, transactinides. All isotopes of these elements are unstable due to radioactive decay. The elements of the actinide series fill the 5f shell, meaning that their electron configuration commonly follows $[Rn]5f^{1-14}6d^17s^2$. There are two important exceptions: thorium has the configuration $[Rn]5f^06d^27s^2$, which rationalizes its tetra valence. Berkelium has an electron configuration $[Rn]5f^96d^07s^2$ and so favours oxidation state +2.

The restricted availability of the actinides with atomic numbers higher than 97 (berkelium) plus their short half-lives result in a relatively low interest in these elements in environmental research.

A broad review about actinides in biological systems, especially in animals and man, is given by P. W. Durbin in *"The Chemistry of Actinide and Transactinide Elements"*.[1] It serves as an excellent source of work published before 2005.

The references given in this chapter are not meant to be complete. We will follow the elements in the order of their position in the periodic table (Table 26.1).

Several actinides show luminescence properties, which enable their study by emission spectroscopic methods. Besides uranium, which is luminescent

RSC Metallobiology Series No. 2
Binding, Transport and Storage of Metal Ions in Biological Cells
Edited by Wolfgang Maret and Anthony Wedd
© The Royal Society of Chemistry 2014
Published by the Royal Society of Chemistry, www.rsc.org

Table 26.1 Actinide isotopes with the longest half-lives and oxidation states.

Atomic number Main emitter			Element symbol Oxidation states	Isotope	Half-life	
89	Ac	227	21.7 y	β⁻	*+3*	
90	Th	232	14 E 09 y	α	*+4*	
91	Pa	231	32 500 y	β⁻	+3; +4; *+5*;	
92	U	238	4.4 E 09 y	α	+3; +4; +5; *+6*	
93	Np	237	2.14 E 06 y	α	+3; +4; *+5*; +6; +7	
94	Pu	244	81 E 06 y	α	+2; +3; +4; *+5*; +6; +7	
95	Am	243	7340 y	α	+2; *+3*; +4: +5; +6	
96	Cm	247	15.6 E 06 y	α	+2; *+3*; +4	
97	Bk	247	1400 y	α	*+2*; +3; +4	
98	Cf	251	900 y	α	+2; *+3*; +4	
99	Es	252	472 d	α; β⁺	+2; *+3*; +4	
100	Fm	257	100.5 d	α	+2; *+3*; +4	
101	Md	258	51.5 d	α	+2; *+3*	
102	No	259	58 min	α; β⁺	+2; *+3*; +4	
103	Lw	262	3.6 h	β⁺	*+3*	

(Main oxidation state in bold and italic face.)

in all environmentally relevant oxidation states, americium and curium, especially, can be determined by time-resolved laser-induced fluorescence spectroscopy. Measurement of luminescence lifetimes of species of these two elements enables determination of their hydration number.[2]

The wide-ranging isotopic variety of actinides can be appreciated in the section of the nuclide table shown in Figure 26.1, which does not present all isotopes. The most stable nuclides, which may be relevant to biological systems, are marked with black borders. More details may be found in special nuclide charts.[3,4]

The availability of transuranium elements in marine environments was surveyed by P. Scoppa in 1984.[5] It was estimated that, by the year 2000, about 288 metric tons of man-made transuranium elements had arrived in marine environments.[5] The highest amount should be ^{237}Np (~194 tons or about 68%) followed by ^{243}Am (38.9 tons or 13.6%).

Radionuclide concentrations of some actinides have been measured in benthic invertebrates (rock jingle, blue mussel and horse mussel) in the islands of the Aleutian Chain.[6] The highest levels detected were for ^{234}U (0.45-0.84 Bq kg^{-1} wet weight). The concentrations are slightly smaller for ^{238}U whereas those for ^{241}Am, ^{239,240}Pu and ^{235}U are about one order of magnitude lower. The concentration was close to the detection limit for ^{236}U. The presence of isotopes ^{241}Am, ^{239,240}Pu and ^{236}U are due to human activities in the environment.

Several studies deal with the rapid determination of actinides in different samples, mainly human urine. By the use of inductively coupled mass spectrometry and alpha spectrometry in combination with a rapid separation technique, nearly all actinide isotopes can be determined in a relatively short period of time.[7,8]

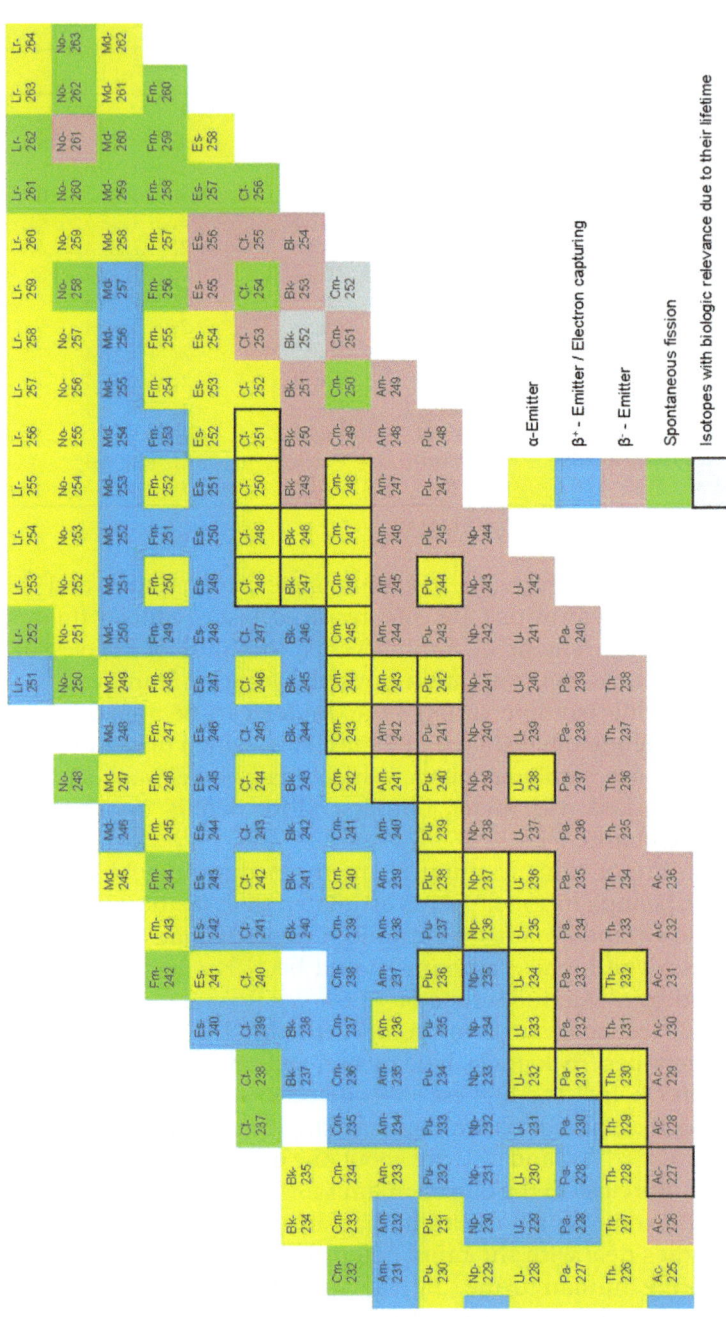

Figure 26.1   Section of the nuclide table.

A bioassay has been developed for rapid analysis of urine.[9] Co-precipitation followed by anion exchange and extraction chromatographic separation and alpha spectrometry were used to determine the concentration of radionuclides. The detection limit is in the order of 20 mBq L^{-1}. Alpha spectrometric methods have also been used for the determination of plutonium isotopes in the urine of workers exposed to radiation.[10] The addition of a ^{236}Pu tracer was used as a calibrant.

26.2 Actinium

Actinium was discovered independently by A. Debierne in 1899 and F. Giesel in 1902. ^{227}Ac is the most stable isotope. It is a beta emitter with a half-life of 21.6 years. As actinium is an element of Group 3, it can be seen as the prototype of the actinide series. Its chemical behaviour is similar to that of lanthanum.

26.3 Thorium

Thorium and the next element protactinium occur naturally in association with uranium. J. Berzelius discovered thorium in 1828 and it is one of the naturally occurring radioactive elements. Up to 25 isotopes are now known with atomic masses ranging from 212 to 236. The longest-living isotope ^{232}Th has a half-life of 1.4×10^{10} years. Thorium exists only in the tetravalent oxidation state and consequently it precipitates readily. This ensures that much less thorium is transferred from soil to plants compared to, for instance, uranium. Only thorium and uranium can be handled in the laboratory in weighable amounts (1000 Bq ^{232}Th with its natural decay products in equilibrium) without additional precautions.[11]

The stability of complexes formed with actinides controls their behaviour in blood. They are transported by plasma proteins.[12] Th(IV) is the most capable of forming strong complexes with transferrin.[†] Similar to iron(III), two thorium(IV) ions bind to the two domains of dimeric transferrin, together with the synergistic bicarbonate ion. Due to the high stability of its hydroxido complexes, it was suggested that Th binds as the monohydroxide. The release of two protons per bound Th imposes a slight decrease in pH upon complex formation.[11] The stability constants for the two sites were estimated to be $10^{33.5}$ and $10^{32.5}$ M^{-1}.[13] These values are comparable to those of the hydroxo species $Th(OH)_3^+$ and $Th(OH)_4$.

26.3.1 Animals

Chronic thorium exposure leads to its bioaccumulation in silver catfish (*Rhamdia quelen*).[14] Fish were exposed to different levels of thorium (~ 25; ~ 70; ~ 210; 610 μg L^{-1}) for 15 days. The response of muscles was analysed

---

[†]See Chapters 10 and 27 for further details of the plasma metallochaperone transferrin.

via the parameters glycogen, glucose, lactate, protein and ammonia. In addition, the levels of lipid peroxidation, catalase and glutathione-*S*-transferase were determined in the gills and in muscle. Gills and skin accumulated the highest levels of Th. However, increasing amounts of Th were found in all organs with increasing contamination. Low Th levels were found to stimulate enzymatic activities while higher levels were inhibitory.

The same species (silver catfish) was subjected to another investigation.[15] The thorium concentrations were in the same range as used before (~ 25; ~ 70; ~ 240; ~ 750 µg L^{-1}) but the duration of exposure was extended to 30 days.[16] The highest contamination was found in the gills and skin and also the muscle. With increasing contamination levels, glutathione-*S*-transferase activity decreased in gills. In the liver, the radioactivity was nearly constant. It was concluded that higher oxidative damage occurs in the gills. However, the authors did not consider chemical speciation. As thorium would be present exclusively in oxidation state +4, the amorphous oxide or precipitated Th(OH)$_4$ is expected to form in these experiments. It appears at present that Th cannot impinge on the cellular redox chemistry but may hamper cellular reactions. This property may also explain the high concentrations of Th observed in the skin where the oxide or hydroxide may have absorbed on the surfaces. Nevertheless, relatively low concentrations of actinides may induce oxidative stress due to their high affinity for several biologically important ligands.

26.3.2 Humans

An increase of Th in faecal samples of a human was observed in Brazil.[17] Initially there was no explanation about the source but it turned out that this person had consumed about 25 g of different types of nuts daily. The analysis of the Brazilian nut species yielded levels of ^{228}Th that may explain the thorium content in the faecal samples. However, additional measurements of urine samples (~ 0.7 mBq kg^{-1}) showed that the intake of ^{228}Th with the nuts cannot explain the high amount of this nuclide in the faecal samples. According to the decay chain of ^{232}Th, this thorium isotope is likely produced by the decay of ingested ^{228}Ra (half-life 5.75 y) *via* ^{228}Ac (half-life 6.13 h). Bio-kinetic calculations support this conclusion.

Human erythrocytes, which serve as a model for cellular membranes, were exposed to Th.[18] Scanning electron microscopy revealed that, depending on the ^{232}Th to cell ratio, the erythrocytes were transformed into equinocytes and/or spherocytes. Atomic force microscopy showed that the roughness of the erythrocyte surface increased significantly. In addition, experiments with erythrocytes that were treated with neuraminidase or were blocked with anti-GpA (gastric parietal cell) antibody suggested that the sialic acid in the membrane as well as the protein glycophorin A play important roles in aggregation and haemolysis induced by Th. The radioactive metal ion also influenced osmotic behaviour, as proven by studies of potassium efflux, osmotic protection and osmotic fragility. Haemolysis studies with inhibitors revealed the role of different ion channels. The presence of non-diffusible cations or anions in the medium showed that Na^+ and Cl^- influx influences the haemolytic effect of thorium.

26.4 Protactinium

K. Fajans and O. H. Gohring discovered protactinium (^{234}Pa) in 1913. The new element was first named brevium. O. Hahn and L. Meitner isolated the longer living isotope ^{231}Pa in 1918. The latter is the only long-living isotope of protactinium and is a member of the U-235 decay chain. Its half-life of 32 500 years makes it an extremely rare naturally occurring element. Pa displays both the pentavalent and tetravalent oxidation states. Until now, examination of the biology of this element has focused only on its radiometric determination. Pa is of no importance to the nuclear fuel cycle.

Only a few studies with Pa have been published and deal mainly with the observed oxidation states and speciation.[19]

New studies of Pa in biological systems have not appeared during the last decade. It is known that about 46% of protactinium fed to rats is stored in the skeleton.[20]

26.5 Uranium

M. H. Klaproth discovered uranium in 1789. The pure metal was isolated in 1841 by E.-M. Peligot. Naturally occurring uranium contains the isotopes ^{238}U (99.2830% by weight), ^{235}U (0.7110%) and ^{234}U (0.0054%). ^{238}U has a half-life of 4.4×10^9 years. The decay chain of ^{238}U is shown in Figure 26.2. The decay chains for the other naturally occurring actinides ^{235}U and ^{232}Th can be found in the literature.[21] The chains demonstrate that the daughter nuclides except ^{206}Pb are also naturally occurring radioactive isotopes.

Uranium forms many minerals; most of them show bright yellow colours and have fluorescent properties. The abundance of uranium on Earth is considered to be in the same order as molybdenum or arsenic.[22] As uranium is a ubiquitous element, its level in human and animal tissues can be about that of chromium.

Uranium exists in four oxidation states; typical colours are presented in Figure 26.3.[22] Dark brown-red uranium(III) is only stable in non-aqueous environments and plays no role in environmental speciation. Dark-green uranium(IV) is formed from uranium(VI) under reducing conditions. In aqueous solutions, slight pink uranium(V) disproportionates rapidly to uranium(IV) and uranium(VI). Hexavalent uranium exists only as the yellow dioxo cation $[U^{VI}O_2]^{2+}$. As the ions of oxidation states VI, IV and III hydrolyse easily, their aqueous solutions have to be acidified.

Depleted uranium (DU) is a by-product in the uranium enrichment process in which natural uranium (0.7% of the fissible isotope ^{235}U) is enriched. The depleted uranium contains about 0.3% ^{235}U and its behaviour in biological environments has been examined lately. New results show that the

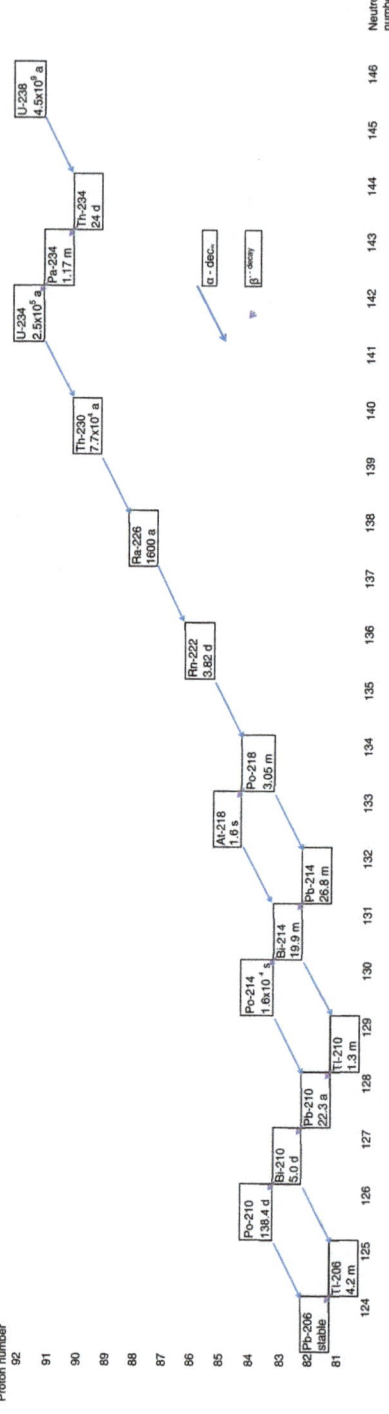

Figure 26.2   Decay chain of ^{238}U. [a = annum (year)]

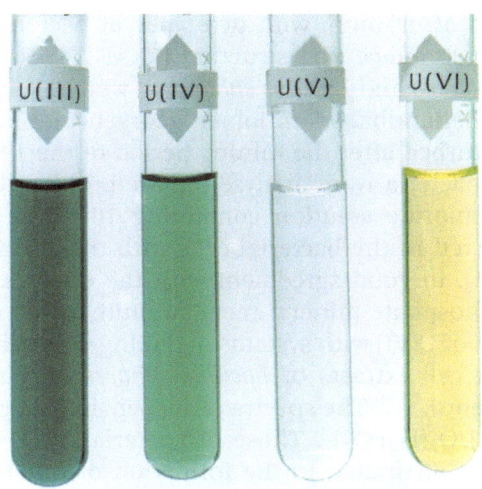

Figure 26.3 Colours of aqueous solutions of uranium in different oxidation states. Photographs by David E. Hobart, courtesy of Los Alamos National Laboratory.

inherent coordination chemistry of uranium leads to changes in the nature of the coordination sphere when DU is transported to different biological compartments.

The uptake and effects of uranium in biological systems have been reviewed.[23] However, its speciation has not been treated in depth. Instead, sequential extraction techniques have been employed to obtain rough estimates of the species formed.

Adenosine triphosphate (ATP) is formed in mitochondria and its energy-transfer functions are mostly intracellular. The binding of uranium and neptunium to ATP has been examined.[24,25] Strong dynamic quenching processes are induced upon complex formation and the decrease in fluorescence intensity can be used for determination of thermodynamic data. The derived stability constants are strongly pH dependent. This indicates that protons are involved in the chemical equilibrium. A formation constant of $\log K_A = -3.80 \pm 0.44$ was found for the 1:1 complex.

Uranium in humans has been studied extensively by determination in urine samples. Only a few selected publications will be cited in this review.

26.5.1 Microorganisms

The uptake of uranium by bacteria isolated from a uranium-containing waste dump has been studied.[26] Isolated *Thiobacillus ferrooxidans* had a slightly higher capability to accumulate uranium than a strain recovered from a coal mine. Reducing bacteria such as *Desulfovibrio desulfuricans* were capable of reducing most of the absorbed U(VI) within 24 h. The rate depended on the pH.

The interaction of uranium with two types of bacteria was examined by extended X-ray absorbance fine structure (EXAFS) spectroscopy.[27] The two bacteria *Bacillus cereus* and *Bacillus sphaericus* were taken from the uranium waste dump "Haberlandhalde" in Johanngeorgenstadt in Saxony, Germany, which was undisturbed after the mining period of the 1950s. Two reference strains of these bacteria were analysed as well. Cell cultures were treated with a sodium chloride solution containing 10^{-4} M uranium(VI). U-O-P bonds were detected in the bacterial cells with the average U-P distance of 0.362 nm (3.62 Å), in good agreement with the equivalent distance in the calcium uranyl-phosphate mineral meta-autunite (0.360 nm).[28]

The interaction of U(VI) with systems including vegetative cells, heat-killed cells, spores and cell extracts of *Bacillus sphaericus* was studied by laser-induced fluorescence.[28,29] The spectra of suspensions of cell extracts was the same as that of $UO_2(H_2PO_4)_2$. Those of bacterial U(VI) surface complexes were different and attributed to the formation of inner sphere complexes with organo-phosphate groups on the cell surface.

Several authors have studied the influence of uranium on the metabolism of bacteria. The biotransformation of uranium citrate by *Clostridium sphenoides* was studied recently.[30] Uranium was reduced to U(IV) only when citric acid was present in excess or when glucose was added. These results are in good agreement with data for *Pseudomonas fluorescens*.[31]

Acidithiobacillus ferrooxidans was isolated from the same uranium waste dump referred to already.[32] It accumulates U(VI), as do the reference bacteria *Pseudomonas stutzeri* and *Pseudomonas migulae*. Each type of bacterium transforms it into phosphate complexes but with different structural parameters.

One of the most studied types of bacterium is *Acidothiobacillus ferrooxidans*. Analysis of the fluorescence spectra of bacteria treated with uranium(VI) revealed three fluorescence lifetimes.[33] It was concluded that one complex is thermodynamically more stable than the others. However, no structural differences could be observed by EXAFS spectroscopy and it was concluded that the same functional groups *are involved* in complex formation.

The response of a bacterial community extracted from a uranium waste dump towards uranium treatment was studied under anaerobic conditions.[34] It could be shown that the U(VI) was reduced to U(IV). This was accompanied within the first four weeks by a strong decrease in bacterial diversity: denitrifying and uranium-resistant bacterial species increased. A further change and an increase in the bacterial diversity was observed after 14 weeks. Models for the bioreduction of uranium, especially under sulfate-reducing conditions, were suggested later.[35] Two different but concurrent processes were established. In the extracellular polymeric substance (EPS) of uranium-reducing bacteria, the reduction process can occur *via* oxidation of Fe(II) (from the iron-nickel-sulfide mineral mackinawite) and/or enzymatic oxidation of acetate to bicarbonate through the Krebs cycle. In addition it was found that, as proposed previously,[34] binding of uranium(IV) to phosphate groups in the biomass stabilized this oxidation state.

Uranium-resistant bacterial strains have been isolated from uranium mines in Portugal.[36] This *Rhodanobacter A2-61* strain tolerates uranium concentrations up to 2 mM. Under aerobic conditions, it was shown that about 24% of the U content of a starting solution (500 μM U) could be removed. Phosphate again played an important role. The authors claim that uranium phosphate was precipitated inside the cells. Structure analysis (X-ray diffraction) supported the formation of a mineral similar to meta-autunite and this is in agreement with X-ray absorption spectral data.[37]

26.5.2 Plants

Many older publications deal with transfer of actinides from soil to plants but speciation studies in plants are very rare. One of the first accounts of uranium speciation in plants determined the nature of uranium complexes formed after uptake of the metal ion by lupines.[38] Two series of experiments were performed with one group grown in soil and the other in hydroponic solution. Speciation was different: uranyl-hydroxo species were the major species in hydroponic solution while carbonate species dominated in the soil pore water. The speciation in several parts of the plant was the same and independent of the species present in the original solution. Uranium is bound mainly to phosphoryl groups in plant compartments, a fact confirmed by both EXAFS and time-resolved laser-induced fluorescence spectroscopy (TRLFS).

It can be concluded that the uptake of radionuclides by plants results in formation of species other than those present in the soil.[39] This review also concluded that uptake generally depends on the concentration of radionuclides in the soil. Some plants (such as black spruce) contained more uranium than was present in the natural background, indicating mechanisms of accumulation.[40]

Uptake by tomato plants growing in uranium-contaminated soil has been examined.[41] The soil was contaminated artificially with several concentrations of uranium. Due to a lack of exact experimental details, only two qualitative conclusions may be drawn:

1) The concentration of uranium decreases in the order root > stem > leaf. This finding would be expected for all plants as the roots have direct contact to the bioavailable uranium.
2) The uranium concentration in the roots is relatively high compared to that in the contaminated soil. This observation means that tomato plants are "uranium accumulators".

Conclusion 1) was, however, not confirmed in another study.[42]

Soil-to-plant transfer factors for uranium have been estimated to be in the wide range from 7.2×10^{-5} to 1×10^{-3}.[43] However, this uptake is not too high compared to another study where a plant originating from uranium sites was examined.[44] Transfer factors for uranium into vegetables were also reported several years ago and were somewhat higher than those quoted above.[45]

Different uptake levels can be expected for actinides in different oxidation states. The uptake of elements of the uranium series was studied at a mine mill.[46] The plants growing on weathered tailings showed the highest uptake for uranium followed by thorium. Unfortunately the data are not comparable to others, because the authors used mixed groups of plants.

Neutron activation analysis has been used to determine uranium levels in plants from a milling site at Köprübasi in Turkey.[47] It was found that the level in all plants except wheat was above 0.6 ppm. The ash of plants collected at a U-deposit showed enrichment of uranium up to 16 ppm.

As organic acids form soluble complexes with actinides, the uptake by plants may increase if such species are present in the soil. Plant uptake of uranium was increased in soil containing acetic, malic and citric acid.[48] The concentration of uranium in shoots from *Brassica juncea* and *Brassica chinensis* increases from 5 mg kg^{-1} to more than 5000 mg kg^{-1} when the soil is treated with citric acid. The authors explain this extremely high uptake by the biodegradation of the citric acid probably to propionic acid.[49] However, the nature of these degradation products was usually not determined.[31,50]

The first study dealing with the speciation of uranium during the growth of plants was carried out in 1998.[51] The pea *Pisum sativum* was investigated under a range of conditions. The speciation of uranium in hydroponics solution (modified Johnson's nutrient solution) was modelled with the program GEOCHEM-PC. The plants were pretreated with a phosphate-containing solution without uranium for 10 days. They were then transferred to nutrient solutions without phosphate but including U(VI) ions and grown for an additional seven days before harvesting. The roots and shoots were rinsed and dried before digestion with nitric acid. The uranium uptake was influenced by pH. At pH 5, when uranium was present as the free uranyl cation, the root concentration was generally higher than that in the shoots. For other species, *Phaseolus acutifolius* and *Beta vulgaris* showed the highest uranium uptake but information on the speciation in these plants is not available.

Uranium accumulation and tolerance in a variety of *Arabidopsis halleri* has been studied.[44] The plant was collected from a former uranium site in Johanngeorgenstadt (Germany). By sequential extraction of soil samples, a correlation was found between low bioavailability of nutrient iron and uptake of non-essential uranium. As reduction of Fe(III) to Fe(II) on the roots is a prerequisite for iron uptake, uranium reduction could take place in the same way. Laser-induced photo-acoustic spectroscopy (LIPAS) confirmed reduction of uranium(VI) to uranium(IV), which is proposed to occur by reduction of uranium(VI) to uranium(V) followed by disproportionation, generating uranium(IV) on the roots.

Arabidopsis plants grown under natural conditions, such as on rock piles, accumulate ~35 mg kg^{-1} uranium (dry weight) in the roots and ~17 mg kg^{-1} in the shoots.[44] Taking into account only the bioavailable uranium in the soil, the soil to plant transfer factor was calculated to be 1.2 for roots and 0.6 for shoots. However, in hydroponically grown plants in the laboratory, uranium uptake was about 100-fold higher for roots and 10-fold higher in shoots when the concentrations are compared to natively grown plants. This dramatic increase in uranium uptake was attributed to iron deficiency in the hydroponic system. The uranium tolerance was estimated by measuring the root elongation and the chlorophyll content as a function of uranium availability. The ratio of chlorophyll a/b isomers in native plants was found to be about seven. Chlorophyll b is more soluble in polar solvents. In the hydroponically grown plants, the ratio decreased drastically due to the lack of iron. Fluorescence measurements of chlorophyll extracted from uranium-containing leaves showed an additional peak that was assigned to a flavonoid. As is the case with other heavy metals, the deficiency of an essential element increases the uptake of non-essential elements.

The influence of uranium in different oxidation states on the cellular glutathione pool in plants has been investigated.[52] Glutathione levels were measured by HPLC methods. Cell cultures of cannola (*Brassica napus*) were grown in a modified medium (Linsmaier and Skoog). Cell cultures were resuspended in a phosphate-free 50% fresh culture medium. After 24 h, the cells were incubated with uranium. The cell suspension was sampled at different times, the cells homogenized and, after precipitation of proteins, the supernatant was concentrated and glutathione (GSH) and glutathione disulfide (GSSG) determined. At 10 μM uranium in the culture medium, the GSH level remained constant but the GSSG level increased immediately. In contrast, at the higher uranium concentration of 50 μM, the concentrations of both forms were drastically reduced. Calculation of the redox potential in the solution using the Nernst equation showed a shift towards more oxidizing conditions at the lower uranium concentrations but the opposite effect at the higher concentrations. It can be concluded that contact with uranium leads to a consumption of reducing equivalents. These redox processes are an important element of cellular defence reactions against oxidative stress. In addition, uranium forms complexes with the carboxylic acid groups of GSH. The disproportionation of initially formed uranium(V) species into uranium(VI) and uranium(IV) may explain why only 37% reduced uranium(IV) was found in the cells, relative to the initial concentration of uranium. The highest uranium(IV) level was detected 2 h after contact of the cells.

26.5.3 Animals

Bioaccumulation of uranium in a freshwater lake system has been investigated.[53] Three fish species (smallmouth bass *Micropterus dolomieu*; yellow perch *Perca flavescens;* bluegill *Lepomis machrochirus*) were sampled. In addition, some invertebrates were collected (families Sphaeriidae, Odonata, Gastropoda and Daphniidae). A significant bioaccumulation of uranium connected to historical mining activities was noticed. Bioaccumulation in invertebrates was about 2–3 orders of magnitude higher than in the fish. The uranium concentration in organs of the fish increased in the order bone > liver > muscle. The $^{15}N/^{14}N$ ratio provides information about trophic level due to a consistent increase in the isotopic ratio between prey and predator. Uranium bioaccumulation correlated negatively with the $^{15}N/^{14}N$ ratio. The authors drew the conclusion that a decrease of the U concentration in the trophic level occurred. The uranium content varied in the species of fish but the relative ratio between the studied organs was constant. Bluegill accumulated the highest and smallmouth bass the lowest amounts of uranium. Thus fish species that consume invertebrates showed higher uranium levels than those that consume other fishes.

The influence of uranium on the life of several fish species in the aquatic environment of a uranium mill has been investigated.[54,55] A study on Northern pike (*Esox lucus*) was divided into two levels of exposure to different elements.[54] Compared to the reference sample, low exposure was characterized by 0.77 µg g^{-1} As, 1.88 µg g^{-1} Be, 16.9 µg g^{-1} Se and 2.85 µg g^{-1} U whereas the element concentrations for high exposure were 4.28, 4.85, 22.9 and 5.68 µg g^{-1}, respectively (data for muscle samples and dry weight). GSH levels in liver increased slightly with increasing exposure while the levels of GSSG remained nearly constant. In contrast, the GSH level in kidney remained constant and the GSSG level for low exposure was somewhat higher. Apparently no significant oxidative stress occurred in northern pike, possibly owing to the fish having developed tolerance against the exposure.

In the same area, spottail shiner (*Notropis hudsonis*) was selected.[55] Low- and high-exposure sites were not included. The highest differences in the dry weight of whole bodies between the reference and exposure sites were for Al (29.22 *vs.* 40.52 µg g^{-1}), B (40.02 *vs.* 230.24 µg g^{-1}) and Se (2.24 *vs.* 17.50 µg g^{-1}). The uranium levels found at both sites were similar (0.03 *vs.* 0.04 µg g^{-1}). The salinities were different (0.23 *vs.* 0.01 mg L^{-1}). Despite a decreased plasma glucose all other factors such as plasma lactate, muscle activities and intramuscular energy storage were comparable. Therefore, it was concluded that the swimming performance for shiner at both sites was similar.

A study of the stress response of the Atlantic salmon to uranium has been performed.[56] The uranium concentrations in the experiments were in the range from 0.25 to 1 mg L^{-1}. In summary, short-term exposure may induce stress responses due to toxicological effects. However, long-term exposure may lead to DNA damage. The authors suggest long-term and multiple endpoint studies for arriving at a better understanding of the mechanisms triggered by the uptake of uranium in animals like fish.

The influence of environmentally relevant uranium concentrations on gene expression in brain, liver, skeleton muscles and gills of zebrafish (*Danio rerio*) has been investigated.[57] The U speciation depended only on the composition of the water in the tanks and adult male animals were exposed to 0, 23 and 130 µg L^{-1} uranyl nitrate with exposure times of 3, 10, 21 and 28 days. The exposure was followed by a period of 8 days when the fish lived in a non-contaminated environment. Then, gene expression and bioaccumulation of uranium were analysed. In the purification time, the bioaccumulation in liver was decreased and a strong expression of genes involved in detoxification was observed. Genes of the ATP-binding cassette transporter family were induced by factors ranging from 4 to 24 in fish exposed to the two contamination levels, consistent with the important role of the liver in detoxification processes. At the highest uranium concentration, an up-regulation of genes in gills by a factor of 21 was observed. This effect was interpreted as an onset of oxidative stress. In the skeleton muscles *coxl*, *atpf51* and *cat* were induced beginning from day 3. This suggested an impact on mitochondrial metabolism and the production of reactive oxygen species. Uranium excretion from brain and skeletal muscles was inefficient. Nevertheless, uranium concentrations in tissues were time-, concentration- and tissue-dependent.

The search for ways to protect against the consequences of exposure to radionuclides has focused on natural substances and metal ions. The influence of *Gingko biloba* against hepatotoxicity and nephrotoxicity has been studied.[58] Mice were treated with uranium-contaminated food for five days. To detect the different influences, six different groups of six mice were formed. Three groups did not receive uranium but the food was mixed with 0, 50, and 150 mg *Gingko biloba* leaf extract per kg body weight, respectively. The other three groups were treated the same way but with the addition of 5 mg uranium per kg body weight. A series of transferases as well as creatinine levels were determined to assess liver and kidney function. Treatment with *Gingko biloba* did show significant protective effects, which increased with increasing concentration of the extract.

The possibility of other metals protecting against the toxicity of uranium has been examined. Uranyl nitrate (10 mg kg^{-1}) was injected into rats. Before and after the injection, the rats were fed with zinc sulfate (10 mg kg^{-1}) once a day for 2 days.[59] They were investigated four days after injection of the zinc salt and 30 days after the uranium application. The data indicated the presence of a protective effect, especially when the zinc was applied before the uranium injection. It was found that zinc pre-treated rats had significantly lower uranium contents in kidney tissues compared to the non-treated group. In addition, rats pre-treated with zinc had a significantly higher survival rate. Pre-treatment decreases the accumulation of uranium in organs and increases its excretion rate. In addition, gene expression levels of metallothionein in the kidney tissues were significantly higher and metallothioneins are known to bind uranium ($\log K_{A1} = 6.5$; $\log K_{A2} = 5.6$[60]). Levels of the oxidative stress marker malondialdehyde also decreased.[59]

The nephrotoxic effects of uranium are well known from studies with animals. Proteomic techniques have been developed for the analysis of changes of urinary proteins upon uranium exposure. A total of 102 proteins in the normal urine of rats were characterized, increasing the known urinary proteome data set.[61] For the uranium study, two groups of rats were treated intravenously with 2.5 and 5 mg kg^{-1} uranyl nitrate, respectively. Urine samples were collected after 24 h. Employing semi-quantitative methods, the authors detected seven proteins with increased levels and seven with decreased levels. The change for three of them could be confirmed with

Western blotting techniques. Some of these proteins were already known to be involved in nephrotoxicity. The results indicate a change of glomerular permeability occurred, resulting in the appearance of albumin and trans-ferrin in the urine. Changes in epidermal growth factor (EGF) and vitamin-D-binding protein point to tubular damage. In addition, a third group of proteins with altered expression levels included some (*e.g.* ceruloplasmin) believed to be associated with metal stress. The particular advantage of this method lies in the possibility to assess uranium toxicity non-invasively.

The direct effect of depleted uranium (DU) on isolated mitochondria of rat kidneys was studied as the putative mechanism of uranium-induced nephrotoxicity is not understood in detail.[62] Uranyl acetate was injected at the levels of 0, 0.5, 1 and 2 mg kg^{-1} body weight of Wistar rats. Blood urea nitrogen and urinary creatinine levels increased. Mitochondria were isolated from the kidneys 24 h after exposition. After the 24 h treatment with uran-ium an increase in oxidative stress accompanied by a change of mitochon-drial membrane potential was observed. When isolated kidney mitochondria were treated with uranyl acetate (50, 100 and 200 μM), the electron transfer chain was inhibited. Such inhibition was associated with an increase in levels of reactive oxygen species, lipid peroxidation, oxidation of glutathione and a decrease of ATP concentration.

To investigate additional ways of protecting the kidneys against the nephrotoxicity of uranium, dietary fish oil (which is rich in ω-3 fatty acids) was used. After a 15-day dietary period, a single nephrotoxic dose of uranyl nitrate (0.5 mg per kg body weight) was given intraperitoneally. Five days after application, serum/urine parameters, enzymes of carbohydrate me-tabolism and brush border membranes, oxidative stress and phosphate transport in the kidneys were analysed. Nephrotoxicity manifested itself by an increase of serum creatinine and urea nitrogen. In addition, the activities of lactate dehydrogenase and NADP-malic enzyme increased. In contrast, the activities of malate-, isocitrate- and glucose-6-phosphate dehydrogenases, glucose-6-phosphatase, fructose-1,6-bisphosphatase decreased and carbo-hydrate metabolism involved in oxidation of NADH or reduction of NADP$^+$ of the brush border membrane were measured. Oxidant/antioxidant bal-ances were changed, as detected by increased activities of lipid peroxidation, superoxide dismutase and glutathione peroxidase and decreased catalase activity. Feeding the rats with fish oil alone increased the activities of en-zymes of glucose metabolism, brush border membrane; both oxidative stress and inorganic phosphorus (Pi) transport was observed. Changes caused by the treatment with uranyl nitrate were prevented by the feeding of the fish oil. Corn oil, in contrast, did not show these effects. Fish oil apparently protects against this kind of nephrotoxicity by enhancing energy metabolism and antioxidant defence.[63]

The differential effects of depleted and enriched uranium on the ster-oidogenic metabolism in rats have been the focus of another investigation.[64] Different groups of rats were treated for nine months with drinking water containing 40 mg L^{-1} U. Uranyl nitrate was used but, under the experimental conditions, other species were formed, of which carbonate species may predominate. The depleted uranium caused no significant change to the production of testicular steroids. But the enriched uranium increased the levels of circulating testosterone by a factor of 2.5. In addition, the expression levels of the nuclear receptors LXR (liver X receptor) and SF-1 (steroidogenic factor 1) and the transcription factor GATA-4 were modified during the treatment with enriched uranium. These results were the first to show different effects between depleted and enriched uranium and to demonstrate that the hazardous effects are caused primarily by the radiotoxicity of the different content of ^{235}U in depleted and enriched uranium samples.

The accumulation of uranium isotopes in several species of deer in northern Poland has been studied.[65] The radioactivity was between 2.5 and 69.4 mBq kg^{-1} wet weight for ^{238}U and 2.9 and 69.2 mBq kg^{-1} for ^{234}U. The activity ratio $^{234}U/^{238}U$ was generally close to 1. The average additional dose rate for consumers in Poland from this source is only about $1.8 \times 10^{-5}\%$ of the effective dose from all natural sources and therefore poses no radiological risk.

The living conditions of Iberian green frogs in an abandoned uranium mine has been investigated.[66] The frogs are living in a contaminated effluent pond. No significant increase in oxidative stress symptoms was detected. Gene expression analysis was performed to address the mechanisms that lead to this tolerance against permanent contamination. An up-regulation was detected for genes that code for the ribosomal protein L17a and for some blood plasma proteins like fibrinogen, haemoglobin and albumin. It was concluded that these proteins may act as protective agents against oxidative stress.

26.5.4 Humans

Due to the use of depleted uranium (DU) in the Gulf War and in the war in the former Yugoslavia, the interest in the behaviour and implications to the human health of this element has increased. This has led to an increased number of publications in this field, which is reflected in this review.

As mentioned already, determination of the uranium content in urine is used to obtain information about uranium accumulation and excretion.

Uranium excretion in urine was used to assess a cohort of Czech uranium miners, their families and the population in the area where the miners live. The levels excreted by the miners was significantly higher than those of the other two groups.[67] A model for excretion using the effective dose rates measured during the miners' work as starting values showed reasonable agreement with ICP-MS measurements of uranium in the urine samples.

It is known that human hair can be used for the estimation of levels of stored radionuclides. This enables an alternative long-term strategy, especially as the urine measurements presumably detect short-term stores. Mass spectrometric methods enable natural, enriched and depleted uranium to be distinguished. Isotope ratio quadrupole inductively coupled mass spectrometry is a sensitive and precise tool for estimation of the isotope composition of uranium samples.[68] The detection limit in human hair is 7.21 $\mu g \ kg^{-1}$. It is possible to determine depleted uranium levels against a background of natural uranium. It is claimed that a 20-fold background is acceptable but higher background levels make possible only screening.

Long-term health effects in connection with DU exposure have been assessed.[69] Uranium concentrations in urine were detected in the range <0.01 to about 65 $\mu g \ g^{-1}$. A connection was found between the amount of uranium bond to creatinine and the uranium content in blood. Low uranium in the urine (<0.1 $\mu g \ g^{-1}$) led to blood concentrations below 0.05 $\mu g \ g^{-1}$, whereas high uranium bond to creatinine (>0.1 $\mu g \ g^{-1}$) was reflected in concentrations above 0.2 $\mu g \ g^{-1}$.

An aboriginal community in Canada was exposed to uranium by drinking water with concentrations between <1 and 845 $\mu g \ L^{-1}$. A non-invasive study using analysis of urine samples for indicators of the function of the kidney and markers for the cell toxicity was undertaken.[70] The study included 54 individuals (between 12 and 73 years old) of a 1480 population size. It was found that a relation exists between the excreted concentrations of uranium in urine with the re-absorptive function of kidneys. The cumulative radiation dose was calculated for a 15-year period. Within this time, a maximum possible uranium intake was 1.761 mg with a calculated dose of 2.1 mSv. This would impose a cancer risk of 13 in 100 000. However, it was concluded that the population size was too small to detect an additional cancer risk from the uranium intake. It was also concluded that the chemical toxicity would have a greater impact on health than the radiation dose.

Uranium in semen has been studied by ICP-MS.[71] Depending on the exposure of the people studied, uranium concentrations were determined to be between <0.8 and 3350 $pg \ g^{-1}$. A significant difference was measured between a group with low uranium exposure (<2.8 $pg \ g^{-1}$) and those individuals with high uranium exposure (concentrations between 2.1 and 3350 $pg \ g^{-1}$). The low exposed group had uranium concentrations in the urine <0.1 $\mu g \ g^{-1}$ and the group with high exposure did excrete uranium in urine with concentrations >0.1 $\mu g \ g^{-1}$.

A search for possible biomarkers of uranium exposure has been performed.[72] Human kidney cells were cultured and treated with uranium. The viability of these cells was determined by measuring the intracellular ATP level. Increasing uranium concentration led to continuous decline in cell viability. In addition osteopontin, a highly phosphorylated sialoprotein that is a prominent component of the mineralized extracellular matrices of bones and teeth, was studied as biomarker for the uranium influence. Cell lysates and supernatants as well as urine samples were assayed with a commercial osteopontin kit. Osteopontin levels decreased with the concentration of added uranyl bicarbonate. Time-dependent experiments showed that osteopontin production in the cells was completely blocked. In addition two groups of human populations with different uranium exposures were involved in this study (together with a non-exposed control group). The first group were uranium miners, exposed by inhalation processes and the second group had drunk uranium-contaminated drill well water. The human populations were categorized according to the uranium level in urine (<5 μg L^{-1}; $5< $ x <10 μg L^{-1}; $10< $ x <30 μg L^{-1} and >30 μg L^{-1}). No significant differences in the osteopontin level relative to the creatinine content could be detected for all groups. It appears that osteopontin cannot be used as a biomarker as the selectivity towards uranium is not sensitive enough.

The fraction of an ingested radionuclide in blood (fl) and the residence time within the different parts of the gastrointestinal tract control the doses delivered to the different target tissues. These can be altered by many factors. For uranium, transfer factors were found to be of the order of 2×10^{-2}.[12] Values of 5×10^{-4} were found for the other actinides. In the case of pentavalent Np and Pu, the factor fl depends on the mass of the ingested radionuclide.

Laser-induced fluorescence spectroscopy has been used for detection of uranium in human urine samples.[73] Human urine is a complex matrix and the measurement of uranium at very low concentrations may be strongly influenced by matrix effects. Therefore original samples were diluted and the uranium was determined as its phosphate complex. The detection limit was determined to be 1.3×10^{-8} M.[73]

Bone as target for the storage of the f-block elements and herein especially uranium has been reviewed recently.[74] It was stated that calcium and phosphate play an important role in the inclusion of actinides into bone minerals. In addition, mechanisms of the interaction of actinides, especially uranium(VI), were suggested. The incorporation of uranyl ions into the bone matrix includes co-precipitation and ion exchange reactions. The main components seem to be the compounds chernikovite ($H_2(UO_2)_2(PO_4)_2{*}nH_2O$) and autunite ($Ca(UO_2)_2(PO_4)_2{*}nH_2O$). Nevertheless, the authors state that the mechanisms are more complicated due to the different oxidation states accessible by the actinides as well as their complexation (speciation) in the liquid phase. It is also stated, correctly, that there are too few studies available presently to make an accurate assessment about the impacts of actinides in this field.

Due to weathering of depleted uranium ammunition in the former Yugoslavia and an increase of bio-available uranium in this region it can be expected that more intense studies in the future will be published.[75]

26.6 Neptunium

Neptunium was the first synthetic transuranium element to be synthesized. ^{239}Np was produced by McMillan and Abelson in 1940 by bombardment of uranium with neutrons. The half-life of ^{237}Np is 2.14×10^6 years. All members of its naturally occurring decay series have decayed. It is possible, however, to detect trace quantities of the element in nature due to nuclear reactions in uranium ores.

Neptunium ions exist in oxidation states between +3 and +7. Np(III) is easily oxidized to Np(IV). The most stable oxidation states are Np(IV) and Np(V), the latter forming the dioxo-cation $[NpO_2]^+$.

In acidic solution, neptunium can be stabilized in four ionic oxidation states: Np^{3+} (pale purple), Np^{4+} (yellow green), $[NpO_2]^+$ (green blue) and $[NpO_2]^{2+}$ (pale pink) (Figure 26.4). The chemical status of Np(VII) remains not completely clear at the moment, it exists only in strong oxidizing and alkaline solutions.

Only studies with synthetic neptunium isotopes have been performed. Changes in oxidation state may also change the nature of complex formation and, consequently, the transportation of actinides in the environment. It is known that microorganisms are capable of reducing neptunium(V).

Early studies concluded that neptunium could be transported in the blood by transferrin and transferred into a cell, presumably by endocytosis. Inside the cell Np is bound first to a cytosolic protein but is later detected as so-called "high-molecular-weight compounds".[12]

Neptunium speciation in citrate media has been studied as this acid is seen to be a key ligand for simulation of biological fluids. Citrate is also an essential cofactor for a range of metalloenzyme substrates.[76] This element lacks sensitive luminescence properties and conventional spectro-photometry was employed. The species $[NpCit]^+$ and $[Np(Cit)_2]^{2-}$ were observed with respective stability constants of log $\beta_{101} = 13.6 \pm 0.3$ and log $\beta_{102} = 25.3 \pm 0.3$. These data are relatively close to the values for plutonium(IV) (log $\beta_{101} = 15.3$; log $\beta_{102} = 30.2$)[77] and thorium(IV) (log $\beta_{101} = 13$ and log $\beta_{102} = 21$).[78]

For pentavalent neptunium the following six species were claimed: $[NpO_2(Cit)]^{2-}$; $[NpO_2(Cit)_2]^{5-}$; $[NpO_2(HCit)]^-$; $[NpO_2(HCit)_2]^{3-}$; $NpO_2(H_2Cit)$

Figure 26.4 Colours of aqueous solutions of neptunium in different oxidation states. Photographs by David E. Hobart, courtesy of Los Alamos National Laboratory.

and $[NpO_2(H_2Cit)_2]^-$. Formation constants were given as $\log K_{101} = 1.6 \pm 0.2$; $\log K_{102} = 0.8 \pm 0.5$; $\log K_{111} = 2.0 \pm 0.2$; $\log K_{112} = 1.2 \pm 0.3$; $\log K_{121} = 1.3 \pm 0.3$ and $\log K_{122} = 0.6 \pm 0.5$, respectively. While speciation calculations indicate that neptunium(IV) exists mainly as two species in citrate solution at a selected pH, neptunium(V) usually is present as multiple species.

The nature of transferrin-actinide(IV) complexes has been compared to the native iron(III) form.[79] It is assumed that, in a plasma sample, the actinides are mainly complexed by transferrin. It was concluded that neptunium(IV) and plutonium(IV) show similarities to the iron(III) form. At pH <7 neptunium did not bind to transferrin but at pH ~7.4, the complexation was quantitative. For plutonium, quantitative binding was at pH ~7.5. No complexation behaviour was detected for thorium(IV) at all pH values. However, this latter conclusion is in direct contrast to those drawn from different data (see discussion above in thorium section).[12] EXAFS studies concluded that the actinide(IV) ions are bound to transferrin *via* one tyrosine residue and one hydroxyl group.

The uptake of neptunium by transferrin has been studied by different techniques.[80] A Np(IV)-transferrin complex was characterized by a combination of IR measurements and EXAFS data. Nitrilotriacetic acid (NTA) was used as the synergistic anion in an attempt to stabilize the complex. Again, the data are consistent with Np(IV) binding to the Fe(III) binding site of transferrin. The addition of NTA means that the transferrin is locked in the open form. Aspartate and histidine residues are not available for binding. The short Np-O-tyrosine distance of 2.34 Å is consistent with the strong basicity of the phenolate group.

The binding of f-block elements to transferrin (Tf) has been evaluated recently.[81] It was concluded that Fe(III) forms the strongest 2 : 1 complex ($\log \beta_2 = 15.99$). Complexes with Th^{4+}, UO_2^{2+}, Pu^{4+} and Cm^{3+} are of the same stoichiometry but are less stable ($\log \beta_2 = 13.79$, 14.17, 12.30 and 13.95, respectively). It was concluded that the ions Th^{4+}, UO_2^{2+} and Cm^{3+} would follow a transferrin receptor (TfR):Tf-mediated uptake pathway, a hypothesis that is yet to be confirmed by cellular uptake studies. Nevertheless, it should be pointed out that the authors have published a different value for the complex Cm_2Tf ($\log \beta_2 = 15.8$).[82] A different value for Pu_2Tf of $\log \beta_2 = 21.25$ has been published.[12,83] In addition, it is proposed that Pu can be acquired by rat cells (PC12) and the transport complex is Pu_CFe_NTF, where C and N are the different binding sites in transferrin.

26.6.1 Microorganisms

The interaction of *Desulfovibrio desulfuricans* with neptunium did not show any pH dependence.[84] For anaerobic bacteria, an increase of the distribution coefficient K_d was found. The results were compared to the interaction of Np(V) with humic acid, which increases with pH. The interpretation was that the complexing sites should predominately be deprotonated functional groups. The K_d values, which describe the concentration ratio of neptunium between the solid phase (bond to the microorganism) and concentration of neptunium in the solution, for humic acid were found to be 100-fold higher than for *Desulfovibrio desulfuricans.*

26.7   Plutonium

Plutonium was discovered in 1940 by Seaborg, McMillan, Kennedy and Wahl. The first isotope ^{238}Pu was produced by deuteron bombardment of uranium. Plutonium can be found in trace amounts in uranium ores.

The most stable isotope is ^{244}Pu with a half-life of 8.1×10^7 years. However, the most important one is ^{239}Pu, which is fissile with a half-life of 24000 years.

Plutonium exhibits oxidation states from $+3$ to $+6$. The redox potentials are similar; several oxidation states can co-exist, depending on the available ligands. Plutonium is one of the most toxic elements.

It exhibits four ionic valence states in aqueous solutions: Pu^{3+} (blue lavender), Pu^{4+} (yellow brown), $[PuO_2]^+$ (pink) and $[PuO_2]^{2+}$ (pink-orange) (Figure 26.5).[85] $[PuO_2]^+$ is unstable in aqueous solutions and disproportionates into Pu^{4+} and $[PuO_2]^{2+}$. Pu^{4+} can then oxidize $[PuO_2]^+$ into $[PuO_2]^{2+}$, itself being reduced to Pu^{3+}. The final products are Pu^{3+} and $[PuO_2]^{2+}$. Plutonium is a very dangerous radiological material. Precautions must also be taken to prevent the unintentional generation of a critical mass.

The ready interconvertability of its five accessible oxidation states make the redox behaviour of plutonium in biological systems an issue of basic interest.

Up to now about 4.2 metric tons of anthropogenic plutonium have been released into the environment.[85] The concentration of the element in aquatic systems is relatively constant and found to be less than 2×10^{-5} Bq L^{-1}

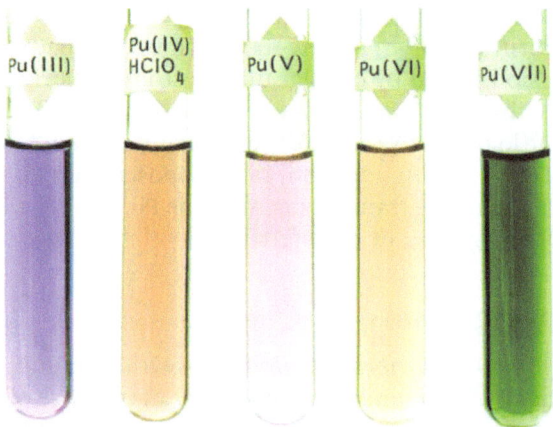

Figure 26.5   Colours of aqueous solutions of plutonium in different oxidation states. Photographs by David E. Hobart, courtesy of Los Alamos National Laboratory.

in lake waters and 4×10^{-5} Bq L^{-1} in ocean water. Assuming that this plutonium comes only from radioactive fallout with a ^{240}Pu/^{239}Pu ratio of 0.18, the chemical concentration of this element in waters can be estimated to be in the order of 4×10^{-17} M. Lake sediments and soils contain plutonium in the order of 2×10^{-16} to 4×10^{-15} moles g^{-1}.

The behaviour and transport of plutonium in the environment has been recently reviewed.[86] However, the interaction with living organisms is described for microorganisms only and mainly from the point of their interaction with colloidal species. Despite this restriction, microorganisms may promote change of oxidation states in plutonium species and excreted organic acids may solubilize plutonium.

26.7.1 Microorganisms

Microorganisms may play an important role in the mobility of actinides. Studies often start with the determination of the uptake of actinides by microorganisms. The association of uranium and plutonium with several bacteria has been studied.[87] It was found that uptake of uranium is higher than for plutonium. However, the oxidation state of the Pu was not determined, so that the data are not comparable. It can only be supposed that uranium was in its hexavalent state, which is highly mobile, whereas plutonium was in oxidation state +4, which is much less mobile.

The uptake of other actinides is of interest for distribution calculations. It was shown that *Pseudomonas stutzeri* and *Bacillus sphaericus*, as representatives for aerobic bacteria, accumulate high amounts of Pu(VI).[88] Lower oxidation states of Pu were detected in this study. A reaction mechanism was developed to explain these results. In the first step, Pu(VI) interacts with the biomass in a reaction with fast kinetics that depends on the concentration of biomass in the suspension. Earlier EXAFS results on the interaction of uranium with biomass indicated that the metal ion was bound to organophosphate groups. In a second reaction step, a slow reduction of bound Pu(VI) occurs. As the product Pu(V) shows a much reduced tendency to interact with the biomass, most of it was found in solution. In a third step, this Pu(V) disproportionates into Pu(VI) and Pu(IV). Both oxidation states react with the bacterial cells. In this way, it could be concluded that the product Pu(IV) forms as a result of disproportionation of Pu(V) and is not by interaction with the biomass.

In the Swedish underground hard rock laboratory, a microorganism has been detected in the past. The interaction of this sulfate-reducing bacterium *Desulfovibrio Äspöensis* with plutonium was studied.[89] By use of extraction methods and UV-Vis spectrophotometry, the amount of surface-adsorbed plutonium was shown to depend on the initial concentration and exposure time. In addition, X-ray absorption near edge spectroscopy (XANES) was used to determine the oxidation state of the bound plutonium and EXAFS was used to explore the coordination sphere of the bound plutonium. Pu(IV) and Pu(IV) polymers were initially bound to the bacterial cell. Nearly all of the reacting Pu(VI) was reduced to Pu(IV) within the first 24 h of contact time. It appeared to be bound to the cell wall weakly as the bound Pu was easily extracted back into solution. However, other indications confirmed that plutonium can penetrate the cell wall.

26.7.2   Plants

Uptake of plutonium by plants has been much less studied than that of uranium, but critical overviews are available.[90,95] This may be due to the problems connected with generating contaminated soil. Two pathways of assimilation are known: direct deposition from radioactive fallout on leaves or by roots. In contrast to other radionuclides, most of the plutonium distributed by radioactive fallout is found in the topsoil and may be re-suspended in the air as dust. The plutonium may be absorbed by the plant *via* leaves. The transfer factor is estimated to be 0.02.[91,92] The transfer factor *via* roots should be several orders of magnitude lower; a value of 0.0005 has been proposed,[93] consistent with previous estimates.[93] The overall estimates of transfer factors range from 1×10^{-7} to 2×10^{-4}.

The order of soil-plant transfer of actinides is proposed to be the following: Np >> Am ~ Cm > Pu.[94] Supporting studies were performed inside a greenhouse over seven vegetation periods. The soil was contaminated artificially with ^{237}Np, ^{238}Pu, ^{241}Am and ^{244}Cm at levels of 1–20 Bq g^{-1} dry mass. The highest transfer factors were measured for grass and corn. The observed transfer of the actinides from the soil to the plants depended on the chemical properties of the particular ion but was also influenced by the properties of the soil. An overview on the distribution of plutonium in soil and plants can be found in the literature.[95]

The conditional stability constant for the Pu(IV)-transferrin complex was determined to be log $\beta = 21.25 \pm 0.75$.[12] The Pu-transferrin complex is not able to facilitate the transport of Pu into the cell as it dissociates near the cell wall.

26.7.3   Animals

The concentration of plutonium isotopes in deer was assessed.[65] The plutonium concentration in many samples was below the detection limit. However, the following ranges of activity concentrations (mBq kg^{-1} w.w.) were provided for ^{238}Pu, $^{239-240}$Pu, respectively: kidney, 0.4 to 1.3 and 0.6 to 2.8; liver, 0.3 to 0.46 and 0.3 to 1.1. The values for muscle were generally lower. From the derived isotope ratios it was concluded that between 44% (red deer kidney) and 91% (red deer muscle) of the plutonium originated from the Chernobyl accident. It was also concluded that there is no radiological risk from consuming deer meat.

It is well known that plutonium is able to form two types of polymeric species (PuO$_2$-like structures and hydroxy-bridged polymers). It has been assumed that these two types will exhibit different behaviour in mammalian cells. PC12 cells from rats were exposed to the two forms for 3 h.[96] After removal of the surface-sorbed plutonium, its localization inside the cells was analysed by synchrotron based X-ray fluorescence microscopy. Mononuclear plutonium was also introduced as complexes of nitrilotriacetic acid or citrate colocalized with the iron within the cells. In contrast, the aged polymeric plutonium was always adsorbed to the cell wall as a large agglomerate and could not enter the cell. In addition, if the polymeric plutonium was formed *in situ*, then it was found mostly as the agglomerate inside the cells.

From the critically evaluated data for Pu equilibrium constants, it was predicted that this element exits from living cells mainly as amorphous PuO_2.[97] An exception was seen for endosomes, where this element may also exit as Pu^{3+}. These predictions were tested by microprobe X-ray absorption near edge structure (μ-XANES) measurements. Cancer cells of the rat (pheochromocytoma, PC12) were incubated *in vitro* for three hours with Pu in different oxidation states. It was found that the Pu was located principally in the cytoplasm and that the nucleus was not affected. It was confirmed that the Pu was present mainly in oxidation state $+4$ and with small amounts (about 10%) in oxidation states $+3$ and $+6$.

26.7.4 Humans

The determination of naturally occurring plutonium requires extremely sensitive methods, as concentrations are very low. A method using accelerator mass spectrometry (AMS) was developed to determine ultra-low traces of plutonium in urine samples.[98] The merits of this method compared to ICP-MS include the absence of isobaric mass interferences, especially if the samples are contaminated with uranium. The detection limits were found to be 0.27 fg L^{-1} ^{239}Pu, 0.28 fg L^{-1} ^{240}Pu and 0.06 fg L^{-1} ^{241}Pu. For determination 1.4 L urine were collected.

Transferrin is the transporter protein (chaperone) of Fe(III) in blood. Its ability to bind and transport metal ions was reviewed, including uranium and plutonium.[99] The conditional stability constants for uranium(VI) are given as log $K_1 = 7.7$ and log $K_2 = 4.6$, respectively[‡] (see also ref. 60). In contrast, no data for Pu were provided. However, it was suggested that plutonium will be transferred *via* ferritin in the bloodstream but cannot enter cells as it will be sorbed onto the cell surface. It is also mentioned that transferrin can bind actinides in all oxidation states from $+2$ to $+6$. The strength of actinide binding was given in the order $Pu^{4+} > Np^{4+} \sim Th^{4+} > UO_2^{2+} > Am^{3+} > Cm^{3+}$. A comparison of the transferrin data with the known stability of albumin complexes was made. It was concluded that, for transport in the bloodstream, albumin complexes may play a more important role for Cm^{3+} and Am^{3+} whereas transferrin complexes may

---

[‡]Differences in the given stability data may be caused by different oxidation states as well as whether or not reactions with protons are included.

have greater impact for Pu^{4+}; Np^{4+}; Th^{4+}; UO_2^{2+}. However, due to the relatively low affinity of these metal ions, it appears that transferrin does not play an important role in incorporation of these metals into cells.

The decorporation of plutonium and its speciation in blood, plus some targeting tissues, have been described.[100] Decorporation in this context means the decontamination of a human body after intake of a radioactive element. The investigators performed a systemic contamination and treated the subject rats with the ligand diethylenetriaminopentaacetic acid (DTPA) very soon after contamination. It was found that the Pu-DTPA complexes are very unstable under *in vivo* conditions. Experiments comparing the data with Pu-citric acid complexes showed that no significant decrease in the plasma plutonium occurred within the first hour. From this it was concluded that complexation of plutonium with DTPA in blood is a slow process and that other compartments are also involved.

26.8   Americium and Curium

The first isotope ^{241}Am was identified by Seaborg, James, Morgan and Ghiorso late in 1944. It was generated by repeated neutron capture reactions of plutonium isotopes in a nuclear reactor. The longest half-life americium isotope is ^{243}Am (7.34×10^3 years, compared to 470 years for ^{241}Am). Americium must be handled with great care to avoid personal contamination.

Curium was discovered by Seaborg, James and Ghiorso in 1944 after helium bombardment of ^{239}Pu. The most stable isotope is ^{247}Cm, with a half-life of 1.56×10^7 years.

These later elements in the series of actinides exist mainly in oxidation state +3 and so their biological effects are expected to be similar. Naturally occurring americium and curium have never been detected. They form trivalent ions in acidic solution. These elements are important components of waste repositories.

Both elements show fluorescence properties. Americium can be excited at about 508 nm, whereas Cm mostly is excited at 395 nm. They have very different fluorescence lifetimes in aqueous media: Am^{3+} about 17 ns; Cm^{3+} about 65 μs. Curium has an extraordinarily high fluorescence emission yield, which enables studies at extremely low concentrations, down to 10^{-11} M. Using Horrock's equation, which establishes a relationship between the fluorescence lifetime and the hydration number of these elements, it can be concluded to coordination changes by complex formation.

Very little information on the speciation of americium in biological systems is available, although a summary of information on soil and uptake by plants is provided in ref. 95.

A TRLFS study estimated the stability of complexes of curium with ATP at trace concentrations of 3×10^{-7} M curium.[101] Three species could be detected. Emission maxima at 598.6, 600.3 and 601 nm were assigned to $[Cm^{III}(H_2ATP)]^+$, $Cm^{III}HATP$ and $[Cm^{III}(ATP)]^-$, respectively, with fluorescence lifetimes of 88 μs, 96 μs and 187 ± 7 μs. The derived stability constants were log $\beta_{121} = 16.86 \pm 0.09$, log $\beta_{111} = 13.23 \pm 0.10$ and log $\beta_{101} = 8.19 \pm 0.16$. The significantly higher lifetime for $[Cm^{III}(ATP)]^-$ was interpreted in terms of a possible ring structure that might be formed as the hydration number decreases, suggested by an increase of the fluorescence lifetime. Cm(III) forms the most stable ATP complexes while those of neptunium(V) are the weakest.

Studies with the Nd(III)- and Sm(III)-transferrin complexes allowed the formation constants of Am(III) and Cm(III) to be estimated as log $K_1 = 6.3 \pm 0.7$ and log $K_1 = 6.5 \pm 0.8$, respectively.[12]

Many studies have examined the different methods of radionuclide determination. A comparison of direct and radiochemical analysis has been carried out.[102] In the case of [241]Am, the agreement between available measurements for the liver are poorer than the data for skeleton and lung. Nevertheless, it is apparent that the highest amounts of [241]Am were found in the skeleton. It was concluded that further series of measurements are necessary for better adjustment and modelling.

As mentioned above, curium shows one of the highest fluorescence efficiencies of the f-block elements and this property can be used to study the interaction of curium(III) species with bacteria. Interaction with *Bacillus subtilis* and *Halobacterium salinarum* shows distribution values K_d from 1000 to 10 000 for the pH range 3.0–5.0.[103] These data are comparable to those obtained for neptunium(V) and *Desulfovibrio desulfuricans*.[84] Changes in the number of ligand water molecules were derived. In the cases of *Bacillus subtilis* and *Halobacterium salinarum*, a respective decrease and increase of the number of bound water molecules were detected with increasing pH. The authors concluded that surface adsorption occurs for *Halobacterium salinarum* but that Cm^{3+} interacts more strongly with the exterior of *Bacillus subtilis*.

The interaction of *Pseudomonas fluorescens* from the Äspo site in Sweden with curium has been studied.[104] The bacteria interacted with curium(III) (0.3 μM) in batch experiments and the resultant curium species were interrogated by TRLFS after extraction. It was concluded that reversible biosorption occurs. Two different Cm-complexes were detected binding to the cell wall. The first one was a phosphate complex (emission max, 599.6 nm) formulated as $[R-O-PO_3H-Cm]^{2+}$. The second was a carboxylate complex (emission max, 601.9 nm) formulated as $[R-COO-Cm]^{2+}$. The surface complexation constants were determined to be log $\beta_{111} = 12.7 \pm 0.6$ and log $\beta_{110} = 6.1 \pm 0.5$, respectively.

The biokinetics of curium and americium were compared in a review in 2008.[105] Those of [242]Cm and [248]Cm appear to be very similar to those for [241]Am. The liver and skeleton are the main organs of deposition in humans and animals. It is reported that Am and Cm are transported very rapidly in the blood, much faster than plutonium. Blood transports the actinide species to the organs of retention, the liver and skeleton. Retention in the liver follows approximately bi-exponential functions. Retention of curium in the skeleton is very long for all species (half-times of retention are reported to be

~1600 days for rats and years for dogs). Both elements appear to form hot areas in the skeleton. However, the binding mechanisms are still not fully understood.

An *in vitro* study of the speciation of curium in urine has been performed.[106] Due to the similarities of the chemistries of curium and europium, the latter lanthanide element has often been used to simulate Cm in biosystems. The very high fluorescence yields of curium and europium allowed studies at extremely low concentrations. The authors added both metal ions to both human urine and a model urine and compared the resultant spectra with those of organic and inorganic components usually found in urine samples. pH was found to be an important parameter. In slightly acidic samples, Cm and Eu existed mainly as citrate complexes. In contrast, in near neutral samples, the speciation was dominated by phosphate- and calcium-containing complexes with citrate complexes also present. Due to the lack of thermodynamic data, especially at higher pH-values, there are some discrepancies between speciation modelling and TRLFS measurements.

This lack of adequate modelling encouraged study of the speciation of curium in biological systems in more detail.[107] TRLFS and FT-IR spectroscopy (for europium) allowed resolution of the following citrate complexes: $Cm^{III}(HCit)$; $[Cm^{III}(H_2Cit)(HCit)]^{2-}$; $[Cm^{III}(HCit)_2]^{3-}$ and $[Cm^{III}(Cit)_2]^{5-}$. The respective stability constants were estimated to be log $\beta_{111} = 21.0 \pm 0.2$, log $\beta_{1(21)(11)} = 43.8 \pm 0.3$, log $\beta_{122} = 38.4 \pm 0.7$. An estimate of log β_{102} could not be determined. In addition indications of the formation of $[Cm^{III}(Cit)]^-$ were seen. Structural parameters for both $Eu^{III}(HCit)$ and $[Eu^{III}(Cit)]^-$ were found by FT-IR spectroscopy and compared to DFT calculations. There is evidence for the binding of the metal to a hydroxylato group in $[Eu^{III}(Cit)]^-$. The authors concluded that in human urine with pH values < 6, curium is present mainly as $[Cm^{III}(H_2Cit)(HCit)]^{2-}$ while europium is present as a mixture of $[Eu^{III}(H_2Cit)(HCit)]^{2-}$ and $EuHCit^0$.

26.9 Higher Actinide Elements (Berkelium)

Thompson, Ghiorso and Seaborg discovered berkelium in 1949. It was synthesized in the cyclotron by bombardment of ^{241}Am with helium ions. ^{247}Bk has the longest half-life of 1400 years. Berkelium should be soluble in dilute mineral acids. It has been found in oxidation states 3+ and 4+. Generation of the latter requires strong oxidizing agents.

No studies of berkelium in biological systems have been reported and none on the higher actinides have been published in the last decade. Californium has been used as a neutron source to irradiate blood lymphocytes.

26.10 Outlook

The behaviour of actinide elements in biological systems is mainly studied by describing transport and storage of the ions. Nevertheless, it can be expected that, in the future, identification of speciation in these processes will play a more important role. The very interesting possibility of visualizing metal ions within living cells has been reviewed.[108] The use of fluorescent sensors in microscopy and spatially resolved mass spectrometric methods has the potential to detect the localization of metal ions within cells and their compartments. Up to now this approach has been applied only to the following metal ions: Zn^{2+}, Cu^+/Cu^{2+}, Fe^{2+}/Fe^{3+}, Mg^{2+}, Ni^{2+}, Cd^{2+}, Pb^{2+} and Hg^{2+}. It is expected that, in the near future, radioactive heavy metals such as the actinides will be detected inside living cells by the use of fluorescent labels.

The synthesis of 5-isothiocyanato-1,10-phenanthroline-2,9-dicarboxylic acid (DCP) and its complex with uranyl ions has been reported.[109] The complex interacts with the polyclonal antibodies 8A11, 10A3 and 12F6 when it is coupled to a carrier protein for injection into mice. Dissociation constants for the respective UO_2(DCP)-antibody complexes were estimated to be 5.5, 2.4 and 0.9 nM. Metal-free DCP is bound to the antibodies with lower affinities (by about three orders of magnitude). Comparison with other metal-DCP complexes showed that these three antibodies allow very sensitive detection of UO_2(DCP). It was concluded that such specific uranium-binding capabilities could be pivotal to monitor and control uranium contamination, supporting the concept of developing immunoassays for UO_2^{2+}.

An additional tool for the localization and characterization of processes involving metals has been highlighted.[110] The authors suggest the use of synchrotron techniques for the determination of metals inside plant cells. XAS-based methods are element-specific as well as sensitive to the environment of the metal ion and are expected to play an important role in the study of the interactions of metal ions with living cells in the future.

In addition, exploration of processes of incorporation of actinide ions into cells, their forms of storage and the excretion processes will all take centre stage in the future.

References

1. P. W. Durbin, in *The Chemistry of Actinide and Transactinide Elements*, ed. L. R. Morss, N. M. Edelstein and J. Fuger, Springer, New York, 3rd edn, vol. 5 , 2006, pp. 3339–3440.
2. R. N. Collins, T. Saito, N. Aoyagi, T. E. Payne, T. Kimura and T. D. Waite, *J. Env. Qual.*, 2011, **50**, 731–741.
3. G. Pfennig, H. Klewe-Nebenius and W. Seelmann-Eggebert, *Karlsruher Nuklidkarte*, Forschungszentrum Karlsruhe, 1998.
4. www.nndc.bnl.gov/nudat2/.
5. P. Scoppa, *Inorg. Chim. Acta*, 1984, **95**, 23–27.
6. J. Burger, M. Gochfeld, C. Jeitner, M. Gray, T. Shukla, S. Shukla and S. Burke, *Environ. Monit. Assess.*, 2007, **128**, 329–341.
7. S. L. Maxwell, *J. Radioanal. Nucl. Chem.*, 2008, **275**, 497–502.
8. S. L. Maxwell and V. D. Jones, *Talanta*, 2009, **80**, 143–150.

9. X. Dai and S. Kramer-Tremblay, *Health Physics*, 2011, **101**, 144–147.

10. R. Kumar, J. R. Yadav, D. D. Rao and L. Chand, *J. Radioanal. Nucl. Chem.*, 2010, **283**, 785–788.

11. Strahlenschutzverordnung, *Verordnung über den Schutz vor Schäden durch ionisierende Strahlen, vom 20 Juli 2001* (BGBl. I S. 1714), Carl Heymanns Verlag KG Köln, Berlin, Bonn, München, 2001.

12. D. M. Taylor, *J. Alloys Compounds*, 1998, **271–273**, 6–10.

13. W. R. Harris, C. J. Carrano, V. L. Pecoraro and K. N. Raymond, *J. Am. Chem. Soc.*, 1981, **103**, 2231–2237.

14. A. Le Du, A. Sabatie-Gogova, A. Morgenstern and G. Montavon, *J. Inorg. Biochem.*, 2012, **109**, 82–89.

15. L. M. Correa, D. Kochhann, A. G. Becker, M. A. Pavanato, S. F. Llesuy, V. L. Loro, A. Raabe, M. F. Mesko, E. M. M. Flores, V. L. Dressler and B. Baldisserotto, *Aquat. Toxicol.*, 2008, **88**, 250–256.

16. D. Kochhann, M. A. Pavanato, S. F. Llesuy, L. M. Correa, A. P. K. Riffel, V. L. Loro, M. F. Mesko, E. M. M. Flores, V. L. Dressler and B. Baldisserotto, *Chemosphere*, 2009, 77, 384–391.

17. R. K. Bull, T. J. Smith and A. W. Phipps, *Rad. Prot. Dos.*, 2006, **121**, 425–428.

18. A. Kumar, M. Ali, B. N. Pandey, P. A. Hassan and K. P. Mishra, *Biochimie*, 2010, **92**, 869–879.

19. C. M. Marquardt, P. J. Panak, C. Apostolidis, A. Morgenstern, C. Walther, R. Klenze and Th. Fanghänel, *Radiochim. Acta*, 2004, **92**, 445–447.

20. J. G. Hamilton, *Medical and Health Physics Quarterly Report UCRL-157*, University of California Radiation Laboratory, 1948, pp. 6–11.

21. K. H. Lieser, *Einführung in die Kernchemie*, VCH Weinheim, 3rd edn, 1991, pp. 107–109.

22. H. J. Rössler, *Lehrbuch der Mineralogie*, Deutscher Verlag für Grundstoffindustrie, Leipzig. 5th edn, 1991, p. 184.

23. D. Ribera, F. Labrot, G. Tisnerat and J. F. Narbonne, *Rev. Environ. Contam. Toxicol.*, 1996, **146**, 53–89.

24. G. Geipel, G. Bernhard, V. Brendler and T. Reich, in *NRC5, 5th International Conference on Nuclear and Radiochemistry*, Extended Abstracts, Pontresina, Switzerland, 2000, pp. 473–476.

25. E. N. Rizkalla, F. Netoux, S. Dabosseignon and M. Pages, *J. Inorg. Biochem.*, 1993, **51**, 701–703.

26. P. Panak, B. C. Hard, K. Pietzsch, S. Kutschke, K. Röske, S. Selenska-Pobell, G. Bernhard and H. Nitsche, *J. Alloys Compounds*, 1998, **271–273**, 262–265.

27. C. Hennig, P. J. Panak, T. Reich, A. Rossberg, J. Raff, S. Selenska-Pobell, W. Matz, J. J. Bucher, G. Bernhard and H. Nitsche, *Radiochim. Acta*, 2001, **89**, 625–631.

28. H. A. Thompson, G. E. Brown and G. A. Parks, *Amer. Mineralogist*, 1997, **82**, 483–489.

29. P. J. Panak, R. Knopp, C. H. Booth and H. Nitsche, *Radiochim. Acta*, 2002, **90**, 779–785.

30. A. C. Francis, G. A. Joshi-Tope, C. J. Dodge and J. B. Gillow, *J. Nucl. Sci. Techn. Suppl.*, 2002, **3**, 935–942.
31. G. A. Joshi-Tope and A. J. Francis, *J. Bacteriol.*, 1995, **177**, 1989–1993.
32. M. Merroun, C. Hennig, A. Rossberg, T. Reich, R. Nicolai, K. H. Heise and S. Selenska-Pobell, in *Uranium in the Aquatic Environment*, ed. B. J. Merkel, Springer Verlag Berlin, Heidelberg, New York, 2002, pp. 505–512.
33. M. Merroun, C. Hennig, A. Rossberg, G. Geipel, T. Reich and S. Selenska-Pobell, *Biochem. Soc. Trans.*, 2002, **30**, 669–672.
34. A. Geissler, M. Merroun, G. Geipel, H. Reuther and S. Selenska-Pobell, *Geobiology*, 2009, **7**, 282–294.
35. J. R. Bargar, K. H. Williams, K. M. Campbell, P. E. Long, J. E. Stubbs, E. I. Suvorova, J. S. Lezama-Pacheco, D. S. Alessi, M. Stylo, S. M. Webb, J. A. Davis, D. E. Giammar, L. Y. Blue and R. Bernier-Latmani, *Proc. Natl Acad. Sci. USA*, 2013, **110**, 4506–4511.
36. T. Sousa, A. P. Chung, A. Pereira, A. P. Piedade and P. V. Morais, *Metallomics*, 2013, **5**, 390–397.
37. M. Nedelkova, M. L. Merroun, A. Rossberg, C. Hennig and S. Selenska-Pobell, *FEMS Microbiol. Ecol.*, 2007, **59**, 694–705.
38. A. Guenther, G. Bernhard, G. Geipel, T. Reich, A. Rossberg and H. Nitsche, *Radiochim. Acta*, 2003, **91**, 319–328.
39. J. J. Mortvedt, *J. Envron. Qual.*, 1994, **23**, 643–648.
40. L. S. Simon and S. A. Ibrahim, in *Radium in the Environment*, International Atomic Energy Agency, Vienna, Austria, 1988.
41. A. Kaur, S. Singh and H. S. Virk, *Nucl. Tracks Radiat. Meas.*, 1988, **15**, 795–799.
42. K. P. Singh, *Curr. Sci.*, 1997, **73**, 532–537.
43. O. Frindik, *Landwirtschaftliche Forschung*, 1986, **39**, 75–83.
44. K. Viehweger and G. Geipel, *Env. Exp. Botany*, 2010, **69**, 39–46.
45. S. Nalezinski and D. Lux, in *BfS Annual Report 1997*, 1998, pp. 37–42.
46. A. Ibrahim and F. W. Whicker, *J. Radioanal. Nucl. Chem. Articles*, 1991, **156**, 253–258.
47. G. Yaprak, N. F. Cam and G. Yener, *J. Radioanal. Nucl. Chem.*, 1998, **238**, 167–173.
48. J. W. Huang, M. J. Blaylock, Y. Kapulnik and B. D. Ensley, *Environ. Sci. Technol.*, 1998, **32**, 2004–2012.
49. T. K. Walker, V. Subramaniam and F. Challenger, *J. Chem. Soc.*, 1927, **200–208**, 3044–3054.
50. A. J. Francis, C. J. Dodge and J. B. Gollow, *Nature*, 1992, **356**, 140–142.
51. S. D. Ebbs, D. J. Brady and L. V. Kochian, *J. Exp. Botany*, 1998, **49**, 1183–1195.
52. K. Viehweger, G. Geipel and G. Bernhard, *Biometals*, 2011, **24**, 1197–1204.
53. L. D. Kraemer and D. Evans, *Aquat. Toxicol.*, 2012, **124–125**, 163–170.
54. J. M. Kelly and D. M. Janz, *Aquat. Toxicol.*, 2009, **92**, 240–249.

55. M. M. Goertzen, D. W. Hauck, J. Phibbs, L. P. Weber and D. M. Janz, *Ecotoxicol. Env. Saf.*, 2012, **75**, 142–150.

56. Y. Song, B. Salbu, L. S. Heier, H. C. Teien, O. C. Lind, D. Oughton, K. Petersen, B. O. Rosseland, L. Skipperud and K. E. Tollefsen, *Aquat. Toxicol.*, 2012, **112**, 62–71.

57. A. Lerebours, P. Gonzalez, C. Adam, V. Camilleri, J. P. Bourdineaud and J. Garnier-Laplace, *Env. Toxicol. Chem.*, 2009, **28**, 1271–1278.

58. K. Yapar, K. Cavusoglu, E. Oruc and E. Yalcin, *J. Med. Food*, 2010, **13**, 179–188.

59. Y. Hao, J. Ren, J. Liu, S. Luo, T. Ma, R. Li and Y. Su, *Basic Clin. Pharmacol. Toxicol.*, 2012, **111**, 402–410.

60. J. Michon, S. Frelon, C. Garnier and F. Coppin, *J. Fluoresc.*, 2010, **20**, 581–590.

61. V. Malard, J. C. Gaillard, F. Berenguer, N. Sage and E. Quemeneur, *Biochim. Biophys. Acta*, 2009, **1794**, 882–892.

62. F. Shaki, M. J. Hosseini, M. Ghazi-Khansari and J. Pourahmad, *Biochim. Biophys. Acta*, 2012, **1820**, 1940–1950.

63. S. Priyamvada, S. A. Khan, M. W. Khan, S. Khan, N. Farooq, F. Khan and A. N. K. Yusufi, *Prostaglandins Leukotrienes Essent. Fatty Acids*, 2010, **82**, 35–44.

64. E. Grignard, Y. Gueguen, S. Grison, J. M. Lobaccaro, P. Gourmelon and M. Souidi, *Int. J. Toxicol.*, 2008, **27**, 323–328.

65. B. Skwarcez, A. Boryło, M. Prucnal and D. I. Strumińska-Parulska, *Pol. J. Environ. Stud.*, 2010, **19**, 771–778.

66. S. M. Marques, S. Chaves, F. Gonçalves and R. Pereira, *Ecotoxicol. Environ. Saf.*, 2013, **87**, 115–119.

67. I. Malátová, V. Bečková, L. Tomášek and J. Hůlka, *Rad. Prot. Dos.*, 2011, **147**, 593–599.

68. S. D'Illio, N. Violante, O. Senofonte, C. Majorani and F. Petrucci, *Anal. Methods*, 2010, **2**, 1184–1190.

69. K. S. Squibb, J. M. Gaitens, S. Engelhardt, J. A. Centeno, H. Xu, P. Gray and M. A. McDiarmid, *J. Occup. Environ. Med.*, 2012, **54**, 724–732.

70. M. L. Zamora, J. M. Zielinski, G. B. Moodie, R. A. Falcomer, W. C. Hunt and K. Capello, *Arch. Environ. Occup. Health*, 2009, **64**, 228–241.

71. T. I. Todorov, J. W. Ehnik, G. Guandalini, H. Xu, D. Hoover, L. Anderson, K. Squibb, M. A. McDiarmid and J. A. Centeno, *J. Trace Elem. Med. Biol.*, 2013, **27**, 2–6.

72. O. Prat, E. Ansoborlo, N. Sage, D. Cavadore, J. Lecoix, P. Kurttio and E. Quemeneur, *Environ. Int.*, 2011, **37**, 657–662.

73. P. Decambox, P. Mauchien and C. Moulin, *Appl. Spectrosc.*, 1991, **45**, 116–121.

74. C. Vidaud, D. Bourgeois and D. Meyer, *Chem. Res. Toxicol.*, 2012, **25**, 1161–1175.

75. W. Schimmack, U. Gerstmann, W. Schultz and G. Geipel, *Radiat. Environ. Biophys.*, 2007, **46**, 221–227.

76. L. Bonin, C. Den Auwer, E. Ansoborlo, G. Cote and P. Moisy, *Radiochim. Acta*, 2007, **95**, 371–379.
77. D. Nebel, *Z. Phys. Chem.*, 1966, **232**, 161–166.
78. D. Raymond, J. R. Dufflied and D. R. Williams, *Inorg. Chim. Acta*, 1987, **140**, 309–314.
79. A. Jeanson, M. Ferrand, H. Funke, C. Hennig, P. Moisy, P. L. Solari, C. Vidaud and C. Den Auwer, *Chemistry*, 2010, **16**, 1378–1387.
80. I. Llorens, C. Den Auwer, Ph. Moisy, E. Ansoborlo, C. Vidaud and H. Funke, *FEBS J.*, 2005, **272**, 1739–1744.
81. G. J.-P. Deblonde, M. Sturzbecher-Hoehne, A. B. Mason and R. J. Abergel, *Metallomics*, 2013, **5**, 619–625.
82. M. Sturzbecher-Hoehne, C. Goujon, G. J. P. Deblonde, A. B. Mason and R. J. Abergel, *J. Am. Chem. Soc.*, 2013, **135**, 2676–2683.
83. M. P. Jensen, D. Gorman-Lewis, B. Aryal, T. Paunesku, S. Vogt, P. G. Rickert, S. Seifert, B. Lai, G. E. Woloschak and L. Soderholm, *Nature Chem. Biol.*, 2011, **7**, 560–565.
84. T. Kobuta, T. Sasaki, O. Tochiyama and A. Kudo, *J. Nucl. Sci. Technol. Suppl.*, 2002, **3**, 946–952.
85. R. W. Perkins and C. W. Thomas, in *Transuranic Elements in the Environment*, ed. W. C. Hansen, Technical Information Center, US Department of Energy, 1980, p. 53.
86. A. B. Kersting, *Inorg. Chem.*, 2013, **52**, 3533–3546.
87. J. B. Gillow, M. Dunn, A. J. Francis, D. A. Lucero and H. W. Papenguth, *Radiochim. Acta*, 2000, **88**, 769–774.
88. P. J. Panak and H. Nitsche, *Radiochim. Acta*, 2001, **89**, 499–503.
89. H. Moll, M. L. Merroun, C. Hennig, A. Rossberg, S. Selenska-Pobell and G. Bernhard, *Radiochim. Acta*, 2006, **94**, 815–824.
90. J. Auer, Verhalten radioaktiver Nuklide im System Pflanze – Boden, in *ZALF Berichte*, 1993, pp. 33–45.
91. K. Bunzl and W. Kracke, *Sci. Total Environ.*, 1987, **63**, 111–118.
92. K. Bunzl and W. Kracke, *J. Radioanal. Nucl. Chem.*, 1989, **138**, 83–88.
93. E. Haunold, O. Horak and M. Gerzabek, *Bodenkultur*, 1987, **38**, 95–99.
94. M. Pimpl, Untersuchungen zum Boden/Pflanzen Transfer von Np-237, Pu-238, Am-241 und Cm-244; KfK-4452, Kernforschungszentrum Karlsruhe, 1998.
95. P. J. Coughtrey, D. Jackson and M. C. Thorne, *Radionuclide Distribution and Transport in Terrestrial and Aquatic Ecosystems. A Critical Review of Data*, Balkema, Rotterdam and Boston, vol. 5 , 1984.
96. B. P. Aryal, D. Gorman-Lewis, T. Paunesku, R. E. Wilson, B. Lai, S. Vogt, G. E. Woloschak and M. P. Jensen, *Int. J. Rad. Biol.*, 2011, **87**, 1023–1032.
97. D. Gorman-Lewis, B. P. Aryal, T. Paunesku, S. Vogt, B. Lai, G. E. Woloschak and M. P. Jensen, *Inorg. Chem.*, 2011, **50**, 7591–7597.
98. X. Dai, M. Christl, S. Kramer-Tremblay and H. A. Synal, *J. Anal. Atom. Spectrom.*, 2012, **27**, 126–130.
99. J. B. Vincent and S. Love, *Biochim. Biophys. Acta*, 2012, **182**, 372–378.

100. A. L. Serandur, O. Gremy, M. Frechou, D. Renault, J. L. Poncy and P. Fritsch, *Radiat. Res.*, 2008, **170**, 208–215.
101. H. Moll, G. Geipel and G. Bernhard, *Inorg. Chim. Acta*, 2005, **358**, 2275–2282.
102. T. P. Lynch, S. Y. Tolmachev and A. C. James, *Rad. Prot. Dos.*, 2009, **134**, 94–101.
103. T. Ozaki, J. B. Gillow, A. J. Francis, T. Kimura, T. Ohnuki and Z. Yoshida, *J. Nucl. Sci. Technol. Suppl.*, 2002, **3**, 950–954.
104. H. Moll, L. Luetke, A. Barkleit and G. Bernhard, *Geomicrobiol. J.*, 2013, **30**, 337–346.
105. F. Menetrier, D. M. Taylor and A. Comte, *Appl. Radiat. Isot.*, 2008, **66**, 632–647.
106. A. Heller, A. Barkleit and G. Bernhard, *Chem. Res. Toxicol.*, 2011, **24**, 193–203.
107. A. Heller, A. Barkleit, H. Foerstendorf, S. Tsushima, K. Heim and G. Bernhard, *Dalton Trans.*, 2012, **41**, 13969–13983.
108. K. M. Dean, Y. Qin and A. E. Palmer, *Biochim. Biophys. Acta*, 2012, **1823**, 1406–1415.
109. R. C. Blake, A. R. Pavlov, M. Khosraviani, H. E. Ensley, G. E. Kiefer, H. Yu, X. Li and D. A. Blake, *Bioconjugate Chem.*, 2004, **15**, 1125–1136.
110. E. Donner, T. Punshon, M. L. Guerinot and E. Lombi, *Anal. Bioanal. Chem.*, 2012, **402**, 3287–3298.

CHAPTER 27

Aluminium

CHRISTOPHER EXLEY

The Birchall Centre, Lennard-Jones Laboratories, Keele University, UK
Email: c.exley@keele.ac.uk

27.1 Introduction

The biological reactivity of aluminium is primarily the bioinorganic chemistry of its free solvated trivalent cation $Al^{3+}_{(aq)}$.[1,2] There are possible exceptions to this where complexes of $Al^{3+}_{(aq)}$, as opposed to the free aqua cation, have biological reactivity. An example might be its complex with ATP and perhaps other nucleotides where $Al^{3+}_{(aq)}$ replaces $Mg^{2+}_{(aq)}$.[3] However, in the main we have to consider the reactions of $Al^{3+}_{(aq)}$ with oxygen donor ligands (and occasionally fluoride) associated with myriad biomolecules. It is the major antagonist to $Mg^{2+}_{(aq)}$ and an effective, occasional competitor for $Ca^{2+}_{(aq)}$ and $Fe^{2+}_{(aq)}/Fe^{3+}_{(aq)}$.[†]

The coordination chemistry of aluminium in relation to biology has been extensively reviewed[4,5] and I will not repeat these valuable data herein. While the reactivity of aluminium supports a wide breadth of potential biochemical reactions, it can also be a double-edged sword in that it often forms strong bonds that are extremely slow to dissociate, meaning that aluminium can be locked away in complexes, polymers and precipitates. This goes some way to explain its propensity to accumulate in structures such as bone[6] and compartments such as cell nuclei.[7] It also explains the predominantly inhibitory action of aluminium in biochemistry though aluminium can also have a stimulatory effect and has been described as biphasic in its actions on

---

[†]The biological chemistries of Mg, Ca and Fe are outlined in Chapters 4, 5, 10 and 11.

RSC Metallobiology Series No. 2
Binding, Transport and Storage of Metal Ions in Biological Cells
Edited by Wolfgang Maret and Anthony Wedd
© The Royal Society of Chemistry 2014
Published by the Royal Society of Chemistry, www.rsc.org

biological systems.[8] It is of importance to emphasize that the bioinorganic chemistry of aluminium, and predominantly $Al^{3+}_{(aq)}$, does not preclude its involvement in essential biochemistry, it is simply its exclusion from biochemical evolution that has created the paradox of its ubiquity and yet non-essentiality.[9] During the past 100 years or so, human activities have interfered substantially in the biogeochemical cycle of aluminium and biota are now experiencing a burgeoning exposure to biologically available aluminium and over the next millennia it is highly likely that essential aluminium biochemistry will emerge.

27.2 The Transport of Aluminium from Extracellular to Intracellular Environments

How might aluminium, as the free cation $Al^{3+}_{(aq)}$ or as a charged or neutral complex, be transported across cell membranes? This question has been addressed in Figure 27.1, where both the generalized forms of

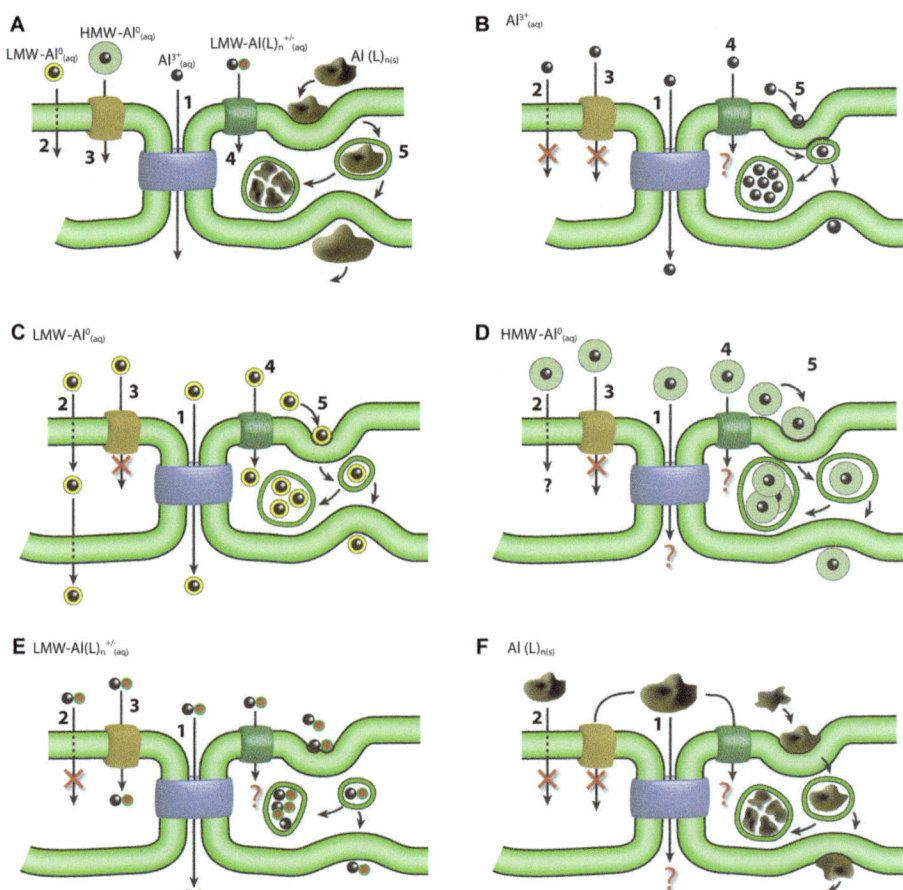

"transportable" aluminium and their potential transport routes have been summarized. The evidence for intracellular aluminium is incontrovertible while that which supports mechanisms for its cellular internalization is equivocal at best. We know that aluminium crosses cell and organelle membranes but we know very little as to how its transmembrane transport is achieved. The functional integrity of the great majority of biological lipid-based membranes is optimal under conditions of circumneutral pH. Such "physiological" conditions are also optimal for binding aluminium and the formation therewith of stable linkages or complexes[1,3,9] and so it is unlikely that any aluminium, as $Al^{3+}_{(aq)}$, is transported transcellularly (*i.e.* across individual cells), neither actively nor passively (Figure 27.1). There are no precedents across all biota for the active or passive transcellular transport of a trivalent metal cation and there are no reasons to expect that $Al^{3+}_{(aq)}$ would be any different. While there is a burgeoning number of transmembrane transporters of mono- and divalent metal cations, none of these have been shown to transport $Al^{3+}_{(aq)}$ successfully and more often as not aluminium has actually been identified as an inhibitor of such transport processes.

There is no unequivocal evidence of any active transcellular transport of $Al^{3+}_{(aq)}$ under circumneutral pH and recent publications purporting such[10,11] are critically flawed in both their experimental methods and their interpretation of data. For example, there are two very powerful reasons why

---

Figure 27.1 The transport of aluminium across biological membranes. (A) There are five major routes by which aluminium could be transported across cell membranes or cell epi-/endothelia: 1. paracellular; 2. transcellular; 3. active transport; 4. channels; 5. adsorptive or receptor-mediated endocytosis. There are five major classes of forms of aluminium that could participate in these transport routes: (i) free solvated trivalent cation ($Al^{3+}_{(aq)}$); (ii) low-molar-mass, neutral, soluble complexes ($LMW-Al^0_{(aq)}$); (iii) high-molar-mass, neutral, soluble complexes ($HMW-Al^0_{(aq)}$); (iv) low-molar-mass, charged, soluble complexes ($LMW-Al(L)_n^{+/-}{}_{(aq)}$); (v) nano- and microparticulates ($Al (L)_{n(s)}$). (B) $Al^{3+}_{(aq)}$ is predicted to move across epi-/endothelia by a paracellular route but its movement into cells is restricted to ion channels under exceptional circumstances and plausibly some degree of receptor-mediated endocytosis of the bound cation. (C) $LMW-Al^0_{(aq)}$ complexes will move across epi-/endothelia by a paracellular route and into cells *via* transcellular diffusion both across membranes and through channels as well as by endocytosis. (D) $HMW-Al^0_{(aq)}$ complexes are probably too large to follow a paracellular route between cells while their transport into cells probably only occurs to any significant degree by an endocytotic mechanism. (E) $LMW-Al(L)_n^{+/-}{}_{(aq)}$ complexes are predicted to pass between cells using a paracellular route and into cells by active transport and receptor-mediated endocytosis. (F) $Al(L)_{n(s)}$ nano- and microparticulates probably only utilize adsorption-mediated endocytosis to enter the cytoplasm of epi-/endothelial cells. All forms of aluminium that enter the cell cytosol by endocytosis may either accumulate in double-membrane vesicles (lysosomes/endosomes) in the cytosol or they may be expelled from the cell by a process of exocytosis.

the data presented in ref. 10 did not identify an aluminium-specific transporter in rice. The first is that they did not present any conclusive evidence of aluminium transport across the cell membrane and its subsequent presence in the cytosol or cytosolic compartments. All of the experimental data that purported to show this were based upon analyses of total aluminium in cell and membrane fractions. It is notoriously difficult to use such methods to confirm the cellular uptake of aluminium. Even when every precaution is taken to ensure that aluminium associated with external surfaces such as membranes is excluded from so-called intracellular compartments, it is extremely difficult to demonstrate statistically significant differences in aluminium concentrations between the various fractions. These are significant methodological issues that need to be identified and corrections demonstrated before such data could be viewed as unequivocal. However, the intracellular accumulation of aluminium could have been demonstrated using light and fluorescence microscopy and even semi-quantitatively using electron microscopy. None of these methods were used in this paper to demonstrate the passage of aluminium across membranes by the putative aluminium-specific transporter. A second major flaw in the research concerned the possible specificity of the transporter for aluminium. Actually no attempt was made in this paper to establish if the transporter also transports other trivalent metals such as, for example, $Fe^{3+}_{(aq)}$. Clearly this was a major oversight as if Nrat1, the putative aluminium-specific transporter, was actually involved in the transport of the metabolically essential Fe(III) then it would also be likely to transport Al(III), albeit probably much less efficiently. If aluminium-specific transporters do exist in some extant organism, then a much more rigorous methodology will be required to provide unequivocal proof of such.

There are circumstances where the passive transport of $Al^{3+}_{(aq)}$ might be predicted. Under acidic conditions, for example, pH ≤ 4.00, the proton $H^{+}_{(aq)}$ begins to compete effectively with $Al^{3+}_{(aq)}$ and, through preventing aluminium from being bound by ligands associated with potential membrane carriers, low-pH milieu may promote the transcellular passage of $Al^{3+}_{(aq)}$ *via* $H^{+}_{(aq)}$-saturated cation channels. Thus low pH and effective competition from protons could protect membrane-associated functional groups from binding $Al^{3+}_{(aq)}$ and allow the cation free passage down a concentration gradient (Figure 27.1). While such a mechanism is currently speculative, there is evidence from research in yeast of intracellular accumulation of aluminium in acidic (pH ≤ 4.00) milieu.[12,13] There have been numerous suggestions that aluminium might "piggy-back" upon other molecules in order to cross membranes through diffusion, ion symports/antiports and active transport; for example, aluminium bound by amino acids or other common extracellular ligands such as citrate.[14–16] In this way, aluminium might masquerade as divalent, monovalent or even anionic species, each with the potential, at least, to be transported by the requisite transport system in the membrane (Figure 27.1C,D,E). However, once again, the direct evidence that any of these mechanisms successfully transport aluminium across cell membranes is equivocal and not unexpectedly so when one considers how binding $Al^{3+}_{(aq)}$, as opposed to other more usual interactions with essential metal ions, is likely to significantly alter the biophysical and biochemical properties of the putative transport vehicle. There is evidence that where aluminium is bound into simple, neutral, low-molar-mass and lipophilic complexes, it is transported across cell membranes by simple diffusion, albeit at very slow rates (Figure 27.1C,D).[17,18]

A consistent feature of the intracellular aluminium load is its presence in lysosomes.[19,20] While the observation of such aluminium-replete vesicles could be explained by the packaging of intracellular aluminium, it could equally implicate receptor- or adsorptive-mediated endocytosis (Figure 27.1) as a mechanism by which extracellular aluminium is transported across the cell membrane.[21,22] Transport *via* this route might be as a direct response to the action of aluminium at the extracellular surface; for example, adsorbed aluminium hydroxide or binding of $Al^{3+}_{(aq)}$ by a membrane protein or receptor, or it might be completely coincidental with aluminium being endocytosed as part of the extracellular medium as an endosome is formed (Figure 27.1). Additionally it has been proposed that aluminium crosses lysosomal membranes as electroneutral hydroxide complexes[23] and a similar mechanism, possibly involving aquaporin-like channels, has also been demonstrated for the transport of aluminium across tonoplast and cell membranes.[24,25] Receptor-mediated endocytosis is implicated in the cellular internalization of aluminium as a transferrin-aluminium complex.[26,27] However, the significance of this route as a mechanism of uptake of aluminium is complicated by evidence that not all transferrin-aluminium complexes (*e.g.* TfAl and $TfAl_2$) are bound by all transferrin receptors.[27,28] In addition, any aluminium-transferrin-receptor complex that might be endocytosed is unlikely to be metabolized further due to the distinct differences in redox chemistry between iron (III) and aluminium (III).[29] Aluminium certainly interferes with both transferrin-dependent and -independent cellular uptake of iron but whether this also results in the significant internalization of aluminium remains to be fully elucidated.[30]

27.3 The Systemic Uptake of Aluminium

Aluminium is found in all body fluids including blood, cerebral spinal fluid, interstitial fluid of the brain, sweat, lymph and urine.[31] Its omnipresence in such systemic compartments demonstrates its propensity to traverse epi-/endothelia through either paracellular (*i.e.* between adjacent cells) or transcellular routes (Figure 27.1). There is little if, arguably, any evidence that the latter play any significant role in the systemic uptake of aluminium. Indeed, there is evidence, for example in gastric mucosa cells, that the uptake of aluminium into epi-/endothelial cells actually limits its further entry into the body.[32] It is not impossible that aluminium crosses epi-/endothelia *via* transcellular mechanisms; it is simply the case that such a mechanism remains to be elucidated for aluminium. All epi-/endothelia have a property defined as residual leakiness, which enables the passive movement of substances across their surfaces and primarily through spaces between adjacent cells. The permeability of these "gaps" is additionally influenced by the presence of a suite of proteins that together constitute what are known as tight junctions. However, even the most impermeable of tight junctions, such as might be found in the endothelial cell layer of the blood-brain barrier, are subject to residual leakiness and so aluminium, particularly if present as neutral complexes of low molar mass, inorganic and organic, will pass across such barriers by a passive mechanism. In addition, any extra-mural factor that directly or indirectly influences the permeability of tight junctions will subsequently also change their permeability to aluminium; for example, as has recently been demonstrated for fatty acids and aluminium in Caco-2 cell monolayers.[33] Aluminium *per se*, probably acting through $Al^{3+}_{(aq)}$ as a pro-oxidant,[34] has been shown to increase the permeability of the blood-brain barrier both to itself and to other substances.[35,36] Similar effects of aluminium on the barrier properties of membranes have been attributed to the structure and function of tight junctions.[37-39] A majority of evidence implicates inter- or paracellular transport as the major, or at least the most rapid, mechanism by which aluminium crosses membranes. Aluminium diffuses freely between cells as well as actively altering paracellular permeability through direct effects upon tight junction structure and function.

27.4   Aluminium Transport in Blood

The form or chemical species of aluminium that enters the blood will significantly influence its subsequent transport and fate.[40] Whole blood is an extremely dynamic medium in terms of both its physical flow or movement throughout the body and the heterogeneity of its composition. The latter will have significant implications for the transport and fate of aluminium and, in the main, has not received adequate research effort. The field has essentially been side-tracked by research on isolated serum (not blood) samples in which it is invariably shown using various fractionation techniques that the majority of aluminium is found in a complex with the iron transport protein transferrin.[41] At chemical equilibrium this is almost certain to be true as transferrin is a major ligand in serum, approximately 30 μM of available binding sites at any one time, and transferrin binds aluminium with some avidity.[42] However, when is blood ever at or even close to chemical equilibrium? At any one moment in time, the form and fate of aluminium in whole blood will be determined by a raft of competitive equilibria with each of these being, in the main, far from equilibrium.[43] Some of the contributors to such equilibria will actually be compartments such as cells, others will be surfaces including membranes, while many will be "soluble" complexes including peptides, proteins, amino acids, organic acids, nucleic acids and inorganic ligands including hydroxide, phosphate, fluoride and silicic acid. The role played by any one contributor at any one moment of time will be influenced by its propensity to bind $Al^{3+}_{(aq)}$, the rate at which this reaction takes place and its concentration relative to other competitive ligands. It is surmised that transferrin is not the major transport ligand for aluminium in blood as its association and dissociation kinetics with $Al^{3+}_{(aq)}$ are not consistent with the relatively rapid excretion of aluminium in the urine *via* the glomerulus of the kidney.[40] It is probable that the rate of excretion can only be explained by aluminium being transported by paracellular routes across glomerular endothelia as neutral low-molecular-mass complexes (Figure 27.1C). There is scant appreciation of the role of non-equilibrium dynamics in the binding and transport of metals and very few investigators have looked to apply such an approach to the understanding of the behaviour of aluminium in blood.[40,43,44] These concepts are not easily accessed and tested through bench-top experiments and a computational model has been used to demonstrate that the approach to equilibrium, whereby aluminium is bound by transferrin, can proceed through significant and biologically relevant intermediate steps in which aluminium is shuttled between low-molar-mass forms such as hydroxide, phosphate and citrate.[44] It is highly likely that these thermodynamically unstable steps are the rate- and route-determining steps in the biological fate of aluminium in blood. It is similarly possible that kinetic constraints also govern the lability and activity of aluminium in all fluid environments, extracellular and intracellular. Together they constitute a form of homeostasis with aluminium being retained both physically and chemically in myriad forms and each form being capable of acting as a sink or source of labile and potentially biologically reactive aluminium. It is always important to emphasize that there is no evolutionarily directed or conserved biology to enable aluminium homeostasis and so this non-essential but highly biologically reactive metal cation is at the whim of the predominant or pre-eminent chemistry of any particular environment.[8,9] This unpredictability makes biologically available aluminium a concern for all forms of life on Earth.[2]

27.5 Aluminium and Bacteria: a Special Case?

There is an assumption within the literature that siderophore-mediated uptake of iron by bacteria also applies to aluminium. There are certainly many examples of iron siderophores that form strong complexes with $Al^{3+}_{(aq)}$[45] and there are many data demonstrating the association of aluminium with bacteria or (centrifuged) pellets of bacteria.[46–49] Many of these studies also purport to show the compartmentalization of aluminium within different fractions relating to extra and intracellular locations.[49,50] However, the classical methods used for discriminating different cellular compartments in bacteria are fraught with problems when they are equally applied to the presence of aluminium in the same locations. Those who have recognized these problems and looked to remedy them have, as often as not, had to conclude that they were unable to confirm unequivocally the presence of aluminium in cytosol and cytosolic compartments. There are tantalizing data from transmission electron microscopy that bacterial inclusion bodies contain aluminium[50] but the details required to confirm such are largely missing.

27.6   Conclusions

There are myriad assumptions, some having achieved the status of dogma, concerning the binding, transport and storage of aluminium in biological systems. There are very few unequivocal mechanisms to describe such processes with the main reason for this being a paucity of appropriate research in the field. The status of a non-essential metal does not lend itself easily to funded research in a highly competitive funding environment while the physicochemical properties of aluminium often preclude its study by conventional means in physiological milieu. Human activities are ensuring that the burden of aluminium in the biotic cycle is burgeoning and this should increase the urgency with which we strive to understand the biological availability, biochemical reactivity and ecotoxicology of the third most abundant element of the Earth's crust. Human exposure to aluminium[51] may accelerate this need even further to enable research to be focused upon the as yet unanswered questions concerning the binding, transport and storage of aluminium in biological systems.

References

1. R. B. Martin, *Clin. Chem.*, 1986, **32**, 1797.
2. C. Exley, *J. Inorg. Biochem.*, 2003, **97**, 1.
3. C. Exley, *Curr. Inorg. Chem.*, 2012, **2**, 3.
4. R. J. P. Williams, *Coord. Chem. Rev.*, **149**, 1.
5. P. Zatta, *Coord. Chem. Rev.*, 2002, **228**, 91.
6. V. Parsons, C. Davies, C. Goode, C. Ogg and J. Siddiqui, *BMJ*, 1971, **4**, 273.
7. H. Matsumoto, S. Morimura and E. Takahashi, *Plant Cell Physiol.*, 1977, **18**, 987.
8. C. Exley and J. D. Birchall, *J. Theor. Biol.*, 1992, **159**, 83.
9. C. Exley, *Trends Biochem. Sci.*, 2009, **34**, 589.
10. J. Xia, N. Yamaji, T. Kasai and J. F. Ma, *Proc. Natl Acad. Sci. USA*, 2010, **107**, 1838.
11. N. VanDuyn, R. Settivari, J. LeVora, S. Y. Zhou, J. Unrine and R. Nass, *J. Neurochem.*, 2013, **124**, 147.
12. M. J. Wu, P. J. O'Doherty, P. A. Murphy, V. Lyons, M. Christophersen, P. J. Rogers, T. D. Bailey and V. J. Higgins, *Int. J. Mol. Sci.*, 2011, **12**, 8119.
13. M. J. Wu, P. A. Murphy, P. J. O'Doherty, S. Mierusznski, M. Jones, C. Kersaitis, P. J. Rogers, T. D. Bailey and V. J. Higgins, *Biometals*, 2012, **25**, 553.
14. R. A. Yokel, D. D. Allen and D. C. Ackley, *J. Inorg. Biochem.*, 1999, **76**, 127.

15. K. Nagasawa, S. Ito, T. Kakuda, K. Nagai, I. Tamai, A. Tsuji and S. Fujimoto, *Toxicol. Lett.*, 2005, **155**, 289.
16. K. Nagasawa, J. Akagi, M. Koma, T. Kakuda, K. Nagai, S. Shimohama and S. Fujimoto, *Life Sci.*, 2006, **79**, 89.
17. G. Berthon, *Coord. Chem. Rev.*, 1996, **149**, 241.
18. Y. Zhou and R. A. Yokel, *Toxicol. Sci.*, 2005, **87**, 15.
19. W. Linss, R. Martin, G. Stein, H. Braunlich and C. Fleck, *Acta Histochem.*, 1991, **90**, 65.
20. K. Kametani and T. Nagata, *Med. Molecular Morphol.*, 2006, **39**, 97.
21. P. Illêŝ, M. Schlicht, J. Pavlovkin, I. Lichtscheidl, F. Baluŝka and M. Oveĉka, *J. Exp. Bot.*, 2006, **57**, 4201.
22. M. Amenós, I. Corrales, C. Poschenrieder, P. Illêŝ, F. Baluŝka and J. Barceló, *Plant Cell Physiol.*, 2009, **50**, 528.
23. M. Bragadin, S. Manente, G. Scutari, M. P. Rigobello and A. Bindoli, *J. Inorg. Biochem.*, 2004, **98**, 1169.
24. G. J. Taylor, J. L. McDonald-Stephens, D. B. Hunter, P. M. Bertsch, D. Elmore, Z. Rengel and R. J. Reid, *Plant Physiol.*, 2000, **123**, 987.
25. T. Negishi, K. Oshima, M. Hattori, M. Kanai, S. Mano, M. Nishimura and K. Yoshida, *PLoS One*, 2012, 7, e43189.
26. A. J. Roskams and J. R. Connor, *Proc. Natl Acad. Sci. USA*, 1990, **87**, 9024.
27. J.-M. El Hage Chahine, M. Hémadi and N.-T. Ha-Duong, *Biochim. Biophys. Acta*, 2012, **1820**, 334.
28. T. Sakajiri, T. Yamamura, T. Kikuchi, K. Ichimura, T. Sawada and H. Yajima, *Biol. Trace Elem. Res.*, 2010, **136**, 279.
29. J. I. Mujika, B. Escribano, E. Akhmatskaya, J. M. Ugalde and X. Lopez, *Biochemistry*, 2012, **51**, 7017.
30. G. Perez, N. Pregi, D. Vittori, C. Di Risio, G. Garbossa and A. Nesse, *Biochim. Biophys. Acta*, 2005, **1745**, 124.
31. C. Exley, in *Molecular and Supramolecular Bioinorganic Chemistry: Applications in Medical Sciences*, ed. A. L. R. Merce, J. Felcman and M. A. L. Recio, Nova Science Publishers Inc., New York, 2008, p. 45.
32. S. Maghraoui, A. Ayadi, J. N. Audinot, A. A. Ben, M. H. Jaafoura, A. El Hili, H. N. Migeon and L. Tekaya, *Microscopy Res. Tech.*, 2012, **75**, 182.
33. B. Aspenstrom-Fagerlund, B. Sundstrom, J. Tallkvist, N. G. Ilback and A. W. Glynn, *Chemico-Biol. Interact.*, 2009, **181**, 272.
34. C. Exley, *Free Radic. Biol. Med.*, 2004, **36**, 380.
35. W. A. Banks and A. J. Kastin, *Neurosci. Biobehav. Rev.*, 1989, **13**, 47.
36. L. Chen, R. A. Yokel, B. Hennig and M. Toborek, *J. Neuroimmune Pharmacol.*, 2008, 3, 286.
37. J. Freda, D. A. Sanchez and H. L. Bergman, *Can. J. Fish. Aquat. Sci.*, 1991, **48**, 2028.
38. M. Sargazi, N. B. Roberts and A. Shenkin, *J. Inorg. Biochem.*, 2001, **87**, 37.
39. Y. Song, Y. Xue, X. Liu, P. Wang and L. Liu, *Neurosci. Lett.*, 2008, **445**, 42.
40. C. Exley, J. Beardmore and G. Rugg, *Int. J. Quant. Chem.*, 2007, **107**, 275.
41. S. Murko, J. Scancar and R. Milacic, *J. Anal. Atom. Spec.*, 2011, **26**, 86.
42. G. A. Trapp, *Life Sci.*, 1983, **33**, 311.

43. J. Beardmore, G. Rugg and C. Exley, *J. Inorg. Biochem.*, 2007, **101**, 1187.
44. J. Beardmore and C. Exley, *J. Inorg. Biochem.*, 2009, **103**, 205.
45. M. A. Santos, *Coord. Chem. Rev.*, 2008, **252**, 1213.
46. X. Hu and G. L. Boyer, *Appl. Environ. Microbiol.*, 1996, **62**, 4044.
47. V. D. Appanna, J. Foucault and N. Legere, *Microbios.*, 1998, **93**, 147.
48. J. Fischer, A. Quentmeier, S. Gansel, V. Sabados and C. G. Friedrich, *Arch. Microbiol.*, 2002, **178**, 554.
49. J. Greenwald, G. Zeder-Lutz, A. Hagege, H. Celia and F. Pattus, *J. Bacteriol.*, 2008, **190**, 6548.
50. V. D. Appanna and R. Hamel, *FEMS Microbiol. Lett.*, 1996, **143**, 223.
51. C. Exley, *Environ. Sci. Proc. Impacts*, 2013, **15**, 1807–1816.

CHAPTER 28

Lead

VIRGINIA M. CANGELOSI AND VINCENT L. PECORARO*

Department of Chemistry, University of Michigan, Ann Arbor,
MI, 48109, USA
*Email: vlpec@umich.edu

28.1 Introduction to Lead in Biology

Lead (Pb) is a non-essential toxic element for which very few organisms have evolved defences. Since antiquity, lead and its compounds have been used in roles that vary from manufacture of pipes (hence the word "plumbing") to additives of gasoline and of wine. Due to its widespread modern use in gasoline, paint and solder, lead has become a common and persistent environmental pollutant. This is reflected in lead body burdens that are 1000 times higher than they were in pre-industrial individuals.[1] Since the phasing out of leaded gasoline in the United States, the average blood lead level (BLL) for children aged 1–5 has dropped from 15.0 µg dL^{-1} (720 nM) in 1976–1980 to 1.84 µg dL^{-1} (89 nM) in 2009–2010.[2,3] Now the most common sources of lead exposure are dust and chips from lead-based paint (banned in the United States in 1978, but not removed in all homes and schools) and contaminated soil (containing both paint dust and deposited lead from gasoline exhaust).[4]

Like with many toxins, children are more significantly affected than adults. Due to the time they spend on the floor and their significant hand-to-mouth activity, children are more likely to be exposed to lead. Additionally, the manifestation of toxicity occurs at lower lead body burdens in children due to their rapidly developing central nervous systems and the ability of their gut to absorb a higher fraction of nutrients.[5] *It is now commonly*

RSC Metallobiology Series No. 2
Binding, Transport and Storage of Metal Ions in Biological Cells
Edited by Wolfgang Maret and Anthony Wedd
© The Royal Society of Chemistry 2014
Published by the Royal Society of Chemistry, www.rsc.org

accepted that no amount of lead exposure is safe. In May 2012, the Centers for Disease Control replaced their 1991 "level of concern" (BLLs $\geq$10 μg dL^{-1} ($\geq$480 nM) for children aged 1–5) with an upper reference interval value BLL of 5 μg dL^{-1} (240 nM) (the 97.5th percentile of BLLs in children aged 1–5 from 2007–2010).[6,7] Even at low levels, lead can cause a multitude of health problems including anaemia, renal impairment, neurological damage, depressive and panic disorders, violent behaviour and decreased IQ resulting in poor school performance.[3,8–10] A pooled analysis of international data from six separate studies on the relationship between BLLs and intelligence revealed that an increase in BLL from <1 to 10 μg dL^{-1} (<48 to 480 nM) resulted in a decline of 6.2 IQ points.[11] Overall, the relationship between lead and morbidity is complex, multi-faceted and difficult to assess.[12] A negative shift in IQ may be documented only at the lower edge of the IQ range while going unnoticed in the general population. Nonetheless, when one considers the size of the population as a whole, a shift of 6.2 IQ points integrated among all individuals is a massive loss.

The symptoms associated with its toxicity reveal that lead is distributed widely throughout the body, affecting the cardiovascular, nervous, skeletal, digestive, endocrine, muscular and reproductive systems. Lead is not an essential element and there are few evolved detoxification pathways. Rather, at the molecular level, Pb(II) ions are capable of targeting a multitude of enzymes, proteins and biomolecules and of disrupting their normal functions. They form strong bonds with sulfur, oxygen and nitrogen ligand atoms and tolerate a wide range of coordination geometries. These properties allow them to replace structural and catalytic metal ions in proteins and to bind opportunistically to a variety of sites. Despite its larger size, Pb(II) (1.19 Å) displaces Zn(II) (0.74 Å) and Ca(II) (0.99 Å) in metalloproteins, but it is certainly not limited to these targets.

This chapter will begin with a brief discussion of the properties of lead and its coordination chemistry. Next, a discussion on ingestion, transport through the gastrointestinal tract and binding in the blood will be presented. A select group of important biochemical targets will be discussed in detail. Finally, a number of bacterial lead-detoxification pathways will be discussed.

28.1.1   The Coordination Chemistry of Pb(II)

Lead is the heaviest Group 14 element with an atomic number of 82, an electronic configuration of [Xe]4f^{14}5d^{10}6s^{2}6p^2 and an atomic mass of 207.2 g mol^{-1}. It exhibits three common oxidation states (0, II, IV) with Pb(II) being most abundant in the environment and the most relevant biologically. In general, lead complexes are thermodynamically stable and kinetically labile in aqueous media.[13] Ligand association and exchange reactions are fast on the physiological time-scale. Ligand exchange rates are affected significantly by pH and by some small molecules (especially buffers) that compete as ligands. One of the few kinetic studies that have been carried out on lead-protein complexes suggests that this is true also for large biomole-cules.[14] The lead and zinc content of certain zinc finger proteins rapidly equilibrate *via* metal-exchange reactions, suggesting that lead binding is under thermodynamic, not kinetic, control. With an electronegativity of 1.8, lead has a high affinity for the electronegative elements S, O, N, Cl and Br. Pb(II) prefers thiolate ligands over oxygen or nitrogen donors. This is re-flected in the much higher affinity for cysteine rather than glutamic acid or histidine. However, for more complex ligands featuring mixed donor groups of varying chelating and steric properties, the preference is much less clear.[13]

28.1.2 Coordination Numbers and Geometries of Lead Complexes

A recent review of crystal structures in the Cambridge Structural Database revealed coordination numbers (CN) ranging from 2 to 12 with the most common being 4 and 6 for Pb(II) and 4 for Pb(IV).[15] However, in the all-sulfur coordination environments found in many of the proteins discussed in this chapter, lead actually avoids four-coordination.[16] The geometries of Pb(II) complexes are determined by the type and number of coordinating ligands. In addition, one of their more interesting features is the potential stereochemical activity of the "lone pair" of electrons. When the ligands are very bulky or the CN is high (9–10), they are distributed in a holo-directed fashion and the lone pair is stereochemically inactive (Figure 28.1A). How-ever, when the coordination number is low (CN = 2–5), ligands are found clustered to one side, reflecting the presence of a stereochemically active lone pair (Figure 28.1B, Figure 28.2).[†] Calculations on four-coordinate Pb(II)-complexes show that a hemi-directed geometry is favoured by >5 kcal mol⁻¹. Pb(II) complexes with intermediate coordination numbers

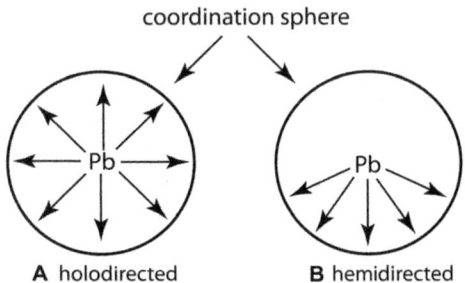

coordination sphere

A holodirected   **B** hemidirected

Figure 28.1 (A) Holo-directed and (B) hemi-directed coordination. Modified with permission from ref. 15. Copyright (1998) American Chemical Society.

[†]There is debate on the degree to which the lone pair on Pb(II) is hybridized.[13] Calculations suggest that the Pb(II) lone pair has both *s* and *p* character in hemi-directed complexes and has purely *s* character in holo-directed complexes.[15]

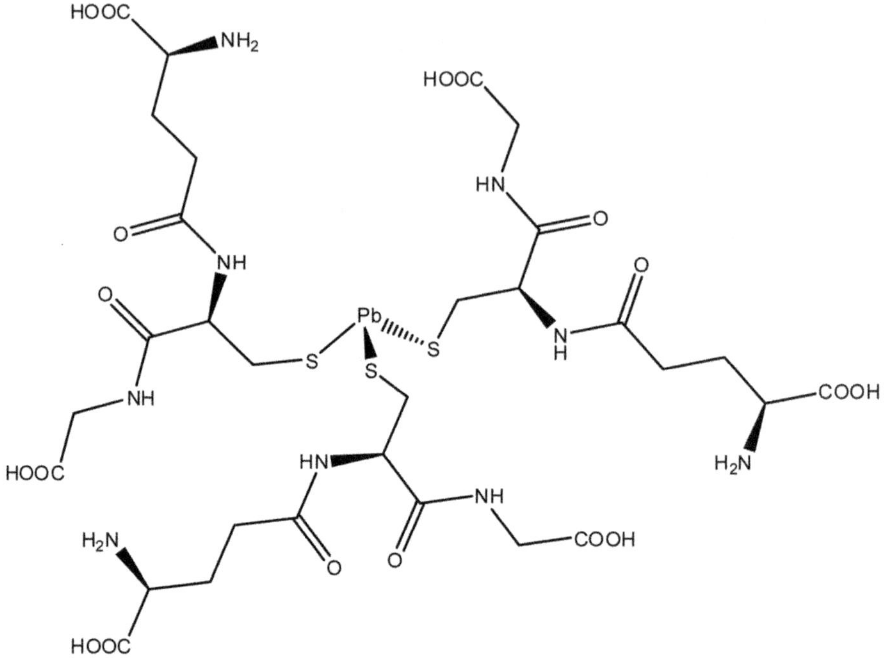

Figure 28.2   Structure of Pb(II)(GS)$_3$$^-$.

(6–8) are observed in both holo- and hemi-directed stereochemistries. With no lone pair to dictate geometry, Pb(IV)-complexes have holo-directed geometries.

28.1.3   Coordination of Pb(II) to Proteins

A quantitative analysis of 48 lead-containing protein crystal structures within the Protein Data Bank in 2008 confirmed that lead can bind to a variety of ligands and sites in proteins.[17] About one-third of the binding sites resulted from displacement of a native metal ion, with the other two-thirds being examples of opportunistic binding to sites not previously inhabited by metal ions. Three-quarters of the sites showed 2–5 ligands coordinated to Pb, but up to 9 ligands were observed. Sites to which lead was observed to bind had a total charge of 0 to −4 (mean negative charge 1.7 ± 1.1). The ligands, in order of frequency, were side-chain glutamate (38.4%), side-chain aspartate (20.3%), water (20.3%), side-chain and small molecule sulfur (7.3%) and side-chain nitrogen atoms (5%) (Figure 28.3). It is important to be mindful that these are the results of analysis of 48 lead-binding sites in crystallized proteins and do not necessarily reflect binding preferences *in vivo*. Rather, the adaptability of Pb(II) to different coordination numbers, geometries and ligands suggests the ion can bind promiscuously to metalloproteins and non-metalloproteins alike.

Bond distances are extremely sensitive to coordination number, donor type and directionality of the ligands. In the above survey, the mean bonding

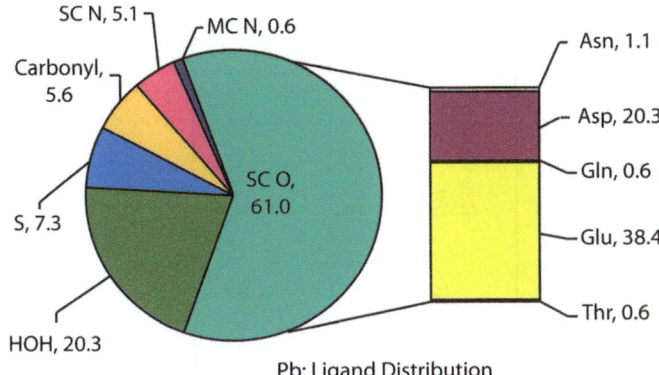

Figure 28.3 Percent distribution of ligands for Pb(II) in protein crystal structures (O = oxygen, N = nitrogen, S = sulfur, HOH = oxygen from water, SC = side-chain, MC = main-chain).
Modified from ref. 17. with permission from Elsevier.

distances were 2.7 ± 0.4 Å for Pb-O (amino acids), 2.8 ± 0.4 Å for Pb-O (water), 2.6 ± 0.4 Å for Pb-N and 3.2 ± 0.3 Å for Pb-S. These distances, however, are averaged for multiple bond types and are not a good predictor of individual bond distances. A more useful analysis that considers lead-ligand bonds for specific proteins, peptides and small molecules is given in Table 28.1. A focus on $Pb^{II}Cys_3$ coordination reveals that Pb-S bond distances range between 2.6 and 2.8 Å, much shorter than the 3.2 Å average distance described above.

28.1.4 Coordination by Sulfur

While Figure 28.3 suggests that oxygen binding is predominant in lead-proteins, sulfur coordination is very important. Protein sites carrying Cys residues and the Cys-carrying tripeptide γ-L-glutamyl-L-cysteinylglycine, glutathione (GSH) are obvious targets.

Spectroscopically, it is difficult to distinguish between Cys_3 and Cys_4 coordination.[18,19] However, in zinc finger proteins, Pb(II) prefers trigonal pyramidal Cys_3 binding sites even when an additional cysteine residue is available.[16] A re-examination of the small-molecule literature revealed that Pb(II) prefers ≤3 or ≥5 ligands when in a sulfur-only environment. [207]Pb NMR demonstrated that only three-coordinate structures were present in similar Zn finger proteins and that GSH also formed a $Pb^{II}S_3$ structure, even in the presence of excess ligand.[20] Similarly, $Pb^{II}Cys_3$ centres were seen in the metalloregulatory protein CadC even when a fourth ligand was available (see Section 28.3.4.1).[21,22] In the same way, δ-aminolevulinic acid dehydratase (ALAD) and the metalloregulatory protein PbrR both bind Pb(II) as $PbCys_3$ with the latter being the only known Pb(II)-specific protein.[23] The

Table 28.1 Pb(II) bond distances observed in proteins.

Name	Coordinating ligands	CN[a]	Bond type	Avg. bond length (Å)[d]	Technique	PDB	Resolution (Å)	Ref.
ALAD	Cys₃	3	Pb-S	2.8	XRD[b]	1QNV	2.5	95
Pb(GS)₃⁻	Cys₃	3	Pb-S	2.65	EXAFS[c]			86
PbrR691	Cys₃ or Cys₄	3 or 4	Pb-S	2.67	EXAFS[c]			18
CP-CCCC	Cys₃	3	Pb-S	2.64	EXAFS[c]			16
Pb(TRIL12C)₃⁻ Pb(TRIL16C)₃⁻	Cys₃	3	Pb-S	2.63	EXAFS[c]			24
CaM (EFII)	Asp₃GluAsnThr	8	Primary Pb-O Secondary Pb-O	2.4 3.0	XRD[b]	1N0Y	1.75	116

[a] CN = Coordination Number.
[b] XRD = X-Ray Diffraction.
[c] EXAFS = Extended X-Ray Absorption Fine Structure.
[d] Excellent reviews have been published on typical bond lengths for Pb(II) complexes with oxygen, sulfur and selenium donors.[183,184]

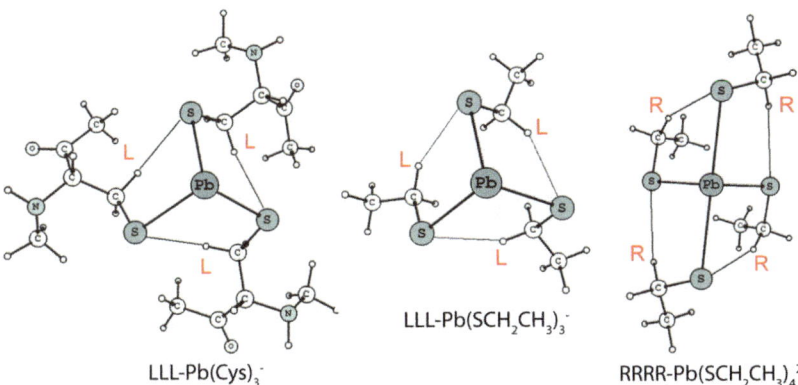

LLL-Pb(Cys)$_3^-$

LLL-Pb(SCH$_2$CH$_3$)$_3^-$

RRRR-Pb(SCH$_2$CH$_3$)$_4^{2-}$

Figure 28.4   Lowest energy DFT-structures of lead-complexes. Thin black lines indicate C-H···S hydrogen bonds.
Adapted with permission from ref. 19. Copyright (2012) American Chemical Society.

discrimination of these Cys$_3$ sites for Pb(II) may be due in part to built-in hydrogen bonding networks that pre-organize the sites for the trigonal pyramidal geometry that is preferred by lead. In DFT calculations carried out on [Pb^{II}(SCH$_2$CH$_3$)$_3$]$^-$, [Pb^{II}(Cys)$_3$]$^-$ and [Pb^{II}(SCH$_2$CH$_3$)$_4$]$^{2-}$, all of the optimized structures showed Pb-S bond lengths of 2.6–2.7 Å plus hydrogen bonding distances of 2.8–3.1 Å between a hydrogen on the C$_\beta$ atom of one ligand and the sulfur atom of an adjacent one (Figure 28.4).[19] However, [Pb^{II}(GS)$_3$]$^-$ and other complexes with peptides providing three thiolate sulfur atoms also obey these structural constraints without the presence of these H-bonds.[24]

28.1.5   *Endo* and *Exo* Orientations

The orientation of the ligands around a Pb(II) ion influences the steric properties of both the ligands and the ion itself. The nomenclature for three-coordinate Pb(SR)$_3$ centres is based on the position of the sulfur-bound carbon and lead atoms relative to the plane of the three sulfur atoms. When the carbon atoms (R groups in Figure 28.5) and lead are on the same side of that plane, the coordination environment is designated as *endo*. When the carbon atoms and lead are on opposite sides, the designation is *exo*. Mixed *endo/exo* coordination environments are possible, but are not discussed here.[25]

DFT calculations indicate that *endo vs. exo* preference depends, at least partially, on the steric environment around the lead ion and its lone pair of electrons. All of the energy-minimized structures reported for Pb^{II}(SCH$_2$CH$_3$)$_3$ and Pb^{II}Cys$_3$ centres were in the *exo* orientation.[19] Similarly, the *exo* isomer of Pb^{II}(SCH$_3$)$_3$ was found to be preferred over the *endo* orientation by 3.8 kcal mol^{-1}.[25] While the steric designation of these

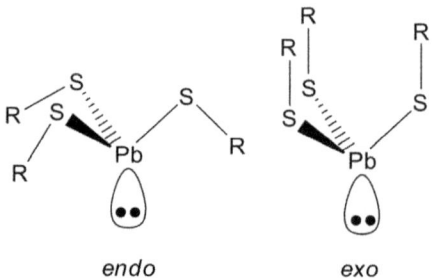

endo   *exo*

Figure 28.5   *Endo* and *exo* Pb^{II}(SR)$_3$ centres.

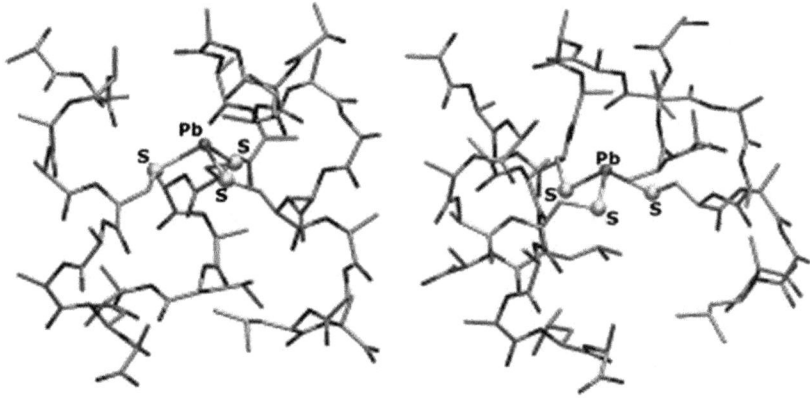

Figure 28.6   *Exo* (left) and *endo* (right) orientations in the optimized [Pb(II)(ACA)$_3$]$^-$ structure in which the lone pair is pointing toward the cavity.
Reprinted with permission from ref. 25. Copyright (2012) WILEY-VCH Verlag GmbH & Co. KGaA, Weinheim.

small-molecule complexes is relatively straightforward, the situation is more complex within protein environments. Calculations on three-helix bundle proteins revealed the importance of steric bulk and the influence of the lone pair. The *endo* orientation was estimated to be 3.3 kcal mol^{-1} lower in energy than the *exo* equivalent for [Pb^{II}(ACA)$_3$]$^-$ where ACA is a 10-amino acid α-helix providing a relatively crowded Cys$_3$-binding site (Figure 28.6B).[25] However, in a model where a cavity was introduced above the binding site (Leu-to-Ala substitutions in ACA), the difference between *endo* and *exo* was small and the lone pair was predicted to occupy the cavity (Figure 28.6A).[25] This illustrates the importance of the steric bulk of the lone pair of electrons in Pb(II) binding sites. Preorganization that accommodates this lone pair could be one way that the metalloregulatory protein PbrR distinguishes Pb(II) from other metal ions and may help explain why Pb(II) binds so well to the active site of δ-aminolevulinic acid dehydratase, normally occupied by Zn(II).

28.1.6 Bioavailability of Lead in Cells

Pb(II) rarely exists as the free ion within mammalian cells. Upon entry, it is likely to be bound by glutathione and ultimately ends up bound to proteins. The combination of fast kinetics and thermodynamically strong Pb-S, Pb-O and Pb-N bonds allows Pb(II) to bind biomolecules adventitiously at a variety of sites. However, in order to disrupt cellular processes, it must be capable of binding tightly to the target molecules and be present in sufficiently high concentrations. Identifying those targets is very challenging. To do so, one must know both the concentration of free Pb(II) and the affinity of Pb(II) for the protein. If competing with another metal, then the concentrations of both free metal ions and their relative affinities for the protein must be known.[26] Unfortunately, very few lead-protein affinities have been determined and the nature of *in vivo* equilibria is complicated by the presence of a variety of other metal ions, proteins and small-molecule ligands. Further, free, or bioavailable, lead concentrations have not yet been experimentally measured and the concentration can vary depending on cell type, location within the cell (*e.g.* within an organelle) and exposure. Within blood plasma, a picomolar free lead concentration has been estimated.[27] This further emphasizes the ability of lead to bind proteins considering a typical whole blood lead concentration of 0.1 μM in a person with a BLL of 2 μg dL^{-1}.[26] At typical transition metal concentrations and picomolar free lead, 1–3 orders of magnitude stronger lead binding and a $K_D \leq 10^{-12}$ M would be required for competition.[26]

28.2 Eukaryotic Cells

28.2.1 Introduction

Because lead is not an essential metal ion, most organisms have not developed specific mechanisms for its binding, transport and storage. Rather, it appears to be transported by an array of mechanisms and to bind to a wide variety of proteins and biomolecules, many of which will be described in detail below. The focus of this section will be on the transport of lead from the gastrointestinal tract into the blood stream and its binding there. The ensuing distribution and storage within bone,[28] organs,[29] muscle[30,31] and brain[27,31–34] that cause many of the toxic effects will be described in lesser detail. Generally, ingestion is followed by absorption into the blood stream where the half-life is several weeks. Subsequently, lead is distributed to the soft tissue and bone where it has half-lives of months and at least 5–10 years, respectively.[35] Because of the very long half-life, bone is the primary storage site in the body. Adults store 80–95% of retained lead in bone whereas children store 70% in bone with the other 30% remaining in soft tissue.[35] Pb(II) can replace Ca(II) ions in the calcium hydroxyapatite portion of bone.[36] Mobilization of lead occurs during periods of high bone matrix turnover such as pregnancy,[37] lactation,[38] menopause[39] and as a result of thyroid disease such as thyrotoxicosis.[40] In addition to being stored within the bone matrix, lead can affect bone cell function both directly and indirectly. This is reflected in changes to circulating levels of hormones, changes in the ability of cells to respond to hormones and to synthesize bone matrix and substitution of lead for calcium in the calcium messenger system.[28] Eventually lead is excreted from the body, with filtration through the kidneys resulting in renal damage.[41]

28.2.2   Uptake into the Intestinal Cells and Release into the Blood

In humans, most lead exposure occurs by inhalation, skin contact or ingestion. Workers with occupational exposure may inhale lead dust. A reasonable model suggests that 35–40% of inhaled lead is deposited in the lungs and, of that, 95% is absorbed into the blood.[42] Organolead compounds such as the former gasoline additive tetraethyllead can be absorbed directly through the skin. The most common route to lead exposure in humans is, however, through ingestion, which may occur through the consumption of contaminated food, breast milk,[43] alternative medicines[44] and occupational exposure. The amount of lead that is absorbed through the gastrointestinal tract depends heavily on the type, amount and timing of the consumption, as well as the nutritional make-up of the food with which it is ingested. For example, in a controlled study, adults who were fasting for 16 hours before ingestion of the sample absorbed 35% of the lead while those who consumed the sample with a meal absorbed only 8%.[45] Generally, adults absorb about 10% of bioavailable ingested lead through the gastrointestinal tract, while children, due to their increased need for nutrients, absorb a much higher amount (50%).[42]

The absorption and distribution of lead occurs in two distinct phases. The first phase is fast and involves its transport into intestinal cells. The second phase is much slower, involving transfer into the circulation.[46] Despite over 50 years of research, the pathways by which lead is absorbed and transferred into the blood are still not understood adequately at the molecular level.[47]

Lead is not an essential nutrient, so it is unlikely that specific mechanisms have evolved for its transport. Rather, it appears to utilize several different pathways that have evolved for the uptake of dietary divalent metal ions from the small intestine (Figure 28.7).[48] For example, Ca^{2+} enters epithelial cells actively through calcium channels, is transported by calcium-binding proteins (CaBP) and then is exported into the blood by an ATPase.[‡] *In vitro* studies on bovine and chick intestinal CaBPs reveal that Pb(II) has a higher affinity than Ca(II) for both proteins (*e.g.* $K_D = 6.2 \times 10^{-8}$ M for Pb(II) and 5×10^{-7} M for Ca(II) for chick CaBP) although it is unknown if this is consequential *in vivo*. In each protein, Pb(II) appears to bind directly to the

---

[‡]Uptake of calcium, iron and zinc are discussed in Chapters 5, 11 and 22, respectively.

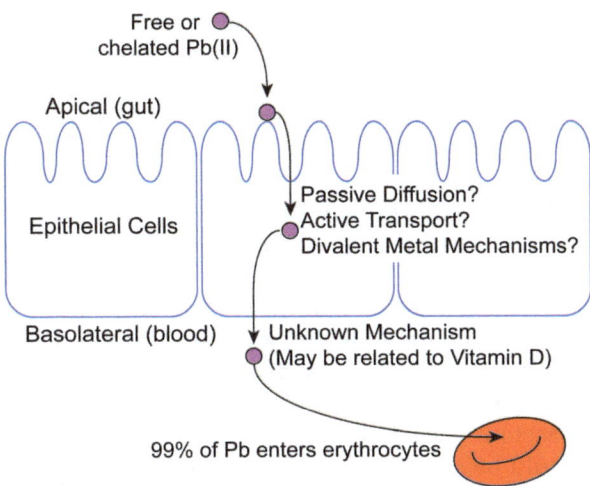

Figure 28.7   Absorption of Pb(II) from the gastrointestinal tract into epithelial cells and transfer to the blood.

calcium-binding site rather than opportunistically to sulfhydryl groups.[49] Lead absorption may also occur through Zn(II) uptake mechanisms, given that Zn(II) inhibits Pb(II) absorption.[50]

Fe(II) enters epithelial cells in the duodenum through the ubiquitous divalent metal transporter DMT1.[51] Iron deficiency increases gastrointestinal lead absorption, possibly *via* coordination to iron-binding proteins in the intestine.[52] Addition of Fe(II) to the mucosal fluid of everted rat duodenal segments did not decrease the amount of lead uptake by cells, but did reduce serosal transfer suggesting that iron and lead compete for certain metal transfer mechanisms within the duodenum.[53] Initial evidence supported the transfer of Pb(II) by DMT1, but it was later shown that lead enters epithelial cells through a DMT1-independent proton-driven mechanism, the details of which are still not fully understood.[54,55] In addition to known ion-transport pathways, lead transport has been shown to occur through a passive diffusion process linked to the movement of water, an active transport requiring oxygen and phosphate bond energy and time- and temperature-dependent transport involving sulfhydryl groups.[50,53,56] In the latter case, the extrusion of lead was found to be partially dependent upon metabolic energy. Anion exchange is not involved in the uptake of lead by kidney epithelial cells.[57]

Most nutrient absorption occurs in one of the three sections of the small intestine: the duodenum, the jejunum and the ileum. The uptake of lead into these sections has been studied using animal models and epithelial cells. Originally it was believed that the lead absorption occurs mainly in the duodenum, the first section of the small intestine in which iron absorption occurs. A comparison of lead transfer into everted sacs of rat small intestine revealed that the duodenum showed the highest level of lead transfer from the lumen through the epithelium and into the serosal space.[53] However, a more recent study on isolated segments of mouse duodenum and ileum shows comparable rates of lead absorption in both sections.[58] This suggests that ileal lead absorption may be substantial considering that food material resides within the ileum about 20 times longer than it does in the duodenum.[59]

Absorption of lead into epithelial cells is a two-step process involving rapid binding to the surface followed by a slower uptake phase. In chicks that ingested lead acetate, 50% of the lead was absorbed into or onto intestinal tissue within the first five minutes.[46] Another study found that the serosal compartment, duodenum and jejunum of rats rapidly absorb lead (>50% absorbed within the first 10 min.) suggesting that lead binds covalently to the tissue surface, likely to surface phosphates.[56] The surface binding of lead acetate was visualized using electron microscopy. Within two minutes, lead could be seen bound to the microvilli of guinea pig intestinal epithelial cells and within 10 to 30 minutes, it had entered the cells and could be observed within the vacuoles (Figure 28.8).[60]

The above studies were carried out with lead acetate, but ingested lead may be absorbed differently if it is chelated prior to ingestion. For example, there is a high turnover of bone matrix in mothers during lactation leading to the mobilization of stored lead into the blood stream. This form enters breast milk and is consumed by nursing children. Almost 90% of the lead was associated with the milk protein casein in rat and bovine milk, in infant formula and in preliminary studies with human milk. Casein was found to have a high affinity and capacity for Pb(II) binding, suggesting that Pb(II) is

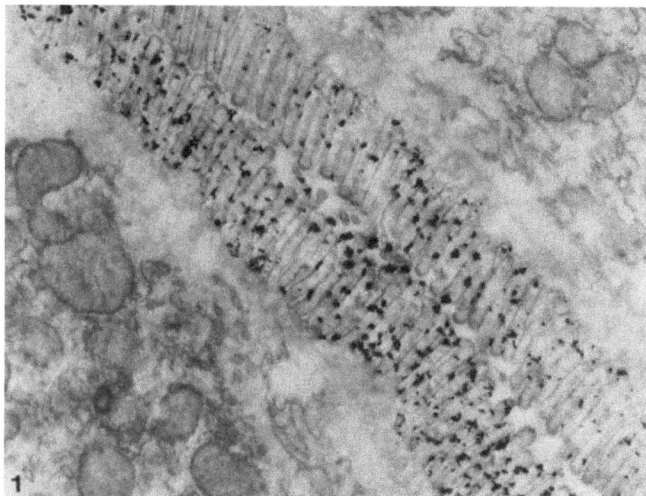

Figure 28.8   Lead particles (dark spots) binding to microvilli of epithelial cells from the small intestine.
Reprinted from ref. 60 with kind permission of Springer Science + Business Media.

replacing casein-bound Ca(II). It is unclear if lead acetate and casein-bound lead are absorbed by the same mechanism.[43]

Overall, the microvilli of epithelial cells within the small intestine bind lead to their surface within minutes. Transport into the cell is slower and probably occurs by a variety of mechanisms intended to facilitate the influx of other divalent cations. Once within the epithelial cells, its chemical state is unclear, as is the mechanism of transport through the cell. The mechanism of efflux into the bloodstream is also unknown but may be associated with the function of vitamin D. Its administration does not enhance the accumulation of lead by intestinal tissue, but does accelerate the transfer of lead through the basal membrane of the epithelium to the blood.[61] Clearly, the absorption of lead in the gastrointestinal tract is complex, apparently multifaceted and still not understood in detail.

28.2.3 Transport through the Cytoplasmic Membrane of Red Blood Cells

Once lead enters the bloodstream, the vast majority (99%) associates with erythrocytes. Only about 1% remains in the plasma where it is bound to albumin (0.4%) and sulfur-containing biomolecules and proteins (0.6%).[62,63] At high blood lead levels (BLLs), the percentage of lead found in the plasma increases. The non-linear relationship between BLLs and plasma lead levels suggests that erythrocytes have a limited capacity to bind lead.[64] Within intact red blood cells, lead binds internally within the cytoplasmic pool, not to the cell exterior as initially thought.[65]

Over 90% of the influx of lead across the cytoplasmic membrane of erythrocytes occurs by passive diffusion through an anion exchange pathway that is blocked by anion transport inhibitors.[57,66,67] HCO_3^- stimulates lead uptake and the mechanism appears to depend on the formation of an anionic lead carbonate complex in solution (but not $PbCO_3$ itself, which is neutral). The human erythrocyte cell membrane consists of almost 20% anion exchange protein AE1 (which exchanges Cl^- and HCO_3^-) so there are many locations at which anion exchange can occur. Passive influx of lead into erythrocytes may also occur through calcium transport mechanisms, as was observed in erythrocyte ghosts (red blood cells with cytoplasmic contents removed).[68] The rate of lead influx is directly proportional to the external lead concentration up to 1.2 μM (saturation may occur at higher concentrations but the insolubility of Pb(II) under the conditions limited the concentration range in the investigation).[66]

The efflux of lead occurs with rates that are similar to those for the influx of lead.[66] Less than 50% of lead efflux occurs by the passive pathways described above with the remainder occurring by an energy-dependent extrusion.[65,69] The active efflux of Pb(II) ions from resealed erythrocyte ghosts is ATP-dependent and appears to be carried out by a calcium pump, but little else is known about the mechanism.[70]

Although this section of the chapter focuses on lead in blood, important work describing the transport pathways utilized by lead to enter cell types other than erythrocytes must be mentioned, especially those involved in the nervous system. The metal enters cells through several different uptake mechanisms, most of which are specific for other metal ions. Although DMT1 is not responsible for transporting Pb(II) into epithelial cells, an *in vitro* model suggests that a protein with properties similar to iron-response element positive (IRE +) DMT1 transports Pb(II) across the blood-brain barrier.[54,71] Further, the expression of DMT1 in rat brain tissue was increased upon exposure to the combination of Pb(II) and Cd(II), but not with either metal on its own.[72] Lead also appears to utilize Ca(II)-transport pathways to enter other cells including L-type Ca(II) channels in bovine chromaffin cells,[73] calcium channels in bovine adrenal medullary cells[74] and channels that are activated by the depletion of intracellular Ca(II) in rat pituitary and glial cells and in human embryonic kidney and brain capillary endothelial cells.[75,76] Pb(II) does not enter cerebellar granule neurons through L-type Ca(II) channels, but rather through *N*-methyl-D-aspartate (NMDA)-receptors.[77]

28.2.4   Coordination Chemistry of Lead Ions

Throughout the body, lead can disrupt many physiological processes by binding to a multitude of proteins and biomolecules. Presented below are some of the better studied and significant biomolecules and proteins involved in lead binding in the blood and the brain.

28.2.4.1   *Lead-glutathione Complexes*

GSH helps to protect against the toxic effects of certain metal ions by co-ordinating them until transfer to proteins.[78–80] *In vitro*, complexes of GSH and Pb(II) transfer the metal ion to both Cys-rich proteins (metallothioneins) and phytochelatins (inducible plant peptides with (γGluCys)$_x$Gly structures).[78] Although the complexation of lead by GSH *in vivo* has not been observed directly, lead-exposed workers and their blood both exhibit decreased levels of GSH and decreased activities of GSH-dependent enzymes compared to controls.[81,82] Lead binds GSH primarily through the sulfur donor with minor interactions with the C-terminal carboxylate.[83–85] Initial studies suggested that Pb(GS)$_2{}^{2-}$ forms with $K_D = 10^{-15}$. However, more recent work suggests that the predominately formed species is [Pb^{II}(GS)$_3$]$^-$ (Figure 28.2).[20,86–88] GSH can facilitate the transport of some metal ions. For example, the P-type ATPase pump ZntA in *Escherichia coli* typically transports Zn(II), but can also transport Pb(II). The highest activity for hydrolysis of ATP was observed for Pb(II) bound to cysteine or GSH, suggesting that metal-thiolate complexes are the true substrates of this pump.[89]

28.2.4.2   Binding of Pb(II) to δ-Aminolevulinic Acid Dehydratase (ALAD)

Anaemia is a common symptom of lead toxicity and it was believed previously that this was due to binding of lead to haem.[90] It is now well established that the anaemia is due to a disruption of the haem biosynthetic pathway inducing haemoglobin deficiency. While lead inhibits ferrochelatase only during extreme, acute toxicity *in vivo*, it indirectly causes an increase in the production of zinc protoporphyrin, which is fluorescent and used as a biomarker for lead exposure.[91] Lead directly inhibits the second step of haem biosynthesis in which the enzyme ALAD catalyses the condensation of two molecules of δ-aminolevulinic acid to produce porphobilinogen. Pb(II) targets the active site of ALAD, displacing the catalytic Zn(II) ion and thereby inhibiting activity at concentrations as low as femtomolar in lead ($K_i = 0.07$ pM).[92] This not only aborts haem synthesis but leads to a build-up of neurotoxic δ-aminolevulinic acid (ALA).[93] The affinity of Pb(II) for S_3-sites is 2–3 orders of magnitude higher than that of Zn(II). A small-molecule model of ALAD with a scorpionate S_3 site (Scheme 28.1) was estimated to have a 500-fold greater affinity for Pb(II) than for Zn(II).[94] Estimates for ALAD itself, however, suggested only a 25-fold greater affinity.[92]

A comparison of the crystal structures of homo-octameric Zn(II)-ALAD and Pb(II)-ALAD reveals that lead occupies the Cys_3-active site with very little disruption of the site itself or of the overall fold of the protein.[95] The biggest difference is that a serine residue (Ser179) that is non-bonding in the Zn-structure is reoriented in the lead-structure so that its backbone carbonyl oxygen atom lies 0.9 Å closer to the metal ion (Figure 28.9). Consistent with the larger size of Pb(II), the Pb-S bonds are an average of 0.54 Å longer than the Zn-S bonds. This causes the ion to protrude slightly further into the substrate-binding cavity and positions its lone pair of electrons in a way that blocks incoming ALA from binding at the required angle.[96] Furthermore, the reaction mechanism involves activation of the ketone group on ALA upon coordination to Zn(II). Pb(II), being a significantly weaker Lewis acid than Zn(II), cannot activate ALA to the same degree.

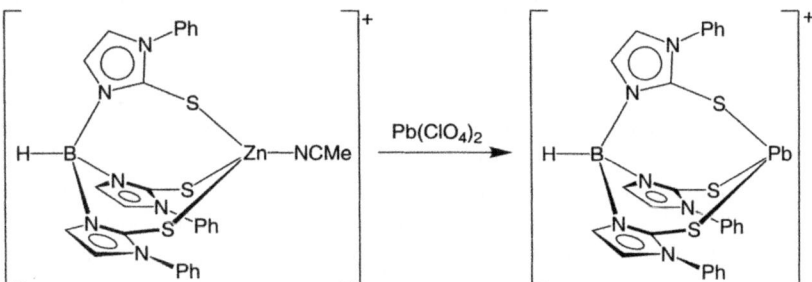

Scheme 28.1   Substitution of Pb(II) into $\{[Tm^{Ph}]Zn(NCMe)\}(ClO_4)$.
Reprinted with permission from ref. 94. Copyright (2000) American Chemical Society.

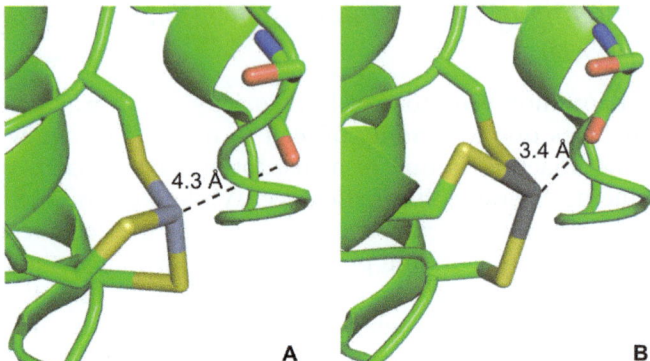

Figure 28.9 Comparison of the active site of ALAD when occupied by A) Zn(II) and
B) Pb(II). Images generated from PDB 1EB3 and 1QNV, respectively.

The crystal structure also contains a second Pb(II) ion within the substrate-binding pocket that is 4.4 Å away from that occupying the active site.[95]
The binding of two ions could increase lead solubility *in vivo*.[17] ALAD is the
main lead-binding site in erythrocytes (and not haemoglobin, as formerly
presumed). In the blood of occupationally exposed workers, 35–81% of detected lead is coordinated to this enzyme.[97] Overall, ALAD appears to have a
lead-binding capacity of ∼40 µg dL^{-1} (1.9 µM) of whole blood with ∼6 lead
ions bound to each octamer, *i.e.* <1 ion per monomer. The 50% decrease in
ALAD activity at BLLs of 15 µg dL^{-1} (720 nM) may be rationalized by the
presence of only 1–2 lead atoms being bound per octamer.[64,97]

28.2.4.3 Binding of Pb(II) to Pyrimidine 5′-Nucleotidase Type I (P5N-I)

Found in erythrocytes, pyrimidine 5′-nucleotidase type I (P5N-I) catalyses the
dephosphorylation of pyrimidine 5′-mononucleotides, a necessary step for
the degradation of ribosomal DNA in maturing erythrocytes. P5N-I requires
Mg(II) for activity and is strongly inhibited by Pb(II).[98,99] Inhibition results in
an increased rate of erythrocyte destruction and decreased haemoglobin
production, despite P5N-I not being part of the haem biosynthetic pathway.[100] In the blood of lead-exposed workers, 12–26% of the lead was found
to be associated with this enzyme.[64] A comparison of the crystal structures of
the native enzyme with its lead-inhibited form reveals that lead binds within
the catalytic site, but away from the cationic cavity occupied by Mg(II)
(Figure 28.10).[101] In fact, the lead atom occupies the same space that contains a water molecule in the magnesium-containing enzyme. The ligands
vary slightly for the two metal ions with the main difference being that Mg(II)
is bound by the backbone carbonyl of Asp51 while, instead, Pb(II) interacts
electrostatically with the carboxylate of Asp242 (monodentate, Pb-O = 2.8 Å).
Inhibition likely results from the combination of the improper positioning

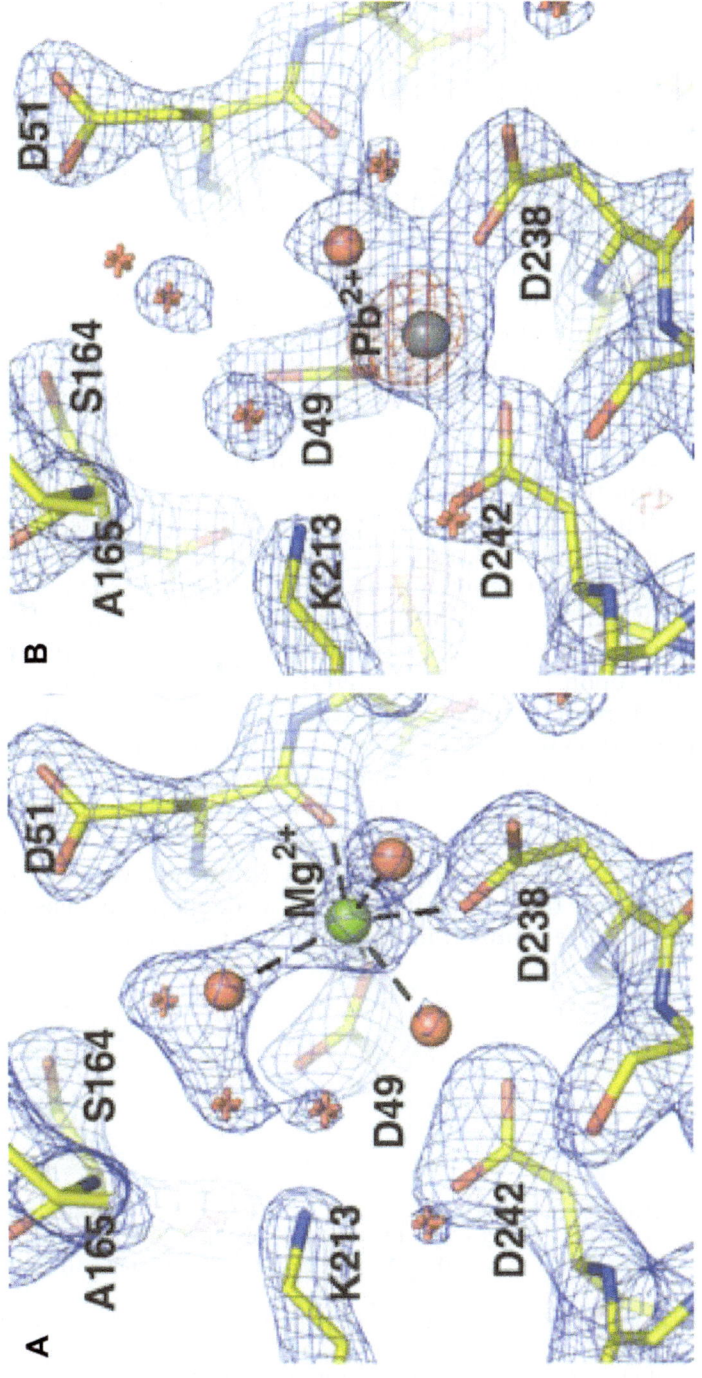

Figure 28.10 Comparison of the active site of P5N-I when occupied by A) Mg(II) and B) Pb(II). This figure was modified from that which was originally published in ref. 101.
Copyright (2006) the American Society for Biochemistry and Molecular Biology.

of the catalytic Asp49 and a lowering of the affinity of phosphate for the cationic cavity upon Pb(II)-binding.

28.2.4.4 Lead Binding to Zinc Finger Proteins

While the disruption of ALAD and P5N-I function helps to explain the anaemia associated with lead poisoning, it does not account for the developmental problems observed in children at relatively low BLLs. Some developmental processes are regulated within cell nuclei by transcription factors and lead is known to concentrate in the nuclei of rat kidney cells.[29,102,103] Zinc finger proteins recognize specific DNA, RNA and protein targets. A zinc finger motif employs Zn(II)-bound to mixed Cys/His sites or Cys sites with four-coordinate tetrahedral geometry to stabilize protein folds. Erythrocyte formation and neurological development depends on zinc finger transcription factors such as nuclear hormone receptors and GATA proteins.[26] These contain Cys_4-binding sites and substitution by Pb(II) would be expected to disrupt these essential processes.

Pb(II) binds tightly to zinc finger consensus peptides containing Cys_4 (CCCC), Cys_3His (CCHC) and Cys_2His_2 (CCHH) binding sites with affinities that increase with the number of cysteine residues (Figure 28.11).[14] Despite the availability of four ligands, Pb(II) is three-coordinate within zinc finger proteins.[20,104] Pb(II), which prefers a trigonal pyramidal geometry, induces improper folding upon substitution of tetrahedral Zn(II). Both CCHC and CCHH bind Zn(II) more tightly than Pb(II), but CCCC binds Pb(II) 30-fold stronger than Zn(II). Only three of the four cysteine residues in CCCC coordinate the Pb(II) ion. Similarly the affinities of Cys_3His sites depend upon the positioning of the cysteine residues. While CCHC prefers Zn(II) over Pb(II), CCCH has a stronger affinity for Pb(II).[16]

The preference for lead binding has been confirmed in chicken and human GATA-type zinc finger proteins (CCCC), where Pb(II) displaces Zn(II) from the Cys_4 sites. These important transcription factors play an essential

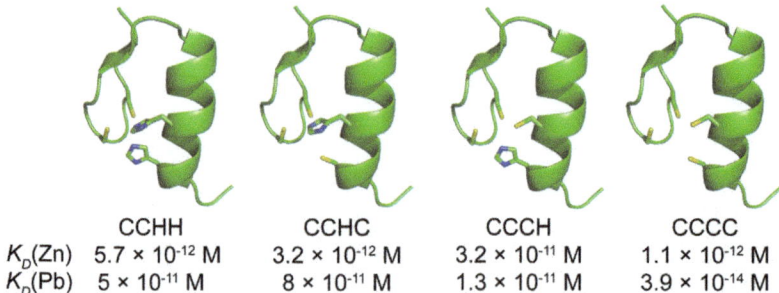

	CCHH	CCHC	CCCH	CCCC
K_D(Zn)	5.7×10^{-12} M	3.2×10^{-12} M	3.2×10^{-11} M	1.1×10^{-12} M
K_D(Pb)	5×10^{-11} M	8×10^{-11} M	1.3×10^{-11} M	3.9×10^{-14} M

Figure 28.11 Models of CCHH, CCHC, CCCH and CCCC consensus zinc finger proteins and Zn(II)- and Pb(II)-binding constants at pH 7.2.[14,16] Models generated by modification of PDB ID: 1MEY; NDB ID: PDTB41 using PyMOL.[185]

role in the regulation of neurological, urogenital and cardiac development. Upon Pb(II)-binding, GATA showed a decreased ability to activate transcription due to a decrease in DNA-binding affinity.[105] This is not surprising if we consider lead concentrations in human tissue. In the tissues of humans that were not occupationally exposed, lead concentrations range from 0.01–2.5 ppm.[106] In those organs with a density similar to that of water, this is equivalent to $\sim$0.05–12 µM lead (although lead levels are significantly higher in those who have been exposed). Considering the 30-fold difference in the dissociation constants of lead and zinc, lead should easily compete with 200 µM zinc for binding.

Although Pb(II) targets Cys_4 sites, binding to Cys_3His and Cys_2His_2 may be possible in cases of lead poisoning. Some Cys_3His-type zinc finger proteins that bind mRNA and regulate post-translational gene expression are found in the cytosol, making them highly susceptible to lead binding.[107] Based on the affinities of Cys_3His type zinc fingers for Pb(II), mRNA-binding zinc finger proteins may be a target for lead binding.[14,16] *In vitro*, Pb(II) has been shown also to compete with Zn(II) for binding to the zinc finger protein human protamine 2 (HP2), which is involved critically in spermatogenesis. Pb(II) binds at two different Cys-containing sites, altering the conformation of the protein and inhibiting HP2-DNA binding. This observation may help to account for the decreased fertility observed in men with occupational lead exposure.[108] Inhibition of DNA binding was also observed for the transcription factors TFIIIA and Sp1 that contain Cys_2His_2 zinc fingers. *In vitro*, micromolar concentrations of lead ($>$5 µM for TFIIIA and $>$10 µM for SP1) inhibit DNA binding, although this may not be significant *in vivo*.[109] The transcription factor Zn(II)-TFIIIA and its DNA adducts both bind Pb(II). Complete dissociation of TFIIIA from the DNA occurred with a four-fold excess of Pb(II) in a 13 µM Zn(II) buffer; however, it is unclear if this is relevant *in vivo*.[110]

28.2.4.5 Binding of Pb(II) to Calmodulin

Calmodulin (CaM) is a Ca(II)-binding signal transducer protein found in both the cytoplasm and within organelles. Expressed in all eukaryotic cells, CaM is involved in calcium signalling pathways for over 100 biological processes involving growth, memory, immune response, inflammation and muscle contraction.§ Calcium binding occurs in each of CaM's four highly conserved EF-hand motifs.[111] These motifs feature helix-loop-helix structures of about 30 amino acids that bind Ca(II) with pentagonal bipyramidal geometry. They are found in over half of all CaBPs, making CaM a good representative of this important class of molecules. As for other CaBPs (Section 28.2.2), Pb(II) has a higher affinity than Ca(II) for CaM[52,110] and is capable of binding at all four Ca(II)-binding sites.[113,114] However, lead-

---

§See Chapter 5 for more detail on the structure and function of calmodulin.

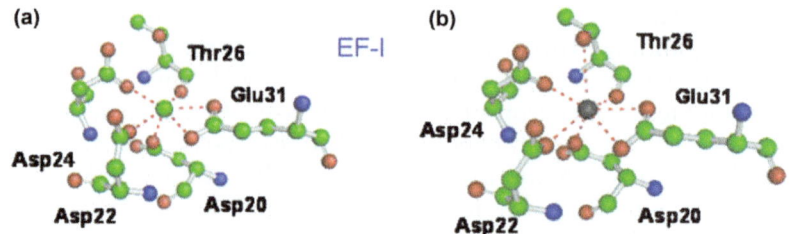

Figure 28.12 Paired binding sites for (a) Ca^{II}-CaM (PDB 1EXR) and (b) Pb^{II}-CaM (PDB 1N0Y).
Reprinted from ref. 17 with permission from Elsevier.

for-calcium displacement does not occur evenly at all sites due to differences in the binding affinities. In the N-terminal domain, Ca(II) binds with $K_D = 11.5\ \mu M$ while Pb(II) binds with $K_D = 1.4\ \mu M$. In the C-terminal domain, $K_D = 2.0\ \mu M$ for Ca(II) and $K_D = 0.73\ \mu M$ for Pb(II).[115]

Crystal structures of Ca^{II}-CaM and Pb^{II}-CaM reveal that binding of lead affects the conformation of the protein, presumably inhibiting its function.[17,116] Overall, however, perturbation of the metal-binding sites is minimal, suggesting that conformational changes between the proteins are not due to the lead-for-calcium ion replacement. Within the metal-binding sites, Ca(II) and Pb(II) share the same general coordination geometry and ligands with only minor differences in bond distances and angles due to the larger size of Pb(II) (Figure 28.12). However, in addition to the four Pb(II) ions bound within the EF-hand motifs, the crystal structure of Pb^{II}-CaM reveals the binding of 10 additional Pb(II) ions to the protein surface (Figure 28.13). Surface binding *in vitro* is a common phenomenon and is not necessarily representative of *in vivo* binding. In this case, however, surface binding occurs with high affinity to areas of high electrostatic potential and causes conformational changes to the protein structure.[17] This opportunistic binding helps to explain the observation that the addition of Pb(II) to CaM initially activates the protein, but causes inhibition at higher lead concentrations.[117] First, Pb(II) seems to bind to the EF-hand sites activating the protein, then opportunistic binding to negative sites on the surface of the protein causes conformational changes that inhibit activity.[17,115]

28.2.4.6 Other Brain Proteins

Neurotoxicity results from the disruption of the normal functions of central nervous system proteins when they coordinate lead, often at calcium-binding sites.[118] Lead binding to voltage-gated calcium channels results in "blocked" channels, thus affecting synaptic function.[31] Synaptotagmin I plays a critical role in the release of neurotransmitters from synaptic vesicles, allowing signalling to occur in neurons. Pb(II) binds more tightly than Ca(II) to the Ca(II)-binding site in synaptotagmin I, but alters its interaction with phospholipids and syntaxin, thus altering function.[34] Protein kinase C

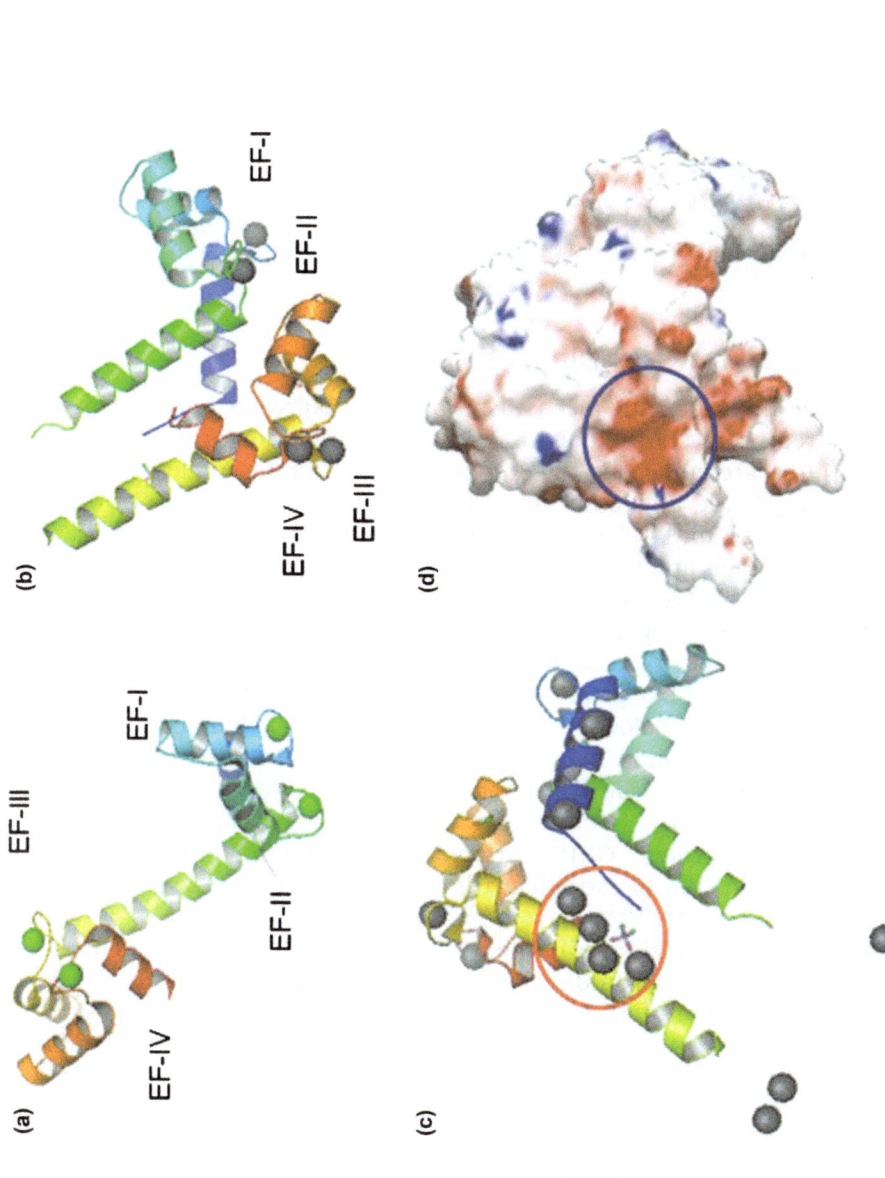

Figure 28.13 Proteins (a) Ca-bound calmodulin (1exr) and (b) Pb-bound calmodulin (1n0y). EF-I through EF-IV are Ca^{2+}-binding sites. (c) Four Pb$_2^+$ ions (circle) opportunistically bound along groove between chains in 1n0y. (d) Electrostatic potential surface map of 1n0y rendered with GRASP, showing region of negative charge density (circle) where Pb^{2+} ions are bound. Reprinted from ref. 17 with permission from Elsevier.

(PKC) regulates cell growth and function by phosphorylating critical regulatory proteins. PKC is a Ca- and Zn-dependent enzyme and is an unusual example of an enzyme that can be both inhibited and activated by a toxic metal. At higher concentrations, lead binds to multiple sites on the enzyme causing inhibition; however, at picomolar concentrations lead actually activates the enzyme.[27,32] When induced by lead, the PKC-dependent expression of new genes may be the source of some of the learning deficits caused by lead.[33]

28.2.4.7 Transport Proteins

Pb(II) is likely transported through the cell by CaBPs or as complexes with glutathione.[78,112]

28.2.5 Regulatory Mechanisms

Lead-specific detoxification mechanisms within eukaryotes have not been fully explored. It is known that a series of low-molar-mass lead-binding proteins (LBPs) are induced in the presence of lead, but it is unclear whether or not these provide protection from its toxic effects. The nature of these proteins and their inter-relationships is debatable.[29] LBPs have been identified in erythrocytes and in several organs (brain, lung, liver and kidney) and have been characterized to various degrees. Over 35% of the lead in human kidneys is coordinated by Acyl-CoA binding protein (9 kDa, $K_D = 24$ nM) and thymosin β4 (5 kDa, $K_D = 5.0$ nM).[119] Along with inducible p32/6.3 and several others, these LBPs protect against lead-induced renal toxicity, which can occur even at BLLs below 5 μg dL^{-1} (240 nM).[29,120] The human brain also contains thymosin β4 and p32/6.3 as well as 12 kDa and 23 kDa LBPs.[120–123] These are thought to provide protection. Lead accumulates in astroglial cells before it reaches sensitive neurons.[124,125] Sequestration is carried out by the glucose-regulated protein GRP78, a molecular chaperone found in the endoplasmic reticulum of astroglia. It is expressed in the presence of lead and chelates it strongly.[126,127]

In order to reach the kidneys or the brain, lead must be transported in the blood where it is mostly coordinated to ALAD and P5N-I. In addition to inhibiting the activity of ALAD (Section 28.2.4.2), lead induces its expression, suggesting that this enzyme may double as an inducible protection against lead toxicity.[128,129] Differences in the ALAD genes affect both enzyme expression and binding of lead.[130] It is possible that the protein resulting from the ALAD-2 gene offers more protection by binding Pb(II) more tightly.[29]

When BLLs increase above 50 μg dL^{-1} (2.4 μM), a LBP with a molar mass of less than 30 kDa overtakes ALAD as the dominant lead-containing species.[131] In addition to these two proteins, a 10 kDa LBP is inducible at BLLs greater than 39 μg dL^{-1} (1.9 μM) and is present in the blood of occupationally exposed workers, but not in the blood of a control group.[132] Lead toxicity symptoms were manifested at much lower BLLs in workers lacking this protein than in workers with large quantities of this protein.[133] Similarly, a metallothionein-like protein (molar mass 6.5 kDa, Cys content 33%) protects against the symptoms of lead poisoning.[134] In one case study, two lead-exposed workers were described. One showed symptoms of lead toxicity and had a BLL of 161 µg dL^{-1} (7.8 µM), while the other was asymptomatic with a BLL of 180 µg dL^{-1} (8.7 µM). In the former, 20% of the lead was bound to the protein while in the latter, 70% was bound suggesting that the protein may provide protection by sequestering lead in a non-bioavailable form.[135] In addition, cytosolic metallothioneins are mildly inducible by lead, although to a lesser degree than by zinc or cadmium ions.[136,137] LBPs ($K_D = 10^{-8}$ M), likely metallothioneins, sequester lead in the kidney and brain and have been shown to help restore ALAD activity.[138]

Overall, a series of metallothioneins and metallothionein-like LBPs are present in the blood cells and many organs. Several of these have been shown to be inducible, giving protection from lead toxicity. This protection results from the strong binding affinities these proteins have for lead, effectively sequestering it in a non-bioavailable form.

28.2.6 Conclusions

Using a variety of pathways that evolved for the uptake and transfer of essential divalent metal ions, ingested lead travels from the small intestine into the bloodstream and then is distributed into tissue and stored in bone. Within cells, Pb(II) appears to be initially coordinated by glutathione and then is transferred to proteins where it binds opportunistically or to binding sites usually occupied by calcium or zinc. Upon binding, lead can affect protein folding and function, the consequences of which are manifested in the symptoms of lead poisoning. Some LBPs are induced in the presence of lead, although their role is not yet understood. Much remains to be learned about the storage, transport and binding of lead in humans. Until release of lead into the environment is ceased and houses, schools and soils of cities are remediated, lead poisoning will remain a serious public health problem. In the meantime, effective treatments must be developed, including specific chelation and removal from the body, but this requires a better understanding of the (often very long) journey that lead takes upon entering the human body.

28.3 Prokaryotic Cells

28.3.1 Introduction

Some bacteria have evolved detoxification pathways to protect against lead. These include those existing in the soil around certain industries that utilize this heavy metal. Protection mechanisms range from extrusion of lead to its precipitation within the cell and even to isolation of the cell by secreting an exopolymer. Each of the most highly studied systems contains a metalloregulatory protein that senses lead and induces expression of a pump for the extrusion of the ion from the cytoplasm. This section will describe the import and export of lead through the cell membrane, its transport within the cell and the mechanisms of protection.

28.3.2   Transport through the Membrane(s)

28.3.2.1   *Why Would a Cell Import Lead?*

The transport of lead into a prokaryotic cell poses an immediate question: why would a bacterial cell import this purely toxic element? If no specific mechanism is in place to allow import of Pb(II), one might assume that the cell is safe from its toxic effects. However, this assumption overlooks both adventitious entry and possible damage to the cell exterior. In fact, heavy metal detoxification for elements such as Pb, Hg or As is often initiated by their transport into the cell. Such accumulation is typically carried out in two phases. First, the negatively charged cell wall rapidly binds cationic metal ions that are then transferred into the cell.[139] This active transport phase apparently protects the outer membrane from damage. Once sequestered within the cell, the ion can be exported (resulting in no overall benefit but costing energy) or stored in a non-toxic form, thereby removing the ion from the external environment. It is likely that there exist a multitude of pathways for Pb(II) transport considering the diverse conditions under which bacteria acquire or develop resistance. However, two major mechanisms that involve P-type ATPases and ionophores have been identified and studied.

28.3.2.2   *P-type ATPases*

These pumps are a superfamily of enzymes that maintain homeostasis by transport of cations into and out of cells using energy provided by the hydrolysis of ATP. One branch transports soft metals and features Cys-Pro-Cys, Cys-Pro-His or Cys-Pro-Ser motifs in the sixth membrane-spanning domain, that presumably form part of binding sites in an ion transport chain in the protein.[140] They also feature characteristic Cys- or His-rich metal binding sites at the polar N-terminus that may contribute to the metal ion transport or serve to regulate it. Soft-metal transporting ATPases are thought to exist in archaea, bacteria and eukarya, including humans.¶ In humans, copper and silver are transported by the P-type ATPases CopA and CopB, human Menkes and Wilson disease-related proteins.[141] In prokaryotes, several P-type ATPases are known for the transport of zinc (ZntA), cadmium (CadA) and arsenic and antimony (ArsAB). Lead is exported by ZntA, CadA and PbrA, the Pb(II)-efflux protein.

¶See Chapters 16–20, 22, 23 and 25 for discussion of P-type ATPases involved in transport of Cu, Ag, Au, Zn, Cd, Sb and Bi.

28.3.2.2.1 ZntA. Even essential transition metals can be toxic in excess, so bacteria have evolved methods to avoid over-accumulation. The efflux of Zn(II) from *Escherichia coli* is accomplished by ZntA, a P-type ATPase.[142,143] As well as its native substrate, ZntA is capable of transporting the toxic metal ions Cd(II) and Pb(II). Pb(II) inhibits the transport of Zn(II) and expression of ZntA confers resistance to Pb(II).[144,145] This ion can also activate the ATPase and increase its rate of phosphorylation.[146] The highest ATPase activities were observed for Pb(II)-Cys and Pb(II)-GSH complexes, suggesting that the thiolate complexes are likely to be the native substrates.[89] The amino terminus of ZntA contains the sequence Gly-X-Asp-Cys-X-X-Cys along with four additional Cys residues, all of which are not essential to the function and specificity of ZntA as a Zn(II) pump.[147,148] However, this motif may increase the rate of metal binding to the transporter, thus increasing its overall catalytic rate. NMR structures of apo- and Zn-bound ZntA fragments reveal a $Zn(II)Cys_2Asp$ binding site.[149] However, this represents only one of many possible binding modes for Pb(II) during efflux across the membrane.

28.3.2.2.2 CadA. This P-type ATPase is found in *Staphylococcus aureus* and is homologous (35% identity) with ZntA.[150] It also exports Cd(II), Zn(II) and Pb(II).[143] Pb(II) induces expression of CadA, inhibits Zn(II) transport and confers resistance to lead.[144]

28.3.2.2.3 PbrA and PbrT. Within *Cupriavidus*[‖] *metallidurans* strain CH34, the *pbr* operon encodes two separate lead transmembrane transport proteins, PbrT and PbrA. Expression of PbrT in the absence of the remainder of the *pbr* operon resulted in hypersensitivity to Pb(II), consistent with a role for PbrT as a Pb(II)-uptake protein.[151] Little else is known about PbrT except that it is an inner-membrane protein and does not appear to be an ATPase. On the other hand, PbrA is a Pb(II) efflux ATPase that counteracts the role of PbrT and exhibits similar properties to those of CadA and ZntA.[145] Its selectivity may arise from two Cys-Pro-Thr-Glu-Glu metal-binding sites that are present instead of the consensus Cys-X-X-Cys sequence observed in other soft-metal-transporting P-type ATPases.[151]

The efflux of Pb(II) ions helps protect the cell from damage by increasing the pH of the external medium. When *C. metallidurans* CH34 is exposed to Pb(II), the external pH increases and external Pb(II) precipitates as hydroxides and carbonates, thus lowering the local lead ion concentration.[151] While this phenomenon has not been studied in detail, it is similar to that observed for the *czc* cadmium-zinc-cobalt resistance system in the same strain of bacterium.[152] In that system, the pH increase seems to result from the czc-mediated antiporter efflux of cations.

[‖] This bacterium was formerly known as *Ralstonia*.

28.3.2.3   Transport via Ionophores

Molecules that transport ions through a lipid bilayer by shielding the charge from the hydrophobic environment are called ionophores. Several that fall within the class of polyether antibiotics and that feature carboxylate functions have been isolated from the *Streptomyces* genus and bind Pb(II) as 1 : 1 complexes (Figure 28.14). Selectivity for transport of Pb(II) over other divalent cations has been demonstrated for ionophores A23187, ionomycin, monensin and nigericin, with selectivity increasing in that order.[153,154] The relative affinities for Pb(II) are an important contributing factor to the observed selectivities. Transport rates for monensin, nigericin and ionomycin have been estimated at 1.0×10^5, 2.4×10^5 and 11.9×10^5 $M^{-1}s^{-1}$ at pH 7, respectively.[155] When used in combination with dimercaptosuccinic acid, the typical treatment for lead poisoning, monensin improves the effectiveness of removal of lead from the femur, brain and heart of rats.[156]

28.3.3   Coordination and Transport within the Cell

As for eukaryotes, Pb(II) is likely coordinated and transported by sulfur-containing proteins and biomolecules, especially glutathione, which reaches concentrations of up to 10 mM in some bacterial cells.[157] In the case of *C. metallidurans* strain CH34, a metallochaperone PbrD is encoded within the *pbr* operon. It is a protein containing ~200 amino acids and features the hypothetical binding site(s) Cys-7X-Cys-Cys-7X-Cys-7X-His-14X-Cys. While it is not strictly required for lead resistance, cells lacking PbrD acquire less lead than do wild-type cells.[151] Inactivation of *pbrD* leads to a significant increase in *pbr* induction and a slight increase in *zntA* induction. This observation fits the hypothesis that PbrD sequesters Pb(II), lowering cytoplasmic concentrations of free lead ions and safely transporting them to efflux ATPases. When PbrD is not present, cytoplasmic lead levels are higher, leading to induction of additional proteins that remove Pb(II) from the cell.[158]

28.3.4   Regulatory Mechanisms

Several specific mechanisms have evolved for lead regulation. The *cad*, *znt* and *pbr* operons are described below, with a focus on the regulatory proteins. A discussion of the corresponding transport proteins was provided in the preceding sections.

28.3.4.1   CadC

In *S. aureus*, regulation of Pb(II) is carried out by the pI258 *cad* operon that consists of the efflux ATPase CadA (see above) and the metalloregulator CadC, both of which are needed for maximal resistance to Cd(II).[159] CadC, a member of the ArsR/SmtB family of transcriptional regulators, suppresses transcription of CadA by binding specifically to the *cad* operator on DNA in the absence of inducing cations.[160] In the presence of Pb(II), Cd(II), Bi(III)

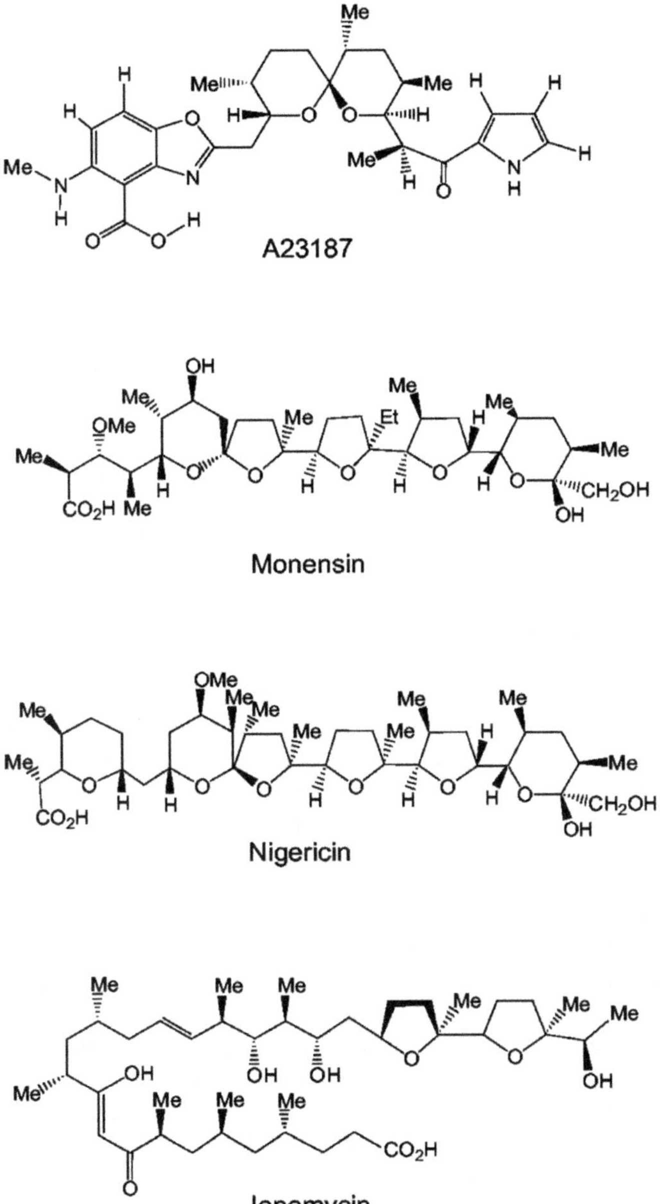

Figure 28.14 Structures of Pb(II)-binding ionophores. This research was originally published in ref. 154.
Copyright (2002) the American Society for Biochemistry and Molecular Biology.

and several other metal ions, CadC undergoes a conformational change that leads to its release, allowing transcription of CadA and the subsequent efflux of the toxic metal ions. Although originally characterized in terms of Cd(II)

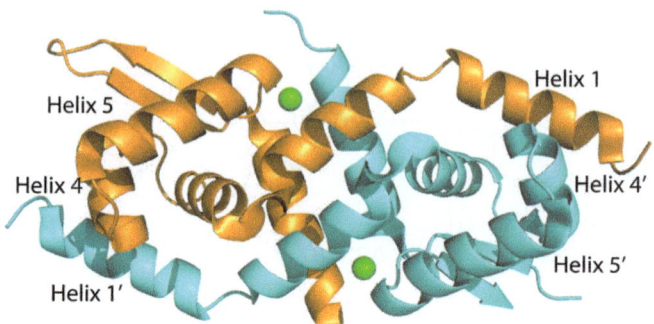

Figure 28.15   CadC dimer structure prepared from PDB 1U2W. Structural Zn(II) ions are shown as green spheres. The inducing metal ions are absent.

regulation, some investigators argue that CadC is primarily a lead regulator due to the fact that it is more sensitive to the binding of Pb(II) than to Cd(II). In modified *E. coli* cells, complete de-repression of *cadCA* occurred at 200 nM Pb(II) and 300 nM Cd(II).[144]

CadC is a 122-residue homodimer that contains two Cys_3/Cys_4 metal binding sites at the interface, both of which are required for de-repression.[21,161] These sites are composed of Cys7 and Cys11 from one monomer and Cys58 and Cys60 from the other.[162,163] Although Cd(II) binds the four Cys residues in a distorted tetrahedral stereochemistry, Cys11 is not required for de-repression and does not coordinate Pb(II). Instead, a stable trigonal pyramidal complex $Pb^{II}Cys_3$ is observed,[22,164] presumably with the lone pair of Pb(II) occupying the apical position. The X-ray crystal structure (1.9 Å resolution) reveals a helix-turn-helix motif in helices 4 and 5 as the putative DNA binding domain (Figure 28.15).[165] Helix 4 from one subunit and the N-terminus from the other (helix 1') provide the metal-binding sites. Transcriptional regulation is induced when metallation causes the N-terminus to be placed in close contact with helix 4, sterically blocking part of the helix-loop-helix DNA binding motif.

Metalloregulation is governed not only by metal site affinity, but also by the kinetics of binding. Pb(II) binding to CadC occurs in two distinct phases.[166] The first is a fast bimolecular encounter between the metal ion and CadC with a rate that is dependent upon the individual ion (Pb(II) > Bi(III) > > Co(II)). The second step is slower with a rate independent of metal type. Cys7 appears to be involved in this slow step, which may effect the conformational change that results in release of CadC from the DNA.

28.3.4.2   ZntR

When *E. coli* is exposed to an excess of Zn(II), the ATPase ZntA exports it from the cell. ZntA also confers lead resistance, with Pb(II) inhibiting the transport of Zn(II).[144] The expression of ZntA is controlled by the transcriptional regulator ZntR.[167] It belongs to the MerR family of transcriptional regulators and responds to Zn(II), Cd(II) or Pb(II) binding.[168] Zn(II) binding is extremely strong (log $K_D = -15.2$ at pH 8.0) but the binding constant for Pb(II) binding is unknown.[169]

The mechanism by which ZntR regulates transcription is very similar to that of MerR, but has only been studied for Zn(II) binding, not for Pb(II). The apo form of the homodimer of ZntR distorts the ZntA-promoter region of DNA when bound. Like MerR, apo-ZntR is thought to bend the DNA toward itself, causing the formation of two kinks in the DNA. This bend keeps RNA polymerase from accessing the full promoter site, thus repressing transcription. The binding of Zn(II) relaxes the bend and causes unwinding of the centre of the operator, making the promoter a better substrate for RNA polymerase. With full access to the operator for RNA polymerase, transcription is activated. Thus, ZntR becomes a transcriptional activator in the presence of Zn(II).[170] Presumably, activation by Pb(II) occurs by the same mechanism.

While no crystal structure yet exists for Pb^{II}-ZntR, much can be learned from the structure of an N-terminally truncated Zn^{II}-bound fragment solved at 1.9 Å resolution.[171] ZntR contains two metal-binding domains, each of which binds two tetrahedral Zn(II) ions (separation 3.6 Å) at the dimeric interface. Each zinc ion is coordinated by two residues from one monomer: Cys114 and Cys124 bind Zn1 while Cys115 and His119 bind Zn2 (Figure 28.16). Cys79 from the other monomer acts as a bridging ligand, and an oxygen atom from a bridging phosphate or sulfate occupies the fourth ligand position on each Zn(II) ion. Neither Cys115 nor His119 are required for induction by Pb(II), suggesting that binding of the ion to these residues is not essential to activate transcription.[172] It is not known how differences in metal size and coordination affect conformational changes to the protein or its ability to bind DNA.

28.3.4.3 *The Lead Resistance Operon pbr*

The first Pb(II)-sensing protein discovered was PbrR, the regulator of lead homeostasis in *C. metallidurans* strain CH34. It controls transcription of the resistance genes *pbrRTUABCD*, which occurs from a divergent promoter with *pbrRTU* from one strand and *prbABCD* from the other.[151] Like ZntR, PbrR is a member of the MerR family of metal-sensing regulatory proteins and shares many characteristics.

However, its mechanism of action is more complex than those for *cad* and *znt*, each of which encode for a metalloregulator and an efflux protein only (Figure 28.17). In addition to a regulator (PbrR) and an efflux ATPase (PbrA), an uptake protein (PbrT), an inner-membrane fusion phosphatase (PbrB/PbrC) and a lead-binding protein (PbrD) are encoded.[145,151,173,174] An inactivated inner membrane permease gene (*pbrU*) may also be present.[173,174] The detailed roles of these proteins are not yet understood. As described in Section 28.3.2.2.3, PbrT transports Pb(II) into the cell to protect the cell

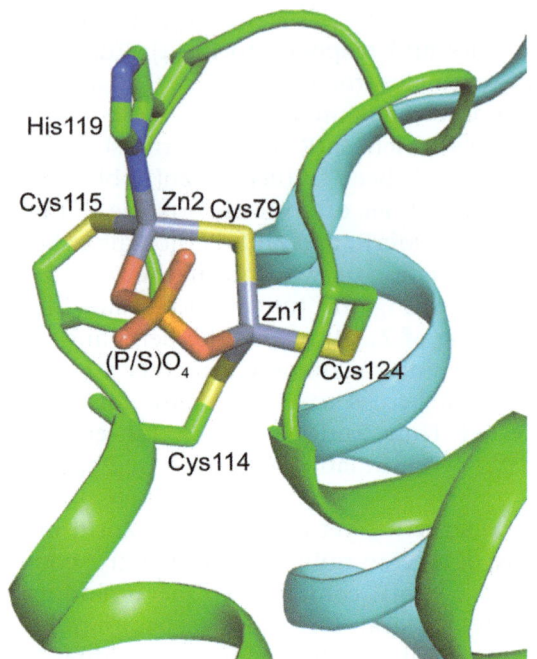

Figure 28.16   ZntR binds two Zn(II) ions in tetrahedral geometry. Prepared from PDB 1Q08.

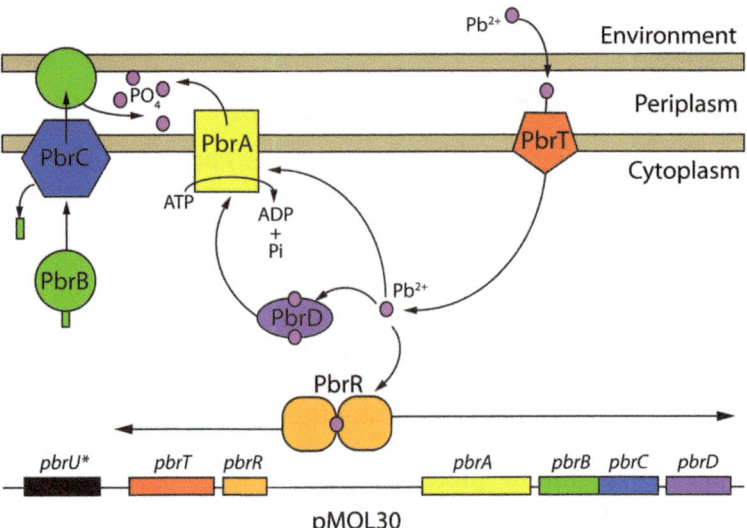

Figure 28.17   Model for *pbr* resistance in *Cupriavidus metallidurans*.

exterior where it is bound by the metallochaperone PbrD. PbrD transports it to efflux ATPases (PbrA, ZntA) that export it from the cytosol to the periplasm. Within the periplasm, the ion is precipitated by phosphates produced by PbrB/PbrC.[145] An alternative mechanism has PbrD transporting the Pb(II) ion to PbrR, signalling for further expression of the Pbr proteins. This relatively complex metalloregulatory system efficiently takes Pb(II) from the external environment and stores it within the cell as inert lead phosphate. With the exception of *pbrU*, each gene is required for maximal resistance.

28.3.4.3.1 PbrR.

In the absence of Pb(II), the homodimeric PbrR binds to the promoter region for the *pbr* operon.[175] Its affinity for the promoter region decreases in the presence of the ion and the mechanism of transcriptional activation is thought to be similar to that of MerR and ZntR. The three cysteine residues 14, 79 and 134 are essential for Pb(II)-induced transcription to occur. Cys14 is found in the helix-turn-helix DNA binding domain. Pb(II) coordination is therefore expected to alter the protein conformation and its interaction with the DNA promoter region, thereby changing the accessibility of the promoter to RNA polymerase. Nearby His, Glu, Lys and Arg residues may bind to Pb(II) as well but it is unclear what role, if any, they play in regulation. Further structural studies are needed.

28.3.4.3.2 PbrR691.

PbrR691 is a PbrR-homologue also found in *C. metallidurans* strain CH34. PbrR691 binds one equivalent of Pb(II) per dimer with $K_D = 0.2$ μM. This is almost 1000-fold higher selectivity than for Zn(II), Cd(II), Hg(II), Co(II), Ni(II) and Cu(II).[23] Sequence alignment with MerR shows conservation of the three metal-binding residues (Cys78, Cys113, Cys122) and spectroscopic studies indicate that three, or possibly four, Cys side-chains are binding Pb(II). Because Pb(II) prefers a hemi-directed geometry at low coordination numbers, this suggests that PbrR691 may discriminate Pb(II) from other ions by presenting a pre-organized hemi-directed Cys_3 binding site.[18] This discrimination allowed the development of PbrR691 into a Pb(II) sensor.[23]

28.3.4.3.3 PbrB/C.

One of the genes controlled by PbrR encodes for the fusion protein PbrBC.[173,174] Removal of the *pbrC* portion of the gene does not affect resistance to lead, suggesting that resistance is coded by the *pbrB* portion.[145] Bound to the inner-membrane within the periplasm, PbrBC is an undecaprenyl pyrophosphate (C55-PP) phosphatase.[145] Its expression leads to the increased production of phosphates, which precipitate Pb(II).** However, this protection only occurs as part of the greater lead-resistance mechanism. In the absence of any P-type ATPase efflux pumps, Pb(II) cannot reach the periplasm resulting in build-up in the cytoplasm and phosphate over-production in the periplasm.[145]

**The chemical formula of the lead phosphate precipitate is not known. The K_{sp} values for lead phosphates range from 10^{-10} for $PbHPO_4$ to 10^{-77} for $Pb_5(PO_4)_3OH$.[176]

28.3.4.4 Other Examples of Lead Defences

While the *pbr* operon in *C. metallidurans* strain CH34 is the most highly studied of those in lead resistant bacteria, it is not the only one that protects by precipitation of Pb(II) as a phosphate. $Pb_3(PO_4)_2$ is nearly insoluble $(K_{sp} = 3.9 \times 10^{-45})$ and does not inhibit the growth of *S. aureus*, even lead-sensitive strains, suggesting that precipitation of lead phosphate salts is a safe way for cells to sequester and detoxify lead.[176,177] In fact, this method has been identified in a number of bacterial families. One species of *Citrobacter* can accumulate up to 36% of its dry weight as $PbHPO_4$ precipitated on the cell surface.[178,179] Lead resistance was also observed in five strains of *S. aureus*. Transmission electron microscopy (TEM) and energy dispersive spectroscopy revealed intracellular crystalline lead phosphate inclusions.[177] Surprisingly, despite precipitation of phosphate in both of these species, lead resistance is not dependent on overall phosphatase activity.[180] The quorum-sensing bacterium *Vibrio harveyi* precipitates lead as $Pb_9(PO_4)_6$. Interestingly, this unusual inorganic salt can only be prepared synthetically at temperatures above 200 °C.[181] TEM showed that lead accumulates in the cytoplasm of *Bacillus megaterium*, but in this case the nature of the species involved was not identified.[182] Finally, some bacteria secrete biopolymers, or exopolymers, into the surrounding environment as an extracellular defence mechanism. In *Pseudomonas marginalis*, lead is sequestered within an exopolymer, preventing its entry to the cells.[182] Exopolymer secretion occurred independently of the presence of metals ions, suggesting that this defence mechanism is not specific for lead, but is a general protection against toxic substances.

28.3.5 Conclusions

Bacteria have evolved defences to protect against heavy metal toxicity. Lead resistance has been observed in a number of bacteria, the most highly studied of which are *E. coli*, *S. aureus* and *C. metallidurans* strain CH34. The respective regulatory operons *znt*, *cad* and *pbr* protect cells by expressing genes for regulatory and efflux proteins, which detoxify Pb(II). In each system, a DNA-bound regulatory protein represses transcription until metal binding induces expression. An efflux ATPase transports Pb(II) out of the cytoplasm, providing protection. The *pbr* system also includes a metallochaperone, and a phosphatase. Comparison of the three systems confirms that consistent strategies for lead resistance emerged in bacteria.

28.4 Future Prospects

This chapter summarizes our current knowledge about the binding, storage and transport of lead in biological systems. We have learned that understanding the symptoms and developing treatments for lead poisoning is complicated by the fact that Pb can bind to many different protein classes.

While we have established a strong foundation for understanding the role of Pb(II) in the human body and bacteria, there is still a great deal to be learned. Fundamentally, the vast majority of studies have focused on "clean systems" such as looking at how Pb binds to individual molecules such as glutathione or zinc fingers in the test tube; however, to understand lead toxicity fully one must consider the interaction of multiple components simultaneously. With the greater understanding of metal associated cellular cascades such as apoptosis, it will be incumbent to examine these types of complex, multicomponent interactions within cells. In many cases, this will require a better understanding of the maintenance of cellular redox poise. Furthermore, the link between lead-bound proteins and RNAs essential in regulatory processes will need to be considered. To achieve all of these objectives thorough measurements of binding affinities, cellular concentrations and detoxification pathways are necessary. Newly expanded techniques such as ^{207}Pb NMR spectroscopy should allow for the better characterization of lead coordination environments in proteins. The development of specific lead sensors can clarify the distribution of lead *in vivo*. Ultimately, the further development of this field will allow for the design of safe, specific chelators for the treatment of lead poisoning, thus mitigating the consequences of exposure.

Acknowledgements

Research by the authors reported in this publication was supported by the National Institutes of Health under Award Numbers F32GM100543 (National Institute of General Medical Sciences) and R01ES012236 (National Institute of Environmental Health Sciences). The content is solely the responsibility of the authors and does not necessarily represent the official views of the National Institutes of Health.

References

1. C. Patterson, J. Ericson, M. Manea-Krichten and H. Shirahata, *Sci. Total Environ.*, 1991, **107**, 205–236.
2. K. R. Mahaffey, J. L. Annest, J. Roberts and R. S. Murphy, *New. Eng. J. Med.*, 1982, **307**, 573–579.
3. N. Zhang, H. W. Baker, M. Tufts, R. E. Raymond, H. Salihu and M. R. Elliott, *Am. J. Public Health*, 2013, **103**, e72–e77.
4. H. W. Mielke, *Am. Sci.*, 1999, **87**, 62–73.
5. H. Needleman, *Annu. Rev. Med.*, 2004, **55**, 209–222.
6. Centers for Disease Control, *Preventing lead poisoning in young children*, Atlanta, GA, US Department of Health and Human Services, CDC, 1991. Available at http://www.cdc.gov/nceh/lead/publications/books/plpyc/contents.htm.
7. Centers for Disease Control Advisory Committee on Childhood Lead Poisoning Prevention, *Low level lead exposure harms children: a renewed*

call for primary prevention, Atlanta, GA, US Department of Health and Human Services, CDC, 2012. Available at http://www.cdc.gov/nceh/lead/acclpp/final_document_030712.pdf.

8. M. F. Bouchard, D. C. Bellinger, J. Weuve, J. Matthews-Bellinger, S. E. Gilman, R. O. Wright, J. Schwartz and M. G. Weisskopf, *Arch. Gen. Psychiat.*, 2009, **66**, 1313–1319.

9. D. C. Bellinger, K. M. Stiles and H. L. Needleman, *Pediatrics*, 1992, **90**, 855–861.

10. H. W. Mielke and S. Zahran, *Environ. Int.*, 2012, **43**, 48–55.

11. B. P. Lanphear, R. Hornung, J. Khoury, K. Yolton, P. Baghurst, D. C. Bellinger, R. L. Canfield, K. N. Dietrich, R. Bornschein, T. Greene, S. J. Rothenberg, H. L. Needleman, L. Schnaas, G. Wasserman, J. Graziano and R. Roberts, *Environ. Health Perspect.*, 2005, **113**, 894–899.

12. D. C. Bellinger, *Neurotoxicol. Teratol.*, 2009, **31**, 267–274.

13. E. S. Claudio, H. A. Godwin and J. S. Magyar, *Prog. Inorg. Chem.*, 2002, **51**, 1–144.

14. J. C. Payne, M. A. ter Horst and H. A. Godwin, *J. Am. Chem. Soc.*, 1999, **121**, 6850–6855.

15. L. Shimoni-Livny, J. P. Glusker and C. W. Bock, *Inorg. Chem.*, 1998, **37**, 1853–1867.

16. J. S. Magyar, T.-C. Weng, C. M. Stern, D. F. Dye, B. W. Rous, J. C. Payne, B. M. Bridgewater, A. Mijovilovich, G. Parkin, J. M. Zaleski, J. E. Penner-Hahn and H. A. Godwin, *J. Am. Chem. Soc.*, 2005, **127**, 9495–9505.

17. M. Kirberger and J. J. Yang, *J. Inorg. Biochem.*, 2008, **102**, 1901–1909.

18. P. R. Chen, E. C. Wasinger, J. Zhao, D. van der Lelie, L. X. Chen and C. He, *J. Am. Chem. Soc.*, 2007, **129**, 12350–12351.

19. A. A. Jarzęcki, *J. Phys. Chem. A*, 2012, **116**, 571–581.

20. K. P. Neupane and V. L. Pecoraro, *J. Inorg. Biochem.*, 2011, **105**, 1030–1034.

21. L. S. Busenlehner, N. J. Cosper, R. A. Scott, B. P. Rosen, M. D. Wong and D. P. Giedroc, *Biochemistry*, 2001, **40**, 4426–4436.

22. J. L. Apuy, L. S. Busenlehner, D. H. Russell and D. P. Giedroc, *Biochemistry*, 2004, **43**, 3824–3834.

23. P. Chen, B. Greenberg, S. Taghavi, C. Romano, D. van der Lelie and C. He, *Angew. Chem. Int. Ed.*, 2005, **44**, 2715–2719.

24. M. Matzapetakis, D. Ghosh, T.-C. Weng, J. E. Penner-Hahn and V. L. Pecoraro, *J. Biol. Inorg. Chem.*, 2006, **11**, 876–890.

25. G. Zampella, K. P. Neupane, L. De Gioia and V. L. Pecoraro, *Chem.-Eur. J.*, 2012, **18**, 2040–2050.

26. H. A. Godwin, *Curr. Opin. Chem. Biol.*, 2001, **5**, 223–227.

27. J. Markovac and G. W. Goldstein, *Nature*, 1988, **344**, 71–73.

28. J. G. Pounds, G. J. Long and J. F. Rosen, *Environ. Health Persp.*, 1991, **91**, 17–32.

29. H. C. Gonick, *J. Toxicol.*, 2011, **2011**, 686050.

30. S. Chao, C.-H. Bu and W. Y. Cheung, *Arch. Toxicol.*, 1990, **64**, 490–496.

31. W. D. Atchison, *J. Bioenerg. Biomembr.*, 2003, **35**, 507–532.

32. J. L. Tomsig and J. B. Suszkiw, *J. Neurochem.*, 1995, **64**, 2667–2673.

33. J. Bressler, K. A. Kim, T. Chakraborti and G. Goldstein, *Neurochem. Res.*, 1999, **24**, 595–600.

34. C. M. L. S. Bouton, L. P. Frelin, C. E. Forde, H. A. Godwin and J. Pevsner, *J. Neurochem.*, 2001, **76**, 1724–1735.

35. L. Patrick, *Altern. Med. Rev.*, 2006, **11**, 2–22.

36. D. E. Ellis, J. Terra, O. Warschkow, M. Jiang, G. B. González, J. S. Okasinski, M. J. Bedzyk, A. M. Rossi and J.-G. Eon, *Phys. Chem. Chem. Phys.*, 2006, **8**, 967–976.

37. J. G. Dorea and C. M. Donangelo, *Clin. Nutr.*, 2006, **25**, 369–376.

38. J. G. Dorea, *Brit. J. Nutr.*, 2004, **92**, 21–40.

39. E. K. Silbergeld, J. Schwartz and K. Mahaffey, *Environ. Res.*, 1988, **47**, 79–94.

40. R. H. Goldman, R. White, S. N. Kales and H. Hu, *Am. J. Ind. Med.*, 1994, **25**, 417–424.

41. E. B. Ekong, B. G. Jaar and V. M. Weaver, *Kidney Int.*, 2006, **70**, 2074–2084.

42. R. W. Leggett, *Environ. Health Perspect.*, 1993, **101**, 598–616.

43. J. R. Beach and S. J. Henning, *Pediatr. Res.*, 1988, **23**, 58–62.

44. S. K. Karri, R. B. Saper and S. N. Kales, *Curr. Drug Saf.*, 2008, **3**, 54–59.

45. M. B. Rabinowitz, J. D. Kipple and G. W. Wetherill, *Am. J. Clin. Nutr.*, 1980, **33**, 1784–1788.

46. H. M. Mykkänen and R. H. Wasserman, *J. Nutr.*, 1981, **111**, 1757–1765.

47. E. J. Martinez-Finley, S. Chakraborty, S. J. B. Fretham and M. Aschner, *Metallomics*, 2012, **4**, 593–605.

48. G. L. Diamond, P. E. Goodrum, S. P. Felter and W. L. Ruoff, *Drug Chem. Toxicol.*, 1998, **21**, 223–251.

49. C. S. Fullmer, S. Edelstein and R. H. Wasserman, *J. Biol. Chem.*, 1985, **260**, 6816–6819.

50. C. M. Dekaney, E. D. Harris, G. R. Bratton and L. A. Jaeger, *Biol. Trace. Elem. Res.*, 1997, **58**, 13–24.

51. H. Gunshin, B. Mackenzie, U. V. Berger, Y. Gunshin, M. F. Romero, W. F. Boron, S. Nussberger, J. L. Gollan and M. A. Hediger, *Nature*, 1997, **388**, 482–488.

52. J. N. Morrison and J. Quarterman, *Biol. Trace. Elem. Res.*, 1987, **14**, 115–126.

53. J. C. Barton, *Am. J. Physiol. Gastrointest. Liver Physiol.*, 1984, **247**, G193–G198.

54. D. I. Bannon, R. Abounader, P. S. J. Lees and J. P. Bressler, *Am. J. Physiol. Cell Physiol.*, 2003, **284**, C44–C50.

55. I. Aduayom and C. Jumarie, *J. Biochem. Mol. Toxic.*, 2005, **19**, 256–265.

56. J. A. Blair, I. P. L. Coleman and M. E. Hilburn, *J. Physiol.*, 1979, **286**, 343–350.

57. D. I. Bannon, L. Olivi and J. Bressler, *Toxicology*, 2000, **147**, 101–107.

58. B. Elsenhans, H. Janser, W. Windisch and K. Schümann, *Toxicology*, 2011, **284**, 7–11.

59. W. Forth and W. Rummel, *Physiol. Rev.*, 1973, **53**, 724–792.

60. K. A. Hussein, S. B. Coghill, G. Milne and D. Hopwood, *Histochemistry*, 1984, **81**, 591–596.
61. H. M. Mykkänen and R. H. Wasserman, *J. Nutr.*, 1982, **112**, 520–527.
62. P. E. DeSilva, *Brit. J. Ind. Med.*, 1981, **38**, 209–217.
63. A. J. A. Al-Modhefer, M. W. B. Bradbury and T. J. B. Simons, *Clin. Sci.*, 1991, **81**, 823–829.
64. I. A. Bergdahl, M. Sheveleva, A. Schütz, V. G. Artamonova and S. Skerfving, *Toxicol. Sci.*, 1998, **46**, 247–253.
65. T. J. B. Simons, *FEBS Lett.*, 1984, **172**, 250–254.
66. T. J. Simons, *J. Physiol.*, 1986, **378**, 267–286.
67. T. J. B. Simons, *J. Physiol.*, 1986, **378**, 287–312.
68. J. V. Calderón-Salinas, M. A. Quintanar-Escorcia, M. T. González-Martínez and C. E. Hernández-Luna, *Hum. Exp. Toxicol.*, 1999, **18**, 327–332.
69. T. J. B. Simons, *Pflüg. Arch.-Eur. J. Phys.*, 1993, **423**, 307–313.
70. T. J. B. Simons, *J. Physiol.*, 1988, **405**, 105–113.
71. Q. Wang, W. Luo, W. Zhang, M. Liu, H. Song and J. Chen, *Toxicol. Vitr.*, 2011, **25**, 991–998.
72. C. Gu, S. Chen, X. Xu, L. Zheng, Y. Li, K. Wu, J. Liu, Z. Qi, D. Han, G. Chen and X. Huo, *Neurochem. Res.*, 2009, **34**, 1150–1156.
73. J. L. Tomsig and J. B. Suszkiw, *Biochim. Biophys. Acta*, 1991, **1069**, 197–200.
74. T. J. Simons and G. Pocock, *J. Neurochem.*, 1987, **48**, 383–389.
75. L. E. Kerper and P. M. Hinkle, *J. Biol. Chem.*, 1997, **272**, 8346–8352.
76. L. E. Kerper and P. M. Hinkle, *Toxicol. Appl. Pharm.*, 1997, **146**, 127–133.
77. M. Mazzolini, S. Traverso and C. Marchetti, *J. Neurochem.*, 2001, **79**, 407–416.
78. R. K. Mehra, V. R. Kodati and R. Abdullah, *Biochem. Biophys. Res. Commun.*, 1995, **215**, 730–736.
79. R. K. Singhal, M. E. Anderson and A. Meister, *FASEB J.*, 1987, **1**, 220–223.
80. J. Yang, R. Swati, T. L. Stemmler and B. P. Rosen, *Biochemistry*, 2010, **49**, 3658–3666.
81. A. Hunaiti, M. Soud and A. Khalil, *Sci. Total Environ.*, 1995, **170**, 95–100.
82. A. A. Hunaiti and M. Soud, *Sci. Total Environ.*, 2000, **248**, 45–50.
83. B. J. Fuhr and D. L. Rabenstein, *J. Am. Chem. Soc.*, 1973, **95**, 6944–6950.
84. L. A. P. Kane-Maguire and P. J. Riley, *J. Coord. Chem.*, 1993, **28**, 105–120.
85. B. K. Singh, R. K. Sharma and B. S. Garg, *J. Therm. Anal. Calorim.*, 2006, **84**, 593–600.
86. V. Mah and F. Jalilehvand, *Inorg. Chem.*, 2012, **51**, 6285–6298.
87. A. M. Corrie, M. D. Walker and D. R. Williams, *J. Chem. Soc. Dalt.*, 1976, 1012–1015.
88. A. M. Corrie and D. R. Williams, *J. Chem. Soc. Dalt.*, 1976, 1068–1072.
89. R. Sharma, C. Rensing, B. P. Rosen and B. Mitra, *J. Biol. Chem.*, 2000, **275**, 3873–3878.
90. I. A. Bergdahl, A. Grubb, A. Schütz, R. J. Desnick, J. G. Wetmur, S. Sassa and S. Skerfving, *Pharmacol. Toxicol.*, 1997, **81**, 153–158.
91. R. F. Labbé, H. J. Vreman and D. K. Stevenson, *Clin. Chem.*, 1999, **45**, 2060–2072.

92. T. J. B. Simons, *Eur. J. Biochem.*, 1995, **234**, 178–183.
93. M. J. Warren, J. B. Cooper, S. P. Wood and P. M. Shoolingin-Jordan, *Trends Biochem. Sci.*, 1998, **23**, 217–221.
94. B. M. Bridgewater and G. Parkin, *J. Am. Chem. Soc.*, 2000, **122**, 7140–7141.
95. P. T. Erskine, E. M. H. Duke, I. J. Tickle, N. M. Senior, M. J. Warren and J. B. Cooper, *Acta Crystallogr. D*, 2000, **D56**, 421–430.
96. G. Parkin, *Chem. Rev.*, 2004, **104**, 699–767.
97. A. Schutz and S. Skerfving, *Scand. J. Work Environ. Health*, 1976, **3**, 176–184.
98. A. Amici, M. Emanuelli, E. Ferretti, N. Raffaelli, S. Ruggieri and G. Magni, *Biochem. J.*, 1994, **304**, 987–992.
99. D. E. Paglia, W. N. Valentine and K. Fink, *J. Clin. Invest.*, 1977, **60**, 1362–1366.
100. Y. Kim, C.-I. Yoo, C. R. Lee, J. H. Lee, H. Lee, S.-R. Kim, S.-H. Chang, W.-J. Lee, C.-H. Hwang and Y. H. Lee, *Ind. Health*, 2002, **40**, 23–27.
101. E. Bitto, C. A. Bingman, G. E. Wesenberg, J. G. McCoy and G. N. Phillips, *J. Biol. Chem.*, 2006, **281**, 20521–20529.
102. R. A. Goyer, D. L. Leonard, J. F. Moore, B. Rhyne and M. R. Krigman, *Arch. Environ. Health*, 1970, **20**, 705–711.
103. R. A. Goyer, P. May, M. M. Cates and M. R. Krigman, *Lab. Invest.*, 1970, **22**, 245–251.
104. A. A. Jarzęcki, *Inorg. Chem.*, 2007, **46**, 7509–7521.
105. A. B. Ghering, L. M. M. Jenkins, B. L. Schenck, S. Deo, R. A. Mayer, M. J. Pikaart, J. G. Omichinski and H. A. Godwin, *J. Am. Chem. Soc.*, 2005, **127**, 3751–3759.
106. P. S. Barry, *Brit. J. Ind. Med.*, 1975, **32**, 119–139.
107. J. L. Michalek, A. N. Besold and S. L. J. Michel, *Dalt. Trans.*, 2011, 12619–12632.
108. B. Quintanilla-Vega, D. J. Hoover, W. Bal, E. K. Silbergeld, M. P. Waalkes and L. D. Anderson, *Chem. Res. Toxicol.*, 2000, **13**, 594–600.
109. J. S. Hanas, J. S. Rodgers, J. A. Bantle and Y.-G. Cheng, *Mol. Pharmacol.*, 1999, **56**, 982–988.
110. D. H. Petering, M. Huang, S. Moteki and C. F. Shaw, III, *Mar. Environ. Res.*, 2000, **50**, 89–92.
111. D. Chin and A. R. Means, *Trends Cell. Biol.*, 2000, **10**, 322–328.
112. E. Habermann, K. Crowell and P. Janicki, *Arch. Toxicol.*, 1983, **54**, 61–70.
113. H. Ouyang and H. J. Vogel, *Biometals*, 1998, **11**, 213–222.
114. J. M. Aramini, T. Hiraoki, M. Yazawa, T. Yuan, M. Zhang and H. J. Vogel, *J. Biol. Inorg. Chem.*, 1996, **1**, 39–48.
115. M. Kirberger, H. C. Wong, J. Jiang and J. J. Yang, *J. Inorg. Biochem.*, 2013, **125**, 40–49.
116. M. A. Wilson and A. T. Brunger, *Acta Crystallogr. D*, 2003, **D59**, 1782–1792.
117. M. Kern, M. Wisniewski, L. Cabell and G. Audesirk, *NeuroToxicology*, 2000, **21**, 353–363.

118. Y. Finkelstein, M. E. Markowitz and J. F. Rosen, *Brain Res. Rev.*, 1998, **27**, 168–176.
119. D. R. Smith, M. W. Kahng, B. Quintanilla-Vega and B. A. Fowler, *Chem.-Biol. Interact.*, 1998, **115**, 39–52.
120. P. M. Egle and K. R. Shelton, *J. Biol. Chem.*, 1986, **261**, 2294–2298.
121. B. Quintanilla-Vega, D. R. Smith, M. W. Kahng, J. M. Hernández, A. Albores and B. A. Fowler, *Chem.-Biol. Interact.*, 1995, **98**, 193–209.
122. P. L. Goering, P. Mistry and B. A. Fowler, *J. Pharmacol. Exp. Ther.*, 1986, **237**, 220–225.
123. G. DuVal and B. A. Fowler, *Biochem. Biophys. Res. Commun.*, 1989, **159**, 177–184.
124. M. W. Bradbury and R. Deane, *NeuroToxicology*, 1993, **14**, 131–136.
125. E. Tiffany-Castiglioni, *Neurotoxicology*, 1993, **14**, 513–536.
126. Y. Qian, E. D. Harris, Y. Zheng and E. Tiffany-Castiglioni, *Toxicol. Appl. Pharm.*, 2000, **163**, 260–266.
127. Y. Qian, Y. Zheng, K. S. Ramos and E. Tiffany-Castiglioni, *NeuroToxicology*, 2005, **26**, 267–275.
128. H. Fujita, K. Sato and S. Sano, *Int. Arch. Occup. Environ. Health*, 1982, **50**, 287–297.
129. C. Boudene, N. Despaux-Pages, E. Comoy and C. Bohuon, *Int. Arch. Occup. Environ. Health*, 1984, **55**, 87–96.
130. S. N. Kelada, E. Shelton, R. B. Kaufmann and M. J. Khoury, *Am. J. Epidemiol.*, 2001, **154**, 1–13.
131. Y. Xie, M. Chiba, A. Shinohara, H. Watanabe and Y. Inaba, *Ind. Health*, 1998, **36**, 234–239.
132. Y. Lolin and P. Ogorman, *Ann. Clin. Biochem.*, 1988, **25**, 688–697.
133. S. R. Raghavan, B. D. Culver and H. C. Gonick, *J. Toxicol. Environ. Health*, 1981, 7, 561–568.
134. H. J. Church, J. P. Day, R. A. Braithwaite and S. S. Brown, *J. Inorg. Biochem.*, 1993, **49**, 55–68.
135. H. J. Church, J. P. Day, R. A. Braithwaite and S. S. Brown, *NeuroToxicology*, 1993, **14**, 359–364.
136. H. Ikebuchi, R. Teshima, K. Suzuki, T. Terao and Y. Yamane, *Biochem. J.*, 1986, **233**, 541–546.
137. T. Maitani, A. Watahiki and K. T. Suzuki, *Toxicol. Appl. Pharm.*, 1986, **83**, 211–217.
138. P. L. Goering and B. A. Fowler, *Arch. Biochem. Biophys.*, 1987, **253**, 48–55.
139. P. R. Norris and D. P. Kelly, *J. Gen. Microbiol.*, 1977, **99**, 317–324.
140. M. Solioz and C. Vulpe, *Trends Biochem. Sci.*, 1996, **21**, 237–241.
141. C. Rensing, M. Ghosh and B. P. Rosen, *J. Bacteriol.*, 1999, **181**, 5891–5897.
142. S. J. Beard, R. Hashim, J. Membrillo-Hernández, M. N. Hughes and R. K. Poole, *Mol. Microbiol.*, 1997, **25**, 883–891.
143. C. Rensing, B. Mitra and B. P. Rosen, *Proc. Natl Acad. Sci. USA*, 1997, **94**, 14326–14331.
144. C. Rensing, Y. Sun, B. Mitra and B. P. Rosen, *J. Biol. Chem.*, 1998, **273**, 32614–32617.

145. A. Hynninen, T. Touzé, L. Pitkänen, D. Mengin-Lecreulx and M. Virta, *Mol. Microbiol.*, 2009, **74**, 384–394.

146. J. Okkeri and T. Haltia, *Biochemistry*, 1999, **38**, 14109–14116.

147. B. Mitra and R. Sharma, *Biochemistry*, 2001, **40**, 7694–7699.

148. Z.-J. Hou, S. Narindrasorasak, B. Bhushan, B. Sarkar and B. Mitra, *J. Biol. Chem.*, 2001, **276**, 40858–40863.

149. L. Banci, I. Bertini, S. Ciofi-Baffoni, L. A. Finney, C. E Outten and T. V. O'Halloran, *J. Mol. Biol.*, 2002, **323**, 883–897.

150. G. Nucifora, *Proc. Natl Acad. Sci. USA*, 1989, **86**, 3544–3548.

151. B. Borremans, J. L. Hobman, A. Provoost, N. L. Brown and D. Van Der Lelie, *J. Bacteriol.*, 2001, **183**, 5651–5658.

152. L. Diels, Q. Dong, D. van der Lelie, W. Baeyens and M. Mergeay, *J. Ind. Microbiol.*, 1995, **14**, 142–53.

153. W. L. Erdahl, C. J. Chapman, R. W. Taylor and D. R. Pfeiffer, *J. Biol. Chem.*, 2000, **275**, 7071–7079.

154. S. A. Hamidinia, O. I. Shimelis, B. Tan, W. L. Erdahl, C. J. Chapman, G. D. Renkes, R. W. Taylor and D. R. Pfeiffer, *J. Biol. Chem.*, 2002, **277**, 38111–38120.

155. S. A. Hamidinia, B. Tan, W. L. Erdahl, C. J. Chapman, R. W. Taylor and D. R. Pfeiffer, *Biochemistry*, 2004, **43**, 15956–15965.

156. S. A. Hamidinia, W. L. Erdahl, C. J. Chapman, G. E. Steinbaugh, R. W. Taylor and D. R. Pfeiffer, *Environ. Health Perspect.*, 2005, **114**, 484–493.

157. R. C. Fahey, W. C. Brown, W. B. Adams and M. B. Worsham, *J. Bacteriol.*, 1978, **133**, 1126–1129.

158. S. Taghavi, C. Lesaulnier, S. Monchy, R. Wattiez, M. Mergeay and D. van der Lelie, *Antonie Van Leeuwenhoek*, 2009, **96**, 171–182.

159. K. P. Yoon and S. Silver, *J. Bacteriol.*, 1991, **173**, 7636–7642.

160. G. Endo and S. Silver, *J. Bacteriol.*, 1995, **177**, 4437–4441.

161. Y. Sun, M. D. Wong and B. P. Rosen, *Mol. Microbiol.*, 2002, **44**, 1323–1329.

162. Y. Sun, M. D. Wong and B. P. Rosen, *J. Biol. Chem.*, 2001, **276**, 14955–14960.

163. M. D. Wong, Y.-F. Lin and B. P. Rosen, *J. Biol. Chem.*, 2002, **277**, 40930–40936.

164. L. S. Busenlehner, T.-C. Weng, J. E. Penner-Hahn and D. P. Giedroc, *J. Mol. Biol.*, 2002, **319**, 685–701.

165. J. Ye, A. Kandegedara, P. Martin and B. P. Rosen, *J. Bacteriol.*, 2005, **187**, 4214–4221.

166. L. S. Busenlehner and D. P. Giedroc, *J. Inorg. Biochem.*, 2006, **100**, 1024–1034.

167. K. R. Brocklehurst, J. L. Hobman, B. Lawley, L. Blank, S. J. Marshall, N. L. Brown and A. P. Morby, *Mol. Microbiol.*, 1999, **31**, 893–902.

168. M. R. Binet and R. K. Poole, *FEBS Lett.*, 2000, **473**, 67–70.

169. Y. Hitomi, C. E. Outten and T. V. O'Halloran, *J. Am. Chem. Soc.*, 2001, **123**, 8614–8615.

170. C. E. Outten, F. W. Outten and T. V. O'Halloran, *J. Biol. Chem.*, 1999, **274**, 37517–37524.

171. A. Changela, K. Chen, Y. Xue, J. Holschen, C. E. Outten, T. V. O'Halloran and A. Mondragón, *Science*, 2003, **301**, 1383–1387.

172. S. Khan, K. R. Brocklehurst, G. W. Jones and A. P. Morby, *Biochem. Bioph. Res. Co.*, 2002, **299**, 438–445.

173. S. Taghavi, C. Lesaulnier, S. Monchy, R. Wattiez, M. Mergeay and D. van der Lelie, *Antonie Van Leeuwenhoek*, 2009, **96**, 171–182.

174. S. Monchy, M. A. Benotmane, P. Janssen, T. Vallaeys, S. Taghavi, D. van der Lelie and M. Mergeay, *J. Bacteriol.*, 2007, **189**, 7417–7425.

175. J. L. Hobman, D. J. Julian and N. L. Brown, *BMC Microbiol.*, 2012, **12**, 109.

176. H. L. Clever, *J. Phys. Chem. Ref. Data*, 1980, **9**, 751–784.

177. H. Levinson, *FEMS Microbiol. Lett.*, 1996, **145**, 421–425.

178. R. M. Aickin and A. C. R. Dean, *Microbios. Lett.*, 1979, **9**, 55–66.

179. R. M. Aickin and A. C. R. Dean, *Microbios. Lett.*, 1979, **9**, 7–15.

180. H. Levinson and I. Mahler, *FEMS Microbiol. Lett.*, 1998, **161**, 135–138.

181. C. E. Mire, J. A. Tourjee, W. F. O. Brien, K. V. Ramanujachary and G. B. Hecht, *Appl. Environ. Microb.*, 2004, **70**, 855–864.

182. T. M. Roane, *Microb. Ecol.*, 1999, **37**, 218–224.

183. R. L. Davidovich, V. Stavila, D. V. Marinin, E. I. Voit and K. H. Whitmire, *Coord. Chem. Rev.*, 2009, **253**, 1316–1352.

184. R. L. Davidovich, V. Stavila and K. H. Whitmire, *Coord. Chem. Rev.*, 2010, **254**, 2193–2226.

185. The PyMOL Molecular Graphics System, Version 1.5.0.4, Schrödinger, LLC.

Subject Index